Practice
"Work the Problems"

NEW

Feature	Description	Benefit	Page
"Assess Your Understanding" contains a variety of problems at the end of each section.			
'Are You Prepared?' Problems	These assess your retention of the prerequisite material you'll need. Answers are given at the end of the section exercises. This feature is related to the Preparing for This Section feature.	Do you always remember what you've learned? Working these problems is the best way to find out. If you get one wrong, you'll know exactly what you need to review, and where to review it!	401
Concepts and Vocabulary	These Fill-in-the-Blank and True/False items assess your understanding of key definitions and concepts in the current section.	Learning math is more than memorization – it's about discovering connections. These problems help you understand the 'big ideas' before diving into skill building.	401
Skill Building	Correlated to section examples, these problems provide straightforward practice, organized by difficulty.	It's important to dig in and develop your skills. These problems provide you with ample practice to do so.	402-403
Applications and Extensions	These problems allow you to apply your skills to real-world problems. These problems also allow you to extend concepts learned in the section.	You will see that the material learned within the section has many uses in everyday life.	403-404
Discussion and Writing	"Discussion and Writing" problems are colored red. These support class discussion, verbalization of mathematical ideas, and writing and research projects.	To verbalize an idea, or to describe it clearly in writing, shows real understanding. These problems nurture that understanding. They're challenging , but you'll get out what you put in.	418
Now Work Problems	Many examples refer you to a related homework problem. These related problems are marked by a pencil and yellow numbers.	If you get stuck while working problems, look for the closest Now Work problem and refer back to the related example to see if it helps.	402

Review
"Study for Quizzes and Tests"

Feature	Description	Benefit	Page
Chapter Reviews at the end of each chapter contain...			
"Things to Know"	A detailed list of important theorems, formulas, and definitions, from the chapter.	Review these and you'll know the most important material in the chapter!	493-494
"You Should be Able To..."	Contains a complete list of objectives by section, with corresponding practice exercises.	Do the recommended exercises and you'll have mastery over the key material. If you get something wrong, review the suggested page numbers and try again.	495
Review Exercises	These provide comprehensive review and practice of key skills, matched to the Learning Objectives for each section.	Practice makes perfect. These problems combine exercises from all sections, giving you a comprehensive review in one place.	496-500
Chapter Test	About 15–20 problems that can be taken as a Chapter Test. Be sure to take the Chapter Test under test conditions—no notes!	Be prepared. Take the sample practice test. This will get you ready for your instructor's test.	500
Chapter Projects	The Chapter Project applies what you've learned in the chapter. Additional projects are available on the Instructor's Resource Center (IRC)	The Project gives you an opportunity to apply what you've learned in the chapter to solve a problem related to the opening article. If your instructor allows, these make excellent opportunities to work in a group, which is often the best way of learning math.	501
Cumulative Review	These problem sets appear at the end of each chapter, beginning with Chapter 2. They combine problems from previous chapters, providing an ongoing cumulative review.	These are really important. They will ensure that you are not forgetting anything as you go. These will go a long way toward keeping you constantly primed for quizzes and tests.	501-502

NEW

COLLEGE ALGEBRA

FOURTH EDITION

COLLEGE ALGEBRA
Enhanced with Graphing Utilities

Michael Sullivan
Chicago State University

Michael Sullivan, III
Joliet Junior College

PEARSON

Prentice
Hall

Upper Saddle River, New Jersey 07458

Library of Congress Cataloging-in-Publication Data

Sullivan, Michael
    College algebra: enhanced with graphing utilities / Michael Sullivan, Michael Sullivan,
      III.—4th ed.
    Includes index.
    ISBN 0-13-149104-0
    1. Algebra—Textbooks.   2. Algebra—Graphic methods.   I. Sullivan, Michael
    2- II. Title.
    QA154.3.S765 2006
    512.0—dc22   2004060075

Editor-in-Chief: Sally Yagan
Acquisitions Editor: Adam Jaworski
Project Manager: Dawn Murrin
Executive Managing Editor: Kathleen Schiaparelli
Senior Managing Editor: Linda Mihatov Behrens
Production Editor: Bob Walters, Prepress Management, Inc.
Assistant Manufacturing Manager/Buyer: Michael Bell
Marketing Manager: Halee Dinsey
Marketing Assistant: Joon Moon
Editorial Assistant/Print Supplements Editor: Christopher Truechan
Art Director: Heather Scott
Interior Designer: Judith Matz-Coniglio
Cover Designer: Susan Anderson-Smith
Art Editor: Thomas Benfatti
Director of Creative Services: Paul Belfanti
Director, Image Resource Center: Melinda Reo
Manager, Rights and Permissions: Zina Arabia
Manager, Visual Research: Beth Brenzel
Image Permission Coordinator: Carolyn Gauntt
Cover Image Permission Manager: Karen Sanatar
Photo Researcher: Melinda Alexander
Art Studio: Artworks:
    Senior Manager: Patricia Burns
    Production Manager: Ronda Whitson
    Manager, Production Technologies: Matt Haas
    Illustrators: Ryan Currier, Mark Landis
    Art Quality Assurance: Timothy Nguyen, Stacy Smith, Pamela Taylor

© 2006, 2003, 2000, 1996 Pearson Education, Inc.
Pearson Prentice Hall
Pearson Education, Inc.
Upper Saddle River, New Jersey 07458

Printed in the United States of America
10 9 8 7 6 5 4 3 2 1

ISBN 0-13-149104-0

Pearson Education LTD., London
Pearson Education Australia PTY, Limited, Sydney
Pearson Education Singapore, Pte. Ltd
Pearson Education North Asia Ltd. Hong Kong
Pearson Education Canada, Ltd., Toronto
Pearson Educación de Mexico, S.A. de C.V.
Pearson Education – Japan, Tokyo
Pearson Education Malaysia, Pte. Ltd

In memory of Mary . . .
Wife and Mother

Contents

Three Distinct Series

Students have different goals, learning styles, and levels of preparation. Instructors have different teaching philosophies, styles, and techniques. Rather than write one series to fit all, the Sullivans have written three distinct series. All share the same goal—to develop a high level of mathematical understanding and an appreciation for the way mathematics can describe the world around us. The manner of reaching that goal, however, differs from series to series.

Contemporary Series, Seventh Edition

The Contemporary Series is the most traditional in approach, yet modern in its treatment of precalculus mathematics. Graphing utility coverage is optional and can be included or excluded based on instructor preference.

College Algebra, College Algebra Essentials, Algebra & Trigonometry, Trigonometry, Precalculus, Precalculus Essentials

Enhanced with Graphing Utilities Series, Fourth Edition

This series provides a more thorough integration of graphing utilities into topics, allowing students to explore mathematical concepts and foreshadow ideas usually studied in later courses. Using technology, the approach to solving certain problems differs from the Contemporary series, while the emphasis on understanding concepts and building strong skills does not.

College Algebra, College Algebra Essentials, Algebra & Trigonometry, Trigonometry, Precalculus

Graphing, Data, and Analysis Series, Third Edition

This series differs from the others, utilizing a functions approach which serves as the organizing principle tying concepts together. Each chapter introduces a new type of function, then develops all concepts pertaining to that particular function. The solutions of equations and inequalities, instead of being developed as stand-alone topics, are developed in the context of the underlying functions. Technology is integrated at the same level as the Enhanced with Graphing Utilities series (described above).

College Algebra, Algebra & Trigonometry, Precalculus

Preface
to the Instructor

As professors at both an urban university and a community college, Michael Sullivan and Michael Sullivan, III are aware of the varied needs of College Algebra students, ranging from those who have little mathematical background and a fear of mathematics courses, to those having a strong mathematical education and a high level of motivation. For some of your students, this will be their last course in mathematics, whereas others will further their mathematical education. This text is written for both groups.

As a teacher, and as an author of precalculus, engineering calculus, finite mathematics, and business calculus texts, Michael Sullivan understands what students must know if they are to be focused and successful in upper level math courses. However, as a father of four, he also understands the realities of college life. His co-author and son, Michael Sullivan, III, believes in the value of technology as a tool for learning that enhances understanding without sacrificing math skills. Together, both authors have taken great pains to ensure that the text contains solid, student-friendly examples and problems, as well as a clear and seamless writing style.

A tremendous benefit of authoring a successful series is the broad-based feedback we receive from teachers and students. Virtually every change in this edition is the result of their thoughtful comments and suggestions. We are sincerely grateful for this support and hope that we have been able to take these ideas and, building upon a successful third edition, make this series an even better tool for learning and teaching. We continue to encourage you to share with us your experiences teaching from this text.

New Features
in the Fourth Edition

Rather than provide a list of new features here, we have placed that information in the end sheet on the front cover of this book. The new features are easy to spot by the "New" in the left column.

This places the new features in their proper context, as building blocks of an overall learning system that has been carefully crafted over the years to help students get the most out of the time they put into studying. Please take the time to review this, and to discuss it with your students at the beginning of your course. Our experience has been that when students utilize these features, they are more successful in the course.

Organizational Changes in the Fourth Edition

- "Integer Exponents" and "Square Roots" now appear earlier in Chapter R in Section R.2. "Radicals" is now combined with "Rational Exponents" in Chapter R.

- "Setting Up Equations: Applications" has been renamed to "Problem Solving: Interest, Mixture, Uniform Motion, Constant Rate Jobs" and now appears in Section 1.6. The easier applications from "Setting Up Equations: Applications" are now in Section 1.2.

- "Synthetic Division" has been moved from the former Chapter 5: "The Zeros of a Polynomial Function" to Chapter R.

- "Quadratic Equations in the Complex Number System" has been moved from former Chapter 5: "The Zeros of a Polynomial Function" to Chapter 1, Section 1.4. This section remains optional and also allows for either early or late coverage of complex numbers and quadratic equations with a negative discriminant.

- Portions of Chapter 2: "Linear and Quadratic Functions" and Chapter 3: "Functions and Their Graphs" have been combined into one chapter, Chapter 2: "Functions and Their Graphs."

- Former Section 2.3: "Quadratic Functions" and former Section 2.4: "Quadratic Models" have been combined into one section, Section 3.1: "Quadratic Functions and Models."

- "Composite Functions," formerly in Chapter 3: "Functions and Their Graphs," now appears as Section 4.1 in Chapter 4: "Exponential and Logarithmic Functions."

- Former Chapter 4: "Polynomial and Rational Functions" and Chapter 5: "The Zeros of a Polynomial Function" have been combined into one chapter, Chapter 3: "Polynomial and Rational Functions."

- Former Section 4.3: "Rational Functions I" has been renamed as Section 3.3: "Properties of Rational Functions." Former Section 4.4: "Rational Functions II: Analyzing Graphs" has been renamed Section 3.4: "The Graph of a Rational Function; Inverse and Joint Variation."
- "Systems of Linear Equations: Two Equations Containing Two Variables" and "Systems of Linear Equations Containing Three Variables" have been combined in one section as "Systems of Linear Equations: Substitution and Elimination."

Using the Fourth Edition Effectively with Your Syllabus

To meet the varied needs of diverse syllabi, this book contains more content than is likely to be covered in a typical College Algebra course. As the chart illustrates, this book has been organized with flexibility of use in mind. Within a given chapter, certain sections are optional (see the detail following the flow chart) and can be omitted without loss of continuity.

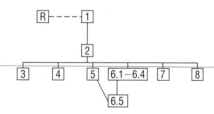

Chapter R Review

This chapter consists of review material. It may be used as the first part of the course or later as a just-in time review when the content is required. Specific references to this chapter occur throughout the book to assist in the review process.

Chapter 1 Graphs, Equations, and Inequalities

This chapter presents an introduction to graphing and the graphing utility. Equations and inequalities are solved algebraically and graphically. The coverage of complex numbers and quadratic equations with a negative discriminant is optional and may be postponed or skipped entirely without loss of continuity.

Chapter 2 Functions and Their Graphs

Perhaps the most important chapter. Section 2.8 is optional.

Chapter 3 Polynomial and Rational Functions

Topic selection depends on your syllabus.

Chapter 4 Exponential and Logarithmic Functions

Sections 4.1–4.6 follow in sequence. Sections 4.7–4.9 are optional.

Chapter 5 Systems of Equations and Inequalities

Sections 5.2–5.6 may be covered in any order. Section 5.7 requires Section 5.6.

Chapter 6 Analytic Geometry

Sections 6.1–6.4 follow in sequence. Section 6.5 requires Section 5.1 and Sections 6.1–6.4.

Chapter 7 Sequences; Induction; The Binomial Theorem

There are three independent parts: Sections 7.1–7.3; Section 7.4; and Section 7.5.

Chapter 8 Counting and Probability

The sections follow in sequence.

Acknowledgments

Textbooks are written by authors, but evolve from an idea to final form through the efforts of many people. It was Don Dellen who first suggested this series. Don is remembered for his extensive contributions to publishing and mathematics.

We would also like to extend thanks to the following individuals for their important assistance and encouragement in the preparation of this edition: From Prentice Hall: Halee Dinsey for her innovative marketing skills; Adam Jaworski for his substantial contributions, ideas, and enthusiasm; Patrice Jones, who remains a huge fan and supporter; Dawn Murrin for her talent and skill in getting supplements out; Bob Walters, who continues to amaze us with his organizational skill as Production Editor; Sally Yagan for her continued support and genuine interest; and the Prentice Hall Sales Team, for their continued confidence in our books. We also extend out thanks to Teri Lovelace and the entire Laurel Technical Services staff, for their dedication to accuracy in checking manuscript and answers. Special thanks to Celeste Hernandez for her help in preparing chapter projects, and to Kevin Bodden for his work in preparing the chapter tests.

Finally, our grateful thanks to the dedicated users and reviewers of our books, whose collective insights form the backbone of each textbook revision. We apologize for any omissions:

James Africh, College of DuPage
Steve Agronsky, Cal Poly State University
Grant Alexander, Joliet Junior College
Dave Anderson, South Suburban College

Joby Milo Anthony, University of Central Florida
James E. Arnold, University of Wisconsin-Milwaukee

Adel Arshaghi, Center for Educational Merit
Carolyn Autray, University of West Georgia
Agnes Azzolino, Middlesex County College
Wilson P Banks, Illinois State University

Sudeshna Basu, Howard University

Dale R. Bedgood, East Texas State University

Beth Beno, South Suburban College

Carolyn Bernath, Tallahassee Community College

William H. Beyer, University of Akron

Annette Blackwelder, Florida State University

Richelle Blair, Lakeland Community College

Kevin Bodden, Lewis and Clark College

Bob Bradshaw, Ohlone College

Trudy Bratten, Grossmont College

Tim Bremer, Broome Community College

Joanne Brunner, Joliet Junior College

Warren Burch, Brevard Community College

Mary Butler, Lincoln Public Schools

Jim Butterbach, Joliet Junior College

William J. Cable, University of Wisconsin-Stevens Point

Lois Calamia, Brookdale Community College

Jim Campbell, Lincoln Public Schools

Roger Carlsen, Moraine Valley Community College

Elena Catoiu, Joliet Junior College

Mathews Chakkanakuzhi, Palomar College

John Collado, South Suburban College

Nelson Collins, Joliet Junior College

Jim Cooper, Joliet Junior College

Denise Corbett, East Carolina University

Theodore C. Coskey, South Seattle Community College

Donna Costello, Plano Senior High School

Paul Crittenden, University of Nebraska at Lincoln

John Davenport, East Texas State University

Faye Dang, Joliet Junior College

Antonio David, Del Mar College

Duane E. Deal, Ball State University

Timothy Deis, University of Wisconsin-Platteville

Vivian Dennis, Eastfield College

Deborah Dillon, R. L. Turner High School

Guesna Dohrman, Tallahassee Community College

Karen R. Dougan, University of Florida

Louise Dyson, Clark College

Paul D. East, Lexington Community College

Don Edmondson, University of Texas-Austin

Erica Egizio, Joliet Junior College

Christopher Ennis, University of Minnesota

Ralph Esparza, Jr., Richland College

Garret J. Etgen, University of Houston

Pete Falzone, Pensacola Junior College

W.A. Ferguson, University of Illinois-Urbana/Champaign

Iris B. Fetta, Clemson University

Mason Flake, student at Edison Community College

Timothy W. Flood, Pittsburg State University

Merle Friel, Humboldt State University

Richard A. Fritz, Moraine Valley Community College

Dewey Furness, Ricke College

Randy Gallaher, Lewis and Clark College

Tina Garn, University of Arizona

Dawit Getachew, Chicago State University

Wayne Gibson, Rancho Santiago College

Robert Gill, University of Minnesota Duluth

Sudhir Kumar Goel, Valdosta State University

Joan Goliday, Sante Fe Community College

Frederic Gooding, Goucher College

Sue Graupner, Lincoln Public Schools

Jennifer L. Grimsley, University of Charleston

Ken Gurganus, University of North Carolina

James E. Hall, University of Wisconsin-Madison

Judy Hall, West Virginia University

Edward R. Hancock, DeVry Institute of Technology

Julia Hassett, DeVry Institute-Dupage

Christopher Hay-Jahans, University of South Dakota

Michah Heibel, Lincoln Public Schools

LaRae Helliwell, San Jose City College

Brother Herron, Brother Rice High School

Robert Hoburg, Western Connecticut State University

Lynda Hollingsworth, Northwest Missouri State University

Charla Holzbog, Denison High School

Lee Hruby, Naperville North High School

Miles Hubbard, St. Cloud State University

Kim Hughes, California State College-San Bernardino

Ron Jamison, Brigham Young University

Richard A. Jensen, Manatee Community College

Sandra G. Johnson, St. Cloud State University

Tuesday Johnson, New Mexico State University

Moana H. Karsteter, Tallahassee Community College

Donna Katula, Joliet Junior College

Arthur Kaufman, College of Staten Island

Thomas Kearns, North Kentucky University

Shelia Kellenbarger, Lincoln Public Schools

Lynne Kowski, Raritan Valley Community College

Keith Kuchar, Manatee Community College

Tor Kwembe, Chicago State University

Linda J. Kyle, Tarrant Country Jr. College

H.E. Lacey, Texas A & M University

Harriet Lamm, Coastal Bend College

James Lapp, Fort Lewis College

Matt Larson, Lincoln Public Schools

Christopher Lattin, Oakton Community College

Julia Ledet, Lousiana State University

Adele LeGere, Oakton Community College

Kevin Leith, University of Houston

JoAnn Lewin, Edison College

Jeff Lewis, Johnson County Community College

Janice C. Lyon, Tallahassee Community College

Virginia McCarthy, Iowa State University

Jean McArthur, Joliet Junior College

Karla McCavit, Albion College

Tom McCollow, DeVry Institute of Technology

Marilyn McCollum, North Carolina State University

Jill McGowan, Howard University

Will McGowant, Howard University

Angela McNulty, Joliet Junior College

Laurence Maher, North Texas State University

Jay A. Malmstrom, Oklahoma City Community College

Rebecca Mann, Apollo High School

Sherry Martina, Naperville North High School

Alec Matheson, Lamar University

Nancy Matthews, University of Oklahoma

James Maxwell, Oklahoma State University-Stillwater

Marsha May, Midwestern State University

Judy Meckley, Joliet Junior College

David Meel, Bowling Green State University

Carolyn Meitler, Concordia University

Samia Metwali, Erie Community College

Rich Meyers, Joliet Junior College

Eldon Miller, University of Mississippi

James Miller, West Virginia University

Michael Miller, Iowa State University

Kathleen Miranda, SUNY at Old Westbury

Thomas Monaghan, Naperville North High School

Craig Morse, Naperville North High School

Samad Mortabit, Metropolitan State University

Pat Mower, Washburn University

A. Muhundan, Manatee Community College

Jane Murphy, Middlesex Community College

Richard Nadel, Florida International University

Gabriel Nagy, Kansas State University

Bill Naegele, South Suburban College

Karla Neal, Lousiana State University

Lawrence E. Newman, Holyoke Community College

James Nymann, University of Texas-El Paso

Mark Omodt, Anoka-Ramsey Community College

Seth F. Oppenheimer, Mississippi State University

Linda Padilla, Joliet Junior College

E. James Peake, Iowa State University

Kelly Pearson, Murray State University

Philip Pina, Florida Atlantic University

Michael Prophet, University of Northern Iowa

Neal C. Raber, University of Akron

Thomas Radin, San Joaquin Delta College

Ken A. Rager, Metropolitan State College

Kenneth D. Reeves, San Antonio College

Elsi Reinhardt, Truckee Meadows Community College

Jane Ringwald, Iowa State University

Stephen Rodi, Austin Community College

William Rogge, Lincoln Northeast High School

Howard L. Rolf, Baylor University
Mike Rosenthal, Florida International
University
Phoebe Rouse, Lousiana State University
Edward Rozema, University of Tennessee at
Chattanooga
Dennis C. Runde, Manatee Community
College
Alan Saleski, Loyola University of Chicago
Susan Sandmeyer, Jamestown Community
College
Linda Schmidt, Greenville Technical College
A.K. Shamma, University of West Florida
Martin Sherry, Lower Columbia College
Tatrana Shubin, San Jose State University
Anita Sikes, Delgado Community College
Timothy Sipka, Alma College
Lori Smellegar, Manatee Community College

Gayle Smith, Loyola Blakefield
John Spellman, Southwest Texas State
University
Rajalakshmi Sriram, Okaloosa-Walton
Community College
Becky Stamper, Western Kentucky University
Judy Staver, Florida Community College-
South
Neil Stephens, Hinsdale South High School
Patrick Stevens, Joliet Junior College
Christopher Terry, Augusta State University
Diane Tesar, South Suburban College
Tommy Thompson, Brookhaven College
Martha K. Tietze, Shawnee Mission Northwest
High School
Richard J.Tondra, Iowa State University
Marvel Townsend, University of Florida
Jim Trudnowski, Carroll College

Robert Tuskey, Joliet Junior College
Richard G.Vinson, University of South
Alabama
Mary Voxman, University of Idaho
Jennifer Walsh, Daytona Beach Community
College
Donna Wandke, Naperville North High School
Darlene Whitkenack, Northern Illinois
University
Christine Wilson, West Virginia University
Brad Wind, Florida International University
Mary Wolyniak, Broome Community College
Canton Woods, Auburn University
Tamara S.Worner, Wayne State College
Terri Wright, New Hampshire Community
Technical College, Manchester
George Zazi, Chicago State University

Michael Sullivan
Michael Sullivan, III

Student Study-Pak

Everything a student needs to succeed in one place. Free, packaged with the book, or available for purchase standalone. Study-Pak contains:

- **Student Solutions Manual**
  Fully worked solutions to odd-numbered exercises.

- **Pearson Tutor Center**
  Tutors provide one-on-one tutoring for any problem with an answer at the back of the book. Students access the Tutor Center via toll-free phone, fax, or email.

- **CD Lecture Series**
  A comprehensive set of CD-ROMs, tied to the textbook, containing short video clips of an instructor working key book examples.

- **Algebra Review**
  Four chapters of Intermediate Algebra review. Perfect for a slower-paced course or for individual review.

- **Chapter Test Prep Video CD**
  Gives student **step-by-step video solutions to every problem in each Chapter Test.** An instructor works through each chapter test problem providing valuable review to enhance mastery of key chapter content. Easy video navigation allows students to instantly access the solution to any problem.

MathXL®

MathXL® is a powerful online homework, tutorial, and assessment system that accompanies your textbook. Instructors can create, edit, and assign online homework and tests using algorithmically generated exercises correlated at the objective level to the textbook. Student work is tracked in an online gradebook. Students can take chapter tests and receive personalized study plans based on their results. The study plan diagnoses weaknesses and links students to tutorial exercises for objectives they need to study. Students can also access video clips from selected exercises. MathXL® is available to qualified adopters. For more information, visit our website at **www.mathxl.com**, or contact your Prentice Hall sales representative for a demonstration.

MyMathLab

MyMathLab is a text-specific, customizable online course for your textbooks. MyMathLab is powered by CourseCompass™—Pearson Education's online teaching and learning environment—and by MathXL®—our online homework, tutorial, and assessment system. MyMathLab gives you the tools you need to deliver all or a portion of your course online, whether your students are in a lab setting or working from home.

MyMathLab provides a rich and flexible set of course materials, featuring free-response exercises that are algorithmically generated for unlimited practice. Students can use online tools such as video lectures and a multimedia textbook to improve their performance. Instructors can use MyMathLab's homework and test managers to select and assign online exercises correlated to the textbook, and can import TestGen tests for added flexibility. The online gradebook—designed specifically for mathematics—automatically tracks students' homework and test results and gives the instructor control over how to calculate final grades. MyMathLab is available to qualified adopters. For more information, visit our website at **www.mymathlab.com** or contact your Prentice Hall sales representative for a product demonstration.

List of Applications

Photo Credits

COLLEGE ALGEBRA

TO THE STUDENT

As you begin, you may feel anxious about the number of theorems, definitions, procedures, and equations. You may wonder if you can learn it all in time. Don't worry, your concerns are normal. This textbook was written with you in mind. If you attend class, work hard, and read and study this book, you will build the knowledge and skills you need to be successful. Here's how you can use the book to your benefit.

READ CAREFULLY

When you get busy, it's easy to skip reading and go right to the problems. Don't . . . the book has a large number of examples and clear explanations to help you break down the mathematics into easy-to-understand steps. Reading will provide you with a clearer understanding, beyond simple memorization. Read it before class (not after) so you can ask questions about anything you didn't understand. You'll be amazed at how much more you'll get out of class if you do this.

GETTING THE MOST FROM THIS BOOK

We've provided a brief guide to getting the most from this book (please refer to the "Prepare for Class," "Practice," and "Review" pages on the inside front cover of this book). Spend fifteen minutes reviewing the guide and familiarizing yourself with the features by flipping to the page numbers provided. Then, as you read, use them. This is the best way to make the most of your textbook.

Please do not hesitate to contact us, through Prentice Hall, with any questions, suggestions or comments that would improve this text. We look forward to hearing from you, and best of luck with all your studies.

Best Wishes!

Michael Sullivan
Michael Sullivan, III

Review

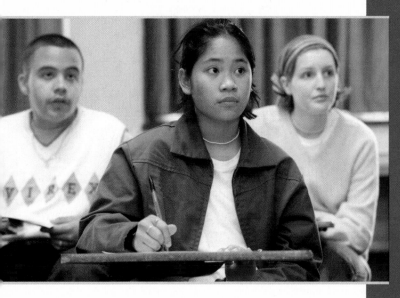

R

A LOOK BACK As the title states, this chapter is a review. Its purpose is to help you recall facts that you learned in an earlier course. The topics chosen for inclusion here represent information that will be useful to prepare for this course.

A LOOK AHEAD Your instructor may decide to cover this chapter or not, depending on the course syllabus. Regardless, as you proceed through the book, references will be made to Chapter R that you can use as a "just in time" review.

OUTLINE

R.1 Real Numbers

PREPARING FOR THIS BOOK *Before getting started, read "To the Students" at the front of this book.*

OBJECTIVES 1 Classify Numbers
2 Approximate Decimals
3 Evaluate Numerical Expressions
4 Work with Properties of Real Numbers

When we want to treat a collection of similar but distinct objects as a whole, we use the idea of a **set**. For example, the set of *digits* consists of the collection of numbers 0, 1, 2, 3, 4, 5, 6, 7, 8, and 9. If we use the symbol D to denote the set of digits, then we can write

$$D = \{0, 1, 2, 3, 4, 5, 6, 7, 8, 9\}$$

In this notation, the braces { } are used to enclose the objects, or **elements**, in the set. This method of denoting a set is called the **roster method**. A second way to denote a set is to use **set-builder notation**, where the set D of digits is written as

$$D = \{\quad x \quad | \quad x \text{ is a digit}\}$$

Read as "D is the set of all x such that x is a digit."

EXAMPLE 1   **Using Set-builder Notation and the Roster Method**

(a) $E = \{x | x \text{ is an even digit}\} = \{0, 2, 4, 6, 8\}$
(b) $O = \{x | x \text{ is an odd digit}\} = \{1, 3, 5, 7, 9\}$ ◄

In listing the elements of a set, we do not list an element more than once because the elements of a set are distinct. Also, the order in which the elements are listed is not relevant. For example, $\{2, 3\}$ and $\{3, 2\}$ both represent the same set.

If every element of a set A is also an element of a set B, then we say that A is a **subset** of B. If two sets A and B have the same elements, then we say that A **equals** B. For example, $\{1, 2, 3\}$ is a subset of $\{1, 2, 3, 4, 5\}$, and $\{1, 2, 3\}$ equals $\{2, 3, 1\}$.

Finally, if a set has no elements, it is called the **empty set**, or the **null set**, and is denoted by the symbol $\varnothing$.

1 Classify Numbers

It is helpful to classify the various kinds of numbers that we deal with as sets. The **counting numbers**, or **natural numbers**, are the numbers in the set $\{1, 2, 3, 4, \dots\}$. (The three dots, called an **ellipsis**, indicate that the pattern continues indefinitely.) As their name implies, these numbers are often used to count things. For example, there are 26 letters in our alphabet; there are 100 cents in a dollar. The **whole numbers** are the numbers in the set $\{0, 1, 2, 3, \dots\}$, that is, the counting numbers together with 0.

The **integers** are the set of numbers $\{\dots, -3, -2, -1, 0, 1, 2, 3, \dots\}$.

These numbers are useful in many situations. For example, if your checking account has $10 in it and you write a check for $15, you can represent the current balance as −$5.

Notice that the set of counting numbers is a subset of the set of whole numbers. Each time we expand a number system, such as from the whole numbers to the integers, we do so in order to be able to handle new, and usually more complicated, problems. The integers allow us to solve problems requiring both positive and negative counting numbers, such as profit/loss, height above/below sea level, temperature above/below 0°F, and so on.

But integers alone are not sufficient for *all* problems. For example, they do not answer the question "What part of a dollar is 38 cents?" To answer such a question, we enlarge our number system to include *rational numbers*. For example, $\frac{38}{100}$ answers the question "What part of a dollar is 38 cents?"

A **rational number** is a number that can be expressed as a quotient $\frac{a}{b}$ of two integers. The integer a is called the **numerator**, and the integer b, which cannot be 0, is called the **denominator**. The rational numbers are the numbers in the set $\{x | x = \frac{a}{b}$, where a, b are integers and $b \neq 0.\}$.

Examples of rational numbers are $\frac{3}{4}, \frac{5}{2}, \frac{0}{4}, -\frac{2}{3}$, and $\frac{100}{3}$. Since $\frac{a}{1} = a$ for any integer a, it follows that the set of integers is a subset of the set of rational numbers.

Rational numbers may be represented as **decimals**. For example, the rational numbers $\frac{3}{4}, \frac{5}{2}, -\frac{2}{3}$, and $\frac{7}{66}$ may be represented as decimals by carrying out the indicated division:

$$\frac{3}{4} = 0.75 \qquad \frac{5}{2} = 2.5 \qquad -\frac{2}{3} = -0.666\ldots = -0.\overline{6} \qquad \frac{7}{66} = 0.1060606\ldots = 0.1\overline{06}$$

Notice that the decimal representations of $\frac{3}{4}$ and $\frac{5}{2}$ terminate, or end. The decimal representations of $-\frac{2}{3}$ and $\frac{7}{66}$ do not terminate, but they do exhibit a pattern of repetition. For $-\frac{2}{3}$, the 6 repeats indefinitely, as indicated by the bar over the 6; for $\frac{7}{66}$, the block 06 repeats indefinitely, as indicated by the bar over the 06. It can be shown that every rational number may be represented by a decimal that either terminates or is nonterminating with a repeating block of digits, and vice versa.

On the other hand, some decimals do not fit into either of these categories. Such decimals represent **irrational numbers**. Every irrational number may be represented by a decimal that neither repeats nor terminates. In other words, irrational numbers cannot be written in the form $\frac{a}{b}$, where a, b are integers and $b \neq 0$.

Irrational numbers occur naturally. For example, consider the isosceles right triangle whose legs are each of length 1. See Figure 1. The length of the hypotenuse is $\sqrt{2}$, an irrational number.

Also, the number that equals the ratio of the circumference C to the diameter d of any circle, denoted by the symbol π (the Greek letter pi), is an irrational number. See Figure 2.

Figure 1

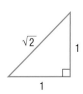

Figure 2

$\pi = \dfrac{C}{d}$

Together, the rational numbers and irrational numbers form the set of **real numbers**.

Figure 3 shows the relationship of various types of numbers.[*]

Figure 3

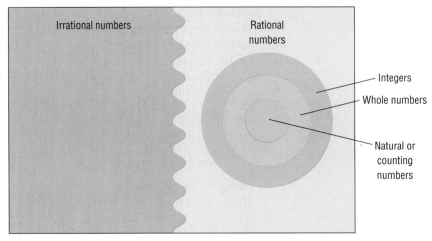

Irrational numbers   Rational numbers   Integers   Whole numbers   Natural or counting numbers

Real numbers

EXAMPLE 2   **Classifying the Numbers in a Set**

List the numbers in the set

$$\left\{-3, \frac{4}{3}, 0.12, \sqrt{2}, \pi, 2.151515\ldots = 2.\overline{15}, 10\right\}$$

that are

(a) Natural numbers   (b) Integers   (c) Rational numbers
(d) Irrational numbers   (e) Real numbers

Solution   (a) 10 is the only natural number.

(b) -3 and 10 are integers.

(c) $-3, \frac{4}{3}, 0.12, 2.151515\ldots = 2.\overline{15}$, and 10 are rational numbers.

(d) $\sqrt{2}$ and π are irrational numbers.

(e) All the numbers listed are real numbers.   ◀

 NOW WORK PROBLEM **11**.

2 Approximate Decimals

Every decimal may be represented by a real number (either rational or irrational), and every real number may be represented by a decimal.

---

[*]The set of real numbers is a subset of the set of complex numbers. We discuss complex numbers in Section 1.4.

In practice, the decimal representation of an irrational number is given as an approximation. For example, using the symbol $\approx$ (read as "approximately equal to"), we can write

$$\sqrt{2} \approx 1.4142 \qquad \pi \approx 3.1416$$

In approximating decimals, we either *round off* or *truncate* to a given number of decimal places.[*] The number of places establishes the location of the *final digit* in the decimal approximation.

> **Truncation:** Drop all the digits that follow the specified final digit in the decimal.
> **Rounding:** Identify the specified final digit in the decimal. If the next digit is 5 or more, add 1 to the final digit; if the next digit is 4 or less, leave the final digit as it is. Then truncate following the final digit.

EXAMPLE 3

Approximating a Decimal to Two Places

Approximate 20.98752 to two decimal places by

(a) Truncating   (b) Rounding

Solution

For 20.98752, the final digit is 8, since it is two decimal places from the decimal point.

(a) To truncate, we remove all digits following the final digit 8. The truncation of 20.98752 to two decimal places is 20.98.

(b) The digit following the final digit 8 is the digit 7. Since 7 is 5 or more, we add 1 to the final digit 8 and truncate. The rounded form of 20.98752 to two decimal places is 20.99. ◄

EXAMPLE 4

Approximating a Decimal to Two and Four Places

Number	Rounded to Two Decimal Places	Rounded to Four Decimal Places	Truncated to Two Decimal Places	Truncated to Four Decimal Places
(a) 3.14159	3.14	3.1416	3.14	3.1415
(b) 0.056128	0.06	0.0561	0.05	0.0561
(c) 893.46125	893.46	893.4613	893.46	893.4612 ◄

NOW WORK PROBLEM **15.**

Calculators

Calculators are finite machines. As a result, they are incapable of displaying decimals that contain a large number of digits. For example, some calculators are capable of displaying only eight digits. When a number requires more than eight digits, the calculator either truncates or rounds. To see how your calculator handles decimals, divide 2 by 3. How many digits do you see? Is the last digit a 6 or a 7? If it is a 6, your calculator truncates; if it is a 7, your calculator rounds.

---

[*]Sometimes we say "correct to a given number of decimal places" instead of "truncate."

There are different kinds of calculators. An **arithmetic** calculator can only add, subtract, multiply, and divide numbers; therefore, this type is not adequate for this course. **Scientific** calculators have all the capabilities of arithmetic calculators and also contain **function keys** labeled ln, log, sin, cos, tan, x^y, inv, and so on. As you proceed through this text, you will discover how to use many of the function keys. **Graphing** calculators have all the capabilities of scientific calculators and contain a screen on which graphs can be displayed.

Figure 4 shows $\frac{2}{3}$ on a TI-84 Plus calculator. How many digits are displayed?

Does a TI-84 Plus round or truncate? What does your calculator do?

Figure 4

3 Evaluate Numerical Expressions

In algebra, we use letters such as x, y, a, b, and c to represent numbers. The symbols used in algebra for the operations of addition, subtraction, multiplication, and division are $+, -, \cdot$, and $/$. The words used to describe the results of these operations are **sum**, **difference**, **product**, and **quotient**. Table 1 summarizes these ideas.

Table 1

Operation	Symbol	Words
Addition	$a + b$	Sum: a plus b
Subtraction	$a - b$	Difference: a minus b
Multiplication	$a \cdot b, (a) \cdot b, a \cdot (b), (a) \cdot (b),$ $ab, (a)b, a(b), (a)(b)$	Product: a times b
Division	a/b or $\dfrac{a}{b}$	Quotient: a divided by b

In algebra, we generally avoid using the multiplication sign $\times$ and the division sign $\div$ so familiar in arithmetic. Notice also that when two expressions are placed next to each other without an operation symbol, as in ab, or in parentheses, as in $(a)(b)$, it is understood that the expressions, called **factors**, are to be multiplied.

We also prefer not to use mixed numbers in algebra. When mixed numbers are used, addition is understood; for example, $2\frac{3}{4}$ means $2 + \frac{3}{4}$. In algebra, use of a mixed number may be confusing because the absence of an operation symbol between two terms is generally taken to mean multiplication. The expression $2\frac{3}{4}$ is therefore written instead as 2.75 or as $\frac{11}{4}$.

The symbol $=$, called an **equal sign** and read as "equals" or "is," is used to express the idea that the number or expression on the left of the equal sign is equivalent to the number or expression on the right.

EXAMPLE 5 **Writing Statements Using Symbols**

(a) The sum of 2 and 7 equals 9. In symbols, this statement is written as $2 + 7 = 9$.

(b) The product of 3 and 5 is 15. In symbols, this statement is written as $3 \cdot 5 = 15$. ◄

NOW WORK PROBLEM 27.

Consider the expression $2 + 3 \cdot 6$. It is not clear whether we should add 2 and 3 to get 5, and then multiply by 6 to get 30; or first multiply 3 and 6 to get 18, and then add 2 to get 20. To avoid this ambiguity, we have the following agreement.

In Words
Multiply first, then add.

We agree that whenever the two operations of addition and multiplication separate three numbers the multiplication operation always will be performed first, followed by the addition operation.

For $2 + 3 \cdot 6$, we have

$$2 + 3 \cdot 6 = 2 + 18 = 20$$

| EXAMPLE 6 | **Finding the Value of an Expression** |

Evaluate each expression.

(a) $3 + 4 \cdot 5$ (b) $8 \cdot 2 + 1$ (c) $2 + 2 \cdot 2$

Solution

(a) $3 + 4 \cdot 5 = 3 + 20 = 23$ (b) $8 \cdot 2 + 1 = 16 + 1 = 17$
            ↑   ↑
      Multiply first   Multiply first

(c) $2 + 2 \cdot 2 = 2 + 4 = 6$ ◄

Figure 5

Figure 5 shows the solutions to Example 6 using a graphing calculator. Notice that the calculator follows the order of operations agreed to.

✏ **NOW WORK PROBLEM 39.**

To first add 3 and 4 and then multiply the result by 5, we use parentheses and write $(3 + 4) \cdot 5$. Whenever parentheses appear in an expression, it means "perform the operations within the parentheses first!"

| EXAMPLE 7 | **Finding the Value of an Expression** |

(a) $(5 + 3) \cdot 4 = 8 \cdot 4 = 32$
(b) $(4 + 5) \cdot (8 - 2) = 9 \cdot 6 = 54$ ◄

When we divide two expressions, as in

$$\frac{2 + 3}{4 + 8}$$

it is understood that the division bar acts like parentheses; that is,

$$\frac{2 + 3}{4 + 8} = \frac{(2 + 3)}{(4 + 8)}$$

The following list gives the rules for the order of operations.

Rules for the Order of Operations

1. Begin with the innermost parentheses and work outward. Remember that in dividing two expressions the numerator and denominator are treated as if they were enclosed in parentheses.
2. Perform multiplications and divisions, working from left to right.
3. Perform additions and subtractions, working from left to right.

| EXAMPLE 8 | **Finding the Value of an Expression** |

Evaluate each expression.

(a) $8 \cdot 2 + 3$ (b) $5 \cdot (3 + 4) + 2$

(c) $\dfrac{2 + 5}{2 + 4 \cdot 7}$ (d) $2 + [4 + 2 \cdot (10 + 6)]$

Solution

(a) $8 \cdot 2 + 3 = 16 + 3 = 19$

  ↑ Multiply first

(b) $5 \cdot (3 + 4) + 2 = 5 \cdot 7 + 2 = 35 + 2 = 37$

  ↑ Parentheses first   ↑ Multiply before adding

(c) $\dfrac{2 + 5}{2 + 4 \cdot 7} = \dfrac{2 + 5}{2 + 28} = \dfrac{7}{30}$

(d) $2 + [4 + 2 \cdot (10 + 6)] = 2 + [4 + 2 \cdot (16)]$
$$= 2 + [4 + 32] = 2 + 36 = 38 \quad \blacktriangleleft$$

Be careful if you use a calculator. For Example 8(c), you need to use parentheses. See Figure 6.[*] If you don't, the calculator will compute the expression

$$2 + \frac{5}{2} + 4 \cdot 7 = 2 + 2.5 + 28 = 32.5$$

giving a wrong answer.

Figure 6

```
(2+5)/(2+4*7)
          .233333333
Ans►Frac
              7/30
```

NOW WORK PROBLEMS **45** AND **53.**

4 **Work with Properties of Real Numbers**

We have used the equal sign to mean that one expression is equivalent to another. Four important properties of equality are listed next. In this list, a, b, and c represent real numbers.

> **1.** The **reflexive property** states that a number always equals itself; that is, $a = a$.
> **2.** The **symmetric property** states that if $a = b$ then $b = a$.
> **3.** The **transitive property** states that if $a = b$ and $b = c$ then $a = c$.
> **4.** The **principle of substitution** states that if $a = b$ then we may substitute b for a in any expression containing a.

Now, let's consider some other properties of real numbers. We begin with an example.

| EXAMPLE 9 | **Commutative Properties** |

(a) $3 + 5 = 8$ (b) $2 \cdot 3 = 6$
     $5 + 3 = 8$ $3 \cdot 2 = 6$
     $3 + 5 = 5 + 3$ $2 \cdot 3 = 3 \cdot 2 \quad \blacktriangleleft$

[*]Notice that we converted the decimal to its fraction form. Consult your manual to see how your calculator does this.

This example illustrates the **Commutative Property** of real numbers, which states that the order in which addition or multiplication takes place will not affect the final result.

Commutative Properties

$$a + b = b + a \qquad \textbf{(1a)}$$
$$a \cdot b = b \cdot a \qquad \textbf{(1b)}$$

Here, and in the properties listed next and on pages 10–12, a, b, and c represent real numbers.

EXAMPLE 10 **Associative Properties**

(a) $2 + (3 + 4) = 2 + 7 = 9$
    $(2 + 3) + 4 = 5 + 4 = 9$
    $2 + (3 + 4) = (2 + 3) + 4$

(b) $2 \cdot (3 \cdot 4) = 2 \cdot 12 = 24$
    $(2 \cdot 3) \cdot 4 = 6 \cdot 4 = 24$
    $2 \cdot (3 \cdot 4) = (2 \cdot 3) \cdot 4$ ◄

The way we add or multiply three real numbers will not affect the final result. Expressions such as $2 + 3 + 4$ and $3 \cdot 4 \cdot 5$ present no ambiguity, even though addition and multiplication are performed on one pair of numbers at a time. This property is called the **Associative Property**.

Associative Properties

$$a + (b + c) = (a + b) + c = a + b + c \qquad \textbf{(2a)}$$
$$a \cdot (b \cdot c) = (a \cdot b) \cdot c = a \cdot b \cdot c \qquad \textbf{(2b)}$$

The next property is perhaps the most important.

Distributive Property

$$a \cdot (b + c) = a \cdot b + a \cdot c \qquad \textbf{(3a)}$$
$$(a + b) \cdot c = a \cdot c + b \cdot c \qquad \textbf{(3b)}$$

The Distributive Property may be used in two different ways.

EXAMPLE 11 **Using the Distributive Property**

(a) $2 \cdot (x + 3) = 2 \cdot x + 2 \cdot 3 = 2x + 6$ Use to remove parentheses.
(b) $3x + 5x = (3 + 5)x = 8x$ Use to combine two expressions.
(c) $(x + 2)(x + 3) = x(x + 3) + 2(x + 3) = (x^2 + 3x) + (2x + 6)$
    $= x^2 + (3x + 2x) + 6 = x^2 + 5x + 6$ ◄

NOW WORK PROBLEM 71.

The real numbers 0 and 1 have unique properties.

EXAMPLE 12	**Identity Properties**

(a) $4 + 0 = 0 + 4 = 4$ (b) $3 \cdot 1 = 1 \cdot 3 = 3$ ◀

The properties of 0 and 1 illustrated in Example 12 are called the **Identity Properties**.

Identity Properties

$$0 + a = a + 0 = a \qquad \text{(4a)}$$
$$a \cdot 1 = 1 \cdot a = a \qquad \text{(4b)}$$

We call 0 the **additive identity** and 1 the **multiplicative identity**.

For each real number a, there is a real number $-a$, called the **additive inverse** of a, having the following property:

Additive Inverse Property

$$a + (-a) = -a + a = 0 \qquad \text{(5a)}$$

EXAMPLE 13	**Finding an Additive Inverse**

(a) The additive inverse of 6 is -6, because $6 + (-6) = 0$.

(b) The additive inverse of -8 is $-(-8) = 8$, because $-8 + 8 = 0$. ◀

The additive inverse of a, that is, $-a$, is often called the *negative* of a or the *opposite* of a. The use of such terms can be dangerous, because they suggest that the additive inverse is a negative number, which may not be the case. For example, the additive inverse of -3, or $-(-3)$, equals 3, a positive number.

For each *nonzero* real number a, there is a real number $\frac{1}{a}$, called the **multiplicative inverse** of a, having the following property:

Multiplicative Inverse Property

$$a \cdot \frac{1}{a} = \frac{1}{a} \cdot a = 1 \qquad \text{if } a \neq 0 \qquad \text{(5b)}$$

The multiplicative inverse $\frac{1}{a}$ of a nonzero real number a is also referred to as the **reciprocal** of a.

EXAMPLE 14	**Finding a Reciprocal**

(a) The reciprocal of 6 is $\frac{1}{6}$, because $6 \cdot \frac{1}{6} = 1$.

(b) The reciprocal of -3 is $\frac{1}{-3}$, because $-3 \cdot \frac{1}{-3} = 1$.

(c) The reciprocal of $\frac{2}{3}$ is $\frac{3}{2}$, because $\frac{2}{3} \cdot \frac{3}{2} = 1$. ◀

With these properties for adding and multiplying real numbers, we can now define the operations of subtraction and division as follows:

The **difference** $a - b$, also read "a less b" or "a minus b," is defined as

$$a - b = a + (-b) \tag{6}$$

To subtract b from a, add the opposite of b to a.

If b is a nonzero real number, the **quotient** $\dfrac{a}{b}$, also read as "a divided by b" or "the ratio of a to b," is defined as

$$\frac{a}{b} = a \cdot \frac{1}{b} \qquad \text{if } b \neq 0 \tag{7}$$

EXAMPLE 15   **Working with Differences and Quotients**

(a) $8 - 5 = 8 + (-5) = 3$ (b) $4 - 9 = 4 + (-9) = -5$ (c) $\dfrac{5}{8} = 5 \cdot \dfrac{1}{8}$ ◀

For any number a, the product of a times 0 is always 0; that is,

In Words
The result of multiplying by zero is zero.

Multiplication by Zero

$$a \cdot 0 = 0 \tag{8}$$

Division Properties

$$\frac{0}{a} = 0 \qquad \frac{a}{a} = 1 \qquad \text{if } a \neq 0 \tag{9}$$

WARNING
Division by 0 is not defined. One reason is to avoid the following difficulty:
$\dfrac{2}{0} = x$ means to find x such that $0 \cdot x = 2$. But $0 \cdot x$ equals 0 for all x, so there is no unique number x such that $\dfrac{2}{0} = x$. ∎

Rules of Signs

$$a(-b) = -(ab) \qquad (-a)b = -(ab) \qquad (-a)(-b) = ab$$
$$-(-a) = a \qquad \frac{a}{-b} = \frac{-a}{b} = -\frac{a}{b} \qquad \frac{-a}{-b} = \frac{a}{b} \tag{10}$$

EXAMPLE 16   **Applying the Rules of Signs**

(a) $2(-3) = -(2 \cdot 3) = -6$ (b) $(-3)(-5) = 3 \cdot 5 = 15$

(c) $\dfrac{3}{-2} = \dfrac{-3}{2} = -\dfrac{3}{2}$ (d) $\dfrac{-4}{-9} = \dfrac{4}{9}$ (e) $\dfrac{x}{-2} = \dfrac{1}{-2} \cdot x = -\dfrac{1}{2}x$ ◀

Cancellation Properties

$$ac = bc \quad \text{implies} \quad a = b \quad \text{if } c \neq 0$$

$$\frac{ac}{bc} = \frac{a}{b} \qquad\qquad \text{if } b \neq 0, c \neq 0 \qquad \textbf{(11)}$$

EXAMPLE 17 **Using the Cancellation Properties**

(a) If $2x = 6$, then

$$
\begin{aligned}
2x &= 6 \\
2x &= 2 \cdot 3 \qquad \text{Factor 6.} \\
x &= 3 \qquad \text{Cancel the 2's.}
\end{aligned}
$$

NOTE
We follow the common practice of using slash marks to indicate cancellations. ∎

(b) $\dfrac{18}{12} = \dfrac{3 \cdot \cancel{6}}{2 \cdot \cancel{6}} = \dfrac{3}{2}$

Cancel the 6's.

◀

Zero-Product Property

$$\text{If } ab = 0, \text{ then } a = 0, \text{ or } b = 0, \text{ or both.} \qquad \textbf{(12)}$$

EXAMPLE 18 **Using the Zero-Product Property**

If $2x = 0$, then either $2 = 0$ or $x = 0$. Since $2 \neq 0$, it follows that $x = 0$. ◀

NOTE
In writing quotients, we shall follow the usual convention and write the quotient in lowest terms; that is, we write it so that any common factors of the numerator and the denominator have been removed using the cancellation properties, equation (11). As examples,

$$\frac{90}{24} = \frac{15 \cdot \cancel{6}}{4 \cdot \cancel{6}} = \frac{15}{4}$$

$$\frac{24x^2}{18x} = \frac{4 \cdot \cancel{6} \cdot x \cdot \cancel{x}}{3 \cdot \cancel{6} \cdot \cancel{x}}$$

$$= \frac{4x}{3} \qquad x \neq 0 \quad ∎$$

Arithmetic of Quotients

$$\frac{a}{b} + \frac{c}{d} = \frac{ad}{bd} + \frac{bc}{bd} = \frac{ad + bc}{bd} \qquad \text{if } b \neq 0, d \neq 0 \qquad \textbf{(13)}$$

$$\frac{a}{b} \cdot \frac{c}{d} = \frac{ac}{bd} \qquad\qquad\qquad \text{if } b \neq 0, d \neq 0 \qquad \textbf{(14)}$$

$$\frac{\dfrac{a}{b}}{\dfrac{c}{d}} = \frac{a}{b} \cdot \frac{d}{c} = \frac{ad}{bc} \qquad\quad \text{if } b \neq 0, c \neq 0, d \neq 0 \qquad \textbf{(15)}$$

EXAMPLE 19 **Adding, Subtracting, Multiplying, and Dividing Quotients**

(a) $\dfrac{2}{3} + \dfrac{5}{2} = \dfrac{2 \cdot 2}{3 \cdot 2} + \dfrac{3 \cdot 5}{3 \cdot 2} = \dfrac{2 \cdot 2 + 3 \cdot 5}{3 \cdot 2} = \dfrac{4 + 15}{6} = \dfrac{19}{6}$

By equation (13)

(b) $\dfrac{3}{5} - \dfrac{2}{3} = \dfrac{3}{5} + \left(-\dfrac{2}{3}\right) = \dfrac{3}{5} + \dfrac{-2}{3}$

         ↑   ↑
By equation (6)   By equation (10)

$\qquad = \dfrac{3 \cdot 3 + 5 \cdot (-2)}{5 \cdot 3} = \dfrac{9 + (-10)}{15} = \dfrac{-1}{15} = -\dfrac{1}{15}$

      ↑
By equation (13)

NOTE
Slanting the cancellation marks in different directions for different factors, as shown in Example 19(c), is a good practice to follow, since it will help in checking for errors. ∎

(c) $\dfrac{8}{3} \cdot \dfrac{15}{4} = \dfrac{8 \cdot 15}{3 \cdot 4} = \dfrac{2 \cdot \cancel{4} \cdot \cancel{3} \cdot 5}{\cancel{3} \cdot \cancel{4} \cdot 1} = \dfrac{2 \cdot 5}{1} = 10$

        ↑   ↑
   By equation (14)   By equation (11)

(d) $\dfrac{\dfrac{3}{5}}{\dfrac{7}{9}} = \dfrac{3}{5} \cdot \dfrac{9}{7} = \dfrac{3 \cdot 9}{5 \cdot 7} = \dfrac{27}{35}$

      ↑   ↑
          By equation (14)

By equation (15)

◀

NOW WORK PROBLEMS 55, 59, AND 69.

Sometimes it is easier to add two quotients using *least common multiples* (LCM). The LCM of two numbers is the smallest number that each has as a common multiple.

EXAMPLE 20 **Finding the Least Common Multiple of Two Numbers**

Find the least common multiple of 15 and 12.

Solution To find the LCM of 15 and 12, we look at multiples of 15 and 12.

$$15, \quad 30, \quad 45, \quad 60, \quad 75, \quad 90, \quad 105, \quad 120, \ldots$$
$$12, \quad 24, \quad 36, \quad 48, \quad 60, \quad 72, \quad 84, \quad 96, \quad 108, \quad 120, \ldots$$

The *common* multiples are in blue. The *least* common multiple is 60.   ◀

EXAMPLE 21 **Using the Least Common Multiple to Add Two Quotients**

Find: $\dfrac{8}{15} + \dfrac{5}{12}$

Solution We use the LCM of the denominators of the quotients and rewrite each quotient using the LCM as a common denominator. The LCM of the denominators (12 and 15) is 60. Rewrite each quotient using 60 as the denominator.

$$\dfrac{8}{15} + \dfrac{5}{12} = \dfrac{8}{15} \cdot \dfrac{4}{4} + \dfrac{5}{12} \cdot \dfrac{5}{5} = \dfrac{32}{60} + \dfrac{25}{60} = \dfrac{32 + 25}{60} = \dfrac{57}{60}$$

◀

NOW WORK PROBLEM 63.

HISTORICAL FEATURE

The real number system has a history that stretches back at least to the ancient Babylonians (1800 BC). It is remarkable how much the ancient Babylonian attitudes resemble our own. As we stated in the text, the fundamental difficulty with irrational numbers is that they cannot be written as quotients of integers or, equivalently, as repeating or terminating decimals. The Babylonians wrote their numbers in a system based on 60 in the same way that we write ours based on 10. They would carry as many places for π as the accuracy of the problem demanded, just as we now use

$$\pi \approx 3\frac{1}{7} \quad \text{or} \quad \pi \approx 3.1416 \quad \text{or} \quad \pi \approx 3.14159$$

$$\text{or} \quad \pi \approx 3.14159265358979$$

depending on how accurate we need to be.

Things were very different for the Greeks, whose number system allowed only rational numbers. When it was discovered that $\sqrt{2}$ was not a rational number, this was regarded as a fundamental flaw in the number concept. So serious was the matter that the Pythagorean Brotherhood (an early mathematical society) is said to have drowned one of its members for revealing this terrible secret.

Greek mathematicians then turned away from the number concept, expressing facts about whole numbers in terms of line segments.

In astronomy, however, Babylonian methods, including the Babylonian number system, continued to be used. Simon Stevin (1548–1620), probably using the Babylonian system as a model, invented the decimal system, complete with rules of calculation, in 1585. [Others, for example, al-Kashi of Samarkand (d. 1429), had made some progress in the same direction.] The decimal system so effectively conceals the difficulties that the need for more logical precision began to be felt only in the early 1800s. Around 1880, Georg Cantor (1845–1918) and Richard Dedekind (1831–1916) gave precise definitions of real numbers. Cantor's definition, although more abstract and precise, has its roots in the decimal (and hence Babylonian) numerical system.

Sets and set theory were a spin-off of the research that went into clarifying the foundations of the real number system. Set theory has developed into a large discipline of its own, and many mathematicians regard it as the foundation upon which modern mathematics is built. Cantor's discoveries that infinite sets can also be counted and that there are different sizes of infinite sets are among the most astounding results of modern mathematics.

Historical Problems

The Babylonian number system was based on 60. In this system 2,30 means $2 + \dfrac{30}{60} = 2.5$, and 4,25,14 means

$$4 + \frac{25}{60} + \frac{14}{60^2} = 4 + \frac{1514}{3600} = 4.42055555\ldots$$

1. What are the following Babylonian numbers when written as fractions and as decimals?
   (a) 2,20   **(b)** 4,52,30   **(c)** 3,8,29,44
2. What are the following numbers in Babylonian notation?
   (a) $1\dfrac{1}{3}$ **(b)** $2\dfrac{5}{6}$

R.1 Assess Your Understanding

Concepts and Vocabulary

1. The numbers in the set $\{x \mid x = \dfrac{a}{b}$, where a, b are integers and $b \neq 0\}$, are called _____ numbers.

2. The value of the expression $4 + 5 \cdot 6 - 3$ is _____.

3. The fact that $2x + 3x = (2 + 3)x$ is a consequence of the _____ Property.

4. "The product of 5 and $x + 3$ equals 6" may be written as _____.

5. *True or False:* Rational numbers have decimals that either terminate or are nonterminating with a repeating block of digits.

6. *True or False:* The Zero-Product Property states that the product of any number and zero equals zero.

7. *True or False:* The least common multiple of 12 and 18 is 6.

8. *True or False:* No real number is both rational and irrational.

Skill Building

In Problems 9–14, list the numbers in each set that are (a) Natural numbers, (b) Integers, (c) Rational numbers, (d) Irrational numbers, (e) Real numbers.

9. $A = \left\{-6, \dfrac{1}{2}, -1.333\ldots = -1.\overline{3}, \pi, 2, 5\right\}$

10. $B = \left\{-\dfrac{5}{3}, 2.060606\ldots = 2.\overline{06}, 1.25, 0, 1, \sqrt{5}\right\}$

11. $C = \left\{0, 1, \dfrac{1}{2}, \dfrac{1}{3}, \dfrac{1}{4}\right\}$

12. $D = \{-1, -1.1, -1.2, -1.3\}$

13. $E = \left\{\sqrt{2}, \pi, \sqrt{2} + 1, \pi + \dfrac{1}{2}\right\}$

14. $F = \left\{-\sqrt{2}, \pi + \sqrt{2}, \dfrac{1}{2} + 10.3\right\}$

In Problems 15–26, approximate each number (a) rounded and (b) truncated to three decimal places.

15. 18.9526 **16.** 25.86134 **17.** 28.65319 **18.** 99.05249 **19.** 0.06291 **20.** 0.05388

21. 9.9985 **22.** 1.0006 **23.** $\dfrac{3}{7}$ **24.** $\dfrac{5}{9}$ **25.** $\dfrac{521}{15}$ **26.** $\dfrac{81}{5}$

In Problems 27–36, write each statement using symbols.

27. The sum of 3 and 2 equals 5.
29. The sum of x and 2 is the product of 3 and 4.
31. The product of 3 and y is the sum of 1 and 2.
33. x minus 2 equals 6.
35. x divided by 2 is 6.

28. The product of 5 and 2 equals 10.
30. The sum of 3 and y is the sum of 2 and 2.
32. The product of 2 and x is the product of 4 and 6.
34. 2 minus y equals 6.
36. 2 divided by x is 6.

In Problems 37–70, evaluate each expression.

37. $9 - 4 + 2$

38. $6 - 4 + 3$

39. $-6 + 4 \cdot 3$

40. $8 - 4 \cdot 2$

41. $4 + 5 - 8$

42. $8 - 3 - 4$

43. $4 + \dfrac{1}{3}$

44. $2 - \dfrac{1}{2}$

45. $6 - [3 \cdot 5 + 2 \cdot (3 - 2)]$

46. $2 \cdot [8 - 3(4 + 2)] - 3$

47. $2 \cdot (3 - 5) + 8 \cdot 2 - 1$

48. $1 - (4 \cdot 3 - 2 + 2)$

49. $10 - [6 - 2 \cdot 2 + (8 - 3)] \cdot 2$

50. $2 - 5 \cdot 4 - [6 \cdot (3 - 4)]$

51. $(5 - 3)\dfrac{1}{2}$

52. $(5 + 4)\dfrac{1}{3}$

53. $\dfrac{4 + 8}{5 - 3}$

54. $\dfrac{2 - 4}{5 - 3}$

55. $\dfrac{3}{5} \cdot \dfrac{10}{21}$

56. $\dfrac{5}{9} \cdot \dfrac{3}{10}$

57. $\dfrac{6}{25} \cdot \dfrac{10}{27}$

58. $\dfrac{21}{25} \cdot \dfrac{100}{3}$

59. $\dfrac{3}{4} + \dfrac{2}{5}$

60. $\dfrac{4}{3} + \dfrac{1}{2}$

61. $\dfrac{5}{6} + \dfrac{9}{5}$

62. $\dfrac{8}{9} + \dfrac{15}{2}$

63. $\dfrac{5}{18} + \dfrac{1}{12}$

64. $\dfrac{2}{15} + \dfrac{8}{9}$

65. $\dfrac{1}{30} - \dfrac{7}{18}$

66. $\dfrac{3}{14} - \dfrac{2}{21}$

67. $\dfrac{3}{20} - \dfrac{2}{15}$

68. $\dfrac{6}{35} - \dfrac{3}{14}$

69. $\dfrac{\frac{5}{18}}{\frac{11}{27}}$

70. $\dfrac{\frac{5}{21}}{\frac{2}{35}}$

In Problems 71–82, use the Distributive Property to remove the parentheses.

71. $6(x + 4)$ **72.** $4(2x - 1)$ **73.** $x(x - 4)$ **74.** $4x(x + 3)$
75. $(x + 2)(x + 4)$ **76.** $(x + 5)(x + 1)$ **77.** $(x - 2)(x + 1)$ **78.** $(x - 4)(x + 1)$
79. $(x - 8)(x - 2)$ **80.** $(x - 4)(x - 2)$ **81.** $(x + 2)(x - 2)$ **82.** $(x - 3)(x + 3)$

Discussion and Writing

83. Explain to a friend how the Distributive Property is used to justify the fact that $2x + 3x = 5x$.

84. Explain to a friend why $2 + 3 \cdot 4 = 14$, whereas $(2 + 3) \cdot 4 = 20$.

85. Explain why $2(3 \cdot 4)$ is not equal to $(2 \cdot 3) \cdot (2 \cdot 4)$.

86. Explain why $\dfrac{4 + 3}{2 + 5}$ is not equal to $\dfrac{4}{2} + \dfrac{3}{5}$.

87. Is subtraction commutative? Support your conclusion with an example.

88. Is subtraction associative? Support your conclusion with an example.

89. Is division commutative? Support your conclusion with an example.

90. Is division associative? Support your conclusion with an example.

91. If $2 = x$, why does $x = 2$?

92. If $x = 5$, why does $x^2 + x = 30$?

93. Are there any real numbers that are both rational and irrational? Are there any real numbers that are neither? Explain your reasoning.

94. Explain why the sum of a rational number and an irrational number must be irrational.

95. What rational number does the repeating decimal $0.9999\ldots$ equal?

R.2 Algebra Review

OBJECTIVES
1 Graph Inequalities
2 Find Distance on the Real Number Line
3 Evaluate Algebraic Expressions
4 Determine the Domain of a Variable
5 Use the Laws of Exponents
6 Evaluate Square Roots
7 Use a Graphing Calculator to Evaluate Exponents
8 Use Scientific Notation

Real numbers can be represented by points on a line called the **real number line**. There is a one-to-one correspondence between real numbers and points on a line. That is, every real number corresponds to a point on the line, and each point on the line has a unique real number associated with it.

Pick a point on the line somewhere in the center, and label it O. This point, called the **origin**, corresponds to the real number 0. See Figure 7. The point one unit to the right of O corresponds to the number 1. The distance between 0 and 1 determines the **scale** of the number line. For example, the point associated with the number 2 is twice as far from O as 1 is. Notice that an arrowhead on the right end of the line indicates the direction in which the numbers increase. Figure 7 also shows the points associated with the irrational numbers $\sqrt{2}$ and π. Points to the left of the origin correspond to the real numbers -1, -2, and so on.

Figure 7
Real number line

The real number associated with a point P is called the **coordinate** of P, and the line whose points have been assigned coordinates is called the **real number line**.

✏ NOW WORK PROBLEM 11.

The real number line consists of three classes of real numbers, as shown in Figure 8.

Figure 8

1. The **negative real numbers** are the coordinates of points to the left of the origin O.
2. The real number **zero** is the coordinate of the origin O.
3. The **positive real numbers** are the coordinates of points to the right of the origin O.

Negative and positive numbers have the following multiplication properties:

Multiplication Properties of Positive and Negative Numbers

1. The product of two positive numbers is a positive number.
2. The product of two negative numbers is a positive number.
3. The product of a positive number and a negative number is a negative number.

1   Graph Inequalities

Figure 9

(a) $a < b$

(b) $a = b$

(c) $a > b$

An important property of the real number line follows from the fact that, given two numbers (points) a and b, either a is to the left of b, a is at the same location as b, or a is to the right of b. See Figure 9.

If a is to the left of b, we say that "a is less than b" and write $a < b$. If a is to the right of b, we say that "a is greater than b" and write $a > b$. If a is at the same location as b, then $a = b$. If a is either less than or equal to b, we write $a \leq b$. Similarly, $a \geq b$ means that a is either greater than or equal to b. Collectively, the symbols $<, >, \leq$, and $\geq$ are called **inequality symbols**.

Note that $a < b$ and $b > a$ mean the same thing. It does not matter whether we write $2 < 3$ or $3 > 2$.

Furthermore, if $a < b$ or if $b > a$, then the difference $b - a$ is positive. Do you see why?

EXAMPLE 1   **Using Inequality Symbols**

(a) $3 < 7$ (b) $-8 > -16$ (c) $-6 < 0$

(d) $-8 < -4$ (e) $4 > -1$ (f) $8 > 0$ ◀

In Example 1(a), we conclude that $3 < 7$ either because 3 is to the left of 7 on the real number line or because the difference, $7 - 3 = 4$, is a positive real number.

Similarly, we conclude in Example 1(b) that $-8 > -16$ either because -8 lies to the right of -16 on the real number line or because the difference, $-8 - (-16) = -8 + 16 = 8$, is a positive real number.

Look again at Example 1. Note that the inequality symbol always points in the direction of the smaller number.

Statements of the form $a < b$ or $b > a$ are called **strict inequalities**, whereas statements of the form $a \leq b$ or $b \geq a$ are called **nonstrict inequalities**. An **inequality** is a statement in which two expressions are related by an inequality symbol. The expressions are referred to as the **sides** of the inequality.

Based on the discussion thus far, we conclude that

$$a > 0 \quad \text{is equivalent to } a \text{ is positive}$$
$$a < 0 \quad \text{is equivalent to } a \text{ is negative}$$

We sometimes read $a > 0$ by saying that "a is positive." If $a \geq 0$, then either $a > 0$ or $a = 0$, and we may read this as "a is nonnegative."

NOW WORK PROBLEMS 15 AND 25.

We shall find it useful in later work to graph inequalities on the real number line.

EXAMPLE 2   **Graphing Inequalities**

(a) On the real number line, graph all numbers x for which $x > 4$.

(b) On the real number line, graph all numbers x for which $x \leq 5$.

Figure 10   **Solution**

-2 -1 0 1 2 3 4 5 6 7

Figure 11

-2 -1 0 1 2 3 4 5 6 7

(a) See Figure 10. Notice that we use a left parenthesis to indicate that the number 4 is not part of the graph.

(b) See Figure 11. Notice that we use a right bracket to indicate that the number 5 is part of the graph.   ◀

NOW WORK PROBLEM 31.

Figure 12

2 Find Distance on the Real Number Line

For this objective we introduce *absolute value*. The absolute value of a number a is the distance from 0 to a on the number line. For example, -4 is 4 units from 0, and 3 is 3 units from 0. See Figure 12. Thus, the absolute value of -4 is 4, and the absolute value of 3 is 3.

A more formal definition of absolute value is given next.

The **absolute value** of a real number a, denoted by the symbol $|a|$, is defined by the rules

$$|a| = a \quad \text{if } a \geq 0 \qquad \text{and} \qquad |a| = -a \quad \text{if } a < 0$$

For example, since $-4 < 0$, the second rule must be used to get $|-4| = -(-4) = 4$.

| EXAMPLE 3 | Computing Absolute Value |

(a) $|8| = 8$ (b) $|0| = 0$ (c) $|-15| = -(-15) = 15$ ◄

Look again at Figure 12. The distance from -4 to 3 is 7 units. This distance is the difference $3 - (-4)$, obtained by subtracting the smaller coordinate from the larger. However, since $|3 - (-4)| = |7| = 7$ and $|-4 - 3| = |-7| = 7$, we can use absolute value to calculate the distance between two points without being concerned about which is smaller.

If P and Q are two points on a real number line with coordinates a and b, respectively, the **distance between P and Q**, denoted by $d(P, Q)$, is

$$d(P, Q) = |b - a|$$

Since $|b - a| = |a - b|$, it follows that $d(P, Q) = d(Q, P)$.

| EXAMPLE 4 | Finding Distance on a Number Line |

Let P, Q, and R be points on a real number line with coordinates $-5, 7$, and -3, respectively. Find the distance

(a) between P and Q (b) between Q and R

Solution See Figure 13.

Figure 13

(a) $d(P, Q) = |7 - (-5)| = |12| = 12$
(b) $d(Q, R) = |-3 - 7| = |-10| = 10$ ◄

NOW WORK PROBLEM **37**.

3 Evaluate Algebraic Expressions

As we said earlier, in algebra we use letters such as $x, y, a, b,$ and c to represent numbers. If the letter used is to represent *any* number from a given set of numbers, it is called a **variable**. A **constant** is either a fixed number, such as 5 or $\sqrt{3}$, or a letter that represents a fixed (possibly unspecified) number.

Constants and variables are combined using the operations of addition, subtraction, multiplication, and division to form *algebraic expressions*. Examples of algebraic expressions include

$$x + 3 \qquad \frac{3}{1 - t} \qquad 7x - 2y$$

To evaluate an algebraic expression, substitute for each variable its numerical value.

EXAMPLE 5	**Evaluating an Algebraic Expression**

Evaluate each expression if $x = 3$ and $y = -1$.

(a) $x + 3y$ \qquad (b) $5xy$ \qquad (c) $\dfrac{3y}{2 - 2x}$ \qquad (d) $|-4x + y|$

Solution

(a) Substitute 3 for x and -1 for y in the expression $x + 3y$.

$$x + 3y = 3 + 3(-1) = 3 + (-3) = 0$$
$$\uparrow$$
$$x = 3, y = -1$$

(b) If $x = 3$ and $y = -1$, then

$$5xy = 5(3)(-1) = -15$$

(c) If $x = 3$ and $y = -1$, then

$$\frac{3y}{2 - 2x} = \frac{3(-1)}{2 - 2(3)} = \frac{-3}{2 - 6} = \frac{-3}{-4} = \frac{3}{4}$$

(d) If $x = 3$ and $y = -1$, then

$$|-4x + y| = |-4(3) + (-1)| = |-12 + (-1)| = |-13| = 13 \qquad \blacktriangleleft$$

A graphing calculator can be used to evaluate an algebraic expression. Figure 14 shows the results of Example 5 using a TI-84 Plus.

Figure 14

(a)

(b)

(c) \qquad\qquad (d)

NOW WORK PROBLEMS 39 AND 47.

4 Determine the Domain of a Variable

In working with expressions or formulas involving variables, the variables may be allowed to take on values from only a certain set of numbers. For example, in the formula for the area A of a circle of radius r, $A = \pi r^2$, the variable r is necessarily restricted to the positive real numbers. In the expression $\dfrac{1}{x}$, the variable x cannot take on the value 0, since division by 0 is not defined.

The set of values that a variable may assume is called the **domain of the variable**.

EXAMPLE 6   **Finding the Domain of a Variable**

The domain of the variable x in the expression

$$\frac{5}{x - 2}$$

is $\{x \mid x \neq 2\}$, since, if $x = 2$, the denominator becomes 0, which is not defined.   ◄

EXAMPLE 7   **Circumference of a Circle**

In the formula for the circumference C of a circle of radius r,

$$C = 2\pi r$$

the domain of the variable r, representing the radius of the circle, is the set of positive real numbers. The domain of the variable C, representing the circumference of the circle, is also the set of positive real numbers.   ◄

In describing the domain of a variable, we may use either set notation or words, whichever is more convenient.

NOW WORK PROBLEM 57.

5 Use the Laws of Exponents

Integer exponents provide a shorthand device for representing repeated multiplications of a real number. For example,

$$3^4 = 3 \cdot 3 \cdot 3 \cdot 3 = 81$$

Additionally, many formulas have exponents. For example,

- The formula for the horsepower rating H of an engine is

$$H = \frac{D^2 N}{2.5}$$

where D is the diameter of a cylinder and N is the number of cylinders.

- A formula for the resistance R of blood flowing in a blood vessel is

$$R = C\frac{L}{r^4}$$

where L is the length of the blood vessel, r is the radius, and C is a positive constant.

If a is a real number and n is a positive integer, then the symbol a^n represents the product of n factors of a. That is,

$$a^n = \underbrace{a \cdot a \cdot \ldots \cdot a}_{n \text{ factors}} \qquad (1)$$

Here it is understood that $a^1 = a$.

Then $a^2 = a \cdot a$, $a^3 = a \cdot a \cdot a$, and so on. In the expression a^n, a is called the **base** and n is called the **exponent**, or **power**. We read a^n as "a raised to the power n" or as "a to the nth power." We usually read a^2 as "a squared" and a^3 as "a cubed."

In working with exponents, the operation of *raising to a power* is performed before any other operation. As examples,

$$4 \cdot 3^2 = 4 \cdot 9 = 36 \qquad 2^2 + 3^2 = 4 + 9 = 13$$
$$-2^4 = -1 \cdot 2^4 = -16 \qquad 5 \cdot 3^2 + 2 \cdot 4 = 5 \cdot 9 + 2 \cdot 4 = 45 + 8 = 53$$

Parentheses are used to indicate operations to be performed first. For example,

$$(-2)^4 = (-2)(-2)(-2)(-2) = 16 \qquad (2 + 3)^2 = 5^2 = 25$$

If $a \neq 0$, we define

$$a^0 = 1 \quad \text{if } a \neq 0$$

If $a \neq 0$ and if n is a positive integer, then we define

$$a^{-n} = \frac{1}{a^n} \quad \text{if } a \neq 0$$

Whenever you encounter a negative exponent, think "reciprocal."

EXAMPLE 8	**Evaluating Expressions Containing Negative Exponents**

(a) $2^{-3} = \dfrac{1}{2^3} = \dfrac{1}{8}$ (b) $x^{-4} = \dfrac{1}{x^4}$ (c) $\left(\dfrac{1}{5}\right)^{-2} = \dfrac{1}{\left(\dfrac{1}{5}\right)^2} = \dfrac{1}{\dfrac{1}{25}} = 25$ ◀

NOW WORK PROBLEMS 75 AND 95.

The following properties, called the **Laws of Exponents**, can be proved using the preceding definitions. In the list, a and b are real numbers, and m and n are integers.

Laws of Exponents

$$a^m a^n = a^{m+n} \qquad (a^m)^n = a^{mn} \qquad (ab)^n = a^n b^n$$
$$\frac{a^m}{a^n} = a^{m-n} = \frac{1}{a^{n-m}}, \quad \text{if } a \neq 0 \qquad \left(\frac{a}{b}\right)^n = \frac{a^n}{b^n}, \quad \text{if } b \neq 0$$

| EXAMPLE 9 | **Using the Laws of Exponents** |

(a) $x^{-3} \cdot x^5 = x^{-3+5} = x^2, \quad x \neq 0$

(b) $(x^{-3})^2 = x^{-3\cdot2} = x^{-6} = \dfrac{1}{x^6}, \quad x \neq 0$

(c) $(2x)^3 = 2^3 \cdot x^3 = 8x^3$

(d) $\left(\dfrac{2}{3}\right)^4 = \dfrac{2^4}{3^4} = \dfrac{16}{81}$ (e) $\dfrac{x^{-2}}{x^{-5}} = x^{-2-(-5)} = x^3, \quad x \neq 0$ ◀

NOW WORK PROBLEM 77.

| EXAMPLE 10 | **Using the Laws of Exponents** |

Write each expression so that all exponents are positive.

(a) $\dfrac{x^5 y^{-2}}{x^3 y}, \quad x \neq 0, \quad y \neq 0$ (b) $\left(\dfrac{x^{-3}}{3y^{-1}}\right)^{-2}, \quad x \neq 0, \quad y \neq 0$

Solution   (a) $\dfrac{x^5 y^{-2}}{x^3 y} = \dfrac{x^5}{x^3} \cdot \dfrac{y^{-2}}{y} = x^{5-3} \cdot y^{-2-1} = x^2 y^{-3} = x^2 \cdot \dfrac{1}{y^3} = \dfrac{x^2}{y^3}$

(b) $\left(\dfrac{x^{-3}}{3y^{-1}}\right)^{-2} = \dfrac{(x^{-3})^{-2}}{(3y^{-1})^{-2}} = \dfrac{x^6}{3^{-2}(y^{-1})^{-2}} = \dfrac{x^6}{\dfrac{1}{9}y^2} = \dfrac{9x^6}{y^2}$ ◀

NOW WORK PROBLEM 87.

6 Evaluate Square Roots

A real number is squared when it is raised to the power 2. The inverse of squaring is finding a **square root**. For example, since $6^2 = 36$ and $(-6)^2 = 36$, the numbers 6 and -6 are square roots of 36.

The symbol $\sqrt{}$, called a **radical sign**, is used to denote the **principal**, or nonnegative, square root. For example, $\sqrt{36} = 6$.

In general, if a is a nonnegative real number, the nonnegative number b, such that $b^2 = a$ is the **principal square root** of a, is denoted by $b = \sqrt{a}$.

The following comments are noteworthy:

1. Negative numbers do not have square roots (in the real number system), because the square of any real number is *nonnegative*. For example, $\sqrt{-4}$ is not a real number, because there is no real number whose square is -4.

2. The principal square root of 0 is 0, since $0^2 = 0$. That is, $\sqrt{0} = 0$.

3. The principal square root of a positive number is positive.

4. If $c \geq 0$, then $(\sqrt{c})^2 = c$. For example, $(\sqrt{2})^2 = 2$ and $(\sqrt{3})^2 = 3$.

| EXAMPLE 11 | **Evaluating Square Roots** |

(a) $\sqrt{64} = 8$ (b) $\sqrt{\dfrac{1}{16}} = \dfrac{1}{4}$ (c) $\left(\sqrt{1.4}\right)^2 = 1.4$

(d) $\sqrt{(-3)^2} = |-3| = 3$ ◄

Examples 11(a) and (b) are examples of square roots of perfect squares, since $64 = 8^2$ and $\dfrac{1}{16} = \left(\dfrac{1}{4}\right)^2$.

Notice the need for the absolute value in Example 11(d). Since $a^2 \geq 0$, the principal square root of a^2 is defined whether $a > 0$ or $a < 0$. However, since the principal square root is nonnegative, we need the absolute value to ensure the nonnegative result.

In general, we have

$$\sqrt{a^2} = |a| \qquad \qquad \textbf{(2)}$$

| EXAMPLE 12 | **Using Equation (2)** |

(a) $\sqrt{(2.3)^2} = |2.3| = 2.3$ (b) $\sqrt{(-2.3)^2} = |-2.3| = 2.3$

(c) $\sqrt{x^2} = |x|$ ◄

━━━ **NOW WORK PROBLEM 83.**

7 Use a Graphing Calculator to Evaluate Exponents

Your calculator has a caret key, $\boxed{\wedge}$, which is used for computations involving exponents.

| EXAMPLE 13 | **Exponents on a Graphing Calculator** |

Evaluate: $(2.3)^5$

Figure 15

Solution   Figure 15 shows the result using a TI-84 Plus graphing calculator.   ◄

Figure 15

```
2.3^5
         64.36343
```

━━━ **NOW WORK PROBLEM 113.**

8 Use Scientific Notation

Measurements of physical quantities can range from very small to very large. For example, the mass of a proton is approximately 0.00000000000000000000000000167 kilogram and the mass of Earth is about 5,980,000,000,000,000,000,000,000 kilograms. These numbers obviously are tedious to write down and difficult to read, so we use exponents to rewrite each.

When a number has been written as the product of a number x, where $1 \leq x < 10$, times a power of 10, it is said to be written in **scientific notation**.

In scientific notation,

$$\text{Mass of a proton} = 1.67 \times 10^{-27} \text{ kilogram}$$
$$\text{Mass of Earth} = 5.98 \times 10^{24} \text{ kilograms}$$

Converting a Decimal to Scientific Notation

To change a positive number into scientific notation:

1. Count the number N of places that the decimal point must be moved in order to arrive at a number x, where $1 \le x < 10$.

2. If the original number is greater than or equal to 1, the scientific notation is $x \times 10^N$. If the original number is between 0 and 1, the scientific notation is $x \times 10^{-N}$.

EXAMPLE 14   **Using Scientific Notation**

Write each number in scientific notation.

(a) 9582   (b) 1.245   (c) 0.285   (d) 0.000561

Solution   (a) The decimal point in 9582 follows the 2. Thus, we count

$$9 \; 5 \; 8 \; 2 \; .$$
$$3 \; 2 \; 1$$

stopping after three moves, because 9.582 is a number between 1 and 10. Since 9582 is greater than 1, we write

$$9582 = 9.582 \times 10^3$$

(b) The decimal point in 1.245 is between the 1 and 2. Since the number is already between 1 and 10, the scientific notation for it is $1.245 \times 10^0 = 1.245$.

(c) The decimal point in 0.285 is between the 0 and the 2. We count

$$0 \; . \; 2 \; 8 \; 5$$
$$1$$

stopping after one move, because 2.85 is a number between 1 and 10. Since 0.285 is between 0 and 1, we write

$$0.285 = 2.85 \times 10^{-1}$$

(d) The decimal point in 0.000561 is moved as follows:

$$0 \; . \; 0 \; 0 \; 0 \; 5 \; 6 \; 1$$
$$1 \; 2 \; 3 \; 4$$

As a result,

$$0.000561 = 5.61 \times 10^{-4} \qquad \blacktriangleleft$$

✏ **NOW WORK PROBLEM 119.**

EXAMPLE 15   **Changing from Scientific Notation to Decimals**

Write each number as a decimal.

(a) 2.1×10^4 (b) 3.26×10^{-5} (c) 1×10^{-2}

Solution (a) $2.1 \times 10^4 = 2$. 1 0 0 0 $\times 10^4 = 21{,}000$

                                     1 2 3 4

(b) $3.26 \times 10^{-5} = 0$ 0 0 0 0 3 . 2 $6 \times 10^{-5} = 0.0000326$

                              5 4 3 2 1

(c) $1 \times 10^{-2} = 0$ 0 1 . $0 \times 10^{-2} = 0.01$

                            2 1

On a graphing calculator, a number such as 3.615×10^{12} is usually displayed as 3.615E12 .

NOW WORK PROBLEM 127.

EXAMPLE 16 | **Using Scientific Notation**

(a) The diameter of the smallest living cell is only about 0.00001 centimeter (cm).[*] Express this number in scientific notation.

(b) The surface area of Earth is about 1.97×10^8 square miles.[†] Express the surface area as a whole number.

Solution (a) 0.00001 cm $= 1 \times 10^{-5}$ cm because the decimal point is moved five places and the number is less than 1.

(b) 1.97×10^8 square miles $= 197{,}000{,}000$ square miles

NOW WORK PROBLEM 151.

[*]Powers of Ten, Philip and Phylis Morrison.
[†]2005 Information Please Almanac.

HISTORICAL FEATURE

The word *algebra* is derived from the Arabic word *al-jabr*. This word is a part of the title of a ninth-century work, "Hisâb al-jabr w'al-muqâbalah," written by Mohammed ibn Mûsâ al-Khowârizmî. The word *al-jabr* means "a restoration," a reference to the fact that, if a number is added to one side of an equation, then it must also be added to the other side in order to "restore" the equality. The title of the work, freely translated, is "The Science of Reduction and Cancellation." Of course, today, algebra has come to mean a great deal more.

R.2 Assess Your Understanding

Concepts and Vocabulary

1. A _____ is a letter used in algebra to represent any number from a given set of numbers.

2. On the real number line, the real number zero is the coordinate of the _____.

3. An inequality of the form $a > b$ is called a _____ inequality.

4. In the expression 2^4, the number 2 is called the _____ and 4 is called the _____.

5. In scientific notation, $1234.5678 =$ _____.

6. *True or False:* The product of two negative real numbers is always greater than zero.

7. *True or False*: The distance between two points on the real number line is always greater than zero.

8. *True or False*: The absolute value of a real number is always greater than zero.

9. *True or False:* When a number is expressed in scientific notation, it is expressed as the product of a number x, $0 \le x < 1$, and a power of 10.

10. *True or False:* To multiply two expressions having the same base, retain the base and multiply the exponents.

Skill Building

11. On the real number line, label the points with coordinates $0, 1, -1, \dfrac{5}{2}, -2.5, \dfrac{3}{4}$, and 0.25.

12. Repeat Problem 11 for the coordinates $0, -2, 2, -1.5, \dfrac{3}{2}, \dfrac{1}{3}$, and $\dfrac{2}{3}$.

In Problems 13–22, replace the question mark by $<$, $>$, or $=$, whichever is correct.

13. $\dfrac{1}{2}\,?\,0$

14. $5\,?\,6$

15. $-1\,?\,-2$

16. $-3\,?\,-\dfrac{5}{2}$

17. $\pi\,?\,3.14$

18. $\sqrt{2}\,?\,1.41$

19. $\dfrac{1}{2}\,?\,0.5$

20. $\dfrac{1}{3}\,?\,0.33$

21. $\dfrac{2}{3}\,?\,0.67$

22. $\dfrac{1}{4}\,?\,0.25$

In Problems 23–28, write each statement as an inequality.

23. x is positive

24. z is negative

25. x is less than 2

26. y is greater than -5

27. x is less than or equal to 1

28. x is greater than or equal to 2

In Problems 29–32, graph the numbers x on the real number line.

29. $x \geq -2$

30. $x < 4$

31. $x > -1$

32. $x \leq 7$

In Problems 33–38, use the real number line given below to compute each distance.

$$
\begin{array}{ccccccccc}
 & A & & B & C & D & & E & \\
\end{array}
$$

A B C D E

$-4\ -3\ -2\ -1\ \ 0\ \ 1\ \ 2\ \ 3\ \ 4\ \ 5\ \ 6$

33. $d(C, D)$

34. $d(C, A)$

35. $d(D, E)$

36. $d(C, E)$

37. $d(A, E)$

38. $d(D, B)$

In Problems 39–46, evaluate each expression if $x = -2$ and $y = 3$. Verify your results using a graphing utility

39. $x + 2y$

40. $3x + y$

41. $5xy + 2$

42. $-2x + xy$

43. $\dfrac{2x}{x - y}$

44. $\dfrac{x + y}{x - y}$

45. $\dfrac{3x + 2y}{2 + y}$

46. $\dfrac{2x - 3}{y}$

In Problems 47–56, find the value of each expression if $x = 3$ and $y = -2$. Verify your results using a graphing utility.

47. $|x + y|$

48. $|x - y|$

49. $|x| + |y|$

50. $|x| - |y|$

51. $\dfrac{|x|}{x}$

52. $\dfrac{|y|}{y}$

53. $|4x - 5y|$

54. $|3x + 2y|$

55. $\|4x| - |5y\|$

56. $3|x| + 2|y|$

In Problems 57–64, determine which value(s) given below, if any, must be excluded from the domain of the variable in each expression.

(a) $x = 3$ (b) $x = 1$ (c) $x = 0$ (d) $x = -1$

57. $\dfrac{x^2 - 1}{x}$

58. $\dfrac{x^2 + 1}{x}$

59. $\dfrac{x}{x^2 - 9}$

60. $\dfrac{x}{x^2 + 9}$

61. $\dfrac{x^2}{x^2 + 1}$

62. $\dfrac{x^3}{x^2 - 1}$

63. $\dfrac{x^2 + 5x - 10}{x^3 - x}$

64. $\dfrac{-9x^2 - x + 1}{x^3 + x}$

In Problems 65–68, determine the domain of the variable x in each expression.

65. $\dfrac{4}{x - 5}$

66. $\dfrac{-6}{x + 4}$

67. $\dfrac{x}{x + 4}$

68. $\dfrac{x - 2}{x - 6}$

In Problems 69–72, use the formula $C = \dfrac{5}{9}(F - 32)$ for converting degrees Fahrenheit into degrees Celsius to find the Celsius measure of each Fahrenheit temperature.

69. $F = 32°$

70. $F = 212°$

71. $F = 77°$

72. $F = -4°$

In Problems 73–84, simplify each expression.

73. $(-4)^2$

74. -4^2

75. 4^{-2}

76. -4^{-2}

77. $3^{-6} \cdot 3^4$

78. $4^{-2} \cdot 4^3$

79. $(3^{-2})^{-1}$

80. $(2^{-1})^{-3}$

81. $\sqrt{25}$

82. $\sqrt{36}$

83. $\sqrt{(-4)^2}$

84. $\sqrt{(-3)^2}$

In Problems 85–94, simplify each expression. Express the answer so that all exponents are positive. Whenever an exponent is 0 or negative, we assume that the base is not 0.

85. $(8x^3)^2$

86. $(-4x^2)^{-1}$

87. $(x^2y^{-1})^2$

88. $(x^{-1}y)^3$

89. $\dfrac{x^2y^3}{xy^4}$

90. $\dfrac{x^{-2}y}{xy^2}$

91. $\dfrac{(-2)^3x^4(yz)^2}{3^2xy^3z}$

92. $\dfrac{4x^{-2}(yz)^{-1}}{2^3x^4y}$

93. $\left(\dfrac{3x^{-1}}{4y^{-1}}\right)^{-2}$

94. $\left(\dfrac{5x^{-2}}{6y^{-2}}\right)^{-3}$

In Problems 95–106, find the value of each expression if $x = 2$ and $y = -1$. Verify your results using a graphing utility.

95. $2xy^{-1}$

96. $-3x^{-1}y$

97. $x^2 + y^2$

98. x^2y^2

99. $(xy)^2$

100. $(x + y)^2$

101. $\sqrt{x^2}$

102. $(\sqrt{x})^2$

103. $\sqrt{x^2 + y^2}$

104. $\sqrt{x^2} + \sqrt{y^2}$

105. x^y

106. y^x

107. Find the value of the expression $2x^3 - 3x^2 + 5x - 4$ if $x = 2$. What is the value if $x = 1$?

108. Find the value of the expression $4x^3 + 3x^2 - x + 2$ if $x = 1$. What is the value if $x = 2$?

109. What is the value of $\dfrac{(666)^4}{(222)^4}$?

110. What is the value of $(0.1)^3(20)^3$?

In Problems 111–118, use a calculator to evaluate each expression. Round your answer to three decimal places.

111. $(8.2)^6$

112. $(3.7)^5$

113. $(6.1)^{-3}$

114. $(2.2)^{-5}$

115. $(-2.8)^6$

116. $-(2.8)^6$

117. $(-8.11)^{-4}$

118. $-(8.11)^{-4}$

In Problems 119–126, write each number in scientific notation.

119. 454.2

120. 32.14

121. 0.013

122. 0.00421

123. 32,155

124. 21,210

125. 0.000423

126. 0.0514

In Problems 127–134, write each number as a decimal.

127. 6.15×10^4

128. 9.7×10^3

129. 1.214×10^{-3}

130. 9.88×10^{-4}

131. 1.1×10^8

132. 4.112×10^2

133. 8.1×10^{-2}

134. 6.453×10^{-1}

Applications and Extensions

In Problems 135–144, express each statement as an equation involving the indicated variables.

135. Area of a Rectangle The area A of a rectangle is the product of its length l and its width w.

136. Perimeter of a Rectangle The perimeter P of a rectangle is twice the sum of its length l and its width w.

137. Circumference of a Circle The circumference C of a circle is the product of π and its diameter d.

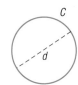

138. Area of a Triangle The area A of a triangle is one-half the product of its base b and its height h.

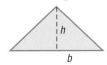

139. Area of an Equilateral Triangle The area A of an equilateral triangle is $\dfrac{\sqrt{3}}{4}$ times the square of the length x of one side.

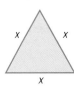

140. Perimeter of an Equilateral Triangle The perimeter P of an equilateral triangle is 3 times the length x of one side.

141. Volume of a Sphere The volume V of a sphere is $\dfrac{4}{3}$ times π times the cube of the radius r.

142. Surface Area of a Sphere The surface area S of a sphere is 4 times π times the square of the radius r.

143. Volume of a Cube   The volume V of a cube is the cube of the length x of a side.

144. Surface Area of a Cube   The surface area S of a cube is 6 times the square of the length x of a side.

145. Manufacturing Cost   The weekly production cost C of manufacturing x watches is given by the formula $C = 4000 + 2x$, where the variable C is in dollars.
(a) What is the cost of producing 1000 watches?
(b) What is the cost of producing 2000 watches?

146. Balancing a Checkbook   At the beginning of the month, Mike had a balance of $210 in his checking account. During the next month, he deposited $80, wrote a check for $120, made another deposit of $25, wrote two checks for $60 and $32, and was assessed a monthly service charge of $5. What was his balance at the end of the month?

147. U.S. Voltage   In the United States, normal household voltage is 115 volts. It is acceptable for the actual voltage x to differ from normal by at most 5 volts. A formula that describes this is
$$|x - 115| \le 5$$
(a) Show that a voltage of 113 volts is acceptable.
(b) Show that a voltage of 109 volts is not acceptable.

148. Foreign Voltage   In other countries, normal household voltage is 220 volts. It is acceptable for the actual voltage x to differ from normal by at most 8 volts. A formula that describes this is
$$|x - 220| \le 8$$
(a) Show that a voltage of 214 volts is acceptable.
(b) Show that a voltage of 209 volts is not acceptable.

149. Making Precision Ball Bearings   The FireBall Company manufactures ball bearings for precision equipment. One of its products is a ball bearing with a stated radius of 3 centimeters (cm). Only ball bearings with a radius within 0.01 cm of this stated radius are acceptable. If x is the radius of a ball bearing, a formula describing this situation is
$$|x - 3| \le 0.01$$
(a) Is a ball bearing of radius $x = 2.999$ acceptable?
(b) Is a ball bearing of radius $x = 2.89$ acceptable?

150. Body Temperature   Normal human body temperature is 98.6°F. A temperature x that differs from normal by at least 1.5°F is considered unhealthy. A formula that describes this is
$$|x - 98.6| \ge 1.5$$
(a) Show that a temperature of 97°F is unhealthy.
(b) Show that a temperature of 100°F is not unhealthy.

151. Distance from Earth to Its Moon   The distance from Earth to the Moon is about 4×10^8 meters.[*] Express this distance as a whole number.

152. Height of Mt. Everest   The height of Mt. Everest is 8848 meters.[*] Express this height in scientific notation.

153. Wavelength of Visible Light   The wavelength of visible light is about 5×10^{-7} meter.[*] Express this wavelength as a decimal.

154. Diameter of an Atom   The diameter of an atom is about 1×10^{-10} meter.[*] Express this diameter as a decimal.

155. Diameter of Copper Wire   The smallest commercial copper wire is about 0.0005 inch in diameter.[†] Express this diameter using scientific notation.

156. Smallest Motor   The smallest motor ever made is less than 0.05 centimeter wide.[†] Express this width using scientific notation.

157. Astronomy   One light-year is defined by astronomers to be the distance that a beam of light will travel in 1 year (365 days). If the speed of light is 186,000 miles per second, how many miles are in a light-year? Express your answer in scientific notation.

158. Astronomy   How long does it take a beam of light to reach Earth from the Sun, when the Sun is 93,000,000 miles from Earth? Express your answer in seconds, using scientific notation.

159. Does $\dfrac{1}{3}$ equal 0.333? If not, which is larger? By how much?

160. Does $\dfrac{2}{3}$ equal 0.666? If not, which is larger? By how much?

Discussion and Writing

161. Is there a positive real number "closest" to 0?

162. I'm thinking of a number! It lies between 1 and 10; its square is rational and lies between 1 and 10. The number is larger than π. Correct to two decimal places (that is, truncated to two decimal places) name the number. Now think of your own number, describe it, and challenge a fellow student to name it.

163. Write a brief paragraph that illustrates the similarities and differences between "less than" ($<$) and "less than or equal to" ($\le$).

164. Give a reason why the statement $5 < 8$ is true.

[*]*Powers of Ten*, Philip and Phylis Morrison.
[†]*2005 Information Please Almanac.*

R.3 Geometry Review

OBJECTIVES   1 Use the Pythagorean Theorem and Its Converse
                2 Know Geometry Formulas

In this section we review some topics studied in geometry that we shall need in later chapters.

1 Use the Pythagorean Theorem and Its Converse

Figure 16

The Pythagorean Theorem is a statement about *right triangles*. A **right triangle** is one that contains a **right angle**, that is, an angle of 90°. The side of the triangle opposite the 90° angle is called the **hypotenuse**; the remaining two sides are called **legs**. In Figure 16 we have used c to represent the length of the hypotenuse and a and b to represent the lengths of the legs. Notice the use of the symbol $\ulcorner$ to show the 90° angle. We now state the Pythagorean Theorem.

Pythagorean Theorem   In a right triangle, the square of the length of the hypotenuse is equal to the sum of the squares of the lengths of the legs. That is, in the right triangle shown in Figure 16,

$$c^2 = a^2 + b^2 \qquad\qquad \textbf{(1)}$$

We prove this result at the end of the section.

EXAMPLE 1   **Finding the Hypotenuse of a Right Triangle**

In a right triangle, one leg is of length 4 and the other is of length 3. What is the length of the hypotenuse?

Solution   Since the triangle is a right triangle, we use the Pythagorean Theorem with $a = 4$ and $b = 3$ to find the length c of the hypotenuse. From equation (1), we have

$$c^2 = a^2 + b^2$$
$$c^2 = 4^2 + 3^2 = 16 + 9 = 25$$
$$c = \sqrt{25} = 5$$
◄

✐━━━━━ **NOW WORK PROBLEM 9.**

The converse of the Pythagorean Theorem is also true.

Converse of the Pythagorean Theorem   In a triangle, if the square of the length of one side equals the sum of the squares of the lengths of the other two sides, then the triangle is a right triangle. The 90° angle is opposite the longest side.

We prove this result at the end of this section.

| EXAMPLE 2 | **Verifying That a Triangle Is a Right Triangle** |

Show that a triangle whose sides are of lengths 5, 12, and 13 is a right triangle. Identify the hypotenuse.

Solution   We square the lengths of the sides.

$$5^2 = 25, \quad 12^2 = 144, \quad 13^2 = 169$$

Notice that the sum of the first two squares (25 and 144) equals the third square (169). Hence, the triangle is a right triangle. The longest side, 13, is the hypotenuse. See Figure 17. ◀

Figure 17

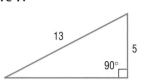

━━━ **NOW WORK PROBLEM 17.**

| EXAMPLE 3 | **Applying the Pythagorean Theorem** |

The tallest inhabited building in the world is the Sears Tower in Chicago. If the observation tower is 1450 feet above ground level, how far can a person standing in the observation tower see (with the aid of a telescope)? Use 3960 miles for the radius of Earth. See Figure 18.

Figure 18

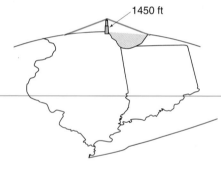

1450 ft

Source: Council on Tall Buildings and Urban Habitat (1997): Sears Tower No. 1 for tallest roof (1450 ft) and tallest occupied floor (1431 ft).

Solution   From the center of Earth, draw two radii: one through the Sears Tower and the other to the farthest point a person can see from the tower. See Figure 19. Apply the Pythagorean Theorem to the right triangle. Since 1 mile = 5280 feet, we have $1450 \text{ feet} = \dfrac{1450}{5280}$ miles. Then,

Figure 19

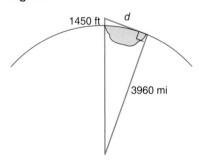

1450 ft d

3960 mi

$$d^2 + (3960)^2 = \left(3960 + \frac{1450}{5280} \right)^2$$

$$d^2 = \left(3960 + \frac{1450}{5280} \right)^2 - (3960)^2 \approx 2175.08$$

$$d \approx 46.64$$

A person can see about 47 miles from the observation tower. ◀

━━━ **NOW WORK PROBLEM 43.**

2   Know Geometry Formulas

Certain formulas from geometry are useful in solving algebra problems. We list some of these formulas next.

For a rectangle of length l and width w,

$$\text{Area} = lw \qquad \text{Perimeter} = 2l + 2w$$

For a triangle with base b and altitude h,

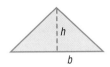

$$\text{Area} = \frac{1}{2}bh$$

For a circle of radius r (diameter $d = 2r$),

$$\text{Area} = \pi r^2 \qquad \text{Circumference} = 2\pi r = \pi d$$

For a closed rectangular box of length l, width w, and height h,

$$\text{Volume} = lwh \qquad \text{Surface area} = 2lw + 2lh + 2wh$$

For a sphere of radius r,

$$\text{Volume} = \frac{4}{3}\pi r^3 \qquad \text{Surface area} = 4\pi r^2$$

For a right circular cylinder of height h and radius r,

$$\text{Volume} = \pi r^2 h \qquad \text{Surface area} = 2\pi r^2 + 2\pi rh$$

NOW WORK PROBLEM **25.**

EXAMPLE 4	**Using Geometry Formulas**

A Christmas tree ornament is in the shape of a semicircle on top of a triangle. How many square centimeters (cm) of copper is required to make the ornament if the height of the triangle is 6 cm and the base is 4 cm?

Figure 20

Solution   See Figure 20. The amount of copper required equals the shaded area. This area is the sum of the area of the triangle and the semicircle. The triangle has height $h = 6$ and base $b = 4$. The semicircle has diameter $d = 4$, so its radius is $r = 2$.

$$\text{Area} = \text{Area of triangle} + \text{Area of semicircle}$$

$$= \frac{1}{2}bh + \frac{1}{2}\pi r^2 = \frac{1}{2}(4)(6) + \frac{1}{2}\pi 2^2 \qquad b = 4;\ h = 6;\ r = 2$$

$$= 12 + 2\pi \approx 18.28 \text{ cm}^2$$

About 18.28 cm^2 of copper is required.   ◀

NOW WORK PROBLEM **39.**

Proof of the Pythagorean Theorem We begin with a square, each side of length $a + b$. In this square, we can form four right triangles, each having legs equal in length to a and b. See Figure 21.

Figure 21

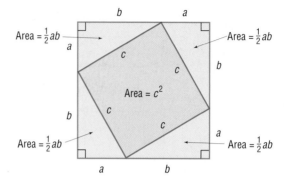

All these triangles are **congruent** (two sides and their included angle are equal). As a result, the hypotenuse of each is the same, say c, and the color shading in Figure 21 indicates a square with an area equal to c^2. The area of the original square with side $a + b$ equals the sum of the areas of the four triangles (each of area $\frac{1}{2}ab$) plus the area of the square with side c. Thus,

$$(a + b)^2 = \frac{1}{2}ab + \frac{1}{2}ab + \frac{1}{2}ab + \frac{1}{2}ab + c^2$$
$$a^2 + 2ab + b^2 = 2ab + c^2$$
$$a^2 + b^2 = c^2$$

Figure 22

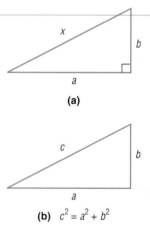

(a)

(b) $c^2 = a^2 + b^2$

The proof is complete. ∎

Proof of the Converse of the Pythagorean Theorem We begin with two triangles: one a right triangle with legs a and b and the other a triangle with sides a, b, and c for which $c^2 = a^2 + b^2$. See Figure 22. By the Pythagorean Theorem, the length x of the third side of the first triangle is

$$x^2 = a^2 + b^2$$

But $c^2 = a^2 + b^2$. Hence,

$$x^2 = c^2$$
$$x = c$$

The two triangles have the same sides and are therefore congruent; hence, corresponding angles are equal. Thus, the angle opposite side c of the second triangle equals 90°.

The proof is complete. ∎

R.3 Assess Your Understanding

Concepts and Vocabulary

1. A _____ triangle is one that contains an angle of 90 degrees. The longest side is called the _____.

2. For a triangle with base b and altitude h, a formula for the area A is _____.

3. The formula for the circumference C of a circle of radius r is _____.

4. *True or False:* In a right triangle, the square of the length of the longest side equals the sum of the squares of the lengths of the other two sides.

5. *True or False:* The triangle with sides of length 6, 8, and 10 is a right triangle.

6. *True or False:* The volume of a sphere of radius r is $\frac{4}{3}\pi r^2$.

Skill Building

In Problems 7–12, the lengths of the legs of a right triangle are given. Find the hypotenuse.

7. $a = 5, \quad b = 12$

8. $a = 6, \quad b = 8$

9. $a = 10, \quad b = 24$

10. $a = 4, \quad b = 3$

11. $a = 7, \quad b = 24$

12. $a = 14, \quad b = 48$

In Problems 13–20, the lengths of the sides of a triangle are given. Determine which are right triangles. For those that are, identify the hypotenuse.

13. $3, 4, 5$

14. $6, 8, 10$

15. $4, 5, 6$

16. $2, 2, 3$

17. $7, 24, 25$

18. $10, 24, 26$

19. $6, 4, 3$

20. $5, 4, 7$

21. Find the area A of a rectangle with length 4 inches and width 2 inches.

22. Find the area A of a rectangle with length 9 centimeters and width 4 centimeters.

23. Find the area A of a triangle with height 4 inches and base 2 inches.

24. Find the area A of a triangle with height 9 centimeters and base 4 centimeters.

25. Find the area A and circumference C of a circle of radius 5 meters.

26. Find the area A and circumference C of a circle of radius 2 feet.

27. Find the volume V and surface area S of a rectangular box with length 8 feet, width 4 feet, and height 7 feet.

28. Find the volume V and surface area S of a rectangular box with length 9 inches, width 4 inches, and height 8 inches.

29. Find the volume V and surface area S of a sphere of radius 4 centimeters.

30. Find the volume V and surface area S of a sphere of radius 3 feet.

31. Find the volume V and surface area S of a right circular cylinder with radius 9 inches and height 8 inches.

32. Find the volume V and surface area S of a right circular cylinder with radius 8 inches and height 9 inches.

Applications and Extensions

In Problems 33–36, find the area of the shaded region.

33.

34.

35.

36.
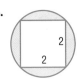

37. How many feet has a wheel with a diameter of 16 inches traveled after four revolutions?

38. How many revolutions will a circular disk with a diameter of 4 feet have completed after it has rolled 20 feet?

39. In the figure shown, $ABCD$ is a square, with each side of length 6 feet. The width of the border (shaded portion) between the outer square $EFGH$ and $ABCD$ is 2 feet. Find the area of the border.

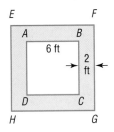

40. Refer to the figure. Square $ABCD$ has an area of 100 square feet; square $BEFG$ has an area of 16 square feet. What is the area of the triangle CGF?

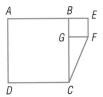

41. **Architecture**   A Norman window consists of a rectangle surmounted by a semicircle. Find the area of the Norman window shown in the illustration. How much wood frame is needed to enclose the window?

6'

4'

42. **Construction**   A circular swimming pool, 20 feet in diameter, is enclosed by a wooden deck that is 3 feet wide. What is the area of the deck? How much fence is required to enclose the deck?

20'   3'

In Problems 43–45, use the facts that the radius of Earth is 3960 miles and 1 mile = 5280 feet.

43. **How Far Can You See?**   The conning tower of the U.S.S. *Silversides*, a World War II submarine now permanently stationed in Muskegon, Michigan, is approximately 20 feet above sea level. How far can you see from the conning tower?

44. **How Far Can You See?**   A person who is 6 feet tall stands on the beach in Fort Lauderdale, Florida, and looks out onto the Atlantic Ocean. Suddenly, a ship appears on the horizon. How far is the ship from shore?

45. **How Far Can You See?**   The deck of a destroyer is 100 feet above sea level. How far can a person see from the deck? How far can a person see from the bridge, which is 150 feet above sea level?

46. Suppose that m and n are positive integers with $m > n$. If $a = m^2 - n^2, b = 2mn$, and $c = m^2 + n^2$, show that $a, b,$ and c are the lengths of the sides of a right triangle. (These formulas can be used to find the sides of a right triangle that are integers, such as 3, 4, 5; 5, 12, 13; and so on. Such triplets of integers are called **Pythagorean triples**.)

Discussion and Writing

47. You have 1000 feet of flexible pool siding and wish to construct a swimming pool. Experiment with rectangular-shaped pools with perimeters of 1000 feet. How do their areas vary? What is the shape of the rectangle with the largest area? Now compute the area enclosed by a circular pool with a perimeter (circumference) of 1000 feet. What would be your choice of shape for the pool? If rectangular, what is your preference for dimensions? Justify your choice. If your only consideration is to have a pool that encloses the most area, what shape should you use?

48. **The Gibb's Hill Lighthouse, Southampton, Bermuda,** in operation since 1846, stands 117 feet high on a hill 245 feet high, so its beam of light is 362 feet above sea level. A brochure states that the light itself can be seen on the horizon about 26 miles distant. Verify the correctness of this information. The brochure further states that ships 40 miles away can see the light and planes flying at 10,000 feet can see it 120 miles away. Verify the accuracy of these statements. What assumption did the brochure make about the height of the ship?

120 miles

40 miles

R.4 Polynomials

> **OBJECTIVES**
> 1 Recognize Monomials
> 2 Recognize Polynomials
> 3 Add and Subtract Polynomials
> 4 Multiply Polynomials
> 5 Know Formulas for Special Products
> 6 Divide Polynomials Using Long Division
> 7 Work with Polynomials in Two Variables

We have described algebra as a generalization of arithmetic in which letters are used to represent real numbers. From now on, we shall use the letters at the end of the alphabet, such as x, y, and z, to represent variables and the letters at the beginning of the alphabet, such as a, b, and c, to represent constants. In the expressions $3x + 5$ and $ax + b$, it is understood that x is a variable and that a and b are constants, even though the constants a and b are unspecified. As you will find out, the context usually makes the intended meaning clear.

1 Recognize Monomials

NOTE
The nonnegative integers are the integers 0, 1, 2, 3, ... ∎

A **monomial** in one variable is the product of a constant times a variable raised to a nonnegative integer power. A monomial is of the form

$$ax^k$$

where a is a constant, x is a variable, and $k \geq 0$ is an integer. The constant a is called the **coefficient** of the monomial. If $a \neq 0$, then k is called the **degree** of the monomial.

EXAMPLE 1 | **Examples of Monomials**

Monomial	Coefficient	Degree	
(a) $6x^2$	6	2	
(b) $-\sqrt{2}x^3$	$-\sqrt{2}$	3	
(c) 3	3	0	Since $3 = 3 \cdot 1 = 3x^0$, $x \neq 0$
(d) $-5x$	-5	1	Since $-5x = -5x^1$
(e) x^4	1	4	Since $x^4 = 1 \cdot x^4$

◀

Now let's look at some expressions that are not monomials.

EXAMPLE 2 | **Examples of Nonmonomial Expressions**

(a) $3x^{1/2}$ is not a monomial, since the exponent of the variable x is $\frac{1}{2}$ and $\frac{1}{2}$ is not a nonnegative integer.

(b) $4x^{-3}$ is not a monomial, since the exponent of the variable x is -3 and -3 is not a nonnegative integer. ◀

NOW WORK PROBLEM **7.**

2 Recognize Polynomials

Two monomials with the same variable raised to the same power are called **like terms**. For example, $2x^4$ and $-5x^4$ are like terms. In contrast, the monomials $2x^3$ and $2x^5$ are not like terms.

We can add or subtract like terms using the Distributive Property. For example,

$$2x^2 + 5x^2 = (2 + 5)x^2 = 7x^2 \quad \text{and} \quad 8x^3 - 5x^3 = (8 - 5)x^3 = 3x^3$$

The sum or difference of two monomials having different degrees is called a **binomial**. The sum or difference of three monomials with three different degrees is called a **trinomial**. For example,

$$x^2 - 2 \text{ is a binomial.}$$
$$x^3 - 3x + 5 \text{ is a trinomial.}$$
$$2x^2 + 5x^2 + 2 = 7x^2 + 2 \text{ is a binomial.}$$

In Words

A polynomial is a sum of monomials.

A **polynomial** in one variable is an algebraic expression of the form

$$a_n x^n + a_{n-1} x^{n-1} + \cdots + a_1 x + a_0 \qquad (1)$$

where $a_n, a_{n-1}, \ldots, a_1, a_0$ are constants,[*] called the **coefficients** of the polynomial, $n \geq 0$ is an integer, and x is a variable. If $a_n \neq 0$, it is called the **leading coefficient**, and n is called the **degree** of the polynomial.

The monomials that make up a polynomial are called its **terms**. If all the coefficients are 0, the polynomial is called the **zero polynomial**, which has no degree.

Polynomials are usually written in **standard form**, beginning with the nonzero term of highest degree and continuing with terms in descending order according to degree. If a power of x is missing, it is because its coefficient is zero.

EXAMPLE 3	Examples of Polynomials

Polynomial	Coefficients	Degree
$-8x^3 + 4x^2 - 6x + 2$	$-8, 4, -6, 2$	3
$3x^2 - 5 = 3x^2 + 0 \cdot x + (-5)$	$3, 0, -5$	2
$8 - 2x + x^2 = 1 \cdot x^2 + (-2)x + 8$	$1, -2, 8$	2
$5x + \sqrt{2} = 5x^1 + \sqrt{2}$	$5, \sqrt{2}$	1
$3 = 3 \cdot 1 = 3 \cdot x^0$	3	0
0	0	No degree ◀

Although we have been using x to represent the variable, letters such as y or z are also commonly used.

$3x^4 - x^2 + 2$ is a polynomial (in x) of degree 4.

$9y^3 - 2y^2 + y - 3$ is a polynomial (in y) of degree 3.

$z^5 + \pi$ is a polynomial (in z) of degree 5.

---

[*]The notation a_n is read as "a sub n." The number n is called a **subscript** and should not be confused with an exponent. We use subscripts to distinguish one constant from another when a large or undetermined number of constants is required.

Algebraic expressions such as

$$\frac{1}{x} \quad \text{and} \quad \frac{x^2 + 1}{x + 5}$$

are not polynomials. The first is not a polynomial because $\dfrac{1}{x} = x^{-1}$ has an exponent that is not a nonnegative integer. Although the second expression is the quotient of two polynomials, the polynomial in the denominator has degree greater than 0, so the expression cannot be a polynomial.

NOW WORK PROBLEM **17.**

3 Add and Subtract Polynomials

Polynomials are added and subtracted by combining like terms.

EXAMPLE 4	**Adding Polynomials**

Find the sum of the polynomials:

$$8x^3 - 2x^2 + 6x - 2 \quad \text{and} \quad 3x^4 - 2x^3 + x^2 + x$$

Solution   We shall find the sum in two ways.

Horizontal Addition: The idea here is to group the like terms and then combine them.

$$(8x^3 - 2x^2 + 6x - 2) + (3x^4 - 2x^3 + x^2 + x)$$
$$= 3x^4 + (8x^3 - 2x^3) + (-2x^2 + x^2) + (6x + x) - 2$$
$$= 3x^4 + 6x^3 - x^2 + 7x - 2$$

Vertical Addition: The idea here is to vertically line up the like terms in each polynomial and then add the coefficients.

$$
\begin{array}{r}
8x^3 - 2x^2 + 6x - 2 \\
(+) \quad 3x^4 - 2x^3 + x^2 + x \\
\hline
3x^4 + 6x^3 - x^2 + 7x - 2
\end{array}
$$

◄

We can subtract two polynomials horizontally or vertically as well.

EXAMPLE 5	**Subtracting Polynomials**

Find the difference: $(3x^4 - 4x^3 + 6x^2 - 1) - (2x^4 - 8x^2 - 6x + 5)$

Solution   *Horizontal Subtraction:*

$$(3x^4 - 4x^3 + 6x^2 - 1) - (2x^4 - 8x^2 - 6x + 5)$$

$$= 3x^4 - 4x^3 + 6x^2 - 1 + \underbrace{(-2x^4 + 8x^2 + 6x - 5)}$$

Be sure to change the sign of each term in the second polynomial.

$$\uparrow$$
$$= (3x^4 - 2x^4) + (-4x^3) + (6x^2 + 8x^2) + 6x + (-1 - 5)$$

Group like terms.

$$= x^4 - 4x^3 + 14x^2 + 6x - 6$$

NOTE
Vertical subtraction will be used when we divide polynomials.   ■

Vertical Subtraction: We line up like terms, change the sign of each coefficient of the second polynomial, and add.

$$
\begin{array}{lccccc}
 & x^4 & x^3 & x^2 & x^1 & x^0 \\
 3x^4 - 4x^3 + 6x^2 & & - 1 =
\end{array}
$$

$$
\begin{array}{l}
 & x^4 & x^3 & x^2 & x^1 & x^0 \\
 3x^4 - 4x^3 + & 6x^2 & & - 1 \\
(-)\ [2x^4 \quad\quad - 8x^2 - 6x + 5] = (+)\ -2x^4 \quad + 8x^2 + 6x - 5 \\
\hline
\quad\quad\quad\quad x^4 - 4x^3 + 14x^2 + 6x - 6 \ \blacktriangleleft
\end{array}
$$

The choice of which of these methods to use for adding and subtracting polynomials is left to you. To save space, we shall most often use the horizontal format.

✏ **NOW WORK PROBLEM 29.**

4   Multiply Polynomials

Two monomials may be multiplied using the Laws of Exponents and the Commutative and Associative Properties. For example,

$$(2x^3) \cdot (5x^4) = (2 \cdot 5) \cdot (x^3 \cdot x^4) = 10x^{3+4} = 10x^7$$

Products of polynomials are found by repeated use of the Distributive Property and the Laws of Exponents. Again, you have a choice of horizontal or vertical format.

EXAMPLE 6   **Multiplying Polynomials**

Find the product: $(2x + 5)(x^2 - x + 2)$

Solution   ***Horizontal Multiplication:***

$$(2x + 5)(x^2 - x + 2) = 2x(x^2 - x + 2) + 5(x^2 - x + 2)$$
↑
Distributive Property

$$= (2x \cdot x^2 - 2x \cdot x + 2x \cdot 2) + (5 \cdot x^2 - 5 \cdot x + 5 \cdot 2)$$
↑
Distributive Property

$$= (2x^3 - 2x^2 + 4x) + (5x^2 - 5x + 10)$$
↑
Law of exponents

$$= 2x^3 + 3x^2 - x + 10$$
↑
Combine like terms

Vertical Multiplication: The idea here is very much like multiplying a two-digit number by a three-digit number.

$$
\begin{array}{rl}
x^2 - x + 2 & \\
2x + 5 & \\
\hline
2x^3 - 2x^2 + 4x & \text{This line is } 2x(x^2 - x + 2). \\
(+)\quad 5x^2 - 5x + 10 & \text{This line is } 5(x^2 - x + 2). \\
\hline
2x^3 + 3x^2 - x + 10 & \text{Sum of the above two lines.} \ \blacktriangleleft
\end{array}
$$

✏ **NOW WORK PROBLEM 41.**

5   Know Formulas for Special Products

Certain products, which we call **special products**, occur frequently in algebra. We can calculate them easily using the **FOIL** (*First*, *Outer*, *Inner*, *Last*) method of multiplying two binomials.

$$(ax + b)(cx + d) = ax(cx + d) + b(cx + d)$$

$$= ax \cdot cx + ax \cdot d + b \cdot cx + b \cdot d$$
$$= acx^2 + adx + bcx + bd$$
$$= acx^2 + (ad + bc)x + bd$$

EXAMPLE 7   **Using FOIL**

(a) $(x - 3)(x + 3) = x^2 + 3x - 3x - 9 = x^2 - 9$

(b) $(x + 2)^2 = (x + 2)(x + 2) = x^2 + 2x + 2x + 4 = x^2 + 4x + 4$

(c) $(x - 3)^2 = (x - 3)(x - 3) = x^2 - 3x - 3x + 9 = x^2 - 6x + 9$

(d) $(x + 3)(x + 1) = x^2 + x + 3x + 3 = x^2 + 4x + 3$

(e) $(2x + 1)(3x + 4) = 6x^2 + 8x + 3x + 4 = 6x^2 + 11x + 4$ ◀

NOW WORK PROBLEMS **47 AND 55.**

Some products have been given special names because of their form. The following special products are based on Examples 7(a), (b), and (c).

Difference of Two Squares

$$(x - a)(x + a) = x^2 - a^2 \qquad \textbf{(2)}$$

Squares of Binomials, or Perfect Squares

$$(x + a)^2 = x^2 + 2ax + a^2 \qquad \textbf{(3a)}$$
$$(x - a)^2 = x^2 - 2ax + a^2 \qquad \textbf{(3b)}$$

EXAMPLE 8   **Using Special Product Formulas**

(a) $(x - 5)(x + 5) = x^2 - 5^2 = x^2 - 25$ Difference of two squares

(b) $(x + 7)^2 = x^2 + 2 \cdot 7 \cdot x + 7^2 = x^2 + 14x + 49$ Square of a binomial

(c) $(2x + 1)^2 = (2x)^2 + 2 \cdot 1 \cdot 2x + 1^2 = 4x^2 + 4x + 1$ Notice that we used 2x in place of x in formula (3a).

(d) $(3x - 4)^2 = (3x)^2 - 2 \cdot 4 \cdot 3x + 4^2 = 9x^2 - 24x + 16$ Replace x by 3x in formula (3b). ◀

NOW WORK PROBLEMS **65, 67, AND 69.**

Let's look at some more examples that lead to general formulas.

EXAMPLE 9

Cubing a Binomial

(a) $(x + 2)^3 = (x + 2)(x + 2)^2 = (x + 2)(x^2 + 4x + 4)$ Formula (3a)
$$= (x^3 + 4x^2 + 4x) + (2x^2 + 8x + 8)$$
$$= x^3 + 6x^2 + 12x + 8$$

(b) $(x - 1)^3 = (x - 1)(x - 1)^2 = (x - 1)(x^2 - 2x + 1)$ Formula (3b)
$$= (x^3 - 2x^2 + x) - (x^2 - 2x + 1)$$
$$= x^3 - 3x^2 + 3x - 1$$ ◀

Cubes of Binomials, or Perfect Cubes

$$(x + a)^3 = x^3 + 3ax^2 + 3a^2x + a^3 \qquad \textbf{(4a)}$$
$$(x - a)^3 = x^3 - 3ax^2 + 3a^2x - a^3 \qquad \textbf{(4b)}$$

━ NOW WORK PROBLEM 85.

EXAMPLE 10

Forming the Difference of Two Cubes

$$(x - 1)(x^2 + x + 1) = x(x^2 + x + 1) - 1(x^2 + x + 1)$$
$$= x^3 + x^2 + x - x^2 - x - 1$$
$$= x^3 - 1$$ ◀

EXAMPLE 11

Forming the Sum of Two Cubes

$$(x + 2)(x^2 - 2x + 4) = x(x^2 - 2x + 4) + 2(x^2 - 2x + 4)$$
$$= x^3 - 2x^2 + 4x + 2x^2 - 4x + 8$$
$$= x^3 + 8$$ ◀

Examples 10 and 11 lead to two more special products.

Difference of Two Cubes

$$(x - a)(x^2 + ax + a^2) = x^3 - a^3 \qquad \textbf{(5)}$$

Sum of Two Cubes

$$(x + a)(x^2 - ax + a^2) = x^3 + a^3 \qquad \textbf{(6)}$$

6 Divide Polynomials Using Long Division

The procedure for dividing two polynomials is similar to the procedure for dividing two integers.

| EXAMPLE 12 | **Dividing Two Integers** |

Divide 842 by 15.

Solution

$$
\begin{array}{r}
56 \quad \leftarrow \text{Quotient} \\
\text{Divisor} \rightarrow \quad 15\overline{)842} \quad \leftarrow \text{Dividend} \\
\underline{75} \quad \leftarrow 5 \cdot 15 \ (\text{Subtract}) \\
92 \\
\underline{90} \quad \leftarrow 6 \cdot 15 \ (\text{Subtract}) \\
2 \quad \leftarrow \text{Remainder}
\end{array}
$$

So, $\dfrac{842}{15} = 56 + \dfrac{2}{15}$.   ◄

In the long division process detailed in Example 12, the number 15 is called the **divisor**, the number 842 is called the **dividend**, the number 56 is called the **quotient**, and the number 2 is called the **remainder**.

To check the answer obtained in a division problem, multiply the quotient by the divisor and add the remainder. The answer should be the dividend.

$$
(\text{Quotient})(\text{Divisor}) + \text{Remainder} = \text{Dividend}
$$

For example, we can check the results obtained in Example 12 as follows:

$$
(56)(15) + 2 = 840 + 2 = 842
$$

To divide two polynomials, we first must write each polynomial in standard form. The process then follows a pattern similar to that of Example 12. The next example illustrates the procedure.

| EXAMPLE 13 | **Dividing Two Polynomials** |

Find the quotient and the remainder when

$$
3x^3 + 4x^2 + x + 7 \quad \text{is divided by} \quad x^2 + 1
$$

Solution

Each polynomial is in standard form. The dividend is $3x^3 + 4x^2 + x + 7$, and the divisor is $x^2 + 1$.

STEP 1: Divide the leading term of the dividend, $3x^3$, by the leading term of the divisor, x^2. Enter the result, $3x$, over the term $3x^3$, as follows:

$$
\begin{array}{r}
3x \\
x^2 + 1 \overline{)3x^3 + 4x^2 + x + 7}
\end{array}
$$

STEP 2: Multiply $3x$ by $x^2 + 1$ and enter the result below the dividend.

$$
\begin{array}{r}
3x \\
x^2 + 1 \overline{)3x^3 + 4x^2 + x + 7} \\
\underline{3x^3 + 3x }
\end{array}
$$

$\leftarrow 3x \cdot (x^2 + 1) = 3x^3 + 3x$

Notice that we align the 3x term under the x to make the next step easier.

STEP 3: Subtract and bring down the remaining terms.

$$
\begin{array}{r}
3x \\
x^2 + 1 \overline{\smash{)}3x^3 + 4x^2 + x + 7} \\
\underline{3x^3 + 3x } \\
4x^2 - 2x + 7
\end{array}
$$

← Subtract (change the signs and add).

← Bring down the $4x^2$ and the 7.

STEP 4: Repeat Steps 1–3 using $4x^2 - 2x + 7$ as the dividend.

$$
\begin{array}{r}
3x + 4 \\
x^2 + 1 \overline{\smash{)}3x^3 + 4x^2 + x + 7} \\
\underline{3x^3 + 3x } \\
4x^2 - 2x + 7 \\
\underline{4x^2 + 4} \\
-2x + 3
\end{array}
$$

← Divide $4x^2$ by x^2 to get 4.

← Multiply $x^2 + 1$ by 4; subtract.

Since x^2 does not divide $-2x$ evenly (that is, the result is not a monomial), the process ends. The quotient is $3x + 4$, and the remainder is $-2x + 3$.

✔ **CHECK:** (Quotient)(Divisor) + Remainder

$$
\begin{aligned}
&= (3x + 4)(x^2 + 1) + (-2x + 3) \\
&= 3x^3 + 4x^2 + 3x + 4 + (-2x + 3) \\
&= 3x^3 + 4x^2 + x + 7 = \text{Dividend}
\end{aligned}
$$

Then

$$
\frac{3x^3 + 4x^2 + x + 7}{x^2 + 1} = 3x + 4 + \frac{-2x + 3}{x^2 + 1}
$$

The next example combines the steps involved in long division.

EXAMPLE 14 | **Dividing Two Polynomials**

Find the quotient and the remainder when

$$
x^4 - 3x^3 + 2x - 5 \quad \text{is divided by} \quad x^2 - x + 1
$$

Solution In setting up this division problem, it is necessary to leave a space for the missing x^2 term in the dividend.

$$
\begin{array}{r}
x^2 - 2x - 3 \\
x^2 - x + 1 \overline{\smash{)}x^4 - 3x^3 + 2x - 5} \\
\underline{x^4 - x^3 + x^2 } \\
-2x^3 - x^2 + 2x - 5 \\
\underline{-2x^3 + 2x^2 - 2x } \\
-3x^2 + 4x - 5 \\
\underline{-3x^2 + 3x - 3} \\
x - 2
\end{array}
$$

Divisor →

Subtract →

Subtract →

Subtract →

← Quotient

← Dividend

← Remainder

✔ **CHECK:** (Quotient)(Divisor) + Remainder

$$
\begin{aligned}
&= (x^2 - 2x - 3)(x^2 - x + 1) + x - 2 \\
&= x^4 - x^3 + x^2 - 2x^3 + 2x^2 - 2x - 3x^2 + 3x - 3 + x - 2 \\
&= x^4 - 3x^3 + 2x - 5 = \text{Dividend}
\end{aligned}
$$

As a result,

$$\frac{x^4 - 3x^3 + 2x - 5}{x^2 - x + 1} = x^2 - 2x - 3 + \frac{x - 2}{x^2 - x + 1}$$ ◄

The process of dividing two polynomials leads to the following result:

Theorem Let Q be a polynomial of positive degree and let P be a polynomial whose degree is greater than or equal to the degree of Q. The remainder after dividing P by Q is either the zero polynomial or a polynomial whose degree is less than the degree of the divisor Q.

NOW WORK PROBLEM **93.**

7 Work with Polynomials in Two Variables

A **monomial in two variables** x and y has the form $ax^n y^m$, where a is a constant, x and y are variables, and n and m are nonnegative integers. The **degree** of a monomial is the sum of the powers of the variables.

For example,

$$2xy^3, \quad x^2y^2, \quad \text{and} \quad x^3y$$

are monomials, each of which has degree 4.

A **polynomial in two variables** x and y is the sum of one or more monomials in two variables. The **degree of a polynomial** in two variables is the highest degree of all the monomials with nonzero coefficients.

EXAMPLE 15 **Examples of Polynomials in Two Variables**

$$3x^2 + 2x^3y + 5 \qquad \pi x^3 - y^2 \qquad x^4 + 4x^3y - xy^3 + y^4$$

Two variables,   Two variables,   Two variables,
degree is 4.   degree is 3.   degree is 4.   ◄

Multiplying polynomials in two variables is handled in the same way as polynomials in one variable.

EXAMPLE 16 **Using a Special Product Formula**

To multiply $(2x - y)^2$, use the Squares of Binomials (3b) formula with $2x$ instead of x and y instead of a.

$$(2x - y)^2 = (2x)^2 - 2 \cdot y \cdot 2x + y^2 = 4x^2 - 4xy + y^2$$ ◄

NOW WORK PROBLEM **79.**

R.4 Assess Your Understanding

Concepts and Vocabulary

1. The polynomial $3x^4 - 2x^3 + 13x^2 - 5$ is of degree _____. The leading coefficient is _____.

2. $(x^2 - 4)(x^2 + 4) = $ _____.

3. $(x - 2)(x^2 + 2x + 4) = $ _____.

4. *True or False:* $4x^{-2}$ is a monomial of degree -2.

5. *True or False:* The degree of the product of two nonzero polynomials equals the sum of their degrees.

6. *True or False:* $(x + a)(x^2 + ax + a) = x^3 + a^3$.

Skill Building

In Problems 7–16, tell whether the expression is a monomial. If it is, name the variable(s) and the coefficient and give the degree of the monomial.

7. $2x^3$ 8. $-4x^2$ 9. $\dfrac{8}{x}$ 10. $-2x^{-3}$ 11. $-2xy^2$

12. $5x^2y^3$ 13. $\dfrac{8x}{y}$ 14. $-\dfrac{2x^2}{y^3}$ 15. $x^2 + y^2$ 16. $3x^2 + 4$

In Problems 17–26, tell whether the expression is a polynomial. If it is, give its degree.

17. $3x^2 - 5$ 18. $1 - 4x$ 19. 5 20. $-\pi$ 21. $3x^2 - \dfrac{5}{x}$

22. $\dfrac{3}{x} + 2$ 23. $2y^3 - \sqrt{2}$ 24. $10z^2 + z$ 25. $\dfrac{x^2 + 5}{x^3 - 1}$ 26. $\dfrac{3x^3 + 2x - 1}{x^2 + x + 1}$

In Problems 27–46, add, subtract, or multiply, as indicated. Express your answer as a single polynomial in standard form.

27. $(x^2 + 4x + 5) + (3x - 3)$ 28. $(x^3 + 3x^2 + 2) + (x^2 - 4x + 4)$

29. $(x^3 - 2x^2 + 5x + 10) - (2x^2 - 4x + 3)$ 30. $(x^2 - 3x - 4) - (x^3 - 3x^2 + x + 5)$

31. $(6x^5 + x^3 + x) + (5x^4 - x^3 + 3x^2)$ 32. $(10x^5 - 8x^2) + (3x^3 - 2x^2 + 6)$

33. $(x^2 - 3x + 1) + 2(3x^2 + x - 4)$ 34. $-2(x^2 + x + 1) + (-5x^2 - x + 2)$

35. $6(x^3 + x^2 - 3) - 4(2x^3 - 3x^2)$ 36. $8(4x^3 - 3x^2 - 1) - 6(4x^3 + 8x - 2)$

37. $(x^2 - x + 2) + (2x^2 - 3x + 5) - (x^2 + 1)$ 38. $(x^2 + 1) - (4x^2 + 5) + (x^2 + x - 2)$

39. $9(y^2 - 3y + 4) - 6(1 - y^2)$ 40. $8(1 - y^3) + 4(1 + y + y^2 + y^3)$

41. $x(x^2 + x - 4)$ 42. $4x^2(x^3 - x + 2)$ 43. $-2x^2(4x^3 + 5)$

44. $5x^3(3x - 4)$ 45. $(x + 1)(x^2 + 2x - 4)$ 46. $(2x - 3)(x^2 + x + 1)$

In Problems 47–64, multiply the polynomials using the FOIL method. Express your answer as a single polynomial in standard form.

47. $(x + 2)(x + 4)$ 48. $(x + 3)(x + 5)$ 49. $(2x + 5)(x + 2)$

50. $(3x + 1)(2x + 1)$ 51. $(x - 4)(x + 2)$ 52. $(x + 4)(x - 2)$

53. $(x - 3)(x - 2)$ 54. $(x - 5)(x - 1)$ 55. $(2x + 3)(x - 2)$

56. $(2x - 4)(3x + 1)$ 57. $(-2x + 3)(x - 4)$ 58. $(-3x - 1)(x + 1)$

59. $(-x - 2)(-2x - 4)$ 60. $(-2x - 3)(3 - x)$ 61. $(x - 2y)(x + y)$

62. $(2x + 3y)(x - y)$ 63. $(-2x - 3y)(3x + 2y)$ 64. $(x - 3y)(-2x + y)$

In Problems 65–88, multiply the polynomials using the special product formulas. Express your answer as a single polynomial in standard form.

65. $(x - 7)(x + 7)$ 66. $(x - 1)(x + 1)$ 67. $(2x + 3)(2x - 3)$ 68. $(3x + 2)(3x - 2)$

69. $(x + 4)^2$ 70. $(x + 5)^2$ 71. $(x - 4)^2$ 72. $(x - 5)^2$

73. $(3x + 4)(3x - 4)$ 74. $(5x - 3)(5x + 3)$ 75. $(2x - 3)^2$ 76. $(3x - 4)^2$

77. $(x + y)(x - y)$ 78. $(x + 3y)(x - 3y)$ 79. $(3x + y)(3x - y)$ 80. $(3x + 4y)(3x - 4y)$

81. $(x + y)^2$ 82. $(x - y)^2$ 83. $(x - 2y)^2$ 84. $(2x + 3y)^2$

85. $(x - 2)^3$ 86. $(x + 1)^3$ 87. $(2x + 1)^3$ 88. $(3x - 2)^3$

In Problems 89–104, find the quotient and the remainder. Check your work by verifying that

$$(Quotient)(Divisor) + Remainder = Dividend$$

89. $4x^3 - 3x^2 + x + 1$ divided by $x + 2$ 90. $3x^3 - x^2 + x - 2$ divided by $x + 2$

91. $4x^3 - 3x^2 + x + 1$ divided by x^2 92. $3x^3 - x^2 + x - 2$ divided by x^2

93. $5x^4 - 3x^2 + x + 1$ divided by $x^2 + 2$ 94. $5x^4 - x^2 + x - 2$ divided by $x^2 + 2$

95. $4x^5 - 3x^2 + x + 1$ divided by $2x^3 - 1$ 96. $3x^5 - x^2 + x - 2$ divided by $3x^3 - 1$

97. $2x^4 - 3x^3 + x + 1$ divided by $2x^2 + x + 1$ 98. $3x^4 - x^3 + x - 2$ divided by $3x^2 + x + 1$

99. $-4x^3 + x^2 - 4$ divided by $x - 1$ 100. $-3x^4 - 2x - 1$ divided by $x - 1$

101. $1 - x^2 + x^4$ divided by $x^2 + x + 1$ 102. $1 - x^2 + x^4$ divided by $x^2 - x + 1$

103. $x^3 - a^3$ divided by $x - a$ 104. $x^5 - a^5$ divided by $x - a$

Discussion and Writing

105. Explain why the degree of the product of two nonzero polynomials equals the sum of their degrees.

106. Explain why the degree of the sum of two polynomials of different degrees equals the larger of their degrees.

107. Give a careful statement about the degree of the sum of two polynomials of the same degree.

108. Do you prefer adding two polynomials using the horizontal method or the vertical method? Write a brief position paper defending your choice.

109. Do you prefer to memorize the rule for the square of a binomial $(x + a)^2$ or to use FOIL to obtain the product? Write a brief position paper defending your choice.

R.5 Factoring Polynomials

OBJECTIVES   1 Factor the Difference of Two Squares and the Sum and the Difference of Two Cubes
2 Factor Perfect Squares
3 Factor a Second-degree Polynomial: $x^2 + Bx + C$
4 Factor by Grouping
5 Factor a Second-degree Polynomial: $Ax^2 + Bx + C$

Consider the following product:

$$(2x + 3)(x - 4) = 2x^2 - 5x - 12$$

The two polynomials on the left side are called **factors** of the polynomial on the right side. Expressing a given polynomial as a product of other polynomials, that is, finding the factors of a polynomial, is called **factoring**.

We shall restrict our discussion here to factoring polynomials in one variable into products of polynomials in one variable, where all coefficients are integers. We call this **factoring over the integers**.

Any polynomial can be written as the product of 1 times itself or as -1 times its additive inverse. If a polynomial cannot be written as the product of two other polynomials (excluding 1 and -1), then the polynomial is said to be **prime**. When a polynomial has been written as a product consisting only of prime factors, it is said to be **factored completely**. Examples of prime polynomials are

$$2, \quad 3, \quad 5, \quad x, \quad x + 1, \quad x - 1, \quad 3x + 4$$

The first factor to look for in a factoring problem is a common monomial factor present in each term of the polynomial. If one is present, use the Distributive Property to factor it out.

EXAMPLE 1	Identifying Common Monomial Factors

Polynomial	Common Monomial Factor	Remaining Factor	Factored Form
$2x + 4$	2	$x + 2$	$2x + 4 = 2(x + 2)$
$3x - 6$	3	$x - 2$	$3x - 6 = 3(x - 2)$
$2x^2 - 4x + 8$	2	$x^2 - 2x + 4$	$2x^2 - 4x + 8 = 2(x^2 - 2x + 4)$
$8x - 12$	4	$2x - 3$	$8x - 12 = 4(2x - 3)$
$x^2 + x$	x	$x + 1$	$x^2 + x = x(x + 1)$
$x^3 - 3x^2$	x^2	$x - 3$	$x^3 - 3x^2 = x^2(x - 3)$
$6x^2 + 9x$	$3x$	$2x + 3$	$6x^2 + 9x = 3x(2x + 3)$

◀

Notice that, once all common monomial factors have been removed from a polynomial, the remaining factor is either a prime polynomial of degree 1 or a polynomial of degree 2 or higher. (Do you see why?)

━━━━━ NOW WORK PROBLEM 5.

1 Factor the Difference of Two Squares and the Sum and Difference of Two Cubes

When you factor a polynomial, first check for common monomial factors. Then see whether you can use one of the special formulas discussed in the previous section.

Difference of Two Squares	$x^2 - a^2 = (x - a)(x + a)$
Perfect Squares	$x^2 + 2ax + a^2 = (x + a)^2$
	$x^2 - 2ax + a^2 = (x - a)^2$
Sum of Two Cubes	$x^3 + a^3 = (x + a)(x^2 - ax + a^2)$
Difference of Two Cubes	$x^3 - a^3 = (x - a)(x^2 + ax + a^2)$

EXAMPLE 2 | **Factoring the Difference of Two Squares**

Factor completely: $x^2 - 4$

Solution | We notice that $x^2 - 4$ is the difference of two squares, x^2 and 2^2.

$$x^2 - 4 = (x - 2)(x + 2)$$ ◀

EXAMPLE 3 | **Factoring the Difference of Two Cubes**

Factor completely: $x^3 - 1$

Solution | Because $x^3 - 1$ is the difference of two cubes, x^3 and 1^3, we find that

$$x^3 - 1 = (x - 1)(x^2 + x + 1)$$ ◀

EXAMPLE 4 | **Factoring the Sum of Two Cubes**

Factor completely: $x^3 + 8$

Solution | Because $x^3 + 8$ is the sum of two cubes, x^3 and 2^3, we have

$$x^3 + 8 = (x + 2)(x^2 - 2x + 4)$$ ◀

EXAMPLE 5 | **Factoring the Difference of Two Squares**

Factor completely: $x^4 - 16$

Solution | Because $x^4 - 16$ is the difference of two squares, $x^4 = (x^2)^2$ and $16 = 4^2$, we have

$$x^4 - 16 = (x^2 - 4)(x^2 + 4)$$

But $x^2 - 4$ is also the difference of two squares. Thus,

$$x^4 - 16 = (x^2 - 4)(x^2 + 4) = (x - 2)(x + 2)(x^2 + 4)$$ ◀

━━━━━ NOW WORK PROBLEMS 15 AND 33.

2 Factor Perfect Squares

When the first term and third term of a trinomial are both positive and are perfect squares, such as x^2, $9x^2$, 1, and 4, check to see whether the trinomial is a perfect square.

| EXAMPLE 6 | **Factoring Perfect Squares** |

Factor completely: $x^2 + 6x + 9$

Solution   The first term, x^2, and the third term, $9 = 3^2$, are perfect squares. Because the middle term $6x$ is twice the product of x and 3, we have a perfect square.

$$x^2 + 6x + 9 = (x + 3)^2 \qquad \blacktriangleleft$$

| EXAMPLE 7 | **Factoring Perfect Squares** |

Factor completely: $9x^2 - 6x + 1$

Solution   The first term, $9x^2 = (3x)^2$, and the third term, $1 = 1^2$, are perfect squares. Because the middle term $-6x$ is -2 times the product of $3x$ and 1, we have a perfect square.

$$9x^2 - 6x + 1 = (3x - 1)^2 \qquad \blacktriangleleft$$

| EXAMPLE 8 | **Factoring Perfect Squares** |

Factor completely: $25x^2 + 30x + 9$

Solution   The first term, $25x^2 = (5x)^2$, and the third term, $9 = 3^2$, are perfect squares. Because the middle term $30x$ is twice the product of $5x$ and 3, we have a perfect square.

$$25x^2 + 30x + 9 = (5x + 3)^2 \qquad \blacktriangleleft$$

NOW WORK PROBLEMS 25 AND 93.

If a trinomial is not a perfect square, it may be possible to factor it using the technique discussed next.

3 Factor a Second-degree Polynomial: $x^2 + Bx + C$

The idea behind factoring a second-degree polynomial like $x^2 + Bx + C$ is to see whether it can be made equal to the product of two, possibly equal, first-degree polynomials.

For example, we know that

$$(x + 3)(x + 4) = x^2 + 7x + 12$$

The factors of $x^2 + 7x + 12$ are $x + 3$ and $x + 4$. Notice the following:

$$x^2 + 7x + 12 = (x + 3)(x + 4)$$

  — 12 is the product of 3 and 4
  — 7 is the sum of 3 and 4

In general, if $x^2 + Bx + C = (x + a)(x + b)$, then $ab = C$ and $a + b = B$.

To factor a second-degree polynomial $x^2 + Bx + C$, find integers whose product is C and whose sum is B. That is, if there are numbers a, b, where $ab = C$ and $a + b = B$, then

$$x^2 + Bx + C = (x + a)(x + b)$$

EXAMPLE 9 | **Factoring Trinomials**

Factor completely: $x^2 + 7x + 10$

Solution | First, determine all integers whose product is 10 and then compute their sums.

Integers whose product is 10	1, 10	−1, −10	2, 5	−2, −5
Sum	11	−11	7	−7

The integers 2 and 5 have a product of 10 and add up to 7, the coefficient of the middle term. As a result, we have

$$x^2 + 7x + 10 = (x + 2)(x + 5) \quad \blacktriangleleft$$

EXAMPLE 10 | **Factoring Trinomials**

Factor completely: $x^2 - 6x + 8$

Solution | First, determine all integers whose product is 8 and then compute each sum.

Integers whose product is 8	1, 8	−1, −8	2, 4	−2, −4
Sum	9	−9	6	−6

Since −6 is the coefficient of the middle term,

$$x^2 - 6x + 8 = (x - 2)(x - 4) \quad \blacktriangleleft$$

EXAMPLE 11 | **Factoring Trinomials**

Factor completely: $x^2 - x - 12$

Solution | First, determine all integers whose product is −12 and then compute each sum.

Integers whose product is −12	1, −12	−1, 12	2, −6	−2, 6	3, −4	−3, 4
Sum	−11	11	−4	4	−1	1

Since −1 is the coefficient of the middle term,

$$x^2 - x - 12 = (x + 3)(x - 4) \quad \blacktriangleleft$$

EXAMPLE 12 | **Factoring Trinomials**

Factor completely: $x^2 + 4x - 12$

Solution | The integers −2 and 6 have a product of −12 and have the sum 4. Thus,

$$x^2 + 4x - 12 = (x - 2)(x + 6) \quad \blacktriangleleft$$

To avoid errors in factoring, always check your answer by multiplying it out to see if the result equals the original expression.

When none of the possibilities works, the polynomial is prime.

EXAMPLE 13   **Identifying Prime Polynomials**

Show that $x^2 + 9$ is prime.

Solution   First, list the integers whose product is 9 and then compute their sums.

Integers whose product is 9	1, 9	−1, −9	3, 3	−3, −3
Sum	10	−10	6	−6

Since the coefficient of the middle term in $x^2 + 9 = x^2 + 0x + 9$ is 0 and none of the sums equals 0, we conclude that $x^2 + 9$ is prime.   ◀

Example 13 demonstrates a more general result:

Theorem   Any polynomial of the form $x^2 + a^2$, a real, is prime.

NOW WORK PROBLEMS 39 AND 77.

4 Factor by Grouping

Sometimes a common factor does not occur in every term of the polynomial, but in each of several groups of terms that together make up the polynomial. When this happens, the common factor can be factored out of each group by means of the Distributive Property. This technique is called **factoring by grouping**.

EXAMPLE 14   **Factoring by Grouping**

Factor completely by grouping: $(x^2 + 2)x + (x^2 + 2) \cdot 3$

Solution   Notice the common factor $x^2 + 2$. By applying the Distributive Property, we have

$$(x^2 + 2)x + (x^2 + 2) \cdot 3 = (x^2 + 2)(x + 3)$$

Since $x^2 + 2$ and $x + 3$ are prime, the factorization is complete.   ◀

The next example shows a factoring problem that occurs in calculus.

EXAMPLE 15   **Factoring by Grouping**

Factor completely by grouping: $3(x - 1)^2(x + 2)^4 + 4(x - 1)^3(x + 2)^3$

Solution   Here, $(x - 1)^2(x + 2)^3$ is a common factor of $3(x - 1)^2(x + 2)^4$ and of $4(x - 1)^3(x + 2)^3$. As a result,

$$3(x - 1)^2(x + 2)^4 + 4(x - 1)^3(x + 2)^3 = (x - 1)^2(x + 2)^3[3(x + 2) + 4(x - 1)]$$
$$= (x - 1)^2(x + 2)^3[3x + 6 + 4x - 4]$$
$$= (x - 1)^2(x + 2)^3(7x + 2)   ◀$$

EXAMPLE 16	Factoring by Grouping

Factor completely by grouping: $x^3 - 4x^2 + 2x - 8$

Solution To see if factoring by grouping will work, group the first two terms and the last two terms. Then look for a common factor in each group. In this example, we can factor x^2 from $x^3 - 4x^2$ and 2 from $2x - 8$. The remaining factor in each case is the same, $x - 4$. This means that factoring by grouping will work, as follows:

$$
\begin{aligned}
x^3 - 4x^2 + 2x - 8 &= (x^3 - 4x^2) + (2x - 8) \\
&= x^2(x - 4) + 2(x - 4) \\
&= (x^2 + 2)(x - 4)
\end{aligned}
$$

Since $x^2 + 2$ and $x - 4$ are prime, the factorization is complete. ◄

✏ **NOW WORK PROBLEMS 51 AND 121.**

5 Factor a Second-degree Polynomial: $Ax^2 + Bx + C$

To factor a second-degree polynomial $Ax^2 + Bx + C$, when $A \neq 1$ and $A, B,$ and C have no common factors, follow these steps:

Steps for Factoring $Ax^2 + Bx + C$, where $A \neq 1$, and $A, B,$ and C Have No Common Factors

STEP 1: Find the value of AC.

STEP 2: Find integers whose product is AC that add up to B. That is, find a and b so that $ab = AC$ and $a + b = B$.

STEP 3: Write $Ax^2 + Bx + C = Ax^2 + ax + bx + C$.

STEP 4: Factor this last expression by grouping.

EXAMPLE 17	Factoring Trinomials

Factor completely: $2x^2 + 5x + 3$

Solution Comparing $2x^2 + 5x + 3$ to $Ax^2 + Bx + C$, we find that $A = 2, B = 5,$ and $C = 3$.

STEP 1: The value of AC is $2 \cdot 3 = 6$.

STEP 2: Determine the integers whose product is $AC = 6$ and compute their sums.

Integers whose product is 6	1, 6	−1, −6	2, 3	−2, −3
Sum	7	−7	5	−5

STEP 3: The integers whose product is 6 that add up to $B = 5$ are 2 and 3.

$$2x^2 + 5x + 3 = 2x^2 + 2x + 3x + 3$$

Step 4: Factor by grouping.

$$2x^2 + 2x + 3x + 3 = (2x^2 + 2x) + (3x + 3)$$
$$= 2x(x + 1) + 3(x + 1)$$
$$= (2x + 3)(x + 1)$$

Thus,

$$2x^2 + 5x + 3 = (2x + 3)(x + 1) \qquad \blacktriangleleft$$

EXAMPLE 18	**Factoring Trinomials**

Factor completely: $2x^2 - x - 6$

Solution   Comparing $2x^2 - x - 6$ to $Ax^2 + Bx + C$, we find that $A = 2, B = -1$, and $C = -6$.

Step 1: The value of AC is $2 \cdot (-6) = -12$.

Step 2: Determine the integers whose product is $AC = -12$ and compute their sums.

Integers whose product is −12	1, −12	−1, 12	2, −6	−2, 6	3, −4	−3, 4
Sum	−11	11	−4	4	−1	1

Step 3: The integers whose product is -12 that add up to $B = -1$ are -4 and 3.

$$2x^2 - x - 6 = 2x^2 - 4x + 3x - 6$$

Step 4: Factor by grouping.

$$2x^2 - 4x + 3x - 6 = (2x^2 - 4x) + (3x - 6)$$
$$= 2x(x - 2) + 3(x - 2)$$
$$= (2x + 3)(x - 2)$$

As a result,

$$2x^2 - x - 6 = (2x + 3)(x - 2) \qquad \blacktriangleleft$$

NOW WORK PROBLEM **57.**

We close this section with a capsule summary.

Summary

Type of Polynomial	Method	Example
Any polynomial	Look for common monomial factors. (Always do this first!)	$6x^2 + 9x = 3x(2x + 3)$
Binomials of degree 2 or higher	Check for a special product: Difference of two squares, $x^2 - a^2$ Difference of two cubes, $x^3 - a^3$ Sum of two cubes, $x^3 + a^3$	$x^2 - 16 = (x - 4)(x + 4)$ $x^3 - 64 = (x - 4)(x^2 + 4x + 16)$ $x^3 + 27 = (x + 3)(x^2 - 3x + 9)$
Trinomials of degree 2	Check for a perfect square, $(x \pm a)^2$ (p. 47) Factoring $x^2 + Bx + C$ (p. 48) Factoring $Ax^2 + Bx + C$ (p. 50)	$x^2 + 8x + 16 = (x + 4)^2$ $x^2 - 10x + 25 = (x - 5)^2$ $x^2 - x - 2 = (x - 2)(x + 1)$ $6x^2 + x - 1 = (2x + 1)(3x - 1)$
Three or more terms	Grouping	$2x^3 - 3x^2 + 4x - 6 = (2x - 3)(x^2 + 2)$

R.5 Assess Your Understanding

Concepts and Vocabulary

1. If factored completely, $3x^3 - 12x =$ _____.

2. If a polynomial cannot be written as the product of two other polynomials (excluding 1 and −1), then the polynomial is said to be _____.

3. *True or False:* The polynomial $x^2 + 4$ is prime.

4. *True or False:* $3x^3 - 2x^2 - 6x + 4 = (3x - 2)(x^2 + 2)$.

Skill Building

In Problems 5–14, factor each polynomial by removing the common monomial factor.

5. $3x + 6$
6. $7x - 14$
7. $ax^2 + a$
8. $ax - a$
9. $x^3 + x^2 + x$
10. $x^3 - x^2 + x$
11. $2x^2 - 2x$
12. $3x^2 - 3x$
13. $3x^2y - 6xy^2 + 12xy$
14. $60x^2y - 48xy^2 + 72x^3y$

In Problems 15–22, factor the difference of two squares.

15. $x^2 - 1$
16. $x^2 - 4$
17. $4x^2 - 1$
18. $9x^2 - 1$
19. $x^2 - 16$
20. $x^2 - 25$
21. $25x^2 - 4$
22. $36x^2 - 9$

In Problems 23–32, factor the perfect squares.

23. $x^2 + 2x + 1$
24. $x^2 - 4x + 4$
25. $x^2 + 4x + 4$
26. $x^2 - 2x + 1$
27. $x^2 - 10x + 25$
28. $x^2 + 10x + 25$
29. $4x^2 + 4x + 1$
30. $9x^2 + 6x + 1$
31. $16x^2 + 8x + 1$
32. $25x^2 + 10x + 1$

In Problems 33–38, factor the sum or difference of two cubes.

33. $x^3 - 27$
34. $x^3 + 125$
35. $x^3 + 27$
36. $27 - 8x^3$
37. $8x^3 + 27$
38. $64 - 27x^3$

In Problems 39–50, factor each polynomial.

39. $x^2 + 5x + 6$
40. $x^2 + 6x + 8$
41. $x^2 + 7x + 6$
42. $x^2 + 9x + 8$
43. $x^2 + 7x + 10$
44. $x^2 + 11x + 10$
45. $x^2 - 10x + 16$
46. $x^2 - 17x + 16$
47. $x^2 - 7x - 8$
48. $x^2 - 2x - 8$
49. $x^2 + 7x - 8$
50. $x^2 + 2x - 8$

In Problems 51–56, factor by grouping.

51. $2x^2 + 4x + 3x + 6$
52. $3x^2 - 3x + 2x - 2$
53. $2x^2 - 4x + x - 2$
54. $3x^2 + 6x - x - 2$
55. $6x^2 + 9x + 4x + 6$
56. $9x^2 - 6x + 3x - 2$

In Problems 57–68, factor each polynomial.

57. $3x^2 + 4x + 1$
58. $2x^2 + 3x + 1$
59. $2z^2 + 5z + 3$
60. $6z^2 + 5z + 1$
61. $3x^2 + 2x - 8$
62. $3x^2 + 10x + 8$
63. $3x^2 - 2x - 8$
64. $3x^2 - 10x + 8$
65. $3x^2 + 14x + 8$
66. $3x^2 - 14x + 8$
67. $3x^2 + 10x - 8$
68. $3x^2 - 10x - 8$

In Problems 69–116, factor completely each polynomial. If the polynomial cannot be factored, say it is prime.

69. $x^2 - 36$
70. $x^2 - 9$
71. $2 - 8x^2$
72. $3 - 27x^2$
73. $x^2 + 7x + 10$
74. $x^2 + 5x + 4$
75. $x^2 - 10x + 21$
76. $x^2 - 6x + 8$
77. $4x^2 - 8x + 32$
78. $3x^2 - 12x + 15$
79. $x^2 + 4x + 16$
80. $x^2 + 12x + 36$
81. $15 + 2x - x^2$
82. $14 + 6x - x^2$
83. $3x^2 - 12x - 36$
84. $x^3 + 8x^2 - 20x$
85. $y^4 + 11y^3 + 30y^2$
86. $3y^3 - 18y^2 - 48y$
87. $4x^2 + 12x + 9$
88. $9x^2 - 12x + 4$
89. $6x^2 + 8x + 2$
90. $8x^2 + 6x - 2$
91. $x^4 - 81$
92. $x^4 - 1$
93. $x^6 - 2x^3 + 1$
94. $x^6 + 2x^3 + 1$
95. $x^7 - x^5$
96. $x^8 - x^5$
97. $16x^2 + 24x + 9$
98. $9x^2 - 24x + 16$
99. $5 + 16x - 16x^2$
100. $5 + 11x - 16x^2$
101. $4y^2 - 16y + 15$
102. $9y^2 + 9y - 4$
103. $1 - 8x^2 - 9x^4$
104. $4 - 14x^2 - 8x^4$
105. $x(x + 3) - 6(x + 3)$
106. $5(3x - 7) + x(3x - 7)$
107. $(x + 2)^2 - 5(x + 2)$
108. $(x - 1)^2 - 2(x - 1)$

109. $(3x - 2)^3 - 27$ **110.** $(5x + 1)^3 - 1$ **111.** $3(x^2 + 10x + 25) - 4(x + 5)$

112. $7(x^2 - 6x + 9) + 5(x - 3)$ **113.** $x^3 + 2x^2 - x - 2$ **114.** $x^3 - 3x^2 - x + 3$

115. $x^4 - x^3 + x - 1$ **116.** $x^4 + x^3 + x + 1$

Applications and Extensions

In Problems 117–126, expressions that occur in calculus are given. Factor completely each expression.

117. $2(3x + 4)^2 + (2x + 3) \cdot 2(3x + 4) \cdot 3$ **118.** $5(2x + 1)^2 + (5x - 6) \cdot 2(2x + 1) \cdot 2$

119. $2x(2x + 5) + x^2 \cdot 2$ **120.** $3x^2(8x - 3) + x^3 \cdot 8$

121. $2(x + 3)(x - 2)^3 + (x + 3)^2 \cdot 3(x - 2)^2$ **122.** $4(x + 5)^3(x - 1)^2 + (x + 5)^4 \cdot 2(x - 1)$

123. $(4x - 3)^2 + x \cdot 2(4x - 3) \cdot 4$ **124.** $3x^2(3x + 4)^2 + x^3 \cdot 2(3x + 4) \cdot 3$

125. $2(3x - 5) \cdot 3(2x + 1)^3 + (3x - 5)^2 \cdot 3(2x + 1)^2 \cdot 2$ **126.** $3(4x + 5)^2 \cdot 4(5x + 1)^2 + (4x + 5)^3 \cdot 2(5x + 1) \cdot 5$

127. Show that $x^2 + 4$ is prime.   **128.** Show that $x^2 + x + 1$ is prime.

Discussion and Writing

129. Make up a polynomial that factors into a perfect square.

130. Explain to a fellow student what you look for first when presented with a factoring problem. What do you do next?

R.6 Synthetic Division

> **OBJECTIVE 1** Divide Polynomials Using Synthetic Division

1 Divide Polynomials Using Synthetic Division

To find the quotient as well as the remainder when a polynomial of degree 1 or higher is divided by $x - c$, a shortened version of long division, called **synthetic division**, makes the task simpler.

To see how synthetic division works, we will use long division to divide the polynomial $2x^3 - x^2 + 3$ by $x - 3$.

$$
\begin{array}{r}
2x^2 + 5x + 15 \quad \leftarrow \text{Quotient} \\
x - 3 \overline{) 2x^3 - x^2 \qquad + 3} \\
\underline{2x^3 - 6x^2} \\
5x^2 \\
\underline{5x^2 - 15x} \\
15x + 3 \\
\underline{15x - 45} \\
48 \quad \leftarrow \text{Remainder}
\end{array}
$$

✔ **CHECK:** (Divisor) $\cdot$ (Quotient) + Remainder

$$= (x - 3)(2x^2 + 5x + 15) + 48$$
$$= 2x^3 + 5x^2 + 15x - 6x^2 - 15x - 45 + 48$$
$$= 2x^3 - x^2 + 3$$

The process of synthetic division arises from rewriting the long division in a more compact form, using simpler notation. For example, in the long division on page 53, the terms in blue are not really necessary because they are identical to the terms directly above them. With these terms removed, we have

$$
\require{enclose}
\begin{array}{r}
2x^2 + 5x\ \ + 15 \\
x - 3 \enclose{longdiv}{2x^3 - \ x^2\qquad\quad + \ 3} \\
\underline{-\ 6x^2\qquad\qquad} \\
5x^2\qquad\qquad \\
\underline{-\ 15x\qquad} \\
15x\qquad \\
\underline{-\ 45} \\
48
\end{array}
$$

Most of the x's that appear in this process can also be removed, provided that we are careful about positioning each coefficient. In this regard, we will need to use 0 as the coefficient of x in the dividend, because that power of x is missing. Now we have

$$
\begin{array}{r}
2x^2 + 5x + 15 \\
x - 3 \enclose{longdiv}{2\ \ - 1\qquad 0\qquad 3} \\
\underline{-\ 6\qquad\qquad} \\
5\qquad\qquad \\
\underline{-\ 15\qquad} \\
15\qquad \\
\underline{-\ 45} \\
48
\end{array}
$$

We can make this display more compact by moving the lines up until the numbers in color align horizontally.

$$
\begin{array}{r}
2x^2 + 5x + 15 \qquad \text{Row 1} \\
x - 3 \enclose{longdiv}{2\ \ - 1\qquad 0\qquad 3}\quad \text{Row 2} \\
\underline{-\ 6\ - 15 - 45}\quad \text{Row 3} \\
\bigcirc\quad 5\quad 15\quad 48\quad\ \text{Row 4}
\end{array}
$$

Because the leading coefficient of the divisor is always 1, we know that the leading coefficient of the dividend will also be the leading coefficient of the quotient. So we place the leading coefficient of the quotient, 2, in the circled position. Now, the first three numbers in row 4 are precisely the coefficients of the quotient, and the last number in row 4 is the remainder. Thus, row 1 is not really needed, so we can compress the process to three rows, where the bottom row contains both the coefficients of the quotient and the remainder.

$$
\begin{array}{r}
x - 3 \enclose{longdiv}{2\ \ - 1\qquad 0\qquad 3}\quad \text{Row 1} \\
\underline{-\ 6 - 15 - 45}\quad \text{Row 2 (subtract)} \\
2\quad 5\quad 15\quad 48\quad\ \text{Row 3}
\end{array}
$$

Recall that the entries in row 3 are obtained by subtracting the entries in row 2 from those in row 1. Rather than subtracting the entries in row 2, we can change the sign of each entry and add. With this modification, our display will look like this:

$$
\begin{array}{r}
x - 3 \enclose{longdiv}{2\ \ - 1\qquad 0\qquad 3}\quad \text{Row 1} \\
\underline{6\quad 15\quad 45}\quad \text{Row 2 (add)} \\
2\quad 5\quad 15\quad 48\quad\ \text{Row 3}
\end{array}
$$

Notice that the entries in row 2 are three times the prior entries in row 3. Our last modification to the display replaces the $x - 3$ by 3. The entries in row 3 give the quotient and the remainder, as shown next.

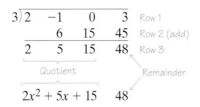

$$2x^2 + 5x + 15 \qquad 48$$

Let's go through an example step by step.

EXAMPLE 1 **Using Synthetic Division to Find the Quotient and Remainder**

Use synthetic division to find the quotient and remainder when

$$x^3 - 4x^2 - 5 \quad \text{is divided by} \quad x - 3$$

Solution **STEP 1:** Write the dividend in descending powers of x. Then copy the coefficients, remembering to insert a 0 for any missing powers of x.

$$1 \quad -4 \quad 0 \quad -5 \quad \text{Row 1}$$

STEP 2: Insert the usual division symbol. In synthetic division, the divisor is of the form $x - c$, and c is the number placed to the left of the division symbol. Here, since the divisor is $x - 3$, we insert 3 to the left of the division symbol.

$$3\overline{)1 \quad -4 \quad 0 \quad -5} \quad \text{Row 1}$$

STEP 3: Bring the 1 down two rows, and enter it in row 3.

$$
\begin{array}{r}
3\overline{)1 \quad -4 \quad 0 \quad -5} \quad \text{Row 1} \\
\downarrow \qquad\qquad\qquad\quad \text{Row 2} \\
\hline
1 \qquad\qquad\qquad\quad \text{Row 3}
\end{array}
$$

STEP 4: Multiply the latest entry in row 3 by 3, and place the result in row 2, one column over to the right.

$$
\begin{array}{r}
3\overline{)1 \quad -4 \quad 0 \quad -5} \quad \text{Row 1} \\
3 \qquad\qquad\qquad \text{Row 2} \\
\hline
1 \qquad\qquad\qquad\qquad \text{Row 3}
\end{array}
$$

STEP 5: Add the entry in row 2 to the entry above it in row 1, and enter the sum in row 3.

$$
\begin{array}{r}
3\overline{)1 \quad -4 \quad 0 \quad -5} \quad \text{Row 1} \\
3 \qquad\qquad\qquad \text{Row 2} \\
\hline
1 \quad -1 \qquad\qquad\quad \text{Row 3}
\end{array}
$$

STEP 6: Repeat Steps 4 and 5 until no more entries are available in row 1.

$$
\begin{array}{r}
3\overline{)1 \quad -4 \quad 0 \quad -5} \quad \text{Row 1} \\
3 \quad -3 \quad -9 \quad \text{Row 2} \\
\hline
1 \quad -1 \quad -3 \quad -14 \quad \text{Row 3}
\end{array}
$$

STEP 7: The final entry in row 3, the -14, is the remainder; the other entries in row 3, the 1, -1, and -3, are the coefficients (in descending order) of a polynomial whose degree is 1 less than that of the dividend. This is the quotient. Thus,

$$\text{Quotient} = x^2 - x - 3 \qquad \text{Remainder} = -14$$

✔ **CHECK:** (Divisor)(Quotient) + Remainder
$$= (x - 3)(x^2 - x - 3) + (-14)$$
$$= (x^3 - x^2 - 3x - 3x^2 + 3x + 9) + (-14)$$
$$= x^3 - 4x^2 - 5 = \text{Dividend} \qquad \blacktriangleleft$$

Let's do an example in which all seven steps are combined.

| EXAMPLE 2 | **Using Synthetic Division to Verify a Factor** |

Use synthetic division to show that $x + 3$ is a factor of
$$2x^5 + 5x^4 - 2x^3 + 2x^2 - 2x + 3$$

Solution The divisor is $x + 3 = x - (-3)$, so we place -3 to the left of the division symbol. Then the row 3 entries will be multiplied by -3, entered in row 2, and added to row 1.

$$
\begin{array}{r|rrrrrr}
-3) & 2 & 5 & -2 & 2 & -2 & 3 \\
    &   & -6 & 3 & -3 & 3 & -3 \\
\hline
    & 2 & -1 & 1 & -1 & 1 & 0 \\
\end{array}
\quad
\begin{array}{l}
\text{Row 1} \\
\text{Row 2} \\
\text{Row 3} \\
\end{array}
$$

Because the remainder is 0, we have

(Divisor)(Quotient) + Remainder
$$= (x + 3)(2x^4 - x^3 + x^2 - x + 1) = 2x^5 + 5x^4 - 2x^3 + 2x^2 - 2x + 3$$

As we see, $x + 3$ is a factor of $2x^5 + 5x^4 - 2x^3 + 2x^2 - 2x + 3$. $\qquad \blacktriangleleft$

As Example 2 illustrates, the remainder after division gives information about whether the divisor is, or is not, a factor. We shall have more to say about this in Chapter 3.

✏ **NOW WORK PROBLEMS 7 AND 17.**

R.6 Assess Your Understanding

Concepts and Vocabulary

1. To check division, the expression that is being divided, the dividend, should equal the product of the _____ and the _____ plus the _____.

2. To divide $2x^3 - 5x + 1$ by $x + 3$ using synthetic division, the first step is to write _____) _____.

3. *True or False:* In using synthetic division, the divisor is always a polynomial of degree 1, whose leading coefficient is 1.

4. *True or False:* $-2)\overline{5 \quad 3 \quad 2 \quad 1}$ means $\dfrac{5x^3 + 3x^2 + 2x + 1}{x + 2} = 5x^2 - 7x + 16 + \dfrac{-31}{x + 2}$
$$ \underline{-10 \quad 14 \quad -32}$$
$$ 5 \quad -7 \quad 16 \quad -31$$

Skill Building

In Problems 5–16, use synthetic division to find the quotient and remainder when:

5. $x^3 - x^2 + 2x + 4$ is divided by $x - 2$

6. $x^3 + 2x^2 - 3x + 1$ is divided by $x + 1$

7. $3x^3 + 2x^2 - x + 3$ is divided by $x - 3$

8. $-4x^3 + 2x^2 - x + 1$ is divided by $x + 2$

9. $x^5 - 4x^3 + x$ is divided by $x + 3$

10. $x^4 + x^2 + 2$ is divided by $x - 2$

11. $4x^6 - 3x^4 + x^2 + 5$ is divided by $x - 1$

12. $x^5 + 5x^3 - 10$ is divided by $x + 1$

13. $0.1x^3 + 0.2x$ is divided by $x + 1.1$

14. $0.1x^2 - 0.2$ is divided by $x + 2.1$

15. $x^5 - 1$ is divided by $x - 1$

16. $x^5 + 1$ is divided by $x + 1$

In Problems 17–26, use synthetic division to determine whether $x - c$ is a factor of the given polynomial.

17. $4x^3 - 3x^2 - 8x + 4; \quad x - 2$

18. $-4x^3 + 5x^2 + 8; \quad x + 3$

19. $3x^4 - 6x^3 - 5x + 10; \quad x - 2$

20. $4x^4 - 15x^2 - 4; \quad x - 2$

21. $3x^6 + 82x^3 + 27; \quad x + 3$

22. $2x^6 - 18x^4 + x^2 - 9; \quad x + 3$

23. $4x^6 - 64x^4 + x^2 - 15; \quad x + 4$

24. $x^6 - 16x^4 + x^2 - 16; \quad x + 4$

25. $2x^4 - x^3 + 2x - 1; \quad x - \dfrac{1}{2}$

26. $3x^4 + x^3 - 3x + 1; \quad x + \dfrac{1}{3}$

Applications and Extensions

27. Find the sum of $a, b, c,$ and d if

$$\frac{x^3 - 2x^2 + 3x + 5}{x + 2} = ax^2 + bx + c + \frac{d}{x + 2}$$

Discussion and Writing

28. When dividing a polynomial by $x - c$, do you prefer to use long division or synthetic division? Does the value of c make a difference to you in choosing? Give reasons.

R.7 Rational Expressions

> **OBJECTIVES** 1 Reduce a Rational Expression to Lowest Terms
>
> 2 Multiply and Divide Rational Expressions
>
> 3 Add and Subtract Rational Expressions
>
> 4 Use the Least Common Multiple Method
>
> 5 Simplify Complex Rational Expressions

1 Reduce a Rational Expression to Lowest Terms

If we form the quotient of two polynomials, the result is called a **rational expression**. Some examples of rational expressions are

(a) $\dfrac{x^3 + 1}{x}$ (b) $\dfrac{3x^2 + x - 2}{x^2 + 5}$ (c) $\dfrac{x}{x^2 - 1}$ (d) $\dfrac{xy^2}{(x - y)^2}$

Expressions (a), (b), and (c) are rational expressions in one variable, x, whereas (d) is a rational expression in two variables, x and y.

Rational expressions are described in the same manner as rational numbers. In expression (a), the polynomial $x^3 + 1$ is called the **numerator**, and x is called the **denominator**. When the numerator and denominator of a rational expression contain no common factors (except 1 and -1), we say that the rational expression is **reduced to lowest terms**, or **simplified**.

The polynomial in the denominator of a rational expression cannot be equal to 0 because division by 0 is not defined. For example, for the expression $\dfrac{x^3 + 1}{x}$, x cannot take on the value 0. The domain of the variable x is $\{x \mid x \neq 0\}$.

A rational expression is reduced to lowest terms by factoring completely the numerator and the denominator and canceling any common factors by using the Cancellation Property:

$$\frac{a\cancel{c}}{b\cancel{c}} = \frac{a}{b} \qquad \text{if } b \neq 0, c \neq 0 \tag{1}$$

| EXAMPLE 1 | **Reducing a Rational Expression to Lowest Terms** |

Reduce to lowest terms: $\dfrac{x^2 + 4x + 4}{x^2 + 3x + 2}$

Solution We begin by factoring the numerator and the denominator.

$$x^2 + 4x + 4 = (x + 2)(x + 2)$$
$$x^2 + 3x + 2 = (x + 2)(x + 1)$$

Since a common factor, $x + 2$, appears, the original expression is not in lowest terms. To reduce it to lowest terms, we use the Cancellation Property:

WARNING
Apply the Cancellation Property only to rational expressions written in factored form. Be sure to cancel only common factors! ■

$$\frac{x^2 + 4x + 4}{x^2 + 3x + 2} = \frac{\cancel{(x + 2)}(x + 2)}{\cancel{(x + 2)}(x + 1)} = \frac{x + 2}{x + 1} \qquad x \neq -2, -1 \qquad \blacktriangleleft$$

| EXAMPLE 2 | **Reducing Rational Expressions to Lowest Terms** |

Reduce each rational expression to lowest terms.

(a) $\dfrac{x^3 - 8}{x^3 - 2x^2}$ (b) $\dfrac{8 - 2x}{x^2 - x - 12}$

Solution (a) $\dfrac{x^3 - 8}{x^3 - 2x^2} = \dfrac{\cancel{(x - 2)}(x^2 + 2x + 4)}{x^2\cancel{(x - 2)}} = \dfrac{x^2 + 2x + 4}{x^2} \qquad x \neq 0, 2$

(b) $\dfrac{8 - 2x}{x^2 - x - 12} = \dfrac{2(4 - x)}{(x - 4)(x + 3)} = \dfrac{2(-1)\cancel{(x - 4)}}{\cancel{(x - 4)}(x + 3)} = \dfrac{-2}{x + 3} \qquad x \neq -3, 4 \qquad \blacktriangleleft$

✏ **NOW WORK PROBLEM 5.**

2 Multiply and Divide Rational Expressions

The rules for multiplying and dividing rational expressions are the same as the rules for multiplying and dividing rational numbers. If $\dfrac{a}{b}$ and $\dfrac{c}{d}$, $b \neq 0, d \neq 0$, are two rational expressions, then

$$\frac{a}{b} \cdot \frac{c}{d} = \frac{ac}{bd} \qquad \text{if } b \neq 0, d \neq 0 \tag{2}$$

$$\frac{\dfrac{a}{b}}{\dfrac{c}{d}} = \frac{a}{b} \cdot \frac{d}{c} = \frac{ad}{bc} \qquad \text{if } b \neq 0, c \neq 0, d \neq 0 \tag{3}$$

In using equations (2) and (3) with rational expressions, be sure first to factor each polynomial completely so that common factors can be canceled. Leave your answer in factored form.

EXAMPLE 3   **Multiplying and Dividing Rational Expressions**

Perform the indicated operation and simplify the result. Leave your answer in factored form.

(a) $\dfrac{x^2 - 2x + 1}{x^3 + x} \cdot \dfrac{4x^2 + 4}{x^2 + x - 2}$

(b) $\dfrac{\dfrac{x + 3}{x^2 - 4}}{\dfrac{x^2 - x - 12}{x^3 - 8}}$

Solution   (a) $\dfrac{x^2 - 2x + 1}{x^3 + x} \cdot \dfrac{4x^2 + 4}{x^2 + x - 2} = \dfrac{(x - 1)^2}{x(x^2 + 1)} \cdot \dfrac{4(x^2 + 1)}{(x + 2)(x - 1)}$

$$= \dfrac{(x - 1)^2(4)\cancel{(x^2 + 1)}}{x\cancel{(x^2 + 1)}(x + 2)\cancel{(x - 1)}}$$

$$= \dfrac{4(x - 1)}{x(x + 2)} \qquad x \neq -2, 0, 1$$

(b) $\dfrac{\dfrac{x + 3}{x^2 - 4}}{\dfrac{x^2 - x - 12}{x^3 - 8}} = \dfrac{x + 3}{x^2 - 4} \cdot \dfrac{x^3 - 8}{x^2 - x - 12}$

$$= \dfrac{x + 3}{(x - 2)(x + 2)} \cdot \dfrac{(x - 2)(x^2 + 2x + 4)}{(x - 4)(x + 3)}$$

$$= \dfrac{\cancel{(x + 3)}\,\cancel{(x - 2)}(x^2 + 2x + 4)}{\cancel{(x - 2)}(x + 2)(x - 4)\cancel{(x + 3)}}$$

$$= \dfrac{x^2 + 2x + 4}{(x + 2)(x - 4)} \qquad x \neq -3, -2, 2, 4$$

◀

─ **NOW WORK PROBLEMS 17 AND 25.**

3   **Add and Subtract Rational Expressions**

The rules for adding and subtracting rational expressions are the same as the rules for adding and subtracting rational numbers. Thus, if the denominators of two rational expressions to be added (or subtracted) are equal, we add (or subtract) the numerators and keep the common denominator.

In Words
To add (or subract) two rational expressions with the same denominator, keep the common denominator and add (or subtract) the numerators.

If $\dfrac{a}{b}$ and $\dfrac{c}{b}$ are two rational expressions, then

$$\dfrac{a}{b} + \dfrac{c}{b} = \dfrac{a + c}{b} \qquad \dfrac{a}{b} - \dfrac{c}{b} = \dfrac{a - c}{b} \qquad \text{if } b \neq 0 \tag{4}$$

EXAMPLE 4   **Adding and Subtracting Rational Expressions with Equal Denominators**

Perform the indicated operation and simplify the result. Leave your answer in factored form.

(a) $\dfrac{2x^2 - 4}{2x + 5} + \dfrac{x + 3}{2x + 5} \qquad x \neq -\dfrac{5}{2}$

(b) $\dfrac{x}{x - 3} - \dfrac{3x + 2}{x - 3} \qquad x \neq 3$

Solution   (a) $\dfrac{2x^2 - 4}{2x + 5} + \dfrac{x + 3}{2x + 5} = \dfrac{(2x^2 - 4) + (x + 3)}{2x + 5}$

$= \dfrac{2x^2 + x - 1}{2x + 5} = \dfrac{(2x - 1)(x + 1)}{2x + 5}$

(b) $\dfrac{x}{x - 3} - \dfrac{3x + 2}{x - 3} = \dfrac{x - (3x + 2)}{x - 3} = \dfrac{x - 3x - 2}{x - 3}$

$= \dfrac{-2x - 2}{x - 3} = \dfrac{-2(x + 1)}{x - 3}$ ◀

| EXAMPLE 5 | **Adding Rational Expressions Whose Denominators Are Additive Inverses of Each Other** |

Perform the indicated operation and simplify the result. Leave your answer in factored form.

$$\dfrac{2x}{x - 3} + \dfrac{5}{3 - x} \qquad x \neq 3$$

Solution   Notice that the denominators of the two rational expressions are different. However, the denominator of the second expression is the additive inverse of the denominator of the first. That is,

$$3 - x = -x + 3 = -1 \cdot (x - 3) = -(x - 3)$$

Then

$$\dfrac{2x}{x - 3} + \dfrac{5}{3 - x} = \dfrac{2x}{x - 3} + \dfrac{5}{-(x - 3)} = \dfrac{2x}{x - 3} + \dfrac{-5}{x - 3}$$

$$\underset{\substack{\uparrow \\ 3 - x = -(x - 3)}}{} \qquad \underset{\substack{\uparrow \\ \frac{a}{-b} = \frac{-a}{b}}}{}$$

$$= \dfrac{2x + (-5)}{x - 3} = \dfrac{2x - 5}{x - 3} \quad ◀$$

✏ **NOW WORK PROBLEMS 37 AND 43.**

If the denominators of two rational expressions to be added or subtracted are not equal, we can use the general formulas for adding and subtracting quotients.

$$\dfrac{a}{b} + \dfrac{c}{d} = \dfrac{a \cdot d}{b \cdot d} + \dfrac{b \cdot c}{b \cdot d} = \dfrac{ad + bc}{bd} \qquad \text{if } b \neq 0, d \neq 0 \qquad \textbf{(5a)}$$

$$\dfrac{a}{b} - \dfrac{c}{d} = \dfrac{a \cdot d}{b \cdot d} - \dfrac{b \cdot c}{b \cdot d} = \dfrac{ad - bc}{bd} \qquad \text{if } b \neq 0, d \neq 0 \qquad \textbf{(5b)}$$

| EXAMPLE 6 | **Adding and Subtracting Rational Expressions with Unequal Denominators** |

Perform the indicated operation and simplify the result. Leave your answer in factored form.

(a) $\dfrac{x - 3}{x + 4} + \dfrac{x}{x - 2} \qquad x \neq -4, 2$ (b) $\dfrac{x^2}{x^2 - 4} - \dfrac{1}{x} \qquad x \neq -2, 0, 2$

Solution   (a) $\dfrac{x-3}{x+4} + \dfrac{x}{x-2} = \dfrac{x-3}{x+4} \cdot \dfrac{x-2}{x-2} + \dfrac{x+4}{x+4} \cdot \dfrac{x}{x-2}$

(5a)

$$= \dfrac{(x-3)(x-2) + (x+4)(x)}{(x+4)(x-2)}$$

$$= \dfrac{x^2 - 5x + 6 + x^2 + 4x}{(x+4)(x-2)} = \dfrac{2x^2 - x + 6}{(x+4)(x-2)}$$

(b) $\dfrac{x^2}{x^2-4} - \dfrac{1}{x} = \dfrac{x^2}{x^2-4} \cdot \dfrac{x}{x} - \dfrac{x^2-4}{x^2-4} \cdot \dfrac{1}{x} = \dfrac{x^2(x) - (x^2-4)(1)}{(x^2-4)(x)}$

(5b)

$$= \dfrac{x^3 - x^2 + 4}{(x-2)(x+2)(x)}$$ ◀

NOW WORK PROBLEM **47.**

4 Use the Least Common Multiple Method

If the denominators of two rational expressions to be added (or subtracted) have common factors, we usually do not use the general rules given by equations (5a) and (5b). Just as with fractions, we apply the **least common multiple (LCM) method**. The LCM method uses the polynomial of least degree that has each denominator polynomial as a factor.

The LCM Method for Adding or Subtracting Rational Expressions

The Least Common Multiple (LCM) Method requires four steps:

STEP 1: Factor completely the polynomial in the denominator of each rational expression.

STEP 2: The LCM of the denominator is the product of each of these factors raised to a power equal to the greatest number of times that the factor occurs in the polynomials.

STEP 3: Write each rational expression using the LCM as the common denominator.

STEP 4: Add or subtract the rational expressions using equation (4).

Let's work an example that only requires Steps 1 and 2.

EXAMPLE 7   ### Finding the Least Common Multiple

Find the least common multiple of the following pair of polynomials:

$$x(x-1)^2(x+1) \quad \text{and} \quad 4(x-1)(x+1)^3$$

Solution   **STEP 1:** The polynomials are already factored completely as

$$x(x-1)^2(x+1) \quad \text{and} \quad 4(x-1)(x+1)^3$$

STEP 2: Start by writing the factors of the left-hand polynomial. (Or you could start with the one on the right.)

$$x(x - 1)^2(x + 1)$$

Now look at the right-hand polynomial. Its first factor, 4, does not appear in our list, so we insert it.

$$4x(x - 1)^2(x + 1)$$

The next factor, $x - 1$, is already in our list, so no change is necessary. The final factor is $(x + 1)^3$. Since our list has $x + 1$ to the first power only, we replace $x + 1$ in the list by $(x + 1)^3$. The LCM is

$$4x(x - 1)^2(x + 1)^3 \qquad \blacktriangleleft$$

Notice that the LCM is, in fact, the polynomial of least degree that contains $x(x - 1)^2(x + 1)$ and $4(x - 1)(x + 1)^3$ as factors.

━━✎━━ **NOW WORK PROBLEM 53.**

EXAMPLE 8	**Using the Least Common Multiple to Add Rational Expressions**

Perform the indicated operation and simplify the result. Leave your answer in factored form.

$$\frac{x}{x^2 + 3x + 2} + \frac{2x - 3}{x^2 - 1} \qquad x \neq -2, -1, 1$$

Solution   **STEP 1:** Factor completely the polynomials in the denominators.

$$x^2 + 3x + 2 = (x + 2)(x + 1)$$
$$x^2 - 1 = (x - 1)(x + 1)$$

STEP 2: The LCM is $(x + 2)(x + 1)(x - 1)$. Do you see why?

STEP 3: Write each rational expression using the LCM as the denominator.

$$\frac{x}{x^2 + 3x + 2} = \frac{x}{(x + 2)(x + 1)} = \frac{x}{(x + 2)(x + 1)} \cdot \frac{x - 1}{x - 1} = \frac{x(x - 1)}{(x + 2)(x + 1)(x - 1)}$$

Multiply numerator and denominator by $x - 1$ to get the LCM in the denominator.

$$\frac{2x - 3}{x^2 - 1} = \frac{2x - 3}{(x - 1)(x + 1)} = \frac{2x - 3}{(x - 1)(x + 1)} \cdot \frac{x + 2}{x + 2} = \frac{(2x - 3)(x + 2)}{(x - 1)(x + 1)(x + 2)}$$

Multiply numerator and denominator by $x + 2$ to get the LCM in the denominator.

STEP 4: Now we can add by using equation (4).

$$\frac{x}{x^2 + 3x + 2} + \frac{2x - 3}{x^2 - 1} = \frac{x(x - 1)}{(x + 2)(x + 1)(x - 1)} + \frac{(2x - 3)(x + 2)}{(x + 2)(x + 1)(x - 1)}$$

$$= \frac{(x^2 - x) + (2x^2 + x - 6)}{(x + 2)(x + 1)(x - 1)}$$

$$= \frac{3x^2 - 6}{(x + 2)(x + 1)(x - 1)} = \frac{3(x^2 - 2)}{(x + 2)(x + 1)(x - 1)} \qquad \blacktriangleleft$$

| EXAMPLE 9 | **Using the Least Common Multiple to Subtract Rational Expressions** |

Perform the indicated operations and simplify the result. Leave your answer in factored form.

$$\frac{3}{x^2 + x} - \frac{x + 4}{x^2 + 2x + 1} \qquad x \ne -1, 0$$

Solution **STEP 1:** Factor completely the polynomials in the denominators.

$$x^2 + x = x(x + 1)$$
$$x^2 + 2x + 1 = (x + 1)^2$$

STEP 2: The LCM is $x(x + 1)^2$.

STEP 3: Write each rational expression using the LCM as the denominator.

$$\frac{3}{x^2 + x} = \frac{3}{x(x + 1)} = \frac{3}{x(x + 1)} \cdot \frac{x + 1}{x + 1} = \frac{3(x + 1)}{x(x + 1)^2}$$

$$\frac{x + 4}{x^2 + 2x + 1} = \frac{x + 4}{(x + 1)^2} = \frac{x + 4}{(x + 1)^2} \cdot \frac{x}{x} = \frac{x(x + 4)}{x(x + 1)^2}$$

STEP 4: Subtract, using equation (4).

$$\frac{3}{x^2 + x} - \frac{x + 4}{x^2 + 2x + 1} = \frac{3(x + 1)}{x(x + 1)^2} - \frac{x(x + 4)}{x(x + 1)^2}$$

$$= \frac{3(x + 1) - x(x + 4)}{x(x + 1)^2}$$

$$= \frac{3x + 3 - x^2 - 4x}{x(x + 1)^2}$$

$$= \frac{-x^2 - x + 3}{x(x + 1)^2} \qquad \blacktriangleleft$$

NOW WORK PROBLEM 63.

5 Simplify Complex Rational Expressions

When sums and/or differences of rational expressions appear as the numerator and/or denominator of a quotient, the quotient is called a **complex rational expression**.[*] For example,

$$\frac{1 + \dfrac{1}{x}}{1 - \dfrac{1}{x}} \qquad \text{and} \qquad \frac{\dfrac{x^2}{x^2 - 4} - 3}{\dfrac{x - 3}{x + 2} - 1}$$

are complex rational expressions. To **simplify** a complex rational expression means to write it as a rational expression reduced to lowest terms. This can be accomplished in either of two ways.

*Some texts use the term **complex fraction**.

Simplifying a Complex Rational Expression

METHOD 1: Treat the numerator and denominator of the complex rational expression separately, performing whatever operations are indicated and simplifying the results. Follow this by simplifying the resulting rational expression.

METHOD 2: Find the LCM of the denominators of all rational expressions that appear in the complex rational expression. Multiply the numerator and denominator of the complex rational expression by the LCM and simplify the result.

We will use both methods in the next example. By carefully studying each method, you can discover situations in which one method may be easier to use than the other.

EXAMPLE 10 **Simplifying a Complex Rational Expression**

Simplify: $\dfrac{\dfrac{1}{2}+\dfrac{3}{x}}{\dfrac{x+3}{4}}$ $x \neq -3, 0$

Solution **Method 1:** First, we perform the indicated operation in the numerator, and then we divide.

$$\frac{\dfrac{1}{2}+\dfrac{3}{x}}{\dfrac{x+3}{4}} = \underset{\substack{\uparrow \\ \text{Rule for adding quotients}}}{\frac{\dfrac{1\cdot x + 2\cdot 3}{2\cdot x}}{\dfrac{x+3}{4}}} = \frac{\dfrac{x+6}{2x}}{\dfrac{x+3}{4}} = \underset{\substack{\uparrow \\ \text{Rule for dividing quotients}}}{\frac{x+6}{2x}\cdot\frac{4}{x+3}}$$

$$= \underset{\substack{\uparrow \\ \text{Rule for multiplying quotients}}}{\frac{(x+6)\cdot 4}{2\cdot x\cdot(x+3)}} = \frac{\cancel{2}\cdot 2\cdot(x+6)}{\cancel{2}\cdot x\cdot(x+3)} = \frac{2(x+6)}{x(x+3)}$$

Method 2: The rational expressions that appear in the complex rational expression are

$$\frac{1}{2},\ \frac{3}{x},\ \frac{x+3}{4}$$

The LCM of their denominators is $4x$. We multiply the numerator and denominator of the complex rational expression by $4x$ and then simplify.

$$\frac{\dfrac{1}{2}+\dfrac{3}{x}}{\dfrac{x+3}{4}} = \underset{\substack{\uparrow \\ \text{Multiply the} \\ \text{numerator and} \\ \text{denominator by } 4x.}}{\frac{4x\cdot\left(\dfrac{1}{2}+\dfrac{3}{x}\right)}{4x\cdot\left(\dfrac{x+3}{4}\right)}} = \underset{\substack{\uparrow \\ \text{Distributive Property} \\ \text{in numerator}}}{\frac{4x\cdot\dfrac{1}{2}+4x\cdot\dfrac{3}{x}}{\dfrac{4x\cdot(x+3)}{4}}}$$

$$= \frac{2\cdot 2x\cdot\dfrac{1}{\cancel{2}}+4\cancel{x}\cdot\dfrac{3}{\cancel{x}}}{\dfrac{\cancel{4}x\cdot(x+3)}{\cancel{4}}} = \underset{\substack{\uparrow \\ \text{Simplify}}}{\frac{2x+12}{x(x+3)}} = \underset{\substack{\uparrow \\ \text{Factor}}}{\frac{2(x+6)}{x(x+3)}}$$

◀

| EXAMPLE 11 | **Simplifying a Complex Rational Expression** |

Simplify: $\dfrac{\dfrac{x^2}{x-4}+2}{\dfrac{2x-2}{x}-1}$

Solution   We will use Method 1.

$$\frac{\dfrac{x^2}{x-4}+2}{\dfrac{2x-2}{x}-1}=\frac{\dfrac{x^2}{x-4}+\dfrac{2(x-4)}{x-4}}{\dfrac{2x-2}{x}-\dfrac{x}{x}}=\frac{\dfrac{x^2+2x-8}{x-4}}{\dfrac{2x-2-x}{x}}$$

$$=\frac{\dfrac{(x+4)(x-2)}{x-4}}{\dfrac{x-2}{x}}=\frac{(x+4)\cancel{(x-2)}}{x-4}\cdot\frac{x}{\cancel{x-2}}$$

$$=\frac{(x+4)\cdot x}{x-4}$$

◀

━━━◖ NOW WORK PROBLEM **73**.

Application

| EXAMPLE 12 | **Solving an Application in Electricity** |

An electrical circuit contains two resistors connected in parallel, as shown in Figure 23. If the resistance of each is R_1 and R_2 ohms, respectively, their combined resistance R is given by the formula

$$R=\frac{1}{\dfrac{1}{R_1}+\dfrac{1}{R_2}}$$

Express R as a rational expression; that is, simplify the right-hand side of this formula. Evaluate the rational expression if $R_1=6$ ohms and $R_2=10$ ohms.

Solution   We will use Method 2. If we consider 1 as the fraction $\dfrac{1}{1}$, then the rational expressions in the complex rational expression are

$$\frac{1}{1},\ \frac{1}{R_1},\ \frac{1}{R_2}$$

Figure 23

The LCM of the denominators is R_1R_2. We multiply the numerator and denominator of the complex rational expression by R_1R_2 and simplify.

$$\frac{1}{\dfrac{1}{R_1}+\dfrac{1}{R_2}}=\frac{1\cdot R_1R_2}{\left(\dfrac{1}{R_1}+\dfrac{1}{R_2}\right)\cdot R_1R_2}=\frac{R_1R_2}{\dfrac{1}{R_1}\cdot R_1R_2+\dfrac{1}{R_2}\cdot R_1R_2}=\frac{R_1R_2}{R_2+R_1}$$

Thus,

$$R=\frac{R_1R_2}{R_2+R_1}$$

If $R_1=6$ and $R_2=10$, then

$$R=\frac{6\cdot 10}{10+6}=\frac{60}{16}=\frac{15}{4}\quad\text{ohms}$$

◀

R.7 Assess Your Understanding

Concepts and Vocabulary

1. When the numerator and denominator of a rational expression contain no common factors (except 1 and −1), the rational expression is in _____ _____.

2. LCM is an abbreviation for _____ _____ _____.

3. *True or False:* The rational expression $\dfrac{2x^3 - 4x}{x - 2}$ is reduced to lowest terms.

4. *True or False:* The LCM of $2x^3 + 6x^2$ and $6x^4 + 4x^3$ is $4x^3(x + 1)$.

Skill Building

In Problems 5–16, reduce each rational expression to lowest terms.

5. $\dfrac{3x + 9}{x^2 - 9}$

6. $\dfrac{4x^2 + 8x}{12x + 24}$

7. $\dfrac{x^2 - 2x}{3x - 6}$

8. $\dfrac{15x^2 + 24x}{3x^2}$

9. $\dfrac{24x^2}{12x^2 - 6x}$

10. $\dfrac{x^2 + 4x + 4}{x^2 - 16}$

11. $\dfrac{y^2 - 25}{2y^2 - 8y - 10}$

12. $\dfrac{3y^2 - y - 2}{3y^2 + 5y + 2}$

13. $\dfrac{x^2 + 4x - 5}{x^2 - 2x + 1}$

14. $\dfrac{x - x^2}{x^2 + x - 2}$

15. $\dfrac{x^2 + 5x - 14}{2 - x}$

16. $\dfrac{2x^2 + 5x - 3}{1 - 2x}$

In Problems 17–34, perform the indicated operation and simplify the result. Leave your answer in factored form.

17. $\dfrac{3x + 6}{5x^2} \cdot \dfrac{x}{x^2 - 4}$

18. $\dfrac{3}{2x} \cdot \dfrac{x^2}{6x + 10}$

19. $\dfrac{4x^2}{x^2 - 16} \cdot \dfrac{x - 4}{2x}$

20. $\dfrac{12}{x^2 - x} \cdot \dfrac{x^2 - 1}{4x - 2}$

21. $\dfrac{4x - 8}{-3x} \cdot \dfrac{12}{12 - 6x}$

22. $\dfrac{6x - 27}{5x} \cdot \dfrac{2}{4x - 18}$

23. $\dfrac{x^2 - 3x - 10}{x^2 + 2x - 35} \cdot \dfrac{x^2 + 4x - 21}{x^2 + 9x + 14}$

24. $\dfrac{x^2 + x - 6}{x^2 + 4x - 5} \cdot \dfrac{x^2 - 25}{x^2 + 2x - 15}$

25. $\dfrac{\dfrac{6x}{x^2 - 4}}{\dfrac{3x - 9}{2x + 4}}$

26. $\dfrac{\dfrac{12x}{5x + 20}}{\dfrac{4x^2}{x^2 - 16}}$

27. $\dfrac{\dfrac{8x}{x^2 - 1}}{\dfrac{10x}{x + 1}}$

28. $\dfrac{\dfrac{x - 2}{4x}}{\dfrac{x^2 - 4x + 4}{12x}}$

29. $\dfrac{\dfrac{4 - x}{4 + x}}{\dfrac{4x}{x^2 - 16}}$

30. $\dfrac{\dfrac{3 + x}{3 - x}}{\dfrac{x^2 - 9}{9x^3}}$

31. $\dfrac{\dfrac{x^2 + 7x + 12}{x^2 - 7x + 12}}{\dfrac{x^2 + x - 12}{x^2 - x - 12}}$

32. $\dfrac{\dfrac{x^2 + 7x + 6}{x^2 + x - 6}}{\dfrac{x^2 + 5x - 6}{x^2 + 5x + 6}}$

33. $\dfrac{\dfrac{2x^2 - x - 28}{3x^2 - x - 2}}{\dfrac{4x^2 + 16x + 7}{3x^2 + 11x + 6}}$

34. $\dfrac{\dfrac{9x^2 + 3x - 2}{12x^2 + 5x - 2}}{\dfrac{9x^2 - 6x + 1}{8x^2 - 10x - 3}}$

In Problems 35–52, perform the indicated operations and simplify the result. Leave your answer in factored form.

35. $\dfrac{x}{2} + \dfrac{5}{2}$

36. $\dfrac{3}{x} - \dfrac{6}{x}$

37. $\dfrac{x^2}{2x - 3} - \dfrac{4}{2x - 3}$

38. $\dfrac{3x^2}{2x - 1} - \dfrac{9}{2x - 1}$

39. $\dfrac{x + 1}{x - 3} + \dfrac{2x - 3}{x - 3}$

40. $\dfrac{2x - 5}{3x + 2} + \dfrac{x + 4}{3x + 2}$

41. $\dfrac{3x + 5}{2x - 1} - \dfrac{2x - 4}{2x - 1}$

42. $\dfrac{5x - 4}{3x + 4} - \dfrac{x + 1}{3x + 4}$

43. $\dfrac{4}{x - 2} + \dfrac{x}{2 - x}$

44. $\dfrac{6}{x - 1} - \dfrac{x}{1 - x}$

45. $\dfrac{4}{x - 1} - \dfrac{2}{x + 2}$

46. $\dfrac{2}{x + 5} - \dfrac{5}{x - 5}$

47. $\dfrac{x}{x + 1} + \dfrac{2x - 3}{x - 1}$

48. $\dfrac{3x}{x - 4} + \dfrac{2x}{x + 3}$

49. $\dfrac{x - 3}{x + 2} - \dfrac{x + 4}{x - 2}$

50. $\dfrac{2x - 3}{x - 1} - \dfrac{2x + 1}{x + 1}$

51. $\dfrac{x}{x^2 - 4} + \dfrac{1}{x}$

52. $\dfrac{x - 1}{x^3} + \dfrac{x}{x^2 + 1}$

In Problems 53–60, find the LCM of the given polynomials.

53. $x^2 - 4,\quad x^2 - x - 2$

54. $x^2 - x - 12,\quad x^2 - 8x + 16$

55. $x^3 - x,\quad x^2 - x$

56. $3x^2 - 27,\quad 2x^2 - x - 15$

57. $4x^3 - 4x^2 + x,\quad 2x^3 - x^2,\quad x^3$

58. $x - 3,\quad x^2 + 3x,\quad x^3 - 9x$

59. $x^3 - x,\quad x^3 - 2x^2 + x,\quad x^3 - 1$

60. $x^2 + 4x + 4,\quad x^3 + 2x^2,\quad (x + 2)^3$

In Problems 61–72, perform the indicated operations and simplify the result. Leave your answer in factored form.

61. $\dfrac{x}{x^2 - 7x + 6} - \dfrac{x}{x^2 - 2x - 24}$

62. $\dfrac{x}{x - 3} - \dfrac{x + 1}{x^2 + 5x - 24}$

63. $\dfrac{4x}{x^2 - 4} - \dfrac{2}{x^2 + x - 6}$

64. $\dfrac{3x}{x - 1} - \dfrac{x - 4}{x^2 - 2x + 1}$

65. $\dfrac{3}{(x - 1)^2(x + 1)} + \dfrac{2}{(x - 1)(x + 1)^2}$

66. $\dfrac{2}{(x + 2)^2(x - 1)} - \dfrac{6}{(x + 2)(x - 1)^2}$

67. $\dfrac{x + 4}{x^2 - x - 2} - \dfrac{2x + 3}{x^2 + 2x - 8}$

68. $\dfrac{2x - 3}{x^2 + 8x + 7} - \dfrac{x - 2}{(x + 1)^2}$

69. $\dfrac{1}{x} - \dfrac{2}{x^2 + x} + \dfrac{3}{x^3 - x^2}$

70. $\dfrac{x}{(x - 1)^2} + \dfrac{2}{x} - \dfrac{x + 1}{x^3 - x^2}$

71. $\dfrac{1}{h}\left(\dfrac{1}{x + h} - \dfrac{1}{x}\right)$

72. $\dfrac{1}{h}\left[\dfrac{1}{(x + h)^2} - \dfrac{1}{x^2}\right]$

In Problems 73–82, perform the indicated operations and simplify the result. Leave your answer in factored form.

73. $\dfrac{1 + \dfrac{1}{x}}{1 - \dfrac{1}{x}}$

74. $\dfrac{4 + \dfrac{1}{x^2}}{3 - \dfrac{1}{x^2}}$

75. $\dfrac{2 - \dfrac{x + 1}{x}}{3 + \dfrac{x - 1}{x + 1}}$

76. $\dfrac{1 - \dfrac{x}{x + 1}}{2 - \dfrac{x - 1}{x}}$

77. $\dfrac{\dfrac{x + 4}{x - 2} - \dfrac{x - 3}{x + 1}}{x + 1}$

78. $\dfrac{\dfrac{x - 2}{x + 1} - \dfrac{x}{x - 2}}{x + 3}$

79. $\dfrac{\dfrac{x - 2}{x + 2} + \dfrac{x - 1}{x + 1}}{\dfrac{x}{x + 1} - \dfrac{2x - 3}{x}}$

80. $\dfrac{\dfrac{2x + 5}{x} - \dfrac{x}{x - 3}}{\dfrac{x^2}{x - 3} - \dfrac{(x + 1)^2}{x + 3}}$

81. $1 - \dfrac{1}{1 - \dfrac{1}{x}}$

82. $1 - \dfrac{1}{1 - \dfrac{1}{1 - x}}$

Applications and Extensions

In Problems 83–90, expressions that occur in calculus are given. Reduce each expression to lowest terms.

83. $\dfrac{(2x + 3)\cdot 3 - (3x - 5)\cdot 2}{(3x - 5)^2}$

84. $\dfrac{(4x + 1)\cdot 5 - (5x - 2)\cdot 4}{(5x - 2)^2}$

85. $\dfrac{x\cdot 2x - (x^2 + 1)\cdot 1}{(x^2 + 1)^2}$

86. $\dfrac{x\cdot 2x - (x^2 - 4)\cdot 1}{(x^2 - 4)^2}$

87. $\dfrac{(3x + 1)\cdot 2x - x^2\cdot 3}{(3x + 1)^2}$

88. $\dfrac{(2x - 5)\cdot 3x^2 - x^3\cdot 2}{(2x - 5)^2}$

89. $\dfrac{(x^2 + 1)\cdot 3 - (3x + 4)\cdot 2x}{(x^2 + 1)^2}$

90. $\dfrac{(x^2 + 9)\cdot 2 - (2x - 5)\cdot 2x}{(x^2 + 9)^2}$

91. The Lensmaker's Equation   The focal length f of a lens with index of refraction n is

$$\dfrac{1}{f} = (n - 1)\left[\dfrac{1}{R_1} + \dfrac{1}{R_2}\right]$$

where R_1 and R_2 are the radii of curvature of the front and back surfaces of the lens. Express f as a rational expression. Evaluate the rational expression for $n = 1.5$, $R_1 = 0.1$ meter, and $R_2 = 0.2$ meter.

92. Electrical Circuits   An electrical circuit contains three resistors connected in parallel. If the resistance of each is $R_1, R_2,$ and R_3 ohms, respectively, their combined resistance R is given by the formula

$$\dfrac{1}{R} = \dfrac{1}{R_1} + \dfrac{1}{R_2} + \dfrac{1}{R_3}$$

Express R as a rational expression. Evaluate R for $R_1 = 5$ ohms, $R_2 = 4$ ohms, and $R_3 = 10$ ohms.

Discussion and Writing

93. The following expressions are called **continued fractions**:

$$1 + \frac{1}{x}, \quad 1 + \cfrac{1}{1 + \cfrac{1}{x}}, \quad 1 + \cfrac{1}{1 + \cfrac{1}{1 + \cfrac{1}{x}}}, \quad 1 + \cfrac{1}{1 + \cfrac{1}{1 + \cfrac{1}{1 + \cfrac{1}{x}}}}, \quad \ldots$$

Each simplifies to an expression of the form

$$\frac{ax + b}{bx + c}$$

Trace the successive values of $a, b,$ and c as you "continue" the fraction. Can you discover the patterns that these values follow? Go to the library and research Fibonacci numbers. Write a report on your findings.

94. Explain to a fellow student when you would use the LCM method to add two rational expressions. Give two examples of adding two rational expressions, one in which you use the LCM and the other in which you do not.

95. Which of the two methods given in the text for simplifying complex rational expressions do you prefer? Write a brief paragraph stating the reasons for your choice.

R.8 *nth* Roots; Rational Exponents

PREPARING FOR THIS SECTION   *Before getting started, review the following:*

• Exponents, Square Roots (Section R.2, pp. 20–23)

Now Work the 'Are You Prepared?' problems on page 74

OBJECTIVES   1 Work with *nth* Roots
2 Simplify Radicals
3 Rationalize Denominators
4 Simplify Expressions with Rational Exponents

---

1 Work with *nth* Roots

The **principal *n*th root of a real number** a, $n \geq 2$ an integer, symbolized by $\sqrt[n]{a}$, is defined as follows:

$$\sqrt[n]{a} = b \quad \text{means} \quad a = b^n$$

where $a \geq 0$ and $b \geq 0$ if n is even and a, b are any real numbers if n is odd.

Notice that if a is negative and n is even then $\sqrt[n]{a}$ is not defined. When it is defined, the principal *n*th root of a number is unique.

The symbol $\sqrt[n]{a}$ for the principal *n*th root of a is called a **radical**; the integer n is called the **index**, and a is called the **radicand**. If the index of a radical is 2, we call $\sqrt[2]{a}$ the **square root** of a and omit the index 2 by simply writing $\sqrt{a}$. If the index is 3, we call $\sqrt[3]{a}$ the **cube root** of a.

EXAMPLE 1   **Simplifying Principal *nth* Roots**

(a) $\sqrt[3]{8} = \sqrt[3]{2^3} = 2$

(b) $\sqrt[3]{-64} = \sqrt[3]{(-4)^3} = -4$

(c) $\sqrt[4]{\dfrac{1}{16}} = \sqrt[4]{\left(\dfrac{1}{2}\right)^4} = \dfrac{1}{2}$

(d) $\sqrt[6]{(-2)^6} = |-2| = 2$ ◀

These are examples of **perfect roots**, since each simplifies to a rational number. Notice the absolute value in Example 1(d). If *n* is even, the principal *n*th root must be nonnegative.

In general, if $n \geq 2$ is an integer and *a* is a real number, we have

$$\sqrt[n]{a^n} = a \qquad \text{if } n \geq 3 \text{ is odd} \tag{1a}$$

$$\sqrt[n]{a^n} = |a| \qquad \text{if } n \geq 2 \text{ is even} \tag{1b}$$

━━━━━━ **NOW WORK PROBLEM 7.**

Radicals provide a way of representing many irrational real numbers. For example, there is no rational number whose square is 2. Using radicals, we can say that $\sqrt{2}$ is the positive number whose square is 2.

| EXAMPLE 2 | **Using a Calculator to Approximate Roots**

Use a calculator to approximate $\sqrt[5]{16}$.

Figure 24 · **Solution** Figure 24 shows the result. ◀

```
5*⁵√16
       1.741101127
```

━━━━━━ **NOW WORK PROBLEM 101.**

2 Simplify Radicals

Let $n \geq 2$ and $m \geq 2$ denote positive integers, and let *a* and *b* represent real numbers. Assuming that all radicals are defined, we have the following properties:

Properties of Radicals

$$\sqrt[n]{ab} = \sqrt[n]{a} \, \sqrt[n]{b} \tag{2a}$$

$$\sqrt[n]{\frac{a}{b}} = \frac{\sqrt[n]{a}}{\sqrt[n]{b}} \tag{2b}$$

$$\sqrt[n]{a^m} = \left(\sqrt[n]{a}\right)^m \tag{2c}$$

When used in reference to radicals, the direction to "simplify" will mean to remove from the radicals any perfect roots that occur as factors. Let's look at some examples of how the preceding rules are applied to simplify radicals.

| EXAMPLE 3 | **Simplifying Radicals**

(a) $\sqrt{32} = \sqrt{16 \cdot 2} = \sqrt{16} \cdot \sqrt{2} = 4\sqrt{2}$

    ↑
    Factor out 16, a perfect square.

(b) $\sqrt[3]{16} = \sqrt[3]{8 \cdot 2} = \sqrt[3]{8} \cdot \sqrt[3]{2} = \sqrt[3]{2^3} \cdot \sqrt[3]{2} = 2\sqrt[3]{2}$

    ↑   ↑
    Factor   (2a)
    out 8,
    a perfect cube.

(c) $\sqrt[3]{-16x^4} = \sqrt[3]{-8 \cdot 2 \cdot x^3 \cdot x} = \sqrt[3]{(-8x^3)(2x)}$

        ↑   ↑

      Factor perfect   Combine perfect

      cubes inside radical. cubes.

$$= \sqrt[3]{(-2x)^3 \cdot 2x} = \sqrt[3]{(-2x)^3} \cdot \sqrt[3]{2x} = -2x\sqrt[3]{2x}$$

                  ↑

                (2a)

(d) $\sqrt[3]{\dfrac{8x^5}{27}} = \sqrt[3]{\dfrac{2^3 x^3 x^2}{3^3}} = \sqrt[3]{\left(\dfrac{2x}{3}\right)^3 \cdot x^2} = \sqrt[3]{\left(\dfrac{2x}{3}\right)^3} \cdot \sqrt[3]{x^2} = \dfrac{2x}{3}\sqrt[3]{x^2}$ ◀

NOW WORK PROBLEMS 11 AND 17.

Two or more radicals can be combined, provided that they have the same index and the same radicand. Such radicals are called **like radicals**.

EXAMPLE 4   **Combining Like Radicals**

(a) $-8\sqrt{12} + \sqrt{3} = -8\sqrt{4 \cdot 3} + \sqrt{3}$
$$= -8 \cdot \sqrt{4}\sqrt{3} + \sqrt{3}$$
$$= -16\sqrt{3} + \sqrt{3} = -15\sqrt{3}$$

(b) $\sqrt[3]{8x^4} + \sqrt[3]{-x} + 4\sqrt[3]{27x} = \sqrt[3]{2^3 x^3 x} + \sqrt[3]{-1 \cdot x} + 4\sqrt[3]{3^3 x}$
$$= \sqrt[3]{(2x)^3} \cdot \sqrt[3]{x} + \sqrt[3]{-1} \cdot \sqrt[3]{x} + 4\sqrt[3]{3^3} \cdot \sqrt[3]{x}$$
$$= 2x\sqrt[3]{x} - 1 \cdot \sqrt[3]{x} + 12\sqrt[3]{x}$$
$$= (2x + 11)\sqrt[3]{x}$$ ◀

NOW WORK PROBLEM 33.

3 **Rationalize Denominators**

When radicals occur in quotients, it is customary to rewrite the quotient so that the denominator contains no radicals. This process is referred to as **rationalizing the denominator**.

The idea is to multiply by an appropriate expression so that the new denominator contains no square roots. For example:

If Denominator Contains the Factor	Multiply by	To Obtain Denominator Free of Radicals
$\sqrt{3}$	$\sqrt{3}$	$\left(\sqrt{3}\right)^2 = 3$
$\sqrt{3} + 1$	$\sqrt{3} - 1$	$\left(\sqrt{3}\right)^2 - 1^2 = 3 - 1 = 2$
$\sqrt{2} - 3$	$\sqrt{2} + 3$	$\left(\sqrt{2}\right)^2 - 3^2 = 2 - 9 = -7$
$\sqrt{5} - \sqrt{3}$	$\sqrt{5} + \sqrt{3}$	$\left(\sqrt{5}\right)^2 - \left(\sqrt{3}\right)^2 = 5 - 3 = 2$
$\sqrt[3]{4}$	$\sqrt[3]{2}$	$\sqrt[3]{4} \cdot \sqrt[3]{2} = \sqrt[3]{8} = 2$

In rationalizing the denominator of a quotient, be sure to multiply both the numerator and the denominator by the expression.

EXAMPLE 5	**Rationalizing Denominators**

Rationalize the denominator of each expression:

(a) $\dfrac{1}{\sqrt{3}}$ (b) $\dfrac{5}{4\sqrt{2}}$ (c) $\dfrac{\sqrt{2}}{\sqrt{3}-\sqrt{2}}$

Solution (a) The denominator contains the factor $\sqrt{3}$, so we multiply the numerator and denominator by $\sqrt{3}$ to obtain

$$\frac{1}{\sqrt{3}} = \frac{1}{\sqrt{3}} \cdot \frac{\sqrt{3}}{\sqrt{3}} = \frac{\sqrt{3}}{\left(\sqrt{3}\right)^2} = \frac{\sqrt{3}}{3}$$

(b) The denominator contains the factor $\sqrt{2}$, so we multiply the numerator and denominator by $\sqrt{2}$ to obtain

$$\frac{5}{4\sqrt{2}} = \frac{5}{4\sqrt{2}} \cdot \frac{\sqrt{2}}{\sqrt{2}} = \frac{5\sqrt{2}}{4\left(\sqrt{2}\right)^2} = \frac{5\sqrt{2}}{4\cdot 2} = \frac{5\sqrt{2}}{8}$$

(c) The denominator contains the factor $\sqrt{3}-\sqrt{2}$, so we multiply the numerator and denominator by $\sqrt{3}+\sqrt{2}$ to obtain

$$\frac{\sqrt{2}}{\sqrt{3}-\sqrt{2}} = \frac{\sqrt{2}}{\sqrt{3}-\sqrt{2}} \cdot \frac{\sqrt{3}+\sqrt{2}}{\sqrt{3}+\sqrt{2}} = \frac{\sqrt{2}\left(\sqrt{3}+\sqrt{2}\right)}{\left(\sqrt{3}\right)^2 - \left(\sqrt{2}\right)^2}$$

$$= \frac{\sqrt{2}\sqrt{3}+\left(\sqrt{2}\right)^2}{3-2} = \sqrt{6}+2 \quad \blacktriangleleft$$

NOW WORK PROBLEM 47.

4 Simplify Expressions with Rational Exponents

Radicals are used to define rational exponents.

If a is a real number and $n \geq 2$ is an integer, then

$$a^{1/n} = \sqrt[n]{a} \tag{3}$$

provided that $\sqrt[n]{a}$ exists.

Note that if n is even and $a < 0$ then $\sqrt[n]{a}$ and $a^{1/n}$ do not exist.

EXAMPLE 6	**Writing Expressions Containing Fractional Exponents as Radicals**

(a) $4^{1/2} = \sqrt{4} = 2$ (b) $8^{1/2} = \sqrt{8} = 2\sqrt{2}$

(c) $(-27)^{1/3} = \sqrt[3]{-27} = -3$ (d) $16^{1/3} = \sqrt[3]{16} = 2\sqrt[3]{2}$ $\quad \blacktriangleleft$

If a is a real number and m and n are integers containing no common factors, with $n \geq 2$, then

$$a^{m/n} = \sqrt[n]{a^m} = \left(\sqrt[n]{a}\right)^m \tag{4}$$

provided that $\sqrt[n]{a}$ exists.

We have two comments about equation (4):

1. The exponent $\dfrac{m}{n}$ must be in lowest terms and n must be positive.

2. In simplifying the rational expression $a^{m/n}$, either $\sqrt[n]{a^m}$ or $(\sqrt[n]{a})^m$ may be used, the choice depending on which is easier to simplify. Generally, taking the root first, as in $(\sqrt[n]{a})^m$, is easier.

EXAMPLE 7

Using Equation (4)

(a) $4^{3/2} = \left(\sqrt{4}\right)^3 = 2^3 = 8$ (b) $(-8)^{4/3} = \left(\sqrt[3]{-8}\right)^4 = (-2)^4 = 16$

(c) $(32)^{-2/5} = \left(\sqrt[5]{32}\right)^{-2} = 2^{-2} = \dfrac{1}{4}$ (d) $25^{6/4} = 25^{3/2} = \left(\sqrt{25}\right)^3 = 5^3 = 125$ ◀

NOW WORK PROBLEM 55.

It can be shown that the Laws of Exponents hold for rational exponents. The next example illustrates using the Law of Exponents to simplify.

EXAMPLE 8

Simplifying Expressions Containing Rational Exponents

Simplify each expression. Express your answer so that only positive exponents occur. Assume that the variables are positive.

(a) $(x^{2/3}y)(x^{-2}y)^{1/2}$ (b) $\left(\dfrac{2x^{1/3}}{y^{2/3}}\right)^{-3}$ (c) $\left(\dfrac{9x^2y^{1/3}}{x^{1/3}y}\right)^{1/2}$

Solution

(a) $(x^{2/3}y)(x^{-2}y)^{1/2} = (x^{2/3}y)\left[(x^{-2})^{1/2}y^{1/2}\right]$

$= x^{2/3}yx^{-1}y^{1/2}$

$= (x^{2/3} \cdot x^{-1})(y \cdot y^{1/2})$

$= x^{-1/3}y^{3/2}$

$= \dfrac{y^{3/2}}{x^{1/3}}$

(b) $\left(\dfrac{2x^{1/3}}{y^{2/3}}\right)^{-3} = \left(\dfrac{y^{2/3}}{2x^{1/3}}\right)^3 = \dfrac{(y^{2/3})^3}{(2x^{1/3})^3} = \dfrac{y^2}{2^3(x^{1/3})^3} = \dfrac{y^2}{8x}$

(c) $\left(\dfrac{9x^2y^{1/3}}{x^{1/3}y}\right)^{1/2} = \left(\dfrac{9x^{2-(1/3)}}{y^{1-(1/3)}}\right)^{1/2} = \left(\dfrac{9x^{5/3}}{y^{2/3}}\right)^{1/2} = \dfrac{9^{1/2}(x^{5/3})^{1/2}}{(y^{2/3})^{1/2}} = \dfrac{3x^{5/6}}{y^{1/3}}$ ◀

NOW WORK PROBLEM 71.

The next two examples illustrate some algebra that you will need to know for certain calculus problems.

EXAMPLE 9	Writing an Expression as a Single Quotient

Write the following expression as a single quotient in which only positive exponents appear.

$$(x^2 + 1)^{1/2} + x \cdot \frac{1}{2}(x^2 + 1)^{-1/2} \cdot 2x$$

Solution

$$(x^2 + 1)^{1/2} + x \cdot \frac{1}{2}(x^2 + 1)^{-1/2} \cdot 2x = (x^2 + 1)^{1/2} + \frac{x^2}{(x^2 + 1)^{1/2}}$$

$$= \frac{(x^2 + 1)^{1/2}(x^2 + 1)^{1/2} + x^2}{(x^2 + 1)^{1/2}}$$

$$= \frac{(x^2 + 1) + x^2}{(x^2 + 1)^{1/2}}$$

$$= \frac{2x^2 + 1}{(x^2 + 1)^{1/2}}$$ ◀

NOW WORK PROBLEM **77**.

EXAMPLE 10	Factoring an Expression Containing Rational Exponents

Factor: $\dfrac{4}{3}x^{1/3}(2x + 1) + 2x^{4/3}$

Solution

We begin by looking for factors that are common to the two terms. Notice that 2 and $x^{1/3}$ are common factors.

$$\frac{4}{3}x^{1/3}(2x + 1) + 2x^{4/3} = 2x^{1/3}\left[\frac{2}{3}(2x + 1) + x\right]$$

$$= \frac{2}{3}x^{1/3}(7x + 2)$$ ◀

NOW WORK PROBLEM **89**.

HISTORICAL FEATURE

The radical sign, $\sqrt{\ }$, was first used in print by Christoff Rudolff in 1525. It is thought to be the manuscript form of the letter *r* (for the Latin word *radix* = *root*), although this is not quite conclusively proved. It took a long time for $\sqrt{\ }$ to become the standard symbol for a square root and much longer to standardize $\sqrt[3]{\ }$, $\sqrt[4]{\ }$, $\sqrt[5]{\ }$, and so on. The indexes of the root were placed in every conceivable position, with

$$\sqrt[3]{8}, \quad \sqrt{③8}, \quad \text{and} \quad \sqrt[\ ]{8}_3$$

all being variants for $\sqrt[3]{8}$. The notation $\sqrt{\ }\sqrt{16}$ was popular for $\sqrt[4]{16}$. By the 1700s, the index had settled where we now put it.

The bar on top of the present radical symbol, as follows,

$$\sqrt{a^2 + 2ab + b^2}$$

is the last survivor of the **vinculum**, a bar placed atop an expression to indicate what we would now indicate with parentheses. For example,

$$\overline{ab + c} = a(b + c)$$

R.8 Assess Your Understanding

'Are You Prepared?'

Answers are given at the end of these exercises. If you get a wrong answer, read the pages in red.

1. $(-3)^2 =$ _____; $-3^2 =$ _____ (pp. 20–23)

2. $\sqrt{16} =$ _____; $\sqrt{(-4)^2} =$ _____ (pp. 20–23)

Concepts and Vocabulary

3. In the symbol $\sqrt[n]{a}$, the integer n is called the _____.

4. *True or False:* $\sqrt[5]{-32} = -2$

5. We call $\sqrt[3]{a}$ the _____ of a.

6. *True or False:* $\sqrt[4]{(-3)^4} = -3$

Skill Building

In Problems 7–42, simplify each expression. Assume that all variables are positive when they appear.

7. $\sqrt[3]{27}$

8. $\sqrt[4]{16}$

9. $\sqrt[3]{-8}$

10. $\sqrt[3]{-1}$

11. $\sqrt{8}$

12. $\sqrt[3]{54}$

13. $\sqrt[3]{-8x^4}$

14. $\sqrt[4]{48x^5}$

15. $\sqrt[4]{x^{12}y^8}$

16. $\sqrt[5]{x^{10}y^5}$

17. $\sqrt[4]{\dfrac{x^9y^7}{xy^3}}$

18. $\sqrt[3]{\dfrac{3xy^2}{81x^4y^2}}$

19. $\sqrt{36x}$

20. $\sqrt{9x^5}$

21. $\sqrt{3x^2}\sqrt{12x}$

22. $\sqrt{5x}\sqrt{20x^3}$

23. $\left(\sqrt{5}\sqrt[3]{9}\right)^2$

24. $\left(\sqrt[3]{3}\sqrt{10}\right)^4$

25. $(3\sqrt{6})(2\sqrt{2})$

26. $(5\sqrt{8})(-3\sqrt{3})$

27. $3\sqrt{2} + 4\sqrt{2}$

28. $6\sqrt{5} - 4\sqrt{5}$

29. $-\sqrt{18} + 2\sqrt{8}$

30. $2\sqrt{12} - 3\sqrt{27}$

31. $(\sqrt{3} + 3)(\sqrt{3} - 1)$

32. $(\sqrt{5} - 2)(\sqrt{5} + 3)$

33. $5\sqrt[3]{2} - 2\sqrt[3]{54}$

34. $9\sqrt[3]{24} - \sqrt[3]{81}$

35. $(\sqrt{x} - 1)^2$

36. $(\sqrt{x} + \sqrt{5})^2$

37. $\sqrt[3]{16x^4} - \sqrt[3]{2x}$

38. $\sqrt[4]{32x} + \sqrt[4]{2x^5}$

39. $\sqrt{8x^3} - 3\sqrt{50x},$

40. $3x\sqrt{9y} + 4\sqrt{25y}$

41. $\sqrt[3]{16x^4y} - 3x\sqrt[3]{2xy} + 5\sqrt[3]{-2xy^4}$

42. $8xy - \sqrt{25x^2y^2} + \sqrt[3]{8x^3y^3}$

In Problems 43–54, rationalize the denominator of each expression. Assume that all variables are positive when they appear.

43. $\dfrac{1}{\sqrt{2}}$

44. $\dfrac{2}{\sqrt{3}}$

45. $\dfrac{-\sqrt{3}}{\sqrt{5}}$

46. $\dfrac{-\sqrt{3}}{\sqrt{8}}$

47. $\dfrac{\sqrt{3}}{5 - \sqrt{2}}$

48. $\dfrac{\sqrt{2}}{\sqrt{7} + 2}$

49. $\dfrac{2 - \sqrt{5}}{2 + 3\sqrt{5}}$

50. $\dfrac{\sqrt{3} - 1}{2\sqrt{3} + 3}$

51. $\dfrac{5}{\sqrt[3]{2}}$

52. $\dfrac{-2}{\sqrt[3]{9}}$

53. $\dfrac{\sqrt{x + h} - \sqrt{x}}{\sqrt{x + h} + \sqrt{x}}$

54. $\dfrac{\sqrt{x + h} + \sqrt{x - h}}{\sqrt{x + h} - \sqrt{x - h}}$

In Problems 55–66, simplify each expression.

55. $8^{2/3}$

56. $4^{3/2}$

57. $(-27)^{1/3}$

58. $16^{3/4}$

59. $16^{3/2}$

60. $25^{3/2}$

61. $9^{-3/2}$

62. $16^{-3/2}$

63. $\left(\dfrac{9}{8}\right)^{3/2}$

64. $\left(\dfrac{27}{8}\right)^{2/3}$

65. $\left(\dfrac{8}{9}\right)^{-3/2}$

66. $\left(\dfrac{8}{27}\right)^{-2/3}$

In Problems 67–74, simplify each expression. Express your answer so that only positive exponents occur. Assume that the variables are positive.

67. $x^{3/4}x^{1/3}x^{-1/2}$

68. $x^{2/3}x^{1/2}x^{-1/4}$

69. $(x^3y^6)^{1/3}$

70. $(x^4y^8)^{3/4}$

71. $(x^2y)^{1/3}(xy^2)^{2/3}$

72. $(xy)^{1/4}(x^2y^2)^{1/2}$

73. $(16x^2y^{-1/3})^{3/4}$

74. $(4x^{-1}y^{1/3})^{3/2}$

Applications and Extensions

In Problems 75–88, expressions that occur in calculus are given. Write each expression as a single quotient in which only positive exponents and/or radicals appear.

75. $\dfrac{x}{(1 + x)^{1/2}} + 2(1 + x)^{1/2}, \quad x > -1$

76. $\dfrac{1 + x}{2x^{1/2}} + x^{1/2}, \quad x > 0$

77. $2x(x^2 + 1)^{1/2} + x^2 \cdot \dfrac{1}{2}(x^2 + 1)^{-1/2} \cdot 2x$

78. $(x + 1)^{1/3} + x \cdot \dfrac{1}{3}(x + 1)^{-2/3}, \quad x \neq -1$

79. $\sqrt{4x + 3} \cdot \dfrac{1}{2\sqrt{x - 5}} + \sqrt{x - 5} \cdot \dfrac{1}{5\sqrt{4x + 3}}$, $x > 5$

80. $\dfrac{\sqrt[3]{8x + 1}}{3\sqrt[3]{(x - 2)^2}} + \dfrac{\sqrt[3]{x - 2}}{24\sqrt[3]{(8x + 1)^2}}$, $x \ne 2, x \ne -\dfrac{1}{8}$

81. $\dfrac{\sqrt{1 + x} - x \cdot \dfrac{1}{2\sqrt{1 + x}}}{1 + x}$, $x > -1$

82. $\dfrac{\sqrt{x^2 + 1} - x \cdot \dfrac{2x}{2\sqrt{x^2 + 1}}}{x^2 + 1}$

83. $\dfrac{(x + 4)^{1/2} - 2x(x + 4)^{-1/2}}{x + 4}$, $x > -4$

84. $\dfrac{(9 - x^2)^{1/2} + x^2(9 - x^2)^{-1/2}}{9 - x^2}$, $-3 < x < 3$

85. $\dfrac{\dfrac{x^2}{(x^2 - 1)^{1/2}} - (x^2 - 1)^{1/2}}{x^2}$, $x < -1$ or $x > 1$

86. $\dfrac{(x^2 + 4)^{1/2} - x^2(x^2 + 4)^{-1/2}}{x^2 + 4}$

87. $\dfrac{\dfrac{1 + x^2}{2\sqrt{x}} - 2x\sqrt{x}}{(1 + x^2)^2}$, $x > 0$

88. $\dfrac{2x(1 - x^2)^{1/3} + \dfrac{2}{3}x^3(1 - x^2)^{-2/3}}{(1 - x^2)^{2/3}}$, $x \ne -1, x \ne 1$

✍ *In Problems 89–98, expressions that occur in calculus are given. Factor each expression. Express your answer so that only positive exponents occur.*

89. $(x + 1)^{3/2} + x \cdot \dfrac{3}{2}(x + 1)^{1/2}$, $x \ge -1$

90. $(x^2 + 4)^{4/3} + x \cdot \dfrac{4}{3}(x^2 + 4)^{1/3} \cdot 2x$

91. $6x^{1/2}(x^2 + x) - 8x^{3/2} - 8x^{1/2}$, $x \ge 0$

92. $6x^{1/2}(2x + 3) + x^{3/2} \cdot 8$, $x \ge 0$

93. $3(x^2 + 4)^{4/3} + x \cdot 4(x^2 + 4)^{1/3} \cdot 2x$

94. $2x(3x + 4)^{4/3} + x^2 \cdot 4(3x + 4)^{1/3}$

95. $4(3x + 5)^{1/3}(2x + 3)^{3/2} + 3(3x + 5)^{4/3}(2x + 3)^{1/2}$, $x \ge -\dfrac{3}{2}$

96. $6(6x + 1)^{1/3}(4x - 3)^{3/2} + 6(6x + 1)^{4/3}(4x - 3)^{1/2}$, $x \ge \dfrac{3}{4}$

97. $3x^{-1/2} + \dfrac{3}{2}x^{1/2}$, $x > 0$

98. $8x^{1/3} - 4x^{-2/3}$, $x \ne 0$

In Problems 99–106, use a calculator to approximate each radical. Round your answer to two decimal places.

99. $\sqrt{2}$

100. $\sqrt{7}$

101. $\sqrt[3]{4}$

102. $\sqrt[3]{-5}$

103. $\dfrac{2 + \sqrt{3}}{3 - \sqrt{5}}$

104. $\dfrac{\sqrt{5} - 2}{\sqrt{2} + 4}$

105. $\dfrac{3\sqrt[3]{5} - \sqrt{2}}{\sqrt{3}}$

106. $\dfrac{2\sqrt{3} - \sqrt[3]{4}}{\sqrt{2}}$

---

107. Calculating the Amount of Gasoline in a Tank   A Shell station stores its gasoline in underground tanks that are right circular cylinders lying on their sides. See the illustration. The volume V of gasoline in the tank (in gallons) is given by the formula

$$V = 40h^2 \sqrt{\dfrac{96}{h} - 0.608}$$

where h is the height of the gasoline (in inches) as measured on a depth stick.
   (a) If $h = 12$ inches, how many gallons of gasoline are in the tank?
   (b) If $h = 1$ inch, how many gallons of gasoline are in the tank?

108. Inclined Planes   The final velocity v of an object in feet per second (ft/sec) after it slides down a frictionless inclined plane of height h feet is

$$v = \sqrt{64h + v_0^2}$$

where v_0 is the initial velocity (in ft/sec) of the object.

   (a) What is the final velocity v of an object that slides down a frictionless inclined plane of height 4 feet? Assume that the initial velocity is 0.
   (b) What is the final velocity v of an object that slides down a frictionless inclined plane of height 16 feet? Assume that the initial velocity is 0.
   (c) What is the final velocity v of an object that slides down a frictionless inclined plane of height 2 feet with an initial velocity of 4 ft/sec?

Problems 109–112 require the following information.
Period of a Pendulum *The period T, in seconds, of a pendulum of length l, in feet, may be approximated using the formula*

$$T = 2\pi\sqrt{\frac{l}{32}}$$

In Problems 109–112, express your answer both as a square root and as a decimal.

109. Find the period T of a pendulum whose length is 64 feet.

110. Find the period T of a pendulum whose length is 16 feet.

111. Find the period T of a pendulum whose length is 8 inches.

112. Find the period T of a pendulum whose length is 4 inches.

Discussion and Writing

113. Give an example to show that $\sqrt{a^2}$ is not equal to a. Use it to explain why $\sqrt{a^2} = |a|$.

'Are you Prepared?' Answers

1. $9; -9$ **2.** $4; 4$

Chapter Review

Things to Know

Classification of numbers (pp. 2–4)

Counting numbers	$1, 2, 3 \ldots$
Whole numbers	$0, 1, 2, 3 \ldots$
Integers	$\ldots, -2, -1, 0, 1, 2, \ldots$
Rational numbers	Quotients of two integers (denominator not equal to 0); terminating or repeating decimals
Irrational numbers	Nonrepeating, nonterminating decimals
Real numbers	Rational or irrational numbers

Properties of real numbers (pp. 8–12)

Commutative properties	$a + b = b + a, \quad a \cdot b = b \cdot a$
Associative properties	$a + (b + c) = (a + b) + c, \quad a \cdot (b \cdot c) = (a \cdot b) \cdot c$
Distributive Property	$a \cdot (b + c) = a \cdot b + a \cdot c$
Identity properties	$a + 0 = a, \quad a \cdot 1 = a$
Inverse properties	$a + (-a) = 0; \quad a \cdot \dfrac{1}{a} = 1, \quad$ where $a \neq 0$
Zero-Product Property	If $ab = 0$, then $a = 0$ or $b = 0$ or both.

Absolute value (p. 18)

$$|a| = a \text{ if } a \geq 0, \quad |a| = -a \text{ if } a < 0$$

Exponents (pp. 20–21)

$$a^n = \underbrace{a \cdot a \cdot \ldots \cdot a}_{n \text{ factors}}, \quad n \text{ a positive integer}; \quad a^0 = 1, \quad a \neq 0; \quad a^{-n} = \frac{1}{a^n} \quad a \neq 0, n \text{ a positive integer}$$

Laws of Exponents (p. 21)

$$a^m \cdot a^n = a^{m+n}, \quad (a^m)^n = a^{mn}, \quad (a \cdot b)^n = a^n \cdot b^n, \quad \frac{a^m}{a^n} = a^{m-n} = \frac{1}{a^{n-m}}, \quad \left(\frac{a}{b}\right)^n = \frac{a^n}{b^n}$$

Pythagorean Theorem (p. 29)

In a right triangle, the square of the length of the hypotenuse is equal to the sum of the squares of the lengths of the legs.

Converse of the Pythagorean Theorem (p. 29)

In a triangle, if the square of the length of one side equals the sum of the squares of the lengths of the other two sides, then the triangle is a right triangle.

Polynomial (p. 36)

Algebraic expression of the form $a_n x^n + a_{n-1} x^{n-1} + \cdots + a_1 x + a_0$, n a nonnegative integer

Special products/factoring formulas (pp. 39–40)

$$(x - a)(x + a) = x^2 - a^2$$
$$(x + a)^2 = x^2 + 2ax + a^2, \quad (x - a)^2 = x^2 - 2ax + a^2$$
$$(x + a)(x + b) = x^2 + (a + b)x + ab$$
$$(ax + b)(cx + d) = acx^2 + (ad + bc)x + bd$$
$$(x + a)^3 = x^3 + 3ax^2 + 3a^2x + a^3, \quad (x - a)^3 = x^3 - 3ax^2 + 3a^2x - a^3$$
$$(x - a)(x^2 + ax + a^2) = x^3 - a^3, \quad (x + a)(x^2 - ax + a^2) = x^3 + a^3$$

Radicals; Rational exponents (pp. 68, 69 and 71)

$\sqrt[n]{a} = b$ means $a = b^n$, where $a \geq 0$ and $b \geq 0$ if $n \geq 2$ is even and a, b are any real numbers if $n \geq 3$ is odd

$$\sqrt[n]{ab} = \sqrt[n]{a}\,\sqrt[n]{b} \qquad \sqrt[n]{\frac{a}{b}} = \frac{\sqrt[n]{a}}{\sqrt[n]{b}} \qquad \sqrt[n]{a^m} = (\sqrt[n]{a})^m \qquad a^{1/n} = \sqrt[n]{a} \qquad a^{m/n} = \sqrt[n]{a^m} = (\sqrt[n]{a})^m$$

Objectives

Section		You should be able to . . .	Review Exercises
R.1	1	Classify numbers (p. 2)	1, 2
	2	Approximate Decimals (p. 4)	3–6
	3	Evaluate numerical expressions (p. 6)	7–10
	4	Work with properties of real numbers (p. 8)	13, 14
R.2	1	Graph inequalities (p. 17)	15, 16
	2	Find distance on the real number line (p. 18)	17, 18
	3	Evaluate algebraic expressions (p. 19)	19–22, 106
	4	Determine the domain of a variable (p. 20)	25, 26
	5	Use the Laws of Exponents (p. 20)	33, 34
	6	Evaluate square roots (p. 22)	11, 23
	7	Use a graphing calculator to evaluate exponents (p. 23)	37
	8	Use scientific notation (p. 23)	40, 103
R.3	1	Use the Pythagorean Theorem and its converse (p. 29)	104, 105, 111
	2	Know geometry formulas (p. 31)	108, 109, 110
R.4	1	Recognize monomials (p. 35)	38
	2	Recognize polynomials (p. 36)	39
	3	Add and subtract polynomials (p. 37)	41, 42
	4	Multiply polynomials (p. 38)	43, 44, 47, 48
	5	Know formulas for special products (p. 39)	45, 46
	6	Divide polynomials using long division (p. 40)	53–58
	7	Work with polynomials in two variables (p. 43)	49–52

Review Exercises

In Problems 1 and 2, list the numbers in the set that are (a) Natural numbers, (b) Integers, (c) Rational numbers, (d) Irrational numbers, (e) Real numbers.

1. $A = \left\{ -10, 0.65, 1.343434\ldots, \sqrt{7}, \dfrac{1}{9} \right\}$

2. $B = \left\{ 0, -5, \dfrac{1}{3}, 0.59, 1.333\ldots, 2\sqrt{2}, \dfrac{\pi}{2} \right\}$

In Problems 3–6, approximate each decimal (a) rounded and (b) truncated to two decimal places.

3. 782.5684　　　**4.** 690.1448　　　**5.** 9.6204　　　**6.** 5.0882

In Problems 7–12, evaluate each expression. Verify your results using a graphing utility.

7. $-6 + 4 \cdot (8 - 3)$　**8.** $\dfrac{5}{18} + \dfrac{1}{12}$　　**9.** $\dfrac{4}{3} \cdot \dfrac{9}{16}$　　**10.** $\dfrac{\dfrac{5}{18}}{\dfrac{11}{27}}$　　**11.** $\sqrt{(-3)^2}$　　**12.** $(-3)^5$

In Problems 13 and 14, use the Distributive Property to remove the parentheses.

13. $4(x - 3)$　　　　　　　　　　**14.** $(x - 2)(3x + 1)$

In Problems 15 and 16, graph the numbers x on the real number line.

15. $x > 3$　　　　　　　　　　　**16.** $x \le 5$

In Problems 17 and 18, let P, Q, and R be points on a real number line with coordinates −2, 3, and 9, respectively.

17. Find the distance between P and Q.　　　　**18.** Find the distance between Q and R.

In Problems 19–24, evaluate each expression if x = −5 and y = 7. Verify your results using a graphing utility.

19. $\dfrac{4x}{x + y}$　　**20.** $|2x - 3y|$　　**21.** $5x^{-1}y^2$　　**22.** $2x^2y^{-2}$　　**23.** $\sqrt{x^2}$　　**24.** $\sqrt[3]{x^3}$

In Problems 25 and 26, determine the domain of the variable.

25. $\dfrac{3}{x - 6}$　　　　　　　　　**26.** $\dfrac{x + 1}{x + 5}$

In Problems 27–36, simplify each expression. All exponents must be positive. Assume that $x > 0$ and $y > 0$, when they appear.

27. $\sqrt{32}$ **28.** $\sqrt{75}$ **29.** $\sqrt[3]{-16}$ **30.** $\sqrt[5]{64}$ **31.** $5\sqrt{8} - 2\sqrt{32}$

32. $4\sqrt{12} + 5\sqrt{27}$ **33.** $\dfrac{(x^2 y)^{-4}}{(xy)^{-3}}$ **34.** $\left(\dfrac{x^2 y^2}{x^{-1}}\right)^2$ **35.** $(25x^{-4/3} y^{-2/3})^{3/2}$ **36.** $(27x^{-3/2} y^{5/2})^{2/3}$

37. Use a calculator to evaluate $(1.5)^4$.

38. Identify the coefficient and degree of the monomial $-5x^3$.

39. Identify the coefficients and degree of the polynomial $3x^5 + 4x^4 - 2x^3 + 5x - 12$.

40. Write 3.275×10^5 as an integer.

In Problems 41–52, perform the indicated operation.

41. $(2x^4 - 8x^3 + 5x - 1) + (6x^3 + x^2 + 4)$ **42.** $(x^3 + 8x^2 - 3x + 4) - (4x^3 - 7x^2 - 2x + 3)$

43. $(2x + 1)(3x - 5)$ **44.** $(2x - 5)(3x^2 + 2)$ **45.** $(4x + 1)(4x - 1)$

46. $(5x + 2)^2$ **47.** $(x + 1)(x + 2)(x - 3)$ **48.** $(x + 1)(x + 3)(x - 5)$

49. $(3x + y)^2$ **50.** $(2x - 3y)(2x + 3y)$ **51.** $(x + 2y)^3$ **52.** $(x - y)(3x + 5y)$

In Problems 53–58, find the quotient and remainder. Check your work by verifying that

$$(Quotient)(Divisor) + Remainder = Dividend$$

53. $3x^3 - x^2 + x + 4$ divided by $x - 3$ **54.** $2x^3 - 3x^2 + x + 1$ divided by $x - 2$

55. $-3x^4 + x^2 + 2$ divided by $x^2 + 1$ **56.** $-4x^3 + x^2 - 2$ divided by $x^2 - 1$

57. $x^5 + 1$ divided by $x + 1$ **58.** $x^5 - 1$ divided by $x - 1$

In Problems 59–74, factor each polynomial completely (over the integers). If the polynomial cannot be factored, say it is prime.

59. $x^2 + 5x - 14$ **60.** $x^2 - 9x + 14$ **61.** $6x^2 - 5x - 6$ **62.** $6x^2 + x - 2$

63. $3x^2 - 15x - 42$ **64.** $2x^3 + 18x^2 + 28x$ **65.** $8x^3 + 1$ **66.** $27x^3 - 8$

67. $2x^3 + 3x^2 - 2x - 3$ **68.** $2x^3 + 3x^2 + 2x + 3$ **69.** $25x^2 - 4$ **70.** $16x^2 - 1$

71. $9x^2 + 1$ **72.** $x^2 - x + 1$ **73.** $x^2 + 8x + 16$ **74.** $4x^2 + 12x + 9$

In Problems 75–84, perform the indicated operation and simplify. Leave your answer in factored form.

75. $\dfrac{2x^2 + 11x + 14}{x^2 - 4}$ **76.** $\dfrac{x^2 - 5x - 14}{4 - x^2}$ **77.** $\dfrac{9x^2 - 1}{x^2 - 9} \cdot \dfrac{3x - 9}{9x^2 + 6x + 1}$ **78.** $\dfrac{x^2 - 25}{x^3 - 4x^2 - 5x} \cdot \dfrac{x^2 + x}{1 - x^2}$

79. $\dfrac{x + 1}{x - 1} - \dfrac{x - 1}{x + 1}$ **80.** $\dfrac{x}{x + 1} - \dfrac{2x}{x + 2}$ **81.** $\dfrac{3x + 4}{x^2 - 4} - \dfrac{2x - 3}{x^2 + 4x + 4}$

82. $\dfrac{x^2}{2x^2 + 5x - 3} + \dfrac{x^2}{2x^2 - 5x + 2}$ **83.** $\dfrac{\dfrac{x^2 - 1}{x^2 - 5x + 6}}{\dfrac{x + 1}{x - 2}}$ **84.** $\dfrac{\dfrac{x + 4}{3}}{\dfrac{1}{4} + \dfrac{2}{x}}$

In Problems 85–88, simplify each expression.

85. $\sqrt{\dfrac{9x^2}{25y^4}}$, $x \geq 0, y > 0$ **86.** $\dfrac{\sqrt{2x}\,\sqrt{5xy^3}}{\sqrt{10y}}$, $x \geq 0, y > 0$ **87.** $\sqrt[3]{27x^4 y^{12}}$ **88.** $\dfrac{\sqrt[4]{243x^5 y}}{\sqrt[4]{3xy^9}}$, $x > 0, y > 0$

In Problems 89–94, rationalize the denominator of each expression.

89. $\dfrac{4}{\sqrt{5}}$ **90.** $\dfrac{-2}{\sqrt{3}}$ **91.** $\dfrac{2}{1 - \sqrt{2}}$ **92.** $\dfrac{-4}{1 + \sqrt{3}}$ **93.** $\dfrac{1 + \sqrt{5}}{1 - \sqrt{5}}$ **94.** $\dfrac{4\sqrt{3} + 2}{2\sqrt{3} + 1}$

△ In Problems 95–100, write each expression as a single quotient in which only positive exponents and/or radicals appear.

95. $(2 + x^2)^{1/2} + x \cdot \frac{1}{2}(2 + x^2)^{-1/2} \cdot 2x$

96. $(x^2 + 4)^{2/3} + x \cdot \frac{2}{3}(x^2 + 4)^{-1/3} \cdot 2x$

97. $\dfrac{(x + 4)^{1/2} \cdot 2x - x^2 \cdot \frac{1}{2}(x + 4)^{-1/2}}{x + 4}, \quad x > -4$

98. $\dfrac{(x^2 + 4)^{1/2} \cdot 2x - x^2 \cdot \frac{1}{2}(x^2 + 4)^{-1/2} \cdot 2x}{x^2 + 4}$

99. $\dfrac{\dfrac{x^2}{\sqrt{x^2 - 1}} - \sqrt{x^2 - 1}}{x^2}$

100. $\dfrac{\dfrac{1 + x^2}{2\sqrt{x}} - 2x\sqrt{x}}{(1 + x^2)^2}$

△ In Problems 101 and 102, factor each expression.

101. $3(x^2 + 4)^{4/3} + x \cdot 4(x^2 + 4)^{1/3} \cdot 2x$

102. $2x(3x + 4)^{4/3} + x^2 \cdot 4(3x + 4)^{1/3}$

---

103. U.S. Population According to the U.S. Census Bureau, the U.S. population in 2000 was 281,421,906. Write the U.S. population in scientific notation.

104. Find the hypotenuse of a right triangle whose legs are of lengths 5 and 8.

105. The lengths of the sides of a triangle are 12, 16, and 20. Is this a right triangle?

106. Manufacturing Cost The weekly production cost C (in dollars) of manufacturing x hand calculators is given by the formula

$$C = 3000 + 6x - \frac{x^2}{1000}$$

(a) What is the cost of producing 1000 hand calculators?
(b) What is the cost of producing 3000 hand calculators?

107. Quarterly Corporate Earnings In the first quarter of its fiscal year, a company posted earnings of $1.20 per share. During the second and third quarters, it posted losses of $0.75 per share and $0.30 per share, respectively. In the fourth quarter, it earned a modest $0.20 per share. What were the annual earnings per share of this company?

108. Design A window consists of a rectangle surmounted by a triangle. Find the area of the window shown in the illustration. How much wood frame is needed to enclose the window?

4 ft   4 ft

6 ft

5 ft

109. Construction A rectangular swimming pool, 20 feet long and 10 feet wide, is enclosed by a wooden deck that is 3 feet wide. What is the area of the deck? How much fence is required to enclose the deck?

20 ft
10 ft
3 ft
3 ft

110. Construction A statue with a circular base of radius 3 feet is enclosed by a circular pond as shown in the illustration. What is the surface area of the pond? How much fence is required to enclose the pond?

2 ft
3 ft

111. How Far Can a Pilot See? On a recent flight to San Francisco, the pilot announced that we were 139 miles from the city, flying at an altitude of 35,000 feet. The pilot claimed that he could see the Golden Gate Bridge and beyond. Was he telling the truth? How far could he see?

112. Use the material in this chapter to create a problem that uses each of the following words:
(a) Simplify
(b) Factor
(c) Reduce

113. A rational number is defined as the quotient of two integers. When written as a decimal, the decimal will either repeat or terminate. By looking at the denominator of the rational number, there is a way to tell in advance whether its decimal representation will repeat or terminate. Make a list of rational numbers and their decimals. See if you can discover the pattern. Confirm your conclusion by consulting books on number theory at the library. Write a brief essay on your findings.

114. The current time is 12 noon CST. What time (CST) will it be 12,997 hours from now?

115. Both $\dfrac{a}{0}$ $(a \neq 0)$ and $\dfrac{0}{0}$ are undefined, but for different reasons. Write a paragraph or two explaining the different reasons.

Chapter Test

1. From the set $A = \left\{ -5, 0, \dfrac{\pi}{4}, \dfrac{4}{3}, 1.5, \sqrt{3}, 2.333\ldots = 2.\overline{3}, 6 \right\}$ list the numbers in the set that are:
(a) Natural numbers, (b) Integers, (c) Rational numbers, (d) Irrational numbers, and (e) Real numbers.

In Problems 2 and 3, approximate each decimal (a) rounded to two decimal places and (b) truncated to two decimal places.

2. -2.537

3. 3.14159

In Problems 4–7, evaluate each expression.

4. $2 - 3(5 - 7)$

5. $\dfrac{2}{5} + \dfrac{1}{5} \cdot \dfrac{4}{3}$

6. $\sqrt{(-2)^2 + 5}$

7. $4 \cdot 81^{-3/4}$

In Problems 8–18, perform the indicated operation and simplify.

8. $(x - 3)(4x + 1)(x - 2)$

9. $(2a - 1)(2a + 1)$

10. $(3y + 2z)(y - 5z)$

11. $\dfrac{4x}{x^2 - 5x - 14} - \dfrac{1}{x + 2}$

12. $(2x^3 - 7x + 9) - (x^2 - 2x - 3)$

13. $(3m + 7n)^2$

14. $2x + 4 + \dfrac{4}{x + 3}$

15. $(2b + 1)(4b^2 - 2b + 1)$

16. $\dfrac{\dfrac{x + 2}{4}}{\dfrac{3}{x} + \dfrac{3}{2}}, \quad x \neq 0$

17. $\dfrac{x^2 - 9}{x + 2} \cdot \dfrac{x^2 - 2x - 8}{x + 3}$

18. $\sqrt{8x^3 y} - 2\sqrt{32xy^3} + 2x\sqrt{2xy} \quad x \geq 0, y \geq 0$

In Problems 19–22, simplify each expression. All exponents must be positive. Assume that all variables are positive.

19. $\dfrac{15a^2 b}{3b^2} \cdot \dfrac{ab^4 c^{-1}}{-10ac}$

20. $\left(\dfrac{x^2 yz^3}{xy^2 z} \right)^{-2}$

21. $\dfrac{x + \dfrac{2}{x + 1}}{x - 1}$

22. $(x^{1/2} y^{3/4})^{2/3} (xy^{-2})^{1/2}$

23. Evaluate $\dfrac{5x + 1}{x^3 - 2x^2}$ for $x = -2$ and $x = 3$. Explain why $x = 0$ must be excluded from the domain of the variable.

24. Rationalize the denominator: $\dfrac{3}{\sqrt{2} - 1}$

In Problems 25–31, factor each polynomial completely (over the integers). If a polynomial cannot be factored, say it is prime.

25. $z^2 - 4z - 21$

26. $2x^2 - 6x + 7$

27. $64w^3 + 125$

28. $16c^4 - 9c^2$

29. $5a^3 - 30a^2 b + 45ab^2$

30. $8x^2 + 2x - 15$

31. $3y^3 - 12y^2 + 2y - 8$

32. Find the quotient and remainder when

$$3x^3 - 2x^2 - 13x + 8$$

is divided by $x + 2$.

33. A rectangular room at Tumble Town is 12 feet longer than it is wide. Mats are placed in the center of the room and cover 640 square feet of the floor. Let x represent the width of the room.

(a) Write an expression for the area of the floor not covered by mats in completely factored form.

(b) How much of the floor is not covered by mats if the width of the room is 40 feet?

34. At a pioneering camp, Nathan and Joe built a lookout tower for their campsite. Guylines were attached to the top of the tower and staked to the ground 15 feet from the base of the tower. If each guyline was 25 feet long, how tall was the tower? Round your answer to two decimal places, if necessary.

35. A circular wading pool has a diameter of 2 meters. If Emma fills the pool to a depth of 45 centimeters, how much water is in the pool? Round your answer to two decimal places, if necessary.

Graphs, Equations, and Inequalities

1

A LOOK BACK In Chapter R, we reviewed skills from Intermediate Algebra.

A LOOK AHEAD In this chapter we make the connection between algebra and geometry through the rectangular coordinate system. The idea of using a system of rectangular coordinates dates back to ancient times, when such a system was used for surveying and city planning. Apollonius of Perga, in 200 BC, used a form of rectangular coordinates in his work on conics, although this use does not stand out as clearly as it does in modern treatments. Sporadic use of rectangular coordinates continued until the 1600s. By that time, algebra had developed sufficiently so that René Descartes (1596–1650) and Pierre de Fermat (1601–1665) could take the crucial step, which was the use of rectangular coordinates to translate geometry problems into algebra problems, and vice versa. This step was important for two reasons. First, it allowed both geometers and algebraists to gain new insights into their subjects, which previously had been regarded as separate, but now were seen to be connected in many important ways. Second, these insights made the development of calculus possible which greatly enlarged the number of areas in which mathematics could be applied and made possible a much deeper understanding of these areas.

With the advent of modern technology, in particular graphing utilities, now not only are we able to visualize the dual roles of algebra and geometry, but we are also able to solve many problems that required advanced methods before this technology.

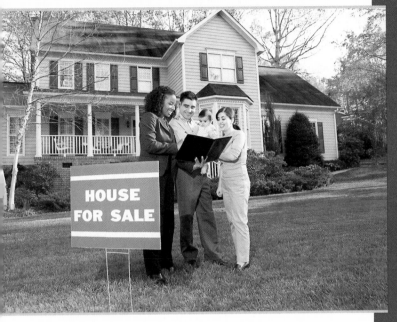

30-year mortgage rates back under 6%

WASHINGTON (Reuters)—Interest rates on 30-year mortgages fell under 6% this week, while rates on 15-year and adjustable mortgages also eased, mortgage finance company Freddie Mac said Thursday.

Thirty-year mortgage rates averaged 5.99% in the week ended Aug. 5 compared with 6.08% a week earlier, Freddie Mac said.

Freddie Mac said 15-year mortgages averaged 5.40%, down from 5.49% last week. One-year adjustable rate mortgages dipped to an average of 4.08% from 4.17% last week.

Freddie Mac is a mortgage finance company chartered by Congress that buys mortgages from lenders and packages them into securities for investors or holds them in its own portfolio.

USA Today, August 5, 2004.

—*See Chapter Project 1.*

OUTLINE

1.1 Rectangular Coordinates; Graphing Utilities; Introduction to Graphing Equations

PREPARING FOR THIS SECTION *Before getting started, review the following:*

• Algebra Review (Section R.2, pp. 16–25)

• Geometry Review (Section R.3, pp. 29–32)

Now work the 'Are You Prepared?' problems on page 97.

OBJECTIVES
1. Use the Distance Formula
2. Use the Midpoint Formula
3. Graph Equations by Hand by Plotting Points
4. Graph Equations Using a Graphing Utility
5. Use a Graphing Utility to Create Tables
6. Find Intercepts from a Graph
7. Use a Graphing Utility to Approximate Intercepts

We locate a point on the real number line by assigning it a single real number, called the *coordinate of the point*. For work in a two-dimensional plane, we locate points by using two numbers.

We begin with two real number lines located in the same plane: one horizontal and the other vertical. We call the horizontal line the **x-axis**, the vertical line the **y-axis**, and the point of intersection the **origin O**. See Figure 1. We assign coordinates to every point on these number lines using a convenient scale. In mathematics, we usually use the same scale on each axis; in applications, a different scale is often used on each axis.

Figure 1

The origin O has a value of 0 on both the x-axis and y-axis. Points on the x-axis to the right of O are associated with positive real numbers, and those to the left of O are associated with negative real numbers. Points on the y-axis above O are associated with positive real numbers, and those below O are associated with negative real numbers. In Figure 1, the x-axis and y-axis are labeled as x and y, respectively, and we have used an arrow at the end of each axis to denote the positive direction.

The coordinate system described here is called a **rectangular** or **Cartesian**[*] **coordinate system**. The plane formed by the x-axis and y-axis is sometimes called the **xy-plane**, and the x-axis and y-axis are referred to as the **coordinate axes**.

Figure 2

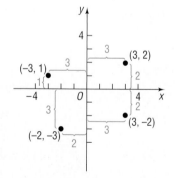

Any point P in the xy-plane can then be located by using an **ordered pair** (x, y) of real numbers. Let x denote the signed distance of P from the y-axis (*signed* means that, if P is to the right of the y-axis, then $x > 0$, and if P is to the left of the y-axis, then $x < 0$); and let y denote the signed distance of P from the x-axis. The ordered pair (x, y), also called the **coordinates** of P, then gives us enough information to locate the point P in the plane.

For example, to locate the point whose coordinates are $(-3, 1)$, go 3 units along the x-axis to the left of O and then go straight up 1 unit. We **plot** this point by placing a dot at this location. See Figure 2, in which the points with coordinates $(-3, 1), (-2, -3), (3, -2)$, and $(3, 2)$ are plotted.

The origin has coordinates $(0, 0)$. Any point on the x-axis has coordinates of the form $(x, 0)$, and any point on the y-axis has coordinates of the form $(0, y)$.

---

[*]Named after René Descartes (1596–1650), a French mathematician, philosopher, and theologian.

Figure 3

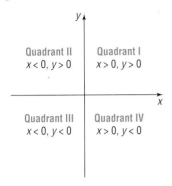

Quadrant II $x < 0, y > 0$	Quadrant I $x > 0, y > 0$
Quadrant III $x < 0, y < 0$	Quadrant IV $x > 0, y < 0$

Figure 4
$y = 2x$

Figure 5

If (x, y) are the coordinates of a point P, then x is called the **x-coordinate**, or **abscissa**, of P and y is the **y-coordinate**, or **ordinate**, of P. We identify the point P by its coordinates (x, y) by writing $P = (x, y)$. Usually, we will simply say "the point (x, y)" rather than "the point whose coordinates are (x, y)."

The coordinate axes divide the xy-plane into four sections called **quadrants**, as shown in Figure 3. In quadrant I, both the x-coordinate and the y-coordinate of all points are positive; in quadrant II, x is negative and y is positive; in quadrant III, both x and y are negative; and in quadrant IV, x is positive and y is negative. Points on the coordinate axes belong to no quadrant.

NOW WORK PROBLEM **11.**

Graphing Utilities

All graphing utilities, that is, all graphing calculators and all computer software graphing packages, graph equations by plotting points on a screen. The screen itself actually consists of small rectangles, called **pixels**. The more pixels the screen has, the better the resolution. Most graphing calculators have 48 pixels per square inch; most computer screens have 32 to 108 pixels per square inch. When a point to be plotted lies inside a pixel, the pixel is turned on (lights up). The graph of an equation is a collection of lighted pixels. Figure 4 shows how the graph of $y = 2x$ looks on a TI-84 Plus graphing calculator.

The screen of a graphing utility will display the coordinate axes of a rectangular coordinate system. However, you must set the scale on each axis. You must also include the smallest and largest values of x and y that you want included in the graph. This is called **setting the viewing rectangle** or **viewing window**. Figure 5 illustrates a typical viewing window.

To select the viewing window, we must give values to the following expressions:

Xmin:	the smallest value of x shown on the viewing window
Xmax:	the largest value of x shown on the viewing window
Xscl:	the number of units per tick mark on the x-axis
Ymin:	the smallest value of y shown on the viewing window
Ymax:	the largest value of y shown on the viewing window
Yscl:	the number of units per tick mark on the y-axis

Figure 6 illustrates these settings and their relation to the Cartesian coordinate system.

Figure 6

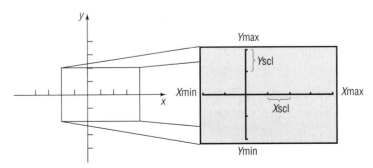

If the scale used on each axis is known, we can determine the minimum and maximum values of x and y shown on the screen by counting the tick marks. Look again at Figure 5. For a scale of 1 on each axis, the minimum and maximum values of x are -10 and 10, respectively; the minimum and maximum values of y are also -10

and 10. If the scale is 2 on each axis, then the minimum and maximum values of x are -20 and 20, respectively; and the minimum and maximum values of y are -20 and 20, respectively.

Conversely, if we know the minimum and maximum values of x and y, we can determine the scales being used by counting the tick marks displayed. We shall follow the practice of showing the minimum and maximum values of x and y in our illustrations so that you will know how the window was set. See Figure 7.

Figure 7

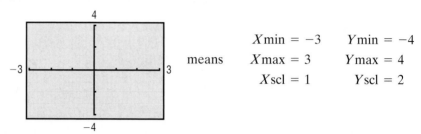

$$X\text{min} = -3 \qquad Y\text{min} = -4$$
$$\text{means} \qquad X\text{max} = 3 \qquad Y\text{max} = 4$$
$$X\text{scl} = 1 \qquad Y\text{scl} = 2$$

EXAMPLE 1	**Finding the Coordinates of a Point Shown on a Graphing Utility Screen**

Find the coordinates of the point shown in Figure 8. Assume the coordinates are integers.

Figure 8

Solution   First we note that the viewing window used in Figure 8 is

$$X\text{min} = -3 \qquad Y\text{min} = -4$$
$$X\text{max} = 3 \qquad Y\text{max} = 4$$
$$X\text{scl} = 1 \qquad Y\text{scl} = 2$$

The point shown is 2 tick units to the left on the horizontal axis (scale $= 1$) and 1 tick up on the vertical scale (scale $= 2$). The coordinates of the point shown are $(-2, 2)$. ◀

 NOW WORK PROBLEMS 15 AND 25.

1   Use the Distance Formula

If the same units of measurement, such as inches, centimeters, and so on, are used for both the x-axis and y-axis, then all distances in the xy-plane can be measured using this unit of measurement.

EXAMPLE 2	**Finding the Distance between Two Points**

Find the distance d between the points $(1, 3)$ and $(5, 6)$.

Solution   First we plot the points $(1, 3)$ and $(5, 6)$ and connect them with a straight line. See Figure 9(a). We are looking for length d. We begin by drawing a horizontal line from

$(1, 3)$ to $(5, 3)$ and a vertical line from $(5, 3)$ to $(5, 6)$, forming a right triangle, as in Figure 9(b). One leg of the triangle is of length 4 (since $|5 - 1| = 4$) and the other is of length 3 (since $|6 - 3| = 3$). By the Pythagorean Theorem, the square of the distance d that we seek is

$$d^2 = 4^2 + 3^2 = 16 + 9 = 25$$
$$d = \sqrt{25} = 5$$

Figure 9

(a)

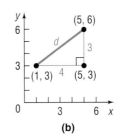
(b)

The **distance formula** provides a straightforward method for computing the distance between two points.

Theorem

Distance Formula

In Words

To compute the distance between two points, find the difference of the x-coordinates, square it, and add this to the square of the difference of the y-coordinates. The square root of this sum is the distance.

The distance between two points $P_1 = (x_1, y_1)$ and $P_2 = (x_2, y_2)$, denoted by $d(P_1, P_2)$, is

$$d(P_1, P_2) = \sqrt{(x_2 - x_1)^2 + (y_2 - y_1)^2} \tag{1}$$

Figure 10 illustrates the theorem.

Proof of the Distance Formula Let (x_1, y_1) denote the coordinates of point P_1, and let (x_2, y_2) denote the coordinates of point P_2. Assume that the line joining P_1 and P_2 is neither horizontal nor vertical. Refer to Figure 11(a). The coordinates of P_3 are (x_2, y_1). The horizontal distance from P_1 to P_3 is the absolute value of the difference of the x-coordinates, $|x_2 - x_1|$. The vertical distance from P_3 to P_2 is the absolute value of the difference of the y-coordinates, $|y_2 - y_1|$. See Figure 11(b). The distance $d(P_1, P_2)$ that we seek is the length of the hypotenuse of the right triangle, so, by the Pythagorean Theorem, it follows that

$$[d(P_1, P_2)]^2 = |x_2 - x_1|^2 + |y_2 - y_1|^2$$
$$= (x_2 - x_1)^2 + (y_2 - y_1)^2$$
$$d(P_1, P_2) = \sqrt{(x_2 - x_1)^2 + (y_2 - y_1)^2}$$

Figure 10

Figure 11

(a)

(b)

Now, if the line joining P_1 and P_2 is horizontal, then the y-coordinate of P_1 equals the y-coordinate of P_2; that is, $y_1 = y_2$. Refer to Figure 12(a). In this case, the distance formula (1) still works, because, for $y_1 = y_2$, it reduces to

$$d(P_1, P_2) = \sqrt{(x_2 - x_1)^2 + 0^2} = \sqrt{(x_2 - x_1)^2} = |x_2 - x_1|$$

Figure 12

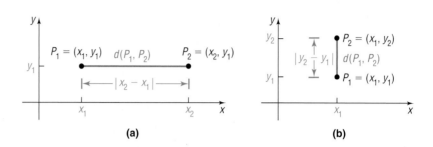

(a)   (b)

A similar argument holds if the line joining P_1 and P_2 is vertical. See Figure 12(b). The distance formula is valid in all cases. ∎

EXAMPLE 3

Finding the Length of a Line Segment

Figure 13

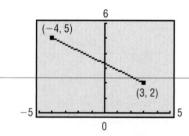

Find the length of the line segment shown in Figure 13.

Solution   The length of the line segment is the distance between the points $P_1 = (x_1, y_1) = (-4, 5)$, and $P_2 = (x_2, y_2) = (3, 2)$. Using the distance formula (1) with $x_1 = -4$, $y_1 = 5$, $x_2 = 3$ and $y_2 = 2$ the length d is

$$d = \sqrt{(x_2 - x_1)^2 + (y_2 - y_1)^2} = \sqrt{[3 - (-4)]^2 + (2 - 5)^2}$$
$$= \sqrt{7^2 + (-3)^2} = \sqrt{49 + 9} = \sqrt{58} \approx 7.62 \qquad \blacktriangleleft$$

NOW WORK PROBLEM **31**.

The distance between two points $P_1 = (x_1, y_1)$ and $P_2 = (x_2, y_2)$ is never a negative number. Furthermore, the distance between two points is 0 only when the points are identical, that is, when $x_1 = x_2$ and $y_1 = y_2$. Also, because $(x_2 - x_1)^2 = (x_1 - x_2)^2$ and $(y_2 - y_1)^2 = (y_1 - y_2)^2$, it makes no difference whether the distance is computed from P_1 to P_2 or from P_2 to P_1; that is, $d(P_1, P_2) = d(P_2, P_1)$.

The introduction to this chapter mentioned that rectangular coordinates enable us to translate geometry problems into algebra problems, and vice versa. The next example shows how algebra (the distance formula) can be used to solve geometry problems.

EXAMPLE 4

Using Algebra to Solve Geometry Problems

Consider the three points $A = (-2, 1)$, $B = (2, 3)$, and $C = (3, 1)$.

(a) Plot each point and form the triangle ABC.

(b) Find the length of each side of the triangle.

(c) Verify that the triangle is a right triangle.

(d) Find the area of the triangle.

Figure 14

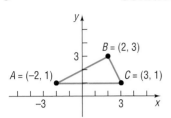

Solution

(a) Points A, B, C and triangle ABC are plotted in Figure 14.

(b) $d(A, B) = \sqrt{[2 - (-2)]^2 + (3 - 1)^2} = \sqrt{16 + 4} = \sqrt{20} = 2\sqrt{5}$

$d(B, C) = \sqrt{(3 - 2)^2 + (1 - 3)^2} = \sqrt{1 + 4} = \sqrt{5}$

$d(A, C) = \sqrt{[3 - (-2)]^2 + (1 - 1)^2} = \sqrt{25 + 0} = 5$

(c) To show that the triangle is a right triangle, we need to show that the sum of the squares of the lengths of two of the sides equals the square of the length of the third side. (Why is this sufficient?) Looking at Figure 14, it seems reasonable to conjecture that the right angle is at vertex B. We shall check to see whether

$$[d(A, B)]^2 + [d(B, C)]^2 = [d(A, C)]^2$$

We find that

$$[d(A, B)]^2 + [d(B, C)]^2 = \left(2\sqrt{5}\right)^2 + \left(\sqrt{5}\right)^2$$
$$= 20 + 5 = 25 = [d(A, C)]^2$$

so it follows from the converse of the Pythagorean Theorem that triangle ABC is a right triangle.

(d) Because the right angle is at vertex B, the sides AB and BC form the base and height of the triangle. Its area is

$$\text{Area} = \frac{1}{2}(\text{Base})(\text{Height}) = \frac{1}{2}\left(2\sqrt{5}\right)\left(\sqrt{5}\right) = 5 \text{ square units}$$ ◀

NOW WORK PROBLEM 49.

Figure 15

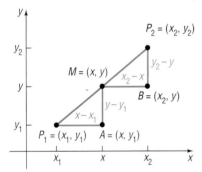

2 **Use the Midpoint Formula**

We now derive a formula for the coordinates of the **midpoint of a line segment**. Let $P_1 = (x_1, y_1)$ and $P_2 = (x_2, y_2)$ be the endpoints of a line segment, and let $M = (x, y)$ be the point on the line segment that is the same distance from P_1 as it is from P_2. See Figure 15. The triangles $P_1 A M$ and $M B P_2$ are congruent.[*] [Do you see why? Angle $AP_1 M =$ angle BMP_2,[†] angle $P_1 M A =$ angle $MP_2 B$, and $d(P_1, M) = d(M, P_2)$ is given. So, we have angle–side–angle.] Hence, corresponding sides are equal in length. That is,

$$x - x_1 = x_2 - x \quad \text{and} \quad y - y_1 = y_2 - y$$
$$2x = x_1 + x_2 \qquad\qquad 2y = y_1 + y_2$$
$$x = \frac{x_1 + x_2}{2} \qquad\qquad y = \frac{y_1 + y_2}{2}$$

Theorem

Midpoint Formula

The midpoint $M = (x, y)$ of the line segment from $P_1 = (x_1, y_1)$ to $P_2 = (x_2, y_2)$ is

$$M = (x, y) = \left(\frac{x_1 + x_2}{2}, \frac{y_1 + y_2}{2}\right) \qquad \textbf{(2)}$$

In Words

To find the midpoint of a line segment, average the x-coordinates and average the y-coordinates of the endpoints.

---

[*]The following statement is a postulate from geometry. Two triangles are congruent if their sides are the same length (SSS), or if two sides and the included angle are the same (SAS), or if two angles and the included sides are the same (ASA).
[†]Another postulate from geometry states that the transversal $\overline{P_1 P_2}$ forms congruent corresponding angles with the parallel line segments $\overline{P_1 A}$ and $\overline{MB}$.

EXAMPLE 5	Finding the Midpoint of a Line Segment

Find the midpoint of a line segment from $P_1 = (-5, 5)$ to $P_2 = (3, 1)$. Plot the points P_1 and P_2 and their midpoint. Check your answer.

Solution We apply the midpoint formula (2) using $x_1 = -5$, $y_1 = 5$, $x_2 = 3$, and $y_2 = 1$. Then the coordinates (x, y) of the midpoint M are

$$x = \frac{x_1 + x_2}{2} = \frac{-5 + 3}{2} = -1 \quad \text{and} \quad y = \frac{y_1 + y_2}{2} = \frac{5 + 1}{2} = 3$$

Figure 16

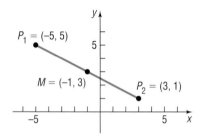

That is, $M = (-1, 3)$. See Figure 16.

✔ **CHECK:** Because M is the midpoint, we check the answer by verifying that $d(P_1, M) = d(M, P_2)$:

$$d(P_1, M) = \sqrt{[-1 - (-5)]^2 + (3 - 5)^2} = \sqrt{16 + 4} = \sqrt{20}$$
$$d(M, P_2) = \sqrt{[3 - (-1)]^2 + (1 - 3)^2} = \sqrt{16 + 4} = \sqrt{20}$$

◀

━ **NOW WORK PROBLEM 55.**

3 Graph Equations by Hand by Plotting Points

An **equation in two variables**, say x and y, is a statement in which two expressions involving x and y are equal. The expressions are called the **sides** of the equation. Since an equation is a statement, it may be true or false, depending on the value of the variables. Any values of x and y that result in a true statement are said to **satisfy** the equation.

For example, the following are all equations in two variables x and y:

$$x^2 + y^2 = 5 \qquad 2x - y = 6 \qquad y = 2x + 5 \qquad x^2 = y$$

The first of these, $x^2 + y^2 = 5$, is satisfied for $x = 1, y = 2$, since $1^2 + 2^2 = 1 + 4 = 5$. Other choices of x and y also satisfy this equation. It is not satisfied for $x = 2$ and $y = 3$, since $2^2 + 3^2 = 4 + 9 = 13 \neq 5$.

The **graph of an equation in two variables** x and y consists of the set of points in the xy-plane whose coordinates (x, y) satisfy the equation.

Graphs play an important role in helping us to visualize the relationships that exist between two variables or quantities. Figure 17 shows the relation between the

Figure 17
SOURCE: T. Rowe Price

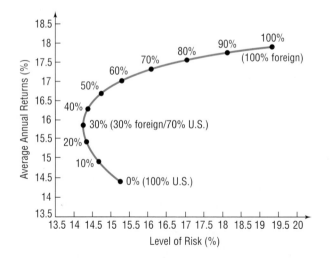

level of risk in a stock portfolio and the average annual rate of return. From the graph, we can see that when 30% of a portfolio of stocks is invested in foreign companies, risk is minimized.

| **EXAMPLE 6** | **Determining Whether a Point Is on the Graph of an Equation** |

Determine if the following points are on the graph of the equation $2x - y = 6$.

(a) $(2, 3)$ \qquad\qquad\qquad (b) $(2, -2)$

Solution   (a) For the point $(2, 3)$, we check to see if $x = 2, y = 3$ satisfies the equation $2x - y = 6$.

$$2x - y = 2(2) - 3 = 4 - 3 = 1 \neq 6$$

The equation is not satisfied, so the point $(2, 3)$ is not on the graph.

(b) For the point $(2, -2)$, we have

$$2x - y = 2(2) - (-2) = 4 + 2 = 6$$

The equation is satisfied, so the point $(2, -2)$ is on the graph.   ◀

 NOW WORK PROBLEM 65.

| **EXAMPLE 7** | **Graphing an Equation by Hand by Plotting Points** |

Graph the equation: $y = 2x + 5$

Solution   We want to find all points (x, y) that satisfy the equation. To locate some of these points (and thus get an idea of the pattern of the graph), we assign some numbers to x and find corresponding values for y.

If	**Then**	**Point on Graph**
$x = 0$	$y = 2(0) + 5 = 5$	$(0, 5)$
$x = 1$	$y = 2(1) + 5 = 7$	$(1, 7)$
$x = -5$	$y = 2(-5) + 5 = -5$	$(-5, -5)$
$x = 10$	$y = 2(10) + 5 = 25$	$(10, 25)$

By plotting these points and then connecting them, we obtain the graph of the equation (a *line*), as shown in Figure 18.

Figure 18

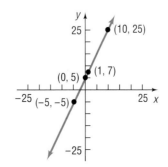

◀

EXAMPLE 8	**Graphing an Equation by Hand by Plotting Points**

Graph the equation: $y = x^2$

Solution   Table 1 provides several points on the graph. In Figure 19 we plot these points and connect them with a smooth curve to obtain the graph (a *parabola*).

Table 1

x	$y = x^2$	(x, y)
-4	16	$(-4, 16)$
-3	9	$(-3, 9)$
-2	4	$(-2, 4)$
-1	1	$(-1, 1)$
0	0	$(0, 0)$
1	1	$(1, 1)$
2	4	$(2, 4)$
3	9	$(3, 9)$
4	16	$(4, 16)$

Figure 19

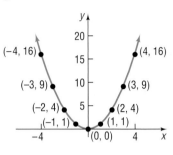

◀

NOW WORK PROBLEM 83.

The graphs of the equations shown in Figures 18 and 19 do not show all points. For example, in Figure 18, the point $(20, 45)$ is a part of the graph of $y = 2x + 5$, but it is not shown. Since the graph of $y = 2x + 5$ could be extended out as far as we please, we use arrows to indicate that the pattern shown continues. It is important when illustrating a graph to present enough of the graph so that any viewer of the illustration will "see" the rest of it as an obvious continuation of what is actually there. This is referred to as a **complete graph**.

One way to obtain a complete graph of an equation is to plot a sufficient number of points on the graph until a pattern becomes evident. Then these points are connected with a smooth curve following the suggested pattern. But how many points are sufficient? Sometimes knowledge about the equation tells us. For example, we will learn in Section 1.8 that, if an equation is of the form $y = mx + b$, then its graph is a line. In this case, two points would suffice to obtain the graph.

One purpose of this book is to investigate the properties of equations in order to decide whether a graph is complete. Sometimes we shall graph equations by plotting a sufficient number of points on the graph until a pattern becomes evident and then connect these points with a smooth curve following the suggested pattern. (Shortly, we shall investigate various techniques that will enable us to graph an equation without plotting so many points.) Other times we shall graph equations using a graphing utility.

4   **Graph Equations Using a Graphing Utility**

From Examples 2 and 3, we see that a graph can be obtained by plotting points in a rectangular coordinate system and connecting them. Graphing utilities perform these same steps when graphing an equation. For example, the TI-84 Plus determines 95 evenly spaced input values,* uses the equation to determine the output val-

---

*These input values depend on the values of Xmin and Xmax. For example, if Xmin $= -10$ and Xmax $= 10$, then the first input value will be -10 and the next input value will be $-10 + (10 - (-10))/94 = -9.7872$, and so on.

ues, plots these points on the screen, and finally (if in the connected mode) draws a line between consecutive points.

To graph an equation in two variables x and y using a graphing utility requires that the equation be written in the form $y = \{\text{expression in } x\}$. If the original equation is not in this form, replace it by equivalent equations until the form $y = \{\text{expression in } x\}$ is obtained. In general, there are four ways to obtain equivalent equations.

Procedures That Result in Equivalent Equations

1. Interchange the two sides of the equation:

$$\text{Replace} \quad 3x + 5 = y \quad \text{by} \quad y = 3x + 5$$

2. Simplify the sides of the equation by combining like terms, eliminating parentheses, and so on:

$$\text{Replace} \quad 2y + 2 + 6 = 2x + 5(x + 1)$$
$$\text{by} \quad 2y + 8 = 7x + 5$$

3. Add or subtract the same expression on both sides of the equation:

$$\text{Replace} \quad y + 3x - 5 = 4$$
$$\text{by} \quad y + 3x - 5 + 5 = 4 + 5$$

4. Multiply or divide both sides of the equation by the same nonzero expression:

$$\text{Replace} \quad 3y = 6 - 2x$$
$$\text{by} \quad \frac{1}{3} \cdot 3y = \frac{1}{3}(6 - 2x)$$

EXAMPLE 9	**Expressing an Equation in the Form $y = \{\text{expression in } x\}$**

Solve for y: $\quad 2y + 3x - 5 = 4$

Solution We replace the original equation by a succession of equivalent equations.

$$2y + 3x - 5 = 4$$
$$2y + 3x - 5 + 5 = 4 + 5 \qquad \text{Add 5 to both sides.}$$
$$2y + 3x = 9 \qquad \text{Simplify.}$$
$$2y + 3x - 3x = 9 - 3x \qquad \text{Subtract 3x from both sides.}$$
$$2y = 9 - 3x \qquad \text{Simplify.}$$
$$\frac{2y}{2} = \frac{9 - 3x}{2} \qquad \text{Divide both sides by 2.}$$
$$y = \frac{9 - 3x}{2} \qquad \text{Simplify.} \qquad \blacktriangleleft$$

Now we are ready to graph equations using a graphing utility. Most graphing utilities require the following steps.

Steps for Graphing an Equation Using a Graphing Utility

STEP 1: Solve the equation for y in terms of x.

STEP 2: Get into the graphing mode of your graphing utility. The screen will usually display $Y =$, prompting you to enter the expression involving x that you found in Step 1. (Consult your manual for the correct way to enter the expression; for example, $y = x^2$ might be entered as x^2 or as $x*x$ or as xx^Y2).

STEP 3: Select the viewing window. Without prior knowledge about the behavior of the graph of the equation, it is common to select the **standard viewing window**[*] initially. The viewing window is then adjusted based on the graph that appears. In this text, the standard viewing window will be

$$X\text{min} = -10 \qquad Y\text{min} = -10$$
$$X\text{max} = 10 \qquad Y\text{max} = 10$$
$$X\text{scl} = 1 \qquad Y\text{scl} = 1$$

STEP 4: Graph.

STEP 5: Adjust the viewing window until a complete graph is obtained.

EXAMPLE 10 **Graphing an Equation on a Graphing Utility**

Graph the equation: $6x^2 + 3y = 36$

STEP 1: We solve for y in terms of x.

$$6x^2 + 3y = 36$$

$$3y = -6x^2 + 36 \qquad \text{Subtract } 6x^2 \text{ from both sides of the equation.}$$

$$y = -2x^2 + 12 \qquad \text{Divide both sides of the equation by 3 and simplify.}$$

STEP 2: From the graphing mode, enter the expression $-2x^2 + 12$ after the prompt $Y =$. Figure 20 shows the expression entered on a TI-84 Plus.

STEP 3: Set the viewing window to the standard viewing window.

STEP 4: Graph. The screen should look like Figure 21.

STEP 5: The graph of $y = -2x^2 + 12$ is not complete. The value of Ymax must be increased so that the top portion of the graph is visible. After increasing the value of Ymax to 12, we obtain the graph in Figure 22.[†] The graph is now complete.

Figure 20

Figure 21

Figure 22

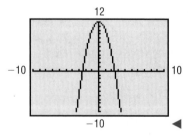

---

[*]Some graphing utilities have a ZOOM-STANDARD feature that automatically sets the viewing window to the standard viewing window and graphs the equation.

[†]Some graphing utilities have a ZOOM-FIT feature that determines the appropriate Ymin and Ymax for a given Xmin and Xmax. Consult your owner's manual for the appropriate keystrokes.

5 Use a Graphing Utility to Create Tables

In addition to graphing equations, graphing utilities can also be used to create a table of values that satisfy the equation. This feature is especially useful in determining an appropriate viewing window when graphing an equation. Many graphing utilities require the following steps to create a table.

Steps for Creating a Table of Values Using a Graphing Utility

STEP 1: Solve the equation for y in terms of x.

STEP 2: Enter the expression in x following the $Y =$ prompt of the graphing utility.

STEP 3: Set up the table. Graphing utilities typically have two modes for creating tables. In the AUTO mode, the user determines a starting point for the table (TblStart) and ΔTbl (pronounced "delta-table"). The ΔTbl feature determines the increment for x. The ASK mode requires the user to enter values of x and then the utility determines the corresponding value of y.

STEP 4: Create the table. The user can scroll within the table if the table was created in AUTO mode.

EXAMPLE 11

Creating a Table Using a Graphing Utility

Create a table that displays the points on the graph of $6x^2 + 3y = 36$ for $x = -3, -2, -1, 0, 1, 2,$ and 3.

Solution

Table 2

X	Y1
-3	-6
-2	4
-1	10
0	12
1	10
2	4
3	-6

Y1 ▤ -2X²+12

STEP 1: We solved the equation for y in Example 10 and obtained $y = -2x^2 + 12$.

STEP 2: Enter the expression in x following the $Y =$ prompt.

STEP 3: We set up the table in the AUTO mode with TblStart $= -3$ and ΔTbl $= 1$.

STEP 4: Create the table. The screen should look like Table 2.   ◀

In looking at Table 2, we notice that $y = 12$ when $x = 0$. This information could have been used to help to create the initial viewing window by letting us know that Ymax needs to be at least 12 in order to get a complete graph.

6 Find Intercepts from a Graph

The points, if any, at which a graph crosses or touches the coordinate axes are called the **intercepts**. See Figure 23. The x-coordinate of a point at which the graph crosses or touches the x-axis is an **x-intercept**, and the y-coordinate of a point at which the graph crosses or touches the y-axis is a **y-intercept**. For a graph to be complete, all its intercepts must be displayed.

Figure 23

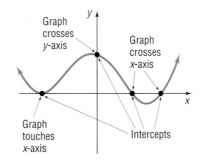

| EXAMPLE 12 | Finding Intercepts from a Graph |

Find the intercepts of the graph in Figure 24. What are its x-intercepts? What are its y-intercepts?

Figure 24

Solution   The intercepts of the graph are the points

$$(-3, 0), \quad (0, 3), \quad \left(\frac{3}{2}, 0\right), \quad \left(0, -\frac{4}{3}\right), \quad (0, -3.5), \quad (4.5, 0)$$

The x-intercepts are -3, $\dfrac{3}{2}$, and 4.5; the y-intercepts are -3.5, $-\dfrac{4}{3}$, and 3.   ◄

In Example 12, you should notice the following usage: If we do not specify the type of intercept (x- versus y-), then we report the intercept as an ordered pair. However, if we specify the type of intercept, then we only report the coordinate of the intercept. For x-intercepts, we report the x-coordinate of the intercept; for y-intercepts, we report the y-coordinate of the intercept.

➤ **NOW WORK PROBLEM 71.**

7 Use a Graphing Utility to Approximate Intercepts

We can use a graphing utility to approximate the intercepts of the graph of an equation, as illustrated in the next example.

| EXAMPLE 13 | Finding Intercepts Using a Graphing Utility |

Use a graphing utility to approximate the intercepts of the equation $y = x^3 - 16$.

Solution   Figure 25(a) shows the graph of $y = x^3 - 16$.

Figure 25

(a)   (b)   (c)

The eVALUEate feature of a TI-84 Plus graphing calculator accepts as input a value of x and determines the value of y. If we let $x = 0$, we find that the y-intercept is -16. See Figure 25(b).

The ZERO feature of a TI-84 Plus is used to find the x-intercept(s). See Figure 25(c). Rounded to two decimal places, the x-intercept is 2.52.   ◄

➤ **NOW WORK PROBLEM 93.**

To find the intercepts algebraically requires the ability to solve equations, the subject of the following four sections.

1.1 Assess Your Understanding

'Are You Prepared?'

Answers are given at the end of these exercises. If you get a wrong answer, read the pages listed in red.

1. On the real number line the origin is assigned the number _____. (p. 16)

2. If -3 and 5 are the coordinates of two points on the real number line, the distance between these points is _____. (p. 18)

3. If 3 and 4 are the legs of a right triangle, the hypotenuse is _____. (p. 29)

4. Use the converse of the Pythagorean Theorem to show that a triangle whose sides are of lengths 11, 60, and 61 is a right triangle. (pp. 29–30)

Concepts and Vocabulary

5. If (x, y) are the coordinates of a point P in the xy-plane, then x is called the _____ of P and y is the _____ of P.

6. The coordinate axes divide the xy-plane into four sections called _____.

7. If three distinct points P, Q, and R all lie on a line and if $d(P, Q) = d(Q, R)$ then Q is called the _____ of the line segment from P to R.

8. *True or False:* The distance between two points is sometimes a negative number.

9. *True or False:* The point $(-1, 4)$ lies in quadrant IV of the Cartesian plane.

10. *True or False:* The midpoint of a line segment is found by averaging the x-coordinates and averaging the y-coordinates of the endpoints.

Skill Building

In Problems 11 and 12, plot each point in the xy-plane. Tell in which quadrant or on what coordinate axis each point lies.

11. (a) $A = (-3, 2)$
   (b) $B = (6, 0)$
   (c) $C = (-2, -2)$
   (d) $D = (6, 5)$
   (e) $E = (0, -3)$
   (f) $F = (6, -3)$

12. (a) $A = (1, 4)$
   (b) $B = (-3, -4)$
   (c) $C = (-3, 4)$
   (d) $D = (4, 1)$
   (e) $E = (0, 1)$
   (f) $F = (-3, 0)$

13. Plot the points $(2, 0)$, $(2, -3)$, $(2, 4)$, $(2, 1)$, and $(2, -1)$. Describe the set of all points of the form $(2, y)$, where y is a real number.

14. Plot the points $(0, 3)$, $(1, 3)$, $(-2, 3)$, $(5, 3)$ and $(-4, 3)$. Describe the set of all points of the form $(x, 3)$, where x is a real number.

In Problems 15–18, determine the coordinates of the points shown. Tell in which quadrant each point lies. Assume the coordinates are integers.

15.

16.

17.

18.

In Problems 19–24, select a setting so that each given point will lie within the viewing window.

19. $(-10, 5)$, $(3, -2)$, $(4, -1)$

20. $(5, 0)$, $(6, 8)$, $(-2, -3)$

21. $(40, 20)$, $(-20, -80)$, $(10, 40)$

22. $(-80, 60)$, $(20, -30)$, $(-20, -40)$

23. $(0, 0)$, $(100, 5)$, $(5, 150)$

24. $(0, -1)$, $(100, 50)$, $(-10, 30)$

In Problems 25–30, determine the viewing window used.

25.

26.

27.

28.

29.

30.

In Problems 31–44, find the distance $d(P_1, P_2)$ between the points P_1 and P_2.

31.

32.

33.

34.

35. $P_1 = (3, -4);$ $P_2 = (5, 4)$
37. $P_1 = (-3, 2);$ $P_2 = (6, 0)$
39. $P_1 = (4, -3);$ $P_2 = (6, 4)$
41. $P_1 = (-0.2, 0.3);$ $P_2 = (2.3, 1.1)$
43. $P_1 = (a, b);$ $P_2 = (0, 0)$

36. $P_1 = (-1, 0);$ $P_2 = (2, 4)$
38. $P_1 = (2, -3);$ $P_2 = (4, 2)$
40. $P_1 = (-4, -3);$ $P_2 = (6, 2)$
42. $P_1 = (1.2, 2.3);$ $P_2 = (-0.3, 1.1)$
44. $P_1 = (a, a);$ $P_2 = (0, 0)$

In Problems 45–48, find the length of the line segment. Assume that the endpoints of each line segment have integer coordinates.

45.

46.

47.

48.
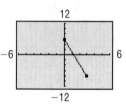

In Problems 49–54, plot each point and form the triangle ABC. Verify that the triangle is a right triangle. Find its area.

49. $A = (-2, 5);$ $B = (1, 3);$ $C = (-1, 0)$
51. $A = (-5, 3);$ $B = (6, 0);$ $C = (5, 5)$
53. $A = (4, -3);$ $B = (0, -3);$ $C = (4, 2)$

50. $A = (-2, 5);$ $B = (12, 3);$ $C = (10, -11)$
52. $A = (-6, 3);$ $B = (3, -5);$ $C = (-1, 5)$
54. $A = (4, -3);$ $B = (4, 1);$ $C = (2, 1)$

In Problems 55–64, find the midpoint of the line segment joining the points P_1 and P_2.

55. $P_1 = (3, -4);$ $P_2 = (5, 4)$
57. $P_1 = (-3, 2);$ $P_2 = (6, 0)$
59. $P_1 = (4, -3);$ $P_2 = (6, 1)$
61. $P_1 = (-0.2, 0.3);$ $P_2 = (2.3, 1.1)$
63. $P_1 = (a, b);$ $P_2 = (0, 0)$

56. $P_1 = (-2, 0);$ $P_2 = (2, 4)$
58. $P_1 = (2, -3);$ $P_2 = (4, 2)$
60. $P_1 = (-4, -3);$ $P_2 = (2, 2)$
62. $P_1 = (1.2, 2.3);$ $P_2 = (-0.3, 1.1)$
64. $P_1 = (a, a);$ $P_2 = (0, 0)$

In Problems 65–70, tell whether the given points are on the graph of the equation.

65. Equation: $y = x^4 - \sqrt{x}$
Points: $(0, 0); (1, 1); (-1, 0)$

66. Equation: $y = x^3 - 2\sqrt{x}$
Points: $(0, 0); (1, 1); (1, -1)$

67. Equation: $y^2 = x^2 + 9$
Points: $(0, 3); (3, 0); (-3, 0)$

68. Equation: $y^3 = x + 1$
Points: $(1, 2); (0, 1); (-1, 0)$

69. Equation: $x^2 + y^2 = 4$
Points: $(0, 2); (-2, 2); \left(\sqrt{2}, \sqrt{2}\right)$

70. Equation: $x^2 + 4y^2 = 4$
Points: $(0, 1); (2, 0); \left(2, \dfrac{1}{2}\right)$

In Problems 71–78, the graph of an equation is given. List the intercepts of the graph.

71.

72.

73.

74.

75.

76.

77.

78.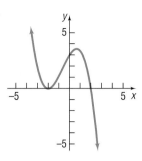

In Problems 79–90, graph each equation by hand by plotting points. Verify your results using a graphing utility.

79. $y = x + 2$ **80.** $y = x - 6$ **81.** $y = 2x + 8$ **82.** $y = 3x - 9$

83. $y = x^2 - 1$ **84.** $y = x^2 - 9$ **85.** $y = -x^2 + 4$ **86.** $y = -x^2 + 1$

87. $2x + 3y = 6$ **88.** $5x + 2y = 10$ **89.** $9x^2 + 4y = 36$ **90.** $4x^2 + y = 4$

In Problems 91–98, graph each equation using a graphing utility. Use a graphing utility to approximate the intercepts rounded to two decimal places. Use the TABLE feature to help to establish the viewing window.

91. $y = 2x - 13$ **92.** $y = -3x + 14$ **93.** $y = 2x^2 - 15$ **94.** $y = -3x^2 + 19$

95. $3x - 2y = 43$ **96.** $4x + 5y = 82$ **97.** $5x^2 + 3y = 37$ **98.** $2x^2 - 3y = 35$

Applications and Extensions

99. The medians of a triangle are the line segments from each vertex to the midpoint of the opposite side (see the figure). Find the lengths of the medians of the triangle with vertices at $A = (0, 0)$, $B = (6, 0)$, and $C = (4, 4)$.

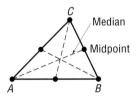

100. An **equilateral triangle** is one in which all three sides are of equal length. If two vertices of an equilateral triangle are $(0, 4)$ and $(0, 0)$, find the third vertex. How many of these triangles are possible?

*In Problems 101–104, find the length of each side of the triangle determined by the three points P_1, P_2, and P_3. State whether the triangle is an isosceles triangle, a right triangle, neither of these, or both. (An **isosceles triangle** is one in which at least two of the sides are of equal length.)*

101. $P_1 = (2, 1)$; $P_2 = (-4, 1)$; $P_3 = (-4, -3)$ **102.** $P_1 = (-1, 4)$; $P_2 = (6, 2)$; $P_3 = (4, -5)$

103. $P_1 = (-2, -1)$; $P_2 = (0, 7)$; $P_3 = (3, 2)$ **104.** $P_1 = (7, 2)$; $P_2 = (-4, 0)$; $P_3 = (4, 6)$

---

105. Baseball A major league baseball "diamond" is actually a square, 90 feet on a side (see the figure). What is the distance directly from home plate to second base (the diagonal of the square)?

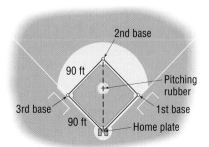

106. Little League Baseball The layout of a Little League playing field is a square, 60 feet on a side.* How far is it directly from home plate to second base (the diagonal of the square)?
*SOURCE: *Little League Baseball, Official Regulations and Playing Rules, 2005.*

107. Baseball Refer to Problem 105. Overlay a rectangular coordinate system on a major league baseball diamond so that the origin is at home plate, the positive x-axis lies in the direction from home plate to first base, and the positive y-axis lies in the direction from home plate to third base.
(a) What are the coordinates of first base, second base, and third base? Use feet as the unit of measurement.
(b) If the right fielder is located at $(310, 15)$, how far is it from there to second base?
(c) If the center fielder is located at $(300, 300)$, how far is it from there to third base?

108. Little League Baseball Refer to Problem 106. Overlay a rectangular coordinate system on a Little League baseball diamond so that the origin is at home plate, the positive *x*-axis lies in the direction from home plate to first base, and the positive *y*-axis lies in the direction from home plate to third base.

(a) What are the coordinates of first base, second base, and third base? Use feet as the unit of measurement.

(b) If the right fielder is located at $(180, 20)$, how far is it from there to second base?

(c) If the center fielder is located at $(220, 220)$, how far is it from there to third base?

109. A Dodge Neon and a Mack truck leave an intersection at the same time. The Neon heads east at an average speed of 30 miles per hour, while the truck heads south at an average speed of 40 miles per hour. Find an expression for their distance apart *d* (in miles) at the end of *t* hours.

110. A hot-air balloon, headed due east at an average speed of 15 miles per hour and at a constant altitude of 100 feet, passes over an intersection (see the figure). Find an expression for its distance *d* (measured in feet) from the intersection *t* seconds later.

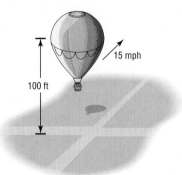

Discussion and Writing

In Problem 111, you may use a graphing utility, but it is not required.

111. (a) Graph $y = \sqrt{x^2}$, $y = x$, $y = |x|$, and $y = (\sqrt{x})^2$, noting which graphs are the same.

(b) Explain why the graphs of $y = \sqrt{x^2}$ and $y = |x|$ are the same.

(c) Explain why the graphs of $y = x$ and $y = (\sqrt{x})^2$ are not the same.

(d) Explain why the graphs of $y = \sqrt{x^2}$ and $y = x$ are not the same.

112. Make up an equation with the points $(2, 0)$, $(4, 0)$, and $(0, 1)$. Compare your equation with a friend's equation. Comment on any similarities.

113. Draw a graph that contains the points $(-2, -1)$, $(0, 1)$, $(1, 3)$, and $(3, 5)$. Compare your graph with those of other students. Are most of the graphs almost straight lines? How many are "curved"? Discuss the various ways that these points might be connected.

114. Explain what is meant by a complete graph.

'Are You Prepared?' Answers

1. 0 **2.** 8 **3.** 5 **4.** $11^2 + 60^2 = 61^2$

1.2 Solving Equations Using a Graphing Utility; Linear and Rational Equations

PREPARING FOR THIS SECTION *Before getting started, review the following:*

• Domain of a Variable (Section R.2, p. 20) • Rational Expressions (Section R.7, pp. 57–65)

Now work the 'Are You Prepared?' problems on page 109.

OBJECTIVES 1 Solve Equations Using a Graphing Utility
2 Solve Linear Equations
3 Solve Rational Equations
4 Solve Problems That Can Be Modeled by Linear Equations

In the last section, we learned how to graph equations in two variables. We use this knowledge, along with a graphing utility, to help us to verify solutions to **equations in one variable**. An **equation in one variable** is a statement in which two expressions, at least one containing the variable, are equal. The expressions are called the **sides** of the equation. Since an equation is a statement, it may be true or false, depending on the value of the variable. Unless otherwise restricted, the admissible values of the variable are those in the domain of the variable. Those admissible values of the variable, if any, that result in a true statement are called **solutions**, or **roots**, of the equation. To **solve an equation** means to find all the solutions of the equation.

For example, the following are all equations in one variable, x:

$$x + 5 = 9 \qquad x^2 + 5x = 2x - 2 \qquad \frac{x^2 - 4}{x + 1} = 0 \qquad x^2 + 9 = 5$$

The first of these statements, $x + 5 = 9$, is true when $x = 4$ and false for any other choice of x. We say that 4 is a solution of the equation $x + 5 = 9$ or that 4 **satisfies** the equation $x + 5 = 9$, because, when we substitute 4 for x, a true statement results.

Sometimes an equation will have more than one solution. For example, the equation

$$\frac{x^2 - 4}{x + 1} = 0$$

has $x = -2$ and $x = 2$ as solutions.

Usually, we will write the solution of an equation in set notation. This set is called the **solution set** of the equation. For example, the solution set of the equation $x^2 - 9 = 0$ is $\{-3, 3\}$.

Unless indicated otherwise, we will limit ourselves to real solutions, that is, solutions that are real numbers. Some equations have no real solution. For example, $x^2 + 9 = 5$ has no real solution, because there is no real number whose square, when added to 9, equals 5.

An equation that is satisfied for every choice of the variable for which both sides are defined is called an **identity**. For example, the equation

$$3x + 5 = x + 3 + 2x + 2$$

is an identity, because this statement is true for any real number x.

In this text, we present two methods for solving equations: algebraic and graphical. We shall see as we proceed through this book that some equations can be solved using algebraic techniques that result in *exact* solutions. For other equations, however, there are no algebraic techniques that lead to an exact solution. For such equations, a graphing utility can often be used to investigate possible solutions. When a graphing utility is used to solve an equation, usually *approximate* solutions are obtained.

One goal of this text is to determine when equations can be solved algebraically. If an algebraic method for solving an equation exists, we shall use it to obtain an exact solution. A graphing utility can then be used to support the algebraic result. However, if an equation must be solved for which no algebraic techniques are available, a graphing utility will be used to obtain approximate solutions.

1   Solve Equations Using a Graphing Utility

When a graphing utility is used to solve an equation, usually *approximate* solutions are obtained. Unless otherwise stated, we shall follow the practice of giving approximate solutions as decimals *rounded to two decimal places*.

The ZERO (or ROOT) feature of a graphing utility can be used to find the solutions of an equation when one side of the equation is 0. In using this feature to solve equations, we make use of the fact that when the graph of an equation crosses or touches the x-axis, then $y = 0$. For this reason, any value of x for which $y = 0$ will be a solution to the equation. That is, solving an equation for x when one side of the equation is 0 is equivalent to finding where the graph of the corresponding equation crosses or touches the x-axis.

| EXAMPLE 1 | Using ZERO (or ROOT) to Approximate Solutions of an Equation |

Find the solution(s) of the equation $x^3 - x + 1 = 0$. Round answers to two decimal places.

Solution The solutions of the equation $x^3 - x + 1 = 0$ are the same as the x-coordinates where the graph of $Y_1 = x^3 - x + 1$ crosses or touches the x-axis. We begin by graphing the equation.

Figure 26 shows the graph.

From the graph there appears to be one x-intercept (solution to the equation) between -2 and -1.

Using the ZERO (or ROOT) feature of our graphing utility, we determine that the x-intercept, and thus the solution to the equation, is $x = -1.32$ rounded to two decimal places. See Figure 27.

Figure 26

Figure 27

NOW WORK PROBLEM **17**.

A second method for solving equations using a graphing utility involves the IN-TERSECT feature of the graphing utility. This feature is used most effectively when one side of the equation is not 0.

| EXAMPLE 2 | Using INTERSECT to Approximate Solutions of an Equation |

Find the solution(s) to the equation $4x^4 - 3 = 2x + 1$. Round answers to two decimal places.

Solution We begin by graphing each side of the equation as follows: graph $Y_1 = 4x^4 - 3$ and $Y_2 = 2x + 1$. See Figure 28.

At the point of intersection of the graphs, the value of the y-coordinate is the same. Thus, the x-coordinate of the point of intersection represents the solution to the equation. Do you see why? The INTERSECT feature on a graphing utility determines the point of intersection of the graphs. Using this feature, we find that the graphs intersect at $(-0.87, -0.73)$ and $(1.12, 3.23)$ rounded to two decimal places. See Figure 29(a) and (b). The solutions of the equation are $x = -0.87$ and $x = 1.12$, rounded to two decimal places.

Figure 28

Figure 29

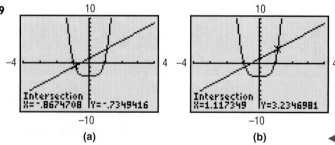

(a)   (b)

NOW WORK PROBLEM **19**.

Summary

Steps for Approximating Solutions of Equations Using ZERO (or ROOT)

STEP 1: Write the equation in the form {expression in x} = 0.

STEP 2: Graph Y_1 = {expression in x}.

STEP 3: Use ZERO (or ROOT) to determine each x-intercept of the graph.

Steps for Approximating Solutions of Equations Using INTERSECT

STEP 1: Graph Y_1 = {expression in x on the left side of the equation}
Y_2 = {expression in x on the right side of the equation}

STEP 2: Use INTERSECT to determine the x-coordinate of each point of intersection.

Solving Equations Algebraically

One method for solving equations algebraically requires that a series of *equivalent equations* be developed from the original equation until an obvious solution results.

Two or more equations that have precisely the same solutions are called **equivalent equations**.

For example, all the following equations are equivalent, because each has only the solution $x = 5$:

$$2x + 3 = 13$$
$$2x = 10$$
$$x = 5$$

The question, though, is "How do I obtain an equivalent equation?" In general, there are five ways to do so.

Procedures That Result in Equivalent Equations

1. Interchange the two sides of the equation:
Replace $\quad 3 = x \quad$ by $\quad x = 3$

2. Simplify the sides of the equation by combining like terms, eliminating parentheses, and so on:
Replace $\quad x + 2 + 6 = 2x + 3(x + 1)$
by $\quad\quad x + 8 = 5x + 3$

3. Add or subtract the same expression on both sides of the equation:
Replace $\quad\quad\quad 3x - 5 = 4$
by $\quad (3x - 5) + 5 = 4 + 5$

4. Multiply or divide both sides of the equation by the same nonzero expression:
Replace $\quad \dfrac{3x}{x - 1} = \dfrac{6}{x - 1} \quad x \neq 1$

by $\quad \dfrac{3x}{x - 1} \cdot (x - 1) = \dfrac{6}{x - 1} \cdot (x - 1)$

5. If one side of the equation is 0 and the other side can be factored, then we may use the Zero-Product Property and set each factor equal to 0:
Replace $\quad\quad\quad x(x - 3) = 0$
by $\quad x = 0 \quad$ or $\quad x - 3 = 0$

WARNING
Squaring both sides of an equation does not necessarily lead to an equivalent equation. ■

Whenever it is possible to solve an equation in your head, do so. For example:

The solution of $2x = 8$ is $x = 4$.

The solution of $3x - 15 = 0$ is $x = 5$.

NOW WORK PROBLEM 9.

We now introduce a specific type of equation that can be solved algebraically to obtain exact solutions.

2 Solve Linear Equations

Linear equations are equations such as

$$3x + 12 = 0, \qquad \frac{3}{4}x - \frac{1}{5} = 0, \qquad 0.62x - 0.3 = 0$$

A **linear equation in one variable** is equivalent to an equation of the form

$$\boxed{ax + b = 0}$$

where a and b are real numbers and $a \neq 0$.

Sometimes a linear equation is called a **first-degree equation**, because the left side is a polynomial in x of degree 1.

EXAMPLE 3	**Solving a Linear Equation**

Solve the equation: $3(x - 2) = 5(x - 1)$

Algebraic Solution

$$3(x - 2) = 5(x - 1)$$
$$3x - 6 = 5x - 5 \qquad \text{Use the Distributive Property.}$$
$$3x - 6 - 5x = 5x - 5 - 5x \qquad \text{Subtract 5x from each side.}$$
$$-2x - 6 = -5 \qquad \text{Simplify.}$$
$$-2x - 6 + 6 = -5 + 6 \qquad \text{Add 6 to each side.}$$
$$-2x = 1 \qquad \text{Simplify.}$$
$$\frac{-2x}{-2} = \frac{1}{-2} \qquad \text{Divide each side by } -2.$$
$$x = -\frac{1}{2} \qquad \text{Simplify.}$$

Graphing Solution

Graph $Y_1 = 3(x - 2)$ and $Y_2 = 5(x - 1)$. See Figure 30. Using INTERSECT, we find the point of intersection to be $(-0.5, -7.5)$. The solution of the equation is $x = -0.5$.

Figure 30

✔ **CHECK:** Let $x = -\frac{1}{2}$ in the expression in x on the left side of the equation and simplify. Let $x = -\frac{1}{2}$ in the expression in x on the right side of the equation and simplify. If the two expressions are equal, the solution checks.

$$3(x - 2) = 3\left(-\frac{1}{2} - 2\right) = 3\left(-\frac{5}{2}\right) = -\frac{15}{2}$$

$$5(x - 1) = 5\left(-\frac{1}{2} - 1\right) = 5\left(-\frac{3}{2}\right) = -\frac{15}{2}$$

Since the two expressions are equal, the solution $x = -\frac{1}{2}$ checks.

The solution set is $\left\{-\frac{1}{2}\right\}$.

◄

NOW WORK PROBLEM 35.

The next example illustrates the solution of an equation that does not appear to be linear, but leads to a linear equation upon simplification.

| **EXAMPLE 4** | **Solving an Equation That Leads to a Linear Equation** |

Solve the equation: $(2x - 1)(x - 1) = (x - 5)(2x - 5)$

Algebraic Solution

$(2x - 1)(x - 1) = (x - 5)(2x - 5)$

$2x^2 - 3x + 1 = 2x^2 - 15x + 25$ Multiply and combine like terms.

$2x^2 - 3x + 1 - 2x^2 = 2x^2 - 15x + 25 - 2x^2$ Subtract $2x^2$ from each side.

$-3x + 1 = -15x + 25$ Simplify.

$-3x + 1 - 1 = -15x + 25 - 1$ Subtract 1 from each side.

$-3x = -15x + 24$ Simplify.

$-3x + 15x = -15x + 24 + 15x$ Add $15x$ to each side.

$12x = 24$ Simplify.

$\dfrac{12x}{12} = \dfrac{24}{12}$ Divide each side by 12.

$x = 2$ Simplify.

Graphing Solution

Graph $Y_1 = (2x - 1)(x - 1)$ and $Y_2 = (x - 5)(2x - 5)$. See Figure 31. Using INTERSECT, we find the point of intersection to be $(2, 3)$. The solution of the equation is $x = 2$.

Figure 31

✔ **CHECK:** $(2x - 1)(x - 1) = (2 \cdot 2 - 1)(2 - 1) = (3)(1) = 3$

$(x - 5)(2x - 5) = (2 - 5)(2 \cdot 2 - 5) = (-3)(-1) = 3$

Since the two expressions are equal, the solution checks. The solution set is $\{2\}$. ◀

 NOW WORK PROBLEM 55.

3   Solve Rational Equations

We now introduce another type of equation, the *rational equation*. A **rational equation** is an equation that contains a rational expression. Examples of rational equations are

$$\frac{3}{x + 1} = \frac{2}{x - 1} + 7 \quad \text{and} \quad \frac{x - 5}{x - 4} = \frac{3}{x + 2}$$

To solve a rational equation, multiply both sides of the equation by the least common multiple of the denominators of the rational expressions that make up the rational equation. We shall only show the algebraic solution to rational equations. We leave it to you to verify the results using a graphing utility.

EXAMPLE 5	Solving a Rational Equation

Solve the equation: $\dfrac{3}{x-2} = \dfrac{1}{x-1} + \dfrac{7}{(x-1)(x-2)}$

Solution First, we note that the domain of the variable is $\{x \mid x \neq 1, x \neq 2\}$. We clear the equation of rational expressions by multiplying both sides by the least common multiple of the denominators of the three rational expressions, $(x-1)(x-2)$.

$$\frac{3}{x-2} = \frac{1}{x-1} + \frac{7}{(x-1)(x-2)}$$

$$(x-1)(x-2)\frac{3}{x-2} = (x-1)(x-2)\left[\frac{1}{x-1} + \frac{7}{(x-1)(x-2)}\right]$$

Multiply both sides by $(x-1)(x-2)$. Cancel on the left.

$$3x - 3 = (x-1)(x-2)\frac{1}{x-1} + (x-1)(x-2)\frac{7}{(x-1)(x-2)}$$

Use the Distributive Property on each side; cancel on the right.

$$3x - 3 = (x-2) + 7$$

$$3x - 3 = x + 5$$

Combine like terms.

$$2x = 8$$

Add 3 to each side; subtract x from each side.

$$x = 4$$

Divide by 2.

✔ **Check:** $\dfrac{3}{x-2} = \dfrac{3}{4-2} = \dfrac{3}{2}$

$$\frac{1}{x-1} + \frac{7}{(x-1)(x-2)} = \frac{1}{4-1} + \frac{7}{(4-1)(4-2)} = \frac{1}{3} + \frac{7}{3\cdot 2} = \frac{2}{6} + \frac{7}{6} = \frac{9}{6} = \frac{3}{2}$$

Since the two expressions are equal, the solution $x = 4$ checks.

The solution set is $\{4\}$. ◄

✎— **NOW WORK PROBLEM 71.**

The next example illustrates a rational equation that has no solution.

EXAMPLE 6	A Rational Equation with No Solution

Solve the equation: $\dfrac{3x}{x-1} + 2 = \dfrac{3}{x-1}$

Solution First, we note that the domain of the variable is $\{x \mid x \neq 1\}$. Since the two rational expressions in the equation have the same denominator, $x - 1$, we can simplify by multiplying both sides by $x - 1$. The resulting equation is equivalent to the original equation, since we are multiplying by $x - 1$, which is not 0 (remember, $x \neq 1$).

$$\frac{3x}{x-1} + 2 = \frac{3}{x-1}$$

$$\left(\frac{3x}{x-1} + 2\right) \cdot (x-1) = \frac{3}{x-1} \cdot (x-1)$$

Multiply both sides by x − 1; cancel on the right.

$$\frac{3x}{x-1} \cdot (x-1) + 2 \cdot (x-1) = 3$$

Use the Distributive Property on the left side; cancel on the left.

$$3x + 2x - 2 = 3$$ Simplify.

$$5x - 2 = 3$$ Combine like terms.

$$5x = 5$$ Add 2 to each side.

$$x = 1$$ Divide both sides by 5.

The solution appears to be 1. But recall that $x = 1$ is not in the domain of the variable. The equation has no solution. The solution set is $\emptyset$.   ◀

━━━ **NOW WORK PROBLEM 61.**

4   Solve Problems That Can Be Modeled by Linear Equations

Although each situation has its own unique features, we can provide an outline of the steps to follow in setting up applied problems.

Steps for Setting Up Applied Problems

STEP 1: Read the problem carefully, perhaps two or three times. Pay particular attention to the question being asked in order to identify what you are looking for. If you can, determine realistic possibilities for the answer.

STEP 2: Assign a letter (variable) to represent what you are looking for, and, if necessary, express any remaining unknown quantities in terms of this variable.

STEP 3: Make a list of all the known facts, and translate them into mathematical expressions. These may take the form of an equation (or, later, an inequality) involving the variable. If possible, draw an appropriately labeled diagram to assist you. Sometimes a table or chart helps.

STEP 4: Solve the equation for the variable, and then answer the question using a complete sentence.

STEP 5: Check the answer with the facts in the problem. If it agrees, congratulations! If it does not agree, try again.

Let's look at two examples.

EXAMPLE 7   **Investments**

A total of $18,000 is invested, some in stocks and some in bonds. If the amount invested in bonds is half that invested in stocks, how much is invested in each category?

Solution   **STEP 1:** We are being asked to find the amount of two investments. These amounts must total $18,000. (Do you see why?)

STEP 2: If we let x equal the amount invested in stocks, then $18,000 - x$ is the amount invested in bonds. Do you see why? Look at Step 3.

STEP 3: We set up a table:

Amount in Stocks	Amount in Bonds	Reason
x	$18{,}000 - x$	Total invested is $18,000

We also know that

Total amount invested in bonds	is	one-half that in stocks
$18{,}000 - x$	$=$	$\frac{1}{2}(x)$

STEP 4:

$$18{,}000 - x = \frac{1}{2}x$$

$$18{,}000 = x + \frac{1}{2}x \qquad \text{Add } x \text{ to both sides.}$$

$$18{,}000 = \frac{3}{2}x \qquad \text{Simplify.}$$

$$\left(\frac{2}{3}\right)18{,}000 = \left(\frac{2}{3}\right)\left(\frac{3}{2}x\right) \qquad \text{Multiply both sides by } \frac{2}{3}.$$

$$12{,}000 = x \qquad \text{Simplify.}$$

So, $12,000 is invested in stocks and $18,000 - $12,000 = $6000 is invested in bonds.

STEP 5: The total invested is $12,000 + $6000 = $18,000, and the amount in bonds, $6000, is half that in stocks, $12,000. ◄

 NOW WORK PROBLEM 95.

EXAMPLE 8	**Determining an Hourly Wage**

Shannon grossed $435 one week by working 52 hours. Her employer pays time-and-a-half for all hours worked in excess of 40 hours. With this information, can you determine Shannon's regular hourly wage?

Solution   **STEP 1:** We are looking for an hourly wage. Our answer will be in dollars per hour.

STEP 2: Let x represent the regular hourly wage; x is measured in dollars per hour.

STEP 3: We set up a table:

	Hours Worked	Hourly Wage	Salary
Regular	40	x	$40x$
Overtime	12	$1.5x$	$12(1.5x) = 18x$

The sum of regular salary plus overtime salary will equal $435. From the table, $40x + 18x = 435$.

STEP 4:

$$40x + 18x = 435$$

$$58x = 435$$

$$x = 7.50$$

Shannon's regular hourly wage is $7.50 per hour.

STEP 5: Forty hours yields a salary of $40(7.50) = $300, and 12 hours of overtime yields a salary of $12(1.5)(7.50) = $135, for a total of $435. ◄

NOW WORK PROBLEM 99.

HISTORICAL FEATURE

Solving equations is among the oldest of mathematical activities, and efforts to systematize this activity determined much of the shape of modern mathematics.

Consider the following problem and its solution using only words: Solve the problem of how many apples Jim has, given that

"Bob's five apples and Jim's apples together make twelve apples" by thinking,

"Jim's apples are all twelve apples less Bob's five apples" and then concluding,

"Jim has seven apples."

The mental steps translated into algebra are

$$5 + x = 12$$
$$x = 12 - 5$$
$$= 7$$

The solution of this problem using only words is the earliest form of algebra. Such problems were solved exactly this way in Babylonia in 1800 BC. We know almost nothing of mathematical work before this date, although most authorities believe the sophistication of the earliest known texts indicates that a long period of previous development must have occurred. The method of writing out equations in words persisted for thousands of years, and although it now seems extremely cumbersome, it was used very effectively by many generations of mathematicians. The Arabs developed a good deal of the theory of cubic equations while writing out all the equations in words. About AD 1500, the tendency to abbreviate words into written equations began to lead in the direction of modern notation; for example, the Latin word *et* (meaning *and*) developed into the plus sign, +. Although the occasional use of letters to represent variables dates back to AD 1200, the practice did not become common until about AD 1600. Development thereafter was rapid, and by 1635 algebraic notation did not differ essentially from what we use now.

1.2 Assess Your Understanding

'Are You Prepared?'

Answers are given at the end of these exercises. If you get a wrong answer, read the pages listed in red.

1. Is $x = 4$ in the domain of $\dfrac{3}{x - 4}$? (p. 20)

2. Find the least common multiple of the denominators of $\dfrac{1}{x - 1}$ and $\dfrac{3}{x + 3}$. (pp. 61–62)

Concepts and Vocabulary

3. Two equations that have the same solution set are called _____.

4. An equation that is satisfied for every value of the variable for which both sides are defined is called a(n) _____.

5. An equation of the form $ax + b = 0$ is called a(n) _____ equation or a(n) _____ equation.

6. *True or False:* The solution of the equation $3x - 8 = 0$ is $\dfrac{3}{8}$.

7. *True or False:* Some equations have no solution.

8. *True or False:* When solving equations using a graphing utility, the solutions are always exact.

Skill Building

In Problems 9–16, mentally solve each equation.

9. $7x = 21$

10. $6x = -24$

11. $3x + 15 = 0$

12. $6x + 18 = 0$

13. $2x - 3 = 0$

14. $3x + 4 = 0$

15. $\dfrac{1}{3}x = \dfrac{5}{12}$

16. $\dfrac{2}{3}x = \dfrac{9}{2}$

In Problems 17–28, use a graphing utility to approximate the real solutions, if any, of each equation rounded to two decimal places. All solutions lie between −10 and 10.

17. $x^3 - 4x + 2 = 0$

18. $x^3 - 8x + 1 = 0$

19. $-2x^4 + 5 = 3x - 2$

20. $-x^4 + 1 = 2x^2 - 3$

21. $x^4 - 2x^3 + 3x - 1 = 0$

22. $3x^4 - x^3 + 4x^2 - 5 = 0$

23. $-x^3 - \dfrac{5}{3}x^2 + \dfrac{7}{2}x + 2 = 0$

24. $-x^4 + 3x^3 + \dfrac{7}{3}x^2 - \dfrac{15}{2}x + 2 = 0$

25. $-\dfrac{2}{3}x^4 - 2x^3 + \dfrac{5}{2}x = -\dfrac{2}{3}x^2 + \dfrac{1}{2}$

26. $\dfrac{1}{4}x^3 - 5x = \dfrac{1}{5}x^2 - 4$

27. $x^4 - 5x^2 + 2x + 11 = 0$

28. $-3x^4 + 8x^2 - 2x - 9 = 0$

In Problems 29–76, solve each equation algebraically. Verify your results using a graphing utility.

29. $3x + 4 = x$

30. $2x + 9 = 5x$

31. $2t - 6 = 3 - t$

32. $5y + 6 = -18 - y$

33. $6 - x = 2x + 9$

34. $3 - 2x = 2 - x$

35. $3 + 2n = 4n + 7$

36. $6 - 2m = 3m + 1$

37. $2(3 + 2x) = 3(x - 4)$

38. $3(2 - x) = 2x - 1$

39. $8x - (3x + 2) = 3x - 10$

40. $7 - (2x - 1) = 10$

41. $\dfrac{3}{2}x + 2 = \dfrac{1}{2} - \dfrac{1}{2}x$

42. $\dfrac{1}{3}x = 2 - \dfrac{2}{3}x$

43. $\dfrac{1}{2}x - 5 = \dfrac{3}{4}x$

44. $1 - \dfrac{1}{2}x = 6$

45. $\dfrac{2}{3}p = \dfrac{1}{2}p - \dfrac{1}{3}$

46. $\dfrac{1}{2} - \dfrac{1}{3}p = \dfrac{4}{3}$

47. $0.9t = 0.4 + 0.1t$

48. $0.9t = 1 + t$

49. $\dfrac{x + 1}{3} + \dfrac{x + 2}{7} = 2$

50. $\dfrac{2x + 1}{3} + 16 = 3x$

51. $\dfrac{2}{y} + \dfrac{4}{y} = 3$

52. $\dfrac{4}{y} - 5 = \dfrac{5}{2y}$

53. $\dfrac{1}{2} + \dfrac{2}{x} = \dfrac{3}{4}$

54. $\dfrac{3}{x} - \dfrac{1}{3} = \dfrac{1}{6}$

55. $(x + 7)(x - 1) = (x + 1)^2$

56. $(x + 2)(x - 3) = (x + 3)^2$

57. $x(2x - 3) = (2x + 1)(x - 4)$

58. $x(1 + 2x) = (2x - 1)(x - 2)$

59. $z(z^2 + 1) = 3 + z^3$

60. $w(4 - w^2) = 8 - w^3$

61. $\dfrac{x}{x - 2} + 3 = \dfrac{2}{x - 2}$

62. $\dfrac{2x}{x + 3} = \dfrac{-6}{x + 3} - 2$

63. $\dfrac{2x}{x^2 - 4} = \dfrac{4}{x^2 - 4} - \dfrac{3}{x + 2}$

64. $\dfrac{x}{x^2 - 9} + \dfrac{4}{x + 3} = \dfrac{3}{x^2 - 9}$

65. $\dfrac{x}{x + 2} = \dfrac{3}{2}$

66. $\dfrac{3x}{x - 1} = 2$

67. $\dfrac{5}{2x - 3} = \dfrac{3}{x + 5}$

68. $\dfrac{-4}{x + 4} = \dfrac{-3}{x + 6}$

69. $\dfrac{6t + 7}{4t - 1} = \dfrac{3t + 8}{2t - 4}$

70. $\dfrac{8w + 5}{10w - 7} = \dfrac{4w - 3}{5w + 7}$

71. $\dfrac{4}{x - 2} = \dfrac{-3}{x + 5} + \dfrac{7}{(x + 5)(x - 2)}$

72. $\dfrac{-4}{2x + 3} + \dfrac{1}{x - 1} = \dfrac{1}{(2x + 3)(x - 1)}$

73. $\dfrac{2}{y + 3} + \dfrac{3}{y - 4} = \dfrac{5}{y + 6}$

74. $\dfrac{5}{5z - 11} + \dfrac{4}{2z - 3} = \dfrac{-3}{5 - z}$

75. $\dfrac{x}{x^2 - 1} - \dfrac{x + 3}{x^2 - x} = \dfrac{-3}{x^2 + x}$

76. $\dfrac{x + 1}{x^2 + 2x} - \dfrac{x + 4}{x^2 + x} = \dfrac{-3}{x^2 + 3x + 2}$

Applications and Extensions

77. If $(a, 2)$ is a point on the graph of $y = 5x + 4$, what is a?

78. If $(2, b)$ is a point on the graph of $y = x^2 + 3x$, what is b?

79. If (a, b) is a point on the graph of $2x + 3y = 6$, write an equation that relates a to b.

80. If $(2, 0)$ and $(0, 5)$ are points on the graph of $y = mx + b$, what are m and b?

In Problems 81–86, solve each equation. The letters a, b, and c are constants.

81. $ax - b = c, \quad a \neq 0$

82. $1 - ax = b, \quad a \neq 0$

83. $\dfrac{x}{a} + \dfrac{x}{b} = c, \quad a \neq 0, b \neq 0, a \neq -b$

84. $\dfrac{a}{x} + \dfrac{b}{x} = c, \quad c \neq 0, \quad a + b \neq 0$

85. $\dfrac{1}{x - a} + \dfrac{1}{x + a} = \dfrac{2}{x - 1}$

86. $\dfrac{b + c}{x + a} = \dfrac{b - c}{x - a}, \quad c \neq 0, a \neq 0$

87. Find the number a for which $x = 4$ is a solution of the equation

$$x + 2a = 16 + ax - 6a$$

88. Find the number b for which $x = 2$ is a solution of the equation

$$x + 2b = x - 4 + 2bx$$

Problems 89–94 list some formulas that occur in applications. Solve each formula for the indicated variable.

89. Electricity $\dfrac{1}{R} = \dfrac{1}{R_1} + \dfrac{1}{R_2}$ for R

90. Finance $A = P(1 + rt)$ for r

91. Mechanics $F = \dfrac{mv^2}{R}$ for R

92. Chemistry $PV = nRT$ for T

93. Mathematics $S = \dfrac{a}{1 - r}$ for r

94. Mechanics $v = -gt + v_0$ for t

95. Finance A total of $20,000 is to be invested, some in bonds and some in certificates of deposit (CDs). If the amount invested in bonds is to exceed that in CDs by $3000, how much will be invested in each type of investment?

96. Finance A total of $10,000 is to be divided between Sean and George, with George to receive $3000 less than Sean. How much will each receive?

97. Finance An inheritance of $900,000 is to be divided among Scott, Alice, and Tricia in the following manner: Alice is to receive $\dfrac{3}{4}$ of what Scott gets, while Tricia gets $\dfrac{1}{2}$ of what Scott gets. How much does each receive?

98. Sharing the Cost of a Pizza Judy and Tom agree to share the cost of an $18 pizza based on how much each ate. If Tom ate $\dfrac{2}{3}$ the amount that Judy ate, how much should each pay?
[**Hint:** Some pizza may be left.]

Tom's portion

Judy's portion

99. Computing Hourly Wages Sandra, who is paid time-and-a-half for hours worked in excess of 40 hours, had gross weekly wages of $442 for 48 hours worked. What is her regular hourly rate?

100. Computing Hourly Wages Leigh is paid time-and-a-half for hours worked in excess of 40 hours and double-time for hours worked on Sunday. If Leigh had gross weekly wages of $342 for working 50 hours, 4 of which were on Sunday, what is her regular hourly rate?

101. Computing Grades Going into the final exam, which will count as two tests, Brooke has test scores of 80, 83, 71, 61, and 95. What score does Brooke need on the final in order to have an average score of 80?

102. Computing Grades Going into the final exam, which will count as two-thirds of the final grade, Mike has test scores of 86, 80, 84, and 90. What score does Mike need on the final in order to earn a B, which requires an average score of 80? What does he need to earn an A, which requires an average of 90?

103. Business: Discount Pricing A builder of tract homes reduced the price of a model by 15%. If the new price is $125,000, what was its original price? How much can be saved by purchasing the model?

104. Business: Discount Pricing A car dealer, at a year-end clearance, reduces the list price of last year's models by 15%. If a certain four-door model has a discounted price of $8000, what was its list price? How much can be saved by purchasing last year's model?

105. Business: Marking up the Price of Books A college book store marks up the price that it pays the publisher for a book by 35%. If the selling price of a book is $94.50, how much did the bookstore pay for this book?

106. Personal Finance: Cost of a Car The suggested list price of a new car is $24,000. The dealer's cost is 85% of list. How much will you pay if the dealer is willing to accept $100 over cost for the car?

107. Business: Theater Attendance The manager of the Coral Theater wants to know whether the majority of its patrons is adults or children. During a week in July, 5200 tickets were sold and the receipts totaled $32,200. The adult admission is $8.50, and the children's admission is $6.00. How many adult patrons were there?

108. Business: Discount Pricing A wool suit, discounted by 30% for a clearance sale, has a price tag of $399. What was the suit's original price?

109. Geometry The perimeter of a rectangle is 60 feet. Find its length and width if the length is 8 feet longer than the width.

110. Geometry The perimeter of a rectangle is 42 meters. Find its length and width if the length is twice the width.

Discussion and Writing

111. One of the steps in the following list contains an error. Identify it and explain what is wrong.

$$x = 2 \tag{1}$$
$$3x - 2x = 2 \tag{2}$$
$$3x = 2x + 2 \tag{3}$$
$$x^2 + 3x = x^2 + 2x + 2 \tag{4}$$
$$x^2 + 3x - 10 = x^2 + 2x - 8 \tag{5}$$
$$(x - 2)(x + 5) = (x - 2)(x + 4) \tag{6}$$
$$x + 5 = x + 4 \tag{7}$$
$$5 = 4 \tag{8}$$

112. The equation

$$\frac{5}{x + 3} + 3 = \frac{8 + x}{x + 3}$$

has no solution, yet when we go through the process of solving it we obtain $x = -3$. Write a brief paragraph to explain what causes this to happen.

113. Make up an equation that has no solution and give it to a fellow student to solve. Ask the fellow student to write a critique of your equation.

114. Explain the difference between the directions "solve," "evaluate," and "simplify." Write an example using each direction with the expression $3(x + 2) - x$.

'Are You Prepared?' Answers

1. No   **2.** $(x - 1)(x + 3)$

1.3 Quadratic Equations

PREPARING FOR THIS SECTION   *Before getting started, review the following:*

• Zero-Product Property (Section R.1, pp. 12)   • Factoring (Section R.5, pp. 45–51)

Now work the 'Are You Prepared?' problems on page 121.

OBJECTIVES
1. Solve Quadratic Equations by Factoring
2. Solve Quadratic Equations Using the Square Root Method
3. Solve Quadratic Equations by Completing the Square
4. Solve Quadratic Equations Using the Quadratic Formula
5. Solve Problems That Can Be Modeled by Quadratic Equations

Quadratic equations are equations such as

$$2x^2 + x + 8 = 0, \qquad 3x^2 - 5x = 0, \qquad x^2 - 9 = 0$$

A general definition is given next.

> A **quadratic equation** is an equation equivalent to one of the form
>
> $$ax^2 + bx + c = 0 \tag{1}$$
>
> where a, b, and c are real numbers and $a \neq 0$.

A quadratic equation written in the form $ax^2 + bx + c = 0$ is said to be in **standard form**.

Sometimes, a quadratic equation is called a **second-degree equation**, because the left side is a polynomial of degree 2. We shall discuss three algebraic ways of solving quadratic equations: by factoring, by completing the square, and by using the quadratic formula.

1 Solve Quadratic Equations by Factoring

When a quadratic equation is written in standard form, $ax^2 + bx + c = 0$, it may be possible to factor the expression on the left side as the product of two first-degree polynomials. Then, by setting each factor equal to 0 and solving the resulting linear equations, we obtain the *exact* solutions of the quadratic equation. This approach leads us to a basic premise in mathematics. Whenever a problem is encountered, use techniques that reduce the problem to one you already know how to solve. In this instance, we are reducing quadratic equations to linear equations using the technique of factoring.

Let's look at an example.

EXAMPLE 1   **Solving a Quadratic Equation by Factoring and by Graphing**

Solve the equation: $x^2 = 12 - x$

Algebraic Solution

We put the equation in standard form by adding $x - 12$ to each side:

$$x^2 = 12 - x$$

$$x^2 + x - 12 = 0$$

The left side of the equation may now be factored as

$$(x + 4)(x - 3) = 0$$

so that

$$x + 4 = 0 \quad \text{or} \quad x - 3 = 0$$

$$x = -4 \qquad\qquad x = 3$$

The solution set is $\{-4, 3\}$.   ◄

Graphing Solution

Graph $Y_1 = x^2$ and $Y_2 = 12 - x$. See Figure 32(a). From the graph it appears that there are two points of intersection: one near -4, the other near 3. Using INTERSECT, the points of intersection are $(-4, 16)$ and $(3, 9)$, so the solutions of the equation are $x = -4$ and $x = 3$. See Figure 32(b) and (c). The solution set is $\{-4, 3\}$.

Figure 32

(a)

(b)

(c)   ◄

✏ **NOW WORK PROBLEM 11.**

When the left side factors into two linear equations with the same solution, the quadratic equation is said to have a **repeated solution**. We also call this solution a **root of multiplicity 2**, or a **double root**.

EXAMPLE 2	Solving a Quadratic Equation by Factoring and by Graphing

Solve the equation: $x^2 - 6x + 9 = 0$

Algebraic Solution

This equation is already in standard form, and the left side can be factored:

$$x^2 - 6x + 9 = 0$$

$$(x - 3)(x - 3) = 0$$

so

$$x = 3 \quad \text{or} \quad x = 3$$

This equation has only the repeated solution 3. The solution set is $\{3\}$. ◀

Graphing Solution

Figure 33 shows the graph of the equation

$$Y_1 = x^2 - 6x + 9$$

From the graph it appears that there is one x-intercept, 3. Using ZERO (or ROOT), we find that the only x-intercept is $x = 3$. The equation has only the repeated solution 3. The solution set is $\{3\}$.

Figure 33

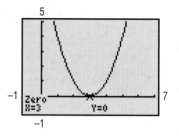

◀

NOW WORK PROBLEM 23.

2 Solve Quadratic Equations Using the Square Root Method

Suppose that we wish to solve the quadratic equation

$$x^2 = p \tag{2}$$

where $p \geq 0$ is a nonnegative number. We proceed as in the earlier examples:

$$x^2 - p = 0 \qquad \text{Put in standard form.}$$

$$(x - \sqrt{p})(x + \sqrt{p}) = 0 \qquad \text{Factor (over the real numbers).}$$

$$x = \sqrt{p} \quad \text{or} \quad x = -\sqrt{p} \qquad \text{Solve.}$$

We have the following result:

If $x^2 = p$ and $p \geq 0$, then $x = \sqrt{p}$ or $x = -\sqrt{p}$. $\tag{3}$

When statement (3) is used, it is called the **Square Root Method**. In statement (3), note that if $p > 0$ the equation $x^2 = p$ has two solutions, $x = \sqrt{p}$ and $x = -\sqrt{p}$. We usually abbreviate these solutions as $x = \pm\sqrt{p}$, read as "x equals plus or minus the square root of p." For example, the two solutions of the equation

$$x^2 = 4$$

are

$$x = \pm\sqrt{4}$$

and, since $\sqrt{4} = 2$, we have

$$x = \pm 2$$

The solution set is $\{-2, 2\}$.

| EXAMPLE 3 | **Solving Quadratic Equations by Using the Square Root Method** |

Solve each equation.

(a) $x^2 = 5$ (b) $(x - 2)^2 = 16$

Solution   (a) $x^2 = 5$

$$x = \pm\sqrt{5} \qquad \text{\textit{Use the Square Root Method.}}$$

$$x = \sqrt{5} \quad \text{or} \quad x = -\sqrt{5}$$

The solution set is $\{-\sqrt{5}, \sqrt{5}\}$.

(b) $(x - 2)^2 = 16$

$$x - 2 = \pm\sqrt{16} \qquad \text{\textit{Use the Square Root Method.}}$$

$$x - 2 = \sqrt{16} \quad \text{or} \quad x - 2 = -\sqrt{16}$$

$$x - 2 = 4 \qquad\qquad x - 2 = -4$$

$$x = 6 \qquad\qquad\quad x = -2$$

The solution set is $\{-2, 6\}$.

✔ **CHECK:** Verify the solutions to Example 3 using a graphing utility. Are the solutions provided by the utility exact?   ◄

NOW WORK PROBLEM 31.

3   Solve Quadratic Equations by Completing the Square

We now introduce the method of **completing the square**. The idea behind this method is to "adjust" the left side of a quadratic equation, $ax^2 + bx + c = 0$, so that it becomes a perfect square, that is, the square of a first-degree polynomial. For example, $x^2 + 6x + 9$ and $x^2 - 4x + 4$ are perfect squares because

$$x^2 + 6x + 9 = (x + 3)^2 \quad \text{and} \quad x^2 - 4x + 4 = (x - 2)^2$$

How do we "adjust" the left side? We do it by adding the appropriate number to create a perfect square. For example, to make $x^2 + 6x$ a perfect square, we add 9.

Let's look at several examples of completing the square when the coefficient of x^2 is 1.

Start	Add	Result
$x^2 + 4x$	4	$x^2 + 4x + 4 = (x + 2)^2$
$x^2 + 12x$	36	$x^2 + 12x + 36 = (x + 6)^2$
$x^2 - 6x$	9	$x^2 - 6x + 9 = (x - 3)^2$
$x^2 + x$	$\dfrac{1}{4}$	$x^2 + x + \dfrac{1}{4} = \left(x + \dfrac{1}{2}\right)^2$

Do you see the pattern? Provided that the coefficient of x^2 is 1, we complete the square by adding the square of one-half of the coefficient of x.

Start	Add	Result
$x^2 + mx$	$\left(\dfrac{m}{2}\right)^2$	$x^2 + mx + \left(\dfrac{m}{2}\right)^2 = \left(x + \dfrac{m}{2}\right)^2$

↗ **NOW WORK PROBLEM 35.**

The next example illustrates how the procedure of completing the square can be used to solve a quadratic equation.

EXAMPLE 4 **Solving a Quadratic Equation by Completing the Square**

Solve by completing the square: $x^2 + 5x + 4 = 0$

Solution We always begin this procedure by rearranging the equation so that the constant is on the right side.

$$x^2 + 5x + 4 = 0$$
$$x^2 + 5x = -4$$

Since the coefficient of x^2 is 1, we can complete the square on the left side by adding $\left(\dfrac{1}{2}\cdot 5\right)^2 = \dfrac{25}{4}$. Of course, in an equation, whatever we add to the left side also must be added to the right side. We add $\dfrac{25}{4}$ to *both* sides.

$$x^2 + 5x + \frac{25}{4} = -4 + \frac{25}{4} \qquad \text{Add } \tfrac{25}{4} \text{ to both sides.}$$

$$\left(x + \frac{5}{2}\right)^2 = \frac{9}{4} \qquad \text{Factor; simplify.}$$

$$x + \frac{5}{2} = \pm\sqrt{\frac{9}{4}} \qquad \text{Use the Square Root Method.}$$

$$x + \frac{5}{2} = \pm\frac{3}{2}$$

$$x = -\frac{5}{2} \pm \frac{3}{2}$$

$$x = -\frac{5}{2} + \frac{3}{2} = -1 \quad \text{or} \quad x = -\frac{5}{2} - \frac{3}{2} = -4$$

The solution set is $\{-4, -1\}$.

✔ **CHECK:** Verify the solution to Example 4 using a graphing utility.   ◀

The solution of the equation in Example 4 can also be obtained by factoring. Rework Example 4 using factoring.

↗ **NOW WORK PROBLEM 39.**

4   Solve Quadratic Equations Using The Quadratic Formula

We can use the method of completing the square to obtain a general formula for solving the quadratic equation

$$ax^2 + bx + c = 0, \qquad a > 0$$

NOTE
There is no loss in generality to assume that $a > 0$, since if $a < 0$ we can multiply both sides by -1 to obtain an equivalent equation with a positive leading coefficient.   ■

As in Example 4, we begin by rearranging the terms as

$$ax^2 + bx = -c$$

Since $a > 0$, we can divide both sides by a to get

$$x^2 + \frac{b}{a}x = -\frac{c}{a}$$

Now the coefficient of x^2 is 1. To complete the square on the left side, add the square of $\frac{1}{2}$ of the coefficient of x; that is, add

$$\left(\frac{1}{2} \cdot \frac{b}{a}\right)^2 = \frac{b^2}{4a^2}$$

to each side. Then

$$x^2 + \frac{b}{a}x + \frac{b^2}{4a^2} = \frac{b^2}{4a^2} - \frac{c}{a}$$

$$\left(x + \frac{b}{2a}\right)^2 = \frac{b^2 - 4ac}{4a^2} \qquad \frac{b^2}{4a^2} - \frac{c}{a} = \frac{b^2}{4a^2} - \frac{4ac}{4a^2} = \frac{b^2 - 4ac}{4a^2} \qquad \textbf{(4)}$$

Provided that $b^2 - 4ac \geq 0$, we now can use the Square Root Method to get

$$x + \frac{b}{2a} = \pm\sqrt{\frac{b^2 - 4ac}{4a^2}}$$

$$x + \frac{b}{2a} = \frac{\pm\sqrt{b^2 - 4ac}}{2a} \qquad \text{The square root of a quotient equals the quotient of the square roots. Also, } \sqrt{4a^2} = 2a \text{ since } a > 0.$$

$$x = -\frac{b}{2a} \pm \frac{\sqrt{b^2 - 4ac}}{2a} \qquad \text{Add } -\frac{b}{2a} \text{ to both sides.}$$

$$= \frac{-b \pm \sqrt{b^2 - 4ac}}{2a} \qquad \text{Combine the quotients on the right.}$$

What if $b^2 - 4ac$ is negative? Then equation (4) states that the left expression (a real number squared) equals the right expression (a negative number). Since this occurrence is impossible for real numbers, we conclude that if $b^2 - 4ac < 0$ the quadratic equation has no *real* solution.[*]

We now state the *quadratic formula*.

---

[*]We consider quadratic equations where $b^2 - 4ac$ is negative in the next section.

Theorem

Quadratic Formula

Consider the quadratic equation

$$ax^2 + bx + c = 0 \qquad a \neq 0$$

If $b^2 - 4ac < 0$, this equation has no real solution.

If $b^2 - 4ac \geq 0$, the real solution(s) of this equation is (are) given by the **quadratic formula**.

$$x = \frac{-b \pm \sqrt{b^2 - 4ac}}{2a}$$

The quantity $b^2 - 4ac$ is called the **discriminant** of the quadratic equation, because its value tells us whether the equation has real solutions. In fact, it also tells us how many solutions to expect.

Discriminant of a Quadratic Equation

For a quadratic equation $ax^2 + bx + c = 0$:

1. If $b^2 - 4ac > 0$, there are two unequal real solutions.

2. If $b^2 - 4ac = 0$, there is a repeated real solution, a root of multiplicity 2.

3. If $b^2 - 4ac < 0$, there is no real solution.

When asked to find the real solutions, if any, of a quadratic equation, always evaluate the discriminant first to see how many real solutions there are.

EXAMPLE 5

Solving a Quadratic Equation by Using the Quadratic Formula and by Graphing

Find the real solutions, if any, of the equation $3x^2 - 5x + 1 = 0$.

Algebraic Solution

The equation is in standard form, so we compare it to $ax^2 + bx + c = 0$ to find $a, b,$ and c.

$$3x^2 - 5x + 1 = 0$$
$$ax^2 + bx + c = 0, \qquad a = 3, b = -5, c = 1$$

With $a = 3, b = -5,$ and $c = 1$, we evaluate the discriminant $b^2 - 4ac$.

$$b^2 - 4ac = (-5)^2 - 4(3)(1) = 25 - 12 = 13$$

Since $b^2 - 4ac > 0$, there are two real solutions.

We use the quadratic formula with $a = 3, b = -5, c = 1$, and $b^2 - 4ac = 13$.

$$x = \frac{-b \pm \sqrt{b^2 - 4ac}}{2a} = \frac{-(-5) \pm \sqrt{13}}{2(3)} = \frac{5 \pm \sqrt{13}}{6}$$

The solution set is $\left\{ \dfrac{5 - \sqrt{13}}{6}, \dfrac{5 + \sqrt{13}}{6} \right\}$. These solutions are exact. ◄

Graphing Solution

Figure 34 shows the graph of the equation

$$Y_1 = 3x^2 - 5x + 1$$

There are two x-intercepts: one between 0 and 1, the other between 1 and 2. The solutions to the equation are 0.23 and 1.43, rounded to two decimal places. These solutions are approximate.

Figure 34

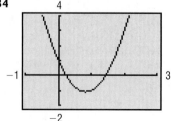

◄

NOW WORK PROBLEMS **49** AND **59.**

EXAMPLE 6	Solving a Quadratic Equation by Using the Quadratic Formula and by Graphing

Find the real solutions, if any, of the equation

$$3x^2 + 2 = 4x$$

Algebraic Solution

The equation, as given, is not in standard form.

$$3x^2 + 2 = 4x$$

$$3x^2 - 4x + 2 = 0 \qquad \text{Subtract } 4x \text{ from both sides to put the equation in standard form.}$$

$$ax^2 + bx + c = 0 \qquad \text{Compare to standard form.}$$

With $a = 3$, $b = -4$, and $c = 2$, we find that

$$b^2 - 4ac = (-4)^2 - 4(3)(2) = 16 - 24$$
$$= -8$$

Since $b^2 - 4ac < 0$, the equation has no real solution. ◄

Graphing Solution

We use the standard form of the equation and graph

$$Y_1 = 3x^2 - 4x + 2$$

See Figure 35. We see that there are no x-intercepts, so the equation has no real solution, as expected.

Figure 35

◄

  NOW WORK PROBLEM 51.

Summary

Procedure for Solving a Quadratic Equation Algebraically

To solve a quadratic equation, first put it in standard form:

$$ax^2 + bx + c = 0$$

Then:

STEP 1: Identify a, b, and c.

STEP 2: Evaluate the discriminant, $b^2 - 4ac$.

STEP 3: (a) If the discriminant is negative, the equation has no real solution.

  (b) If the discriminant is nonnegative, determine whether the left side can be factored. If you can easily spot factors, use the factoring method to solve the equation. Otherwise, use the quadratic formula or the method of completing the square.

---

5   Solve Problems That Can Be Modeled by Quadratic Equations

Many applied problems require the solution of a quadratic equation. Let's look at one that you will probably see again in a slightly different form if you study calculus.

EXAMPLE 7	**Constructing a Box**

From each corner of a square piece of sheet metal, remove a square of side 9 centimeters. Turn up the edges to form an open box. If the box is to hold 144 cubic centimeters (cm^3), what should be the dimensions of the piece of sheet metal?

Solution   We use Figure 36 as a guide. We have labeled by x the length of a side of the square piece of sheet metal. The box will be of height 9 centimeters, and its square base will have $x - 18$ as the length of a side. The volume (Length $\times$ Width $\times$ Height) of the box is therefore

$$(x - 18)(x - 18) \cdot 9 = 9(x - 18)^2$$

Figure 36

Since the volume of the box is to be 144 cm^3, we have

$$9(x - 18)^2 = 144$$

$$(x - 18)^2 = 16 \qquad \textit{Divide each side by 9.}$$

$$x - 18 = \pm 4 \qquad \textit{Use the Square Root Method.}$$

$$x = 18 \pm 4$$

$$x = 22 \quad \text{or} \quad x = 14$$

We discard the solution $x = 14$ (do you see why?) and conclude that the sheet metal should be 22 centimeters by 22 centimeters.

✔ **CHECK:** If we begin with a piece of sheet metal 22 centimeters by 22 centimeters, cut out a 9 centimeter square from each corner, and fold up the edges, we get a box whose dimensions are 9 by 4 by 4, with volume $9 \times 4 \times 4 = 144$ cm^3, as required. ◄

NOW WORK PROBLEM 85.

HISTORICAL FEATURE

Problems using quadratic equations are found in the oldest known mathematical literature. Babylonians and Egyptians were solving such problems before 1800 BC. Euclid solved quadratic equations geometrically in his *Data* (300 BC), and the Hindus and Arabs gave rules for solving any quadratic equation with real roots. Because negative numbers were not freely used before AD 1500, there were several different types of quadratic equations, each with its own rule. Thomas Harriot (1560–1621) introduced the method of factoring to obtain solutions, and François Viète (1540–1603) introduced a method that is essentially completing the square.

Until modern times it was usual to neglect the negative roots (if there were any), and equations involving square roots of negative quantities were regarded as unsolvable until the 1500s.

Historical Problems

1. *One of al-Khwăiizmi solutions* We solve $x^2 + 12x = 85$ by drawing the square shown. The area of the four white rectangles and the yellow square is $x^2 + 12x$. We then set this expression equal to 85 to get the equation $x^2 + 12x = 85$. If we add the four blue squares, we will have a larger square of known area. Complete the solution.

2. *Viète's method* We solve $x^2 + 12x - 85 = 0$ by letting $x = u + z$. Then

$$(u + z)^2 + 12(u + z) - 85 = 0$$
$$u^2 + (2z + 12)u + (z^2 + 12z - 85) = 0$$

Now select z so that $2z + 12 = 0$ and finish the solution.

3. *Another method to get the quadratic formula* Look at equation (4) on page 117. Rewrite the right side as $\left(\dfrac{\sqrt{b^2 - 4ac}}{2a}\right)^2$ and then

subtract it from each side. The right side is now 0 and the left side is a difference of two squares. If you factor this difference of two squares, you will easily be able to get the quadratic formula, and, moreover, the quadratic expression is factored, which is sometimes useful.

1.3 Assess Your Understanding

'Are You Prepared?'

Answers are given at the end of these exercises. If you get a wrong answer, read the pages listed in red.

1. Factor: $x^2 - 5x - 6$ (pp. 47–49)

2. Factor: $2x^2 - x - 3$ (pp. 50–51)

3. The solution set of the equation $(x - 3)(3x + 5) = 0$ is _____. (p. 12)

4. *True or False:* $\sqrt{x^2} = |x|$. (pp. 22–23)

Concepts and Vocabulary

5. When a quadratic equation has a repeated solution, it is called a(n) _____ root or a root of _____ _____.

6. The quantity $b^2 - 4ac$ is called the _____ of a quadratic equation. If it is _____, the equation has no real solution.

7. Quadratic equations are sometimes called _____ -degree equations.

8. *True or False:* To complete the square in the expression $x^2 + 8x$, add 16.

9. *True or False:* Quadratic equations always have two real solutions.

10. *True or False:* A quadratic equation that is in the form $ax^2 + bx + c = 0$ is said to be in standard form.

Skill Building

In Problems 11–28, solve each equation by factoring. Verify your solution using a graphing utility.

11. $x^2 - 9x = 0$

12. $x^2 + 4x = 0$

13. $x^2 - 25 = 0$

14. $x^2 - 9 = 0$

15. $z^2 + z - 6 = 0$

16. $v^2 + 7v + 6 = 0$

17. $2x^2 - 5x - 3 = 0$

18. $3x^2 + 5x + 2 = 0$

19. $3t^2 - 48 = 0$

20. $2y^2 - 50 = 0$

21. $x(x + 8) + 12 = 0$

22. $x(x - 4) = 12$

23. $4x^2 + 9 = 12x$

24. $25x^2 + 16 = 40x$

25. $2x^2 - x = 15$

26. $6x^2 + x = 2$

27. $\dfrac{4(x - 2)}{x - 3} + \dfrac{3}{x} = \dfrac{-3}{x(x - 3)}$

28. $\dfrac{5}{x + 4} = 4 + \dfrac{3}{x - 2}$

In Problems 29–34, solve each equation by the Square Root Method. Verify your solution using a graphing utility.

29. $x^2 = 25$

30. $x^2 = 36$

31. $(x - 1)^2 = 4$

32. $(x + 2)^2 = 1$

33. $(2x + 3)^2 = 9$

34. $(3x - 2)^2 = 4$

In Problems 35–38, what number should be added to complete the square of each expression?

35. $x^2 + 8x$

36. $x^2 - 4x$

37. $x^2 - \dfrac{1}{2}x$

38. $x^2 + \dfrac{1}{3}x$

In Problems 39–44, solve each equation by completing the square. Verify your solution using a graphing utility.

39. $x^2 + 4x = 21$

40. $x^2 - 6x = 13$

41. $x^2 - \dfrac{1}{2}x - \dfrac{3}{16} = 0$

42. $x^2 + \dfrac{2}{3}x - \dfrac{1}{3} = 0$

43. $3x^2 + x - \dfrac{1}{2} = 0$

44. $2x^2 - 3x - 1 = 0$

In Problems 45–62, find the real solutions, if any, of each equation. Use the quadratic formula. Verify your solution using a graphing utility.

45. $x^2 - 4x + 2 = 0$

46. $x^2 + 4x + 2 = 0$

47. $x^2 - 4x - 1 = 0$

48. $x^2 + 6x + 1 = 0$

49. $2x^2 - 5x + 3 = 0$

50. $2x^2 + 5x + 3 = 0$

51. $4y^2 - y + 2 = 0$

52. $4t^2 + t + 1 = 0$

53. $4x^2 = 1 - 2x$

54. $2x^2 = 1 - 2x$

55. $4x^2 = 9x + 2$

56. $5x = 4x^2 + 5$

57. $9t^2 - 6t + 1 = 0$

58. $4u^2 - 6u + 9 = 0$

59. $\dfrac{3}{4}x^2 - \dfrac{1}{4}x - \dfrac{1}{2} = 0$

60. $\dfrac{2}{3}x^2 - x - 3 = 0$

61. $4 - \dfrac{1}{x} - \dfrac{2}{x^2} = 0$

62. $4 + \dfrac{1}{x} - \dfrac{1}{x^2} = 0$

In Problems 63–74, find the exact real solutions, if any, of each equation. Use any method. Verify your solution using a graphing utility.

63. $x^2 - 5 = 0$

64. $x^2 - 6 = 0$

65. $16x^2 - 8x + 1 = 0$

66. $4x^2 - 12x + 9 = 0$

67. $10x^2 - 19x - 15 = 0$

68. $6x^2 + 7x - 20 = 0$

69. $2 + z = 6z^2$

70. $2 = y + 6y^2$

71. $x^2 + \sqrt{2}x = \dfrac{1}{2}$

72. $\dfrac{1}{2}x^2 = \sqrt{2}x + 1$

73. $x^2 + x = 4$

74. $x^2 + x = 1$

In Problems 75–80, determine whether each quadratic equation has two unequal real solutions, a repeated real solution, or no real solution, without solving the equation. [**Hint:** *Use the discriminant.*]

75. $2x^2 - 6x + 7 = 0$

76. $x^2 + 4x + 7 = 0$

77. $9x^2 - 30x + 25 = 0$

78. $25x^2 - 20x + 4 = 0$

79. $3x^2 + 5x - 8 = 0$

80. $2x^2 - 3x - 7 = 0$

Applications and Extensions

81. Dimensions of a Window The area of the opening of a rectangular window is to be 143 square feet. If the length is to be 2 feet more than the width, what are the dimensions?

82. Dimensions of a Window The area of a rectangular window is to be 306 square centimeters. If the length exceeds the width by 1 centimeter, what are the dimensions?

83. Geometry Find the dimensions of a rectangle whose perimeter is 26 meters and whose area is 40 square meters.

84. Watering a Field An adjustable water sprinkler that sprays water in a circular pattern is placed at the center of a square field whose area is 1250 square feet (see the figure). What is the shortest radius setting that can be used if the field is to be completely enclosed within the circle?

85. Constructing a Box   An open box is to be constructed from a square piece of sheet metal by removing a square of side 1 foot from each corner and turning up the edges. If the box is to hold 4 cubic feet, what should be the dimensions of the sheet metal?

86. Constructing a Box   Rework Problem 85 if the piece of sheet metal is a rectangle whose length is twice its width.

87. Physics   A ball is thrown vertically upward from the top of a building 96 feet tall with an initial velocity of 80 feet per second. The distance s (in feet) of the ball from the ground after t seconds is $s = 96 + 80t - 16t^2$.
  (a) After how many seconds does the ball strike the ground?
  (b) After how many seconds will the ball pass the top of the building on its way down?

88. Physics   An object is propelled vertically upward with an initial velocity of 20 meters per second. The distance s (in meters) of the object from the ground after t seconds is $s = -4.9t^2 + 20t$.
  (a) When will the object be 15 meters above the ground?
  (b) When will it strike the ground?
  (c) Will the object reach a height of 100 meters?

89. Reducing the Size of a Candy Bar   A jumbo chocolate bar with a rectangular shape measures 12 centimeters in length, 7 centimeters in width, and 3 centimeters in thickness. Due to escalating costs of cocoa, management decides to reduce the volume of the bar by 10%. To accomplish this reduction, management decides that the new bar should have the same 3-centimeter thickness, but the length and width of each should be reduced an equal number of centimeters. What should be the dimensions of the new candy bar?

90. Reducing the Size of a Candy Bar   Rework Problem 89 if the reduction is to be 20%.

91. Constructing a Border around a Pool   A circular pool measures 10 feet across. One cubic yard of concrete is to be used to create a circular border of uniform width around the pool. If the border is to have a depth of 3 inches, how wide will the border be? (1 cubic yard = 27 cubic feet) See the illustration.

92. Constructing a Border around a Pool   Rework Problem 91 if the depth of the border is 4 inches.

93. Constructing a Border around a Garden   A landscaper, who just completed a rectangular flower garden measuring 6 feet by 10 feet, orders 1 cubic yard of premixed cement, all of which is to be used to create a border of uniform width around the garden. If the border is to have a depth of 3 inches, how wide will the border be?

94. Dimensions of a Patio   A contractor orders 8 cubic yards of premixed cement, all of which is to be used to pour a patio that will be 4 inches thick. If the length of the patio is specified to be twice the width, what will be the patio dimensions? (1 cubic yard = 27 cubic feet)

95. The sum of the consecutive integers $1, 2, 3, \ldots, n$ is given by the formula $\frac{1}{2}n(n + 1)$. How many consecutive integers, starting with 1, must be added to get a sum of 666?

96. Geometry   If a polygon of n sides has $\frac{1}{2}n(n - 3)$ diagonals, how many sides will a polygon with 65 diagonals have? Is there a polygon with 80 diagonals?

97. Show that the sum of the roots of a quadratic equation is $-\frac{b}{a}$.

98. Show that the product of the roots of a quadratic equation is $\frac{c}{a}$.

99. Find k such that the equation $kx^2 + x + k = 0$ has a repeated real solution.

100. Find k such that the equation $x^2 - kx + 4 = 0$ has a repeated real solution.

101. Show that the real solutions of the equation $ax^2 + bx + c = 0$ are the negatives of the real solutions of the equation $ax^2 - bx + c = 0$. Assume that $b^2 - 4ac \geq 0$.

102. Show that the real solutions of the equation $ax^2 + bx + c = 0$ are the reciprocals of the real solutions of the equation $cx^2 + bx + a = 0$. Assume that $b^2 - 4ac \geq 0$.

Discussion and Writing

103. Which of the following pairs of equations are equivalent? Explain.
 (a) $x^2 = 9$; $x = 3$
 (b) $x = \sqrt{9}$; $x = 3$
 (c) $(x - 1)(x - 2) = (x - 1)^2$; $x - 2 = x - 1$

104. Describe three ways that you might solve a quadratic equation. State your preferred method; explain why you chose it.

105. Explain the benefits of evaluating the discriminant of a quadratic equation before attempting to solve it.

106. Create three quadratic equations: one having two distinct solutions, one having no real solution, and one having exactly one real solution.

107. The word *quadratic* seems to imply four (*quad*), yet a quadratic equation is an equation that involves a polynomial of degree 2. Investigate the origin of the term *quadratic* as it is used in the expression *quadratic equation*. Write a brief essay on your findings.

'Are You Prepared?' Answers

 1. $(x - 6)(x + 1)$ **2.** $(2x - 3)(x + 1)$ **3.** $\left\{-\dfrac{5}{3}, 3\right\}$ **4.** True

1.4 Complex Numbers; Quadratic Equations in the Complex Number System

PREPARING FOR THIS SECTION   *Before getting started, review the following:*

• Classification of Numbers (Review, Section R.1, pp. 2–4)

• Rationalizing Denominators (Section R.8, pp. 70–71)

Now work the 'Are You Prepared?' problems on page 131.

OBJECTIVES   1 Add, Subtract, Multiply, and Divide Complex Numbers
   2 Solve Quadratic Equations with a Negative Discriminant

One property of a real number is that its square is nonnegative (greater than or equal to 0). For example, there is no real number x for which

$$x^2 = -1$$

To remedy this situation, we introduce a number called the **imaginary unit**, which we denote by **i** and whose square is -1; that is,

$$i^2 = -1$$

This should not surprise you. If our universe were to consist only of integers, there would be no number x for which $2x = 1$. This unfortunate circumstance was remedied by introducing numbers such as $\dfrac{1}{2}$ and $\dfrac{2}{3}$, the *rational numbers*. If our universe were to consist only of rational numbers, there would be no x whose square equals 2. That is, there would be no number x for which $x^2 = 2$. To remedy this, we introduced numbers such as $\sqrt{2}$ and $\sqrt[3]{5}$, the *irrational numbers*. The *real numbers*, you will recall, consist of the rational numbers and the irrational numbers. Now, if our universe were to consist only of real numbers, then there would be no number x whose square is -1. To remedy this, we introduce a number i, whose square is -1.

In the progression outlined, each time that we encountered a situation that was unsuitable, we introduced a new number system to remedy this situation. And each new number system contained the earlier number system as a subset. The number system that results from introducing the number i is called the **complex number system**.

> **Complex numbers** are numbers of the form $a + bi$, where a and b are real numbers. The real number a is called the **real part** of the number $a + bi$; the real number b is called the **imaginary part** of $a + bi$.

For example, the complex number $-5 + 6i$ has the real part -5 and the imaginary part 6.

When a complex number is written in the form $a + bi$, where a and b are real numbers, we say it is in **standard form**. However, if the imaginary part of a complex number is negative, such as in the complex number $3 + (-2)i$, we agree to write it instead in the form $3 - 2i$.

Also, the complex number $a + 0i$ is usually written merely as a. This serves to remind us that the real numbers are a subset of the complex numbers. The complex number $0 + bi$ is usually written as bi. Sometimes the complex number bi is called a **pure imaginary number**.

1 Add, Subtract, Multiply, and Divide Complex Numbers

Equality, addition, subtraction, and multiplication of complex numbers are defined so as to preserve the familiar rules of algebra for real numbers. Thus, two complex numbers are equal if and only if their real parts are equal and their imaginary parts are equal. That is,

Equality of Complex Numbers

$$a + bi = c + di \quad \text{if and only if} \quad a = c \text{ and } b = d \qquad \textbf{(1)}$$

Two complex numbers are added by forming the complex number whose real part is the sum of the real parts and whose imaginary part is the sum of the imaginary parts. That is,

Sum of Complex Numbers

$$(a + bi) + (c + di) = (a + c) + (b + d)i \qquad \textbf{(2)}$$

To subtract two complex numbers, we use this rule:

Difference of Complex Numbers

$$(a + bi) - (c + di) = (a - c) + (b - d)i \qquad \textbf{(3)}$$

| EXAMPLE 1 | **Adding and Subtracting Complex Numbers** |

(a) $(3 + 5i) + (-2 + 3i) = [3 + (-2)] + (5 + 3)i = 1 + 8i$

(b) $(6 + 4i) - (3 + 6i) = (6 - 3) + (4 - 6)i = 3 + (-2)i = 3 - 2i$ ◀

Figure 37

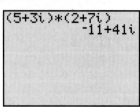

Some graphing calculators have the capability of handling complex numbers. For example, Figure 37 shows the results of Example 1 using a TI-84 Plus graphing calculator.

━━━ **NOW WORK PROBLEM 13.**

Products of complex numbers are calculated as illustrated in Example 2.

| EXAMPLE 2 | **Multiplying Complex Numbers** |

$$
\begin{aligned}
(5 + 3i) \cdot (2 + 7i) &= 5 \cdot (2 + 7i) + 3i(2 + 7i) & \text{Distributive Property}\\
&= 10 + 35i + 6i + 21i^2 & \text{Distributive Property}\\
&= 10 + 41i + 21(-1) & i^2 = -1\\
&= -11 + 41i &
\end{aligned}
$$
◀

Based on the procedure of Example 2, we define the **product** of two complex numbers by the following formula:

Product of Complex Numbers

$$(a + bi) \cdot (c + di) = (ac - bd) + (ad + bc)i \qquad \textbf{(4)}$$

Do not bother to memorize formula (4). Instead, whenever it is necessary to multiply two complex numbers, follow the usual rules for multiplying two binomials, as in Example 2, remembering that $i^2 = -1$. For example,

$$(2i)(2i) = 4i^2 = -4$$
$$(2 + i)(1 - i) = 2 - 2i + i - i^2 = 3 - i$$

Figure 38

(5+3i)*(2+7i)
                 -11+41i

Graphing calculators may also be used to multiply complex numbers. Figure 38 shows the result obtained in Example 2 using a TI-84 Plus graphing calculator.

━━━ **NOW WORK PROBLEM 19.**

Algebraic properties for addition and multiplication, such as the Commutative, Associative, and Distributive Properties, hold for complex numbers. However, the property that every nonzero complex number has a multiplicative inverse, or reciprocal, requires a closer look.

Conjugates

If $z = a + bi$ is a complex number, then its **conjugate**, denoted by $\bar{z}$, is defined as

$$\bar{z} = \overline{a + bi} = a - bi$$

For example, $\overline{2 + 3i} = 2 - 3i$ and $\overline{-6 - 2i} = -6 + 2i$.

| EXAMPLE 3 | **Multiplying a Complex Number by Its Conjugate** |

Find the product of the complex number $z = 3 + 4i$ and its conjugate $\bar{z}$.

Solution   Since $\bar{z} = 3 - 4i$, we have

$$z\bar{z} = (3 + 4i)(3 - 4i) = 9 - 12i + 12i - 16i^2 = 9 + 16 = 25 \qquad ◀$$

The result obtained in Example 3 has an important generalization.

Theorem   The product of a complex number and its conjugate is a nonnegative real number. That is, if $z = a + bi$, then

$$\boxed{z\bar{z} = a^2 + b^2} \qquad \textbf{(5)}$$

Proof   If $z = a + bi$, then

$$z\bar{z} = (a + bi)(a - bi) = a^2 - (bi)^2 = a^2 - b^2 i^2 = a^2 + b^2 \qquad ■$$

To express the reciprocal of a nonzero complex number z in standard form, multiply the numerator and denominator of $\dfrac{1}{z}$ by its conjugate $\bar{z}$. That is, if $z = a + bi$ is a nonzero complex number, then

$$\frac{1}{a + bi} = \frac{1}{z} = \frac{1}{z} \cdot \frac{\bar{z}}{\bar{z}} = \underset{\underset{\text{Use (5).}}{\uparrow}}{\frac{\bar{z}}{z\bar{z}}} = \frac{a - bi}{a^2 + b^2} = \frac{a}{a^2 + b^2} - \frac{b}{a^2 + b^2} i$$

| EXAMPLE 4 | **Writing the Reciprocal of a Complex Number in Standard Form** |

Write $\dfrac{1}{3 + 4i}$ in standard form $a + bi$; that is, find the reciprocal of $3 + 4i$.

Figure 39

```
1/(3+4i)▶Frac
       3/25-4/25i
```

Solution   The idea is to multiply the numerator and denominator by the conjugate of $3 + 4i$, that is, the complex number $3 - 4i$. The result is

$$\frac{1}{3 + 4i} = \frac{1}{3 + 4i} \cdot \frac{3 - 4i}{3 - 4i} = \frac{3 - 4i}{9 + 16} = \frac{3}{25} - \frac{4}{25} i \qquad ◀$$

A graphing calculator can be used to verify the result of Example 4. See Figure 39.
To express the quotient of two complex numbers in standard form, we multiply the numerator and denominator of the quotient by the conjugate of the denominator.

| EXAMPLE 5 | **Writing the Quotient of Complex Numbers in Standard Form** |

Write each of the following in standard form.

(a) $\dfrac{1 + 4i}{5 - 12i}$ (b) $\dfrac{2 - 3i}{4 - 3i}$

Solution   (a) $\dfrac{1 + 4i}{5 - 12i} = \dfrac{1 + 4i}{5 - 12i} \cdot \dfrac{5 + 12i}{5 + 12i} = \dfrac{5 + 12i + 20i + 48i^2}{25 + 144}$

$$= \frac{-43 + 32i}{169} = \frac{-43}{169} + \frac{32}{169} i$$

(b) $\dfrac{2 - 3i}{4 - 3i} = \dfrac{2 - 3i}{4 - 3i} \cdot \dfrac{4 + 3i}{4 + 3i} = \dfrac{8 + 6i - 12i - 9i^2}{16 + 9}$

$\phantom{(b) \dfrac{2-3i}{4-3i}} = \dfrac{17 - 6i}{25} = \dfrac{17}{25} - \dfrac{6}{25}i$ ◀

— **NOW WORK PROBLEM 27.**

EXAMPLE 6 | **Writing Other Expressions in Standard Form**

If $z = 2 - 3i$ and $w = 5 + 2i$, write each of the following expressions in standard form.

(a) $\dfrac{z}{w}$ (b) $\overline{z + w}$ (c) $z + \overline{z}$

Solution   (a) $\dfrac{z}{w} = \dfrac{z \cdot \overline{w}}{w \cdot \overline{w}} = \dfrac{(2 - 3i)(5 - 2i)}{(5 + 2i)(5 - 2i)} = \dfrac{10 - 4i - 15i + 6i^2}{25 + 4}$

$\phantom{(a) \dfrac{z}{w}} = \dfrac{4 - 19i}{29} = \dfrac{4}{29} - \dfrac{19}{29}i$

(b) $\overline{z + w} = \overline{(2 - 3i) + (5 + 2i)} = \overline{7 - i} = 7 + i$

(c) $z + \overline{z} = (2 - 3i) + (2 + 3i) = 4$ ◀

The conjugate of a complex number has certain general properties that we shall find useful later.

For a real number $a = a + 0i$, the conjugate is $\overline{a} = \overline{a + 0i} = a - 0i = a$. That is,

Theorem   The conjugate of a real number is the real number itself.

Other properties that are direct consequences of the definition of the conjugate are given next. In each statement, z and w represent complex numbers.

Theorem   The conjugate of the conjugate of a complex number is the complex number itself.

$$\overline{(\overline{z})} = z \qquad\qquad\qquad (6)$$

The conjugate of the sum of two complex numbers equals the sum of their conjugates.

$$\overline{z + w} = \overline{z} + \overline{w} \qquad\qquad\qquad (7)$$

The conjugate of the product of two complex numbers equals the product of their conjugates.

$$\overline{z \cdot w} = \overline{z} \cdot \overline{w} \qquad\qquad\qquad (8)$$

We leave the proofs of equations (6), (7), and (8) as exercises. See Problems 86–88.

Powers of i

The powers of i follow a pattern that is useful to know.

$$i^1 = i \qquad\qquad i^5 = i^4 \cdot i = 1 \cdot i = i$$
$$i^2 = -1 \qquad\qquad i^6 = i^4 \cdot i^2 = -1$$
$$i^3 = i^2 \cdot i = -i \qquad\qquad i^7 = i^4 \cdot i^3 = -i$$
$$i^4 = i^2 \cdot i^2 = (-1)(-1) = 1 \qquad i^8 = i^4 \cdot i^4 = 1$$

And so on. The powers of i repeat with every fourth power.

EXAMPLE 7 **Evaluating Powers of i**

(a) $i^{27} = i^{24} \cdot i^3 = (i^4)^6 \cdot i^3 = 1^6 \cdot i^3 = -i$

(b) $i^{101} = i^{100} \cdot i^1 = (i^4)^{25} \cdot i = 1^{25} \cdot i = i$ ◀

EXAMPLE 8 **Writing the Power of a Complex Number in Standard Form**

Write $(2 + i)^3$ in standard form.

Solution We use the special product formula for $(x + a)^3$.

$$(x + a)^3 = x^3 + 3ax^2 + 3a^2x + a^3$$

Using this special product formula,

$$(2 + i)^3 = 2^3 + 3 \cdot i \cdot 2^2 + 3 \cdot i^2 \cdot 2 + i^3$$
$$= 8 + 12i + 6(-1) + (-i)$$
$$= 2 + 11i \qquad ◀$$

NOW WORK PROBLEMS **33** AND **41**.

2 Solve Quadratic Equations with a Negative Discriminant

Quadratic equations with a negative discriminant have no real number solution. However, if we extend our number system to allow complex numbers, quadratic equations will always have a solution. Since the solution to a quadratic equation involves the square root of the discriminant, we begin with a discussion of square roots of negative numbers.

If N is a positive real number, we define the **principal square root of** $-N$, denoted by $\sqrt{-N}$, as

$$\boxed{\sqrt{-N} = \sqrt{N}\,i}$$

where i is the imaginary unit and $i^2 = -1$.

EXAMPLE 9 **Evaluating the Square Root of a Negative Number**

(a) $\sqrt{-1} = \sqrt{1}\,i = i$ 　　　　　　(b) $\sqrt{-4} = \sqrt{4}\,i = 2i$

(c) $\sqrt{-8} = \sqrt{8}\,i = 2\sqrt{2}\,i$ ◀

| **EXAMPLE 10** | **Solving Equations** |

Solve each equation in the complex number system.

(a) $x^2 = 4$ (b) $x^2 = -9$

WARNING
When working with square roots of negative numbers, do not set the square root of a product equal to the product of the square roots (which can be done with positive numbers). To see why, look at this calculation: We know that $\sqrt{100} = 10$. However, it is also true that $100 = (-25)(-4)$, so

$$10 = \sqrt{100}$$
$$= \sqrt{(-25)(-4)}$$
$$\neq \sqrt{-25}\,\sqrt{-4}$$

because $\sqrt{-25} \cdot \sqrt{-4}$

$$= \left(\sqrt{25}i\right)\left(\sqrt{4}i\right)$$
$$= (5i)(2i)$$
$$= 10i^2 = -10 \qquad\blacksquare$$

Solution   (a) $x^2 = 4$

$$x = \pm\sqrt{4} = \pm 2$$

The equation has the solution set $\{-2, 2\}$.

(b) $x^2 = -9$

$$x = \pm\sqrt{-9} = \pm\sqrt{9}i = \pm 3i$$

The equation has the solution set $\{-3i, 3i\}$.   ◄

NOW WORK PROBLEM **53**.

Because we have defined the square root of a negative number, we can now re-state the quadratic formula without restriction.

Theorem

In the complex number system, the solutions of the quadratic equation $ax^2 + bx + c = 0$, where $a, b,$ and c are real numbers and $a \neq 0$, are given by the formula

$$x = \frac{-b \pm \sqrt{b^2 - 4ac}}{2a} \qquad\qquad\textbf{(9)}$$

| **EXAMPLE 11** | **Solving Quadratic Equations in the Complex Number System** |

Solve the equation $x^2 - 4x + 8 = 0$ in the complex number system.

Solution   Here $a = 1, b = -4, c = 8,$ and $b^2 - 4ac = 16 - 4(1)(8) = -16$. Using equation (9), we find that

$$x = \frac{-(-4) \pm \sqrt{-16}}{2(1)} = \frac{4 \pm \sqrt{16}i}{2} = \frac{4 \pm 4i}{2} = 2 \pm 2i$$

The equation has the solution set $\{2 - 2i, 2 + 2i\}$.

✔ CHECK:

$$2 + 2i:\quad (2 + 2i)^2 - 4(2 + 2i) + 8 = 4 + 8i + 4i^2 - 8 - 8i + 8$$
$$= 4 - 4 = 0$$

$$2 - 2i:\quad (2 - 2i)^2 - 4(2 - 2i) + 8 = 4 - 8i + 4i^2 - 8 + 8i + 8$$
$$= 4 - 4 = 0 \qquad\qquad ◄$$

Figure 40 shows the check of the solution using a TI-84 Plus graphing calculator.

Figure 40

NOW WORK PROBLEM **59**.

The discriminant $b^2 - 4ac$ of a quadratic equation still serves as a way to determine the character of the solutions.

Character of the Solutions of a Quadratic Equation

In the complex number system, consider a quadratic equation $ax^2 + bx + c = 0$ with real coefficients.

1. If $b^2 - 4ac > 0$, the equation has two unequal real solutions.
2. If $b^2 - 4ac = 0$, the equation has a repeated real solution, a double root.
3. If $b^2 - 4ac < 0$, the equation has two complex solutions that are not real. The solutions are conjugates of each other.

The third conclusion in the display is a consequence of the fact that if $b^2 - 4ac = -N < 0$ then, by the quadratic formula, the solutions are

$$x = \frac{-b + \sqrt{b^2 - 4ac}}{2a} = \frac{-b + \sqrt{-N}}{2a} = \frac{-b + \sqrt{N}i}{2a} = \frac{-b}{2a} + \frac{\sqrt{N}}{2a}i$$

and

$$x = \frac{-b - \sqrt{b^2 - 4ac}}{2a} = \frac{-b - \sqrt{-N}}{2a} = \frac{-b - \sqrt{N}i}{2a} = \frac{-b}{2a} - \frac{\sqrt{N}}{2a}i$$

which are conjugates of each other.

EXAMPLE 12

Determining the Character of the Solutions of a Quadratic Equation

Without solving, determine the character of the solutions of each equation.

(a) $3x^2 + 4x + 5 = 0$ (b) $2x^2 + 4x + 1 = 0$

(c) $9x^2 - 6x + 1 = 0$

Solution

(a) Here $a = 3, b = 4$, and $c = 5$, so $b^2 - 4ac = 16 - 4(3)(5) = -44$. The solutions are two complex numbers that are not real and are conjugates of each other.

(b) Here $a = 2, b = 4$, and $c = 1$, so $b^2 - 4ac = 16 - 8 = 8$. The solutions are two unequal real numbers.

(c) Here $a = 9, b = -6$, and $c = 1$, so $b^2 - 4ac = 36 - 4(9)(1) = 0$. The solution is a repeated real number, that is, a double root. ◄

NOW WORK PROBLEM 73.

1.4 Assess Your Understanding

'Are You Prepared?'

Answers are given at the end of these exercises. If you get a wrong answer, read the pages listed in red.

1. Name the integers and the rational numbers in the set $\left(-3, 0, \sqrt{2}, \dfrac{6}{5}, \pi\right)$. (pp. 2–4)

2. *True or False*: Rational numbers and irrational numbers are in the set of real numbers. (pp. 2–4)

3. Rationalize the denominator of $\dfrac{3}{2 + \sqrt{3}}$. (pp. 70–71)

Concepts and Vocabulary

4. In the complex number $5 + 2i$, the number 5 is called the _____ part; the number 2 is called the _____ part; the number i is called the _____ _____.

5. The equation $x^2 = -4$ has the solution set _____.

6. *True or False*: The conjugate of $2 + 5i$ is $-2 - 5i$.

7. *True or False*: All real numbers are complex numbers.

8. *True or False*: If $2 - 3i$ is a solution of a quadratic equation with real coefficients, then $-2 + 3i$ is also a solution.

Skill Building

In Problems 9–46, write each expression in the standard form $a + bi$. Verify your results using a graphing utility.

9. $(2 - 3i) + (6 + 8i)$

10. $(4 + 5i) + (-8 + 2i)$

11. $(-3 + 2i) - (4 - 4i)$

12. $(3 - 4i) - (-3 - 4i)$

13. $(2 - 5i) - (8 + 6i)$

14. $(-8 + 4i) - (2 - 2i)$

15. $3(2 - 6i)$

16. $-4(2 + 8i)$

17. $2i(2 - 3i)$

18. $3i(-3 + 4i)$

19. $(3 - 4i)(2 + i)$

20. $(5 + 3i)(2 - i)$

21. $(-6 + i)(-6 - i)$

22. $(-3 + i)(3 + i)$

23. $\dfrac{10}{3 - 4i}$

24. $\dfrac{13}{5 - 12i}$

25. $\dfrac{2 + i}{i}$

26. $\dfrac{2 - i}{-2i}$

27. $\dfrac{6 - i}{1 + i}$

28. $\dfrac{2 + 3i}{1 - i}$

29. $\left(\dfrac{1}{2} + \dfrac{\sqrt{3}}{2}i\right)^2$

30. $\left(\dfrac{\sqrt{3}}{2} - \dfrac{1}{2}i\right)^2$

31. $(1 + i)^2$

32. $(1 - i)^2$

33. i^{23}

34. i^{14}

35. i^{-15}

36. i^{-23}

37. $i^6 - 5$

38. $4 + i^3$

39. $6i^3 - 4i^5$

40. $4i^3 - 2i^2 + 1$

41. $(1 + i)^3$

42. $(3i)^4 + 1$

43. $i^7(1 + i^2)$

44. $2i^4(1 + i^2)$

45. $i^6 + i^4 + i^2 + 1$

46. $i^7 + i^5 + i^3 + i$

In Problems 47–52, perform the indicated operations and express your answer in the form $a + bi$.

47. $\sqrt{-4}$

48. $\sqrt{-9}$

49. $\sqrt{-25}$

50. $\sqrt{-64}$

51. $\sqrt{(3 + 4i)(4i - 3)}$

52. $\sqrt{(4 + 3i)(3i - 4)}$

In Problems 53–72, solve each equation in the complex number system. Check your results using a graphing utility.

53. $x^2 + 4 = 0$

54. $x^2 - 4 = 0$

55. $x^2 - 16 = 0$

56. $x^2 + 25 = 0$

57. $x^2 - 6x + 13 = 0$

58. $x^2 + 4x + 8 = 0$

59. $x^2 - 6x + 10 = 0$

60. $x^2 - 2x + 5 = 0$

61. $8x^2 - 4x + 1 = 0$

62. $10x^2 + 6x + 1 = 0$

63. $5x^2 + 2x + 1 = 0$

64. $13x^2 + 6x + 1 = 0$

65. $x^2 + x + 1 = 0$

66. $x^2 - x + 1 = 0$

67. $x^3 - 8 = 0$

68. $x^3 + 27 = 0$

69. $x^4 - 16 = 0$

70. $x^4 - 1 = 0$

71. $x^4 + 13x^2 + 36 = 0$

72. $x^4 + 3x^2 - 4 = 0$

In Problems 73–78, without solving, determine the character of the solutions of each equation. Verify your answer using a graphing utility.

73. $3x^2 - 3x + 4 = 0$

74. $2x^2 - 4x + 1 = 0$

75. $2x^2 + 3x - 4 = 0$

76. $x^2 + 2x + 6 = 0$

77. $9x^2 - 12x + 4 = 0$

78. $4x^2 + 12x + 9 = 0$

Applications and Extensions

79. $2 + 3i$ is a solution of a quadratic equation with real coefficients. Find the other solution.

80. $4 - i$ is a solution of a quadratic equation with real coefficients. Find the other solution.

In Problems 81–84, z = 3 − 4i and w = 8 + 3i. Write each expression in the standard form a + bi.

81. $z + \overline{z}$ **82.** $w - \overline{w}$ **83.** $z\overline{z}$ **84.** $\overline{z - w}$

---

85. Use $z = a + bi$ to show that $z + \overline{z} = 2a$ and that $z - \overline{z} = 2bi$.

86. Use $z = a + bi$ to show that $(\overline{\overline{z}}) = z$.

87. Use $z = a + bi$ and $w = c + di$ to show that $\overline{z + w} = \overline{z} + \overline{w}$.

88. Use $z = a + bi$ and $w = c + di$ to show that $\overline{z \cdot w} = \overline{z} \cdot \overline{w}$.

Discussion and Writing

89. Explain to a friend how you would add two complex numbers and how you would multiply two complex numbers. Explain any differences in the two explanations.

90. Write a brief paragraph that compares the method used to rationalize denominators and the method used to write the quotient of two complex numbers in standard form.

91. Use an Internet search engine to investigate the origins of complex numbers. Write a paragraph describing what you find and present it to the class.

'Are you Prepared?' Answers

1. Integers: $\{-3, 0\}$; rational numbers: $\left\{-3, 0, \dfrac{6}{5}\right\}$ **2.** True   **3.** $3(2 - \sqrt{3})$

1.5 Radical Equations; Equations Quadratic in Form; Absolute Value Equations; Factorable Equations

PREPARING FOR THIS SECTION *Before getting started, review the following:*

- Absolute Value (Section R.2, p. 18)
- Square Roots (Section R.2, pp. 22–23)
- Factoring Polynomials (Section R.5, pp. 45–51)
- n^{th} Roots; Rational Exponents (Section R.8, pp. 68–73)

Now work the 'Are You Prepared?' problems on page 139.

OBJECTIVES
1. Solve Radical Equations
2. Solve Equations Quadratic in Form
3. Solve Absolute Value Equations
4. Solve Equations by Factoring

---

1 Solve Radical Equations

When the variable in an equation occurs in a square root, cube root, and so on, that is, when it occurs in a radical, the equation is called a **radical equation**. Sometimes a suitable operation will change a radical equation to one that is linear or quadratic. A commonly used procedure is to isolate the most complicated radical on one side of the equation and then eliminate it by raising each side to a power equal to the index of the radical. Care must be taken, however, because apparent solutions that are not, in fact, solutions of the original equation may result. These are called **extraneous solutions**. In radical equations, extraneous solutions may occur when the index of the radical is even. Therefore, we need to check all answers when working with radical equations.

| **EXAMPLE 1** | **Solving a Radical Equation** |

Find the real solutions of the equation: $\sqrt[3]{2x - 4} - 2 = 0$

Algebraic Solution

The equation contains a radical whose index is 3. We isolate it on the left side:

$$\sqrt[3]{2x - 4} - 2 = 0$$
$$\sqrt[3]{2x - 4} = 2$$

Now raise each side to the third power (the index of the radical is 3) and solve.

$$\left(\sqrt[3]{2x - 4}\right)^3 = 2^3 \qquad \text{Raise each side to the power 3.}$$
$$2x - 4 = 8 \qquad \text{Simplify.}$$
$$2x = 12 \qquad \text{Add 4 to both sides.}$$
$$x = 6 \qquad \text{Divide both sides by 2.}$$

Graphing Solution

Figure 41 shows the graph of the equation $Y_1 = \sqrt[3]{2x - 4} - 2$. From the graph, we see one x-intercept near 6. Using ZERO (or ROOT), we find that the x-intercept is 6. The only solution is $x = 6$.

Figure 41

✔ **Check:** $\sqrt[3]{2(6) - 4} - 2 = \sqrt[3]{12 - 4} - 2 = \sqrt[3]{8} - 2 = 2 - 2 = 0$

The solution set is $\{6\}$.   ◄

— **NOW WORK PROBLEM 13.**

| **EXAMPLE 2** | **Solving a Radical Equation** |

Find the real solutions of the equation: $\sqrt{x - 1} = x - 7$

Algebraic Solution

We square both sides since the index of a square root is 2.

$$\sqrt{x - 1} = x - 7$$
$$\left(\sqrt{x - 1}\right)^2 = (x - 7)^2 \qquad \text{Square both sides.}$$
$$x - 1 = x^2 - 14x + 49 \qquad \text{Remove parentheses.}$$
$$x^2 - 15x + 50 = 0 \qquad \text{Put in standard form.}$$
$$(x - 10)(x - 5) = 0 \qquad \text{Factor.}$$
$$x = 10 \quad \text{or} \quad x = 5 \qquad \text{Apply the Zero-Product Property and solve.}$$

Graphing Solution

Graph $Y_1 = \sqrt{x - 1}$ and $Y_2 = x - 7$. See Figure 42. From the graph, there is one point of intersection. Using INTERSECT, the point of intersection is $(10, 3)$, so the solution is $x = 10$.

Figure 42

✔ **Check:** $x = 10$: $\sqrt{x - 1} = \sqrt{10 - 1} = \sqrt{9} = 3$ and $x - 7 = 10 - 7 = 3$

$x = 5$: $\sqrt{x - 1} = \sqrt{5 - 1} = \sqrt{4} = 2$ and $x - 7 = 5 - 7 = -2$

The apparent solution $x = 5$ is extraneous; the only solution of the equation is $x = 10$. The solution set is $\{10\}$.   ◄

— **NOW WORK PROBLEM 25.**

Sometimes, we need to raise each side to a power more than once in order to solve a radical equation algebraically.

EXAMPLE 3   **Solving a Radical Equation**

Find the real solutions of the equation: $\sqrt{2x+3} - \sqrt{x+2} = 2$

Algebraic Solution

First, we choose to isolate the more complicated radical expression (in this case, $\sqrt{2x+3}$) on the left side:

$$\sqrt{2x+3} = \sqrt{x+2} + 2$$

Now square both sides (the index of the radical is 2):

$$\left(\sqrt{2x+3}\right)^2 = \left(\sqrt{x+2} + 2\right)^2$$

$2x + 3 = \left(\sqrt{x+2}\right)^2 + 4\sqrt{x+2} + 4$ Remove parentheses.

$2x + 3 = x + 2 + 4\sqrt{x+2} + 4$ Simplify.

$2x + 3 = x + 6 + 4\sqrt{x+2}$ Combine like terms.

Because the equation still contains a radical, we isolate the remaining radical on the right side and again square both sides.

$x - 3 = 4\sqrt{x+2}$ Isolate the radical on the right side.

$(x-3)^2 = 16(x+2)$ Square both sides.

$x^2 - 6x + 9 = 16x + 32$ Remove parentheses.

$x^2 - 22x - 23 = 0$ Put in standard form.

$(x-23)(x+1) = 0$ Factor.

$x = 23$ or $x = -1$

Graphing Solution

Graph $Y_1 = \sqrt{2x+3} - \sqrt{x+2}$ and $Y_2 = 2$. See Figure 43. From the graph there is one point of intersection. Using INTERSECT, the point of intersection is $(23, 2)$, so the solution is $x = 23$.

Figure 43

✔ **CHECK:** $x = 23$: $\sqrt{2(23)+3} - \sqrt{23+2} = \sqrt{49} - \sqrt{25} = 7 - 5 = 2$

$x = -1$: $\sqrt{2(-1)+3} - \sqrt{-1+2} = \sqrt{1} - \sqrt{1} = 1 - 1 = 0$

The solution set of the equation is $\{23\}$.   ◄

━━━━ **NOW WORK PROBLEM 35.**

2 Solve Equations Quadratic in Form

The equation $x^4 + x^2 - 12 = 0$ is not quadratic in x, but it is quadratic in x^2. That is, if we let $u = x^2$, we get $u^2 + u - 12 = 0$, a quadratic equation. This equation can be solved for u and, in turn, by using $u = x^2$, we can find the solutions x of the original equation.

In general, if an appropriate substitution u transforms an equation into one of the form

$$au^2 + bu + c = 0, \quad a \neq 0$$

then the original equation is called an **equation of the quadratic type** or an **equation quadratic in form**.

The difficulty of solving such an equation lies in the determination that the equation is, in fact, quadratic in form. After you are told an equation is quadratic in form, it is easy enough to see it, but some practice is needed to enable you to recognize them on your own.

EXAMPLE 4	**Solving Equations That Are Quadratic in Form**

Find the real solutions of the equation: $(x + 2)^2 + 11(x + 2) - 12 = 0$

Solution For this equation, let $u = x + 2$. Then $u^2 = (x + 2)^2$, and the original equation,

$$(x + 2)^2 + 11(x + 2) - 12 = 0$$

becomes

$$u^2 + 11u - 12 = 0 \qquad \text{Let } u = x + 2.$$
$$(u + 12)(u - 1) = 0 \qquad \text{Factor.}$$
$$u = -12 \quad \text{or} \quad u = 1 \qquad \text{Solve.}$$

But we want to solve for x. Because $u = x + 2$, we have

$$x + 2 = -12 \quad \text{or} \quad x + 2 = 1$$
$$x = -14 \qquad\qquad x = -1$$

✔ CHECK: $x = -14$: $(-14 + 2)^2 + 11(-14 + 2) - 12$
$$= (-12)^2 + 11(-12) - 12 = 144 - 132 - 12 = 0$$
$x = -1$: $(-1 + 2)^2 + 11(-1 + 2) - 12 = 1 + 11 - 12 = 0$

The original equation has the solution set $\{-14, -1\}$. ◄

✔ CHECK: Verify the solution of Example 4 using a graphing utility.

EXAMPLE 5	**Solving Equations That Are Quadratic in Form**

Find the real solutions of the equation: $x + 2\sqrt{x} - 3 = 0$

Solution For the equation $x + 2\sqrt{x} - 3 = 0$, let $u = \sqrt{x}$. Then $u^2 = x$, and the original equation,

$$x + 2\sqrt{x} - 3 = 0$$

becomes

$$u^2 + 2u - 3 = 0 \qquad \text{Let } u = \sqrt{x}.$$
$$(u + 3)(u - 1) = 0 \qquad \text{Factor.}$$
$$u = -3 \quad \text{or} \quad u = 1 \qquad \text{Solve.}$$

Since $u = \sqrt{x}$, we have $\sqrt{x} = -3$ or $\sqrt{x} = 1$. The first of these, $\sqrt{x} = -3$, has no real solution, since the square root of a real number is never negative. The second, $\sqrt{x} = 1$, has the solution $x = 1$.

✔ CHECK: $1 + 2\sqrt{1} - 3 = 1 + 2 - 3 = 0$

The solution set of the original equation is $\{1\}$. ◄

✔ CHECK: Verify the solution to Example 5 using a graphing utility.

Another method for solving Example 5 would be to treat it as a radical equation. Solve it this way for practice.

The idea should now be clear. If an equation contains an expression and that same expression squared, make a substitution for the expression. You may get a quadratic equation.

 NOW WORK PROBLEM 59.

3 Solve Absolute Value Equations

Recall that, on the real number line, the absolute value of a equals the distance from the origin to the point whose coordinate is a. For example, there are two points whose distance from the origin is 5 units, -5 and 5. Thus the equation $|x| = 5$ will have the solution set $\{-5, 5\}$. This leads to the following result:

Equations Involving Absolute Value

If a is a positive real number and if u is any algebraic expression, then

$$|u| = a \quad \text{is equivalent to} \quad u = a \text{ or } u = -a \qquad (1)$$

| EXAMPLE 6 | Solving an Equation Involving Absolute Value |

Solve the equation $|x + 4| = 13$.

Algebraic Solution

This follows the form of equation (1), where $u = x + 4$. There are two possibilities.

$$x + 4 = 13 \quad \text{or} \quad x + 4 = -13$$
$$x = 9 \qquad\qquad x = -17$$

The solution set is $\{-17, 9\}$.   ◀

Graphing Solution

For this equation, graph $Y_1 = |x + 4|$ and $Y_2 = 13$ on the same screen and find their point(s) of intersection, if any. See Figure 44. Using the INTERSECT command (twice), we find the points of intersection to be $(-17, 13)$ and $(9, 13)$. The solution set is $\{-17, 9\}$.

Figure 44

20 $Y_1 = |x + 4|$

$Y_2 = 13$

-25 25

0

◀

NOW WORK PROBLEM 81.

4 Solve Equations by Factoring

We have already solved certain quadratic equations using factoring. Let's look at examples of other kinds of equations that can be solved by factoring.

EXAMPLE 7	Solving Equations by Factoring

Solve the equation: $x^4 = 4x^2$

Algebraic Solution

We begin by collecting all terms on one side. This results in 0 on one side and an expression to be factored on the other.

$$x^4 = 4x^2$$
$$x^4 - 4x^2 = 0$$
$$x^2(x^2 - 4) = 0 \qquad \text{Factor.}$$
$$x^2 = 0 \quad \text{or} \qquad x^2 - 4 = 0 \qquad \text{Apply the Zero-Product Property.}$$
$$x^2 = 4$$
$$x = 0 \quad \text{or} \quad x = -2 \quad \text{or} \quad x = 2 \qquad \text{Use the Square Root Method.}$$

The solution set is $\{-2, 0, 2\}$.

Graphing Solution

Graph $Y_1 = x^4$ and $Y_2 = 4x^2$ on the same screen and find their point(s) of intersection, if any. See Figure 45. Using the INTER-SECT command (three times), we find the points of intersection to be $(-2, 16)$, $(0, 0)$, and $(2, 16)$. The solution set is $\{-2, 0, 2\}$.

Figure 45

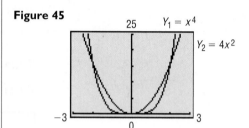

✔ **CHECK:** $x = -2$: $(-2)^4 = 16$ and $4(-2)^2 = 16$ So -2 is a solution.

$x = 0$: $0^4 = 0$ and $4 \cdot 0^2 = 0$ So 0 is a solution.

$x = 2$: $2^4 = 16$ and $4 \cdot 2^2 = 16$ So 2 is a solution. ◄

EXAMPLE 8	Solving Equations by Factoring

Solve the equation: $x^3 - x^2 - 4x + 4 = 0$

Algebraic Solution

Do you recall the method of factoring by grouping? (If not, review pp. 49–50.) We group the terms of $x^3 - x^2 - 4x + 4 = 0$ as follows:

$$(x^3 - x^2) - (4x - 4) = 0$$

Factor out x^2 from the first grouping and 4 from the second.

$$x^2(x - 1) - 4(x - 1) = 0$$

This reveals the common factor $(x - 1)$, so we have

$$(x^2 - 4)(x - 1) = 0$$
$$(x - 2)(x + 2)(x - 1) = 0 \qquad \text{Factor again.}$$
$$x - 2 = 0 \quad \text{or} \quad x + 2 = 0 \quad \text{or} \quad x - 1 = 0 \qquad \text{Set each factor equal to 0.}$$
$$x = 2 \qquad\qquad x = -2 \qquad\qquad x = 1 \qquad \text{Solve.}$$

The solution set is $\{-2, 1, 2\}$.

Graphing Solution

Graph $Y_1 = x^3 - x^2 - 4x + 4$. See Figure 46. Using ZERO (three times), the values of x for which $y = 0$ are -2, 1, and 2. The solution set is $\{-2, 1, 2\}$.

Figure 46

✔ **CHECK:**

$x = -2$: $(-2)^3 - (-2)^2 - 4(-2) + 4 = -8 - 4 + 8 + 4 = 0$ -2 is a solution.

$x = 1$: $1^3 - 1^2 - 4(1) + 4 = 1 - 1 - 4 + 4 = 0$ 1 is a solution.

$x = 2$: $2^3 - 2^2 - 4(2) + 4 = 8 - 4 - 8 + 4 = 0$ 2 is a solution. ◄

NOW WORK PROBLEM 107.

1.5 Assess Your Understanding

'Are You Prepared?'

Answers are given at the end of these exercises. If you get a wrong answer, read the pages listed in red.

1. *True or False:* The principal square root of any nonnegative real number is always nonnegative. (pp. 22–23)

2. $\sqrt[3]{-8} =$ _____ (p. 68)

3. Factor $6x^3 - 2x^2$ (pp. 45–46)

4. Factor $x^3 + 4x^2 - 9x - 36$ (pp. 49–50)

5. $|a| =$ _____ if $a < 0$. (p. 18)

Concepts and Vocabulary

6. When an apparent solution does not satisfy the original equation, it is called a(n) _____ solution.

7. If u is an expression involving x, the equation $au^2 + bu + c = 0$, $a \neq 0$, is called a(n) equation _____ _____ _____.

8. *True or False:* Radical equations sometimes have no real solution.

Skill Building

In Problems 9–46, find the real solutions of each equation. Verify your results using a graphing utility.

9. $\sqrt{x} = 3$

10. $\sqrt{z} = 10$

11. $\sqrt{y + 3} = 5$

12. $\sqrt{t - 3} = 7$

13. $\sqrt{2t - 1} = 1$

14. $\sqrt{3t + 4} = 2$

15. $\sqrt{3t + 4} = -6$

16. $\sqrt{5t + 3} = -2$

17. $\sqrt[3]{1 - 2x} - 3 = 0$

18. $\sqrt[3]{1 - 2x} - 1 = 0$

19. $\sqrt[4]{5x - 4} = 2$

20. $\sqrt[5]{2x - 3} = -1$

21. $\sqrt[5]{x^2 + 2x} = -1$

22. $\sqrt[4]{x^2 + 16} = \sqrt{5}$

23. $x = 8\sqrt{x}$

24. $x = 3\sqrt{x}$

25. $\sqrt{15 - 2x} = x$

26. $\sqrt{12 - x} = x$

27. $x = 2\sqrt{x - 1}$

28. $x = 2\sqrt{-x - 1}$

29. $\sqrt{x^2 - x - 4} = x + 2$

30. $\sqrt{3 - x + x^2} = x - 2$

31. $3 + \sqrt{3x + 1} = x$

32. $2 + \sqrt{12 - 2x} = x$

33. $\sqrt{2x + 3} - \sqrt{x + 1} = 1$

34. $\sqrt{3x + 7} + \sqrt{x + 2} = 1$

35. $\sqrt{3x + 1} - \sqrt{x - 1} = 2$

36. $\sqrt{3x - 5} - \sqrt{x + 7} = 2$

37. $\sqrt{3 - 2\sqrt{x}} = \sqrt{x}$

38. $\sqrt{10 + 3\sqrt{x}} = \sqrt{x}$

39. $(3x + 1)^{1/2} = 4$

40. $(3x - 5)^{1/2} = 2$

41. $(5x - 2)^{1/3} = 2$

42. $(2x + 1)^{1/3} = -1$

43. $(x^2 + 9)^{1/2} = 5$

44. $(x^2 - 16)^{1/2} = 9$

45. $x^{3/2} - 3x^{1/2} = 0$

46. $x^{3/4} - 9x^{1/4} = 0$

In Problems 47–78, find the real solutions of each equation. Verify your results using a graphing utility.

47. $t^4 - 16 = 0$

48. $y^4 - 4 = 0$

49. $x^4 - 5x^2 + 4 = 0$

50. $x^4 - 10x^2 + 25 = 0$

51. $3x^4 - 2x^2 - 1 = 0$

52. $2x^4 - 5x^2 - 12 = 0$

53. $x^6 + 7x^3 - 8 = 0$

54. $x^6 - 7x^3 - 8 = 0$

55. $(x + 2)^2 + 7(x + 2) + 12 = 0$

56. $(2x + 5)^2 - (2x + 5) - 6 = 0$

57. $(3x + 4)^2 - 6(3x + 4) + 9 = 0$

58. $(2 - x)^2 + (2 - x) - 20 = 0$

59. $2(s + 1)^2 - 5(s + 1) = 3$

60. $3(1 - y)^2 + 5(1 - y) + 2 = 0$

61. $x - 4\sqrt{x} = 0$

62. $x + 8\sqrt{x} = 0$

63. $x + \sqrt{x} = 20$

64. $x + \sqrt{x} = 6$

65. $t^{1/2} - 2t^{1/4} + 1 = 0$

66. $z^{1/2} - 4z^{1/4} + 4 = 0$

67. $4x^{1/2} - 9x^{1/4} + 4 = 0$

68. $x^{1/2} - 3x^{1/4} + 2 = 0$

69. $\sqrt[4]{5x^2 - 6} = x$

70. $\sqrt[4]{4 - 5x^2} = x$

71. $\dfrac{1}{(x + 1)^2} = \dfrac{1}{x + 1} + 2$

72. $\dfrac{1}{(x - 1)^2} + \dfrac{1}{x - 1} = 12$

73. $3x^{-2} - 7x^{-1} - 6 = 0$

74. $2x^{-2} - 3x^{-1} - 4 = 0$

75. $2x^{2/3} - 5x^{1/3} - 3 = 0$

76. $3x^{4/3} + 5x^{2/3} - 2 = 0$

77. $\left(\dfrac{v}{v + 1}\right)^2 + \dfrac{2v}{v + 1} = 8$

78. $\left(\dfrac{y}{y - 1}\right)^2 = 6\left(\dfrac{y}{y - 1}\right) + 7$

In Problems 79–100, solve each equation. Verify your results using a graphing utility.

79. $|x| = 6$

80. $|x| = 12$

81. $|2x + 3| = 5$

82. $|3x - 1| = 2$

83. $|1 - 4t| + 8 = 13$

84. $|1 - 2z| + 6 = 9$

85. $|-2x| = 8$

86. $|-x| = 1$

87. $4 - |2x| = 3$ **88.** $5 - \left|\frac{1}{2}x\right| = 3$ **89.** $\frac{2}{3}|x| = 9$ **90.** $\frac{3}{4}|x| = 9$

91. $\left|\frac{x}{3} + \frac{2}{5}\right| = 2$ **92.** $\left|\frac{x}{2} - \frac{1}{3}\right| = 1$ **93.** $|u - 2| = -\frac{1}{2}$ **94.** $|2 - v| = -1$

95. $|x^2 - 9| = 0$ **96.** $|x^2 - 16| = 0$ **97.** $|x^2 - 2x| = 3$ **98.** $|x^2 + x| = 12$

99. $|x^2 + x - 1| = 1$ **100.** $|x^2 + 3x - 2| = 2$

In Problems 101–114, find the real solutions of each equation by factoring. Verify your results using a graphing utility.

101. $x^3 - 9x = 0$ **102.** $x^4 - x^2 = 0$ **103.** $4x^3 = 3x^2$

104. $x^5 = 4x^3$ **105.** $x^3 + x^2 - 20x = 0$ **106.** $x^3 + 6x^2 - 7x = 0$

107. $x^3 + x^2 - x - 1 = 0$ **108.** $x^3 + 4x^2 - x - 4 = 0$ **109.** $x^3 - 3x^2 - 4x + 12 = 0$

110. $x^3 - 3x^2 - x + 3 = 0$ **111.** $2x^3 + 4 = x^2 + 8x$ **112.** $3x^3 + 4x^2 = 27x + 36$

113. $5x^3 + 45x = 2x^2 + 18$ **114.** $3x^3 + 12x = 5x^2 + 20$

In Problems 115–120, find the real solutions of each equation. Use a calculator to express any solutions rounded to two decimal places.

115. $x - 4x^{1/2} + 2 = 0$ **116.** $x^{2/3} + 4x^{1/3} + 2 = 0$ **117.** $x^4 + \sqrt{3}x^2 - 3 = 0$

118. $x^4 + \sqrt{2}x^2 - 2 = 0$ **119.** $\pi(1 + t)^2 = \pi + 1 + t$ **120.** $\pi(1 + r)^2 = 2 + \pi(1 + r)$

Applications and Extensions

121. If $k = \dfrac{x + 3}{x - 3}$ and $k^2 - k = 12$, find x.

122. If $k = \dfrac{x + 3}{x - 4}$ and $k^2 - 3k = 28$, find x.

123. Find all points having an x-coordinate of 2 whose distance from the point $(-2, -1)$ is 5.

124. Find all points having a y-coordinate of -3 whose distance from the point $(1, 2)$ is 13.

125. Find all points on the x-axis that are 5 units from the point $(4, -3)$.

126. Find all points on the y-axis that are 5 units from the point $(4, 4)$.

127. Physics: Using Sound to Measure Distance The distance to the surface of the water in a well can sometimes be found by dropping an object into the well and measuring the time elapsed until a sound is heard. If t_1 is the time (measured in seconds) that it takes for the object to strike the water, then t_1 will obey the equation $s = 16t_1^2$, where s is the distance (measured in feet). It follows that $t_1 = \dfrac{\sqrt{s}}{4}$. Suppose that t_2 is the time that it takes for the sound of the impact to reach your ears. Because sound travels at a speed of approximately 1100 feet per second, the time t_2 for the sound to travel the distance s will be $t_2 = \dfrac{s}{1100}$. See the illustration. Now $t_1 + t_2$ is the total time that elapses from the moment that the object is dropped to the moment that a sound is heard. We have the equation

$$\text{Total time elapsed} = \frac{\sqrt{s}}{4} + \frac{s}{1100}$$

Find the distance to the water's surface if the total time elapsed from dropping a rock to hearing it hit water is 4 seconds.

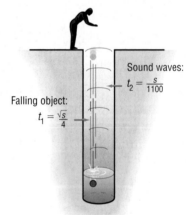

Sound waves:
$t_2 = \frac{s}{1100}$

Falling object:
$t_1 = \frac{\sqrt{s}}{4}$

Discussion and Writing

128. Make up a radical equation that has no solution.

129. Make up a radical equation that has an extraneous solution.

130. Discuss the step in the solving process for radical equations that leads to the possibility of extraneous solutions. Why is there no such possibility for linear and quadratic equations?

131. In your own words, explain what it means for an equation to be quadratic in form.

132. The equation $|x| = -2$ has no real solution. Why?

'Are You Prepared?' Answers

   1. True   **2.** -2 **3.** $2x^2(3x - 1)$ **4.** $(x - 3)(x + 3)(x + 4)$ **5.** $-a$

1.6 Problem Solving: Interest, Mixture, Uniform Motion, Constant Rate Jobs

OBJECTIVES   1 Translate Verbal Descriptions into Mathematical Expressions
2 Solve Interest Problems
3 Solve Mixture Problems
4 Solve Uniform Motion Problems
5 Solve Constant Rate Job Problems

Applied (word) problems do not come in the form "Solve the equation. . . . " Instead, they supply information using words, a verbal description of the real problem. So, to solve applied problems, we must be able to translate the verbal description into the language of mathematics. We do this by using variables to represent unknown quantities and then finding relationships (such as equations) that involve these variables. The process of doing all this is called **mathematical modeling**.

Any solution to the mathematical problem must be checked against the mathematical problem, the verbal description, and the real problem. See Figure 47 for an illustration of the **modeling process**.

Figure 47

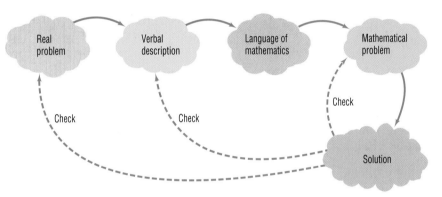

1 Translate Verbal Descriptions into Mathematical Expressions

Let's look at a few examples that will help you to translate certain words into mathematical symbols.

EXAMPLE 1   **Translating Verbal Descriptions into Mathematical Expressions**

(a) For uniform motion, the velocity of an object equals the distance traveled divided by the time required.

Translation: If v is the velocity, s the distance, and t the time, then $v = \dfrac{s}{t}$.

(b) Let x denote a number.

The number 5 times as large as x is $5x$.

The number 3 less than x is $x - 3$.

The number that exceeds x by 4 is $x + 4$.

The number that, when added to x, gives 5 is $5 - x$.   ◀

NOW WORK PROBLEM 7.

Always check the units used to measure the variables of an applied problem. In Example 1(a), if v is measured in miles per hour, then the distance s must be expressed in miles and the time t must be expressed in hours. It is a good practice to check units to be sure that they are consistent and make sense.

The steps to follow for setting up applied problems, given earlier, is repeated here.

Steps for Setting Up Applied Problems

STEP 1: Read the problem carefully, perhaps two or three times. Pay particular attention to the question being asked in order to identify what you are looking for. If you can, determine realistic possibilities for the answer.

STEP 2: Assign a letter (variable) to represent what you are looking for, and, if necessary, express any remaining unknown quantities in terms of this variable.

STEP 3: Make a list of all the known facts, and translate them into mathematical expressions. These may take the form of an equation (or, later, an inequality) involving the variable. If possible, draw an appropriately labeled diagram to assist you. Sometimes a table or chart helps.

STEP 4: Solve the equation for the variable, and then answer the question.

STEP 5: Check the answer with the facts in the problem. If it agrees, congratulations! If it does not agree, try again.

---

2 Solve Interest Problems

Interest is money paid for the use of money. The total amount borrowed (whether by an individual from a bank in the form of a loan or by a bank from an individual in the form of a savings account) is called the **principal**. The **rate of interest**, expressed as a percent, is the amount charged for the use of the principal for a given period of time, usually on a yearly (that is, per annum) basis.

Simple Interest Formula

If a principal of P dollars is borrowed for a period of t years at a per annum interest rate r, expressed as a decimal, the interest I charged is

$$I = Prt \qquad (1)$$

Interest charged according to formula (1) is called **simple interest**. When using formula (1), be sure to express r as a decimal.

EXAMPLE 2 | **Finance: Computing Interest on a Loan**

Suppose that Juanita borrows $500 for 6 months at the simple interest rate of 9% per annum. What is the interest that Juanita will be charged on this loan? How much does Juanita owe after 6 months?

Solution   The rate of interest is given per annum, so the actual time that the money is borrowed must be expressed in years. The interest charged would be the principal, $500, times the rate of interest (9% = 0.09) times the time in years, $\frac{1}{2}$:

$$\text{Interest charged} = I = Prt = (500)(0.09)\left(\frac{1}{2}\right) = \$22.50$$

After 6 months, Juanita will owe what she borrowed plus the interest: $500 + $22.50 = $522.50. ◀

| **EXAMPLE 3** | **Financial Planning** |

Candy has $70,000 to invest and requires an overall rate of return of 9%. She can invest in a safe, government-insured certificate of deposit, but it only pays 8%. To obtain 9%, she agrees to invest some of her money in noninsured corporate bonds paying 12%. How much should be placed in each investment to achieve her goal?

Solution   **STEP 1:** The question is asking for two dollar amounts: the principal to invest in the corporate bonds and the principal to invest in the certificate of deposit.

STEP 2: We let x represent the amount (in dollars) to be invested in the bonds. Then $70{,}000 - x$ is the amount that will be invested in the certificate. (Do you see why?)

STEP 3: We set up a table:

	Principal ($)	Rate	Time (yr)	Interest ($)
Bonds	x	12% = 0.12	1	0.12x
Certificate	70,000 − x	8% = 0.08	1	0.08(70,000 − x)
Total	70,000	9% = 0.09	1	0.09(70,000) = 6300

Since the total interest from the investments is equal to $0.09(70{,}000) = 6300$, we must have the equation

$$0.12x + 0.08(70{,}000 - x) = 6300$$

(Note that the units are consistent: the unit is dollars on each side.)

STEP 4:
$$0.12x + 5600 - 0.08x = 6300$$
$$0.04x = 700$$
$$x = 17{,}500$$

Candy should place $17,500 in the bonds and $70,000 − $17,500 = $52,500 in the certificate.

STEP 5: The interest on the bonds after 1 year is 0.12($17,500) = $2100; the interest on the certificate after 1 year is 0.08($52,500) = $4200. The total annual interest is $6300, the required amount. ◀

NOW WORK PROBLEM 17.

3 Solve Mixture Problems

Oil refineries sometimes produce gasoline that is a blend of two or more types of fuel; bakeries occasionally blend two or more types of flour for their bread. These problems are referred to as **mixture problems** because they combine two or more quantities to form a mixture.

| EXAMPLE 4 | **Blending Coffees** |

The manager of a Starbucks store decides to experiment with a new blend of coffee. She will mix some B grade Colombian coffee that sells for $5 per pound with some A grade Arabica coffee that sells for $10 per pound to get 100 pounds of the new blend. The selling price of the new blend is to be $7 per pound, and there is to be no difference in revenue from selling the new blend versus selling the other types. How many pounds of the B grade Colombian and A grade Arabica coffees are required?

Solution Let x represent the number of pounds of the B grade Colombian coffee. Then $100 - x$ equals the number of pounds of the A grade Arabica coffee. See Figure 48.

Figure 48

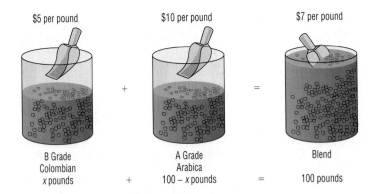

Since there is to be no difference in revenue between selling the A and B grades separately versus the blend, we have

$$\left\{\begin{matrix}\text{Price per pound}\\\text{of B grade}\end{matrix}\right\}\left\{\begin{matrix}\text{\# Pounds}\\\text{B grade}\end{matrix}\right\} + \left\{\begin{matrix}\text{Price per pound}\\\text{of A grade}\end{matrix}\right\}\left\{\begin{matrix}\text{\# Pounds}\\\text{A grade}\end{matrix}\right\} = \left\{\begin{matrix}\text{Price per pound}\\\text{of blend}\end{matrix}\right\}\left\{\begin{matrix}\text{\# Pounds}\\\text{blend}\end{matrix}\right\}$$

$$\$5 \quad\cdot\quad x \quad+\quad \$10 \quad\cdot\ (100 - x) = \quad \$7 \quad\cdot\quad 100$$

We have the equation

$$5x + 10(100 - x) = 7 \cdot (100)$$
$$5x + 1000 - 10x = 700$$
$$-5x = -300$$
$$x = 60$$

The manager should blend 60 pounds of B grade Colombian coffee with $100 - 60 = 40$ pounds of A grade Arabica coffee to get the desired blend.

✔ CHECK: The 60 pounds of B grade coffee would sell for $(\$5)(60) = \$300$, and the 40 pounds of A grade coffee would sell for $(\$10)(40) = \$400$; the total revenue, $700, equals the revenue obtained from selling the blend, as desired. ◀

NOW WORK PROBLEM 21.

4 **Solve Uniform Motion Problems**

Objects that move at a constant velocity are said to be in **uniform motion**. When the average velocity of an object is known, it can be interpreted as its constant velocity. For example, a bicyclist traveling at an average velocity of 25 miles per hour is in uniform motion.

Uniform Motion Formula

If an object moves at an average velocity v, the distance s covered in time t is given by the formula

$$s = vt \tag{2}$$

That is, Distance = Velocity · Time.

| EXAMPLE 5 | **Physics: Uniform Motion** |

Tanya, who is a long-distance runner, runs at an average velocity of 8 miles per hour (mi/hr). Two hours after Tanya leaves your house, you leave in your Honda and follow the same route. If your average velocity is 40 mi/hr, how long will it be before you catch up to Tanya? How far will each of you be from your home?

Solution Refer to Figure 49. We use t to represent the time (in hours) that it takes the Honda to catch up to Tanya. When this occurs, the total time elapsed for Tanya is $t + 2$ hours.

Figure 49

Set up the following table:

	Velocity mi/hr	Time hr	Distance mi
Tanya	8	$t + 2$	$8(t + 2)$
Honda	40	t	$40t$

Since the distance traveled is the same, we are led to the following equation:

$$8(t + 2) = 40t$$
$$8t + 16 = 40t$$
$$32t = 16$$
$$t = \frac{1}{2} \text{ hour}$$

It will take the Honda $\frac{1}{2}$ hour to catch up to Tanya. Each will have gone 20 miles.

✔ **CHECK:** In 2.5 hours, Tanya travels a distance of $(2.5)(8) = 20$ miles. In $\frac{1}{2}$ hour, the Honda travels a distance of $\left(\frac{1}{2}\right)(40) = 20$ miles. ◄

EXAMPLE 6	**Physics: Uniform Motion**

A motorboat heads upstream a distance of 24 miles on a river whose current is running at 3 miles per hour (mi/hr). The trip up and back takes 6 hours. Assuming that the motorboat maintained a constant speed relative to the water, what was its speed?

Figure 50

24 miles

$v - 3$ mi/hr

$v + 3$ mi/hr

Solution See Figure 50. We use v to represent the constant speed of the motorboat relative to the water. Then the true speed going upstream is $v - 3$ mi/hr, and the true speed going downstream is $v + 3$ mi/hr. Since Distance = Velocity × Time, then Time $= \dfrac{\text{Distance}}{\text{Velocity}}$. We set up a table.

	Velocity mi/hr	Distance mi	Time $= \dfrac{\text{Distance}}{\text{Velocity}}$ hr
Upstream	$v - 3$	24	$\dfrac{24}{v - 3}$
Downstream	$v + 3$	24	$\dfrac{24}{v + 3}$

Since the total time up and back is 6 hours, we have

$$\frac{24}{v - 3} + \frac{24}{v + 3} = 6$$

$$\frac{24(v + 3) + 24(v - 3)}{(v - 3)(v + 3)} = 6 \qquad \text{Add the quotients on the left.}$$

$$\frac{48v}{v^2 - 9} = 6 \qquad \text{Simplify.}$$

$$48v = 6(v^2 - 9)$$

$$6v^2 - 48v - 54 = 0 \qquad \text{Place in standard form.}$$

$$v^2 - 8v - 9 = 0 \qquad \text{Divide by 6.}$$

$$(v - 9)(v + 1) = 0 \qquad \text{Factor.}$$

$$v = 9 \quad \text{or} \quad v = -1 \qquad \text{Apply the Zero-Product Property and solve.}$$

We discard the solution $v = -1$ mi/hr, so the speed of the motorboat relative to the water is 9 mi/hr. ◀

 NOW WORK PROBLEM 27.

5 Solve Constant Rate Job Problems

This section involves jobs that are performed at a **constant rate**. Our assumption is that, if a job can be done in t units of time, $\dfrac{1}{t}$ of the job is done in 1 unit of time. Let's look at an example.

EXAMPLE 7	**Working Together to Do a Job**

At 10 AM Danny is asked by his father to weed the garden. From past experience, Danny knows that this will take him 4 hours, working alone. His older brother, Mike, when it is his turn to do this job, requires 6 hours. Since Mike wants to go golfing with Danny and has a reservation for 1 PM, he agrees to help Danny. Assuming no gain or loss of efficiency, when will they finish if they work together? Can they make the golf date?

Solution

Table 3

	Hours to Do Job	Part of Job Done in 1 Hour
Danny	4	$\dfrac{1}{4}$
Mike	6	$\dfrac{1}{6}$
Together	t	$\dfrac{1}{t}$

We set up Table 3. In 1 hour, Danny does $\dfrac{1}{4}$ of the job, and in 1 hour, Mike does $\dfrac{1}{6}$ of the job. Let t be the time (in hours) that it takes them to do the job together. In 1 hour, then, $\dfrac{1}{t}$ of the job is completed. We reason as follows:

$$\begin{pmatrix} \text{Part done by Danny} \\ \text{in 1 hour} \end{pmatrix} + \begin{pmatrix} \text{Part done by Mike} \\ \text{in 1 hour} \end{pmatrix} = \begin{pmatrix} \text{Part done together} \\ \text{in 1 hour} \end{pmatrix}$$

From Table 3,

$$\frac{1}{4} + \frac{1}{6} = \frac{1}{t}$$
$$\frac{3 + 2}{12} = \frac{1}{t}$$
$$\frac{5}{12} = \frac{1}{t}$$
$$5t = 12$$
$$t = \frac{12}{5}$$

Working together, the job can be done in $\dfrac{12}{5}$ hours, or 2 hours, 24 minutes. They should make the golf date, since they will finish at 12:24 PM. ◀

NOW WORK PROBLEM **29.**

1.6 Assess Your Understanding

Concepts and Vocabulary

1. Using variables to represent unknown quantities and then finding relationships that involve these variables is referred to as _____ _____.

2. The money paid for the use of money is _____.

3. Objects that move at a constant velocity are said to be in _____ _____.

4. *True or False:* The amount charged for the use of principal for a given period of time is called the rate of interest.

5. *True or False:* If an object moves at an average velocity v, the distance s covered in time t is given by the formula $s = vt$.

6. Suppose that you want to mix two coffees in order to obtain 100 pounds of the blend. If x represents the number of pounds of coffee A, write an algebraic expression that represents the number of pounds of coffee B.

Applications and Extensions

In Problems 7–16, translate each sentence into a mathematical equation. Be sure to identify the meaning of all symbols.

7. **Geometry**   The area of a circle is the product of the number π and the square of the radius.

8. **Geometry**   The circumference of a circle is the product of the number π and twice the radius.

9. **Geometry**   The area of a square is the square of the length of a side.

10. **Geometry**   The perimeter of a square is four times the length of a side.

11. **Physics**   Force equals the product of mass and acceleration.

12. **Physics**   Pressure is force per unit area.

13. **Physics**   Work equals force times distance.

14. Physics Kinetic energy is one-half the product of the mass and the square of the velocity.

15. Business The total variable cost of manufacturing x dishwashers is $150 per dishwasher times the number of dishwashers manufactured.

16. Business The total revenue derived from selling x dishwashers is $250 per dishwasher times the number of dishwashers sold.

17. Financial Planning Betsy, a recent retiree, requires $6000 per year in extra income. She has $50,000 to invest and can invest in B-rated bonds paying 15% per year or in a certificate of deposit (CD) paying 7% per year. How much money should be invested in each to realize exactly $6000 in interest per year?

18. Financial Planning After 2 years, Betsy (see Problem 17) finds that she will now require $7000 per year. Assuming that the remaining information is the same, how should the money be reinvested?

19. Banking A bank loaned out $12,000, part of it at the rate of 8% per year and the rest at the rate of 18% per year. If the interest received totaled $1000, how much was loaned at 8%?

20. Banking Wendy, a loan officer at a bank, has $1,000,000 to lend and is required to obtain an average return of 18% per year. If she can lend at the rate of 19% or at the rate of 16%, how much can she lend at the 16% rate and still meet her requirement?

21. Blending Teas The manager of a store that specializes in selling tea decides to experiment with a new blend. She will mix some Earl Grey tea that sells for $5 per pound with some Orange Pekoe tea that sells for $3 per pound to get 100 pounds of the new blend. The selling price of the new blend is to be $4.50 per pound, and there is to be no difference in revenue from selling the new blend versus selling the other types. How many pounds of the Earl Grey tea and Orange Pekoe tea are required?

22. Business: Blending Coffee A coffee manufacturer wants to market a new blend of coffee that sells for $3.90 per pound by mixing two coffees that sell for $2.75 and $5 per pound, respectively. What amounts of each coffee should be blended to obtain the desired mixture?

[**Hint:** Assume that the total weight of the desired blend is 100 pounds.]

23. Business: Mixing Nuts A nut store normally sells cashews for $4.00 per pound and peanuts for $1.50 per pound. But at the end of the month the peanuts had not sold well, so, in order to sell 60 pounds of peanuts, the manager decided to mix the 60 pounds of peanuts with some cashews and sell the mixture for $2.50 per pound. How many pounds of cashews should be mixed with the peanuts to ensure no change in the profit?

24. Business: Mixing Candy A candy store sells boxes of candy containing caramels and cremes. Each box sells for $12.50 and holds 30 pieces of candy (all pieces are the same size). If the caramels cost $0.25 to produce and the cremes cost $0.45 to produce, how many of each should be in a box to make a profit of $3?

25. Physics: Uniform Motion A motorboat can maintain a constant speed of 16 miles per hour relative to the water. The boat makes a trip upstream to a certain point in 20 minutes; the return trip takes 15 minutes. What is the speed of the current? (See the figure.)

26. Physics: Uniform Motion A motorboat heads upstream on a river that has a current of 3 miles per hour. The trip upstream takes 5 hours, and the return trip takes 2.5 hours. What is the speed of the motorboat? (Assume that the motorboat maintains a constant speed relative to the water.)

27. Physics: Uniform Motion A motorboat maintained a constant speed of 15 miles per hour relative to the water in going 10 miles upstream and then returning. The total time for the trip was 1.5 hours. Use this information to find the speed of the current.

28. Physics: Uniform Motion Two cars enter the Florida Turnpike at Commercial Boulevard at 8:00 AM, each heading for Wildwood. One car's average speed is 10 miles per hour more than the other's. The faster car arrives at Wildwood at 11:00 AM, $\frac{1}{2}$ hour before the other car. What is the average speed of each car? How far did each travel?

29. Working Together on a Job Trent can deliver his newspapers in 30 minutes. It takes Lois 20 minutes to do the same route. How long would it take them to deliver the newspapers if they work together?

30. Working Together on a Job Patrice, by himself, can paint four rooms in 10 hours. If he hires April to help, they can do the same job together in 6 hours. If he lets April work alone, how long will it take her to paint four rooms?

31. Enclosing a Garden A gardener has 46 feet of fencing to be used to enclose a rectangular garden that has a border 2 feet wide surrounding it (see the figure).
   (a) If the length of the garden is to be twice its width, what will be the dimensions of the garden?
   (b) What is the area of the garden?
   (c) If the length and width of the garden are to be the same, what would be the dimensions of the garden?
   (d) What would be the area of the square garden?

32. **Construction**   A pond is enclosed by a wooden deck that is 3 feet wide. The fence surrounding the deck is 100 feet long.
    (a) If the pond is square, what are its dimensions?
    (b) If the pond is rectangular and the length of the pond is to be three times its width, what are its dimensions?
    (c) If the pond is circular, what is its diameter?
    (d) Which pond has the most area?

33. **Football**   A tight end can run the 100-yard dash in 12 seconds. A defensive back can do it in 10 seconds. The tight end catches a pass at his own 20-yard line with the defensive back at the 15-yard line. (See the figure.) If no other players are nearby, at what yard line will the defensive back catch up to the tight end?

    [**Hint:** At time $t = 0$, the defensive back is 5 yards behind the tight end.]

34. **Computing Business Expense**   Therese, an outside salesperson, uses her car for both business and pleasure. Last year, she traveled 30,000 miles, using 900 gallons of gasoline. Her car gets 40 miles per gallon on the highway and 25 in the city. She can deduct all highway travel, but no city travel, on her taxes. How many miles should Therese be allowed as a business expense?

35. **Mixing Water and Antifreeze**   How much water should be added to 1 gallon of pure antifreeze to obtain a solution that is 60% antifreeze?

36. **Mixing Water and Antifreeze**   The cooling system of a certain foreign-made car has a capacity of 15 liters. If the system is filled with a mixture that is 40% antifreeze, how much of this mixture should be drained and replaced by pure antifreeze so that the system is filled with a solution that is 60% antifreeze?

37. **Chemistry: Salt Solutions**   How much water must be evaporated from 32 ounces of a 4% salt solution to make a 6% salt solution?

38. **Chemistry: Salt Solutions**   How much water must be evaporated from 240 gallons of a 3% salt solution to produce a 5% salt solution?

39. **Purity of Gold**   The purity of gold is measured in karats, with pure gold being 24 karats. Other purities of gold are expressed as proportional parts of pure gold. Thus, 18-karat gold is $\frac{18}{24}$, or 75% pure gold; 12-karat gold is $\frac{12}{24}$, or 50% pure gold; and so on. How much 12-karat gold should be mixed with pure gold to obtain 60 grams of 16-karat gold?

40. **Chemistry: Sugar Molecules**   A sugar molecule has twice as many atoms of hydrogen as it does oxygen and one more atom of carbon than oxygen. If a sugar molecule has a total of 45 atoms, how many are oxygen? How many are hydrogen?

41. **Running a Race**   Mike can run the mile in 6 minutes, and Dan can run the mile in 9 minutes. If Mike gives Dan a head start of 1 minute, how far from the start will Mike pass Dan? (See the figure.) How long does it take?

42. **Range of an Airplane**   An air rescue plane averages 300 miles per hour in still air. It carries enough fuel for 5 hours of flying time. If, upon takeoff, it encounters a head wind of 30 mi/hr, how far can it fly and return safely? (Assume that the wind remains constant.)

43. **Emptying Oil Tankers**   An oil tanker can be emptied by the main pump in 4 hours. An auxiliary pump can empty the tanker in 9 hours. If the main pump is started at 9 AM, when should the auxiliary pump be started so that the tanker is emptied by noon?

44. **Cement Mix**   A 20-pound bag of Economy brand cement mix contains 25% cement and 75% sand. How much pure cement must be added to produce a cement mix that is 40% cement?

45. **Emptying a Tub**   A bathroom tub will fill in 15 minutes with both faucets open and the stopper in place. With both faucets closed and the stopper removed, the tub will empty in 20 minutes. How long will it take for the tub to fill if both faucets are open and the stopper is removed?

46. **Using Two Pumps**   A 5-horsepower (hp) pump can empty a pool in 5 hours. A smaller, 2-hp pump empties the same pool in 8 hours. The pumps are used together to begin emptying this pool. After 2 hours, the 2-hp pump breaks down. How long will it take the larger pump to empty the pool?

47. **Business: Large Order Discounts**   A company charges $200 for each box of tools on orders of 150 or fewer boxes. If a customer orders x boxes in excess of 150, the cost for each box ordered is reduced by x dollars. If a customer's bill came to $30,625, how many boxes were ordered?

48. Constructing a Coffee Can A 39-ounce can of Hills Bros.® coffee requires 188.5 square inches of aluminum. If its height is 7 inches, what is its radius? (The surface area S of a right cylinder is $S = 2\pi r^2 + 2\pi rh$, where r is the radius and h is the height.)

49. Comparing Olympic Heroes In the 1984 Olympics, Carl Lewis of the United States won the gold medal in the 100-meter race with a time of 9.99 seconds. In the 1896 Olympics, Thomas Burke, also of the United States, won the gold medal in the 100-meter race in 12.0 seconds. If they ran in the same race repeating their respective times, by how many meters would Lewis beat Burke?

Discussion and Writing

50. Critical Thinking You are the manager of a clothing store and have just purchased 100 dress shirts for $20.00 each. After 1 month of selling the shirts at the regular price, you plan to have a sale giving 40% off the original selling price. However, you still want to make a profit of $4 on each shirt at the sale price. What should you price the shirts at initially to ensure this? If, instead of 40% off at the sale, you give 50% off, by how much is your profit reduced?

51. Critical Thinking Make up a word problem that requires solving a linear equation as part of its solution. Exchange problems with a friend. Write a critique of your friend's problem.

52. Critical Thinking Without solving, explain what is wrong with the following mixture problem: How many liters of 25% ethanol should be added to 20 liters of 48% ethanol to obtain a solution of 58% ethanol? Now go through an algebraic solution. What happens?

53. Computing Average Speed In going from Chicago to Atlanta, a car averages 45 miles per hour, and in going from Atlanta to Miami, it averages 55 miles per hour. If Atlanta is halfway between Chicago and Miami, what is the average speed from Chicago to Miami? Discuss an intuitive solution. Write a paragraph defending your intuitive solution. Then solve the problem algebraically. Is your intuitive solution the same as the algebraic one? If not, find the flaw.

54. Speed of a Plane On a recent flight from Phoenix to Kansas City, a distance of 919 nautical miles, the plane arrived 20 minutes early. On leaving the aircraft, I asked the captain, "What was our tail wind?" He replied, "I don't know, but our ground speed was 550 knots." How can you determine if enough information is provided to find the tail wind? If possible, find the tail wind. (1 knot = 1 nautical mile per hour)

1.7 Solving Inequalities

PREPARING FOR THIS SECTION *Before getting started, review the following:*

• Inequalities (Section R.2, p. 17)

• Absolute Value (Section R.2, p. 18)

Now work the 'Are You Prepared?' problems on page 159.

OBJECTIVES 1 Use Interval Notation
2 Use Properties of Inequalities
3 Solve Linear Inequalities Algebraically and Graphically
4 Solve Combined Inequalities Algebraically and Graphically
5 Solve Absolute Value Inequalities Algebraically and Graphically

Suppose that a and b are two real numbers and $a < b$. We shall use the notation $a < x < b$ to mean that x is a number *between* a and b. Thus, the expression $a < x < b$ is equivalent to the two inequalities $a < x$ and $x < b$. Similarly, the expression $a \leq x \leq b$ is equivalent to the two inequalities $a \leq x$ and $x \leq b$. The remaining two possibilities, $a \leq x < b$ and $a < x \leq b$, are defined similarly.

Although it is acceptable to write $3 \geq x \geq 2$, it is preferable to reverse the inequality symbols and write instead $2 \leq x \leq 3$ so that, as you read from left to right, the values go from smaller to larger.

A statement such as $2 \leq x \leq 1$ is false because there is no number x for which $2 \leq x$ and $x \leq 1$. Finally, we never mix inequality symbols, as in $2 \leq x \geq 3$.

1 Use Interval Notation

Let a and b represent two real numbers with $a < b$:

A **closed interval**, denoted by $[a, b]$, consists of all real numbers x for which $a \leq x \leq b$.
An **open interval**, denoted by (a, b), consists of all real numbers x for which $a < x < b$.
The **half-open**, or **half-closed, intervals** are $(a, b]$, consisting of all real numbers x for which $a < x \leq b$, and $[a, b)$, consisting of all real numbers x for which $a \leq x < b$.

In each of these definitions, a is called the **left endpoint** and b the **right endpoint** of the interval.

The symbol ∞ (read as "infinity") is not a real number but a notational device used to indicate unboundedness in the positive direction. The symbol $-\infty$ (read as "minus infinity" or "negative infinity") also is not a real number, but a notational device used to indicate unboundedness in the negative direction. Using the symbols ∞ and $-\infty$, we can define five other kinds of intervals:

$[a, \infty)$	consists of all real numbers x for which $x \geq a$
(a, ∞)	consists of all real numbers x for which $x > a$
$(-\infty, a]$	consists of all real numbers x for which $x \leq a$
$(-\infty, a)$	consists of all real numbers x for which $x < a$
$(-\infty, \infty)$	consists of all real numbers x

Note that ∞ and $-\infty$ are never included as endpoints since they are not real numbers.

Table 4 summarizes interval notation, corresponding inequality notation, and their graphs.

Table 4

Interval	Inequality	Graph
The open interval (a, b)	$a < x < b$	
The closed interval $[a, b]$	$a \leq x \leq b$	
The half-open interval $[a, b)$	$a \leq x < b$	
The half-open interval $(a, b]$	$a < x \leq b$	
The interval $[a, \infty)$	$x \geq a$	
The interval (a, ∞)	$x > a$	
The interval $(-\infty, a]$	$x \leq a$	
The interval $(-\infty, a)$	$x < a$	
The interval $(-\infty, \infty)$	All real numbers	

EXAMPLE 1	**Writing Inequalities Using Interval Notation**

Write each inequality using interval notation.

(a) $1 \leq x \leq 3$ (b) $-4 < x < 0$ (c) $x > 5$ (d) $x \leq 1$

Solution (a) $1 \leq x \leq 3$ describes all numbers x between 1 and 3, inclusive. In interval notation, we write $[1, 3]$.

(b) In interval notation, $-4 < x < 0$ is written $(-4, 0)$.

(c) $x > 5$ consists of all numbers x greater than 5. In interval notation, we write $(5, \infty)$.

(d) In interval notation, $x \leq 1$ is written $(-\infty, 1]$. ◀

EXAMPLE 2	**Writing Intervals Using Inequality Notation**

Write each interval as an inequality involving x.

(a) $[1, 4)$ (b) $(2, \infty)$ (c) $[2, 3]$ (d) $(-\infty, -3]$

Solution (a) $[1, 4)$ consists of all numbers x for which $1 \leq x < 4$.

(b) $(2, \infty)$ consists of all numbers x for which $x > 2$.

(c) $[2, 3]$ consists of all numbers x for which $2 \leq x \leq 3$.

(d) $(-\infty, -3]$ consists of all numbers x for which $x \leq -3$. ◀

━━ **NOW WORK PROBLEMS 11, 23, AND 31.**

2 Use Properties of Inequalities

The product of two positive real numbers is positive, the product of two negative real numbers is positive, and the product of 0 and 0 is 0. For any real number a, the value of a^2 is 0 or positive; that is, a^2 is nonnegative. This is called the **nonnegative property**.

In Words
The square of a real number is never negative.

Nonnegative Property

For any real number a,

$$a^2 \geq 0 \tag{1}$$

If we add the same number to both sides of an inequality, we obtain an equivalent inequality. For example, since $3 < 5$, then $3 + 4 < 5 + 4$ or $7 < 9$. This is called the **addition property** of inequalities.

Addition Property of Inequalities

For real numbers $a, b,$ and c,

$$\text{if } a < b, \text{ then } a + c < b + c \tag{2a}$$
$$\text{if } a > b, \text{ then } a + c > b + c \tag{2b}$$

The addition property states that the sense, or direction, of an inequality remains unchanged if the same number is added to each side. Figure 51 illustrates the addition property (2a). In Figure 51(a), we see that a lies to the left of b. If c is positive, then $a + c$ and $b + c$ each lie c units to the right of a and b, respectively. Consequently, $a + c$ must lie to the left of $b + c$; that is, $a + c < b + c$. Figure 51(b) illustrates the situation if c is negative.

Figure 51

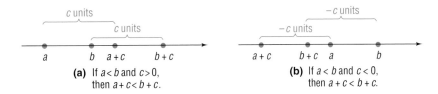

(a) If $a < b$ and $c > 0$,
then $a + c < b + c$.

(b) If $a < b$ and $c < 0$,
then $a + c < b + c$.

Draw an illustration similar to Figure 51 that illustrates the addition property (2b).

EXAMPLE 3   **Addition Property of Inequalities**

(a) If $x < -5$, then $x + 5 < -5 + 5$ or $x + 5 < 0$.
(b) If $x > 2$, then $x + (-2) > 2 + (-2)$ or $x - 2 > 0$. ◀

- - NOW WORK PROBLEM 39.

We will use two examples to arrive at our next property.

EXAMPLE 4   **Multiplying an Inequality by a Positive Number**

Express as an inequality the result of multiplying each side of the inequality $3 < 7$ by 2.

Solution   We begin with

$$3 < 7$$

Multiplying each side by 2 yields the numbers 6 and 14, so we have

$$6 < 14$$ ◀

EXAMPLE 5   **Multiplying an Inequality by a Negative Number**

Express as an inequality the result of multiplying each side of the inequality $9 > 2$ by -4.

Solution   We begin with

$$9 > 2$$

Multiplying each side by -4 yields the numbers -36 and -8, so we have

$$-36 < -8$$ ◀

Note that the effect of multiplying both sides of $9 > 2$ by the negative number -4 is that the direction of the inequality symbol is reversed.

Examples 4 and 5 illustrate the following general **multiplication properties** for inequalities:

Multiplication Properties for Inequalities

For real numbers a, b, and c,

$$\text{if } a < b \text{ and if } c > 0, \text{ then } ac < bc.$$
$$\text{if } a < b \text{ and if } c < 0, \text{ then } ac > bc.$$
$$\text{(3a)}$$
$$\text{if } a > b \text{ and if } c > 0, \text{ then } ac > bc.$$
$$\text{if } a > b \text{ and if } c < 0, \text{ then } ac < bc.$$
$$\text{(3b)}$$

The multiplication properties state that the sense, or direction, of an inequality *remains the same* if each side is multiplied by a *positive* real number, whereas the direction is *reversed* if each side is multiplied by a *negative* real number.

EXAMPLE 6 **Multiplication Property of Inequalities**

(a) If $2x < 6$, then $\dfrac{1}{2}(2x) < \dfrac{1}{2}(6)$ or $x < 3$.

(b) If $\dfrac{x}{-3} > 12$, then $-3\left(\dfrac{x}{-3}\right) < -3(12)$ or $x < -36$.

(c) If $-4x > -8$, then $\dfrac{-4x}{-4} < \dfrac{-8}{-4}$ or $x < 2$.

(d) If $-x < 8$, then $(-1)(-x) > (-1)(8)$ or $x > -8$. ◀

NOW WORK PROBLEM 45.

3 Solve Linear Inequalities Algebraically and Graphically

An **inequality in one variable** is a statement involving two expressions, at least one containing the variable, separated by one of the inequality symbols, $<$, $\leq$, $>$, or $\geq$. To **solve an inequality** means to find all values of the variable for which the statement is true. These values are called **solutions** of the inequality.

For example, the following are all inequalities involving one variable, x:

$$x + 5 < 8, \qquad 2x - 3 \geq 4, \qquad x^2 - 1 \leq 3, \qquad \frac{x+1}{x-2} > 0$$

Two inequalities having exactly the same solution set are called **equivalent inequalities**. As with equations, one method for solving an inequality is to replace it by a series of equivalent inequalities until an inequality with an obvious solution, such as $x < 3$, is obtained. We obtain equivalent inequalities by applying some of the same operations as those used to find equivalent equations. The addition property and the multiplication properties form the basis for the following procedures.

Procedures That Leave the Inequality Symbol Unchanged

1. Simplify both sides of the inequality by combining like terms and eliminating parentheses:

Replace $x + 2 + 6 > 2x + 5(x + 1)$

by $x + 8 > 7x + 5$

2. Add or subtract the same expression on both sides of the inequality:

Replace $3x - 5 < 4$

by $(3x - 5) + 5 < 4 + 5$

3. Multiply or divide both sides of the inequality by the same *positive* expression:

Replace $4x > 16$ by $\dfrac{4x}{4} > \dfrac{16}{4}$

Procedures That Reverse the Sense or Direction of the Inequality Symbol

1. Interchange the two sides of the inequality:

Replace $3 < x$ by $x > 3$

2. Multiply or divide both sides of the inequality by the same *negative* expression:

Replace $-2x > 6$ by $\dfrac{-2x}{-2} < \dfrac{6}{-2}$

To solve an inequality using a graphing utility, we follow these steps:

Steps for Solving Inequalities Graphically

STEP 1: Write the inequality in one of the following forms:

$$Y_1 < Y_2, \qquad Y_1 > Y_2, \qquad Y_1 \le Y_2, \qquad Y_1 \ge Y_2$$

STEP 2: Graph Y_1 and Y_2 on the same screen.

STEP 3: If the inequality is of the form $Y_1 < Y_2$, determine on what interval Y_1 is below Y_2.

If the inequality is of the form $Y_1 > Y_2$, determine on what interval Y_1 is above Y_2.

If the inequality is not strict ($\le$ or $\ge$), include the x-coordinates of the points of intersection in the solution.

As the examples that follow illustrate, we solve inequalities using many of the same steps that we would use to solve equations. In writing the solution of an inequality, we may use either set notation or interval notation, whichever is more convenient.

| EXAMPLE 7 | Solving an Inequality |

Solve the inequality $4x + 7 \geq 2x - 3$, and graph the solution set.

Algebraic Solution

$$4x + 7 \geq 2x - 3$$

$4x + 7 - 7 \geq 2x - 3 - 7$ Subtract 7 from both sides.

$4x \geq 2x - 10$ Simplify.

$4x - 2x \geq 2x - 10 - 2x$ Subtract 2x from both sides.

$2x \geq -10$ Simplify.

$\dfrac{2x}{2} \geq \dfrac{-10}{2}$ Divide both sides by 2. (The direction of the inequality symbol is unchanged.)

$x \geq -5$ Simplify.

The solution set is $\{x \mid x \geq -5\}$ or, using interval notation, all numbers in the interval $[-5, \infty)$.

Figure 53

See Figure 53 for the graph of the solution set.

Graphing Solution

We graph $Y_1 = 4x + 7$ and $Y_2 = 2x - 3$ on the same screen. See Figure 52. Using the INTERSECT command, we find that Y_1 and Y_2 intersect at $x = -5$. The graph of Y_1 is above that of Y_2, $Y_1 > Y_2$, to the right of the point of intersection. Since the inequality is not strict, the solution set is $\{x \mid x \geq -5\}$ or, using interval notation, $[-5, \infty)$.

Figure 52

NOW WORK PROBLEM 53.

4 Solve Combined Inequalities Algebraically and Graphically

Now let's look at how to solve combined inequalities.

| EXAMPLE 8 | Solving Combined Inequalities |

Solve the inequality $-5 < 3x - 2 < 1$ and draw a graph to illustrate the solution.

Algebraic Solution

Recall that the inequality

$$-5 < 3x - 2 < 1$$

is equivalent to the two inequalities

$$-5 < 3x - 2 \quad \text{and} \quad 3x - 2 < 1$$

We will solve each of these inequalities separately.

$-5 < 3x - 2$ $3x - 2 < 1$

$-5 + 2 < 3x - 2 + 2$ Add 2 to both sides. $3x - 2 + 2 < 1 + 2$

$-3 < 3x$ Simplify. $3x < 3$

$\dfrac{-3}{3} < \dfrac{3x}{3}$ Divide both sides by 3. $\dfrac{3x}{3} < \dfrac{3}{3}$

$-1 < x$ Simplify. $x < 1$

The solution set of the original pair of inequalities consists of all x for which

$$-1 < x \quad \text{and} \quad x < 1$$

This may be written more compactly as $\{x \mid -1 < x < 1\}$. In interval notation, the solution is $(-1, 1)$.

Graphing Solution

To solve a combined inequality, we graph each part: $Y_1 = -5$, $Y_2 = 3x - 2$, and $Y_3 = 1$. We seek the values of x for which the graph of Y_2 is between the graphs of Y_1 and Y_3. See Figure 54. The point of intersection of Y_1 and Y_2 is $(-1, -5)$, and the point of intersection of Y_2 and Y_3 is $(1, 1)$. The inequality is true for all values of x between these two intersection points. Since the inequality is strict, the solution set is $\{x \mid -1 < x < 1\}$ or, using interval notation, $(-1, 1)$.

Figure 54

Figure 55

See Figure 55 for the graph of the solution set. ◄

We observe in the algebraic solution of Example 8 that the two inequalities that were solved required exactly the same steps. A shortcut to solving the original inequality algebraically is to deal with the two inequalities at the same time, as follows:

$$-5 < \quad 3x - 2 < 1$$
$$-5 + 2 < 3x - 2 + 2 < 1 + 2 \qquad \text{Add 2 to each part.}$$
$$-3 < \quad 3x \quad < 3 \qquad \text{Simplify.}$$
$$\frac{-3}{3} < \quad \frac{3x}{3} \quad < \frac{3}{3} \qquad \text{Divide each part by 3.}$$
$$-1 < \quad x \quad < 1 \qquad \text{Simplify.}$$

NOW WORK PROBLEM 73.

5 Solve Absolute Value Inequalities Algebraically and Graphically

Let's look at an inequality involving absolute value.

EXAMPLE 9	**Solving an Inequality Involving Absolute Value**

Solve the inequality: $|x| < 4$

Algebraic Solution

We are looking for all points whose coordinate x is a distance less than 4 units from the origin. See Figure 56 for an illustration.

Figure 56

Less than 4 units from origin

Because any x between -4 and 4 satisfies the condition $|x| < 4$, the solution set consists of all numbers x for which $-4 < x < 4$, that is, all x in the interval $(-4, 4)$. ◄

Graphing Solution

We graph $Y_1 = |x|$ and $Y_2 = 4$ on the same screen. See Figure 57. Using the INTERSECT command (twice), we find that Y_1 and Y_2 intersect at $x = -4$ and at $x = 4$. The graph of Y_1 is below that of Y_2, $Y_1 < Y_2$, between the points of intersection. Since the inequality is strict, the solution set is $\{x \mid -4 < x < 4\}$ or, using interval notation, $(-4, 4)$.

Figure 57

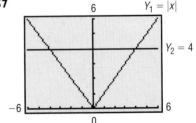

We are led to the following results:

Inequalities Involving Absolute Value

If a is any positive number and if u is any algebraic expression, then

$\|u\| < a$	is equivalent to	$-a < u < a$	**(4)**
$\|u\| \le a$	is equivalent to	$-a \le u \le a$	**(5)**

In other words, $|u| < a$ is equivalent to $-a < u$ and $u < a$.

EXAMPLE 10 Solving an Inequality Involving Absolute Value

Solve the inequality $|2x + 4| \leq 3$, and graph the solution set.

Algebraic Solution

$$|2x + 4| \leq 3$$

This follows the form of statement (5); the expression $u = 2x + 4$ is inside the absolute value bars.

$$-3 \leq 2x + 4 \leq 3$$

Apply statement (5).

$$-3 - 4 \leq 2x + 4 - 4 \leq 3 - 4$$

Subtract 4 from each part.

$$-7 \leq 2x \leq -1$$

Simplify.

$$\frac{-7}{2} \leq \frac{2x}{2} \leq \frac{-1}{2}$$

Divide each part by 2.

$$-\frac{7}{2} \leq x \leq -\frac{1}{2}$$

Simplify.

The solution set is $\left\{ x \middle| -\frac{7}{2} \leq x \leq -\frac{1}{2} \right\}$, that is, all x in the interval $\left[-\frac{7}{2}, -\frac{1}{2} \right]$.

Figure 59

See Figure 59 for a graph of the solution set.

Graphing Solution

We graph $Y_1 = |2x + 4|$ and $Y_2 = 3$ on the same screen. See Figure 58. Using the IN-TERSECT command (twice), we find that Y_1 and Y_2 intersect at $x = -3.5$ and at $x = -0.5$. The graph of Y_1 is below that of Y_2, $Y_1 < Y_2$, between the points of intersection. Since the inequality is not strict, the solution set is $\{x | -3.5 \leq x \leq -0.5\}$ or, using interval notation, $[-3.5, -0.5]$.

Figure 58

$Y_1 = |2x + 4|$

$Y_2 = 3$

NOW WORK PROBLEM **91**.

EXAMPLE 11 Solving an Inequality Involving Absolute Value

Solve the inequality $|x| > 3$.

Algebraic Solution

We are looking for all points whose coordinate x is a distance greater than 3 units from the origin. Figure 60 illustrates the situation.

Figure 60

We conclude that any x less than -3 or greater than 3 satisfies the condition $|x| > 3$. Consequently, the solution set consists of all numbers x for which $x < -3$ or $x > 3$, that is, all x in the intervals $(-\infty, -3)$ or $(3, \infty)$. ◄

Graphing Solution

We graph $Y_1 = |x|$ and $Y_2 = 3$ on the same screen. See Figure 61. Using the INTERSECT command (twice), we find that Y_1 and Y_2 intersect at $x = -3$ and at $x = 3$. The graph of Y_1 is above that of Y_2, $Y_1 > Y_2$, to the left of $x = -3$ and to the right of $x = 3$. Since the inequality is strict, the solution set is $\{x | x < -3 \text{ or } x > 3\}$. Using interval notation, the solution is $(-\infty, -3)$ or $(3, \infty)$.

Figure 61

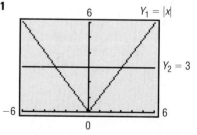

$Y_1 = |x|$

$Y_2 = 3$

WARNING
A common error to be avoided is to attempt to write the solution $x < 1$ or $x > 4$ as $1 > x > 4$, which is incorrect, since there are no numbers x for which $1 > x$ and $x > 4$. Another common error is to "mix" the symbols and write $1 < x > 4$, which makes no sense. ∎

Inequalities Involving Absolute Value

If a is any positive number and u is any algebraic expression, then

$\lvert u \rvert > a$	is equivalent to	$u < -a$ or $u > a$ **(6)**
$\lvert u \rvert \geq a$	is equivalent to	$u \leq -a$ or $u \geq a$ **(7)**

EXAMPLE 12	Solving an Inequality Involving Absolute Value

Solve the inequality $\lvert 2x - 5 \rvert > 3$, and graph the solution set.

Algebraic Solution

$\lvert 2x - 5 \rvert > 3$ This follows the form of statement (6); the expression $u = 2x - 5$ is inside the absolute value bars.

$$2x - 5 < -3 \quad \text{or} \quad 2x - 5 > 3 \qquad \text{Apply statement (6).}$$

$$2x - 5 + 5 < -3 + 5 \quad \text{or} \quad 2x - 5 + 5 > 3 + 5 \quad \text{Add 5 to each part.}$$

$$2x < 2 \quad \text{or} \quad 2x > 8 \qquad \text{Simplify.}$$

$$\frac{2x}{2} < \frac{2}{2} \quad \text{or} \quad \frac{2x}{2} > \frac{8}{2} \qquad \text{Divide each part by 2.}$$

$$x < 1 \quad \text{or} \quad x > 4 \qquad \text{Simplify.}$$

The solution set is $\{x \mid x < 1 \text{ or } x > 4\}$, that is, all x in the intervals $(-\infty, 1)$ or $(4, \infty)$.

Graphing Solution

We graph $Y_1 = \lvert 2x - 5 \rvert$ and $Y_2 = 3$ on the same screen. See Figure 62. Using the IN-TERSECT command (twice), we find that Y_1 and Y_2 intersect at $x = 1$ and at $x = 4$. The graph of Y_1 is above that of Y_2, $Y_1 > Y_2$, to the left of $x = 1$ and to the right of $x = 4$. Since the inequality is strict, the solution set is $\{x \mid x < 1 \text{ or } x > 4\}$. Using interval notation, the solution is $(-\infty, 1)$ or $(4, \infty)$.

Figure 62

Figure 63

See Figure 63 for a graph of the solution set. ◀

NOW WORK PROBLEM 95.

1.7 Assess Your Understanding

'Are You Prepared?'

Answers are given at the end of these exercises. If you get a wrong answer, read the pages listed in red.

1. Graph the inequality: $x \geq -2$. (p. 17)

2. *True or False:* The absolute value of a negative number is positive. (p. 18)

Concepts and Vocabulary

3. If each side of an inequality is multiplied by a(n) _____ number, then the sense of the inequality symbol is reversed.

4. A(n) _____ _____, denoted $[a, b]$, consists of all real numbers x for which $a \leq x \leq b$.

5. The _____ _____ state that the sense, or direction, of an inequality remains the same if each side is multiplied by a positive number, while the direction is reversed if each side is multiplied by a negative number.

In Problems 6–9, determine if the statement is True or False if $a < b$ and $c < 0$.

6. $a + c < b + c$ **7.** $a - c < b - c$ **8.** $ac > bc$ **9.** $\dfrac{a}{c} < \dfrac{b}{c}$

10. *True or False:* The square of any real number is always nonnegative.

Skill Building

In Problems 11–16, express the graph shown in color using interval notation. Also express each as an inequality involving x.

11.

12.

13.

14.

15.

16.

In Problems 17–22, an inequality is given. Write the inequality obtained by:
(a) *Adding 3 to each side of the given inequality.*   (b) *Subtracting 5 from each side of the given inequality.*
(c) *Multiplying each side of the given inequality by 3.*   (d) *Multiplying each side of the given inequality by −2.*

17. $3 < 5$ **18.** $2 > 1$ **19.** $4 > -3$

20. $-3 > -5$ **21.** $2x + 1 < 2$ **22.** $1 - 2x > 5$

In Problems 23–30, write each inequality using interval notation, and illustrate each inequality using the real number line.

23. $0 \le x \le 4$ **24.** $-1 < x < 5$ **25.** $4 \le x < 6$ **26.** $-2 < x < 0$

27. $x \ge 4$ **28.** $x \le 5$ **29.** $x < -4$ **30.** $x > 1$

In Problems 31–38, write each interval as an inequality involving x, and illustrate each inequality using the real number line.

31. $[2, 5]$ **32.** $(1, 2)$ **33.** $(-3, -2)$ **34.** $[0, 1)$

35. $[4, \infty)$ **36.** $(-\infty, 2]$ **37.** $(-\infty, -3)$ **38.** $(-8, \infty)$

In Problems 39–52, fill in the blank with the correct inequality symbol.

39. If $x < 5$, then $x - 5$ _____ 0.

40. If $x < -4$, then $x + 4$ _____ 0.

41. If $x > -4$, then $x + 4$ _____ 0.

42. If $x > 6$, then $x - 6$ _____ 0.

43. If $x \ge -4$, then $3x$ _____ -12.

44. If $x \le 3$, then $2x$ _____ 6.

45. If $x > 6$, then $-2x$ _____ -12.

46. If $x > -2$, then $-4x$ _____ 8.

47. If $x \ge 5$, then $-4x$ _____ -20.

48. If $x \le -4$, then $-3x$ _____ 12.

49. If $2x < 6$, then x _____ 3.

50. If $3x \le 12$, then x _____ 4.

51. If $-\dfrac{1}{2}x \le 3$, then x _____ -6.

52. If $-\dfrac{1}{4}x > 1$, then x _____ -4.

In Problems 53–106, solve each inequality algebraically. Express your answer using set notation or interval notation. Graph the solution set. Verify your results using a graphing utility.

53. $x + 1 < 5$ **54.** $x - 6 < 1$ **55.** $1 - 2x \le 3$

56. $2 - 3x \le 5$ **57.** $3x - 7 > 2$ **58.** $2x + 5 > 1$

59. $3x - 1 \ge 3 + x$ **60.** $2x - 2 \ge 3 + x$ **61.** $-2(x + 3) < 8$

62. $-3(1 - x) < 12$ **63.** $4 - 3(1 - x) \le 3$ **64.** $8 - 4(2 - x) \le -2x$

65. $\dfrac{1}{2}(x - 4) > x + 8$ **66.** $3x + 4 > \dfrac{1}{3}(x - 2)$ **67.** $\dfrac{x}{2} \ge 1 - \dfrac{x}{4}$

68. $\dfrac{x}{3} \ge 2 + \dfrac{x}{6}$ **69.** $0 \le 2x - 6 \le 4$ **70.** $4 \le 2x + 2 \le 10$

71. $-5 \le 4 - 3x \le 2$ **72.** $-3 \le 3 - 2x \le 9$ **73.** $-3 < \dfrac{2x - 1}{4} < 0$

74. $0 < \dfrac{3x + 2}{2} < 4$

75. $1 < 1 - \dfrac{1}{2}x < 4$

76. $0 < 1 - \dfrac{1}{3}x < 1$

77. $(x + 2)(x - 3) > (x - 1)(x + 1)$

78. $(x - 1)(x + 1) > (x - 3)(x + 4)$

79. $x(4x + 3) \le (2x + 1)^2$

80. $x(9x - 5) \le (3x - 1)^2$

81. $\dfrac{1}{2} \le \dfrac{x + 1}{3} < \dfrac{3}{4}$

82. $\dfrac{1}{3} < \dfrac{x + 1}{2} \le \dfrac{2}{3}$

83. $|x| < 6$

84. $|x| < 9$

85. $|x| > 4$

86. $|x| > 1$

87. $|2x| < 8$

88. $|3x| < 15$

89. $|3x| > 12$

90. $|2x| > 6$

91. $|x - 2| + 2 < 3$

92. $|x + 4| + 3 < 5$

93. $|3t - 2| \le 4$

94. $|2u + 5| \le 7$

95. $|x - 3| \ge 2$

96. $|x + 4| \ge 2$

97. $|1 - 4x| - 7 < -2$

98. $|1 - 2x| - 4 < -1$

99. $|1 - 2x| > |-3|$

100. $|2 - 3x| > |-1|$

101. $|2x + 1| < -1$

102. $|3x - 4| \ge 0$

103. $-3 < x + 5 < 2x$

104. $2 < x - 3 < 2x$

105. $x + 2 < 2x - 1 < 5x$

106. $2x - 1 < 3x + 5 < 5x - 7$

Applications and Extensions

107. Express the fact that x differs from 2 by less than $\dfrac{1}{2}$ as an inequality involving an absolute value. Solve for x.

108. Express the fact that x differs from -1 by less than 1 as an inequality involving an absolute value. Solve for x.

109. Express the fact that x differs from -3 by more than 2 as an inequality involving an absolute value. Solve for x.

110. Express the fact that x differs from 2 by more than 3 as an inequality involving an absolute value. Solve for x.

111. A young adult may be defined as someone older than 21, but less than 30 years of age. Express this statement using inequalities.

112. Middle-aged may be defined as being 40 or more and less than 60. Express this statement using inequalities.

113. Body Temperature Normal human body temperature is 98.6°F. If a temperature x that differs from normal by at least 1.5° is considered unhealthy, write the condition for an unhealthy temperature x as an inequality involving an absolute value, and solve for x.

114. Household Voltage In the United States, normal household voltage is 115 volts. However, it is not uncommon for actual voltage to differ from normal voltage by at most 5 volts. Express this situation as an inequality involving an absolute value. Use x as the actual voltage and solve for x.

115. Life Expectancy Metropolitan Life Insurance Co. reported that an average 25-year-old male in 2000 could expect to live at least 50.6 more years, and an average 25-year-old female in 2000 could expect to live at least 55.4 more years.

(a) To what age can an average 25-year-old male expect to live? Express your answer as an inequality.

(b) To what age can an average 25-year-old female expect to live? Express your answer as an inequality.

(c) Who can expect to live longer, a male or a female? By how many years?

116. General Chemistry For a certain ideal gas, the volume V (in cubic centimeters) equals 20 times the temperature T in kelvins (K). If the temperature varies from 353 to 393 K, inclusive, what is the corresponding range of the volume of the gas?

117. Real Estate A real estate agent agrees to sell a large apartment complex according to the following commission schedule: $45,000 plus 25% of the selling price in excess of $900,000. Assuming that the complex will sell at some price between $900,000 and $1,100,000, inclusive, over what range does the agent's commission vary? How does the commission vary as a percent of selling price?

118. Sales Commission A used car salesperson is paid a commission of $25 plus 40% of the selling price in excess of owner's cost. The owner claims that used cars typically sell for at least owner's cost plus $70 and at most owner's cost plus $300. For each sale made, over what range can the salesperson expect the commission to vary?

119. Federal Tax Withholding The percentage method of withholding for federal income tax (2004) states that a single person whose weekly wages, after subtracting withholding allowances, are over $592, but not over $1317, shall have $74.35 plus 25% of the excess over $592 withheld. Over what range does the amount withheld vary if the weekly wages vary from $600 to $700, inclusive?
SOURCE: *Employer's Tax Guide*. Department of the Treasury, Internal Revenue Service, 2004.

120. Federal Tax Withholding Rework Problem 119 if the weekly wages vary from $800 to $900, inclusive.

121. Electricity Rates Commonwealth Edison Company's summer charge for electricity is $8.275¢ per kilowatt-hour. In addition, each monthly bill contains a customer charge of $10.07. If last summer's bills ranged from a low of $65.96 to a high of $217.02, over what range did usage vary (in kilowatt-hours)?
SOURCE: Commonwealth Edison Co., Chicago, Illinois, 2004.

122. Water Bills The Village of Oak Lawn charges homeowners $27.18 per quarter-year plus $1.90 per 1000 gallons for water usage in excess of 12,000 gallons. In 2004, one homeowner's quarterly bill ranged from a high of $43.33 to a low of $30.03. Over what range did water usage vary?
SOURCE: Village of Oak Lawn, Illinois, 2004.

123. Markup of a New Car The markup over dealer's cost of a new car ranges from 12% to 18%. If the sticker price is $8800, over what range will the dealer's cost vary?

124. IQ Tests A standard intelligence test has an average score of 100. According to statistical theory, of the people who take the test, the 2.5% with the highest scores will have scores of more than 1.96σ above the average, where σ (sigma, a number called the *standard deviation*) depends on the nature of the test. If $\sigma = 12$ for this test and there is (in principle) no upper limit to the score possible on the test, write the interval of possible test scores of the people in the top 2.5%.

125. Computing Grades In your Economics 101 class, you have scores of 68, 82, 87, and 89 on the first four of five tests. To get a grade of B, the average of the first five test scores must

be greater than or equal to 80 and less than 90. Solve an inequality to find the range of the score that you need on the last test to get a B.

What do I need to get a B?

126. Computing Grades Repeat Problem 125 if the fifth test counts double.

127. A car that averages 25 miles per gallon has a tank that holds 20 gallons of gasoline. After a trip that covered at least 300 miles, the car ran out of gasoline. What is the range of the amount of gasoline (in gallons) that was in the tank at the start of the trip?

128. Repeat Problem 127 if the same car runs out of gasoline after a trip of no more than 250 miles.

129. Arithmetic Mean If $a < b$, show that $a < \dfrac{a+b}{2} < b$. The number $\dfrac{a+b}{2}$ is called the **arithmetic mean** of a and b.

130. Refer to Problem 129. Show that the arithmetic mean of a and b is equidistant from a and b.

131. Geometric Mean If $0 < a < b$, show that $a < \sqrt{ab} < b$. The number $\sqrt{ab}$ is called the **geometric mean** of a and b.

132. Refer to Problems 129 and 131. Show that the geometric mean of a and b is less than the arithmetic mean of a and b.

133. Harmonic Mean For $0 < a < b$, let h be defined by

$$\frac{1}{h} = \frac{1}{2}\left(\frac{1}{a} + \frac{1}{b}\right)$$

Show that $a < h < b$. The number h is called the **harmonic mean** of a and b.

134. Refer to Problems 129, 131, and 133. Show that the harmonic mean of a and b equals the geometric mean squared divided by the arithmetic mean.

Discussion and Writing

135. Make up an inequality that has no solution. Make up one that has exactly one solution.

136. How would you explain to a fellow student the underlying reason for the multiplication properties for inequalities (page 154); that is, the sense or direction of an inequality remains the same if each side is multiplied by a positive real number, while the direction is reversed if each side is multiplied by a negative real number.

137. The inequality $x^2 + 1 < -5$ has no solution. Explain why.

138. Do you prefer to use inequality notation or interval notation to express the solution to an inequality? Give your reasons. Are there particular circumstances when you prefer one to the other? Cite examples.

'Are You Prepared?' Answers

1.
$$\begin{array}{ccc} \hline -4 & -2 & 0 \end{array}$$

2. True

1.8 Lines

OBJECTIVES 1 Calculate and Interpret the Slope of a Line
2 Graph Lines Given a Point and the Slope
3 Find the Equation of a Vertical Line
4 Use the Point–Slope Form of a Line; Identify Horizontal Lines
5 Find the Equation of a Line Given Two Points
6 Write the Equation of a Line in Slope–Intercept Form
7 Identify the Slope and y-Intercept of a Line from Its Equation
8 Graph Lines Written in General Form Using Intercepts
9 Find Equations of Parallel Lines
10 Find Equations of Perpendicular Lines

In this section we study a certain type of equation that contains two variables, called a *linear equation*, and its graph, a *line*.

1 **Calculate and Interpret the Slope of a Line**

Consider the staircase illustrated in Figure 64. Each step contains exactly the same horizontal **run** and the same vertical **rise**. The ratio of the rise to the run, called the *slope*, is a numerical measure of the steepness of the staircase. For example, if the run is increased and the rise remains the same, the staircase becomes less steep. If the run is kept the same, but the rise is increased, the staircase becomes more steep. This important characteristic of a line is best defined using rectangular coordinates.

Figure 64

Let $P = (x_1, y_1)$ and $Q = (x_2, y_2)$ be two distinct points. If $x_1 \neq x_2$, the **slope m** of the nonvertical line L containing P and Q is defined by the formula

$$m = \frac{y_2 - y_1}{x_2 - x_1}, \qquad x_1 \neq x_2 \qquad\qquad \textbf{(1)}$$

If $x_1 = x_2$, L is a **vertical line** and the slope m of L is **undefined** (since this results in division by 0).

Figure 65(a) provides an illustration of the slope of a nonvertical line; Figure 65(b) illustrates a vertical line.

Figure 65

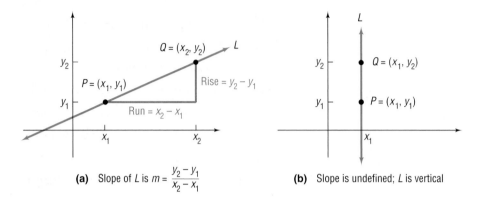

(a) Slope of L is $m = \dfrac{y_2 - y_1}{x_2 - x_1}$

(b) Slope is undefined; L is vertical

As Figure 65(a) illustrates, the slope m of a nonvertical line may be viewed as

$$m = \frac{y_2 - y_1}{x_2 - x_1} = \frac{\text{Rise}}{\text{Run}}$$

We can also express the slope m of a nonvertical line as

$$m = \frac{y_2 - y_1}{x_2 - x_1} = \frac{\text{Change in } y}{\text{Change in } x} = \frac{\Delta y}{\Delta x}$$

That is, the slope m of a nonvertical line L measures the amount that y changes as x changes from x_1 to x_2. This is called the **average rate of change** of y with respect to x.

Two comments about computing the slope of a nonvertical line may prove helpful:

1. Any two distinct points on the line can be used to compute the slope of the line. (See Figure 66 for justification.)

Figure 66
Triangles ABC and PQR are similar (equal angles). Hence, ratios of corresponding sides are proportional so that

Slope using P and $Q = \dfrac{y_2 - y_1}{x_2 - x_1}$

$= \dfrac{d(B, C)}{d(A, C)} = $ Slope using A and B

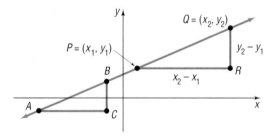

2. The slope of a line may be computed from $P = (x_1, y_1)$ to $Q = (x_2, y_2)$ or from Q to P because

$$\frac{y_2 - y_1}{x_2 - x_1} = \frac{y_1 - y_2}{x_1 - x_2}$$

EXAMPLE 1

Finding and Interpreting the Slope of a Line Containing Two Points

The slope m of the line containing the points $(1, 2)$ and $(5, -3)$ may be computed as

$$m = \frac{-3 - 2}{5 - 1} = \frac{-5}{4} = -\frac{5}{4} \quad \text{or as} \quad m = \frac{2 - (-3)}{1 - 5} = \frac{5}{-4} = -\frac{5}{4}$$

For every 4-unit change in x, y will change by -5 units. That is, if x increases by 4 units, then y will decrease by 5 units. The average rate of change of y with respect to x is $-\dfrac{5}{4}$.   ◀

NOW WORK PROBLEMS 7 AND 13.

Square Screens

To get an undistorted view of slope, the same scale must be used on each axis. However, most graphing utilities have a rectangular screen. Because of this, using the same interval for both x and y will result in a distorted view. For example, Figure 67 shows the graph of the line $y = x$ connecting the points $(-4, -4)$ and $(4, 4)$. We expect the line to bisect the first and third quadrants, but it doesn't. We need to adjust the selections for Xmin, Xmax, Ymin, and Ymax so that a **square screen** results. On most graphing utilities, this is accomplished by setting the ratio of x to y at $3:2$.*

Figure 68 shows the graph of the line $y = x$ on a square screen using a TI-84 Plus. Notice that the line now bisects the first and third quadrants. Compare this illustration to Figure 67.

To get a better idea of the meaning of the slope m of a line, consider the following:

Figure 67

Figure 68

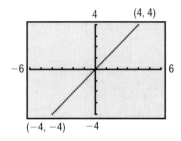

──── Seeing the Concept ────

On the same square screen, graph the following equations:

$Y_1 = 0$ *Slope of line is 0.*

$Y_2 = \dfrac{1}{4}x$ *Slope of line is $\dfrac{1}{4}$.*

$Y_3 = \dfrac{1}{2}x$ *Slope of line is $\dfrac{1}{2}$.*

$Y_4 = x$ *Slope of line is 1.*

$Y_5 = 2x$ *Slope of line is 2.*

$Y_6 = 6x$ *Slope of line is 6.*

See Figure 69.

Figure 69

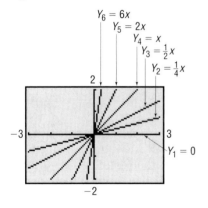

──── Seeing the Concept ────

On the same square screen, graph the following equations:

$Y_1 = 0$ *Slope of line is 0.*

$Y_2 = -\dfrac{1}{4}x$ *Slope of line is $-\dfrac{1}{4}$.*

$Y_3 = -\dfrac{1}{2}x$ *Slope of line is $-\dfrac{1}{2}$.*

$Y_4 = -x$ *Slope of line is -1.*

$Y_5 = -2x$ *Slope of line is -2.*

$Y_6 = -6x$ *Slope of line is -6.*

See Figure 70.

Figure 70

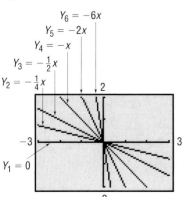

*Most graphing utilities have a feature that automatically squares the viewing window. Consult your owner's manual for the appropriate keystrokes.

Figures 69 and 70 illustrate the following facts:

1. When the slope of a line is positive, the line slants upward from left to right.

2. When the slope of a line is negative, the line slants downward from left to right.

3. When the slope is 0, the line is horizontal.

Figures 69 and 70 also illustrate that the closer the line is to the vertical position, the greater the magnitude of the slope.

2 Graph Lines Given a Point and the Slope

The next example illustrates how the slope of a line can be used to graph the line.

EXAMPLE 2 **Graphing a Line Given a Point and a Slope**

Draw a graph of the line that contains the point $(3, 2)$ and has a slope of:

(a) $\dfrac{3}{4}$ (b) $-\dfrac{4}{5}$

Solution (a) Slope $= \dfrac{\text{Rise}}{\text{Run}}$. The fact that the slope is $\dfrac{3}{4}$ means that for every horizontal movement (run) of 4 units to the right there will be a vertical movement (rise) of 3 units. If we start at the given point $(3, 2)$ and move 4 units to the right and 3 units up, we reach the point $(7, 5)$. By drawing the line through this point and the point $(3, 2)$, we have the graph. See Figure 71.

(b) The fact that the slope is

$$-\frac{4}{5} = \frac{-4}{5} = \frac{\text{Rise}}{\text{Run}}$$

means that for every horizontal movement of 5 units to the right there will be a corresponding vertical movement of -4 units (a downward movement). If we start at the given point $(3, 2)$ and move 5 units to the right and then 4 units down, we arrive at the point $(8, -2)$. By drawing the line through these points, we have the graph. See Figure 72.

Figure 71

Slope $= \dfrac{3}{4}$

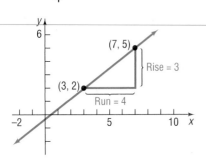

Figure 72

Slope $= -\dfrac{4}{5}$

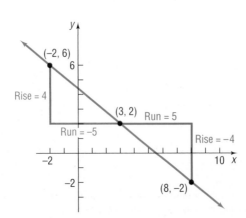

Alternatively, we can set

$$-\frac{4}{5} = \frac{4}{-5} = \frac{\text{Rise}}{\text{Run}}$$

so that for every horizontal movement of -5 units (a movement to the left) there will be a corresponding vertical movement of 4 units (upward). This approach brings us to the point $(-2, 6)$, which is also on the graph shown in Figure 72. ◀

NOW WORK PROBLEM 19.

3 Find the Equation of a Vertical Line

Now that we have discussed the slope of a line, we are ready to derive equations of lines. As we shall see, there are several forms of the equation of a line. Let's start with an example.

| EXAMPLE 3 | **Graphing a Line** |

Graph the equation: $x = 3$

Solution To graph $x = 3$ by hand, recall that we are looking for all points (x, y) in the plane for which $x = 3$. No matter what y-coordinate is used, the corresponding x-coordinate always equals 3. Consequently, the graph of the equation $x = 3$ is a vertical line with x-intercept 3 and undefined slope. See Figure 73(a).

To use a graphing utility, we need to express the equation in the form $y = \{\text{expression in } x\}$. But $x = 3$ cannot be put into this form, so an alternative method must be used. Consult your manual to determine the methodology required to draw vertical lines. Figure 73(b) shows the graph that you should obtain.

Figure 73
$x = 3$

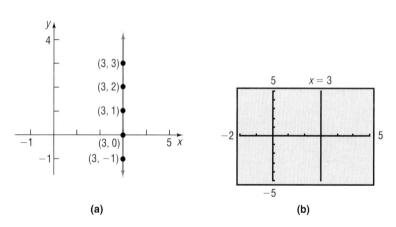

(a)   (b)

As suggested by Example 3, we have the following result:

Theorem **Equation of a Vertical Line**

A vertical line is given by an equation of the form

$$x = a$$

where a is the x-intercept.

Figure 74

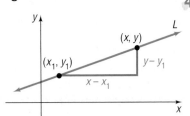

4 Use the Point–Slope Form of a Line; Identify Horizontal Lines

Now let L be a nonvertical line with slope m and containing the point (x_1, y_1). See Figure 74. For any other point (x, y) on L, we have

$$m = \frac{y - y_1}{x - x_1} \quad \text{or} \quad y - y_1 = m(x - x_1)$$

Theorem

Point–Slope Form of an Equation of a Line

An equation of a nonvertical line of slope m that contains the point (x_1, y_1) is

$$y - y_1 = m(x - x_1) \qquad \textbf{(2)}$$

EXAMPLE 4

Using the Point–Slope Form of a Line

An equation of the line with slope 4 and containing the point $(1, 2)$ can be found by using the point–slope form with $m = 4$, $x_1 = 1$, and $y_1 = 2$.

$$y - y_1 = m(x - x_1)$$
$$y - 2 = 4(x - 1) \qquad m = 4, x_1 = 1, y_1 = 2$$
$$y = 4x - 2$$

See Figure 75 for the graph.

Figure 75

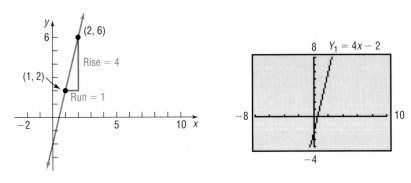

EXAMPLE 5

Finding the Equation of a Horizontal Line

Find an equation of the horizontal line containing the point $(3, 2)$.

Solution

Because all the y-values are equal on a horizontal line, the slope of a horizontal line is 0. To get an equation, we use the point–slope form with $m = 0$, $x_1 = 3$, and $y_1 = 2$.

$$y - y_1 = m(x - x_1)$$
$$y - 2 = 0 \cdot (x - 3) \qquad m = 0, x_1 = 3, \text{and } y_1 = 2$$
$$y - 2 = 0$$
$$y = 2$$

See Figure 76 for the graph.

Figure 76
$y = 2$

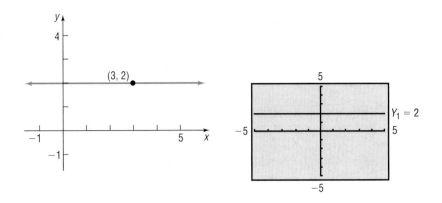

As suggested by Example 5, we have the following result:

Theorem   **Equation of a Horizontal Line**

A horizontal line is given by an equation of the form

$$y = b$$

where b is the y-intercept.

5   Find the Equation of a Line Given Two Points

We use the slope formula and the point–slope form of a line to find the equation given two points.

| **EXAMPLE 6** | **Finding an Equation of a Line Given Two Points** |

Find an equation of the line L containing the points $(2, 3)$ and $(-4, 5)$. Graph the line L.

Solution   We first compute the slope of the line.

$$m = \frac{5 - 3}{-4 - 2} = \frac{2}{-6} = -\frac{1}{3}$$

Figure 77

We use the point $(2, 3)$ and the slope $m = -\frac{1}{3}$ to get the point–slope form of the equation of the line.

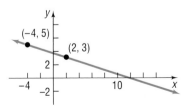

$$y - 3 = -\frac{1}{3}(x - 2)$$

See Figure 77 for the graph.   ◄

In the solution to Example 6, we could have used the other point, $(-4, 5)$, instead of the point $(2, 3)$. The equation that results, although it looks different, is equivalent to the equation that we obtained in the example. (Try it for yourself.)

✏ **NOW WORK PROBLEM 33.**

6 Write the Equation of a Line in Slope–Intercept Form

Another useful equation of a line is obtained when the slope m and y-intercept b are known. In this event, we know both the slope m of the line and a point $(0, b)$ on the line; thus, we may use the point–slope form, equation (2), to obtain the following equation:

$$y - b = m(x - 0) \quad \text{or} \quad y = mx + b$$

Theorem

Slope–Intercept Form of an Equation of a Line

An equation of a line L with slope m and y-intercept b is

$$y = mx + b \tag{3}$$

Figure 78 $y = mx + 2$

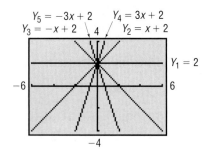

$Y_5 = -3x + 2 \quad Y_4 = 3x + 2$
$Y_3 = -x + 2 \quad Y_2 = x + 2$
$Y_1 = 2$

───── **Seeing the Concept** ─────

To see the role that the slope m plays, graph the following lines on the same square screen.

$$Y_1 = 2$$
$$Y_2 = x + 2$$
$$Y_3 = -x + 2$$
$$Y_4 = 3x + 2$$
$$Y_5 = -3x + 2$$

See Figure 78. What do you conclude about the lines $y = mx + 2$?

Figure 79 $y = 2x + b$

$Y_2 = 2x + 1$
$Y_4 = 2x + 4$
$Y_1 = 2x$
$Y_3 = 2x - 1$
$Y_5 = 2x - 4$

───── Seeing the Concept ─────

To see the role of the y-intercept b, graph the following lines on the same square screen.

$$Y_1 = 2x$$
$$Y_2 = 2x + 1$$
$$Y_3 = 2x - 1$$
$$Y_4 = 2x + 4$$
$$Y_5 = 2x - 4$$

See Figure 79. What do you conclude about the lines $y = 2x + b$?

7 Identify the Slope and y-Intercept of a Line from Its Equation

When the equation of a line is written in slope–intercept form, it is easy to find the slope m and y-intercept b of the line. For example, suppose that the equation of a line is

$$y = -2x + 3$$

Compare it to $y = mx + b$.

$$y = -2x + 3$$
$$\uparrow \quad \uparrow$$
$$y = \quad mx + b$$

The slope of this line is -2 and its y-intercept is 3.

───── **NOW WORK PROBLEM 65.**

| **EXAMPLE 7** | **Finding the Slope and y-Intercept** |

Find the slope m and y-intercept b of the equation $2x + 4y = 8$. Graph the equation.

Solution To obtain the slope and y-intercept, we transform the equation into its slope–intercept form by solving for y.

$$2x + 4y = 8$$
$$4y = -2x + 8$$
$$y = -\frac{1}{2}x + 2 \qquad {\color{gray} y = mx + b}$$

Figure 80
$2x + 4y = 8$

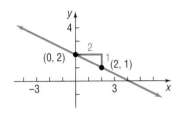

The coefficient of x, $-\dfrac{1}{2}$, is the slope, and the y-intercept is 2. We can graph the line using the fact that the y-intercept is 2 and the slope is $-\dfrac{1}{2}$. Then, starting at the point $(0, 2)$, go to the right 2 units and then down 1 unit to the point $(2, 1)$. See Figure 80. ◄

NOW WORK PROBLEM 71.

8   Graph Lines Written in General Form Using Intercepts

The form of the equation of the line in Example 7, $2x + 4y = 8$, is called the *general form*.

> The equation of a line L is in **general form** when it is written as
>
> $$Ax + By = C \qquad\qquad (4)$$
>
> where A, B, and C are real numbers and A and B are not both 0.

When we want to graph a linear equation that is written in general form, we can solve the equation for y and write the equation in slope–intercept form as we did in Example 7. Another approach to graphing the equation would be to find its intercepts. Remember, the intercepts of the graph of an equation are the points where the graph crosses or touches a coordinate axis. In Section 1.1, we learned how to identify the intercepts from the graph of an equation and how to approximate intercepts using a graphing utility. Now that we have discussed solving equations in one variable, we are prepared to find intercepts algebraically from the equation. The intercepts of the graph of an equation can be found algebraically by using the fact that points on the x-axis have y-coordinates equal to 0 and points on the y-axis have x-coordinates equal to 0.

Procedure for Finding Intercepts

- To find the x-intercept(s), if any, of the graph of an equation, let $y = 0$ in the equation and solve for x.
- To find the y-intercept(s), if any, of the graph of an equation, let $x = 0$ in the equation and solve for y.

| **EXAMPLE 8** | **Graphing a Linear Equation Using Its Intercepts** |

Graph the linear equation $2x + 4y = 8$ by finding its intercepts.

Solution   To obtain the x-intercept, let $y = 0$ in the equation and solve for x.

$$2x + 4y = 8$$
$$2x + 4(0) = 8 \qquad \text{Let } y = 0.$$
$$2x = 8$$
$$x = 4 \qquad \text{Divide both sides by 2.}$$

The x-intercept is 4 and the point $(4, 0)$ is on the graph of the equation.
To obtain the y-intercept, let $x = 0$ in the equation and solve for y.

$$2x + 4y = 8$$
$$2(0) + 4y = 8 \qquad \text{Let } x = 0.$$
$$4y = 8$$
$$y = 2 \qquad \text{Divide both sides by 4.}$$

Figure 81

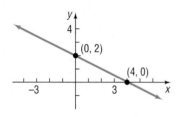

The y-intercept is 2 and the point $(0, 2)$ is on the graph of the equation.
We plot the points $(4, 0)$ and $(0, 2)$ in a Cartesian plane and draw a line through the points. See Figure 81.   ◀

NOW WORK PROBLEM **85.**

Every line has an equation that is equivalent to an equation written in general form. For example, a vertical line whose equation is

$$x = a$$

can be written in the general form

$$1 \cdot x + 0 \cdot y = a \qquad A = 1, B = 0, C = a$$

A horizontal line whose equation is

$$y = b$$

can be written in the general form

$$0 \cdot x + 1 \cdot y = b \qquad A = 0, B = 1, C = b$$

Figure 82
The distinct lines are parallel if and only if their slopes are equal.

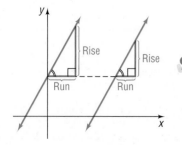

Lines that are neither vertical nor horizontal have general equations of the form

$$Ax + By = C \qquad A \neq 0 \text{ and } B \neq 0$$

Because the equation of every line can be written in general form, any equation equivalent to equation (4) is called a **linear equation**.

9   Find Equations of Parallel Lines

When two lines (in the plane) do not intersect (that is, they have no points in common), they are said to be **parallel**. Look at Figure 82. There we have drawn two lines and have constructed two right triangles by drawing sides parallel to the coordinate axes. These lines are parallel if and only if the right triangles are similar. (Do you see why? Two angles are equal.) And the triangles are similar if and only if the ratios of corresponding sides are equal.

This suggests the following result:

Theorem   **Criterion for Parallel Lines**

Two nonvertical lines are parallel if and only if their slopes are equal and they have different y-intercepts.

The use of the words "if and only if" in the preceding theorem means that actually two statements are being made, one the converse of the other.

If two nonvertical lines are parallel, then their slopes are equal and they have different y-intercepts.

If two nonvertical lines have equal slopes and they have different y-intercepts, then they are parallel.

EXAMPLE 9   **Showing That Two Lines Are Parallel**

Show that the lines given by the following equations are parallel:

$$L_1: \quad 2x + 3y = 6, \qquad L_2: \quad 4x + 6y = 0$$

Solution   To determine whether these lines have equal slopes and different y-intercepts, we write each equation in slope–intercept form:

$$L_1: \quad 2x + 3y = 6 \qquad\qquad L_2: \quad 4x + 6y = 0$$

$$3y = -2x + 6 \qquad\qquad\qquad 6y = -4x$$

$$y = -\frac{2}{3}x + 2 \qquad\qquad\qquad y = -\frac{2}{3}x$$

$$\text{Slope} = -\frac{2}{3}; \ y\text{-intercept} = 2 \qquad \text{Slope} = -\frac{2}{3}; \ y\text{-intercept} = 0$$

Because these lines have the same slope, $-\dfrac{2}{3}$, but different y-intercepts, the lines are parallel. See Figure 83.

Figure 83
Parallel lines

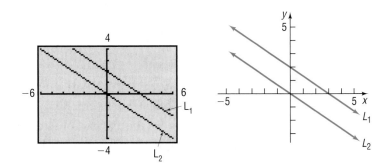

| **EXAMPLE 10** | **Finding a Line That Is Parallel to a Given Line** |

Find an equation for the line that contains the point $(2, -3)$ and is parallel to the line $2x + y = 6$.

Solution Since the two lines are to be parallel, the slope of the line that we seek equals the slope of the line $2x + y = 6$. We begin by writing the equation of the line $2x + y = 6$ in slope–intercept form.

$$2x + y = 6$$
$$y = -2x + 6$$

The slope is -2. Since the line that we seek contains the point $(2, -3)$, we use the point–slope form to obtain

$y - y_1 = m(x - x_1)$	Point–slope form
$y - (-3) = -2(x - 2)$	$m = -2, x_1 = 2, y_1 = -3$
$y + 3 = -2x + 4$	Simplify
$y = -2x + 1$	Slope–intercept form
$2x + y = 1$	General form

This line is parallel to the line $2x + y = 6$ and contains the point $(2, -3)$. See Figure 84.

Figure 84

$Y_1 = -2x + 1$ $Y_2 = -2x + 6$

$2x + y = 6$

$(2, -3)$

$2x + y = 1$

──── **NOW WORK PROBLEM 53.**

Figure 85
Perpendicular lines

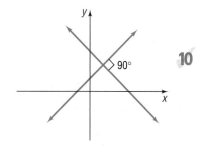

10 Find Equations of Perpendicular Lines

When two lines intersect at a right angle (90°), they are said to be **perpendicular**. See Figure 85.

The following result gives a condition, in terms of their slopes, for two lines to be perpendicular.

Theorem **Criterion for Perpendicular Lines**

Two nonvertical lines are perpendicular if and only if the product of their slopes is -1.

Here we shall prove the "only if" part of the statement:

If two nonvertical lines are perpendicular, then the product of their slopes is -1.

You are asked to prove the "if" part of the theorem; that is:

If two nonvertical lines have slopes whose product is -1, then the lines are perpendicular.

Figure 86

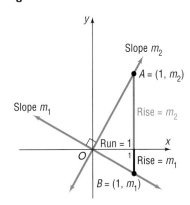

Proof Let m_1 and m_2 denote the slopes of the two lines. There is no loss in generality (that is, neither the angle nor the slopes are affected) if we situate the lines so that they meet at the origin. See Figure 86. The point $A = (1, m_2)$ is on the line having slope m_2, and the point $B = (1, m_1)$ is on the line having slope m_1. (Do you see why this must be true?)

Suppose that the lines are perpendicular. Then triangle OAB is a right triangle. As a result of the Pythagorean Theorem, it follows that

$$[d(O, A)]^2 + [d(O, B)]^2 = [d(A, B)]^2 \qquad \textbf{(5)}$$

By the distance formula, we can write the squares of these distances as

$$[d(O, A)]^2 = (1 - 0)^2 + (m_2 - 0)^2 = 1 + m_2^2$$

$$[d(O, B)]^2 = (1 - 0)^2 + (m_1 - 0)^2 = 1 + m_1^2$$

$$[d(A, B)]^2 = (1 - 1)^2 + (m_2 - m_1)^2 = m_2^2 - 2m_1m_2 + m_1^2$$

Using these facts in equation (5), we get

$$(1 + m_2^2) + (1 + m_1^2) = m_2^2 - 2m_1m_2 + m_1^2$$

which, upon simplification, can be written as

$$m_1m_2 = -1$$

If the lines are perpendicular, the product of their slopes is -1. ∎

You may find it easier to remember the condition for two nonvertical lines to be perpendicular by observing that the equality $m_1m_2 = -1$ means that m_1 and m_2 are negative reciprocals of each other; that is, either $m_1 = -\dfrac{1}{m_2}$ or $m_2 = -\dfrac{1}{m_1}$.

EXAMPLE 11	**Finding the Slope of a Line Perpendicular to Another Line**

If a line has slope $\dfrac{3}{2}$, any line having slope $-\dfrac{2}{3}$ is perpendicular to it. ◀

EXAMPLE 12	**Finding the Equation of a Line Perpendicular to a Given Line**

Find an equation of the line that contains the point $(1, -2)$ and is perpendicular to the line $x + 3y = 6$. Graph the two lines.

Solution We first write the equation of the given line in slope–intercept form to find its slope.

$$x + 3y = 6$$
$$3y = -x + 6 \qquad \text{\textit{Proceed to solve for y.}}$$
$$y = -\frac{1}{3}x + 2 \qquad \text{\textit{Place in the form y = mx + b.}}$$

The given line has slope $-\dfrac{1}{3}$. Any line perpendicular to this line will have slope 3.

Because we require the point $(1, -2)$ to be on this line with slope 3, we use the point–slope form of the equation of a line.

$$y - y_1 = m(x - x_1) \qquad \text{\textit{Point–slope form}}$$
$$y - (-2) = 3(x - 1) \qquad \text{\textit{m = 3, x}}_1\text{\textit{ = 1, y}}_1\text{\textit{ = -2.}}$$

To obtain other forms of the equations, we proceed as follows:

$$y + 2 = 3(x - 1)$$
$$y + 2 = 3x - 3 \qquad \text{Simplify}$$
$$y = 3x - 5 \qquad \text{Slope–intercept form}$$
$$3x - y = 5 \qquad \text{General form}$$

Figure 87 shows the graphs.

Figure 87

WARNING
Be sure to use a square screen
when you graph perpendicular lines.
Otherwise, the angle between the
two lines will appear distorted. ■

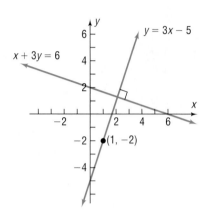

NOW WORK PROBLEM 59.

1.8 Assess Your Understanding

Concepts and Vocabulary

1. The slope of a vertical line is _____; the slope of a horizontal line is _____.

2. Two nonvertical lines have slopes m_1 and m_2, respectively. The lines are parallel if _____ and the _____ are unequal; the lines are perpendicular if _____.

3. A horizontal line is given by an equation of the form _____, where b is the _____.

4. *True or False*: Vertical lines have an undefined slope.

5. *True or False*: The slope of the line $2y = 3x + 5$ is 3.

6. *True or False*: Perpendicular lines have slopes that are reciprocals of one another.

Skill Building

In Problems 7–10, (a) find the slope of the line and (b) interpret the slope.

7.

8.

9.

10.

In Problems 11–18, plot each pair of points and determine the slope of the line containing them. Graph the line.

11. $(2, 3); (4, 0)$ **12.** $(4, 2); (3, 4)$ **13.** $(-2, 3); (2, 1)$ **14.** $(-1, 1); (2, 3)$

15. $(-3, -1); (2, -1)$ **16.** $(4, 2); (-5, 2)$ **17.** $(-1, 2); (-1, -2)$ **18.** $(2, 0); (2, 2)$

In Problems 19–26, graph, by hand, the line containing the point P and having slope m.

19. $P = (1, 2); m = 3$ **20.** $P = (2, 1); m = 4$ **21.** $P = (2, 4); m = -\dfrac{3}{4}$

22. $P = (1, 3); m = -\dfrac{2}{5}$ **23.** $P = (-1, 3); m = 0$ **24.** $P = (2, -4); m = 0$

25. $P = (0, 3)$; slope undefined   **26.** $P = (-2, 0)$; slope undefined

In Problems 27–32, the slope and a point on a line are given. Use this information to locate three additional points on the line. Answers may vary.

[Hint: *It is not necessary to find the equation of the line. See Example 2.]*

27. Slope 4; point $(1, 2)$ **28.** Slope 2; point $(-2, 3)$ **29.** Slope $-\dfrac{3}{2}$; point $(2, -4)$

30. Slope $\dfrac{4}{3}$; point $(-3, 2)$ **31.** Slope -2; point $(-2, -3)$ **32.** Slope -1; point $(4, 1)$

In Problems 33–40, find an equation of the line L.

33.   **34.**   **35.**   **36.**

37.   **38.**   **39.**   **40.**

L is parallel to $y = 2x$ L is parallel to $y = -x$ L is perpendicular to $y = 2x$ L is perpendicular to $y = -x$

In Problems 41–64, find an equation for the line with the given properties. Express your answer using either the general form or the slope–intercept form of the equation of a line, whichever you prefer.

41. Slope $= 3$; containing the point $(-2, 3)$ **42.** Slope $= 2$; containing the point $(4, -3)$

43. Slope $= -\dfrac{2}{3}$; containing the point $(1, -1)$ **44.** Slope $= \dfrac{1}{2}$; containing the point $(3, 1)$

45. Containing the points $(1, 3)$ and $(-1, 2)$ **46.** Containing the points $(-3, 4)$ and $(2, 5)$

47. Slope $= -3$; y-intercept $= 3$ **48.** Slope $= -2$; y-intercept $= -2$

49. x-intercept $= 2$; y-intercept $= -1$ **50.** x-intercept $= -4$; y-intercept $= 4$

51. Slope undefined; containing the point $(2, 4)$ **52.** Slope undefined; containing the point $(3, 8)$

53. Parallel to the line $y = 2x$; containing the point $(-1, 2)$ **54.** Parallel to the line $y = -3x$; containing the point $(-1, 2)$

55. Parallel to the line $2x - y = -2$; containing the point $(0, 0)$ **56.** Parallel to the line $x - 2y = -5$; containing the point $(0, 0)$

57. Parallel to the line $x = 5$; containing the point $(4, 2)$ **58.** Parallel to the line $y = 5$; containing the point $(4, 2)$

59. Perpendicular to the line $y = \dfrac{1}{2}x + 4$;
containing the point $(1, -2)$

60. Perpendicular to the line $y = 2x - 3$;
containing the point $(1, -2)$

61. Perpendicular to the line $2x + y = 2$;
containing the point $(-3, 0)$

62. Perpendicular to the line $x - 2y = -5$;
containing the point $(0, 4)$

63. Perpendicular to the line $x = 8$; containing the point $(3, 4)$ **64.** Perpendicular to the line $y = 8$; containing the point $(3, 4)$

In Problems 65–84, find the slope and y-intercept of each line. Graph the line by hand. Check your graph using a graphing utility.

65. $y = 2x + 3$ **66.** $y = -3x + 4$ **67.** $\frac{1}{2}y = x - 1$ **68.** $\frac{1}{3}x + y = 2$ **69.** $y = \frac{1}{2}x + 2$

70. $y = 2x + \frac{1}{2}$ **71.** $x + 2y = 4$ **72.** $-x + 3y = 6$ **73.** $2x - 3y = 6$ **74.** $3x + 2y = 6$

75. $x + y = 1$ **76.** $x - y = 2$ **77.** $x = -4$ **78.** $y = -1$ **79.** $y = 5$

80. $x = 2$ **81.** $y - x = 0$ **82.** $x + y = 0$ **83.** $2y - 3x = 0$ **84.** $3x + 2y = 0$

In Problems 85–94, (a) find the intercepts of the graph of each equation and (b) graph the equation.

85. $2x + 3y = 6$ **86.** $3x - 2y = 6$ **87.** $-4x + 5y = 40$ **88.** $6x - 4y = 24$

89. $7x + 2y = 21$ **90.** $5x + 3y = 18$ **91.** $\frac{1}{2}x + \frac{1}{3}y = 1$ **92.** $x - \frac{2}{3}y = 4$

93. $0.2x - 0.5y = 1$ **94.** $-0.3x + 0.4y = 1.2$

95. Find an equation of the *x*-axis.   **96.** Find an equation of the *y*-axis.

In Problems 97–100, match each graph with the correct equation:

(a) $y = x$ (b) $y = 2x$ (c) $y = \frac{x}{2}$ (d) $y = 4x$

97.

98.

99.

100.
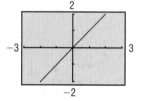

In Problems 101–104, write an equation of each line. Express your answer using either the general form or the slope–intercept form of the equation of a line, whichever you prefer.

101.

102.

103.

104.
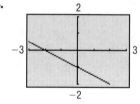

Applications and Extensions

105. Truck Rentals A truck rental company rents a moving truck for one day by charging $29 plus $0.07 per mile. Write a linear equation that relates the cost *C*, in dollars, of renting the truck to the number *x* of miles driven. What is the cost of renting the truck if the truck is driven 110 miles? 230 miles?

106. Cost Equation The **fixed costs** of operating a business are the costs incurred regardless of the level of production. Fixed costs include rent, fixed salaries, and costs of buying machinery. The **variable costs** of operating a business are the costs that change with the level of output. Variable costs include raw materials, hourly wages, and electricity. Suppose that a manufacturer of jeans has fixed costs of $500 and variable costs of $8 for each pair of jeans manufactured. Write a linear equation that relates the cost *C*, in dollars, of manufacturing the jeans to the number *x* pairs of jeans manufactured. What is the cost of manufacturing 400 pairs of jeans? 740 pairs?

107. Cost of Sunday Home Delivery The cost to the *Chicago Tribune* for Sunday home delivery is approximately $0.53 per newspaper with fixed costs of $1,070,000. Write an equation that relates the cost *C* and the number *x* of copies delivered. Source: *Chicago Tribune*, 2002.

108. Wages of a Car Salesperson Dan receives $375 per week for selling new and used cars at a car dealership in Oak Lawn, Illinois. In addition, he receives 5% of the profit on any sales that he generates. Write an equation that relates Dan's weekly salary *S* when he has sales that generate a profit of *x* dollars.

109. Electricity Rates in Illinois Commonwealth Edison Company supplies electricity to residential customers for a monthly customer charge of $7.58 plus 8.275 cents per kilowatt-hour for up to 400 kilowatt-hours.

(a) Write an equation that relates the monthly charge C, in dollars, to the number x of kilowatt-hours used in a month, $0 \leq x \leq 400$.
(b) Graph this equation.
(c) What is the monthly charge for using 100 kilowatt-hours?
(d) What is the monthly charge for using 300 kilowatt-hours?
(e) Interpret the slope of the line.
SOURCE: Commonwealth Edison Company, December 2002.

110. Electricity Rates in Florida Florida Power & Light Company supplies electricity to residential customers for a monthly customer charge of $5.25 plus 6.787 cents per kilowatt-hour for up to 750 kilowatt-hours.
(a) Write an equation that relates the monthly charge C, in dollars, to the number x of kilowatt-hours used in a month, $0 \leq x \leq 750$.
(b) Graph this equation.
(c) What is the monthly charge for using 200 kilowatt-hours?

(d) What is the monthly charge for using 500 kilowatt-hours?
(e) Interpret the slope of the line.
SOURCE: Florida Power & Light Company, January 2003.

111. Measuring Temperature The relationship between Celsius (°C) and Fahrenheit (°F) degrees of measuring temperature is linear. Find an equation relating °C and °F if 0°C corresponds to 32°F and 100°C corresponds to 212°F. Use the equation to find the Celsius measure of 70°F.

112. Measuring Temperature The Kelvin (K) scale for measuring temperature is obtained by adding 273 to the Celsius temperature.
(a) Write an equation relating K and °C.
(b) Write an equation relating K and °F (see Problem 111).

113. Product Promotion A cereal company finds that the number of people who will buy one of its products in the first month that it is introduced is linearly related to the amount of money it spends on advertising. If it spends $40,000 on advertising, then 100,000 boxes of cereal will be sold, and if it spends $60,000, then 200,000 boxes will be sold.
(a) Write an equation describing the relation between the amount A spent on advertising and the number x of boxes sold.
(b) How much advertising is needed to sell 300,000 boxes of cereal?
(c) Interpret the slope.

114. Show that the line containing the points (a, b) and (b, a), $a \neq b$, is perpendicular to the line $y = x$. Also show that the midpoint of (a, b) and (b, a) lies on the line $y = x$.

115. The equation $2x - y = C$ defines a **family of lines**, one line for each value of C. On one set of coordinate axes, graph the members of the family when $C = -4$, $C = 0$, and $C = 2$. Can you draw a conclusion from the graph about each member of the family?

116. Rework Problem 115 for the family of lines $Cx + y = -4$.

Discussion and Writing

117. Which of the following equations might have the graph shown? (More than one answer is possible.)
(a) $2x + 3y = 6$ (e) $x - y = -1$
(b) $-2x + 3y = 6$ (f) $y = 3x - 5$
(c) $3x - 4y = -12$ (g) $y = 2x + 3$
(d) $x - y = 1$ (h) $y = -3x + 3$

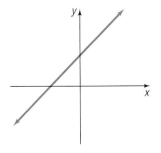

118. Which of the following equations might have the graph shown? (More than one answer is possible.)
(a) $2x + 3y = 6$ (e) $x - y = -1$
(b) $2x - 3y = 6$ (f) $y = -2x - 1$
(c) $3x + 4y = 12$ (g) $y = -\dfrac{1}{2}x + 10$
(d) $x - y = 1$ (h) $y = x + 4$

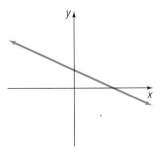

119. The figure shows the graph of two parallel lines. Which of the following pairs of equations might have such a graph?

(a) $x - 2y = 3$
$x + 2y = 7$

(b) $x + y = 2$
$x + y = -1$

(c) $x - y = -2$
$x - y = 1$

(d) $x - y = -2$
$2x - 2y = -4$

(e) $x + 2y = 2$
$x + 2y = -1$

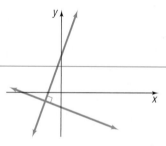

120. The figure shows the graph of two perpendicular lines. Which of the following pairs of equations might have such a graph?

(a) $y - 2x = 2$
$y + 2x = -1$

(b) $y - 2x = 0$
$2y + x = 0$

(c) $2y - x = 2$
$2y + x = -2$

(d) $y - 2x = 2$
$x + 2y = -1$

(e) $2x + y = -2$
$2y + x = -2$

121. The accepted symbol used to denote the slope of a line is the letter m. Investigate the origin of this symbolism. Begin by consulting a French dictionary and looking up the French word *monter*. Write a brief essay on your findings.

122. The term *grade* is used to describe the inclination of a road. How does this term relate to the notion of slope of a line? Is a 4% grade very steep? Investigate the grades of some mountainous roads and determine their slopes. Write a brief essay on your findings.

123. Carpentry Carpenters use the term *pitch* to describe the steepness of staircases and roofs. How does pitch relate to slope? Investigate typical pitches used for stairs and for roofs. Write a brief essay on your findings.

124. Can the equation of every line be written in slope–intercept form? Why?

125. Does every line have exactly one x-intercept and one y-intercept? Are there any lines that have no intercepts?

126. What can you say about two lines that have equal slopes and equal y-intercepts?

127. What can you say about two lines with the same x-intercept and the same y-intercept? Assume that the x-intercept is not 0.

128. If two lines have the same slope, but different x-intercepts, can they have the same y-intercept?

129. If two lines have the same y-intercept, but different slopes, can they have the same x-intercept?

1.9 Circles

OBJECTIVES 1 Write the Standard Form of the Equation of a Circle
2 Graph a Circle by Hand and by Using a Graphing Utility
3 Work with the General Form of the Equation of a Circle

1 Write the Standard Form of the Equation of a Circle

One advantage of a coordinate system is that it enables us to translate a geometric statement into an algebraic statement, and vice versa. Consider, for example, the following geometric statement that defines a circle.

A **circle** is a set of points in the xy-plane that are a fixed distance r from a fixed point (h, k). The fixed distance r is called the **radius**, and the fixed point (h, k) is called the **center** of the circle.

Figure 88

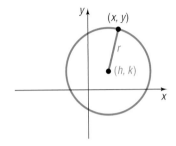

Figure 88 shows the graph of a circle. To find the equation, we let (x, y) represent the coordinates of any point on a circle with radius r and center (h, k). Then the distance between the points (x, y) and (h, k) must always equal r. That is, by the distance formula

$$\sqrt{(x - h)^2 + (y - k)^2} = r$$

or, equivalently,

$$(x - h)^2 + (y - k)^2 = r^2$$

The **standard form of an equation of a circle** with radius r and center (h, k) is

$$(x - h)^2 + (y - k)^2 = r^2 \qquad \textbf{(1)}$$

The standard form of an equation of a circle of radius r with center at the origin (0, 0) is

$$x^2 + y^2 = r^2$$

If the radius r = 1, the circle whose center is at the origin is called the **unit circle** and has the equation

$$x^2 + y^2 = 1$$

See Figure 89.

Figure 89
Unit circle $x^2 + y^2 = 1$

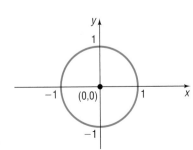

EXAMPLE 1	Writing the Standard Form of the Equation of a Circle

Write the standard form of the equation of the circle with radius 5 and center $(-3, 6)$.

Solution   Using the form of equation (1) and substituting the values $r = 5$, $h = -3$, and $k = 6$, we have

$$(x - h)^2 + (y - k)^2 = r^2$$
$$(x + 3)^2 + (y - 6)^2 = 25$$

◀

━ NOW WORK PROBLEM 5.

2 Graph a Circle by Hand and by Using a Graphing Utility

The graph of any equation of the form (1) is that of a circle with radius r and center (h, k).

EXAMPLE 2	Graphing a Circle by Hand and by Using a Graphing Utility

Graph the equation: $(x + 3)^2 + (y - 2)^2 = 16$

Solution   The graph of the equation is a circle. To graph the equation by hand, we first compare the given equation to the standard form of the equation of a circle. The comparison yields information about the circle.

$$(x + 3)^2 + (y - 2)^2 = 16$$
$$(x - (-3))^2 + (y - 2)^2 = 4^2$$
$$\uparrow \qquad\qquad \uparrow \qquad \uparrow$$
$$(x - h)^2 + (y - k)^2 = r^2$$

Figure 90

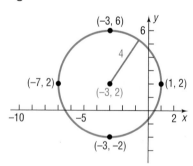

We see that $h = -3$, $k = 2$, and $r = 4$. The circle has center $(-3, 2)$ and a radius of 4 units. To graph this circle, we first plot the center $(-3, 2)$. Since the radius is 4, we can locate four points on the circle by plotting points 4 units to the left, to the right, up, and down from the center. These four points can then be used as guides to obtain the graph. See Figure 90.

To graph a circle on a graphing utility, we must write the equation in the form $y = \{$expression involving $x\}$.* We must solve for y in the equation

$$(x + 3)^2 + (y - 2)^2 = 16$$
$$(y - 2)^2 = 16 - (x + 3)^2 \qquad \text{Subtract } (x + 3)^2 \text{ from both sides.}$$
$$y - 2 = \pm\sqrt{16 - (x + 3)^2} \qquad \text{Use the Square Root Method.}$$
$$y = 2 \pm \sqrt{16 - (x + 3)^2} \qquad \text{Add 2 to both sides.}$$

---

*Some graphing utilities (e.g., TI-83, TI-84, and TI-86) have a CIRCLE function that allows the user to enter only the coordinates of the center of the circle and its radius to graph the circle.

Figure 91

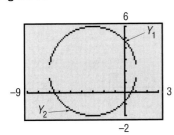

To graph the circle, we graph the top half

$$Y_1 = 2 + \sqrt{16 - (x + 3)^2}$$

and the bottom half

$$Y_2 = 2 - \sqrt{16 - (x + 3)^2}$$

Also, be sure to use a square screen. Otherwise, the circle will appear distorted. Figure 91 shows the graph on a TI-84 Plus.   ◄

NOW WORK PROBLEMS 21(a) AND (b).

EXAMPLE 3	**Finding the Intercepts of a Circle**

For the circle $(x + 3)^2 + (y - 2)^2 = 16$, find the intercepts, if any, of its graph.

Solution This is the equation discussed and graphed in Example 2. To find the x-intercepts, if any, let $y = 0$ and solve for x. Then

$$(x + 3)^2 + (y - 2)^2 = 16$$
$$(x + 3)^2 + (0 - 2)^2 = 16 \qquad \text{\small y = 0}$$
$$(x + 3)^2 + 4 = 16 \qquad \text{\small Simplify.}$$
$$(x + 3)^2 = 12 \qquad \text{\small Simplify.}$$
$$x + 3 = \pm\sqrt{12} \qquad \text{\small Apply the Square Root Method.}$$
$$x = -3 \pm 2\sqrt{3} \qquad \text{\small Solve for x.}$$

The x-intercepts are $-3 - 2\sqrt{3} \approx -6.46$ and $-3 + 2\sqrt{3} \approx 0.46$.

To find the y-intercepts, if any, we let $x = 0$ and solve for y. Then

$$(x + 3)^2 + (y - 2)^2 = 16$$
$$(0 + 3)^2 + (y - 2)^2 = 16 \qquad \text{\small x = 0}$$
$$9 + (y - 2)^2 = 16$$
$$(y - 2)^2 = 7$$
$$y - 2 = \pm\sqrt{7} \qquad \text{\small Apply the Square Root Method.}$$
$$y = 2 \pm \sqrt{7} \qquad \text{\small Solve for y.}$$

The y-intercepts are $2 - \sqrt{7} \approx -0.65$ and $2 + \sqrt{7} \approx 4.65$.

Look back at Figure 90 to verify the approximate locations of the intercepts. ◄

NOW WORK PROBLEM 21(c).

3 **Work with the General Form of the Equation of a Circle**

If we eliminate the parentheses from the standard form of the equation of the circle given in Example 3, we get

$$(x + 3)^2 + (y - 2)^2 = 16$$
$$x^2 + 6x + 9 + y^2 - 4y + 4 = 16$$

which we find, upon simplifying, is equivalent to

$$x^2 + y^2 + 6x - 4y - 3 = 0$$

It can be shown that any equation of the form

$$x^2 + y^2 + ax + by + c = 0$$

has a graph that is a circle, or a point, or has no graph at all. For example, the graph of the equation $x^2 + y^2 = 0$ is the single point $(0, 0)$. The equation $x^2 + y^2 + 5 = 0$, or $x^2 + y^2 = -5$, has no graph, because sums of squares of real numbers are never negative. When its graph is a circle, the equation

$$x^2 + y^2 + ax + by + c = 0$$

is referred to as the **general form of the equation of a circle**.

If an equation of a circle is in the general form, we use the method of completing the square to put the equation in standard form so that we can identify its center and radius.

| **EXAMPLE 4** | **Graphing a Circle Whose Equation Is in General Form** |

Graph the equation $x^2 + y^2 + 4x - 6y + 12 = 0$

Solution We complete the square in both x and y to put the equation in standard form. Group the expression involving x, group the expression involving y, and put the constant on the right side of the equation. The result is

$$(x^2 + 4x) + (y^2 - 6y) = -12$$

Next, complete the square of each expression in parentheses. Remember that any number added on the left side of the equation must be added on the right.

$$(x^2 + 4x + 4) + (y^2 - 6y + 9) = -12 + 4 + 9$$
$$\underbrace{\left(\tfrac{4}{2}\right)^2 = 4} \qquad \underbrace{\left(\tfrac{-6}{2}\right)^2 = 9}$$

$$(x + 2)^2 + (y - 3)^2 = 1 \quad \text{Factor}$$

We recognize this equation as the standard form of the equation of a circle with radius 1 and center $(-2, 3)$. To graph the equation by hand, use the center $(-2, 3)$ and the radius 1. See Figure 92 (a).

To graph the equation using a graphing utility, we need to solve for y.

$$(y - 3)^2 = 1 - (x + 2)^2$$
$$y - 3 = \pm\sqrt{1 - (x + 2)^2} \qquad \text{Use the Square Root Method.}$$
$$y = 3 \pm \sqrt{1 - (x + 2)^2} \qquad \text{Add 3 to both sides.}$$

Figure 92(b) illustrates the graph.

Figure 92

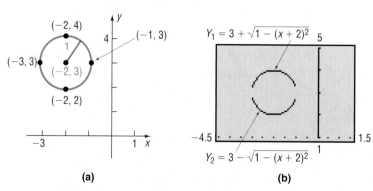

(a)

(b)

NOW WORK PROBLEM 25.

| EXAMPLE 5 | Finding the General Equation of a Circle |

Find the general equation of the circle whose center is $(1, -2)$ and whose graph contains the point $(4, -2)$.

Solution   To find the equation of a circle, we need to know its center and its radius. Here, we know that the center is $(1, -2)$. Since the point $(4, -2)$ is on the graph, the radius r will equal the distance from $(4, -2)$ to the center $(1, -2)$. See Figure 93. Thus,

Figure 93

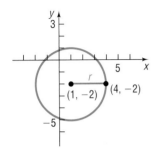

$$r = \sqrt{(4 - 1)^2 + [-2 - (-2)]^2}$$
$$= \sqrt{9} = 3$$

The standard form of the equation of the circle is

$$(x - 1)^2 + (y + 2)^2 = 9$$

Eliminating the parentheses and rearranging terms, we get the general equation

$$x^2 + y^2 - 2x + 4y - 4 = 0$$ ◄

Overview

The discussion in Sections 1.8 and 1.9 about circles and lines dealt with two main types of problems that can be generalized as follows:

1. Given an equation, classify it and graph it.
2. Given a graph, or information about a graph, find its equation.

This text deals with both types of problems. We shall study various equations, classify them, and graph them. Although the second type of problem is usually more difficult to solve than the first, in many instances a graphing utility can be used to solve such problems.

1.9 Assess Your Understanding

Concepts and Vocabulary

1. *True or False:* Every equation of the form
$$x^2 + y^2 + ax + by + c = 0$$
has a circle as its graph.

2. For a circle, the _____ is the distance from the center to any point on the circle.

3. *True or False:* The radius of the circle $x^2 + y^2 = 9$ is 3.

4. *True or False:* The center of the circle
$$(x + 3)^2 + (y - 2)^2 = 13$$
is $(3, -2)$.

Skill Building

In Problems 5–8, find the center and radius of each circle. Write the standard form of the equation.

5.

6.

7.

8.
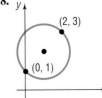

In Problems 9–18, write the standard form of the equation and the general form of the equation of each circle of radius r and center (h, k). *By hand, graph each circle.*

9. $r = 2$; $(h, k) = (0, 0)$ **10.** $r = 3$; $(h, k) = (0, 0)$ **11.** $r = 2$; $(h, k) = (0, 2)$ **12.** $r = 3$; $(h, k) = (1, 0)$

13. $r = 5$; $(h, k) = (4, -3)$ **14.** $r = 4$; $(h, k) = (2, -3)$ **15.** $r = 4$; $(h, k) = (-2, 1)$ **16.** $r = 7$; $(h, k) = (-5, -2)$

17. $r = \frac{1}{2}$; $(h, k) = \left(\frac{1}{2}, 0\right)$ **18.** $r = \frac{1}{2}$; $(h, k) = \left(0, -\frac{1}{2}\right)$

In Problems 19–30, (a) find the center (h, k) *and radius r of each circle; (b) graph each circle; (c) find the intercepts, if any, of the graphs.*

19. $x^2 + y^2 = 4$

20. $x^2 + (y - 1)^2 = 1$

21. $2(x - 3)^2 + 2y^2 = 8$

22. $3(x + 1)^2 + 3(y - 1)^2 = 6$

23. $x^2 + y^2 - 2x + 4y - 4 = 0$

24. $x^2 + y^2 + 4x + 2y - 20 = 0$

25. $x^2 + y^2 + 4x - 4y - 1 = 0$

26. $x^2 + y^2 - 6x + 2y + 9 = 0$

27. $x^2 + y^2 - x + 2y + 1 = 0$

28. $x^2 + y^2 + x + y - \frac{1}{2} = 0$

29. $2x^2 + 2y^2 - 12x + 8y - 24 = 0$

30. $2x^2 + 2y^2 + 8x + 7 = 0$

In Problems 31–36, find the general form of the equation of each circle.

31. Center at the origin and containing the point $(-2, 3)$

32. Center $(1, 0)$ and containing the point $(-3, 2)$

33. Center $(2, 3)$ and tangent to the x-axis

34. Center $(-3, 1)$ and tangent to the y-axis

35. With endpoints of a diameter at $(1, 4)$ and $(-3, 2)$

36. With endpoints of a diameter at $(4, 3)$ and $(0, 1)$

In Problems 37–40, match each graph with the correct equation.
(a) $(x - 3)^2 + (y + 3)^2 = 9$ (b) $(x + 1)^2 + (y - 2)^2 = 4$ (c) $(x - 1)^2 + (y + 2)^2 = 4$ (d) $(x + 3)^2 + (y - 3)^2 = 9$

37.

38.

39.

40.

In Problems 41–44, find the standard form of the equation of each circle. Assume that the center has integer coordinates and that the radius is an integer.

41.

42.

43.

44.

Applications and Extensions

45. Weather Satellites Earth is represented on a map of a portion of the solar system so that its surface is the circle with equation $x^2 + y^2 + 2x + 4y - 4091 = 0$. A weather satellite circles 0.6 unit above Earth with the center of its circular orbit at the center of Earth. Find the equation for the orbit of the satellite on this map.

46. The **tangent line** to a circle may be defined as the line that intersects the circle in a single point, called the **point of tangency** (see the figure).

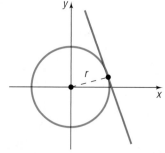

If the equation of the circle is $x^2 + y^2 = r^2$ and the equation of the tangent line is $y = mx + b$, show that:

(a) $r^2(1 + m^2) = b^2$

   [**Hint:** The quadratic equation $x^2 + (mx + b)^2 = r^2$ has exactly one solution.]

(b) The point of tangency is $\left(\dfrac{-r^2m}{b}, \dfrac{r^2}{b} \right)$.

(c) The tangent line is perpendicular to the line containing the center of the circle and the point of tangency.

47. The Greek method for finding the equation of the tangent line to a circle used the fact that at any point on a circle the lines containing the center and the tangent line are perpendicular (see Problem 46). Use this method to find an equation of the tangent line to the circle $x^2 + y^2 = 9$ at the point $(1, 2\sqrt{2})$.

48. Use the Greek method described in Problem 47 to find an equation of the tangent line to the circle $x^2 + y^2 - 4x + 6y + 4 = 0$ at the point $(3, 2\sqrt{2} - 3)$.

49. Refer to Problem 46. The line $x - 2y + 4 = 0$ is tangent to a circle at $(0, 2)$. The line $y = 2x - 7$ is tangent to the same circle at $(3, -1)$. Find the center of the circle.

50. Find an equation of the line containing the centers of the two circles

$$x^2 + y^2 - 4x + 6y + 4 = 0 \quad \text{and}$$

$$x^2 + y^2 + 6x + 4y + 9 = 0$$

51. If a circle of radius 2 is made to roll along the x-axis, what is an equation for the path of the center of the circle?

Discussion and Writing

52. Which of the following equations might have the graph shown? (More than one answer is possible.)

(a) $(x - 2)^2 + (y + 3)^2 = 13$

(b) $(x - 2)^2 + (y - 2)^2 = 8$

(c) $(x - 2)^2 + (y - 3)^2 = 13$

(d) $(x + 2)^2 + (y - 2)^2 = 8$

(e) $x^2 + y^2 - 4x - 9y = 0$

(f) $x^2 + y^2 + 4x - 2y = 0$

(g) $x^2 + y^2 - 9x - 4y = 0$

(h) $x^2 + y^2 - 4x - 4y = 4$

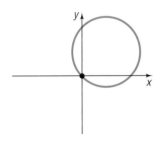

53. Which of the following equations might have the graph shown? (More than one answer is possible.)

(a) $(x - 2)^2 + y^2 = 3$

(b) $(x + 2)^2 + y^2 = 3$

(c) $x^2 + (y - 2)^2 = 3$

(d) $(x + 2)^2 + y^2 = 4$

(e) $x^2 + y^2 + 10x + 16 = 0$

(f) $x^2 + y^2 + 10x - 2y = 1$

(g) $x^2 + y^2 + 9x + 10 = 0$

(h) $x^2 + y^2 - 9x - 10 = 0$

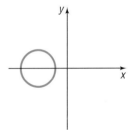

54. Explain how the center and radius of a circle can be used to graph a circle by hand.

55. Explain how the center and radius of a circle can be used to establish an initial viewing window.

56. If the circumference of a circle is 6π, what is its radius?

Chapter Review

Things to Know

Formulas

Distance formula (p. 87)	$d = \sqrt{(x_2 - x_1)^2 + (y_2 - y_1)^2}$
Midpoint formula (p. 89)	$(x, y) = \left(\dfrac{x_1 + x_2}{2}, \dfrac{y_1 + y_2}{2} \right)$
Slope (p. 163)	$m = \dfrac{y_2 - y_1}{x_2 - x_1}$, if $x_1 \neq x_2$; undefined if $x_1 = x_2$
Parallel lines (p. 173)	Equal slopes ($m_1 = m_2$) with different y-intercepts
Perpendicular lines (p. 174)	Product of slopes is -1 ($m_1 \cdot m_2 = -1$)

Quadratic equation and quadratic formula

The real solutions of the equation $ax^2 + bx + c = 0$, $a \neq 0$, are given by $x = \dfrac{-b \pm \sqrt{b^2 - 4ac}}{2a}$, provided $b^2 - 4ac \geq 0$. If $b^2 - 4ac < 0$, there are no real solutions. (p. 118)

In the complex number system, the solutions of the equation $ax^2 + bx + c = 0$, $a \neq 0$, are given by $x = \dfrac{-b \pm \sqrt{b^2 - 4ac}}{2a}$. (p. 130)

Discriminant (pp. 118 and 131)

If $b^2 - 4ac > 0$, there are two distinct real solutions.

If $b^2 - 4ac = 0$, there is one repeated real solution.

If $b^2 - 4ac < 0$, there are no real solutions, but there are two distinct complex solutions that are not real; the complex solutions are conjugates of each other.

Equations

Vertical line (p. 167)	$x = a$
Point–slope form of the equation of a line (p. 168)	$y - y_1 = m(x - x_1)$; m is the slope of the line, (x_1, y_1) is a point on the line
Horizontal line (p. 169)	$y = b$
Slope–intercept form of the equation of a line (p. 170)	$y = mx + b$; m is the slope of the line, b is the y-intercept
General form of the equation of a line (p. 171)	$Ax + By = C$; A, B not both 0
Standard form of the equation of a circle (p. 181)	$(x - h)^2 + (y - k)^2 = r^2$; r is the radius of the circle, (h, k) is the center of the circle
Equation of the unit circle (p. 181)	$x^2 + y^2 = 1$
General form of the equation of a circle (p. 184)	$x^2 + y^2 + ax + by + c = 0$

Interval notation (p. 151)

$[a, b]$	$\{x \mid a \leq x \leq b\}$	$[a, \infty)$	$\{x \mid x \geq a\}$
$[a, b)$	$\{x \mid a \leq x < b\}$	(a, ∞)	$\{x \mid x > a\}$
$(a, b]$	$\{x \mid a < x \leq b\}$	$(-\infty, a]$	$\{x \mid x \leq a\}$
(a, b)	$\{x \mid a < x < b\}$	$(-\infty, a)$	$\{x \mid x < a\}$
		$(-\infty, \infty)$	All real numbers

Properties of inequalities

Addition property (p. 152)	If $a < b$, then $a + c < b + c$. If $a > b$, then $a + c > b + c$.

Multiplication properties (p. 154)	(a) If $a < b$ and if $c > 0$, then $ac < bc$.
	     If $a < b$ and if $c < 0$, then $ac > bc$.
	
	 (b) If $a > b$ and if $c > 0$, then $ac > bc$
	     If $a > b$ and if $c < 0$, then $ac < bc$.

Absolute value

If $|u| = a$, $a > 0$, then $u = -a$ or $u = a$. (p. 137)

If $|u| \leq a$, $a > 0$, then $-a \leq u \leq a$. (p. 157)

If $|u| \geq a$, $a > 0$, then $u \leq -a$ or $u \geq a$. (p. 159)

Objectives

Section		You should be able to	Review Exercises
1.1	1	Use the distance formula (p. 86)	91(a)–96(a); 127, 128, 132
	2	Use the midpoint formula (p. 89)	91(b)–96(b)
	3	Graph equations by hand by plotting points (p. 90)	113–118
	4	Graph equations using a graphing utility (p. 92)	97, 113–118
	5	Use a graphing utility to create tables (p. 95)	97
	6	Find intercepts from a graph (p. 95)	98
	7	Use a graphing utility to approximate intercepts (p. 96)	113–118
1.2	1	Solve equations using a graphing utility (p. 101)	51–54
	2	Solve linear equations (p. 103)	1–6
	3	Solve rational equations (p. 105)	7, 8, 36, 47, 48
	4	Solve problems that can be modeled by linear equations (p. 107)	154
1.3	1	Solve quadratic equations by factoring (p. 113)	10, 13, 33–35
	2	Solve quadratic equations using the Square Root Method (p. 114)	49, 50
	3	Solve quadratic equations by completing the square (p. 115)	9, 10, 13–16, 19, 33–35
	4	Solve quadratic equations using the quadratic formula (p. 117)	9, 10, 13–16, 19, 33–35
	5	Solve problems that can be modeled by quadratic equations (p. 119)	138, 144, 151, 152, 156, 157
1.4	1	Add, subtract, multiply, and divide complex numbers (p. 125)	73–82
	2	Solve quadratic equations with a negative discriminant (p. 129)	83–90
1.5	1	Solve radical equations (p. 133)	17, 18, 23–28, 37, 38
	2	Solve equations quadratic in form (p. 135)	21, 22, 31, 32
	3	Solve absolute value equations (p. 137)	39–42
	4	Solve equations by factoring (p. 137)	43–46
1.6	1	Translate verbal descriptions into mathematical expressions (p. 141)	133, 134
	2	Solve interest problems (p. 142)	135, 136
	3	Solve mixture problems (p. 143)	147–150
	4	Solve uniform motion problems (p. 144)	137, 139–143, 153
	5	Solve constant rate job problems (p. 146)	145, 146, 155, 158
1.7	1	Use interval notation (p. 151)	55–68
	2	Use properties of inequalities (p. 152)	55–68
	3	Solve linear inequalities algebraically and graphically (p. 154)	55–56
	4	Solve combined inequalities algebraically and graphically (p. 156)	57–60
	5	Solve absolute value inequalities algebraically and graphically (p. 157)	61–68
1.8	1	Calculate and interpret the slope of a line (p. 163)	91(c), (d)–96(c), (d)
	2	Graph lines given a point and the slope (p. 166)	99–108
	3	Find the equation of a vertical line (p. 167)	101
	4	Use the point–slope form of a line; identify horizontal lines (p. 168)	99–108
	5	Find the equation of a line given two points (p. 169)	104

Review Exercises

In Problems 1–46, find all the real solutions, if any, of each equation. (Where they appear, a, b, m, and n are positive constants.)

1. $2 - \dfrac{x}{3} = 8$

2. $\dfrac{x}{4} - 2 = 4$

3. $-2(5 - 3x) + 8 = 4 + 5x$

4. $(6 - 3x) - 2(1 + x) = 6x$

5. $\dfrac{3x}{4} - \dfrac{x}{3} = \dfrac{1}{12}$

6. $\dfrac{4 - 2x}{3} + \dfrac{1}{6} = 2x$

7. $\dfrac{x}{x - 1} = \dfrac{6}{5}, \quad x \neq 1$

8. $\dfrac{4x - 5}{3 - 7x} = 2, \quad x \neq \dfrac{3}{7}$

9. $x(1 - x) = 6$

10. $x(1 + x) = 6$

11. $\dfrac{1}{2}\left(x - \dfrac{1}{3}\right) = \dfrac{3}{4} - \dfrac{x}{6}$

12. $\dfrac{1 - 3x}{4} = \dfrac{x + 6}{3} + \dfrac{1}{2}$

13. $(x - 1)(2x + 3) = 3$

14. $x(2 - x) = 3(x - 4)$

15. $2x + 3 = 4x^2$

16. $1 + 6x = 4x^2$

17. $\sqrt[3]{x^2 - 1} = 2$

18. $\sqrt{1 + x^3} = 3$

19. $x(x + 1) + 2 = 0$

20. $3x^2 - x + 1 = 0$

21. $x^4 - 5x^2 + 4 = 0$

22. $3x^4 + 4x^2 + 1 = 0$

23. $\sqrt{2x - 3} + x = 3$

24. $\sqrt{2x - 1} = x - 2$

25. $\sqrt[4]{2x + 3} = 2$

26. $\sqrt[5]{3x + 1} = -1$

27. $\sqrt{x + 1} + \sqrt{x - 1} = \sqrt{2x + 1}$

28. $\sqrt{2x - 1} - \sqrt{x - 5} = 3$

29. $2x^{1/2} - 3 = 0$

30. $3x^{1/4} - 2 = 0$

31. $x^{-6} - 7x^{-3} - 8 = 0$

32. $6x^{-1} - 5x^{-1/2} + 1 = 0$

33. $x^2 + m^2 = 2mx + (nx)^2, \quad n \neq 1, n \neq -1$

34. $b^2x^2 + 2ax = x^2 + a^2, \quad b \neq 1, b \neq -1$

35. $10a^2x^2 - 2abx - 36b^2 = 0$

36. $\dfrac{1}{x - m} + \dfrac{1}{x - n} = \dfrac{2}{x}, \quad x \neq 0, x \neq m, x \neq n$

37. $\sqrt{x^2 + 3x + 7} - \sqrt{x^2 - 3x + 9} + 2 = 0$

38. $\sqrt{x^2 + 3x + 7} - \sqrt{x^2 + 3x + 9} = 2$

39. $|2x + 3| = 7$

40. $|3x - 1| = 5$

41. $|2 - 3x| + 2 = 9$

42. $|1 - 2x| + 1 = 4$

43. $2x^3 = 3x^2$

44. $5x^4 = 9x^3$

45. $2x^3 + 5x^2 - 8x - 20 = 0$

46. $3x^3 + 5x^2 - 3x - 5 = 0$

47. $\dfrac{1}{x - 1} + \dfrac{3}{x + 2} = \dfrac{11}{x^2 + x - 2}$

48. $\dfrac{5}{x + 5} - \dfrac{2}{x - 3} = \dfrac{-1}{x^2 + 2x - 15}$

49. $(x - 2)^2 = 9$

50. $(2x + 1)^2 = 25$

In Problems 51–54, use a graphing utility to approximate the solutions of each equation rounded to two decimal places. All solutions lie between −10 and 10.

51. $x^3 - 5x + 3 = 0$

52. $-x^3 + 3x + 1 = 0$

53. $x^4 - 3 = 2x + 1$

54. $-x^4 + 7 = x^2 - 2$

In Problems 55–68, solve each inequality. Express your answer using set notation or interval notation. Graph the solution set.

55. $\dfrac{2x-3}{5} + 2 \le \dfrac{x}{2}$

56. $\dfrac{5-x}{3} \le 6x - 4$

57. $-9 \le \dfrac{2x+3}{-4} \le 7$

58. $-4 < \dfrac{2x-2}{3} < 6$

59. $2 < \dfrac{3-3x}{12} < 6$

60. $-3 \le \dfrac{5-3x}{2} \le 6$

61. $|3x + 4| < \dfrac{1}{2}$

62. $|1 - 2x| < \dfrac{1}{3}$

63. $|2x - 5| \ge 9$

64. $|3x + 1| \ge 10$

65. $2 + |2 - 3x| \le 4$

66. $\dfrac{1}{2} + \left|\dfrac{2x-1}{3}\right| \le 1$

67. $1 - |2 - 3x| < -4$

68. $1 - \left|\dfrac{2x-1}{3}\right| < -2$

In Problems 69–72, what number should be added to complete the square of each expression?

69. $x^2 + 6x$

70. $x^2 - 10x$

71. $x^2 - \dfrac{4}{3}x$

72. $x^2 + \dfrac{4}{5}x$

In Problems 73–82, use the complex number system and write each expression in the standard form $a + bi$.

73. $(6 + 3i) - (2 - 4i)$

74. $(8 - 3i) + (-6 + 2i)$

75. $4(3 - i) + 3(-5 + 2i)$

76. $2(1 + i) - 3(2 - 3i)$

77. $\dfrac{3}{3 + i}$

78. $\dfrac{4}{2 - i}$

79. i^{50}

80. i^{29}

81. $(2 + 3i)^3$

82. $(3 - 2i)^3$

In Problems 83–90, solve each equation in the complex number system.

83. $x^2 + x + 1 = 0$

84. $x^2 - x + 1 = 0$

85. $2x^2 + x - 2 = 0$

86. $3x^2 - 2x - 1 = 0$

87. $x^2 + 3 = x$

88. $2x^2 + 1 = 2x$

89. $x(1 - x) = 6$

90. $x(1 + x) = 2$

In Problems 91–96, find the following for each pair of points:
(a) *The distance between the points*
(b) *The midpoint of the line segment connecting the points*
(c) *The slope of the line containing the points*
(d) *Then interpret the slope found in part (c)*

91. $(0, 0)$; $(4, 2)$

92. $(0, 0)$; $(-4, 6)$

93. $(1, -1)$; $(-2, 3)$

94. $(-2, 2)$; $(1, 4)$

95. $(4, -4)$; $(4, 8)$

96. $(-3, 4)$; $(2, 4)$

97. Graph $y = -x^2 + 15$ using a graphing utility. Create a table of values to determine a good initial viewing window.

98. List the intercepts of the following graph.

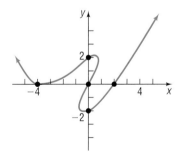

In Problems 99–108, find an equation of the line having the given characteristics. Express your answer using either the general form or the slope–intercept form of the equation of a line, whichever you prefer. Graph the line.

99. Slope $= -2$; containing the point $(3, -1)$

100. Slope $= 0$; containing the point $(-5, 4)$

101. Slope undefined; containing the point $(-3, 4)$

102. x-intercept $= 2$; containing the point $(4, -5)$

103. y-intercept $= -2$; containing the point $(5, -3)$

104. Containing the points $(3, -4)$ and $(2, 1)$

105. Parallel to the line $2x - 3y = -4$; containing the point $(-5, 3)$

106. Parallel to the line $x + y = 2$; containing the point $(1, -3)$

107. Perpendicular to the line $x + y = 2$; containing the point $(4, -3)$

108. Perpendicular to the line $3x - y = -4$; containing the point $(-2, 4)$

In Problems 109–112, graph each line by finding its intercepts.

109. $4x + 6y = 36$ **110.** $7x - 3y = 42$ **111.** $\frac{1}{2}x + \frac{5}{2}y = 10$ **112.** $-\frac{2}{3}x + \frac{1}{2}y = 2$

In Problems 113–118, graph each equation by hand by plotting points. Verify your results using a graphing utility. Approximate the intercepts using a graphing utility and label them on the graph.

113. $2x - 3y = 6$ **114.** $x + 2y = 4$ **115.** $y = x^2 - 9$ **116.** $y = x^2 + 4$

117. $x^2 + 2y = 16$ **118.** $2x^2 - 4y = 24$

In Problems 119–122, find the standard form of the equation of the circle whose center and radius are given.

119. $(h, k) = (-2, 3); r = 4$ **120.** $(h, k) = (3, 4); r = 4$ **121.** $(h, k) = (-1, -2); r = 1$ **122.** $(h, k) = (2, -4); r = 3$

In Problems 123–126, find the center and radius of each circle. Graph each circle by hand. Determine the intercepts of the graph of each circle.

123. $x^2 + y^2 - 2x + 4y - 4 = 0$ **124.** $x^2 + y^2 + 4x - 4y - 1 = 0$

125. $3x^2 + 3y^2 - 6x + 12y = 0$ **126.** $2x^2 + 2y^2 - 4x = 0$

---

127. Show that the points $A = (3, 4), B = (1, 1)$, and $C = (-2, 3)$ are the vertices of an isosceles triangle.

128. Show that the points $A = (-2, 0), B = (-4, 4)$, and $C = (8, 5)$ are the vertices of a right triangle in two ways:
(a) By using the converse of the Pythagorean Theorem.
(b) By using the slopes of the lines joining the vertices.

129. Show that the points $A = (2, 5), B = (6, 1)$, and $C = (8, -1)$ lie on a straight line by using slopes.

130. Show that the points $A = (1, 5), B = (2, 4)$, and $C = (-3, 5)$ lie on a circle with center $(-1, 2)$. What is the radius of this circle?

131. The endpoints of the diameter of a circle are $(-3, 2)$ and $(5, -6)$. Find the center and radius of the circle. Write the general equation of this circle.

132. Find two numbers y such that the distance from $(-3, 2)$ to $(5, y)$ is 10.

133. Translate the following statement into a mathematical expression: The perimeter p of a rectangle is the sum of two times the length l and two times the width w.

134. Translate the following statement into a mathematical expression: The total cost C of manufacturing x bicycles in one day is $50,000 plus $95 times the number of bicycles manufactured.

135. Banking A bank lends out $9000 at 7% simple interest. At the end of 1 year, how much interest is owed on the loan?

136. Financial Planning Steve, a recent retiree, requires $5000 per year in extra income. He has $70,000 to invest and can invest in A-rated bonds paying 8% per year or in a certificate of deposit (CD) paying 5% per year. How much money should be invested in each to realize exactly $5000 in interest per year?

137. Lightning and Thunder A flash of lightning is seen, and the resulting thunderclap is heard 3 seconds later. If the speed of sound averages 1100 feet per second, how far away is the storm?

138. Physics: Intensity of Light The intensity I (in candlepower) of a certain light source obeys the equation $I = \dfrac{900}{x^2}$, where x is the distance (in meters) from the light. Over what range of distances can an object be placed from this light source so that the range of intensity of light is from 1600 to 3600 candlepower, inclusive?

139. Extent of Search and Rescue A search plane has a cruising speed of 250 miles per hour and carries enough fuel for at most 5 hours of flying. If there is a wind that averages 30 miles per hour and the direction of the search is with the wind one way and against it the other, how far can the search plane travel before it has to turn back?

140. Extent of Search and Rescue If the search plane described in Problem 139 is able to add a supplementary fuel tank that allows for an additional 2 hours of flying, how much farther can the plane extend its search?

141. Rescue at Sea A life raft, set adrift from a sinking ship 150 miles offshore, travels directly toward a Coast Guard station at the rate of 5 miles per hour. At the time that the raft is set adrift, a rescue helicopter is dispatched from the Coast Guard station. If the helicopter's average speed is 90 miles per hour, how long will it take the helicopter to reach the life raft?

142. Physics: Uniform Motion Two bees leave two locations 150 meters apart and fly, without stopping, back and forth between these two locations at average speeds of 3 meters per second and 5 meters per second, respectively. How long is it until the bees meet for the first time? How long is it until they meet for the second time?

143. Physics: Uniform Motion A Metra commuter train leaves Union Station in Chicago at 12 noon. Two hours later, an Amtrak train leaves on the same track, traveling at an average speed that is 50 miles per hour faster than the Metra train. At 3 PM the Amtrak train is 10 miles behind the commuter train. How fast is each going?

144. Physics An object is thrown down from the top of a building 1280 feet tall with an initial velocity of 32 feet per second. The distance s (in feet) of the object from the ground after t seconds is $s = 1280 - 32t - 16t^2$.
(a) When will the object strike ground?
(b) What is the height of the object after 4 seconds?

145. Working Together to Get a Job Done Clarissa and Shawna, working together, can paint the exterior of a house in 6 days. Clarissa by herself can complete this job in 5 days less than Shawna. How long will it take Clarissa to complete the job by herself?

146. Emptying a Tank Two pumps of different sizes, working together, can empty a fuel tank in 5 hours. The larger pump can empty this tank in 4 hours less than the smaller. If the larger pump is out of order, how long will it take the smaller one to do the job alone?

147. Chemistry: Mixing Acids A laboratory has 60 cubic centimeters (cm^3) of a solution that is 40% HCl acid. How many cubic centimeters of a 15% solution of HCl acid should be mixed with the 60 cm^3 of 40% acid to obtain a solution of 25% HCl? How much of the 25% solution is there?

148. Business: Blending Coffee A coffee house has 20 pounds of a coffee that sells for $4 per pound. How many pounds of a coffee that sells for $8 per pound should be mixed with the 20 pounds of $4-per-pound coffee to obtain a blend that will sell for $5 per pound? How much of the $5-per-pound coffee is there to sell?

149. Chemistry: Salt Solutions How much water should be added to 64 ounces of a 10% salt solution to make a 2% salt solution?

150. Chemistry: Salt Solutions How much water must be evaporated from 64 ounces of a 2% salt solution to make a 10% salt solution?

151. Geometry The hypotenuse of a right triangle measures 13 centimeters. Find the lengths of the legs if their sum is 17 centimeters.

152. Geometry The diagonal of a rectangle measures 10 inches. If the length is 2 inches more than the width, find the dimensions of the rectangle.

---

153. Physics: Uniform Motion A man is walking at an average speed of 4 miles per hour alongside a railroad track. A freight train, going in the same direction at an average speed of 30 miles per hour, requires 5 seconds to pass the man. How long is the freight train? Give your answer in feet.

154. Framing a Painting An artist has 50 inches of oak trim to frame a painting. The frame is to have a border 3 inches wide surrounding the painting.
(a) If the painting is square, what are its dimensions? What are the dimensions of the frame?
(b) If the painting is rectangular with a length twice its width, what are the dimensions of the painting? What are the dimensions of the frame?

155. Using Two Pumps An 8-horsepower (hp) pump can fill a tank in 8 hours. A smaller, 3-hp pump fills the same tank in 12 hours. The pumps are used together to begin filling this tank. After 4 hours, the 8-hp pump breaks down. How long will it take the smaller pump to fill the tank?

156. Pleasing Proportion One formula stating the relationship between the length l and width w of a rectangle of "pleasing proportion" is $l^2 = w(l + w)$. How should a 4 foot by 8 foot sheet of plasterboard be cut so that the result is a rectangle of "pleasing proportion" with a width of 4 feet?

157. **Business: Determining the Cost of a Charter** A group of 20 senior citizens can charter a bus for a one-day excursion trip for $15 per person. The charter company agrees to reduce the price of each ticket by 10¢ for each additional passenger in excess of 20 who goes on the trip, up to a maximum of 44 passengers (the capacity of the bus). If the final bill from the charter company was $482.40, how many seniors went on the trip, and how much did each pay?

158. **Utilizing Copying Machines** A new copying machine can do a certain job in 1 hour less than an older copier. Together they can do this job in 72 minutes. How long would it take the older copier by itself to do the job?

159. **Evening Up a Race** In a 100-meter race, Todd crosses the finish line 5 meters ahead of Scott. To even things up, Todd suggests to Scott that they race again, this time with Todd lining up 5 meters behind the start.
   (a) Assuming that Todd and Scott run at the same pace as before, does the second race end in a tie?
   (b) If not, who wins?

(c) By how many meters does he win?
(d) How far back should Todd start so that the race ends in a tie?

After running the race a second time, Scott, to even things up, suggests to Todd that he (Scott) line up 5 meters in front of the start.
   (e) Assuming again that they run at the same pace as in the first race, does the third race result in a tie?
   (f) If not, who wins?
   (g) By how many meters?
   (h) How far up should Scott start so that the race ends in a tie?

160. Explain the differences among the following three problems. Are there any similarities in their solution?
   (a) Write the expression as a single quotient:

$$\frac{x}{x - 2} + \frac{x}{x^2 - 4}$$

   (b) Solve: $\dfrac{x}{x - 2} + \dfrac{x}{x^2 - 4} = 0$

   (c) Solve: $\dfrac{x}{x - 2} + \dfrac{x}{x^2 - 4} < 0$

Chapter Test

1. Suppose the points $(-2, -3)$ and $(4, 5)$ are the endpoints of a diameter of a circle.
   (a) Find the distance between the two points.
   (b) Find the midpoint of the line segment connecting the two points.
   (c) Find the equation of the circle, in standard form, containing the two points.
   (d) Graph the circle by hand.

In Problems 2–9, solve each equation algebraically. Answers should be exact.

2. $2x^2 + 6x = x - 3$

3. $x + 1 = \sqrt{x + 7}$

4. $2 - \dfrac{3}{m} = \dfrac{2}{m + 2}$

5. $5x - 8 = -4(x - 1) + 6$

6. $5|3 - 2b| - 7 = 8$

7. $x^4 + x^2 = 3x^2 + 8$

8. $x^2 - 4x + 7 = 0$

9. $2x^2 + x - 1 = x(x + 7) + 2$

In Problems 10–12, graph each equation by hand by plotting points. Use a graphing utility to approximate intercepts and label them on the graph.

10. $y - 14 = (x + 1)^2$

11. $2x - 7y = 21$

12. $y = x^3 - 3x + 2$

13. Given the linear equation $10x - 6y = 20$, do the following:
   (a) Find the slope of the line.
   (b) Find the intercepts of the graph of the line.
   (c) Graph the equation using the intercepts.
   (d) Find the equation of the line that is perpendicular to the given line and contains the point $(-1, 5)$.
   (e) Find the equation of the line that is parallel to the given line and contains the point $(-1, 5)$.

In Problems 14–16, use a graphing utility to approximate the real solutions of each equation rounded to two decimal places. All solutions lie between −10 and 10.

14. $2x^3 - x^2 - 2x + 1 = 0$

15. $x^4 - 5x^2 - 8 = 0$

16. $-x^3 + 7x - 2 = x^2 + 3x - 3$

In Problems 17–19, solve each inequality. Express your answer in set-builder notation or interval notation. Graph the solution set.

17. $\dfrac{2x + 3}{4} < -2$

18. $|2x + 3| - 4 \geq 3$

19. $-7 < 3 - 5x \leq 8$

In Problems 20–22, write each expression in the standard form $a + bi$.

20. $2(3 - 7i) - (4 + 11i)$

21. $(3 + 10i)(8 + i)$

22. $\dfrac{2 + i}{5 - 3i}$

---

23. Jamie is a cashier at a local supermarket. On average, she can check out a customer in 5 minutes. Scott, a cashier trainee, takes an average of 8.5 minutes to check out customers. How long would it take Jamie and Scott to check out 65 customers if they work together at different registers? Round to two decimal places if necessary.

24. A health food retailer sells a mixture of dried cherries, cranberries, and pecans for $15.00 per pound and dried banana chips for $2.25 per pound. The retailer decides to create a new mix by adding banana chips to the original mix. How many pounds of banana chips must be mixed with 40 pounds of the original mix to obtain a new mixture that sells for $10.25 per pound with no loss in revenue? Round to two decimal places if necessary.

25. A Mountainsmith Auspex 4000 backpack originally sold for $275.00 but is advertised at 42% off. What is the sale price of the backpack?

26. Glenn invests $10,000 in a Certificate of Deposit (CD) that pays simple interest of 4% per annum. How much interest will Glenn earn after 3 months?

Chapter Projects

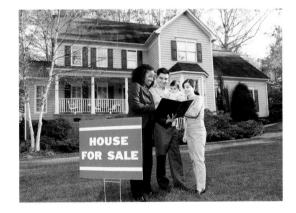

1. **Home Mortgages** While you may not be in the market for a home right now, it is probably an event that will occur for you within the next few years. Formula (1) gives the monthly payment P required to pay off a loan L at an annual rate of interest r, expressed as a decimal, but usually given as a percent. The time t, measured in months, is the du-

ration of the loan, so a 15-year mortgage requires $t = 12 \times 15 = 180$ monthly payments.

$$P = L\left[\frac{\dfrac{r}{12}}{1 - \left(1 + \dfrac{r}{12}\right)^{-t}}\right] \qquad \begin{array}{l} P = \text{monthly payment} \\ L = \text{loan amount} \\ r = \text{annual rate of interest,} \\ \quad \text{expressed as a decimal} \\ t = \text{length of loan, in months} \end{array} \quad \textbf{(1)}$$

On July 8, 2003, average interest rates were:
(a) 5.52% on 30-year mortgages
(b) 4.85% on 15-year mortgages

1. For each rate, calculate the monthly payment for a loan of $200,000.

2. Then compute the total amount paid over the term of each loan.

3. Calculate the interest paid on each loan.

On August 5, 2004, average interest rates were:
(a) 5.99% on 30-year mortgages
(b) 5.40% on 15-year mortgages

4. For each rate, calculate the monthly payment for a loan of $200,000.

5. Then compute the total amount paid over the term of each loan.

6. Calculate the interest paid on each loan.

7. Solve equation (1) for L.

8. If you can afford to pay $1000 per month for a mortgage payment, calculate the amount you can borrow on August 5, 2004.

   (a) on a 30-year mortgage.

   (b) on a 15-year mortgage.

9. Check with your local lending institution for current rates for 30-year and 15-year mortages. Calculate how much you can borrow with a $1000 per month payment.

10. Repeat Problem 9 if you can afford a payment of $1300 per month.

11. Do you think that the interest rate plays an important role in determining how much you can afford to pay for a house?

12. Comment on the two types of mortgages: 30-year and 15-year. Which would you take? Why?

The following projects are available on the Instructor's Resource Center (IRC):

2. **Project at Motorola**   *Mobile Phone Usage*

3. **Economics**   *Isocost Lines*

Functions and Their Graphs

2

A LOOK BACK Up to now, our discussion has focused on equations. We have solved equations containing one variable and developed techniques for graphing equations containing two variables.

A LOOK AHEAD In this chapter, we look at a special type of equation involving two variables called a *function*. This chapter deals with what a function is, how to graph functions, properties of functions, and how functions are used in applications. The word function apparently was introduced by René Descartes in 1637. For him, a function simply meant any positive integral power of a variable *x*. Gottfried Wilhelm Leibniz (1646–1716), who always emphasized the geometric side of mathematics, used the word function to denote any quantity associated with a curve, such as the coordinates of a point on the curve. Leonhard Euler (1707–1783) employed the word to mean any equation or formula involving variables and constants. His idea of a function is similar to the one most often seen in courses that precede calculus. Later, the use of functions in investigating heat flow equations led to a very broad definition, due to Lejeune Dirichlet (1805–1859), which describes a function as a rule or correspondence between two sets. It is his definition that we use here.

Portable Phones Power Up

Latest Census Bureau Data Show Rise in Cell Phones

Jan. 24—Cell phone ownership jumped from just over 5.2 million in 1990 to nearly 110 million in 2000—a more than twenty-fold increase in just 10 years, new figures show. Driving the phenomenal growth are several factors—one of which is the decreasing costs. The data note that the average monthly cell phone bill has been almost cut in half—from $81 to just over $45—over the past decade. Survey data, compiled by the Cellular Telecommunications and Internet Association (CTIA) in Washington, D.C., and published in the U.S. Census Bureau's *Statistical Abstract of the United States: 2001* report, point to the rapid adoption for cell phone technology as a key reason behind the growth.

  "The cell phone industry has shown remarkable growth over the decade," says Glenn King, chief of the statistical compendia branch of the Commerce Department that produces the abstract. Noting that there were only 55 million users in 1997, "It's doubled in the last three years alone," he says. With cost of service coming down, cell phone service becomes accessible to everyone," says Charles Golvin, a senior analyst with Forrester Research. "People can be in touch anytime and anywhere they want to."

By Paul Eng, abcNEWS.com

—See Chapter Project 1.

OUTLINE

2.1 Symmetry; Graphing Key Equations

PREPARING FOR THIS SECTION   *Before getting started, review the following:*

• Introduction to Graphing Equations (Section 1.1, pp. 90–96)   • Intercepts (Section 1.8, pp. 171–172)

Now work the 'Are You Prepared?' problems on page 202.

OBJECTIVES   1 Test an Equation for Symmetry with Respect to (a) the *x*-Axis, (b) the *y*-Axis, and (c) the Origin

2 Know How to Graph Key Equations

In Chapter 1, we saw the role that intercepts play in obtaining key points on the graph of an equation. Recall, to find the *x*-intercepts of an equation, if any, we let $y = 0$ and solve for *x*. To find the *y*-intercepts of an equation, if any, we let $x = 0$ and solve for *y*. Another helpful tool for graphing equations by hand involves *symmetry*, particularly symmetry with respect to the *x*-axis, the *y*-axis, and the origin.

1 Test an Equation for Symmetry

A graph is said to be **symmetric with respect to the *x*-axis** if, for every point (x, y) on the graph, the point $(x, -y)$ is also on the graph.

A graph is said to be **symmetric with respect to the *y*-axis** if, for every point (x, y) on the graph, the point $(-x, y)$ is also on the graph.

A graph is said to be **symmetric with respect to the origin** if, for every point (x, y) on the graph, the point $(-x, -y)$ is also on the graph.

Figure 1 illustrates the definition. Notice that, when a graph is symmetric with respect to the *x*-axis, the part of the graph above the *x*-axis is a reflection or mirror image of the part below it, and vice versa. And when a graph is symmetric with respect to the *y*-axis, the part of the graph to the right of the *y*-axis is a reflection of the part to the left of it, and vice versa. Symmetry with respect to the origin may be viewed in two ways:

1. As a reflection about the *y*-axis, followed by a reflection about the *x*-axis

2. As a projection along a line through the origin so that the distances from the origin are equal

Figure 1

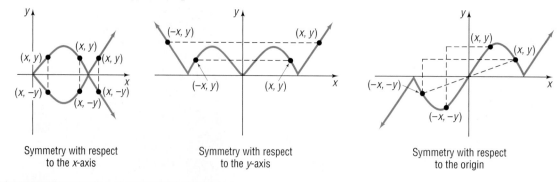

Symmetry with respect to the *x*-axis

Symmetry with respect to the *y*-axis

Symmetry with respect to the origin

EXAMPLE 1   **Symmetric Points**

(a) If a graph is symmetric with respect to the *x*-axis and the point $(4, 2)$ is on the graph, then the point $(4, -2)$ is also on the graph.

(b) If a graph is symmetric with respect to the y-axis and the point $(4, 2)$ is on the graph, then the point $(-4, 2)$ is also on the graph.

(c) If a graph is symmetric with respect to the origin and the point $(4, 2)$ is on the graph, then the point $(-4, -2)$ is also on the graph. ◀

NOW WORK PROBLEM **7**.

When the graph of an equation is symmetric with respect to the x-axis, the y-axis, or the origin, the number of points that you need to plot in order to see the pattern is reduced. For example, if the graph of an equation is symmetric with respect to the y-axis, then, once points to the right of the y-axis are plotted, an equal number of points on the graph can be obtained by reflecting them about the y-axis. Because of this, before we graph an equation, we first want to determine whether it has any symmetry. The following tests are used for this purpose.

Tests for Symmetry

To test the graph of an equation for symmetry with respect to the

x-Axis   Replace y by $-y$ in the equation. If an equivalent equation results, the graph of the equation is symmetric with respect to the x-axis.

y-Axis   Replace x by $-x$ in the equation. If an equivalent equation results, the graph of the equation is symmetric with respect to the y-axis.

Origin   Replace x by $-x$ and y by $-y$ in the equation. If an equivalent equation results, the graph of the equation is symmetric with respect to the origin.

| EXAMPLE 2 | **Finding Intercepts and Testing an Equation for Symmetry** |

For the equation $y = \dfrac{x^2 - 4}{x^2 + 1}$ (a) find the intercepts and (b) test for symmetry.

Solution (a) To obtain the x-intercept(s), let $y = 0$ in the equation and solve for x.

$$\frac{x^2 - 4}{x^2 + 1} = 0 \qquad \text{Let } y = 0$$

$$x^2 - 4 = 0 \qquad \text{Multiply both sides by } x^2 + 1$$

$$x = -2 \quad \text{or} \quad x = 2 \qquad \text{Factor, use the Square Root Method and simplify}$$

To obtain the y-intercept(s), let $x = 0$ in the equation and solve for y.

$$\frac{0^2 - 4}{0^2 + 1} = \frac{-4}{1} = -4$$

The x-intercepts are -2 and 2; the y-intercept is -4.

(b) We now test the equation for symmetry with respect to the x-axis, the y-axis, and the origin.

x-Axis: To test for symmetry with respect to the x-axis, replace y by $-y$. Since $-y = \dfrac{x^2 - 4}{x^2 + 1}$ is not equivalent to $y = \dfrac{x^2 - 4}{x^2 + 1}$, the graph of the equation is not symmetric with respect to the x-axis.

y-Axis: To test for symmetry with respect to the y-axis, replace x by $-x$. Since $y = \dfrac{(-x)^2 - 4}{(-x)^2 + 1} = \dfrac{x^2 - 4}{x^2 + 1}$ is equivalent to $y = \dfrac{x^2 - 4}{x^2 + 1}$, the graph of the equation is symmetric with respect to the y-axis.

Origin: To test for symmetry with respect to the origin, replace x by $-x$ and y by $-y$.

$$-y = \frac{(-x)^2 - 4}{(-x)^2 + 1} \qquad \text{Replace x by } -x \text{ and y by } -y.$$

$$-y = \frac{x^2 - 4}{x^2 + 1} \qquad \text{Simplify.}$$

Figure 2

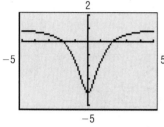

Since the result is not equivalent to the original equation, the graph of the equation $y = \frac{x^2 - 4}{x^2 + 1}$ is not symmetric with respect to the origin. ◄

—— Seeing the Concept ——

Figure 2 shows the graph of $y = \frac{x^2 - 4}{x^2 + 1}$ using a graphing utility. Do you see the symmetry with respect to the y-axis?

NOW WORK PROBLEM 33.

2 Know How to Graph Key Equations

The next three examples use intercepts, symmetry, and point plotting to obtain the graphs of key equations. It is important to know the graphs of these key equations because we use them later in the chapter. The first of these is $y = x^3$.

EXAMPLE 3 | ### Graphing the Equation $y = x^3$ by Finding Intercepts and Checking for Symmetry

Graph the equation $y = x^3$ by hand by plotting points. Find any intercepts and check for symmetry first.

Solution First, we seek the intercepts. When $x = 0$, then $y = 0$; and when $y = 0$, then $x = 0$. The origin $(0, 0)$ is the only intercept. Now we test for symmetry.

x-Axis: Replace y by $-y$. Since $-y = x^3$ is not equivalent to $y = x^3$, the graph is not symmetric with respect to the x-axis.

y-Axis: Replace x by $-x$. Since $y = (-x)^3 = -x^3$ is not equivalent to $y = x^3$, the graph is not symmetric with respect to the y-axis.

Origin: Replace x by $-x$ and y by $-y$. Since $-y = (-x)^3 = -x^3$ is equivalent to $y = x^3$ (multiply both sides by -1), the graph is symmetric with respect to the origin.

To graph by hand, we use the equation to obtain several points on the graph. Because of the symmetry, we only need to locate points on the graph for which $x \geq 0$. See Table 1. Points on the graph could also be obtained using the TABLE feature on a graphing utility. See Table 2. Do you see the symmetry with respect to the origin from the table? Figure 3 shows the graph.

Figure 3

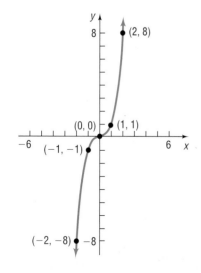

Table 1

x	$y = x^3$	(x, y)
0	0	(0, 0)
1	1	(1, 1)
2	8	(2, 8)
3	27	(3, 27)

Table 2

X	Y1
-3	-27
-2	-8
-1	-1
0	0
1	1
2	8
3	27

Y1■X^3

◄

| EXAMPLE 4 | **Graphing the Equation $x = y^2$** |

Graph the equation $x = y^2$. Find any intercepts and check for symmetry first.

Solution

The lone intercept is $(0, 0)$. The graph is symmetric with respect to the x-axis since $x = (-y)^2$ is equivalent to $x = y^2$. The graph is not symmetric with respect to the y-axis or the origin.

To graph $x = y^2$ by hand, we use the equation to obtain several points on the graph. Because the equation is solved for x, it is easier to assign values to y and use the equation to determine the corresponding values of x. See Table 3. Because of the symmetry, we can restrict ourselves to points whose y-coordinates are positive. We then use the symmetry to find additional points on the graph. For example, since $(1, 1)$ is on the graph, so is $(1, -1)$. Since $(4, 2)$ is on the graph, so is $(4, -2)$, and so on. We plot these points and connect them with a smooth curve to obtain Figure 4.

To graph the equation $x = y^2$ using a graphing utility, we must write the equation in the form $y = \{\text{expression in } x\}$. We proceed to solve for y.

$$x = y^2$$
$$y^2 = x$$
$$y = \pm\sqrt{x} \qquad \text{Square Root Method}$$

To graph $x = y^2$, we need to graph both $Y_1 = \sqrt{x}$ and $Y_2 = -\sqrt{x}$ on the same screen. Figure 5 shows the result. Table 4 shows various values of y for a given value of x when $Y_1 = \sqrt{x}$ and $Y_2 = -\sqrt{x}$. Notice that when $x < 0$ we get an error. Can you explain why?

Table 3

y	$x = y^2$	(x, y)
0	0	$(0, 0)$
1	1	$(1, 1)$
2	4	$(4, 2)$
3	9	$(9, 3)$

Figure 4

Figure 5

Table 4

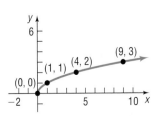

◀

Look again at either Figure 4 or 5. If we restrict y so that $y \geq 0$, the equation $x = y^2$, $y \geq 0$, may be written as $y = \sqrt{x}$. The portion of the graph of $x = y^2$ in quadrant I is the graph of $y = \sqrt{x}$. See Figure 6.

Figure 6

EXAMPLE 5   **Graphing the Equation** $y = \dfrac{1}{x}$

Graph the equation $y = \dfrac{1}{x}$. Find any intercepts and check for symmetry first.

Solution   We check for intercepts first. If we let $x = 0$, we obtain a 0 in the denominator, which is not defined. We conclude that there is no y-intercept. If we let $y = 0$, we get the equation $\dfrac{1}{x} = 0$, which has no solution. We conclude that there is no x-intercept.

The graph of $y = \dfrac{1}{x}$ does not cross or touch the coordinate axes.

Next we check for symmetry.

x-Axis: Replacing y by $-y$ yields $-y = \dfrac{1}{x}$, which is not equivalent to $y = \dfrac{1}{x}$.

y-Axis: Replacing x by $-x$ yields $y = \dfrac{1}{-x} = -\dfrac{1}{x}$, which is not equivalent to $y = \dfrac{1}{x}$.

Origin: Replacing x by $-x$ and y by $-y$ yields $-y = -\dfrac{1}{x}$, which is equivalent to $y = \dfrac{1}{x}$. The graph is symmetric only with respect to the origin.

We can use the equation to form Table 5 and obtain some points on the graph. Because of symmetry, we only find points (x, y) for which x is positive. From Table 5 we infer that, if x is a large and positive number, then $y = \dfrac{1}{x}$ is a positive number close to 0. We also infer that if x is a positive number close to 0 then $y = \dfrac{1}{x}$ is a large and positive number. Armed with this information, we can graph the equation.

Figure 7 illustrates some of these points and the graph of $y = \dfrac{1}{x}$. Observe how the absence of intercepts and the existence of symmetry with respect to the origin were utilized. Figure 8 confirms our algebraic analysis using a TI-84 Plus.

Table 5

x	$y = \dfrac{1}{x}$	(x, y)
$\dfrac{1}{10}$	10	$\left(\dfrac{1}{10}, 10\right)$
$\dfrac{1}{3}$	3	$\left(\dfrac{1}{3}, 3\right)$
$\dfrac{1}{2}$	2	$\left(\dfrac{1}{2}, 2\right)$
1	1	$(1, 1)$
2	$\dfrac{1}{2}$	$\left(2, \dfrac{1}{2}\right)$
3	$\dfrac{1}{3}$	$\left(3, \dfrac{1}{3}\right)$
10	$\dfrac{1}{10}$	$\left(10, \dfrac{1}{10}\right)$

Figure 7

Figure 8

2.1 Assess Your Understanding

'Are You Prepared?'

Answers are given at the end of these exercises. If you get a wrong answer, read the pages listed in red.

1. Graph $y = x^2 - 1$ by plotting points. (pp. 91–92)

2. Find the intercepts of $3x - 5y = 15$. (pp. 171–172)

Concepts and Vocabulary

3. If for every point (x, y) on the graph of an equation, the point $(-x, y)$ is also on the graph, then the graph is symmetric with respect to the _____.

4. The graph of the equation $y = x^3 - 5x$ is symmetric with respect to the _____.

5. *True or False:* The graph of the equation $y = x^4 + x^2 + 1$ is symmetric with respect to the y-axis.

6. *True or False:* The graph of a circle whose center is at the origin will have symmetry with respect to the x-axis, the y-axis, and the origin.

Skill Building

In Problems 7–16, plot each point. Then plot the point that is symmetric to it with respect to (a) the x-axis; (b) the y-axis; (c) the origin.

7. $(3, 4)$ **8.** $(5, 3)$ **9.** $(-2, 1)$ **10.** $(4, -2)$ **11.** $(5, -2)$

12. $(-1, -1)$ **13.** $(-3, -4)$ **14.** $(4, 0)$ **15.** $(0, -3)$ **16.** $(-3, 0)$

In Problems 17–24, the graph of an equation is given. Find the intercepts. Indicate whether the graph is symmetric with respect to the x-axis, the y-axis, or the origin.

17.

18.

19.

20.

21.

22.

23.

24.

In Problems 25–28, draw a complete graph so that it has the type of symmetry indicated.

25. y-axis

26. x-axis

27. Origin

28. y-axis
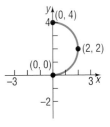

In Problems 29–32, draw a quick sketch of the equation by hand. Be sure to label all intercepts.

29. $y = x^3$ **30.** $x = y^2$ **31.** $y = \sqrt{x}$ **32.** $y = \dfrac{1}{x}$

In Problems 33–48, (a) find the intercepts of each equation and (b) test each equation for symmetry with respect to the x-axis, the y-axis, and the origin. Verify your results using a graphing utility.

33. $y = 3x$ **34.** $y = -5x$ **35.** $y = x^2 - 9$ **36.** $y = 4 - x^2$

37. $x = y^2 - 4$ **38.** $x = y^2 - 1$ **39.** $y = x^3 - 8x$ **40.** $y = -2x^3 + 8x$

41. $x^2 + y^2 = 9$ **42.** $x^2 + y^2 = 25$ **43.** $y + 4 = x^2 - 3x$ **44.** $y - 8 = x^2 + 6x$

45. $y + 4 = |x|$ **46.** $y = |x - 4|$ **47.** $y = \dfrac{x^2 - 9}{x}$ **48.** $y = \dfrac{x^2 - 4}{x}$

Applications and Extensions

49. Given that the point $(1, 2)$ is on the graph of an equation that is symmetric with respect to the origin, what other point is on the graph?

50. If the graph of an equation is symmetric with respect to the y-axis and 6 is an x-intercept of this graph, name another x-intercept.

51. If the graph of an equation is symmetric with respect to the origin and -4 is an x-intercept of this graph, name another x-intercept.

Discussion and Writing

52. An equation is being tested for symmetry with respect to the x-axis, the y-axis, and the origin. Explain why, if two of these symmetries are present, the remaining one must also be present.

53. Describe three types of symmetry that the graph of an equation could have.

'Are You Prepared?' Answers

1.

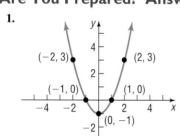

2. $(0, -3); (5, 0)$

2.2 Functions

PREPARING FOR THIS SECTION *Before getting started, review the following:*

- Intervals (Section 1.7, pp. 151–152)
- Evaluating Algebraic Expressions, Domain of a Variable (Review, Section R.2, pp. 19–20)
- Solving Inequalities (Section 1.7, pp. 154–157)

↘ Now work the 'Are You Prepared?' problems on page 216.

OBJECTIVES   1 Determine Whether a Relation Represents a Function
  2 Find the Value of a Function
  3 Find the Domain of a Function
  4 Form the Sum, Difference, Product, and Quotient of Two Functions

Figure 9

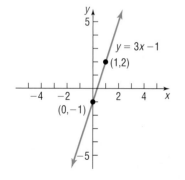

1 Determine Whether a Relation Represents a Function

We often see situations where one variable is somehow linked to the value of another variable. For example, an individual's level of education is linked to annual income. Engine size is linked to gas mileage. When the value of one variable is related to the value of a second variable, we have a *relation*. A **relation** is a correspondence between two sets. If x and y are two elements in these sets and if a relation exists between x and y, then we say that x **corresponds** to y or that y **depends on** x, and we write $x \rightarrow y$.

We have a number of ways to express relations between two sets. For example, the equation $y = 3x - 1$ shows a relation between x and y. It says that if we take some number x multiply it by 3 and then subtract 1, we obtain the corresponding value of y. In this sense, x serves as the **input** to the relation and y is the **output** of the relation. We can also express this relation as a graph as shown in Figure 9.

Not only can relations be expressed through an equation or graph, but we can also express relations through a technique called *mapping*. A **map** illustrates a relation by using a set of inputs and drawing arrows to the corresponding element in the set of outputs. Ordered pairs can be used to represent $x \rightarrow y$ as (x, y). We illustrate these two concepts in Example 1.

EXAMPLE 1	**Maps and Ordered Pairs as Relations**

Figure 10 shows a relation between states and the number of representatives each has in the House of Representatives. The relation might be named "number of representatives."

Figure 10

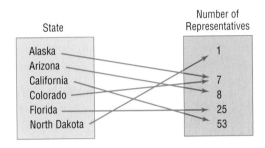

In this relation, Alaska corresponds to 7, Arizona corresponds to 8, and so on. Using ordered pairs, this relation would be expressed as

$$\{(\text{Alaska}, 7), (\text{Arizona}, 8), (\text{California}, 53), (\text{Colorado}, 7), (\text{Florida}, 25), (\text{North Dakota}, 1)\} \quad \blacktriangleleft$$

We now present what is one of the most important concepts in algebra—the *function*. A function is a special type of relation. To understand the idea behind a function, let's revisit the relation presented in Example 1. If we were to ask "How many representatives does Alaska have" you would respond "7." In other words, each input "state" corresponds to a single output "number of representatives."

Let's consider a second relation where we have a correspondence between four people and their phone numbers. See Figure 11. Notice that Maureen has two telephone numbers; therefore, if asked, "What is Maureen's phone number?", you cannot assign a single number to her.

Figure 11

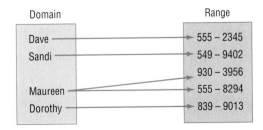

Let's look at one more relation. Figure 12 is a relation that shows a correspondence between "animals" and "life expectancy." If asked to determine the average life expectancy of a dog, we would all respond "11 years." If asked to determine the average life expectancy of a rabbit, we would all respond "7 years."

Figure 12

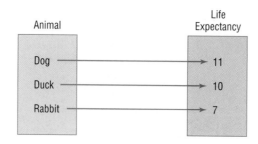

Notice that the relations presented in Figures 10 and 12 have something in common. What is it? The common link between these two relations is that each input corresponds to only one output. This leads to the definition of a *function*.

Figure 13

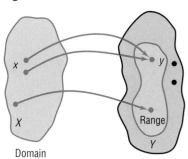

Let X and Y be two nonempty sets.[*] A **function** from X into Y is a relation that associates with each element of X exactly one element of Y.

The set X is called the **domain** of the function. For each element x in X, the corresponding element y in Y is called the **value** of the function at x, or the **image** of x. The set of all images of the elements in the domain is called the **range** of the function. See Figure 13.

Since there may be some elements in Y that are not the image of some x in X, it follows that the range of a function may be a subset of Y, as shown in Figure 13.

Not all relations between two sets are functions. The next example shows how to determine whether a relation is a function or not.

EXAMPLE 2   **Determining Whether a Relation Represents a Function**

Determine whether the following relations represent functions. If the relation is a function, then state its domain and range.

(a) See Figure 14. For this relation, the domain represents the level of education and the range represents the unemployment rate.

Figure 14
[SOURCE: *Statistical Abstract of the United States, 2003*]

(b) See Figure 15. For this relation, the domain represents the number of calories in a sandwich from a fast-food restaurant and the range represents the fat content (in grams).

---

[*]The sets X and Y will usually be sets of real numbers, in which case a (real) function results. The two sets can also be sets of complex numbers, and then we have defined a complex function. In the broad definition (due to Lejeune Dirichlet), X and Y can be any two sets.

Figure 15
[SOURCE: Each company's Web site]

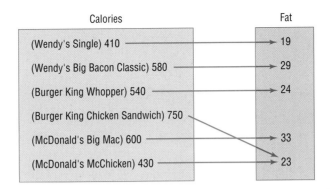

Calories

(Wendy's Single) 410
(Wendy's Big Bacon Classic) 580
(Burger King Whopper) 540
(Burger King Chicken Sandwich) 750
(McDonald's Big Mac) 600
(McDonald's McChicken) 430

Fat

19
29
24
33
23

(c) See Figure 16. For this relation, the domain represents the weight of pear-cut diamonds and the range represents their price.

Figure 16
[SOURCE: diamonds.com]

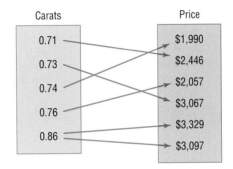

Carats

0.71
0.73
0.74
0.76
0.86

Price

$1,990
$2,446
$2,057
$3,067
$3,329
$3,097

Solution

(a) The relation is a function because each element in the domain corresponds to exactly one element in the range. The domain of the relation is {No High School Diploma, High School Diploma, Some College, College Graduate} and the range of the relation is {7.7%, 5.9%, 5.4%, 3.4%}.

(b) The relation is a function because each element in the domain corresponds to exactly one element in the range. Notice that it is okay for more than one element in the domain to correspond to the same element in the range (McDonald's and Burger King's chicken sandwich both have 23 grams of fat). The domain of the relation is {410, 580, 540, 750, 600, 430}. The range of the relation is {19, 29, 24, 33, 23}.

(c) The relation is not a function because each element in the domain does not correspond to exactly one element in the range. If a 0.86 carat diamond is chosen from the domain, a single price cannot be assigned to it. ◄

NOW WORK PROBLEM 15.

In Words

For a function, no input has more than one output.

The idea behind a function is its predictability. If the input is known, we can use the function to determine the output. With "nonfunctions," we don't have this predictability. Look back at Figure 15. The inputs are {410, 580, 540, 750, 600, 430}. The correspondence is "number of fat grams" and the outputs are {19, 29, 24, 33, 23}. If asked "How many fat grams in a 410-calorie sandwich?" we can use the correspondence to answer "19." Now consider Figure 16. If asked "What is the price of a 0.86 carat diamond?" we could not give a single response because two outputs result from the single input "0.86." For this reason, the relation in Figure 16 is not a function.

In Words

For a function, the domain is the set of inputs, and the range is the set of outputs.

We may also think of a function as a set of ordered pairs (x, y) in which no two ordered pairs have the same first element, but different second elements. The set of all first elements x is the domain of the function, and the set of all second elements y is its range. Each element x in the domain corresponds to exactly one element y in the range.

EXAMPLE 3 **Determining Whether a Relation Represents a Function**

Determine whether each relation represents a function. If it is a function, state the domain and range.

(a) $\{(1, 4), (2, 5), (3, 6), (4, 7)\}$

(b) $\{(1, 4), (2, 4), (3, 5), (6, 10)\}$

(c) $\{(-3, 9), (-2, 4), (0, 0), (1, 1), (-3, 8)\}$

Solution (a) This relation is a function because there are no ordered pairs with the same first element and different second elements. The domain of this function is $\{1, 2, 3, 4\}$, and its range is $\{4, 5, 6, 7\}$.

(b) This relation is a function because there are no ordered pairs with the same first element and different second elements. The domain of this function is $\{1, 2, 3, 6\}$, and its range is $\{4, 5, 10\}$.

(c) This relation is not a function because there are two ordered pairs, $(-3, 9)$ and $(-3, 8)$, that have the same first element, but different second elements. ◀

In Example 3(b), notice that 1 and 2 in the domain each have the same image in the range. This does not violate the definition of a function; two different first elements can have the same second element. A violation of the definition occurs when two ordered pairs have the same first element and different second elements, as in Example 3(c).

NOW WORK PROBLEM 19.

Up to now we have shown how to identify when a relation is a function for relations defined by mappings [Example 2] and ordered pairs [Example 3]. We know that relations can also be expressed as equations. We discuss next the circumstances under which equations are functions.

To determine whether an equation, where y depends on x, is a function, it is often easiest to solve the equation for y. If any value of x in the domain corresponds to more than one y, the equation does not define a function; otherwise, it does define a function.

EXAMPLE 4 **Determining Whether an Equation Is a Function**

Determine if the equation $y = 2x - 5$ defines y as a function of x.

Solution The equation tells us to take an input x, multiply it by 2, and then subtract 5. For any input x, these operations yield only one output y. For example, if $x = 1$, then $y = 2(1) - 5 = -3$. If $x = 3$, then $y = 2(3) - 5 = 1$. For this reason, the equation is a function. ◀

EXAMPLE 5	**Determining Whether an Equation Is a Function**

Determine if the equation $x^2 + y^2 = 1$ defines y as a function of x.

Solution   To determine whether the equation $x^2 + y^2 = 1$, which defines the unit circle, is a function, we need to solve the equation for y.

$$x^2 + y^2 = 1$$
$$y^2 = 1 - x^2$$

$$y = \pm\sqrt{1 - x^2} \qquad \text{Square Root Method}$$

For values of x between -1 and 1, two values of y result. For example, if $x = 0$, then $y = \pm1$, so two different outputs result from the same input. This means that the equation $x^2 + y^2 = 1$ does not define a function.   ◀

NOW WORK PROBLEM 33.

2   Find the Value of a Function

Functions are often denoted by letters such as f, F, g, G, and others. If f is a function, then for each number x in its domain the corresponding image in the range is designated by the symbol $f(x)$, read as "f of x" or as "f at x." We refer to $f(x)$ as the **value of f at the number x**; $f(x)$ is the number that results when x is given and the function f is applied; $f(x)$ does *not* mean "f times x." For example, the function given in Example 4 may be written as $y = f(x) = 2x - 5$. Then $f\left(\dfrac{3}{2}\right) = -2$.

Figure 17 illustrates some other functions. Notice that, in every function illustrated, for each x in the domain there is one value in the range.

Figure 17

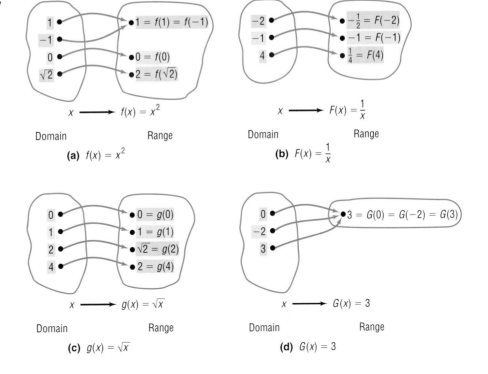

(a) $f(x) = x^2$

(b) $F(x) = \dfrac{1}{x}$

(c) $g(x) = \sqrt{x}$

(d) $G(x) = 3$

Figure 18

Input x

Output
$y = f(x)$

Sometimes it is helpful to think of a function f as a machine that receives as input a number from the domain, manipulates it, and outputs the value. See Figure 18.

The restrictions on this input/output machine are as follows:

1. It only accepts numbers from the domain of the function.

2. For each input, there is exactly one output (which may be repeated for different inputs).

For a function $y = f(x)$, the variable x is called the **independent variable**, because it can be assigned any of the permissible numbers from the domain. The variable y is called the **dependent variable**, because its value depends on x.

Any symbol can be used to represent the independent and dependent variables. For example, if f is the *cube function*, then f can be given by $f(x) = x^3$ or $f(t) = t^3$ or $f(z) = z^3$. All three functions are the same. Each tells us to cube the independent variable. In practice, the symbols used for the independent and dependent variables are based on common usage, such as using C for cost in business.

The independent variable is also called the **argument** of the function. Thinking of the independent variable as an argument can sometimes make it easier to find the value of a function. For example, if f is the function defined by $f(x) = x^3$, then f tells us to cube the argument. Thus, $f(2)$ means to cube 2, $f(a)$ means to cube the number a, and $f(x + h)$ means to cube the quantity $x + h$.

EXAMPLE 6   **Finding Values of a Function**

For the function f defined by $f(x) = 2x^2 - 3x$, evaluate

(a) $f(3)$ (b) $f(x) + f(3)$ (c) $f(-x)$

(d) $-f(x)$ (e) $f(x + 3)$ (f) $\dfrac{f(x + h) - f(x)}{h}$, $h \neq 0$

Solution   (a) We substitute 3 for x in the equation for f to get

$$f(3) = 2(3)^2 - 3(3) = 18 - 9 = 9$$

(b) $f(x) + f(3) = (2x^2 - 3x) + (9) = 2x^2 - 3x + 9$

(c) We substitute $-x$ for x in the equation for f.

$$f(-x) = 2(-x)^2 - 3(-x) = 2x^2 + 3x$$

(d) $-f(x) = -(2x^2 - 3x) = -2x^2 + 3x$

(e) $f(x + 3) = 2(x + 3)^2 - 3(x + 3)$ Notice the use of parentheses here.

$$= 2(x^2 + 6x + 9) - 3x - 9$$

$$= 2x^2 + 12x + 18 - 3x - 9$$

$$= 2x^2 + 9x + 9$$

(f) $\dfrac{f(x + h) - f(x)}{h} = \dfrac{[2(x + h)^2 - 3(x + h)] - [2x^2 - 3x]}{h}$

$\uparrow$

$f(x + h) = 2(x + h)^2 - 3(x + h)$

$= \dfrac{2(x^2 + 2xh + h^2) - 3x - 3h - 2x^2 + 3x}{h}$ Simplify.

$= \dfrac{2x^2 + 4xh + 2h^2 - 3h - 2x^2}{h}$ Simplify.

$= \dfrac{4xh + 2h^2 - 3h}{h}$ Simplify.

$= \dfrac{h(4x + 2h - 3)}{h}$ Factor out h.

$= 4x + 2h - 3$ Cancel h.   ◄

Notice in this example that $f(x + 3) \neq f(x) + f(3)$ and $f(-x) \neq -f(x)$. The expression in part (f) is called the **difference quotient** of f, an important expression in calculus.

NOW WORK PROBLEMS 39 AND 73.

Most calculators have special keys that enable you to find the value of certain commonly used functions. For example, you should be able to find the square function $f(x) = x^2$, the square root function $f(x) = \sqrt{x}$, the reciprocal function $f(x) = \dfrac{1}{x} = x^{-1}$, and many others that will be discussed later in this book (such as ln x and log x). Verify the results of Example 7, which follows, on your calculator.

EXAMPLE 7   **Finding Values of a Function on a Calculator**

(a) $f(x) = x^2$; $f(1.234) = 1.522756$

(b) $F(x) = \dfrac{1}{x}$; $F(1.234) = 0.8103727715$

(c) $g(x) = \sqrt{x}$; $g(1.234) = 1.110855526$ ◄

NOTE   Graphing calculators can be used to evaluate any function that you wish. Figure 19 shows the result obtained in Example 6(a) on a TI-84 Plus graphing calculator with the function to be evaluated, $f(x) = 2x^2 - 3x$, in Y_1.

Figure 19

Implicit Form of a Function

In general, when a function f is defined by an equation in x and y, we say that the function f is given **implicitly**. If it is possible to solve the equation for y in terms of x, then we write $y = f(x)$ and say that the function is given **explicitly**. For example,

Implicit Form	Explicit Form
$3x + y = 5$	$y = f(x) = -3x + 5$
$x^2 - y = 6$	$y = f(x) = x^2 - 6$
$xy = 4$	$y = f(x) = \dfrac{4}{x}$

We list next a summary of some important facts to remember about a function f.

Summary

Important Facts About Functions

(a) For each x in the domain of f, there is exactly one image $f(x)$ in the range; however, an element in the range can result from more than one x in the domain.
(b) f is the symbol that we use to denote the function. It is symbolic of the equation that we use to get from an x in the domain to $f(x)$ in the range.
(c) If $y = f(x)$, then x is called the independent variable or argument of f, and y is called the dependent variable or the value of f at x.

3 Find the Domain of a Function

Often the domain of a function f is not specified; instead, only the equation defining the function is given. In such cases, we agree that the **domain of f** is the largest set of real numbers for which the value $f(x)$ is a real number. The domain of a function f is the same as the domain of the variable x in the expression $f(x)$.

EXAMPLE 8 **Finding the Domain of a Function**

Find the domain of each of the following functions:

(a) $f(x) = x^2 + 5x$ (b) $g(x) = \dfrac{3x}{x^2 - 4}$ (c) $h(t) = \sqrt{4 - 3t}$

Solution
(a) The function tells us to square a number and then add five times the number. Since these operations can be performed on any real number, we conclude that the domain of f is the set of all real numbers.
(b) The function g tells us to divide $3x$ by $x^2 - 4$. Since division by 0 is not defined, the denominator $x^2 - 4$ can never be 0, so x can never equal -2 or 2. The domain of the function g is $\{x \mid x \neq -2, x \neq 2\}$.
(c) The function h tells us to take the square root of $4 - 3t$. But only nonnegative numbers have real square roots, so the expression under the square root must be nonnegative (greater than or equal to zero). This requires that

$$4 - 3t \geq 0$$
$$-3t \geq -4$$
$$t \leq \frac{4}{3}$$

The domain of h is $\left\{t \mid t \leq \dfrac{4}{3}\right\}$ or the interval $\left(-\infty, \dfrac{4}{3}\right]$.   ◀

NOW WORK PROBLEM 51.

If x is in the domain of a function f, we shall say that **f is defined at x**, or **$f(x)$ exists**. If x is not in the domain of f, we say that **f is not defined at x**, or **$f(x)$ does not exist**. For example, if $f(x) = \dfrac{x}{x^2 - 1}$, then $f(0)$ exists, but $f(1)$ and $f(-1)$ do not exist. (Do you see why?)

We have not said much about finding the range of a function. The reason is that when a function is defined by an equation it is often difficult to find the range.[*] Therefore, we shall usually be content to find just the domain of a function when only the rule for the function is given. We shall express the domain of a function using inequalities, interval notation, set notation, or words, whichever is most convenient.

When we use functions in applications, the domain may be restricted by physical or geometric considerations. For example, the domain of the function f defined by $f(x) = x^2$ is the set of all real numbers. However, if f is used to obtain the area of a square when the length x of a side is known, then we must restrict the domain of f to the positive real numbers, since the length of a side can never be 0 or negative.

| **EXAMPLE 9** | **Finding the Domain in an Application** |

Express the area of a circle as a function of its radius. Find the domain.

Figure 20

Solution   See Figure 20. We know that the formula for the area A of a circle of radius r is $A = \pi r^2$. If we use r to represent the independent variable and A to represent the dependent variable, the function expressing this relationship is

$$A(r) = \pi r^2$$

In this setting, the domain is $\{r \mid r > 0\}$. (Do you see why?)   ◀

Observe in the solution to Example 9 that we used the symbol A in two ways: It is used to name the function, and it is used to symbolize the dependent variable. This double use is common in applications and should not cause any difficulty.

NOW WORK PROBLEM 85.

4   Form the Sum, Difference, Product, and Quotient of Two Functions

Next we introduce some operations on functions. We shall see that functions, like numbers, can be added, subtracted, multiplied, and divided. For example, if $f(x) = x^2 + 9$ and $g(x) = 3x + 5$, then

$$f(x) + g(x) = (x^2 + 9) + (3x + 5) = x^2 + 3x + 14$$

The new function $y = x^2 + 3x + 14$ is called the *sum function $f + g$*. Similarly,

$$f(x) \cdot g(x) = (x^2 + 9)(3x + 5) = 3x^3 + 5x^2 + 27x + 45$$

The new function $y = 3x^3 + 5x^2 + 27x + 45$ is called the *product function $f \cdot g$*.

---

[*]In Section 4.2 we discuss a way to find the range for a special class of functions.

The general definitions are given next.

If f and g are functions:

The **sum** $f + g$ is the function defined by

$$(f + g)(x) = f(x) + g(x)$$

The domain of $f + g$ consists of the numbers x that are in the domains of both f and g.

The **difference** $f - g$ is the function defined by

$$(f - g)(x) = f(x) - g(x)$$

The domain of $f - g$ consists of the numbers x that are in the domains of both f and g.

The product $f \cdot g$ is the function defined by

$$(f \cdot g)(x) = f(x) \cdot g(x)$$

The domain of $f \cdot g$ consists of the numbers x that are in the domains of both f and g.

The **quotient** $\dfrac{f}{g}$ is the function defined by

$$\left(\frac{f}{g}\right)(x) = \frac{f(x)}{g(x)} \qquad g(x) \neq 0$$

The domain of $\dfrac{f}{g}$ consists of the numbers x for which $g(x) \neq 0$ that are in the domains of both f and g.

EXAMPLE 10 | **Operations on Functions**

Let f and g be two functions defined as

$$f(x) = \frac{1}{x + 2} \quad \text{and} \quad g(x) = \frac{x}{x - 1}$$

Find the following, and determine the domain in each case.

(a) $(f + g)(x)$ (b) $(f - g)(x)$ (c) $(f \cdot g)(x)$ (d) $\left(\dfrac{f}{g}\right)(x)$

Solution The domain of f is $\{x \mid x \neq -2\}$ and the domain of g is $\{x \mid x \neq 1\}$.

(a) $(f + g)(x) = f(x) + g(x) = \dfrac{1}{x + 2} + \dfrac{x}{x - 1} = \dfrac{x - 1}{(x + 2)(x - 1)} + \dfrac{x(x + 2)}{(x + 2)(x - 1)} = \dfrac{x^2 + 3x - 1}{(x + 2)(x - 1)}$

The domain of $f + g$ consists of those numbers x that are in the domains of both f and g. Therefore, the domain of $f + g$ is $\{x \mid x \neq -2, x \neq 1\}$.

(b) $(f - g)(x) = f(x) - g(x) = \dfrac{1}{x + 2} - \dfrac{x}{x - 1} = \dfrac{x - 1}{(x + 2)(x - 1)} - \dfrac{x(x + 2)}{(x + 2)(x - 1)} = \dfrac{-(x^2 + x + 1)}{(x + 2)(x - 1)}$

The domain of $f - g$ consists of those numbers x that are in the domains of both f and g. Therefore, the domain of $f - g$ is $\{x \mid x \neq -2, x \neq 1\}$.

(c) $(f \cdot g)(x) = f(x) \cdot g(x) = \dfrac{1}{x + 2} \cdot \dfrac{x}{x - 1} = \dfrac{x}{(x + 2)(x - 1)}$

The domain of $f \cdot g$ consists of those numbers x that are in the domains of both f and g. Therefore, the domain of $f \cdot g$ is $\{x \mid x \neq -2, x \neq 1\}$.

(d) $\left(\dfrac{f}{g}\right)(x) = \dfrac{f(x)}{g(x)} = \dfrac{\dfrac{1}{x + 2}}{\dfrac{x}{x - 1}} = \dfrac{1}{x + 2} \cdot \dfrac{x - 1}{x} = \dfrac{x - 1}{x(x + 2)}$

The domain of $\dfrac{f}{g}$ consists of the numbers x for which $g(x) \neq 0$ that are in the domains of both f and g. Since $g(x) = 0$ when $x = 0$, we exclude 0 as well as -2 and 1. The domain of $\dfrac{f}{g}$ is $\{x \mid x \neq -2, x \neq 0, x \neq 1\}$. ◀

NOW WORK PROBLEM **61**.

In calculus, it is sometimes helpful to view a complicated function as the sum, difference, product, or quotient of simpler functions. For example,

$F(x) = x^2 + \sqrt{x}$ is the sum of $f(x) = x^2$ and $g(x) = \sqrt{x}$.

$H(x) = \dfrac{x^2 - 1}{x^2 + 1}$ is the quotient of $f(x) = x^2 - 1$ and $g(x) = x^2 + 1$.

Summary

We list here some of the important vocabulary introduced in this section, with a brief description of each term.

Function
A relation between two sets of real numbers so that each number x in the first set, the domain, has corresponding to it exactly one number y in the second set.
A set of ordered pairs (x, y) or $(x, f(x))$ in which no first element is paired with two different second elements.
The range is the set of y values of the function for the x values in the domain.
A function f may be defined implicitly by an equation involving x and y or explicitly by writing $y = f(x)$.

Unspecified domain
If a function f is defined by an equation and no domain is specified, then the domain will be taken to be the largest set of real numbers for which the equation defines a real number.

Function notation
$y = f(x)$
f is a symbol for the function.
x is the independent variable or argument.
y is the dependent variable.
$f(x)$ is the value of the function at x, or the image of x.

2.2 Assess Your Understanding

'Are You Prepared?'

Answers are given at the end of these exercises. If you get a wrong answer, read the pages listed in red.

1. The inequality $-1 < x < 3$ can be written in interval notation as _____. (pp. 151–152)

2. If $x = -2$, the value of the expression $3x^2 - 5x + \dfrac{1}{x}$ is _____. (p. 19)

3. The domain of the variable in the expression $\dfrac{x-3}{x+4}$ is _____. (p. 20)

4. Solve the inequality: $3 - 2x > 5$. Graph the solution set. (pp. 154–156)

Concepts and Vocabulary

5. If f is a function defined by the equation $y = f(x)$, then x is called the _____ variable and y is the _____ variable.

6. The set of all images of the elements in the domain of a function is called the _____.

7. If the domain of f is all real numbers in the interval $[0, 7]$ and the domain of g is all real numbers in the interval $[-2, 5]$, the domain of $f + g$ is all real numbers in the interval _____.

8. The domain of $\dfrac{f}{g}$ consists of numbers x for which $g(x)$ _____ 0 that are in the domains of both _____ and _____.

9. If $f(x) = x + 1$ and $g(x) = x^3$, then _____ $= x^3 - (x + 1)$.

10. *True or False:* Every relation is a function.

11. *True or False:* The domain of $(f \cdot g)(x)$ consists of the numbers x that are in the domains of both f and g.

12. *True or False:* The independent variable is sometimes referred to as the argument of the function.

13. *True or False:* If no domain is specified for a function f, then the domain of f is taken to be the set of real numbers.

14. *True or False:* The domain of the function $f(x) = \dfrac{x^2 - 4}{x}$ is $\{x \,|\, x \neq \pm 2\}$.

Skill Building

In Problems 15–26, determine whether each relation represents a function. For each function, state the domain and range.

15.

16.

17.

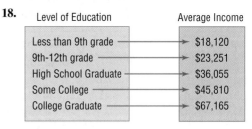

18.

Level of Education	Average Income
Less than 9th grade	$18,120
9th-12th grade	$23,251
High School Graduate	$36,055
Some College	$45,810
College Graduate	$67,165

19. $\{(2, 6), (-3, 6), (4, 9), (2, 10)\}$

21. $\{(1, 3), (2, 3), (3, 3), (4, 3)\}$

23. $\{(-2, 4), (-2, 6), (0, 3), (3, 7)\}$

25. $\{(-2, 4), (-1, 1), (0, 0), (1, 1)\}$

20. $\{(-2, 5), (-1, 3), (3, 7), (4, 12)\}$

22. $\{(0, -2), (1, 3), (2, 3), (3, 7)\}$

24. $\{(-4, 4), (-3, 3), (-2, 2), (-1, 1), (-4, 0)\}$

26. $\{(-2, 16), (-1, 4), (0, 3), (1, 4)\}$

In Problems 27–38, determine whether the equation defines y as a function of x.

27. $y = x^2$

28. $y = x^3$

29. $y = \dfrac{1}{x}$

30. $y = |x|$

31. $y^2 = 4 - x^2$

32. $y = \pm\sqrt{1 - 2x}$

33. $x = y^2$

34. $x + y^2 = 1$

35. $y = 2x^2 - 3x + 4$ **36.** $y = \dfrac{3x - 1}{x + 2}$ **37.** $2x^2 + 3y^2 = 1$ **38.** $x^2 - 4y^2 = 1$

In Problems 39–46, find the following values for each function:

   (a) $f(0)$ (b) $f(1)$ (c) $f(-1)$ (d) $f(-x)$ (e) $-f(x)$ (f) $f(x + 1)$ (g) $f(2x)$ (h) $f(x + h)$

39. $f(x) = 3x^2 + 2x - 4$ **40.** $f(x) = -2x^2 + x - 1$ **41.** $f(x) = \dfrac{x}{x^2 + 1}$ **42.** $f(x) = \dfrac{x^2 - 1}{x + 4}$

43. $f(x) = |x| + 4$ **44.** $f(x) = \sqrt{x^2 + x}$ **45.** $f(x) = \dfrac{2x + 1}{3x - 5}$ **46.** $f(x) = 1 - \dfrac{1}{(x + 2)^2}$

In Problems 47–60, find the domain of each function.

47. $f(x) = -5x + 4$ **48.** $f(x) = x^2 + 2$ **49.** $f(x) = \dfrac{x}{x^2 + 1}$ **50.** $f(x) = \dfrac{x^2}{x^2 + 1}$

51. $g(x) = \dfrac{x}{x^2 - 16}$ **52.** $h(x) = \dfrac{2x}{x^2 - 4}$ **53.** $F(x) = \dfrac{x - 2}{x^3 + x}$ **54.** $G(x) = \dfrac{x + 4}{x^3 - 4x}$

55. $h(x) = \sqrt{3x - 12}$ **56.** $G(x) = \sqrt{1 - x}$ **57.** $f(x) = \dfrac{4}{\sqrt{x - 9}}$ **58.** $f(x) = \dfrac{x}{\sqrt{x - 4}}$

59. $p(x) = \sqrt{\dfrac{2}{x - 1}}$ **60.** $q(x) = \sqrt{-x - 2}$

In Problems 61–70, for the given functions f and g, find the following functions and state the domain of each.

   (a) $f + g$ (b) $f - g$ (c) $f \cdot g$ (d) $\dfrac{f}{g}$

61. $f(x) = 3x + 4$; $g(x) = 2x - 3$
63. $f(x) = x - 1$; $g(x) = 2x^2$
65. $f(x) = \sqrt{x}$; $g(x) = 3x - 5$
67. $f(x) = 1 + \dfrac{1}{x}$; $g(x) = \dfrac{1}{x}$
69. $f(x) = \dfrac{2x + 3}{3x - 2}$; $g(x) = \dfrac{4x}{3x - 2}$

62. $f(x) = 2x + 1$; $g(x) = 3x - 2$
64. $f(x) = 2x^2 + 3$; $g(x) = 4x^3 + 1$
66. $f(x) = |x|$; $g(x) = x$
68. $f(x) = \sqrt{x - 2}$; $g(x) = \sqrt{4 - x}$
70. $f(x) = \sqrt{x + 1}$; $g(x) = \dfrac{2}{x}$

71. Given $f(x) = 3x + 1$ and $(f + g)(x) = 6 - \dfrac{1}{2}x$, find the function g.

72. Given $f(x) = \dfrac{1}{x}$ and $\left(\dfrac{f}{g}\right)(x) = \dfrac{x + 1}{x^2 - x}$, find the function g.

In Problems 73–78, find the difference quotient of f, that is, find $\dfrac{f(x + h) - f(x)}{h}$, $h \neq 0$, for each function. Be sure to simplify.

73. $f(x) = 4x + 3$ **74.** $f(x) = -3x + 1$ **75.** $f(x) = x^2 - x + 4$

76. $f(x) = x^2 + 5x - 1$ **77.** $f(x) = x^3 - 2$ **78.** $f(x) = \dfrac{1}{x + 3}$

Applications and Extensions

79. If $f(x) = 2x^3 + Ax^2 + 4x - 5$ and $f(2) = 5$, what is the value of A?

80. If $f(x) = 3x^2 - Bx + 4$ and $f(-1) = 12$, what is the value of B?

81. If $f(x) = \dfrac{3x + 8}{2x - A}$ and $f(0) = 2$, what is the value of A?

82. If $f(x) = \dfrac{2x - B}{3x + 4}$ and $f(2) = \dfrac{1}{2}$, what is the value of B?

83. If $f(x) = \dfrac{2x - A}{x - 3}$ and $f(4) = 0$, what is the value of A? Where is f not defined?

84. If $f(x) = \dfrac{x - B}{x - A}$, $f(2) = 0$, and $f(1)$ is undefined, what are the values of A and B?

85. Geometry   Express the area A of a rectangle as a function of the length x if the length of the rectangle is twice its width.

86. Geometry   Express the area A of an isosceles right triangle as a function of the length x of one of the two equal sides.

87. Constructing Functions   Express the gross salary G of a person who earns \$10 per hour as a function of the number x of hours worked.

88. Constructing Functions Tiffany, a commissioned salesperson, earns $100 base pay plus $10 per item sold. Express her gross salary G as a function of the number x of items sold.

89. Effect of Gravity on Earth If a rock falls from a height of 20 meters on Earth, the height H (in meters) after x seconds is approximately

$$H(x) = 20 - 4.9x^2$$

(a) What is the height of the rock when $x = 1$ second? $x = 1.1$ seconds? $x = 1.2$ seconds? $x = 1.3$ seconds?
(b) When is the height of the rock 15 meters? When is it 10 meters? When is it 5 meters?
(c) When does the rock strike the ground?

90. Effect of Gravity on Jupiter If a rock falls from a height of 20 meters on the planet Jupiter, its height H (in meters) after x seconds is approximately

$$H(x) = 20 - 13x^2$$

(a) What is the height of the rock when $x = 1$ second? $x = 1.1$ seconds? $x = 1.2$ seconds?
(b) When is the height of the rock 15 meters? When is it 10 meters? When is it 5 meters?
(c) When does the rock strike the ground?

91. Cost of Trans-Atlantic Travel A Boeing 747 crosses the Atlantic Ocean (3000 miles) with an airspeed of 500 miles per hour. The cost C (in dollars) per passenger is given by

$$C(x) = 100 + \frac{x}{10} + \frac{36,000}{x}$$

where x is the ground speed (airspeed $\pm$ wind).
(a) What is the cost per passenger for quiescent (no wind) conditions?
(b) What is the cost per passenger with a head wind of 50 miles per hour?
(c) What is the cost per passenger with a tail wind of 100 miles per hour?
(d) What is the cost per passenger with a head wind of 100 miles per hour?

92. Cross-sectional Area The cross-sectional area of a beam cut from a log with radius 1 foot is given by the function $A(x) = 4x\sqrt{1 - x^2}$, where x represents the length, in feet, of half the base of the beam. See the figure. Determine the cross-sectional area of the beam if the length of half the base of the beam is as follows:
(a) One-third of a foot
(b) One-half of a foot
(c) Two-thirds of a foot

$A(x) = 4x\sqrt{1 - x^2}$

93. Economics The **participation rate** is the number of people in the labor force divided by the civilian population (excludes military). Let $L(x)$ represent the size of the labor force in year x and $P(x)$ represent the civilian population in year x. Determine a function that represents the participation rate R as a function of x.

94. Crimes Suppose that $V(x)$ represents the number of violent crimes committed in year x and $P(x)$ represents the number of property crimes committed in year x. Determine a function T that represents the combined total of violent crimes and property crimes in year x.

95. Health Care Suppose that $P(x)$ represents the percentage of income spent on health care in year x and $I(x)$ represents income in year x. Determine a function H that represents total health care expenditures in year x.

96. Income Tax Suppose that $I(x)$ represents the income of an individual in year x before taxes and $T(x)$ represents the individual's tax bill in year x. Determine a function N that represents the individual's net income (income after taxes) in year x.

97. Some functions f have the property that $f(a + b) = f(a) + f(b)$ for all real numbers a and b. Which of the following functions have this property?
(a) $h(x) = 2x$ (b) $g(x) = x^2$
(c) $F(x) = 5x - 2$ (d) $G(x) = \dfrac{1}{x}$

Discussion and Writing

98. Are the functions $f(x) = x - 1$ and $g(x) = \dfrac{x^2 - 1}{x + 1}$ the same? Explain.

99. Investigate when, historically, the use of the function notation $y = f(x)$ first appeared.

'Are You Prepared?' Answers

1. $(-1, 3)$ 2. 21.5 3. $\{x | x \neq -4\}$ 4. $\{x | x < -1\}$

2.3 The Graph of a Function

PREPARING FOR THIS SECTION   *Before getting started, review the following:*

- Graphs of Equations (Section 1.1, pp. 90–92)
- Intercepts (Section 1.1, pp. 95–96; Section 1.9, p. 183)

Now work the 'Are You Prepared?' problems on page 223.

OBJECTIVES   1   Identify the Graph of a Function
2   Obtain Information from or about the Graph of a Function

In applications, a graph often demonstrates more clearly the relationship between two variables than, say, an equation or table would. For example, Table 6 shows the average price of gasoline, adjusted for inflation, for the years 1978–2004. If we plot these data and then connect the points, we obtain Figure 21.

Table 6

Year	Price	Year	Price
1978	1.5207	1992	1.3460
1979	1.9263	1993	1.4622
1980	2.4197	1994	1.4166
1981	2.4258	1995	1.4111
1982	2.1136	1996	1.4804
1983	1.8210	1997	1.4697
1984	1.7732	1998	1.2685
1985	1.6998	1999	1.4629
1986	1.3043	2000	1.7469
1987	1.2831	2001	1.6821
1988	1.2674	2002	1.5339
1989	1.3134	2003	1.8308
1990	1.4160	2004	2.0047
1991	1.4490		

SOURCE: Statistical Abstract of the United States.

Figure 21
Price of Gasoline Adjusted
for Inflation, 1978–2004

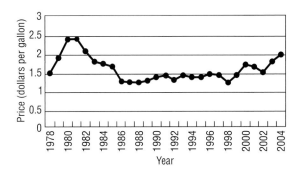

NOTE

When we select a viewing window to graph a function on a calculator, the values of Xmin, Xmax give the domain that we wish to view, while Ymin, Ymax give the range that we wish to view. These settings usually do not represent the actual domain and range of the function.   ■

We can see from the graph that the price of gasoline (adjusted for inflation) rose rapidly from 1978 to 1981 and was falling from 2000 to 2002. The graph also shows that the highest price occurred in 1981. To learn information such as this from an equation requires that some calculations be made.

Look again at Figure 21. The graph shows that for each date on the horizontal axis there is only one price on the vertical axis. The graph represents a function, although the exact rule for getting from date to price is not given.

When a function is defined by an equation in x and y, the **graph of the function** is the graph of the equation, that is, the set of points (x, y) in the xy-plane that satisfies the equation.

1   Identify the Graph of a Function

Not every collection of points in the xy-plane represents the graph of a function. Remember, for a function, each number x in the domain has exactly one image y in the range. This means that the graph of a function cannot contain two points with the same x-coordinate and different y-coordinates. Therefore, the graph of a function must satisfy the following **vertical-line test**.

Theorem

Vertical-line Test

A set of points in the xy-plane is the graph of a function if and only if every vertical line intersects the graph in at most one point.

In other words, if any vertical line intersects a graph at more than one point, the graph is not the graph of a function.

EXAMPLE 1

Identifying the Graph of a Function

Which of the graphs in Figure 22 are graphs of functions?

Figure 22

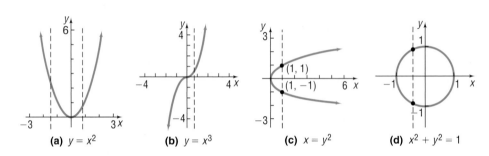

(a) $y = x^2$ (b) $y = x^3$ (c) $x = y^2$ (d) $x^2 + y^2 = 1$

Solution

The graphs in Figures 22(a) and 22(b) are graphs of functions, because every vertical line intersects each graph in at most one point. The graphs in Figures 22(c) and 22(d) are not graphs of functions, because there is a vertical line that intersects each graph in more than one point. ◀

 NOW WORK PROBLEM 15.

2 **Obtain Information from or about the Graph of a Function**

If (x, y) is a point on the graph of a function f, then y is the value of f at x; that is, $y = f(x)$. The next example illustrates how to obtain information about a function if its graph is given.

EXAMPLE 2

Obtaining Information from the Graph of a Function

Let f be the function whose graph is given in Figure 23. (The graph of f might represent the distance that the bob of a pendulum is from its *at-rest* position. Negative values of y mean that the pendulum is to the left of the at-rest position, and positive values of y mean that the pendulum is to the right of the at-rest position.)

Figure 23

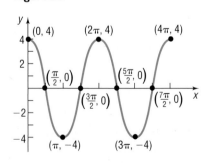

(a) What are $f(0)$, $f\left(\dfrac{3\pi}{2}\right)$, and $f(3\pi)$?

(b) What is the domain of f?

(c) What is the range of f?

(d) List the intercepts. (Recall that these are the points, if any, where the graph crosses or touches the coordinate axes.)

(e) How often does the line $y = 2$ intersect the graph?

(f) For what values of x does $f(x) = -4$?

(g) For what values of x is $f(x) > 0$?

Solution   (a) Since $(0, 4)$ is on the graph of f, the y-coordinate 4 is the value of f at the x-coordinate 0; that is, $f(0) = 4$. In a similar way, we find that when $x = \dfrac{3\pi}{2}$ then $y = 0$, so $f\left(\dfrac{3\pi}{2}\right) = 0$. When $x = 3\pi$, then $y = -4$, so $f(3\pi) = -4$.

(b) To determine the domain of f, we notice that the points on the graph of f will have x-coordinates between 0 and 4π, inclusive; and for each number x between 0 and 4π, there is a point $(x, f(x))$ on the graph. The domain of f is $\{x | 0 \le x \le 4\pi\}$ or the interval $[0, 4\pi]$.

(c) The points on the graph all have y-coordinates between -4 and 4, inclusive; and for each such number y, there is at least one number x in the domain. The range of f is $\{y | -4 \le y \le 4\}$ or the interval $[-4, 4]$.

(d) The intercepts are

$$(0, 4), \left(\frac{\pi}{2}, 0\right), \left(\frac{3\pi}{2}, 0\right), \left(\frac{5\pi}{2}, 0\right), \quad \text{and} \quad \left(\frac{7\pi}{2}, 0\right)$$

(e) If we draw the horizontal line $y = 2$ on the graph in Figure 23, then we find that it intersects the graph four times.

(f) Since $(\pi, -4)$ and $(3\pi, -4)$ are the only points on the graph for which $y = f(x) = -4$, we have $f(x) = -4$ when $x = \pi$ and $x = 3\pi$.

(g) To determine where $f(x) > 0$, we look at Figure 23 and determine the x-values for which the y-coordinate is positive. This occurs on the intervals $\left[0, \dfrac{\pi}{2}\right), \left(\dfrac{3\pi}{2}, \dfrac{5\pi}{2}\right)$, and $\left(\dfrac{7\pi}{2}, 4\pi\right]$. Using inequality notation, $f(x) > 0$ for $0 \le x < \dfrac{\pi}{2}, \dfrac{3\pi}{2} < x < \dfrac{5\pi}{2}$, and $\dfrac{7\pi}{2} < x \le 4\pi$.   ◀

When the graph of a function is given, its domain may be viewed as the shadow created by the graph on the x-axis by vertical beams of light. Its range can be viewed as the shadow created by the graph on the y-axis by horizontal beams of light. Try this technique with the graph given in Figure 23.

NOW WORK PROBLEMS 9 AND 13.

| **EXAMPLE 3** | **Obtaining Information about the Graph of a Function** |

Consider the function: $f(x) = \dfrac{x}{x + 2}$

(a) Is the point $\left(1, \dfrac{1}{2}\right)$ on the graph of f?

(b) If $x = 2$, what is $f(x)$? What point is on the graph of f?

(c) If $f(x) = 2$, what is x? What point is on the graph of f?

Solution (a) When $x = 1$, then

$$f(x) = \frac{x}{x + 2}$$

$$f(1) = \frac{1}{1 + 2} = \frac{1}{3}$$

The point $\left(1, \frac{1}{3}\right)$ is on the graph of f; the point $\left(1, \frac{1}{2}\right)$ is not.

(b) If $x = 2$, then

$$f(x) = \frac{x}{x + 2}$$

$$f(2) = \frac{2}{2 + 2} = \frac{2}{4} = \frac{1}{2}$$

The point $\left(2, \frac{1}{2}\right)$ is on the graph of f.

(c) If $f(x) = 2$, then

$$f(x) = 2$$

$$\frac{x}{x + 2} = 2$$

$$x = 2(x + 2) \qquad \text{Multiply both sides by x + 2.}$$

$$x = 2x + 4 \qquad \text{Remove parentheses.}$$

$$x = -4 \qquad \text{Solve for x.}$$

If $f(x) = 2$, then $x = -4$. The point $(-4, 2)$ is on the graph of f. ◄

NOW WORK PROBLEM 25.

| EXAMPLE 4 | **Average Cost Function** |

The average cost $\overline{C}$ of manufacturing x computers per day is given by the function

$$\overline{C}(x) = 0.56x^2 - 34.39x + 1212.57 + \frac{20,000}{x}$$

Determine the average cost of manufacturing:

(a) 30 computers in a day
(b) 40 computers in a day
(c) 50 computers in a day
(d) Graph the function $\overline{C} = \overline{C}(x), 0 < x \le 80$.
(e) Create a TABLE with TblStart = 1 and ΔTbl = 1. Which value of x minimizes the average cost?

Solution (a) The average cost of manufacturing $x = 30$ computers is

$$\overline{C}(30) = 0.56(30)^2 - 34.39(30) + 1212.57 + \frac{20,000}{30} = \$1351.54$$

(b) The average cost of manufacturing $x = 40$ computers is

$$\overline{C}(40) = 0.56(40)^2 - 34.39(40) + 1212.57 + \frac{20{,}000}{40} = \$1232.97$$

(c) The average cost of manufacturing $x = 50$ computers is

$$\overline{C}(50) = 0.56(50)^2 - 34.39(50) + 1212.57 + \frac{20{,}000}{50} = \$1293.07$$

(d) See Figure 24 for the graph of $\overline{C} = \overline{C}(x)$.

(e) With the function $\overline{C} = \overline{C}(x)$ in Y_1, we create Table 7. We scroll down until we find a value of x for which Y_1 is smallest. Table 8 shows that manufacturing $x = 41$ computers minimizes the average cost at \$1231.74 per computer.

Figure 24

Table 7

X	Y1
1	21179
2	11146
3	7781.1
4	6084
5	5054.6
6	4359.7
7	3856.4

Y1◾.56X²-34.39X...

Table 8

X	Y1
38	1240.7
39	1235.9
40	1233
41	1231.7
42	1232.2
43	1234.4
44	1238.1

Y1=1231.74487805   ◄

✏ **NOW WORK PROBLEM 29.**

Summary

Graph of a function   The collection of points (x, y) that satisfies the equation $y = f(x)$.
A collection of points is the graph of a function provided that every vertical line intersects the graph in at most one point (vertical-line test).

2.3 Assess Your Understanding

'Are You Prepared?'

Answers are given at the end of these exercises. If you get a wrong answer, read the pages listed in red.

1. The intercepts of the equation $x^2 + 4y^2 = 16$ are _____.
(pp. 171–172 and 183)

2. *True or False:* The point $(-2, -6)$ is on the graph of the equation $x = 2y - 2$. (pp. 90–92)

Concepts and Vocabulary

3. A set of points in the xy-plane is the graph of a function if and only if every _____ line intersects the graph in at most one point.

4. If the point $(5, -3)$ is a point on the graph of f, then $f(\_\_\_) = \_\_\_$.

5. Find a so that the point $(-1, 2)$ is on the graph of $f(x) = ax^2 + 4$.

6. *True or False:* A function can have more than one y-intercept.

7. *True or False:* The graph of a function $y = f(x)$ always crosses the y-axis.

8. *True or False:* The y-intercept of the graph of the function $y = f(x)$, whose domain is all real numbers, is $f(0)$.

Skill Building

9. Use the given graph of the function f to answer parts (a)–(n).

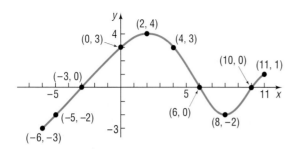

(a) Find $f(0)$ and $f(-6)$.
(b) Find $f(6)$ and $f(11)$.
(c) Is $f(3)$ positive or negative?
(d) Is $f(-4)$ positive or negative?
(e) For what numbers x is $f(x) = 0$?
(f) For what numbers x is $f(x) > 0$?
(g) What is the domain of f?
(h) What is the range of f?
(i) What are the x-intercepts?
(j) What is the y-intercept?

(k) How often does the line $y = \dfrac{1}{2}$ intersect the graph?

(l) How often does the line $x = 5$ intersect the graph?
(m) For what values of x does $f(x) = 3$?
(n) For what values of x does $f(x) = -2$?

10. Use the given graph of the function f to answer parts (a)–(n).

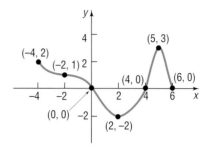

(a) Find $f(0)$ and $f(6)$.
(b) Find $f(2)$ and $f(-2)$.
(c) Is $f(3)$ positive or negative?
(d) Is $f(-1)$ positive or negative?
(e) For what numbers x is $f(x) = 0$?
(f) For what numbers x is $f(x) < 0$?
(g) What is the domain of f?
(h) What is the range of f?
(i) What are the x-intercepts?
(j) What is the y-intercept?
(k) How often does the line $y = -1$ intersect the graph?
(l) How often does the line $x = 1$ intersect the graph?
(m) For what value of x does $f(x) = 3$?
(n) For what value of x does $f(x) = -2$?

In Problems 11–22, determine whether the graph is that of a function by using the vertical-line test. If it is, use the graph to find:
  (a) *Its domain and range*
  (b) *The intercepts, if any*
  (c) *Any symmetry with respect to the x-axis, the y-axis, or the origin*

11.

12.

13.

14.

15.

16.

17.

18.

19.

20.

21.

22.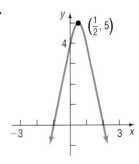

In Problems 23–28, answer the questions about the given function.

23. $f(x) = 2x^2 - x - 1$
  (a) Is the point $(-1, 2)$ on the graph of f?
  (b) If $x = -2$, what is $f(x)$? What point is on the graph of f?
  (c) If $f(x) = -1$, what is x? What point(s) are on the graph of f?
  (d) What is the domain of f?
  (e) List the x-intercepts, if any, of the graph of f.
  (f) List the y-intercept, if there is one, of the graph of f.

24. $f(x) = -3x^2 + 5x$
  (a) Is the point $(-1, 2)$ on the graph of f?
  (b) If $x = -2$, what is $f(x)$? What point is on the graph of f?
  (c) If $f(x) = -2$, what is x? What point(s) are on the graph of f?
  (d) What is the domain of f?
  (e) List the x-intercepts, if any, of the graph of f.
  (f) List the y-intercept, if there is one, of the graph of f.

25. $f(x) = \dfrac{x + 2}{x - 6}$
  (a) Is the point $(3, 14)$ on the graph of f?
  (b) If $x = 4$, what is $f(x)$? What point is on the graph of f?
  (c) If $f(x) = 2$, what is x? What point(s) are on the graph of f?
  (d) What is the domain of f?
  (e) List the x-intercepts, if any, of the graph of f.
  (f) List the y-intercept, if there is one, of the graph of f.

26. $f(x) = \dfrac{x^2 + 2}{x + 4}$
  (a) Is the point $\left(1, \dfrac{3}{5}\right)$ on the graph of f?

(b) If $x = 0$, what is $f(x)$? What point is on the graph of f?
(c) If $f(x) = \dfrac{1}{2}$, what is x? What point(s) are on the graph of f?
(d) What is the domain of f?
(e) List the x-intercepts, if any, of the graph of f.
(f) List the y-intercept, if there is one, of the graph of f.

27. $f(x) = \dfrac{2x^2}{x^4 + 1}$
  (a) Is the point $(-1, 1)$ on the graph of f?
  (b) If $x = 2$, what is $f(x)$? What point is on the graph of f?
  (c) If $f(x) = 1$, what is x? What point(s) are on the graph of f?
  (d) What is the domain of f?
  (e) List the x-intercepts, if any, of the graph of f.
  (f) List the y-intercept, if there is one, of the graph of f.

28. $f(x) = \dfrac{2x}{x - 2}$
  (a) Is the point $\left(\dfrac{1}{2}, -\dfrac{2}{3}\right)$ on the graph of f?
  (b) If $x = 4$, what is $f(x)$? What point is on the graph of f?
  (c) If $f(x) = 1$, what is x? What point(s) are on the graph of f?
  (d) What is the domain of f?
  (e) List the x-intercepts, if any, of the graph of f.
  (f) List the y-intercept, if there is one, of the graph of f.

Applications and Extensions

29. Motion of a Golf Ball A golf ball is hit with an initial velocity of 130 feet per second at an inclination of 45° to the horizontal. In physics, it is established that the height h of the golf ball is given by the function

$$h(x) = \frac{-32x^2}{130^2} + x$$

where x is the horizontal distance that the golf ball has traveled. (continued on page 226)

(a) Determine the height of the golf ball after it has traveled 100 feet.
(b) What is the height after it has traveled 300 feet?
(c) What is the height after it has traveled 500 feet?
(d) How far was the golf ball hit?
(e) Graph the function $h = h(x)$.
(f) Use a graphing utility to determine the distance that the ball has traveled when the height of the ball is 90 feet.
(g) Create a TABLE with TblStart = 0 and ΔTbl = 25. To the nearest 25 feet, how far does the ball travel before it reaches a maximum height? What is the maximum height?
(h) Adjust the value of ΔTbl until you determine the distance, to within 1 foot, that the ball travels before it reaches a maximum height.

30. Cross-sectional Area   The cross-sectional area of a beam cut from a log with radius 1 foot is given by the function $A(x) = 4x\sqrt{1 - x^2}$, where x represents the length, in feet, of half the base of the beam. See the figure.

$$A(x) = 4x\sqrt{1 - x^2}$$

(a) Find the domain of A.
(b) Graph the function $A = A(x)$.

(c) Create a TABLE with TblStart = 0 and ΔTbl = 0.1 for $0 \le x \le 1$. Which value of x maximizes the cross-sectional area? What should be the length of the base of the beam to maximize the cross-sectional area?

31. Cost of Trans-Atlantic Travel   A Boeing 747 crosses the Atlantic Ocean (3000 miles) with an airspeed of 500 miles per hour. The cost C (in dollars) per passenger is given by

$$C(x) = 100 + \frac{x}{10} + \frac{36{,}000}{x}$$

where x is the ground speed (airspeed $\pm$ wind).
(a) Graph the function $C = C(x)$.
(b) Create a TABLE with TblStart = 0 and ΔTbl = 50.
(c) To the nearest 50 miles per hour, what ground speed minimizes the cost per passenger?

32. Effect of Elevation on Weight   If an object weighs m pounds at sea level, then its weight W (in pounds) at a height of h miles above sea level is given approximately by

$$W(h) = m\left(\frac{4000}{4000 + h}\right)^2$$

(a) If Amy weighs 120 pounds at sea level, how much will she weigh on Pike's Peak, which is 14,110 feet above sea level?
(b) Use a graphing utility to graph the function $W = W(h)$. Use $m = 120$ pounds.
(c) Create a Table with TblStart = 0 and ΔTbl = 0.5 to see how the weight W varies as h changes from 0 to 5 miles.
(d) At what height will Amy weigh 119.95 pounds?
(e) Does your answer to part (d) seem reasonable? Explain.

Discussion and Writing

33. Describe how you would proceed to find the domain and range of a function if you were given its graph. How would your strategy change if you were given the equation defining the function instead of its graph?

34. How many x-intercepts can the graph of a function have?

35. How many y-intercepts can the graph of a function have?

36. Is a graph that consists of a single point the graph of a function? Can you write the equation of such a function?

37. Match each of the following functions with the graphs that best describes the situation.
(a) The cost of building a house as a function of its square footage
(b) The height of an egg dropped from a 300-foot building as a function of time
(c) The height of a human as a function of time
(d) The demand for Big Macs as a function of price
(e) The height of a child on a swing as a function of time

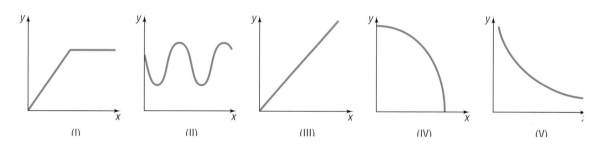

(I)   (II)   (III)   (IV)   (V)

38. Match each of the following functions with the graph that best describes the situation.
(a) The temperature of a bowl of soup as a function of time
(b) The number of hours of daylight per day over a 2-year period
(c) The population of Florida as a function of time
(d) The distance of a car traveling at a constant velocity as a function of time
(e) The height of a golf ball hit with a 7-iron as a function of time

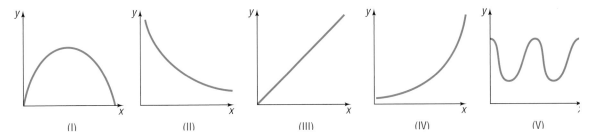

(I)   (II)   (III)   (IV)   (V)

---

39. Consider the following scenario: Barbara decides to take a walk. She leaves home, walks 2 blocks in 5 minutes at a constant speed, and realizes that she forgot to lock the door. So Barbara runs home in 1 minute. While at her doorstep, it takes her 1 minute to find her keys and lock the door. Barbara walks 5 blocks in 15 minutes and then decides to jog home. It takes her 7 minutes to get home. Draw a graph of Barbara's distance from home (in blocks) as a function of time.

40. Consider the following scenario: Jayne enjoys riding her bicycle through the woods. At the forest preserve, she gets on her bicycle and rides up a 2000-foot incline in 10 minutes. She then travels down the incline in 3 minutes. The next 5000 feet is level terrain and she covers the distance in 20 minutes. She rests for 15 minutes. Jayne then travels 10,000 feet in 30 minutes. Draw a graph of Jayne's distance traveled (in feet) as a function of time.

41. The following sketch represents the distance d (in miles) that Kevin is from home as a function of time t (in hours). Answer the questions based on the graph. In parts (a)–(g), how many hours elapsed and how far was Kevin from home during this time?

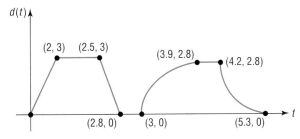

(a) From $t = 0$ to $t = 2$
(b) From $t = 2$ to $t = 2.5$
(c) From $t = 2.5$ to $t = 2.8$

(d) From $t = 2.8$ to $t = 3$
(e) From $t = 3$ to $t = 3.9$
(f) From $t = 3.9$ to $t = 4.2$
(g) From $t = 4.2$ to $t = 5.3$
(h) What is the farthest distance that Kevin is from home?
(i) How many times did Kevin return home?

42. The following sketch represents the speed v (in miles per hour) of Michael's car as a function of time t (in minutes).

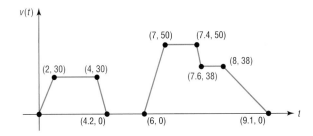

(a) Over what interval of time is Michael traveling fastest?
(b) Over what interval(s) of time is Michael's speed zero?
(c) What is Michael's speed between 0 and 2 minutes?
(d) What is Michael's speed between 4.2 and 6 minutes?
(e) What is Michael's speed between 7 and 7.4 minutes?
(f) When is Michael's speed constant?

43. Draw the graph of a function whose domain is $\{x \mid -3 \le x \le 8, \ x \ne 5\}$ and whose range is $\{y \mid -1 \le y \le 2, \ y \ne 0\}$. What point(s) in the rectangle $-3 \le x \le 8, -1 \le y \le 2$ cannot be on the graph? Compare your graph with those of other students. What differences do you see?

44. Is there a function whose graph is symmetric with respect to the x-axis? Explain.

'Are You Prepared?' Answers

1. $(-4, 0), (4, 0), (0, -2), (0, 2)$ **2.** False

2.4 Properties of Functions

PREPARING FOR THIS SECTION   *Before getting started, review the following:*

- Intervals (Section 1.7, pp. 151–152)
- Intercepts (Section 1.1, pp. 95–96 and Section 1.8, pp. 171–172).
- Slope of a Line (Section 1.8, pp. 163–165)
- Point–Slope Form of a Line (Section 1.8, p. 168)

Now work the 'Are You Prepared?' problems on page 236.

OBJECTIVES
1 Determine Even and Odd Functions from a Graph
2 Identify Even and Odd Functions from the Equation
3 Use a Graph to Determine Where a Function Is Increasing, Decreasing, or Constant
4 Use a Graph to Locate Local Maxima and Local Minima
5 Use a Graphing Utility to Approximate Local Maxima and Local Minima and to Determine Where a Function Is Increasing or Decreasing
6 Find the Average Rate of Change of a Function

It is easiest to obtain the graph of a function $y = f(x)$ by knowing certain properties that the function has and the impact of these properties on the way that the graph will look.

1 Determine Even and Odd Functions from a Graph

The words *even* and *odd*, when applied to a function f, describe the symmetry that exists for the graph of the function.

A function f is even if and only if, whenever the point (x, y) is on the graph of f then the point $(-x, y)$ is also on the graph. Using function notation, we define an even function as follows:

A function f is **even** if, for every number x in its domain, the number $-x$ is also in the domain and

$$f(-x) = f(x)$$

A function f is odd if and only if, whenever the point (x, y) is on the graph of f then the point $(-x, -y)$ is also on the graph. Using function notation, we define an odd function as follows:

A function f is **odd** if, for every number x in its domain, the number $-x$ is also in the domain and

$$f(-x) = -f(x)$$

Refer to Section 2.1, where the tests for symmetry are listed. The following results are then evident.

Theorem

A function is even if and only if its graph is symmetric with respect to the y-axis. A function is odd if and only if its graph is symmetric with respect to the origin.

| EXAMPLE 1 | **Determining Even and Odd Functions from the Graph** |

Determine whether each graph given in Figure 25 is the graph of an even function, an odd function, or a function that is neither even nor odd.

Figure 25

(a)   (b)   (c)

Solution

(a) The graph in Figure 25(a) is that of an even function, because the graph is symmetric with respect to the y-axis.

(b) The function whose graph is given in Figure 25(b) is neither even nor odd, because the graph is neither symmetric with respect to the y-axis nor symmetric with respect to the origin.

(c) The function whose graph is given in Figure 25(c) is odd, because its graph is symmetric with respect to the origin.   ◄

━━━━━━━━━━▶ **NOW WORK PROBLEMS 21(a), (b), AND (d).**

2 Identify Even and Odd Functions from the Equation

A graphing utility can be used to conjecture whether a function is even, odd, or neither. As stated, when the graph of an even function contains the point (x, y), it must also contain the point $(-x, y)$. Therefore, if the graph shows evidence of symmetry with respect to the y-axis, we would conjecture that the function is even. In addition, if the graph shows evidence of symmetry with respect to the origin, we would conjecture that the function is odd.

In the next example, we use a graphing utility to conjecture whether a function is even, odd, or neither. Then we verify our conjecture algebraically.

| EXAMPLE 2 | **Identifying Even and Odd Functions** |

Use a graphing utility to conjecture whether each of the following functions is even, odd, or neither. Verify the conjecture algebraically. Then state whether the graph is symmetric with respect to the y-axis or with respect to the origin.

(a) $f(x) = x^2 - 5$ (b) $g(x) = x^3 - 1$ (c) $h(x) = 5x^3 - x$

Solution

(a) Graph the function. See Figure 26. It appears that the graph is symmetric with respect to the y-axis. We conjecture that the function is even.

Figure 26

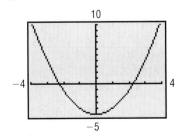

To algebraically verify the conjecture, we replace x by $-x$ in $f(x) = x^2 - 5$. Then

$$f(-x) = (-x)^2 - 5 = x^2 - 5 = f(x)$$

Since $f(-x) = f(x)$, we conclude that f is an even function, and the graph is symmetric with respect to the y-axis.

(b) Graph the function. See Figure 27 on page 230. It appears that there is no symmetry. We conjecture that the function is neither even nor odd.

To algebraically verify that the function is not even, we find $g(-x)$ and compare the result with $g(x)$.

$$g(-x) = (-x)^3 - 1 = -x^3 - 1; \quad g(x) = x^3 - 1$$

Figure 27

Figure 28

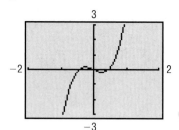

Since $g(-x) \neq g(x)$, the function is not even.

To algebraically verify that the function is not odd, we find $-g(x)$ and compare the result with $g(-x)$.

$$-g(x) = -(x^3 - 1) = -x^3 + 1; \quad g(-x) = -x^3 - 1$$

Since $g(-x) \neq -g(x)$, the function is not odd. The graph is not symmetric with respect to the y-axis nor is it symmetric with respect to the origin.

(c) Graph the function. See Figure 28. It appears that there is symmetry with respect to the origin. We conjecture that the function is odd.

To algebraically verify the conjecture, we replace x by $-x$ in $h(x) = 5x^3 - x$. Then

$$h(-x) = 5(-x)^3 - (-x) = -5x^3 + x = -(5x^3 - x) = -h(x)$$

Since $h(-x) = -h(x)$, h is an odd function, and the graph of h is symmetric with respect to the origin. ◀

━━━ **NOW WORK PROBLEM 33.**

 3 **Use a Graph to Determine Where a Function Is Increasing, Decreasing, or Constant**

Consider the graph given in Figure 29. If you look from left to right along the graph of the function, you will notice that parts of the graph are rising, parts are falling, and parts are horizontal. In such cases, the function is described as *increasing*, *decreasing*, or *constant*, respectively.

Figure 29

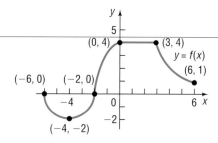

EXAMPLE 3

Determining Where a Function Is Increasing, Decreasing, or Constant from Its Graph

Where is the function in Figure 29 increasing? Where is it decreasing? Where is it constant?

Solution

To answer the question of where a function is increasing, where it is decreasing, and where it is constant, we use strict inequalities involving the independent variable x, or we use open intervals[*] of x-coordinates. The graph in Figure 29 is rising (increasing) from the point $(-4, -2)$ to the point $(0, 4)$, so we conclude that it is increasing on the open interval $(-4, 0)$ or for $-4 < x < 0$. The graph is falling (decreasing) from the point $(-6, 0)$ to the point $(-4, -2)$ and from the point $(3, 4)$ to the point $(6, 1)$. We conclude that the graph is decreasing on the open intervals $(-6, -4)$ and $(3, 6)$ or for $-6 < x < -4$ and $3 < x < 6$. The graph is constant on the open interval $(0, 3)$ or for $0 < x < 3$. ◀

More precise definitions follow:

[*]The open interval (a, b) consists of all real numbers x for which $a < x < b$.

A function f is **increasing** on an open interval I if, for any choice of x_1 and x_2 in I, with $x_1 < x_2$, we have $f(x_1) < f(x_2)$.

A function f is **decreasing** on an open interval I if, for any choice of x_1 and x_2 in I, with $x_1 < x_2$, we have $f(x_1) > f(x_2)$.

A function f is **constant** on an interval I if, for all choices of x in I, the values $f(x)$ are equal.

Figure 30 illustrates the definitions. The graph of an increasing function goes up from left to right, the graph of a decreasing function goes down from left to right, and the graph of a constant function remains at a fixed height.

Figure 30

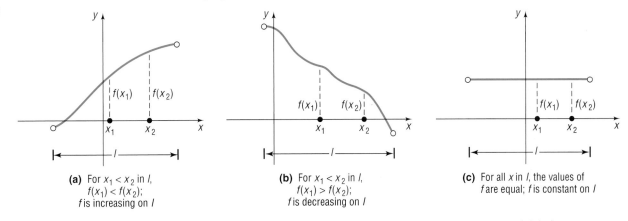

(a) For $x_1 < x_2$ in I,
$f(x_1) < f(x_2)$;
f is increasing on I

(b) For $x_1 < x_2$ in I,
$f(x_1) > f(x_2)$;
f is decreasing on I

(c) For all x in I, the values of
f are equal; f is constant on I

 NOW WORK PROBLEMS 11, 13, 15, AND 21(c).

4 Use a Graph to Locate Local Maxima and Local Minima

When the graph of a function is increasing to the left of $x = c$ and decreasing to the right of $x = c$, then at c the value of f is largest. This value is called a *local maximum* of f. See Figure 31(a).

When the graph of a function is decreasing to the left of $x = c$ and is increasing to the right of $x = c$, then at c the value of f is the smallest. This value is called a *local minimum* of f. See Figure 31(b).

Figure 31

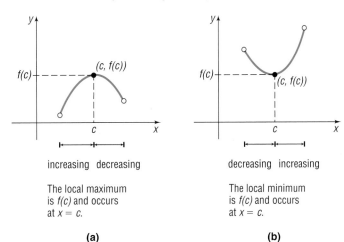

increasing decreasing

The local maximum
is $f(c)$ and occurs
at $x = c$.

(a)

decreasing increasing

The local minimum
is $f(c)$ and occurs
at $x = c$.

(b)

A function f has a **local maximum** at c if there is an open interval I containing c so that, for all $x \neq c$ in I, $f(x) \leq f(c)$. We call $f(c)$ a **local maximum of f**.

A function f has a **local minimum** at c if there is an open interval I containing c so that, for all $x \neq c$ in I, $f(x) \geq f(c)$. We call $f(c)$ a **local minimum of f**.

If f has a local maximum at c, then the value of f at c is greater than or equal to the values of f near c. If f has a local minimum at c, then the value of f at c is less than or equal to the values of f near c. The word *local* is used to suggest that it is only near c that the value $f(c)$ is largest or smallest.

| EXAMPLE 4 | **Finding Local Maxima and Local Minima from the Graph of a Function and Determining Where the Function Is Increasing, Decreasing, or Constant** |

Figure 32

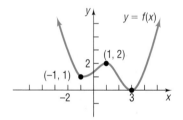

Figure 32 shows the graph of a function f.

(a) At what number(s), if any, does f have a local maximum?
(b) What are the local maxima?
(c) At what number(s), if any, does f have a local minimum?
(d) What are the local minima?
(e) List the intervals on which f is increasing. List the intervals on which f is decreasing.

Solution The domain of f is the set of real numbers.

(a) f has a local maximum at 1, since for all x close to 1, $x \neq 1$, we have $f(x) \leq f(1)$.

(b) The local maximum is $f(1) = 2$.
(c) f has a local minimum at -1 and at 3.
(d) The local minima are $f(-1) = 1$ and $f(3) = 0$.
(e) The function whose graph is given in Figure 32 is increasing for all values of x between -1 and 1 and for all values of x greater than 3. That is, the function is increasing on the intervals $(-1, 1)$ and $(3, \infty)$ or for $-1 < x < 1$ and $x > 3$. The function is decreasing for all values of x less than -1 and for all values of x between 1 and 3. That is, the function is decreasing on the intervals $(-\infty, -1)$ and $(1, 3)$ or for $x < -1$ and $1 < x < 3$. ◄

NOW WORK PROBLEMS 17 AND 19.

5 Use a Graphing Utility to Approximate Local Maxima and Local Minima and to Determine Where a Function Is Increasing or Decreasing

To locate the exact value at which a function f has a local maximum or a local minimum usually requires calculus. However, a graphing utility may be used to approximate these values by using the MAXIMUM and MINIMUM features.

| EXAMPLE 5 | **Using a Graphing Utility to Approximate Local Maxima and Minima and to Determine Where a Function Is Increasing or Decreasing** |

(a) Use a graphing utility to graph $f(x) = 6x^3 - 12x + 5$ for $-2 < x < 2$. Approximate where f has a local maximum and where f has a local minimum.
(b) Determine where f is increasing and where it is decreasing.

Solution

(a) Graphing utilities have a feature that finds the maximum or minimum point of a graph within a given interval. Graph the function f for $-2 < x < 2$. Using MAXIMUM, we find that the local maximum is 11.53 and it occurs at $x = -0.82$, rounded to two decimal places. See Figure 33(a). Using MINIMUM, we find that the local minimum is -1.53 and it occurs at $x = 0.82$, rounded to two decimal places. See Figure 33(b).

Figure 33

(a)   (b)

(b) Looking at Figures 33(a) and (b), we see that the graph of f is increasing from $x = -2$ to $x = -0.82$ and from $x = 0.82$ to $x = 2$, so f is increasing on the intervals $(-2, -0.82)$ and $(0.82, 2)$ or for $-2 < x < -0.82$ and $0.82 < x < 2$. The graph is decreasing from $x = -0.82$ to $x = 0.82$, so f is decreasing on the interval $(-0.82, 0.82)$ or for $-0.82 < x < 0.82$.   ◀

 NOW WORK PROBLEM 45.

6 Find the Average Rate of Change of a Function

In Section 1.8 we said that the slope of a line could be interpreted as the average rate of change. Often we are interested in the rate at which functions change. To find the average rate of change of a function between any two points on its graph, we calculate the slope of the line containing the two points.

If c is in the domain of a function $y = f(x)$, the **average rate of change of f** from c to x is defined as

$$\text{Average rate of change} = \frac{\Delta y}{\Delta x} = \frac{f(x) - f(c)}{x - c}, \qquad x \neq c \qquad \textbf{(1)}$$

In calculus, this expression is called the **difference quotient** of f at c.

The symbol Δy in (1) is the "change in y" and Δx is the "change in x." The average rate of change of f is the change in y divided by the change in x.

EXAMPLE 6	**Finding the Average Rate of Change**

Find the average rate of change of $f(x) = 3x^2$:

(a) From 1 to 3   (b) From 1 to 5   (c) From 1 to 7

Solution

(a) The average rate of change of $f(x) = 3x^2$ from 1 to 3 is

$$\frac{\Delta y}{\Delta x} = \frac{f(3) - f(1)}{3 - 1} = \frac{27 - 3}{3 - 1} = \frac{24}{2} = 12$$

Figure 34

(b) The average rate of change of $f(x) = 3x^2$ from 1 to 5 is

$$\frac{\Delta y}{\Delta x} = \frac{f(5) - f(1)}{5 - 1} = \frac{75 - 3}{5 - 1} = \frac{72}{4} = 18$$

(c) The average rate of change of $f(x) = 3x^2$ from 1 to 7 is

$$\frac{\Delta y}{\Delta x} = \frac{f(7) - f(1)}{7 - 1} = \frac{147 - 3}{7 - 1} = \frac{144}{6} = 24$$ ◀

See Figure 34 for a graph of $f(x) = 3x^2$. The function f is increasing for $x > 0$. The fact that the average rates of change are getting larger indicates that the graph is getting steeper, that is, it is increasing at an increasing rate.

NOW WORK PROBLEM 53.

EXAMPLE 7   **Average Rate of Change of a Population**

A strain of E-coli Beu 397-recA441 is placed into a nutrient broth at 30° Celsius and allowed to grow. The data shown in Table 9 are collected. The population is measured in grams and the time in hours. Since population P depends on time t and each input corresponds to exactly one output, we can say that population is a function of time, so $P(t)$ represents the population at time t. For example, $P(2.5) = 0.18$. Figure 35 shows a graph of the function.

(a) Find the average rate of change of the population from 0 to 2.5 hours.
(b) Find the average rate of change of the population from 4.5 to 6 hours.
(c) What is happening to the average rate of change as time passes?

Table 9

Time (hours), x	Population (grams), y
0	0.09
2.5	0.18
3.5	0.26
4.5	0.35
6	0.50

Figure 35

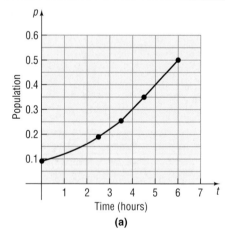

(a)

Solution   (a) Average rate of change $= \dfrac{\text{Population at 2.5 hours} - \text{Population at 0 hours}}{2.5 - 0}$

$$= \frac{P(2.5) - P(0)}{2.5 - 0}$$

$$= \frac{0.18 - 0.09}{2.5}$$

$$= 0.036 \text{ gram per hour}$$

On average, the population is increasing at a rate of 0.036 gram per hour from 0 to 2.5 hours.

(b) Average rate of change $= \dfrac{\text{Population at 6 hours} - \text{Population at 4.5 hours}}{6 - 4.5}$

$$= \frac{P(6) - P(4.5)}{6 - 4.5}$$

$$= \frac{0.5 - 0.35}{1.5}$$

$$= 0.1 \text{ gram per hour}$$

On average, the population is increasing at a rate of 0.1 gram per hour from 4.5 to 6 hours.

(c) Not only is the size of the population increasing over time, but the rate of increase is also increasing. We say that the population is increasing at an increasing rate. ◀

NOW WORK PROBLEM 69.

The Secant Line

The average rate of change of a function has an important geometric interpretation. Look at the graph of $y = f(x)$ in Figure 36. We have labeled two points on the graph: $(c, f(c))$ and $(x, f(x))$. The line containing these two points is called the **secant line**; its slope is

$$m_{\text{sec}} = \frac{f(x) - f(c)}{x - c}$$

Figure 36

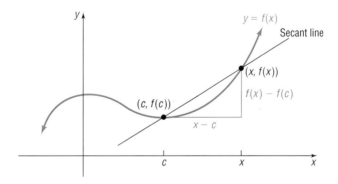

Theorem **Slope of the Secant Line**

The average rate of change of a function equals the slope of the secant line containing two points on its graph.

EXAMPLE 8 **Finding the Equation of a Secant Line**

Suppose that $g(x) = 3x^2 - 2x + 3$.

(a) Find the average rate of change of g from -2 to x.

(b) Use the result of part (a) to find the average rate of change of g from -2 to 1. Interpret this result.

(c) Find an equation of the secant line containing $(-2, g(-2))$ and $(1, g(1))$.

(d) Using a graphing utility, draw the graph of g and the secant line obtained in part (c) on the same screen.

Solution   (a) The average rate of change of $g(x) = 3x^2 - 2x + 3$ from -2 to x is

$$\text{Average rate of change} = \frac{g(x) - g(-2)}{x - (-2)}$$

$$= \frac{(3x^2 - 2x + 3) - 19}{x + 2} \qquad g(-2) = 3(-2)^2 - 2(-2) + 3 = 19$$

$$= \frac{3x^2 - 2x - 16}{x + 2}$$

$$= \frac{(3x - 8)(x + 2)}{x + 2} \qquad \text{Factor.}$$

$$= 3x - 8 \qquad \text{Cancel like factors.}$$

(b) The average rate of change of g from -2 to x is given by $3x - 8$. Therefore, the average rate of change of g from -2 to 1 is $3(1) - 8 = -5$. The slope of the secant line joining $(-2, g(-2))$ and $(1, g(1))$ is -5.

(c) We use the point–slope form to find an equation of the secant line.

$$y - y_1 = m_{\sec}(x - x_1) \qquad \text{Point–slope form of the secant line}$$

$$y - 4 = -5(x - 1) \qquad x_1 = 1, \, y_1 = g(1) = 4, \, m_{\sec} = -5$$

$$y - 4 = -5x + 5 \qquad \text{Simplify}$$

$$y = -5x + 9 \qquad \text{Slope–intercept form of the secant line}$$

Figure 37

(d) Figure 37 shows the graph of g along with the secant line $y = -5x + 9$. ◀

NOW WORK PROBLEM 59.

2.4 Assess Your Understanding

'Are You Prepared?'

Answers are given at the end of these exercises. If you get a wrong answer, read the pages listed in red.

1. The interval $(2, 5)$ can be written as the inequality _____. (pp. 151–152)

2. The slope of the line containing the points $(-2, 3)$ and $(3, 8)$ is _____. (pp. 163–165)

3. Write $x > 3$ using interval notation. (pp. 151–152)

4. Write the point–slope form of the line with slope 5 containing the point $(3, -2)$. (p. 168)

5. The intercepts of the equation $y = x^2 - 9$ are _____. (pp. 171–172 and 183)

Concepts and Vocabulary

6. A function f is _____ on an open interval I if, for any choice of x_1 and x_2 in I, with $x_1 < x_2$, we have $f(x_1) < f(x_2)$.

7. An _____ function f is one for which $f(-x) = f(x)$ for every x in the domain of f; an _____ function f is one for which $f(-x) = -f(x)$ for every x in the domain of f.

8. *True or False:* A function f is decreasing on an open interval I if, for any choice of x_1 and x_2 in I, with $x_1 < x_2$, we have $f(x_1) > f(x_2)$.

9. *True or False:* A function f has a local maximum at c if there is an open interval I containing c so that, for all $x \neq c$ in I, $f(x) \leq f(c)$.

10. *True or False:* Even functions have graphs that are symmetric with respect to the origin.

Skill Building

In Problems 11–20, use the given graph of the function f.

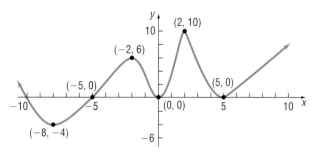

11. Is *f* increasing on the interval (−8, −2)?

12. Is *f* decreasing on the interval (−8, −4)?

13. Is *f* increasing on the interval (2, 10)?

14. Is *f* decreasing on the interval (2, 5)?

15. List the interval(s) on which *f* is increasing.

16. List the interval(s) on which *f* is decreasing.

17. Is there a local maximum at 2? If yes, what is it?

18. Is there a local maximum at 5? If yes, what is it?

19. List the numbers at which *f* has a local maximum. What are these local maxima?

20. List the numbers at which *f* has a local minimum. What are these local minima?

In Problems 21–28, the graph of a function is given. Use the graph to find:
(a) *The intercepts, if any*
(b) *Its domain and range*
(c) *The intervals on which it is increasing, decreasing, or constant*
(d) *Whether it is even, odd, or neither*

21.

22.

23.

24.

25.

26.

27.

28.

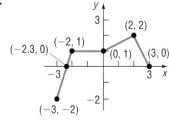

In Problems 29–32, the graph of a function f is given. Use the graph to find:
  (a) *The numbers, if any, at which f has a local maximum. What are these local maxima?*
  (b) *The numbers, if any, at which f has a local minimum. What are these local minima?*

29.

30.

31.

32.
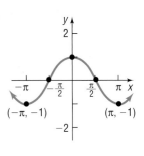

In Problems 33–44, determine algebraically whether each function is even, odd, or neither.

33. $f(x) = 4x^3$

34. $f(x) = 2x^4 - x^2$

35. $g(x) = -3x^2 - 5$

36. $h(x) = 3x^3 + 5$

37. $F(x) = \sqrt[3]{x}$

38. $G(x) = \sqrt{x}$

39. $f(x) = x + |x|$

40. $f(x) = \sqrt[3]{2x^2 + 1}$

41. $g(x) = \dfrac{1}{x^2}$

42. $h(x) = \dfrac{x}{x^2 - 1}$

43. $h(x) = \dfrac{-x^3}{3x^2 - 9}$

44. $F(x) = \dfrac{2x}{|x|}$

In Problems 45–52, use a graphing utility to graph each function over the indicated interval and approximate any local maxima and local minima. Determine where the function is increasing and where it is decreasing. Round answers to two decimal places.

45. $f(x) = x^3 - 3x + 2 \quad (-2, 2)$

46. $f(x) = x^3 - 3x^2 + 5 \quad (-1, 3)$

47. $f(x) = x^5 - x^3 \quad (-2, 2)$

48. $f(x) = x^4 - x^2 \quad (-2, 2)$

49. $f(x) = -0.2x^3 - 0.6x^2 + 4x - 6 \quad (-6, 4)$

50. $f(x) = -0.4x^3 + 0.6x^2 + 3x - 2 \quad (-4, 5)$

51. $f(x) = 0.25x^4 + 0.3x^3 - 0.9x^2 + 3 \quad (-3, 2)$

52. $f(x) = -0.4x^4 - 0.5x^3 + 0.8x^2 - 2 \quad (-3, 2)$

53. Find the average rate of change of $f(x) = -2x^2 + 4$
  (a) From 0 to 2
  (b) From 1 to 3
  (c) From 1 to 4

54. Find the average rate of change of $f(x) = -x^3 + 1$
  (a) From 0 to 2
  (b) From 1 to 3
  (c) From −1 to 1

55. Find the average rate of change of $g(x) = x^3 - 2x + 1$
  (a) From −3 to −2
  (b) From −1 to 1
  (c) From 1 to 3

56. Find the average rate of change of $h(x) = x^2 - 2x + 3$
  (a) From −1 to 1
  (b) From 0 to 2
  (c) From 2 to 5

57. $f(x) = 5x - 2$
  (a) Find the average rate of change from 1 to x.
  (b) Use the result of part (a) to find the average rate of change of f from 1 to 3. Interpret this result.
  (c) Find an equation of the secant line containing $(1, f(1))$ and $(3, f(3))$.
  (d) Using a graphing utility, draw the graph of f and the secant line on the same screen.

58. $f(x) = -4x + 1$
  (a) Find the average rate of change from 2 to x.
  (b) Use the result of part (a) to find the average rate of change of f from 2 to 5. Interpret this result.
  (c) Find an equation of the secant line containing $(2, f(2))$ and $(5, f(5))$.
  (d) Using a graphing utility, draw the graph of f and the secant line on the same screen.

59. $g(x) = x^2 - 2$
  (a) Find the average rate of change from −2 to x.
  (b) Use the result of part (a) to find the average rate of change of g from −2 to 1. Interpret this result.
  (c) Find an equation of the secant line containing $(-2, g(-2))$ and $(1, g(1))$.
  (d) Using a graphing utility, draw the graph of g and the secant line on the same screen.

60. $g(x) = x^2 + 1$
  (a) Find the average rate of change from −1 to x.
  (b) Use the result of part (a) to find the average rate of change of g from −1 to 2. Interpret this result.
  (c) Find an equation of the secant line containing $(-1, g(-1))$ and $(2, g(2))$.
  (d) Using a graphing utility, draw the graph of g and the secant line on the same screen.

61. $h(x) = x^2 - 2x$
  (a) Find the average rate of change from 2 to x.
  (b) Use the result of part (a) to find the average rate of change of h from 2 to 4. Interpret this result.
  (c) Find an equation of the secant line containing $(2, h(2))$ and $(4, h(4))$.
  (d) Using a graphing utility, draw the graph of h and the secant line on the same screen.

62. $h(x) = -2x^2 + x$
  (a) Find the average rate of change from 0 to x.
  (b) Use the result of part (a) to find the average rate of change of h from 0 to 3. Interpret this result.
  (c) Find an equation of the secant line containing $(0, h(0))$ and $(3, h(3))$.
  (d) Using a graphing utility, draw the graph of h and the secant line on the same screen.

Applications and Extensions

63. Constructing an Open Box An open box with a square base is to be made from a square piece of cardboard 24 inches on a side by cutting out a square from each corner and turning up the sides (see the figure).

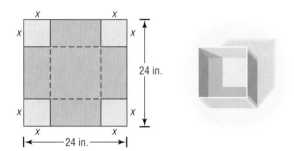

  (a) Express the volume V of the box as a function of the length x of the side of the square cut from each corner.
  (b) What is the volume if a 3-inch square is cut out?
  (c) What is the volume if a 10-inch square is cut out?
  (d) Graph $V = V(x)$. For what value of x is V largest?

64. Constructing an Open Box An open box with a square base is required to have a volume of 10 cubic feet.
  (a) Express the amount A of material used to make such a box as a function of the length x of a side of the square base.
  (b) How much material is required for a base 1 foot by 1 foot?
  (c) How much material is required for a base 2 feet by 2 feet?
  (d) Graph $A = A(x)$. For what value of x is A smallest?

65. Maximum Height of a Ball The height s of a ball (in feet) thrown with an initial velocity of 80 feet per second from an initial height of 6 feet is given as a function of the time t (in seconds) by

$$s(t) = -16t^2 + 80t + 6$$

  (a) Graph $s = s(t)$.
  (b) Determine the time at which height is maximum.
  (c) What is the maximum height?

66. Maximum Height of a Ball On July 1, 2004, the Cassini probe became the first spacecraft to orbit the planet Saturn. Although Saturn is about 764 times the size of Earth, it has a very similar gravitational force. The height s of an object thrown upward from Saturn's surface with an initial velocity of 100 feet per second is given as a function of time t (in seconds) by $s(t) = -17.28t^2 + 100t$.
  (a) Use a graphing utility to graph $s = s(t)$.
  (b) Determine the time at which height is a maximum.
  (c) What is the maximum height?
  (d) The same object thrown from the surface of Earth would have a height given by $s(t) = -16t^2 + 100t$. Determine the maximum height of the object on Earth and compare this to your result from part (c).

67. Minimum Average Cost The average cost of producing x riding lawn mowers per hour is given by

$$\overline{C}(x) = 0.3x^2 + 21x - 251 + \frac{2500}{x}$$

  (a) Graph $\overline{C} = \overline{C}(x)$
  (b) Determine the number of riding lawn mowers to produce in order to minimize average cost.
  (c) What is the minimum average cost?

68. Medicine Concentration The concentration C of a medication in the bloodstream t hours after being administered is given by

$$C(t) = -0.002x^4 + 0.039t^3 - 0.285t^2 + 0.766t + 0.085.$$

  (a) After how many hours will the concentration be highest?
  (b) A woman nursing a child must wait until the concentration is below 0.5 before she can feed him. After taking the medication, how long must she wait before feeding her child?

69. Revenue from Selling Bikes The following data represent the total revenue that would be received from selling x bicycles at Tunney's Bicycle Shop.

Number of Bicycles, x	Total Revenue, R (Dollars)
0	0
25	28,000
60	45,000
102	53,400
150	59,160
190	62,360
223	64,835
249	66,525

(a) Plot the points (x, R) in a rectangular coordinate system.
(b) Draw a line through the points $(0, 0)$ and $(25, 28000)$ on the graph found in part (a).
(c) Find the average rate of change of revenue from 0 to 25 bicycles.
(d) Interpret the average rate of change found in part (c).
(e) Draw a line through the points $(190, 62360)$ and $(223, 64835)$ on the graph found in part (a).
(f) Find the average rate of change of revenue from 190 to 223 bicycles.
(g) Interpret the average rate of change found in part (f).
(h) What is happening to the average rate of change of revenue as the number of bicycles increases?

70. Cost of Manufacturing Bikes The following data represent the monthly cost of producing bicycles at Tunney's Bicycle Shop.

Number of Bicycles, x	Total Cost of Production, C (Dollars)
0	24,000
25	27,750
60	31,500
102	35,250
150	39,000
190	42,750
223	46,500
249	50,250

(a) Plot the points (x, C) in a rectangular coordinate system.
(b) Draw a line through the points $(0, 24000)$ and $(25, 27750)$ on the graph found in part (a).

(c) Find the average rate of change of the cost from 0 to 25 bicycles.
(d) Interpret the average rate of change found in part (c).
(e) Draw a line through the points $(190, 42750)$ and $(223, 46500)$ on the graph found in part (a).
(f) Find the average rate of change of the cost from 190 to 223 bicycles.
(g) Interpret the average rate of change found in part (f).
(h) What is happening to the average rate of change of cost as the number of bicycles increases?

71. For the function $f(x) = x^2$, compute each average rate of change:
(a) From 0 to 1
(b) From 0 to 0.5
(c) From 0 to 0.1
(d) From 0 to 0.01
(e) From 0 to 0.001
(f) Graph each of the secant lines.
(g) What do you think is happening to the secant lines?
(h) What is happening to the slopes of the secant lines? Is there some number that they are getting closer to? What is that number?

72. For the function $f(x) = x^2$, compute each average rate of change:
(a) From 1 to 2
(b) From 1 to 1.5
(c) From 1 to 1.1
(d) From 1 to 1.01
(e) From 1 to 1.001
(f) Graph each of the secant lines.
(g) What do you think is happening to the secant lines?
(h) What is happening to the slopes of the secant lines? Is there some number that they are getting closer to? What is that number?

Problems 73–80 require the following discussion of a secant line. The slope of the secant line containing the two points $(x, f(x))$ and $(x + h, f(x + h))$ on the graph of a function $y = f(x)$ may be given as

$$m_{\text{sec}} = \frac{f(x + h) - f(x)}{(x + h) - x} = \frac{f(x + h) - f(x)}{h}, \qquad h \neq 0$$

In calculus, this expression is called the **difference quotient of f**.
(a) *Express the slope of the secant line of each function in terms of x and h. Be sure to simplify your answer.*
(b) *Find m_{sec} for $h = 0.5, 0.1$, and 0.01 at $x = 1$. What value does m_{sec} approach as h approaches 0?*
(c) *Find the equation for the secant line at $x = 1$ with $h = 0.01$.*
(d) *Graph f and the secant line found in part (c) on the same viewing window.*

73. $f(x) = 2x + 5$

74. $f(x) = -3x + 2$

75. $f(x) = x^2 + 2x$

76. $f(x) = 2x^2 + x$

77. $f(x) = 2x^2 - 3x + 1$

78. $f(x) = -x^2 + 3x - 2$

79. $f(x) = \dfrac{1}{x}$

80. $f(x) = \dfrac{1}{x^2}$

Discussion and Writing

81. Draw the graph of a function that has the following characteristics: domain: all real numbers; range: all real numbers; intercepts: $(0, -3)$ and $(3, 0)$; a local maximum of -2 is at -1; a local minimum of -6 is at 2. Compare your graph with others. Comment on any differences.

82. Redo Problem 81 with the following additional information: increasing on $(-\infty, -1), (2, \infty)$; decreasing on $(-1, 2)$. Again compare your graph with others and comment on any differences.

83. How many x-intercepts can a function defined on an interval have if it is increasing on that interval? Explain.

84. Suppose that a friend of yours does not understand the idea of increasing and decreasing functions. Provide an explanation complete with graphs that clarifies the idea.

85. Can a function be both even and odd? Explain.

86. Using a graphing utility, graph $y = 5$ on the interval $(-3, 3)$. Use MAXIMUM to find the local maxima on $(-3, 3)$. Comment on the result provided by the calculator.

'Are You Prepared?' Answers

1. $2 < x < 5$ **2.** 1 **3.** $(3, \infty)$ **4.** $y + 2 = 5(x - 3)$ **5.** $(-3, 0), (3, 0), (0, -9)$

2.5 Linear Functions and Models

PREPARING FOR THIS SECTION   *Before getting started, review the following:*

- Lines (Section 1.8, pp. 163–172)
- Linear Equations (Section 1.2, pp. 103–105)

✎ Now work the 'Are You Prepared?' problems on page 249.

OBJECTIVES
1. Graph Linear Functions
2. Work with Applications of Linear Functions
3. Draw and Interpret Scatter Diagrams
4. Distinguish between Linear and Nonlinear Relations
5. Use a Graphing Utility to Find the Line of Best Fit
6. Construct a Linear Model Using Direct Variation

1   Graph Linear Functions

In Section 1.8, we introduced lines. In particular, we discussed the slope–intercept form of the equation of a line $y = mx + b$. When we write the slope–intercept form of a line using function notation, we have a *linear function*.

A **linear function** is a function of the form

$$f(x) = mx + b$$

The graph of a linear function is a line with slope m and y-intercept b.

EXAMPLE 1	**Graphing a Linear Function**

Figure 38

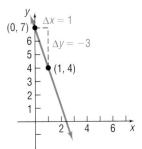

Graph the linear function: $f(x) = -3x + 7$

Solution   This is a linear function with slope $m = -3$ and y-intercept 7. To graph this function, we start by plotting the point $(0, 7)$, the y-intercept, and use the slope to find an additional point by moving right 1 unit and down 3 units. See Figure 38.   ◀

Alternatively, we could have found an additional point by evaluating the function at some $x \neq 0$. For $x = 1$, we find $f(1) = -3(1) + 7 = 4$ and obtain the point $(1, 4)$ on the graph.

An important property of a linear function is that the average rate of change between any two values is always the same; that is, it is a constant. Moreover, this constant is the slope of the line.

Theorem

Average Rate of Change of Linear Functions

Linear functions have a constant average rate of change. The constant average rate of change of $f(x) = mx + b$ is

$$\frac{\Delta y}{\Delta x} = m$$

Proof The average rate of change of $f(x) = mx + b$ from c to x is

$$\text{Average rate of change} = \frac{f(x) - f(c)}{x - c}$$
$$= \frac{(mx + b) - (mc + b)}{x - c}$$
$$= \frac{mx - mc}{x - c}$$
$$= \frac{m(x - c)}{x - c}$$
$$= m \qquad \blacksquare$$

Based on the theorem just proved, the average rate of change of $f(x) = 3x - 1$ is 3. The average rate of change of $g(x) = -\frac{2}{5}x + 5$ is $-\frac{2}{5}$.

 NOW WORK PROBLEM **11**.

2 Work with Applications of Linear Functions

There are many applications of linear functions. Let's look at one from accounting.

EXAMPLE 2

Straight-line Depreciation

Book value is the value of an asset that a company uses to create its balance sheet. Some companies depreciate their assets using straight-line depreciation so that the value of the asset declines by a fixed amount each year. The amount of the decline depends on the useful life that the company places on the asset. Suppose that a company just purchased a fleet of new cars for its sales force at a cost of $28,000 per car. The company chooses to depreciate each vehicle using the straight-line method over 7 years. This means that each car will depreciate by $\frac{\$28{,}000}{7} = \$4000$ per year.

(a) Write a linear function that expresses the book value V of each car as a function of its age, x.

(b) Graph the linear function.

(c) What is the book value of each car after 3 years?

(d) Interpret the slope.

(e) When will the book value of each car be $8000?

[**Hint:** Solve the equation $V(x) = 8000$.]

Solution   (a) If we let $V(x)$ represent the value of each car after x years, then $V(0)$ represents the original value of each car, so $V(0) = \$28{,}000$. The y-intercept of the linear function is $\$28{,}000$. Because each car depreciates by $\$4000$ per year, the slope of the linear function is $-\$4000$. The linear function that represents the value of each car after x years is

$$V(x) = -4000x + 28{,}000$$

(b) Figure 39 shows the graph of $V = V(x)$.

(c) The book value of each car after 3 years is

$$V(3) = -4000(3) + 28{,}000$$
$$= \$16{,}000$$

(d) Since the slope of $V(x) = -4000x + 28{,}000$ is -4000, the average rate of change of book value is $-\$4000$ per year. So, for each additional year that passes, the book value of each car decreases by $\$4000$.

(e) To find when the book value is $\$8000$, we solve the equation

$$V(x) = 8000$$
$$-4000x + 28{,}000 = 8000$$
$$-4000x = -20{,}000 \qquad \text{\small Subtract \$28,000 from each side.}$$
$$x = \frac{-20{,}000}{-4000} = 5 \qquad \text{\small Divide both sides by } -4000.$$

Each car will have a book value of $\$8000$ after 5 years.   ◀

Figure 39

Book value ($)

28,000
24,000
20,000
16,000
12,000
8000
4000

1 2 3 4 5 6 7
Age of vehicle (years)

 NOW WORK PROBLEM 39.

Next we look at an application from economics.

EXAMPLE 3	**Supply and Demand**

The quantity supplied of a good is the amount of a product that a company is willing to make available for sale at a given price. The quantity demanded of a good is the amount of a product that consumers are willing to purchase at a given price. Suppose that the quantity supplied, S, and quantity demanded, D, of cellular telephones each month are given by the following functions:

$$S(p) = 60p - 900$$
$$D(p) = -15p + 2850$$

where p is the price (in dollars) of the telephone.

(a) The **equilibrium price** of a product is defined as the price at which quantity supplied equals quantity demanded. That is, the equilibrium price is the price at which $S(p) = D(p)$. Find the equilibrium price of cellular telephones. What is the equilibrium quantity, the amount demanded (or supplied), at the equilibrium price?

(b) Determine the prices for which quantity supplied is greater than quantity demanded. That is, solve the inequality $S(p) > D(p)$.

(c) Graph $S = S(p)$, $D = D(p)$ and label the equilibrium price.

Solution   (a) We solve the equation $S(p) = D(p)$.

$$60p - 900 = -15p + 2850 \qquad \text{\small{$S(p) = D(p)$}}$$
$$60p = -15p + 3750 \qquad \text{\small{Add 900 to each side.}}$$
$$75p = 3750 \qquad \text{\small{Add 15p to each side.}}$$
$$p = \$50 \qquad \text{\small{Divide each side by 75.}}$$

The equilibrium price of cellular phones is $50. To find the equilibrium quantity, we evaluate either $S(p)$ or $D(p)$ at $p = 50$.

$$S(50) = 60(50) - 900 = 2100$$

The equilibrium quantity is 2100 cellular phones. At a price of $50 per phone, the company will sell 2100 phones each month and have no shortages or excess inventory.

(b) We solve the inequality $S(p) > D(p)$.

$$60p - 900 > -15p + 2850 \qquad \text{\small{$S(p) > D(p)$}}$$
$$60p > -15p + 3750 \qquad \text{\small{Add 900 to each side.}}$$
$$75p > 3750 \qquad \text{\small{Add 15p to each side.}}$$
$$p > \$50 \qquad \text{\small{Divide each side by 75.}}$$

If the company charges more than $50 per phone, then quantity supplied will exceed quantity demanded. In this case the company will have excess phones in inventory.

(c) Figure 40 shows the graphs of $S = S(p)$ and $D = D(p)$.

Figure 40

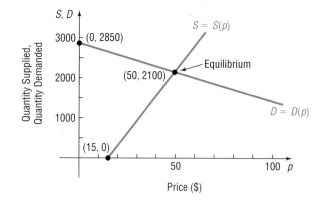

NOW WORK PROBLEM 35.

3   Draw and Interpret Scatter Diagrams

Linear functions can also be created by *fitting* a linear function to data. The first step in finding this relation is to plot the ordered pairs using rectangular coordinates. The resulting graph is called a **scatter diagram**. When drawing a scatter diagram, the independent variable is plotted on the horizontal axis and the dependent variable is plotted on the vertical axis.

| EXAMPLE 4 | Drawing and Interpreting a Scatter Diagram |

In baseball, the on-base percentage for a team represents the percentage of time that the players safely reach base. The data given in Table 10 represent the number of runs scored y and the on-base percentage x for teams in the National League during the 2003 baseball season.

Table 10

Team	On-Base Percentage, x	Runs Scored, y	(x, y)
Atlanta	34.9	907	(0.349, 907)
St. Louis	35.0	876	(0.350, 876)
Colorado	34.4	853	(0.344, 853)
Houston	33.6	805	(0.336, 805)
Philadelphia	34.3	791	(0.343, 791)
San Francisco	33.8	755	(0.338, 755)
Pittsburgh	33.8	753	(0.338, 753)
Florida	33.3	751	(0.333, 751)
Chicago Cubs	32.3	724	(0.323, 724)
Arizona	33.0	717	(0.330, 717)
Milwaukee	32.9	714	(0.329, 714)
Montreal	32.6	711	(0.326, 711)
Cincinnati	31.8	694	(0.318, 694)
San Diego	33.3	678	(0.333, 678)
NY Mets	31.4	642	(0.314, 642)
Los Angeles	30.3	574	(0.303, 574)

SOURCE: espn.com

(a) Draw a scatter diagram of the data by hand, treating on-base percentage as the independent variable.

(b) Use a graphing utility to draw a scatter diagram.

(c) Describe what happens as the on-base percentage increases.

Solution

(a) To draw a scatter diagram by hand, we plot the ordered pairs listed in Table 10, with the on-base percentage as the x-coordinate and the runs scored as the y-coordinate. See Figure 41(a). Notice that the points in the scatter diagram are not connected.

(b) Figure 41(b) shows a scatter diagram using a TI-84 Plus graphing calculator.

Figure 41

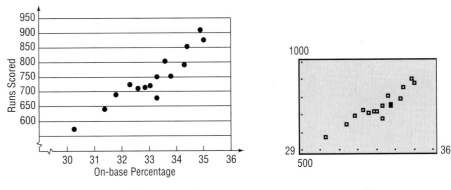

(a)   (b)

(c) We see from the scatter diagrams that, as the on-base percentage increases, the number of runs scored also increases. ◄

NOW WORK PROBLEM 25(a).

4 Distinguish between Linear and Nonlinear Relations

Scatter diagrams are used to help us to see the type of relation that exists between two variables. In this text, we will discuss a variety of different relations that may exist between two variables. For now, we concentrate on distinguishing between linear and nonlinear relations. See Figure 42.

Figure 42

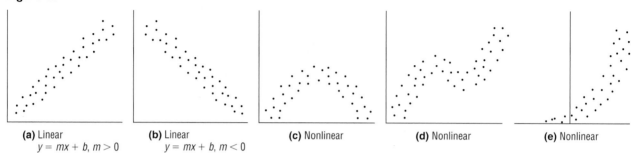

(a) Linear
$y = mx + b, m > 0$

(b) Linear
$y = mx + b, m < 0$

(c) Nonlinear

(d) Nonlinear

(e) Nonlinear

EXAMPLE 5   Distinguishing between Linear and Nonlinear Relations

Determine whether the relation between the two variables in Figure 43 is linear or nonlinear.

Figure 43

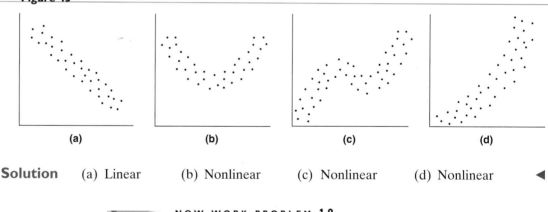

(a)

(b)

(c)

(d)

Solution   (a) Linear   (b) Nonlinear   (c) Nonlinear   (d) Nonlinear   ◄

NOW WORK PROBLEM 19.

In this section we will study data whose scatter diagrams imply that a linear relation exists between the two variables. Nonlinear data will be discussed in later chapters.

5 Use a Graphing Utility to Find the Line of Best Fit

Suppose that the scatter diagram of a set of data appears to be linearly related as in Figure 42(a) or (b). We might wish to find an equation of a line that relates the two variables. One way to obtain an equation for such data is to draw a line through two points on the scatter diagram and determine the equation of the line.

| EXAMPLE 6 | **Finding an Equation for Linearly Related Data** |

Using the data in Table 10 from Example 4:

(a) Select two points and find an equation of the line containing the points.

(b) Graph the line on the scatter diagram obtained in Example 4(b).

Solution

(a) Select two points, say $(32.6, 711)$ and $(34.9, 907)$. The slope of the line joining the points $(32.6, 711)$ and $(34.9, 907)$ is

$$m = \frac{907 - 711}{34.9 - 32.6} = \frac{196}{2.3} = 85.22$$

rounded to two decimal places. The equation of the line with slope 85.22 and passing through $(32.6, 711)$ is found using the point–slope form with $m = 85.22$, $x_1 = 32.6$, and $y_1 = 711$.

$$y - y_1 = m(x - x_1)$$
$$y - 711 = 85.22(x - 32.6)$$
$$y - 711 = 85.22x - 2778.17$$
$$y = 85.22x - 2067.17$$

Figure 44

(b) Figure 44 shows the scatter diagram with the graph of the line found in part (a). ◀

Select two other points and complete the solution. Add the line obtained to Figure 44.

NOW WORK PROBLEMS **25(b)** AND **(c)**.

The line obtained in Example 6 depends on the selection of points, which will vary from person to person. So the line that we found might be different from the line you found. Although the line we found in Example 6 appears to fit the data well, there may be a line that "fits it better." Do you think your line fits the data better? Is there a line of *best fit*? As it turns out, there is a method for finding the line that best fits linearly related data (called the *line of best fit*).[*]

| EXAMPLE 7 | **Finding the Line of Best Fit** |

Using the data in Table 10 from Example 4:

(a) Find the line of best fit using a graphing utility.

(b) Graph the line of best fit on the scatter diagram obtained in Example 4(b).

(c) Interpret the slope.

(d) Use the line of best fit to predict the number of runs a team will score if their on-base percentage is 34.1.

[*]We shall not discuss the underlying mathematics of lines of best fit in this book.

Solution   (a) Graphing utilities contain built-in programs that find the line of best fit for a collection of points in a scatter diagram. Upon executing the LINear REGression program, we obtain the results shown in Figure 45. The output that the utility provides shows us the equation $y = ax + b$, where a is the slope of the line and b is the y-intercept. The line of best fit that relates on-base percentage to runs scored may be expressed as the line $y = 62.17x - 1315.53$.

(b) Figure 46 shows the graph of the line of best fit, along with the scatter diagram.

Figure 45

Figure 46

(c) The slope of the line of best fit is 62.17, which means that, for every 1% increase in the on-base percentage, runs scored increase 62.17.

(d) Letting $x = 34.1$ in the equation of the line of best fit, we obtain $y = 62.17(34.1) - 1315.53 \approx 804$ runs. ◄

━━━ **NOW WORK PROBLEMS 25(d) AND (e).**

Does the line of best fit appear to be a good fit? In other words, does the line appear to accurately describe the relation between on-base percentage and runs scored?

And just how "good" is this line of best fit? The answers are given by what is called the *correlation coefficient*. Look again at Figure 45. The last line of output is $r = 0.924$. This number, called the **correlation coefficient,** r, $-1 \leq r \leq 1$, is a measure of the strength of the *linear relation* that exists between two variables. The closer that $|r|$ is to 1, the more perfect the linear relationship is. If r is close to 0, there is little or no *linear* relationship between the variables. A negative value of r, $r < 0$, indicates that as x increases y decreases; a positive value of r, $r > 0$, indicates that as x increases y does also. The data given in Table 10, having a correlation coefficient of 0.924, are indicative of a strong linear relationship with positive slope.

6 Construct a Linear Model Using Direct Variation

When a mathematical model is developed for a real-world problem, it often involves relationships between quantities that are expressed in terms of proportionality.

Force is proportional to acceleration.

For an ideal gas held at a constant temperature, pressure and volume are inversely proportional.

The force of attraction between two heavenly bodies is inversely proportional to the square of the distance between them.

Revenue is directly proportional to sales.

Each of these statements illustrates the idea of **variation**, or how one quantity varies in relation to another quantity. Quantities may vary *directly, inversely,* or *jointly*. We discuss direct variation here.

Let x and y denote two quantities. Then y **varies directly** with x, or y is **directly proportional to** x, if there is a nonzero number k such that

$$y = kx$$

The number k is called the **constant of proportionality**.

If y varies directly with x, then y is a linear function of x. The graph in Figure 47 illustrates the relationship between y and x if y varies directly with x and $k > 0$, $x \geq 0$. Note that the constant of proportionality is, in fact, the slope of the line.

If we know that two quantities vary directly, then knowing the value of each quantity in one instance enables us to write a formula that is true in all cases.

Figure 47
$y = kx, k > 0, x \geq 0$

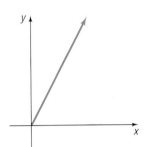

EXAMPLE 8   **Mortgage Payments**

The monthly payment p on a mortgage varies directly with the amount borrowed B. If the monthly payment on a 30-year mortgage is $6.65 for every $1000 borrowed, find a formula that relates the monthly payment p to the amount borrowed B for a mortgage with the same terms. Then find the monthly payment p when the amount borrowed B is $120,000.

Solution   Because p varies directly with B, we know that

$$p = kB$$

for some constant k. Because $p = 6.65$ when $B = 1000$, it follows that

$$6.65 = k(1000)$$

$$k = 0.00665$$

So we have

$$p = 0.00665B$$

We see that p is a linear function of B: $p(B) = 0.00665B$. In particular, when $B = \$120,000$, we find that

$$p(120,000) = 0.00665(\$120,000) = \$798$$

Figure 48 illustrates the relationship between the monthly payment p and the amount borrowed B.   ◀

Figure 48

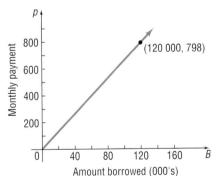

(120 000, 798)

Monthly payment
Amount borrowed (000's)

NOW WORK PROBLEM 45.

2.5   Assess Your Understanding

'Are You Prepared?'

Answers are given at the end of these exercises. If you get a wrong answer, read the pages listed in red.

1. Graph $y = 2x - 3$. (p. 171)

2. Find the slope of the line joining the points $(2, 5)$ and $(-1, 3)$. (pp. 163–165)

3. Find an equation of the line whose slope is -3 and contains the point $(-1, 5)$. (p. 168)

4. Solve $6x - 900 = -15x + 2850$.

Concepts and Vocabulary

5. For the graph of the linear function $f(x) = mx + b$, m is the _____ and b is the _____.

6. A _____ _____ is used to help us to see the type of relation, if any, that may exist between two variables.

7. If x and y are two quantities, then y is directly proportional to x if there is a nonzero number k such that _____.

8. *True or False:* The average rate of change of a linear function $f(x) = mx + b$ is the constant m.

9. *True or False:* The correlation coefficient is a measure of the strength of a linear relation between two variables and must lie between -1 and 1, inclusive.

10. *True or False:* If y varies directly with x, then y is a linear function of x.

Skill Building

For Problems 11–18, graph each linear function by hand. Determine the average rate of change of each function.

11. $f(x) = 2x + 3$

12. $g(x) = 5x - 4$

13. $h(x) = -3x + 4$

14. $p(x) = -x + 6$

15. $f(x) = \dfrac{1}{4}x - 3$

16. $h(x) = -\dfrac{2}{3}x + 4$

17. $F(x) = 4$

18. $G(x) = -2$

In Problems 19–24, examine the scatter diagram and determine whether the type of relation, if any, that may exist is linear or nonlinear.

19.

20.

21.

22.

23.

24.
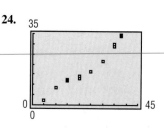

In Problems 25–30:
  (a) *Draw a scatter diagram by hand and using a graphing utility.*
  (b) *Select two points from the scatter diagram and find the equation of the line containing the points selected.*
  (c) *Graph the line found in part (b) on the scatter diagram.*
  (d) *Use a graphing utility to find the line of best fit.*
  (e) *Use a graphing utility to graph the line of best fit on the scatter diagram.*

25.

x	3	4	5	6	7	8	9
y	4	6	7	10	12	14	16

26.

x	3	5	7	9	11	13
y	0	2	3	6	9	11

27.

x	-2	-1	0	1	2
y	-4	0	1	4	5

28.

x	-2	-1	0	1	2
y	7	6	3	2	0

29.

x	-20	-17	-15	-14	-10
y	100	120	118	130	140

30.

x	-30	-27	-25	-20	-14
y	10	12	13	13	18

Applications and Extensions

31. Car Rentals The cost C, in dollars, of renting a moving truck for a day is given by the function $C(x) = 0.25x + 35$, where x is the number of miles driven.
  (a) What is the cost if you drive $x = 40$ miles?
  (b) If the cost of renting the moving truck is $80, how many miles did you drive?
  (c) Suppose that you want to keep the cost below $100. What is the maximum number of miles that you can drive?

32. Phone Charges The monthly cost C, in dollars, of a certain cellular phone plan is given by the function $C(x) = 0.38x + 5$, where x is the number of minutes used on the phone.
  (a) What is the cost if you talk on the phone for $x = 50$ minutes?
  (b) Suppose that your monthly bill is $29.32. How many minutes did you use the phone?

(c) Suppose that you budget yourself $60 per month for the phone. What is the maximum number of minutes that you can talk?

33. Disability Benefits The average monthly benefit B, in dollars, for individuals on disability is given by the function $B(t) = 19.25t + 585.72$, where t is the number of years since 1990.

    (a) What was the average monthly benefit in 2000 $(t = 10)$?

    (b) In what year will the average monthly benefit be $893.72?

    (c) In what year will the average monthly benefit exceed $1000?

34. Health Expenditures The total private health expenditures H, in billions of dollars, is given by the function $H(t) = 26t + 411$, where t is the number of years since 1990.

    (a) What was the total private health expenditures in 2000 $(t = 10)$?

    (b) In what year will total private health expenditures be $879 billion?

    (c) In what year will total private health expenditures exceed $1 trillion ($1000 billion)?

35. Supply and Demand Suppose that the quantity supplied S and quantity demanded D of T-shirts at a concert are given by the following functions:

$$S(p) = -200 + 50p$$
$$D(p) = 1000 - 25p$$

where p is the price in dollars. The equilibrium price of a market is defined as the price at which quantity supplied equals quantity demanded $(S = D)$.

    (a) Find the equilibrium price for T-shirts at this concert. What is the equilibrium quantity?

    (b) Determine the prices for which quantity demanded is greater than quantity supplied.

    (c) What do you think will eventually happen to the price of T-shirts if quantity demanded is greater than quantity supplied?

36. Supply and Demand Suppose that the quantity supplied S and quantity demanded D of hot dogs at a baseball game are given by the following functions:

$$S(p) = -2000 + 3000p$$
$$D(p) = 10,000 - 1000p$$

where p is the price in dollars. The equilibrium price of a market is defined as the price at which quantity supplied equals quantity demanded $(S = D)$.

    (a) Find the equilibrium price for hot dogs at the baseball game. What is the equilibrium quantity?

    (b) Determine the prices for which quantity demanded is less than quantity supplied.

    (c) What do you think will eventually happen to the price of hot dogs if quantity demanded is less than quantity supplied?

The point at which a company's profits equal zero is called the company's **break-even point**. *For Problems 37 and 38, let R represent a company's revenue, let C represent the company's costs, and let x represent the number of units produced and sold each day.*

(a) *Find the firm's break-even point; that is, find x so that R = C.*

(b) *Find the values of x such that R(x) > C(x). This represents the number of units that the company must sell to earn a profit.*

37. $R(x) = 8x$
$C(x) = 4.5x + 17,500$

38. $R(x) = 12x$
$C(x) = 10x + 15,000$

---

39. Straight-line Depreciation Suppose that a company has just purchased a new computer for $3000. The company chooses to depreciate the computer using the straight-line method over 3 years.

    (a) Write a linear function that expresses the book value of the computer as a function of its age.

    (b) Graph the linear function.

    (c) What is the book value of the computer after 2 years?

    (d) When will the computer be worth $2000?

40. Straight-line Depreciation Suppose that a company has just purchased a new machine for its manufacturing facility for $120,000. The company chooses to depreciate the machine using the straight-line method over 10 years.

    (a) Write a linear function that expresses the book value of the machine as a function of its age.

    (b) Graph the linear function.

    (c) What is the book value of the machine after 4 years?

    (d) When will the machine be worth $60,000?

41. Cost Function The simplest cost function is the linear cost function $C(x) = mx + b$, where C is the cost of producing x units. Here the y-intercept b represents the fixed costs of operating a business and the slope m represents the variable cost of producing each unit. Suppose that a small bicycle manufacturer has daily fixed costs of $1800 and each bicycle costs $90 to manufacture.

    (a) Write a linear function that expresses the cost C of manufacturing x bicycles in a day.

    (b) Graph the linear function.

    (c) What is the cost of manufacturing 14 bicycles in a day?

    (d) How many bicycles could be manufactured for $3780?

42. Cost Function Refer to Problem 41. Suppose that the landlord of the building increases the bicycle manufacturer's rent by $100 per month.
(a) Assuming that the manufacturer is open for business 20 days per month, what are the new daily fixed costs?
(b) Write a linear function that expresses the cost C of manufacturing x bicycles in a day with the higher rent.
(c) Graph the linear function.
(d) What is the cost of manufacturing 14 bicycles in a day?
(e) How many bicycles could be manufactured for $3780?

43. Truck Rentals A truck rental company rents a truck for one day by charging $29 plus $0.07 per mile.
(a) Write a linear function that relates the cost C, in dollars, of renting the truck to the number x of miles driven.
(b) What is the cost of renting the truck if the truck is driven 110 miles? 230 miles?

44. Long Distance A phone company offers a long distance package by charging $5 plus $0.05 per minute.
(a) Write a linear function that relates the cost C in dollars of talking x minutes.
(b) What is the cost of talking 105 minutes? 180 minutes?

45. Mortgage Payments The monthly payment p on a mortgage varies directly with the amount borrowed B. If the monthly payment on a 30-year mortgage is $6.49 for every $1000 borrowed, find a linear function that relates the monthly payment p to the amount borrowed B for a mortgage with the same terms. Then find the monthly payment p when the amount borrowed B is $145,000.

46. Mortgage Payments The monthly payment p on a mortgage varies directly with the amount borrowed B. If the monthly payment on a 15-year mortgage is $8.99 for every $1000 borrowed, find a linear function that relates the monthly payment p to the amount borrowed B for a mortgage with the same terms. Then find the monthly payment p when the amount borrowed B is $175,000.

47. Revenue Function At the corner Shell station, the revenue R varies directly with the number g of gallons of gasoline sold. If the revenue is $23.40 when the number of gallons sold is 12, find a linear function that relates revenue R to the number g of gallons of gasoline. Then find the revenue R when the number of gallons of gasoline sold is 10.5.

48. Cost Function The cost C of chocolate-covered almonds varies directly with the number A of pounds of almonds purchased. If the cost is $23.75 when the number of pounds of chocolate-covered almonds purchased is 5, find a linear function that relates the cost C to the number A of pounds of almonds purchased. Then find the cost C when the number of pounds of almonds purchased is 3.5.

49. Sand Castles Researchers at Bournemouth University discovered that the amount of water W varies directly with the amount of sand S that is needed to make the perfect sand castle. If 15 gallons of sand is mixed with 1.875 gallons of water, a perfect sand castle can be built. How many gallons of water should be mixed with 40 gallons of sand to create a perfect sand castle?

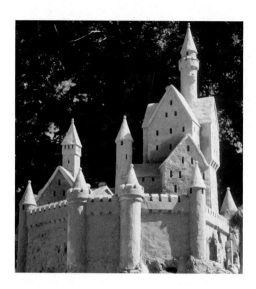

50. Physics: Falling Objects The velocity v of a falling object is directly proportional to the time t of the fall. If, after 2 seconds, the velocity of the object is 64 feet per second, what will its velocity be after 3 seconds?

51. Per Capita Disposable Income versus Consumption An economist wishes to estimate a linear function that relates per capita consumption expenditures C and disposable income I. Both C and I are measured in dollars. The following data represent the per capita disposable income (income after taxes) and per capita consumption in the United States for 1995 to 2003.

Year	Per Capita Disposable Income (I)	Per Capita Consumption (C)
1995	20,316	19,061
1996	21,127	19,938
1997	21,871	20,807
1998	22,212	21,385
1999	23,968	22,491
2000	25,472	23,863
2001	26,175	24,690
2002	27,259	25,622
2003	28,227	26,650

Source: U.S. Department of Commerce

Let I represent the independent variable and C the dependent variable.
(a) Use a graphing utility to draw a scatter diagram.
(b) Use a graphing utility to find the line of best fit to the data. Express the solution using function notation.
(c) Interpret the slope. The slope of this line is called the **marginal propensity to consume**.
(d) Predict the per capita consumption in a year when disposable income is $28,750.
(e) What is the income in a year when consumption is $26,900?

52. Height versus Head Circumference   A pediatrician wanted to estimate a linear function that relates a child's height, H to their head circumference, C. She randomly selects nine children from her practice, measures their height and head circumference, and obtains the data shown below. Let H represent the independent variable and C the dependent variable.

(a) Use a graphing utility to draw a scatter diagram.
(b) Use a graphing utility to find the line of best fit to the data. Express the solution using function notation.
(c) Interpret the slope.
(d) Predict the head circumference of a child that is 26 inches tall.
(e) What is the height of a child whose head circumference is 17.4 inches?

Height, H (inches)	Head Circumference, C (inches)
25.25	16.4
25.75	16.9
25	16.9
27.75	17.6
26.5	17.3
27	17.5
26.75	17.3
26.75	17.5
27.5	17.5

Source: Denise Slucki, Student at Joliet Junior College

53. Gestation Period versus Life Expectancy   A researcher would like to estimate the linear function relating the gestation period of an animal, G, and its life expectancy, L. She collects the following data.

Animal	Gestation (or incubation) Period, G (days)	Life Expectancy, L (years)
Cat	63	11
Chicken	22	7.5
Dog	63	11
Duck	28	10
Goat	151	12
Lion	108	10
Parakeet	18	8
Pig	115	10
Rabbit	31	7
Squirrel	44	9

Source: Time Almanac 2000

Let G represent the independent variable and L the dependent variable.
(a) Use a graphing utility to draw a scatter diagram.
(b) Use a graphing utility to find the line of best fit to the data. Express the solution using function notation.
(c) Interpret the slope.

(d) Predict the life expectancy of an animal whose gestation period is 89 days.

54. Mortgage Qualification   The amount of money that a lending institution will allow you to borrow mainly depends on the interest rate and your annual income. The following data represent the annual income, I, required by a bank in order to lend L dollars at an interest rate of 7.5% for 30 years.

Annual Income, I ($)	Loan Amount, L ($)
15,000	44,600
20,000	59,500
25,000	74,500
30,000	89,400
35,000	104,300
40,000	119,200
45,000	134,100
50,000	149,000
55,000	163,900
60,000	178,800
65,000	193,700
70,000	208,600

Source: Information Please Almanac, 1999

Let I represent the independent variable and L the dependent variable.
(a) Use a graphing utility to draw a scatter diagram of the data.
(b) Use a graphing utility to find the line of best fit to the data.
(c) Graph the line of best fit on the scatter diagram drawn in part (a).
(d) Interpret the slope of the line of best fit.
(e) Determine the loan amount that an individual would qualify for if her income is $42,000.

55. Demand for Jeans   The marketing manager at Levi-Strauss wishes to find a function that relates the demand D for men's jeans and p, the price of the jeans. The following data were obtained based on a price history of the jeans.

Price ($/Pair), p	Demand (Pairs of Jeans Sold per Day), D
20	60
22	57
23	56
23	53
27	52
29	49
30	44

(a) Does the relation defined by the set of ordered pairs (p, D) represent a function?
(b) Draw a scatter diagram of the data.
(c) Using a graphing utility, find the line of best fit relating price and quantity demanded.

(d) Interpret the slope.
(e) Express the relationship found in part (c) using function notation.
(f) What is the domain of the function?
(g) How many jeans will be demanded if the price is $28 a pair?

56. Advertising and Sales Revenue   A marketing firm wishes to find a function that relates the sales *S* of a product and *A*,

	Advertising Expenditures, *A*	Sales, *S*
	20	335
	22	339
	22.5	338
	24	343
	24	341
	27	350
	28.3	351

the amount spent on advertising the product. The data are obtained from past experience. Advertising and sales are measured in thousands of dollars.
(a) Does the relation defined by the set of ordered pairs (*A*, *S*) represent a function?
(b) Draw a scatter diagram of the data.
(c) Using a graphing utility, find the line of best fit relating advertising expenditures and sales.
(d) Interpret the slope.
(e) Express the relationship found in part (c) using function notation.
(f) What is the domain of the function?
(g) Predict sales if advertising expenditures are $25,000.

Discussion and Writing

57. Maternal Age versus Down Syndrome   A biologist would like to know how the age of the mother affects the incidence rate of Down syndrome. The following data represent the age of the mother and the incidence rate of Down syndrome per 1000 pregnancies.

Age of Mother, *x*	Incidence of Down Syndrome, *y*
33	2.4
34	3.1
35	4
36	5
37	6.7
38	8.3
39	10
40	13.3
41	16.7
42	22.2
43	28.6
44	33.3
45	50

Source: Hook, E.B., *Journal of the American Medical Association*, 249, 2034-2038, 1983

Draw a scatter diagram treating age of the mother as the independent variable. Explain why it would not make sense to find the line of best fit for these data.

58. Harley Davidson, Inc., Stock   The following data represent the closing year-end price of Harley Davidson, Inc., stock for the years 1990 to 2003. We let *x* = 1 represent 1990, *x* = 2 represent 1991, and so on.

Year, *x*	Closing Price, *y*
1990 (*x* = 1)	1.1609
1991 (*x* = 2)	2.6988
1992 (*x* = 3)	4.5381
1993 (*x* = 4)	5.3379
1994 (*x* = 5)	6.8032
1995 (*x* = 6)	7.0328
1996 (*x* = 7)	11.5585
1997 (*x* = 8)	13.4799
1998 (*x* = 9)	23.5424
1999 (*x* = 10)	31.9342
2000 (*x* = 11)	39.7277
2001 (*x* = 12)	54.31
2002 (*x* = 13)	46.20
2003 (*x* = 14)	47.53

Source: Yahoo! Finance

Draw a scatter diagram of the data treating the year, *x*, as the independent variable. Explain why it would not make sense to find the line of best fit for these data.

59. Under what circumstances is a linear function $f(x) = mx + b$ odd? Can a linear function ever be even?

60. Find the line of best fit for the ordered pairs $(1, 5)$ and $(3, 8)$. What is the correlation coefficient for these data? Why is this result reasonable?

61. What does a correlation coefficient of 0 imply?

62. What would it mean if a teacher states that a student's grade in the class is directly proportional to the amount of time that the student studies?

'Are You Prepared?' Answers

1.

2. $\dfrac{2}{3}$ **3.** $y = -3x + 2$ **4.** $\{50\}$

2.6 Library of Functions; Piecewise-defined Functions

PREPARING FOR THIS SECTION   *Before getting started, review the following:*

- Intercepts (Section 1.1, pp. 95–96; Section 1.8, pp. 171–172)
- Graphs of Key Equations (Section 2.1; Example 3, p. 200, Example 4, p. 200, Example 5, p. 202)

Now work the 'Are You Prepared?' problems on page 262.

OBJECTIVES   1 Graph the Functions Listed in the Library of Functions
2 Graph Piecewise-defined Functions

We now introduce a few more functions to add to our list of important functions. We begin with the *square root function*.

In Section 2.1 we graphed the equation $y = \sqrt{x}$. Figure 49 shows a graph of $f(x) = \sqrt{x}$. Based on the graph, we have the following properties:

Figure 49

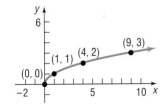

Properties of $f(x) = \sqrt{x}$

1. The x-intercept of the graph of $f(x) = \sqrt{x}$ is 0. The y-intercept of the graph of $f(x) = \sqrt{x}$ is also 0.

2. The function is neither even nor odd.

3. It is increasing on the interval $(0, \infty)$.

4. It has a local minimum of 0 at $x = 0$.

EXAMPLE 1	**Graphing the Cube Root Function**

(a) Determine whether $f(x) = \sqrt[3]{x}$ is even, odd, or neither. State whether the graph of f is symmetric with respect to the y-axis or symmetric with respect to the origin.

(b) Determine the intercepts, if any, of the graph of $f(x) = \sqrt[3]{x}$.

(c) Graph $f(x) = \sqrt[3]{x}$.

Solution   (a) Because

$$f(-x) = \sqrt[3]{-x} = -\sqrt[3]{x} = -f(x)$$

the function is odd. The graph of f is symmetric with respect to the origin.

(b) The y-intercept is $f(0) = \sqrt[3]{0} = 0$. The x-intercept is found by solving the equation $f(x) = 0$.

$$\begin{array}{ll} f(x) = 0 & \\ \sqrt[3]{x} = 0 & \quad f(x) = \sqrt[3]{x} \\ x = 0 & \quad \textit{Cube both sides of the equation.} \end{array}$$

The x-intercept is also 0.

(c) We use the function to form Table 11 and obtain some points on the graph. Because of the symmetry with respect to the origin, we only need to find points (x, y) for which $x \geq 0$. Figure 50 shows the graph of $f(x) = \sqrt[3]{x}$.

Table 11

x	$y = f(x) = \sqrt[3]{x}$	(x, y)
0	0	$(0, 0)$
$\dfrac{1}{8}$	$\dfrac{1}{2}$	$\left(\dfrac{1}{8}, \dfrac{1}{2}\right)$
1	1	$(1, 1)$
2	$\sqrt[3]{2} \approx 1.26$	$(2, \sqrt[3]{2})$
8	2	$(8, 2)$

Figure 50

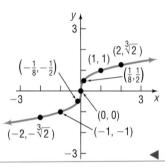

From the results of Example 1 and Figure 50, we have the following properties of the cube root function.

Properties of $f(x) = \sqrt[3]{x}$

1. The x-intercept of the graph of $f(x) = \sqrt[3]{x}$ is 0. The y-intercept of the graph of $f(x) = \sqrt[3]{x}$ is also 0.

2. The function is odd.

3. It is increasing on the interval $(-\infty, \infty)$.

4. It does not have a local minimum or a local maximum.

EXAMPLE 2   **Graphing the Absolute Value Function**

(a) Determine whether $f(x) = |x|$ is even, odd, or neither. State whether the graph of f is symmetric with respect to the y-axis or symmetric with respect to the origin.

(b) Determine the intercepts, if any, of the graph of $f(x) = |x|$.

(c) Graph $f(x) = |x|$.

Solution   (a) Because

$$f(-x) = |-x| = |x| = f(x)$$

the function is even. The graph of f is symmetric with respect to the y-axis.

(b) The y-intercept is $f(0) = |0| = 0$. The x-intercept is found by solving the equation $f(x) = |x| = 0$. So the x-intercept is 0.

(c) We use the function to form Table 12 and obtain some points on the graph. Because of the symmetry with respect to the y-axis, we only need to find points (x, y) for which $x \geq 0$. Figure 51 shows the graph of $f(x) = |x|$.

Table 12

x	y = f(x) = |x|	(x, y)
0	0	(0, 0)
1	1	(1, 1)
2	2	(2, 2)
3	3	(3, 3)

Figure 51

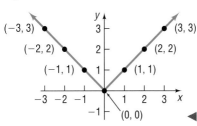

From the results of Example 2 and Figure 51, we have the following properties of the absolute value function.

Properties of $f(x) = |x|$

1. The x-intercept of the graph of $f(x) = |x|$ is 0. The y-intercept of the graph of $f(x) = |x|$ is also 0.

2. The function is even.

3. It is decreasing on the interval $(-\infty, 0)$. It is increasing on the interval $(0, \infty)$.

4. It has a local minimum of 0 at $x = 0$.

────── Seeing the Concept ──────

Graph $y = |x|$ on a square screen and compare what you see with Figure 51. Note that some graphing calculators use abs(x) for absolute value.

1 Graph the Functions Listed in the Library of Functions

We now provide a summary of the key functions that we have encountered. In going through this list, pay special attention to the properties of each function, particularly to the shape of each graph. Knowing these graphs will lay the foundation for later graphing techniques.

Linear Function

$$f(x) = mx + b, \qquad m \text{ and } b \text{ are real numbers}$$

Figure 52
Linear Function

y↑ $f(x) = mx + b, m > 0$

(0, b)

x

See Figure 52.

The domain of a **linear function** is the set of all real numbers. The graph of this function is a nonvertical line with slope m and y-intercept b. A linear function is increasing if $m > 0$, decreasing if $m < 0$, and constant if $m = 0$.

Figure 53
Constant Function

Constant Function

$$f(x) = b, \quad b \text{ is a real number}$$

See Figure 53.

A **constant function** is a special linear function ($m = 0$). Its domain is the set of all real numbers; its range is the set consisting of a single number b. Its graph is a horizontal line whose y-intercept is b. The constant function is an even function whose graph is constant over its domain.

Figure 54
Identity Function

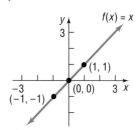

Identity Function

$$f(x) = x$$

See Figure 54.

The **identity function** is also a special linear function. Its domain and range are the set of all real numbers. Its graph is a line whose slope is $m = 1$ and whose y-intercept is 0. The line consists of all points for which the x-coordinate equals the y-coordinate. The identity function is an odd function that is increasing over its domain. Note that the graph bisects quadrants I and III.

Figure 55
Square Function

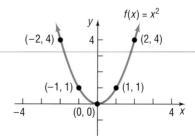

Square Function

$$f(x) = x^2$$

See Figure 55.

The domain of the **square function** f is the set of all real numbers; its range is the set of nonnegative real numbers. The graph of this function is a parabola whose intercept is at $(0, 0)$. The square function is an even function that is decreasing on the interval $(-\infty, 0)$ and increasing on the interval $(0, \infty)$.

Figure 56
Cube Function

Cube Function

$$f(x) = x^3$$

See Figure 56.

The domain and the range of the **cube function** is the set of all real numbers. The intercept of the graph is at $(0, 0)$. The cube function is odd and is increasing on the interval $(-\infty, \infty)$.

Figure 57
Square Root Function

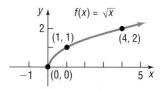

Square Root Function

$$f(x) = \sqrt{x}$$

See Figure 57.

The domain and the range of the **square root function** is the set of nonnegative real numbers. The intercept of the graph is at $(0, 0)$. The square root function is neither even nor odd and is increasing on the interval $(0, \infty)$.

Figure 58
Cube Root Function

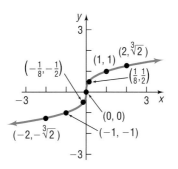

Cube Root Function

$$f(x) = \sqrt[3]{x}$$

See Figure 58.

The domain and the range of the **cube root function** is the set of all real numbers. The intercept of the graph is at $(0, 0)$. The cube root function is an odd function that is increasing on the interval $(-\infty, \infty)$.

Figure 59
Reciprocal Function

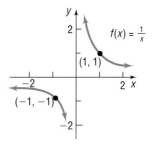

Reciprocal Function

$$f(x) = \frac{1}{x}$$

Refer to Example 5, page 202, for a discussion of the equation $y = \dfrac{1}{x}$. See Figure 59.

The domain and the range of the **reciprocal function** is the set of all nonzero real numbers. The graph has no intercepts. The reciprocal function is decreasing on the intervals $(-\infty, 0)$ and $(0, \infty)$ and is an odd function.

Absolute Value Function

$$f(x) = |x|$$

Figure 60
Absolute Value Function

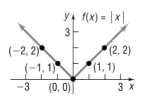

See Figure 60.

The domain of the **absolute value function** is the set of all real numbers; its range is the set of nonnegative real numbers. The intercept of the graph is at $(0, 0)$. If $x \geq 0$, then $f(x) = x$, and the graph of f is part of the line $y = x$; if $x < 0$, then $f(x) = -x$, and the graph of f is part of the line $y = -x$. The absolute value function is an even function; it is decreasing on the interval $(-\infty, 0)$ and increasing on the interval $(0, \infty)$.

The notation $\text{int}(x)$ stands for the largest integer less than or equal to x. For example,

$$\text{int}(1) = 1, \quad \text{int}(2.5) = 2, \quad \text{int}\left(\frac{1}{2}\right) = 0, \quad \text{int}\left(-\frac{3}{4}\right) = -1, \quad \text{int}(\pi) = 3$$

This type of correspondence occurs frequently enough in mathematics that we give it a name.

Greatest Integer Function

$$f(x) = \text{int}(x)^* = \text{greatest integer less than or equal to } x$$

*Some books use the notation $f(x) = [\![x]\!]$ instead of $\text{int}(x)$.

Table 13

x	y = f(x) = int(x)	(x, y)
−1	−1	(−1, −1)
−$\frac{1}{2}$	−1	$\left(-\frac{1}{2}, -1\right)$
−$\frac{1}{4}$	−1	$\left(-\frac{1}{4}, -1\right)$
0	0	(0, 0)
$\frac{1}{4}$	0	$\left(\frac{1}{4}, 0\right)$
$\frac{1}{2}$	0	$\left(\frac{1}{2}, 0\right)$
$\frac{3}{4}$	0	$\left(\frac{3}{4}, 0\right)$

We obtain the graph of $f(x) = \text{int}(x)$ by plotting several points. See Table 13. For values of x, $-1 \leq x < 0$, the value of $f(x) = \text{int}(x)$ is -1; for values of x, $0 \leq x < 1$, the value of f is 0. See Figure 61 for the graph.

Figure 61
Greatest Integer Function

The domain of the **greatest integer function** is the set of all real numbers; its range is the set of integers. The y-intercept of the graph is 0. The x-intercepts lie in the interval $[0, 1)$. The greatest integer function is neither even nor odd. It is constant on every interval of the form $[k, k + 1)$, for k an integer. In Figure 61, we use a solid dot to indicate, for example, that at $x = 1$ the value of f is $f(1) = 1$; we use an open circle to illustrate that the function does not assume the value of 0 at $x = 1$.

From the graph of the greatest integer function, we can see why it also called a **step function**. At $x = 0$, $x = \pm 1$, $x = \pm 2$, and so on, this function exhibits what is called a *discontinuity*; that is, at integer values, the graph suddenly "steps" from one value to another without taking on any of the intermediate values. For example, to the immediate left of $x = 3$, the y-coordinates are 2, and to the immediate right of $x = 3$, the y-coordinates are 3.

Figure 62 shows the graph of $f(x) = \text{int}(x)$ on a TI-84 Plus.

NOTE

When graphing a function using a graphing utility, you can choose either the **connected mode**, in which points plotted on the screen are connected, making the graph appear without any breaks, or the **dot mode**, in which only the points plotted appear. When graphing the greatest integer function with a graphing utility, it is necessary to be in the **dot mode**. This is to prevent the utility from "connecting the dots" when f(x) changes from one integer value to the next. See Figure 62. ∎

Figure 62
$f(x) = \text{int}(x)$

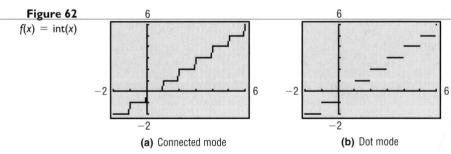

(a) Connected mode   (b) Dot mode

The functions that we have discussed so far are basic. Whenever you encounter one of them, you should see a mental picture of its graph. For example, if you encounter the function $f(x) = x^2$, you should see in your mind's eye a picture like Figure 55.

━━━ **NOW WORK PROBLEMS 9 THROUGH 16.**

2 Graph Piecewise-defined Functions

Sometimes a function is defined differently on different parts of its domain. For example, the absolute value function $f(x) = |x|$ is actually defined by two equations: $f(x) = x$ if $x \geq 0$ and $f(x) = -x$ if $x < 0$. For convenience, we generally combine these equations into one expression as

$$f(x) = |x| = \begin{cases} x & \text{if } x \geq 0 \\ -x & \text{if } x < 0 \end{cases}$$

When functions are defined by more than one equation, they are called **piecewise-defined** functions.

Let's look at another example of a piecewise-defined function.

| EXAMPLE 3 | **Analyzing a Piecewise-defined Function** |

The function f is defined as

$$f(x) = \begin{cases} -x + 1 & \text{if } -1 \le x < 1 \\ 2 & \text{if } x = 1 \\ x^2 & \text{if } x > 1 \end{cases}$$

(a) Find $f(0)$, $f(1)$, and $f(2)$.

(b) Determine the domain of f.

(c) Graph f by hand.

(d) Use the graph to find the range of f.

Solution

(a) To find $f(0)$, we observe that when $x = 0$ the equation for f is given by $f(x) = -x + 1$. So we have

$$f(0) = -0 + 1 = 1$$

When $x = 1$, the equation for f is $f(x) = 2$. Thus,

$$f(1) = 2$$

When $x = 2$, the equation for f is $f(x) = x^2$. So

$$f(2) = 2^2 = 4$$

Figure 63

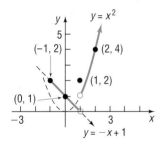

(b) To find the domain of f, we look at its definition. We conclude that the domain of f is $\{x \mid x \ge -1\}$, or the interval $[-1, \infty)$.

(c) To graph f by hand, we graph "each piece." First we graph the line $y = -x + 1$ and keep only the part for which $-1 \le x < 1$. Then we plot the point $(1, 2)$ because, when $x = 1$, $f(x) = 2$. Finally, we graph the parabola $y = x^2$ and keep only the part for which $x > 1$. See Figure 63.

(d) From the graph, we conclude that the range of f is $\{y \mid y > 0\}$, or the interval $(0, \infty)$. ◀

To graph a piecewise-defined function on a graphing calculator, we use the TEST menu to enter inequalities that allow us to restrict the domain function. For example, to graph the function in Example 3 using a TI-84 Plus graphing calculator, we would enter the function in Y_1 as shown in Figure 64(a). We then graph the function and obtain the result in Figure 64(b). When graphing piecewise-defined functions on a graphing calculator, you should use dot mode so that the calculator does not attempt to connect the "pieces" of the function.

Figure 64

(a)

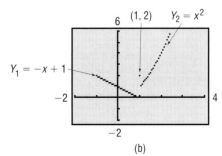

(b)

NOW WORK PROBLEM 29.

| **EXAMPLE 4** | **Cost of Electricity** |

In May 2004, Commonwealth Edison Company supplied electricity to residences for a monthly customer charge of $7.13 plus 8.275¢ per kilowatt-hour (kWhr) for the first 400 kWhr supplied in the month and 6.208¢ per kWhr for all usage over 400 kWhr in the month.

(a) What is the charge for using 300 kWhr in a month?

(b) What is the charge for using 700 kWhr in a month?

(c) If C is the monthly charge for x kWhr, express C as a function of x.

SOURCE: Commonwealth Edison Co., Chicago, Illinois, 2004.

Solution

(a) For 300 kWhr, the charge is $7.13 plus 8.275¢ = $0.08275 per kWhr. That is,

$$\text{Charge} = \$7.13 + \$0.08275(300) = \$31.96$$

(b) For 700 kWhr, the charge is $7.13 plus 8.275¢ per kWhr for the first 400 kWhr plus 6.208¢ per kWhr for the 300 kWhr in excess of 400. That is,

$$\text{Charge} = \$7.13 + \$0.08275(400) + \$0.06208(300) = \$58.85$$

(c) If $0 \leq x \leq 400$, the monthly charge C (in dollars) can be found by multiplying x times $0.08275 and adding the monthly customer charge of $7.13. So, if $0 \leq x \leq 400$, then $C(x) = 0.08275x + 7.13$. For $x > 400$, the charge is $0.08275(400) + 7.13 + 0.06208(x - 400)$, since $x - 400$ equals the usage in excess of 400 kWhr, which costs $0.06208 per kWhr. That is, if $x > 400$, then

$$C(x) = 0.08275(400) + 7.13 + 0.06208(x - 400)$$
$$= 40.23 + 0.06208(x - 400)$$
$$= 0.06208x + 15.40$$

The rule for computing C follows two equations:

Figure 65

$$C(x) = \begin{cases} 0.08275x + 7.13 & \text{if } 0 \leq x \leq 400 \\ 0.06208x + 15.40 & \text{if } x > 400 \end{cases}$$

See Figure 65 for the graph.   ◀

2.6 Assess Your Understanding

'Are You Prepared?'

Answers are given at the end of these exercises. If you get a wrong answer, read the pages listed in red.

1. Sketch the graph of $y = \sqrt{x}$. (pp. 200–201)

2. Sketch the graph of $y = \dfrac{1}{x}$. (p. 202)

3. List the intercepts of the equation $y = x^3 - 8$. (pp. 171–172)

Concepts and Vocabulary

4. The graph of $f(x) = mx + b$ is decreasing if m is _____ than zero.

5. When functions are defined by more than one equation they are called _____ functions.

6. *True or False:* The cube function is odd and is increasing on the interval $(-\infty, \infty)$.

7. *True or False:* The cube root function is odd and is decreasing on the interval $(-\infty, \infty)$.

8. *True or False:* The domain and the range of the reciprocal function are the set of all real numbers.

Skill Building

In Problems 9–16, match each graph to its function.

   A. *Constant function*   B. *Linear function*   C. *Square function*
   D. *Cube function*   E. *Square root function*   F. *Reciprocal function*
   G. *Absolute value function*   H. *Cube root function*

9.   **10.**   **11.**   **12.**

13.   **14.**   **15.** 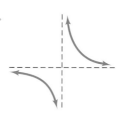   **16.**

In Problems 17–24, sketch the graph of each function. Be sure to label three points on the graph.

17. $f(x) = x$ **18.** $f(x) = x^2$ **19.** $f(x) = x^3$ **20.** $f(x) = \sqrt{x}$

21. $f(x) = \dfrac{1}{x}$ **22.** $f(x) = |x|$ **23.** $f(x) = \sqrt[3]{x}$ **24.** $f(x) = 3$

25. If $f(x) = \begin{cases} x^2 & \text{if } x < 0 \\ 2 & \text{if } x = 0 \\ 2x + 1 & \text{if } x > 0 \end{cases}$

     find: (a) $f(-2)$ (b) $f(0)$ (c) $f(2)$

26. If $f(x) = \begin{cases} x^3 & \text{if } x < 0 \\ 3x + 2 & \text{if } x \geq 0 \end{cases}$

     find: (a) $f(-1)$ (b) $f(0)$ (c) $f(1)$

27. If $f(x) = \text{int}(2x)$, find (a) $f(1.2)$, (b) $f(1.6)$, (c) $f(-1.8)$.

28. If $f(x) = \text{int}\left(\dfrac{x}{2}\right)$, find (a) $f(1.2)$, (b) $f(1.6)$, (c) $f(-1.8)$.

In Problems 29–40:

   (a) *Find the domain of each function.*   (b) *Locate any intercepts.*
   (c) *Graph each function.*   (d) *Based on the graph, find the range.*

29. $f(x) = \begin{cases} 2x & \text{if } x \neq 0 \\ 1 & \text{if } x = 0 \end{cases}$
            30. $f(x) = \begin{cases} 3x & \text{if } x \neq 0 \\ 4 & \text{if } x = 0 \end{cases}$

31. $f(x) = \begin{cases} -2x + 3 & x < 1 \\ 3x - 2 & x \geq 1 \end{cases}$
        32. $f(x) = \begin{cases} x + 3 & x < -2 \\ -2x - 3 & x \geq -2 \end{cases}$

33. $f(x) = \begin{cases} x + 3 & -2 \leq x < 1 \\ 5 & x = 1 \\ -x + 2 & x > 1 \end{cases}$
    34. $f(x) = \begin{cases} 2x + 5 & -3 \leq x < 0 \\ -3 & x = 0 \\ -5x & x > 0 \end{cases}$

35. $f(x) = \begin{cases} 1 + x & \text{if } x < 0 \\ x^2 & \text{if } x \geq 0 \end{cases}$
          36. $f(x) = \begin{cases} \dfrac{1}{x} & \text{if } x < 0 \\ \sqrt[3]{x} & \text{if } x \geq 0 \end{cases}$

37. $f(x) = \begin{cases} |x| & \text{if } -2 \leq x < 0 \\ 1 & \text{if } x = 0 \\ x^3 & \text{if } x > 0 \end{cases}$
     38. $f(x) = \begin{cases} 3 + x & \text{if } -3 \leq x < 0 \\ 3 & \text{if } x = 0 \\ \sqrt{x} & \text{if } x > 0 \end{cases}$

39. $f(x) = 2\,\text{int}(x)$
                   40. $f(x) = \text{int}(2x)$

In Problems 41–44, the graph of a piecewise-defined function is given. Write a definition for each function.

41.

42.

43.

44.
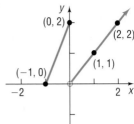

Applications and Extensions

45. Cell Phone Service Sprint PCS offers a monthly cellular phone plan for $35. It includes 300 anytime minutes plus $0.40 per minute for additional minutes. The following function is used to compute the monthly cost for a subscriber:

$$C(x) = \begin{cases} 35 & \text{if } 0 < x \le 300 \\ 0.40x - 85 & \text{if } x > 300 \end{cases}$$

where x is the number of anytime minutes used. Compute the monthly cost of the cellular phone for use of the following anytime minutes:
(a) 200   (b) 365   (c) 301
SOURCE: Sprint PCS.

46. Parking at O'Hare International Airport The short-term parking (no more than 24 hours) fee F (in dollars) for parking x hours at O'Hare International Airport's main parking garage can be modeled by the function

$$F(x) = \begin{cases} 3 & \text{if } 0 < x \le 3 \\ 5 \, \text{int}(x+1) + 1 & \text{if } 3 < x < 9 \\ 50 & \text{if } 9 \le x \le 24 \end{cases}$$

Determine the fee for parking in the short-term parking garage for
(a) 2 hours   (b) 7 hours
(c) 15 hours   (d) 8 hours and 24 minutes
SOURCE: O'Hare International Airport.

47. Cost of Natural Gas In May 2003, the Peoples Gas Company had the following rate schedule for natural gas usage in single-family residences:

Monthly service charge	$9.45
Per therm service charge	
1st 50 therms	$0.36375/therm
Over 50 therms	$0.11445/therm
Gas charge	$0.6338/therm

(a) What is the charge for using 50 therms in a month?
(b) What is the charge for using 500 therms in a month?
(c) Construct a function that relates the monthly charge C for x therms of gas.
(d) Graph this function.
SOURCE: The Peoples Gas Company, Chicago, Illinois, 2003.

48. Cost of Natural Gas In May 2003, Nicor Gas had the following rate schedule for natural gas usage in single-family residences:

Monthly customer charge	$6.45
Distribution charge	
1st 20 therms	$0.2012/therm
Next 30 therms	$0.1117/therm
Over 50 therms	$0.0374/therm
Gas supply charge	$0.7268/therm

(a) What is the charge for using 40 therms in a month?
(b) What is the charge for using 202 therms in a month?
(c) Construct a function that gives the monthly charge C for x therms of gas.
(d) Graph this function.
SOURCE: Nicor Gas, Aurora, Illinois, 2003.

49. Federal Income Tax Two 2004 Tax Rate Schedules are given in the accompanying table. If x equals taxable income and y equals the tax due, construct a function $y = f(x)$ for Schedule X.

REVISED 2004 TAX RATE SCHEDULES											
Schedule X—						**Schedule Y-1—**					
	If Taxable Income		The Tax Is				If Taxable Income			The Tax Is	
	Is over	But Not over	This Amount	Plus This %	Of the Excess over		Is over	But Not over	This Amount	Plus Of the This %	Excess over
Single	$0	$7,150	$0.00	10%	$0.00	**Married**	$0	$14,300	$0.00	10%	$0.00
	$7,150	$29,050	$715.00	15%	$7,150	**Filing**	$14,300	$58,100	$1,430.00	15%	$14,300
	$29,050	$70,350	$4000.00	25%	$29,050	**Jointly**	$58,100	$117,250	$8,000.00	25%	$58,100
	$70,350	$146,750	$14,325.00	28%	$70,350	**or**	$117,250	$178,650	$22,787.50	28%	$117,250
	$146,750	$319,100	$35,717.00	33%	$146,750	**Qualifying**	$178,650	$319,100	$39,979.50	33%	$178,650
	$319,100	—	$92,592.50	35%	$319,100	**Widow(er)**	$319,100	—	$86,328.00	35%	$319,100

SOURCE: Internal Revenue Service

50. Federal Income Tax Refer to the revised 2004 tax rate schedules. If x equals taxable income and y equals the tax due, construct a function $y = f(x)$ for Schedule Y-1.

51. Cost of Transporting Goods A trucking company transports goods between Chicago and New York, a distance of 960 miles. The company's policy is to charge, for each pound, $0.50 per mile for the first 100 miles, $0.40 per mile for the next 300 miles, $0.25 per mile for the next 400 miles, and no charge for the remaining 160 miles.
   (a) Graph the relationship between the cost of transportation in dollars and mileage over the entire 960-mile route.
   (b) Find the cost as a function of mileage for hauls between 100 and 400 miles from Chicago.
   (c) Find the cost as a function of mileage for hauls between 400 and 800 miles from Chicago.

52. Car Rental Costs An economy car rented in Florida from National Car Rental® on a weekly basis costs $95 per week. Extra days cost $24 per day until the day rate exceeds the weekly rate, in which case the weekly rate applies. Also, any part of a day used counts as a full day. Find the cost C of renting an economy car as a piecewise-defined function of the number x of days used, where $7 \leq x \leq 14$. Graph this function.

53. Minimum Payments for Credit Cards Holders of credit cards issued by banks, department stores, oil companies, and so on, receive bills each month that state minimum amounts that must be paid by a certain due date. The minimum due depends on the total amount owed. One such credit card company uses the following rules: For a bill of less than $10, the entire amount is due. For a bill of at least $10 but less than $500, the minimum due is $10. A minimum of $30 is due on a bill of at least $500 but less than $1000, a minimum of $50 is due on a bill of at least $1000 but less than $1500, and a minimum of $70 is due on bills of $1500 or more. Find the function f that describes the minimum payment due on a bill of x dollars. Graph f.

54. Interest Payments for Credit Cards Refer to Problem 53. The card holder may pay any amount between the minimum due and the total owed. The organization issuing the card charges the card holder interest of 1.5% per month for the first $1000 owed and 1% per month on any unpaid balance over $1000. Find the function g that gives the amount of interest charged per month on a balance of x dollars. Graph g.

55. Wind Chill The wind chill factor represents the equivalent air temperature at a standard wind speed that would produce the same heat loss as the given temperature and wind speed. One formula for computing the equivalent temperature is

$$W = \begin{cases} t & 0 \leq v < 1.79 \\ 33 - \dfrac{(10.45 + 10\sqrt{v} - v)(33 - t)}{22.04} & 1.79 \leq v \leq 20 \\ 33 - 1.5958(33 - t) & v > 20 \end{cases}$$

where v represents the wind speed (in meters per second) and t represents the air temperature (°C). Compute the wind chill for the following:
   (a) An air temperature of 10°C and a wind speed of 1 meter per second (m/sec)
   (b) An air temperature of 10°C and a wind speed of 5 m/sec
   (c) An air temperature of 10°C and a wind speed of 15 m/sec
   (d) An air temperature of 10°C and a wind speed of 25 m/sec
   (e) Explain the physical meaning of the equation corresponding to $0 \leq v < 1.79$.
   (f) Explain the physical meaning of the equation corresponding to $v > 20$.

56. Wind Chill Redo Problem 55(a)–(d) for an air temperature of -10°C.

Discussion and Writing

57. Exploration Graph $y = x^2$. Then on the same screen graph $y = x^2 + 2$, followed by $y = x^2 + 4$, followed by $y = x^2 - 2$. What pattern do you observe? Can you predict the graph of $y = x^2 - 4$? Of $y = x^2 + 5$?

58. Exploration Graph $y = x^2$. Then on the same screen graph $y = (x - 2)^2$, followed by $y = (x - 4)^2$, followed by $y = (x + 2)^2$. What pattern do you observe? Can you predict the graph of $y = (x + 4)^2$? Of $y = (x - 5)^2$?

59. Exploration Graph $y = |x|$. Then on the same screen graph $y = 2|x|$, followed by $y = 4|x|$, followed by $y = \frac{1}{2}|x|$.

What pattern do you observe? Can you predict the graph of $y = \frac{1}{4}|x|$? Of $y = 5|x|$?

60. Exploration Graph $y = x^2$. Then on the same screen graph $y = -x^2$. What pattern do you observe? Now try $y = |x|$ and $y = -|x|$. What do you conclude?

61. Exploration Graph $y = \sqrt{x}$. Then on the same screen graph $y = \sqrt{-x}$. What pattern do you observe? Now try $y = 2x + 1$ and $y = 2(-x) + 1$. What do you conclude?

62. Exploration Graph $y = x^3$. Then on the same screen graph $y = (x - 1)^3 + 2$. Could you have predicted the result?

63. Exploration Graph $y = x^2$, $y = x^4$, and $y = x^6$ on the same screen. What do you notice is the same about each graph? What do you notice that is different?

64. Exploration Graph $y = x^3$, $y = x^5$, and $y = x^7$ on the same screen. What do you notice is the same about each graph? What do you notice that is different?

65. Consider the equation

$$y = \begin{cases} 1 & \text{if } x \text{ is rational} \\ 0 & \text{if } x \text{ is irrational} \end{cases}$$

Is this a function? What is its domain? What is its range? What is its y-intercept, if any? What are its x-intercepts, if any? Is it even, odd, or neither? How would you describe its graph?

66. Define some functions that pass through $(0, 0)$ and $(1, 1)$ and are increasing for $x \geq 0$. Begin your list with $y = \sqrt{x}$, $y = x$, and $y = x^2$. Can you propose a general result about such functions?

'Are You Prepared' Answers

1.

2.

3. $(0, -8)$, $(2, 0)$

2.7 Graphing Techniques: Transformations

> **OBJECTIVES**
> 1 Graph Functions Using Vertical and Horizontal Shifts
> 2 Graph Functions Using Compressions and Stretches
> 3 Graph Functions Using Reflections about the x-Axis or y-Axis

At this stage, if you were asked to graph any of the functions defined by $y = x$, $y = x^2$, $y = x^3$, $y = \sqrt{x}$, $y = \sqrt[3]{x}$, $y = |x|$, or $y = \dfrac{1}{x}$, your response should be, "Yes, I recognize these functions and know the general shapes of their graphs." (If this is not your answer, review the previous section, Figures 54 through 60).

Sometimes we are asked to graph a function that is "almost" like one that we already know how to graph. In this section, we look at some of these functions and develop techniques for graphing them. Collectively, these techniques are referred to as **transformations**.

1 Graph Functions Using Vertical and Horizontal Shifts

——— Exploration ———

On the same screen, graph each of the following functions:

$$Y_1 = x^2$$
$$Y_2 = x^2 + 2$$
$$Y_3 = x^2 - 2$$

What do you observe?

Result Figure 66 illustrates the graphs. You should have observed a general pattern. With $Y_1 = x^2$ on the screen, the graph of $Y_2 = x^2 + 2$ is identical to that of $Y_1 = x^2$, except that it is shifted vertically up 2 units. The graph of $Y_3 = x^2 - 2$ is identical to that of $Y_1 = x^2$, except that it is shifted vertically down 2 units.

Figure 66

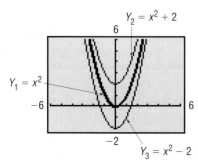

We are led to the following conclusion:

If a real number k is added to the right side of a function $y = f(x)$, the graph of the new function $y = f(x) + k$ is the graph of f **shifted vertically up** k units (if $k > 0$) or **down** $|k|$ units (if $k < 0$).

Let's look at an example.

EXAMPLE 1	**Vertical Shift Down**

Use the graph of $f(x) = x^2$ to obtain the graph of $h(x) = x^2 - 4$.

Solution   Table 14 lists some points on the graphs of $f = Y_1$ and $h = Y_2$. Notice that each y-coordinate of h is 4 units less than the corresponding y-coordinate of f.
The graph of h is identical to that of f, except that it is shifted down 4 units. See Figure 67.

Table 14

Figure 67

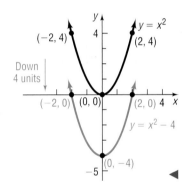

NOW WORK PROBLEM 35.

──── Exploration ────

On the same screen, graph each of the following functions:

$$Y_1 = x^2$$
$$Y_2 = (x - 3)^2$$
$$Y_3 = (x + 2)^2$$

What do you observe?

Figure 68

$Y_1 = x^2$

$Y_3 = (x + 2)^2$ $Y_2 = (x - 3)^2$

Result   Figure 68 illustrates the graphs. You should have observed the following pattern. With the graph of $Y_1 = x^2$ on the screen, the graph of $Y_2 = (x - 3)^2$ is identical to that of $Y_1 = x^2$, except it is shifted horizontally to the right 3 units. The graph of $Y_3 = (x + 2)^2$ is identical to that of $Y_1 = x^2$, except it is shifted horizontally to the left 2 units.

We are led to the following conclusion.

If the argument x of a function f is replaced by $x - h, h > 0$, the graph of the new function $y = f(x - h)$ is the graph of f **shifted horizontally right** h units.
If the argument x of a function f is repaced by $x + h, h > 0$, the graph of the new function $y = f(x + h)$ is the graph of f **shifted horizontally left** h units..

NOW WORK PROBLEM 39.

Vertical and horizontal shifts are sometimes combined.

| EXAMPLE 2 | **Combining Vertical and Horizontal Shifts** |

Graph the function $f(x) = (x + 3)^2 - 5$.

Solution We graph f in steps. First, we note that the rule for f is basically a square function, so we begin with the graph of $y = x^2$ as shown in Figure 69(a). To get the graph of $y = (x + 3)^2$, we shift the graph of $y = x^2$ horizontally 3 units to the left. See Figure 69(b). Finally, to get the graph of $y = (x + 3)^2 - 5$, we shift the graph of $y = (x + 3)^2$ vertically down 5 units. See Figure 69(c). Note the points plotted on each graph. Using key points can be helpful in keeping track of the transformation that has taken place.

Figure 69

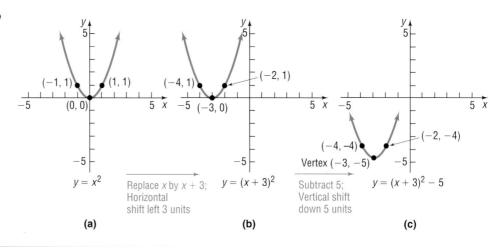

(a) $y = x^2$ Replace x by $x + 3$; Horizontal shift left 3 units   (b) $y = (x + 3)^2$ Subtract 5; Vertical shift down 5 units   (c) $y = (x + 3)^2 - 5$

✔ **CHECK:** Graph $Y_1 = f(x) = (x + 3)^2 - 5$ and compare the graph to Figure 69(c). ◀

In Example 2, if the vertical shift had been done first, followed by the horizontal shift, the final graph would have been the same. (Try it for yourself.)

 NOW WORK PROBLEMS 41 AND 77.

2 Graph Functions Using Compressions and Stretches

─────── **Exploration** ───────

On the same screen, graph each of the following functions:

$$Y_1 = |x|$$

$$Y_2 = 2|x|$$

$$Y_3 = \frac{1}{2}|x|$$

Figure 70

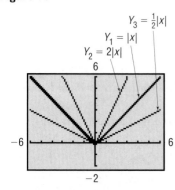

Result Figure 70 illustrates the graphs. You should have observed the following pattern. The graph of $Y_2 = 2|x|$ can be obtained from the graph of $Y_1 = |x|$ by multiplying each y-coordinate of $Y_1 = |x|$ by 2. This is sometimes referred to as a vertical *stretch* using a factor of 2.

The graph of $Y_3 = \frac{1}{2}|x|$ can be obtained from the graph of $Y_1 = |x|$ by multiplying each y-coordinate by $\frac{1}{2}$. This is sometimes referred to as a vertical *compression* using a factor of $\frac{1}{2}$.

Look at Tables 15 and 16, where $Y_1 = |x|$, $Y_2 = 2|x|$, and $Y_3 = \dfrac{1}{2}|x|$. Notice that the values for Y_2 in Table 15 are two times the values of Y_1 for each x-value. Therefore, the graph of Y_2 will be vertically *stretched* by a factor of 2. Likewise, the values of Y_3 in Table 16 are half the values of Y_1 for each x-value. Therefore, the graph of Y_3 will be vertically *compressed* by a factor of $\dfrac{1}{2}$.

Table 15

X	Y₁	**Y₂**
-2	2	4
-1	1	2
0	0	0
1	1	2
2	2	4
3	3	6
4	4	8

Y₂∎2abs(X)

Table 16

X	Y₁	**Y₃**
-2	2	1
-1	1	.5
0	0	0
1	1	.5
2	2	1
3	3	1.5
4	4	2

Y₃∎.5abs(X)

When the right side of a function $y = f(x)$ is multiplied by a positive number a, the graph of the new function $y = af(x)$ is obtained by multiplying each y-coordinate on the graph of $y = f(x)$ by a. The new graph is a **vertically compressed** (if $0 < a < 1$) or a **vertically stretched** (if $a > 1$) version of the graph of $y = f(x)$.

NOW WORK PROBLEM 43.

What happens if the argument x of a function $y = f(x)$ is multiplied by a positive number a, creating a new function $y = f(ax)$? To find the answer, we look at the following Exploration.

──── Exploration ────

On the same screen, graph each of the following functions:

$$Y_1 = f(x) = x^2$$
$$Y_2 = f(4x) = (4x)^2$$
$$Y_3 = f\left(\frac{1}{2}x\right) = \left(\frac{1}{2}x\right)^2$$

Result   You should have obtained the graphs shown in Figure 71. Look at Table 17(a). Notice that (1, 1), (4, 16), and (16, 256) are points on the graph of $Y_1 = x^2$. Also, (0.25, 1), (1, 16), and (4, 256) are points on the graph of $Y_2 = (4x)^2$. For each y-coordinate, the x-coordinate on the graph of Y_2 is $\dfrac{1}{4}$ of the x-coordinate on Y_1. We conclude the graph of $Y_2 = (4x)^2$ is obtained by multiplying the x-coordinate of each point on the graph of $Y_1 = x^2$ by $\dfrac{1}{4}$. The graph of $Y_2 = (4x)^2$ is the graph of $Y_1 = x^2$ compressed horizontally.

Figure 71

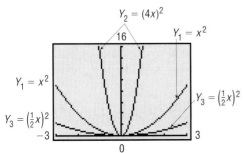

$Y_2 = (4x)^2$
16
$Y_1 = x^2$
$Y_1 = x^2$
$Y_3 = \left(\frac{1}{2}x\right)^2$
$Y_3 = \left(\frac{1}{2}x\right)^2$
−3
3
0

Table 17

X	Y₁	**Y₂**
0	0	0
.25	.0625	1
1	1	16
4	16	256
16	256	4096

Y₂∎(4X)²

(a)

X	Y₁	**Y₃**
0	0	0
.5	.25	.0625
1	1	.25
2	4	1
4	16	4
8	64	16
16	256	64

Y₃∎(X/2)²

(b)

Look at Table 17(b). Notice that (0.5, 0.25), (1, 1), (2, 4), and (4, 16) are points on the graph of $Y_1 = x^2$. Also, (1, 0.25), (2, 1), (4, 4), and (8, 16) are points on the graph of $Y_3 = \left(\frac{1}{2}x\right)^2$. For each y-coordinate, the x-coordinate on the graph of Y_3 is 2 times the x-coordinate on Y_1. We conclude the graph of $Y_3 = \left(\frac{1}{2}x\right)^2$ is obtained by multiplying the x-coordinate of each point on the graph of $Y_1 = x^2$ by 2. The graph of $Y_3 = \left(\frac{1}{2}x\right)^2$ is the graph of $Y_1 = x^2$ stretched horizontally.

> If the argument x of a function $y = f(x)$ is multiplied by a positive number a, the graph of the new function $y = f(ax)$ is obtained by multiplying each x-coordinate of $y = f(x)$ by $\frac{1}{a}$. A **horizontal compression** results if $a > 1$, and a **horizontal stretch** occurs if $0 < a < 1$.

Let's look at an example.

EXAMPLE 3	**Graphing Using Stretches and Compressions**

The graph of $y = f(x)$ is given in Figure 72. Use this graph to find the graphs of

(a) $y = 2f(x)$ (b) $y = f(3x)$

Figure 72

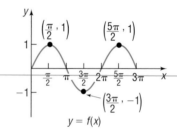

$y = f(x)$

Solution

(a) The graph of $y = 2f(x)$ is obtained by multiplying each y-coordinate of $y = f(x)$ by 2. See Figure 73(a).

(b) The graph of $y = f(3x)$ is obtained from the graph of $y = f(x)$ by multiplying each x-coordinate of $y = f(x)$ by $\frac{1}{3}$. See Figure 73(b).

Figure 73

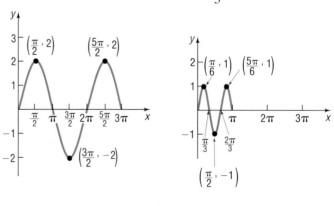

       (a) $y = 2f(x)$ **(b)** $y = f(3x)$

NOW WORK PROBLEMS 65(e) AND (g).

3 Graph Functions Using Reflections about the *x*-Axis or *y*-Axis

─── Exploration ───────────────────────────────

Reflection about the *x*-axis:
(a) Graph $Y_1 = x^2$, followed by $Y_2 = -x^2$.
(b) Graph $Y_1 = |x|$, followed by $Y_2 = -|x|$.
(c) Graph $Y_1 = x^2 - 4$, followed by $Y_2 = -(x^2 - 4) = -x^2 + 4$.

Result   See Tables 18(a), (b), and (c) and Figures 74(a), (b), and (c). In each instance, Y_2 is the reflection about the *x*-axis of Y_1.

Table 18

X	Y₁	**Y2**
-3	9	-9
-2	4	-4
-1	1	-1
0	0	0
1	1	-1
2	4	-4
3	9	-9

Y₂**∎**-X²

(a)

X	Y₁	**Y2**
-3	3	-3
-2	2	-2
-1	1	-1
0	0	0
1	1	-1
2	2	-2
3	3	-3

Y₂**∎**-abs(X)

(b)

X	Y₁	**Y2**
-3	5	-5
-2	0	0
-1	-3	3
0	-4	4
1	-3	3
2	0	0
3	5	-5

Y₂**∎**-X²+4

(c)

Figure 74

(a)

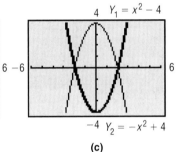

(b)   **(c)**

When the right side of the function $y = f(x)$ is multiplied by -1, the graph of the new function $y = -f(x)$ is the **reflection about the *x*-axis** of the graph of the function $y = f(x)$.

✏── NOW WORK PROBLEM **47**.

─── Exploration ───────────────────────────────

Reflection about the *y*-axis:
(a) Graph $Y_1 = \sqrt{x}$, followed by $Y_2 = \sqrt{-x}$.
(b) Graph $Y_1 = x + 1$, followed by $Y_2 = -x + 1$.
(c) Graph $Y_1 = x^4 + x$, followed by $Y_2 = (-x)^4 + (-x) = x^4 - x$.

Result   See Tables 19(a), (b), and (c) and Figures 75(a), (b), and (c). In each instance, Y_2 is the reflection about the *y*-axis of Y_1.

Table 19

X	Y₁	**Y2**
-3	ERROR	1.7321
-2	ERROR	1.4142
-1	ERROR	1
0	0	0
1	1	ERROR
2	1.4142	ERROR
3	1.7321	ERROR

Y₂**∎**√(-X)

(a)

X	Y₁	**Y2**
-3	-2	4
-2	-1	3
-1	0	2
0	1	1
1	2	0
2	3	-1
3	4	-2

Y₂**∎**-X+1

(b)

X	Y₁	**Y2**
-3	78	84
-2	14	18
-1	0	2
0	0	0
1	2	0
2	18	14
3	84	78

Y₂**∎**X^4-X

(c)

Figure 75

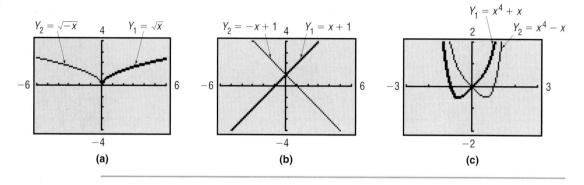

(a) (b) (c)

When the graph of the function $y = f(x)$ is known, the graph of the new function $y = f(-x)$ is the **reflection about the y-axis** of the graph of the function $y = f(x)$.

Summary of Graphing Techniques

Table 20 summarizes the graphing procedures that we have just discussed.

Table 20

To Graph:	Draw the Graph of f and:	Functional Change to f(x)
Vertical shifts		
$y = f(x) + k,\quad k > 0$	Raise the graph of f by k units.	Add k to $f(x)$.
$y = f(x) - k,\quad k > 0$	Lower the graph of f by k units.	Subtract k from $f(x)$.
Horizontal shifts		
$y = f(x + h),\quad h > 0$	Shift the graph of f to the left h units.	Replace x by $x + h$.
$y = f(x - h),\quad h > 0$	Shift the graph of f to the right h units.	Replace x by $x - h$.
Compressing or stretching		
$y = af(x),\quad a > 0$	Multiply each y-coordinate of $y = f(x)$ by a. Stretch the graph of f vertically if $a > 1$. Compress the graph of f vertically if $0 < a < 1$.	Multiply $f(x)$ by a.
$y = f(ax),\quad a > 0$	Multiply each x-coordinate of $y = f(x)$ by $\dfrac{1}{a}$. Stretch the graph of f horizontally if $0 < a < 1$. Compress the graph of f horizontally if $a > 1$.	Replace x by ax.
Reflection about the x-axis		
$y = -f(x)$	Reflect the graph of f about the x-axis.	Multiply $f(x)$ by -1.
Reflection about the y-axis		
$y = f(-x)$	Reflect the graph of f about the y-axis.	Replace x by $-x$.

The examples that follow combine some of the procedures outlined in this section to get the required graph.

EXAMPLE 4

Determining the Function Obtained from a Series of Transformations

Find the function that is finally graphed after the following three transformations are applied to the graph of $y = |x|$.

1. Shift left 2 units.

2. Shift up 3 units.

3. Reflect about the y-axis.

Solution

1. Shift left 2 units: Replace x by $x + 2$. $y = |x + 2|$

2. Shift up 3 units: Add 3. $y = |x + 2| + 3$

3. Reflect about the y-axis: Replace x by $-x$. $y = |-x + 2| + 3$ ◀

━━━━━━ NOW WORK PROBLEM 27.

| **EXAMPLE 5** | **Combining Graphing Procedures** |

Graph the function: $f(x) = \dfrac{3}{x - 2} + 1$

Solution

We use the following steps to obtain the graph of f:

STEP 1: $y = \dfrac{1}{x}$ Reciprocal function.

STEP 2: $y = \dfrac{3}{x}$ Multiply by 3; vertical stretch of the graph of $y = \dfrac{1}{x}$ by a factor of 3.

STEP 3: $y = \dfrac{3}{x - 2}$ Replace x by $x - 2$; horizontal shift to the right 2 units.

STEP 4: $y = \dfrac{3}{x - 2} + 1$ Add 1; vertical shift up 1 unit.

See Figure 76.

Figure 76

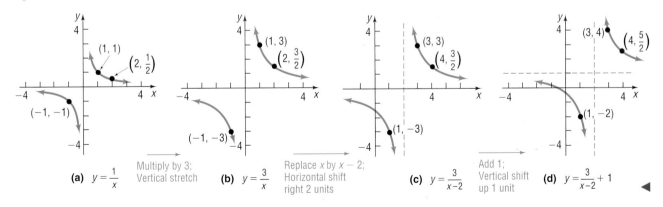

(a) $y = \dfrac{1}{x}$ Multiply by 3; Vertical stretch

(b) $y = \dfrac{3}{x}$ Replace x by $x - 2$; Horizontal shift right 2 units

(c) $y = \dfrac{3}{x-2}$ Add 1; Vertical shift up 1 unit

(d) $y = \dfrac{3}{x-2} + 1$ ◀

Other orderings of the steps shown in Example 5 would also result in the graph of f. For example, try this one:

STEP 1: $y = \dfrac{1}{x}$ Reciprocal function

STEP 2: $y = \dfrac{1}{x - 2}$ Replace x by $x - 2$; horizontal shift to the right 2 units.

STEP 3: $y = \dfrac{3}{x-2}$ Multiply by 3; vertical stretch of the graph of $y = \dfrac{1}{x-2}$ by a factor of 3.

STEP 4: $y = \dfrac{3}{x-2} + 1$ Add 1; vertical shift up 1 unit.

| EXAMPLE 6 | **Combining Graphing Procedures** |

Graph the function: $f(x) = \sqrt{1-x} + 2$

Solution   We use the following steps to obtain the graph of $y = \sqrt{1-x} + 2$:

STEP 1: $y = \sqrt{x}$ Square root function

STEP 2: $y = \sqrt{x+1}$ Replace x by $x + 1$; horizontal shift left 1 unit.

STEP 3: $y = \sqrt{-x+1} = \sqrt{1-x}$ Replace x by $-x$; reflect about y-axis.

STEP 4: $y = \sqrt{1-x} + 2$ Add 2; vertical shift up 2 units.

See Figure 77.

Figure 77

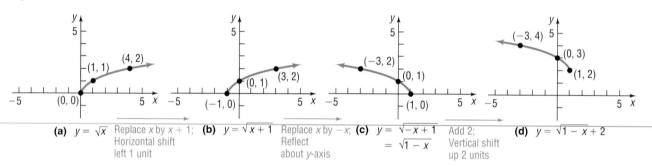

(a) $y = \sqrt{x}$ Replace x by $x + 1$; **(b)** $y = \sqrt{x+1}$ Replace x by $-x$; **(c)** $y = \sqrt{-x+1}$ Add 2; **(d)** $y = \sqrt{1-x} + 2$
          Horizontal shift   Reflect $= \sqrt{1-x}$ Vertical shift
          left 1 unit   about y-axis   up 2 units

✔ **CHECK:** Graph $Y_1 = f(x) = \sqrt{1-x} + 2$ and compare the graph to Figure 77(d).   ◀

NOW WORK PROBLEM 57.

2.7 Assess Your Understanding

Concepts and Vocabulary

1. Suppose that the graph of a function f is known. Then the graph of $y = f(x - 2)$ may be obtained by a(n) _____ shift of the graph of f to the _____ a distance of 2 units.

2. Suppose that the graph of a function f is known. Then the graph of $y = f(-x)$ may be obtained by a reflection about the _____-axis of the graph of the function $y = f(x)$.

3. Suppose that the x-intercepts of the graph of $y = f(x)$ are -2, 1, and 5. The x-intercepts of $y = f(x + 3)$ are _____, _____, and _____.

4. *True or False:* The graph of $y = -f(x)$ is the reflection about the x-axis of the graph of $y = f(x)$.

5. *True or False:* To obtain the graph of $y = f(x + 2) - 3$, shift the graph of $y = f(x)$ horizontally to the right 2 units and vertically down 3 units.

6. *True or False:* Suppose that the x-intercepts of the graph of $y = f(x)$ are -3 and 2. Then the x-intercepts of the graph of $y = 2f(x)$ are -3 and 2.

Skill Building

In Problems 7–18, match each graph to one of the following functions without using a graphing utility.

A. $y = x^2 + 2$ B. $y = -x^2 + 2$ C. $y = |x| + 2$ D. $y = -|x| + 2$

E. $y = (x - 2)^2$ F. $y = -(x + 2)^2$ G. $y = |x - 2|$ H. $y = -|x + 2|$

I. $y = 2x^2$ J. $y = -2x^2$ K. $y = 2|x|$ L. $y = -2|x|$

7.

8.

9.

10.

11.

12.

13.

14.

15.

16.

17.

18.
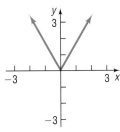

In Problems 19–26, write the function whose graph is the graph of $y = x^3$, but is:

19. Shifted to the right 4 units

20. Shifted to the left 4 units

21. Shifted up 4 units

22. Shifted down 4 units

23. Reflected about the y-axis

24. Reflected about the x-axis

25. Vertically stretched by a factor of 4

26. Horizontally stretched by a factor of 4

In Problems 27–30, find the function that is finally graphed after the following transformations are applied to the graph of $y = \sqrt{x}$.

27. (1) Shift up 2 units
    (2) Reflect about the x-axis
    (3) Reflect about the y-axis

28. (1) Reflect about the x-axis
    (2) Shift right 3 units
    (3) Shift down 2 units

29. (1) Reflect about the x-axis
    (2) Shift up 2 units
    (3) Shift left 3 units

30. (1) Shift up 2 units
    (2) Reflect about the y-axis
    (3) Shift left 3 units

31. If $(3, 0)$ is a point on the graph of $y = f(x)$, which of the following must be on the graph of $y = -f(x)$?
    (a) $(0, 3)$ (b) $(0, -3)$
    (c) $(3, 0)$ (d) $(-3, 0)$

32. If $(3, 0)$ is a point on the graph of $y = f(x)$, which of the following must be on the graph of $y = f(-x)$?
    (a) $(0, 3)$ (b) $(0, -3)$
    (c) $(3, 0)$ (d) $(-3, 0)$

33. If $(0, 3)$ is a point on the graph of $y = f(x)$, which of the following must be on the graph of $y = 2f(x)$?
    (a) $(0, 3)$ (b) $(0, 2)$
    (c) $(0, 6)$ (d) $(6, 0)$

34. If $(3, 0)$ is a point on the graph of $y = f(x)$, which of the following must be on the graph of $y = \dfrac{1}{2} f(x)$?
    (a) $(3, 0)$ (b) $\left(\dfrac{3}{2}, 0 \right)$
    (c) $\left(0, \dfrac{3}{2} \right)$ (d) $\left(\dfrac{1}{2}, 0 \right)$

In Problems 35–64, graph each function using the techniques of shifting, compressing, stretching, and/or reflecting. Start with the graph of the basic function (for example, $y = x^2$) and show all stages. Verify your results using a graphing utility.

35. $f(x) = x^2 - 1$

36. $f(x) = x^2 + 4$

37. $g(x) = x^3 + 1$

38. $g(x) = x^3 - 1$

39. $h(x) = \sqrt{x - 2}$

40. $h(x) = \sqrt{x + 1}$

41. $f(x) = (x - 1)^3 + 2$

42. $f(x) = (x + 2)^3 - 3$

43. $g(x) = 4\sqrt{x}$

44. $g(x) = \frac{1}{2}\sqrt{x}$

45. $h(x) = \frac{1}{2x}$

46. $h(x) = 3\sqrt[3]{x}$

47. $f(x) = -\sqrt[3]{x}$

48. $f(x) = -\sqrt{x}$

49. $g(x) = |-x|$

50. $g(x) = \sqrt[3]{-x}$

51. $h(x) = -x^3 + 2$

52. $h(x) = \frac{1}{-x} + 2$

53. $f(x) = 2(x + 1)^2 - 3$

54. $f(x) = 3(x - 2)^2 + 1$

55. $g(x) = \sqrt{x - 2} + 1$

56. $g(x) = |x + 1| - 3$

57. $h(x) = \sqrt{-x} - 2$

58. $h(x) = \frac{4}{x} + 2$

59. $f(x) = -(x + 1)^3 - 1$

60. $f(x) = -4\sqrt{x - 1}$

61. $g(x) = 2|1 - x|$

62. $g(x) = 4\sqrt{2 - x}$

63. $h(x) = 2\,\text{int}(x - 1)$

64. $h(x) = \text{int}(-x)$

In Problems 65–68, the graph of a function f is illustrated. Use the graph of f as the first step toward graphing each of the following functions:

(a) $F(x) = f(x) + 3$

(b) $G(x) = f(x + 2)$

(c) $P(x) = -f(x)$

(d) $H(x) = f(x + 1) - 2$

(e) $Q(x) = \frac{1}{2}f(x)$

(f) $g(x) = f(-x)$

(g) $h(x) = f(2x)$

65.

66.

67.

68.

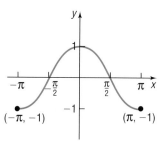

Applications and Extensions

69. Suppose that the x-intercepts of the graph of $y = f(x)$ are -5 and 3.

(a) What are the x-intercepts of the graph of $y = f(x + 2)$?

(b) What are the x-intercepts of the graph of $y = f(x - 2)$?

(c) What are the x-intercepts of the graph of $y = 4f(x)$?

(d) What are the x-intercepts of the graph of $y = f(-x)$?

70. Suppose that the x-intercepts of the graph of $y = f(x)$ are -8 and 1.

(a) What are the x-intercepts of the graph of $y = f(x + 4)$?

(b) What are the x-intercepts of the graph of $y = f(x - 3)$?

(c) What are the x-intercepts of the graph of $y = 2f(x)$?

(d) What are the x-intercepts of the graph of $y = f(-x)$?

71. Suppose that the function $y = f(x)$ is increasing on the interval $(-1, 5)$.

(a) Over what interval is the graph of $y = f(x + 2)$ increasing?

(b) Over what interval is the graph of $y = f(x - 5)$ increasing?

(c) What can be said about the graph of $y = -f(x)$?

(d) What can be said about the graph of $y = f(-x)$?

72. Suppose that the function $y = f(x)$ is decreasing on the interval $(-2, 7)$.
   (a) Over what interval is the graph of $y = f(x + 2)$ decreasing?
   (b) Over what interval is the graph of $y = f(x - 5)$ decreasing?
   (c) What can be said about the graph of $y = -f(x)$?
   (d) What can be said about the graph of $y = f(-x)$?

73. The graph of a function f is illustrated in the figure.
   (a) Draw the graph of $y = |f(x)|$.
   (b) Draw the graph of $y = f(|x|)$.

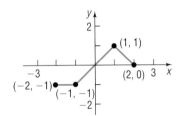

74. The graph of a function f is illustrated in the figure.
   (a) Draw the graph of $y = |f(x)|$.
   (b) Draw the graph of $y = f(|x|)$.

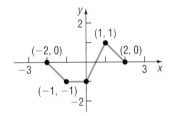

In Problems 75–84, complete the square of each quadratic expression. Then graph each function by hand using the technique of shifting. Verify your results using a graphing utility. (If necessary, refer to Section 1.3 to review completing the square.)

75. $f(x) = x^2 + 2x$ **76.** $f(x) = x^2 - 6x$

77. $f(x) = x^2 - 8x + 1$ **78.** $f(x) = x^2 + 4x + 2$

79. $f(x) = x^2 + x + 1$ **80.** $f(x) = x^2 - x + 1$

81. $f(x) = 2x^2 - 12x + 19$

82. $f(x) = 3x^2 + 6x + 1$

83. $f(x) = -3x^2 - 12x - 17$

84. $f(x) = -2x^2 - 12x - 13$

85. The equation $y = (x - c)^2$ defines a *family of parabolas*, one parabola for each value of c. On one set of coordinate axes, graph the members of the family for $c = 0$, $c = 3$, and $c = -2$.

86. Repeat Problem 85 for the family of parabolas $y = x^2 + c$.

87. **Temperature Measurements** The relationship between the Celsius (°C) and Fahrenheit (°F) scales for measuring temperature is given by the equation

$$F = \frac{9}{5}C + 32$$

The relationship between the Celsius (°C) and Kelvin (K) scales is $K = C + 273$. Graph the equation $F = \frac{9}{5}C + 32$ using degrees Fahrenheit on the y-axis and degrees Celsius on the x-axis. Use the techniques introduced in this section to obtain the graph showing the relationship between Kelvin and Fahrenheit temperatures.

88. **Period of a Pendulum** The period T (in seconds) of a simple pendulum is a function of its length l (in feet) defined by the equation

$$T = 2\pi\sqrt{\frac{l}{g}}$$

where $g \approx 32.2$ feet per second per second is the acceleration of gravity.
   (a) Use a graphing utility to graph the function $T = T(l)$.
   (b) Now graph the functions $T = T(l + 1)$, $T = T(l + 2)$, and $T = T(l + 3)$.
   (c) Discuss how adding to the length l changes the period T.
   (d) Now graph the functions $T = T(2l)$, $T = T(3l)$, and $T = T(4l)$.
   (e) Discuss how multiplying the length l by factors of 2, 3, and 4 changes the period T.

89. **Cigar Company Profits** The daily profits of a cigar company from selling x cigars are given by

$$p(x) = -0.05x^2 + 100x - 2000$$

The government wishes to impose a tax on cigars (sometimes called a *sin tax*) that gives the company the option of either paying a flat tax of $10,000 per day or a tax of 10% on profits. As chief financial officer (CFO) of the company, you need to decide which tax is the better option for the company.
   (a) On the same screen, graph $Y_1 = p(x) - 10,000$ and $Y_2 = (1 - 0.10)p(x)$.
   (b) Based on the graph, which option would you select? Why?
   (c) Using the terminology learned in this section, describe each graph in terms of the graph of $p(x)$.
   (d) Suppose that the government offered the options of a flat tax of $4800 or a tax of 10% on profits. Which would you select? Why?

Discussion and Writing

90. Suppose that the graph of a function f is known. Explain how the graph of $y = 4f(x)$ differs from the graph of $y = f(4x)$.

91. Exploration

    (a) Use a graphing utility to graph $y = x + 1$ and $y = |x + 1|$.

    (b) Graph $y = 4 - x^2$ and $y = |4 - x^2|$.

    (c) Graph $y = x^3 + x$ and $y = |x^3 + x|$.

    (d) What do you conclude about the relationship between the graphs of $y = f(x)$ and $y = |f(x)|$?

92. Exploration

    (a) Use a graphing utility to graph $y = x + 1$ and $y = |x| + 1$.

    (b) Graph $y = 4 - x^2$ and $y = 4 - |x|^2$.

    (c) Graph $y = x^3 + x$ and $y = |x|^3 + |x|$.

    (d) What do you conclude about the relationship between the graphs of $y = f(x)$ and $y = f(|x|)$?

2.8 Mathematical Models: Constructing Functions

> **OBJECTIVE** 1 Construct and Analyze Functions

1 Construct and Analyze Functions

Real-world problems often result in mathematical models that involve functions. These functions need to be constructed or built based on the information given. In constructing functions, we must be able to translate the verbal description into the language of mathematics. We do this by assigning symbols to represent the independent and dependent variables and then finding the function or rule that relates these variables.

EXAMPLE 1   **Area of a Rectangle with Fixed Perimeter**

The perimeter of a rectangle is 50 feet. Express its area A as a function of the length l of a side.

Figure 78

Solution   Consult Figure 78. If the length of the rectangle is l and if w is its width, then the sum of the lengths of the sides is the perimeter, 50.

$$l + w + l + w = 50$$
$$2l + 2w = 50$$
$$l + w = 25$$
$$w = 25 - l$$

The area A is length times width, so

$$A = lw = l(25 - l)$$

The area A as a function of l is

$$A(l) = l(25 - l) \qquad \blacktriangleleft$$

Note that we use the symbol A as the dependent variable and also as the name of the function that relates the length l to the area A. This double usage is common in applications and should cause no difficulties.

EXAMPLE 2   **Economics: Demand Equations**

In economics, revenue R, in dollars, is defined as the amount of money received from the sale of a product and is equal to the unit selling price p, in dollars, of the product times the number x of units actually sold. That is,

$$R = xp$$

In economics, the Law of Demand states that p and x are related: As one increases, the other decreases. Suppose that p and x are related by the following **demand equation**:

$$p = -\frac{1}{10}x + 20, \qquad 0 \le x \le 200$$

Express the revenue R as a function of the number x of units sold.

Solution   Since $R = xp$ and $p = -\frac{1}{10}x + 20$, it follows that

$$R(x) = xp = x\left(-\frac{1}{10}x + 20\right) = -\frac{1}{10}x^2 + 20x$$ ◀

NOW WORK PROBLEM 3.

| EXAMPLE 3 | **Finding the Distance from the Origin to a Point on a Graph** |

Let $P = (x, y)$ be a point on the graph of $y = x^2 - 1$.

(a) Express the distance d from P to the origin O as a function of x.

(b) What is d if $x = 0$?

(c) What is d if $x = 1$?

(d) What is d if $x = \dfrac{\sqrt{2}}{2}$?

(e) Use a graphing utility to graph the function $d = d(x)$, $x \ge 0$. Rounded to two decimal places, find the value(s) of x at which d has a local minimum. [This gives the point(s) on the graph of $y = x^2 - 1$ closest to the origin.]

Figure 79

Solution   (a) Figure 79 illustrates the graph. The distance d from P to O is

$$d = \sqrt{(x-0)^2 + (y-0)^2} = \sqrt{x^2 + y^2}$$

Since P is a point on the graph of $y = x^2 - 1$, we substitute $x^2 - 1$ for y. Then

$$d(x) = \sqrt{x^2 + (x^2 - 1)^2} = \sqrt{x^4 - x^2 + 1}$$

We have expressed the distance d as a function of x.

(b) If $x = 0$, the distance d is
$$d(0) = \sqrt{1} = 1$$

(c) If $x = 1$, the distance d is
$$d(1) = \sqrt{1 - 1 + 1} = 1$$

(d) If $x = \dfrac{\sqrt{2}}{2}$, the distance d is

Figure 80

$$d\left(\frac{\sqrt{2}}{2}\right) = \sqrt{\left(\frac{\sqrt{2}}{2}\right)^4 - \left(\frac{\sqrt{2}}{2}\right)^2 + 1} = \sqrt{\frac{1}{4} - \frac{1}{2} + 1} = \frac{\sqrt{3}}{2}$$

(e) Figure 80 shows the graph of $Y_1 = \sqrt{x^4 - x^2 + 1}$. Using the MINIMUM feature on a graphing utility, we find that when $x \approx 0.71$ the value of d is smallest. The local minimum is $d \approx 0.87$ rounded to two decimal places. [By symmetry, it follows that when $x \approx -0.71$ the value of d is also a local minimum.]   ◀

NOW WORK PROBLEM 9.

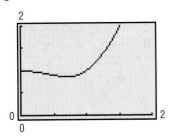

| EXAMPLE 4 | **Filling a Swimming Pool** |

A rectangular swimming pool 20 meters long and 10 meters wide is 4 meters deep at one end and 1 meter deep at the other. Figure 81 illustrates a cross-sectional view of the pool. Water is being pumped into the pool to a height of 3 meters at the deep end.

Figure 81

(a) Find a function that expresses the volume V of water in the pool as a function of the height x of the water at the deep end.

(b) Find the volume when the height is 1 meter.

(c) Find the volume when the height is 2 meters.

(d) At what height is the volume 20 cubic meters? 100 cubic meters?

Solution   (a) Let L denote the distance (in meters) measured at water level from the deep end to the short end. Notice that L and x form the sides of a triangle that is similar to the triangle whose sides are 20 meters by 3 meters. This means L and x are related by the equation

$$\frac{L}{x} = \frac{20}{3} \quad \text{or} \quad L = \frac{20x}{3}, \qquad 0 \le x \le 3$$

The volume V of water in the pool at any time is

$$V = \left(\begin{array}{c}\text{cross-sectional}\\\text{triangular area}\end{array}\right)(\text{width}) = \left(\frac{1}{2}Lx\right)(10) \quad \text{cubic meters}$$

Since $L = \dfrac{20x}{3}$, we have

$$V(x) = \left(\frac{1}{2} \cdot \frac{20x}{3} \cdot x\right)(10) = \frac{100}{3}x^2 \quad \text{cubic meters}$$

(b) When the height x of the water is 1 meter, the volume $V = V(x)$ is

$$V(1) = \frac{100}{3} \cdot 1^2 = 33\frac{1}{3} \quad \text{cubic meters}$$

(c) When the height x of the water is 2 meters, the volume $V = V(x)$ is

$$V(2) = \frac{100}{3} \cdot 2^2 = \frac{400}{3} = 133\frac{1}{3} \quad \text{cubic meters}$$

(d) By solving the equation $\dfrac{100}{3}x^2 = 20$, we find that when $x \approx 0.77$ meter, the volume is 20 cubic meters. By solving the equation $\dfrac{100}{3}x^2 = 100$, we find that when $x \approx 1.73$ meters, the volume is 100 cubic meters.   ◀

Figure 82

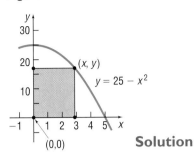

| **EXAMPLE 5** | **Area of a Rectangle** |

A rectangle has one corner on the graph of $y = 25 - x^2$, another at the origin, a third on the positive y-axis, and the fourth on the positive x-axis. See Figure 82.

(a) Express the area A of the rectangle as a function of x.

(b) What is the domain of A?

(c) Graph $A = A(x)$.

(d) For what value of x is the area largest?

Solution (a) The area A of the rectangle is $A = xy$, where $y = 25 - x^2$. Substituting this expression for y, we obtain $A(x) = x(25 - x^2) = 25x - x^3$.

(b) Since x represents a side of the rectangle, we have $x > 0$. In addition, the area must be positive, so $y = 25 - x^2 > 0$, which implies that $x^2 < 25$, so $-5 < x < 5$. Combining these restrictions, we have the domain of A as $\{x | 0 < x < 5\}$ or $(0, 5)$ using interval notation.

(c) See Figure 83 for the graph of $A = A(x)$.

(d) Using MAXIMUM, we find that the area is a maximum of 48.11 at $x = 2.89$, each rounded to two decimal places. See Figure 84.

Figure 83

Figure 84

─ NOW WORK PROBLEM 15.

| **EXAMPLE 6** | **Making a Playpen**＊ |

A manufacturer of children's playpens makes a square model that can be opened at one corner and attached at right angles to a wall or, perhaps, the side of a house. If each side is 3 feet in length, the open configuration doubles the available area in which the child can play from 9 square feet to 18 square feet. See Figure 85.

Now suppose that we place hinges at the outer corners to allow for a configuration like the one shown in Figure 86.

Figure 85

Figure 86

＊Adapted from Proceedings, S*ummer Conference for College Teachers on Applied Mathematics* (University of Missouri, Rolla), 1971.

(a) Express the area A of this configuration as a function of the distance x between the two parallel sides.

(b) Find the domain of A.

(c) Find A if $x = 5$.

(d) Graph $A = A(x)$. For what value of x is the area largest? What is the maximum area?

Solution

(a) Refer to Figure 86. The area A that we seek consists of the area of a rectangle (with width 3 and length x) and the area of an isosceles triangle (with base x and two equal sides of length 3). The height h of the triangle may be found using the Pythagorean Theorem.

$$h^2 + \left(\frac{x}{2}\right)^2 = 3^2$$

$$h^2 = 3^2 - \left(\frac{x}{2}\right)^2 = 9 - \frac{x^2}{4} = \frac{36 - x^2}{4}$$

$$h = \frac{1}{2}\sqrt{36 - x^2}$$

The area A enclosed by the playpen is

$$A = \text{area of rectangle} + \text{area of triangle} = 3x + \frac{1}{2}x\left(\frac{1}{2}\sqrt{36 - x^2}\right)$$

$$A(x) = 3x + \frac{x\sqrt{36 - x^2}}{4}$$

Now the area A is expressed as a function of x.

(b) To find the domain of A, we note first that $x > 0$, since x is a length. Also, the expression under the square root must be positive, so

$$36 - x^2 > 0$$
$$x^2 < 36$$
$$-6 < x < 6$$

Figure 87

Combining these restrictions, we find that the domain of A is $0 < x < 6$, or $(0, 6)$ using interval notation.

(c) If $x = 5$, the area is

$$A(5) = 3(5) + \frac{5}{4}\sqrt{36 - (5)^2} \approx 19.15 \text{ square feet}$$

If the width of the playpen is 5 feet, its area is 19.15 square feet.

(d) See Figure 87. The maximum area is about 19.82 square feet, obtained when x is about 5.58 feet.   ◀

2.8 Assess Your Understanding

Applications and Extensions

1. **Volume of a Cylinder** The volume V of a right circular cylinder of height h and radius r is $V = \pi r^2 h$. If the height is twice the radius, express the volume V as a function of r.

2. **Volume of a Cone** The volume V of a right circular cone is $V = \frac{1}{3}\pi r^2 h$. If the height is twice the radius, express the volume V as a function of r.

3. Demand Equation   The price p, in dollars, and the quantity x sold of a certain product obey the demand equation

$$p = -\frac{1}{6}x + 100, \qquad 0 \le x \le 600$$

(a) Express the revenue R as a function of x. (Remember, $R = xp$.)
(b) What is the revenue if 200 units are sold?
(c) Graph the revenue function using a graphing utility.
(d) What quantity x maximizes revenue? What is the maximum revenue?
(e) What price should the company charge to maximize revenue?

4. Demand Equation   The price p, in dollars, and the quantity x sold of a certain product obey the demand equation

$$p = -\frac{1}{3}x + 100, \qquad 0 \le x \le 300$$

(a) Express the revenue R as a function of x.
(b) What is the revenue if 100 units are sold?
(c) Graph the revenue function using a graphing utility.
(d) What quantity x maximizes revenue? What is the maximum revenue?
(e) What price should the company charge to maximize revenue?

5. Demand Equation   The price p, in dollars, and the quantity x sold of a certain product obey the demand equation

$$x = -5p + 100, \qquad 0 \le p \le 20$$

(a) Express the revenue R as a function of x.
(b) What is the revenue if 15 units are sold?
(c) Graph the revenue function using a graphing utility.
(d) What quantity x maximizes revenue? What is the maximum revenue?
(e) What price should the company charge to maximize revenue?

6. Demand Equation   The price p, in dollars, and the quantity x sold of a certain product obey the demand equation

$$x = -20p + 500, \qquad 0 \le p \le 25$$

(a) Express the revenue R as a function of x.
(b) What is the revenue if 20 units are sold?
(c) Graph the revenue function using a graphing utility.
(d) What quantity x maximizes revenue? What is the maximum revenue?
(e) What price should the company charge to maximize revenue?

7. Enclosing a Rectangular Field   David has available 400 yards of fencing and wishes to enclose a rectangular area.
(a) Express the area A of the rectangle as a function of the width x of the rectangle.
(b) What is the domain of A?
(c) Graph $A = A(x)$ using a graphing utility. For what value of x is the area largest?

8. Enclosing a Rectangular Field along a River   Beth has 3000 feet of fencing available to enclose a rectangular field. One side of the field lies along a river, so only three sides require fencing.
(a) Express the area A of the rectangle as a function of x, where x is the length of the side parallel to the river.
(b) Graph $A = A(x)$ using a graphing utility. For what value of x is the area largest?

9. Let $P = (x, y)$ be a point on the graph of $y = x^2 - 8$.
(a) Express the distance d from P to the origin as a function of x.
(b) What is d if $x = 0$?
(c) What is d if $x = 1$?
(d) Use a graphing utility to graph $d = d(x)$.
(e) For what values of x is d smallest?

10. Let $P = (x, y)$ be a point on the graph of $y = x^2 - 8$.
(a) Express the distance d from P to the point $(0, -1)$ as a function of x.
(b) What is d if $x = 0$?
(c) What is d if $x = -1$?
(d) Use a graphing utility to graph $d = d(x)$.
(e) For what values of x is d smallest?

11. Let $P = (x, y)$ be a point on the graph of $y = \sqrt{x}$.
(a) Express the distance d from P to the point $(1, 0)$ as a function of x.
(b) Use a graphing utility to graph $d = d(x)$.
(c) For what values of x is d smallest?

12. Let $P = (x, y)$ be a point on the graph of $y = \dfrac{1}{x}$.

(a) Express the distance d from P to the origin as a function of x.
(b) Use a graphing utility to graph $d = d(x)$.
(c) For what values of x is d smallest?

13. A right triangle has one vertex on the graph of $y = x^3, x > 0$, at (x, y), another at the origin, and the third on the positive y-axis at $(0, y)$, as shown in the figure. Express the area A of the triangle as a function of x.

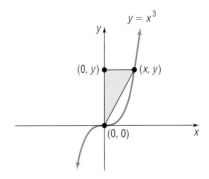

14. A right triangle has one vertex on the graph of $y = 9 - x^2, x > 0$, at (x, y), another at the origin, and the third on the positive x-axis at $(x, 0)$. Express the area A of the triangle as a function of x.

15. A rectangle has one corner on the graph of $y = 16 - x^2$, another at the origin, a third on the positive y-axis, and the fourth on the positive x-axis (see the figure).

(a) Express the area A of the rectangle as a function of x.
(b) What is the domain of A?
(c) Graph $A = A(x)$. For what value of x is A largest?

16. A rectangle is inscribed in a semicircle of radius 2 (see the figure). Let $P = (x, y)$ be the point in quadrant I that is a vertex of the rectangle and is on the circle.

(a) Express the area A of the rectangle as a function of x.
(b) Express the perimeter p of the rectangle as a function of x.
(c) Graph $A = A(x)$. For what value of x is A largest?
(d) Graph $p = p(x)$. For what value of x is p largest?

17. A rectangle is inscribed in a circle of radius 2 (see the figure). Let $P = (x, y)$ be the point in quadrant I that is a vertex of the rectangle and is on the circle.

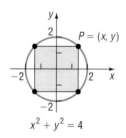

(a) Express the area A of the rectangle as a function of x.
(b) Express the perimeter p of the rectangle as a function of x.
(c) Graph $A = A(x)$. For what value of x is A largest?
(d) Graph $p = p(x)$. For what value of x is p largest?

18. A circle of radius r is inscribed in a square (see the figure).

(a) Express the area A of the square as a function of the radius r of the circle.
(b) Express the perimeter p of the square as a function of r.

19. A wire 10 meters long is to be cut into two pieces. One piece will be shaped as a square, and the other piece will be shaped as a circle (see the figure).

(a) Express the total area A enclosed by the pieces of wire as a function of the length x of a side of the square.
(b) What is the domain of A?
(c) Graph $A = A(x)$. For what value of x is A smallest?

20. A wire 10 meters long is to be cut into two pieces. One piece will be shaped as an equilateral triangle, and the other piece will be shaped as a circle.
(a) Express the total area A enclosed by the pieces of wire as a function of the length x of a side of the equilateral triangle.
(b) What is the domain of A?
(c) Graph $A = A(x)$. For what value of x is A smallest?

21. A wire of length x is bent into the shape of a circle.
(a) Express the circumference of the circle as a function of x.
(b) Express the area of the circle as a function of x.

22. A wire of length x is bent into the shape of a square.
(a) Express the perimeter of the square as a function of x.
(b) Express the area of the square as a function of x.

23. A semicircle of radius r is inscribed in a rectangle so that the diameter of the semicircle is the length of the rectangle (see the figure).

(a) Express the area A of the rectangle as a function of the radius r of the semicircle.
(b) Express the perimeter p of the rectangle as a function of r.

24. An equilateral triangle is inscribed in a circle of radius r. See the figure. Express the circumference C of the circle as a function of the length x of a side of the triangle.

[**Hint:** First show that $r^2 = \dfrac{x^2}{3}$.]

25. An equilateral triangle is inscribed in a circle of radius r. See the figure in Problem 24. Express the area A within the circle, but outside the triangle, as a function of the length x of a side of the triangle.

26. Two cars leave an intersection at the same time. One is headed south at a constant speed of 30 miles per hour, and the other is headed west at a constant speed of 40 miles per hour (see the figure). Express the distance d between the cars as a function of the time t.

[**Hint:** At $t = 0$, the cars leave the intersection.]

27. Two cars are approaching an intersection. One is 2 miles south of the intersection and is moving at a constant speed of 30 miles per hour. At the same time, the other car is 3 miles east of the intersection and is moving at a constant speed of 40 miles per hour.
 (a) Express the distance d between the cars as a function of time t.
   [**Hint:** At $t = 0$, the cars are 2 miles south and 3 miles east of the intersection, respectively.]
 (b) Use a graphing utility to graph $d = d(t)$. For what value of t is d smallest?

28. Inscribing a Cylinder in a Sphere   Inscribe a right circular cylinder of height h and radius r in a sphere of fixed radius R. See the illustration. Express the volume V of the cylinder as a function of h.

[**Hint:** $V = \pi r^2 h$. Note also the right triangle.]

Sphere

29. Inscribing a Cylinder in a Cone   Inscribe a right circular cylinder of height h and radius r in a cone of fixed radius R and fixed height H. See the illustration. Express the volume V of the cylinder as a function of r.

[**Hint:** $V = \pi r^2 h$. Note also the similar triangles.]

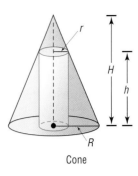

Cone

30. Installing Cable TV   MetroMedia Cable is asked to provide service to a customer whose house is located 2 miles from the road along which the cable is buried. The nearest connection box for the cable is located 5 miles down the road (see the figure).

 (a) If the installation cost is \$10 per mile along the road and \$14 per mile off the road, express the total cost C of installation as a function of the distance x (in miles) from the connection box to the point where the cable installation turns off the road. Give the domain.
 (b) Compute the cost if $x = 1$ mile.
 (c) Compute the cost if $x = 3$ miles.
 (d) Graph the function $C = C(x)$. Use TRACE to see how the cost C varies as x changes from 0 to 5.
 (e) What value of x results in the least cost?

31. Time Required to Go from an Island to a Town   An island is 2 miles from the nearest point P on a straight shoreline. A town is 12 miles down the shore from P. See the illustration.

 (a) If a person can row a boat at an average speed of 3 miles per hour and the same person can walk 5 miles per

hour, express the time T that it takes to go from the island to town as a function of the distance x from P to
where the person lands the boat.

(b) What is the domain of T?

(c) How long will it take to travel from the island to town if
the person lands the boat 4 miles from P?

(d) How long will it take if the person lands the boat 8 miles
from P?

32. Water is poured into a container in the shape of a right circular cone with radius 4 feet and height 16 feet (see the fig-

ure). Express the volume V of the water in the cone as a
function of the height h of the water.

[Hint: The volume V of a cone of radius r and height h is

$$V = \frac{1}{3}\pi r^2 h.]$$

Chapter Review

Library of Functions

Linear function (p. 257)

$f(x) = mx + b$

Graph is a line with slope m and y-intercept b.

Constant function (p. 258)

$f(x) = b$

Graph is a horizontal line with y-intercept b.

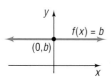

Identity function (p. 258)

$f(x) = x$

Graph is a line with slope 1 and y-intercept 0.

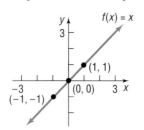

Square function (p. 258)

$f(x) = x^2$

Graph is a parabola with intercept at $(0, 0)$.

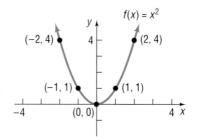

Cube function (p. 258)

$f(x) = x^3$

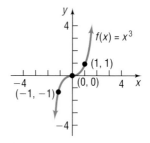

Square root function (p. 259)

$f(x) = \sqrt{x}$

Cube root function (p. 259)

$f(x) = \sqrt[3]{x}$

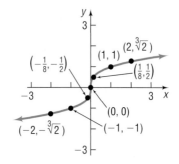

Reciprocal function (p. 259)

$f(x) = \dfrac{1}{x}$

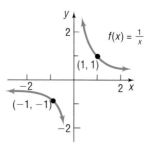

Absolute value function (p. 259)

$f(x) = |x|$

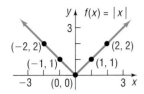

Greatest integer function (p. 260)

$f(x) = \text{int}(x)$

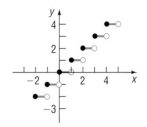

Things to Know

Function (pp. 206–210)	A relation between two sets of real numbers so that each number x in the first set, the domain, has corresponding to it exactly one number y in the second set. The range is the set of y values of the function for the x values in the domain.
	x is the independent variable; y is the dependent variable.
	A function can also be characterized as a set of ordered pairs (x, y) in which no first element is paired with two different second elements.
Function notation (pp. 209–212)	$y = f(x)$
	f is a symbol for the function.
	x is the argument, or independent variable.
	y is the dependent variable.
	$f(x)$ is the value of the function at x, or the image of x.
	A function f may be defined implicitly by an equation involving x and y or explicitly by writing $y = f(x)$.
Difference quotient of f (p. 211 and p. 240)	$\dfrac{f(x + h) - f(x)}{h}, \quad h \neq 0$
Domain (p. 206 and p. 212)	If unspecified, the domain of a function f is the largest set of real numbers for which $f(x)$ is a real number.
Vertical-line test (p. 220)	A set of points in the plane is the graph of a function if and only if every vertical line intersects the graph in at most one point.
Even function f (p. 228)	$f(-x) = f(x)$ for every x in the domain ($-x$ must also be in the domain).
Odd function f (p. 228)	$f(-x) = -f(x)$ for every x in the domain ($-x$ must also be in the domain).
Increasing function (p. 230)	A function f is increasing on an open interval I if, for any choice of x_1 and x_2 in I, with $x_1 < x_2$, we have $f(x_1) < f(x_2)$.
Decreasing function (p. 230)	A function f is decreasing on an open interval I if, for any choice of x_1 and x_2 in I, with $x_1 < x_2$, we have $f(x_1) > f(x_2)$.
Constant function (p. 230)	A function f is constant on an interval I if, for all choices of x in I, the values of $f(x)$ are equal.
Local maximum (p. 231)	A function f has a local maximum at c if there is an open interval I containing c so that, for all $x \neq c$ in I, $f(x) \leq f(c)$.
Local minimum (p. 231)	A function f has a local minimum at c if there is an open interval I containing c so that, for all $x \neq c$ in I, $f(x) \geq f(c)$.
Average rate of change of a function (p. 233)	The average rate of change of f from c to x is $$\frac{\Delta y}{\Delta x} = \frac{f(x) - f(c)}{x - c}, \qquad x \neq c$$
Linear function (p. 241)	$f(x) = mx + b$
	Graph is a line with slope m and y-intercept b.

Objectives

Section		You should be able to . . .	Review Exercises
2.1	1	Test an equation for symmetry with respect to (a) the x-axis, (b) the y-axis, and (c) the origin (p. 198)	1–8
	2	Know how to graph key equations (p. 200)	9, 10
2.2	1	Determine whether a relation represents a function (p. 204)	11, 12, 91(a)
	2	Find the value of a function (p. 209)	13–18, 33, 34
	3	Find the domain of a function (p. 212)	19–26, 91(f)
	4	Form the sum, difference, product, and quotient of two functions (p. 213)	27–32
2.3	1	Identify the graph of a function (p. 219)	57, 58
	2	Obtain information from or about the graph of a function (p. 220)	35–38
2.4	1	Determine even and odd functions from a graph (p. 228)	37(e), 38(e)
	2	Identify even and odd functions from the equation (p. 229)	39–46
	3	Use a graph to determine where a function is increasing, decreasing, or constant (p. 230)	37(b), 38(b)
	4	Use a graph to locate local maxima and local minima (p. 231)	37(c), 38(c)
	5	Use a graphing utility to approximate local maxima and local minima and to determine where a function is increasing or decreasing (p. 232)	47–50, 97
	6	Find the average rate of change of a function (p. 233)	51–56
2.5	1	Graph linear functions (p. 241)	59–62
	2	Work with applications of linear functions (p. 242)	85, 86, 89
	3	Draw and interpret scatter diagrams (p. 244)	91(b)
	4	Distinguish between linear and nonlinear relations (p. 246)	91(b)
	5	Use a graphing utility to find the line of best fit (p. 246)	91(c)
	6	Construct a linear model using direct variation (p. 248)	92, 93
2.6	1	Graph the functions listed in the library of functions (p. 257)	63, 64
	2	Graph piecewise-defined functions (p. 260)	77–80
2.7	1	Graph functions using vertical and horizontal shifts (p. 266)	35(f), 36(f), 36(g) 65, 66, 69–76
	2	Graph functions using compressions and stretches (p. 268)	35(g), 36(h), 67, 68, 75, 76
	3	Graph functions using reflections about the x-axis or y-axis (p. 271)	35(h), 67, 71, 72, 76
2.8	1	Construct and analyze functions (p. 278)	88, 90, 94–96, 98–102

Review Exercises

In Problems 1–8, test each equation for symmetry with respect to the x-axis, the y-axis, and the origin.

1. $2x = 3y^2$ **2.** $y = 5x$ **3.** $x^2 + 4y^2 = 16$ **4.** $9x^2 - y^2 = 9$

5. $y = x^4 + 2x^2 + 1$ **6.** $y = x^3 - x$ **7.** $x^2 + x + y^2 + 2y = 0$ **8.** $x^2 + 4x + y^2 - 2y = 0$

9. Sketch a graph of $y = x^3$. **10.** Sketch a graph of $y = \sqrt{x}$.

In Problems 11 and 12, determine whether each relation represents a function. For each function, state the domain and range.

11. $\{(-1, 0), (2, 3), (4, 0)\}$ **12.** $\{(4, -1), (2, 1), (4, 2)\}$

In Problems 13–18, find the following for each function:

(a) $f(2)$ (b) $f(-2)$ (c) $f(-x)$ (d) $-f(x)$ (e) $f(x - 2)$ (f) $f(2x)$

13. $f(x) = \dfrac{3x}{x^2 - 1}$ **14.** $f(x) = \dfrac{x^2}{x + 1}$ **15.** $f(x) = \sqrt{x^2 - 4}$

16. $f(x) = |x^2 - 4|$ **17.** $f(x) = \dfrac{x^2 - 4}{x^2}$ **18.** $f(x) = \dfrac{x^3}{x^2 - 9}$

In Problems 19–26, find the domain of each function.

19. $f(x) = \dfrac{x}{x^2 - 9}$

20. $f(x) = \dfrac{3x^2}{x - 2}$

21. $f(x) = \sqrt{2 - x}$

22. $f(x) = \sqrt{x + 2}$

23. $h(x) = \dfrac{\sqrt{x}}{|x|}$

24. $g(x) = \dfrac{|x|}{x}$

25. $f(x) = \dfrac{x}{x^2 + 2x - 3}$

26. $F(x) = \dfrac{1}{x^2 - 3x - 4}$

In Problems 27–32, find $f + g$, $f - g$, $f \cdot g$, and $\dfrac{f}{g}$ for each pair of functions. State the domain of each.

27. $f(x) = 2 - x$; $g(x) = 3x + 1$

28. $f(x) = 2x - 1$; $g(x) = 2x + 1$

29. $f(x) = 3x^2 + x + 1$; $g(x) = 3x$

30. $f(x) = 3x$; $g(x) = 1 + x + x^2$

31. $f(x) = \dfrac{x + 1}{x - 1}$; $g(x) = \dfrac{1}{x}$

32. $f(x) = \dfrac{1}{x - 3}$; $g(x) = \dfrac{3}{x}$

In Problems 33 and 34, find the difference quotient of each function f; that is, find

$$\frac{f(x + h) - f(x)}{h}, \qquad h \neq 0$$

33. $f(x) = -2x^2 + x + 1$

34. $f(x) = 3x^2 - 2x + 4$

35. Using the graph of the function f shown:
   (a) Find the domain and the range of f.
   (b) List the intercepts.
   (c) Find $f(-2)$.
   (d) For what values of x does $f(x) = -3$?
   (e) Solve $f(x) > 0$.
   (f) Graph $y = f(x - 3)$.
   (g) Graph $y = f\!\left(\dfrac{1}{2}x\right)$.
   (h) Graph $y = -f(x)$.

36. Using the graph of the function g shown:
   (a) Find the domain and the range of g.
   (b) Find $g(-1)$.
   (c) List the intercepts.
   (d) For what value of x does $g(x) = -3$?
   (e) Solve $g(x) > 0$.
   (f) Graph $y = g(x - 2)$.
   (g) Graph $y = g(x) + 1$.
   (h) Graph $y = 2g(x)$.

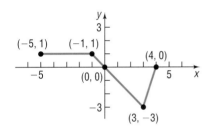

In Problems 37 and 38, use the graph of the function f to find:
(a) *The domain and the range of f*
(b) *The intervals on which f is increasing, decreasing, or constant*
(c) *The local minima and local maxima*
(d) *Whether the graph is symmetric with respect to the x-axis, the y-axis, or the origin*
(e) *Whether the function is even, odd, or neither*
(f) *The intercepts, if any*

37.

38.

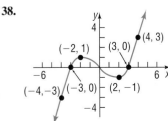

In Problems 39–46, determine (algebraically) whether the given function is even, odd, or neither.

39. $f(x) = x^3 - 4x$

40. $g(x) = \dfrac{4 + x^2}{1 + x^4}$

41. $h(x) = \dfrac{1}{x^4} + \dfrac{1}{x^2} + 1$

42. $F(x) = \sqrt{1 - x^3}$

43. $G(x) = 1 - x + x^3$

44. $H(x) = 1 + x + x^2$

45. $f(x) = \dfrac{x}{1 + x^2}$

46. $g(x) = \dfrac{1 + x^2}{x^3}$

In Problems 47–50, use a graphing utility to graph each function over the indicated interval. Approximate any local maxima and local minima. Determine where the function is increasing and where it is decreasing.

47. $f(x) = 2x^3 - 5x + 1$ $(-3, 3)$

48. $f(x) = -x^3 + 3x - 5$ $(-3, 3)$

49. $f(x) = 2x^4 - 5x^3 + 2x + 1$ $(-2, 3)$

50. $f(x) = -x^4 + 3x^3 - 4x + 3$ $(-2, 3)$

In Problems 51 and 52, find the average rate of change of f:
(a) *From 1 to 2*
(b) *From 0 to 1*
(c) *From 2 to 4*

51. $f(x) = 8x^2 - x$

52. $f(x) = 2x^3 + x$

In Problems 53–56, find the average rate of change from 2 to x for each function f. Be sure to simplify.

53. $f(x) = 2 - 5x$

54. $f(x) = 2x^2 + 7$

55. $f(x) = 3x - 4x^2$

56. $f(x) = x^2 - 3x + 2$

In Problems 57 and 58, tell which of the following graphs are graphs of functions.

57.

(a)   (b)

58.

(a)   (b)

In Problems 59–62, graph each linear function.

59. $f(x) = 2x - 5$

60. $g(x) = -4x + 7$

61. $h(x) = \dfrac{4}{5}x - 6$

62. $F(x) = -\dfrac{1}{3}x + 1$

In Problems 63 and 64, sketch the graph of each function. Be sure to label at least three points.

63. $f(x) = |x|$ **64.** $f(x) = \sqrt[3]{x}$

In Problems 65–76, graph each function using the techniques of shifting, compressing or stretching, and reflections. Identify any intercepts on the graph. State the domain and, based on the graph, find the range.

65. $F(x) = |x| - 4$

66. $f(x) = |x| + 4$

67. $g(x) = -2|x|$

68. $g(x) = \dfrac{1}{2}|x|$

69. $h(x) = \sqrt{x - 1}$

70. $h(x) = \sqrt{x} - 1$

71. $f(x) = \sqrt{1 - x}$

72. $f(x) = -\sqrt{x + 3}$

73. $h(x) = (x - 1)^2 + 2$

74. $h(x) = (x + 2)^2 - 3$

75. $g(x) = 3(x - 1)^3 + 1$

76. $g(x) = -2(x + 2)^3 - 8$

In Problems 77–80,
   (a) *Find the domain of each function.*
   (b) *Locate any intercepts.*
   (c) *Graph each function.*
   (d) *Based on the graph, find the range.*

77. $f(x) = \begin{cases} 3x & -2 < x \le 1 \\ x + 1 & x > 1 \end{cases}$

78. $f(x) = \begin{cases} x - 1 & -3 < x < 0 \\ 3x - 1 & x \ge 0 \end{cases}$

79. $f(x) = \begin{cases} x & -4 \le x < 0 \\ 1 & x = 0 \\ 3x & x > 0 \end{cases}$

80. $f(x) = \begin{cases} x^2 & -2 \le x \le 2 \\ 2x - 1 & x > 2 \end{cases}$

81. Given that f is a linear function, $f(4) = -5$, and $f(0) = 3$, write the equation that defines f.

82. Given that g is a linear function with slope $= -4$ and $g(-2) = 2$, write the equation that defines g.

83. A function f is defined by

$$f(x) = \frac{Ax + 5}{6x - 2}$$

If $f(1) = 4$, find A.

84. A function g is defined by

$$g(x) = \frac{A}{x} + \frac{8}{x^2}$$

If $g(-1) = 0$, find A.

85. Temperature Conversion   The temperature T of the air is approximately a linear function of the altitude h for altitudes within 10,000 meters of the surface of Earth. If the surface temperature is 30°C and the temperature at 10,000 meters is 5°C, find the function $T = T(h)$.

86. Speed as a Function of Time   The speed v (in feet per second) of a car is a linear function of the time t (in seconds) for $10 \le t \le 30$. If after each second the speed of the car has increased by 5 feet per second and if after 20 seconds the speed is 80 feet per second, how fast is the car going after 30 seconds? Find the function $v = v(t)$.

87. Spheres   The volume V of a sphere of radius r is $V = \frac{4}{3}\pi r^3$; the surface area S of this sphere is $S = 4\pi r^2$. Express the volume V as a function of the surface area S. If the surface area doubles, how does the volume change?

88. Page Design   A page with dimensions of $8\frac{1}{2}$ inches by 11 inches has a border of uniform width x surrounding the printed matter of the page, as shown in the figure.

$8\frac{1}{2}$ in.

11 in.

(a) Write a formula for the area A of the printed part of the page as a function of the width x of the border.
(b) Give the domain and the range of A.
(c) Find the area of the printed the page for borders of widths 1 inch, 1.2 inches, and 1.5 inches.
(d) Graph the function $A = A(x)$.

(e) Use TRACE to determine what margin should be used to obtain an area of 70 square inches and of 50 square inches.

89. Strength of a Beam   The strength of a rectangular wooden beam is proportional to the product of the width and the cube of its depth (see the figure). If the beam is to be cut from a log in the shape of a cylinder of radius 3 feet, express the strength S of the beam as a function of the width x. What is the domain of S?

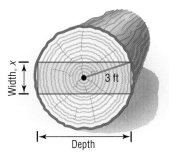

Width, x

3 ft

Depth

90. Material Needed to Make a Drum   A steel drum in the shape of a right circular cylinder is required to have a volume of 100 cubic feet.
(a) Express the amount A of material required to make the drum as a function of the radius r of the cylinder.
(b) How much material is required if the drum is of radius 3 feet?
(c) How much material is required if the drum is of radius 4 feet?
(d) How much material is required if the drum is of radius 5 feet?
(e) Graph $A = A(r)$. For what value of r is A smallest?

91. High School versus College GPA   An administrator at Southern Illinois University wants to find a function that relates a student's college grade point average G to the high school grade point average x. She randomly selects eight students and obtains the following data:

High School GPA, x	College GPA, G
2.73	2.43
2.92	2.97
3.45	3.63
3.78	3.81
2.56	2.83
2.98	2.81
3.67	3.45
3.10	2.93

(a) Does the relation defined by the set of ordered pairs (x, G) represent a function?
(b) Draw a scatter diagram of the data. Are the data linear?
(c) Using a graphing utility, find the line of best fit relating high school GPA and college GPA.

(d) Interpret the slope.

(e) Express the relationship found in part (c) using function notation.

(f) What is the domain of the function?

(g) Predict a student's college GPA if her high school GPA is 3.23.

92. Mortgage Payments   The monthly payment p on a mortgage varies directly with the amount borrowed B. If the monthly payment on a 30-year mortgage is $854.00 when $130,000 is borrowed, find a linear function that relates the monthly payment p to the amount borrowed B for a mortgage with the same terms. Then find the monthly payment p when the amount borrowed B is $165,000.

93. Revenue Function   At the corner Esso station, the revenue R varies directly with the number g of gallons of gasoline sold. If the revenue is $28.89 when the number of gallons sold is 13.5, find a linear function that relates revenue R to the number g of gallons of gasoline. Then find the revenue R when the number of gallons of gasoline sold is 11.2.

94. Landscaping   A landscape engineer has 200 feet of border to enclose a rectangular pond. What dimensions will result in the largest pond?

95. Geometry   Find the length and width of a rectangle whose perimeter is 20 feet and whose area is 16 square feet.

96. A rectangle has one vertex on the line $y = 10 - x, x > 0$, another at the origin, one on the positive x-axis, and one on the positive y-axis. Find the largest area A that can be enclosed by the rectangle.

97. Minimizing Marginal Cost   The marginal cost of a product can be thought of as the cost of producing one additional unit of output. For example, if the marginal cost of producing the 50th product is $6.20, then it cost $6.20 to increase production from 49 to 50 units of output. Callaway Golf Company has determined that the marginal cost C of manufacturing x Big Bertha golf clubs may be expressed by the quadratic function

$$C(x) = 4.9x^2 - 617.4x + 19,600$$

(a) How many clubs should be manufactured to minimize the marginal cost?

(b) At this level of production, what is the marginal cost?

98. Find the point on the line $y = x$ that is closest to the point $(3, 1)$.

99. Find the point on the line $y = x + 1$ that is closest to the point $(4, 1)$.

100. A rectangle has one vertex on the graph of $y = 10 - x^2, x > 0$, another at the origin, one on the positive x-axis, and one on the positive y-axis. Find the largest area A that can be enclosed by the rectangle.

101. Constructing a Closed Box   A closed box with a square base is required to have a volume of 10 cubic feet.

(a) Express the amount A of material used to make such a box as a function of the length x of a side of the square base.

(b) How much material is required for a base 1 foot by 1 foot?

(c) How much material is required for a base 2 feet by 2 feet?

(d) Graph $A = A(x)$. For what value of x is A smallest?

102. Cost of a Drum   A drum in the shape of a right circular cylinder is required to have a volume of 500 cubic centimeters. The top and bottom are made of material that costs 6¢ per square centimeter; the sides are made of material that costs 4¢ per square centimeter.

(a) Express the total cost C of the material as a function of the radius r of the cylinder.

(b) What is the cost if the radius is 4 cm?

(c) What is the cost if the radius is 8 cm?

(d) Graph $C = C(r)$. For what value of r is the cost C least?

Chapter Test

1. Test the equation $5x - 2y^2 = 7$ for symmetry with respect to the x-axis, the y-axis, and the origin.

2. Determine whether each relation represents a function. For each function, state the domain and range.
   (a) $\{(2, 5), (4, 6), (6, 7), (8, 8)\}$
   (b) $\{(1, 3), (4, -2), (-3, 5), (1, 7)\}$
   (c)

   (d)

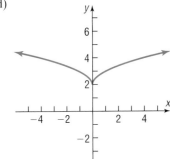

In Problems 3–5, find the domain of each function and evaluate each function at $x = -1$.

3. $f(x) = \sqrt{4 - 5x}$

4. $g(x) = \dfrac{x + 2}{|x + 2|}$

5. $h(x) = \dfrac{x - 4}{x^2 + 5x - 36}$

6. Using the graph of the function f below:
   (a) Find the domain and the range of f.
   (b) List the intercepts.
   (c) Find $f(1)$.
   (d) For what value(s) of x does $f(x) = -3$?
   (e) Solve $f(x) < 0$

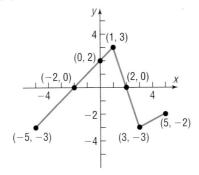

7. Use a graphing utility to graph the function $f(x) = -x^4 + 2x^3 + 4x^2 - 2$ on the interval $(-5, 5)$. Approximate any local maxima and local minima rounded to two decimal places. Determine where the function is increasing and where it is decreasing.

8. Consider the function $g(x) = \begin{cases} 2x + 1 & x < -1 \\ x - 4 & x \geq -1 \end{cases}$.
   (a) Graph the function.
   (b) List the intercepts.
   (c) Find $g(-5)$.
   (d) Find $g(2)$.

9. For the function $f(x) = 3x^2 - 2x + 4$, find the average rate of change from 3 to x.

10. For the functions $f(x) = 2x^2 + 1$ and $g(x) = 3x - 2$, find the following and simplify:
   (a) $f - g$
   (b) $f \cdot g$
   (c) $f(x + h) - f(x)$

11. Graph each function using the techniques of shifting, compressing or stretching, and reflections. Start with the graph of the basic function and show all stages.
   (a) $h(x) = -2(x + 1)^3 + 3$
   (b) $g(x) = |x + 4| + 2$

12. Consider the two data sets:

Set A:

x	0	1	2	3	4	5	6	7
y	−5	−9	−8	−6	−9	−6	−3	4

Set B:

x	0	1	2	3	4	5	6	7
y	−4	−5	−1	0	4	4	7	9

   (a) Use a graphing utility to make a scatter diagram for each set of data, and determine which of the two data sets has a stronger linear relationship.
   (b) Use a graphing utility to find the line of best fit for the data set you selected in part (a).

13. The variable interest rate on a student loan changes each July 1 based on the bank prime loan rate. For the years 1992–2004, this rate can be approximated by the model $r(x) = -0.115x^2 + 1.183x + 5.623$, where x is the number of years since 1992 and r is the interest rate as a percent.

(SOURCE: U.S. Federal Reserve)
   (a) Use a graphing utility to estimate the highest rate during this time period. During which year was the interest rate the highest?
   (b) Use the model to estimate the rate in 2010. Does this value seem reasonable?

14. A community skating rink is in the shape of a rectangle with semicircles attached at the ends. The length of the rectangle is 20 feet less than twice the width. The thickness of the ice is 0.75 inch.

(a) Write the ice volume, V, as a function of the width, x.

(b) How much ice is in the rink if the width is 90 feet?

Chapter Projects

1. Cell Phone Service This month when you were paying your bills, you noticed that your original service contract for your cell phone service had expired. Since your cell number is portable to another company now, you decided to investigate different companies as well. Noting that you travel outside your local calling area on a regular basis, you decide to stick with only those plans that have roaming charges included. Here is what you found:

	Anytime minutes included	Charge for each extra minute	Mobile-to-mobile minutes	National Long Distance	Nights and Weekends
Company A:					
Plan A1: $49.99	600	$0.45	Unlimited	Included	Unlimited (after 9 pm)
Plan A2: $59.99	900	$0.35	Unlimited	Included	Unlimited (after 7 pm)
Company B:					
Plan B1: $39.99	450 with rollover	$0.45	Unlimited	Included	3000 min (after 9 pm)
Plan B2: $49.99	600 with rollover	$0.40	Unlimited	Included	unlimited (after 9 pm)
Company C:					
Plan C1: $45.00	300	$0.40	Unlimited	Included	Unlimited (after 9 pm)
Plan C2: $60.00	700	$0.40	Unlimited	Included	Unlimited (after 9 pm)

Each plan requires a two-year contract.

(a) Determine the total cost of each plan for the life of the contract, assuming that you stay within the allotted anytime minutes provided by each contract.

(b) If you expect to use 600 anytime minutes and 2500 night and weekend minutes per month, which plan provides the best deal? If you expect to use 600 anytime minutes and 3500 night and weekend minutes, which plan provides the best deal?

(c) Ignoring any night and weekend usage, if you expect to use 425 anytime minutes each month, which option provides the best deal? What if you use 750 anytime minutes per month?

(d) Each monthly charge includes a specific number of peak time minutes in the monthly fee. Write a function for each option, where C is the monthly cost and x is the number of anytime minutes used.

(e) Graph each of the functions from part (d).

(f) For each of the companies A, B, and C, determine the average price per minute for each plan, based on no extra minutes used. For each company, which plan is better?

(g) Now, looking at the three plans that you found to be the best for Companies A, B, and C, in part (f), which of those three seems to be the best deal?

(h) Based upon your own cell phone usage, which plan would be the best for you?

SOURCE: Based on rates from the websites of the companies: ATTWireless, Cingular, and Sprint PCS, for area code 76201 on October 4, 2004. *(www.attwireless.com, www.cingular.com, www.sprint.com)*

The following projects are available on the Instructor's Resource Center (IRC):

2. **Project at Motorola**   *Pricing Wireless Service*

3. **Cost of Cable**

4. **Oil Spill**

Cumulative Review

In Problems 1–8, find the real solutions of each equation.

1. $-5x + 4 = 0$

2. $x^2 - 7x + 12 = 0$

3. $3x^2 - 5x - 2 = 0$

4. $4x^2 + 4x + 1 = 0$

5. $4x^2 - 2x + 4 = 0$

6. $\sqrt[3]{1 - x} = 2$

7. $\sqrt[5]{1 - x} = 2$

8. $|2 - 3x| = 1$

9. In the complex number system, solve $4x^2 - 2x + 4 = 0$.

10. Solve the inequality $-2 < 3x - 5 < 7$. Graph the solution set.

In Problems 11–14, graph each equation.

11. $-3x + 4y = 12$

12. $y = 3x + 12$

13. $x^2 + y^2 + 2x - 4y + 4 = 0$

14. $y = (x + 1)^2 - 3$

15. For the graph of the function f below:

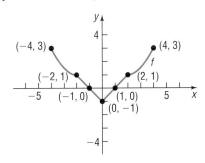

(a) Find the domain and the range of f.
(b) Find the intercepts.
(c) Is the graph of f symmetric with respect to the x-axis, the y-axis, or the origin?
(d) Find $f(2)$.
(e) For what value(s) of x is $f(x) = 3$?
(f) Solve $f(x) < 0$.
(g) Graph $y = f(x) + 2$.
(h) Graph $y = f(-x)$.
(i) Graph $y = 2f(x)$.
(j) Is f even, odd, or neither?
(k) Find the interval(s) on which f is increasing.

(l) Find the interval(s) on which f is decreasing.
(m) Find the local maxima and local minima.
(n) Find the average rate of change of f from 1 to 4.

16. Find the distance between the points $P = (-1, 3)$ and $Q = (4, -2)$.

17. Which of the following points are on the graph of $y = x^3 - 3x + 1$?
   (a) $(-2, -1)$ (b) $(2, 3)$ (c) $(3, 1)$

18. Determine the intercepts of $y = 3x^2 + 14x - 5$.

19. Use a graphing utility to find the solution(s) of the equation $x^4 - 3x^3 + 4x - 1 = 0$.

   Hint: All solutions are between $x = -10$ and $x = 10$.

20. Find the equation of the line perpendicular to the line $y = 2x + 1$ and containing the point $(3, 5)$. Express your answer in slope–intercept form and graph both lines.

21. Determine whether the following relation represents a function: $\{(-3, 8), (1, 3), (2, 5), (3, 8)\}$.

22. For the function f defined by $f(x) = x^2 - 4x + 1$, evaluate:
   (a) $f(2)$ (b) $f(x) + f(2)$
   (c) $f(-x)$ (d) $-f(x)$
   (e) $f(x + 2)$
   (f) $\dfrac{f(x + h) - f(x)}{h}$, $h \neq 0$

23. Find the domain of $h(z) = \dfrac{3z - 1}{z^2 - 6z - 7}$.

24. Determine whether the following graph is the graph of a function.

25. Consider the function $f(x) = \dfrac{x}{x + 4}$.

   (a) Is the point $\left(1, \dfrac{1}{4}\right)$ on the graph of f?
   (b) If $x = -2$, what is $f(x)$? What point is on the graph of f?
   (c) If $f(x) = 2$, what is x? What point is on the graph of f?

Polynomial and Rational Functions

3

A LOOK BACK In Chapter 2, we began our discussion of functions. We defined domain and range and independent and dependent variables; we found the value of a function and graphed functions. We continued our study of functions by listing the properties that a function might have, like being even or odd, and we created a library of functions, naming key functions and listing their properties, including their graphs.

A LOOK AHEAD In this chapter, we look at two general classes of functions: polynomial functions and rational functions, and examine their properties. Polynomial functions are arguably the simplest expressions in algebra. For this reason, they are often used to approximate other, more complicated, functions. Rational functions are simply ratios of polynomial functions.

We also introduce methods that can be used to determine the zeros of polynomial functions that are not factored completely. We end the chapter with a discussion of complex zeros.

Ragweed plus bad air will equal fall misery

A hot, dry, smoggy summer is adding up to an insufferable fall for allergy sufferers.

Experts say people with allergies to ragweed, who normally sneeze and wheeze through September, will be dabbing their eyes and nose well into October, even though pollen counts are actually no worse than usual.

The difference this year is that a summer of heavy smog has made the pollen in the air feel much thicker, making it more irritating.

"Smog plays an extremely important role in allergy suffering," says Frances Coates, owner of Ottawa-based Aerobiology Research, which measures pollen levels. "It holds particles in the air a lot longer."

Low rainfall delayed the ragweed season until the end of August. It usually hits early in the month.

SOURCE: Adapted by Louise Surette from *Toronto Star*, September 8, 2001.

See Chapter Project 1.

OUTLINE

3.1 Quadratic Functions and Models

PREPARING FOR THIS SECTION *Before getting started, review the following:*

- Intercepts (Section 1.1, pp. 95–96; Section 1.8, pp. 171–172)
- Quadratic Equations (Section 1.3, pp. 112–119)

- Completing the Square (Section 1.3, pp. 115–116)
- Graphing Techniques: Transformations (Section 2.7, pp. 266–274)

Now work the 'Are You Prepared?' problems on page 311.

OBJECTIVES 1 Graph a Quadratic Function Using Transformations

2 Identify the Vertex and Axis of Symmetry of a Quadratic Function

3 Graph a Quadratic Function Using Its Vertex, Axis, and Intercepts

4 Use the Maximum or Minimum Value of a Quadratic Function to Solve Applied Problems

5 Use a Graphing Utility to Find the Quadratic Function of Best Fit to Data

A *quadratic function* is a function that is defined by a second-degree polynomial in one variable.

A **quadratic function** is a function of the form

$$f(x) = ax^2 + bx + c \qquad \textbf{(1)}$$

where a, b, and c are real numbers and $a \neq 0$. The domain of a quadratic function is the set of all real numbers.

Many applications require a knowledge of quadratic functions. For example, suppose that Texas Instruments collects the data shown in Table 1, which relate the number of calculators sold at the price p (in dollars) per calculator. Since the price of a product determines the quantity that will be purchased, we treat price as the independent variable. The relationship between the number x of calculators sold and the price p per calculator may be approximated by the linear equation

$$x = 21{,}000 - 150p$$

Table 1

Price per Calculator, p (Dollars)	Number of Calculators, x
60	11,100
65	10,115
70	9,652
75	8,731
80	8,087
85	7,205
90	6,439

Then the revenue R derived from selling x calculators at the price p per calculator is equal to the unit selling price p of the product times the number x of units actually sold. That is,

$$R = xp$$
$$R(p) = (21{,}000 - 150p)p$$
$$= -150p^2 + 21{,}000p$$

So, the revenue R is a quadratic function of the price p. Figure 1 illustrates the graph of this revenue function, whose domain is $0 \le p \le 140$, since both x and p must be non-negative. Later in this section we shall determine the price p that maximizes revenue.

A second situation in which a quadratic function appears involves the motion of a projectile. Based on Newton's second law of motion (force equals mass times acceleration, $F = ma$), it can be shown that, ignoring air resistance, the path of a projectile propelled upward at an inclination to the horizontal is the graph of a quadratic function. See Figure 2 for an illustration. Later in this section we shall analyze the path of a projectile.

Figure 1

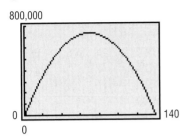

Figure 2
Path of a cannonball

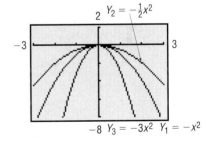

1   Graph a Quadratic Function Using Transformations

We know how to graph the quadratic function $f(x) = x^2$. Figure 3 shows the graph of three functions of the form $f(x) = ax^2, a > 0$, for $a = 1$, $a = \dfrac{1}{2}$, and $a = 3$. Notice that the larger the value of a, the "narrower" the graph is, and the smaller the value of a, the "wider" the graph is.

Figure 3

Figure 4

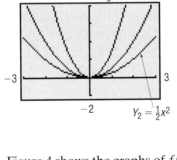

Figure 4 shows the graphs of $f(x) = ax^2$ for $a < 0$. Notice that these graphs are reflections about the x-axis of the graphs in Figure 3. Based on the results of these two figures, we can draw some general conclusions about the graph of $f(x) = ax^2$. First, as $|a|$ increases, the graph becomes "taller" (a vertical stretch), and as $|a|$ gets closer to zero, the graph gets "shorter" (a vertical compression). Second, if a is positive, then the graph opens "up," and if a is negative, then the graph opens "down."

The graphs in Figures 3 and 4 are typical of the graphs of all quadratic functions, which we call **parabolas**.* Refer to Figure 5, where two parabolas are pictured. The one on the left **opens up** and has a lowest point; the one on the right **opens down** and has a highest point. The lowest or highest point of a parabola is called the **vertex**.

Figure 5
Graphs of a quadratic function,
$f(x) = ax^2 + bx + c, a \ne 0$

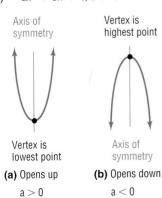

*We shall study parabolas using a geometric definition later in this book.

The vertical line passing through the vertex in each parabola in Figure 5 is called the **axis of symmetry** (usually abbreviated to **axis**) of the parabola. Because the parabola is symmetric about its axis, the axis of symmetry of a parabola can be used to find additional points on the parabola when graphing by hand.

The parabolas shown in Figure 5 are the graphs of a quadratic function $f(x) = ax^2 + bx + c, a \neq 0$. Notice that the coordinate axes are not included in the figure. Depending on the values of a, b, and c, the axes could be placed anywhere. The important fact is that the shape of the graph of a quadratic function will look like one of the parabolas in Figure 5.

In the following example, we use techniques from Section 2.7 to graph a quadratic function $f(x) = ax^2 + bx + c, a \neq 0$. In so doing, we shall complete the square and write the function f in the form $f(x) = a(x - h)^2 + k$.

EXAMPLE 1	**Graphing a Quadratic Function Using Transformations**

Graph the function $f(x) = 2x^2 + 8x + 5$. Find the vertex and axis of symmetry.

Solution We begin by completing the square on the right side.

$$f(x) = 2x^2 + 8x + 5$$

$$= 2(x^2 + 4x) + 5 \qquad \text{Factor out the 2 from } 2x^2 + 8x.$$

$$= 2(x^2 + 4x + 4) + 5 - 8 \qquad \text{Complete the square of } 2(x^2 + 4x). \text{ Notice that the}$$
$$\text{factor of 2 requires that 8 be added and subtracted.}$$

$$= 2(x + 2)^2 - 3 \tag{2}$$

The graph of f can be obtained in three stages, as shown in Figure 6. Now compare this graph to the graph in Figure 5(a). The graph of $f(x) = 2x^2 + 8x + 5$ is a parabola that opens up and has its vertex (lowest point) at $(-2, -3)$. Its axis of symmetry is the line $x = -2$.

Figure 6

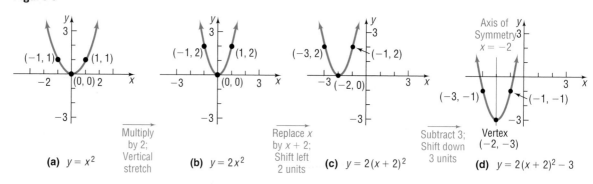

(a) $y = x^2$ Multiply by 2; Vertical stretch   (b) $y = 2x^2$ Replace x by $x + 2$; Shift left 2 units   (c) $y = 2(x + 2)^2$ Subtract 3; Shift down 3 units   (d) $y = 2(x + 2)^2 - 3$

✔ **CHECK:** Use a graphing utility to graph $f(x) = 2x^2 + 8x + 5$ and use the MINIMUM command to locate its vertex. ◄

NOW WORK PROBLEM **27.**

The method used in Example 1 can be used to graph any quadratic function $f(x) = ax^2 + bx + c, a \neq 0$, as follows:

$$f(x) = ax^2 + bx + c$$

$$= a\left(x^2 + \frac{b}{a}x\right) + c \qquad \text{Factor out } a \text{ from } ax^2 + bx.$$

$$= a\left(x^2 + \frac{b}{a}x + \frac{b^2}{4a^2}\right) + c - a\left(\frac{b^2}{4a^2}\right) \qquad \text{Complete the square by adding and sub-}$$
$$\text{tracting } a\left(\frac{b^2}{4a^2}\right). \text{ Look closely at this step!}$$

$$= a\left(x + \frac{b}{2a}\right)^2 + c - \frac{b^2}{4a} \qquad \text{Factor}$$

$$= a\left(x + \frac{b}{2a}\right)^2 + \frac{4ac - b^2}{4a} \qquad c - \frac{b^2}{4a} = c \cdot \frac{4a}{4a} - \frac{b^2}{4a} = \frac{4ac - b^2}{4a}$$

Based on these results, we conclude the following:

$$\text{If } h = -\frac{b}{2a} \quad \text{and} \quad k = \frac{4ac - b^2}{4a}, \quad \text{then}$$

$$\boxed{f(x) = ax^2 + bx + c = a(x - h)^2 + k} \qquad \textbf{(3)}$$

The graph of $f(x) = a(x - h)^2 + k$ is the parabola $y = ax^2$ shifted horizontally h units (replace x by $x - h$) and vertically k units (add k). As a result, the vertex is at (h, k), and the graph opens up if $a > 0$ and down if $a < 0$. The axis of symmetry is the vertical line $x = h$.

For example, compare equation (3) with equation (2) of Example 1.

$$f(x) = 2(x + 2)^2 - 3$$
$$= 2(x - (-2))^2 - 3$$
$$= a(x - h)^2 + k$$

We conclude that $a = 2$, so the graph opens up. Also, we find that $h = -2$ and $k = -3$, so its vertex is at $(-2, -3)$.

2   Identify the Vertex and Axis of Symmetry of a Quadratic Function

We do not need to complete the square to obtain the vertex. In almost every case, it is easier to obtain the vertex of a quadratic function f by remembering that its x-coordinate is $h = -\dfrac{b}{2a}$. The y-coordinate can then be found by evaluating f at $-\dfrac{b}{2a}$ to find $k = f\left(-\dfrac{b}{2a}\right)$.

We summarize these remarks as follows:

Properties of the Graph of a Quadratic Function

$$f(x) = ax^2 + bx + c, \quad a \neq 0$$

$$\text{Vertex} = \left(-\frac{b}{2a}, f\left(-\frac{b}{2a}\right)\right) \qquad \text{Axis of symmetry: the line } x = -\frac{b}{2a} \qquad \textbf{(4)}$$

Parabola opens up if $a > 0$; the vertex is a minimum point.

Parabola opens down if $a < 0$; the vertex is a maximum point.

EXAMPLE 2	Locating the Vertex without Graphing

Without graphing, locate the vertex and axis of symmetry of the parabola defined by $f(x) = -3x^2 + 6x + 1$. Does it open up or down?

Solution   For this quadratic function, $a = -3$, $b = 6$, and $c = 1$. The x-coordinate of the vertex is

$$h = -\frac{b}{2a} = -\frac{6}{-6} = 1$$

The y-coordinate of the vertex is

$$k = f\left(-\frac{b}{2a}\right) = f(1) = -3 + 6 + 1 = 4$$

The vertex is located at the point $(1, 4)$. The axis of symmetry is the line $x = 1$. Because $a = -3 < 0$, the parabola opens down. ◀

3 Graph a Quadratic Function Using Its Vertex, Axis, and Intercepts

The information we gathered in Example 2, together with the location of the intercepts, usually provides enough information to graph $f(x) = ax^2 + bx + c$, $a \neq 0$, by hand.

The y-intercept is the value of f at $x = 0$; that is, $f(0) = c$.

The x-intercepts, if there are any, are found by solving the quadratic equation

$$ax^2 + bx + c = 0$$

This equation has two, one, or no real solutions, depending on whether the discriminant $b^2 - 4ac$ is positive, 0, or negative. The graph of f has x-intercepts, as follows:

The x-intercepts of a Quadratic Function

1. If the discriminant $b^2 - 4ac > 0$, the graph of $f(x) = ax^2 + bx + c$ has two distinct x-intercepts and so will cross the x-axis in two places.

2. If the discriminant $b^2 - 4ac = 0$, the graph of $f(x) = ax^2 + bx + c$ has one x-intercept and touches the x-axis at its vertex.

3. If the discriminant $b^2 - 4ac < 0$, the graph of $f(x) = ax^2 + bx + c$ has no x-intercept and so will not cross or touch the x-axis.

Figure 7 illustrates these possibilities for parabolas that open up.

Figure 7
$f(x) = ax^2 + bx + c, a > 0$

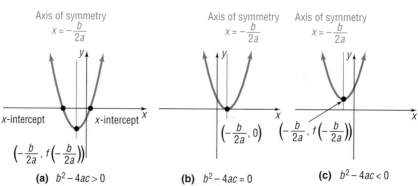

(a) $b^2 - 4ac > 0$
Two x-intercepts

(b) $b^2 - 4ac = 0$
One x-intercept

(c) $b^2 - 4ac < 0$
No x-intercepts

| EXAMPLE 3 | **Graphing a Quadratic Function by Hand Using Its Vertex, Axis, and Intercepts** |

Use the information from Example 2 and the locations of the intercepts to graph $f(x) = -3x^2 + 6x + 1$. Determine the domain and the range of f. Determine where f is increasing and where it is decreasing.

Solution   In Example 2, we found the vertex to be at $(1, 4)$ and the axis of symmetry to be $x = 1$. The y-intercept is found by letting $x = 0$. The y-intercept is $f(0) = 1$. The x-intercepts are found by solving the equation $f(x) = 0$. This results in the equation

$$-3x^2 + 6x + 1 = 0 \qquad a = -3, b = 6, c = 1$$

The discriminant $b^2 - 4ac = (6)^2 - 4(-3)(1) = 36 + 12 = 48 > 0$, so the equation has two real solutions and the graph has two x-intercepts. Using the quadratic formula, we find that

$$x = \frac{-b + \sqrt{b^2 - 4ac}}{2a} = \frac{-6 + \sqrt{48}}{-6} = \frac{-6 + 4\sqrt{3}}{-6} \approx -0.15$$

and

$$x = \frac{-b - \sqrt{b^2 - 4ac}}{2a} = \frac{-6 - \sqrt{48}}{-6} = \frac{-6 - 4\sqrt{3}}{-6} \approx 2.15$$

The x-intercepts are approximately -0.15 and 2.15.

The graph is illustrated in Figure 8. Notice how we used the y-intercept and the axis of symmetry, $x = 1$, to obtain the additional point $(2, 1)$ on the graph.

The domain of f is the set of all real numbers. Based on the graph, the range of f is the interval $(-\infty, 4]$. The function f is increasing on the interval $(-\infty, 1)$ and decreasing on the interval $(1, \infty)$.

✔ **CHECK:** Graph $f(x) = -3x^2 + 6x + 1$ using a graphing utility. Use ZERO (or ROOT) to locate the two x-intercepts. Use MAXIMUM to locate the vertex.   ◀

Figure 8

Axis of symmetry

━▷ **NOW WORK PROBLEM 35.**

If the graph of a quadratic function has only one x-intercept or no x-intercepts, it is usually necessary to plot an additional point to obtain the graph by hand.

| EXAMPLE 4 | **Graphing a Quadratic Function by Hand Using Its Vertex, Axis, and Intercepts** |

Graph $f(x) = x^2 - 6x + 9$ by determining whether the graph opens up or down and by finding its vertex, axis of symmetry, y-intercept, and x-intercepts, if any. Determine the domain and the range of f. Determine where f is increasing and where it is decreasing.

Solution   For $f(x) = x^2 - 6x + 9$, we have $a = 1$, $b = -6$, and $c = 9$. Since $a = 1 > 0$, the parabola opens up. The x-coordinate of the vertex is

$$h = -\frac{b}{2a} = -\frac{-6}{2(1)} = 3$$

Figure 9

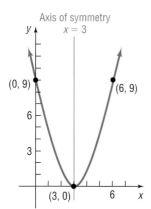

Axis of symmetry
$x = 3$

The y-coordinate of the vertex is

$$k = f(3) = (3)^2 - 6(3) + 9 = 0$$

So, the vertex is at $(3, 0)$. The axis of symmetry is the line $x = 3$. The y-intercept is $f(0) = 9$. Since the vertex $(3, 0)$ lies on the x-axis, the graph touches the x-axis at the x-intercept. By using the axis of symmetry and the y-intercept at $(0, 9)$, we can locate the additional point $(6, 9)$ on the graph. See Figure 9.

The domain of f is the set of all real numbers. Based on the graph, the range of f is the interval $[0, \infty)$. The function f is decreasing on the interval $(-\infty, 3)$ and increasing on the interval $(3, \infty)$. ◀

Graph the function in Example 4 by completing the square and using transformations. Which method do you prefer?

─ **NOW WORK PROBLEM 43.**

| **EXAMPLE 5** | **Graphing a Quadratic Function by Hand Using Its Vertex, Axis, and Intercepts** |

Graph $f(x) = 2x^2 + x + 1$ by determining whether the graph opens up or down and by finding its vertex, axis of symmetry, y-intercept, and x-intercepts, if any. Determine the domain and the range of f. Determine where f is increasing and where it is decreasing.

Solution For $f(x) = 2x^2 + x + 1$, we have $a = 2, b = 1$, and $c = 1$. Since $a = 2 > 0$, the parabola opens up. The x-coordinate of the vertex is

$$h = -\frac{b}{2a} = -\frac{1}{4}$$

The y-coordinate of the vertex is

$$k = f\left(-\frac{1}{4}\right) = 2\left(\frac{1}{16}\right) + \left(-\frac{1}{4}\right) + 1 = \frac{7}{8}$$

NOTE

In Example 5, since the vertex is above the x-axis and the parabola opens up, we can conclude the graph of the quadratic function will have no x-intercepts. ■

So, the vertex is at $\left(-\frac{1}{4}, \frac{7}{8}\right)$. The axis of symmetry is the line $x = -\frac{1}{4}$. The y-intercept is $f(0) = 1$. The x-intercept(s), if any, obey the equation $2x^2 + x + 1 = 0$. Since the discriminant $b^2 - 4ac = (1)^2 - 4(2)(1) = -7 < 0$, this equation has no real solutions, and therefore the graph has no x-intercepts. We use the point $(0, 1)$ and the axis of symmetry $x = -\frac{1}{4}$ to locate the additional point $\left(-\frac{1}{2}, 1\right)$ on the graph. See Figure 10.

The domain of f is the set of all real numbers. Based on the graph, the range of f is the interval $\left[\frac{7}{8}, \infty\right)$. The function f is decreasing on the interval $\left(-\infty, -\frac{1}{4}\right)$ and increasing on the interval $\left(-\frac{1}{4}, \infty\right)$. ◀

Figure 10

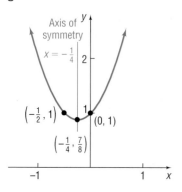

Axis of symmetry
$x = -\frac{1}{4}$

$\left(-\frac{1}{2}, 1\right)$ $(0, 1)$

$\left(-\frac{1}{4}, \frac{7}{8}\right)$

─ **NOW WORK PROBLEM 47.**

Given the vertex (h, k) and one additional point on the graph of a quadratic function $f(x) = ax^2 + bx + c, a \neq 0$, we can use

$$f(x) = a(x - h)^2 + k \qquad\qquad \textbf{(5)}$$

to obtain the quadratic function.

EXAMPLE 6   **Finding the Quadratic Function Given Its Vertex and One Other Point**

Determine the quadratic function whose vertex is $(1, -5)$ and whose y-intercept is -3. The graph of the parabola is shown in Figure 11.

Figure 11

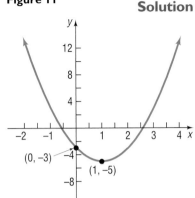

Solution   The vertex is $(1, -5)$, so $h = 1$ and $k = -5$. Substitute these values into equation (5).

$$f(x) = a(x - h)^2 + k \qquad \text{Equation (5)}$$
$$f(x) = a(x - 1)^2 - 5 \qquad h = 1, k = -5$$

To determine the value of a, we use the fact that $f(0) = -3$ (the y-intercept).

$$f(x) = a(x - 1)^2 - 5$$
$$-3 = a(0 - 1)^2 - 5 \qquad x = 0, y = f(0) = -3$$
$$-3 = a - 5$$
$$a = 2$$

The quadratic function whose graph is shown in Figure 11 is

$$f(x) = a(x - h)^2 + k = 2(x - 1)^2 - 5 = 2x^2 - 4x - 3$$

✔ **CHECK:** Figure 12 shows the graph $f(x) = 2x^2 - 4x - 3$ using a graphing utility.

Figure 12

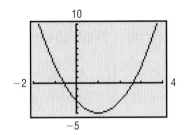

NOW WORK PROBLEM 53.

Summary

Steps for Graphing a Quadratic Function $f(x) = ax^2 + bx + c, a \neq 0$, by Hand.

Option 1

STEP 1: Complete the square in x to write the quadratic function in the form $f(x) = a(x - h)^2 + k$.

STEP 2: Graph the function in stages using transformations.

Option 2

STEP 1: Determine the vertex $\left(-\dfrac{b}{2a}, f\left(-\dfrac{b}{2a} \right) \right)$.

STEP 2: Determine the axis of symmetry, $x = -\dfrac{b}{2a}$.

STEP 3: Determine the y-intercept, $f(0)$.

STEP 4: (a) If $b^2 - 4ac > 0$, then the graph of the quadratic function has two x-intercepts, which are found by solving the equation $ax^2 + bx + c = 0$.
(b) If $b^2 - 4ac = 0$, the vertex is the x-intercept.
(c) If $b^2 - 4ac < 0$, there are no x-intercepts.

STEP 5: Determine an additional point by using the y-intercept and the axis of symmetry.

STEP 6: Plot the points and draw the graph.

4 Use the Maximum or Minimum Value of a Quadratic Function to Solve Applied Problems

When a mathematical model leads to a quadratic function, the properties of this quadratic function can provide important information about the model. For example, for a quadratic revenue function, we can find the maximum revenue; for a quadratic cost function, we can find the minimum cost.

To see why, recall that the graph of a quadratic function

$$f(x) = ax^2 + bx + c, \quad a \neq 0$$

is a parabola with vertex at $\left(-\dfrac{b}{2a}, f\left(-\dfrac{b}{2a} \right) \right)$. This vertex is the highest point on the graph if $a < 0$ and the lowest point on the graph if $a > 0$. If the vertex is the highest point $(a < 0)$, then $f\left(-\dfrac{b}{2a} \right)$ is the **maximum value** of f. If the vertex is the lowest point $(a > 0)$, then $f\left(-\dfrac{b}{2a} \right)$ is the **minimum value** of f.

This property of the graph of a quadratic function enables us to answer questions involving optimization (finding maximum or minimum values) in models involving quadratic functions.

| EXAMPLE 7 | **Finding the Maximum or Minimum Value of a Quadratic Function** |

Determine whether the quadratic function

$$f(x) = x^2 - 4x - 5$$

has a maximum or minimum value. Then find the maximum or minimum value.

Solution We compare $f(x) = x^2 - 4x - 5$ to $f(x) = ax^2 + bx + c$. We conclude that $a = 1, b = -4$, and $c = -5$. Since $a > 0$, the graph of f opens up, so the vertex is a minimum point. The minimum value occurs at

$$x = -\frac{b}{2a} = -\frac{-4}{2(1)} = \frac{4}{2} = 2$$

$$\uparrow$$
$$a = 1, b = -4$$

The minimum value is

$$f\left(-\frac{b}{2a} \right) = f(2) = 2^2 - 4(2) - 5 = 4 - 8 - 5 = -9$$

Table 2

X	Y1
-1	0
0	-5
1	-8
2	-9
3	-8
4	-5
5	0

Y1■X²−4X−5

✔ **CHECK:** We can support the algebraic solution using the TABLE feature on a graphing utility. Create Table 2 by letting $Y_1 = x^2 - 4x - 5$. From the table we see that the smallest value of y occurs when $x = 2$, leading us to investigate the number 2 further. The symmetry about $x = 2$ confirms that 2 is the x-coordinate of the vertex of the parabola. We conclude that the minimum value is -9 and occurs at $x = 2$. ◀

NOW WORK PROBLEM 61.

| EXAMPLE 8 | **Maximizing Revenue** |

The marketing department at Texas Instruments has found that, when certain calculators are sold at a price of p dollars per unit, the revenue R (in dollars) as a function of the price p is

$$R(p) = -150p^2 + 21,000p$$

What unit price should be established to maximize revenue? If this price is charged, what is the maximum revenue?

Solution   The revenue R is

$$R(p) = -150p^2 + 21,000p \qquad R(p) = ap^2 + bp + c$$

The function R is a quadratic function with $a = -150, b = 21,000$, and $c = 0$. Because $a < 0$, the vertex is the highest point on the parabola. The revenue R is therefore a maximum when the price p is

$$p = -\frac{b}{2a} = -\frac{21,000}{2(-150)} = -\frac{21,000}{-300} = \$70.00$$

$$a = -150, b = 21,000$$

The maximum revenue R is

$$R(70) = -150(70)^2 + 21,000(70) = \$735,000$$

See Figure 13 for an illustration.

Figure 13

NOW WORK PROBLEM 71.

| EXAMPLE 9 | **Maximizing the Area Enclosed by a Fence** |

A farmer has 2000 yards of fence to enclose a rectangular field. What are the dimensions of the rectangle that encloses the most area?

Figure 14   **Solution**   Figure 14 illustrates the situation. The available fence represents the perimeter of the rectangle. If x is the length and w is the width, then

$$2x + 2w = 2000 \qquad \textbf{(6)}$$

The area A of the rectangle is

$$A = xw$$

To express A in terms of a single variable, we solve equation (6) for w and substitute the result in $A = xw$. Then A involves only the variable x. [You could also solve equation (6) for x and express A in terms of w alone. Try it!]

$$2x + 2w = 2000 \qquad \text{\color{gray}Equation (6)}$$
$$2w = 2000 - 2x \qquad \text{\color{gray}Solve for w.}$$
$$w = \frac{2000 - 2x}{2} = 1000 - x$$

Then the area A is

$$A = xw = x(1000 - x) = -x^2 + 1000x$$

Now, A is a quadratic function of x.

$$A(x) = -x^2 + 1000x \qquad \text{\color{gray}$a = -1, b = 1000, c = 0$}$$

Figure 15

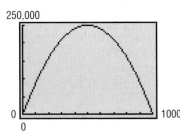

Figure 15 shows a graph of $A(x) = -x^2 + 1000x$ using a graphing utility. Since $a < 0$, the vertex is a maximum point on the graph of A. The maximum value occurs at

$$x = -\frac{b}{2a} = -\frac{1000}{2(-1)} = 500$$

The maximum value of A is

$$A\left(-\frac{b}{2a}\right) = A(500) = -500^2 + 1000(500) = -250,000 + 500,000 = 250,000$$

The largest rectangle that can be enclosed by 2000 yards of fence has an area of 250,000 square yards. Its dimensions are 500 yards by 500 yards.   ◀

NOW WORK PROBLEM 77.

EXAMPLE 10   **Analyzing the Motion of a Projectile**

A projectile is fired from a cliff 500 feet above the water at an inclination of $45°$ to the horizontal, with a muzzle velocity of 400 feet per second. In physics, it is established that the height h of the projectile above the water is given by

$$h(x) = \frac{-32x^2}{(400)^2} + x + 500$$

where x is the horizontal distance of the projectile from the base of the cliff. See Figure 16.

Figure 16

(a) Find the maximum height of the projectile.

(b) How far from the base of the cliff will the projectile strike the water?

Solution   (a)   The height of the projectile is given by a quadratic function.

$$h(x) = \frac{-32x^2}{(400)^2} + x + 500 = \frac{-1}{5000}x^2 + x + 500$$

We are looking for the maximum value of h. Since $a < 0$, the maximum value is obtained at the vertex. We compute

$$x = -\frac{b}{2a} = -\frac{1}{2\left(-\dfrac{1}{5000}\right)} = \frac{5000}{2} = 2500$$

The maximum height of the projectile is

$$h(2500) = \frac{-1}{5000}(2500)^2 + 2500 + 500 = -1250 + 2500 + 500 = 1750 \text{ ft}$$

(b)   The projectile will strike the water when the height is zero. To find the distance x traveled, we need to solve the equation

$$h(x) = \frac{-1}{5000}x^2 + x + 500 = 0$$

— Seeing the Concept —

Graph

$$h(x) = \frac{-1}{5000}x^2 + x + 500,$$

$$0 \le x \le 5500$$

Use MAXIMUM to find the maximum height of the projectile, and use ROOT or ZERO to find the distance from the base of the cliff to where it strikes the water. Compare your results with those obtained in Example 10.

We find the discriminant first.

$$b^2 - 4ac = 1^2 - 4\left(\frac{-1}{5000}\right)(500) = 1.4$$

then,

$$x = \frac{-b \pm \sqrt{b^2 - 4ac}}{2a} = \frac{-1 \pm \sqrt{1.4}}{2\left(-\dfrac{1}{5000}\right)} \approx \begin{cases} -458 \\ 5458 \end{cases}$$

We discard the negative solution and find that the projectile will strike the water at a distance of about 5458 feet from the base of the cliff.   ◀

NOW WORK PROBLEM 81.

EXAMPLE 11   **The Golden Gate Bridge**

The Golden Gate Bridge, a suspension bridge, spans the entrance to San Francisco Bay. Its 746-foot-tall towers are 4200 feet apart. The bridge is suspended from two huge cables more than 3 feet in diameter; the 90-foot-wide roadway is 220 feet above the water. The cables are parabolic in shape[*] and touch the road surface at the center of the bridge. Find the height of the cable at a distance of 1000 feet from the center.

Solution   See Figure 17. We begin by choosing the placement of the coordinate axes so that the x-axis coincides with the road surface and the origin coincides with the center of the bridge. As a result, the twin towers will be vertical (height $746 - 220 = 526$ feet above the road) and located 2100 feet from the center. Also, the cable, which has the

Figure 17

[*]A cable suspended from two towers is in the shape of a **catenary**, but when a horizontal roadway is suspended from the cable, the cable takes the shape of a parabola.

shape of a parabola, will extend from the towers, open up, and have its vertex at $(0, 0)$. The choice of placement of the axes enables us to identify the equation of the parabola as $y = ax^2$, $a > 0$. We can also see that the points $(-2100, 526)$ and $(2100, 526)$ are on the graph.

Based on these facts, we can find the value of a in $y = ax^2$.

$$y = ax^2$$
$$526 = a(2100)^2 \qquad x = 2100, y = 526$$
$$a = \frac{526}{(2100)^2}$$

The equation of the parabola is therefore

$$y = \frac{526}{(2100)^2} x^2$$

The height of the cable when $x = 1000$ is

$$y = \frac{526}{(2100)^2} (1000)^2 \approx 119.3 \text{ feet}$$

The cable is 119.3 feet high at a distance of 1000 feet from the center of the bridge. ◀

NOW WORK PROBLEM 83.

5 Use a Graphing Utility to Find the Quadratic Function of Best Fit to Data

In Section 2.5, we found the line of best fit for data that appeared to be linearly related. It was noted that data may also follow a nonlinear relation. Figures 18(a) and (b) show scatter diagrams of data that follow a quadratic relation.

Figure 18

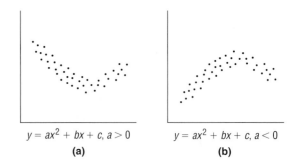

$$y = ax^2 + bx + c, \, a > 0 \qquad\qquad y = ax^2 + bx + c, \, a < 0$$
(a)   (b)

EXAMPLE 12 **Fitting a Quadratic Function to Data**

A farmer collected the data given in Table 3, which shows crop yields Y for various amounts of fertilizer used, x.

(a) With a graphing utility, draw a scatter diagram of the data. Comment on the type of relation that may exist between the two variables.

(b) Use a graphing utility to find the quadratic function of best fit to these data.

(c) Use the function found in part (b) to determine the optimal amount of fertilizer to apply.

Table 3

Plot	Fertilizer, x (Pounds/100 ft²)	Yield (Bushels)
1	0	4
2	0	6
3	5	10
4	5	7
5	10	12
6	10	10
7	15	15
8	15	17
9	20	18
10	20	21
11	25	20
12	25	21
13	30	21
14	30	22
15	35	21
16	35	20
17	40	19
18	40	19

(d) Use the function found in part (b) to predict crop yield when the optimal amount of fertilizer is applied.

(e) Draw the quadratic function of best fit on the scatter diagram.

Solution

(a) Figure 19 shows the scatter diagram, from which it appears that the data follow a quadratic relation, with $a < 0$.

(b) Upon executing the QUADratic REGression program, we obtain the results shown in Figure 20. The output that the utility provides shows us the equation $y = ax^2 + bx + c$. The quadratic function of best fit is

$$Y(x) = -0.0171x^2 + 1.0765x + 3.8939$$

where x represents the amount of fertilizer used and Y represents crop yield.

(c) Based on the quadratic function of best fit, the optimal amount of fertilizer to apply is

$$x = -\frac{b}{2a} = -\frac{1.0765}{2(-0.0171)} \approx 31.5 \text{ pounds of fertilizer per 100 square feet}$$

(d) We evaluate the function $Y(x)$ for $x = 31.5$.

$$Y(31.5) = -0.0171(31.5)^2 + 1.0765(31.5) + 3.8939 \approx 20.8 \text{ bushels}$$

If we apply 31.5 pounds of fertilizer per 100 square feet, the crop yield will be 20.8 bushels according to the quadratic function of best fit.

(e) Figure 21 shows the graph of the quadratic function found in part (b) drawn on the scatter diagram.

Figure 19

Figure 20

Figure 21

Look again at Figure 20. Notice that the output given by the graphing calculator does not include r, the correlation coefficient. Recall that the correlation coefficient is a measure of the strength of a *linear* relation that exists between two variables. The graphing calculator does not provide an indication of how well the function fits the data in terms of r since a quadratic function cannot be expressed as a linear function.

NOW WORK PROBLEM **101.**

3.1 Assess Your Understanding

'Are You Prepared?'

Answers are given at the end of these exercises. If you get a wrong answer, read the pages listed in red.

1. List the intercepts of the equation $y = x^2 - 9$. (pp. 171–172 and 183)

2. Solve the equation $2x^2 + 7x - 4 = 0$. (pp. 113–114 and 117–119)

3. To complete the square of $x^2 - 5x$, you add the number _____. (pp. 115–116)

4. To graph $y = (x - 4)^2$, you shift the graph of $y = x^2$ to the _____ a distance of _____ units. (pp. 266–274)

Concepts and Vocabulary

5. The graph of a quadratic function is called a(n) _____.

6. The vertical line passing through the vertex of a parabola is called the _____.

7. The x-coordinate of the vertex of $f(x) = ax^2 + bx + c$, $a \neq 0$, is _____.

8. *True or False:* The graph of $f(x) = 2x^2 + 3x - 4$ opens up.

9. *True or False:* The y-coordinate of the vertex of $f(x) = -x^2 + 4x + 5$ is $f(2)$.

10. *True or False:* If the discriminant $b^2 - 4ac = 0$, the graph of $f(x) = ax^2 + bx + c, a \neq 0$, will touch the x-axis at its vertex.

Skill Building

In Problems 11–18, match each graph to one the following functions without using a graphing utility.

11. $f(x) = x^2 - 1$

12. $f(x) = -x^2 - 1$

13. $f(x) = x^2 - 2x + 1$

14. $f(x) = x^2 + 2x + 1$

15. $f(x) = x^2 - 2x + 2$

16. $f(x) = x^2 + 2x$

17. $f(x) = x^2 - 2x$

18. $f(x) = x^2 + 2x + 2$

A.

B.

C.

D.

E.

F.

G.

H.
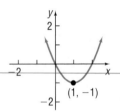

In Problems 19–34, graph the function f by starting with the graph of $y = x^2$ and using transformations (shifting, compressing, stretching, and/or reflection). Verify your results using a graphing utility.

[Hint: *If necessary, write f in the form $f(x) = a(x - h)^2 + k$.]*

19. $f(x) = \dfrac{1}{4}x^2$

20. $f(x) = 2x^2$

21. $f(x) = \dfrac{1}{4}x^2 - 2$

22. $f(x) = 2x^2 - 3$

23. $f(x) = \dfrac{1}{4}x^2 + 2$

24. $f(x) = 2x^2 + 4$

25. $f(x) = \dfrac{1}{4}x^2 + 1$

26. $f(x) = -2x^2 - 2$

27. $f(x) = x^2 + 4x + 2$

28. $f(x) = x^2 - 6x - 1$

29. $f(x) = 2x^2 - 4x + 1$

30. $f(x) = 3x^2 + 6x$

31. $f(x) = -x^2 - 2x$

32. $f(x) = -2x^2 + 6x + 2$

33. $f(x) = \dfrac{1}{2}x^2 + x - 1$

34. $f(x) = \dfrac{2}{3}x^2 + \dfrac{4}{3}x - 1$

In Problems 35–52, graph each quadratic function by determining whether its graph opens up or down and by finding its vertex, axis of symmetry, y-intercept, and x-intercepts, if any. Determine the domain and the range of the function. Determine where the function is increasing and where it is decreasing. Verify your results using a graphing utility.

35. $f(x) = x^2 + 2x$

36. $f(x) = x^2 - 4x$

37. $f(x) = -x^2 - 6x$

38. $f(x) = -x^2 + 4x$

39. $f(x) = 2x^2 - 8x$

40. $f(x) = 3x^2 + 18x$

41. $f(x) = x^2 + 2x - 8$

42. $f(x) = x^2 - 2x - 3$

43. $f(x) = x^2 + 2x + 1$

44. $f(x) = x^2 + 6x + 9$

45. $f(x) = 2x^2 - x + 2$

46. $f(x) = 4x^2 - 2x + 1$

47. $f(x) = -2x^2 + 2x - 3$

48. $f(x) = -3x^2 + 3x - 2$

49. $f(x) = 3x^2 + 6x + 2$

50. $f(x) = 2x^2 + 5x + 3$

51. $f(x) = -4x^2 - 6x + 2$

52. $f(x) = 3x^2 - 8x + 2$

In Problems 53–58, determine the quadratic function whose graph is given.

53.

54.

55.

56.

57.

58.

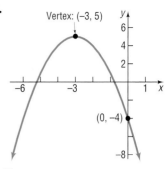

In Problems 59–66, determine, without graphing, whether the given quadratic function has a maximum value or a minimum value and then find the value. Use the TABLE feature on a graphing utility to support your answer.

59. $f(x) = 2x^2 + 12x$ **60.** $f(x) = -2x^2 + 12x$ **61.** $f(x) = 2x^2 + 12x - 3$ **62.** $f(x) = 4x^2 - 8x + 3$

63. $f(x) = -x^2 + 10x - 4$ **64.** $f(x) = -2x^2 + 8x + 3$ **65.** $f(x) = -3x^2 + 12x + 1$ **66.** $f(x) = 4x^2 - 4x$

Applications and Extensions

67. The graph of the function $f(x) = ax^2 + bx + c$ has vertex at $(0, 2)$ and passes through the point $(1, 8)$. Find $a, b,$ and c.

68. The graph of the function $f(x) = ax^2 + bx + c$ has vertex at $(1, 4)$ and passes through the point $(-1, -8)$. Find $a, b,$ and c.

Answer Problems 69 and 70 using the following: A quadratic function of the form $f(x) = ax^2 + bx + c$ with $b^2 - 4ac > 0$ may also be written in the form $f(x) = a(x - r_1)(x - r_2)$, where r_1 and r_2 are the x-intercepts of the graph of the quadratic function.

69. (a) Find a quadratic function whose x-intercepts are -3 and 1 with $a = 1$; $a = 2$; $a = -2$; $a = 5$.
   (b) How does the value of a affect the intercepts?
   (c) How does the value of a affect the axis of symmetry?
   (d) How does the value of a affect the vertex?
   (e) Compare the x-coordinate of the vertex with the midpoint of the x-intercepts. What might you conclude?

70. (a) Find a quadratic function whose x-intercepts are -5 and 3 with $a = 1$; $a = 2$; $a = -2$; $a = 5$.
   (b) How does the value of a affect the intercepts?
   (c) How does the value of a affect the axis of symmetry?
   (d) How does the value of a affect the vertex?
   (e) Compare the x-coordinate of the vertex with the midpoint of the x-intercepts. What might you conclude?

71. Maximizing Revenue   Suppose that the manufacturer of a gas clothes dryer has found that, when the unit price is p dollars, the revenue R (in dollars) is

$$R(p) = -4p^2 + 4000p$$

What unit price for the dryer should be established to maximize revenue? What is the maximum revenue?

72. Maximizing Revenue   The John Deere company has found that the revenue from sales of heavy-duty tractors is a function of the unit price p (in dollars) that it charges. If the revenue R is

$$R(p) = -\frac{1}{2}p^2 + 1900p$$

what unit price p (in dollars) should be charged to maximize revenue? What is the maximum revenue?

73. Demand Equation   The price p (in dollars) and the quantity x sold of a certain product obey the demand equation

$$p = -\frac{1}{6}x + 100, \qquad 0 \le x \le 600$$

(a) Express the revenue R as a function of x. (Remember, $R = xp$.)
(b) What is the revenue if 200 units are sold?
(c) What quantity x maximizes revenue? What is the maximum revenue?
(d) What price should the company charge to maximize revenue?

74. Demand Equation The price p (in dollars) and the quantity x sold of a certain product obey the demand equation

$$p = -\frac{1}{3}x + 100, \qquad 0 \le x \le 300$$

(a) Express the revenue R as a function of x.
(b) What is the revenue if 100 units are sold?
(c) What quantity x maximizes revenue? What is the maximum revenue?
(d) What price should the company charge to maximize revenue?

75. Demand Equation The price p (in dollars) and the quantity x sold of a certain product obey the demand equation

$$x = -5p + 100, \qquad 0 \le p \le 20$$

(a) Express the revenue R as a function of x.
(b) What is the revenue if 15 units are sold?
(c) What quantity x maximizes revenue? What is the maximum revenue?
(d) What price should the company charge to maximize revenue?

76. Demand Equation The price p (in dollars) and the quantity x sold of a certain product obey the demand equation

$$x = -20p + 500, \qquad 0 \le p \le 25$$

(a) Express the revenue R as a function of x.
(b) What is the revenue if 20 units are sold?
(c) What quantity x maximizes revenue? What is the maximum revenue?
(d) What price should the company charge to maximize revenue?

77. Enclosing a Rectangular Field David has available 400 yards of fencing and wishes to enclose a rectangular area.
(a) Express the area A of the rectangle as a function of the width w of the rectangle.
(b) For what value of w is the area largest?
(c) What is the maximum area?

78. Enclosing a Rectangular Field Beth has 3000 feet of fencing available to enclose a rectangular field.
(a) Express the area A of the rectangle as a function of x where x is the length of the rectangle.
(b) For what value of x is the area largest?
(c) What is the maximum area?

79. Enclosing the Most Area with a Fence A farmer with 4000 meters of fencing wants to enclose a rectangular plot that borders on a river. If the farmer does not fence the side along the river, what is the largest area that can be enclosed? (See the figure.)

$4000 - 2x$

80. Enclosing the Most Area with a Fence A farmer with 2000 meters of fencing wants to enclose a rectangular plot that borders on a straight highway. If the farmer does not fence the side along the highway, what is the largest area that can be enclosed?

81. Analyzing the Motion of a Projectile A projectile is fired from a cliff 200 feet above the water at an inclination of 45° to the horizontal, with a muzzle velocity of 50 feet per second. The height h of the projectile above the water is given by

$$h(x) = \frac{-32x^2}{(50)^2} + x + 200$$

where x is the horizontal distance of the projectile from the base of the cliff.
(a) How far from the base of the cliff is the height of the projectile a maximum?
(b) Find the maximum height of the projectile.
(c) How far from the base of the cliff will the projectile strike the water?
(d) Using a graphing utility, graph the function h, $0 \le x \le 200$.
(e) When the height of the projectile is 100 feet above the water, how far is it from the cliff?

82. Analyzing the Motion of a Projectile A projectile is fired at an inclination of 45° to the horizontal, with a muzzle velocity of 100 feet per second. The height h of the projectile is given by

$$h(x) = \frac{-32x^2}{(100)^2} + x$$

where x is the horizontal distance of the projectile from the firing point.
(a) How far from the firing point is the height of the projectile a maximum?
(b) Find the maximum height of the projectile.
(c) How far from the firing point will the projectile strike the ground?
(d) Using a graphing utility, graph the function h, $0 \le x \le 350$.
(e) When the height of the projectile is 50 feet above the ground, how far has it traveled horizontally?

83. Suspension Bridge A suspension bridge with weight uniformly distributed along its length has twin towers that extend 75 meters above the road surface and are 400 meters apart. The cables are parabolic in shape and are suspended from the tops of the towers. The cables touch the road surface at the center of the bridge. Find the height of the cables at a point 100 meters from the center. (Assume that the road is level.)

84. Architecture A parabolic arch has a span of 120 feet and a maximum height of 25 feet. Choose suitable rectangular coordinate axes and find the equation of the parabola. Then calculate the height of the arch at points 10 feet, 20 feet, and 40 feet from the center.

85. Constructing Rain Gutters A rain gutter is to be made of aluminum sheets that are 12 inches wide by turning up the edges 90°. What depth will provide maximum cross-sectional area and hence allow the most water to flow?

86. Norman Windows A Norman window has the shape of a rectangle surmounted by a semicircle of diameter equal to the width of the rectangle (see the figure). If the perimeter of the window is 20 feet, what dimensions will admit the most light (maximize the area)?

[**Hint:** Circumference of a circle $= 2\pi r$; area of a circle $= \pi r^2$, where r is the radius of the circle.]

87. Constructing a Stadium A track and field playing area is in the shape of a rectangle with semicircles at each end (see the figure). The inside perimeter of the track is to be 1500 meters. What should the dimensions of the rectangle be so that the area of the rectangle is a maximum?

88. Architecture A special window has the shape of a rectangle surmounted by an equilateral triangle (see the figure). If the perimeter of the window is 16 feet, what dimensions will admit the most light?

[**Hint:** Area of an equilateral triangle $= \left(\dfrac{\sqrt{3}}{4}\right)x^2$, where x is the length of a side of the triangle.]

89. Chemical Reactions A self-catalytic chemical reaction results in the formation of a compound that causes the formation ratio to increase. If the reaction rate V is given by

$$V(x) = kx(a - x), \qquad 0 \le x \le a$$

where k is a positive constant, a is the initial amount of the compound, and x is the variable amount of the compound, for what value of x is the reaction rate a maximum?

90. Calculus: Simpson's Rule The figure shows the graph of $y = ax^2 + bx + c$. Suppose that the points $(-h, y_0)$, $(0, y_1)$, and (h, y_2) are on the graph. It can be shown that the area enclosed by the parabola, the x-axis, and the lines $x = -h$ and $x = h$ is

$$\text{Area} = \frac{h}{3}(2ah^2 + 6c)$$

Show that this area may also be given by

$$\text{Area} = \frac{h}{3}(y_0 + 4y_1 + y_2)$$

91. Use the result obtained in Problem 90 to find the area enclosed by $f(x) = -5x^2 + 8$, the x-axis, and the lines $x = -1$ and $x = 1$.

92. Use the result obtained in Problem 90 to find the area enclosed by $f(x) = 2x^2 + 8$, the x-axis, and the lines $x = -2$ and $x = 2$.

93. Use the result obtained in Problem 90 to find the area enclosed by $f(x) = x^2 + 3x + 5$, the x-axis, and the lines $x = -4$ and $x = 4$.

94. Use the result obtained in Problem 90 to find the area enclosed by $f(x) = -x^2 + x + 4$, the x-axis, and the lines $x = -1$ and $x = 1$.

95. A rectangle has one vertex on the line $y = 10 - x$, $x > 0$, another at the origin, one on the positive x-axis, and one on the positive y-axis. Find the largest area A that can be enclosed by the rectangle.

96. Let $f(x) = ax^2 + bx + c$, where a, b, and c are odd integers. If x is an integer, show that $f(x)$ must be an odd integer.

[Hint: x is either an even integer or an odd integer.]

97. Hunting The function $H(x) = -1.01x^2 + 114.3x + 451.0$ models the number of individuals who engage in hunting activities whose annual income is x thousand dollars.

SOURCE: Based on data obtained from the National Sporting Goods Association.

(a) What is the income level for which there are the most hunters? Approximately how many hunters earn this amount?

(b) Using a graphing utility, graph $H = H(x)$. Are the number of hunters increasing or decreasing for individuals earning between $20,000 and $40,000?

98. Advanced Degrees The function $P(x) = -0.008x^2 + 0.868x - 11.884$ models the percentage of the U.S. population in March 2000 whose age is given by x that has earned an advanced degree (more than a bachelor's degree).

SOURCE: Based on data obtained from the U.S. Census Bureau.

(a) What is the age for which the highest percentage of Americans have earned an advanced degree? What is the highest percentage?

(b) Using a graphing utility, graph $P = P(x)$. Is the percentage of Americans that have earned an advanced degree increasing or decreasing for individuals between the ages of 40 and 50?

99. Male Murder Victims The function $M(x) = 0.76x^2 - 107.00x + 3854.18$ models the number of male murder victims who are x years of age ($20 \le x < 90$).

SOURCE: Based on data obtained from the Federal Bureau of Investigation.

(a) Use the model to approximate the number of male murder victims who are $x = 23$ years of age.

(b) At what age is the number of male murder victims 1456?

(c) Using a graphing utility, graph $M = M(x)$.

(d) Based on the graph drawn in part (c), describe what happens to the number of male murder victims as age increases.

100. Health Care Expenditures The function $H(x) = 0.004x^2 - 0.197x + 5.406$ models the percentage of total income that an individual who is x years of age spends on health care.

SOURCE: Based on data obtained from the Bureau of Labor Statistics.

(a) Use the model to approximate the percentage of total income that an individual who is $x = 45$ years of age spends on health care.

(b) At what age is the percentage of income spent on health care 10%?

(c) Using a graphing utility, graph $H = H(x)$.

(d) Based on the graph drawn in part (c), describe what happens to the percentage of income spent on health care as individuals age.

101. Life Cycle Hypothesis An individual's income varies with his or her age. The following table shows the median income I of individuals of different age groups within the United States for 2001. For each age group, let the class midpoint represent the independent variable, x. For the class "65 years and older," we will assume that the class midpoint is 69.5.

Age	Class Midpoint, x	Median Income, I
15–24 years	19.5	$9,301
25–34 years	29.5	$30,510
35–44 years	39.5	$38,340
45–54 years	49.5	$41,104
55–64 years	59.5	$35,637
65 years and older	69.5	$19,688

SOURCE: U.S. Census Bureau

(a) Draw a scatter diagram of the data. Comment on the type of relation that may exist between the two variables.

(b) Use a graphing utility to find the quadratic function of best fit to these data.

(c) Use the function found in part (b) to determine the age at which an individual can expect to earn the most income.

(d) Use the function found in part (b) to predict the peak income earned.

(e) With a graphing utility, graph the quadratic function of best fit on the scatter diagram.

102. Life Cycle Hypothesis An individual's income varies with his or her age. The following table shows the median income I of individuals of different age groups within the United States for 2000. For each age group, the class midpoint represents the independent variable, x. For the age group "65 years and older," we will assume that the class midpoint is 69.5.

Age	Class Midpoint, x	Median Income, I
15–24 years	19.5	$9,548
25–34 years	29.5	$30,633
35–44 years	39.5	$37,088
45–54 years	49.5	$41,072
55–64 years	59.5	$34,414
65 years and older	69.5	$19,167

SOURCE: U.S. Census Bureau

(a) Draw a scatter diagram of the data. Comment on the type of relation that may exist between the two variables.
(b) Use a graphing utility to find the quadratic function of best fit to these data.
(c) Use the function found in part (b) to determine the age at which an individual can expect to earn the most income.
(d) Use the function found in part (b) to predict the peak income earned.
(e) With a graphing utility, graph the quadratic function of best fit on the scatter diagram.
(f) Compare the results of parts (c) and (d) in 2000 to those of 2001 in Problem 101.

103. **Height of a Ball** A shot-putter throws a ball at an inclination of 45° to the horizontal. The following data represent the height of the ball h at the instant that it has traveled x feet horizontally.

Distance, x	Height, h
20	25
40	40
60	55
80	65
100	71
120	77
140	77
160	75
180	71
200	64

(a) Draw a scatter diagram of the data. Comment on the type of relation that may exist between the two variables.
(b) Use a graphing utility to find the quadratic function of best fit to these data.

(c) Use the function found in part (b) to determine how far the ball will travel before it reaches its maximum height.
(d) Use the function found in part (b) to find the maximum height of the ball.
(e) With a graphing utility, graph the quadratic function of best fit on the scatter diagram.

104. **Miles per Gallon** An engineer collects the following data showing the speed s of a Ford Taurus and its average miles per gallon, M.

Speed, s	Miles per Gallon, M
30	18
35	20
40	23
40	25
45	25
50	28
55	30
60	29
65	26
65	25
70	25

(a) Draw a scatter diagram of the data. Comment on the type of relation that may exist between the two variables.
(b) Use a graphing utility to find the quadratic function of best fit to these data.
(c) Use the function found in part (b) to determine the speed that maximizes miles per gallon.
(d) Use the function found in part (b) to predict miles per gallon for a speed of 63 miles per hour.
(e) With a graphing utility, graph the quadratic function of best fit on the scatter diagram.

Discussion and Writing

105. Make up a quadratic function that opens down and has only one x-intercept. Compare yours with others in the class. What are the similarities? What are the differences?

106. On one set of coordinate axes, graph the family of parabolas $f(x) = x^2 + 2x + c$ for $c = -3, c = 0$, and $c = 1$. Describe the characteristics of a member of this family.

107. On one set of coordinate axes, graph the family of parabolas $f(x) = x^2 + bx + 1$ for $b = -4, b = 0$, and $b = 4$. Describe the general characteristics of this family.

108. State the circumstances that cause the graph of a quadratic function $f(x) = ax^2 + bx + c$ to have no x-intercepts.

109. Why does the graph of a quadratic function open up if $a > 0$ and down if $a < 0$?

110. Refer to Example 8 on page 307. Notice that if the price charged for the calculators is $0 or $140 the revenue is $0. It is easy to explain why revenue would be $0 if the price charged is $0, but how can revenue be $0 if the price charged is $140?

'Are You Prepared?' Answers

1. $(0, -9), (-3, 0), (3, 0)$ 2. $\left\{-4, \frac{1}{2}\right\}$ 3. $\frac{25}{4}$ 4. right; 4

3.2 Polynomial Functions and Models

PREPARING FOR THIS SECTION *Before getting started, review the following:*

- Polynomials (Section R.4, pp. 36–37)
- Using a Graphing Utility to Approximate Local Maxima and Local Minima (Section 2.4, pp. 232–233)
- Graphing Techniques: Transformations (Section 2.7, pp. 266–274)
- Intercepts (Section 1.8, pp. 171–172; Section 1.9, p. 183)

 Now work the 'Are You Prepared?' problems on page 330.

OBJECTIVES
1 Identify Polynomial Functions and Their Degree
2 Graph Polynomial Functions Using Transformations
3 Identify the Zeros of a Polynomial Function and Their Multiplicity
4 Analyze the Graph of a Polynomial Function
5 Find the Cubic Function of Best Fit to Data

1 Identify Polynomial Functions and Their Degree

Polynomial functions are among the simplest expressions in algebra. They are easy to evaluate: only addition and repeated multiplication are required. Because of this, they are often used to approximate other, more complicated functions. In this section, we investigate properties of this important class of functions.

A **polynomial function** is a function of the form

$$f(x) = a_n x^n + a_{n-1} x^{n-1} + \cdots + a_1 x + a_0 \qquad \textbf{(1)}$$

where $a_n, a_{n-1}, \ldots, a_1, a_0$ are real numbers and n is a nonnegative integer. The domain is the set of all real numbers.

A polynomial function is a function whose rule is given by a polynomial in one variable. The **degree** of a polynomial function is the degree of the polynomial in one variable, that is, the largest power of x that appears.

EXAMPLE 1

Identifying Polynomial Functions

Determine which of the following are polynomial functions. For those that are, state the degree; for those that are not, tell why not.

(a) $f(x) = 2 - 3x^4$ (b) $g(x) = \sqrt{x}$ (c) $h(x) = \dfrac{x^2 - 2}{x^3 - 1}$

(d) $F(x) = 0$ (e) $G(x) = 8$ (f) $H(x) = -2x^3(x - 1)^2$

Solution

(a) f is a polynomial function of degree 4.

(b) g is not a polynomial function because $g(x) = \sqrt{x} = x^{\frac{1}{2}}$, so the variable x is raised to the $\dfrac{1}{2}$ power, which is not a nonnegative integer.

(c) h is not a polynomial function. It is the ratio of two polynomials, and the polynomial in the denominator is of positive degree.

(d) F is the zero polynomial function; it is not assigned a degree.

(e) G is a nonzero constant function. It is a polynomial function of degree 0 since $G(x) = 8 = 8x^0$.

(f) $H(x) = -2x^3(x - 1)^2 = -2x^3(x^2 - 2x + 1) = -2x^5 + 4x^4 - 2x^3$. So H is a polynomial function of degree 5. Do you see how to find the degree of H without multiplying out?   ◀

⬤───── **NOW WORK PROBLEMS 11 AND 15.**

We have already discussed in detail polynomial functions of degrees 0, 1, and 2. See Table 4 for a summary of the properties of the graphs of these polynomial functions.

Table 4

Degree	Form	Name	Graph
No degree	$f(x) = 0$	Zero function	The x-axis
0	$f(x) = a_0,\quad a_0 \neq 0$	Constant function	Horizontal line with y-intercept a_0
1	$f(x) = a_1 x + a_0,\quad a_1 \neq 0$	Linear function	Nonvertical, nonhorizontal line with slope a_1 and y-intercept a_0
2	$f(x) = a_2 x^2 + a_1 x + a_0,\quad a_2 \neq 0$	Quadratic function	Parabola: Graph opens up if $a_2 > 0$; graph opens down if $a_2 < 0$

One objective of this section is to analyze the graph of a polynomial function. If you take a course in calculus, you will learn that the graph of every polynomial function is both smooth and continuous. By **smooth**, we mean that the graph contains no sharp corners or cusps; by **continuous**, we mean that the graph has no gaps or holes and can be drawn without lifting pencil from paper. See Figures 22(a) and (b).

Figure 22

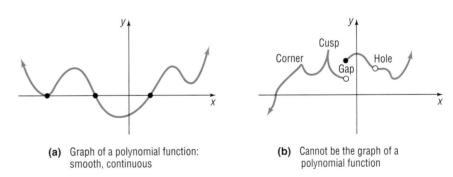

(a) Graph of a polynomial function: smooth, continuous

(b) Cannot be the graph of a polynomial function

2   Graph Polynomial Functions Using Transformations

We begin the analysis of the graph of a polynomial function by discussing *power functions*, a special kind of polynomial function.

In Words

A power function is a function that is defined by a single monomial.

A **power function of degree** n is a function of the form

$$f(x) = ax^n \tag{2}$$

where a is a real number, $a \neq 0$, and $n > 0$ is an integer.

Examples of power functions are

$$f(x) = 3x \qquad f(x) = -5x^2 \qquad f(x) = 8x^3 \qquad f(x) = -5x^4$$
$$\text{degree 1} \qquad\quad \text{degree 2} \qquad\quad\; \text{degree 3} \qquad\quad\; \text{degree 4}$$

The graph of a power function of degree 1, $f(x) = ax$, is a straight line, with slope a, that passes through the origin. The graph of a power function of degree 2, $f(x) = ax^2$, is a parabola, with vertex at the origin, that opens up if $a > 0$ and down if $a < 0$.

If we know how to graph a power function of the form $f(x) = x^n$, then a compression or stretch and, perhaps, a reflection about the x-axis will enable us to obtain the graph of $g(x) = ax^n$. Consequently, we shall concentrate on graphing power functions of the form $f(x) = x^n$.

We begin with power functions of even degree of the form $f(x) = x^n$, $n \geq 2$ and n even.

—————— Exploration ——————

Using your graphing utility and the viewing window $-2 \leq x \leq 2$, $-4 \leq y \leq 16$, graph the function $Y_1 = f(x) = x^4$. On the same screen, graph $Y_2 = g(x) = x^8$. Now, also on the same screen, graph $Y_3 = h(x) = x^{12}$. What do you notice about the graphs as the magnitude of the exponent increases? Repeat this procedure for the viewing window $-1 \leq x \leq 1$, $0 \leq y \leq 1$. What do you notice?

Result See Figures 23(a) and (b).

Figure 23

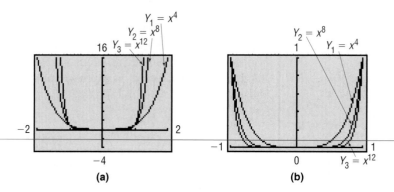

(a)

(b)

Table 5

X	Y₂	Y₃
-1	1	1
1	1	1
.5	.00391	2.4E-4
.1	1E-8	1E-12
.01	1E-16	1E-24
.001	1E-24	1E-36
0	0	0

Y₂⊟X^8

The domain of $f(x) = x^n$, $n \geq 2$ and n even, is the set of all real numbers, and the range is the set of nonnegative real numbers. Such a power function is an even function (do you see why?), so its graph is symmetric with respect to the y-axis. Its graph always contains the origin $(0, 0)$ and the points $(-1, 1)$ and $(1, 1)$.

For large n, it appears that the graph coincides with the x-axis near the origin, but it does not; the graph actually touches the x-axis only at the origin. See Table 5, where $Y_2 = x^8$ and $Y_3 = x^{12}$. For x close to 0, the values of y are positive and close to 0. Also, for large n, it may appear that for $x < -1$ or for $x > 1$ the graph is vertical, but it is not; it is only increasing very rapidly. If you TRACE along one of the graphs, these distinctions will be clear.

To summarize:

Properties of Power Functions, $f(x) = x^n$, n Is an Even Integer

1. The graph is symmetric with respect to the y-axis, so f is even.

2. The domain is the set of all real numbers. The range is the set of nonnegative real numbers.

3. The graph always contains the points $(0, 0)$, $(1, 1)$, and $(-1, 1)$.

4. As the exponent n increases in magnitude, the graph becomes more vertical when $x < -1$ or $x > 1$; but for x near the origin, the graph tends to flatten out and lie closer to the x-axis.

NOTE

Don't forget how graphing calculators express scientific notation. In Table 5, $-1E-8$ means -1×10^{-8}. ∎

Now we consider power functions of odd degree of the form $f(x) = x^n$, *n odd.*

─── **Exploration** ───────────────────

Using your graphing utility and the viewing window $-2 \le x \le 2$, $-16 \le y \le 16$, graph the function $Y_1 = f(x) = x^3$. On the same screen, graph $Y_2 = g(x) = x^7$ and $Y_3 = h(x) = x^{11}$. What do you notice about the graphs as the magnitude of the exponent increases? Repeat this procedure for the viewing window $-1 \le x \le 1$, $-1 \le y \le 1$. What do you notice?

Result   The graphs on your screen should look like Figures 24(a) and (b).

Figure 24

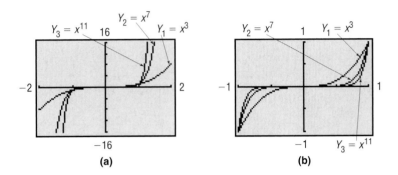

(a)   (b)

The domain and the range of $f(x) = x^n$, $n \ge 3$ and n odd, are the set of real numbers. Such a power function is an odd function (do you see why?), so its graph is symmetric with respect to the origin. Its graph always contains the origin $(0, 0)$ and the points $(-1, -1)$ and $(1, 1)$.

It appears that the graph coincides with the x-axis near the origin, but it does not; the graph actually crosses the x-axis only at the origin. Also, it appears that as x increases the graph is vertical, but it is not; it is increasing very rapidly. TRACE along the graphs to verify these distinctions.

To summarize:

Properties of Power Functions, $f(x) = x^n$, n Is an Odd Integer

1. The graph is symmetric with respect to the origin, so f is odd.

2. The domain and the range are the set of all real numbers.

3. The graph always contains the points $(0, 0)$, $(1, 1)$, and $(-1, -1)$.

4. As the exponent n increases in magnitude, the graph becomes more vertical when $x < -1$ or $x > 1$, but for x near the origin, the graph tends to flatten out and lie closer to the x-axis.

The methods of shifting, compression, stretching, and reflection studied in Section 2.7, when used with the facts just presented, will enable us to graph polynomial functions that are transformations of power functions.

EXAMPLE 2	**Graphing Polynomial Functions Using Transformations**

Graph: $f(x) = 1 - x^5$

Solution Figure 25 shows the required stages.

Figure 25

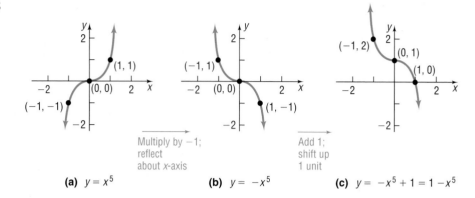

(a) $y = x^5$

(b) $y = -x^5$

(c) $y = -x^5 + 1 = 1 - x^5$

✔ **CHECK:** Verify the result of Example 2 by graphing $Y_1 = f(x) = 1 - x^5$. ◀

EXAMPLE 3 **Graphing Polynomial Functions Using Transformations**

Graph: $f(x) = \dfrac{1}{2}(x - 1)^4$

Solution Figure 26 shows the required stages.

Figure 26

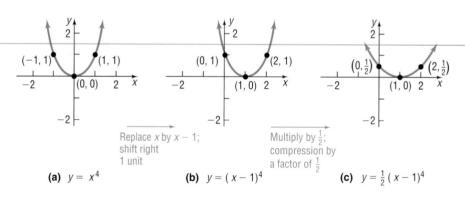

(a) $y = x^4$

(b) $y = (x - 1)^4$

(c) $y = \dfrac{1}{2}(x - 1)^4$

✔ **CHECK:** Verify the result of Example 3 by graphing

$$Y_1 = f(x) = \frac{1}{2}(x - 1)^4$$

◀

NOW WORK PROBLEMS 23 AND 27.

3 Identify the Zeros of a Polynomial Function and Their Multiplicity

Figure 27 shows the graph of a polynomial function with four *x*-intercepts. Notice that at the *x*-intercepts the graph must either cross the *x*-axis or touch the *x*-axis. Consequently, between consecutive *x*-intercepts the graph is either above the *x*-axis or below the *x*-axis. We will make use of this property of the graph of a polynomial shortly.

Figure 27

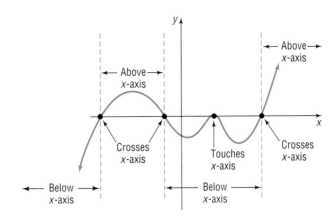

If a polynomial function f is factored completely, it is easy to solve the equation $f(x) = 0$ and locate the x-intercepts of the graph. For example, if $f(x) = (x - 1)^2(x + 3)$, then the solutions of the equation

$$f(x) = (x - 1)^2(x + 3) = 0$$

are identified as 1 and -3. Based on this result, we make the following observations:

If f is a polynomial function and r is a real number for which $f(r) = 0$, then r is called a (real) **zero of f**, or **root of f**. If r is a (real) zero of f, then

(a) r is an x-intercept of the graph of f.

(b) $(x - r)$ is a factor of f.

So the real zeros of a function are the x-intercepts of its graph, and they are found by solving the equation $f(x) = 0$.

| EXAMPLE 4 | **Finding a Polynomial from Its Zeros** |

(a) Find a polynomial of degree 3 whose zeros are $-3, 2,$ and 5.

(b) Graph the polynomial found in part (a) to verify your result.

Solution   (a) If r is a zero of a polynomial f, then $x - r$ is a factor of f. This means that $x - (-3) = x + 3, x - 2,$ and $x - 5$ are factors of f. As a result, any polynomial of the form

$$f(x) = a(x + 3)(x - 2)(x - 5)$$

where a is any nonzero real number, qualifies.

(b) The value of a causes a stretch, compression, or reflection, but does not affect the x-intercepts. We choose to graph f with $a = 1$.

$$f(x) = (x + 3)(x - 2)(x - 5) = x^3 - 4x^2 - 11x + 30$$

Figure 28 shows the graph of f. Notice that the x-intercepts are $-3, 2,$ and 5. ◀

Figure 28

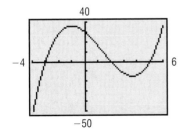

— Seeing the Concept —

Graph the function found in Example 4 for $a = 2$ and $a = -1$. Does the value of a affect the zeros of f? How does the value of a affect the graph of f?

NOW WORK PROBLEM **37**.

If the same factor $x - r$ occurs more than once, then r is called a **repeated**, or **multiple, zero of f**. More precisely, we have the following definition.

If $(x - r)^m$ is a factor of a polynomial f and $(x - r)^{m+1}$ is not a factor of f, then r is called a **zero of multiplicity m of f**.

EXAMPLE 5	Identifying Zeros and Their Multiplicities

For the polynomial

$$f(x) = 5(x - 2)(x + 3)^2\left(x - \frac{1}{2}\right)^4$$

2 is a zero of multiplicity 1 because the exponent on the factor $x - 2$ is 1.

-3 is a zero of multiplicity 2 because the exponent on the factor $x + 3$ is 2.

$\dfrac{1}{2}$ is a zero of multiplicity 4 because the exponent on the factor $x - \frac{1}{2}$ is 4. ◀

NOW WORK PROBLEM 45(a).

In Example 5 notice that, if you add the multiplicities $(1 + 2 + 4 = 7)$, you obtain the degree of the polynomial.

Suppose that it is possible to factor completely a polynomial function and, as a result, locate all the x-intercepts of its graph (the real zeros of the function). The following example illustrates the role that the multiplicity of an x-intercept plays.

EXAMPLE 6	Investigating the Role of Multiplicity

For the polynomial $f(x) = x^2(x - 2)$:

(a) Find the x- and y-intercepts of the graph of f.

(b) Using a graphing utility, graph the polynomial.

(c) For each x-intercept, determine whether it is of odd or even multiplicity.

Solution

(a) The y-intercept is $f(0) = 0^2(0 - 2) = 0$. The x-intercepts satisfy the equation

$$f(x) = x^2(x - 2) = 0$$

from which we find that

$$x^2 = 0 \quad \text{or} \quad x - 2 = 0$$
$$x = 0 \quad \text{or} \quad x = 2$$

The x-intercepts are 0 and 2.

(b) See Figure 29 for the graph of f.

(c) We can see from the factored form of f that 0 is a zero or root of multiplicity 2, and 2 is a zero or root of multiplicity 1; so 0 is of even multiplicity and 2 is of odd multiplicity. ◀

We can use a TABLE to further analyze the graph. See Table 6. The sign of $f(x)$ is the same on each side of $x = 0$ and the graph of f just touches the x-axis at $x = 0$ (a zero of even multiplicity). The sign of $f(x)$ changes from one side of $x = 2$ to the other and the graph of f crosses the x-axis at $x = 2$ (a zero of odd multiplicity). These observations suggest the following result:

Figure 29

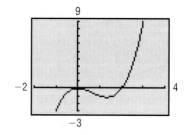

Table 6

X	Y1
-2	-16
-1	-3
0	0
1	-1
2	0
3	9
4	32

$Y_1 \boxminus X^2(X-2)$

If r Is a Zero of Even Multiplicity

Sign of $f(x)$ does not change from one side
of r to the other side of r.

Graph **touches**
x-axis at r.

If r Is a Zero of Odd Multiplicity

Sign of $f(x)$ changes from one side
of r to the other side of r.

Graph **crosses**
x-axis at r.

✎ NOW WORK PROBLEM **45(b)**.

Figure 30

Points on the graph where the graph changes from an increasing function to a decreasing function, or vice versa, are called **turning points**.*

Look at Figure 30. The graph of $f(x) = x^2(x - 2) = x^3 - 2x^2$, has a turning point at $(0, 0)$. After utilizing MINIMUM, we find that the graph also has a turning point at $(1.33, -1.19)$, rounded to two decimal places.

———— Exploration ————————————————————

Graph $Y_1 = x^3$, $Y_2 = x^3 - x$, and $Y_3 = x^3 + 3x^2 + 4$. How many turning points do you see? How does the number of turning points relate to the degree? Graph $Y_1 = x^4$, $Y_2 = x^4 - \frac{4}{3}x^3$, and $Y_3 = x^4 - 2x^2$. How many turning points do you see? How does the number of turning points compare to the degree?

The following theorem from calculus supplies the answer.

Theorem

If f is a polynomial function of degree n, then f has at most $n - 1$ turning points.

One last remark about Figure 30. Notice that the graph of $f(x) = x^2(x - 2)$ looks somewhat like the graph of $y = x^3$. In fact, for very large values of x, either positive or negative, there is little difference.

———— Exploration ————————————————————

For each pair of functions Y_1 and Y_2 given in parts (a), (b), and (c), graph Y_1 and Y_2 on the same viewing window. Create a TABLE or TRACE for large positive and large negative values of x. What do you notice about the graphs of Y_1 and Y_2 as x becomes very large and positive or very large and negative?

(a) $Y_1 = x^2(x - 2)$; $Y_2 = x^3$

(b) $Y_1 = x^4 - 3x^3 + 7x - 3$; $Y_2 = x^4$

(c) $Y_1 = -2x^3 + 4x^2 - 8x + 10$; $Y_2 = -2x^3$

---

*Graphing utilities can be used to approximate turning points. Calculus is needed to find the exact location of turning points.

The behavior of the graph of a function for large values of x, either positive or negative, is referred to as its **end behavior**.

Theorem

End Behavior

For large values of x, either positive or negative, that is, for large $|x|$, the graph of the polynomial

$$f(x) = a_n x^n + a_{n-1} x^{n-1} + \cdots + a_1 x + a_0$$

resembles the graph of the power function

$$y = a_n x^n$$

Look back at Figures 23 and 24. Based on this theorem and the previous discussion on power functions, the end behavior of a polynomial can be of only four types. See Figure 31.

Figure 31
End Behavior

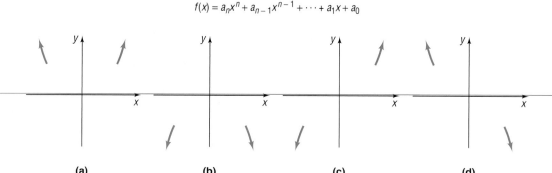

$$f(x) = a_n x^n + a_{n-1} x^{n-1} + \cdots + a_1 x + a_0$$

(a)	(b)	(c)	(d)
$n \geq 2$ even; $a_n > 0$	$n \geq 2$ even; $a_n < 0$	$n \geq 3$ odd; $a_n > 0$	$n \geq 3$ odd; $a_n < 0$

For example, consider the polynomial function $f(x) = -2x^4 + x^3 + 4x^2 - 7x + 1$. The graph of f will resemble the graph of the power function $y = -2x^4$ for large $|x|$. The graph of f will look like Figure 31(b) for large $|x|$.

NOW WORK PROBLEM **45(c)**.

Summary

Graph of a Polynomial Function $f(x) = a_n x^n + a_{n-1} x^{n-1} + \cdots + a_1 x + a_0,$ $a_n \neq 0$

Degree of the polynomial f: n

Maximum number of turning points: $n - 1$

At a zero of even multiplicity: The graph of f touches the x-axis.

At a zero of odd multiplicity: The graph of f crosses the x-axis.

Between zeros, the graph of f is either above or below the x-axis.

End behavior: For large $|x|$, the graph of f behaves like the graph of $y = a_n x^n$.

4   Analyze the Graph of a Polynomial Function

| EXAMPLE 7 | **Analyzing the Graph of a Polynomial Function** |

For the polynomial: $f(x) = x^4 - 2x^3 - 8x^2$

(a) Find the degree of the polynomial. Determine the end behavior; that is, find the power function that the graph of f resembles for large values of $|x|$.

(b) Find the x- and y-intercepts of the graph of f.

(c) Determine whether the graph crosses or touches the x-axis at each x-intercept.

(d) Use a graphing utility to graph f.

(e) Determine the number of turning points on the graph of f. Approximate the turning points, if any exist, rounded to two decimal places.

(f) Use the information obtained in parts (a) to (e) to draw a complete graph of f by hand.

(g) Find the domain of f. Use the graph to find the range of f.

(h) Use the graph to determine where f is increasing and where f is decreasing.

Solution   (a) The polynomial is of degree 4. End behavior: the graph of f resembles that of the power function $y = x^4$ for large values of $|x|$.

(b) The y-intercept is $f(0) = 0$. To find the x-intercepts, if any, we first factor f.

$$f(x) = x^4 - 2x^3 - 8x^2 = x^2(x^2 - 2x - 8) = x^2(x - 4)(x + 2)$$

We find the x-intercepts by solving the equation

$$f(x) = x^2(x - 4)(x + 2) = 0$$

So

$$x^2 = 0 \quad \text{or} \quad x - 4 = 0 \quad \text{or} \quad x + 2 = 0$$
$$x = 0 \quad \text{or} \qquad x = 4 \quad \text{or} \qquad x = -2$$

The x-intercepts are $-2, 0,$ and 4.

(c) The x-intercept 0 is a zero of multiplicity 2, so the graph of f will touch the x-axis at 0; -2 and 4 are zeros of multiplicity 1, so the graph of f will cross the x-axis at -2 and 4.

(d) See Figure 32 for the graph of f.

(e) From the graph of f shown in Figure 32, we can see that f has three turning points. Using MAXIMUM, one turning point is at $(0, 0)$. Using MINIMUM, the two remaining turning points are at $(-1.39, -6.35)$ and $(2.89, -45.33)$, rounded to two decimal places.

(f) Figure 33 shows a graph of f using the information obtained in parts (a) to (e).

(g) The domain of f is the set of all real numbers. The range of f is the interval $[-45.33, \infty)$.

(h) Based on the graph, f is decreasing on the intervals $(-\infty, -1.39)$ and $(0, 2.89)$ and is increasing on the intervals $(-1.39, 0)$ and $(2.89, \infty)$.   ◀

NOW WORK PROBLEM 69.

For polynomial functions that have noninteger coefficients and for polynomials that are not easily factored, we utilize the graphing utility early in the analysis of the graph. This is because the amount of information that can be obtained from algebraic analysis is limited.

Figure 32

Figure 33

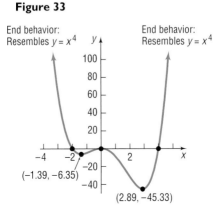

End behavior: Resembles $y = x^4$
End behavior: Resembles $y = x^4$

$(-1.39, -6.35)$

$(2.89, -45.33)$

EXAMPLE 8	Using a Graphing Utility to Analyze the Graph of a Polynomial Function

For the polynomial $f(x) = x^3 + 2.48x^2 - 4.3155x + 2.484406$:

(a) Find the degree of the polynomial. Determine the end behavior: that is, find the power function that the graph of f resembles for large values of $|x|$.

(b) Graph f using a graphing utility.

(c) Find the x- and y-intercepts of the graph.

(d) Use a TABLE to find points on the graph around each x-intercept.

(e) Determine the local maxima and local minima, if any exist, rounded to two decimal places. That is, locate any turning points.

(f) Use the information obtained in parts (a)–(e) to draw a complete graph of f by hand. Be sure to label the intercepts, turning points, and the points obtained in part (d).

(g) Find the domain of f. Use the graph to find the range of f.

(h) Use the graph to determine where f is increasing and where f is decreasing.

Solution

(a) The degree of the polynomial is 3. End behavior: the graph of f resembles that of the power function $y = x^3$ for large values of $|x|$.

Figure 34

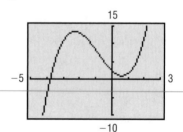

(b) See Figure 34 for the graph of f.

(c) The y-intercept is $f(0) = 2.484406$. In Example 7 we could easily factor $f(x)$ to find the x-intercepts. However, it is not readily apparent how $f(x)$ factors in this example. Therefore, we use a graphing utility's ZERO (or ROOT) feature and find the lone x-intercept to be -3.79, rounded to two decimal places.

(d) Table 7 shows values of x around the x-intercept. The points $(-4, -4.57)$ and $(-2, 13.04)$ are on the graph.

Table 7

X	Y1
-4	-4.574
-2	13.035

Y1◼X^3+2.48X²-4...

(e) From the graph we see that it has two turning points: one between -3 and -2 the other between 0 and 1. Rounded to two decimal places, the local maximum is 13.36 and occurs at $x = -2.28$; the local minimum is 1 and occurs at $x = 0.63$. The turning points are $(-2.28, 13.36)$ and $(0.63, 1)$.

(f) Figure 35 shows a graph of f drawn by hand using the information obtained in parts (a) to (e).

Figure 35

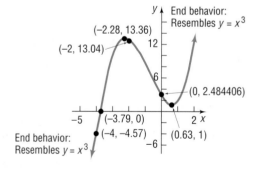

(g) The domain and the range of f are the set of all real numbers.

(h) Based on the graph, f is decreasing on the interval $(-2.28, 0.63)$ and is increasing on the intervals $(-\infty, -2.28)$ and $(0.63, \infty)$. ◄

NOW WORK PROBLEM 81.

Summary

Steps for Graphing a Polynomial by Hand

To analyze the graph of a polynomial function $y = f(x)$, follow these steps:

STEP 1: End behavior: find the power function that the graph of f resembles for large values of x.

STEP 2: (a) Find the x-intercepts, if any, by solving the equation $f(x) = 0$.
(b) Find the y-intercept by letting $x = 0$ and finding the value of $f(0)$.

STEP 3: Determine whether the graph of f crosses or touches the x-axis at each x-intercept.

STEP 4: Use a graphing utility to graph f. Determine the number of turning points on the graph of f. Approximate any turning points rounded to two decimal places.

STEP 5: Use the information obtained in Steps 1 to 4 to draw a complete graph of f by hand.

5 Find the Cubic Function of Best Fit to Data

In Section 2.5 we found the line of best fit from data and in Section 3.1 we found the quadratic function of best fit. It is also possible to find polynomial functions of best fit. However, most statisticians do not recommend finding polynomials of best fit of degree higher than 3.

Data that follow a cubic relation should look like Figure 36(a) or (b).

Figure 36

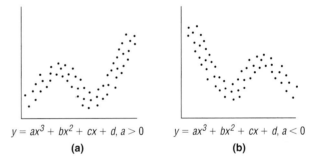

$$y = ax^3 + bx^2 + cx + d, a > 0$$
(a)

$$y = ax^3 + bx^2 + cx + d, a < 0$$
(b)

| **EXAMPLE 9** | **A Cubic Function of Best Fit** |

Table 8

The data in Table 8 represent the average miles per gallon of vans, pickups, and sport utility vehicles (SUVs) in the United States for 1991–2001, where 1 represents 1991, 2 represents 1992, and so on.

(a) Draw a scatter diagram of the data using the year x as the independent variable and average miles per gallon M as the dependent variable. Comment on the type of relation that may exist between the two variables x and M.

(b) Using a graphing utility, find the cubic function of best fit $M = M(x)$ to these data.

(c) Graph the cubic function of best fit on your scatter diagram.

(d) Use the function found in part (b) to predict the average miles per gallon in 2002 $(x = 12)$.

Year, x	Average Miles per Gallon, M
1991, 1	17.0
1992, 2	17.3
1993, 3	17.4
1994, 4	17.3
1995, 5	17.3
1996, 6	17.2
1997, 7	17.2
1998, 8	17.2
1999, 9	17.0
2000, 10	17.4
2001, 11	17.6

SOURCE: United States Federal Highway Administration

Solution   (a) Figure 37 shows the scatter diagram. A cubic relation may exist between the two variables.

(b) Upon executing the CUBIC REGression program, we obtain the results shown in Figure 38. The output the utility provides shows us the equation $y = ax^3 + bx^2 + cx + d$. The cubic function of best fit to the data is $M(x) = 0.0058x^3 - 0.1007x^2 + 0.4950x + 16.6258$.

(c) Figure 39 shows the graph of the cubic function of best fit on the scatter diagram. The function fits the data reasonably well.

Figure 37

Figure 38

Figure 39

(d) We evaluate the function $M(x)$ at $x = 12$.
$$M(12) = 0.0058(12)^3 - 0.1007(12)^2 + 0.4950(12) + 16.6258 \approx 18.1$$

The model predicts that the average miles per gallon of vans, pickups, and SUVs will be 18.1 miles per gallon in 2002.   ◀

3.2 Assess Your Understanding

'Are You Prepared?'

Answers are given at the end of these exercises. If you get a wrong answer, read the pages listed in red.

1. The intercepts of the equation $9x^2 + 4y = 36$ are _____. (pp. 171–172 and 183)

2. *True or False:* The expression $4x^3 - 3.6x^2 - \sqrt{2}$ is a polynomial. (pp. 36–37)

3. To graph $y = x^2 - 4$, you would shift the graph of $y = x^2$ _____ a distance of _____ units. (pp. 266–268)

4. Use a graphing utility to approximate (rounded to two decimal places) the local maxima and local minima of $f(x) = x^3 - 2x^2 - 4x + 5$, for $-10 < x < 10$ (pp. 232–233)

Concepts and Vocabulary

5. The graph of every polynomial function is both _____ and _____.

6. A number r for which $f(r) = 0$ is called a(n) _____ of the function f.

7. If r is a zero of even multiplicity for a function f, the graph of f _____ the x-axis at r.

8. *True or False:* The graph of $f(x) = x^2(x - 3)(x + 4)$ has exactly three x-intercepts.

9. *True or False:* The x-intercepts of the graph of a polynomial function are called turning points.

10. *True or False:* End behavior: the graph of the function $f(x) = 3x^4 - 6x^2 + 2x + 5$ resembles $y = x^4$ for large values of $|x|$.

Skill Building

In Problems 11–22, determine which functions are polynomial functions. For those that are, state the degree. For those that are not, tell why not.

11. $f(x) = 4x + x^3$

12. $f(x) = 5x^2 + 4x^4$

13. $g(x) = \dfrac{1 - x^2}{2}$

14. $h(x) = 3 - \dfrac{1}{2}x$

15. $f(x) = 1 - \dfrac{1}{x}$

16. $f(x) = x(x - 1)$

17. $g(x) = x^{3/2} - x^2 + 2$

18. $h(x) = \sqrt{x}\,(\sqrt{x} - 1)$

19. $F(x) = 5x^4 - \pi x^3 + \dfrac{1}{2}$

20. $F(x) = \dfrac{x^2 - 5}{x^3}$

21. $G(x) = 2(x - 1)^2(x^2 + 1)$

22. $G(x) = -3x^2(x + 2)^3$

In Problems 23–36, use transformations of the graph of $y = x^4$ or $y = x^5$ to graph each function.

23. $f(x) = (x + 1)^4$

24. $f(x) = (x - 2)^5$

25. $f(x) = x^5 - 3$

26. $f(x) = x^4 + 2$

27. $f(x) = \dfrac{1}{2}x^4$

28. $f(x) = 3x^5$

29. $f(x) = -x^5$

30. $f(x) = -x^4$

31. $f(x) = (x - 1)^5 + 2$

32. $f(x) = (x + 2)^4 - 3$

33. $f(x) = 2(x + 1)^4 + 1$

34. $f(x) = \dfrac{1}{2}(x - 1)^5 - 2$

35. $f(x) = 4 - (x - 2)^5$

36. $f(x) = 3 - (x + 2)^4$

In Problems 37–44, form a polynomial whose zeros and degree are given.

37. Zeros: $-1, 1, 3;$ degree 3

38. Zeros: $-2, 2, 3;$ degree 3

39. Zeros: $-3, 0, 4;$ degree 3

40. Zeros: $-4, 0, 2;$ degree 3

41. Zeros: $-4, -1, 2, 3;$ degree 4

42. Zeros: $-3, -1, 2, 5;$ degree 4

43. Zeros: -1, multiplicity 1;   3, multiplicity 2;   degree 3

44. Zeros: -2, multiplicity 2;   4, multiplicity 1;   degree 3

In Problems 45–56, for each polynomial function: (a) List each real zero and its multiplicity; (b) Determine whether the graph crosses or touches the x-axis at each x-intercept. (c) Find the power function that the graph of f resembles for large values of $|x|$.

45. $f(x) = 3(x - 7)(x + 3)^2$

46. $f(x) = 4(x + 4)(x + 3)^3$

47. $f(x) = 4(x^2 + 1)(x - 2)^3$

48. $f(x) = 2(x - 3)(x + 4)^3$

49. $f(x) = -2\left(x + \dfrac{1}{2}\right)^2(x^2 + 4)^2$

50. $f(x) = \left(x - \dfrac{1}{3}\right)^2(x - 1)^3$

51. $f(x) = (x - 5)^3(x + 4)^2$

52. $f(x) = \left(x + \sqrt{3}\right)^2(x - 2)^4$

53. $f(x) = 3(x^2 + 8)(x^2 + 9)^2$

54. $f(x) = -2(x^2 + 3)^3$

55. $f(x) = -2x^2(x^2 - 2)$

56. $f(x) = 4x(x^2 - 3)$

In Problems 57–80, for each polynomial function f:
  (a) *Find the degree of the polynomial. Determine the end behavior: that is, find the power function that the graph of f resembles for large values of $|x|$.*
  (b) *Find the x- and y-intercepts of the graph.*
  (c) *Determine whether the graph crosses or touches the x-axis at each x-intercept.*
  (d) *Graph f using a graphing utility.*
  (e) *Determine the local maxima and local minima, if any exist, rounded to two decimal places. That is, locate any turning points.*
  (f) *Use the information obtained in parts (a)–(e) to draw a complete graph of f by hand. Be sure to label the intercepts and turning points.*
  (g) *Find the domain of f. Use the graph to find the range of f.*
  (h) *Use the graph to determine where f is increasing and where f is decreasing.*

57. $f(x) = (x - 1)^2$

58. $f(x) = (x - 2)^3$

59. $f(x) = x^2(x - 3)$

60. $f(x) = x(x + 2)^2$

61. $f(x) = (x + 4)(x - 2)^2$

62. $f(x) = (x - 1)(x + 3)^2$

63. $f(x) = -2(x + 2)(x - 2)^3$

64. $f(x) = -\dfrac{1}{2}(x + 4)(x - 1)^3$

65. $f(x) = (x + 1)(x - 2)(x + 4)$

66. $f(x) = (x - 1)(x + 4)(x - 3)$

67. $f(x) = 4x - x^3$

68. $f(x) = x - x^3$

69. $f(x) = x^2(x - 2)(x + 2)$

70. $f(x) = x^2(x - 3)(x + 4)$

71. $f(x) = (x + 1)^2(x - 2)^2$

72. $f(x) = (x + 1)^3(x - 3)$

73. $f(x) = x^2(x - 3)(x + 1)$

74. $f(x) = x^2(x - 3)(x - 1)$

75. $f(x) = (x + 2)^2(x - 4)^2$

76. $f(x) = (x - 2)^2(x + 2)(x + 4)$

77. $f(x) = x^2(x - 2)(x^2 + 3)$

78. $f(x) = x^2(x^2 + 1)(x + 4)$

79. $f(x) = -x^2(x^2 - 1)(x + 1)$

80. $f(x) = -x^2(x^2 - 4)(x - 5)$

In Problems 81–90, for each polynomial function f:

(a) *Find the degree of the polynomial. Determine the end behavior: that is, find the power function that the graph of f resembles for large values of |x|.*

(b) *Graph f using a graphing utility.*

(c) *Find the x- and y-intercepts of the graph.*

(d) *Use a TABLE to find points on the graph around each x-intercept.*

(e) *Determine the local maxima and local minima, if any exist, rounded to two decimal places. That is, locate any turning points.*

(f) *Use the information obtained in parts (a)–(e) to draw a complete graph of f by hand. Be sure to label the intercepts, turning points, and the points obtained in part (d).*

(g) *Find the domain of f. Use the graph to find the range of f.*

(h) *Use the graph to determine where f is increasing and where f is decreasing.*

81. $f(x) = x^3 + 0.2x^2 - 1.5876x - 0.31752$ **82.** $f(x) = x^3 - 0.8x^2 - 4.6656x + 3.73248$

83. $f(x) = x^3 + 2.56x^2 - 3.31x + 0.89$ **84.** $f(x) = x^3 - 2.91x^2 - 7.668x - 3.8151$

85. $f(x) = x^4 - 2.5x^2 + 0.5625$ **86.** $f(x) = x^4 - 18.5x^2 + 50.2619$

87. $f(x) = 2x^4 - \pi x^3 + \sqrt{5}x - 4$ **88.** $f(x) = -1.2x^4 + 0.5x^2 - \sqrt{3}x + 2$

89. $f(x) = -2x^5 - \sqrt{2}x^2 - x - \sqrt{2}$ **90.** $f(x) = \pi x^5 + \pi x^4 + \sqrt{3}x + 1$

Applications and Extensions

In Problems 91–94, construct a polynomial function that might have this graph. (More than one answer may be possible.)

91.

92.

93.

94.

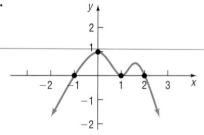

95. Cost of Manufacturing The following data represent the cost C (in thousands of dollars) of manufacturing Chevy Cavaliers and the number x of Cavaliers produced.

Number of Cavaliers Produced, x	Cost, C
0	10
1	23
2	31
3	38
4	43
5	50
6	59
7	70
8	85
9	105
10	135

(a) Draw a scatter diagram of the data using x as the independent variable and C as the dependent variable. Comment on the type of relation that may exist between the two variables C and x.

(b) Find the average rate of change of cost from four to five Cavaliers.

(c) What is the average rate of change in cost from eight to nine Cavaliers?

(d) Use a graphing utility to find the cubic function of best fit C = C(x).

(e) Graph the cubic function of best fit on the scatter diagram.

(f) Use the function found in part (d) to predict the cost of manufacturing 11 Cavaliers.

(g) Interpret the y-intercept.

96. Cost of Printing The following data represent the weekly cost C (in thousands of dollars) of printing textbooks and the number x (in thousands of units) of texts printed.

Number of Text Books, x	Cost, C
0	100
5	128.1
10	144
13	153.5
17	161.2
18	162.6
20	166.3
23	178.9
25	190.2
27	221.8

(a) Draw a scatter diagram of the data using x as the independent variable and C as the dependent variable. Comment on the type of relation that may exist between the two variables C and x.
(b) Find the average rate of change in cost from 10,000 to 13,000 textbooks.
(c) What is the average rate of change in the cost of producing from 18,000 to 20,000 textbooks?
(d) Use a graphing utility to find the cubic function of best fit $C = C(x)$.
(e) Graph the cubic function of best fit on the scatter diagram.
(f) Use the function found in part (d) to predict the cost of printing 22,000 texts per week.
(g) Interpret the y-intercept.

97. **Motor Vehicle Thefts** The following data represent the number T (in thousands) of motor vehicle thefts in the United States for the years 1989–2001, where $x = 1$ represents 1989, $x = 2$ represents 1990, and so on.

Year, x	Motor Vehicle Thefts, T
1989, 1	1565
1990, 2	1636
1991, 3	1662
1992, 4	1611
1993, 5	1563
1994, 6	1539
1995, 7	1472
1996, 8	1394
1997, 9	1354
1998, 10	1243
1999, 11	1152
2000, 12	1160
2001, 13	1226

Source: U.S. Federal Bureau of Investigation

(a) Draw a scatter diagram of the data using x as the independent variable and T as the dependent variable. Comment on the type of relation that may exist between the two variables T and x.
(b) Use a graphing utility to find the cubic function of best fit $T = T(x)$.
(c) Graph the cubic function of best fit on the scatter diagram.
(d) Use the function found in part (b) to predict the number of motor vehicle thefts in 2005.
(e) Do you think the function found in part (b) will be useful in predicting motor vehicle thefts in 2010? Why?

98. **Average Annual Miles** The following data represent the average number of miles driven (in thousands) annually by vans, pickups, and sport utility vehicles (SUVs) for the years 1993–2001, where $x = 1$ represents 1993, $x = 2$ represents 1994, and so on.

Year, x	Average Miles Driven, M
1993, 1	12.4
1994, 2	12.2
1995, 3	12.0
1996, 4	11.8
1997, 5	12.1
1998, 6	12.2
1999, 7	12.0
2000, 8	11.7
2001, 9	11.1

Source: United States Federal Highway Administration

(a) Draw a scatter diagram of the data using x as the independent variable and M as the dependent variable. Comment on the type of relation that may exist between the two variables M and x.
(b) Use a graphing utility to find the cubic function of best fit $M = M(x)$.
(c) Graph the cubic function of best fit on the scatter diagram.
(d) Use the function found in part (b) to predict the average miles driven by vans, pickups, and SUVs in 2005.
(e) Do you think the function found in part (b) will be useful in predicting the average miles driven by vans, pickups, and SUVs in 2010? Why?

Discussion and Writing

99. Can the graph of a polynomial function have no *y*-intercept? Can it have no *x*-intercepts? Explain.

100. Write a few paragraphs that provide a general strategy for graphing a polynomial function. Be sure to mention the following: degree, intercepts, end behavior, and turning points.

101. Make up a polynomial that has the following characteristics: crosses the *x*-axis at −1 and 4, touches the *x*-axis at 0 and 2, and is above the *x*-axis between 0 and 2. Give your polynomial to a fellow classmate and ask for a written critique of your polynomial.

102. Make up two polynomials, not of the same degree, with the following characteristics: crosses the *x*-axis at −2, touches the *x*-axis at 1, and is above the *x*-axis between −2 and 1. Give your polynomials to a fellow classmate and ask for a written critique of your polynomials.

103. The graph of a polynomial function is always smooth and continuous. Name a function studied earlier that is smooth and not continuous. Name one that is continuous, but not smooth.

104. Which of the following statements are true regarding the graph of the cubic polynomial $f(x) = x^3 + bx^2 + cx + d$? (Give reasons for your conclusions.)
(a) It intersects the *y*-axis in one and only one point.
(b) It intersects the *x*-axis in at most three points.
(c) It intersects the *x*-axis at least once.
(d) For $|x|$ very large, it behaves like the graph of $y = x^3$.
(e) It is symmetric with respect to the origin.
(f) It passes through the origin.

'Are You Prepared?' Answers

1. $(-2, 0), (2, 0), (0, 9)$ **2.** True   **3.** down; 4   **4.** Local maximum of 6.48 at $x = -0.67$; local minimum of -3 at $x = 2$

3.3 Properties of Rational Functions

PREPARING FOR THIS SECTION *Before getting started, review the following:*

- Rational Expressions (Section R.7, pp. 57–65)
- Polynomial Division (Section R.4, pp. 40–43)

- Graph of $f(x) = \dfrac{1}{x}$ (Section 2.1, Example 5, p. 202)
- Graphing Techniques: Transformations (Section 2.7, pp. 266–274)

Now work the 'Are You Prepared?' problems on page 343.

OBJECTIVES   1 Find the Domain of a Rational Function
2 Find the Vertical Asymptotes of a Rational Function
3 Find the Horizontal or Oblique Asymptotes of a Rational Function

Ratios of integers are called *rational numbers*. Similarly, ratios of polynomial functions are called *rational functions*.

A **rational function** is a function of the form

$$R(x) = \frac{p(x)}{q(x)}$$

where *p* and *q* are polynomial functions and *q* is not the zero polynomial. The domain is the set of all real numbers except those for which the denominator *q* is 0.

1   Find the Domain of a Rational Function

EXAMPLE 1

Finding the Domain of a Rational Function

(a) The domain of $R(x) = \dfrac{2x^2 - 4}{x + 5}$ is the set of all real numbers x except -5; that is, $\{x | x \neq -5\}$.

(b) The domain of $R(x) = \dfrac{1}{x^2 - 4}$ is the set of all real numbers x except -2 and 2, that is, $\{x | x \neq -2, x \neq 2\}$.

(c) The domain of $R(x) = \dfrac{x^3}{x^2 + 1}$ is the set of all real numbers.

(d) The domain of $R(x) = \dfrac{-x^2 + 2}{3}$ is the set of all real numbers.

(e) The domain of $R(x) = \dfrac{x^2 - 1}{x - 1}$ is the set of all real numbers x except 1, that is, $\{x | x \neq 1\}$.   ◄

It is important to observe that the functions

$$R(x) = \frac{x^2 - 1}{x - 1} \quad \text{and} \quad f(x) = x + 1$$

are not equal, since the domain of R is $\{x | x \neq 1\}$ and the domain of f is the set of all real numbers.

━━━━━ **NOW WORK PROBLEM 13.**

If $R(x) = \dfrac{p(x)}{q(x)}$ is a rational function and if p and q have no common factors, then the rational function R is said to be in **lowest terms**. For a rational function $R(x) = \dfrac{p(x)}{q(x)}$ in lowest terms, the zeros, if any, of the numerator are the x-intercepts of the graph of R and so will play a major role in the graph of R. The zeros of the denominator of R [that is, the numbers x, if any, for which $q(x) = 0$], although not in the domain of R, also play a major role in the graph of R. We will discuss this role shortly.

We have already discussed the properties of the rational function $f(x) = \dfrac{1}{x}$. (Refer to Example 5, page 202.) The next rational function that we take up is $H(x) = \dfrac{1}{x^2}$.

EXAMPLE 2

Graphing $y = \dfrac{1}{x^2}$

Analyze the graph of $H(x) = \dfrac{1}{x^2}$.

Solution

The domain of $H(x) = \dfrac{1}{x^2}$ consists of all real numbers x except 0. The graph has no y-intercept, because x can never equal 0. The graph has no x-intercept because the equation $H(x) = 0$ has no solution. Therefore, the graph of H will not cross either coordinate axis.

Because

$$H(-x) = \frac{1}{(-x)^2} = \frac{1}{x^2} = H(x)$$

H is an even function, so its graph is symmetric with respect to the y-axis.

See Figure 40. Notice that the graph confirms the conclusions just reached. But what happens to the graph as the values of x get closer and closer to 0? We use a TABLE to answer the question. See Table 9. The first four rows show that as x approaches 0 the values of $H(x)$ become larger and larger positive numbers. When this happens, we say that $H(x)$ is **unbounded in the positive direction**. We symbolize this by writing $H(x) \rightarrow \infty$ [read as "$H(x)$ **approaches infinity**"]. In calculus, **limits** are used to convey these ideas. There we use the symbolism $\lim\limits_{x \to 0} H(x) = \infty$, read as "the limit of $H(x)$ as x approaches 0 is infinity" to mean that $H(x) \rightarrow \infty$ as $x \rightarrow 0$.

Look at the last three rows of Table 9. As $x \rightarrow \infty$, the values of $H(x)$ approach 0 (the end behavior of the graph). This is symbolized in calculus by writing $\lim\limits_{x \to \infty} H(x) = 0$.

Figure 41 shows the graph of $H(x) = \dfrac{1}{x^2}$ drawn by hand.

Figure 40

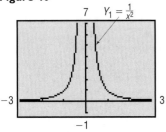

Table 9

X	Y1
.1	100
.01	10000
.001	1E6
1E-4	1E8
10	.01
100	1E-4
1000	1E-6

Y1◻1/X²

Figure 41

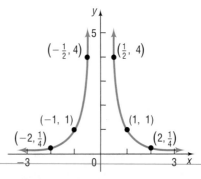

Sometimes transformations (shifting, compressing, stretching, and reflection) can be used to graph a rational function.

EXAMPLE 3   **Using Transformations to Graph a Rational Function**

Graph the rational function: $R(x) = \dfrac{1}{(x - 2)^2} + 1$

Solution   First, notice that the domain of R is the set of all real numbers except $x = 2$. To graph R, we start with the graph of $y = \dfrac{1}{x^2}$. See Figure 42 for the stages.

Figure 42

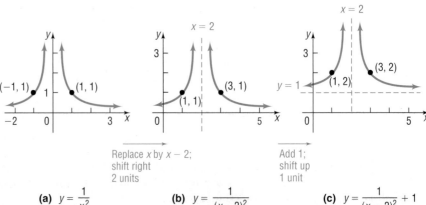

Replace x by $x - 2$;
shift right
2 units

Add 1;
shift up
1 unit

(a) $y = \dfrac{1}{x^2}$

(b) $y = \dfrac{1}{(x - 2)^2}$

(c) $y = \dfrac{1}{(x - 2)^2} + 1$

✔ **Check:** Graph $Y_1 = \dfrac{1}{(x-2)^2} + 1$ using a graphing utility to verify the graph obtained in Figure 42(c). ◀

N O W   W O R K   P R O B L E M   **31.**

Asymptotes

Notice that the y-axis in Figure 42(a) is transformed into the vertical line $x = 2$ in Figure 42(c), and the x-axis in Figure 42(a) is transformed into the horizontal line $y = 1$ in Figure 42(c). The **Exploration** that follows will help us analyze the role of these lines.

—— *Exploration* ——

(a) Using a graphing utility and the TABLE feature, evaluate the function $H(x) = \dfrac{1}{(x-2)^2} + 1$ at
    $x = 10, 100, 1000,$ and $10{,}000$. What happens to the values of H as x becomes unbounded in
    the positive direction, symbolized by $\lim\limits_{x\to\infty} H(x)$?

(b) Evaluate H at $x = -10, -100, -1000,$ and $-10{,}000$. What happens to the values of H as x
    becomes unbounded in the negative direction, symbolized by $\lim\limits_{x\to-\infty} H(x)$?

(c) Evaluate H at $x = 1.5, 1.9, 1.99, 1.999,$ and 1.9999. What happens to the values of H as x
    approaches 2, $x < 2$, symbolized by $\lim\limits_{x\to 2^-} H(x)$?

(d) Evaluate H at $x = 2.5, 2.1, 2.01, 2.001,$ and 2.0001. What happens to the values of H as x
    approaches 2, $x > 2$, symbolized by $\lim\limits_{x\to 2^+} H(x)$?

Result

(a) Table 10 shows the values of $Y_1 = H(x)$ as x approaches ∞. Notice that the values of H are
    approaching 1, so $\lim\limits_{x\to\infty} H(x) = 1$.

(b) Table 11 shows the values of $Y_1 = H(x)$ as x approaches $-\infty$. Again the values of H are
    approaching 1, so $\lim\limits_{x\to-\infty} H(x) = 1$.

(c) From Table 12 we see that, as x approaches 2, $x < 2$, the values of H are increasing without
    bound, so $\lim\limits_{x\to 2^-} H(x) = \infty$.

(d) Finally, Table 13 reveals that, as x approaches 2, $x > 2$, the values of H are increasing without
    bound, so $\lim\limits_{x\to 2^+} H(x) = \infty$.

Table 10

X	Y1
10	1.0156
100	1.0001
1000	1
10000	1

$Y_1 \boxminus 1/(X-2)^2+1$

Table 11

X	Y1
-10	1.0069
-100	1.0001
-1000	1
-10000	1

$Y_1 \boxminus 1/(X-2)^2+1$

Table 12

X	Y1
1.5	5
1.9	101
1.99	10001
1.999	1E6
1.9999	1E8

$Y_1 \boxminus 1/(X-2)^2+1$

Table 13

X	Y1
2.5	5
2.1	101
2.01	10001
2.001	1E6
2.0001	1E8

$Y_1 \boxminus 1/(X-2)^2+1$

The results of the Exploration reveal an important property of rational functions. The vertical line $x = 2$ and the horizontal line $y = 1$ are called *asymptotes* of the graph of H, which we define as follows:

Let R denote a function:

If, as $x \to -\infty$ or as $x \to \infty$, the values of $R(x)$ approach some fixed number L, then the line $y = L$ is a **horizontal asymptote** of the graph of R.

If, as x approaches some number c, the values $|R(x)| \to \infty$, then the line $x = c$ is a **vertical asymptote** of the graph of R. The graph of R never intersects a vertical asymptote.

Even though the asymptotes of a function are not part of the graph of the function, they provide information about how the graph looks. Figure 43 illustrates some of the possibilities.

Figure 43

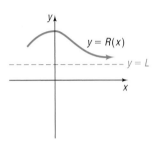

(a) End behavior:
As $x \to \infty$, the values of $R(x)$ approach L [$\lim\limits_{x \to \infty} R(x) = L$]. That is, the points on the graph of R are getting closer to the line $y = L$; $y = L$ is a horizontal asymptote.

(b) End behavior:
As $x \to -\infty$, the values of $R(x)$ approach L [$\lim\limits_{x \to -\infty} R(x) = L$]. That is, the points on the graph of R are getting closer to the line $y = L$; $y = L$ is a horizontal asymptote.

(c) As x approaches c, the values of $|R(x)| \to \infty$ [$\lim\limits_{x \to c^-} R(x) = \infty$; $\lim\limits_{x \to c^+} R(x) = \infty$]. That is, the points on the graph of R are getting closer to the line $x = c$; $x = c$ is a vertical asymptote.

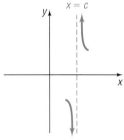

(d) As x approaches c, the values of $|R(x)| \to \infty$ [$\lim\limits_{x \to c^-} R(x) = -\infty$; $\lim\limits_{x \to c^+} R(x) = \infty$]. That is, the points on the graph of R are getting closer to the line $x = c$; $x = c$ is a vertical asymptote.

Figure 44
Oblique asymptote

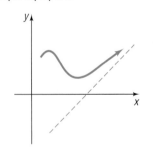

A horizontal asymptote, when it occurs, describes a certain behavior of the graph as $x \to \infty$ or as $x \to -\infty$, that is, its end behavior. **The graph of a function may intersect a horizontal asymptote**.

A vertical asymptote, when it occurs, describes a certain behavior of the graph when x is close to some number c. **The graph of the function will never intersect a vertical asymptote**.

If an asymptote is neither horizontal nor vertical, it is called **oblique**. Figure 44 shows an oblique asymptote. An oblique asymptote, when it occurs, describes the end behavior of the graph. **The graph of a function may intersect an oblique asymptote**.

2 Find the Vertical Asymptotes of a Rational Function

The vertical asymptotes, if any, of a rational function $R(x) = \dfrac{p(x)}{q(x)}$, in lowest terms, are located at the zeros of the denominator of $q(x)$. Suppose that r is a zero, so $x - r$ is a factor. Now, as x approaches r, symbolized as $x \to r$, the values of $x - r$ approach 0, causing the ratio to become unbounded, that is, causing $|R(x)| \to \infty$. Based on the definition, we conclude that the line $x = r$ is a vertical asymptote.

Theorem

Locating Vertical Asymptotes

WARNING
If a rational function is not in lowest terms, an application of this theorem will result in an incorrect listing of vertical asymptotes. ∎

A rational function $R(x) = \dfrac{p(x)}{q(x)}$, *in lowest terms*, will have a vertical asymptote $x = r$ if r is a real zero of the *denominator* q. That is, if $x - r$ is a factor of the denominator q of a rational function $R(x) = \dfrac{p(x)}{q(x)}$, in lowest terms, then R will have the vertical asymptote $x = r$.

| EXAMPLE 4 | **Finding Vertical Asymptotes** |

Find the vertical asymptotes, if any, of the graph of each rational function.

(a) $R(x) = \dfrac{x}{x^2 - 4}$ (b) $F(x) = \dfrac{x + 3}{x - 1}$

(c) $H(x) = \dfrac{x^2}{x^2 + 1}$ (d) $G(x) = \dfrac{x^2 - 9}{x^2 + 4x - 21}$

Solution

(a) R is in lowest terms and the zeros of the denominator $x^2 - 4$ are -2 and 2. Hence, the lines $x = -2$ and $x = 2$ are the vertical asymptotes of the graph of R.

(b) F is in lowest terms and the only zero of the denominator is 1. The line $x = 1$ is the only vertical asymptote of the graph of F.

(c) H is in lowest terms and the denominator has no real zeros because the discriminant of $x^2 + 1$ is $b^2 - 4ac = 0^2 - 4 \cdot 1 \cdot 1 = -4$. The graph of H has no vertical asymptotes.

(d) Factor $G(x)$ to determine if it is in lowest terms.

$$G(x) = \frac{x^2 - 9}{x^2 + 4x - 21} = \frac{(x + 3)(x - 3)}{(x + 7)(x - 3)} = \frac{x + 3}{x + 7}, \qquad x \neq 3$$

The only zero of the denominator of $G(x)$ in lowest terms is -7. Hence, the line $x = -7$ is the only vertical asymptote of the graph of G.   ◄

WARNING

In Example 4(a), the vertical asymptotes are $x = -2$ and $x = 2$. Do not say that the vertical asymptotes are -2 and 2.   ∎

As Example 4 points out, rational functions can have no vertical asymptotes, one vertical asymptote, or more than one vertical asymptote. However, the graph of a rational function will never intersect any of its vertical asymptotes. (Do you know why?)

────── **Exploration** ──────────────────────────────

Graph each of the following rational functions:

$$R(x) = \frac{1}{x - 1} \quad R(x) = \frac{1}{(x - 1)^2} \quad R(x) = \frac{1}{(x - 1)^3} \quad R(x) = \frac{1}{(x - 1)^4}$$

Each has the vertical asymptote $x = 1$. What happens to the value of $R(x)$ as x approaches 1 from the right side of the vertical asymptote; that is, what is $\lim_{x \to 1^+} R(x)$? What happens to the value of $R(x)$ as x approaches 1 from the left side of the vertical asymptote; that is, what is $\lim_{x \to 1^-} R(x)$? How does the multiplicity of the zero in the denominator affect the graph of R?

──

NOW WORK PROBLEM 45. (Find the vertical asymptotes, if any.)

3 **Find the Horizontal or Oblique Asymptotes of a Rational Function**

The procedure for finding horizontal and oblique asymptotes is somewhat more involved. To find such asymptotes, we need to know how the values of a function behave as $x \to -\infty$ or as $x \to \infty$.

If a rational function $R(x)$ is **proper**, that is, if the degree of the numerator is less than the degree of the denominator, then as $x \to -\infty$ or as $x \to \infty$ the value of $R(x)$ approaches 0. Consequently, the line $y = 0$ (the x-axis) is a horizontal asymptote of the graph.

Theorem   If a rational function is proper, the line $y = 0$ is a horizontal asymptote of its graph.

| EXAMPLE 5 | **Finding Horizontal Asymptotes** |

Find the horizontal asymptotes, if any, of the graph of

$$R(x) = \frac{x - 12}{4x^2 + x + 1}$$

Solution   Since the degree of the numerator, 1, is less than the degree of the denominator, 2, the rational function R is proper. We conclude that the line $y = 0$ is a horizontal asymptote of the graph of R.   ◄

To see why $y = 0$ is a horizontal asymptote of the function R in Example 5, we need to investigate the behavior of R as $x \to -\infty$ and $x \to \infty$. When $|x|$ is unbounded, the numerator of R, which is $x - 12$, can be approximated by the power function $y = x$, while the denominator of R, which is $4x^2 + x + 1$, can be approximated by the power function $y = 4x^2$. Applying these ideas to $R(x)$, we find

$$R(x) = \frac{x - 12}{4x^2 + x + 1} \underset{\underset{\text{For } |x| \text{ unbounded}}{\uparrow}}{\approx} \frac{x}{4x^2} = \frac{1}{4x} \underset{\underset{\text{As } x \to -\infty \text{ or } x \to \infty}{\uparrow}}{\to} 0$$

This shows that the line $y = 0$ is a horizontal asymptote of the graph of R.

We verify these results in Tables 14(a) and (b). Notice, as $x \to -\infty$ [Table 14(a)] or $x \to \infty$ [Table 14(b)], respectively, that the values of $R(x)$ approach 0.

Table 14

X	Y1
-10	-.0563
-100	-.0028
-1000	-3E-4
-10000	-3E-5
-1E5	-3E-6
-1E6	-3E-7
-1E7	-3E-8

Y1目(X-12)/(4X²+...

(a)

X	Y1
10	-.0049
100	.00219
1000	2.5E-4
10000	2.5E-5
100000	2.5E-6
1E6	2.5E-7
1E7	2.5E-8

Y1目(X-12)/(4X²+...

(b)

If a rational function $R(x) = \dfrac{p(x)}{q(x)}$ is **improper**, that is, if the degree of the numerator is greater than or equal to the degree of the denominator, we must use long division to write the rational function as the sum of a polynomial $f(x)$ plus a proper rational function $\dfrac{r(x)}{q(x)}$. That is, we write

$$R(x) = \frac{p(x)}{q(x)} = f(x) + \frac{r(x)}{q(x)}$$

where $f(x)$ is a polynomial and $\dfrac{r(x)}{q(x)}$ is a proper rational function. Since $\dfrac{r(x)}{q(x)}$ is proper, then $\dfrac{r(x)}{q(x)} \to 0$ as $x \to -\infty$ or as $x \to \infty$. As a result,

$$R(x) = \frac{p(x)}{q(x)} \to f(x), \qquad \text{as } x \to -\infty \text{ or as } x \to \infty$$

The possibilities are listed next.

1. If $f(x) = b$, a constant, then the line $y = b$ is a horizontal asymptote of the graph of R.

2. If $f(x) = ax + b, a \neq 0$, then the line $y = ax + b$ is an oblique asymptote of the graph of R.

3. In all other cases, the graph of R approaches the graph of f, and there are no horizontal or oblique asymptotes.

The following examples demonstrate these conclusions.

EXAMPLE 6 **Finding Horizontal or Oblique Asymptotes**

Find the horizontal or oblique asymptotes, if any, of the graph of

$$H(x) = \frac{3x^4 - x^2}{x^3 - x^2 + 1}$$

Solution Since the degree of the numerator, 4, is larger than the degree of the denominator, 3, the rational function H is improper. To find any horizontal or oblique asymptotes, we use long division.

$$
\begin{array}{r}
3x + 3 \\
x^3 - x^2 + 1 \overline{)3x^4 \qquad\quad - x^2} \\
\underline{3x^4 - 3x^3 \qquad\quad + 3x} \\
3x^3 - x^2 - 3x \\
\underline{3x^3 - 3x^2 \qquad\quad + 3} \\
2x^2 - 3x - 3
\end{array}
$$

As a result,

$$H(x) = \frac{3x^4 - x^2}{x^3 - x^2 + 1} = 3x + 3 + \frac{2x^2 - 3x - 3}{x^3 - x^2 + 1}$$

As $x \to -\infty$ or as $x \to \infty$,

$$\frac{2x^2 - 3x - 3}{x^3 - x^2 + 1} \approx \frac{2x^2}{x^3} = \frac{2}{x} \to 0$$

As $x \to -\infty$ or as $x \to \infty$, we have $H(x) \to 3x + 3$. We conclude that the graph of the rational function H has an oblique asymptote $y = 3x + 3$.

We verify these results in Tables 15(a) and (b) with $Y_1 = H(x)$ and $Y_2 = 3x + 3$. As $x \to -\infty$ or $x \to \infty$, the difference in the values between Y_1 and Y_2 is indistinguishable.

Table 15

X	Y1	Y2
-10	-27.21	-27
-100	-297	-297
-1000	-2997	-2997
-10000	-29997	-29997
-1E5	-3E5	-3E5
-1E6	-3E6	-3E6
-1E7	-3E7	-3E7

Y1 ▤ (3X^4−X²)/(X...

(a)

X	Y1	Y2
10	33.185	33
100	303.02	303
1000	3003	3003
10000	30003	30003
100000	300003	300003
1E6	3E6	3E6
1E7	3E7	3E7

Y1 ▤ (3X^4−X²)/(X...

(b)

◄

| EXAMPLE 7 | **Finding Horizontal or Oblique Asymptotes** |

Find the horizontal or oblique asymptotes, if any, of the graph of

$$R(x) = \frac{8x^2 - x + 2}{4x^2 - 1}$$

Solution
Since the degree of the numerator, 2, equals the degree of the denominator, 2, the rational function R is improper. To find any horizontal or oblique asymptotes, we use long division.

$$
\begin{array}{r}
2 \\
4x^2 - 1 \overline{)8x^2 - x + 2} \\
\underline{8x^2 \qquad - 2} \\
-x + 4
\end{array}
$$

As a result,

$$R(x) = \frac{8x^2 - x + 2}{4x^2 - 1} = 2 + \frac{-x + 4}{4x^2 - 1}$$

Then, as $x \to -\infty$ or as $x \to \infty$,

$$\frac{-x + 4}{4x^2 - 1} \approx \frac{-x}{4x^2} = \frac{-1}{4x} \to 0$$

As $x \to -\infty$ or as $x \to \infty$, we have $R(x) \to 2$. We conclude that $y = 2$ is a horizontal asymptote of the graph.

✔ **CHECK:** Verify the results of Example 7 by creating a TABLE with $Y_1 = R(x)$ and $Y_2 = 2$. ◀

In Example 7, we note that the quotient 2 obtained by long division is the quotient of the leading coefficients of the numerator polynomial and the denominator polynomial $\left(\dfrac{8}{4}\right)$. This means that we can avoid the long division process for rational functions whose numerator and denominator *are of the same degree* and conclude that the quotient of the leading coefficients will give us the horizontal asymptote.

NOW WORK PROBLEM 41.

| EXAMPLE 8 | **Finding Horizontal or Oblique Asymptotes** |

Find the horizontal or oblique asymptotes, if any, of the graph of

$$G(x) = \frac{2x^5 - x^3 + 2}{x^3 - 1}$$

Solution
Since the degree of the numerator, 5, is larger than the degree of the denominator, 3, the rational function G is improper. To find any horizontal or oblique asymptotes, we use long division.

$$
\begin{array}{r}
2x^2 - 1 \\
x^3 - 1 \overline{)2x^5 - x^3 \qquad\quad + 2} \\
\underline{2x^5 \qquad\quad - 2x^2} \\
-x^3 + 2x^2 + 2 \\
\underline{-x^3 \qquad\quad + 1} \\
2x^2 + 1
\end{array}
$$

As a result,

$$G(x) = \frac{2x^5 - x^3 + 2}{x^3 - 1} = 2x^2 - 1 + \frac{2x^2 + 1}{x^3 - 1}$$

Then, as $x \to -\infty$ or as $x \to \infty$,

$$\frac{2x^2 + 1}{x^3 - 1} \approx \frac{2x^2}{x^3} = \frac{2}{x} \to 0$$

As $x \to -\infty$ or as $x \to \infty$, we have $G(x) \to 2x^2 - 1$. We conclude that, for large values of $|x|$, the graph of G approaches the graph of $y = 2x^2 - 1$. That is, the graph of G will look like the graph of $y = 2x^2 - 1$ as $x \to -\infty$ or $x \to \infty$. Since $y = 2x^2 - 1$ is not a linear function, G has no horizontal or oblique asymptotes. ◀

We now summarize the procedure for finding horizontal and oblique asymptotes.

Summary

Finding Horizontal and Oblique Asymptotes of a Rational Function R

Consider the rational function

$$R(x) = \frac{p(x)}{q(x)} = \frac{a_n x^n + a_{n-1} x^{n-1} + \cdots + a_1 x + a_0}{b_m x^m + b_{m-1} x^{m-1} + \cdots + b_1 x + b_0}$$

in which the degree of the numerator is n and the degree of the denominator is m.

1. If $n < m$, then R is a proper rational function, and the graph of R will have the horizontal asymptote $y = 0$ (the x-axis).

2. If $n \geq m$, then R is improper. Here long division is used.

 (a) If $n = m$, the quotient obtained will be a number $\dfrac{a_n}{b_m}$, and the line $y = \dfrac{a_n}{b_m}$ is a horizontal asymptote.

 (b) If $n = m + 1$, the quotient obtained is of the form $ax + b$ (a polynomial of degree 1), and the line $y = ax + b$ is an oblique asymptote.

 (c) If $n > m + 1$, the quotient obtained is a polynomial of degree 2 or higher, and R has neither a horizontal nor an oblique asymptote. In this case, for $|x|$ unbounded, the graph of R will behave like the graph of the quotient.

NOTE The graph of a rational function either has one horizontal or one oblique asymptote or else has no horizontal and no oblique asymptote. ■

3.3 Assess Your Understanding

'Are You Prepared?'

Answers are given at the end of these exercises. If you get a wrong answer, read the pages listed in red.

1. *True or False:* The quotient of two polynomial expressions is a rational expression. (p. 57)

2. What is the quotient and remainder when $3x^3 - 6x^2 + 3x - 4$ is divided by $x^2 - 1$? (pp. 40–43)

3. Graph $y = \dfrac{1}{x}$. (p. 202)

4. Graph $y = 2(x + 1)^2 - 3$ using transformations. (pp. 266–274)

Concepts and Vocabulary

5. The line _____ is a horizontal asymptote of $R(x) = \dfrac{x^3 - 1}{x^3 + 1}$.

6. The line _____ is a vertical asymptote of $R(x) = \dfrac{x - 1}{x + 1}$.

7. For a rational function R, if the degree of the numerator is less than the degree of the denominator, then R is _____.

8. *True or False:* The domain of every rational function is the set of all real numbers.

9. *True or False:* If an asymptote is neither horizontal nor vertical, it is called oblique.

10. *True or False:* If the degree of the numerator of a rational function equals the degree of the denominator, then the ratio of the leading coefficients gives rise to the horizontal asymptote.

Skill Building

In Problems 11–22, find the domain of each rational function.

11. $R(x) = \dfrac{4x}{x - 3}$

12. $R(x) = \dfrac{5x^2}{3 + x}$

13. $H(x) = \dfrac{-4x^2}{(x - 2)(x + 4)}$

14. $G(x) = \dfrac{6}{(x + 3)(4 - x)}$

15. $F(x) = \dfrac{3x(x - 1)}{2x^2 - 5x - 3}$

16. $Q(x) = \dfrac{-x(1 - x)}{3x^2 + 5x - 2}$

17. $R(x) = \dfrac{x}{x^3 - 8}$

18. $R(x) = \dfrac{x}{x^4 - 1}$

19. $H(x) = \dfrac{3x^2 + x}{x^2 + 4}$

20. $G(x) = \dfrac{x - 3}{x^4 + 1}$

21. $R(x) = \dfrac{3(x^2 - x - 6)}{4(x^2 - 9)}$

22. $F(x) = \dfrac{-2(x^2 - 4)}{3(x^2 + 4x + 4)}$

In Problems 23–28, use the graph shown to find:
  (a) *The domain and range of each function*
  (d) *Vertical asymptotes, if any*
  (b) *The intercepts, if any*
  (e) *Oblique asymptotes, if any*
  (c) *Horizontal asymptotes, if any*

23.

24.

25.

26.

27.

28.

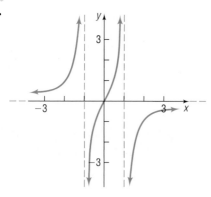

In Problems 29–40, graph each rational function using transformations. Verify your result using a graphing utility.

29. $F(x) = 2 + \dfrac{1}{x}$

30. $Q(x) = 3 + \dfrac{1}{x^2}$

31. $R(x) = \dfrac{1}{(x - 1)^2}$

32. $R(x) = \dfrac{3}{x}$

33. $H(x) = \dfrac{-2}{x + 1}$

34. $G(x) = \dfrac{2}{(x + 2)^2}$

35. $R(x) = \dfrac{-1}{x^2 + 4x + 4}$

36. $R(x) = \dfrac{1}{x - 1} + 1$

37. $G(x) = 1 + \dfrac{2}{(x - 3)^2}$

38. $F(x) = 2 - \dfrac{1}{x + 1}$

39. $R(x) = \dfrac{x^2 - 4}{x^2}$

40. $R(x) = \dfrac{x - 4}{x}$

In Problems 41–52, find the vertical, horizontal, and oblique asymptotes, if any, of each rational function.

41. $R(x) = \dfrac{3x}{x + 4}$

42. $R(x) = \dfrac{3x + 5}{x - 6}$

43. $H(x) = \dfrac{x^3 - 8}{x^2 - 5x + 6}$

44. $G(x) = \dfrac{-x^2 + 1}{x^2 - 5x + 6}$

45. $T(x) = \dfrac{x^3}{x^4 - 1}$

46. $P(x) = \dfrac{4x^5}{x^3 - 1}$

47. $Q(x) = \dfrac{5 - x^2}{3x^4}$

48. $F(x) = \dfrac{-2x^2 + 1}{2x^3 + 4x^2}$

49. $R(x) = \dfrac{3x^4 + 4}{x^3 + 3x}$

50. $R(x) = \dfrac{6x^2 + x + 12}{3x^2 - 5x - 2}$

51. $G(x) = \dfrac{x^3 - 1}{x - x^2}$

52. $F(x) = \dfrac{x - 1}{x - x^3}$

Applications and Extensions

53. Gravity   In physics, it is established that the acceleration due to gravity, g (in meters/sec^2), at a height h meters above sea level is given by

$$g(h) = \frac{3.99 \times 10^{14}}{(6.374 \times 10^6 + h)^2}$$

where 6.374×10^6 is the radius of Earth in meters.
(a) What is the acceleration due to gravity at sea level?
(b) The Sears Tower in Chicago, Illinois, is 443 meters tall. What is the acceleration due to gravity at the top of the Sears Tower?
(c) The peak of Mount Everest is 8848 meters above sea level. What is the acceleration due to gravity on the peak of Mount Everest?
(d) Find the horizontal asymptote of $g(h)$.
(e) Solve $g(h) = 0$. How do you interpret your answer?

54. Population Model   A rare species of insect was discovered in the Amazon Rain Forest. To protect the species, environmentalists declare the insect endangered and transplant the insect into a protected area. The population P of the insect t months after being transplanted is given below.

$$P(t) = \frac{50(1 + 0.5t)}{(2 + 0.01t)}$$

(a) How many insects were discovered? In other words, what was the population when $t = 0$?
(b) What will the population be after 5 years?
(c) Determine the horizontal asymptote of $P(t)$. What is the largest population that the protected area can sustain?

Discussion and Writing

55. If the graph of a rational function R has the vertical asymptote $x = 4$, then the factor $x - 4$ must be present in the denominator of R. Explain why.

56. If the graph of a rational function R has the horizontal asymptote $y = 2$, then the degree of the numerator of R equals the degree of the denominator of R. Explain why.

57. Can the graph of a rational function have both a horizontal and an oblique asymptote? Explain.

58. Make up a rational function that has $y = 2x + 1$ as an oblique asymptote. Explain the methodology that you used.

'Are You Prepared?' Answers

1. True   **2.** Quotient: $3x - 6$; Remainder: $6x - 10$ **3.**

4.

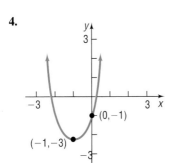

3.4 The Graph of a Rational Function; Inverse and Joint Variation

PREPARING FOR THIS SECTION   *Before getting started, review the following:*

• Finding Intercepts (Section 1.8, pp. 171–172)   • Even or Odd Functions (Section 2.4, pp. 228–230)

Now work the 'Are You Prepared?' problems on page 355.

OBJECTIVES 1 Analyze the Graph of a Rational Function
2 Solve Applied Problems Involving Rational Functions
3 Construct a Model Using Inverse Variation
4 Construct a Model Using Joint or Combined Variation

1 Analyze the Graph of a Rational Function

Graphing utilities make the task of graphing rational functions less time consuming. However, the results of algebraic analysis must be taken into account before drawing conclusions based on the graph provided by the utility. We will use the information collected in the last section in conjunction with the graphing utility to analyze the graph of a rational function $R(x) = \dfrac{p(x)}{q(x)}$. The analysis will require the following steps:

> **Analyzing the Graph of a Rational Function**
>
> **STEP 1:** Find the domain of the rational function.
>
> **STEP 2:** Write R in lowest terms.
>
> **STEP 3:** Locate the intercepts of the graph. The x-intercepts, if any, of $R(x) = \dfrac{p(x)}{q(x)}$ in lowest terms, satisfy the equation $p(x) = 0$. The y-intercept, if there is one, is $R(0)$.
>
> **STEP 4:** Test for symmetry. Replace x by $-x$ in $R(x)$. If $R(-x) = R(x)$, there is symmetry with respect to the y-axis; if $R(-x) = -R(x)$, there is symmetry with respect to the origin.
>
> **STEP 5:** Locate the vertical asymptotes. The vertical asymptotes, if any, of $R(x) = \dfrac{p(x)}{q(x)}$ in lowest terms are found by identifying the real zeros of $q(x)$. Each zero of the denominator gives rise to a vertical asymptote.
>
> **STEP 6:** Locate the horizontal or oblique asymptotes, if any, using the procedure given in Section 3.3. Determine points, if any, at which the graph of R intersects these asymptotes.
>
> **STEP 7:** Graph R using a graphing utility.
>
> **STEP 8:** Use the results obtained in Steps 1 through 7 to graph R by hand.

EXAMPLE 1   **Analyzing the Graph of a Rational Function**

Analyze the graph of the rational function: $R(x) = \dfrac{x - 1}{x^2 - 4}$

Solution   First, we factor both the numerator and the denominator of R.

$$R(x) = \frac{x - 1}{(x + 2)(x - 2)}$$

STEP 1: The domain of R is $\{x | x \neq -2, x \neq 2\}$.

STEP 2: R is in lowest terms.

STEP 3: We locate the x-intercepts by finding the zeros of the numerator. By inspection, 1 is the only x-intercept. The y-intercept is $R(0) = \frac{1}{4}$.

STEP 4: Because

$$R(-x) = \frac{-x - 1}{x^2 - 4} = \frac{-(x + 1)}{x^2 - 4}$$

we conclude that R is neither even nor odd. There is no symmetry with respect to the y-axis or the origin.

Figure 45

Connected mode
(a)

Dot mode
(b)

STEP 5: We locate the vertical asymptotes by finding the zeros of the denominator. Since R is in lowest terms, the graph of R has two vertical asymptotes: the lines $x = -2$ and $x = 2$.

STEP 6: The degree of the numerator is less than the degree of the denominator, so R is proper and the line $y = 0$ (the x-axis) is a horizontal asymptote of the graph. To determine if the graph of R intersects the horizontal asymptote, we solve the equation $R(x) = 0$:

$$\frac{x - 1}{x^2 - 4} = 0$$

$$x - 1 = 0$$

$$x = 1$$

The only solution is $x = 1$, so the graph of R intersects the horizontal asymptote at $(1, 0)$.

STEP 7: The analysis just completed helps us to set the viewing window to obtain a complete graph. Figure 45(a) shows the graph of $R(x) = \frac{x - 1}{x^2 - 4}$ in connected mode, and Figure 45(b) shows it in dot mode. Notice in Figure 45(a) that the graph has vertical lines at $x = -2$ and $x = 2$. This is due to the fact that, when the graphing utility is in connected mode, it will "connect the dots" between consecutive pixels. We know that the graph of R does not cross the lines $x = -2$ and $x = 2$, since R is not defined at $x = -2$ or $x = 2$. When graphing rational functions, dot mode should be used if extraneous vertical lines appear. You should confirm that all the algebraic conclusions that we arrived at in Steps 1 through 6 are part of the graph. For example, the graph has vertical asymptotes at $x = -2$ and $x = 2$, and the graph has a horizontal asymptote at $y = 0$. The y-intercept is $\frac{1}{4}$ and the x-intercept is 1.

Figure 46

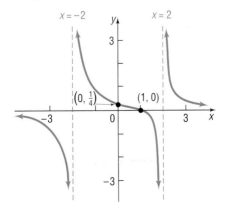

STEP 8: Using the information gathered in Steps 1 through 7, we obtain the graph of R shown in Figure 46.   ◄

━━━━━━ **NOW WORK PROBLEM 7.**

| EXAMPLE 2 | **Analyzing the Graph of a Rational Function** |

Analyze the graph of the rational function: $R(x) = \dfrac{x^2 - 1}{x}$

Solution   **STEP 1:** The domain of R is $\{x \mid x \neq 0\}$.

STEP 2: R is in lowest terms.

STEP 3: The graph has two x-intercepts: -1 and 1. There is no y-intercept, since x cannot equal 0.

STEP 4: Since $R(-x) = -R(x)$, the function is odd and the graph is symmetric with respect to the origin.

STEP 5: Since R is in lowest terms, the graph of $R(x)$ has the line $x = 0$ (the y-axis) as a vertical asymptote.

STEP 6: Since the degree of the numerator, 2, is larger than the degree of the denominator, 1, the rational function R is improper. To find any horizontal or oblique asymptotes, we use long division.

$$\begin{array}{r} x \\ x \overline{)\, x^2 - 1} \\ \underline{x^2 } \\ -1 \end{array}$$

NOTE
Because the determinator of the rational function is a monomial, we can also find the oblique asymptote as follows:

$\dfrac{x^2 - 1}{x} = \dfrac{x^2}{x} - \dfrac{1}{x} = x - \dfrac{1}{x}$

Since $\dfrac{1}{x} \to 0$ as $x \to \infty$, $y = x$ is

an oblique asymptote. ∎

The quotient is x, so the line $y = x$ is an oblique asymptote of the graph. To determine whether the graph of R intersects the asymptote $y = x$, we solve the equation $R(x) = x$.

$$\begin{aligned} R(x) = \frac{x^2 - 1}{x} &= x \\ x^2 - 1 &= x^2 \\ -1 &= 0 \qquad \textit{Impossible.} \end{aligned}$$

We conclude that the equation $\dfrac{x^2 - 1}{x} = x$ has no solution, so the graph of $R(x)$ does not intersect the line $y = x$.

STEP 7: See Figure 47. We see from the graph that there is no y-intercept and there are two x-intercepts, -1 and 1. The symmetry with respect to the origin is also evident. We can also see that there is a vertical asymptote at $x = 0$. Finally, it is not necessary to graph this function in dot mode since no extraneous vertical lines are present.

STEP 8: Using the information gathered in Steps 1 through 7, we obtain the graph of R shown in Figure 48. Notice how the oblique asymptote is used as a guide in graphing the rational function by hand.

Figure 47

Figure 48

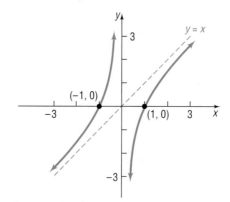

NOW WORK PROBLEM **15.**

| EXAMPLE 3 | **Analyzing the Graph of a Rational Function** |

Analyze the graph of the rational function: $R(x) = \dfrac{3x^2 - 3x}{x^2 + x - 12}$

Solution   We factor R to get

$$R(x) = \frac{3x(x - 1)}{(x + 4)(x - 3)}$$

STEP 1: The domain of R is $\{x \,|\, x \neq -4,\ x \neq 3\}$.

STEP 2: R is in lowest terms.

STEP 3: The graph has two x-intercepts: 0 and 1. The y-intercept is $R(0) = 0$.

STEP 4: Because

$$R(-x) = \frac{-3x(-x - 1)}{(-x + 4)(-x - 3)} = \frac{3x(x + 1)}{(x - 4)(x + 3)}$$

we conclude that R is neither even nor odd. There is no symmetry with respect to the y-axis or the origin.

STEP 5: Since R is in lowest terms, the graph of R has two vertical asymptotes: $x = -4$ and $x = 3$.

STEP 6: Since the degree of the numerator equals the degree of the denominator, the graph has a horizontal asymptote. To find it, we form the quotient of the leading coefficient of the numerator, 3, and the leading coefficient of the denominator, 1. The graph of R has the horizontal asymptote $y = 3$. To find out whether the graph of R intersects the asymptote, we solve the equation $R(x) = 3$.

$$R(x) = \frac{3x^2 - 3x}{x^2 + x - 12} = 3$$
$$3x^2 - 3x = 3x^2 + 3x - 36$$
$$-6x = -36$$
$$x = 6$$

The graph intersects the line $y = 3$ at $x = 6$, and $(6, 3)$ is a point on the graph of R.

STEP 7: Figure 49(a) shows the graph of R in connected mode. Notice the extraneous vertical lines at $x = -4$ and $x = 3$ (the vertical asymptotes). As a result, we also graph R in dot mode. See Figure 49(b).

STEP 8: Figure 49 does not display the graph between the two x-intercepts, 0 and 1. However, because the zeros in the numerator, 0 and 1, are of odd multiplicity (both are multiplicity 1), we know that the graph of R crosses the x-axis at 0 and 1. Therefore, the graph of R is above the x-axis for $0 < x < 1$. To see this part better, we graph R for $-1 \leq x \leq 2$ in Figure 50. Using MAXIMUM, we approximate the turning point to be $(0.52, 0.07)$, rounded to two decimal places.

Figure 49

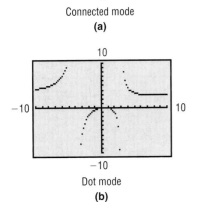

Connected mode
(a)

Dot mode
(b)

Figure 50

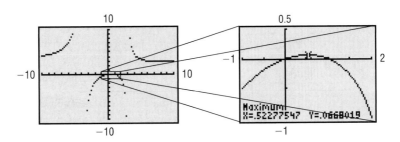

Figure 49 also does not display the graph of R crossing the horizontal asymptote at $(6, 3)$. To see this part better, we graph R for $4 \le x \le 60$ in Figure 51. Using MINIMUM, we approximate the turning point to be $(11.48, 2.75)$, rounded to two decimal places.

Figure 51

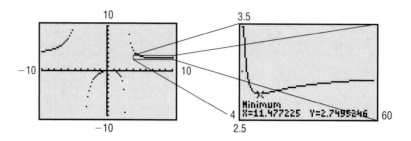

Using this information along with the information gathered in Steps 1 through 7, we obtain the graph of R shown in Figure 52.

Figure 52

EXAMPLE 4	**Analyzing the Graph of a Rational Function with a Hole**

Analyze the graph of the rational function: $R(x) = \dfrac{2x^2 - 5x + 2}{x^2 - 4}$

Solution   We factor R and obtain

$$R(x) = \frac{(2x - 1)(x - 2)}{(x + 2)(x - 2)}$$

STEP 1: The domain of R is $\{x \mid x \ne -2, x \ne 2\}$.

STEP 2: In lowest terms,

$$R(x) = \frac{2x - 1}{x + 2}, \qquad x \ne 2$$

STEP 3: The graph has one x-intercept: 0.5. The y-intercept is $R(0) = -0.5$.

STEP 4: Because

$$R(-x) = \frac{2x^2 + 5x + 2}{x^2 - 4}$$

we conclude that R is neither even nor odd. There is no symmetry with respect to the y-axis or the origin.

STEP 5: Since $x + 2$ is the only factor of the denominator of $R(x)$ *in lowest terms* the graph has one vertical asymptote, $x = -2$. However, the rational function is undefined at both $x = 2$ and $x = -2$.

STEP 6: Since the degree of the numerator equals the degree of the denominator, the graph has a horizontal asymptote. To find it, we form the quotient of the leading coefficient of the numerator, 2, and the leading coefficient of the denominator, 1. The graph of R has the horizontal asymptote $y = 2$. To find whether the graph of R intersects the asymptote, we solve the equation $R(x) = 2$.

$$R(x) = \frac{2x - 1}{x + 2} = 2$$

$$2x - 1 = 2(x + 2)$$

$$2x - 1 = 2x + 4$$

$$-1 = 4 \qquad \text{\textit{Impossible}}$$

The graph does not intersect the line $y = 2$.

Figure 53

STEP 7: Figure 53 shows the graph of $R(x)$. Notice that the graph has one vertical asymptote at $x = -2$. Also, the function appears to be continuous at $x = 2$.

STEP 8: The analysis presented thus far does not explain the behavior of the graph at $x = 2$. We use the TABLE feature of our graphing utility to determine the behavior of the graph of R as x approaches 2. See Table 16. From the table, we conclude that the value of R approaches 0.75 as x approaches 2. This result is further verified by evaluating R in lowest terms at $x = 2$. We conclude that there is a hole in the graph at $(2, 0.75)$. Using the information gathered in Steps 1 through 7, we obtain the graph of R shown in Figure 54.

Table 16

X	Y1
1.99	.74687
1.999	.74969
1.9999	.74997
2	ERROR
2.0001	.75003
2.001	.75031
2.01	.75312

Y1 ◼ (2X²−5X+2)/(...

Figure 54

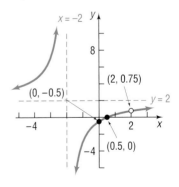

As Example 4 shows, the values excluded from the domain of a rational function give rise to either vertical asymptotes or holes.

NOW WORK PROBLEM 33.

| EXAMPLE 5 | Constructing a Rational Function from Its Graph |

Make up a rational function that might have the graph shown in Figure 55.

Figure 55

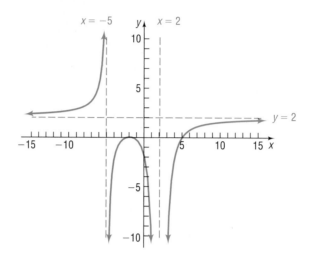

Solution

The numerator of a rational function $R(x) = \dfrac{p(x)}{q(x)}$ in lowest terms determines the x-intercepts of its graph. The graph shown in Figure 55 has x-intercepts -2 (even multiplicity; graph touches the x-axis) and 5 (odd multiplicity; graph crosses the x-axis). So one possibility for the numerator is $p(x) = (x + 2)^2(x - 5)$.

The denominator of a rational function in lowest terms determines the vertical asymptotes of its graph. The vertical asymptotes of the graph are $x = -5$ and $x = 2$. Since $R(x)$ approaches ∞ from the left of $x = -5$ and $R(x)$ approaches $-\infty$ from the right of $x = -5$, we know that $(x + 5)$ is a factor of odd multiplicity in $q(x)$. Also, $R(x)$ approaches $-\infty$ from both sides of $x = 2$, so $(x - 2)$ is a factor of even multiplicity in $q(x)$. A possibility for the denominator is $q(x) = (x + 5)(x - 2)^2$.

Figure 56

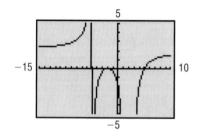

So far we have $R(x) = \dfrac{(x + 2)^2(x - 5)}{(x + 5)(x - 2)^2}$. However, the horizontal asymptote of the graph given in Figure 55, is $y = 2$, so we know that the degree of the numerator must equal the degree in the denominator and the quotient of leading coefficients must be $\dfrac{2}{1}$. This leads to $R(x) = \dfrac{2(x + 2)^2(x - 5)}{(x + 5)(x - 2)^2}$. Figure 56 shows the graph of R drawn on a graphing utility. Since Figure 56 looks similar to Figure 55, we have found a rational function R for the graph in Figure 55. ◄

✎ **NOW WORK PROBLEM 51.**

2 Solve Applied Problems Involving Rational Functions

| EXAMPLE 6 | Finding the Least Cost of a Can |

Reynolds Metal Company manufactures aluminum cans in the shape of a cylinder with a capacity of 500 cubic centimeters $\left(\dfrac{1}{2} \text{ liter} \right)$. The top and bottom of the can are made of a special aluminum alloy that costs 0.05¢ per square centimeter. The sides of the can are made of material that costs 0.02¢ per square centimeter.

(a) Express the cost of material for the can as a function of the radius r of the can.
(b) Use a graphing utility to graph the function $C = C(r)$.
(c) What value of r will result in the least cost?
(d) What is this least cost?

Solution

Figure 57

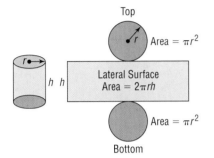

Top
r Area $= \pi r^2$

Lateral Surface
Area $= 2\pi rh$

h h

Area $= \pi r^2$

Bottom

(a) Figure 57 illustrates the situation. Notice that the material required to produce a cylindrical can of height h and radius r consists of a rectangle of area $2\pi rh$ and two circles, each of area πr^2. The total cost C (in cents) of manufacturing the can is therefore

$$C = \text{Cost of top and bottom} + \text{Cost of side}$$
$$= \underbrace{(2\pi r^2 \text{ cm}^2)}_{\substack{\text{Total area}\\\text{of top and}\\\text{bottom}}} \underbrace{(0.05¢/\text{cm}^2)}_{\text{Cost/unit area}} + \underbrace{(2\pi rh \text{ cm}^2)}_{\substack{\text{Total area}\\\text{of side}}} \underbrace{(0.02¢/\text{cm}^2)}_{\text{Cost/unit area}}$$
$$= 0.10\pi r^2 + 0.04\pi rh$$

But we have the additional restriction that the height h and radius r must be chosen so that the volume V of the can is 500 cubic centimeters. Since $V = \pi r^2 h$, we have

$$500 = \pi r^2 h \quad \text{or} \quad h = \frac{500}{\pi r^2}$$

Substituting this expression for h, the cost C, in cents, as a function of the radius r, is

Figure 58

60

0
0
10

$$C(r) = 0.10\pi r^2 + 0.04\pi r \frac{500}{\pi r^2} = 0.10\pi r^2 + \frac{20}{r} = \frac{0.10\pi r^3 + 20}{r}$$

(b) See Figure 58 for the graph of $C(r)$.
(c) Using the MINIMUM command, the cost is least for a radius of about 3.17 centimeters.
(d) The least cost is $C(3.17) \approx 9.47¢$. ◀

━━━ **NOW WORK PROBLEM 61.**

Figure 59

$y = \dfrac{k}{x}, k > 0, x > 0$

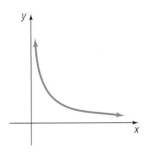

y

x

3 **Construct a Model Using Inverse Variation**

In Section 2.5, we discussed direct variation. We now study **inverse variation**.

Let x and y denote two quantities. Then y **varies inversely** with x, or y is **inversely proportional** to x, if there is a nonzero constant k such that

$$y = \frac{k}{x}$$

The graph in Figure 59 illustrates the relationship between y and x if y varies inversely with x and $k > 0, x > 0$.

| EXAMPLE 7 | Maximum Weight That Can Be Supported by a Piece of Pine |

Figure 60

See Figure 60. The maximum weight W that can be safely supported by a 2-inch by 4-inch piece of pine varies inversely with its length l. Experiments indicate that the maximum weight that a 10-foot-long pine 2-by-4 can support is 500 pounds. Write a general formula relating the maximum weight W (in pounds) to the length l (in feet). Find the maximum weight W that can be safely supported by a length of 25 feet.

Solution Because W varies inversely with l, we know that

$$W = \frac{k}{l}$$

for some constant k. Because $W = 500$ when $l = 10$, we have

$$500 = \frac{k}{10}$$
$$k = 5000$$

Thus, in all cases,

$$W(l) = \frac{5000}{l}$$

In particular, the maximum weight W that can be safely supported by a piece of pine 25 feet in length is

$$W(25) = \frac{5000}{25} = 200 \text{ pounds}$$

Figure 61

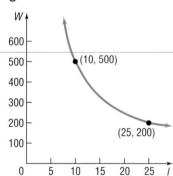

Figure 61 illustrates the relationship between the weight W and the length l. ◄

In direct or inverse variation, the quantities that vary may be raised to powers. For example, in the early seventeenth century, Johannes Kepler (1571–1630) discovered that the square of the period of revolution T of a planet around the Sun varies directly with the cube of its mean distance a from the Sun. That is, $T^2 = ka^3$, where k is the constant of proportionality.

✏ **NOW WORK PROBLEM 63.**

4 Construct a Model Using Joint or Combined Variation

When a variable quantity Q is proportional to the product of two or more other variables, we say that Q **varies jointly** with these quantities. Finally, combinations of direct and/or inverse variation may occur. This is usually referred to as **combined variation**.

Let's look at an example.

| EXAMPLE 8 | Loss of Heat Through a Wall |

The loss of heat through a wall varies jointly with the area of the wall and the difference between the inside and outside temperatures and varies inversely with the thickness of the wall. Write an equation that relates these quantities.

Solution We begin by assigning symbols to represent the quantities:

$$L = \text{Heat loss} \qquad T = \text{Temperature difference}$$
$$A = \text{Area of wall} \qquad d = \text{Thickness of wall}$$

Then

$$L = k\frac{AT}{d}$$

where k is the constant of proportionality.   ◀

NOW WORK PROBLEM 69.

EXAMPLE 9   **Force of the Wind on a Window**

Figure 62

See Figure 62. The force F of the wind on a flat surface positioned at a right angle to the direction of the wind varies jointly with the area A of the surface and the square of the speed v of the wind. A wind of 30 miles per hour blowing on a window measuring 4 feet by 5 feet has a force of 150 pounds. What is the force on a window measuring 3 feet by 4 feet caused by a wind of 50 miles per hour?

Solution Since F varies jointly with A and v^2, we have

$$F = kAv^2$$

where k is the constant of proportionality. We are told that $F = 150$ when $A = 4 \cdot 5 = 20$ and $v = 30$. Then we have

$$150 = k(20)(900) \qquad \text{\small$F = kAv^2$, F = 150, A = 20, v = 30}$$
$$k = \frac{1}{120}$$

The general formula is therefore

$$F = \frac{1}{120}Av^2$$

For a wind of 50 miles per hour blowing on a window whose area is $A = 3 \cdot 4 = 12$ square feet, the force F is

$$F = \frac{1}{120}(12)(2500) = 250 \text{ pounds}$$

◀

3.4 Assess Your Understanding

'Are You Prepared?'

Answers are given at the end of these exercises. If you get a wrong answer, read the pages listed in red.

1. Find the intercepts of $-3x + 5y = 30$. (p. 171–172)

2. Determine whether $f(x) = \dfrac{x^2 + 1}{x^4 - 1}$ is even, odd, or neither. Is the graph of f symmetric with respect to the y-axis or origin? (pp. 229–230)

Concepts and Vocabulary

3. If the numerator and the denominator of a rational function have no common factors, the rational function is _____.

4. *True or False:* The graph of a polynomial function sometimes has a hole.

5. *True or False:* The graph of a rational function never intersects a horizontal asymptote.

6. *True or False:* The graph of a rational function sometimes has a hole.

Skill Building

In Problems 7–44, follow Steps 1 through 8 on page 346 to analyze the graph of each function.

7. $R(x) = \dfrac{x + 1}{x(x + 4)}$

8. $R(x) = \dfrac{x}{(x - 1)(x + 2)}$

9. $R(x) = \dfrac{3x + 3}{2x + 4}$

10. $R(x) = \dfrac{2x + 4}{x - 1}$

11. $R(x) = \dfrac{3}{x^2 - 4}$

12. $R(x) = \dfrac{6}{x^2 - x - 6}$

13. $P(x) = \dfrac{x^4 + x^2 + 1}{x^2 - 1}$

14. $Q(x) = \dfrac{x^4 - 1}{x^2 - 4}$

15. $H(x) = \dfrac{x^3 - 1}{x^2 - 9}$

16. $G(x) = \dfrac{x^3 + 1}{x^2 + 2x}$

17. $R(x) = \dfrac{x^2}{x^2 + x - 6}$

18. $R(x) = \dfrac{x^2 + x - 12}{x^2 - 4}$

19. $G(x) = \dfrac{x}{x^2 - 4}$

20. $G(x) = \dfrac{3x}{x^2 - 1}$

21. $R(x) = \dfrac{3}{(x - 1)(x^2 - 4)}$

22. $R(x) = \dfrac{-4}{(x + 1)(x^2 - 9)}$

23. $H(x) = \dfrac{4(x^2 - 1)}{x^4 - 16}$

24. $H(x) = \dfrac{x^2 + 4}{x^4 - 1}$

25. $F(x) = \dfrac{x^2 - 3x - 4}{x + 2}$

26. $F(x) = \dfrac{x^2 + 3x + 2}{x - 1}$

27. $R(x) = \dfrac{x^2 + x - 12}{x - 4}$

28. $R(x) = \dfrac{x^2 - x - 12}{x + 5}$

29. $F(x) = \dfrac{x^2 + x - 12}{x + 2}$

30. $G(x) = \dfrac{x^2 - x - 12}{x + 1}$

31. $R(x) = \dfrac{x(x - 1)^2}{(x + 3)^3}$

32. $R(x) = \dfrac{(x - 1)(x + 2)(x - 3)}{x(x - 4)^2}$

33. $R(x) = \dfrac{x^2 + x - 12}{x^2 - x - 6}$

34. $R(x) = \dfrac{x^2 + 3x - 10}{x^2 + 8x + 15}$

35. $R(x) = \dfrac{6x^2 - 7x - 3}{2x^2 - 7x + 6}$

36. $R(x) = \dfrac{8x^2 + 26x + 15}{2x^2 - x - 15}$

37. $R(x) = \dfrac{x^2 + 5x + 6}{x + 3}$

38. $R(x) = \dfrac{x^2 + x - 30}{x + 6}$

39. $f(x) = x + \dfrac{1}{x}$

40. $f(x) = 2x + \dfrac{9}{x}$

41. $f(x) = x^2 + \dfrac{1}{x}$

42. $f(x) = 2x^2 + \dfrac{9}{x}$

43. $f(x) = x + \dfrac{1}{x^3}$

44. $f(x) = 2x + \dfrac{9}{x^3}$

In Problems 45–50, graph each function and use MINIMUM to obtain the minimum value, rounded to two decimal places.

45. $f(x) = x + \dfrac{1}{x}, \quad x > 0$

46. $f(x) = 2x + \dfrac{9}{x}, \quad x > 0$

47. $f(x) = x^2 + \dfrac{1}{x}, \quad x > 0$

48. $f(x) = 2x^2 + \dfrac{9}{x}, \quad x > 0$

49. $f(x) = x + \dfrac{1}{x^3}, \quad x > 0$

50. $f(x) = 2x + \dfrac{9}{x^3}, \quad x > 0$

In Problems 51–54, find a rational function that might have this graph. (More than one answer might be possible.)

51.

52.

53.

54.

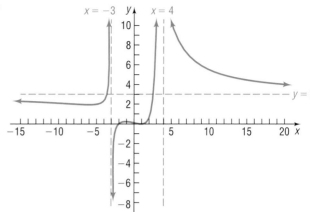

Applications and Extensions

55. Drug Concentration   The concentration C of a certain drug in a patient's bloodstream t hours after injection is given by

$$C(t) = \frac{t}{2t^2 + 1}$$

(a) Find the horizontal asymptote of $C(t)$. What happens to the concentration of the drug as t increases?
(b) Using a graphing utility, graph $C = C(t)$.
(c) Determine the time at which the concentration is highest.

56. Drug Concentration   The concentration C of a certain drug in a patient's bloodstream t minutes after injection is given by

$$C(t) = \frac{50t}{t^2 + 25}$$

(a) Find the horizontal asymptote of $C(t)$. What happens to the concentration of the drug as t increases?
(b) Using a graphing utility, graph $C = C(t)$.
(c) Determine the time at which the concentration is highest.

57. Average Cost   In Problem 95, Exercise 3.2, the cost function C (in thousands of dollars) for manufacturing x Chevy Cavaliers was determined to be

$$C(x) = 0.2x^3 - 2.3x^2 + 14.3x + 10.2$$

Economists define the **average cost function** as

$$\overline{C}(x) = \frac{C(x)}{x}$$

(a) Find the average cost function.
(b) What is the average cost of producing six Cavaliers?
(c) What is the average cost of producing nine Cavaliers?
(d) Using a graphing utility, graph the average cost function.
(e) Using a graphing utility, find the number of Cavaliers that should be produced to minimize the average cost.
(f) What is the minimum average cost?

58. Average Cost   In Problem 96, Exercise 3.2, the cost function C (in thousands of dollars) for printing x textbooks (in thousands of units) was found to be

$$C(x) = 0.015x^3 - 0.595x^2 + 9.15x + 98.43$$

(a) Find the average cost function (refer to Problem 57).
(b) What is the average cost of printing 13,000 textbooks per week?
(c) What is the average cost of printing 25,000 textbooks per week?
(d) Using your graphing utility, graph the average cost function.
(e) Using your graphing utility, find the number of textbooks that should be printed to minimize the average cost.
(f) What is the minimum average cost?

59. Minimizing Surface Area   United Parcel Service has contracted you to design a closed box with a square base that has a volume of 10,000 cubic inches. See the illustration.

(a) Find a function for the surface area of the box.
(b) Using a graphing utility, graph the function found in part (a).
(c) What is the minimum amount of cardboard that can be used to construct the box?
(d) What are the dimensions of the box that minimize the surface area?
(e) Why might UPS be interested in designing a box that minimizes the surface area?

60. Minimizing Surface Area   United Parcel Service has contracted you to design a closed box with a square base that has a volume of 5000 cubic inches. See the illustration.

(a) Find a function for the surface area of the box.
(b) Using a graphing utility, graph the function found in part (a).
(c) What is the minimum amount of cardboard that can be used to construct the box?
(d) What are the dimensions of the box that minimize the surface area?
(e) Why might UPS be interested in designing a box that minimizes the surface area?

61. **Cost of a Can**   A can in the shape of a right circular cylinder is required to have a volume of 500 cubic centimeters. The top and bottom are made of material that costs 6¢ per square centimeter, while the sides are made of material that costs 4¢ per square centimeter.
(a) Express the total cost C of the material as a function of the radius r of the cylinder. (Refer to Figure 57.)
(b) Using a graphing utility graph $C = C(r)$. For what value of r is the cost C a minimum?

62. **Material Needed to Make a Drum**   A steel drum in the shape of a right circular cylinder is required to have a volume of 100 cubic feet.

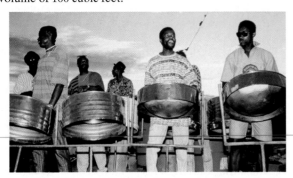

(a) Express the amount A of material required to make the drum as a function of the radius r of the cylinder.
(b) How much material is required if the drum's radius is 3 feet?
(c) How much material is required if the drum's radius is 4 feet?
(d) How much material is required if the drum's radius is 5 feet?
(e) Using a graphing utility graph $A = A(r)$. For what value of r is A smallest?

63. **Demand**   Suppose that the demand D for candy at the movie theater is inversely related to the price p.
(a) When the price of candy is $2.75 per bag, the theater sells 156 bags of candy on a typical Saturday. Express the demand for candy as a function of its price.
(b) Determine the number of bags of candy that will be sold if the price is raised to $3 a bag.

64. **Driving to School**   The time t that it takes to get to school varies inversely with your average speed s.
(a) Suppose that it takes you 40 minutes to get to school when your average speed is 30 miles per hour. Express the driving time to school as a function of average speed.
(b) Suppose that your average speed to school is 40 miles per hour. How long will it take you to get to school?

65. **Pressure**   The volume of a gas V held at a constant temperature in a closed container varies inversely with its pressure P. If the volume of a gas is 600 cubic centimeters (cm^3) when the pressure is 150 millimeters of mercury (mm Hg), find the volume when the pressure is 200 mm Hg.

66. **Resistance**   The current i in a circuit is inversely proportional to its resistance R measured in ohms. Suppose that when the current in a circuit is 30 amperes the resistance is 8 ohms. Find the current in the same circuit when the resistance is 10 ohms.

67. **Weight**   The weight of an object above the surface of Earth varies inversely with the square of the distance from the center of Earth. If Maria weighs 125 pounds when she is on the surface of Earth (3960 miles from the center), determine Maria's weight if she is at the top of Mount McKinley (3.8 miles from the surface of Earth).

68. **Intensity of Light**   The intensity I of light (measured in foot-candles) varies inversely with the square of the distance from the bulb. Suppose that the intensity of a 100-watt light bulb at a distance of 2 meters is 0.075 foot-candle. Determine the intensity of the bulb at a distance of 5 meters.

69. **Geometry**   The volume V of a right circular cylinder varies jointly with the square of its radius r and its height h. The constant of proportionality is π. (See the figure.) Write an equation for V.

70. **Geometry**   The volume V of a right circular cone varies jointly with the square of its radius r and its height h. The constant of proportionality is $\dfrac{\pi}{3}$. (See the figure.) Write an equation for V.

71. **Horsepower**   The horsepower (hp) that a shaft can safely transmit varies jointly with its speed (in revolutions per minute, rpm) and the cube of its diameter. If a shaft of a certain material 2 inches in diameter can transmit 36 hp at 75 rpm, what diameter must the shaft have to transmit 45 hp at 125 rpm?

72. Chemistry: Gas Laws The volume V of an ideal gas varies directly with the temperature T and inversely with the pressure P. Write an equation relating V, T, and P using k as the constant of proportionality. If a cylinder contains oxygen at a temperature of 300 K and a pressure of 15 atmospheres in a volume of 100 liters, what is the constant of proportionality k? If a piston is lowered into the cylinder, decreasing the volume occupied by the gas to 80 liters and raising the temperature to 310 K, what is the gas pressure?

73. Physics: Kinetic Energy The kinetic energy K of a moving object varies jointly with its mass m and the square of its velocity v. If an object weighing 25 pounds and moving with a velocity of 100 feet per second has a kinetic energy of 400 foot-pounds, find its kinetic energy when the velocity is 150 feet per second.

74. Electrical Resistance of a Wire The electrical resistance of a wire varies directly with the length of the wire and inversely with the square of the diameter of the wire. If a wire 432 feet long and 4 millimeters in diameter has a resistance

of 1.24 ohms, find the length of a wire of the same material whose resistance is 1.44 ohms and whose diameter is 3 millimeters.

75. Measuring the Stress of Materials The stress in the material of a pipe subject to internal pressure varies jointly with the internal pressure and the internal diameter of the pipe and inversely with the thickness of the pipe. The stress is 100 pounds per square inch when the diameter is 5 inches, the thickness is 0.75 inch, and the internal pressure is 25 pounds per square inch. Find the stress when the internal pressure is 40 pounds per square inch if the diameter is 8 inches and the thickness is 0.50 inch.

76. Safe Load for a Beam The maximum safe load for a horizontal rectangular beam varies jointly with the width of the beam and the square of the thickness of the beam and inversely with its length. If an 8-foot beam will support up to 750 pounds when the beam is 4 inches wide and 2 inches thick, what is the maximum safe load in a similar beam 10 feet long, 6 inches wide, and 2 inches thick?

Discussion and Writing

77. Graph each of the following functions:

$$y = \frac{x^2 - 1}{x - 1} \qquad y = \frac{x^3 - 1}{x - 1}$$

$$y = \frac{x^4 - 1}{x - 1} \qquad y = \frac{x^5 - 1}{x - 1}$$

Is $x = 1$ a vertical asymptote? Why not? What is happening for $x = 1$? What do you conjecture about $y = \dfrac{x^n - 1}{x - 1}$, $n \geq 1$ an integer, for $x = 1$?

78. Graph each of the following functions:

$$y = \frac{x^2}{x - 1} \qquad y = \frac{x^4}{x - 1} \qquad y = \frac{x^6}{x - 1} \qquad y = \frac{x^8}{x - 1}$$

What similarities do you see? What differences?

79. Write a few paragraphs that provide a general strategy for graphing a rational function. Be sure to mention the following: proper, improper, intercepts, and asymptotes.

80. Create a rational function that has the following characteristics: crosses the x-axis at 2; touches the x-axis at -1; one

vertical asymptote at $x = -5$ and another at $x = 6$; and one horizontal asymptote, $y = 3$. Compare yours to a fellow classmate's. How do they differ? What are their similarities?

81. Create a rational function that has the following characteristics: crosses the x-axis at 3; touches the x-axis at -2; one vertical asymptote, $x = 1$; and one horizontal asymptote, $y = 2$. Give your rational function to a fellow classmate and ask for a written critique of your rational function.

82. The formula on page 354 attributed to Johannes Kepler is one of the famous three Keplerian Laws of Planetary Motion. Go to the library and research these laws. Write a brief paper about these laws and Kepler's place in history.

83. Using a situation that has not been discussed in the text, write a real-world problem that you think involves two variables that vary directly. Exchange your problem with another student's to solve and critique.

84. Using a situation that has not been discussed in the text, write a real-world problem that you think involves two variables that vary inversely. Exchange your problem with another student's to solve and critique.

85. Using a situation that has not been discussed in the text, write a real-world problem that you think involves three variables that vary jointly. Exchange your problem with another student's to solve and critique.

'Are You Prepared?' Answers

1. $(-10, 0)$, $(0, 6)$ **2.** Even; the graph of f is symmetric with respect to the y-axis.

3.5 Polynomial and Rational Inequalities

PREPARING FOR THIS SECTION *Before getting started, review the following:*

• Solving Inequalities (Section 1.7, pp. 150–159)

     Now work the 'Are You Prepared?' problem on page 365.

OBJECTIVES 1 Solve Polynomial Inequalities Algebraically and Graphically

             2 Solve Rational Inequalities Algebraically and Graphically

In this section we solve inequalities that involve polynomials of degree 2 and higher, as well as some that involve rational expressions. To solve such inequalities, we use the information obtained in the previous three sections about the graph of polynomial and rational functions. The general idea follows:

Suppose that the polynomial or rational inequality is in one of the forms

$$f(x) < 0 \qquad f(x) > 0 \qquad f(x) \le 0 \qquad f(x) \ge 0$$

Locate the zeros of f if f is a polynomial function, and locate the zeros of the numerator and the denominator if f is a rational function. If we use these zeros to divide the real number line into intervals, then we know that on each interval the graph of f is either above the x-axis $[f(x) > 0]$ or below the x-axis $[f(x) < 0]$. In other words, we have found the solution of the inequality.

The following steps provide more detail.

Steps for Solving Polynomial and Rational Inequalities Algebraically

STEP 1: Write the inequality so that a polynomial or rational expression f is on the left side and zero is on the right side in one of the following forms:

$$f(x) > 0 \qquad f(x) \ge 0 \qquad f(x) < 0 \qquad f(x) \le 0$$

For rational expressions, be sure that the left side is written as a single quotient.

STEP 2: Determine the numbers at which the expression f on the left side equals zero and, if the expression is rational, the numbers at which the expression f on the left side is undefined.

STEP 3: Use the numbers found in Step 2 to separate the real number line into intervals.

STEP 4: Select a number in each interval and evaluate f at the number.
(a) If the value of f is positive, then $f(x) > 0$ for all numbers x in the interval.
(b) If the value of f is negative, then $f(x) < 0$ for all numbers x in the interval.
If the inequality is not strict ($\ge$ or $\le$), include the solutions of $f(x) = 0$ in the solution set, but be careful not to include values of x where the expression is undefined.

1 Solve Polynomial Inequalities Algebraically and Graphically

| EXAMPLE 1 | Solving a Polynomial Inequality |

Solve the inequality $x^2 \leq 4x + 12$, and graph the solution set.

Algebraic Solution

STEP 1: Rearrange the inequality so that 0 is on the right side.

$$x^2 \leq 4x + 12$$

$$x^2 - 4x - 12 \leq 0 \qquad \text{Subtract } 4x + 12 \text{ from both sides of the inequality.}$$

This inequality is equivalent to the one we wish to solve.

STEP 2: Find the zeros of $f(x) = x^2 - 4x - 12$ by solving the equation $x^2 - 4x - 12 = 0$.

$$x^2 - 4x - 12 = 0$$

$$(x + 2)(x - 6) = 0 \qquad \text{Factor.}$$

$$x = -2 \quad \text{or} \quad x = 6$$

STEP 3: We use the zeros of f to separate the real number line into three intervals.

$$(-\infty, -2) \qquad (-2, 6) \qquad (6, \infty)$$

STEP 4: We select a number in each interval found in Step 3 and evaluate $f(x) = x^2 - 4x - 12$ at each number to determine if $f(x)$ is positive or negative. See Table 17.

Graphing Solution

We graph $Y_1 = x^2$ and $Y_2 = 4x + 12$ on the same screen. See Figure 63. Using the IN-TERSECT command, we find that Y_1 and Y_2 intersect at $x = -2$ and at $x = 6$. The graph of Y_1 is below that of Y_2, $Y_1 < Y_2$, between the points of intersection. Since the inequality is not strict, the solution set is $\{x \mid -2 \leq x \leq 6\}$ or, using interval notation, $[-2, 6]$.

Figure 63

Table 17

Interval	$(-\infty, -2)$	$(-2, 6)$	$(6, \infty)$
Number Chosen	-3	0	7
Value of f	$f(-3) = 9$	$f(0) = -12$	$f(7) = 9$
Conclusion	Positive	Negative	Positive

Since we want to know where $f(x)$ is negative, we conclude that $f(x) < 0$ for all x such that $-2 < x < 6$. However, because the original inequality is not strict, numbers x that satisfy the equation $x^2 = 4x + 12$ are also solutions of the inequality $x^2 \leq 4x + 12$. Thus, we include -2 and 6. The solution set of the given inequality is $\{x \mid -2 \leq x \leq 6\}$ or, using interval notation, $[-2, 6]$.

Figure 64

Figure 64 shows the graph of the solution set.

NOW WORK PROBLEMS 5 AND 9.

| **EXAMPLE 2** | **Solving a Polynomial Inequality** |

Solve the inequality $x^4 > x$, and graph the solution set.

Algebraic Solution

STEP 1: Rearrange the inequality so that 0 is on the right side.

$$x^4 > x$$
$$x^4 - x > 0 \qquad \text{Subtract x from both sides of the inequality.}$$

This inequality is equivalent to the one we wish to solve.

STEP 2: Find the zeros of $f(x) = x^4 - x$ by solving $x^4 - x = 0$.

$$x^4 - x = 0$$
$$x(x^3 - 1) = 0 \qquad \text{Factor out x.}$$
$$x(x - 1)(x^2 + x + 1) = 0 \qquad \text{Factor the difference of two cubes.}$$
$$x = 0 \quad \text{or} \quad x - 1 = 0 \quad \text{or} \quad x^2 + x + 1 = 0 \qquad \text{Set each factor equal to zero and solve.}$$
$$x = 0 \quad \text{or} \quad x = 1$$

The equation $x^2 + x + 1 = 0$ has no real solutions. (Do you see why?)

STEP 3: We use the zeros to separate the real number line into three intervals:

$$(-\infty, 0) \qquad (0, 1) \qquad (1, \infty)$$

STEP 4: We select a test number in each interval found in Step 3 and evaluate $f(x) = x^4 - x$ at each number to determine if $f(x)$ is positive or negative. See Table 18.

Graphing Solution

We graph $Y_1 = x^4$ and $Y_2 = x$ on the same screen. See Figure 65. Using the INTERSECT command, we find that Y_1 and Y_2 intersect at $x = 0$ and at $x = 1$. The graph of Y_1 is above that of Y_2, $Y_1 > Y_2$, to the left of $x = 0$ and to the right of $x = 1$. Since the inequality is strict, the solution set is $\{x \mid x < 0 \text{ or } x > 1\}$ or, using interval notation, $(-\infty, 0)$ or $(1, \infty)$.

Figure 65

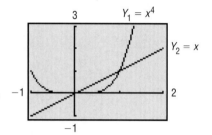

Table 18

Interval	$(-\infty, 0)$	$(0, 1)$	$(1, \infty)$
Number Chosen	-1	$\dfrac{1}{2}$	2
Value of f	$f(-1) = 2$	$f\left(\dfrac{1}{2}\right) = -\dfrac{7}{16}$	$f(2) = 14$
Conclusion	Positive	Negative	Positive

Since we want to know where $f(x)$ is positive, we conclude that $f(x) > 0$ for all numbers x for which $x < 0$ or $x > 1$. Because the original inequality is strict, numbers x that satisfy the equation $x^4 = x$ are not solutions. The solution set of the given inequality is $\{x \mid x < 0 \text{ or } x > 1\}$ or, using interval notation, $(-\infty, 0)$ or $(1, \infty)$.

Figure 66

Figure 66 shows the graph of the solution set.

NOW WORK PROBLEM 21.

2 Solve Rational Inequalities Algebraically and Graphically

| EXAMPLE 3 | Solving a Rational Inequality |

Solve the inequality $\dfrac{4x + 5}{x + 2} \geq 3$, and graph the solution set.

Algebraic Solution

STEP 1: We first note that the domain of the variable consists of all real numbers except -2. We rearrange terms so that 0 is on the right side.

$$\frac{4x + 5}{x + 2} - 3 \geq 0 \qquad \textit{Subtract 3 from both sides of the inequality.}$$

STEP 2: For $f(x) = \dfrac{4x + 5}{x + 2} - 3$, find the zeros of f and the values of x at which f is undefined. To find these numbers, we must express f as a single quotient.

$$
\begin{aligned}
f(x) &= \frac{4x + 5}{x + 2} - 3 \qquad &\textit{Least Common Denominator: } x + 2 \\
&= \frac{4x + 5}{x + 2} - 3 \cdot \frac{x + 2}{x + 2} \qquad &\textit{Multiply 3 by } \tfrac{x + 2}{x + 2} \\
&= \frac{4x + 5 - 3x - 6}{x + 2} \qquad &\textit{Write as a single quotient.} \\
&= \frac{x - 1}{x + 2} \qquad &\textit{Combine like terms.}
\end{aligned}
$$

The zero of f is 1. Also, f is undefined for $x = -2$.

STEP 3: We use the numbers found in Step 2 to separate the real number line into three intervals.

$$(-\infty, -2) \qquad (-2, 1) \qquad (1, \infty)$$

STEP 4: We select a number in each interval found in Step 3 and evaluate $f(x) = \dfrac{4x + 5}{x + 2} - 3$ at each number to determine if $f(x)$ is positive or negative. See Table 19.

Graphing Solution

We first note that the domain of the variable consists of all real numbers except -2. We graph $Y_1 = \dfrac{4x + 5}{x + 2}$ and $Y_2 = 3$ on the same screen. See Figure 67. Using the INTERSECT command, we find that Y_1 and Y_2 intersect at $x = 1$. The graph of Y_1 is above that of Y_2, $Y_1 > Y_2$, to the left of $x = -2$ and to the right of $x = 1$. Since the inequality is not strict, the solution set is $\{x | x < -2 \text{ or } x \geq 1\}$. Using interval notation, the solution set is $(-\infty, -2)$ or $[1, \infty)$.

Figure 67

Table 19

		-2		1	

Interval	$(\infty, -2)$	$(-2, 1)$	$(1, \infty)$
Number Chosen	-3	0	2
Value of f	$f(-3) = 4$	$f(0) = -\dfrac{1}{2}$	$f(2) = \dfrac{1}{4}$
Conclusion	Positive	Negative	Positive

We want to know where $f(x)$ is positive. We conclude that $f(x) > 0$ for all x such that $x < -2$ or $x > 1$. Because the original inequality is not strict, numbers x that satisfy the equation $\dfrac{x - 1}{x + 2} = 0$ are also solutions of the inequality. Since $\dfrac{x - 1}{x + 2} = 0$ only if $x = 1$, we conclude that the solution set is $\{x | x < -2 \text{ or } x \geq 1\}$ or, using interval notation, $(-\infty, -2)$ or $[1, \infty)$.

Figure 68

Figure 68 shows the graph of the solution set.

NOW WORK PROBLEM 35.

EXAMPLE 4

Minimum Sales Requirements

Table 20

Pounds of Cookies (in Hundreds), x	Profit, P
0	−20,000
50	−5990
75	412
120	10,932
200	26,583
270	36,948
340	44,381
420	49,638
525	49,225
610	44,381
700	34,220

Tami is considering leaving her $30,000 a year job and buying a cookie company. According to the financial records of the firm, the relationship between pounds of cookies sold and profit is as exhibited by Table 20.

(a) Draw a scatter diagram of the data in Table 20 with the pounds of cookies sold as the independent variable.

(b) Use a graphing utility to find the quadratic function of best fit.

(c) Use the function found in part (b) to determine the number of pounds of cookies that Tami must sell for the profits to exceed $30,000 a year and therefore make it worthwhile for her to quit her job.

(d) Using the function found in part (b), determine the number of pounds of cookies that Tami should sell to maximize profits.

(e) Using the function found in part (b), determine the maximum profit that Tami can expect to earn.

Solution

(a) Figure 69 shows the scatter diagram.

(b) Using a graphing utility, the quadratic function of best fit is

$$P(x) = -0.31x^2 + 295.86x - 20,042.52$$

where P is the profit achieved from selling x pounds (in hundreds) of cookies. See Figure 70.

(c) Since we want profit to exceed $30,000 we need to solve the inequality

$$-0.31x^2 + 295.86x - 20,042.52 > 30,000$$

We graph

$$Y_1 = P(x) = -0.31x^2 + 295.86x - 20,042.52 \quad \text{and} \quad Y_2 = 30,000$$

on the same screen. See Figure 71.

Figure 69

Figure 70

Figure 71

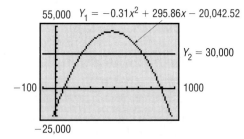

Using the INTERSECT command, we find that Y_1 and Y_2 intersect at $x = 220$ and at $x = 735$. The graph of Y_1 is above that of Y_2, $Y_1 > Y_2$, between the points of intersection. Tami must sell between 220 hundred or 22,000 and 735 hundred or 73,500 pounds of cookies to make it worthwhile to quit her job.

(d) The function $P(x)$ given in part (b) is a quadratic function whose graph opens down. The vertex is therefore the highest point. The x-coordinate of the vertex is

$$x = -\frac{b}{2a} = -\frac{295.86}{2(-0.31)} = 477$$

To maximize profit, 477 hundred (47,700) pounds of cookies must be sold.

(e) The maximum profit is

$$P(477) = -0.31(477)^2 + 295.86(477) - 20,042.52 = \$50,549 \qquad \blacktriangleleft$$

3.5 Assess Your Understanding

'Are You Prepared?'

Answer is given at the end of these exercises. If you get a wrong answer, read the pages listed in red.

1. Solve the inequality: $3 - 4x > 5$. Graph the solution set. (pp. 150–156)

Concepts and Vocabulary

2. *True or False:* A test number for the interval $-5 < x < 1$ is 0.

Skill Building

In Problems 3–56, solve each inequality algebraically. Verify your results using a graphing utility.

3. $(x - 5)(x + 2) < 0$

4. $(x - 5)(x + 2) > 0$

5. $x^2 - 4x > 0$

6. $x^2 + 8x > 0$

7. $x^2 - 9 < 0$

8. $x^2 - 1 < 0$

9. $x^2 + x \geq 12$

10. $x^2 + 7x \leq -12$

11. $2x^2 > 5x + 3$

12. $6x^2 > 6 + 5x$

13. $x(x - 7) < 8$

14. $x(x + 1) < 20$

15. $4x^2 + 9 < 6x$

16. $25x^2 + 16 < 40x$

17. $6(x^2 - 1) \geq 5x$

18. $2(2x^2 - 3x) \geq -9$

19. $(x - 1)(x^2 + x + 1) > 0$

20. $(x + 2)(x^2 - x + 1) > 0$

21. $(x - 1)(x - 2)(x - 3) < 0$

22. $(x + 1)(x + 2)(x + 3) < 0$

23. $x^3 - 2x^2 - 3x \geq 0$

24. $x^3 + 2x^2 - 3x \geq 0$

25. $x^4 > x^2$

26. $x^4 < 4x^2$

27. $x^3 > x^2$

28. $x^3 < 3x^2$

29. $x^4 > 1$

30. $x^3 > 1$

31. $x^2 - 7x - 8 < 0$

32. $x^2 + 12x + 32 \geq 0$

33. $x^4 - 3x^2 - 4 > 0$

34. $x^4 - 5x^2 + 6 < 0$

35. $\dfrac{x + 1}{x - 1} > 0$

36. $\dfrac{x - 3}{x + 1} > 0$

37. $\dfrac{(x - 1)(x + 1)}{x} < 0$

38. $\dfrac{(x - 3)(x + 2)}{x - 1} < 0$

39. $\dfrac{(x - 2)^2}{x^2 - 1} \geq 0$

40. $\dfrac{(x + 5)^2}{x^2 - 4} \geq 0$

41. $6x - 5 < \dfrac{6}{x}$

42. $x + \dfrac{12}{x} < 7$

43. $\dfrac{x + 4}{x - 2} \leq 1$

44. $\dfrac{x + 2}{x - 4} \geq 1$

45. $\dfrac{3x - 5}{x + 2} \leq 2$

46. $\dfrac{x - 4}{2x + 4} \geq 1$

47. $\dfrac{1}{x - 2} < \dfrac{2}{3x - 9}$

48. $\dfrac{5}{x - 3} > \dfrac{3}{x + 1}$

49. $\dfrac{2x + 5}{x + 1} > \dfrac{x + 1}{x - 1}$

50. $\dfrac{1}{x + 2} > \dfrac{3}{x + 1}$

51. $\dfrac{x^2(3 + x)(x + 4)}{(x + 5)(x - 1)} > 0$

52. $\dfrac{x(x^2 + 1)(x - 2)}{(x - 1)(x + 1)} > 0$

53. $\dfrac{2x^2 - x - 1}{x - 4} \leq 0$

54. $\dfrac{3x^2 + 2x - 1}{x + 2} > 0$

55. $\dfrac{x^2 + 3x - 1}{x + 3} > 0$

56. $\dfrac{x^2 - 5x + 3}{x - 5} < 0$

Applications and Extensions

57. For what positive numbers will the cube of a number exceed four times its square?

58. For what positive numbers will the square of a number exceed twice the number?

59. What is the domain of the function $f(x) = \sqrt{x^2 - 16}$?

60. What is the domain of the function $f(x) = \sqrt{x^3 - 3x^2}$?

61. What is the domain of the function $R(x) = \sqrt{\dfrac{x - 2}{x + 4}}$?

62. What is the domain of the function $R(x) = \sqrt{\dfrac{x - 1}{x + 4}}$?

63. Physics A ball is thrown vertically upward with an initial velocity of 80 feet per second. The distance s (in feet) of the ball from the ground after t seconds is $s = 80t - 16t^2$. See the figure.

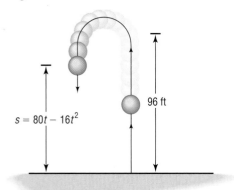

$s = 80t - 16t^2$

96 ft

(a) For what time interval is the ball more than 96 feet above the ground?
(b) Using a graphing utility, graph the relation between s and t.
(c) What is the maximum height of the ball?
(d) After how many seconds does the ball reach the maximum height?

64. Physics A ball is thrown vertically upward with an initial velocity of 96 feet per second. The distance s (in feet) of the ball from the ground after t seconds is $s = 96t - 16t^2$.
(a) For what time interval is the ball more than 112 feet above the ground?
(b) Using a graphing utility, graph the relation between s and t.
(c) What is the maximum height of the ball?
(d) After how many seconds does the ball reach the maximum height?

65. Business The monthly revenue achieved by selling x wristwatches is figured to be $x(40 - 0.2x)$ dollars. The wholesale cost of each watch is $32.
(a) How many watches must be sold each month to achieve a profit (revenue − cost) of at least $50?
(b) Using a graphing utility, graph the revenue function.
(c) What is the maximum revenue that this firm could earn?

(d) How many wristwatches should the firm sell to maximize revenue?
(e) Using a graphing utility, graph the profit function.
(f) What is the maximum profit that this firm can earn?
(g) How many watches should the firm sell to maximize profit?
(h) Provide a reasonable explanation as to why the answers found in parts (d) and (g) differ. Is the shape of the revenue function reasonable in your opinion? Why?

66. Business The monthly revenue achieved by selling x boxes of candy is figured to be $x(5 - 0.05x)$ dollars. The wholesale cost of each box of candy is $1.50.
(a) How many boxes must be sold each month to achieve a profit of at least $60?
(b) Using a graphing utility, graph the revenue function.
(c) What is the maximum revenue that this firm could earn?
(d) How many boxes of candy should the firm sell to maximize revenue?
(e) Using a graphing utility, graph the profit function.
(f) What is the maximum profit that this firm can earn?
(g) How many boxes of candy should the firm sell to maximize profit?
(h) Provide a reasonable explanation as to why the answers found in parts (d) and (g) differ. Is the shape of the revenue function reasonable in your opinion? Why?

67. Cost of Manufacturing In Problem 95 of Section 3.2, a cubic function of best fit relating the cost C of manufacturing x Chevy Cavaliers was found. Budget constraints will not allow Chevy to spend more than $97,000. Determine the number of Cavaliers that could be produced.

68. Cost of Printing In Problem 96 of Section 3.2, a cubic function of best fit relating the cost C of printing x textbooks in a week was found. Budget constraints will not allow the printer to spend more than $170,000 per week. Determine the number of textbooks that could be printed in a week.

69. Prove that, if a, b are real numbers and $a \geq 0, b \geq 0$, then

$$a \leq b \quad \text{is equivalent to} \quad \sqrt{a} \leq \sqrt{b}$$

[**Hint:** $b - a = \left(\sqrt{b} - \sqrt{a}\right)\left(\sqrt{b} + \sqrt{a}\right)$.]

Discussion and Writing

70. Make up an inequality that has no solution. Make up one that has exactly one solution.

71. The inequality $x^2 + 1 < -5$ has no solution. Explain why.

'Are You Prepared?' Answer

1. $\left\{x \mid x < -\dfrac{1}{2}\right\}$

$\xleftarrow{\hspace{1cm}}$ −2 −1 −$\frac{1}{2}$ 0 1

3.6 The Real Zeros of a Polynomial Function

PREPARING FOR THIS SECTION   *Before getting started, review the following:*

- Classification of Numbers (Section R.1, pp. 2–4)
- Factoring Polynomials (Section R.5, pp. 45–51)
- Synthetic Division (Section R.6, pp. 53–56)

- Polynomial Division (Section R.4, pp. 40–43)
- Quadratic Formula (Section 1.3, pp. 117–119)

 Now work the 'Are You Prepared?' problems on page 378.

OBJECTIVES   1 Use the Remainder and Factor Theorems
2 Use the Rational Zeros Theorem
3 Find the Real Zeros of a Polynomial Function
4 Solve Polynomial Equations
5 Use the Theorem for Bounds on Zeros
6 Use the Intermediate Value Theorem

In this section, we discuss techniques that can be used to find the real zeros of a polynomial function. Recall that if r is a real zero of a polynomial function f then $f(r) = 0$, r is an x-intercept of the graph of f, and r is a solution of the equation $f(x) = 0$. For polynomial and rational functions, we have seen the importance of the zeros for graphing. In most cases, however, the zeros of a polynomial function are difficult to find using algebraic methods. No nice formulas like the quadratic formula are available to help us find zeros for polynomials of degree 3 or higher. Formulas do exist for solving any third- or fourth-degree polynomial equation, but they are somewhat complicated. No general formulas exist for polynomial equations of degree 5 or higher. Refer to the Historical Feature at the end of this section for more information.

1 Use the Remainder and Factor Theorems

When we divide one polynomial (the dividend) by another (the divisor), we obtain a quotient polynomial and a remainder, the remainder being either the zero polynomial or a polynomial whose degree is less than the degree of the divisor. To check our work, we verify that

(Quotient)(Divisor) + Remainder = Dividend

This checking routine is the basis for a famous theorem called the **division algorithm**[*] **for polynomials**, which we now state without proof.

Theorem   **Division Algorithm for Polynomials**

If $f(x)$ and $g(x)$ denote polynomial functions and if $g(x)$ is a polynomial whose degree is greater than zero, then there are unique polynomial functions $q(x)$ and $r(x)$ such that

$$\frac{f(x)}{g(x)} = q(x) + \frac{r(x)}{g(x)} \quad \text{or} \quad f(x) = q(x)g(x) + r(x) \qquad (1)$$

$$\underset{\text{dividend}}{\uparrow} \qquad \underset{\text{quotient}}{\uparrow}\ \underset{\text{divisor}}{\uparrow} \qquad \underset{\text{remainder}}{\uparrow}$$

where $r(x)$ is either the zero polynomial or a polynomial of degree less than that of $g(x)$.

[*]A systematic process in which certain steps are repeated a finite number of times is called an **algorithm**. For example, long division is an algorithm.

In equation (1), $f(x)$ is the **dividend**, $g(x)$ is the **divisor**, $q(x)$ is the **quotient**, and $r(x)$ is the **remainder**.

If the divisor $g(x)$ is a first-degree polynomial of the form

$$g(x) = x - c, \qquad c \text{ a real number}$$

then the remainder $r(x)$ is either the zero polynomial or a polynomial of degree 0. As a result, for such divisors, the remainder is some number, say R, and we may write

$$f(x) = (x - c)q(x) + R \tag{2}$$

This equation is an identity in x and is true for all real numbers x. Suppose that $x = c$. Then equation (2) becomes

$$f(c) = (c - c)q(c) + R$$
$$f(c) = R$$

Substitute $f(c)$ for R in equation (2) to obtain

$$f(x) = (x - c)q(x) + f(c) \tag{3}$$

We have now proved the **Remainder Theorem**.

Remainder Theorem

Let f be a polynomial function. If $f(x)$ is divided by $x - c$, then the remainder is $f(c)$.

EXAMPLE 1   **Using the Remainder Theorem**

Find the remainder if $f(x) = x^3 - 4x^2 - 5$ is divided by

(a) $x - 3$ (b) $x + 2$

Solution

(a) We could use long division or synthetic division, but it is easier to use the Remainder Theorem, which says that the remainder is $f(3)$.
$$f(3) = (3)^3 - 4(3)^2 - 5 = 27 - 36 - 5 = -14$$
The remainder is -14.

(b) To find the remainder when $f(x)$ is divided by $x + 2 = x - (-2)$, we evaluate $f(-2)$.
$$f(-2) = (-2)^3 - 4(-2)^2 - 5 = -8 - 16 - 5 = -29$$
The remainder is -29. ◀

Figure 72

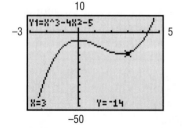

Compare the method used in Example 1(a) with the method used in Example 1 on page 55 (synthetic division). Which method do you prefer? Give reasons.

NOTE   A graphing utility provides another way to find the value of a function, using the eVALUEate feature. Consult your manual for details. See Figure 72 for the results of Example 1(a). ■

An important and useful consequence of the Remainder Theorem is the **Factor Theorem**.

Factor Theorem

Let f be a polynomial function. Then $x - c$ is a factor of $f(x)$ if and only if $f(c) = 0$.

The Factor Theorem actually consists of two separate statements:

---

1. If $f(c) = 0$, then $x - c$ is a factor of $f(x)$.
2. If $x - c$ is a factor of $f(x)$, then $f(c) = 0$.

The proof requires two parts.

Proof

1. Suppose that $f(c) = 0$. Then, by equation (3), we have

$$f(x) = (x - c)q(x) + f(c)$$
$$= (x - c)q(x) + 0$$
$$= (x - c)q(x)$$

   for some polynomial $q(x)$. That is, $x - c$ is a factor of $f(x)$.

2. Suppose that $x - c$ is a factor of $f(x)$. Then there is a polynomial function q such that

$$f(x) = (x - c)q(x)$$

   Replacing x by c, we find that

$$f(c) = (c - c)q(c) = 0 \cdot q(c) = 0$$

   This completes the proof. ∎

One use of the Factor Theorem is to determine whether a polynomial has a particular factor.

EXAMPLE 2

Using the Factor Theorem

Use the Factor Theorem to determine whether the function

$$f(x) = 2x^3 - x^2 + 2x - 3$$

has the factor

(a) $x - 1$ (b) $x + 2$

Solution

The Factor Theorem states that if $f(c) = 0$ then $x - c$ is a factor.

(a) Because $x - 1$ is of the form $x - c$ with $c = 1$, we find the value of $f(1)$. We choose to use substitution.

$$f(1) = 2(1)^3 - (1)^2 + 2(1) - 3 = 2 - 1 + 2 - 3 = 0$$

See also Figure 73(a). By the Factor Theorem, $x - 1$ is a factor of $f(x)$.

(b) We first need to write $x + 2$ in the form $x - c$. Since $x + 2 = x - (-2)$, we find the value of $f(-2)$. See Figure 73(b). Because $f(-2) = -27 \neq 0$, we conclude from the Factor Theorem that $x - (-2) = x + 2$ is not a factor of $f(x)$.

Figure 73

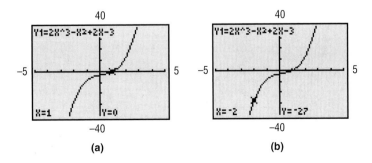

(a)   (b)

◀

In Example 2, we found that $x - 1$ was a factor of f. To write f in factored form, we can use long division or synthetic division. Using synthetic division, we find that

$$
\begin{array}{r|rrrr}
1) & 2 & -1 & 2 & -3 \\
  &   & 2 & 1 & 3 \\
\hline
  & 2 & 1 & 3 & 0
\end{array}
$$

The quotient is $q(x) = 2x^2 + x + 3$ with a remainder of 0, as expected. We can write f in factored form as

$$f(x) = 2x^3 - x^2 + 2x - 3 = (x - 1)(2x^2 + x + 3)$$

NOW WORK PROBLEM 11.

The next theorem concerns the number of real zeros that a polynomial function may have. In counting the zeros of a polynomial, we count each zero as many times as its multiplicity.

Theorem

Number of Real Zeros

A polynomial function of degree n, $n \geq 1$, has at most n real zeros.

Proof The proof is based on the Factor Theorem. If r is a zero of a polynomial function f, then $f(r) = 0$ and, hence, $x - r$ is a factor of $f(x)$. Each zero corresponds to a factor of degree 1. Because f cannot have more first-degree factors than its degree, the result follows. ∎

2 Use the Rational Zeros Theorem

The next result, called the **Rational Zeros Theorem**, provides information about the rational zeros of a polynomial *with integer coefficients*.

Theorem

Rational Zeros Theorem

Let f be a polynomial function of degree 1 or higher of the form

$$\boxed{f(x) = a_n x^n + a_{n-1} x^{n-1} + \cdots + a_1 x + a_0, \qquad a_n \neq 0, a_0 \neq 0}$$

where each coefficient is an integer. If $\dfrac{p}{q}$, in lowest terms, is a rational zero of f, then p must be a factor of a_0, and q must be a factor of a_n.

EXAMPLE 3

Listing Potential Rational Zeros

List the potential rational zeros of

$$f(x) = 2x^3 + 11x^2 - 7x - 6$$

Solution Because f has integer coefficients, we may use the Rational Zeros Theorem. First, we list all the integers p that are factors of $a_0 = -6$ and all the integers q that are factors of the leading coefficient $a_3 = 2$.

$$p: \quad \pm 1, \pm 2, \pm 3, \pm 6$$
$$q: \quad \pm 1, \pm 2$$

Now we form all possible ratios $\dfrac{p}{q}$.

$$\frac{p}{q}: \quad \pm 1, \pm 2, \pm 3, \pm 6, \pm \frac{1}{2}, \pm \frac{3}{2}$$

If f has a rational zero, it will be found in this list, which contains 12 possibilities. ◀

 NOW WORK PROBLEM 21.

Be sure that you understand what the Rational Zeros Theorem says: For a polynomial with integer coefficients, *if* there is a rational zero, it is one of those listed. The Rational Zeros Theorem does not say that if a rational number is in the list of potential rational zeros, then it is a zero. It may be the case that the function does not have any rational zeros.

The Rational Zeros Theorem provides a list of potential rational zeros of a function f. If we graph f, we can get a better sense of the location of the x-intercepts and test to see if they are rational. We can also use the potential rational zeros to select our initial viewing window to graph f and then adjust the window based on the results. The graphs shown throughout the text will be those obtained after setting the final viewing window.

3 Find the Real Zeros of a Polynomial Function

| EXAMPLE 4 | **Finding the Rational Zeros of a Polynomial Function** |

Continue working with Example 3 to find the rational zeros of
$$f(x) = 2x^3 + 11x^2 - 7x - 6$$

Solution We gather all the information that we can about the zeros.

STEP 1: Since f is a polynomial of degree 3, there are at most three real zeros.

STEP 2: We list the potential rational zeros obtained in Example 3:
$$\pm 1, \pm 2, \pm 3, \pm 6, \pm \frac{1}{2}, \pm \frac{3}{2}.$$

Figure 74

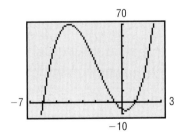

STEP 3: We could, of course, use the Factor Theorem to test each potential rational zero to see if the value of f there is zero. This is not very efficient. The graph of f will tell us approximately where the real zeros are. So we only need to test those rational zeros that are nearby. Figure 74 shows the graph of f. We see that f has three zeros: one near -6, one between -1 and 0, and one near 1. From our original list of potential rational zeros, we will test -6, using eVALUEate.

STEP 4: Because $f(-6) = 0$, we know that -6 is a zero and $x + 6$ is a factor of f. We can use long division or synthetic division to factor f.
$$f(x) = 2x^3 + 11x^2 - 7x - 6$$
$$= (x + 6)(2x^2 - x - 1)$$

Now any solution of the equation $2x^2 - x - 1 = 0$ will be a zero of f. Because of this, we call the equation $2x^2 - x - 1 = 0$ a **depressed equation** of f. Since the degree of the depressed equation of f is less than that of the original polynomial, we work with the depressed equation to find the zeros of f.

The depressed equation $2x^2 - x - 1 = 0$ is a quadratic equation with discriminant $b^2 - 4ac = (-1)^2 - 4(2)(-1) = 9 > 0$. The equation has two real solutions, which can be found by factoring.

$$2x^2 - x - 1 = (2x + 1)(x - 1) = 0$$

$$2x + 1 = 0 \quad \text{or} \quad x - 1 = 0$$

$$x = -\frac{1}{2} \quad \text{or} \quad x = 1$$

The zeros of f are -6, $-\frac{1}{2}$, and 1.

Because $f(x) = (x + 6)(2x^2 - x - 1)$, we completely factor f as follows:

$$f(x) = 2x^3 + 11x^2 - 7x - 6 = (x + 6)(2x^2 - x - 1) = (x + 6)(2x + 1)(x - 1)$$

Notice that the three zeros of f are among those given in the list of potential rational zeros in Example 3. ◀

Steps for Finding the Real Zeros of a Polynomial Function

STEP 1: Use the degree of the polynomial to determine the maximum number of zeros.

STEP 2: If the polynomial has integer coefficients, use the Rational Zeros Theorem to identify those rational numbers that potentially can be zeros.

STEP 3: Using a graphing utility, graph the polynomial function.

STEP 4: (a) Use eVALUEate, substitution, synthetic division, or long division to test a potential rational zero based on the graph.

(b) Each time that a zero (and thus a factor) is found, repeat Step 4 on the depressed equation. In attempting to find the zeros, remember to use (if possible) the factoring techniques that you already know (special products, factoring by grouping, and so on).

| EXAMPLE 5 | · **Finding the Real Zeros of a Polynomial Function** |

Find the real zeros of $f(x) = x^5 - x^4 - 4x^3 + 8x^2 - 32x + 48$. Write f in factored form.

Solution

STEP 1: There are at most five real zeros.

STEP 2: To obtain the list of potential rational zeros, we write the factors p of $a_0 = 48$ and the factors q of the leading coefficient $a_5 = 1$.

$$p: \quad \pm 1, \pm 2, \pm 3, \pm 4, \pm 6, \pm 8, \pm 12, \pm 16, \pm 24, \pm 48$$
$$q: \quad \pm 1$$

The potential rational zeros consist of all possible quotients $\frac{p}{q}$:

$$\frac{p}{q}: \quad \pm 1, \pm 2, \pm 3, \pm 4, \pm 6, \pm 8, \pm 12, \pm 16, \pm 24, \pm 48$$

Figure 75

STEP 3: Figure 75 shows the graph of f. The graph has the characteristics that we expect of the given polynomial of degree 5: no more than four turning points, y-intercept 48, and it behaves like $y = x^5$ for large $|x|$.

STEP 4: Since -3 appears to be a zero and -3 is a potential rational zero, we eVALUEate f at -3 and find that $f(-3) = 0$. We use synthetic division to factor f.

$$
\begin{array}{r|rrrrr}
-3) & 1 & -1 & -4 & 8 & -32 & 48 \\
   &   & -3 & 12 & -24 & 48 & -48 \\
\hline
   & 1 & -4 & 8 & -16 & 16 & 0
\end{array}
$$

We can factor f as

$$f(x) = x^5 - x^4 - 4x^3 + 8x^2 - 32x + 48 = (x + 3)(x^4 - 4x^3 + 8x^2 - 16x + 16)$$

We now work with the first depressed equation:

$$q_1(x) = x^4 - 4x^3 + 8x^2 - 16x + 16 = 0$$

Repeat Step 4: In looking back at Figure 75, it appears that 2 might be a zero of even multiplicity. We check the potential rational zero 2 and find that $f(2) = 0$. Using synthetic division,

$$
\begin{array}{r|rrrrr}
2) & 1 & -4 & 8 & -16 & 16 \\
   &   & 2 & -4 & 8 & -16 \\
\hline
   & 1 & -2 & 4 & -8 & 0
\end{array}
$$

Now we can factor f as

$$f(x) = (x + 3)(x - 2)(x^3 - 2x^2 + 4x - 8)$$

Repeat Step 4: The depressed equation $q_2(x) = x^3 - 2x^2 + 4x - 8 = 0$ can be factored by grouping.

$$x^3 - 2x^2 + 4x - 8 = (x^3 - 2x^2) + (4x - 8) = x^2(x - 2) + 4(x - 2) = (x - 2)(x^2 + 4) = 0$$
$$x - 2 = 0 \quad \text{or} \quad x^2 + 4 = 0$$
$$x = 2$$

Since $x^2 + 4 = 0$ has no real solutions, the real zeros of f are -3 and 2, with 2 being a zero of multiplicity 2. The factored form of f is

$$f(x) = x^5 - x^4 - 4x^3 + 8x^2 - 32x + 48$$
$$= (x + 3)(x - 2)^2(x^2 + 4) \qquad \blacktriangleleft$$

NOW WORK PROBLEM **39.**

4 Solve Polynomial Equations

EXAMPLE 6 Solving a Polynomial Equation

Solve the equation: $x^5 - x^4 - 4x^3 + 8x^2 - 32x + 48 = 0$

Solution The solutions of this equation are the zeros of the polynomial function

$$f(x) = x^5 - x^4 - 4x^3 + 8x^2 - 32x + 48$$

Using the result of Example 5, the real zeros are -3 and 2. These are the real solutions of the equation $x^5 - x^4 - 4x^3 + 8x^2 - 32x + 48 = 0$. $\qquad \blacktriangleleft$

NOW WORK PROBLEM **63.**

In Example 5, the quadratic factor $x^2 + 4$ that appears in the factored form of $f(x)$ is called *irreducible*, because the polynomial $x^2 + 4$ cannot be factored over the real numbers. In general, we say that a quadratic factor $ax^2 + bx + c$ is **irreducible** if it cannot be factored over the real numbers, that is, if it is prime over the real numbers.

Refer back to Examples 4 and 5. The polynomial function of Example 4 has three real zeros, and its factored form contains three linear factors. The polynomial function of Example 5 has two distinct real zeros, and its factored form contains two distinct linear factors and one irreducible quadratic factor.

Theorem   Every polynomial function (with real coefficients) can be uniquely factored into a product of linear factors and/or irreducible quadratic factors.

We shall prove this result in Section 3.7, and, in fact, we shall draw several additional conclusions about the zeros of a polynomial function. One conclusion is worth noting now. If a polynomial (with real coefficients) is of odd degree, then it must contain at least one linear factor. (Do you see why?) This means that it must have at least one real zero.

COROLLARY   A polynomial function (with real coefficients) of odd degree has at least one real zero.

5 Use the Theorem for Bounds on Zeros

One challenge in using a graphing utility is to set the viewing window so that a complete graph is obtained. The next theorem is a tool that can be used to find bounds on the zeros. This will assure that the function does not have any zeros outside these bounds. Then using these bounds to set Xmin and Xmax assures that all the x-intercepts appear in the viewing window.

A number M is a **bound** on the zeros of a polynomial if every zero r lies between $-M$ and M, inclusive. That is, M is a bound to the zeros of a polynomial f if

$$-M \leq \text{any zero of } f \leq M$$

Theorem   **Bounds on Zeros**

Let f denote a polynomial function whose leading coefficient is 1.

$$f(x) = x^n + a_{n-1}x^{n-1} + \cdots + a_1 x + a_0$$

A bound M on the zeros of f is the smaller of the two numbers

$$\text{Max}\{1, |a_0| + |a_1| + \cdots + |a_{n-1}|\}, \quad 1 + \text{Max}\{|a_0|, |a_1|, \ldots, |a_{n-1}|\} \quad \textbf{(4)}$$

where Max { } means "choose the largest entry in { }."

An example will help to make the theorem clear.

| EXAMPLE 7 | **Using the Theorem for Finding Bounds on Zeros** |

Find a bound to the zeros of each polynomial.

(a) $f(x) = x^5 + 3x^3 - 9x^2 + 5$ (b) $g(x) = 4x^5 - 2x^3 + 2x^2 + 1$

Solution   (a) The leading coefficient of f is 1.

$$f(x) = x^5 + 3x^3 - 9x^2 + 5 \qquad a_4 = 0, a_3 = 3, a_2 = -9, a_1 = 0, a_0 = 5$$

We evaluate the expressions in formula (4).

$$\text{Max}\{1, |a_0| + |a_1| + \cdots + |a_{n-1}|\} = \text{Max}\{1, |5| + |0| + |-9| + |3| + |0|\}$$
$$= \text{Max}\{1, 17\} = 17$$
$$1 + \text{Max}\{|a_0|, |a_1|, \ldots, |a_{n-1}|\} = 1 + \text{Max}\{|5|, |0|, |-9|, |3|, |0|\}$$
$$= 1 + 9 = 10$$

The smaller of the two numbers, 10, is the bound. Every zero of f lies between -10 and 10.

(b) First we write g so that its leading coefficient is 1.

$$g(x) = 4x^5 - 2x^3 + 2x^2 + 1 = 4\left(x^5 - \frac{1}{2}x^3 + \frac{1}{2}x^2 + \frac{1}{4}\right)$$

Next we evaluate the two expressions in formula (4) with $a_4 = 0$, $a_3 = -\dfrac{1}{2}$, $a_2 = \dfrac{1}{2}$, $a_1 = 0$, and $a_0 = \dfrac{1}{4}$.

$$\text{Max}\{1, |a_0| + |a_1| + \cdots + |a_{n-1}|\} = \text{Max}\left\{1, \left|\frac{1}{4}\right| + |0| + \left|\frac{1}{2}\right| + \left|-\frac{1}{2}\right| + |0|\right\}$$
$$= \text{Max}\left\{1, \frac{5}{4}\right\} = \frac{5}{4}$$
$$1 + \text{Max}\{|a_0|, |a_1|, \ldots, |a_{n-1}|\} = 1 + \text{Max}\left\{\left|\frac{1}{4}\right|, |0|, \left|\frac{1}{2}\right|, \left|-\frac{1}{2}\right|, |0|\right\}$$
$$= 1 + \frac{1}{2} = \frac{3}{2}$$

The smaller of the two numbers, $\dfrac{5}{4}$, is the bound. Every zero of g lies between $-\dfrac{5}{4}$ and $\dfrac{5}{4}$. ◀

| EXAMPLE 8 | **Obtaining Graphs Using Bounds on Zeros** |

Obtain a graph for each polynomial.

(a) $f(x) = x^5 + 3x^3 - 9x^2 + 5$ (b) $g(x) = 4x^5 - 2x^3 + 2x^2 + 1$

Solution

(a) Based on Example 7(a), every zero lies between -10 and 10. Using $X\min = -10$ and $X\max = 10$, we graph $Y_1 = f(x) = x^5 + 3x^3 - 9x^2 + 5$. Figure 76(a) shows the graph obtained using ZOOM-FIT. Figure 76(b) shows the graph after adjusting the viewing window to improve the graph.

Figure 76

(a)   (b)

Figure 77

(b) Based on Example 7(b), every zero lies between $-\dfrac{5}{4}$ and $\dfrac{5}{4}$. Using $X\min = -\dfrac{5}{4}$ and $X\max = \dfrac{5}{4}$, we graph $Y_1 = g(x) = 4x^5 - 2x^3 + 2x^2 + 1$. Figure 77 shows the graph after using ZOOM-FIT. Here no adjustment of the viewing window is needed. ◀

✏ **NOW WORK PROBLEM 33.**

The next example shows how to proceed when some of the coefficients of the polynomial are not integers.

| **EXAMPLE 9** | **Finding the Zeros of a Polynomial** |

Find all the real zeros of the polynomial function

$$f(x) = x^5 - 1.8x^4 - 17.79x^3 + 31.672x^2 + 37.95x - 8.7121$$

Figure 78

(a)

(b)

Solution

STEP 1: There are at most five real zeros.

STEP 2: Since there are noninteger coefficients, the Rational Zeros Theorem does not apply.

STEP 3: We determine the bounds of f. The leading coefficient of f is 1 with $a_4 = -1.8$, $a_3 = -17.79$, $a_2 = 31.672$, $a_1 = 37.95$, and $a_0 = -8.7121$. We evaluate the expressions using formula (4).

$$\text{Max}\{1, |-8.7121| + |37.95| + |31.672| + |-17.79| + |-1.8|\} = \text{Max}\{1, 97.9241\}$$
$$= 97.9241$$

$$1 + \text{Max}\{|-8.7121|, |37.95|, |31.672|, |-17.79|, |-1.8|\} = 1 + 37.95$$
$$= 38.95$$

The smaller of the two numbers, 38.95, is the bound. Every real zero of f lies between -38.95 and 38.95. Figure 78(a) shows the graph of f with $X\min = -38.95$ and $X\max = 38.95$. Figure 78(b) shows a graph of f after adjusting the viewing window to improve the graph.

STEP 4: From Figure 78(b), we see that f appears to have four x-intercepts: one near -4, one near -1, one between 0 and 1, and one near 3. The x-intercept near 3 might be a zero of even multiplicity since the graph seems to touch the x-axis at that point.

We use the Factor Theorem to determine if -4 and -1 are zeros. Since $f(-4) = f(-1) = 0$, we know that -4 and -1 are zeros. Using ZERO (or ROOT), we find that the remaining zeros are 0.20 and 3.30, rounded to two decimal places.

There are no real zeros on the graph that have not already been identified. So, either 3.30 is a zero of multiplicity 2 or there are two distinct zeros, each of which is 3.30, rounded to two decimal places. (Example 10 will explain how to determine which is true.) ◀

6   Use the Intermediate Value Theorem

The Intermediate Value Theorem requires that the function be *continuous*. Although calculus is needed to explain the meaning precisely, the *idea* of a continuous function is easy to understand. Very basically, a function f is continuous when its graph can be drawn without lifting pencil from paper, that is, when the graph contains no holes or jumps or gaps. For example, every polynomial function is continuous.

Intermediate Value Theorem

Let f denote a continuous function. If $a < b$ and if $f(a)$ and $f(b)$ are of opposite sign, then f has at least one zero between a and b.

Although the proof of this result requires advanced methods in calculus, it is easy to "see" why the result is true. Look at Figure 79.

Figure 79

If $f(a) < 0$ and $f(b) > 0$ and if f is continuous, there is at least one zero between a and b.

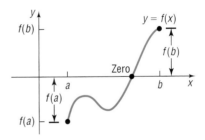

The Intermediate Value Theorem together with the TABLE feature of a graphing utility provides a basis for finding zeros.

EXAMPLE 10   **Using the Intermediate Value Theorem and a Graphing Utility to Locate Zeros**

Continue working with Example 9 to determine whether there is a repeated zero or two distinct zeros near 3.30.

Table 21

X	Y1
3.27	.08567
3.28	.03829
3.29	.00956
3.3	-1E-4
3.31	.0097
3.32	.03936
3.33	.08931

Y1◻X^5-1.8X^4-1...

Solution   We use the TABLE feature of a graphing utility. See Table 21. Since $f(3.29) = 0.00956 > 0$ and $f(3.30) = -0.0001 < 0$, by the Intermediate Value Theorem there is a zero between 3.29 and 3.30. Similarly, since $f(3.30) = -0.0001 < 0$ and $f(3.31) = 0.0097 > 0$, there is another zero between 3.30 and 3.31. Now we know that the five zeros of f are distinct. ◀

NOW WORK PROBLEM 75.

HISTORICAL FEATURE

Tartaglia
(1500–1557)

Formulas for the solution of third- and fourth-degree polynomial equations exist, and, while not very practical, they do have an interesting history.

In the 1500s in Italy, mathematical contests were a popular pastime, and persons possessing methods for solving problems kept them secret. (Solutions that were published were already common knowledge.) Niccolo of Brescia (1500–1557), commonly referred to as Tartaglia ("the stammerer"), had the secret for solving cubic (third-degree) equations, which gave him a decided advantage in the contests. See the Historical Problems.

Girolamo Cardano (1501–1576) found out that Tartaglia had the secret, and, being interested in cubics, he requested it from Tartaglia. The reluctant Tartaglia hesitated for some time, but finally, swearing Cardano to secrecy with midnight oaths by candlelight, told him the secret. Cardano then published the solution in his book *Ars Magna* (1545), giving Tartaglia the credit but rather compromising the secrecy. Tartaglia exploded into bitter recriminations, and each wrote pamphlets that reflected on the other's mathematics, moral character, and ancestry. See the Historical Problems.

The quartic (fourth-degree) equation was solved by Cardano's student Lodovico Ferrari, and this solution also was included, with credit and this time with permission, in the *Ars Magna*.

Attempts were made to solve the fifth-degree equation in similar ways, all of which failed. In the early 1800s, P. Ruffini, Niels Abel, and Evariste Galois all found ways to show that it is not possible to solve fifth-degree equations by formula, but the proofs required the introduction of new methods. Galois's methods eventually developed into a large part of modern algebra.

Historical Problems

Problems 1–8 develop the Tartaglia–Cardano solution of the cubic equation and show why it is not altogether practical.

1. Show that the general cubic equation $y^3 + by^2 + cy + d = 0$

   can be transformed into an equation of the form $x^3 + px + q = 0$ by using the substitution $y = x - \dfrac{b}{3}$.

2. In the equation $x^3 + px + q = 0$, replace x by $H + K$. Let $3HK = -p$, and show that $H^3 + K^3 = -q$.
   [**Hint:** $3H^2K + 3HK^2 = 3HKx$.]

3. Based on Problem 2, we have the two equations

   $$3HK = -p \quad \text{and} \quad H^3 + K^3 = -q$$

   Solve for K in $3HK = -p$ and substitute into $H^3 + K^3 = -q$. Then show that

   $$H = \sqrt[3]{\dfrac{-q}{2} + \sqrt{\dfrac{q^2}{4} + \dfrac{p^3}{27}}}$$

   [**Hint:** Look for an equation that is quadratic in form.]

4. Use the solution for H from Problem 3 and the equation $H^3 + K^3 = -q$ to show that

   $$K = \sqrt[3]{\dfrac{-q}{2} - \sqrt{\dfrac{q^2}{4} + \dfrac{p^3}{27}}}$$

5. Use the results from Problems 2–4 to show that the solution of $x^3 + px + q = 0$ is

   $$x = \sqrt[3]{\dfrac{-q}{2} + \sqrt{\dfrac{q^2}{4} + \dfrac{p^3}{27}}} + \sqrt[3]{\dfrac{-q}{2} - \sqrt{\dfrac{q^2}{4} + \dfrac{p^3}{27}}}$$

6. Use the result of Problem 5 to solve the equation $x^3 - 6x - 9 = 0$.

7. Use a calculator and the result of Problem 5 to solve the equation $x^3 + 3x - 14 = 0$.

8. Use the methods of this chapter to solve the equation $x^3 + 3x - 14 = 0$.

3.6 Assess Your Understanding

'Are You Prepared?'

Answers are given at the end of these exercises. If you get a wrong answer, read the pages listed in red.

1. In the set $\left\{-2, -\sqrt{2}, 0, \dfrac{1}{2}, 4.5, \pi\right\}$, name the numbers that are integers. Which are rational numbers? (pp. 2–4)

2. Factor the expression $6x^2 + x - 2$. (pp. 50–51)

3. Find the quotient and remainder if $3x^4 - 5x^3 + 7x - 4$ is divided by $x - 3$. (pp. 40–43 and pp. 53–56)

4. Solve the equation $x^2 + x - 3 = 0$. (pp. 112–119)

Concepts and Vocabulary

5. In the process of polynomial division, (Divisor)(Quotient) + _____ = _____.

6. When a polynomial function f is divided by $x - c$, the remainder is _____.

7. If a function f, whose domain is all real numbers, is even and if 4 is a zero of f, then _____ is also a zero.

8. *True or False:* Every polynomial function of degree 3 with real coefficients has exactly three real zeros.

9. *True or False:* The only potential rational zeros of $f(x) = 2x^5 - x^3 + x^2 - x + 1$ are $\pm 1, \pm 2$.

10. *True or False:* If f is a polynomial function of degree 4 and if $f(2) = 5$, then

$$\frac{f(x)}{x - 2} = p(x) + \frac{5}{x - 2}$$

where $p(x)$ is a polynomial of degree 3.

Skill Building

In Problems 11–20, use the Factor Theorem to determine whether $x - c$ is a factor of f. If it is, write f in factored form, that is, write f in the form $f(x) = (x - c)$ (quotient).

11. $f(x) = 4x^3 - 3x^2 - 8x + 4$; $c = 3$

12. $f(x) = -4x^3 + 5x^2 + 8$; $c = -2$

13. $f(x) = 3x^4 - 6x^3 - 5x + 10$; $c = 1$

14. $f(x) = 4x^4 - 15x^2 - 4$; $c = 2$

15. $f(x) = 3x^6 + 2x^3 - 176$; $c = -2$

16. $f(x) = 2x^6 - 18x^4 + x^2 - 9$; $c = -3$

17. $f(x) = 4x^6 - 64x^4 + x^2 - 16$; $c = 4$

18. $f(x) = x^6 - 16x^4 + x^2 - 16$; $c = -4$

19. $f(x) = 2x^4 - x^3 + 2x - 1$; $c = -\dfrac{1}{2}$

20. $f(x) = 3x^4 + x^3 - 3x + 1$; $c = -\dfrac{1}{3}$

In Problems 21–32, tell the maximum number of real zeros that each polynomial function may have. Then list the potential rational zeros of each polynomial function. Do not attempt to find the zeros.

21. $f(x) = 3x^4 - 3x^3 + x^2 - x + 1$

22. $f(x) = x^5 - x^4 + 2x^2 + 3$

23. $f(x) = x^5 - 6x^2 + 9x - 3$

24. $f(x) = 2x^5 - x^4 - x^2 + 1$

25. $f(x) = -4x^3 - x^2 + x + 2$

26. $f(x) = 6x^4 - x^2 + 2$

27. $f(x) = 3x^4 - x^2 + 2$

28. $f(x) = -4x^3 + x^2 + x + 2$

29. $f(x) = 2x^5 - x^3 + 2x^2 + 4$

30. $f(x) = 3x^5 - x^2 + 2x + 3$

31. $f(x) = 6x^4 + 2x^3 - x^2 + 2$

32. $f(x) = -6x^3 - x^2 + x + 3$

In Problems 33–38, find the bounds to the zeros of each polynomial function. Use the bounds to obtain a complete graph of f.

33. $f(x) = 2x^3 + x^2 - 1$

34. $f(x) = 3x^3 - 2x^2 + x + 4$

35. $f(x) = x^3 - 5x^2 - 11x + 11$

36. $f(x) = 2x^3 - x^2 - 11x - 6$

37. $f(x) = x^4 + 3x^3 - 5x^2 + 9$

38. $f(x) = 4x^4 - 12x^3 + 27x^2 - 54x + 81$

In Problems 39–56, find the real zeros of f. Use the real zeros to factor f.

39. $f(x) = x^3 + 2x^2 - 5x - 6$

40. $f(x) = x^3 + 8x^2 + 11x - 20$

41. $f(x) = 2x^3 - 13x^2 + 24x - 9$

42. $f(x) = 2x^3 - 5x^2 - 4x + 12$

43. $f(x) = 3x^3 + 4x^2 + 4x + 1$

44. $f(x) = 3x^3 - 7x^2 + 12x - 28$

45. $f(x) = x^3 - 8x^2 + 17x - 6$

46. $f(x) = x^3 + 6x^2 + 6x - 4$

47. $f(x) = x^4 + x^3 - 3x^2 - x + 2$

48. $f(x) = x^4 - x^3 - 6x^2 + 4x + 8$

49. $f(x) = 2x^4 + 17x^3 + 35x^2 - 9x - 45$

50. $f(x) = 4x^4 - 15x^3 - 8x^2 + 15x + 4$

51. $f(x) = 2x^4 - 3x^3 - 21x^2 - 2x + 24$

52. $f(x) = 2x^4 + 11x^3 - 5x^2 - 43x + 35$

53. $f(x) = 4x^4 + 7x^2 - 2$

54. $f(x) = 4x^4 + 15x^2 - 4$

55. $f(x) = 4x^5 - 8x^4 - x + 2$

56. $f(x) = 4x^5 + 12x^4 - x - 3$

In Problems 57–62, find the real zeros of f rounded to two decimal places.

57. $f(x) = x^3 + 3.2x^2 - 16.83x - 5.31$

58. $f(x) = x^3 + 3.2x^2 - 7.25x - 6.3$

59. $f(x) = x^4 - 1.4x^3 - 33.71x^2 + 23.94x + 292.41$

60. $f(x) = x^4 + 1.2x^3 - 7.46x^2 - 4.692x + 15.2881$

61. $f(x) = x^3 + 19.5x^2 - 1021x + 1000.5$

62. $f(x) = x^3 + 42.2x^2 - 664.8x + 1490.4$

In Problems 63–72, find the real solutions of each equation.

63. $x^4 - x^3 + 2x^2 - 4x - 8 = 0$

64. $2x^3 + 3x^2 + 2x + 3 = 0$

65. $3x^3 + 4x^2 - 7x + 2 = 0$

66. $2x^3 - 3x^2 - 3x - 5 = 0$

67. $3x^3 - x^2 - 15x + 5 = 0$

68. $2x^3 - 11x^2 + 10x + 8 = 0$

69. $x^4 + 4x^3 + 2x^2 - x + 6 = 0$

70. $x^4 - 2x^3 + 10x^2 - 18x + 9 = 0$

71. $x^3 - \dfrac{2}{3}x^2 + \dfrac{8}{3}x + 1 = 0$

72. $x^3 - \dfrac{2}{3}x^2 + 3x - 2 = 0$

In Problems 73–78, use the Intermediate Value Theorem to show that each function has a zero in the given interval. Approximate the zero rounded to two decimal places.

73. $f(x) = 8x^4 - 2x^2 + 5x - 1; \quad [0, 1]$

74. $f(x) = x^4 + 8x^3 - x^2 + 2; \quad [-1, 0]$

75. $f(x) = 2x^3 + 6x^2 - 8x + 2; \quad [-5, -4]$

76. $f(x) = 3x^3 - 10x + 9; \quad [-3, -2]$

77. $f(x) = x^5 - x^4 + 7x^3 - 7x^2 - 18x + 18; \quad [1.4, 1.5]$

78. $f(x) = x^5 - 3x^4 - 2x^3 + 6x^2 + x + 2; \quad [1.7, 1.8]$

Applications and Extensions

79. Cost of Manufacturing In Problem 95 of Section 3.2, you found the cost function for manufacturing Chevy Cavaliers. Use the methods learned in this section to determine how many Cavaliers can be manufactured at a total cost of $3,000,000. In other words, solve the equation $C(x) = 3000$.

80. Cost of Printing In Problem 96 of Section 3.2, you found the cost function for printing textbooks. Use the methods learned in this section to determine how many textbooks can be printed at a total cost of $200,000. In other words, solve the equation $C(x) = 200$.

81. Find k such that $f(x) = x^3 - kx^2 + kx + 2$ has the factor $x - 2$.

82. Find k such that $f(x) = x^4 - kx^3 + kx^2 + 1$ has the factor $x + 2$.

83. What is the remainder when $f(x) = 2x^{20} - 8x^{10} + x - 2$ is divided by $x - 1$?

84. What is the remainder when
$f(x) = -3x^{17} + x^9 - x^5 + 2x$ is divided by $x + 1$?

85. One solution of the equation $x^3 - 8x^2 + 16x - 3 = 0$ is 3. Find the sum of the remaining solutions.

86. One solution of the equation $x^3 + 5x^2 + 5x - 2 = 0$ is -2. Find the sum of the remaining solutions.

87. What is the length of the edge of a cube if, after a slice 1 inch thick is cut from one side, the volume remaining is 294 cubic inches?

88. What is the length of the edge of a cube if its volume could be doubled by an increase of 6 centimeters in one edge, an increase of 12 centimeters in a second edge, and a decrease of 4 centimeters in the third edge?

89. Use the Factor Theorem to prove that $x - c$ is a factor of $x^n - c^n$ for any positive integer n.

90. Use the Factor Theorem to prove that $x + c$ is a factor of $x^n + c^n$ if $n \geq 1$ is an odd integer.

91. Let $f(x)$ be a polynomial function whose coefficients are integers. Suppose that r is a real zero of f and that the leading coefficient of f is 1. Use the Rational Zeros Theorem to show that r is either an integer or an irrational number.

92. Prove the Rational Zeros Theorem.

[**Hint:** Let $\dfrac{p}{q}$, where p and q have no common factors except 1 and -1, be a zero of the polynomial function $f(x) = a_n x^n + a_{n-1} x^{n-1} + \cdots + a_1 x + a_0$, whose coefficients are all integers. Show that $a_n p^n + a_{n-1} p^{n-1} q + \cdots + a_1 pq^{n-1} + a_0 q^n = 0$. Now, because p is a factor of the first n terms of this equation, p must also be a factor of the term $a_0 q^n$. Since p is not a factor of q (why?), p must be a factor of a_0. Similarly, q must be a factor of a_n.]

Discussion and Writing

93. Is $\dfrac{1}{3}$ a zero of $f(x) = 2x^3 + 3x^2 - 6x + 7$? Explain.

94. Is $\dfrac{1}{3}$ a zero of $f(x) = 4x^3 - 5x^2 - 3x + 1$? Explain.

95. Is $\dfrac{3}{5}$ a zero of $f(x) = 2x^6 - 5x^4 + x^3 - x + 1$? Explain.

96. Is $\dfrac{2}{3}$ a zero of $f(x) = x^7 + 6x^5 - x^4 + x + 2$? Explain.

'Are You Prepared?' Answers

1. Integers: $\{-2, 0\}$; Rational $\left\{-2, 0, \dfrac{1}{2}, 4.5\right\}$

2. $(3x + 2)(2x - 1)$

3. Quotient: $3x^3 + 4x^2 + 12x + 43$; Remainder: 125

4. $\left\{\dfrac{-1 - \sqrt{13}}{2}, \dfrac{-1 + \sqrt{13}}{2}\right\}$

3.7 Complex Zeros; Fundamental Theorem of Algebra

PREPARING FOR THIS SECTION   *Before getting started, review the following:*

- Complex Numbers (Section 1.4, pp. 124–129)
- Quadratic Equations with a Negative Discriminant (Section 1.4, pp. 129–131)

 Now work the 'Are You Prepared?' problems on page 385.

OBJECTIVES 1 Use the Conjugate Pairs Theorem
2 Find a Polynomial Function with Specified Zeros
3 Find the Complex Zeros of a Polynomial

In Section 3.6 we found the **real** zeros of a polynomial function. In this section we will find the **complex** zeros of a polynomial function. Finding the complex zeros of a function requires finding all zeros of the form $a + bi$. These zeros will be real if $b = 0$.

A variable in the complex number system is referred to as a **complex variable**.

A **complex polynomial function** f of degree n is a function of the form

$$f(x) = a_n x^n + a_{n-1} x^{n-1} + \cdots + a_1 x + a_0 \qquad \textbf{(1)}$$

where $a_n, a_{n-1}, \ldots, a_1, a_0$ are complex numbers, $a_n \neq 0$, n is a nonnegative integer, and x is a complex variable. As before, a_n is called the **leading coefficient** of f. A complex number r is called a **(complex) zero** of f if $f(r) = 0$.

We have learned that some quadratic equations have no real solutions, but that in the complex number system every quadratic equation has a solution, either real or complex. The next result, proved by Karl Friedrich Gauss (1777–1855) when he was 22 years old,[*] gives an extension to complex polynomials. In fact, this result is so important and useful that it has become known as the **Fundamental Theorem of Algebra**.

Fundamental Theorem of Algebra

Every complex polynomial function $f(x)$ of degree $n \geq 1$ has at least one complex zero.

We shall not prove this result, as the proof is beyond the scope of this book. However, using the Fundamental Theorem of Algebra and the Factor Theorem, we can prove the following result:

Theorem

Every complex polynomial function $f(x)$ of degree $n \geq 1$ can be factored into n linear factors (not necessarily distinct) of the form

$$f(x) = a_n (x - r_1)(x - r_2) \cdots \cdots (x - r_n) \qquad \textbf{(2)}$$

where $a_n, r_1, r_2, \ldots, r_n$ are complex numbers. That is, every complex polynomial function of degree $n \geq 1$ has exactly n (not necessarily distinct) zeros.

_____

[*]In all, Gauss gave four different proofs of this theorem, the first one in 1799 being the subject of his doctoral dissertation.

Proof Let

$$f(x) = a_n x^n + a_{n-1} x^{n-1} + \cdots + a_1 x + a_0$$

By the Fundamental Theorem of Algebra, f has at least one zero, say r_1. Then, by the Factor Theorem, $x - r_1$ is a factor, and

$$f(x) = (x - r_1)q_1(x)$$

where $q_1(x)$ is a complex polynomial of degree $n - 1$ whose leading coefficient is a_n. Again by the Fundamental Theorem of Algebra, the complex polynomial $q_1(x)$ has at least one zero, say r_2. By the Factor Theorem, $q_1(x)$ has the factor $x - r_2$, so

$$q_1(x) = (x - r_2)q_2(x)$$

where $q_2(x)$ is a complex polynomial of degree $n - 2$ whose leading coefficient is a_n. Consequently,

$$f(x) = (x - r_1)(x - r_2)q_2(x)$$

Repeating this argument n times, we finally arrive at

$$f(x) = (x - r_1)(x - r_2) \cdot \cdots \cdot (x - r_n)q_n(x)$$

where $q_n(x)$ is a complex polynomial of degree $n - n = 0$ whose leading coefficient is a_n. Thus, $q_n(x) = a_n x^0 = a_n$, and so

$$f(x) = a_n(x - r_1)(x - r_2) \cdot \cdots \cdot (x - r_n)$$

We conclude that every complex polynomial function $f(x)$ of degree $n \geq 1$ has exactly n (not necessarily distinct) zeros. ∎

1 Use the Conjugate Pairs Theorem

We can use the Fundamental Theorem of Algebra to obtain valuable information about the complex zeros of polynomials whose coefficients are real numbers.

Conjugate Pairs Theorem

Let $f(x)$ be a polynomial whose coefficients are real numbers. If $r = a + bi$ is a zero of f, then the complex conjugate $\bar{r} = a - bi$ is also a zero of f.

In other words, for polynomials whose coefficients are real numbers, the zeros occur in conjugate pairs.

Proof Let

$$f(x) = a_n x^n + a_{n-1} x^{n-1} + \cdots + a_1 x + a_0$$

where $a_n, a_{n-1}, \ldots, a_1, a_0$ are real numbers and $a_n \neq 0$. If $r = a + bi$ is a zero of f, then $f(r) = f(a + bi) = 0$, so

$$a_n r^n + a_{n-1} r^{n-1} + \cdots + a_1 r + a_0 = 0$$

We take the conjugate of both sides to get

$$\overline{a_n r^n + a_{n-1} r^{n-1} + \cdots + a_1 r + a_0} = \bar{0}$$

$$\overline{a_n r^n} + \overline{a_{n-1} r^{n-1}} + \cdots + \overline{a_1 r} + \overline{a_0} = \bar{0} \qquad \text{The conjugate of a sum equals the sum of the conjugates (see Section 1.4).}$$

$$\overline{a_n} (\bar{r})^n + \overline{a_{n-1}} (\bar{r})^{n-1} + \cdots + \overline{a_1} \bar{r} + \overline{a_0} = \bar{0} \qquad \text{The conjugate of a product equals the product of the conjugates.}$$

$$a_n (\bar{r})^n + a_{n-1} (\bar{r})^{n-1} + \cdots + a_1 \bar{r} + a_0 = 0 \qquad \text{The conjugate of a real number equals the real number.}$$

This last equation states that $f(\bar{r}) = 0$; that is, $\bar{r} = a - bi$ is a zero of f. ∎

The value of this result should be clear. If we know that, $3 + 4i$ is a zero of a polynomial with real coefficients, then we know that $3 - 4i$ is also a zero. This result has an important corollary.

COROLLARY   A polynomial f of odd degree with real coefficients has at least one real zero.

Proof   Because complex zeros occur as conjugate pairs in a polynomial with real coefficients, there will always be an even number of zeros that are not real numbers. Consequently, since f is of odd degree, one of its zeros has to be a real number. ■

For example, the polynomial $f(x) = x^5 - 3x^4 + 4x^3 - 5$ has at least one zero that is a real number, since f is of degree 5 (odd) and has real coefficients.

EXAMPLE 1   **Using the Conjugate Pairs Theorem**

A polynomial f of degree 5 whose coefficients are real numbers has the zeros $1, 5i$, and $1 + i$. Find the remaining two zeros.

Solution   Since complex zeros appear as conjugate pairs, it follows that $-5i$, the conjugate of $5i$, and $1 - i$, the conjugate of $1 + i$, are the two remaining zeros. ◄

NOW WORK PROBLEM 7.

2 Find a Polynomial Function with Specified Zeros

EXAMPLE 2   **Finding a Polynomial Function Whose Zeros Are Given**

(a) Find a polynomial f of degree 4 whose coefficients are real numbers and that has the zeros $1, 1$, and $-4 + i$.

(b) Graph the polynomial found in part (a) to verify your result.

Solution   (a) Since $-4 + i$ is a zero, by the Conjugate Pairs Theorem, $-4 - i$ must also be a zero of f. Because of the Factor Theorem, if $f(c) = 0$, then $x - c$ is a factor of $f(x)$. So we can now write f as

$$f(x) = a(x - 1)(x - 1)[x - (-4 + i)][x - (-4 - i)]$$

where a is any real number. If we let $a = 1$, we obtain

$$\begin{aligned} f(x) &= (x - 1)(x - 1)[x - (-4 + i)][x - (-4 - i)] \\ &= (x^2 - 2x + 1)[x^2 - (-4 + i)x - (-4 - i)x + (-4 + i)(-4 - i)] \\ &= (x^2 - 2x + 1)(x^2 + 4x - ix + 4x + ix + 16 + 4i - 4i - i^2) \\ &= (x^2 - 2x + 1)(x^2 + 8x + 17) \\ &= x^4 + 8x^3 + 17x^2 - 2x^3 - 16x^2 - 34x + x^2 + 8x + 17 \\ &= x^4 + 6x^3 + 2x^2 - 26x + 17 \end{aligned}$$

Figure 80

(b) A quick analysis of the polynomial f tells us what to expect:

At most three turning points.
For large $|x|$, the graph will behave like $y = x^4$.
A repeated real zero at 1 so that the graph will touch the x-axis at 1.
The only x-intercept is at 1.

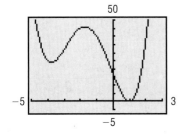

Figure 80 shows the complete graph. (Do you see why? The graph has exactly three turning points and the degree of the polynomial is 4.) ◄

—— **Exploration** ——

Graph the function found in Example 2 for $a = 2$ and $a = -1$. Does the value of a affect the zeros of f? How does the value of a affect the graph of f?

Now we can prove the theorem we conjectured earlier in Section 3.6.

Theorem

Every polynomial function with real coefficients can be uniquely factored over the real numbers into a product of linear factors and/or irreducible quadratic factors.

Proof Every complex polynomial f of degree n has exactly n zeros and can be factored into a product of n linear factors. If its coefficients are real, then those zeros that are complex numbers will always occur as conjugate pairs. As a result, if $r = a + bi$ is a complex zero, then so is $\bar{r} = a - bi$. Consequently, when the linear factors $x - r$ and $x - \bar{r}$ of f are multiplied, we have

$$(x - r)(x - \bar{r}) = x^2 - (r + \bar{r})x + r\bar{r} = x^2 - 2ax + a^2 + b^2$$

This second-degree polynomial has real coefficients and is irreducible (over the real numbers). Thus, the factors of f are either linear or irreducible quadratic factors. ∎

3 **Find the Complex Zeros of a Polynomial**

| **EXAMPLE 3** | **Finding the Complex Zeros of a Polynomial** |

Find the complex zeros of the polynomial function

$$f(x) = 3x^4 + 5x^3 + 25x^2 + 45x - 18$$

Solution **STEP 1:** The degree of f is 4. So f will have four complex zeros.

STEP 2: The Rational Zeros Theorem provides information about the potential rational zeros of polynomials with integer coefficients. For this polynomial (which has integer coefficients), the potential rational zeros are

$$\pm\frac{1}{3}, \pm\frac{2}{3}, \pm1, \pm2, \pm3, \pm6, \pm9, \pm18$$

Figure 81

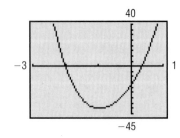

STEP 3: Figure 81 shows the graph of f. The graph has the characteristics that we expect of this polynomial of degree 4: It behaves like $y = 3x^4$ for large $|x|$ and has y-intercept -18. There are x-intercepts near -2 and between 0 and 1.

STEP 4: Because $f(-2) = 0$, we known that -2 is a zero and $x + 2$ is a factor of f. We can use long division or synthetic division to factor f:

$$f(x) = (x + 2)(3x^3 - x^2 + 27x - 9)$$

From the graph of f and the list of potential rational zeros, it appears that $\frac{1}{3}$ may also be a zero of f. Since $f\left(\frac{1}{3}\right) = 0$, we know that $\frac{1}{3}$ is a zero of f. We use synthetic division on the depressed equation of f to factor.

$$
\begin{array}{r|rrrr}
\frac{1}{3}) & 3 & -1 & 27 & -9 \\
           &   & 1 & 0 & 9 \\
\hline
           & 3 & 0 & 27 & 0
\end{array}
$$

Using the bottom row of the synthetic division, we find

$$f(x) = (x + 2)\left(x - \frac{1}{3}\right)(3x^2 + 27) = 3(x + 2)\left(x - \frac{1}{3}\right)(x^2 + 9)$$

The factor $x^2 + 9$ does not have any real zeros; its complex zeros are $\pm 3i$. The complex zeros of $f(x) = 3x^4 + 5x^3 + 25x^2 + 45x - 18$ are $-2, \dfrac{1}{3}, 3i, -3i$.

◀

━━━ **NOW WORK PROBLEM 33.**

3.7 Assess Your Understanding

'Are You Prepared?'

Answers are given at the end of these exercises. If you get a wrong answer, read the pages listed in red.

1. Find the sum and the product of the complex numbers $3 - 2i$ and $-3 + 5i$. (pp. 125–126)

2. In the complex number system, solve the equation $x^2 + 2x + 2 = 0$. (pp. 129–131)

Concepts and Vocabulary

3. Every polynomial function of odd degree with real coefficients will have at least _____ real zero(s).

4. If $3 + 4i$ is a zero of a polynomial function of degree 5 with real coefficients, then so is _____.

5. *True or False:* A polynomial function of degree n with real coefficients has exactly n complex zeros. At most n of them are real zeros.

6. *True or False:* A polynomial function of degree 4 with real coefficients could have $-3, 2 + i, 2 - i$, and $-3 + 5i$ as its zeros.

Skill Building

In Problems 7–16, information is given about a polynomial $f(x)$ whose coefficients are real numbers. Find the remaining zeros of f.

7. Degree 3; zeros: $3, 4 - i$

8. Degree 3; zeros: $4, 3 + i$

9. Degree 4; zeros: $i, 1 + i$

10. Degree 4; zeros: $1, 2, 2 + i$

11. Degree 5; zeros: $1, i, 2i$

12. Degree 5; zeros: $0, 1, 2, i$

13. Degree 4; zeros: $i, 2, -2$

14. Degree 4; zeros: $2 - i, -i$

15. Degree 6; zeros: $2, 2 + i, -3 - i, 0$

16. Degree 6; zeros: $i, 3 - 2i, -2 + i$

In Problems 17–22, form a polynomial $f(x)$ with real coefficients having the given degree and zeros.

17. Degree 4; zeros: $3 + 2i$; 4, multiplicity 2

18. Degree 4; zeros: $i, 1 + 2i$

19. Degree 5; zeros: $2; -i; 1 + i$

20. Degree 6; zeros: $i, 4 - i; 2 + i$

21. Degree 4; zeros: 3, multiplicity 2; $-i$

22. Degree 5; zeros: 1, multiplicity 3; $1 + i$

In Problems 23–30, use the given zero to find the remaining zeros of each function.

23. $f(x) = x^3 - 4x^2 + 4x - 16$; zero: $2i$

24. $g(x) = x^3 + 3x^2 + 25x + 75$; zero: $-5i$

25. $f(x) = 2x^4 + 5x^3 + 5x^2 + 20x - 12$; zero: $-2i$

26. $h(x) = 3x^4 + 5x^3 + 25x^2 + 45x - 18$; zero: $3i$

27. $h(x) = x^4 - 9x^3 + 21x^2 + 21x - 130$; zero: $3 - 2i$

28. $f(x) = x^4 - 7x^3 + 14x^2 - 38x - 60$; zero: $1 + 3i$

29. $h(x) = 3x^5 + 2x^4 + 15x^3 + 10x^2 - 528x - 352$; zero: $-4i$

30. $g(x) = 2x^5 - 3x^4 - 5x^3 - 15x^2 - 207x + 108$; zero: $3i$

In Problems 31–40, find the complex zeros of each polynomial function. Write f in factored form.

31. $f(x) = x^3 - 1$

32. $f(x) = x^4 - 1$

33. $f(x) = x^3 - 8x^2 + 25x - 26$

34. $f(x) = x^3 + 13x^2 + 57x + 85$

35. $f(x) = x^4 + 5x^2 + 4$

36. $f(x) = x^4 + 13x^2 + 36$

37. $f(x) = x^4 + 2x^3 + 22x^2 + 50x - 75$

38. $f(x) = x^4 + 3x^3 - 19x^2 + 27x - 252$

39. $f(x) = 3x^4 - x^3 - 9x^2 + 159x - 52$

40. $f(x) = 2x^4 + x^3 - 35x^2 - 113x + 65$

Discussion and Writing

In Problems 41 and 42, explain why the facts given are contradictory.

41. $f(x)$ is a polynomial of degree 3 whose coefficients are real numbers; its zeros are $4 + i, 4 - i$, and $2 + i$.

42. $f(x)$ is a polynomial of degree 3 whose coefficients are real numbers; its zeros are $2, i$, and $3 + i$.

43. $f(x)$ is a polynomial of degree 4 whose coefficients are real numbers; three of its zeros are $2, 1 + 2i$, and $1 - 2i$. Explain why the remaining zero must be a real number.

44. $f(x)$ is a polynomial of degree 4 whose coefficients are real numbers; two of its zeros are -3 and $4 - i$. Explain why one of the remaining zeros must be a real number. Write down one of the missing zeros.

'Are You Prepared?' Answers

1. Sum: $3i$; product: $1 + 21i$ **2.** $\{-1 - i, -1 + i\}$

Chapter Review

Things to Know

Quadratic function (pp. 298–302)

$f(x) = ax^2 + bx + c$

Graph is a parabola that opens up if $a > 0$ and opens down if $a < 0$.

Vertex: $\left(-\dfrac{b}{2a}, f\left(-\dfrac{b}{2a}\right)\right)$

Axis of symmetry: $x = -\dfrac{b}{2a}$

y-intercept: $f(0)$

x-intercept(s): If any, found by finding the real solutions of the equation $ax^2 + bx + c = 0$.

Power function (pp. 319–321)

$f(x) = x^n, \quad n \geq 2$ even

Domain: all real numbers   Range: nonnegative real numbers

Passes through $(-1, 1), (0, 0), (1, 1)$

Even function

Decreasing on $(-\infty, 0)$, increasing on $(0, \infty)$

$f(x) = x^n, \quad n \geq 3$ odd

Domain: all real numbers   Range: all real numbers

Passes through $(-1, -1), (0, 0), (1, 1)$

Odd function

Increasing on $(-\infty, \infty)$

Polynomial function (p. 318 and pp. 322–326)

$f(x) = a_n x^n + a_{n-1}x^{n-1} + \cdots$
$\quad + a_1 x + a_0, \quad a_n \neq 0$

Domain: all real numbers

At most $n - 1$ turning points

End behavior: Behaves like $y = a_n x^n$ for large $|x|$

Zeros of a polynomial function f (p. 323)

Numbers for which $f(x) = 0$; the real zeros of f are the x-intercepts of the graph of f.

Rational function (p. 334)

$$R(x) = \frac{p(x)}{q(x)}$$

p, q are polynomial functions.

Domain: $\{x | q(x) \neq 0\}$

Vertical asymptotes: With $R(x)$ in lowest terms, if $q(r) = 0$, then $x = r$ is a vertical asymptote.

Horizontal or oblique asymptotes: See the summary on page 343.

Inverse Variation (p. 353)

Let x and y denote two quantities. Then y varies inversely with x, or y is inversely proportional to x, if there is a nonzero constant k such that $y = \dfrac{k}{x}$.

Remainder Theorem (p. 368)

If a polynomial $f(x)$ is divided by $x - c$, then the remainder is $f(c)$.

Factor Theorem (p. 368)

$x - c$ is a factor of a polynomial $f(x)$ if and only if $f(c) = 0$.

Rational Zeros Theorem (p. 370)

Let f be a polynomial function of degree 1 or higher of the form

$$f(x) = a_n x^n + a_{n-1}x^{n-1} + \cdots + a_1 x + a_0, \quad a_n \neq 0, a_0 \neq 0$$

where each coefficient is an integer. If $\dfrac{p}{q}$, in lowest terms, is a rational zero of f, then p must be a factor of a_0 and q must be a factor of a_n.

Intermediate Value Theorem (p. 377)

Let f denote a continuous function. If $a < b$ and $f(a)$ and $f(b)$ are of opposite sign, then f has at least one zero between a and b.

Fundamental Theorem of Algebra (p. 381)

Every complex polynomial function $f(x)$ of degree $n \geq 1$ has at least one complex zero.

Conjugate Pairs Theorem (p. 382)

Let $f(x)$ be a polynomial whose coefficients are real numbers. If $r = a + bi$ is a zero of f, then its complex conjugate $\bar{r} = a - bi$ is also a zero of f.

Objectives

Section		You should be able to . . .	Review Exercises
3.1	1	Graph a quadratic function using transformations (p. 299)	1–6
	2	Identify the vertex and axis of symmetry of a quadratic function (p. 301)	7–16
	3	Graph a quadratic function using its vertex, axis, and intercepts (p. 303)	7–16
	4	Use the maximum or minimum value of a quadratic function to solve applied problems (p. 306)	115–122
	5	Use a graphing utility to find the quadratic function of best fit to data (p. 310)	125
3.2	1	Identify polynomial functions and their degree (p. 318)	23–26
	2	Graph polynomial functions using transformations (p. 319)	1–6, 27–32
	3	Identify the zeros of a polynomial function and their multiplicity (p. 322)	33–40
	4	Analyze the graph of a polynomial function (p. 327)	33–40
	5	Find the cubic function of best fit to data (p. 329)	126
3.3	1	Find the domain of a rational function (p. 335)	41–44
	2	Find the vertical asymptotes of a rational function (p. 338)	41–44
	3	Find the horizontal or oblique asymptotes of a rational function (p. 339)	41–44
3.4	1	Analyze the graph of a rational function (p. 346)	45–56
	2	Solve applied problems involving rational functions (p. 352)	127
	3	Construct a model using inverse variation (p. 353)	123

Review Exercises

In Problems 1–6, graph each function using transformations (shifting, compressing, stretching, and reflection). Verify your result using a graphing utility.

1. $f(x) = (x - 2)^2 + 2$

2. $f(x) = (x + 1)^2 - 4$

3. $f(x) = -(x - 4)^2$

4. $f(x) = (x - 1)^2 - 3$

5. $f(x) = 2(x + 1)^2 + 4$

6. $f(x) = -3(x + 2)^2 + 1$

In Problems 7–16, graph each quadratic function by determining whether its graph opens up or down and by finding its vertex, axis of symmetry, y-intercept, and x-intercepts, if any.

7. $f(x) = (x - 2)^2 + 2$

8. $f(x) = (x + 1)^2 - 4$

9. $f(x) = \frac{1}{4}x^2 - 16$

10. $f(x) = -\frac{1}{2}x^2 + 2$

11. $f(x) = -4x^2 + 4x$

12. $f(x) = 9x^2 - 6x + 3$

13. $f(x) = \frac{9}{2}x^2 + 3x + 1$

14. $f(x) = -x^2 + x + \frac{1}{2}$

15. $f(x) = 3x^2 + 4x - 1$

16. $f(x) = -2x^2 - x + 4$

In Problems 17–22, determine whether the given quadratic function has a maximum value or a minimum value, and then find the value.

17. $f(x) = 3x^2 - 6x + 4$

18. $f(x) = 2x^2 + 8x + 5$

19. $f(x) = -x^2 + 8x - 4$

20. $f(x) = -x^2 - 10x - 3$

21. $f(x) = -3x^2 + 12x + 4$

22. $f(x) = -2x^2 + 4$

In Problems 23–26, determine which functions are polynomial functions. For those that are, state the degree. For those that are not, tell why not.

23. $f(x) = 4x^5 - 3x^2 + 5x - 2$

24. $f(x) = \frac{3x^5}{2x + 1}$

25. $f(x) = 3x^2 + 5x^{1/2} - 1$

26. $f(x) = 3$

In Problems 27–32, graph each function using transformations (shifting, compressing, stretching, and reflection). Show all the stages. Verify your result using a graphing utility.

27. $f(x) = (x + 2)^3$

28. $f(x) = -x^3 + 3$

29. $f(x) = -(x - 1)^4$

30. $f(x) = (x - 1)^4 - 2$

31. $f(x) = 2(x + 1)^4 + 2$

32. $f(x) = (1 - x)^3$

In Problems 33–40, for each polynomial function f:
 (a) *Find the x- and y-intercepts of the graph of f.*
 (b) *Determine whether the graph crosses or touches the x-axis at each x-intercept.*
 (c) *End behavior: Find the power function that the graph of f resembles for large values of |x|.*
 (d) *Use a graphing utility to graph f.*
 (e) *Determine the number of turning points on the graph of f. Approximate the turning points if any exist, rounded to two decimal places.*
 (f) *Use the information obtained in parts (a) to (e) to draw a complete graph of f by hand.*
 (g) *Find the domain of f. Use the graph to find the range of f.*
 (h) *Use the graph to determine where f is increasing and where f is decreasing.*

33. $f(x) = x(x + 2)(x + 4)$

34. $f(x) = x(x - 2)(x - 4)$

35. $f(x) = (x - 2)^2(x + 4)$

36. $f(x) = (x - 2)(x + 4)^2$ **37.** $f(x) = x^3 - 4x^2$ **38.** $f(x) = x^3 + 4x$

39. $f(x) = (x - 1)^2(x + 3)(x + 1)$ **40.** $f(x) = (x - 4)(x + 2)^2(x - 2)$

In Problems 41–44, find the domain of each rational function. Find any horizontal, vertical, or oblique asymptotes.

41. $R(x) = \dfrac{x + 2}{x^2 - 9}$ **42.** $R(x) = \dfrac{x^2 + 4}{x - 2}$ **43.** $R(x) = \dfrac{x^2 + 3x + 2}{(x + 2)^2}$ **44.** $R(x) = \dfrac{x^3}{x^3 - 1}$

In Problems 45–56, discuss each rational function following the eight steps on page 346.

45. $R(x) = \dfrac{2x - 6}{x}$ **46.** $R(x) = \dfrac{4 - x}{x}$ **47.** $H(x) = \dfrac{x + 2}{x(x - 2)}$ **48.** $H(x) = \dfrac{x}{x^2 - 1}$

49. $R(x) = \dfrac{x^2 + x - 6}{x^2 - x - 6}$ **50.** $R(x) = \dfrac{x^2 - 6x + 9}{x^2}$ **51.** $F(x) = \dfrac{x^3}{x^2 - 4}$ **52.** $F(x) = \dfrac{3x^3}{(x - 1)^2}$

53. $R(x) = \dfrac{2x^4}{(x - 1)^2}$ **54.** $R(x) = \dfrac{x^4}{x^2 - 9}$ **55.** $G(x) = \dfrac{x^2 - 4}{x^2 - x - 2}$ **56.** $F(x) = \dfrac{(x - 1)^2}{x^2 - 1}$

In Problems 57–66, solve each inequality (a) algebraically and (b) graphically.

57. $2x^2 + 5x - 12 < 0$ **58.** $3x^2 - 2x - 1 \geq 0$ **59.** $\dfrac{6}{x + 3} \geq 1$ **60.** $\dfrac{-2}{1 - 3x} < 1$

61. $\dfrac{2x - 6}{1 - x} < 2$ **62.** $\dfrac{3 - 2x}{2x + 5} \geq 2$ **63.** $\dfrac{(x - 2)(x - 1)}{x - 3} > 0$ **64.** $\dfrac{x + 1}{x(x - 5)} \leq 0$

65. $\dfrac{x^2 - 8x + 12}{x^2 - 16} > 0$ **66.** $\dfrac{x(x^2 + x - 2)}{x^2 + 9x + 20} \leq 0$

In Problems 67–70, find the remainder R when f(x) is divided by g(x). Is g a factor of f?

67. $f(x) = 8x^3 - 3x^2 + x + 4$; $g(x) = x - 1$ **68.** $f(x) = 2x^3 + 8x^2 - 5x + 5$; $g(x) = x - 2$

69. $f(x) = x^4 - 2x^3 + 15x - 2$; $g(x) = x + 2$ **70.** $f(x) = x^4 - x^2 + 2x + 2$; $g(x) = x + 1$

71. Find the value of $f(x) = 12x^6 - 8x^4 + 1$ at $x = 4$. **72.** Find the value of $f(x) = -16x^3 + 18x^2 - x + 2$ at $x = -2$.

In Problems 73 and 74, tell the maximum number of real zeros that each polynomial function may have. Then list the potential rational zeros of each polynomial function. Do not attempt to find the zeros.

73. $f(x) = 2x^8 - x^7 + 8x^4 - 2x^3 + x + 3$ **74.** $f(x) = -6x^5 + x^4 + 5x^3 + x + 1$

In Problems 75–80, find all the real zeros of each polynomial function.

75. $f(x) = x^3 - 3x^2 - 6x + 8$ **76.** $f(x) = x^3 - x^2 - 10x - 8$

77. $f(x) = 4x^3 + 4x^2 - 7x + 2$ **78.** $f(x) = 4x^3 - 4x^2 - 7x - 2$

79. $f(x) = x^4 - 4x^3 + 9x^2 - 20x + 20$ **80.** $f(x) = x^4 + 6x^3 + 11x^2 + 12x + 18$

In Problems 81–84, determine the real zeros of the polynomial function. Approximate all irrational zeros rounded to two decimal places.

81. $f(x) = 2x^3 - 11.84x^2 - 9.116x + 82.46$ **82.** $f(x) = 12x^3 + 39.8x^2 - 4.4x - 3.4$

83. $g(x) = 15x^4 - 21.5x^3 - 1718.3x^2 + 5308x + 3796.8$ **84.** $g(x) = 3x^4 + 67.93x^3 + 486.265x^2 + 1121.32x + 412.195$

In Problems 85–88, find the real solutions of each equation.

85. $2x^4 + 2x^3 - 11x^2 + x - 6 = 0$ **86.** $3x^4 + 3x^3 - 17x^2 + x - 6 = 0$

87. $2x^4 + 7x^3 + x^2 - 7x - 3 = 0$ **88.** $2x^4 + 7x^3 - 5x^2 - 28x - 12 = 0$

In Problems 89–92, find bounds to the zeros of each polynomial function. Obtain a complete graph of f.

89. $f(x) = x^3 - x^2 - 4x + 2$ **90.** $f(x) = x^3 + x^2 - 10x - 5$

91. $f(x) = 2x^3 - 7x^2 - 10x + 35$ **92.** $f(x) = 3x^3 - 7x^2 - 6x + 14$

In Problems 93–96, use the Intermediate Value Theorem to show that each polynomial has a zero in the given interval. Approximate the zero rounded to two decimal places.

93. $f(x) = 3x^3 - x - 1$; $[0, 1]$

94. $f(x) = 2x^3 - x^2 - 3$; $[1, 2]$

95. $f(x) = 8x^4 - 4x^3 - 2x - 1$; $[0, 1]$

96. $f(x) = 3x^4 + 4x^3 - 8x - 2$; $[1, 2]$

In Problems 97–100, information is given about a complex polynomial $f(x)$ whose coefficients are real numbers. Find the remaining zeros of f. Write a polynomial function whose zeros are given.

97. Degree 3;   zeros: $4 + i, 6$

98. Degree 3;   zeros: $3 + 4i, 5$

99. Degree 4;   zeros: $i, 1 + i$

100. Degree 4;   zeros: $1, 2, 1 + i$

In Problems 101–114, solve each equation in the complex number system.

101. $x^2 + x + 1 = 0$

102. $x^2 - x + 1 = 0$

103. $2x^2 + x - 2 = 0$

104. $3x^2 - 2x - 1 = 0$

105. $x^2 + 3 = x$

106. $2x^2 + 1 = 2x$

107. $x(1 - x) = 6$

108. $x(1 + x) = 2$

109. $x^4 + 2x^2 - 8 = 0$

110. $x^4 + 8x^2 - 9 = 0$

111. $x^3 - x^2 - 8x + 12 = 0$

112. $x^3 - 3x^2 - 4x + 12 = 0$

113. $3x^4 - 4x^3 + 4x^2 - 4x + 1 = 0$

114. $x^4 + 4x^3 + 2x^2 - 8x - 8 = 0$

---

115. Find the point on the line $y = x$ that is closest to the point $(3, 1)$.

[**Hint:** Find the minimum value of the function $f(x) = d^2$, where d is the distance from $(3, 1)$ to a point on the line.]

116. Landscaping   A landscape engineer has 200 feet of border to enclose a rectangular pond. What dimensions will result in the largest pond?

117. Enclosing the Most Area with a Fence   A farmer with 10,000 meters of fencing wants to enclose a rectangular field and then divide it into two plots with a fence parallel to one of the sides (see the figure). What is the largest area that can be enclosed?

118. A rectangle has one vertex on the line $y = 8 - 2x$, $x > 0$, another at the origin, one on the positive x-axis, and one on the positive y-axis. Find the largest area A that can be enclosed by the rectangle.

119. Architecture   A special window in the shape of a rectangle with semicircles at each end is to be constructed so that the outside dimensions are 100 feet in length. See the illustration. Find the dimensions that maximizes the area of the rectangle.

120. Parabolic Arch Bridges   A horizontal bridge is in the shape of a parabolic arch. Given the information shown in the figure, what is the height h of the arch 2 feet from shore?

121. Minimizing Marginal Cost   The marginal cost of a product can be thought of as the cost of producing one additional unit of output. For example, if the marginal cost of producing the 50th product is $6.20, then it cost $6.20 to increase production from 49 to 50 units of output. Callaway Golf Company has determined that the marginal cost C of manufacturing x Big Bertha golf clubs may be expressed by the quadratic function

$$C(x) = 4.9x^2 - 617.4x + 19,600$$

(a) How many clubs should be manufactured to minimize the marginal cost?

(b) At this level of production, what is the marginal cost?

122. Violent Crimes   The function

$$V(t) = -10.0t^2 + 39.2t + 1862.6$$

models the number V (in thousands) of violent crimes committed in the United States t years after 1990. So $t = 0$ represents 1990, $t = 1$ represents 1991, and so on.
(a) Determine the year in which the most violent crimes were committed.
(b) Approximately how many violent crimes were committed during this year?
(c) Using a graphing utility, graph $V = V(t)$. Were the number of violent crimes increasing or decreasing during the years 1994 to 1998?

SOURCE: Based on data obtained from the Federal Bureau of Investigation.

123. **Weight of a Body** The weight of a body varies inversely with the square of its distance from the center of Earth. Assuming that the radius of Earth is 3960 miles, how much would a man weigh at an altitude of 1 mile above Earth's surface if he weighs 200 pounds on Earth's surface?

124. **Resistance due to a Conductor** The resistance (in ohms) of a circular conductor varies directly with the length of the conductor and inversely with the square of the radius of the conductor. If 50 feet of wire with a radius of 6×10^{-3} inch has a resistance of 10 ohms, what would be the resistance of 100 feet of the same wire if the radius is increased to 7×10^{-3} inch?

125. **Advertising** A small manufacturing firm collected the following data on advertising expenditures A (in thousands of dollars) and total revenue R (in thousands of dollars).

Advertising	Total Revenue
20	$6101
22	$6222
25	$6350
25	$6378
27	$6453
28	$6423
29	$6360
31	$6231

(a) Draw a scatter diagram of the data. Comment on the type of relation that may exist between the two variables.
(b) Use a graphing utility to find the quadratic function of best fit to these data.
(c) Use the function found in part (b) to determine the optimal level of advertising for this firm.
(d) Use the function found in part (b) to find the revenue that the firm can expect if it uses the optimal level of advertising.
(e) With a graphing utility, graph the quadratic function of best fit on the scatter diagram.

126. **AIDS Cases in the United States** The following data represent the cumulative number of reported AIDS cases in the United States for 1990–1997.

Year, t	Number of AIDS Cases, A
1990, 1	193,878
1991, 2	251,638
1992, 3	326,648
1993, 4	399,613
1994, 5	457,280
1995, 6	528,215
1996, 7	594,760
1997, 8	653,253

SOURCE: U.S. Center for Disease Control and Prevention

(a) Draw a scatter diagram of the data.
(b) The cubic function of best fit to these data is

$$A(t) = -212t^3 + 2429t^2 + 59{,}569t + 130{,}003$$

Use this function to predict the cumulative number of AIDS cases reported in the United States in 2000.
(c) Use a graphing utility to verify that the function given in part (b) is the cubic function of best fit.
(d) With a graphing utility, draw a scatter diagram of the data and then graph the cubic function of best fit on the scatter diagram.
(e) Do you think the function given in part (b) will be useful in predicting the number of AIDS cases in 2005?

127. **Making a Can** A can in the shape of a right circular cylinder is required to have a volume of 250 cubic centimeters.
(a) Express the amount A of material to make the can as a function of the radius r of the cylinder.
(b) How much material is required if the can is of radius 3 centimeters?
(c) How much material is required if the can is of radius 5 centimeters?
(d) Graph $A = A(r)$. For what value of r is A smallest?

128. Design a polynomial function with the following characteristics: degree 6; four real zeros, one of multiplicity 3; y-intercept 3; behaves like $y = -5x^6$ for large values of $|x|$. Is this polynomial unique? Compare your polynomial with those of other students. What terms will be the same as everyone else's? Add some more characteristics, such as symmetry or naming the real zeros. How does this modify the polynomial?

129. Design a rational function with the following characteristics: three real zeros, one of multiplicity 2; y-intercept 1; vertical asymptotes $x = -2$ and $x = 3$; oblique asymptote $y = 2x + 1$. Is this rational function unique? Compare yours with those of other students. What will be the same as everyone else's? Add some more characteristics, such as symmetry or naming the real zeros. How does this modify the rational function?

130. The illustration shows the graph of a polynomial function.
  (a) Is the degree of the polynomial even or odd?
  (b) Is the leading coefficient positive or negative?
  (c) Is the function even, odd, or neither?
  (d) Why is x^2 necessarily a factor of the polynomial?
  (e) What is the minimum degree of the polynomial?
  (f) Formulate five different polynomials whose graphs could look like the one shown. Compare yours to those of other students. What similarities do you see? What differences?

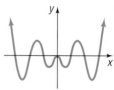

Chapter Test

1. Graph $f(x) = (x - 3)^4 - 2$.

2. $f(x) = 3x^2 - 12x + 4$
  (a) Determine if the function has a maximum or a minimum and explain how you know.
  (b) Algebraically, determine the vertex of the graph.
  (c) Determine the axis of symmetry of the graph.
  (d) Algebraically, determine the intercepts of the graph.
  (e) Graph the function by hand.

3. For the polynomial function $g(x) = 2x^3 + 5x^2 - 28x - 15$,
  (a) Determine the maximum number of real zeros that the function may have.
  (b) Find bounds to the zeros of the function.
  (c) List the potential rational zeros.
  (d) Determine the real zeros of g. Factor g over the reals.

4. Find the complex zeros of $f(x) = x^3 - 4x^2 + 25x - 100$

5. Solve $3x^3 + 2x - 1 = 8x^2 - 4$ in the complex number system.

In Problems 6 and 7, find the domain of each function. Find any horizontal, vertical, or oblique asymptotes.

6. $g(x) = \dfrac{2x^2 - 14x + 24}{x^2 + 6x - 40}$

7. $r(x) = \dfrac{x^2 + 2x - 3}{x + 1}$

8. Sketch the graph of the function in Problem 7. Label all intercepts, vertical asymptotes, horizontal asymptotes, and oblique asymptotes. Verify your results using a graphing utility.

In Problems 9 and 10, write a function that meets the given conditions.

9. Fourth-degree polynomial with real coefficients; zeros: $-2, 0, 3 + i$

10. Rational function; asymptotes: $y = 2, x = 4$; domain: $\{x \mid x \neq 4, x \neq 9\}$

11. Use the Intermediate Value Theorem to show that the function $f(x) = -2x^2 - 3x + 8$ has at least one real zero on the interval $[0, 4]$.

In Problems 12 and 13, solve each inequality algebraically. Write the solution set in interval notation. Verify your results using a graphing utility.

12. $3x^2 - x - 4 \geq x^2 - 3x + 8$

13. $\dfrac{x + 2}{x - 3} < 2$

14. Given the polynomial function $f(x) = -2(x - 1)^2(x + 2)$, do the following:
  (a) Find the x- and y-intercepts of the graph of f.
  (b) Determine whether the graph crosses or touches the x-axis at each x-intercept.
  (c) Find the power function that the graph of f resembles for large values of $|x|$.
  (d) Using results of parts (a)–(c), describe what the graph of f should look like. Verify this with a graphing utility.
  (e) Approximate the turning points of f rounded to two decimal places.
  (f) By hand, use the information in parts (a)–(e) to draw a complete graph of f.

15. The table gives the average home game attendance A for the St. Louis Cardinals (in thousands) during the years 1995–2003, where x is the number of years since 1995.

SOURCE: www.baseball-almanac.com

x	0	1	2	3	4	5	6	7	8
A	24.570	32.774	32.519	39.453	40.197	41.191	38.390	37.182	35.930

  (a) Use a graphing utility to find the quadratic function of best fit to the data. Round coefficients to three decimal places.
  (b) Use the function found in part (a) to estimate the average home game attendance for the St. Louis Cardinals in 2004.

Chapter Projects

1. **Weed Pollen**   Given in the table below is the Weed pollen count for Lexington, Kentucky, from August 4, 2004, to October 6, 2004, the prime season for weed pollen.

   (a) Let the independent variable D represent the date, where $D = 1$ on 8/4, $D = 6$ on 8/9, $D = 9$ on 8/12... and $D = 64$ on 10/6. Let the dependent variable P represent the number of grains per cubic meter of weed pollen. Draw a scatter diagram of the data using your graphing utility, and by hand on graph paper.

   (b) In the hand-drawn scatter diagram, sketch a smooth curve that fits the data. Try to have the curve pass through as many of the points as closely as possible. How many turning points are there? What does this tell you about the degree of the polynomial that could be fit to the data? If the ends of the graph on the left and on the right are extended downward, how many real zeros are indicated by the graph? How many complex? Why?

   (c) Use a graphing utility to determine the quartic function of the best fit. Graph it on your graphing utility. Does it look like what you expected from your hand-drawn graph? Explain.

   (d) Use a graphing utility to determine the cubic function of best fit. Graph it on your graphing utility. Does it look like what you expected from your hand-drawn graph? Explain.

   (e) Use a graphing utility to determine the quadratic function of best fit. Graph it on your graphing utility. Does it look like what you expected from your hand-drawn graph? Explain.

   (f) Which of the three functions seems to fit the best? Explain your reasoning.

   (g) Investigate weed pollen counts at other times of the year. (Go to the National Allergy Bureau at www.aaaai.org.) Would the polynomial you found in part (c) work for other 2-month periods? Why or why not? What about for the same period of time in other years?

Date	Grains per cubic meter	Date	Grains per cubic meter
8/4	18	9/8	45
8/9	14	9/10	93
8/12	12	9/13	14
8/13	16	9/15	10
8/17	37	9/17	11
8/18	30	9/20	5
8/23	53	9/22	3
8/24	22	9/27	4
8/25	47	9/29	6
8/26	7	10/1	4
8/31	214	10/4	4
9/2	31	10/5	7
9/3	44	10/6	7
9/6	50		

SOURCE: National Allergy Bureau at the American Academy of Allergy, Asthma and Immunology website (*www.aaaai.org*)

The following projects are available at the Instructor's Resource Center (IRC):

2. **Project at Motorola**   *How Many Cellphones Can I Make?*

3. **First and Second Differences**

4. **Cannons**

5. **Maclaurin Series**

6. **Theory of Equations**

7. **CBL Experiment**

Cumulative Review

1. Find the distance between the points $P = (1, 3)$ and $Q = (-4, 2)$.

2. Solve the inequality $x^2 \geq x$ and graph the solution set.

3. Solve the inequality $x^2 - 3x < 4$ and graph the solution set.

4. Find a linear function with slope -3 that contains the point $(-1, 4)$. Graph the function.

5. Find the equation of the line parallel to the line $y = 2x + 1$ and containing the point $(3, 5)$. Express your answer in slope–intercept form and graph the line.

6. Graph the equation $y = x^3$.

7. Does the relation $\{(3, 6), (1, 3), (2, 5), (3, 8)\}$ represent a function? Why or why not?

8. Solve the equation $x^3 - 6x^2 + 8x = 0$.

9. Solve the inequality $3x + 2 \leq 5x - 1$ and graph the solution set.

10. Find the center and radius of the circle $x^2 + 4x + y^2 - 2y - 4 = 0$. Graph the circle.

11. For the equation $y = x^3 - 9x$, determine the intercepts and test for symmetry.

12. Find an equation of the line perpendicular to $3x - 2y = 7$ that contains the point $(1, 5)$.

13. Is the following graph the graph of a function? Why or why not?

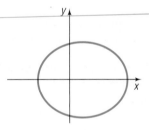

14. For the function $f(x) = x^2 + 5x - 2$, find
    (a) $f(3)$
    (b) $f(-x)$
    (c) $-f(x)$
    (d) $f(3x)$
    (e) $\dfrac{f(x + h) - f(x)}{h}, h \neq 0$

15. Answer the following questions regarding the function $f(x) = \dfrac{x + 5}{x - 1}$.
    (a) What is the domain of f?
    (b) Is the point $(2, 6)$ on the graph of f?
    (c) If $x = 3$, what is $f(x)$? What point is on the graph of f?
    (d) If $f(x) = 9$, what is x? What point is on the graph of f?

16. Graph the function $f(x) = -3x + 7$.

17. Graph $f(x) = 2x^2 - 4x + 1$ by determining whether its graph opens up or down and by finding its vertex, axis of symmetry, y-intercept, and x-intercepts, if any.

18. Find the average rate of change of $f(x) = x^2 + 3x + 1$ from 1 to x. Use this result to find the slope of the secant line containing $(1, f(1))$ and $(2, f(2))$.

19. In parts (a) to (f) use the following graph.

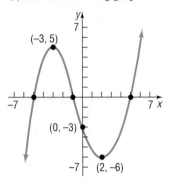

(a) Determine the intercepts.
(b) Based on the graph, tell whether the graph is symmetric with respect to the x-axis, the y-axis, and/or the origin.
(c) Based on the graph, tell whether the function is even, odd, or neither.
(d) List the intervals on which f is increasing. List the intervals on which f is decreasing.
(e) List the numbers, if any, at which f has a local maximum. What are these local maxima?
(f) List the numbers, if any, at which f has a local minimum. What are these local minima?

20. Determine algebraically whether the function
$$f(x) = \frac{5x}{x^2 - 9} \text{ is even, odd, or neither.}$$

21. For the function $f(x) = \begin{cases} 2x + 1 & \text{if } -3 < x < 2 \\ -3x + 4 & \text{if } x \geq 2 \end{cases}$
    (a) Find the domain of f.
    (b) Locate any intercepts.
    (c) Graph the function.
    (d) Based on the graph, find the range.

22. Graph the function $f(x) = -3(x + 1)^2 + 5$ using transformations.

23. Suppose that $f(x) = x^2 - 5x + 1$ and $g(x) = -4x - 7$.
    (a) Find $f + g$ and state its domain.
    (b) Find $\dfrac{f}{g}$ and state its domain.

24. **Demand Equation** The price p (in dollars) and the quantity x sold of a certain product obey the demand equation
$$p = -\frac{1}{10}x + 150, \qquad 0 \leq x \leq 1500$$
    (a) Express the revenue R as a function of x.
    (b) What is the revenue if 100 units are sold?
    (c) What quantity x maximizes revenue? What is the maximum revenue?
    (d) What price should the company charge to maximize revenue?

Exponential and Logarithmic Functions

4

A LOOK BACK Until now, our study of functions has concentrated on polynomial and rational functions. These functions belong to the class of **algebraic functions**, that is, functions that can be expressed in terms of sums, differences, products, quotients, powers, or roots of polynomials. Functions that are not algebraic are termed **transcendental** (they transcend, or go beyond, algebraic functions).

A LOOK AHEAD In this chapter, we study two transcendental functions: The exponential function and the logarithmic function. These functions occur frequently in a wide variety of applications, such as biology, chemistry, economics, and psychology.

The chapter begins with a discussion of composite functions.

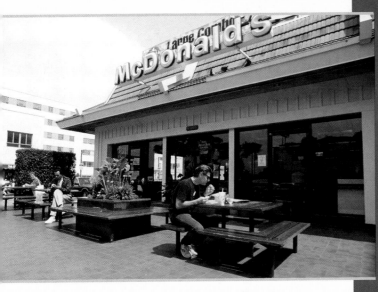

The McDonald's Scalding Coffee Case

April 3, 1996

There is a lot of hype about the McDonald's scalding coffee case. No one is in favor of frivolous cases or outlandish results; however, it is important to understand some points that were not reported in most of the stories about the case. McDonald's coffee was not only hot, it was scalding, capable of almost instantaneous destruction of skin, flesh, and muscle.

Plaintiff's expert, a scholar in thermodynamics applied to human skin burns, testified that liquids at 180 degrees will cause a full thickness burn to human skin in two to seven seconds. Other testimony showed that, as the temperature decreases toward 155 degrees, the extent of the burn relative to that temperature decreases exponentially. Thus, if (the) spill had involved coffee at 155 degrees, the liquid would have cooled and given her time to avoid a serious burn.

ATLA fact sheet © 1995, 1996 Consumer Attorneys of CA. Used with permission of the Association of Trial Lawyers of America.

—See Chapter Project 1.

OUTLINE

4.1 Composite Functions

PREPARING FOR THIS SECTION *Before getting started, review the following:*

• Finding Values of a Function (Section 2.2, pp. 209–211)   • Domain of a Function (Section 2.2, pp. 206–208 and 212–213)

Now work the 'Are You Prepared?' problems on page 401.

OBJECTIVES 1 Form a Composite Function
2 Find the Domain of a Composite Function

1 Form a Composite Function

Suppose that an oil tanker is leaking oil and we want to be able to determine the area of the circular oil patch around the ship. See Figure 1. It is determined that the oil is leaking from the tanker in such a way that the radius of the circular oil patch around the ship is increasing at a rate of 3 feet per minute. Therefore, the radius r of the oil patch at any time t, in minutes, is given by $r(t) = 3t$. So after 20 minutes, the radius of the oil patch is $r(20) = 3(20) = 60$ feet. The area A of a circle as a function of the radius r is given by $A(r) = \pi r^2$. The area of the circular patch of oil after 20 minutes is $A(60) = \pi(60)^2 = 3600\pi$ square feet.

Notice that $60 = r(20)$, so $A(60) = A(r(20))$. The argument of the function A is itself a function! In general, we can find the area of the oil patch as a function of time t by evaluating $A(r(t))$. The function $A(r(t))$ is a special type of function called a *composite function*.

Figure 1

As another example, consider the function $y = (2x + 3)^2$. If we write $y = f(u) = u^2$ and $u = g(x) = 2x + 3$, then, by a substitution process, we can obtain the original function: $y = f(u) = f(g(x)) = (2x + 3)^2$.

In general, suppose that f and g are two functions and that x is a number in the domain of g. By evaluating g at x, we get $g(x)$. If $g(x)$ is in the domain of f, then we may evaluate f at $g(x)$ and thereby obtain the expression $f(g(x))$. The correspondence from x to $f(g(x))$ is called a *composite function* $f \circ g$.

Given two functions f and g, the **composite function**, denoted by $f \circ g$ (read as "f composed with g"), is defined by

$$(f \circ g)(x) = f(g(x))$$

The domain of $f \circ g$ is the set of all numbers x in the domain of g such that $g(x)$ is in the domain of f.

Look carefully at Figure 2. Only those x's in the domain of g for which $g(x)$ is in the domain of f can be in the domain of $f \circ g$. The reason is that if $g(x)$ is not in the domain of f then $f(g(x))$ is not defined. Because of this, the domain of $f \circ g$ is a subset of the domain of g; the range of $f \circ g$ is a subset of the range of f.

Figure 2

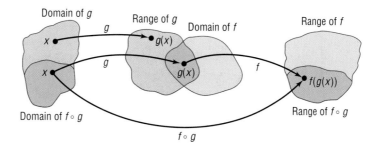

Figure 3 provides a second illustration of the definition. Here, x is the input to the function g, yielding $g(x)$. Then $g(x)$ is the input to the function f, yielding $f(g(x))$. Notice that the "inside" function g in $f(g(x))$ is done first.

Figure 3

Let's look at some examples.

EXAMPLE 1

Evaluating a Composite Function

Suppose that $f(x) = 2x^2 - 3$ and $g(x) = 4x$. Find:

(a) $(f \circ g)(1)$ (b) $(g \circ f)(1)$ (c) $(f \circ f)(-2)$ (d) $(g \circ g)(-1)$

Solution   (a) $(f \circ g)(1) = f(g(1)) = f(4) = 2 \cdot 4^2 - 3 = 29$

$g(x) = 4x$ $f(x) = 2x^2 - 3$
$g(1) = 4$

(b) $(g \circ f)(1) = g(f(1)) = g(-1) = 4 \cdot (-1) = -4$

$$\underset{\substack{\uparrow \\ f(x) = 2x^2 - 3 \\ f(1) = -1}}{} \underset{\substack{\uparrow \\ g(x) = 4x}}{}$$

(c) $(f \circ f)(-2) = f(f(-2)) = f(5) = 2 \cdot 5^2 - 3 = 47$

$$\underset{\substack{\uparrow \\ f(-2) = 2(-2)^2 - 3 = 5}}{}$$

(d) $(g \circ g)(-1) = g(g(-1)) = g(-4) = 4 \cdot (-4) = -16$

$$\underset{\substack{\uparrow \\ g(-1) = -4}}{}$$ ◀

Figure 4

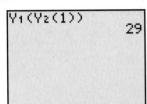

Graphing calculators can be used to evaluate composite functions.* Let $Y_1 = f(x) = 2x^2 - 3$ and $Y_2 = g(x) = 4x$. Then, using a TI-84 Plus graphing calculator, $(f \circ g)(1)$ would be found as shown in Figure 4. Notice that this is the result obtained in Example 1(a).

✏️── **NOW WORK PROBLEM 11.**

2 **Find the Domain of a Composite Function**

EXAMPLE 2 **Finding a Composite Function and Its Domain**

Suppose that $f(x) = x^2 + 3x - 1$ and $g(x) = 2x + 3$.

Find: (a) $f \circ g$ (b) $g \circ f$

Then find the domain of each composite function.

Solution The domain of f and the domain of g are all real numbers.

(a) $(f \circ g)(x) = f(g(x)) = f(2x + 3) = (2x + 3)^2 + 3(2x + 3) - 1$

$$\underset{\substack{\uparrow \\ f(x) = x^2 + 3x - 1}}{}$$

$$= 4x^2 + 12x + 9 + 6x + 9 - 1 = 4x^2 + 18x + 17$$

Since the domains of both f and g are all real numbers, the domain of $f \circ g$ is all real numbers.

(b) $(g \circ f)(x) = g(f(x)) = g(x^2 + 3x - 1) = 2(x^2 + 3x - 1) + 3$

$$\underset{\substack{\uparrow \\ g(x) = 2x + 3}}{}$$

$$= 2x^2 + 6x - 2 + 3 = 2x^2 + 6x + 1$$

Since the domains of both f and g are all real numbers, the domain of $f \circ g$ is all real numbers. ◀

Look back at Figure 2. In determining the domain of the composite function $(f \circ g)(x) = f(g(x))$, keep the following two thoughts in mind about the input x.

1. $g(x)$ must be defined so any x not in the domain of g must be excluded.

2. $f(g(x))$ must be defined so any x for which $g(x)$ is not in the domain of f must be excluded.

*Consult your owner's manual for the appropriate keystrokes.

| EXAMPLE 3 | Finding the Domain of $f \circ g$ |

Find the domain of $(f \circ g)(x)$ if $f(x) = \dfrac{1}{x + 2}$ and $g(x) = \dfrac{4}{x - 1}$

Solution For $(f \circ g)(x) = f(g(x))$, we first note that the domain of g is $\{x | x \neq 1\}$, so we exclude 1 from the domain of $f \circ g$. Next, we note that the domain of f is $\{x | x \neq -2\}$, which means that $g(x)$ cannot equal -2. We solve the equation $g(x) = -2$ to determine what values of x to exclude.

$$\frac{4}{x - 1} = -2 \qquad \qquad g(x) = -2$$
$$4 = -2(x - 1)$$
$$4 = -2x + 2$$
$$2x = -2$$
$$x = -1$$

We also exclude -1 from the domain of $f \circ g$. The domain of $f \circ g$ is $\{x | x \neq -1, x \neq 1\}$.

✔ **CHECK:** For $x = 1$, $g(x) = \dfrac{4}{x - 1}$ is not defined, so $(f \circ g)(x) = f(g(x))$ is not defined.

For $x = -1$, $g(-1) = \dfrac{4}{-2} = -2$, and $(f \circ g)(-1) = f(g(-1)) = f(-2)$ is not defined. ◀

—✏— **NOW WORK PROBLEM 21.**

| EXAMPLE 4 | Finding a Composite Function and Its Domain |

Suppose that $f(x) = \dfrac{1}{x + 2}$ and $g(x) = \dfrac{4}{x - 1}$

Find: (a) $f \circ g$ (b) $f \circ f$

Then find the domain of each composite function.

Solution The domain of f is $\{x | x \neq -2\}$ and the domain of g is $\{x | x \neq 1\}$.

(a) $(f \circ g)(x) = f(g(x)) = f\left(\dfrac{4}{x - 1}\right) = \dfrac{1}{\underset{\underset{f(x) = \frac{1}{x+2}}{\uparrow}}{\dfrac{4}{x - 1} + 2}} = \dfrac{x - 1}{4 + 2(x - 1)} \underset{\underset{\text{Multiply by } \frac{x-1}{x-1}.}{\uparrow}}{} = \dfrac{x - 1}{2x + 2} = \dfrac{x - 1}{2(x + 1)}$

In Example 3, we found the domain of $f \circ g$ to be $\{x | x \neq -1, x \neq 1\}$.

We could also find the domain of $f \circ g$ by first looking at the domain of g: $\{x | x \neq 1\}$. We exclude 1 from the domain of $f \circ g$ as a result. Then we look at $f \circ g$ and notice that x cannot equal -1, since $x = -1$ results in division by 0. So we also exclude -1 from the domain of $f \circ g$. Therefore, the domain of $f \circ g$ is $\{x | x \neq -1, x \neq 1\}$.

(b) $(f \circ f)(x) = f(f(x)) = f\left(\dfrac{1}{x + 2}\right) = \dfrac{1}{\underset{\underset{f(x) = \frac{1}{x+2}}{\uparrow}}{\dfrac{1}{x + 2} + 2}} = \dfrac{x + 2}{1 + 2(x + 2)} \underset{\underset{\text{Multiply by } \frac{x+2}{x+2}.}{\uparrow}}{} = \dfrac{x + 2}{2x + 5}$

The domain of $f \circ f$ consists of those x in the domain of f, $\{x | x \neq -2\}$, for which

$$f(x) = \frac{1}{x + 2} \neq -2 \qquad \frac{1}{x + 2} = -2$$

$$1 = -2(x + 2)$$
$$1 = -2x - 4$$
$$2x = -5$$
$$x = -\frac{5}{2}$$

or, equivalently,

$$x \neq -\frac{5}{2}$$

The domain of $f \circ f$ is $\left\{ x \Big| x \neq -\dfrac{5}{2}, x \neq -2 \right\}$.

We could also find the domain of $f \circ f$ by recognizing that -2 is not in the domain of f and so should be excluded from the domain of $f \circ f$. Then, looking at $f \circ f$, we see that x cannot equal $-\dfrac{5}{2}$. Do you see why? Therefore, the domain of $f \circ f$ is $\left\{ x \Big| x \neq -\dfrac{5}{2}, x \neq -2 \right\}$. ◀

▱━ **NOW WORK PROBLEMS 33 AND 35.**

Look back at Example 2, which illustrates that, in general, $f \circ g \neq g \circ f$. Sometimes $f \circ g$ does equal $g \circ f$, as shown in the next example.

| EXAMPLE 5 | **Showing That Two Composite Functions Are Equal** |

If $f(x) = 3x - 4$ and $g(x) = \dfrac{1}{3}(x + 4)$, show that

$$(f \circ g)(x) = (g \circ f)(x) = x$$

for every x in the domain of $f \circ g$ and $g \circ f$.

Solution

$(f \circ g)(x) = f(g(x))$

$\qquad = f\left(\dfrac{x + 4}{3}\right) \qquad\qquad g(x) = \dfrac{1}{3}(x + 4) = \dfrac{x + 4}{3}$

$\qquad = 3\left(\dfrac{x + 4}{3}\right) - 4 \qquad$ Substitute $g(x)$ into the rule for f, $f(x) = 3x - 4$.

$\qquad = x + 4 - 4 = x$

$(g \circ f)(x) = g(f(x))$

$\qquad = g(3x - 4) \qquad\qquad f(x) = 3x - 4$

$\qquad = \dfrac{1}{3}[(3x - 4) + 4] \qquad$ Substitute $f(x)$ into the rule for g, $g(x) = \dfrac{1}{3}(x + 4)$.

$\qquad = \dfrac{1}{3}(3x) = x$

Thus, $(f \circ g)(x) = (g \circ f)(x) = x$. ◀

─ **Seeing the Concept** ─

Using a graphing calculator, let $Y_1 = f(x) = 3x - 4$, $Y_2 = g(x) = \dfrac{1}{3}(x + 4)$, $Y_3 = f \circ g$, and $Y_4 = g \circ f$. Using the viewing window $-3 \leq x \leq 3, -2 \leq y \leq 2$, graph only Y_3 and Y_4. What do you see? TRACE to verify that $Y_3 = Y_4$.

In Section 4.2, we shall see that there is an important relationship between functions f and g for which $(f \circ g)(x) = (g \circ f)(x) = x$.

▱━ **NOW WORK PROBLEM 45.**

Calculus Application

Some techniques in calculus require that we be able to determine the components of a composite function. For example, the function $H(x) = \sqrt{x + 1}$ is the composition of the functions f and g, where $f(x) = \sqrt{x}$ and $g(x) = x + 1$, because $H(x) = (f \circ g)(x) = f(g(x)) = f(x + 1) = \sqrt{x + 1}$.

EXAMPLE 6   **Finding the Components of a Composite Function**

Find functions f and g such that $f \circ g = H$ if $H(x) = (x^2 + 1)^{50}$.

Solution   The function H takes $x^2 + 1$ and raises it to the power 50. A natural way to decompose H is to raise the function $g(x) = x^2 + 1$ to the power 50. If we let $f(x) = x^{50}$ and $g(x) = x^2 + 1$, then

$$(f \circ g)(x) = f(g(x))$$
$$= f(x^2 + 1)$$
$$= (x^2 + 1)^{50} = H(x)$$

Figure 5

See Figure 5.   ◄

Other functions f and g may be found for which $f \circ g = H$ in Example 6. For example, if $f(x) = x^2$ and $g(x) = (x^2 + 1)^{25}$, then

$$(f \circ g)(x) = f(g(x)) = f((x^2 + 1)^{25}) = [(x^2 + 1)^{25}]^2 = (x^2 + 1)^{50}$$

Although the functions f and g found as a solution to Example 6 are not unique, there is usually a "natural" selection for f and g that comes to mind first.

EXAMPLE 7   **Finding the Components of a Composite Function**

Find functions f and g such that $f \circ g = H$ if $H(x) = \dfrac{1}{x + 1}$.

Solution   Here H is the reciprocal of $g(x) = x + 1$. If we let $f(x) = \dfrac{1}{x}$ and $g(x) = x + 1$, we find that

$$(f \circ g)(x) = f(g(x)) = f(x + 1) = \frac{1}{x + 1} = H(x)$$ ◄

NOW WORK PROBLEM 53.

4.1 Assess Your Understanding

'Are You Prepared?'

Answers are given at the end of these exercises. If you get a wrong answer, read the pages listed in red.

1. Find $f(3)$ if $f(x) = -4x^2 + 5x$. (p. 209)

2. Find $f(3x)$ if $f(x) = 4 - 2x^2$. (p. 209)

3. Find the domain of the function $f(x) = \dfrac{x^2 - 1}{x^2 - 4}$. (pp. 212–213)

Concepts and Vocabulary

4. If $f(x) = x + 1$ and $g(x) = x^3$, then _____ $= (x + 1)^3$.

5. *True or False:* $f(g(x)) = f(x) \cdot g(x)$.

6. *True or False:* The domain of the composite function $(f \circ g)(x)$ is the same as the domain of $g(x)$.

Skill Building

In Problems 7 and 8, evaluate each expression using the values given in the table.

7.

x	−3	−2	−1	0	1	2	3
f(x)	−7	−5	−3	−1	3	5	5
g(x)	8	3	0	−1	0	3	8

(a) $(f \circ g)(1)$ (b) $(f \circ g)(-1)$ (c) $(g \circ f)(-1)$ (d) $(g \circ f)(0)$ (e) $(g \circ g)(-2)$ (f) $(f \circ f)(-1)$

8.

x	−3	−2	−1	0	1	2	3
f(x)	11	9	7	5	3	1	−1
g(x)	−8	−3	0	1	0	−3	−8

(a) $(f \circ g)(1)$ (b) $(f \circ g)(2)$ (c) $(g \circ f)(2)$ (d) $(g \circ f)(3)$ (e) $(g \circ g)(1)$ (f) $(f \circ f)(3)$

In Problems 9 and 10, evaluate each expression using the graphs of $y = f(x)$ and $y = g(x)$ shown below.

9. (a) $g(f(-1))$ (b) $g(f(0))$
      (c) $f(g(-1))$ (d) $f(g(4))$

10. (a) $g(f(1))$ (b) $g(f(5))$
       (c) $f(g(0))$ (d) $f(g(2))$

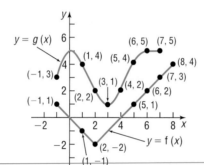

In Problems 11–20, for the given functions f and g, find:
     (a) $(f \circ g)(4)$ (b) $(g \circ f)(2)$ (c) $(f \circ f)(1)$ (d) $(g \circ g)(0)$

11. $f(x) = 2x; \quad g(x) = 3x^2 + 1$

12. $f(x) = 3x + 2; \quad g(x) = 2x^2 - 1$

13. $f(x) = 4x^2 - 3; \quad g(x) = 3 - \dfrac{1}{2}x^2$

14. $f(x) = 2x^2; \quad g(x) = 1 - 3x^2$

15. $f(x) = \sqrt{x}; \quad g(x) = 2x$

16. $f(x) = \sqrt{x + 1}; \quad g(x) = 3x$

17. $f(x) = |x|; \quad g(x) = \dfrac{1}{x^2 + 1}$

18. $f(x) = |x - 2|; \quad g(x) = \dfrac{3}{x^2 + 2}$

19. $f(x) = \dfrac{3}{x + 1}; \quad g(x) = \sqrt[3]{x}$

20. $f(x) = x^{3/2}; \quad g(x) = \dfrac{2}{x + 1}$

In Problems 21–28, find the domain of the composite function $f \circ g$.

21. $f(x) = \dfrac{3}{x - 1}; \quad g(x) = \dfrac{2}{x}$

22. $f(x) = \dfrac{1}{x + 3}; \quad g(x) = \dfrac{-2}{x}$

23. $f(x) = \dfrac{x}{x - 1}; \quad g(x) = \dfrac{-4}{x}$

24. $f(x) = \dfrac{x}{x + 3}; \quad g(x) = \dfrac{2}{x}$

25. $f(x) = \sqrt{x}; \quad g(x) = 2x + 3$

26. $f(x) = x - 2; \quad g(x) = \sqrt{1 - x}$

27. $f(x) = x^2 + 1; \quad g(x) = \sqrt{x - 1}$

28. $f(x) = x^2 + 4; \quad g(x) = \sqrt{x - 2}$

In Problems 29–44, for the given functions f and g, find:

(a) $f \circ g$ (b) $g \circ f$ (c) $f \circ f$ (d) $g \circ g$

State the domain of each composite function.

29. $f(x) = 2x + 3; \quad g(x) = 3x$

30. $f(x) = -x; \quad g(x) = 2x - 4$

31. $f(x) = 3x + 1; \quad g(x) = x^2$

32. $f(x) = x + 1; \quad g(x) = x^2 + 4$

33. $f(x) = x^2; \quad g(x) = x^2 + 4$

34. $f(x) = x^2 + 1; \quad g(x) = 2x^2 + 3$

35. $f(x) = \dfrac{3}{x - 1}; \quad g(x) = \dfrac{2}{x}$

36. $f(x) = \dfrac{1}{x + 3}; \quad g(x) = \dfrac{-2}{x}$

37. $f(x) = \dfrac{x}{x - 1}; \quad g(x) = \dfrac{-4}{x}$

38. $f(x) = \dfrac{x}{x + 3}; \quad g(x) = \dfrac{2}{x}$

39. $f(x) = \sqrt{x}; \quad g(x) = 2x + 3$

40. $f(x) = \sqrt{x - 2}; \quad g(x) = 1 - 2x$

41. $f(x) = x^2 + 1; \quad g(x) = \sqrt{x - 1}$

42. $f(x) = x^2 + 4; \quad g(x) = \sqrt{x - 2}$

43. $f(x) = ax + b; \quad g(x) = cx + d$

44. $f(x) = \dfrac{ax + b}{cx + d}; \quad g(x) = mx$

In Problems 45–52, show that $(f \circ g)(x) = (g \circ f)(x) = x$.

45. $f(x) = 2x; \quad g(x) = \dfrac{1}{2}x$

46. $f(x) = 4x; \quad g(x) = \dfrac{1}{4}x$

47. $f(x) = x^3; \quad g(x) = \sqrt[3]{x}$

48. $f(x) = x + 5; \quad g(x) = x - 5$

49. $f(x) = 2x - 6; \quad g(x) = \dfrac{1}{2}(x + 6)$

50. $f(x) = 4 - 3x; \quad g(x) = \dfrac{1}{3}(4 - x)$

51. $f(x) = ax + b; \quad g(x) = \dfrac{1}{a}(x - b), \quad a \neq 0$

52. $f(x) = \dfrac{1}{x}; \quad g(x) = \dfrac{1}{x}$

In Problems 53–58, find functions f and g so that $f \circ g = H$.

53. $H(x) = (2x + 3)^4$

54. $H(x) = (1 + x^2)^3$

55. $H(x) = \sqrt{x^2 + 1}$

56. $H(x) = \sqrt{1 - x^2}$

57. $H(x) = |2x + 1|$

58. $H(x) = |2x^2 + 3|$

Applications and Extensions

59. If $f(x) = 2x^3 - 3x^2 + 4x - 1$ and $g(x) = 2$, find $(f \circ g)(x)$ and $(g \circ f)(x)$.

60. If $f(x) = \dfrac{x + 1}{x - 1}$, find $(f \circ f)(x)$.

61. If $f(x) = 2x^2 + 5$ and $g(x) = 3x + a$, find a so that the graph of $f \circ g$ crosses the y-axis at 23.

62. If $f(x) = 3x^2 - 7$ and $g(x) = 2x + a$, find a so that the graph of $f \circ g$ crosses the y-axis at 68.

63. Surface Area of a Balloon   The surface area S (in square meters) of a hot-air balloon is given by

$$S(r) = 4\pi r^2$$

where r is the radius of the balloon (in meters). If the radius r is increasing with time t (in seconds) according to the formula $r(t) = \dfrac{2}{3}t^3, t \geq 0$, find the surface area S of the balloon as a function of the time t.

64. Volume of a Balloon   The volume V (in cubic meters) of the hot-air balloon described in Problem 63 is given by $V(r) = \dfrac{4}{3}\pi r^3$. If the radius r is the same function of t as in Problem 63, find the volume V as a function of the time t.

65. Automobile Production   The number N of cars produced at a certain factory in 1 day after t hours of operation is given by $N(t) = 100t - 5t^2, 0 \leq t \leq 10$. If the cost C (in dollars) of producing N cars is $C(N) = 15{,}000 + 8000N$, find the cost C as a function of the time t of operation of the factory.

66. Environmental Concerns   The spread of oil leaking from a tanker is in the shape of a circle. If the radius r (in feet) of the spread after t hours is $r(t) = 200\sqrt{t}$, find the area A of the oil slick as a function of the time t.

67. Production Cost The price p, in dollars, of a certain product and the quantity x sold obey the demand equation

$$p = -\frac{1}{4}x + 100, \qquad 0 \le x \le 400$$

Suppose that the cost C, in dollars, of producing x units is

$$C = \frac{\sqrt{x}}{25} + 600$$

Assuming that all items produced are sold, find the cost C as a function of the price p.

[**Hint:** Solve for x in the demand equation and then form the composite.]

68. Cost of a Commodity The price p, in dollars, of a certain commodity and the quantity x sold obey the demand equation

$$p = -\frac{1}{5}x + 200, \qquad 0 \le x \le 1000$$

Suppose that the cost C, in dollars, of producing x units is

$$C = \frac{\sqrt{x}}{10} + 400$$

Assuming that all items produced are sold, find the cost C as a function of the price p.

69. Volume of a Cylinder The volume V of a right circular cylinder of height h and radius r is $V = \pi r^2 h$. If the height is twice the radius, express the volume V as a function of r.

70. Volume of a Cone The volume V of a right circular cone is $V = \frac{1}{3}\pi r^2 h$. If the height is twice the radius, express the volume V as a function of r.

71. Foreign Exchange Traders often buy foreign currency in hope of making money when the currency's value changes. For example, on October 20, 2003, one U.S. dollar could purchase 0.857118 Euros and one Euro could purchase 128.6054 yen. Let $f(x)$ represent the number of Euros you can buy with x dollars and let $g(x)$ represent the number of yen you can buy with x Euros.
  (a) Find a function that relates dollars to Euros.
  (b) Find a function that relates Euros to yen.
  (c) Use the results of parts (a) and (b) to find a function that relates dollars to yen. That is, find $g(f(x))$.
  (d) What is $g(f(1000))$?

72. If f and g are odd functions, show that the composite function $f \circ g$ is also odd.

73. If f is an odd function and g is an even function, show that the composite functions $f \circ g$ and $g \circ f$ are also even.

'Are You Prepared?' Answers

1. -21 **2.** $4 - 18x^2$ **3.** $\{x \mid x \ne -2, x \ne 2\}$

4.2 One-to-One Functions; Inverse Functions

PREPARING FOR THIS SECTION *Before getting started, review the following:*

• Functions (Section 2.2, pp. 204–215)
• Increasing/Decreasing Functions (Section 2.4, pp. 230–231)

Now work the 'Are You Prepared?' problems on page 415.

OBJECTIVES 1 Determine Whether a Function Is One-to-One
     2 Determine the Inverse of a Function Defined by a Map or an Ordered Pair
     3 Obtain the Graph of the Inverse Function from the Graph of the Function
     4 Find the Inverse of a Function Defined by an Equation

1 **Determine Whether a Function Is One-to-One**

In Section 2.2, we presented four different ways to represent a function: as (1) a map, (2) a set of ordered pairs, (3) a graph, and (4) an equation. For example, Figures 6 and 7 illustrate two different functions represented as mappings. The function in

Figure 6

Figure 7

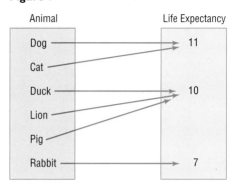

Figure 6 shows the correspondence between states and their population. The function in Figure 7 shows a correspondence between animals and life expectancy.

Suppose we asked a group of people to name the state that has a population of 761,063 based on the function in Figure 6. Everyone in the group would respond South Dakota. Now, if we asked the same group of people to name the animal whose life expectancy is 11 years based on the function in Figure 7, some would respond dog, while others would respond cat. What is the difference between the functions in Figures 6 and 7? In Figure 6, we can see that each element in the domain corresponds to exactly one element in the range. In Figure 7, this is not the case: two different elements in the domain correspond to the same element in the range. We give functions such as the one in Figure 6 a special name.

In Words

A function is not one-to-one if two different inputs correspond to the same output.

A function is **one-to-one** if any two different inputs in the domain correspond to two different outputs in the range. That is, if x_1 and x_2 are two different inputs of a function f, then $f(x_1) \neq f(x_2)$.

Put another way, a function f is one-to-one if no y in the range is the image of more than one x in the domain. A function is not one-to-one if two different elements in the domain correspond to the same element in the range. So, the function in Figure 7 is not one-to-one because two different elements in the domain, dog and cat, both correspond to 11. Figure 8 illustrates the distinction of one-to-one functions, non–one-to-one functions, and nonfunctions.

Figure 8

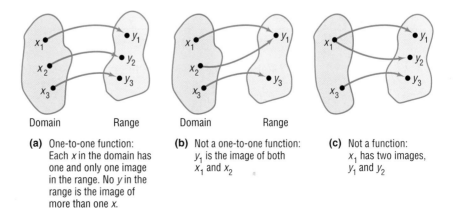

(a) One-to-one function: Each x in the domain has one and only one image in the range. No y in the range is the image of more than one x.

(b) Not a one-to-one function: y_1 is the image of both x_1 and x_2.

(c) Not a function: x_1 has two images, y_1 and y_2.

| EXAMPLE 1 | Determining Whether a Function Is One-to-One |

Determine whether the following functions are one-to-one.

(a) For the following function, the domain represents the age of five males and the range represents their HDL (good) cholesterol (mg/dL).

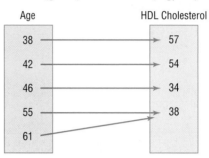

(b) $\{(-2, 6), (-1, 3), (0, 2), (1, 5), (2, 8)\}$

Solution   (a) The function is not one-to-one because there are two different inputs, 55 and 61, that correspond to the same output, 38.

(b) The function is one-to-one because there are no two distinct inputs that correspond to the same output.   ◄

NOW WORK PROBLEMS 9 AND 13.

If the graph of a function f is known, there is a simple test, called the **horizontal-line test**, to determine whether f is one-to-one.

Figure 9

$f(x_1) = f(x_2) = h$ and $x_1 \neq x_2$; f is not a one-to-one function.

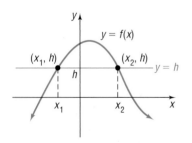

Theorem   **Horizontal-line Test**

If every horizontal line intersects the graph of a function f in at most one point, then f is one-to-one.

The reason that this test works can be seen in Figure 9, where the horizontal line $y = h$ intersects the graph at two distinct points, (x_1, h) and (x_2, h). Since h is the image of both x_1 and x_2, $x_1 \neq x_2$, f is not one-to-one. Based on Figure 9, we can state the horizontal-line test in another way: If the graph of any horizontal line intersects the graph of a function f at more than one point, then f is not one-to-one.

| EXAMPLE 2 | Using the Horizontal-line Test |

For each function, use the graph to determine whether the function is one-to-one.

(a) $f(x) = x^2$ (b) $g(x) = x^3$

Solution   (a) Figure 10(a) illustrates the horizontal-line test for $f(x) = x^2$. The horizontal line $y = 1$ intersects the graph of f twice, at $(1, 1)$ and at $(-1, 1)$, so f is not one-to-one.

(b) Figure 10(b) illustrates the horizontal-line test for $g(x) = x^3$. Because every horizontal line will intersect the graph of g exactly once, it follows that g is one-to-one.

Figure 10

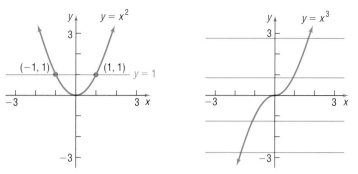

(a) A horizontal line intersects the graph twice; f is not one-to-one

(b) Every horizontal line intersects the graph exactly once; g is one-to-one   ◀

NOW WORK PROBLEM 17.

Let's look more closely at the one-to-one function $g(x) = x^3$. This function is an increasing function. Because an increasing (or decreasing) function will always have different y-values for unequal x-values, it follows that a function that is increasing (or decreasing) over its domain is also a one-to-one function.

Theorem

A function that is increasing on an interval I is a one-to-one function on I.

A function that is decreasing on an interval I is a one-to-one function on I.

2   Determine the Inverse of a Function Defined by a Map or an Ordered Pair

In Section 2.2, we said that a function f can be thought of as a machine that receives an input, say x, from the domain, manipulates it, and outputs the value $f(x)$. The **inverse function of f** receives as input $f(x)$, manipulates it, and outputs x. For example, if $f(x) = 2x + 5$, then $f(3) = 11$. Here, the input is 3 and the output is 11. The inverse of f would "undo" what f does and receive as input 11 and provide as output 3. So, if a function takes inputs and multiplies them by 2, the inverse would take inputs and divide them by 2. For the inverse of a function f to itself be a function, f must be one-to-one.

Recall that we have a variety of ways of representing functions. We will discuss how to find inverses for all four representations of functions: (1) maps, (2) sets of ordered pairs (3) graphs, and (4) equations. We begin with finding inverses of functions represented by maps or sets of ordered pairs.

| EXAMPLE 3 | **Finding the Inverse of a Function Defined by a Map** |

Find the inverse of the following function. Let the domain of the function represent certain states, and let the range represent the state's population. State the domain and the range of the inverse function.

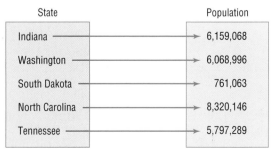

Solution The function is one-to-one, so the inverse will be a function. To find the inverse function, we interchange the elements in the domain with the elements in the range. For
example, the function receives as input Indiana and outputs 6,159,068. So, the inverse receives as input 6,159,068 and outputs Indiana. The inverse function is shown
next.

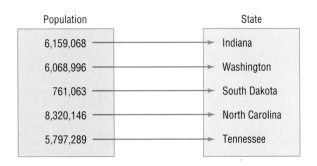

The domain of the inverse function is $\{6\,159\,068, 6\,068\,996, 761\,063, 8\,320\,146,$
$5\,797\,289\}$. The range of the inverse function is $\{$Indiana, Washington, South
Dakota, North Carolina, Tennessee$\}$. ◀

If the function f is a set of ordered pairs (x, y), then the inverse of f is the set of
ordered pairs (y, x).

| EXAMPLE 4 | **Finding the Inverse of a Function Defined By a Set of Ordered Pairs** |

Find the inverse of the following one-to-one function:

$$\{(-3, -27), (-2, -8), (-1, -1), (0, 0), (1, 1), (2, 8), (3, 27)\}$$

Solution The inverse of the given function is found by interchanging the entries in each
ordered pair and so is given by

$$\{(-27, -3), (-8, -2), (-1, -1), (0, 0), (1, 1), (8, 2), (27, 3)\}$$ ◀

NOW WORK PROBLEMS 23 AND 27.

Figure 11

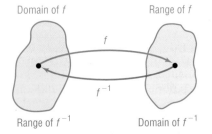

Remember, if f is a one-to-one function, its inverse is a function. Then, to each
x in the domain of f, there is exactly one y in the range (because f is a function); and
to each y in the range of f, there is exactly one x in the domain (because f is one-to-
one). The correspondence from the range of f back to the domain of f is called the
inverse function of f and is denoted by the symbol f^{-1}. Figure 11 illustrates this
definition.

Based on Figure 11, two facts are now apparent about a function f and its
inverse f^{-1}.

WARNING
Be careful! f^{-1} is a symbol for the
inverse function of f. The -1 used
in f^{-1} is not an exponent. That is,
f^{-1} does not mean the reciprocal
of f; $f^{-1}(x)$ is not equal to $\dfrac{1}{f(x)}$. ∎

> Domain of f = Range of f^{-1} Range of f = Domain of f^{-1}

Look again at Figure 11 to visualize the relationship. If we start with x, apply
f, and then apply f^{-1}, we get x back again. If we start with x, apply f^{-1}, and then
apply f, we get the number x back again. To put it simply, what f does, f^{-1} undoes,
and vice versa. See the illustration that follows.

$$\boxed{\text{Input } x} \xrightarrow{\text{Apply } f} \boxed{f(x)} \xrightarrow{\text{Apply } f^{-1}} \boxed{f^{-1}(f(x)) = x}$$

$$\boxed{\text{Input } x} \xrightarrow{\text{Apply } f^{-1}} \boxed{f^{-1}(x)} \xrightarrow{\text{Apply } f} \boxed{f(f^{-1}(x)) = x}$$

In other words,

$$f^{-1}(f(x)) = x \quad \text{where } x \text{ is in the domain of } f$$
$$f(f^{-1}(x)) = x \quad \text{where } x \text{ is in the domain of } f^{-1}$$

Consider the function $f(x) = 2x$, which multiplies the argument x by 2. The inverse function f^{-1} undoes whatever f does. So the inverse function of f is $f^{-1}(x) = \dfrac{1}{2}x$, which divides the argument by 2. For example, $f(3) = 2(3) = 6$ and

Figure 12

$f^{-1}(2x) = \frac{1}{2}(2x) = x$

$f^{-1}(6) = \dfrac{1}{2}(6) = 3$, so f^{-1} undoes what f did. We can verify this by showing that

$$f^{-1}(f(x)) = f^{-1}(2x) = \frac{1}{2}(2x) = x \quad \text{and} \quad f(f^{-1}(x)) = f\left(\frac{1}{2}x\right) = 2\left(\frac{1}{2}x\right) = x$$

See Figure 12.

EXAMPLE 5

Verifying Inverse Functions

— **Exploration** —

Simultaneously graph $Y_1 = x$, $Y_2 = x^3$, and $Y_3 = \sqrt[3]{x}$ on a square screen with $-3 \le x \le 3$. What do you observe about the graphs of $Y_2 = x^3$, its inverse $Y_3 = \sqrt[3]{x}$, and the line $Y_1 = x$?

Repeat this experiment by simultaneously graphing $Y_1 = x$, $Y_2 = 2x + 3$, and $Y_3 = \dfrac{1}{2}(x - 3)$ on a square screen with $-6 \le x \le 3$. Do you see the symmetry of the graph of Y_2 and its inverse Y_3 with respect to the line $Y_1 = x$?

(a) We verify that the inverse of $g(x) = x^3$ is $g^{-1}(x) = \sqrt[3]{x}$ by showing that

$$g^{-1}(g(x)) = g^{-1}(x^3) = \sqrt[3]{x^3} = x \quad \text{for all } x \text{ in the domain of } g$$
$$g(g^{-1}(x)) = g(\sqrt[3]{x}) = (\sqrt[3]{x})^3 = x \quad \text{for all } x \text{ in the domain of } g^{-1}.$$

(b) We verify that the inverse of $f(x) = 2x + 3$ is $f^{-1}(x) = \dfrac{1}{2}(x - 3)$ by showing that

$$f^{-1}(f(x)) = f^{-1}(2x + 3) = \frac{1}{2}[(2x + 3) - 3] = \frac{1}{2}(2x) = x \quad \substack{\text{for all } x \text{ in the} \\ \text{domain of } f}$$

$$f(f^{-1}(x)) = f\left(\frac{1}{2}(x - 3)\right) = 2\left[\frac{1}{2}(x - 3)\right] + 3 = (x - 3) + 3 = x \quad \substack{\text{for all } x \text{ in the} \\ \text{domain of } f^{-1}.} \quad \blacktriangleleft$$

EXAMPLE 6

Verifying Inverse Functions

Verify that the inverse of $f(x) = \dfrac{1}{x - 1}$ is $f^{-1}(x) = \dfrac{1}{x} + 1$. For what values of x is $f^{-1}(f(x)) = x$? For what values of x is $f(f^{-1}(x)) = x$?

Solution

The domain of f is $\{x \mid x \ne 1\}$ and the domain of f^{-1} is $\{x \mid x \ne 0\}$ Now,

$$f^{-1}(f(x)) = f^{-1}\left(\frac{1}{x - 1}\right) = \frac{1}{\dfrac{1}{x - 1}} + 1 = x - 1 + 1 = x \quad \text{provided } x \ne 1.$$

$$f(f^{-1}(x)) = f\left(\frac{1}{x} + 1\right) = \frac{1}{\dfrac{1}{x} + 1 - 1} = \frac{1}{\dfrac{1}{x}} = x \quad \text{provided } x \ne 0$$
$\blacktriangleleft$

➤ **NOW WORK PROBLEM 37.**

3 Obtain the Graph of the Inverse Function from the Graph of the Function

For the functions in Example 5(b), we list points on the graph of $f = Y_1$ and on the graph of $f^{-1} = Y_2$ in Table 1.

We notice that whenever (a, b) is on the graph of f then (b, a) is on the graph of f^{-1}. Figure 13 shows these points plotted. Also shown is the graph of $y = x$, which you should observe is a line of symmetry of the points.

Figure 13

Table 1

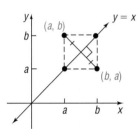

Figure 14

Suppose that (a, b) is a point on the graph of a one-to-one function f defined by $y = f(x)$. Then $b = f(a)$. This means that $a = f^{-1}(b)$, so (b, a) is a point on the graph of the inverse function f^{-1}. The relationship between the point (a, b) on f and the point (b, a) on f^{-1} is shown in Figure 14. The line segment containing (a, b) and (b, a) is perpendicular to the line $y = x$ and is bisected by the line $y = x$. (Do you see why?) It follows that the point (b, a) on f^{-1} is the reflection about the line $y = x$ of the point (a, b) on f.

Theorem

The graph of a function f and the graph of its inverse f^{-1} are symmetric with respect to the line $y = x$.

Figure 15 illustrates this result. Notice that, once the graph of f is known, the graph of f^{-1} may be obtained by reflecting the graph of f about the line $y = x$.

Figure 15

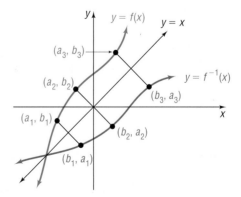

EXAMPLE 7 **Graphing the Inverse Function**

The graph in Figure 16(a) is that of a one-to-one function $y = f(x)$. Draw the graph of its inverse.

Solution   We begin by adding the graph of $y = x$ to Figure 16(a). Since the points $(-2, -1), (-1, 0)$, and $(2, 1)$ are on the graph of f, we know that the points $(-1, -2), (0, -1)$, and $(1, 2)$ must be on the graph of f^{-1}. Keeping in mind that the graph of f^{-1} is the reflection about the line $y = x$ of the graph of f, we can draw f^{-1}. See Figure 16(b).

Figure 16

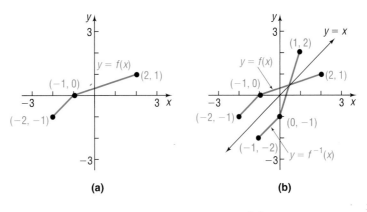

(a)   (b)

NOW WORK PROBLEM **31.**

4 Find the Inverse of a Function Defined by an Equation

The fact that the graphs of a one-to-one function f and its inverse function f^{-1} are symmetric with respect to the line $y = x$ tells us more. It says that we can obtain f^{-1} by interchanging the roles of x and y in f. Look again at Figure 15. If f is defined by the equation

$$y = f(x)$$

then f^{-1} is defined by the equation

$$x = f(y)$$

The equation $x = f(y)$ defines f^{-1} *implicitly*. If we can solve this equation for y, we will have the *explicit* form of f^{-1}, that is,

$$y = f^{-1}(x)$$

Let's use this procedure to find the inverse of $f(x) = 2x + 3$. (Since f is a linear function and is increasing, we know that f is one-to-one and so has an inverse function.)

| EXAMPLE 8 |

Finding the Inverse Function

Find the inverse of $f(x) = 2x + 3$. Also find the domain and range of f and f^{-1}. Graph f and f^{-1} on the same coordinate axes.

Solution   In the equation $y = 2x + 3$, interchange the variables x and y. The result,

$$x = 2y + 3$$

is an equation that defines the inverse f^{-1} implicitly. To find the explicit form, we solve for y.

$$2y + 3 = x$$
$$2y = x - 3$$
$$y = \frac{1}{2}(x - 3)$$

The explicit form of the inverse f^{-1} is therefore

$$f^{-1}(x) = \frac{1}{2}(x - 3)$$

which we verified in Example 5(c).

Next we find

$$\text{Domain of } f = \text{Range of } f^{-1} = (-\infty, \infty)$$
$$\text{Range of } f = \text{Domain of } f^{-1} = (-\infty, \infty)$$

The graphs of $Y_1 = f(x) = 2x + 3$ and its inverse $Y_2 = f^{-1}(x) = \frac{1}{2}(x - 3)$ are shown in Figure 17. Note the symmetry of the graphs with respect to the line $Y_3 = x$.

Figure 17

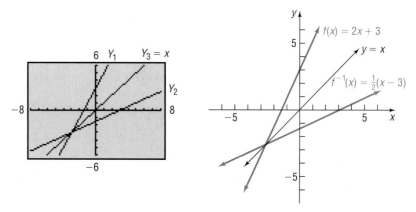

We now outline the steps to follow for finding the inverse of a one-to-one function.

Procedure for Finding the Inverse of a One-to-One Function

STEP 1: In $y = f(x)$, interchange the variables x and y to obtain

$$x = f(y)$$

This equation defines the inverse function f^{-1} implicitly.

STEP 2: If possible, solve the implicit equation for y in terms of x to obtain the explicit form of f^{-1}

$$y = f^{-1}(x)$$

STEP 3: Check the result by showing that

$$f^{-1}(f(x)) = x \quad \text{and} \quad f(f^{-1}(x)) = x$$

EXAMPLE 9 **Finding the Inverse Function**

The function

$$f(x) = \frac{2x + 1}{x - 1}, \qquad x \neq 1$$

is one-to-one. Find its inverse and check the result.

Solution   **STEP 1:** Interchange the variables x and y in

$$y = \frac{2x + 1}{x - 1}$$

to obtain

$$x = \frac{2y + 1}{y - 1}$$

STEP 2: Solve for y.

$$x = \frac{2y + 1}{y - 1}$$

$$x(y - 1) = 2y + 1 \qquad \text{Multiply both sides by } x - 1.$$

$$xy - x = 2y + 1 \qquad \text{Apply the Distributive Property.}$$

$$xy - 2y = x + 1 \qquad \text{Subtract 2x from both sides; add x to both sides.}$$

$$(x - 2)y = x + 1 \qquad \text{Factor.}$$

$$y = \frac{x + 1}{x - 2} \qquad \text{Divide by } x - 2.$$

The inverse is

$$f^{-1}(x) = \frac{x + 1}{x - 2}, \qquad x \neq 2 \qquad \text{Replace y by } f^{-1}(x).$$

STEP 3: ✔ **CHECK:**

$$f^{-1}(f(x)) = f^{-1}\left(\frac{2x + 1}{x - 1}\right) = \frac{\dfrac{2x + 1}{x - 1} + 1}{\dfrac{2x + 1}{x - 1} - 2} = \frac{2x + 1 + x - 1}{2x + 1 - 2(x - 1)} = \frac{3x}{3} = x, \quad x \neq 1$$

$$f(f^{-1}(x)) = f\left(\frac{x + 1}{x - 2}\right) = \frac{2\left(\dfrac{x + 1}{x - 2}\right) + 1}{\dfrac{x + 1}{x - 2} - 1} = \frac{2(x + 1) + x - 2}{x + 1 - (x - 2)} = \frac{3x}{3} = x, \quad x \neq 2$$

◀

──────── **Exploration** ──────────────────────────────

In Example 9, we found that, if $f(x) = \dfrac{2x + 1}{x - 1}$, then $f^{-1}(x) = \dfrac{x + 1}{x - 2}$. Compare the vertical and horizontal asymptotes of f and f^{-1}. What did you find? Are you surprised?

Result   You should have determined that the vertical asymptote of f is $x = 1$ and the horizontal asymptote is $y = 2$. The vertical asymptote of f^{-1} is $x = 2$, and the horizontal asymptote is $y = 1$.

──

        ✏──   **NOW WORK PROBLEM 49.**

We said in Chapter 2 that finding the range of a function f is not easy. However, if f is one-to-one, we can find its range by finding the domain of the inverse function f^{-1}.

EXAMPLE 10 **Finding the Range of a Function**

Find the domain and range of

$$f(x) = \frac{2x + 1}{x - 1}$$

Solution The domain of f is $\{x \mid x \neq 1\}$. To find the range of f, we first find the inverse f^{-1}. Based on Example 9, we have

$$f^{-1}(x) = \frac{x + 1}{x - 2}$$

The domain of f^{-1} is $\{x \mid x \neq 2\}$, so the range of f is $\{y \mid y \neq 2\}$. ◄

NOW WORK PROBLEM 63.

If a function is not one-to-one, then its inverse is not a function. Sometimes, though, an appropriate restriction on the domain of such a function will yield a new function that is one-to-one. Let's look at an example of this common practice.

EXAMPLE 11 **Finding the Inverse of a Domain-restricted Function**

Find the inverse of $y = f(x) = x^2$ if $x \geq 0$.

Solution The function $y = x^2$ is not one-to-one. [Refer to Example 2(a).] However, if we restrict the domain of this function to $x \geq 0$, as indicated, we have a new function that is increasing and therefore is one-to-one. As a result, the function defined by $y = f(x) = x^2, x \geq 0$, has an inverse function, f^{-1}.

We follow the steps given previously to find f^{-1}.

STEP 1: In the equation $y = x^2$, $x \geq 0$, interchange the variables x and y. The result is

$$x = y^2, \quad y \geq 0$$

This equation defines (implicitly) the inverse function.

STEP 2: We solve for y to get the explicit form of the inverse. Since $y \geq 0$, only one solution for y is obtained, namely, $y = \sqrt{x}$. So $f^{-1}(x) = \sqrt{x}$.

STEP 3: ✔ **CHECK:** $f^{-1}(f(x)) = f^{-1}(x^2) = \sqrt{x^2} = |x| = x$, since $x \geq 0$
$$f(f^{-1}(x)) = f(\sqrt{x}) = (\sqrt{x})^2 = x.$$ ◄

Figure 18 illustrates the graphs of $Y_1 = f(x) = x^2, x \geq 0$, and $Y_2 = f^{-1}(x) = \sqrt{x}$.

Figure 18

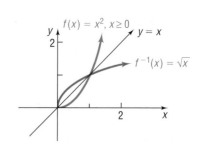

Summary

1. If a function f is one-to-one, then it has an inverse function f^{-1}.
2. Domain of f = Range of f^{-1}; Range of f = Domain of f^{-1}.
3. To verify that f^{-1} is the inverse of f, show that $f^{-1}(f(x)) = x$ for every x in the domain of f and $f(f^{-1}(x)) = x$ for every x in the domain of f^{-1}.
4. The graphs of f and f^{-1} are symmetric with respect to the line $y = x$.
5. To find the range of a one-to-one function f, find the domain of the inverse function f^{-1}.

4.2 Assess Your Understanding

'Are You Prepared?'

Answers are given at the end of these exercises. If you get a wrong answer, read the pages listed in red.

1. Is the set of ordered pairs $\{(1, 3), (2, 3), (-1, 2)\}$ a function? Why or why not? (pp. 204–209)
2. Where is the function $f(x) = x^2$ increasing? Where is it decreasing? (pp. 230–231)
3. What is the domain of $f(x) = \dfrac{x + 5}{x^2 + 3x - 18}$? (pp. 212–213)

Concepts and Vocabulary

4. If every horizontal line intersects the graph of a function f at no more than one point, then f is a(n) _____ function.
5. If f^{-1} denotes the inverse of a function f, then the graphs of f and f^{-1} are symmetric with respect to the line _____.
6. If the domain of a one-to-one function f is $[4, \infty)$, then the range of its inverse, f^{-1}, is _____.
7. *True or False:* If f and g are inverse functions, then the domain of f is the same as the domain of g.
8. *True or False:* If f and g are inverse functions, then their graphs are symmetric with respect to the line $y = x$.

Skill Building

In Problems 9–16, determine whether the function is one-to-one.

9.

10.

11.

12.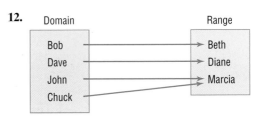

13. $\{(2, 6), (-3, 6), (4, 9), (1, 10)\}$

14. $\{(-2, 5), (-1, 3), (3, 7), (4, 12)\}$

15. $\{(0, 0), (1, 1), (2, 16), (3, 81)\}$

16. $\{(1, 2), (2, 8), (3, 18), (4, 32)\}$

In Problems 17–22, the graph of a function f is given. Use the horizontal-line test to determine whether f is one-to-one.

17.

18.

19.

20.

21.

22.

In Problems 23–30, find the inverse of each one-to-one function. State the domain and the range of each inverse function.

23.

Location	Annual Rainfall (inches)
Mt Waialeale, Hawaii	460.00
Monrovia, Liberia	202.01
Pago Pago, American Samoa	196.46
Moulmein, Burma	191.02
Lae, Papua New Guinea	182.87

SOURCE: *Information Please Almanac*

24.

Title	Domestic Gross
Star Wars	$460,998,007
Star Wars: Episode One – The Phantom Menace	$431,088,295
E.T. the Extra Terrestrial	$399,804,539
Jurassic Park	$357,067,947
Forrest Gump	$329,694,499

SOURCE: *Information Please Almanac*

25.

Age	Monthly Cost of Life Insurance
30	$7.09
40	$8.40
45	$11.29

SOURCE: *eterm.com*

26.

State	Unemployment Rate
Virginia	11%
Nevada	5.5%
Tennessee	5.1%
Texas	6.3%

SOURCE: *United States Statistical Abstract*

27. $\{(-3, 5), (-2, 9), (-1, 2), (0, 11), (1, -5)\}$

28. $\{(-2, 2), (-1, 6), (0, 8), (1, -3), (2, 9)\}$

29. $\{(-2, 1), (-3, 2), (-10, 0), (1, 9), (2, 4)\}$

30. $\{(-2, -8), (-1, -1), (0, 0), (1, 1), (2, 8)\}$

In Problems 31–36, the graph of a one-to-one function f is given. Draw the graph of the inverse function f^{-1}. For convenience (and as a hint), the graph of y = x is also given.

31.

32.

33.

34.

35.

36.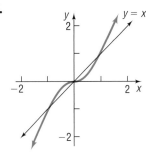

In Problems 37–46, verify that the functions f and g are inverses of each other by showing that $f(g(x)) = x$ and $g(f(x)) = x$. Give any values of x that need to be excluded.

37. $f(x) = 3x + 4;$ $g(x) = \dfrac{1}{3}(x - 4)$

38. $f(x) = 3 - 2x;$ $g(x) = -\dfrac{1}{2}(x - 3)$

39. $f(x) = 4x - 8;$ $g(x) = \dfrac{x}{4} + 2$

40. $f(x) = 2x + 6;$ $g(x) = \dfrac{1}{2}x - 3$

41. $f(x) = x^3 - 8;$ $g(x) = \sqrt[3]{x + 8}$

42. $f(x) = (x - 2)^2,\, x \geq 2;$ $g(x) = \sqrt{x} + 2$

43. $f(x) = \dfrac{1}{x};$ $g(x) = \dfrac{1}{x}$

44. $f(x) = x;$ $g(x) = x$

45. $f(x) = \dfrac{2x + 3}{x + 4};$ $g(x) = \dfrac{4x - 3}{2 - x}$

46. $f(x) = \dfrac{x - 5}{2x + 3};$ $g(x) = \dfrac{3x + 5}{1 - 2x}$

In Problems 47–58, the function f is one-to-one. Find its inverse and check your answer. State the domain and the range of f and f^{-1}. Graph f, f^{-1}, and y = x on the same coordinate axes.

47. $f(x) = 3x$

48. $f(x) = -4x$

49. $f(x) = 4x + 2$

50. $f(x) = 1 - 3x$

51. $f(x) = x^3 - 1$

52. $f(x) = x^3 + 1$

53. $f(x) = x^2 + 4,\quad x \geq 0$

54. $f(x) = x^2 + 9,\quad x \geq 0$

55. $f(x) = \dfrac{4}{x}$

56. $f(x) = -\dfrac{3}{x}$

57. $f(x) = \dfrac{1}{x - 2}$

58. $f(x) = \dfrac{4}{x + 2}$

In Problems 59–70, the function f is one-to-one. Find its inverse and check your answer. State the domain of f and find its range using f^{-1}.

59. $f(x) = \dfrac{2}{3 + x}$

60. $f(x) = \dfrac{4}{2 - x}$

61. $f(x) = \dfrac{3x}{x + 2}$

62. $f(x) = \dfrac{-2x}{x - 1}$

63. $f(x) = \dfrac{2x}{3x - 1}$

64. $f(x) = \dfrac{3x + 1}{-x}$

65. $f(x) = \dfrac{3x + 4}{2x - 3}$

66. $f(x) = \dfrac{2x - 3}{x + 4}$

67. $f(x) = \dfrac{2x + 3}{x + 2}$

68. $f(x) = \dfrac{-3x - 4}{x - 2}$

69. $f(x) = \dfrac{x^2 - 4}{2x^2},\quad x > 0$

70. $f(x) = \dfrac{x^2 + 3}{3x^2},\quad x > 0$

Applications and Extensions

71. Use the graph of $y = f(x)$ given in Problem 31 to evaluate the following:
(a) $f(-1)$ (b) $f(1)$ (c) $f^{-1}(1)$ (d) $f^{-1}(2)$

72. Use the graph of $y = f(x)$ given in Problem 32 to evaluate the following:
(a) $f(2)$ (b) $f(1)$ (c) $f^{-1}(0)$ (d) $f^{-1}(-1)$

73. Find the inverse of the linear function

$$f(x) = mx + b, \quad m \neq 0$$

74. Find the inverse of the function

$$f(x) = \sqrt{r^2 - x^2}, \quad 0 \leq x \leq r$$

75. A function f has an inverse function. If the graph of f lies in quadrant I, in which quadrant does the graph of f^{-1} lie?

76. A function f has an inverse function. If the graph of f lies in quadrant II, in which quadrant does the graph of f^{-1} lie?

77. The function $f(x) = |x|$ is not one-to-one. Find a suitable restriction on the domain of f so that the new function that results is one-to-one. Then find the inverse of f.

78. The function $f(x) = x^4$ is not one-to-one. Find a suitable restriction on the domain of f so that the new function that results is one-to-one. Then find the inverse of f.

79. Height versus Head Circumference The head circumference C of a child is related to the height H of the child (both in inches) through the function

$$H(C) = 2.15C - 10.53$$

(a) Express the head circumference C as a function of height H.

(b) Predict the head circumference of a child who is 26 inches tall.

80. Temperature Conversion To convert from x degrees Celsius to y degrees Fahrenheit, we use the formula $y = f(x) = \dfrac{9}{5}x + 32$. To convert from x degrees Fahrenheit to y degrees Celsius, we use the formula $y = g(x) = \dfrac{5}{9}(x - 32)$. Show that f and g are inverse functions.

81. Demand for Corn The demand for corn obeys the equation $p(x) = 300 - 50x$, where p is the price per bushel (in dollars) and x is the number of bushels produced (in millions). Express the production amount x as a function of the price p.

82. Period of a Pendulum The period T (in seconds) of a simple pendulum is a function of its length l (in feet), given by $T(l) = 2\pi\sqrt{\dfrac{l}{g}}$, where $g \approx 32.2$ feet per second per second is the acceleration of gravity. Express the length l as a function of the period T.

83. Given

$$f(x) = \frac{ax + b}{cx + d}$$

find $f^{-1}(x)$. If $c \neq 0$, under what conditions on $a, b, c,$ and d is $f = f^{-1}$?

Discussion and Writing

84. Can a one-to-one function and its inverse be equal? What must be true about the graph of f for this to happen? Give some examples to support your conclusion.

85. Draw the graph of a one-to-one function that contains the points $(-2, -3)$, $(0, 0)$, and $(1, 5)$. Now draw the graph of its inverse. Compare your graph to those of other students. Discuss any similarities. What differences do you see?

86. Give an example of a function whose domain is the set of real numbers and that is neither increasing nor decreasing on its domain, but is one-to-one.

[**Hint:** Use a piecewise-defined function.]

87. If a function f is even, can it be one-to-one? Explain.

88. Is every odd function one-to-one? Explain.

89. If the graph of a function and its inverse intersect, where must this necessarily occur? Can they intersect anywhere else? Must they intersect?

90. Suppose $C(g)$ represents the cost C of manufacturing g cars. Explain what $C^{-1}(800,000)$ represents.

'Are You Prepared?' Answers

1. Yes; for each input x there is one output y. **2.** Increasing on $(0, \infty)$; decreasing on $(-\infty, 0)$. **3.** $\{x | x \neq -6, x \neq 3\}$

4.3 Exponential Functions

PREPARING FOR THIS SECTION   *Before getting started, review the following:*

- Exponents (Section R.2, pp. 20–22, and Section R.8, pp. 71–73)
- Graphing Techniques: Transformations (Section 2.7, pp. 266–274)

- Average Rate of Change (Section 2.4, pp. 233–236)
- Solving Equations (Section 1.2, pp. 100–107 and Section 1.3, pp. 112–119)
- Horizontal Asymptotes (Section 3.3, pp. 339–343)

 Now work the 'Are You Prepared?' problems on page 430.

> **OBJECTIVES** 1 Evaluate Exponential Functions
> 2 Graph Exponential Functions
> 3 Define the Number *e*
> 4 Solve Exponential Equations

---

1 Evaluate Exponential Functions

In Section R.8, we give a definition for raising a real number *a* to a rational power. Based on that discussion, we gave meaning to expressions of the form

$$a^r$$

where the base *a* is a positive real number and the exponent *r* is a rational number.

 But what is the meaning of a^x, where the base *a* is a positive real number and the exponent *x* is an irrational number? Although a rigorous definition requires methods discussed in calculus, the basis for the definition is easy to follow: Select a rational number *r* that is formed by truncating (removing) all but a finite number of digits from the irrational number *x*. Then it is reasonable to expect that

$$a^x \approx a^r$$

For example, take the irrational number $\pi = 3.14159\ldots$. Then an approximation to a^π is

$$a^\pi \approx a^{3.14}$$

where the digits after the hundredths position have been removed from the value for π. A better approximation would be

$$a^\pi \approx a^{3.14159}$$

where the digits after the hundred-thousandths position have been removed. Continuing in this way, we can obtain approximations to a^π to any desired degree of accuracy.

Most calculators have an $\boxed{x^y}$ key or a caret key $\boxed{\wedge}$ for working with exponents. To evaluate expressions of the form a^x, enter the base *a*, then press the $\boxed{x^y}$ key (or the $\boxed{\wedge}$ key), enter the exponent *x*, and press $\boxed{=}$ (or $\boxed{\text{enter}}$).

| **EXAMPLE 1** | **Using a Calculator to Evaluate Powers of 2** |

Using a calculator, evaluate:

(a) $2^{1.4}$ (b) $2^{1.41}$ (c) $2^{1.414}$ (d) $2^{1.4142}$ (e) $2^{\sqrt{2}}$

Figure 19

$2^1.4$
2.639015822

Solution

Figure 19 shows the solution to part (a) using a TI-84 Plus graphing calculator.

(a) $2^{1.4} \approx 2.639015822$ (b) $2^{1.41} \approx 2.657371628$

(c) $2^{1.414} \approx 2.66474965$ (d) $2^{1.4142} \approx 2.665119089$

(e) $2^{\sqrt{2}} \approx 2.665144143$ ◀

━━ **NOW WORK PROBLEM 11.**

It can be shown that the familiar laws for rational exponents hold for real exponents.

Theorem **Laws of Exponents**

If s, t, a, and b are real numbers with $a > 0$ and $b > 0$, then

$$a^s \cdot a^t = a^{s+t} \qquad (a^s)^t = a^{st} \qquad (ab)^s = a^s \cdot b^s$$

$$1^s = 1 \qquad a^{-s} = \frac{1}{a^s} = \left(\frac{1}{a}\right)^s \qquad a^0 = 1 \qquad \textbf{(1)}$$

We are now ready for the following definition:

An **exponential function** is a function of the form

$$f(x) = a^x$$

where a is a positive real number $(a > 0)$ and $a \neq 1$. The domain of f is the set of all real numbers.

CAUTION It is important to distinguish a power function, $g(x) = x^n$, $n \geq 2$, an integer, from an exponential function, $f(x) = a^x$, $a \neq 1$, a real. In a power function, the base is a variable and the exponent is a constant. In an exponential function, the base is a constant and the exponent is a variable. ■

We exclude the base $a = 1$ because this function is simply the constant function $f(x) = 1^x = 1$. We also need to exclude bases that are negative; otherwise, we would have to exclude many values of x from the domain, such as $x = \dfrac{1}{2}$ and $x = \dfrac{3}{4}$.

[Recall that $(-2)^{1/2} = \sqrt{-2}$, $(-3)^{3/4} = \sqrt[4]{(-3)^3} = \sqrt[4]{-27}$, and so on, are not defined in the set of real numbers.]

Some examples of exponential functions are

$$f(x) = 2^x, \qquad F(x) = \left(\frac{1}{3}\right)^x$$

Notice that in each example the base is a constant and the exponent is a variable.

You may wonder what role the base a plays in the exponential function $f(x) = a^x$. We use the following Exploration to find out.

────── Exploration ──────────────────────────────

(a) Evaluate $f(x) = 2^x$ at $x = -2, -1, 0, 1, 2,$ and 3.
(b) Evaluate $g(x) = 3x + 2$ at $x = -2, -1, 0, 1, 2,$ and 3.
(c) Comment on the pattern that exists in the values of f and g.

Result (a) Table 2 shows the values of $f(x) = 2^x$ for $x = -2, -1, 0, 1, 2,$ and 3.
        (b) Table 3 shows the values of $g(x) = 3x + 2$ for $x = -2, -1, 0, 1, 2,$ and 3.

Table 2

x	$f(x) = 2^x$
-2	$f(-2) = 2^{-2} = \dfrac{1}{2^2} = \dfrac{1}{4}$
-1	$\dfrac{1}{2}$
0	1
1	2
2	4
3	8

Table 3

x	$g(x) = 3x + 2$
-2	$g(-2) = 3(-2) + 2 = -4$
-1	-1
0	2
1	5
2	8
3	11

(c) In Table 2 we notice that each value of the exponential function $f(x) = a^x = 2^x$ could be found by multiplying the previous value of the function by the base, $a = 2$. For example,

$$f(-1) = 2 \cdot f(-2) = 2 \cdot \frac{1}{4} = \frac{1}{2}, \quad f(0) = 2 \cdot f(-1) = 2 \cdot \frac{1}{2} = 1, \quad f(1) = 2 \cdot f(0) = 2 \cdot 1 = 2$$

and so on.

Put another way, we see that the ratio of consecutive outputs is constant for unit increases in the input. The constant equals the value of the base a of the exponential function. For example, for the function $f(x) = 2^x$, we notice that

$$\frac{f(-1)}{f(-2)} = \frac{\frac{1}{2}}{\frac{1}{4}} = 2, \quad \frac{f(1)}{f(0)} = \frac{2}{1} = 2, \quad \frac{f(x+1)}{f(x)} = \frac{2^{x+1}}{2^x} = 2$$

and so on.

From Table 3 we see that the function $g(x) = 3x + 2$ does not have the ratio of consecutive outputs that are constant because it is not exponential. For example,

$$\frac{g(-1)}{g(-2)} = \frac{-1}{-4} = \frac{1}{4} \neq \frac{g(1)}{g(0)} = \frac{5}{2}$$

Instead, because $g(x) = 3x + 2$ is a linear function, for unit increases in the input, the outputs increase by a fixed amount equal to the value of the slope, 3.

───

The results of the Exploration lead to the following result.

Theorem

For an exponential function $f(x) = a^x, a > 0, a \neq 1$, if x is any real number, then

$$\frac{f(x + 1)}{f(x)} = a$$

In Words

For an exponential function $f(x) = a^x$, for 1-unit changes in the input x, the ratio of consecutive outputs is the constant a.

Proof

$$\frac{f(x + 1)}{f(x)} = \frac{a^{x+1}}{a^x} = a^{x+1-x} = a^1 = a \qquad \blacksquare$$

NOW WORK PROBLEM 21.

2 Graph Exponential Functions

First, we graph the exponential function $f(x) = 2^x$.

EXAMPLE 2 | **Graphing an Exponential Function**

Graph the exponential function: $f(x) = 2^x$

Table 4

X	Y1
-10	9.8E-4
-3	.125
-1	.5
0	1
1	2
3	8
10	1024

Y1🔲2^X

Solution The domain of $f(x) = 2^x$ is the set of all real numbers. We begin by locating some points on the graph of $f(x) = 2^x$, as listed in Table 4.

Since $2^x > 0$ for all x, the range of f is $(0, \infty)$. From this, we conclude that the graph has no x-intercepts, and, in fact, the graph will lie above the x-axis for all x. As Table 4 indicates, the y-intercept is 1. Table 4 also indicates that as $x \to -\infty$ the value of $f(x) = 2^x$ gets closer and closer to 0. We conclude that the x-axis ($y = 0$) is a horizontal asymptote to the graph as $x \to -\infty$. This gives us the end behavior for x large and negative.

To determine the end behavior for x large and positive, look again at Table 4. As $x \to \infty$, $f(x) = 2^x$ grows very quickly, causing the graph of $f(x) = 2^x$ to rise very rapidly. It is apparent that f is an increasing function and hence is one-to-one.

Figure 20 shows the graph of $f(x) = 2^x$. Notice that all the conclusions given earlier are confirmed by the graph.

Figure 20

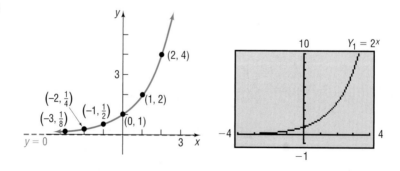

As we shall see, graphs that look like the one in Figure 20 occur very frequently in a variety of situations. For example, look at the graph in Figure 21, which illus-

Figure 21

trates the closing price of a share of Harley Davidson stock. Investors might conclude from this graph that the price of Harley Davidson stock is *behaving exponentially*; that is, the graph exhibits rapid, or exponential, growth.

We shall have more to say about situations that lead to exponential growth later in this chapter. For now, we continue to seek properties of the exponential functions.

The graph of $f(x) = 2^x$ in Figure 20 is typical of all exponential functions that have a base larger than 1. Such functions are increasing functions and hence are one-to-one. Their graphs lie above the x-axis, pass through the point $(0, 1)$, and thereafter rise rapidly as $x \to \infty$. As $x \to -\infty$, the x-axis ($y = 0$) is a horizontal asymptote. There are no vertical asymptotes. Finally, the graphs are smooth and continuous, with no corners or gaps.

Figure 22 illustrates the graphs of two more exponential functions whose bases are larger than 1. Notice that for the larger base the graph is steeper when $x > 0$. Figure 23 shows that when $x < 0$ the graph of the equation with the larger base is closer to the x-axis.

Figure 22

Figure 23

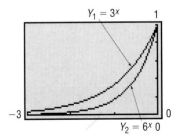

The following display summarizes the information that we have about $f(x) = a^x, a > 1$.

Properties of the Exponential Function $f(x) = a^x$, $a > 1$

1. The domain is the set of all real numbers; the range is the set of positive real numbers.

2. There are no x-intercepts; the y-intercept is 1.

3. The x-axis ($y = 0$) is a horizontal asymptote as $x \rightarrow -\infty$.

4. $f(x) = a^x$, $a > 1$, is an increasing function and is one-to-one.

5. The graph of f contains the points $(0, 1)$, $(1, a)$, and $\left(-1, \dfrac{1}{a}\right)$.

6. The graph of f is smooth and continuous, with no corners or gaps. See Figure 24.

Figure 24
$f(x) = a^x$, $a > 1$

Now we consider $f(x) = a^x$ when $0 < a < 1$.

EXAMPLE 3

Graphing an Exponential Function

Graph the exponential function: $f(x) = \left(\dfrac{1}{2}\right)^x$

Solution

The domain of $f(x) = \left(\dfrac{1}{2}\right)^x$ consists of all real numbers. As before, we locate some points on the graph, as listed in Table 5. Since $\left(\dfrac{1}{2}\right)^x > 0$ for all x, the range of f is the interval $(0, \infty)$. The graph lies above the x-axis and so has no x-intercepts.

The y-intercept is 1. As $x \rightarrow -\infty$, $f(x) = \left(\dfrac{1}{2}\right)^x$ grows very quickly. As $x \rightarrow \infty$, the values of $f(x)$ approach 0. The x-axis ($y = 0$) is a horizontal asymptote as $x \rightarrow \infty$. It is apparent that f is a decreasing function and hence is one-to-one. Figure 25 illustrates the graph.

Table 5

X	Y1
-10	1024
-3	8
-1	2
0	1
1	.5
3	.125
10	9.8E-4

Y1 ▣ (1/2)^X

Figure 25

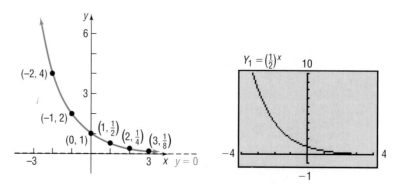

— **Seeing the Concept** —
Using a graphing utility, simultaneously graph:

(a) $Y_1 = 3^x, Y_2 = \left(\dfrac{1}{3}\right)^x$

(b) $Y_1 = 6^x, Y_2 = \left(\dfrac{1}{6}\right)^x$

Conclude that the graph of $Y_2 = \left(\dfrac{1}{a}\right)^x$, for $a > 0$, is the reflection about the y-axis of the graph of $Y_1 = a^x$.

We could have obtained the graph of $y = \left(\dfrac{1}{2}\right)^x$ from the graph of $y = 2^x$ using transformations. The graph of $y = \left(\dfrac{1}{2}\right)^x = 2^{-x}$ is a reflection about the y-axis of the graph of $y = 2^x$. Compare Figures 20 and 25.

The graph of $f(x) = \left(\dfrac{1}{2}\right)^x$ in Figure 25 is typical of all exponential functions that have a base between 0 and 1. Such functions are decreasing and one-to-one. Their graphs lie above the x-axis and pass through the point $(0, 1)$. The graphs rise rapidly as $x \to -\infty$. As $x \to \infty$, the x-axis $(y = 0)$ is a horizontal asymptote. There are no vertical asymptotes. Finally, the graphs are smooth and continuous, with no corners or gaps.

Figure 26 illustrates the graphs of two more exponential functions whose bases are between 0 and 1. Notice that the smaller base results in a graph that is steeper when $x < 0$. Figure 27 shows that when $x > 0$ the graph of the equation with the smaller base is closer to the x-axis.

Figure 26

Figure 27

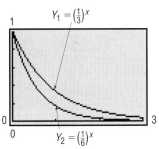

The following display summarizes the information that we have about the function $f(x) = a^x, 0 < a < 1$.

Properties of the Exponential Function $f(x) = a^x, 0 < a < 1$

1. The domain is the set of all real numbers; the range is the set of positive real numbers.
2. There are no x-intercepts; the y-intercept is 1.
3. The x-axis $(y = 0)$ is a horizontal asymptote as $x \to \infty$.
4. $f(x) = a^x, 0 < a < 1$, is a decreasing function and is one-to-one.
5. The graph of f contains the points $(0, 1)$, $(1, a)$, and $\left(-1, \dfrac{1}{a}\right)$.
6. The graph of f is smooth and continuous, with no corners or gaps. See Figure 28.

Figure 28
$f(x) = a^x, 0 < a < 1$

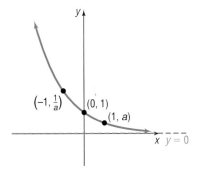

| EXAMPLE 4 | **Graphing Exponential Functions Using Transformations** |

Graph $f(x) = 2^{-x} - 3$ and determine the domain, range, and horizontal asymptote of f.

Solution   We begin with the graph of $y = 2^x$. Figure 29 shows the stages.

Figure 29

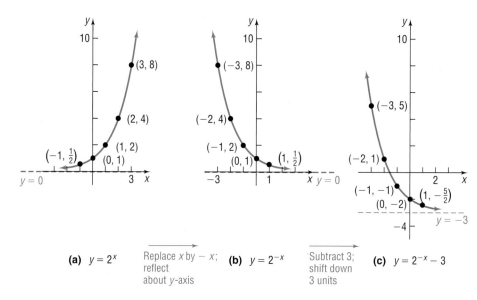

(a) $y = 2^x$ Replace x by $-x$;
reflect
about y-axis

(b) $y = 2^{-x}$ Subtract 3;
shift down
3 units

(c) $y = 2^{-x} - 3$

As Figure 29(c) illustrates, the domain of $f(x) = 2^{-x} - 3$ is the interval $(-\infty, \infty)$ and the range is the interval $(-3, \infty)$. The horizontal asymptote of f is the line $y = -3$.

✔ **CHECK:** Graph $Y_1 = 2^{-x} - 3$ to verify the graph obtained in Figure 29(c). ◀

─── **NOW WORK PROBLEM 37.**

3   Define the Number e

As we shall see shortly, many problems that occur in nature require the use of an exponential function whose base is a certain irrational number, symbolized by the letter e.

Let's look at one way of arriving at this important number e.

The **number e** is defined as the number that the expression

$$\left(1 + \frac{1}{n}\right)^n \tag{2}$$

approaches as $n \to \infty$. In calculus, this is expressed using limit notation as

$$e = \lim_{n \to \infty}\left(1 + \frac{1}{n}\right)^n$$

Table 6 illustrates what happens to the defining expression (2) as n takes on increasingly large values. The last number in the right column in the table is correct to nine decimal places and is the same as the entry given for e on your calculator (if expressed correctly to nine decimal places).

Table 6

n	$\dfrac{1}{n}$	$1 + \dfrac{1}{n}$	$\left(1 + \dfrac{1}{n}\right)^n$
1	1	2	2
2	0.5	1.5	2.25
5	0.2	1.2	2.48832
10	0.1	1.1	2.59374246
100	0.01	1.01	2.704813829
1,000	0.001	1.001	2.716923932
10,000	0.0001	1.0001	2.718145927
100,000	0.00001	1.00001	2.718268237
1,000,000	0.000001	1.000001	2.718280469
1,000,000,000	10^{-9}	$1 + 10^{-9}$	2.718281827

Table 7

X	Y1
-2	.13534
-1	.36788
0	1
1	2.7183
2	7.3891

Y1■e^(X)

The exponential function $f(x) = e^x$, whose base is the number e, occurs with such frequency in applications that it is usually referred to as *the* exponential function. Graphing calculators have the key $\boxed{e^x}$ or $\boxed{\exp(x)}$, which may be used to evaluate the exponential function for a given value of x. Use your calculator to find e^x for $x = -2$, $x = -1$, $x = 0$, $x = 1$, and $x = 2$, as we have done to create Table 7. The graph of the exponential function $f(x) = e^x$ is given in Figures 30 (a) and 30 (b). Since $2 < e < 3$, the graph of $y = e^x$ is increasing and lies between the graphs of $y = 2^x$ and $y = 3^x$. [See Figure 30(c)].

Figure 30
$y = e^x$

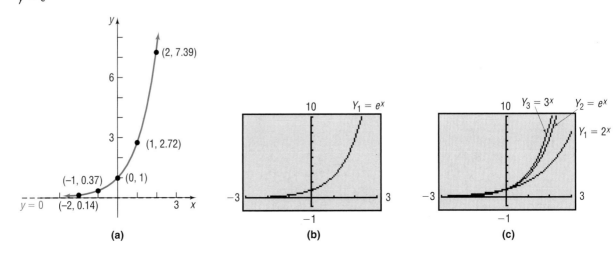

(a)

(b)

(c)

EXAMPLE 5   **Graphing Exponential Functions Using Transformations**

Graph $f(x) = -e^{x-3}$ and determine the domain, the range, and horizontal asymptote of f.

Solution   We begin with the graph of $y = e^x$. Figure 31 shows the stages.

Figure 31

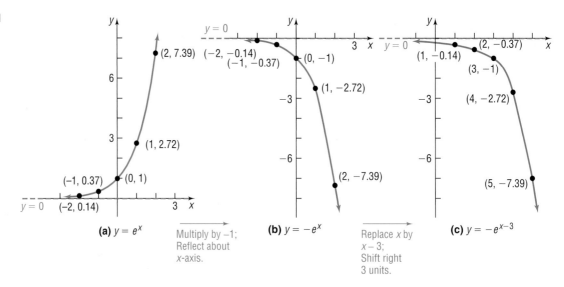

(a) $y = e^x$

Multiply by -1;
Reflect about
x-axis.

(b) $y = -e^x$

Replace x by
$x - 3$;
Shift right
3 units.

(c) $y = -e^{x-3}$

As Figure 31(c) illustrates, the domain of $f(x) = -e^{x-3}$ is the interval $(-\infty, \infty)$, and the range is the interval $(-\infty, 0)$. The horizontal asymptote is the line $y = 0$.

✔ **CHECK:** Graph $Y_1 = -e^{x-3}$ to verify the graph obtained in Figure 31(c). ◄

NOW WORK PROBLEM 45.

4 **Solve Exponential Equations**

Equations that involve terms of the form a^x, $a > 0$, $a \neq 1$, are often referred to as **exponential equations**. Such equations can sometimes be solved by appropriately applying the Laws of Exponents and property (3) below.

$$\text{If} \quad a^u = a^v, \quad \text{then} \quad u = v \qquad \textbf{(3)}$$

Property (3) is a consequence of the fact that exponential functions are one-to-one. To use property (3), each side of the equality must be written with the same base.

EXAMPLE 6   **Solving an Exponential Equation**

Solve: $3^{x+1} = 81$

Solution   Since $81 = 3^4$, we can write the equation as

$$3^{x+1} = 81 = 3^4$$

Figure 32

Now we have the same base, 3, on each side, so we can apply property (3) to obtain

$$x + 1 = 4$$
$$x = 3$$

✔ **CHECK:** We verify the solution by graphing $Y_1 = 3^{x+1}$ and $Y_2 = 81$ to determine where the graphs intersect. See Figure 32. The graphs intersect at $x = 3$. The solution set is $\{3\}$. ◀

 NOW WORK PROBLEM 53.

| EXAMPLE 7 | **Solving an Exponential Equation** |

Solve: $e^{-x^2} = (e^x)^2 \cdot \dfrac{1}{e^3}$

Solution   We use the Laws of Exponents first to get the base e on the right side.

$$(e^x)^2 \cdot \frac{1}{e^3} = e^{2x} \cdot e^{-3} = e^{2x-3}$$

As a result,

$$
\begin{aligned}
e^{-x^2} &= e^{2x-3} \\
-x^2 &= 2x - 3 && \text{Apply Property (3).} \\
x^2 + 2x - 3 &= 0 && \text{Place the quadratic equation in standard form.} \\
(x + 3)(x - 1) &= 0 && \text{Factor.} \\
x = -3 \quad &\text{or} \quad x = 1 && \text{Use the Zero-Product Property.}
\end{aligned}
$$

The solution set is $\{-3, 1\}$.

You should verify these solutions using a graphing utility. ◀

Many applications involve the exponential functions. Let's look at one.

| EXAMPLE 8 | **Exponential Probability** |

Between 9:00 PM and 10:00 PM cars arrive at Burger King's drive-thru at the rate of 12 cars per hour (0.2 car per minute). The following formula from statistics can be used to determine the probability that a car will arrive within t minutes of 9:00 PM.

$$F(t) = 1 - e^{-0.2t}$$

(a) Determine the probability that a car will arrive within 5 minutes of 9 PM (that is, before 9:05 PM).

(b) Determine the probability that a car will arrive within 30 minutes of 9 PM (before 9:30 PM).

(c) Graph F using your graphing utility.

(d) What value does F approach as t becomes unbounded in the positive direction?

Figure 33

```
1-e^(-.2*5)
        .6321205588
```

Solution   (a) The probability that a car will arrive within 5 minutes is found by evaluating $F(t)$ at $t = 5$.

$$F(5) = 1 - e^{-0.2(5)}$$

We evaluate this expression in Figure 33. We conclude that there is a 63% probability that a car will arrive within 5 minutes.

(b) The probability that a car will arrive within 30 minutes is found by evaluating $F(t)$ at $t = 30$.

Figure 34

$$F(30) = 1 - e^{-0.2(30)} \approx 0.9975$$

There is a 99.75% probability that a car will arrive within 30 minutes.

(c) See Figure 34 for the graph of F.

(d) As time passes, the probability that a car will arrive increases. The value that F approaches can be found by letting $t \to \infty$. Since $e^{-0.2t} = \dfrac{1}{e^{0.2t}}$, it follows that $e^{-0.2t} \to 0$ as $t \to \infty$. Thus, F approaches 1 as t gets large. ◀

NOW WORK PROBLEM 81.

Summary

Properties of the Exponential Function

$f(x) = a^x, \quad a > 1$ Domain: the interval $(-\infty, \infty)$; Range: the interval $(0, \infty)$
x-intercepts: none; y-intercept: 1
Horizontal asymptote: x-axis $(y = 0)$ as $x \to -\infty$
Increasing; one-to-one; smooth; continuous
See Figure 24 for a typical graph.

$f(x) = a^x, \quad 0 < a < 1$ Domain: the interval $(-\infty, \infty)$; Range: the interval $(0, \infty)$.
x-intercepts: none; y-intercept: 1
Horizontal asymptote: x-axis $(y = 0)$ as $x \to \infty$
Decreasing; one-to-one; smooth; continuous
See Figure 28 for a typical graph.

If $a^u = a^v$, then $u = v$.

4.3 Assess Your Understanding

'Are You Prepared?'

Answers are given at the end of these exercises. If you get a wrong answer, read the pages listed in red.

1. $4^3 =$ _____; $8^{2/3} =$ _____; $3^{-2} =$ _____. (pp. 20–22 and pp. 71–73)

2. Solve: $3x^2 + 5x - 2 = 0$. (pp. 112–119)

3. *True or False:* To graph $y = (x - 2)^3$, shift the graph of $y = x^3$ to the left 2 units. (pp. 266–274)

4. Find the average rate of change of $f(x) = 3x - 5$ from $x = 0$ to $x = c$. (pp. 233–236)

5. *True or False:* The function $f(x) = \dfrac{2x}{x - 3}$ has $y = 2$ as a horizontal asymptote. (pp. 339–343)

Concepts and Vocabulary

6. The graph of every exponential function $f(x) = a^x, a > 0, a \neq 1$, passes through three points: _____, _____, and _____.

7. If the graph of the exponential function $f(x) = a^x, a > 0, a \neq 1$, is decreasing, then a must be less than _____.

8. If $3^x = 3^4$, then $x =$ _____.

9. *True or False:* The graphs of $y = 3^x$ and $y = \left(\dfrac{1}{3}\right)^x$ are identical.

10. *True or False:* The range of the exponential function $f(x) = a^x, a > 0, a \neq 1$, is the set of all real numbers.

Skill Building

In Problems 11–20, approximate each number using a calculator. Express your answer rounded to three decimal places.

11. (a) $3^{2.2}$ (b) $3^{2.23}$ (c) $3^{2.236}$ (d) $3^{\sqrt{5}}$ 12. (a) $5^{1.7}$ (b) $5^{1.73}$ (c) $5^{1.732}$ (d) $5^{\sqrt{3}}$

13. (a) $2^{3.14}$ (b) $2^{3.141}$ (c) $2^{3.1415}$ (d) 2^{π} 14. (a) $2^{2.7}$ (b) $2^{2.71}$ (c) $2^{2.718}$ (d) 2^{e}

15. (a) $3.1^{2.7}$ (b) $3.14^{2.71}$ (c) $3.141^{2.718}$ (d) π^{e} 16. (a) $2.7^{3.1}$ (b) $2.71^{3.14}$ (c) $2.718^{3.141}$ (d) e^{π}

17. $e^{1.2}$ 18. $e^{-1.3}$ 19. $e^{-0.85}$ 20. $e^{2.1}$

In Problems 21–28, determine whether the given function is exponential or not. For those that are exponential functions, identify the value of the base a. [**Hint:** *Look at the ratio of consecutive values.*]

21.

x	f(x)
−1	3
0	6
1	12
2	18
3	30

22.

x	g(x)
−1	2
0	5
1	8
2	11
3	14

23.

x	H(x)
−1	$\frac{1}{4}$
0	1
1	4
2	16
3	64

24.

x	F(x)
−1	$\frac{2}{3}$
0	1
1	$\frac{3}{2}$
2	$\frac{9}{4}$
3	$\frac{27}{8}$

25.

x	f(x)
−1	$\frac{3}{2}$
0	3
1	6
2	12
3	24

26.

x	g(x)
−1	6
0	1
1	0
2	3
3	10

27.

x	H(x)
−1	2
0	4
1	6
2	8
3	10

28.

x	F(x)
−1	$\frac{1}{2}$
0	$\frac{1}{4}$
1	$\frac{1}{8}$
2	$\frac{1}{16}$
3	$\frac{1}{32}$

In Problems 29–36, the graph of an exponential function is given. Match each graph to one of the following functions.

A. $y = 3^{x}$ B. $y = 3^{-x}$ C. $y = -3^{x}$ D. $y = -3^{-x}$

E. $y = 3^{x} - 1$ F. $y = 3^{x-1}$ G. $y = 3^{1-x}$ H. $y = 1 - 3^{x}$

29.

30.

31.

32.

33.

34.

35.

36.

In Problems 37–44, use transformations to graph each function. Determine the domain, range, and horizontal asymptote of each function.

37. $f(x) = 2^x + 1$ **38.** $f(x) = 2^{x+2}$ **39.** $f(x) = 3^{-x} - 2$ **40.** $f(x) = -3^x + 1$

41. $f(x) = 2 + 3(4^x)$ **42.** $f(x) = 1 - 3(2^x)$ **43.** $f(x) = 2 + 3^{x/2}$ **44.** $f(x) = 1 - 2^{-x/3}$

In Problems 45–52, begin with the graph of $y = e^x$ (Figure 30 (a)) and use transformations to graph each function. Determine the domain, range, and horizontal asymptote of each function.

45. $f(x) = e^{-x}$ **46.** $f(x) = -e^x$ **47.** $f(x) = e^{x+2}$ **48.** $f(x) = e^x - 1$

49. $f(x) = 5 - e^{-x}$ **50.** $f(x) = 9 - 3e^{-x}$ **51.** $f(x) = 2 - e^{-x/2}$ **52.** $f(x) = 7 - 3e^{2x}$

In Problems 53–66, solve each equation.

53. $2^{2x+1} = 4$ **54.** $5^{1-2x} = \dfrac{1}{5}$ **55.** $3^{x^3} = 9^x$ **56.** $4^{x^2} = 2^x$ **57.** $8^{x^2-2x} = \dfrac{1}{2}$

58. $9^{-x} = \dfrac{1}{3}$ **59.** $2^x \cdot 8^{-x} = 4^x$ **60.** $\left(\dfrac{1}{2}\right)^{1-x} = 4$ **61.** $\left(\dfrac{1}{5}\right)^{2-x} = 25$ **62.** $4^x - 2^x = 0$

63. $4^x = 8$ **64.** $9^{2x} = 27$ **65.** $e^{x^2} = (e^{3x}) \cdot \dfrac{1}{e^2}$ **66.** $(e^4)^x \cdot e^{x^2} = e^{12}$

67. If $4^x = 7$, what does 4^{-2x} equal?   **68.** If $2^x = 3$, what does 4^{-x} equal?

69. If $3^{-x} = 2$, what does 3^{2x} equal?   **70.** If $5^{-x} = 3$, what does 5^{3x} equal?

In Problems 71–74, determine the exponential function whose graph is given.

71.

72.

73.

74.

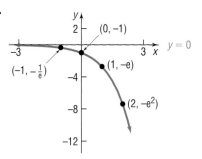

Applications and Extensions

75. Optics If a single pane of glass obliterates 3% of the light passing through it, then the percent p of light that passes through n successive panes is given approximately by the function

$$p(n) = 100(0.97)^n$$

(a) What percent of light will pass through 10 panes?
(b) What percent of light will pass through 25 panes?

76. Atmospheric Pressure The atmospheric pressure p on a balloon or plane decreases with increasing height. This pressure, measured in millimeters of mercury, is related to the height h (in kilometers) above sea level by the function

$$p(h) = 760e^{-0.145h}$$

(a) Find the atmospheric pressure at a height of 2 kilometers (over a mile).
(b) What is it at a height of 10 kilometers (over 30,000 feet)?

77. **Depreciation** The price p of a Honda Civic DX Sedan that is x years old is given by

$$p(x) = 16,630(0.90)^x$$

(a) How much does a 3-year-old Civic DX Sedan cost?
(b) How much does a 9-year-old Civic DX Sedan cost?

78. **Healing of Wounds** The normal healing of wounds can be modeled by an exponential function. If A_0 represents the original area of the wound and if A equals the area of the wound, then the function

$$A(n) = A_0 e^{-0.35n}$$

describes the area of a wound after n days following an injury when no infection is present to retard the healing. Suppose that a wound initially had an area of 100 square millimeters.

(a) If healing is taking place, how large will the area of the wound be after 3 days?
(b) How large will it be after 10 days?

79. **Drug Medication** The function

$$D(h) = 5e^{-0.4h}$$

can be used to find the number of milligrams D of a certain drug that is in a patient's bloodstream h hours after the drug has been administered. How many milligrams will be present after 1 hour? After 6 hours?

80. **Spreading of Rumors** A model for the number N of people in a college community who have heard a certain rumor is

$$N = P(1 - e^{-0.15d})$$

where P is the total population of the community and d is the number of days that have elapsed since the rumor began. In a community of 1000 students, how many students will have heard the rumor after 3 days?

81. **Exponential Probability** Between 12:00 PM and 1:00 PM, cars arrive at Citibank's drive-thru at the rate of 6 cars per hour (0.1 car per minute). The following formula from probability can be used to determine the probability that a car will arrive within t minutes of 12:00 PM:

$$F(t) = 1 - e^{-0.1t}$$

(a) Determine the probability that a car will arrive within 10 minutes of 12:00 PM (that is, before 12:10 PM).
(b) Determine the probability that a car will arrive within 40 minutes of 12:00 PM (before 12:40 PM).
(c) What value does F approach as t becomes unbounded in the positive direction?
(d) Graph F using a graphing utility.
(e) Using TRACE, determine how many minutes are needed for the probability to reach 50%.

82. **Exponential Probability** Between 5:00 PM and 6:00 PM, cars arrive at Jiffy Lube at the rate of 9 cars per hour (0.15 car per minute). The following formula from probability can be used to determine the probability that a car will arrive within t minutes of 5:00 PM:

$$F(t) = 1 - e^{-0.15t}$$

(a) Determine the probability that a car will arrive within 15 minutes of 5:00 PM (that is, before 5:15 PM).
(b) Determine the probability that a car will arrive within 30 minutes of 5:00 PM (before 5:30 PM).
(c) What value does F approach as t becomes unbounded in the positive direction?
(d) Graph F using a graphing utility.
(e) Using TRACE, determine how many minutes are needed for the probability to reach 60%.

83. **Poisson Probability** Between 5:00 PM and 6:00 PM, cars arrive at McDonald's drive-thru at the rate of 20 cars per hour. The following formula from probability can be used to determine the probability that x cars will arrive between 5:00 PM and 6:00 PM.

$$P(x) = \frac{20^x e^{-20}}{x!}$$

where

$$x! = x \cdot (x - 1) \cdot (x - 2) \cdot \cdots \cdot 3 \cdot 2 \cdot 1$$

(a) Determine the probability that $x = 15$ cars will arrive between 5:00 PM and 6:00 PM.
(b) Determine the probability that $x = 20$ cars will arrive between 5:00 PM and 6:00 PM.

84. **Poisson Probability** People enter a line for the *Demon Roller Coaster* at the rate of 4 per minute. The following formula from probability can be used to determine the probability that x people will arrive within the next minute.

$$P(x) = \frac{4^x e^{-4}}{x!}$$

where

$$x! = x \cdot (x - 1) \cdot (x - 2) \cdot \cdots \cdot 3 \cdot 2 \cdot 1$$

(a) Determine the probability that $x = 5$ people will arrive within the next minute.
(b) Determine the probability that $x = 8$ people will arrive within the next minute.

85. **Relative Humidity** The relative humidity is the ratio (expressed as a percent) of the amount of water vapor in the air to the maximum amount that it can hold at a specific temperature. The relative humidity, R, is found using the following formula:

$$R = 10^{\left(\frac{4221}{T+459.4} - \frac{4221}{D+459.4} + 2\right)}$$

where T is the air temperature (in °F) and D is the dew point temperature (in °F).

(a) Determine the relative humidity if the air temperature is 50° Fahrenheit and the dew point temperature is 41° Fahrenheit.
(b) Determine the relative humidity if the air temperature is 68° Fahrenheit and the dew point temperature is 59° Fahrenheit.
(c) What is the relative humidity if the air temperature and the dew point temperature are the same?

86. Learning Curve   Suppose that a student has 500 vocabulary words to learn. If the student learns 15 words after 5 minutes, the function

$$L(t) = 500(1 - e^{-0.0061t})$$

approximates the number of words L that the student will learn after t minutes.

(a) How many words will the student learn after 30 minutes?

(b) How many words will the student learn after 60 minutes?

87. Current in a RL Circuit   The equation governing the amount of current I (in amperes) after time t (in seconds) in a single RL circuit consisting of a resistance R (in ohms), an inductance L (in henrys), and an electromotive force E (in volts) is

$$I = \frac{E}{R}[1 - e^{-(R/L)t}]$$

(a) If $E = 120$ volts, $R = 10$ ohms, and $L = 5$ henrys, how much current I_1 is flowing after 0.3 second? After 0.5 second? After 1 second?

(b) What is the maximum current?

(c) Graph this function $I = I_1(t)$, measuring I along the y-axis and t along the x-axis.

(d) If $E = 120$ volts, $R = 5$ ohms, and $L = 10$ henrys, how much current I_2 is flowing after 0.3 second? After 0.5 second? After 1 second?

(e) What is the maximum current?

(f) Graph this function $I = I_2(t)$ on the same coordinate axes as $I_1(t)$.

88. Current in a RC Circuit   The equation governing the amount of current I (in amperes) after time t (in microseconds) in a single RC circuit consisting of a resistance R (in ohms), a capacitance C (in microfarads), and an electromotive force E (in volts) is

$$I = \frac{E}{R}e^{-t/(RC)}$$

(a) If $E = 120$ volts, $R = 2000$ ohms, and $C = 1.0$ microfarad, how much current I_1 is flowing initially $(t = 0)$? After 1000 microseconds? After 3000 microseconds?

(b) What is the maximum current?

(c) Graph this function $I = I_1(t)$, measuring I along the y-axis and t along the x-axis.

(d) If $E = 120$ volts, $R = 1000$ ohms, and $C = 2.0$ microfarads, how much current I_2 is flowing initially? After 1000 microseconds? After 3000 microseconds?

(e) What is the maximum current?

(f) Graph this function $I = I_2(t)$ on the same coordinate axes as $I_1(t)$.

89. Another Formula for e   Use a calculator to compute the values of

$$2 + \frac{1}{2!} + \frac{1}{3!} + \cdots + \frac{1}{n!}$$

for $n = 4, 6, 8,$ and 10. Compare each result with e.

[**Hint:** $1! = 1, 2! = 2 \cdot 1, 3! = 3 \cdot 2 \cdot 1,$
$n! = n(n - 1) \cdots \cdot (3)(2)(1).$]

90. Another Formula for e   Use a calculator to compute the various values of the expression. Compare the values to e.

$$2 + \cfrac{1}{1 + \cfrac{1}{2 + \cfrac{2}{3 + \cfrac{3}{4 + \cfrac{4}{\text{etc.}}}}}}$$

91. Difference Quotient   If $f(x) = a^x$, show that

$$\frac{f(x + h) - f(x)}{h} = a^x \cdot \frac{a^h - 1}{h}$$

92. If $f(x) = a^x$, show that $f(A + B) = f(A) \cdot f(B)$.

93. If $f(x) = a^x$, show that $f(-x) = \dfrac{1}{f(x)}$.

94. If $f(x) = a^x$, show that $f(\alpha x) = [f(x)]^\alpha$.

Problems 95 and 96 provide definitions for two other transcendental functions.

95. The **hyperbolic sine function**, designated by sinh x, is defined as

$$\sinh x = \frac{1}{2}(e^x - e^{-x})$$

(a) Show that $f(x) = \sinh x$ is an odd function.

(b) Graph $f(x) = \sinh x$ using a graphing utility.

96. The **hyperbolic cosine function**, designated by cosh x, is defined as

$$\cosh x = \frac{1}{2}(e^x + e^{-x})$$

(a) Show that $f(x) = \cosh x$ is an even function.

(b) Graph $f(x) = \cosh x$ using a graphing utility.

(c) Refer to Problem 95. Show that, for every x,

$$(\cosh x)^2 - (\sinh x)^2 = 1$$

97. **Historical Problem** Pierre de Fermat (1601–1665) conjectured that the function

$$f(x) = 2^{(2^x)} + 1$$

for $x = 1, 2, 3, \ldots$, would always have a value equal to a prime number. But Leonhard Euler (1707–1783) showed that this formula fails for $x = 5$. Use a calculator to determine the prime numbers produced by f for $x = 1, 2, 3, 4$. Then show that $f(5) = 641 \times 6{,}700{,}417$, which is not prime.

Discussion and Writing

98. The bacteria in a 4-liter container double every minute. After 60 minutes the container is full. How long did it take to fill half the container?

99. Explain in your own words what the number e is. Provide at least two applications that require the use of this number.

100. Do you think that there is a power function that increases more rapidly than an exponential function whose base is greater than 1? Explain.

101. As the base a of an exponential function $f(x) = a^x$, $a > 1$, increases, what happens to the behavior of its graph for $x > 0$? What happens to the behavior of its graph for $x < 0$?

102. The graphs of $y = a^{-x}$ and $y = \left(\dfrac{1}{a}\right)^x$ are identical. Why?

'Are You Prepared?' Answers

1. $64; 4; \dfrac{1}{9}$ **2.** $\left\{-2, \dfrac{1}{3}\right\}$ **3.** False **4.** 3 **5.** True

4.4 Logarithmic Functions

PREPARING FOR THIS SECTION *Before getting started, review the following:*

- Solving Inequalities (Section 1.7, pp. 150–157)
- Polynomial and Rational Inequalities (Section 3.5, pp. 360–363)

Now work the 'Are You Prepared?' problems on page 444.

OBJECTIVES 1 Change Exponential Expressions to Logarithmic Expressions and Logarithmic Expressions to Exponential Expressions
2 Evaluate Logarithmic Expressions
3 Determine the Domain of a Logarithmic Function
4 Graph Logarithmic Functions
5 Solve Logarithmic Equations

Recall that a one-to-one function $y = f(x)$ has an inverse function that is defined (implicitly) by the equation $x = f(y)$. In particular, the exponential function $y = f(x) = a^x$, $a > 0$, $a \neq 1$, is one-to-one and hence has an inverse function that is defined implicitly by the equation

$$x = a^y, \qquad a > 0, \qquad a \neq 1$$

This inverse function is so important that it is given a name, the *logarithmic function*.

The **logarithmic function to the base a**, where $a > 0$ and $a \neq 1$, is denoted by $y = \log_a x$ (read as "y is the logarithm to the base a of x") and is defined by

$$y = \log_a x \quad \text{if and only if} \quad x = a^y$$

The domain of the logarithmic function $y = \log_a x$ is $x > 0$.

A *logarithm* is merely a name for a certain exponent.

| EXAMPLE 1 | **Relating Logarithms to Exponents** |

(a) If $y = \log_3 x$, then $x = 3^y$. For example, $4 = \log_3 81$ is equivalent to $81 = 3^4$.

(b) If $y = \log_5 x$, then $x = 5^y$. For example, $-1 = \log_5\left(\dfrac{1}{5}\right)$ is equivalent to $\dfrac{1}{5} = 5^{-1}$.

◀

1 Change Exponential Expressions to Logarithmic Expressions and Logarithmic Expressions to Exponential Expressions

We can use the definition of a logarithm to convert from exponential form to logarithmic form, and vice versa, as the following two examples illustrate.

| EXAMPLE 2 | **Changing Exponential Expressions to Logarithmic Expressions** |

Change each exponential expression to an equivalent expression involving a logarithm.

(a) $1.2^3 = m$ (b) $e^b = 9$ (c) $a^4 = 24$

Solution We use the fact that $y = \log_a x$ and $x = a^y$, $a > 0$, $a \neq 1$, are equivalent.

(a) If $1.2^3 = m$, then $3 = \log_{1.2} m$.

(b) If $e^b = 9$, then $b = \log_e 9$.

(c) If $a^4 = 24$, then $4 = \log_a 24$.

◀

━━━━━ NOW WORK PROBLEM 9.

| EXAMPLE 3 | **Changing Logarithmic Expressions to Exponential Expressions** |

Change each logarithmic expression to an equivalent expression involving an exponent.

(a) $\log_a 4 = 5$ (b) $\log_e b = -3$ (c) $\log_3 5 = c$

Solution (a) If $\log_a 4 = 5$, then $a^5 = 4$.

(b) If $\log_e b = -3$, then $e^{-3} = b$.

(c) If $\log_3 5 = c$, then $3^c = 5$.

◀

━━━━━ NOW WORK PROBLEM 21.

2 Evaluate Logarithmic Expressions

To find the exact value of a logarithm, we write the logarithm in exponential notation and use the fact that if $a^u = a^v$ then $u = v$.

| EXAMPLE 4 | **Finding the Exact Value of a Logarithmic Expression** |

Find the exact value of:

(a) $\log_2 16$ (b) $\log_3 \dfrac{1}{27}$

Solution   (a) $y = \log_2 16$

$$2^y = 16 \qquad \text{Change to exponential form.}$$
$$2^y = 2^4 \qquad 16 = 2^4$$
$$y = 4 \qquad \text{Equate exponents.}$$

Therefore, $\log_2 16 = 4$.

(b) $y = \log_3 \dfrac{1}{27}$

$$3^y = \frac{1}{27} \qquad \text{Change to exponential form.}$$

$$3^y = 3^{-3} \qquad \frac{1}{27} = \frac{1}{3^3} = 3^{-3}$$

$$y = -3 \qquad \text{Equate exponents.}$$

Therefore, $\log_3 \dfrac{1}{27} = -3$.   ◀

NOW WORK PROBLEM 33.

3   Determine the Domain of a Logarithmic Function

The logarithmic function $y = \log_a x$ has been defined as the inverse of the exponential function $y = a^x$. That is, if $f(x) = a^x$, then $f^{-1}(x) = \log_a x$. Based on the discussion given in Section 4.2 on inverse functions, for a function f and its inverse f^{-1}, we have

$$\text{Domain of } f^{-1} = \text{Range of } f \quad \text{and} \quad \text{Range of } f^{-1} = \text{Domain of } f$$

Consequently, it follows that

> Domain of the logarithmic function = Range of the exponential function = $(0, \infty)$
>
> Range of the logarithmic function = Domain of the exponential function = $(-\infty, \infty)$

In the next box, we summarize some properties of the logarithmic function:

> $$y = \log_a x \quad (\text{defining equation: } x = a^y)$$
> $$\text{Domain: } 0 < x < \infty \qquad \text{Range: } -\infty < y < \infty$$

The domain of a logarithmic function consists of the *positive* real numbers, so the argument of a logarithmic function must be greater than zero.

EXAMPLE 5   **Finding the Domain of a Logarithmic Function**

Find the domain of each logarithmic function.

(a) $F(x) = \log_2(x + 3)$ (b) $g(x) = \log_5\left(\dfrac{1 + x}{1 - x}\right)$

(c) $h(x) = \log_{1/2}|x|$

Solution   (a) The domain of F consists of all x for which $x + 3 > 0$, that is, $x > -3$. Using interval notation, the domain of f is $(-3, \infty)$.

(b) The domain of g is restricted to

$$\frac{1 + x}{1 - x} > 0$$

Solving this inequality, we find that the domain of g consists of all x between -1 and 1, that is, $-1 < x < 1$ or, using interval notation, $(-1, 1)$.

(c) Since $|x| > 0$, provided that $x \neq 0$, the domain of h consists of all real numbers except zero or, using interval notation, $(-\infty, 0)$ or $(0, \infty)$. ◀

NOW WORK PROBLEMS 47 AND 53.

4 Graph Logarithmic Functions

Since exponential functions and logarithmic functions are inverses of each other, the graph of the logarithmic function $y = \log_a x$ is the reflection about the line $y = x$ of the graph of the exponential function $y = a^x$, as shown in Figure 35.

Figure 35

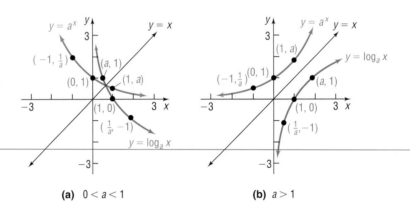

(a) $0 < a < 1$ (b) $a > 1$

For example, to graph $y = \log_2 x$, graph $y = 2^x$ and reflect it about the line $y = x$. See Figure 36. To graph $y = \log_{1/3} x$, graph $y = \left(\frac{1}{3}\right)^x$ and reflect it about the line $y = x$. See Figure 37.

Figure 36

Figure 37

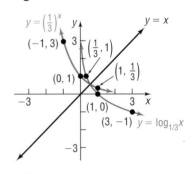

NOW WORK PROBLEM 63.

Properties of the Logarithmic Function $f(x) = \log_a x$

1. The domain is the set of positive real numbers; the range is the set of all real numbers.

2. The x-intercept of the graph is 1. There is no y-intercept.

3. The y-axis ($x = 0$) is a vertical asymptote of the graph.

4. A logarithmic function is decreasing if $0 < a < 1$ and increasing if $a > 1$.

5. The graph of f contains the points $(1, 0)$, $(a, 1)$, and $\left(\dfrac{1}{a}, -1\right)$.

6. The graph is smooth and continuous, with no corners or gaps.

If the base of a logarithmic function is the number e, then we have the **natural logarithm function**. This function occurs so frequently in applications that it is given a special symbol, **ln** (from the Latin, *logarithmus naturalis*). That is,

$$y = \ln x \quad \text{if and only if} \quad x = e^y \qquad \textbf{(1)}$$

Since $y = \ln x$ and the exponential function $y = e^x$ are inverse functions, we can obtain the graph of $y = \ln x$ by reflecting the graph of $y = e^x$ about the line $y = x$. See Figure 38.

Figure 38

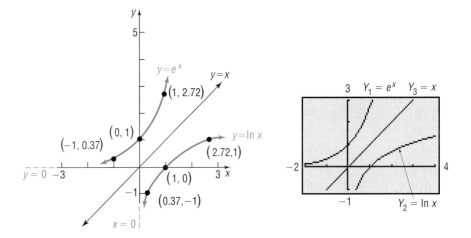

Table 8 displays other points on the graph of $f(x) = \ln x$. Notice for $x \leq 0$ that we obtain an error message. Do you recall why?

Table 8

X	Y2
-1	ERROR
0.5	-.6931
1	0
2	.69315
2.7183	1
3	1.0986

`Y2⊟ln(X)`

| **EXAMPLE 6** | **Graphing Logarithmic Functions Using Transformations** |

Graph $f(x) = -\ln(x + 2)$ by starting with the graph of $y = \ln x$. Determine the domain, range, and vertical asymptote.

Solution   The domain consists of all x for which

$$x + 2 > 0 \quad \text{or} \quad x > -2$$

To obtain the graph of $y = -\ln(x + 2)$, we use the steps illustrated in Figure 39.

Figure 39

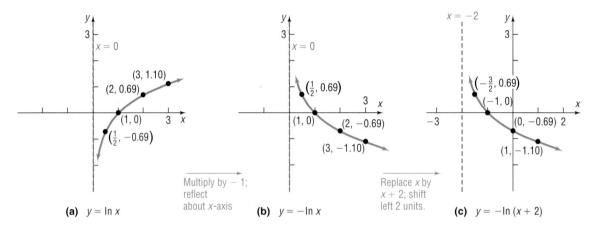

(a) $y = \ln x$ Multiply by -1; reflect about x-axis   **(b)** $y = -\ln x$ Replace x by $x + 2$; shift left 2 units.   **(c)** $y = -\ln(x + 2)$

The range of $f(x) = -\ln(x + 2)$ is the interval $(-\infty, \infty)$, and the vertical asymptote is $x = -2$. [Do you see why? The original asymptote $(x = 0)$ is shifted to the left 2 units.]

✔ **CHECK:** Graph $Y_1 = -\ln(x + 2)$ using a graphing utility to verify Figure 39(c).

 NOW WORK PROBLEM 75.

If the base of a logarithmic function is the number 10, then we have the **common logarithm function**. If the base a of the logarithmic function is not indicated, it is understood to be 10. That is,

$$y = \log x \quad \text{if and only if} \quad x = 10^y$$

Since $y = \log x$ and the exponential function $y = 10^x$ are inverse functions, we can obtain the graph of $y = \log x$ by reflecting the graph of $y = 10^x$ about the line $y = x$. See Figure 40.

Figure 40

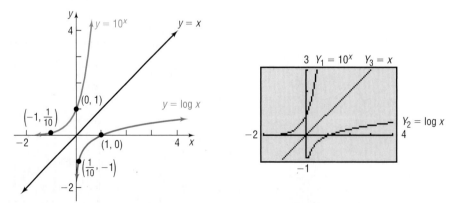

| EXAMPLE 7 | **Graphing Logarithmic Functions Using Transformations** |

Graph $f(x) = 3 \log(x - 1)$. Determine the domain, range, and vertical asymptote of f.

Solution The domain consists of all x for which

$$x - 1 > 0 \quad \text{or} \quad x > 1$$

To obtain the graph of $y = 3 \log(x - 1)$, we use the steps illustrated in Figure 41.

Figure 41

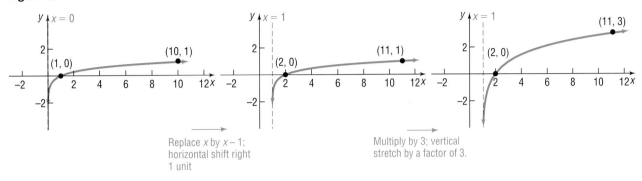

(a) $y = \log x$ Replace x by $x - 1$; horizontal shift right 1 unit   **(b)** $y = \log (x - 1)$ Multiply by 3; vertical stretch by a factor of 3.   **(c)** $y = 3 \log (x - 1)$

The range of $f(x) = 3 \log(x - 1)$ is the interval $(-\infty, \infty)$, and the vertical asymptote is $x = 1$.

✔ **CHECK:** Graph $Y_1 = 3 \log(x - 1)$ using a graphing utility to verify Figure 41(c). ◄

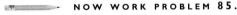 **NOW WORK PROBLEM 85.**

5 Solve Logarithmic Equations

Equations that contain logarithms are called **logarithmic equations**. Care must be taken when solving logarithmic equations algebraically. Be sure to check each apparent solution in the original equation and discard any that are extraneous. In the expression $\log_a M$, remember that a and M are positive and $a \neq 1$.

Some logarithmic equations can be solved by changing from a logarithmic expression to an exponential expression.

| EXAMPLE 8 | **Solving a Logarithmic Equation** |

Solve: (a) $\log_3(4x - 7) = 2$ (b) $\log_x 64 = 2$

Solution (a) We can obtain an exact solution by changing the logarithm to exponential form.

$$\log_3(4x - 7) = 2$$
$$4x - 7 = 3^2 \quad \textit{Change to exponential form.}$$
$$4x - 7 = 9$$
$$4x = 16$$
$$x = 4$$

✔ **CHECK:** $\log_3(4x - 7) = \log_3(16 - 7) = \log_3 9 = 2$ $3^2 = 9$

(b) We can obtain an exact solution by changing the logarithm to exponential form.

$$\log_x 64 = 2$$

$$x^2 = 64 \qquad \textit{Change to exponential form.}$$

$$x = \pm\sqrt{64} = \pm 8 \qquad \textit{Square Root Method.}$$

The base of a logarithm is always positive. As a result, we discard -8; the only solution is 8.

✔ **CHECK:** $\log_8 64 = 2$ $8^2 = 64$ ◀

| **EXAMPLE 9** | **Using Logarithms to Solve Exponential Equations** |

Solve: $e^{2x} = 5$

Solution   We can obtain an exact solution by changing the exponential equation to logarithmic form.

$$e^{2x} = 5$$

$$\ln 5 = 2x \qquad \textit{Change to a logarithmic expression using (1).}$$

$$x = \frac{\ln 5}{2} \qquad \textit{Exact solution.}$$

$$\approx 0.805 \qquad \textit{Approximate solution.}$$ ◀

✏ **NOW WORK PROBLEMS 91 AND 103.**

| **EXAMPLE 10** | **Alcohol and Driving** |

The concentration of alcohol in a person's blood is measurable. Recent medical research suggests that the risk R (given as a percent) of having an accident while driving a car can be modeled by the equation

$$R = 6e^{kx}$$

where x is the variable concentration of alcohol in the blood and k is a constant.

(a) Suppose that a concentration of alcohol in the blood of 0.04 results in a 10% risk ($R = 10$) of an accident. Find the constant k in the equation. Graph $R = 6e^{kx}$ using this value of k.

(b) Using this value of k, what is the risk if the concentration is 0.17?

(c) Using the same value of k, what concentration of alcohol corresponds to a risk of 100%?

(d) If the law asserts that anyone with a risk of having an accident of 20% or more should not have driving privileges, at what concentration of alcohol in the blood should a driver be arrested and charged with a DUI (Driving Under the Influence)?

Solution (a) For a concentration of alcohol in the blood of 0.04 and a risk of 10%, we let $x = 0.04$ and $R = 10$ in the equation and solve for k.

$$R = 6e^{kx}$$

$$10 = 6e^{k(0.04)} \qquad \text{\textit{R = 10; x = 0.04}}$$

$$\frac{10}{6} = e^{0.04k} \qquad \text{\textit{Divide both sides by 6.}}$$

$$0.04k = \ln\frac{10}{6} = 0.5108256 \qquad \text{\textit{Change to a logarithmic expression.}}$$

$$k = 12.77 \qquad \text{\textit{Solve for k.}}$$

See Figure 42 for the graph of $R = 6e^{12.77x}$.

Figure 42

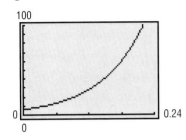

(b) Using $k = 12.77$ and $x = 0.17$ in the equation, we find the risk R to be

$$R = 6e^{kx} = 6e^{(12.77)(0.17)} = 52.6$$

For a concentration of alcohol in the blood of 0.17, the risk of an accident is about 52.6%.

(c) Using $k = 12.77$ and $R = 100$ in the equation, we find the concentration x of alcohol in the blood to be

$$R = 6e^{kx}$$

$$100 = 6e^{12.77x} \qquad \text{\textit{R = 100; k = 12.77}}$$

$$\frac{100}{6} = e^{12.77x} \qquad \text{\textit{Divide both sides by 6.}}$$

$$12.77x = \ln\frac{100}{6} = 2.8134 \qquad \text{\textit{Change to a logarithmic expression.}}$$

$$x = 0.22 \qquad \text{\textit{Solve for x.}}$$

For a concentration of alcohol in the blood of 0.22, the risk of an accident is 100%.

(d) Using $k = 12.77$ and $R = 20$ in the equation, we find the concentration x of alcohol in the blood to be

$$R = 6e^{kx}$$

$$20 = 6e^{12.77x}$$

$$\frac{20}{6} = e^{12.77x}$$

$$12.77x = \ln\frac{20}{6} = 1.204$$

$$x = 0.094$$

A driver with a concentration of alcohol in the blood of 0.094 or more (9.4%) should be arrested and charged with DUI. ◀

NOTE Most states use 0.08 or 0.10 as the blood alcohol content at which a DUI citation is given. ∎

Summary

Properties of the Logarithmic Function

$f(x) = \log_a x, \quad a > 1$
($y = \log_a x$ means $x = a^y$)

Domain: the interval $(0, \infty)$; Range: the interval $(-\infty, \infty)$
x-intercept: 1; y-intercept: none; vertical asymptote: $x = 0$ (y-axis); increasing; one-to-one
See Figure 43(a) for a typical graph.

$f(x) = \log_a x, \quad 0 < a < 1$
($y = \log_a x$ means $x = a^y$)

Domain: the interval $(0, \infty)$; Range: the interval $(-\infty, \infty)$
x-intercept: 1; y-intercept: none; vertical asymptote: $x = 0$ (y-axis); decreasing; one-to-one
See Figure 43(b) for a typical graph.

Figure 43

(a) $a > 1$ **(b)** $0 < a < 1$

4.4 Assess Your Understanding

'Are You Prepared?'

Answers are given at the end of these exercises. If you get a wrong answer, read the pages listed in red.

1. Solve the inequality: $3x - 7 \leq 8 - 2x$ (pp. 150–157)
2. Solve the inequality: $x^2 - x - 6 > 0$ (pp. 360–362)
3. Solve the inequality: $\dfrac{x-1}{x+4} > 0$ (p. 363)

Concepts and Vocabulary

4. The domain of the logarithmic function $f(x) = \log_a x$ is
_____.

5. The graph of every logarithmic function $f(x) = \log_a x$, $a > 0, a \neq 1$, passes through three points: _____, _____, and
_____.

6. If the graph of a logarithmic function $f(x) = \log_a x$, $a > 0, a \neq 1$, is increasing, then its base must be larger than _____.

7. *True or False:* If $y = \log_a x$, then $y = a^x$.

8. *True or False:* The graph of $f(x) = \log_a x, a > 0, a \neq 1$, has an x-intercept equal to 1 and no y-intercept.

Skill Building

In Problems 9–20, change each exponential expression to an equivalent expression involving a logarithm.

9. $9 = 3^2$
10. $16 = 4^2$
11. $a^2 = 1.6$
12. $a^3 = 2.1$
13. $1.1^2 = M$
14. $2.2^3 = N$
15. $2^x = 7.2$
16. $3^x = 4.6$
17. $x^{\sqrt{2}} = \pi$
18. $x^\pi = e$
19. $e^x = 8$
20. $e^{2.2} = M$

In Problems 21–32, change each logarithmic expression to an equivalent expression involving an exponent.

21. $\log_2 8 = 3$

22. $\log_3\left(\dfrac{1}{9}\right) = -2$

23. $\log_a 3 = 6$

24. $\log_b 4 = 2$

25. $\log_3 2 = x$

26. $\log_2 6 = x$

27. $\log_2 M = 1.3$

28. $\log_3 N = 2.1$

29. $\log_{\sqrt{2}} \pi = x$

30. $\log_\pi x = \dfrac{1}{2}$

31. $\ln 4 = x$

32. $\ln x = 4$

In Problems 33–44, find the exact value of each logarithm without using a calculator.

33. $\log_2 1$

34. $\log_8 8$

35. $\log_5 25$

36. $\log_3\left(\dfrac{1}{9}\right)$

37. $\log_{1/2} 16$

38. $\log_{1/3} 9$

39. $\log_{10} \sqrt{10}$

40. $\log_5 \sqrt[3]{25}$

41. $\log_{\sqrt{2}} 4$

42. $\log_{\sqrt{3}} 9$

43. $\ln \sqrt{e}$

44. $\ln e^3$

In Problems 45–56, find the domain of each function.

45. $f(x) = \ln(x - 3)$

46. $g(x) = \ln(x - 1)$

47. $F(x) = \log_2 x^2$

48. $H(x) = \log_5 x^3$

49. $f(x) = 3 - 2\log_4 \dfrac{x}{2}$

50. $g(x) = 8 + 5\ln(2x)$

51. $f(x) = \ln\left(\dfrac{1}{x + 1}\right)$

52. $g(x) = \ln\left(\dfrac{1}{x - 5}\right)$

53. $g(x) = \log_5\left(\dfrac{x + 1}{x}\right)$

54. $h(x) = \log_3\left(\dfrac{x}{x - 1}\right)$

55. $f(x) = \sqrt{\ln x}$

56. $g(x) = \dfrac{1}{\ln x}$

In Problems 57–60, use a calculator to evaluate each expression. Round your answer to three decimal places.

57. $\ln \dfrac{5}{3}$

58. $\dfrac{\ln 5}{3}$

59. $\dfrac{\ln \dfrac{10}{3}}{0.04}$

60. $\dfrac{\ln \dfrac{2}{3}}{-0.1}$

61. Find a so that the graph of $f(x) = \log_a x$ contains the point $(2, 2)$.

62. Find a so that the graph of $f(x) = \log_a x$ contains the point $\left(\dfrac{1}{2}, -4\right)$.

In Problems 63–66, graph each logarithmic function by hand.

63. $y = \log_3 x$

64. $y = \log_{1/2} x$

65. $y = \log_{1/4} x$

66. $y = \log_5 x$

In Problems 67–74, the graph of a logarithmic function is given. Match each graph to one of the following functions:

A. $y = \log_3 x$

B. $y = \log_3(-x)$

C. $y = -\log_3 x$

D. $y = -\log_3(-x)$

E. $y = \log_3 x - 1$

F. $y = \log_3(x - 1)$

G. $y = \log_3(1 - x)$

H. $y = 1 - \log_3 x$

67.

68.

69.

70.

71.

72.

73.

74.

In Problems 75–90, use transformations to graph each function. Determine the domain, range, and vertical asymptote of each function.

75. $f(x) = \ln(x + 4)$ **76.** $f(x) = \ln(x - 3)$ **77.** $f(x) = 2 + \ln x$ **78.** $f(x) = -\ln(-x)$

79. $g(x) = \ln(2x)$ **80.** $h(x) = \ln\left(\dfrac{1}{2}x\right)$ **81.** $f(x) = 3 \ln x$ **82.** $f(x) = -2 \ln x$

83. $f(x) = \log(x - 4)$ **84.** $f(x) = \log(x + 5)$ **85.** $h(x) = 4 \log x$ **86.** $g(x) = -3 \log x$

87. $F(x) = \log(2x)$ **88.** $G(x) = \log(5x)$ **89.** $h(x) = 3 + \log(x + 2)$ **90.** $g(x) = 2 - \log(x + 1)$

In Problems 91–110, solve each equation.

91. $\log_3 x = 2$ **92.** $\log_5 x = 3$ **93.** $\log_2(2x + 1) = 3$ **94.** $\log_3(3x - 2) = 2$

95. $\log_x 4 = 2$ **96.** $\log_x\left(\dfrac{1}{8}\right) = 3$ **97.** $\ln e^x = 5$ **98.** $\ln e^{-2x} = 8$

99. $\log_4 64 = x$ **100.** $\log_5 625 = x$ **101.** $\log_3 243 = 2x + 1$ **102.** $\log_6 36 = 5x + 3$

103. $e^{3x} = 10$ **104.** $e^{-2x} = \dfrac{1}{3}$ **105.** $e^{2x+5} = 8$ **106.** $e^{-2x+1} = 13$

107. $\log_3(x^2 + 1) = 2$ **108.** $\log_5(x^2 + x + 4) = 2$ **109.** $\log_2 8^x = -3$ **110.** $\log_3 3^x = -1$

Applications and Extensions

In Problems 111–114, (a) graph each function and state its domain, range, and asymptote; (b) determine the inverse function; (c) use the graph obtained in (a) to graph the inverse and state its domain, range, and asymptote.

111. $f(x) = 2^x$ **112.** $f(x) = 5^x$ **113.** $f(x) = 2^{x+3}$ **114.** $f(x) = 5^x - 2$

115. Chemistry The pH of a chemical solution is given by the formula

$$pH = -\log_{10}[H^+]$$

where $[H^+]$ is the concentration of hydrogen ions in moles per liter. Values of pH range from 0 (acidic) to 14 (alkaline).
(a) What is the pH of a solution for which $[H^+]$ is 0.1?
(b) What is the pH of a solution for which $[H^+]$ is 0.01?
(c) What is the pH of a solution for which $[H^+]$ is 0.001?
(d) What happens to pH as the hydrogen ion concentration decreases?
(e) Determine the hydrogen ion concentration of an orange (pH = 3.5).
(f) Determine the hydrogen ion concentration of human blood (pH = 7.4).

116. Diversity Index Shannon's diversity index is a measure of the diversity of a population. The diversity index is given by the formula

$$H = -(p_1 \log p_1 + p_2 \log p_2 + \cdots + p_n \log p_n)$$

where p_1 is the proportion of the population that is species 1, p_2 is the proportion of the population that is species 2, and so on.
(a) According to the U.S. Census Bureau, the distribution of race in the United States in 2000 was as follows:

Race	Proportion
American Indian or Native Alaskan	0.014
Asian	0.041
Black or African American	0.128
Hispanic	0.124
Native Hawaiian or Pacific Islander	0.003
White	0.690.

Source: U.S. Census Bureau

Compute the diversity index of the United States in 2000.
(b) The largest value of the diversity index is given by $H_{max} = \log(S)$, where S is the number of categories of race. Compute H_{max}.
(c) The evenness ratio is given by $E_H = \dfrac{H}{H_{max}}$, where $0 \le E_H \le 1$. If $E_H = 1$, there is complete evenness. Compute the evenness ratio for the United States.
(d) Obtain the distribution of race for the United States in 1990 from the Census Bureau. Compute Shannon's diversity index. Is the United States becoming more diverse? Why?

117. Atmospheric Pressure The atmospheric pressure p on a balloon or an aircraft decreases with increasing height. This pressure, measured in millimeters of mercury, is related to the height h (in kilometers) above sea level by the formula

$$p = 760e^{-0.145h}$$

(a) Find the height of an aircraft if the atmospheric pressure is 320 millimeters of mercury.
(b) Find the height of a mountain if the atmospheric pressure is 667 millimeters of mercury.

118. Healing of Wounds The normal healing of wounds can be modeled by an exponential function. If A_0 represents the original area of the wound and if A equals the area of the wound, then the formula

$$A = A_0 e^{-0.35n}$$

describes the area of a wound after n days following an injury when no infection is present to retard the healing. Suppose that a wound initially had an area of 100 square millimeters.

(a) If healing is taking place, after how many days will the wound be one-half its original size?
(b) How long before the wound is 10% of its original size?

119. Exponential Probability Between 12:00 PM and 1:00 PM, cars arrive at Citibank's drive-thru at the rate of 6 cars per hour (0.1 car per minute). The following formula from statistics can be used to determine the probability that a car will arrive within t minutes of 12:00 PM.

$$F(t) = 1 - e^{-0.1t}$$

(a) Determine how many minutes are needed for the probability to reach 50%.
(b) Determine how many minutes are needed for the probability to reach 80%.
(c) Is it possible for the probability to equal 100%? Explain.

120. Exponential Probability Between 5:00 PM and 6:00 PM, cars arrive at Jiffy Lube at the rate of 9 cars per hour (0.15 car per minute). The following formula from statistics can be used to determine the probability that a car will arrive within t minutes of 5:00 PM.

$$F(t) = 1 - e^{-0.15t}$$

(a) Determine how many minutes are needed for the probability to reach 50%.

(b) Determine how many minutes are needed for the probability to reach 80%.

121. Drug Medication The formula

$$D = 5e^{-0.4h}$$

can be used to find the number of milligrams D of a certain drug that is in a patient's bloodstream h hours after the drug has been administered. When the number of milligrams reaches 2, the drug is to be administered again. What is the time between injections?

122. Spreading of Rumors A model for the number N of people in a college community who have heard a certain rumor is

$$N = P(1 - e^{-0.15d})$$

where P is the total population of the community and d is the number of days that have elapsed since the rumor began. In a community of 1000 students, how many days will elapse before 450 students have heard the rumor?

123. Current in a RL Circuit The equation governing the amount of current I (in amperes) after time t (in seconds) in a simple RL circuit consisting of a resistance R (in ohms), an inductance L (in henrys), and an electromotive force E (in volts) is

$$I = \frac{E}{R}[1 - e^{-(R/L)t}]$$

If $E = 12$ volts, $R = 10$ ohms, and $L = 5$ henrys, how long does it take to obtain a current of 0.5 ampere? Of 1.0 ampere? Graph the equation.

124. Learning Curve Psychologists sometimes use the function

$$L(t) = A(1 - e^{-kt})$$

to measure the amount L learned at time t. The number A represents the amount to be learned, and the number k measures the rate of learning. Suppose that a student has an amount A of 200 vocabulary words to learn. A psychologist determines that the student learned 20 vocabulary words after 5 minutes.

(a) Determine the rate of learning k.
(b) Approximately how many words will the student have learned after 10 minutes?
(c) After 15 minutes?
(d) How long does it take for the student to learn 180 words?

Loudness of Sound *Problems 125–128 use the following discussion: The **loudness** $L(x)$, measured in decibels, of a sound of intensity x, measured in watts per square meter, is defined as $L(x) = 10 \log \dfrac{x}{I_0}$, where $I_0 = 10^{-12}$ watt per square meter is the least intense sound that a human ear can detect. Determine the loudness, in decibels, of each of the following sounds.*

125. Normal conversation: intensity of $x = 10^{-7}$ watt per square meter.

126. Heavy city traffic: intensity of $x = 10^{-3}$ watt per square meter.

127. Amplified rock music: intensity of 10^{-1} watt per square meter.

128. Diesel truck traveling 40 miles per hour 50 feet away: intensity 10 times that of a passenger car traveling 50 miles per hour 50 feet away whose loudness is 70 decibels.

*Problems 129 and 130 use the following discussion: The **Richter scale** is one way of converting seismographic readings into numbers that provide an easy reference for measuring the magnitude M of an earthquake. All earthquakes are compared to a **zero-level earthquake** whose seismographic reading measures 0.001 millimeter at a distance of 100 kilometers from the epicenter. An earthquake whose seismographic reading measures x millimeters has **magnitude** M(x), given by*

$$M(x) = \log\left(\frac{x}{x_0}\right)$$

where $x_0 = 10^{-3}$ is the reading of a zero-level earthquake the same distance from its epicenter. Determine the magnitude of the following earthquakes.

129. Magnitude of an Earthquake   Mexico City in 1985: seismographic reading of 125,892 millimeters 100 kilometers from the center.

130. Magnitude of an Earthquake   San Francisco in 1906: seismographic reading of 7943 millimeters 100 kilometers from the center.

131. Alcohol and Driving   The concentration of alcohol in a person's blood is measurable. Suppose that the risk R (given as a percent) of having an accident while driving a car can be modeled by the equation

$$R = 3e^{kx}$$

where x is the variable concentration of alcohol in the blood and k is a constant.

(a) Suppose that a concentration of alcohol in the blood of 0.06 results in a 10% risk ($R = 10$) of an accident. Find the constant k in the equation.

(b) Using this value of k, what is the risk if the concentration is 0.17?

(c) Using the same value of k, what concentration of alcohol corresponds to a risk of 100%?

(d) If the law asserts that anyone with a risk of having an accident of 15% or more should not have driving privileges, at what concentration of alcohol in the blood should a driver be arrested and charged with a DUI?

(e) Compare this situation with that of Example 10. If you were a lawmaker, which situation would you support? Give your reasons.

Discussion and Writing

132. Is there any function of the form $y = x^\alpha, 0 < \alpha < 1$, that increases more slowly than a logarithmic function whose base is greater than 1? Explain.

133. In the definition of the logarithmic function, the base a is not allowed to equal 1. Why?

134. Critical Thinking   In buying a new car, one consideration might be how well the price of the car holds up over time. Different makes of cars have different depreciation rates. One way to compute a depreciation rate for a car is given here. Suppose that the current prices of a certain Mercedes automobile are as follows:

	Age in Years				
New	**1**	**2**	**3**	**4**	**5**
$38,000	$36,600	$32,400	$28,750	$25,400	$21,200

Use the formula New = Old(e^{Rt}) to find R, the annual depreciation rate, for a specific time t. When might be the best time to trade in the car? Consult the NADA ("blue") book and compare two like models that you are interested in. Which has the better depreciation rate?

'Are You Prepared?' Answers

1. $x \le 3$ **2.** $x < -2$ or $x > 3$ **3.** $x < -4$ or $x > 1$

4.5 Properties of Logarithms

OBJECTIVES   1 Work with the Properties of Logarithms
2 Write a Logarithmic Expression as a Sum or Difference of Logarithms
3 Write a Logarithmic Expression as a Single Logarithm
4 Evaluate Logarithms Whose Base Is Neither 10 nor e
5 Graph Logarithmic Functions Whose Base Is Neither 10 nor e

1 Work with the Properties of Logarithms

Logarithms have some very useful properties that can be derived directly from the definition and the laws of exponents.

EXAMPLE 1	**Establishing Properties of Logarithms**

(a) Show that $\log_a 1 = 0$.   (b) Show that $\log_a a = 1$.

Solution   (a) This fact was established when we graphed $y = \log_a x$ (see Figure 35). To show the result algebraically, let $y = \log_a 1$. Then

$$y = \log_a 1$$
$$a^y = 1 \qquad \text{\textit{Change to an exponent.}}$$
$$a^y = a^0 \qquad \text{\textit{$a^0 = 1$}}$$
$$y = 0 \qquad \text{\textit{Solve for y.}}$$
$$\log_a 1 = 0 \qquad \text{\textit{$y = \log_a 1$}}$$

(b) Let $y = \log_a a$. Then

$$y = \log_a a$$
$$a^y = a \qquad \text{\textit{Change to an exponent.}}$$
$$a^y = a^1 \qquad \text{\textit{$a^1 = a$}}$$
$$y = 1 \qquad \text{\textit{Solve for y.}}$$
$$\log_a a = 1 \qquad \text{\textit{$y = \log_a a$}} \qquad \blacktriangleleft$$

To summarize:

$$\log_a 1 = 0 \qquad \log_a a = 1$$

Theorem   **Properties of Logarithms**

In the properties given next, M and a are positive real numbers, with $a \neq 1$, and r is any real number.

The number $\log_a M$ is the exponent to which a must be raised to obtain M. That is,

$$a^{\log_a M} = M \tag{1}$$

The logarithm to the base a of a raised to a power equals that power. That is,

$$\log_a a^r = r \tag{2}$$

The proof uses the fact that $y = a^x$ and $y = \log_a x$ are inverses.

Proof of Property (1)   For inverse functions,

$$f(f^{-1}(x)) = x \quad \text{for every } x \text{ in the domain of } f^{-1}$$

Using $f(x) = a^x$ and $f^{-1}(x) = \log_a x$, we find

$$f(f^{-1}(x)) = a^{\log_a x} = x \quad \text{for} \qquad x > 0$$

Now let $x = M$ to obtain $a^{\log_a M} = M$, where $M > 0$. ■

Proof of Property (2)   For inverse functions,

$$f^{-1}(f(x)) = x \quad \text{for all } x \text{ in the domain of } f.$$

Using $f(x) = a^x$ and $f^{-1}(x) = \log_a x$, we find

$$f^{-1}(f(x)) = \log_a a^x = x \quad \text{for all real numbers } x.$$

Now let $x = r$ to obtain $\log_a a^r = r$, where r is any real number.   ∎

EXAMPLE 2

Using Properties (1) and (2)

(a) $2^{\log_2 \pi} = \pi$ (b) $\log_{0.2} 0.2^{-\sqrt{2}} = -\sqrt{2}$ (c) $\ln e^{kt} = kt$ ◀

NOW WORK PROBLEM 9.

Other useful properties of logarithms are given next.

Theorem

Properties of Logarithms

In the following properties, M, N, and a are positive real numbers, with $a \neq 1$, and r is any real number.

The Log of a Product Equals the Sum of the Logs

$$\log_a(MN) = \log_a M + \log_a N \tag{3}$$

The Log of a Quotient Equals the Difference of the Logs

$$\log_a\left(\frac{M}{N}\right) = \log_a M - \log_a N \tag{4}$$

The Log of a Power Equals the Product of the Power and the Log

$$\log_a M^r = r \log_a M \tag{5}$$

We shall derive properties (3) and (5) and leave the derivation of property (4) as an exercise (see Problem 101).

Proof of Property (3)   Let $A = \log_a M$ and let $B = \log_a N$. These expressions are equivalent to the exponential expressions

$$a^A = M \quad \text{and} \quad a^B = N$$

Now

$$\log_a(MN) = \log_a(a^A a^B) = \log_a a^{A+B} \qquad \text{Law of Exponents}$$

$$= A + B \qquad \text{Property (2) of logarithms}$$

$$= \log_a M + \log_a N \qquad \qquad ∎$$

Proof of Property (5)   Let $A = \log_a M$. This expression is equivalent to

$$a^A = M$$

Now

$$\log_a M^r = \log_a(a^A)^r = \log_a a^{rA} \qquad \text{Law of Exponents}$$
$$= rA \qquad\qquad \text{Property (2) of logarithms}$$
$$= r \log_a M$$

∎

NOW WORK PROBLEM 13.

2 Write a Logarithmic Expression as a Sum or Difference of Logarithms

Logarithms can be used to transform products into sums, quotients into differences, and powers into factors. Such transformations prove useful in certain types of calculus problems.

EXAMPLE 3	**Writing a Logarithmic Expression as a Sum of Logarithms**

Write $\log_a\left(x\sqrt{x^2+1}\right)$, $x > 0$, as a sum of logarithms. Express all powers as factors.

Solution
$$\log_a\left(x\sqrt{x^2+1}\right) = \log_a x + \log_a \sqrt{x^2+1} \qquad \text{Property (3)}$$
$$= \log_a x + \log_a(x^2+1)^{1/2}$$
$$= \log_a x + \frac{1}{2}\log_a(x^2+1) \qquad \text{Property (5)}$$

◀

EXAMPLE 4	**Writing a Logarithmic Expression as a Difference of Logarithms**

Write

$$\ln \frac{x^2}{(x-1)^3}, \qquad x > 1$$

as a difference of logarithms. Express all powers as factors.

Solution
$$\ln \frac{x^2}{(x-1)^3} = \ln x^2 - \ln(x-1)^3 = 2\ln x - 3\ln(x-1)$$
$$\qquad\qquad\qquad\uparrow \qquad\qquad\qquad\uparrow$$
$$\qquad\text{Property (4)}\qquad\qquad\text{Property (5)}$$

◀

EXAMPLE 5	**Writing a Logarithmic Expression as a Sum and Difference of Logarithms**

Write

$$\log_a \frac{\sqrt{x^2+1}}{x^3(x+1)^4}, \qquad x > 0$$

as a sum and difference of logarithms. Express all powers as factors.

Solution
$$\log_a \frac{\sqrt{x^2+1}}{x^3(x+1)^4} = \log_a \sqrt{x^2+1} - \log_a[x^3(x+1)^4] \qquad \text{Property (4)}$$
$$= \log_a \sqrt{x^2+1} - [\log_a x^3 + \log_a(x+1)^4] \qquad \text{Property (3)}$$
$$= \log_a(x^2+1)^{1/2} - \log_a x^3 - \log_a(x+1)^4$$
$$= \frac{1}{2}\log_a(x^2+1) - 3\log_a x - 4\log_a(x+1) \qquad \text{Property (5)}$$

◀

CAUTION
In using properties (3) through (5), be careful about the values that the variable may assume. For example, the domain of the variable for $\log_a x$ is $x > 0$ and for $\log_a(x-1)$ it is $x > 1$. If we add these functions, the domain is $x > 1$. That is, the equality

$$\log_a x + \log_a(x-1) = \log_a[x(x-1)]$$

is true only for $x > 1$.   ∎

NOW WORK PROBLEM 45.

3 Write a Logarithmic Expression as a Single Logarithm

Another use of properties (3) through (5) is to write sums and/or differences of logarithms with the same base as a single logarithm. This skill will be needed to solve certain logarithmic equations discussed in the next section.

| EXAMPLE 6 | **Writing Expressions as a Single Logarithm** |

Write each of the following as a single logarithm.

(a) $\log_a 7 + 4\log_a 3$ (b) $\dfrac{2}{3}\ln 8 - \ln(3^4 - 8)$

(c) $\log_a x + \log_a 9 + \log_a(x^2 + 1) - \log_a 5$

Solution (a) $\log_a 7 + 4\log_a 3 = \log_a 7 + \log_a 3^4$ Property (5)

$$= \log_a 7 + \log_a 81$$
$$= \log_a(7 \cdot 81)$$ Property (3)
$$= \log_a 567$$

(b) $\dfrac{2}{3}\ln 8 - \ln(3^4 - 8) = \ln 8^{2/3} - \ln(81 - 8)$ Property (5)

$$= \ln 4 - \ln 73$$
$$= \ln\left(\dfrac{4}{73}\right)$$ Property (4)

(c) $\log_a x + \log_a 9 + \log_a(x^2 + 1) - \log_a 5 = \log_a(9x) + \log_a(x^2 + 1) - \log_a 5$

$$= \log_a[9x(x^2 + 1)] - \log_a 5$$
$$= \log_a\left[\dfrac{9x(x^2 + 1)}{5}\right]$$ ◄

WARNING   A common error made by some students is to express the logarithm of a sum as the sum of logarithms.

$\log_a(M + N)$ is not equal to $\log_a M + \log_a N$

Correct statement $\log_a(MN) = \log_a M + \log_a N$ Property (3)

Another common error is to express the difference of logarithms as the quotient of logarithms.

$\log_a M - \log_a N$ is not equal to $\dfrac{\log_a M}{\log_a N}$

Correct statement $\log_a M - \log_a N = \log_a\left(\dfrac{M}{N}\right)$ Property (4)

A third common error is to express a logarithm raised to a power as the product of the power times the logarithm.

$(\log_a M)^r$ is not equal to $r\log_a M$

Correct statement $\log_a M^r = r\log_a M$ Property (5)   ■

NOW WORK PROBLEM **51**.

Two other properties of logarithms that we need to know are consequences of the fact that the logarithmic function $y = \log_a x$ is one-to-one.

Theorem

Properties of Logarithms

In the following properties, M, N, and a are positive real numbers, with $a \neq 1$.

If $M = N$, then $\log_a M = \log_a N$.	**(6)**
If $\log_a M = \log_a N$, then $M = N$.	**(7)**

When property (6) is used, we start with the equation $M = N$ and say "take the logarithm of both sides" to obtain $\log_a M = \log_a N$.

Properties (6) and (7) are useful for solving *exponential and logarithmic equations*, a topic discussed in the next section.

4 Evaluate Logarithms Whose Base Is Neither 10 nor e

Logarithms to the base 10, common logarithms, were used to facilitate arithmetic computations before the widespread use of calculators. (See the Historical Feature at the end of this section.) Natural logarithms, that is, logarithms whose base is the number e, remain very important because they arise frequently in the study of natural phenomena.

Common logarithms are usually abbreviated by writing **log**, with the base understood to be 10, just as natural logarithms are abbreviated by **ln**, with the base understood to be e.

Most calculators have both $\boxed{\log}$ and $\boxed{\ln}$ keys to calculate the common logarithm and natural logarithm of a number. Let's look at an example to see how to approximate logarithms having a base other than 10 or e.

EXAMPLE 7	**Approximating Logarithms Whose Base Is Neither 10 nor e**

Approximate $\log_2 7$. Round the answer to four decimal places.

Solution   Let $y = \log_2 7$. Then $2^y = 7$, so

$$2^y = 7$$
$$\ln 2^y = \ln 7 \qquad \text{Property (6)}$$
$$y \ln 2 = \ln 7 \qquad \text{Property (5)}$$
$$y = \frac{\ln 7}{\ln 2} \qquad \text{Exact solution}$$
$$y \approx 2.8074 \qquad \text{Approximate solution rounded to four decimal places} \qquad \blacktriangleleft$$

Example 7 shows how to approximate a logarithm whose base is 2 by changing to logarithms involving the base e. In general, we use the **Change-of-Base Formula**.

Theorem

Change-of-Base Formula

If $a \neq 1$, $b \neq 1$, and M are positive real numbers, then

$$\log_a M = \frac{\log_b M}{\log_b a} \qquad \qquad \textbf{(8)}$$

Proof We derive this formula as follows: Let $y = \log_a M$. Then

$$a^y = M$$

$$\log_b a^y = \log_b M \qquad \text{Property (6)}$$

$$y \log_b a = \log_b M \qquad \text{Property (5)}$$

$$y = \frac{\log_b M}{\log_b a} \qquad \text{Solve for y.}$$

$$\log_a M = \frac{\log_b M}{\log_b a} \qquad y = \log_a M$$

∎

Since calculators have keys only for $\boxed{\log}$ and $\boxed{\ln}$, in practice, the Change-of-Base Formula uses either $b = 10$ or $b = e$. That is,

$$\log_a M = \frac{\log M}{\log a} \quad \text{and} \quad \log_a M = \frac{\ln M}{\ln a} \qquad \textbf{(9)}$$

EXAMPLE 8 **Using the Change-of-Base Formula**

Approximate: (a) $\log_5 89$ (b) $\log_{\sqrt{2}} \sqrt{5}$
Round answers to four decimal places.

Solution (a) $\log_5 89 = \dfrac{\log 89}{\log 5} \approx \dfrac{1.949390007}{0.6989700043} \approx 2.7889$

or

$$\log_5 89 = \frac{\ln 89}{\ln 5} \approx \frac{4.48863637}{1.609437912} \approx 2.7889$$

(b) $\log_{\sqrt{2}} \sqrt{5} = \dfrac{\log \sqrt{5}}{\log \sqrt{2}} = \dfrac{\dfrac{1}{2}\log 5}{\dfrac{1}{2}\log 2} \approx 2.3219$

or

$$\log_{\sqrt{2}} \sqrt{5} = \frac{\ln \sqrt{5}}{\ln \sqrt{2}} = \frac{\dfrac{1}{2}\ln 5}{\dfrac{1}{2}\ln 2} \approx 2.3219$$

◀

NOW WORK PROBLEMS 17 AND 65.

5 Graph Logarithmic Functions Whose Base Is Neither 10 nor e

We also use the Change-of-Base Formula to graph logarithmic functions whose base is neither 10 nor e.

| EXAMPLE 9 | **Graphing a Logarithmic Function Whose Base Is Neither 10 nor e** |

Use a graphing utility to graph $y = \log_2 x$.

Figure 44

Solution Since graphing utilities only have logarithms with the base 10 or the base e, we need to use the Change-of-Base Formula to express $y = \log_2 x$ in terms of logarithms with base 10 or base e. We can graph either $y = \dfrac{\ln x}{\ln 2}$ or $y = \dfrac{\log x}{\log 2}$ to obtain the graph of $y = \log_2 x$. See Figure 44.

✔ **CHECK:** Verify that $y = \dfrac{\ln x}{\ln 2}$ and $y = \dfrac{\log x}{\log 2}$ result in the same graph by graphing each on the same screen.   ◄

━ NOW WORK PROBLEM 73.

Summary
Properties of Logarithms
In the list that follows, $a > 0$, $a \ne 1$, and $b > 0$, $b \ne 1$; also, $M > 0$ and $N > 0$.

Definition $y = \log_a x$ means $x = a^y$

Properties of logarithms

$\log_a 1 = 0; \quad \log_a a = 1$ $\log_a M^r = r \log_a M$

$a^{\log_a M} = M; \quad \log_a a^r = r$ If $M = N$, then $\log_a M = \log_a N$.

$\log_a(MN) = \log_a M + \log_a N$ If $\log_a M = \log_a N$, then $M = N$.

$\log_a\left(\dfrac{M}{N}\right) = \log_a M - \log_a N$

Change-of-Base Formula $\log_a M = \dfrac{\log_b M}{\log_b a}$

HISTORICAL FEATURE

John Napier
(1550–1617)

Logarithms were invented about 1590 by John Napier (1550–1617) and Joost Bürgi (1552–1632), working independently. Napier, whose work had the greater influence, was a Scottish lord, a secretive man whose neighbors were inclined to believe him to be in league with the devil. His approach to logarithms was very different from ours; it was based on the relationship between arithmetic and geometric sequences, discussed in a later chapter, and not on the inverse function relationship of logarithms to exponential functions (described in Section 4.4). Napier's tables, pub-

lished in 1614, listed what would now be called *natural logarithms* of sines and were rather difficult to use. A London professor, Henry Briggs, became interested in the tables and visited Napier. In their conversations, they developed the idea of common logarithms, which were published in 1617. Their importance for calculation was immediately recognized, and by 1650 they were being printed as far away as China. They remained an important calculation tool until the advent of the inexpensive handheld calculator about 1972, which has decreased their calculational, but not their theoretical, importance.

A side effect of the invention of logarithms was the popularization of the decimal system of notation for real numbers.

4.5 Assess Your Understanding

Concepts and Vocabulary

1. The logarithm of a product equals the _____ of the logarithms.

2. If $\log_8 M = \dfrac{\log_5 7}{\log_5 8}$, then $M =$ _____.

3. $\log_a M^r =$ _____.

4. *True or False:* $\ln(x + 3) - \ln(2x) = \dfrac{\ln(x + 3)}{\ln(2x)}$

5. *True or False:* $\log_2(3x^4) = 4 \log_2(3x)$

6. *True or False:* $\log_2 16 = \dfrac{\ln 16}{\ln 2}$

Skill Building

In Problems 7–22, use properties of logarithms to find the exact value of each expression. Do not use a calculator.

7. $\log_3 3^{71}$

8. $\log_2 2^{-13}$

9. $\ln e^{-4}$

10. $\ln e^{\sqrt{2}}$

11. $2^{\log_2 7}$

12. $e^{\ln 8}$

13. $\log_8 2 + \log_8 4$

14. $\log_6 9 + \log_6 4$

15. $\log_6 18 - \log_6 3$

16. $\log_8 16 - \log_8 2$

17. $\log_2 6 \cdot \log_6 4$

18. $\log_3 8 \cdot \log_8 9$

19. $3^{\log_3 5 - \log_3 4}$

20. $5^{\log_5 6 + \log_5 7}$

21. $e^{\log_{e^2} 16}$

22. $e^{\log_{e^2} 9}$

In Problems 23–30, suppose that $\ln 2 = a$ and $\ln 3 = b$. Use properties of logarithms to write each logarithm in terms of a and b.

23. $\ln 6$

24. $\ln \dfrac{2}{3}$

25. $\ln 1.5$

26. $\ln 0.5$

27. $\ln 8$

28. $\ln 27$

29. $\ln \sqrt[5]{6}$

30. $\ln \sqrt[4]{\dfrac{2}{3}}$

In Problems 31–50, write each expression as a sum and/or difference of logarithms. Express powers as factors.

31. $\log_5(25x)$

32. $\log_3 \dfrac{x}{9}$

33. $\log_2 z^3$

34. $\log_7(x^5)$

35. $\ln(ex)$

36. $\ln \dfrac{e}{x}$

37. $\ln(xe^x)$

38. $\ln \dfrac{x}{e^x}$

39. $\log_a(u^2 v^3), \quad u > 0, v > 0$ **40.** $\log_2\left(\dfrac{a}{b^2}\right), \quad a > 0, b > 0$

41. $\ln\left(x^2\sqrt{1-x}\right), \quad 0 < x < 1$ **42.** $\ln\left(x\sqrt{1+x^2}\right), \quad x > 0$

43. $\log_2\left(\dfrac{x^3}{x-3}\right), \quad x > 3$ **44.** $\log_5\left(\dfrac{\sqrt[3]{x^2+1}}{x^2-1}\right), \quad x > 1$ **45.** $\log\left[\dfrac{x(x+2)}{(x+3)^2}\right], \quad x > 0$ **46.** $\log\left[\dfrac{x^3\sqrt{x+1}}{(x-2)^2}\right], \quad x > 2$

47. $\ln\left[\dfrac{x^2-x-2}{(x+4)^2}\right]^{1/3}, \quad x > 2$ **48.** $\ln\left[\dfrac{(x-4)^2}{x^2-1}\right]^{2/3}, \quad x > 4$

49. $\ln \dfrac{5x\sqrt{1+3x}}{(x-4)^3}, \quad x > 4$ **50.** $\ln\left[\dfrac{5x^2\sqrt[3]{1-x}}{4(x+1)^2}\right], \quad 0 < x < 1$

In Problems 51–64, write each expression as a single logarithm.

51. $3\log_5 u + 4\log_5 v$

52. $2\log_3 u - \log_3 v$

53. $\log_3 \sqrt{x} - \log_3 x^3$

54. $\log_2\left(\dfrac{1}{x}\right) + \log_2\left(\dfrac{1}{x^2}\right)$

55. $\log_4(x^2 - 1) - 5\log_4(x+1)$

56. $\log(x^2 + 3x + 2) - 2\log(x+1)$

57. $\ln\left(\dfrac{x}{x-1}\right) + \ln\left(\dfrac{x+1}{x}\right) - \ln(x^2-1)$ **58.** $\log\left(\dfrac{x^2+2x-3}{x^2-4}\right) - \log\left(\dfrac{x^2+7x+6}{x+2}\right)$ **59.** $8\log_2\sqrt{3x-2} - \log_2\left(\dfrac{4}{x}\right) + \log_2 4$

60. $21\log_3\sqrt[3]{x} + \log_3(9x^2) - \log_3 9$ **61.** $2\log_a(5x^3) - \dfrac{1}{2}\log_a(2x+3)$ **62.** $\dfrac{1}{3}\log(x^3+1) + \dfrac{1}{2}\log(x^2+1)$

63. $2\log_2(x+1) - \log_2(x+3) - \log_2(x-1)$ **64.** $3\log_5(3x+1) - 2\log_5(2x-1) - \log_5 x$

In Problems 65–72, use the Change-of-Base Formula and a calculator to evaluate each logarithm. Round your answer to three decimal places.

65. $\log_3 21$

66. $\log_5 18$

67. $\log_{1/3} 71$

68. $\log_{1/2} 15$

69. $\log_{\sqrt{2}} 7$

70. $\log_{\sqrt{5}} 8$

71. $\log_\pi e$

72. $\log_\pi \sqrt{2}$

In Problems 73–78, graph each function using a graphing utility and the Change-of-Base Formula.

73. $y = \log_4 x$

74. $y = \log_5 x$

75. $y = \log_2(x+2)$

76. $y = \log_4(x-3)$

77. $y = \log_{x-1}(x+1)$

78. $y = \log_{x+2}(x-2)$

Applications and Extensions

In Problems 79–88, express y as a function of x. The constant C is a positive number.

79. $\ln y = \ln x + \ln C$

80. $\ln y = \ln(x+C)$

81. $\ln y = \ln x + \ln(x + 1) + \ln C$

82. $\ln y = 2 \ln x - \ln(x + 1) + \ln C$

83. $\ln y = 3x + \ln C$

84. $\ln y = -2x + \ln C$

85. $\ln(y - 3) = -4x + \ln C$

86. $\ln(y + 4) = 5x + \ln C$

87. $3 \ln y = \dfrac{1}{2} \ln(2x + 1) - \dfrac{1}{3} \ln(x + 4) + \ln C$

88. $2 \ln y = -\dfrac{1}{2} \ln x + \dfrac{1}{3} \ln(x^2 + 1) + \ln C$

89. Find the value of $\log_2 3 \cdot \log_3 4 \cdot \log_4 5 \cdot \log_5 6 \cdot \log_6 7 \cdot \log_7 8$.

90. Find the value of $\log_2 4 \cdot \log_4 6 \cdot \log_6 8$.

91. Find the value of $\log_2 3 \cdot \log_3 4 \cdot \cdots \cdot \log_n(n + 1) \cdot \log_{n+1} 2$.

92. Find the value of $\log_2 2 \cdot \log_2 4 \cdot \cdots \cdot \log_2 2^n$.

93. Show that $\log_a\left(x + \sqrt{x^2 - 1}\right) + \log_a\left(x - \sqrt{x^2 - 1}\right) = 0$.

94. Show that $\log_a\left(\sqrt{x} + \sqrt{x - 1}\right) + \log_a\left(\sqrt{x} - \sqrt{x - 1}\right) = 0$.

95. Show that $\ln(1 + e^{2x}) = 2x + \ln(1 + e^{-2x})$.

96. Difference Quotient If $f(x) = \log_a x$, show that $\dfrac{f(x + h) - f(x)}{h} = \log_a\left(1 + \dfrac{h}{x}\right)^{1/h}$, $h \neq 0$.

97. If $f(x) = \log_a x$, show that $-f(x) = \log_{1/a} x$.

98. If $f(x) = \log_a x$, show that $f(AB) = f(A) + f(B)$.

99. If $f(x) = \log_a x$, show that $f\left(\dfrac{1}{x}\right) = -f(x)$.

100. If $f(x) = \log_a x$, show that $f(x^{\alpha}) = \alpha f(x)$.

101. Show that $\log_a\left(\dfrac{M}{N}\right) = \log_a M - \log_a N$, where a, M, and N are positive real numbers, with $a \neq 1$.

102. Show that $\log_a\left(\dfrac{1}{N}\right) = -\log_a N$, where a and N are positive real numbers, with $a \neq 1$.

Discussion and Writing

103. Graph $Y_1 = \log(x^2)$ and $Y_2 = 2\log(x)$ using a graphing utility. Are they equivalent? What might account for any differences in the two functions?

104. Write an example that illustrates why $\log_2(x + y) \neq \log_2 x + \log_2 y$.

105. Write an example that illustrates why $(\log_a x)^r \neq r \log_a x$.

106. Does $3^{\log_3 -5} = -5$? Why or Why not?

4.6 Logarithmic and Exponential Equations

PREPARING FOR THIS SECTION *Before getting started, review the following:*

• Solving Equations Using a Graphing Utility (Section 1.2, pp. 101–103) • Solving Quadratic Equations (Section 1.3, pp. 112–119)

Now work the 'Are You Prepared?' problems on page 461.

OBJECTIVES 1 Solve Logarithmic Equations Using the Properties of Logarithms

2 Solve Exponential Equations

3 Solve Logarithmic and Exponential Equations Using a Graphing Utility

1 Solve Logarithmic Equations Using the Properties of Logarithms

In Section 4.4 we solved logarithmic equations by changing a logarithm to exponential form. Often, however, some manipulation of the equation (usually using the properties of logarithms) is required before we can change to exponential form.

Our practice will be to solve equations, whenever possible, by finding exact solutions using algebraic methods and exact or approximate solutions using a graphing utility. When algebraic methods cannot be used, approximate solutions will be obtained using a graphing utility. The reader is encouraged to pay particular attention to the form of equations for which exact solutions are possible.

| EXAMPLE 1 | Solving a Logarithmic Equation |

Solve: $2 \log_5 x = \log_5 9$

Algebraic Solution

Because each logarithm is to the same base, 5, we can obtain an exact solution as follows:

$$2 \log_5 x = \log_5 9$$
$$\log_5 x^2 = \log_5 9 \quad \text{{log}}_a M^r = r \log_a M$$
$$x^2 = 9 \quad \text{If } \log_a M = \log_a N, \text{ then } M = N.$$
$$x = 3 \quad \text{or} \quad \cancel{x = -3} \quad \text{Recall that logarithms of nega-}$$

tive numbers are not defined, so, in the expression $2 \log_5 x$, x must be positive. Therefore, -3 is extraneous and we discard it.

✔ **CHECK:** $2 \log_5 3 \overset{?}{=} \log_5 9$
$$\log_5 3^2 \overset{?}{=} \log_5 9 \quad r \log_a M = \log_a M^r$$
$$\log_5 9 = \log_5 9$$

Graphing Solution

To solve the equation using a graphing utility, graph $Y_1 = 2 \log_5 x = \dfrac{2 \log x}{\log 5}$ and

$Y_2 = \log_5 9 = \dfrac{\log 9}{\log 5}$, and determine the point of intersection. See Figure 45.

Figure 45

The solution set is $\{3\}$.

➤ **NOW WORK PROBLEM 9.**

| EXAMPLE 2 | Solving a Logarithmic Equation |

Solve: $\log_4(x + 3) + \log_4(2 - x) = 1$

Algebraic Solution

To obtain an exact solution, we need to express the left side as a single logarithm. Then we will change the expression to exponential form.

$$\log_4(x + 3) + \log_4(2 - x) = 1$$
$$\log_4[(x + 3)(2 - x)] = 1 \quad \log_a M + \log_a N = \log_a(MN)$$
$$(x + 3)(2 - x) = 4^1 = 4 \quad \text{Change to an exponential expression.}$$
$$-x^2 - x + 6 = 4 \quad \text{Simplify.}$$
$$x^2 + x - 2 = 0 \quad \text{Place the quadratic equation in standard form.}$$
$$(x + 2)(x - 1) = 0 \quad \text{Factor.}$$
$$x = -2 \quad \text{or} \quad x = 1 \quad \text{Zero-Product Property}$$

Since the arguments of each logarithmic expression in the equation are positive for both $x = -2$ and $x = 1$, neither is extraneous. We leave the check to you.

Graphing Solution

Graph $Y_1 = \log_4(x + 3) + \log_4(2 - x) = \dfrac{\log(x + 3)}{\log 4} + \dfrac{\log(2 - x)}{\log 4}$ and $Y_2 = 1$ and determine the points of intersection. See Figure 46.

Figure 46

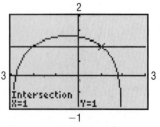

The solution set is $\{-2, 1\}$.

➤ **NOW WORK PROBLEM 13.**

2 Solve Exponential Equations

In Sections 4.3 and 4.4, we solved certain exponential equations algebraically by expressing each side of the equation with the same base. However, many exponential equations cannot be rewritten so each side has the same base. In such cases, properties of logarithms along with algebraic techniques can sometimes be used to obtain a solution.

| EXAMPLE 3 | Solving an Exponential Equation |

Solve: $4^x - 2^x - 12 = 0$

Algebraic Solution

We note that $4^x = (2^2)^x = 2^{2x} = (2^x)^2$, so the equation is actually quadratic in form, and we can rewrite it as

$$(2^x)^2 - 2^x - 12 = 0 \qquad \text{Let } u = 2^x; \text{ then } u^2 - u - 12 = 0.$$

Now we can factor as usual.

$$(2^x - 4)(2^x + 3) = 0 \qquad (u-4)(u+3) = 0$$
$$2^x - 4 = 0 \quad \text{or} \quad 2^x + 3 = 0 \qquad u - 4 = 0 \quad \text{or} \quad u + 3 = 0$$
$$2^x = 4 \qquad\qquad 2^x = -3 \qquad u = 2^x = 4 \qquad u = 2^x = -3$$

The equation on the left has the solution $x = 2$, since $2^x = 4 = 2^2$; the equation on the right has no solution, since $2^x > 0$ for all x. The only solution is 2.

Graphing Solution

Graph $Y_1 = 4^x - 2^x - 12$ and determine the x-intercept. See Figure 47.

Figure 47

The solution set is $\{2\}$. ◀

In the algebraic solution of the previous example, we were able to write the exponential expression using the same base after utilizing some algebra, obtaining an exact solution to the equation. When this is not possible, logarithms can sometimes be used to obtain the solution.

| EXAMPLE 4 | Solving an Exponential Equation |

Solve: $2^x = 5$

Algebraic Solution

Since 5 cannot be written as an integral power of 2, we write the exponential equation as the equivalent logarithmic equation.

$$2^x = 5$$
$$x = \log_2 5 = \frac{\ln 5}{\ln 2}$$
$$\uparrow$$

Change-of-Base Formula (9), Section 4.5

Alternatively, we can solve the equation $2^x = 5$ by taking the natural logarithm (or common logarithm) of each side. Taking the natural logarithm,

$$2^x = 5$$
$$\ln 2^x = \ln 5 \qquad \text{If } M = N, \text{ then } \ln M = \ln N.$$
$$x \ln 2 = \ln 5 \qquad \ln M^r = r \ln M$$
$$x = \frac{\ln 5}{\ln 2} \qquad \text{Exact solution}$$
$$\approx 2.322 \qquad \text{Approximate solution}$$
◀

Graphing Solution

Graph $Y_1 = 2^x$ and $Y_2 = 5$ and determine the x-coordinate of the point of intersection. See Figure 48.

Figure 48

The approximate solution, rounded to three decimal places, is 2.322. ◀

 ━━━ **NOW WORK PROBLEM 21.**

| EXAMPLE 5 | Solving an Exponential Equation |

Solve: $8 \cdot 3^x = 5$

Algebraic Solution

$$8 \cdot 3^x = 5$$

$$3^x = \frac{5}{8} \qquad \text{Solve for } 3^x.$$

$$x = \log_3\left(\frac{5}{8}\right) = \frac{\ln\left(\frac{5}{8}\right)}{\ln 3} \qquad \text{Exact solution}$$

$$\approx -0.428 \qquad \text{Approximate solution} \quad \blacktriangleleft$$

Graphing Solution

Graph $Y_1 = 8 \cdot 3^x$ and $Y_2 = 5$ and determine the x-coordinate of the point of intersection. See Figure 49.

Figure 49

The approximate solution, rounded to three decimal places, is –0.428. $\quad \blacktriangleleft$

| EXAMPLE 6 | Solving an Exponential Equation |

Solve: $5^{x-2} = 3^{3x+2}$

Algebraic Solution

Because the bases are different, we first apply Property (6), Section 4.5 (take the natural logarithm of each side), and then use appropriate properties of logarithms. The result is an equation in x that we can solve.

$$5^{x-2} = 3^{3x+2}$$

$$\ln 5^{x-2} = \ln 3^{3x+2} \qquad \text{If } M = N, \ln M = \ln N.$$

$$(x - 2)\ln 5 = (3x + 2)\ln 3 \qquad \ln M^r = r \ln M$$

$$(\ln 5)x - 2\ln 5 = (3\ln 3)x + 2\ln 3 \qquad \text{Distribute.}$$

$$(\ln 5)x - (3\ln 3)x = 2\ln 3 + 2\ln 5 \qquad \text{Place terms involving } x \text{ on the left.}$$

$$(\ln 5 - 3\ln 3)x = 2(\ln 3 + \ln 5) \qquad \text{Factor.}$$

$$x = \frac{2(\ln 3 + \ln 5)}{\ln 5 - 3\ln 3} \qquad \text{Exact solution}$$

$$\approx -3.212 \qquad \text{Approximate solution} \quad \blacktriangleleft$$

Graphing Solution

Graph $Y_1 = 5^{x-2}$ and $Y_2 = 3^{3x+2}$ and determine the x-coordinate of the point of intersection. See Figure 50.

Figure 50

The approximate solution, rounded to three decimal places, is –3.212. $\quad \blacktriangleleft$

 NOW WORK PROBLEM 29.

3 Solve Logarithmic and Exponential Equations Using a Graphing Utility

The algebraic techniques introduced in this section to obtain exact solutions apply only to certain types of logarithmic and exponential equations. Solutions for other types are usually studied in calculus, using numerical methods. For such types, we can use a graphing utility to approximate the solution.

EXAMPLE 7	**Solving Equations Using a Graphing Utility**

Figure 51

Solve: $x + e^x = 2$
Express the solution(s) rounded to two decimal places.

Solution The solution is found by graphing $Y_1 = x + e^x$ and $Y_2 = 2$. Y_1 is an increasing function (do you know why?), and so there is only one point of intersection for Y_1 and Y_2. Figure 51 shows the graphs of Y_1 and Y_2. Using the INTERSECT command, the solution is 0.44 rounded to two decimal places. ◀

NOW WORK PROBLEM 53.

4.6 Assess Your Understanding

'Are You Prepared?'

Answers are given at the end of these exercises. If you get a wrong answer, read the pages listed in red.

1. Solve $x^2 - 7x - 30 = 0$. (pp. 113–114)

2. Solve $(x + 3)^2 - 4(x + 3) + 3 = 0$. (pp. 135–137)

3. Approximate the solution(s) to $x^3 = x^2 - 5$ using a graphing utility. (pp. 102–103)

4. Approximate the solution(s) to $x^3 - 2x + 2 = 0$ using a graphing utility. (pp. 102–103)

Skill Building

In Problems 5–50, solve each equation. Verify your solution using a graphing utility.

5. $\log_4(x + 2) = \log_4 8$

6. $\log_5(2x + 3) = \log_5 3$

7. $\frac{1}{2}\log_3 x = 2\log_3 2$

8. $-2\log_4 x = \log_4 9$

9. $2\log_5 x = 3\log_5 4$

10. $3\log_2 x = -\log_2 27$

11. $3\log_2(x - 1) + \log_2 4 = 5$

12. $2\log_3(x + 4) - \log_3 9 = 2$

13. $\log x + \log(x + 15) = 2$

14. $\log_4 x + \log_4(x - 3) = 1$

15. $\ln x + \ln(x + 2) = 4$

16. $\ln(x + 1) - \ln x = 2$

17. $2^{2x} + 2^x - 12 = 0$

18. $3^{2x} + 3^x - 2 = 0$

19. $3^{2x} + 3^{x+1} - 4 = 0$

20. $2^{2x} + 2^{x+2} - 12 = 0$

21. $2^x = 10$

22. $3^x = 14$

23. $8^{-x} = 1.2$

24. $2^{-x} = 1.5$

25. $3^{1-2x} = 4^x$

26. $2^{x+1} = 5^{1-2x}$

27. $\left(\frac{3}{5}\right)^x = 7^{1-x}$

28. $\left(\frac{4}{3}\right)^{1-x} = 5^x$

29. $1.2^x = (0.5)^{-x}$

30. $0.3^{1+x} = 1.7^{2x-1}$

31. $\pi^{1-x} = e^x$

32. $e^{x+3} = \pi^x$

33. $5(2^{3x}) = 8$

34. $0.3(4^{0.2x}) = 0.2$

35. $\log_a(x - 1) - \log_a(x + 6) = \log_a(x - 2) - \log_a(x + 3)$

36. $\log_a x + \log_a(x - 2) = \log_a(x + 4)$

37. $\log_{1/3}(x^2 + x) - \log_{1/3}(x^2 - x) = -1$

38. $\log_4(x^2 - 9) - \log_4(x + 3) = 3$

39. $\log_2(x + 1) - \log_4 x = 1$
[**Hint:** Change $\log_4 x$ to base 2.]

40. $\log_2(3x + 2) - \log_4 x = 3$

41. $\log_{16} x + \log_4 x + \log_2 x = 7$

42. $\log_9 x + 3\log_3 x = 14$

43. $\left(\sqrt[3]{2}\right)^{2-x} = 2^{x^2}$

44. $\log_2 x^{\log_2 x} = 4$

45. $\dfrac{e^x + e^{-x}}{2} = 1$

46. $\dfrac{e^x + e^{-x}}{2} = 3$

47. $\dfrac{e^x - e^{-x}}{2} = 2$

48. $\dfrac{e^x - e^{-x}}{2} = -2$

[**Hint:** Multiply each side by e^x.]

49. $\log_5 x + \log_3 x = 1$
[**Hint:** Use the Change-of-Base Formula and factor $\log x$.]

50. $\log_2 x + \log_6 x = 3$

In Problems 51–64, use a graphing utility to solve each equation. Express your answer rounded to two decimal places.

51. $\log_5(x + 1) - \log_4(x - 2) = 1$

52. $\log_2(x - 1) - \log_6(x + 2) = 2$

53. $e^x = -x$

54. $e^{2x} = x + 2$

55. $e^x = x^2$

56. $e^x = x^3$

57. $\ln x = -x$

58. $\ln(2x) = -x + 2$

59. $\ln x = x^3 - 1$

60. $\ln x = -x^2$

61. $e^x + \ln x = 4$

62. $e^x - \ln x = 4$

63. $e^{-x} = \ln x$

64. $e^{-x} = -\ln x$

Applications and Extensions

65. A Population Model The population of the United States in 2000 was 282 million people. In addition, the population of the United States was growing at a rate of 1.1% per year. Assuming that this growth rate continues, the model $P(t) = 282(1.011)^{t-2000}$ represents the population P (in millions of people) in year t.

    (a) According to this model, when will the population of the United States be 303 million people?

    (b) According to this model, when will the population of the United States be 355 million people?

    SOURCE: *Statistical Abstract of the United States*

66. A Population Model The population of the world in 2000 was 6.08 billion people. In addition, the population of the world was growing at a rate of 1.26% per year. Assuming that this growth rate continues, the model $P(t) = 6.08(1.0126)^{t-2000}$ represents the population P (in millions of people) in year t.

    (a) According to this model, when will the population of the world be 9.17 billion people?

    (b) According to this model, when will the population of the world be 11.55 billion people?

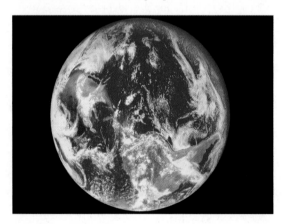

SOURCE: *Statistical Abstract of the United States*

67. Depreciation The value V of a Chevy Cavalier Coupe that is t years old can be modeled by $V(t) = 14,512(0.82)^t$.

    (a) According to the model, when will the car be worth $9000?

    (b) According to the model, when will the car be worth $4000?

    (c) According to the model, when will the car be worth $2000?

SOURCE: *Kelley Blue Book*

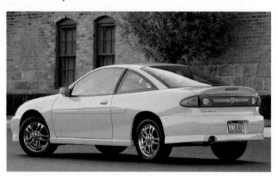

68. Depreciation The value V of a Dodge Stratus that is t years old can be modeled by $V(t) = 19,282(0.84)^t$.

    (a) According to the model, when will the car be worth $15,000?

    (b) According to the model, when will the car be worth $8000?

    (c) According to the model, when will the car be worth $2000?

SOURCE: *Kelley Blue Book*

Discussion and Writing

69. Fill in reasons for each step in the following two solutions.

Solve: $\log_3(x - 1)^2 = 2$

Solution A	**Solution B**
$\log_3(x - 1)^2 = 2$	$\log_3(x - 1)^2 = 2$
$(x - 1)^2 = 3^2 = 9$ _____	$2\log_3(x - 1) = 2$ _____
$(x - 1) = \pm 3$ _____	$\log_3(x - 1) = 1$ _____
$x - 1 = -3$ or $x - 1 = 3$ _____	$x - 1 = 3^1 = 3$ _____
$x = -2$ or $x = 4$ _____	$x = 4$ _____

Both solutions given in Solution A check. Explain what caused the solution $x = -2$ to be lost in Solution B.

'Are You Prepared?' Answers

1. $\{-3, 10\}$ **2.** $\{-2, 0\}$ **3.** $\{-1.43\}$ **4.** $\{-1.77\}$

4.7 Compound Interest

PREPARING FOR THIS SECTION *Before getting started, review the following:*

- Simple Interest (Section 1.6, pp. 142–143)

Now work the 'Are You Prepared?' problems on page 470.

OBJECTIVES
1 Determine the Future Value of a Lump Sum of Money
2 Calculate Effective Rates of Return
3 Determine the Present Value of a Lump Sum of Money
4 Determine the Time Required to Double or Triple a Lump Sum of Money

1 Determine the Future Value of a Lump Sum of Money

Interest is money paid for the use of money. The total amount borrowed (whether by an individual from a bank in the form of a loan or by a bank from an individual in the form of a savings account) is called the **principal**. The **rate of interest**, expressed as a percent, is the amount charged for the use of the principal for a given period of time, usually on a yearly (that is, per annum) basis.

Simple Interest Formula

If a principal of P dollars is borrowed for a period of t years at a per annum interest rate r, expressed as a decimal, the interest I charged is

$$I = Prt \tag{1}$$

Interest charged according to formula (1) is called **simple interest**.

In working with problems involving interest, we define the term **payment period** as follows:

Annually:	Once per year	**Monthly:**	12 times per year
Semiannually:	Twice per year	**Daily:**	365 times per year[*]
Quarterly:	Four times per year		

When the interest due at the end of a payment period is added to the principal so that the interest computed at the end of the next payment period is based on this new principal amount (old principal + interest), the interest is said to have been **compounded**. **Compound interest** is interest paid on principal and previously earned interest.

EXAMPLE 1 | **Computing Compound Interest**

A credit union pays interest of 8% per annum compounded quarterly on a certain savings plan. If $1000 is deposited in such a plan and the interest is left to accumulate, how much is in the account after 1 year?

---

[*]Most banks use a 360-day "year." Why do you think they do?

Solution   We use the simple interest formula, $I = Prt$. The principal P is $1000 and the rate of interest is $8\% = 0.08$. After the first quarter of a year, the time t is $\frac{1}{4}$ year, so the interest earned is

$$I = Prt = (\$1000)(0.08)\left(\frac{1}{4}\right) = \$20$$

The new principal is $P + I = \$1000 + \$20 = \$1020$. At the end of the second quarter, the interest on this principal is

$$I = (\$1020)(0.08)\left(\frac{1}{4}\right) = \$20.40$$

At the end of the third quarter, the interest on the new principal of $\$1020 + \$20.40 = \$1040.40$ is

$$I = (\$1040.40)(0.08)\left(\frac{1}{4}\right) = \$20.81$$

Finally, after the fourth quarter, the interest is

$$I = (\$1061.21)(0.08)\left(\frac{1}{4}\right) = \$21.22$$

After 1 year the account contains $\$1061.21 + \$21.22 = \$1082.43$.   ◄

The pattern of the calculations performed in Example 1 leads to a general formula for compound interest. To fix our ideas, let P represent the principal to be invested at a per annum interest rate r that is compounded n times per year, so the time of each compounding period is $\frac{1}{n}$ years. (For computing purposes, r is expressed as a decimal.) The interest earned after each compounding period is given by formula (1).

$$\text{Interest} = \text{principal} \times \text{rate} \times \text{time} = P \cdot r \cdot \frac{1}{n} = P \cdot \left(\frac{r}{n}\right)$$

The amount A after one compounding period is

$$A = P + I = P + P \cdot \left(\frac{r}{n}\right) = P \cdot \left(1 + \frac{r}{n}\right)$$

After two compounding periods, the amount A, based on the new principal $P \cdot \left(1 + \frac{r}{n}\right)$, is

$$A = \underbrace{P \cdot \left(1 + \frac{r}{n}\right)}_{\substack{\text{New} \\ \text{principal}}} + \underbrace{P \cdot \left(1 + \frac{r}{n}\right)\left(\frac{r}{n}\right)}_{\substack{\text{Interest on} \\ \text{new principal}}} = P \cdot \left(1 + \frac{r}{n}\right)\left(1 + \frac{r}{n}\right) = P \cdot \left(1 + \frac{r}{n}\right)^2$$

After three compounding periods, the amount A is

$$A = P \cdot \left(1 + \frac{r}{n}\right)^2 + P \cdot \left(1 + \frac{r}{n}\right)^2\left(\frac{r}{n}\right) = P \cdot \left(1 + \frac{r}{n}\right)^2 \cdot \left(1 + \frac{r}{n}\right) = P \cdot \left(1 + \frac{r}{n}\right)^3$$

Continuing this way, after n compounding periods (1 year), the amount A is

$$A = P \cdot \left(1 + \frac{r}{n}\right)^n$$

Because t years will contain $n \cdot t$ compounding periods, after t years we have

$$A = P \cdot \left(1 + \frac{r}{n}\right)^{nt}$$

Theorem

Compound Interest Formula

The amount A after t years due to a principal P invested at an annual interest rate r compounded n times per year is

$$A = P \cdot \left(1 + \frac{r}{n}\right)^{nt} \tag{2}$$

— **Exploration** ———
To see the effects of compounding interest monthly on an initial deposit of \$1, graph $Y_1 = \left(1 + \dfrac{r}{12}\right)^{12x}$ with $r = 0.06$ and $r = 0.12$ for $0 \le x \le 30$. What is the future value of \$1 in 30 years when the interest rate per annum is $r = 0.06$ (6%)? What is the future value of \$1 in 30 years when the interest rate per annum is $r = 0.12$ (12%)? Does doubling the interest rate double the future value?

For example, to rework Example 1, we would use $P = \$1000$, $r = 0.08$, $n = 4$ (quarterly compounding), and $t = 1$ year to obtain

$$A = P \cdot \left(1 + \frac{r}{n}\right)^{nt} = 1000\left(1 + \frac{0.08}{4}\right)^{4\cdot1} = \$1082.43$$

In equation (2), the amount A is typically referred to as the **future value** of the account, while P is called the **present value**.

NOW WORK PROBLEM 3.

EXAMPLE 2

Comparing Investments Using Different Compounding Periods

Investing \$1000 at an annual rate of 10% compounded annually, semiannually, quarterly, monthly, and daily will yield the following amounts after 1 year:

Annual compounding ($n = 1$): $A = P \cdot (1 + r)$
$$= (\$1000)(1 + 0.10) = \$1100.00$$

Semiannual compounding ($n = 2$): $A = P \cdot \left(1 + \dfrac{r}{2}\right)^2$
$$= (\$1000)(1 + 0.05)^2 = \$1102.50$$

Quarterly compounding ($n = 4$): $A = P \cdot \left(1 + \dfrac{r}{4}\right)^4$
$$= (\$1000)(1 + 0.025)^4 = \$1103.81$$

Monthly compounding ($n = 12$): $A = P \cdot \left(1 + \dfrac{r}{12}\right)^{12}$
$$= (\$1000)(1 + 0.00833)^{12} = \$1104.71$$

Daily compounding ($n = 365$): $A = P \cdot \left(1 + \dfrac{r}{365}\right)^{365}$
$$= (\$1000)(1 + 0.000274)^{365} = \$1105.16 \ \blacktriangleleft$$

From Example 2 we can see that the effect of compounding more frequently is that the amount after 1 year is higher: \$1000 compounded 4 times a year at 10% results in \$1103.81; \$1000 compounded 12 times a year at 10% results in \$1104.71; and \$1000 compounded 365 times a year at 10% results in \$1105.16. This leads to the following question: What would happen to the amount after 1 year if the number of times that the interest is compounded were increased without bound?

Let's find the answer. Suppose that P is the principal, r is the per annum interest rate, and n is the number of times that the interest is compounded each year. The amount after 1 year is

$$A = P \cdot \left(1 + \frac{r}{n}\right)^n$$

Rewrite this expression as follows:

$$A = P \cdot \left(1 + \frac{r}{n}\right)^n = P \cdot \left(1 + \frac{1}{\frac{n}{r}}\right)^n = P \cdot \left[\left(1 + \frac{1}{\frac{n}{r}}\right)^{n/r}\right]^r = P \cdot \left[\left(1 + \frac{1}{h}\right)^h\right]^r \qquad \text{(3)}$$

$$\uparrow$$
$$h = \frac{n}{r}$$

Now suppose that the number n of times that the interest is compounded per year gets larger and larger; that is, suppose that $n \to \infty$. Then $h = \dfrac{n}{r} \to \infty$, and the expression in brackets equals e. [Refer to equation (2), page 426.] That is, $A \to Pe^r$.

Table 9 compares $\left(1 + \dfrac{r}{n}\right)^n$, for large values of n, to e^r for $r = 0.05$, $r = 0.10$, $r = 0.15$, and $r = 1$. The larger that n gets, the closer $\left(1 + \dfrac{r}{n}\right)^n$ gets to e^r.

No matter how frequent the compounding, the amount after 1 year has the definite ceiling Pe^r.

Table 9

	$\left(1 + \frac{r}{n}\right)^n$			
	$n = 100$	$n = 1000$	$n = 10{,}000$	e^r
$r = 0.05$	1.0512580	1.0512698	1.051271	1.0512711
$r = 0.10$	1.1051157	1.1051654	1.1051704	1.1051709
$r = 0.15$	1.1617037	1.1618212	1.1618329	1.1618342
$r = 1$	2.7048138	2.7169239	2.7181459	2.7182818

When interest is compounded so that the amount after 1 year is Pe^r, we say the interest is **compounded continuously**.

Theorem

Continuous Compounding

The amount A after t years due to a principal P invested at an annual interest rate r compounded continuously is

$$A = Pe^{rt} \qquad \text{(4)}$$

EXAMPLE 3

Using Continuous Compounding

The amount A that results from investing a principal P of $1000 at an annual rate r of 10% compounded continuously for a time t of 1 year is

$$A = \$1000e^{0.10} = (\$1000)(1.10517) = \$1105.17$$

◀

NOW WORK PROBLEM 11.

2 Calculate Effective Rates of Return

The **effective rate of interest** is the equivalent annual simple rate of interest that would yield the same amount as compounding after 1 year. For example, based on Example 3, a principal of $1000 will result in $1105.17 at a rate of 10% compounded continuously. To get this same amount using a simple rate of interest would require that interest of $1105.17 − $1000.00 = $105.17 be earned on the principal. Since $105.17 is 10.517% of $1000, a simple rate of interest of 10.517% is needed to equal 10% compounded continuously. The effective rate of interest of 10% compounded continuously is 10.517%.

Based on the results of Examples 2 and 3, we find the following comparisons:

	Annual Rate	Effective Rate
Annual compounding	10%	10%
Semiannual compounding	10%	10.25%
Quarterly compounding	10%	10.381%
Monthly compounding	10%	10.471%
Daily compounding	10%	10.516%
Continuous compounding	10%	10.517%

EXAMPLE 4	Computing the Effective Rate of Interest

On January 2, 2004, $2000 is placed in an Individual Retirement Account (IRA) that will pay interest of 7% per annum compounded continuously.

(a) What will the IRA be worth on January 1, 2024?

(b) What is the effective rate of interest?

Solution (a) The amount A after 20 years is

$$A = Pe^{rt} = \$2000e^{(0.07)(20)} = \$8110.40$$

(b) First, we compute the interest earned on $2000 at $r = 7\%$ compounded continuously for 1 year.

$$A = \$2000e^{0.07(1)}$$

$$= \$2145.02$$

So the interest earned is $2145.02 − $2000.00 = $145.02. Use the simple interest formula $I = Prt$, with $I = \$145.02$, $P = \$2000$, and $t = 1$, and solve for r, the effective rate of interest.

$$\$145.02 = \$2000 \cdot r \cdot 1$$

$$r = \frac{\$145.02}{\$2000} = 0.07251$$

The effective rate of interest is 7.251%. ◀

--- **Exploration** ---

For the IRA described in Example 4, how long will it be until $A = \$4000$? $6000?

[**Hint:** Graph $Y_1 = 2000e^{0.07x}$ and $Y_2 = 4000$. Use INTERSECT to find x.]

NOW WORK PROBLEM 23.

3 Determine the Present Value of a Lump Sum of Money

When people engaged in finance speak of the "time value of money," they are usually referring to the *present value* of money. The **present value** of A dollars to be received at a future date is the principal that you would need to invest now so that it will grow to A dollars in the specified time period. The present value of money to be received at a future date is always less than the amount to be received, since the amount to be received will equal the present value (money invested now) *plus* the interest accrued over the time period.

We use the compound interest formula (2) to get a formula for present value. If P is the present value of A dollars to be received after t years at a per annum interest rate r compounded n times per year, then, by formula (2),

$$A = P \cdot \left(1 + \frac{r}{n}\right)^{nt}$$

To solve for P, we divide both sides by $\left(1 + \frac{r}{n}\right)^{nt}$. The result is

$$\frac{A}{\left(1 + \frac{r}{n}\right)^{nt}} = P \quad \text{or} \quad P = A \cdot \left(1 + \frac{r}{n}\right)^{-nt}$$

Theorem

Present Value Formulas

The present value P of A dollars to be received after t years, assuming a per annum interest rate r compounded n times per year, is

$$P = A \cdot \left(1 + \frac{r}{n}\right)^{-nt} \tag{5}$$

If the interest is compounded continuously, then

$$P = Ae^{-rt} \tag{6}$$

To prove (6), solve formula (4) for P.

EXAMPLE 5

Computing the Value of a Zero-Coupon Bond

A zero-coupon (noninterest-bearing) bond can be redeemed in 10 years for $1000. How much should you be willing to pay for it now if you want a return of

(a) 8% compounded monthly?

(b) 7% compounded continuously?

Solution

(a) We are seeking the present value of $1000. We use formula (5) with $A = \$1000$, $n = 12$, $r = 0.08$, and $t = 10$.

$$P = A \cdot \left(1 + \frac{r}{n}\right)^{-nt} = \$1000\left(1 + \frac{0.08}{12}\right)^{-12(10)} = \$450.52$$

For a return of 8% compounded monthly, you should pay $450.52 for the bond.

(b) Here we use formula (6) with $A = \$1000$, $r = 0.07$, and $t = 10$.

$$P = Ae^{-rt} = \$1000e^{-(0.07)(10)} = \$496.59$$

For a return of 7% compounded continuously, you should pay \$496.59 for the bond. ◀

NOW WORK PROBLEM **13**.

| EXAMPLE 6 | **Rate of Interest Required to Double an Investment** |

What annual rate of interest compounded annually should you seek if you want to double your investment in 5 years?

Solution If P is the principal and we want P to double, the amount A will be $2P$. We use the compound interest formula with $n = 1$ and $t = 5$ to find r.

$$A = P \cdot \left(1 + \frac{r}{n}\right)^{nt}$$

$2P = P \cdot (1 + r)^5$ $A = 2P, n = 1, t = 5$

$2 = (1 + r)^5$ Cancel the P's.

$1 + r = \sqrt[5]{2}$ Take the fifth root of each side.

$r = \sqrt[5]{2} - 1 \approx 1.148698 - 1 = 0.148698$

The annual rate of interest needed to double the principal in 5 years is 14.87%. ◀

NOW WORK PROBLEM **25**.

---

4 Determine the Time Required to Double or Triple a Lump Sum of Money

---

| EXAMPLE 7 | **Doubling and Tripling Time for an Investment** |

(a) How long will it take for an investment to double in value if it earns 5% compounded continuously?

(b) How long will it take to triple at this rate?

Solution (a) If P is the initial investment and we want P to double, the amount A will be $2P$. We use formula (4) for continuously compounded interest with $r = 0.05$. Then

$$A = Pe^{rt}$$

$2P = Pe^{0.05t}$ $A = 2P, r = 0.05$

$2 = e^{0.05t}$ Cancel the P's.

$0.05t = \ln 2$ Rewrite as a logarithm.

$t = \dfrac{\ln 2}{0.05} \approx 13.86$ Solve for t.

It will take about 14 years to double the investment.

(b) To triple the investment, we set $A = 3P$ in formula (4).

$$A = Pe^{rt}$$
$$3P = Pe^{0.05t} \qquad A = 3P, r = 0.05$$
$$3 = e^{0.05t} \qquad \text{Cancel the P's.}$$
$$0.05t = \ln 3 \qquad \text{Rewrite as a logarithm.}$$
$$t = \frac{\ln 3}{0.05} \approx 21.97 \qquad \text{Solve for t.}$$

It will take about 22 years to triple the investment. ◄

NOW WORK PROBLEM 31.

4.7 Assess Your Understanding

'Are You Prepared?'

Answers are given at the end of these exercises. If you get a wrong answer, read the pages listed in red.

1. What is the interest due if $500 is borrowed for 6 months at a simple interest rate of 6% per annum? (pp. 142–143)

2. If you borrow $5000 and, after 9 months, pay off the loan in the amount of $5500, what per annum rate of interest was charged? (pp. 142–143)

Skill Building

In Problems 3–12, find the amount that results from each investment.

3. $100 invested at 4% compounded quarterly after a period of 2 years

4. $50 invested at 6% compounded monthly after a period of 3 years

5. $500 invested at 8% compounded quarterly after a period of $2\frac{1}{2}$ years

6. $300 invested at 12% compounded monthly after a period of $1\frac{1}{2}$ years

7. $600 invested at 5% compounded daily after a period of 3 years

8. $700 invested at 6% compounded daily after a period of 2 years

9. $10 invested at 11% compounded continuously after a period of 2 years

10. $40 invested at 7% compounded continuously after a period of 3 years

11. $100 invested at 10% compounded continuously after a period of $2\frac{1}{4}$ years

12. $100 invested at 12% compounded continuously after a period of $3\frac{3}{4}$ years

In Problems 13–22, find the principal needed now to get each amount; that is, find the present value.

13. To get $100 after 2 years at 6% compounded monthly

14. To get $75 after 3 years at 8% compounded quarterly

15. To get $1000 after $2\frac{1}{2}$ years at 6% compounded daily

16. To get $800 after $3\frac{1}{2}$ years at 7% compounded monthly

17. To get $600 after 2 years at 4% compounded quarterly

18. To get $300 after 4 years at 3% compounded daily

19. To get $80 after $3\frac{1}{4}$ years at 9% compounded continuously

20. To get $800 after $2\frac{1}{2}$ years at 8% compounded continuously

21. To get $400 after 1 year at 10% compounded continuously

22. To get $1000 after 1 year at 12% compounded continuously

23. Find the effective rate of interest for $5\frac{1}{4}$% compounded quarterly.

24. What interest rate compounded quarterly will give an effective interest rate of 7%?

25. What rate of interest compounded annually is required to double an investment in 3 years?

26. What rate of interest compounded annually is required to double an investment in 10 years?

In Problems 27–30, which of the two rates would yield the larger amount in 1 year?
[**Hint:** *Start with a principal of $10,000 in each instance.*]

27. 6% compounded quarterly or $6\frac{1}{4}$% compounded annually

28. 9% compounded quarterly or $9\frac{1}{4}$% compounded annually

29. 9% compounded monthly or 8.8% compounded daily

30. 8% compounded semiannually or 7.9% compounded daily

31. How long does it take for an investment to double in value if it is invested at 8% per annum compounded monthly? Compounded continuously?

32. How long does it take for an investment to double in value if it is invested at 10% per annum compounded monthly? Compounded continuously?

Applications and Extensions

33. If Tanisha has $100 to invest at 8% per annum compounded monthly, how long will it be before she has $150? If the compounding is continuous, how long will it be?

34. If Angela has $100 to invest at 10% per annum compounded monthly, how long will it be before she has $175? If the compounding is continuous, how long will it be?

35. How many years will it take for an initial investment of $10,000 to grow to $25,000? Assume a rate of interest of 6% compounded continuously.

36. How many years will it take for an initial investment of $25,000 to grow to $80,000? Assume a rate of interest of 7% compounded continuously.

37. What will a $90,000 house cost 5 years from now if the inflation rate over that period averages 3% compounded annually?

38. Sears charges 1.25% per month on the unpaid balance for customers with charge accounts (interest is compounded monthly). A customer charges $200 and does not pay her bill for 6 months. What is the bill at that time?

39. Jerome will be buying a used car for $15,000 in 3 years. How much money should he ask his parents for now so that, if he invests it at 5% compounded continuously, he will have enough to buy the car?

40. John will require $3000 in 6 months to pay off a loan that has no prepayment privileges. If he has the $3000 now, how much of it should he save in an account paying 3% compounded monthly so that in 6 months he will have exactly $3000?

41. George is contemplating the purchase of 100 shares of a stock selling for $15 per share. The stock pays no dividends. The history of the stock indicates that it should grow at an annual rate of 15% per year. How much will the 100 shares of stock be worth in 5 years?

42. Tracy is contemplating the purchase of 100 shares of a stock selling for $15 per share. The stock pays no dividends. Her broker says that the stock will be worth $20 per share in 2 years. What is the annual rate of return on this investment?

43. A business purchased for $650,000 in 1994 is sold in 1997 for $850,000. What is the annual rate of return for this investment?

44. Tanya has just inherited a diamond ring appraised at $5000. If diamonds have appreciated in value at an annual rate of 8%, what was the value of the ring 10 years ago when the ring was purchased?

45. Jim places $1000 in a bank account that pays 5.6% compounded continuously. After 1 year, will he have enough money to buy a computer system that costs $1060? If another bank will pay Jim 5.9% compounded monthly, is this a better deal?

46. On January 1, Kim places $1000 in a certificate of deposit that pays 6.8% compounded continuously and matures in 3 months. Then Kim places the $1000 and the interest in a passbook account that pays 5.25% compounded monthly. How much does Kim have in the passbook account on May 1?

47. Will invests $2000 in a bond trust that pays 9% interest compounded semiannually. His friend Henry invests $2000 in a certificate of deposit that pays $8\frac{1}{2}$% compounded continuously. Who has more money after 20 years, Will or Henry?

48. Suppose that April has access to an investment that will pay 10% interest compounded continuously. Which is better: To be given $1000 now so that she can take advantage of this investment opportunity or to be given $1325 after 3 years?

49. Colleen and Bill have just purchased a house for $150,000, with the seller holding a second mortgage of $50,000. They promise to pay the seller $50,000 plus all accrued interest 5 years from now. The seller offers them three interest options on the second mortgage:
(a) Simple interest at 12% per annum
(b) $11\frac{1}{2}$% interest compounded monthly
(c) $11\frac{1}{4}$% interest compounded continuously
Which option is best; that is, which results in the least interest on the loan?

50. The First National Bank advertises that it pays interest on savings accounts at the rate of 4.25% compounded daily. Find the effective rate if the bank uses (a) 360 days or (b) 365 days in determining the daily rate.

Problems 51–54 involve zero-coupon bonds. A zero-coupon bond is a bond that is sold now at a discount and will pay its face value at the time when it matures; no interest payments are made.

51. A zero-coupon bond can be redeemed in 20 years for $10,000. How much should you be willing to pay for it now if you want a return of:
(a) 10% compounded monthly?
(b) 10% compounded continuously?

52. A child's grandparents are considering buying a $40,000 face value zero-coupon bond at birth so that she will have enough money for her college education 17 years later. If they want a rate of return of 8% compounded annually, what should they pay for the bond?

53. How much should a $10,000 face value zero-coupon bond, maturing in 10 years, be sold for now if its rate of return is to be 8% compounded annually?

54. If Pat pays $12,485.52 for a $25,000 face value zero-coupon bond that matures in 8 years, what is his annual rate of return?

55. Time to Double or Triple an Investment The formula

$$t = \frac{\ln m}{n \ln\left(1 + \dfrac{r}{n}\right)}$$

can be used to find the number of years t required to multiply an investment m times when r is the per annum interest rate compounded n times a year.
(a) How many years will it take to double the value of an IRA that compounds annually at the rate of 12%?

(b) How many years will it take to triple the value of a savings account that compounds quarterly at an annual rate of 6%?
(c) Give a derivation of this formula.

56. Time to Reach an Investment Goal The formula

$$t = \frac{\ln A - \ln P}{r}$$

can be used to find the number of years t required for an investment P to grow to a value A when compounded continuously at an annual rate r.
(a) How long will it take to increase an initial investment of $1000 to $8000 at an annual rate of 10%?
(b) What annual rate is required to increase the value of a $2000 IRA to $30,000 in 35 years?
(c) Give a derivation of this formula.

Discussion and Writing

57. Explain in your own words what the term *compound interest* means. What does *continuous compounding* mean?

58. Explain in your own words the meaning of *present value*.

59. Critical Thinking You have just contracted to buy a house and will seek financing in the amount of $100,000. You go to several banks. Bank 1 will lend you $100,000 at the rate of 8.75% amortized over 30 years with a loan origination fee of 1.75%. Bank 2 will lend you $100,000 at the rate of 8.375% amortized over 15 years with a loan origination fee of 1.5%. Bank 3 will lend you $100,000 at the rate of 9.125% amortized over 30 years with no loan origination fee. Bank 4 will lend you $100,000 at the rate of 8.625% amortized over 15 years with no loan origination fee. Which loan would you take? Why? Be sure to have sound reasons for your choice.

Use the information in the table to assist you. If the amount of the monthly payment does not matter to you, which loan would you take? Again, have sound reasons for your choice. Compare your final decision with others in the class. Discuss.

	Monthly Payment	Loan Origination Fee
Bank 1	$786.70	$1,750.00
Bank 2	$977.42	$1,500.00
Bank 3	$813.63	$0.00
Bank 4	$990.68	$0.00

'Are You Prepared?' Answers

1. $15   **2.** 13.33%

4.8 Exponential Growth and Decay; Newton's Law; Logistic Growth and Decay

> **OBJECTIVES** 1 Find Equations of Populations That Obey the Law of Uninhibited Growth
> 2 Find Equations of Populations That Obey the Law of Decay
> 3 Use Newton's Law of Cooling
> 4 Use Logistic Models

1 Find Equations of Populations That Obey the Law of Uninhibited Growth

Many natural phenomena have been found to follow the law that an amount A varies with time t according to

$$A = A_0 e^{kt} \qquad \textbf{(1)}$$

Here A_0 is the original amount $(t = 0)$ and $k \neq 0$ is a constant.

If $k > 0$, then equation (1) states that the amount A is increasing over time; if $k < 0$, the amount A is decreasing over time. In either case, when an amount A varies over time according to equation (1), it is said to follow the **exponential law** or the **law of uninhibited growth** $(k > 0)$ **or decay** $(k < 0)$. See Figure 52.

Figure 52

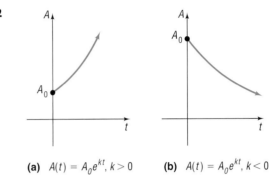

(a) $A(t) = A_0 e^{kt}, k > 0$ (b) $A(t) = A_0 e^{kt}, k < 0$

For example, we saw in Section 4.7 that continuously compounded interest follows the law of uninhibited growth. In this section we shall look at three additional phenomena that follow the exponential law.

Cell division is the growth process of many living organisms, such as amoebas, plants, and human skin cells. Based on an ideal situation in which no cells die and no by-products are produced, the number of cells present at a given time follows the law of uninhibited growth. Actually, however, after enough time has passed, growth at an exponential rate will cease due to the influence of factors such as lack of living space and dwindling food supply. The law of uninhibited growth accurately reflects only the early stages of the cell division process.

The cell division process begins with a culture containing N_0 cells. Each cell in the culture grows for a certain period of time and then divides into two identical cells. We assume that the time needed for each cell to divide in two is constant and does not change as the number of cells increases. These new cells then grow, and eventually each divides in two, and so on.

Uninhibited Growth of Cells

A model that gives the number N of cells in a culture after a time t has passed (in the early stages of growth) is

$$N(t) = N_0 e^{kt}, \qquad k > 0 \qquad \textbf{(2)}$$

where N_0 is the initial number of cells and k is a positive constant that represents the growth rate of the cells.

In using formula (2) to model the growth of cells, we are using a function that yields positive real numbers, even though we are counting the number of cells, which must be an integer. This is a common practice in many applications.

| EXAMPLE 1 | **Bacterial Growth** |

A colony of bacteria grows according to the law of uninhibited growth according to the function $N(t) = 100e^{0.045t}$, where N is measured in grams and t is measured in days.

(a) Determine the initial amount of bacteria.

(b) What is the growth rate of the bacteria?

(c) Graph the function using a graphing utility.

(d) What is the population after 5 days?

(e) How long will it take for the population to reach 140 grams?

(f) What is the doubling time for the population?

Solution

(a) The initial amount of bacteria, N_0, is obtained when $t = 0$, so

$$N_0 = N(0) = 100e^{0.045(0)} = 100 \text{ grams.}$$

(b) Compare $N(t) = 100e^{0.045t}$ to $N(t) = 100e^{kt}$. The value of k, 0.045, indicates a growth rate of 4.5%.

(c) Figure 53 shows the graph of $N(t) = 100e^{0.045t}$.

(d) The population after 5 days is $N(5) = 100e^{0.045(5)} = 125.2$ grams.

(e) To find how long it takes for the population to reach 140 grams, we solve the equation $N(t) = 140$.

Figure 53

$$100e^{0.045t} = 140$$

$$e^{0.045t} = 1.4 \qquad \text{\textit{Divide both sides of the equation by 100.}}$$

$$0.045t = \ln 1.4 \qquad \text{\textit{Rewrite as a logarithm.}}$$

$$t = \frac{\ln 1.4}{0.045} \qquad \text{\textit{Divide both sides of the equation by 0.045.}}$$

$$\approx 7.5 \text{ days}$$

(f) The population doubles when $N(t) = 200$ grams, so we find the doubling time by solving the equation $200 = 100e^{0.045t}$ for t.

$$200 = 100e^{0.045t}$$

$$2 = e^{0.045t} \qquad \text{\textit{Divide both sides of the equation by 100.}}$$

$$\ln 2 = 0.045t \qquad \text{\textit{Rewrite as a logarithm.}}$$

$$t = \frac{\ln 2}{0.045} \qquad \text{\textit{Divide both sides of the equation by 0.045.}}$$

$$\approx 15.4 \text{ days}$$

The population doubles approximately every 15.4 days. ◀

NOW WORK PROBLEM 1.

| EXAMPLE 2 | **Bacterial Growth** |

A colony of bacteria increases according to the law of uninhibited growth.

(a) If the number of bacteria doubles in 3 hours, find the function that gives the number of cells in the culture.

(b) How long will it take for the size of the colony to triple?

(c) How long will it take for the population to double a second time (that is, increase four times)?

Solution (a) Using formula (2), the number N of cells at a time t is

$$N(t) = N_0 e^{kt}$$

where N_0 is the initial number of bacteria present and k is a positive number. We first seek the number k. The number of cells doubles in 3 hours, so we have

$$N(3) = 2N_0$$

But $N(3) = N_0 e^{k(3)}$, so

$$N_0 e^{k(3)} = 2N_0$$
$$e^{3k} = 2 \qquad \text{Divide both sides by } N_0.$$
$$3k = \ln 2 \qquad \text{Write the exponential equation as a logarithm.}$$
$$k = \frac{1}{3} \ln 2$$

Formula (2) for this growth process is therefore

$$N(t) = N_0 e^{\left(\frac{1}{3}\ln 2\right)t}$$

(b) The time t needed for the size of the colony to triple requires that $N = 3N_0$. We substitute $3N_0$ for N to get

$$3N_0 = N_0 e^{\left(\frac{1}{3}\ln 2\right)t}$$
$$3 = e^{\left(\frac{1}{3}\ln 2\right)t}$$
$$\left(\frac{1}{3}\ln 2\right)t = \ln 3$$
$$t = \frac{3\ln 3}{\ln 2} \approx 4.755 \text{ hours}$$

It will take about 4.755 hours or 4 hours, 45 minutes for the size of the colony to triple.

(c) If a population doubles in 3 hours, it will double a second time in 3 more hours, for a total time of 6 hours. ◄

2 Find Equations of Populations That Obey the Law of Decay

Radioactive materials follow the law of uninhibited decay.

Uninhibited Radioactive Decay

The amount A of a radioactive material present at time t is given by

$$A(t) = A_0 e^{kt}, \qquad k < 0 \qquad \text{(3)}$$

where A_0 is the original amount of radioactive material and k is a negative number that represents the rate of decay.

All radioactive substances have a specific **half-life**, which is the time required for half of the radioactive substance to decay. In **carbon dating**, we use the fact that all living organisms contain two kinds of carbon, carbon 12 (a stable carbon) and carbon 14 (a radioactive carbon with a half-life of 5600 years). While an organism is living, the ratio of carbon 12 to carbon 14 is constant. But when an organism dies, the original amount of carbon 12 present remains unchanged, whereas the amount of carbon 14 begins to decrease. This change in the amount of carbon 14 present relative to the amount of carbon 12 present makes it possible to calculate when an organism died.

| EXAMPLE 3 | **Estimating the Age of Ancient Tools** |

Traces of burned wood along with ancient stone tools in an archeological dig in Chile were found to contain approximately 1.67% of the original amount of carbon 14.

(a) If the half-life of carbon 14 is 5600 years, approximately when was the tree cut and burned?

(b) Using a graphing utility, graph the relation between the percentage of carbon 14 remaining and time.

(c) Determine the time that elapses until half of the carbon 14 remains. This answer should equal the half-life of carbon 14.

(d) Use a graphing utility to verify the answer found in part (a).

Solution (a) Using formula (3), the amount A of carbon 14 present at time t is

$$A(t) = A_0 e^{kt}$$

where A_0 is the original amount of carbon 14 present and k is a negative number. We first seek the number k. To find it, we use the fact that after 5600 years half of the original amount of carbon 14 remains, so $A(5600) = \frac{1}{2}A_0$. Then,

$$\frac{1}{2}A_0 = A_0 e^{k(5600)}$$

$$\frac{1}{2} = e^{5600k} \qquad \text{Divide both sides of the equation by } A_0.$$

$$5600k = \ln\frac{1}{2} \qquad \text{Rewrite as a logarithm.}$$

$$k = \frac{1}{5600}\ln\frac{1}{2} \approx -0.000124$$

Formula (3) therefore becomes

$$A(t) = A_0 e^{\frac{\ln\frac{1}{2}}{5600}t}$$

If the amount A of carbon 14 now present is 1.67% of the original amount, it follows that

$$0.0167A_0 = A_0 e^{\frac{\ln\frac{1}{2}}{5600}t}$$

$$0.0167 = e^{\frac{\ln\frac{1}{2}}{5600}t} \qquad \text{Divide both sides of the equation by } A_0.$$

$$\frac{\ln\frac{1}{2}}{5600}t = \ln 0.0167 \qquad \text{Rewrite as a logarithm.}$$

$$t = \frac{5600}{\ln\frac{1}{2}}\ln 0.0167 \approx 33{,}062 \text{ years}$$

The tree was cut and burned about 33,062 years ago. Some archeologists use this conclusion to argue that humans lived in the Americas 33,000 years ago, much earlier than is generally accepted.

(b) Figure 54 shows the graph of $y = e^{\frac{\ln \frac{1}{2}}{5600}x}$, where y is the fraction of carbon 14 present and x is the time.

(c) By graphing $Y_1 = 0.5$ and $Y_2 = e^{\frac{\ln \frac{1}{2}}{5600}x}$, where x is time, and using INTERSECT, we find that it takes 5600 years until half the carbon 14 remains. The half-life of carbon 14 is 5600 years.

(d) By graphing $Y_1 = 0.0167$ and $Y_2 = e^{\frac{\ln \frac{1}{2}}{5600}x}$, where x is time, and using INTERSECT, we find that it takes 33,062 years until 1.67% of the carbon 14 remains.   ◀

Figure 54

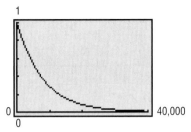

━━━━ **NOW WORK PROBLEM 3.**

3 Use Newton's Law of Cooling

Newton's Law of Cooling[*] states that the temperature of a heated object decreases exponentially over time toward the temperature of the surrounding medium.

Newton's Law of Cooling

The temperature u of a heated object at a given time t can be modeled by the following function:

$$u(t) = T + (u_0 - T)e^{kt}, \qquad k < 0 \qquad \textbf{(4)}$$

where T is the constant temperature of the surrounding medium, u_0 is the initial temperature of the heated object, and k is a negative constant.

EXAMPLE 4

Using Newton's Law of Cooling

An object is heated to 100°C (degrees Celsius) and is then allowed to cool in a room whose air temperature is 30°C.

(a) If the temperature of the object is 80°C after 5 minutes, when will its temperature be 50°C?

(b) Using a graphing utility, graph the relation found between the temperature and time.

(c) Using a graphing utility, verify that after 18.6 minutes the temperature is 50°C.

(d) Using a graphing utility, determine the elapsed time before the object is 35°C.

(e) What do you notice about the temperature as time passes?

Solution   (a) Using formula (4) with $T = 30$ and $u_0 = 100$, the temperature (in degrees Celsius) of the object at time t (in minutes) is

$$u(t) = 30 + (100 - 30)e^{kt} = 30 + 70e^{kt}$$

[*]Named after Sir Isaac Newton (1642–1727), one of the cofounders of calculus.

where k is a negative constant. To find k, we use the fact that $u = 80$ when $t = 5$. Then

$$80 = 30 + 70e^{k(5)}$$
$$50 = 70e^{5k}$$
$$e^{5k} = \frac{50}{70}$$
$$5k = \ln\frac{5}{7}$$
$$k = \frac{1}{5}\ln\frac{5}{7} \approx -0.0673$$

Formula (4) therefore becomes

$$u(t) = 30 + 70e^{\frac{\ln\frac{5}{7}}{5}t}$$

We want to find t when $u = 50°C$, so

$$50 = 30 + 70e^{\frac{\ln\frac{5}{7}}{5}t}$$
$$20 = 70e^{\frac{\ln\frac{5}{7}}{5}t}$$
$$e^{\frac{\ln\frac{5}{7}}{5}t} = \frac{20}{70}$$
$$\frac{\ln\frac{5}{7}}{5}t = \ln\frac{2}{7}$$
$$t = \frac{5}{\ln\frac{5}{7}}\ln\frac{2}{7} \approx 18.6 \text{ minutes}$$

The temperature of the object will be 50°C after about 18.6 minutes or 18 minutes, 37 seconds.

(b) Figure 55 shows the graph of $y = 30 + 70e^{\frac{\ln\frac{5}{7}}{5}x}$, where y is the temperature and x is the time.

(c) By graphing $Y_1 = 50$ and $Y_2 = 30 + 70e^{\frac{\ln\frac{5}{7}}{5}x}$, where x is time, and using INTERSECT, we find that it takes $x = 18.6$ minutes (18 minutes, 37 seconds) for the temperature to cool to 50°C.

(d) By graphing $Y_1 = 35$ and $Y_2 = 30 + 70e^{\frac{\ln\frac{5}{7}}{5}x}$, where x is time, and using INTERSECT, we find that it takes $x = 39.22$ minutes (39 minutes, 13 seconds) for the temperature to cool to 35°C.

(e) As x increases, the value of $e^{\frac{\ln\frac{5}{7}}{5}x}$ approaches zero, so the value of y, the temperature of the object, approaches 30°C, the air temperature of the room.   ◄

Figure 55

NOW WORK PROBLEM 13.

4 Use Logistic Models

The exponential growth model $A(t) = A_0e^{kt}$, $k > 0$, assumes uninhibited growth, meaning that the value of the function grows without limit. Recall that we stated that cell division could be modeled using this function, assuming that no cells die and no by-products are produced. However, cell division would eventually be limited by factors such as living space and food supply. The **logistic model** can describe situations where the growth or decay of the dependent variable is limited. The logistic model is given next.

Logistic Model

In a logistic growth model, the population P after time t obeys the equation

$$P(t) = \frac{c}{1 + ae^{-bt}} \tag{5}$$

where a, b, and c are constants with $c > 0$. The model is a growth model if $b > 0$; the model is a decay model if $b < 0$.

The number c is called the **carrying capacity** (for growth models) because the value $P(t)$ approaches c as t approaches infinity; that is, $\lim\limits_{t \to \infty} P(t) = c$. The number $|b|$ is the growth rate for $b > 0$ and the decay rate for $b < 0$. Figure 56(a) shows the graph of a typical logistic growth function, and Figure 56(b) shows the graph of a typical logistic decay function.

Figure 56

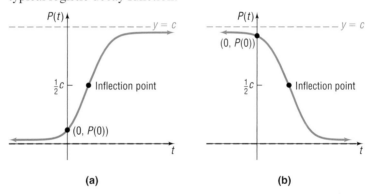

(a)   (b)

Based on the figures, we have the following properties of logistic growth functions.

Properties of the Logistic Growth Function, Equation (5)

1. The domain is the set of all real numbers. The range is the interval $(0, c)$, where c is the carrying capacity.
2. There are no x-intercepts; the y-intercept is $P(0)$.
3. There are two horizontal asymptotes: $y = 0$ and $y = c$.
4. $P(t)$ is an increasing function if $b > 0$ and a decreasing function if $b < 0$.
5. There is an **inflection point** where $P(t)$ equals $\dfrac{1}{2}$ of the carrying capacity. The inflection point is the point on the graph where the graph changes from being curved upward to curved downward for growth functions and the point where the graph changes from being curved downward to curved upward for decay functions.
6. The graph is smooth and continuous, with no corners or gaps.

| EXAMPLE 5 | Fruit Fly Population |

Fruit flies are placed in a half-pint milk bottle with a banana (for food) and yeast plants (for food and to provide a stimulus to lay eggs). Suppose that the fruit fly population after t days is given by

$$P(t) = \frac{230}{1 + 56.5e^{-0.37t}}$$

(a) State the carrying capacity and the growth rate.

(b) Determine the initial population.

(c) Use a graphing utility to graph $P(t)$.

(d) What is the population after 5 days?

(e) How long does it take for the population to reach 180?

(f) How long does it take for the population to reach one-half of the carrying capacity?

Solution

(a) As $t \to \infty$, $e^{-0.37t} \to 0$ and $P(t) \to 230/1$. The carrying capacity of the half-pint bottle is 230 fruit flies. The growth rate is $|b| = |0.37| = 37\%$.

(b) To find the initial number of fruit flies in the half-pint bottle, we evaluate $P(0)$.

$$P(0) = \frac{230}{1 + 56.5e^{-0.37(0)}}$$

$$= \frac{230}{1 + 56.5}$$

$$= 4$$

So initially there were 4 fruit flies in the half-pint bottle.

(c) See Figure 57 for the graph of $P(t)$.

Figure 57

(d) To find the number of fruit flies in the half-pint bottle after 5 days, we evaluate $P(5)$.

$$P(5) = \frac{230}{1 + 56.5e^{-0.37(5)}} \approx 23 \text{ fruit flies}$$

After 5 days, there are approximately 23 fruit flies in the bottle.

(e) To determine when the population of fruit flies will be 180, we solve the equation

$$\frac{230}{1 + 56.5e^{-0.37t}} = 180$$

$$230 = 180(1 + 56.5e^{-0.37t})$$

$$1.2778 = 1 + 56.5e^{-0.37t} \qquad \text{Divide both sides by 180.}$$

$$0.2778 = 56.5e^{-0.37t} \qquad \text{Subtract 1 from both sides.}$$

$$0.0049 = e^{-0.37t} \qquad \text{Divide both sides by 56.5.}$$

$$\ln(0.0049) = -0.37t \qquad \text{Rewrite as a logarithmic expression.}$$

$$t \approx 14.4 \text{ days} \qquad \text{Divide both sides by } -0.37.$$

It will take approximately 14.4 days (14 days, 9 hours) for the population to reach 180 fruit flies.

We could also solve this problem by graphing $Y_1 = \dfrac{230}{1 + 56.5e^{-0.37t}}$ and $Y_2 = 180$ and using INTERSECT. See Figure 58.

(f) One-half of the carrying capacity is 115 fruit flies. We solve $P(t) = 115$ by graphing $Y_1 = \dfrac{230}{1 + 56.5e^{-0.37t}}$ and $Y_2 = 115$ and using INTERSECT. See Figure 59. The population will reach one-half of the carrying capacity in about 10.9 days (10 days, 22 hours).

Figure 58

Figure 59

◀

Look back at Figure 59. Notice the point where the graph reaches 115 fruit flies (one-half of the carrying capacity): the graph changes from being curved upward to being curved downward. Using the language of calculus, we say the graph changes from increasing at an increasing rate to increasing at a decreasing rate. For any logistic growth function, when the population reaches one-half the carrying capacity, the population growth starts to slow down.

— **Exploration** —

On the same viewing rectangle, graph

$$Y_1 = \frac{500}{1 + 24e^{-0.03t}} \text{ and } Y_2 = \frac{500}{1 + 24e^{-0.08t}}.$$

What effect does the growth rate $|b|$ have on the logistic growth function?

NOW WORK PROBLEM 21.

| EXAMPLE 6 | **Wood Products** |

The EFISCEN wood product model classifies wood products according to their life-span. There are four classifications; short (1 year), medium short (4 years), medium long (16 years), and long (50 years). Based on data obtained from the European Forest Institute, the percentage of remaining wood products after t years for wood products with long life-spans (such as those used in the building industry) is given by

$$P(t) = \frac{100.3952}{1 + 0.0316e^{0.0581t}}$$

(a) What is the decay rate?

(b) Use a graphing utility to graph $P(t)$.

(c) What is the percentage of remaining wood products after 10 years?

(d) How long does it take for the percentage of remaining wood products to reach 50 percent?

(e) Explain why the numerator given in the model is reasonable.

Solution

Figure 60

(a) The decay rate is $|b| = |-0.0581| = 5.81\%$.

(b) The graph of $P(t)$ is given in Figure 60.

(c) Evaluate $P(10)$.

$$P(10) = \frac{100.3952}{1 + 0.0316e^{0.0581(10)}} \approx 95.0$$

So 95% of wood products remain after 10 years.

(d) Solve the equation $P(t) = 50$.

$$\frac{100.3952}{1 + 0.0316e^{0.0581t}} = 50$$

$$100.3952 = 50(1 + 0.0316e^{0.0581t})$$

$$2.0079 = 1 + 0.0316e^{0.0581t} \qquad \text{Divide both sides by 50.}$$

$$1.0079 = 0.0316e^{0.0581t} \qquad \text{Subtract 1 from both sides.}$$

$$31.8956 = e^{0.0581t} \qquad \text{Divide both sides by 0.0316.}$$

$$\ln(31.8956) = 0.0581t \qquad \text{Rewrite as a logarithmic expression.}$$

$$t \approx 59.6 \text{ years} \qquad \text{Divide both sides by 0.0581.}$$

It will take approximately 59.6 years for the percentage of wood products remaining to reach 50%.

(e) The numerator of 100.3952 is reasonable because the maximum percentage of wood products remaining that is possible is 100%. ◀

NOW WORK PROBLEM 27.

4.8 Assess Your Understanding

Applications and Extensions

1. **Growth of an Insect Population** The size P of a certain insect population at time t (in days) obeys the function $P(t) = 500e^{0.02t}$.
   (a) Determine the number of insects at $t = 0$ days.
   (b) What is the growth rate of the insect population?
   (c) Graph the function using a graphing utility.
   (d) What is the population after 10 days?
   (e) When will the insect population reach 800?
   (f) When will the insect population double?

2. **Growth of Bacteria** The number N of bacteria present in a culture at time t (in hours) obeys the law of uninhibited growth $N(t) = 1000e^{0.01t}$.
   (a) Determine the number of bacteria at $t = 0$ hours.
   (b) What is the growth rate of the bacteria?
   (c) Graph the function using a graphing utility.
   (d) What is the population after 4 hours?
   (e) When will the number of bacteria reach 1700?
   (f) When will the number of bacteria double?

3. **Radioactive Decay** Strontium 90 is a radioactive material that decays according to the function $A(t) = A_0 e^{-0.0244t}$, where A_0 is the initial amount present and A is the amount present at time t (in years). Assume that a scientist has a sample of 500 grams of strontium 90.
   (a) What is the decay rate of strontium 90?

   (b) Graph the function using a graphing utility.
   (c) How much strontium 90 is left after 10 years?
   (d) When will 400 grams of strontium 90 be left?
   (e) What is the half-life of strontium 90?

4. **Radioactive Decay** Iodine 131 is a radioactive material that decays according to the function $A(t) = A_0 e^{-0.087t}$, where A_0 is the initial amount present and A is the amount present at time t (in days). Assume that a scientist has a sample of 100 grams of iodine 131.
   (a) What is the decay rate of iodine 131?
   (b) Graph the function using a graphing utility.
   (c) How much iodine 131 is left after 9 days?
   (d) When will 70 grams of iodine 131 be left?
   (e) What is the half-life of iodine 131?

5. **Growth of a Colony of Mosquitoes** The population of a colony of mosquitoes obeys the law of uninhibited growth. If there are 1000 mosquitoes initially and there are 1800 after 1 day, what is the size of the colony after 3 days? How long is it until there are 10,000 mosquitoes?

6. **Bacterial Growth** A culture of bacteria obeys the law of uninhibited growth. If 500 bacteria are present initially and there are 800 after 1 hour, how many will be present in the culture after 5 hours? How long is it until there are 20,000 bacteria?

7. **Population Growth** The population of a southern city follows the exponential law. If the population doubled in size over an 18-month period and the current population is 10,000, what will the population be 2 years from now?

8. **Population Decline** The population of a midwestern city follows the exponential law. If the population decreased from 900,000 to 800,000 from 2003 to 2005, what will the population be in 2007?

9. **Radioactive Decay** The half-life of radium is 1690 years. If 10 grams are present now, how much will be present in 50 years?

10. **Radioactive Decay** The half-life of radioactive potassium is 1.3 billion years. If 10 grams are present now, how much will be present in 100 years? In 1000 years?

11. **Estimating the Age of a Tree** A piece of charcoal is found to contain 30% of the carbon 14 that it originally had.
    (a) When did the tree from which the charcoal came die? Use 5600 years as the half-life of carbon 14.
    (b) Using a graphing utility, graph the relation between the percentage of carbon 14 remaining and time.
    (c) Using INTERSECT, determine the time that elapses until half of the carbon 14 remains.
    (d) Verify the answer found in part (a).

12. **Estimating the Age of a Fossil** A fossilized leaf contains 70% of its normal amount of carbon 14.
    (a) How old is the fossil?
    (b) Using a graphing utility, graph the relation between the percentage of carbon 14 remaining and time.
    (c) Using INTERSECT, determine the time that elapses until half of the carbon 14 remains.
    (d) Verify the answer found in part (a).

13. **Cooling Time of a Pizza** A pizza baked at 450°F is removed from the oven at 5:00 PM into a room that is a constant 70°F. After 5 minutes, the pizza is at 300°F.
    (a) At what time can you begin eating the pizza if you want its temperature to be 135°F?
    (b) Using a graphing utility, graph the relation between temperature and time.
    (c) Using INTERSECT, determine the time that needs to elapse before the pizza is 160°F.
    (d) TRACE the function for large values of time. What do you notice about y, the temperature?

14. **Newton's Law of Cooling** A thermometer reading 72°F is placed in a refrigerator where the temperature is a constant 38°F.
    (a) If the thermometer reads 60°F after 2 minutes, what will it read after 7 minutes?

(b) How long will it take before the thermometer reads 39°F?
(c) Using a graphing utility, graph the relation between temperature and time.
(d) Using INTERSECT, determine the time needed to elapse before the thermometer reads 45°F.
(e) TRACE the function for large values of time. What do you notice about y, the temperature?

15. **Newton's Law of Heating** A thermometer reading 8°C is brought into a room with a constant temperature of 35°C.
    (a) If the thermometer reads 15°C after 3 minutes, what will it read after being in the room for 5 minutes? For 10 minutes?
    (b) Graph the relation between temperature and time. TRACE to verify that your answers are correct.

    [**Hint:** You need to construct a formula similar to equation (4).]

16. **Thawing Time of a Steak** A frozen steak has a temperature of 28°F. It is placed in a room with a constant temperature of 70°F. After 10 minutes, the temperature of the steak has risen to 35°F. What will the temperature of the steak be after 30 minutes? How long will it take the steak to thaw to a temperature of 45°F? [See the hint given for Problem 15.] Graph the relation between temperature and time. TRACE to verify that your answer is correct.

17. **Decomposition of Salt in Water** Salt (NaCl) decomposes in water into sodium (NA^+) and chloride (Cl^-) ions according to the law of uninhibited decay. If the initial amount of salt is 25 kilograms and, after 10 hours, 15 kilograms of salt is left, how much salt is left after 1 day? How long does it take until $\frac{1}{2}$ kilogram of salt is left?

18. **Voltage of a Conductor** The voltage of a certain conductor decreases over time according to the law of uninhibited decay. If the initial voltage is 40 volts, and 2 seconds later it is 10 volts, what is the voltage after 5 seconds?

19. **Radioactivity from Chernobyl** After the release of radioactive material into the atmosphere from a nuclear power plant at Chernobyl (Ukraine) in 1986, the hay in Austria was contaminated by iodine 131 (half-life 8 days). If it is all right to feed the hay to cows when 10% of the iodine 131 remains, how long do the farmers need to wait to use this hay?

20. **Pig Roasts** The hotel Bora-Bora is having a pig roast. At noon, the chef put the pig in a large earthen oven. The pig's original temperature was 75°F. At 2:00 PM the chef checked the pig's temperature and was upset because it had reached only 100°F. If the oven's temperature remains a constant 325°F, at what time may the hotel serve its guests, assuming that pork is done when it reaches 175°F?

21. **Proportion of the Population That Owns a VCR** The logistic growth model

$$P(t) = \frac{0.9}{1 + 6e^{-0.32t}}$$

relates the proportion of U.S. households that own a VCR to the year. Let $t = 0$ represent 1984, $t = 1$ represent 1985, and so on.

(a) Determine the maximum proportion of households that will own a VCR.

(b) What proportion of households owned a VCR in 1984 $(t = 0)$?

(c) Use a graphing utility to graph $P(t)$.

(d) What proportion of households owned a VCR in 1999 $(t = 15)$?

(e) When will 0.8 (80%) of U.S. households own a VCR?

(f) How long will it be before 0.45 (45%) of the population owns a VCR?

22. **Market Penetration of Intel's Coprocessor** The logistic growth model

$$P(t) = \frac{0.90}{1 + 3.5e^{-0.339t}}$$

relates the proportion of new personal computers (PCs) sold at Best Buy that have Intel's latest coprocessor t months after it has been introduced.

(a) Determine the maximum proportion of PCs sold at Best Buy that will have Intel's latest coprocessor.

(b) What proportion of computers sold at Best Buy will have Intel's latest coprocessor when it is first introduced $(t = 0)$?

(c) Use a graphing utility to graph $P(t)$.

(d) What proportion of PCs sold will have Intel's latest coprocessor $t = 4$ months after it is introduced?

(e) When will 0.75 (75%) of PCs sold at Best Buy have Intel's latest coprocessor?

(f) How long will it be before 0.45 (45%) of the PCs sold by Best Buy have Intel's latest coprocessor?

23. **Population of a Bacteria Culture** The logistic growth model

$$P(t) = \frac{1000}{1 + 32.33e^{-0.439t}}$$

represents the population (in grams) of a bacterium after t hours.

(a) Determine the carrying capacity of the environment.

(b) What is the growth rate of the bacteria?

(c) Determine the initial population size.

(d) Use a graphing utility to graph $P(t)$.

(e) What is the population after 9 hours?

(f) When will the population be 700 grams?

(g) How long does it take for the population to reach one-half of the carrying capacity?

24. **Population of an Endangered Species** Often environmentalists capture an endangered species and transport the species to a controlled environment where the species can produce offspring and regenerate its population. Suppose that six American bald eagles are captured, transported to Montana, and set free. Based on experience, the environmentalists expect the population to grow according to the model

$$P(t) = \frac{500}{1 + 83.33e^{-0.162t}}$$

where t is measured in years.

(a) Determine the carrying capacity of the environment.

(b) What is the growth rate of the bald eagle?

(c) Use a graphing utility to graph $P(t)$.

(d) What is the population after 3 years?

(e) When will the population be 300 eagles?

(f) How long does it take for the population to reach one-half of the carrying capacity?

25. **The *Challenger* Disaster** After the *Challenger* disaster in 1986, a study was made of the 23 launches that preceded the fatal flight. A mathematical model was developed involving the relationship between the Fahrenheit temperature x around the O-rings and the number y of eroded or leaky primary O-rings. The model stated that

$$y = \frac{6}{1 + e^{-(5.085 - 0.1156x)}}$$

where the number 6 indicates the 6 primary O-rings on the spacecraft.

(a) What is the predicted number of eroded or leaky primary O-rings at a temperature of 100°F?

(b) What is the predicted number of eroded or leaky primary O-rings at a temperature of 60°F?

(c) What is the predicted number of eroded or leaky primary O-rings at a temperature of 30°F?

(d) Graph the equation. At what temperature is the predicted number of eroded or leaky O-rings 1? 3? 5?

SOURCE: Linda Tappin, "Analyzing Data Relating to the *Challenger* Disaster," *Mathematics Teacher*, Vol. 87, No. 6, September 1994, pp. 423–426.

26. Word Users   According to a survey by Olsten Staffing Services, the percentage of companies reporting usage of Microsoft Word t years since 1984 is given by

$$P(t) = \frac{99.744}{1 + 3.014e^{-0.799t}}$$

(a) What is the growth rate in the percentage of Microsoft Word users?

(b) Use a graphing utility to graph $P(t)$.

(c) What was the percentage of Microsoft Word users in 1990?

(d) During what year did the percentage of Microsoft Word users reach 90%?

(e) Explain why the numerator given in the model is reasonable? What does it imply?

27. Home Computers   The logistic model

$$P(t) = \frac{95.4993}{1 + 0.0405e^{0.1968t}}$$

represents the percentage of households that do not own a personal computer t years since 1984.

(a) Evaluate and interpret $P(0)$.

(b) Use a graphing utility to graph $P(t)$.

(c) What percentage of households did not own a personal computer in 1995?

(d) In what year will the percentage of households that do not own a personal computer reach 20%?

SOURCE: U.S. Department of Commerce

28. Farmers   The logistic model

$$W(t) = \frac{14{,}656.248}{1 + 0.059e^{0.057t}}$$

represents the number of farm workers in the United States t years after 1910.

(a) Evaluate and interpret $W(0)$.

(b) Use a graphing utility to graph $W(t)$.

(c) How many farm workers were there in the United States in 1990?

(d) When did the number of farm workers in the United States reach 10,000?

(e) According to this model, what happens to the number of farm workers in the United States as t approaches ∞? Based on this result, do you think that it is reasonable to use this model to predict the number of farm workers in the United States in 2060? Why?

SOURCE: U.S. Department of Agriculture

29. Birthdays   The logistic model

$$P(n) = \frac{113.3198}{1 + 0.115e^{0.0912n}}$$

models the probability that, in a room of n people, no two people share the same birthday.

(a) Use a graphing utility to graph $P(n)$.

(b) In a room of $n = 15$ people, what is the probability that no two share the same birthday?

(c) How many people must be in a room before the probability that no two people share the same birthday is 10%?

(d) What happens to the probability as n increases? Explain what this result means.

30. Bank Failures   The logistic model

$$F(t) = \frac{130.118}{1 + 0.00022e^{3.042t}}$$

models the number of bank failures in the United States t years since 1991.

(a) Evaluate and interpret $F(0)$.

(b) Use a graphing utility to graph $F(t)$.

(c) How many bank failures were there in 1995 ($t = 4$)?

(d) In what year did the number of bank failures reach 10?

(e) What does the model imply will happen to the number of bank failures as time passes?

SOURCE: Federal Deposit Insurance Corporation

4.9   Building Exponential, Logarithmic, and Logistic Models from Data

PREPARING FOR THIS SECTION   *Before getting started, review the following:*

• Scatter Diagrams; Linear Curve Fitting (Section 2.5, pp. 244–248)   • Quadratic Models (Section 3.1, pp. 310–311)

OBJECTIVES   **1** Use a Graphing Utility to Fit an Exponential Function to Data
2 Use a Graphing Utility to Fit a Logarithmic Function to Data
3 Use a Graphing Utility to Fit a Logistic Function to Data

In Section 2.5 we discussed how to find the linear function of best fit $(y = ax + b)$, and in Section 3.1 we discussed how to find the quadratic function of best fit $(y = ax^2 + bx + c)$.

In this section we will discuss how to use a graphing utility to find equations of best fit that describe the relation between two variables when the relation is thought to be exponential $(y = ab^x)$, logarithmic $(y = a + b \ln x)$, or logistic $\left(y = \dfrac{c}{1 + ae^{-bx}}\right)$. As before, we draw a scatter diagram of the data to help to determine the appropriate model to use.

Figure 61 shows scatter diagrams that will typically be observed for the three models. Below each scatter diagram are any restrictions on the values of the parameters.

Figure 61

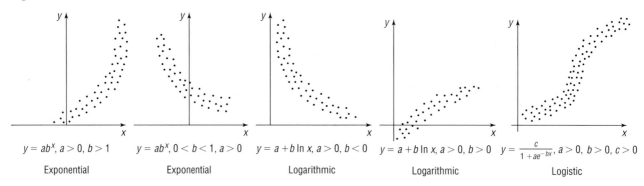

$y = ab^x, a > 0, b > 1$
Exponential

$y = ab^x, 0 < b < 1, a > 0$
Exponential

$y = a + b \ln x, a > 0, b < 0$
Logarithmic

$y = a + b \ln x, a > 0, b > 0$
Logarithmic

$y = \dfrac{c}{1 + ae^{-bx}}, a > 0, b > 0, c > 0$
Logistic

Most graphing utilities have REGression options that fit data to a specific type of curve. Once the data have been entered and a scatter diagram obtained, the type of curve that you want to fit to the data is selected. Then that REGression option is used to obtain the curve of *best fit* of the type selected.

The correlation coefficient r will appear only if the model can be written as a linear expression. As it turns out, r will appear for the linear, power, exponential, and logarithmic models, since these models can be written as a linear expression. Remember, the closer $|r|$ is to 1, the better the fit.

Let's look at some examples.

1 ## Use a Graphing Utility to Fit an Exponential Function to Data

We saw in Section 4.7 that the future value of money behaves exponentially, and we saw in Section 4.8 that growth and decay models also behave exponentially. The next example shows how data can lead to an exponential model.

| EXAMPLE 1 | **Fitting an Exponential Function to Data** |

Beth is interested in finding a function that explains the closing price of Harley Davidson stock at the end of each year. She obtains the data shown in Table 10.

(a) Using a graphing utility, draw a scatter diagram with year as the independent variable.

(b) Using a graphing utility, fit an exponential function to the data.

(c) Express the function found in part (b) in the form $A = A_0 e^{kt}$.

(d) Graph the exponential function found in part (b) or (c) on the scatter diagram.

(e) Using the solution to part (b) or (c), predict the closing price of Harley Davidson stock at the end of 2004.

(f) Interpret the value of k found in part (c).

Table 10

Year, x	Closing Price, y
1987 (x = 1)	0.392
1988 (x = 2)	0.7652
1989 (x = 3)	1.1835
1990 (x = 4)	1.1609
1991 (x = 5)	2.6988
1992 (x = 6)	4.5381
1993 (x = 7)	5.3379
1994 (x = 8)	6.8032
1995 (x = 9)	7.0328
1996 (x = 10)	11.5585
1997 (x = 11)	13.4799
1998 (x = 12)	23.5424
1999 (x = 13)	31.9342
2000 (x = 14)	39.7277
2001 (x = 15)	54.31
2002 (x = 16)	46.20
2003 (x = 17)	47.53

SOURCE: *http://finance.yahoo.com*

Solution

(a) Enter the data into the graphing utility, letting 1 represent 1987, 2 represent 1988, and so on. We obtain the scatter diagram shown in Figure 62.

(b) A graphing utility fits the data in Figure 62 to an exponential function of the form $y = ab^x$ by using the EXPonential REGression option. From Figure 63 we find $y = ab^x = 0.47547(1.3582)^x$.

Figure 62

Figure 63

ExpReg
y=a*b^x
a=.4754667252
b=1.358201358
r²=.9660173956
r=.9828618395

(c) To express $y = ab^x$ in the form $A = A_0 e^{kt}$, where $x = t$ and $y = A$, we proceed as follows:

$$ab^x = A_0 e^{kt}, \qquad x = t$$

When $x = t = 0$, we find $a = A_0$. This leads to

$$a = A_0, \qquad b^x = e^{kt}$$
$$b^x = (e^k)^t$$
$$b = e^k \qquad x = t$$

Since $y = ab^x = 0.47547(1.3582)^x$, we find that $a = 0.47547$ and $b = 1.3582$:

$$a = A_0 = 0.47547 \quad \text{and} \quad b = 1.3582 = e^k$$

We want to find k, so we rewrite $1.3582 = e^k$ as a logarithm and obtain

$$k = \ln(1.3582) \approx 0.3062$$

As a result, $A = A_0 e^{kt} = 0.47547 e^{0.3062t}$.

(d) See Figure 64 for the graph of the exponential function of best fit.

(e) Let $t = 18$ (end of 2004) in the function found in part (c). The predicted closing price of Harley Davidson stock at the end of 2004 is

$$A = 0.47547 e^{0.3062(18)} = \$117.70$$

(f) The value of k represents the annual interest rate compounded continuously.

$$A = A_0 e^{kt} = 0.47547 e^{0.3062t}$$
$$= Pe^{rt} \qquad \text{Equation (4), Section 4.7}$$

The price of Harley Davidson stock has grown at an annual rate of 30.62% (compounded continuously) between 1987 and 2003.   ◀

Figure 64

━━━━ **NOW WORK PROBLEM 1.**

2 Use a Graphing Utility to Fit a Logarithmic Function to Data

Many relations between variables do not follow an exponential model; instead, the independent variable is related to the dependent variable using a logarithmic model.

| EXAMPLE 2 | Fitting a Logarithmic Function to Data |

Jodi, a meteorologist, is interested in finding a function that explains the relation between the height of a weather balloon (in kilometers) and the atmospheric pressure (measured in millimeters of mercury) on the balloon. She collects the data shown in Table 11.

Table 11

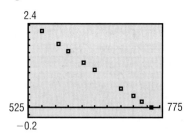

Atmospheric Pressure, p	Height, h
760	0
740	0.184
725	0.328
700	0.565
650	1.079
630	1.291
600	1.634
580	1.862
550	2.235

(a) Using a graphing utility, draw a scatter diagram of the data with atmospheric pressure as the independent variable.

(b) It is known that the relation between atmospheric pressure and height follows a logarithmic model. Using a graphing utility, fit a logarithmic function to the data.

(c) Draw the logarithmic function found in part (b) on the scatter diagram.

(d) Use the function found in part (b) to predict the height of the weather balloon if the atmospheric pressure is 560 millimeters of mercury.

Solution

(a) After entering the data into the graphing utility, we obtain the scatter diagram shown in Figure 65.

(b) A graphing utility fits the data in Figure 65 to a logarithmic function of the form $y = a + b \ln x$ by using the Logarithm REGression option. See Figure 66. The logarithmic function of best fit to the data is

$$h(p) = 45.7863 - 6.9025 \ln p$$

where h is the height of the weather balloon and p is the atmospheric pressure. Notice that $|r|$ is close to 1, indicating a good fit.

(c) Figure 67 shows the graph of $h(p) = 45.7863 - 6.9025 \ln p$ on the scatter diagram.

Figure 65

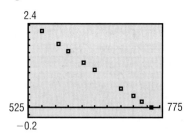

2.4

525 ┄┄┄┄┄┄ 775

−0.2

Figure 66

LnReg
y=a+blnx
a=45.78632064
b=-6.902524299
r=-.9999946336

Figure 67

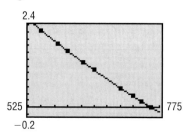

2.4

525 ┄┄┄┄┄┄ 775

−0.2

(d) Using the function found in part (b), Jodi predicts the height of the weather balloon when the atmospheric pressure is 560 to be

$$h(560) = 45.7863 - 6.9025 \ln 560$$

$$\approx 2.108 \text{ kilometers}$$ ◀

✏━━━ **NOW WORK PROBLEM 7.**

3 Use a Graphing Utility to Fit a Logistic Function to Data

Logistic growth models can be used to model situations for which the value of the dependent variable is limited. Many real-world situations conform to this scenario. For example, the population of the human race is limited by the availability of natural resources such as food and shelter. When the value of the dependent variable is limited, a logistic growth model is often appropriate.

EXAMPLE 3	**Fitting a Logistic Function to Data**

The data in Table 12 represent the amount of yeast biomass in a culture after t hours.

Table 12

Time (in hours)	Yeast Biomass	Time (in hours)	Yeast Biomass
0	9.6	10	513.3
1	18.3	11	559.7
2	29.0	12	594.8
3	47.2	13	629.4
4	71.1	14	640.8
5	119.1	15	651.1
6	174.6	16	655.9
7	257.3	17	659.6
8	350.7	18	661.8
9	441.0		

SOURCE: Tor Carlson (Über Geschwindigkeit und Grösse der Hefevermehrung in Würze, Biochemische Zeitschrift, Bd. 57, pp. 313–334, 1913)

(a) Using a graphing utility, draw a scatter diagram of the data with time as the independent variable.

(b) Using a graphing utility, fit a logistic function to the data.

(c) Using a graphing utility, graph the function found in part (b) on the scatter diagram.

(d) What is the predicted carrying capacity of the culture?

(e) Use the function found in part (b) to predict the population of the culture at $t = 19$ hours.

Solution (a) See Figure 68 for a scatter diagram of the data.

(b) A graphing utility fits a logistic growth model of the form $y = \dfrac{c}{1 + ae^{-bx}}$ by using the LOGISTIC regression option. See Figure 69. The logistic function of best fit to the data is

$$y = \frac{663.0}{1 + 71.6e^{-0.5470x}}$$

where y is the amount of yeast biomass in the culture and x is the time.

Figure 68

Figure 69

Figure 70

(c) See Figure 70 for the graph of the logistic function of best fit.

(d) Based on the logistic growth function found in part (b), the carrying capacity of the culture is 663.

(e) Using the logistic growth function found in part (b), the predicted amount of yeast biomass at $t = 19$ hours is

$$y = \frac{663.0}{1 + 71.6e^{-0.5470(19)}} = 661.5$$

◀

NOW WORK PROBLEM 9.

4.9 Assess Your Understanding

Applications and Extensions

1. **Biology** A strain of E-coli Beu 397-recA441 is placed into a nutrient broth at 30°Celsius and allowed to grow. The following data are collected. Theory states that the number of bacteria in the petri dish will initially grow according to the law of uninhibited growth. The population is measured using an optical device in which the amount of light that passes through the petri dish is measured.

Time (hours), x	Population, y
0	0.09
2.5	0.18
3.5	0.26
4.5	0.35
6	0.50

SOURCE: Dr. Polly Lavery, Joliet Junior College

(a) Draw a scatter diagram treating time as the predictor variable.
(b) Using a graphing utility, fit an exponential function to the data.
(c) Express the function found in part (b) in the form $N(t) = N_0 e^{kt}$.
(d) Graph the exponential function found in part (b) or (c) on the scatter diagram.
(e) Use the exponential function from part (b) or (c) to predict the population at $x = 7$ hours.
(f) Use the exponential function from part (b) or (c) to predict when the population will reach 0.75.

2. **Biology** A strain of E-coli SC18del-recA718 is placed into a nutrient broth at 30°Celsius and allowed to grow. The following data are collected. Theory states that the number of bacteria in the petri dish will initially grow according to the law of uninhibited growth. The population is measured using an optical device in which the amount of light that passes through the petri dish is measured.

Time (hours), x	Population, y
2.5	0.175
3.5	0.38
4.5	0.63
4.75	0.76
5.25	1.20

SOURCE: Dr. Polly Lavery, Joliet Junior College

(a) Draw a scatter diagram treating time as the predictor variable.
(b) Using a graphing utility, fit an exponential function to the data.
(c) Express the function found in part (b) in the form $N(t) = N_0 e^{kt}$.
(d) Graph the exponential function found in part (b) or (c) on the scatter diagram.
(e) Use the exponential function from part (b) or (c) to predict the population at $x = 6$ hours.
(f) Use the exponential function from part (b) or (c) to predict when the population will reach 2.1.

3. **Chemistry**   A chemist has a 100-gram sample of a radioactive material. He records the amount of radioactive material every week for 6 weeks and obtains the following data:

Week	Weight (in Grams)
0	100.0
1	88.3
2	75.9
3	69.4
4	59.1
5	51.8
6	45.5

(a) Using a graphing utility, draw a scatter diagram with week as the independent variable.
(b) Using a graphing utility, fit an exponential function to the data.
(c) Express the function found in part (b) in the form $A(t) = A_0 e^{kt}$.
(d) Graph the exponential function found in part (b) or (c) on the scatter diagram.
(e) From the result found in part (b), determine the half-life of the radioactive material.
(f) How much radioactive material will be left after 50 weeks?
(g) When will there be 20 grams of radioactive material?

4. **Chemistry**   A chemist has a 1000-gram sample of a radioactive material. She records the amount of radioactive material remaining in the sample every day for a week and obtains the following data:

Day	Weight (in Grams)
0	1000.0
1	897.1
2	802.5
3	719.8
4	651.1
5	583.4
6	521.7
7	468.3

(a) Using a graphing utility, draw a scatter diagram with day as the independent variable.
(b) Using a graphing utility, fit an exponential function to the data.
(c) Express the function found in part (b) in the form $A(t) = A_0 e^{kt}$.
(d) Graph the exponential function found in part (b) or (c) on the scatter diagram.

(e) From the result found in part (b), find the half-life of the radioactive material.
(f) How much radioactive material will be left after 20 days?
(g) When will there be 200 grams of radioactive material?

5. **Finance**   The following data represent the amount of money an investor has in an investment account each year for 10 years. She wishes to determine the average annual rate of return on her investment.

Year	Value of Account
1994	$10,000
1995	$10,573
1996	$11,260
1997	$11,733
1998	$12,424
1999	$13,269
2000	$13,968
2001	$14,823
2002	$15,297
2003	$16,539

(a) Using a graphing utility, draw a scatter diagram with time as the independent variable and the value of the account as the dependent variable.
(b) Using a graphing utility, fit an exponential function to the data.
(c) Based on the answer in part (b), what was the average annual rate of return from this account over the past 10 years?
(d) If the investor plans on retiring in 2021, what will the predicted value of this account be?
(e) When will the account be worth $50,000?

6. **Finance**   The following data show the amount of money an investor has in an investment account each year for 7 years. He wishes to determine the average annual rate of return on his investment.

Year	Value of Account
1997	$20,000
1998	$21,516
1999	$23,355
2000	$24,885
2001	$27,434
2002	$30,053
2003	$32,622

(a) Using a graphing utility, draw a scatter diagram with time as the independent variable and the value of the account as the dependent variable.

(b) Using a graphing utility, fit an exponential function to the data.

(c) Based on the answer to part (b), what was the average annual rate of return from this account over the past 7 years?

(d) If the investor plans on retiring in 2020, what will the predicted value of his account be?

(e) When will the account be worth $80,000?

7. **Economics and Marketing** The following data represent the price and quantity demanded in 2005 for IBM personal computers.

Price ($/Computer)	Quantity Demanded
2300	152
2000	159
1700	164
1500	171
1300	176
1200	180
1000	189

(a) Using a graphing utility, draw a scatter diagram of the data with price as the dependent variable.

(b) Using a graphing utility, fit a logarithmic function to the data.

(c) Using a graphing utility, draw the logarithmic function found in part (b) on the scatter diagram.

(d) Use the function found in part (b) to predict the number of IBM personal computers that would be demanded if the price were $1650.

8. **Economics and Marketing** The following data represent the price and quantity supplied in 2005 for IBM personal computers.

Price ($/Computer)	Quantity Supplied
2300	180
2000	173
1700	160
1500	150
1300	137
1200	130
1000	113

(a) Using a graphing utility, draw a scatter diagram of the data with price as the dependent variable.

(b) Using a graphing utility, fit a logarithmic function to the data.

(c) Using a graphing utility, draw the logarithmic function found in part (b) on the scatter diagram.

(d) Use the function found in part (b) to predict the number of IBM personal computers that would be supplied if the price were $1650.

9. **Population Model** The following data represent the population of the United States. An ecologist is interested in finding a function that describes the population of the United States.

Year	Population
1900	76,212,168
1910	92,228,496
1920	106,021,537
1930	123,202,624
1940	132,164,569
1950	151,325,798
1960	179,323,175
1970	203,302,031
1980	226,542,203
1990	248,709,873
2000	281,421,906

Source: U.S. Census Bureau

(a) Using a graphing utility, draw a scatter diagram of the data using the year as the independent variable and population as the dependent variable.

(b) Using a graphing utility, fit a logistic function to the data.

(c) Using a graphing utility, draw the function found in part (b) on the scatter diagram.

(d) Based on the function found in part (b), what is the carrying capacity of the United States?

(e) Use the function found in part (b) to predict the population of the United States in 2004.

(f) When will the United States population be 300,000,000?

(g) Compare actual U.S. Census figures to the prediction found in part (e).

10. **Population Model** The following data represent the world population. An ecologist is interested in finding a function that describes the world population.

Year	Population (in Billions)
1993	5.531
1994	5.611
1995	5.691
1996	5.769
1997	5.847
1998	5.925
1999	6.003
2000	6.080
2001	6.157

Source: U.S. Census Bureau

(a) Using a graphing utility, draw a scatter diagram of the data using year as the independent variable and population as the dependent variable.

(b) Using a graphing utility, fit a logistic function to the data.

(c) Using a graphing utility, draw the function found in part (b) on the scatter diagram.

(d) Based on the function found in part (b), what is the carrying capacity of the world?

(e) Use the function found in part (b) to predict the population of the world in 2004.

(f) When will world population be 7 billion?

(g) Compare actual U.S. Census figures to the prediction found in part (e).

11. Population Model The following data represent the population of Illinois. An urban economist is interested in finding a model that describes the population of Illinois.

Year	Population
1900	4,821,550
1910	5,638,591
1920	6,485,280
1930	7,630,654
1940	7,897,241
1950	8,712,176
1960	10,081,158
1970	11,110,285
1980	11,427,409
1990	11,430,602
2000	12,419,293

SOURCE: U.S. Census Bureau

(a) Using a graphing utility, draw a scatter diagram of the data using year as the independent variable and population as the dependent variable.

(b) Using a graphing utility, fit a logistic function to the data.

(c) Using a graphing utility, draw the function found in part (b) on the scatter diagram.

(d) Based on the function found in part (b), what is the carrying capacity of Illinois?

(e) Use the function found in part (b) to predict the population of Illinois in 2010.

12. Population Model The following data represent the population of Pennsylvania. An urban economist is interested in finding a model that describes the population of Pennsylvania.

(a) Using a graphing utility, draw a scatter diagram of the data using year as the independent variable and population as the dependent variable.

(b) Using a graphing utility, fit a logistic function to the data.

(c) Using a graphing utility, draw the function found in part (b) on the scatter diagram.

(d) Based on the function found in part (b), what is the carrying capacity of Pennsylvania?

(e) Use the function found in part (b) to predict the population of Pennsylvania in 2010.

Year	Population
1900	6,302,115
1910	7,665,111
1920	8,720,017
1930	9,631,350
1940	9,900,180
1950	10,498,012
1960	11,319,366
1970	11,800,766
1980	11,864,720
1990	11,881,643
2000	12,281,054

SOURCE: U.S. Census Bureau

Chapter Review

Things to Know

Composite Function (p. 397)	$(f \circ g)(x) = f(g(x))$
One-to-one function f (p. 405)	A function whose inverse is also a function For any choice of elements x_1, x_2 in the domain of f, if $x_1 \neq x_2$, then $f(x_1) \neq f(x_2)$.
Horizontal-line test (p. 406)	If every horizontal line intersects the graph of a function f in at most one point, then f is one-to-one.
Inverse function f^{-1} of f (pp. 407–411)	Domain of f = Range of f^{-1}; Range of f = Domain of f^{-1} $f^{-1}(f(x)) = x$ for all x in the domain of f and $f(f^{-1}(x)) = x$ for all x in the domain of f^{-1}. Graphs of f and f^{-1} are symmetric with respect to the line $y = x$.

Properties of the exponential function (pp. 424 and 425)	$f(x) = a^x,\quad a > 1$	Domain: the interval $(-\infty, \infty)$
		Range: the interval $(0, \infty)$
		x-intercepts: none; y-intercept: 1
		Horizontal asymptote: x-axis $(y = 0)$ as $x \to -\infty$
		Increasing; one-to-one; smooth; continuous
		See Figure 24 for a typical graph.
	$f(x) = a^x,\quad 0 < a < 1$	Domain: the interval $(-\infty, \infty)$
		Range: the interval $(0, \infty)$
		x-intercepts: none; y-intercept: 1
		Horizontal asymptote: x-axis $(y = 0)$ as $x \to \infty$
		Decreasing; one-to-one; smooth; continuous
		See Figure 28 for a typical graph.

Number e (p. 426)   Value approached by the expression $\left(1 + \dfrac{1}{n}\right)^n$ as $n \to \infty$; that is, $\displaystyle\lim_{n\to\infty}\left(1 + \dfrac{1}{n}\right)^n = e$

Property of exponents (p. 428)   If $a^u = a^v$, then $u = v$.

Properties of the logarithmic functions (pp. 435–439)	$f(x) = \log_a x,\quad a > 1$	Domain: the interval $(0, \infty)$
	($y = \log_a x$ means $x = a^y$)	Range: the interval $(-\infty, \infty)$
		x-intercept: 1; y-intercept; none
		Vertical asymptote: $x = 0$ (y-axis)
		Increasing; one-to-one; smooth; continuous
		See Figure 35(b) for a typical graph.
	$f(x) = \log_a x,\quad 0 < a < 1$	Domain: the interval $(0, \infty)$
	($y = \log_a x$ means $x = a^y$)	Range: the interval $(-\infty, \infty)$
		x-intercept: 1; y-intercept; none
		Vertical asymptote: $x = 0$ (y-axis)
		Decreasing; one-to-one; smooth; continuous
		See Figure 35(a) for a typical graph.

Natural logarithm (p. 439) $y = \ln x$ means $x = e^y$.

Properties of logarithms (pp. 449–450, 453)
$\log_a 1 = 0 \qquad \log_a a = 1 \qquad a^{\log_a M} = M \qquad \log_a a^r = r$

$\log_a(MN) = \log_a M + \log_a N \qquad \log_a\!\left(\dfrac{M}{N}\right) = \log_a M - \log_a N$

$\log_a M^r = r \log_a M$

If $M = N$, then $\log_a M = \log_a N$.

If $\log_a M = \log_a N$, then $M = N$.

Formulas

Change-of-Base Formula (p. 453) $\log_a M = \dfrac{\log_b M}{\log_b a}$

Compound Interest Formula (p. 465) $A = P \cdot \left(1 + \dfrac{r}{n}\right)^{nt}$

Continuous compounding (p. 466) $A = Pe^{rt}$

Present Value Formulas (p. 468) $P = A \cdot \left(1 + \dfrac{r}{n}\right)^{-nt} \quad$ or $\quad P = Ae^{-rt}$

Growth and decay (p. 473) $\qquad A(t) = A_0 e^{kt}$

Newton's Law of Cooling (p. 477) $\qquad u(t) = T + (u_0 - T)e^{kt}, \quad k < 0$

Logistic model (p. 479) $\qquad P(t) = \dfrac{c}{1 + ae^{-bt}}$

Objectives

Review Exercises

In Problems 1–6, for the given functions f and g find:

(a) $(f \circ g)(2)$ (b) $(g \circ f)(-2)$ (c) $(f \circ f)(4)$ (d) $(g \circ g)(-1)$

1. $f(x) = 3x - 5$; $g(x) = 1 - 2x^2$

2. $f(x) = 4 - x$; $g(x) = 1 + x^2$

3. $f(x) = \sqrt{x + 2}$; $g(x) = 2x^2 + 1$

4. $f(x) = 1 - 3x^2$; $g(x) = \sqrt{4 - x}$

5. $f(x) = e^x$; $g(x) = 3x - 2$

6. $f(x) = \dfrac{2}{1 + 2x^2}$; $g(x) = 3x$

In Problems 7–12, find $f \circ g$, $g \circ f$, $f \circ f$, and $g \circ g$ for each pair of functions. State the domain of each composite function.

7. $f(x) = 2 - x$; $g(x) = 3x + 1$

8. $f(x) = 2x - 1$; $g(x) = 2x + 1$

9. $f(x) = 3x^2 + x + 1$; $g(x) = |3x|$

10. $f(x) = \sqrt{3x}$; $g(x) = 1 + x + x^2$

11. $f(x) = \dfrac{x + 1}{x - 1}$; $g(x) = \dfrac{1}{x}$

12. $f(x) = \sqrt{x - 3}$; $g(x) = \dfrac{3}{x}$

In Problems 13 and 14, (a) verify that the function is one-to-one, and (b) find the inverse of the given function.

13. $\{(1, 2), (3, 5), (5, 8), (6, 10)\}$

14. $\{(-1, 4), (0, 2), (1, 5), (3, 7)\}$

In Problems 15 and 16, state why the graph of the function is one-to-one. Then draw the graph of the inverse function f^{-1}. For convenience (and as a hint), the graph of $y = x$ is also given.

15.

16.

In Problems 17–22, the function f is one-to-one. Find the inverse of each function and check your answer. Find the domain and the range of f and f^{-1}.

17. $f(x) = \dfrac{2x + 3}{5x - 2}$

18. $f(x) = \dfrac{2 - x}{3 + x}$

19. $f(x) = \dfrac{1}{x - 1}$

20. $f(x) = \sqrt{x - 2}$

21. $f(x) = \dfrac{3}{x^{1/3}}$

22. $f(x) = x^{1/3} + 1$

In Problems 23 and 24, $f(x) = 3^x$ and $g(x) = \log_3 x$.

23. Evaluate: (a) $f(4)$ (b) $g(9)$ (c) $f(-2)$ (d) $g\left(\dfrac{1}{27}\right)$

24. Evaluate: (a) $f(1)$ (b) $g(81)$ (c) $f(-4)$ (d) $g\left(\dfrac{1}{243}\right)$

In Problems 25 and 26, convert each exponential expression to an equivalent expression involving a logarithm. In Problems 27 and 28, convert each logarithmic expression to an equivalent expression involving an exponent.

25. $5^2 = z$

26. $a^5 = m$

27. $\log_5 u = 13$

28. $\log_a 4 = 3$

In Problems 29–32, find the domain of each logarithmic function.

29. $f(x) = \log(3x - 2)$ **30.** $F(x) = \log_5(2x + 1)$ **31.** $H(x) = \log_2(x^2 - 3x + 2)$ **32.** $F(x) = \ln(x^2 - 9)$

In Problems 33–38, evaluate each expression. Do not use a calculator.

33. $\log_2\left(\dfrac{1}{8}\right)$ **34.** $\log_3 81$ **35.** $\ln e^{\sqrt{2}}$

36. $e^{\ln 0.1}$ **37.** $2^{\log_2 0.4}$ **38.** $\log_2 2^{\sqrt{3}}$

In Problems 39–44, write each expression as the sum and/or difference of logarithms. Express powers as factors.

39. $\log_3\left(\dfrac{uv^2}{w}\right),\quad u > 0, v > 0, w > 0$ **40.** $\log_2\left(a^2\sqrt{b}\right)^4,\quad a > 0, b > 0$ **41.** $\log\left(x^2\sqrt{x^3 + 1}\right),\quad x > 0$

42. $\log_5\left(\dfrac{x^2 + 2x + 1}{x^2}\right),\quad x > 0$ **43.** $\ln\left(\dfrac{x\sqrt[3]{x^2 + 1}}{x - 3}\right),\quad x > 3$ **44.** $\ln\left(\dfrac{2x + 3}{x^2 - 3x + 2}\right)^2,\quad x > 2$

In Problems 45–50, write each expression as a single logarithm.

45. $3\log_4 x^2 + \dfrac{1}{2}\log_4 \sqrt{x}$ **46.** $-2\log_3\left(\dfrac{1}{x}\right) + \dfrac{1}{3}\log_3 \sqrt{x}$

47. $\ln\left(\dfrac{x - 1}{x}\right) + \ln\left(\dfrac{x}{x + 1}\right) - \ln(x^2 - 1)$ **48.** $\log(x^2 - 9) - \log(x^2 + 7x + 12)$

49. $2\log 2 + 3\log x - \dfrac{1}{2}[\log(x + 3) + \log(x - 2)]$ **50.** $\dfrac{1}{2}\ln(x^2 + 1) - 4\ln\dfrac{1}{2} - \dfrac{1}{2}[\ln(x - 4) + \ln x]$

In Problems 51 and 52, use the Change-of-Base Formula and a calculator to evaluate each logarithm. Round your answer to three decimal places.

51. $\log_4 19$ **52.** $\log_2 21$

In Problems 53 and 54, graph each function using a graphing utility and the Change-of-Base Formula.

53. $y = \log_3 x$ **54.** $y = \log_7 x$

In Problems 55–64, use transformations to graph each function. Determine the domain, range, and any asymptotes.

55. $f(x) = 2^{x-3}$ **56.** $f(x) = -2^x + 3$ **57.** $f(x) = \dfrac{1}{2}(3^{-x})$ **58.** $f(x) = 1 + 3^{2x}$

59. $f(x) = 1 - e^x$ **60.** $f(x) = 3 + \ln x$ **61.** $f(x) = \dfrac{1}{2}\ln x$ **62.** $f(x) = 3e^x$

63. $f(x) = 3 - e^{-x}$ **64.** $f(x) = 4 - \ln(-x)$

In Problems 65–84, solve each equation.

65. $4^{1-2x} = 2$ **66.** $8^{6+3x} = 4$ **67.** $3^{x^2+x} = \sqrt{3}$ **68.** $4^{x-x^2} = \dfrac{1}{2}$

69. $\log_x 64 = -3$ **70.** $\log_{\sqrt{2}} x = -6$ **71.** $5^x = 3^{x+2}$ **72.** $5^{x+2} = 7^{x-2}$

73. $9^{2x} = 27^{3x-4}$ **74.** $25^{2x} = 5^{x^2-12}$ **75.** $\log_3 \sqrt{x - 2} = 2$ **76.** $2^{x+1} \cdot 8^{-x} = 4$

77. $8 = 4^{x^2} \cdot 2^{5x}$ **78.** $2^x \cdot 5 = 10^x$ **79.** $\log_6(x + 3) + \log_6(x + 4) = 1$

80. $\log(7x - 12) = 2\log x$ **81.** $e^{1-x} = 5$ **82.** $e^{1-2x} = 4$

83. $2^{3x} = 3^{2x+1}$ **84.** $2^{x^3} = 3^{x^2}$

In Problems 85 and 86, use the following result: If x is the atmospheric pressure (measured in millimeters of mercury), then the formula for the altitude h(x) (measured in meters above sea level) is

$$h(x) = (30T + 8000) \log\left(\frac{P_0}{x}\right)$$

where T is the temperature (in degrees Celsius) and P_0 is the atmospheric pressure at sea level, which is approximately 760 millimeters of mercury.

85. Finding the Altitude of an Airplane At what height is a Piper Cub whose instruments record an outside temperature of 0°C and a barometric pressure of 300 millimeters of mercury?

86. Finding the Height of a Mountain How high is a mountain if instruments placed on its peak record a temperature of 5°C and a barometric pressure of 500 millimeters of mercury?

87. Amplifying Sound An amplifier's power output P (in watts) is related to its decibel voltage gain d by the formula $P = 25e^{0.1d}$.

(a) Find the power output for a decibel voltage gain of 4 decibels.
(b) For a power output of 50 watts, what is the decibel voltage gain?

88. Limiting Magnitude of a Telescope A telescope is limited in its usefulness by the brightness of the star that it is aimed at and by the diameter of its lens. One measure of a star's brightness is its *magnitude*; the dimmer the star, the larger its magnitude. A formula for the limiting magnitude L of a telescope, that is, the magnitude of the dimmest star that it can be used to view, is given by

$$L = 9 + 5.1 \log d$$

where d is the diameter (in inches) of the lens.
(a) What is the limiting magnitude of a 3.5-inch telescope?
(b) What diameter is required to view a star of magnitude 14?

89. Salvage Value The number of years n for a piece of machinery to depreciate to a known salvage value can be found using the formula

$$n = \frac{\log s - \log i}{\log(1 - d)}$$

where s is the salvage value of the machinery, i is its initial value, and d is the annual rate of depreciation.
(a) How many years will it take for a piece of machinery to decline in value from $90,000 to $10,000 if the annual rate of depreciation is 0.20 (20%)?
(b) How many years will it take for a piece of machinery to lose half of its value if the annual rate of depreciation is 15%?

90. Funding a College Education A child's grandparents purchase a $10,000 bond fund that matures in 18 years to be used for her college education. The bond fund pays 4% interest compounded semiannually. How much will the bond fund be worth at maturity? What is the effective rate of interest? How long would it take the bond to double in value under these terms?

91. Funding a College Education A child's grandparents wish to purchase a bond that matures in 18 years to be used for her college education. The bond pays 4% interest compounded semiannually. How much should they pay so that the bond will be worth $85,000 at maturity?

92. Funding an IRA First Colonial Bankshares Corporation advertised the following IRA investment plans.

Target IRA Plans

For each $5000 Maturity Value Desired	
Deposit:	At a Term of:
$620.17	20 Years
$1045.02	15 Years
$1760.92	10 Years
$2967.26	5 Years

(a) Assuming continuous compounding, what was the annual rate of interest that they offered?
(b) First Colonial Bankshares claims that $4000 invested today will have a value of over $32,000 in 20 years. Use the answer found in part (a) to find the actual value of $4000 in 20 years. Assume continuous compounding.

93. Estimating the Date That a Prehistoric Man Died The bones of a prehistoric man found in the desert of New Mexico contain approximately 5% of the original amount of carbon 14. If the half-life of carbon 14 is 5600 years, approximately how long ago did the man die?

94. Temperature of a Skillet A skillet is removed from an oven whose temperature is 450°F and placed in a room whose temperature is 70°F. After 5 minutes, the temperature of the skillet is 400°F. How long will it be until its temperature is 150°F?

95. World Population The growth rate of the world's population in 2003 was $k = 1.16\% = 0.0116$. The population of the world in 2003 was 6,302,486,693. Letting $t = 0$ represent 2003, use the uninhibited growth model to predict the world's population in the year 2010.

SOURCE: U.S. Census Bureau.

96. Radioactive Decay The half-life of radioactive cobalt is 5.27 years. If 100 grams of radioactive cobalt is present now, how much will be present in 20 years? In 40 years?

97. Logistic Growth The logistic growth model

$$P(t) = \frac{0.8}{1 + 1.67e^{-0.16t}}$$

represents the proportion of new cars with a Global Positioning System (GPS). Let $t = 0$ represent 2003, $t = 1$ represent 2004, and so on.

(a) What proportion of new cars in 2003 had a GPS?
(b) Determine the maximum proportion of new cars that have a GPS.
(c) Using a graphing utility, graph $P(t)$.
(d) When will 75% of new cars have a GPS?

98. CBL Experiment The following data were collected by placing a temperature probe in a portable heater, removing the probe, and then recording temperature over time.

Time (sec.)	Temperature (°F)
0	165.07
1	164.77
2	163.99
3	163.22
4	162.82
5	161.96
6	161.20
7	160.45
8	159.35
9	158.61
10	157.89
11	156.83
12	156.11
13	155.08
14	154.40
15	153.72

According to Newton's Law of Cooling, these data should follow an exponential model.
(a) Using a graphing utility, draw a scatter diagram for the data.
(b) Using a graphing utility, fit an exponential function to the data.

(c) Graph the exponential function found in part (b) on the scatter diagram.
(d) Predict how long it will take for the probe to reach a temperature of 110°F.

99. Wind Chill Factor The following data represent the wind speed (mph) and wind chill factor at an air temperature of 15°F.

Wind Speed (mph)	Wind Chill Factor (°F)
5	7
10	3
15	0
20	−2
25	−4
30	−5
35	−7

SOURCE: U.S. National Weather Service

(a) Using a graphing utility, draw a scatter diagram with wind speed as the independent variable.
(b) Using a graphing utility, fit a logarithmic function to the data.
(c) Using a graphing utility, draw the logarithmic function found in part (b) on the scatter diagram.
(d) Use the function found in part (b) to predict the wind chill factor if the air temperature is 15°F and the wind speed is 23 mph.

100. Spreading of a Disease Jack and Diane live in a small town of 50 people. Unfortunately, Jack and Diane both have a cold. Those who come in contact with someone who has this cold will themselves catch the cold. The following data represent the number of people in the small town who have caught the cold after t days.

Days, t	Number of People with Cold, C
0	2
1	4
2	8
3	14
4	22
5	30
6	37
7	42
8	44

(a) Using a graphing utility, draw a scatter diagram of the data. Comment on the type of relation that appears to exist between the days and number of people with a cold.

(b) Using a graphing utility, fit a logistic function to the data.

(c) Graph the function found in part (b) on the scatter diagram.

(d) According to the function found in part (b), what is the maximum number of people who will catch the cold? In reality, what is the maximum number of people who could catch the cold?

(e) Sometime between the second and third day, 10 people in the town had a cold. According to the model found in part (b), when did 10 people have a cold?

(f) How long will it take for 46 people to catch the cold?

Chapter Test

1. Given $f(x) = \dfrac{x+2}{x-2}$ and $g(x) = 2x + 5$, find:

(a) $f \circ g$ and state its domain (b) $(g \circ f)(-2)$
(c) $(f \circ g)(-2)$

2. Determine whether the function is one-to-one.

(a) $y = 4x^2 + 3$ (b) $y = \sqrt{x + 3} - 5$

3. Find the inverse of $f(x) = \dfrac{2}{3x - 5}$ and check your answer. State the domain and range of f and f^{-1}.

4. If the point $(3, -5)$ is on the graph of a one-to-one function f, what point must be on the graph of f^{-1}?

In Problems 5–7, find the unknown value without using a calculator.

5. $3^x = 243$ **6.** $\log_b 16 = 2$ **7.** $\log_5 x = 4$

In Problems 8–11, use a calculator to evaluate each expression. Round your answer to three decimal places.

8. $e^3 + 2$ **9.** $\log 20$

10. $\log_3 21$ **11.** $\ln 133$

In Problems 12 and 13, use transformations to graph each function. Determine the domain, range, and any asymptotes.

12. $f(x) = 4^{x+1} - 2$ **13.** $g(x) = 1 - \log_5(x - 2)$

In Problems 14–19, solve each equation.

14. $5^{x+2} = 125$ **15.** $\log(x + 9) = 2$

16. $8 - 2e^{-x} = 4$ **17.** $\log(x^2 + 3) = \log(x + 6)$

18. $7^{x+3} = e^x$ **19.** $\log_2(x - 4) + \log_2(x + 4) = 3$

20. Write $\log_2\left(\dfrac{4x^3}{x^2 - 3x - 18}\right)$ as the sum and/or difference of logarithms. Express powers as factors.

21. A 50-mg sample of a radioactive substance decays to 34 mg after 30 days. How long will it take for there to be 2 mg remaining?

22. The average cost of college at 4-year private colleges was $19,710 in 2003–2004. This was a 6% increase from the previous year.

(a) If the cost of college increases by 6% each year, what will be the average cost of college at 4-year private colleges in 2013–2014?

(b) College savings plans allow individuals to put money aside now to help pay for college later. If one such plan offers a rate of 5% compounded continuously, how much would Angie have needed to put in a college savings plan in 2003 in order to pay for 1 year of the cost of college at a 4-year private college in 2013?

23. The decibel level, D, of sound is given by the equation $D = 10 \log\left(\dfrac{I}{I_0}\right)$, where I is the intensity of the sound and $I_0 = 10^{-12}$ watts per square meter.

(a) If the shout of a single person measures 80 decibels, how loud will the sound be if two people shout at the same time? That is, how loud would the sound be if the intensity doubled?

(b) The pain threshold for sound is 125 decibels. If the Athens Olympic Stadium 2004 (Olympiako Stadio Athinas 'Spyros Louis') can seat 74,400 people, how many people in the crowd need to shout at the same time in order for the resulting sound level to meet or exceed the pain threshold? (Ignore any possible sound dampening.)

24. The table shows the estimated number of U.S. cell phone subscribers (in millions), y, from 1985 to 2003. Use a graphing utility to make a scatter diagram of the data. Fit a logistic model to the data and use the model to predict the number of U.S. cell phone subscribers in 2007. Let $x =$ the number of years since 1985.

SOURCE: Cellular Telecommunications & Internet Association

x	0	2	4	6	8	10	12	14	16	18
y	0.34	1.23	3.51	7.56	16.01	33.76	55.31	86.05	128.37	158.72

Chapter Projects

1. **Hot Coffee** A fast-food restaurant wants a special container to hold coffee. The restaurant wishes the container to quickly cool the coffee from 200° to 130°F and keep the liquid between 110° and 130°F as long as possible. The restaurant has three containers to select from.

   1. The CentiKeeper Company has a container that reduces the temperature of a liquid from 200°F to 100°F in 30 minutes by maintaining a constant temperature of 70°F.

2. The TempControl Company has a container that reduces the temperature of a liquid from 200°F to 110°F in 25 minutes by maintaining a constant temperature of 60°F.

3. The Hot'n'Cold Company has a container that reduces the temperature of a liquid from 200°F to 120°F in 20 minutes by maintaining a constant temperature of 65°F.

You need to recommend which container the restaurant should purchase.

(a) Use Newton's Law of Cooling to find a function relating the temperature of the liquid over time for each container.

(b) How long does it take each container to lower the coffee temperature from 200° to 130°F?

(c) How long will the coffee temperature remain between 110° and 130°F? This temperature is considered the optimal drinking temperature.

(d) Graph each function using a graphing utility.

(e) Which company would you recommend to the restaurant? Why?

(f) How might the cost of the container affect your decision?

The following projects are available on the Instructor's Resource Center (IRC):

2. **Project at Motorola**   *Thermal Fatigue of Solder Connections*
3. **Depreciation of a New Car**
4. **CBL Experiment**

Cumulative Review

1. Is the following graph the graph of a function? If it is, is the function one-to-one?

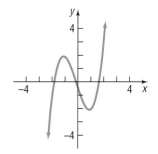

2. For the function $f(x) = 2x^2 - 3x + 1$, find the following:
  (a) $f(3)$ (b) $f(-x)$ (c) $f(x + h)$

3. Determine which of the following points are on the graph of $x^2 + y^2 = 1$.

  (a) $\left(\dfrac{1}{2}, \dfrac{1}{2}\right)$ (b) $\left(\dfrac{1}{2}, \dfrac{\sqrt{3}}{2}\right)$

4. Solve the equation $3(x - 2) = 4(x + 5)$.

5. Graph the line $2x - 4y = 16$.

6. (a) Graph the quadratic function $f(x) = -x^2 + 2x - 3$ by determining whether its graph opens up or down and by finding its vertex, axis of symmetry, y-intercept, and x-intercept(s), if any.

  (b) Solve $f(x) \le 0$.

7. Determine the quadratic function whose graph is given in the figure.

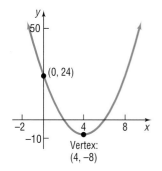

Vertex:
$(4, -8)$

8. Graph $f(x) = 3(x + 1)^3 - 2$ using transformations.

9. Given that $f(x) = x^2 + 2$ and $g(x) = \dfrac{2}{x - 3}$, find $f(g(x))$ and state its domain. What is $f(g(5))$?

10. For the polynomial function $f(x) = 4x^3 + 9x^2 - 30x - 8$:
   (a) Find the real zeros of f.
   (b) Determine the intercepts of the graph of f.
   (c) Use a graphing utility to approximate the local maxima and local minima.
   (d) By hand, draw a complete graph of f. Be sure to label the intercepts and turning points.

11. For the function $g(x) = 3^x + 2$:
   (a) Graph g using transformations. State the domain, range, and horizontal asymptote of g.
   (b) Determine the inverse of g. State the domain, range, and vertical asymptote of g^{-1}.
   (c) On the same graph as g, graph g^{-1}.

12. Solve the equation $4^{x-3} = 8^{2x}$.

13. Solve the equation: $\log_3(x + 1) + \log_3(2x - 3) = \log_9 9$

14. Suppose that $f(x) = \log_3(x + 2)$. Solve:
   (a) $f(x) = 0$.
   (b) $f(x) > 0$.
   (c) $f(x) = 3$.

15. Data Analysis   The following data represent the percent of all drivers that have been stopped by the police for any reason within the past year by age. The median age represents the midpoint of the upper and lower limit for the age range.

Age Range	Median Age, x	Percentage Stopped, y
16–19	17.5	18.2
20–29	24.5	16.8
30–39	34.5	11.3
40–49	44.5	9.4
50–59	54.5	7.7
≥60	69.5	3.8

   (a) Using your graphing utility, draw a scatter diagram of the data treating median age, x, as the independent variable.
   (b) Determine a model that you feel best describes the relation between median age and percentage stopped. You may choose from among linear, quadratic, cubic, power, exponential, logarithmic, or logistic models.
   (c) Provide a justification for the model that you selected in part (b).

Systems of Equations and Inequalities

5

A LOOK BACK In Chapters 1, 3, and 4, we solved equations and inequalities involving a single variable.

A LOOK AHEAD In this chapter we take up the problem of solving equations and inequalities containing two or more variables. There are various ways to solve such problems:

The *method of substitution* for solving equations in several unknowns goes back to ancient times.

The *method of elimination*, although it had existed for centuries, was put into systematic order by Karl Friedrich Gauss (1777–1855) and by Camille Jordan (1838–1922).

The theory of *matrices* was developed in 1857 by Arthur Cayley (1821–1895), although only later were matrices used as we use them in this chapter. Matrices have become a very flexible instrument, useful in almost all areas of mathematics.

The method of *determinants* was invented by Takakazu Seki Kôwa (1642–1708) in 1683 in Japan and by Gottfried Wilhelm von Leibniz (1646–1716) in 1693 in Germany. *Cramer's Rule* is named after Gabriel Cramer (1704–1752) of Switzerland, who popularized the use of determinants for solving linear systems.

Section 5.5, on *partial fraction decomposition*, provides an application of systems of equations. This particular application is one that is used in integral calculus.

Section 5.7 introduces *linear programming*, a modern application of linear inequalities. This topic is particularly useful for students interested in operations research.

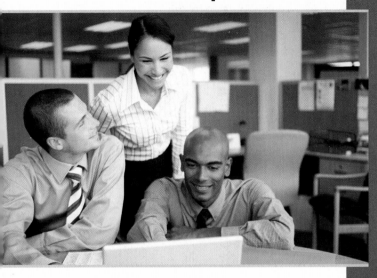

Economic Outcomes

Annual Earnings of Young Adults

For both males and females, earnings increase with education: full-time workers with at least a bachelor's degree have higher median earnings than those with less education. For example, in 2002, male college graduates earned 65 percent more than male high school completers. Females with a bachelor's or higher degree earned 71 percent more than female high school completers. Males and females who dropped out of high school earned 23 and 27 percent less, respectively, than male and female high school completers.

The median earnings of young adults who have at least a bachelor's degree declined in the 1970s relative to their counterparts who were high school completers, before increasing between 1980 and 2002. Males with a bachelor's degree or higher had earnings 19 percent higher than male high school completers in 1980 and had earnings 65 percent higher in 2002. Among females, those with at least a bachelor's degree had earnings 34 percent higher than female high school completers in 1980, compared with earnings 71 percent higher in 2002.

—See Chapter Project 1.

OUTLINE

5.1 Systems of Linear Equations: Substitution and Elimination

PREPARING FOR THIS SECTION *Before getting started, review the following:*

• Linear Equations (Section 1.2, pp. 103–105)   • Lines (Section 1.8, pp 163–176)

 Now work the 'Are You Prepared?' problems on page 515.

OBJECTIVES
1. Solve Systems of Equations by Substitution
2. Solve Systems of Equations by Elimination
3. Identify Inconsistent Systems of Equations Containing Two Variables
4. Express the Solution of a System of Dependent Equations Containing Two Variables
5. Solve Systems of Three Equations Containing Three Variables
6. Identify Inconsistent Systems of Equations Containing Three Variables
7. Express the Solution of a System of Dependent Equations Containing Three Variables

We begin with an example.

EXAMPLE 1   **Movie Theater Ticket Sales**

A movie theater sells tickets for $8.00 each, with seniors receiving a discount of $2.00. One evening the theater took in $3580 in revenue. If x represents the number of tickets sold at $8.00 and y the number of tickets sold at the discounted price of $6.00, write an equation that relates these variables.

Solution   Each nondiscounted ticket brings in $8.00, so x tickets will bring in $8x$ dollars. Similarly, y discounted tickets bring in $6y$ dollars. Since the total brought in is $3580, we must have

$$8x + 6y = 3580 \qquad \blacktriangleleft$$

In Example 1, suppose that we also know that 525 tickets were sold that evening. Then we have another equation relating the variables x and y.

$$x + y = 525$$

The two equations

$$8x + 6y = 3580$$
$$x + y = 525$$

form a *system* of equations.

In general, a **system of equations** is a collection of two or more equations, each containing one or more variables. Example 2 gives some illustrations of systems of equations.

| EXAMPLE 2 | **Examples of Systems of Equations** |

(a) $\begin{cases} 2x + y = 5 & \text{(1)} \\ -4x + 6y = -2 & \text{(2)} \end{cases}$ (1) Two equations containing two variables, x and y

(b) $\begin{cases} x + y^2 = 5 & \text{(1)} \\ 2x + y = 4 & \text{(2)} \end{cases}$ (1) Two equations containing two variables, x and y

(c) $\begin{cases} x + y + z = 6 & \text{(1)} \\ 3x - 2y + 4z = 9 & \text{(2)} \\ x - y - z = 0 & \text{(3)} \end{cases}$ (1) Three equations containing three variables, x, y, and z

(d) $\begin{cases} x + y + z = 5 & \text{(1)} \\ x - y = 2 & \text{(2)} \end{cases}$ (1) Two equations containing three variables, x, y, and z

(e) $\begin{cases} x + y + z = 6 & \text{(1)} \\ 2x + 2z = 4 & \text{(2)} \\ y + z = 2 & \text{(3)} \\ x = 4 & \text{(4)} \end{cases}$ (1) Four equations containing three variables, x, y, and z

◀

We use a brace, as shown, to remind us that we are dealing with a system of equations. We also will find it convenient to number each equation in the system.

A **solution** of a system of equations consists of values for the variables that are solutions of each equation of the system. To **solve** a system of equations means to find all solutions of the system.

For example, $x = 2$, $y = 1$ is a solution of the system in Example 2(a), because

$$\begin{cases} 2x + y = 5 & \text{(1)} \\ -4x + 6y = -2 & \text{(2)} \end{cases} \qquad \begin{cases} 2(2) + 1 = 4 + 1 = 5 \\ -4(2) + 6(1) = -8 + 6 = -2 \end{cases}$$

A solution of the system in Example 2(b) is $x = 1$, $y = 2$, because

$$\begin{cases} x + y^2 = 5 & \text{(1)} \\ 2x + y = 4 & \text{(2)} \end{cases} \qquad \begin{cases} 1 + 2^2 = 1 + 4 = 5 \\ 2(1) + 2 = 2 + 2 = 4 \end{cases}$$

Another solution of the system in Example 2(b) is $x = \dfrac{11}{4}$, $y = -\dfrac{3}{2}$, which you can check for yourself.

A solution of the system in Example 2(c) is $x = 3$, $y = 2$, $z = 1$, because

$$\begin{cases} x + y + z = 6 & \text{(1)} \\ 3x - 2y + 4z = 9 & \text{(2)} \\ x - y - z = 0 & \text{(3)} \end{cases} \qquad \begin{cases} 3 + 2 + 1 = 6 & \text{(1)} \\ 3(3) - 2(2) + 4(1) = 9 - 4 + 4 = 9 & \text{(2)} \\ 3 - 2 - 1 = 0 & \text{(3)} \end{cases}$$

Note that $x = 3$, $y = 3$, $z = 0$ is not a solution of the system in Example 2(c).

$$\begin{cases} x + y + z = 6 & \text{(1)} \\ 3x - 2y + 4z = 9 & \text{(2)} \\ x - y - z = 0 & \text{(3)} \end{cases} \qquad \begin{cases} 3 + 3 + 0 = 6 & \text{(1)} \\ 3(3) - 2(3) + 4(0) = 3 \neq 9 & \text{(2)} \\ 3 - 3 - 0 = 0 & \text{(3)} \end{cases}$$

Although these values satisfy equations (1) and (3), they do not satisfy equation (2). Any solution of the system must satisfy *each* equation of the system.

NOW WORK PROBLEM 9.

When a system of equations has at least one solution, it is said to be **consistent**; otherwise, it is called **inconsistent**.

An equation in n variables is said to be **linear** if it is equivalent to an equation of the form

$$a_1x_1 + a_2x_2 + \cdots + a_nx_n = b$$

where $x_1, x_2, \ldots, x_n$ are n distinct variables, $a_1, a_2, \ldots, a_n, b$ are constants, and at least one of the a's is not 0.

Some examples of linear equations are

$$2x + 3y = 2 \qquad 5x - 2y + 3z = 10 \qquad 8x + 8y - 2z + 5w = 0$$

If each equation in a system of equations is linear, then we have a **system of linear equations**. The systems in Examples 2(a), (c), (d), and (e) are linear, whereas the system in Example 2(b) is nonlinear. In this chapter we shall solve linear systems in Sections 5.1 to 5.3. We discuss nonlinear systems in a later chapter.

We begin by discussing a system of two linear equations containing two variables. We can view the problem of solving such a system as a geometry problem. The graph of each equation in such a system is a line. So a system of two equations containing two variables represents a pair of lines. The lines either (1) intersect or (2) are parallel or (3) are **coincident** (that is, identical).

1. If the lines intersect, then the system of equations has one solution, given by the point of intersection. The system is **consistent** and the equations are **independent**. See Figure 1(a).

2. If the lines are parallel, then the system of equations has no solution, because the lines never intersect. The system is **inconsistent**. See Figure 1(b).

3. If the lines are coincident, then the system of equations has infinitely many solutions, represented by the totality of points on the line. The system is **consistent** and the equations are **dependent**. See Figure 1(c).

Figure 1

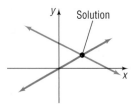
(a) Intersecting lines; system has one solution

(b) Parallel lines; system has no solution

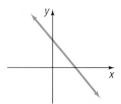
(c) Coincident lines; system has infinitely many solutions

EXAMPLE 3   **Solving a System of Linear Equations Using a Graphing Utility**

Solve: $\begin{cases} 2x + y = 5 & (1) \\ -4x + 6y = 12 & (2) \end{cases}$

Solution

First, we solve each equation for y. This is equivalent to writing each equation in slope–intercept form. Equation (1) in slope–intercept form is $Y_1 = -2x + 5$. Equation (2) in slope–intercept form is $Y_2 = \frac{2}{3}x + 2$. Figure 2 shows the graphs using a graphing utility. From the graph in Figure 2, we see that the lines intersect, so the system is consistent and the equations are independent. Using INTERSECT, we obtain the solution $(1.125, 2.75)$.   ◀

Figure 2

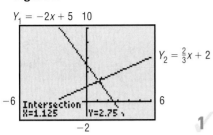

1 Solve Systems of Equations by Substitution

Sometimes to obtain exact solutions, we must use algebraic methods. A number of methods are available to us for solving systems of linear equations algebraically. In this section, we introduce two methods: *substitution* and *elimination*. We illustrate the **method of substitution** by solving the system given in Example 3.

| EXAMPLE 4 | **Solving a System of Linear Equations by Substitution** |

Solve: $\begin{cases} 2x + y = 5 & (1) \\ -4x + 6y = 12 & (2) \end{cases}$

Solution

We solve the first equation for y, obtaining

$$2x + y = 5 \qquad (1)$$
$$y = -2x + 5 \qquad \text{Subtract 2x from each side of (1).}$$

We substitute this result for y into the second equation. The result is an equation containing just the variable x, which we can then solve.

$$-4x + 6y = 12 \qquad (2)$$
$$-4x + 6(-2x + 5) = 12 \qquad \text{Substitute } y = -2x + 5 \text{ in (2).}$$
$$-4x - 12x + 30 = 12 \qquad \text{Remove parentheses.}$$
$$-16x = -18 \qquad \text{Combine like terms and subtract 30 from both sides.}$$
$$x = \frac{-18}{-16} = \frac{9}{8} \qquad \text{Divide each side by } -16.$$

Once we know that $x = \frac{9}{8}$, we can easily find the value of y by **back-substitution**, that is, by substituting $\frac{9}{8}$ for x in one of the original equations. We will use the first equation.

$$2x + y = 5 \qquad (1)$$
$$y = -2x + 5 \qquad \text{Subtract 2x from each side.}$$
$$y = -2\left(\frac{9}{8}\right) + 5 \qquad \text{Substitute } x = \frac{9}{8} \text{ in (1).}$$
$$= \frac{-9}{4} + \frac{20}{4} = \frac{11}{4}$$

The solution of the system is $x = \frac{9}{8} = 1.125$, $y = \frac{11}{4} = 2.75$.

✔ **CHECK:** $\begin{cases} 2x + y = 5: & 2\left(\dfrac{9}{8}\right) + \dfrac{11}{4} = \dfrac{9}{4} + \dfrac{11}{4} = \dfrac{20}{4} = 5 \\[2ex] -4x + 6y = 12: & -4\left(\dfrac{9}{8}\right) + 6\left(\dfrac{11}{4}\right) = -\dfrac{9}{2} + \dfrac{33}{2} = \dfrac{24}{2} = 12 \end{cases}$ ◀

The method used to solve the system in Example 4 is called **substitution**. The steps to be used are outlined next.

Steps for Solving by Substitution

STEP 1: Pick one of the equations and solve for one of the variables in terms of the remaining variables.

STEP 2: Substitute the result into the remaining equations.

STEP 3: If one equation in one variable results, solve this equation. Otherwise repeat Steps 1 and 2 until a single equation with one variable remains.

STEP 4: Find the values of the remaining variables by back-substitution.

STEP 5: Check the solution found.

NOW USE SUBSTITUTION TO WORK PROBLEM **19**.

2 Solve Systems of Equations by Elimination

A second method for solving a system of linear equations is the *method of elimination*. This method is usually preferred over substitution if substitution leads to fractions or if the system contains more than two variables. Elimination also provides the necessary motivation for solving systems using matrices (the subject of Section 5.2).

The idea behind the method of elimination is to replace the original system of equations by an equivalent system so that adding two of the equations eliminates a variable. The rules for obtaining equivalent equations are the same as those studied earlier. However, we may also interchange any two equations of the system and/or replace any equation in the system by the sum (or difference) of that equation and a nonzero multiple of any other equation in the system.

In Words

When using elimination, we want to get the coefficients of one of the variables to be negatives of one another.

Rules for Obtaining an Equivalent System of Equations

1. Interchange any two equations of the system.

2. Multiply (or divide) each side of an equation by the same nonzero constant.

3. Replace any equation in the system by the sum (or difference) of that equation and a nonzero multiple of any other equation in the system.

An example will give you the idea. As you work through the example, pay particular attention to the pattern being followed.

EXAMPLE 5 **Solving a System of Linear Equations by Elimination**

Solve: $\begin{cases} 2x + 3y = 1 & (1) \\ -x + y = -3 & (2) \end{cases}$

Solution   We multiply each side of equation (2) by 2 so that the coefficients of x in the two equations are negatives of one another. The result is the equivalent system

$$\begin{cases} 2x + 3y = 1 & (1) \\ -2x + 2y = -6 & (2) \end{cases}$$

If we add equations (1) and (2) we obtain an equation containing just the variable y, which we can solve.

$$\begin{cases} 2x + 3y = 1 & (1) \\ -2x + 2y = -6 & (2) \\ \hline \quad\quad\;\; 5y = -5 & \text{Add (1) and (2).} \\ \quad\quad\;\;\; y = -1 & \text{Solve for } y. \end{cases}$$

We back-substitute this value for y in equation (1) and simplify.

$$2x + 3y = 1 \quad (1)$$
$$2x + 3(-1) = 1 \quad \text{Substitute } y = -1 \text{ in (1).}$$
$$2x = 4 \quad \text{Simplify.}$$
$$x = 2 \quad \text{Solve for } x.$$

The solution of the original system is $x = 2$, $y = -1$. We leave it to you to check the solution.   ◄

The procedure used in Example 5 is called the **method of elimination**. Notice the pattern of the solution. First, we eliminated the variable x from the second equation. Then we back-substituted; that is, we substituted the value found for y back into the first equation to find x.

NOW USE ELIMINATION TO WORK PROBLEM 19.

Let's return to the movie theater example Example 1.

EXAMPLE 6	**Movie Theater Ticket Sales**

A movie theater sells tickets for $8.00 each, with seniors receiving a discount of $2.00. One evening the theater sold 525 tickets and took in $3580 in revenue. How many of each type of ticket were sold?

Solution   If x represents the number of tickets sold at $8.00 and y the number of tickets sold at the discounted price of $6.00, then the given information results in the system of equations

$$\begin{cases} 8x + 6y = 3580 & (1) \\ x + y = 525 & (2) \end{cases}$$

We use the method of elimination. First, multiply the second equation by -6, and then add the equations.

$$\begin{cases} 8x + 6y = 3580 \\ -6x - 6y = -3150 \\ \hline \quad\;\; 2x = 430 \quad \text{Add the equations.} \\ \quad\;\;\; x = 215 \end{cases}$$

Since $x + y = 525$, then $y = 525 - x = 525 - 215 = 310$. So 215 nondiscounted tickets and 310 senior discount tickets were sold.   ◄

3 Identify Inconsistent Systems of Equations Containing Two Variables

The previous examples dealt with consistent systems of equations that had a single solution. The next two examples deal with two other possibilities that may occur, the first being a system that has no solution.

EXAMPLE 7 | **An Inconsistent System of Linear Equations**

Solve: $\begin{cases} 2x + y = 5 & (1) \\ 4x + 2y = 8 & (2) \end{cases}$

Solution We choose to use the method of substitution and solve equation (1) for y.

$$2x + y = 5 \qquad (1)$$
$$y = -2x + 5 \qquad \text{Subtract 2x from each side.}$$

Now substitute $y = -2x + 5$ for y in equation (2) and solve for x.

$$4x + 2y = 8 \qquad (2)$$
$$4x + 2(-2x + 5) = 8 \qquad \text{Substitute } y = -2x + 5 \text{ in (2).}$$
$$4x - 4x + 10 = 8 \qquad \text{Remove parentheses.}$$
$$0 \cdot x = -2 \qquad \text{Subtract 10 from both sides.}$$

This equation has no solution. We conclude that the system itself has no solution and is therefore inconsistent. ◀

Figure 3

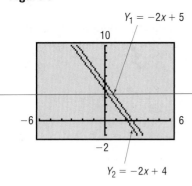

$Y_1 = -2x + 5$

$Y_2 = -2x + 4$

Figure 3 illustrates the pair of lines whose equations form the system in Example 7. Notice that the graphs of the two equations are lines, each with slope -2; one has a y-intercept of 5, the other a y-intercept of 4. The lines are parallel and have no point of intersection. This geometric statement is equivalent to the algebraic statement that the system has no solution.

4 Express the Solution of a System of Dependent Equations Containing Two Variables

---

EXAMPLE 8 | **Solving a System of Dependent Equations**

Solve: $\begin{cases} 2x + y = 4 & (1) \\ -6x - 3y = -12 & (2) \end{cases}$

Solution We choose to use the method of elimination.

$\begin{cases} 2x + y = 4 & (1) \\ -6x - 3y = -12 & (2) \end{cases}$

$\begin{cases} 6x + 3y = 12 & (1) \quad \text{Multiply each side of equation (1) by 3.} \\ -6x - 3y = -12 & (2) \end{cases}$

$\begin{cases} 6x + 3y = 12 & (1) \\ 0 = 0 & (2) \quad \text{Replace equation (2) by the sum of equations (1) and (2).} \end{cases}$

The original system is equivalent to a system containing one equation, so the equations are dependent. This means that any values of x and y for which $6x + 3y = 12$ or, equivalently, $2x + y = 4$ are solutions. For example, $x = 2, y = 0; x = 0, y = 4; x = -2, y = 8; x = 4, y = -4$; and so on, are solutions. There are, in fact, infinitely many values of x and y for which $2x + y = 4$, so the original system has infinitely many solutions. We will write the solution of the original system either as

$$y = -2x + 4$$

where x can be any real number, or as

$$x = -\frac{1}{2}y + 2$$

where y can be any real number.   ◀

Figure 4

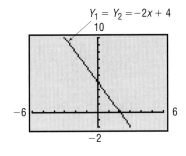

$Y_1 = Y_2 = -2x + 4$

Figure 4 illustrates the situation presented in Example 8. Notice that the graphs of the two equations are lines, each with slope -2 and each with y-intercept 4. The lines are coincident. Notice also that equation (2) in the original system is -3 times equation (1), indicating that the two equations are dependent.

For the system in Example 8, we can write down some of the infinite number of solutions by assigning values to x and then finding $y = -2x + 4$.

If $x = -2$, then $y = -2(-2) + 4 = 8$.

If $x = 0$, then $y = 4$.

If $x = 2$, then $y = 0$.

The ordered pairs $(-2, 8), (0, 4)$, and $(2, 0)$ are three of the points on the line in Figure 4.

✏️ **NOW WORK PROBLEMS 25 AND 29.**

5   Solve Systems of Three Equations Containing Three Variables

Just as with a system of two linear equations containing two variables, a system of three linear equations containing three variables also has either (1) exactly one solution (a consistent system with independent equations), or (2) no solution (an inconsistent system), or (3) infinitely many solutions (a consistent system with dependent equations).

We can view the problem of solving a system of three linear equations containing three variables as a geometry problem. The graph of each equation in such a system is a plane in space. A system of three linear equations containing three variables represents three planes in space. Figure 5 illustrates some of the possibilities.

Figure 5

(a) Consistent system;
one solution

(b) Consistent system;
infinite number of solutions

(c) Inconsistent system;
no solution

Recall that a **solution** to a system of equations consists of values for the variables that are solutions of each equation of the system. For example, $x = 3$, $y = -1$, $z = -5$ is a solution to the system of equations

$$\begin{cases} x + y + z = -3 & (1) \\ 2x - 3y + 6z = -21 & (2) \\ -3x + 5y = -14 & (3) \end{cases} \qquad \begin{array}{l} 3 + (-1) + (-5) = -3 \\ 2(3) - 3(-1) + 6(-5) = 6 + 3 - 30 = -21 \\ -3(3) + 5(-1) = -9 - 5 = -14 \end{array}$$

because these values of the variables are solutions of each equation.

Typically, when solving a system of three linear equations containing three variables, we use the method of elimination. Recall that the idea behind the method of elimination is to form equivalent equations so that adding two of the equations eliminates a variable.

Let's see how elimination works on a system of three equations containing three variables.

EXAMPLE 9   **Solving a System of Three Linear Equations with Three Variables**

Use the method of elimination to solve the system of equations.

$$\begin{cases} x + y - z = -1 & (1) \\ 4x - 3y + 2z = 16 & (2) \\ 2x - 2y - 3z = 5 & (3) \end{cases}$$

Solution   For a system of three equations, we attempt to eliminate one variable at a time, using pairs of equations until an equation with a single variable remains. Our plan of attack on this system will be to use equation (1) to eliminate the variable x from equations (2) and (3).

We begin by multiplying each side of equation (1) by -4 and adding the result to equation (2). (Do you see why? The coefficients of x are now negatives of one another.) We also multiply equation (1) by -2 and add the result to equation (3). Notice that these two procedures result in the removal of the variable x from equations (2) and (3).

$$\begin{array}{l} x + y - z = -1 \;\; (1) \quad \text{Multiply by} -4 \\ 4x - 3y + 2z = 16 \;\; (2) \end{array} \qquad \begin{array}{l} -4x - 4y + 4z = 4 \;\; (1) \\ \underline{4x - 3y + 2z = 16 \;\; (2)} \\ -7y + 6z = 20 \quad \text{Add.} \end{array}$$

$$\begin{array}{l} x + y - z = -1 \;\; (1) \quad \text{Multiply by} -2 \\ 2x - 2y - 3z = 5 \;\; (3) \end{array} \qquad \begin{array}{l} -2x - 2y + 2z = 2 \;\; (1) \\ \underline{2x - 2y - 3z = 5 \;\; (3)} \\ -4y - z = 7 \quad \text{Add.} \end{array}$$

$$\begin{cases} x + y - z = -1 & (1) \\ -7y + 6z = 20 & (2) \\ -4y - z = 7 & (3) \end{cases}$$

We now concentrate on equations (2) and (3), treating them as a system of two equations containing two variables. It is easiest to eliminate z. We multiply each side of equation (3) by 6 and add equations (2) and (3). The result is the new equation (3).

$$\begin{array}{l} -7y + 6z = 20 \;\; (2) \\ -4y - z = 7 \;\; (3) \quad \text{Multiply by 6} \end{array} \qquad \begin{array}{l} -7y + 6z = 20 \;\; (2) \\ \underline{-24y - 6z = 42 \;\; (3)} \\ -31y = 62 \quad \text{Add.} \end{array} \qquad \begin{cases} x + y - z = -1 & (1) \\ -7y + 6z = 20 & (2) \\ -31y = 62 & (3) \end{cases}$$

We now solve equation (3) for y by dividing both sides of the equation by -31.

$$\begin{cases} x + y - z = -1 & (1) \\ -7y + 6z = 20 & (2) \\ y = -2 & (3) \end{cases}$$

Back-substitute $y = -2$ in equation (2) and solve for z.

$$\begin{aligned} -7y + 6z &= 20 & \text{(2)} \\ -7(-2) + 6z &= 20 & \text{Substitute } y = -2 \text{ in (2).} \\ 6z &= 6 & \text{Subtract 14 from both sides of the equation.} \\ z &= 1 & \text{Divide both sides of the equation by 6.} \end{aligned}$$

Finally, we back-substitute $y = -2$ and $z = 1$ in equation (1) and solve for x.

$$\begin{aligned} x + y - z &= -1 & \text{(1)} \\ x + (-2) - 1 &= -1 & \text{Substitute } y = -2 \text{ and } z = 1 \text{ in (1).} \\ x - 3 &= -1 & \text{Simplify.} \\ x &= 2 & \text{Add 3 to both sides.} \end{aligned}$$

The solution of the original system is $x = 2$, $y = -2$, $z = 1$. You should verify this solution. ◀

Look back over the solution given in Example 9. Note the pattern of removing one of the variables from two of the equations, followed by solving this system of two equations and two unknowns. Although which variables to remove is your choice, the methodology remains the same for all systems.

 NOW WORK PROBLEM 43.

6   Identify Inconsistent Systems of Equations Containing Three Variables

| EXAMPLE 10 | **An Inconsistent System of Linear Equations** |

Solve: $\begin{cases} 2x + y - z = -2 & (1) \\ x + 2y - z = -9 & (2) \\ x - 4y + z = 1 & (3) \end{cases}$

Solution   Our plan of attack is the same as in Example 9. However, in this system, it seems easiest to eliminate the variable z first. Do you see why?

Multiply each side of equation (1) by -1 and add the result to equation (2). Add equations (2) and (3).

$$\begin{array}{rl} -2x - y + z = 2 & \text{(1) Multiply by } -1. \\ x + 2y - z = -9 & \text{(2)} \\ \hline -x + y = -7 & \text{Add.} \end{array}$$

$$\begin{cases} 2x + y - z = -2 & (1) \\ -x + y = -7 & (2) \\ 2x - 2y = -8 & (3) \end{cases}$$

$$\begin{array}{rl} x + 2y - z = -9 & \text{(2)} \\ x - 4y + z = 1 & \text{(3)} \\ \hline 2x - 2y = -8 & \text{Add.} \end{array}$$

We now concentrate on equations (2) and (3), treating them as a system of two equations containing two variables. Multiply each side of equation (2) by 2 and add the result to equation (3).

$$-x + y = -7 \quad \text{(2)} \quad \text{Multiply by 2.} \qquad -2x + 2y = -14 \quad \text{(2)}$$
$$2x - 2y = -8 \quad \text{(3)} \qquad\qquad\qquad \underline{\phantom{-2x+{}}2x - 2y = -8} \quad \text{(3)}$$
$$0 = -22 \quad \text{Add.}$$

$$\begin{cases} 2x + y - z = -2 & \text{(1)} \\ -x + y \phantom{{}- z} = -7 & \text{(2)} \\ 0 = -22 & \text{(3)} \end{cases}$$

Equation (3) has no solution and the system is inconsistent. ◄

7 Express the Solution of a System of Dependent Equations Containing Three Variables

EXAMPLE 11   **Solving a System of Dependent Equations**

Solve: $\begin{cases} x - 2y - z = 8 & \text{(1)} \\ 2x - 3y + z = 23 & \text{(2)} \\ 4x - 5y + 5z = 53 & \text{(3)} \end{cases}$

Solution   Our plan is to eliminate x from equations (2) and (3). Multiply each side of equation (1) by -2 and add the result to equation (2). Also, multiply each side of equation (1) by -4 and add the result to equation (3).

$$x - 2y - z = 8 \quad \text{(1)} \quad \text{Multiply by } -2. \qquad -2x + 4y + 2z = -16 \quad \text{(1)}$$
$$2x - 3y + z = 23 \quad \text{(2)} \qquad\qquad\qquad \underline{\phantom{-2x+{}}2x - 3y + z = 23} \quad \text{(2)}$$
$$y + 3z = 7 \quad \text{Add.}$$

$$x - 2y - z = 8 \quad \text{(1)} \quad \text{Multiply by } -4. \qquad -4x + 8y + 4z = -32 \quad \text{(1)}$$
$$4x - 5y + 5z = 53 \quad \text{(3)} \qquad\qquad\qquad \underline{\phantom{-4x+{}}4x - 5y + 5z = 53} \quad \text{(2)}$$
$$3y + 9z = 21 \quad \text{Add.}$$

$$\begin{cases} x - 2y - z = 8 & \text{(1)} \\ y + 3z = 7 & \text{(2)} \\ 3y + 9z = 21 & \text{(3)} \end{cases}$$

Treat equations (2) and (3) as a system of two equations containing two variables, and eliminate the variable y by multiplying both sides of equation (2) by -3 and adding the result to equation (3).

$$y + 3z = 7 \quad \text{Multiply by } -3. \qquad -3y - 9z = -21$$
$$3y + 9z = 21 \qquad\qquad\qquad \underline{\phantom{-3y+{}}3y + 9z = 21} \quad \text{Add.}$$
$$0 = 0$$

$$\begin{cases} x - 2y - z = 8 & \text{(1)} \\ y + 3z = 7 & \text{(2)} \\ 0 = 0 & \text{(3)} \end{cases}$$

The original system is equivalent to a system containing two equations, so the equations are dependent and the system has infinitely many solutions. If we solve equation (2) for y, we can express y in terms of z as $y = -3z + 7$. Substitute this expression into equation (1) to determine x in terms of z.

$$\begin{aligned} x - 2y - z &= 8 & \text{(1)} \\ x - 2(-3z + 7) - z &= 8 & \text{Substitute } y = -3z + 7 \text{ in (1).} \\ x + 6z - 14 - z &= 8 & \text{Remove parentheses.} \\ x + 5z &= 22 & \text{Combine like terms.} \\ x &= -5z + 22 & \text{Solve for } x. \end{aligned}$$

We will write the solution to the system as

$$\begin{cases} x = -5z + 22 \\ y = -3z + 7 \end{cases}$$

where z can be any real number.

This way of writing the solution makes it easier to find specific solutions of the system. To find specific solutions, choose any value of z and use the equations $x = -5z + 22$ and $y = -3z + 7$ to determine x and y. For example, if $z = 0$, then $x = 22$ and $y = 7$, and if $z = 1$, then $x = 17$ and $y = 4$. ◀

⬤━━ **NOW WORK PROBLEM 45.**

Two points in the Cartesian plane determine a unique line. Given three noncollinear points, we can find the (unique) quadratic function whose graph contains these three points.

EXAMPLE 12	**Curve Fitting**

Find real numbers a, b, and c so that the graph of the quadratic function $y = ax^2 + bx + c$ contains the points $(-1, -4)$, $(1, 6)$, and $(3, 0)$.

Solution   We require that the three points satisfy the equation $y = ax^2 + bx + c$.

For the point $(-1, -4)$ we have: $-4 = a(-1)^2 + b(-1) + c$ $-4 = a - b + c$
For the point $(1, 6)$ we have: $6 = a(1)^2 + b(1) + c$ $6 = a + b + c$
For the point $(3, 0)$ we have: $0 = a(3)^2 + b(3) + c$ $0 = 9a + 3b + c$

We wish to determine a, b, and c so that each equation is satisfied. That is, we want to solve the following system of three equations containing three variables:

$$\begin{cases} a - b + c = -4 & (1) \\ a + b + c = 6 & (2) \\ 9a + 3b + c = 0 & (3) \end{cases}$$

Figure 6

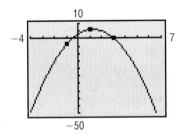

Solving this system of equations, we obtain $a = -2, b = 5$, and $c = 3$. So the quadratic function whose graph contains the points $(-1, -4)$, $(1, 6)$, and $(3, 0)$ is

$$y = -2x^2 + 5x + 3 \qquad y = ax^2 + bx + c, \quad a = -2, b = 5, c = 3$$

Figure 6 shows the graph of the function along with the three points. ◀

5.1 Assess Your Understanding

'Are You Prepared?'

Answers are given at the end of these exercises. If you get a wrong answer, read the pages listed in red.

1. Solve the equation: $3x + 4 = 8 - x$. (pp. 103–104)
2. (a) Graph the line: $3x + 4y = 12$. (pp. 170–172)

(b) What is the slope of a line parallel to this line? (pp. 172–174)

Concepts and Vocabulary

3. If a system of equations has no solution, it is said to be _____.

4. If a system of equations has one or more solutions, the system is said to be _____.

5. *True or False:* A system of two linear equations containing two variables always has at least one solution.

6. *True or False:* A solution of a system of equations consists of values for the variables that are solutions of each equation of the system.

Skill Building

In Problems 7–16, verify that the values of the variables listed are solutions of the system of equations.

7. $\begin{cases} 2x - y = 5 \\ 5x + 2y = 8 \end{cases}$
$x = 2, y = -1$

8. $\begin{cases} 3x + 2y = 2 \\ x - 7y = -30 \end{cases}$
$x = -2, y = 4$

9. $\begin{cases} 3x - 4y = 4 \\ \dfrac{1}{2}x - 3y = -\dfrac{1}{2} \end{cases}$
$x = 2, y = \dfrac{1}{2}$

10. $\begin{cases} 2x + \dfrac{1}{2}y = 0 \\ 3x - 4y = -\dfrac{19}{2} \end{cases}$
$x = -\dfrac{1}{2}, y = 2$

11. $\begin{cases} x - y = 3 \\ \dfrac{1}{2}x + y = 3 \end{cases}$
$x = 4, y = 1$

12. $\begin{cases} x - y = 3 \\ -3x + y = 1 \end{cases}$
$x = -2, y = -5$

13. $\begin{cases} 3x + 3y + 2z = 4 \\ x - y - z = 0 \\ 2y - 3z = -8 \end{cases}$
$x = 1, y = -1, z = 2$

14. $\begin{cases} 4x - z = 7 \\ 8x + 5y - z = 0 \\ -x - y + 5z = 6 \end{cases}$
$x = 2, y = -3, z = 1$

15. $\begin{cases} 3x + 3y + 2z = 4 \\ x - 3y + z = 10 \\ 5x - 2y - 3z = 8 \end{cases}$
$x = 2, y = -2, z = 2$

16. $\begin{cases} 4x - 5z = 6 \\ 5y - z = -17 \\ -x - 6y + 5z = 24 \end{cases}$
$x = 4, y = -3, z = 2$

In Problems 17–54, solve each system of equations. If the system has no solution, say that it is inconsistent.

17. $\begin{cases} x + y = 8 \\ x - y = 4 \end{cases}$

18. $\begin{cases} x + 2y = 5 \\ x + y = 3 \end{cases}$

19. $\begin{cases} 5x - y = 13 \\ 2x + 3y = 12 \end{cases}$

20. $\begin{cases} x + 3y = 5 \\ 2x - 3y = -8 \end{cases}$

21. $\begin{cases} 3x = 24 \\ x + 2y = 0 \end{cases}$

22. $\begin{cases} 4x + 5y = -3 \\ -2y = -4 \end{cases}$

23. $\begin{cases} 3x - 6y = 2 \\ 5x + 4y = 1 \end{cases}$

24. $\begin{cases} 2x + 4y = \dfrac{2}{3} \\ 3x - 5y = -10 \end{cases}$

25. $\begin{cases} 2x + y = 1 \\ 4x + 2y = 3 \end{cases}$

26. $\begin{cases} x - y = 5 \\ -3x + 3y = 2 \end{cases}$

27. $\begin{cases} 2x - y = 0 \\ 3x + 2y = 7 \end{cases}$

28. $\begin{cases} 3x + 3y = -1 \\ 4x + y = \dfrac{8}{3} \end{cases}$

29. $\begin{cases} x + 2y = 4 \\ 2x + 4y = 8 \end{cases}$

30. $\begin{cases} 3x - y = 7 \\ 9x - 3y = 21 \end{cases}$

31. $\begin{cases} 2x - 3y = -1 \\ 10x + y = 11 \end{cases}$

32. $\begin{cases} 3x - 2y = 0 \\ 5x + 10y = 4 \end{cases}$

33. $\begin{cases} 2x + 3y = 6 \\ x - y = \dfrac{1}{2} \end{cases}$

34. $\begin{cases} \dfrac{1}{2}x + y = -2 \\ x - 2y = 8 \end{cases}$

35. $\begin{cases} \dfrac{1}{2}x + \dfrac{1}{3}y = 3 \\ \dfrac{1}{4}x - \dfrac{2}{3}y = -1 \end{cases}$

36. $\begin{cases} \dfrac{1}{3}x - \dfrac{3}{2}y = -5 \\ \dfrac{3}{4}x + \dfrac{1}{3}y = 11 \end{cases}$

37. $\begin{cases} 3x - 5y = 3 \\ 15x + 5y = 21 \end{cases}$

38. $\begin{cases} 2x - y = -1 \\ x + \dfrac{1}{2}y = \dfrac{3}{2} \end{cases}$

39. $\begin{cases} \dfrac{1}{x} + \dfrac{1}{y} = 8 \\ \dfrac{3}{x} - \dfrac{5}{y} = 0 \end{cases}$

40. $\begin{cases} \dfrac{4}{x} - \dfrac{3}{y} = 0 \\ \dfrac{6}{x} + \dfrac{3}{2y} = 2 \end{cases}$

[**Hint:** Let $u = \dfrac{1}{x}$ and $v = \dfrac{1}{y}$, and solve for u and v.
Then $x = \dfrac{1}{u}$ and $y = \dfrac{1}{v}$.]

41. $\begin{cases} x - y = 6 \\ 2x - 3z = 16 \\ 2y + z = 4 \end{cases}$

42. $\begin{cases} 2x + y = -4 \\ -2y + 4z = 0 \\ 3x - 2z = -11 \end{cases}$

43. $\begin{cases} x - 2y + 3z = 7 \\ 2x + y + z = 4 \\ -3x + 2y - 2z = -10 \end{cases}$

44. $\begin{cases} 2x + y - 3z = 0 \\ -2x + 2y + z = -7 \\ 3x - 4y - 3z = 7 \end{cases}$

45. $\begin{cases} x - y - z = 1 \\ 2x + 3y + z = 2 \\ 3x + 2y = 0 \end{cases}$

46. $\begin{cases} 2x - 3y - z = 0 \\ -x + 2y + z = 5 \\ 3x - 4y - z = 1 \end{cases}$

47. $\begin{cases} x - y - z = 1 \\ -x + 2y - 3z = -4 \\ 3x - 2y - 7z = 0 \end{cases}$

48. $\begin{cases} 2x - 3y - z = 0 \\ 3x + 2y + 2z = 2 \\ x + 5y + 3z = 2 \end{cases}$

49. $\begin{cases} 2x - 2y + 3z = 6 \\ 4x - 3y + 2z = 0 \\ -2x + 3y - 7z = 1 \end{cases}$

50. $\begin{cases} 3x - 2y + 2z = 6 \\ 7x - 3y + 2z = -1 \\ 2x - 3y + 4z = 0 \end{cases}$

51. $\begin{cases} x + y - z = 6 \\ 3x - 2y + z = -5 \\ x + 3y - 2z = 14 \end{cases}$

52. $\begin{cases} x - y + z = -4 \\ 2x - 3y + 4z = -15 \\ 5x + y - 2z = 12 \end{cases}$

53. $\begin{cases} x + 2y - z = -3 \\ 2x - 4y + z = -7 \\ -2x + 2y - 3z = 4 \end{cases}$

54. $\begin{cases} x + 4y - 3z = -8 \\ 3x - y + 3z = 12 \\ x + y + 6z = 1 \end{cases}$

Applications and Extensions

55. The perimeter of a rectangular floor is 90 feet. Find the dimensions of the floor if the length is twice the width.

56. The length of fence required to enclose a rectangular field is 3000 meters. What are the dimensions of the field if it is known that the difference between its length and width is 50 meters?

57. Cost of Fast Food   Four large cheeseburgers and two chocolate shakes cost a total of $7.90. Two shakes cost 15¢ more than one cheeseburger. What is the cost of a cheeseburger? A shake?

58. Movie Theater Tickets   A movie theater charges $9.00 for adults and $7.00 for senior citizens. On a day when 325 people paid an admission, the total receipts were $2495. How many who paid were adults? How many were seniors?

59. Mixing Nuts   A store sells cashews for $5.00 per pound and peanuts for $1.50 per pound. The manager decides to mix 30 pounds of peanuts with some cashews and sell the mixture for $3.00 per pound. How many pounds of cashews should be mixed with the peanuts so that the mixture will produce the same revenue as would selling the nuts separately?

60. Financial Planning   A recently retired couple need $12,000 per year to supplement their Social Security. They have $150,000 to invest to obtain this income. They have decided on two investment options: AA bonds yielding 10% per annum and a Bank Certificate yielding 5%.
  (a) How much should be invested in each to realize exactly $12,000?
  (b) If, after two years, the couple requires $14,000 per year in income, how should they reallocate their investment to achieve the new amount?

61. Computing Wind Speed   With a tail wind, a small Piper aircraft can fly 600 miles in 3 hours. Against this same wind, the Piper can fly the same distance in 4 hours. Find the average wind speed and the average airspeed of the Piper.

62. Computing Wind Speed   The average airspeed of a single-engine aircraft is 150 miles per hour. If the aircraft flew the same distance in 2 hours with the wind as it flew in 3 hours against the wind, what was the wind speed?

63. Restaurant Management   A restaurant manager wants to purchase 200 sets of dishes. One design costs $25 per set, while another costs $45 per set. If she only has $7400 to spend, how many of each design should be ordered?

64. Cost of Fast Food   One group of people purchased 10 hot dogs and 5 soft drinks at a cost of $12.50. A second bought 7 hot dogs and 4 soft drinks at a cost of $9.00. What is the cost of a single hot dog? A single soft drink?

We paid $12.50.
How much is one hot dog?
How much is one cola?

We paid $9.00.
How much is one hot dog?
How much is one cola?

65. Computing a Refund   The grocery store we use does not mark prices on its goods. My wife went to this store, bought three 1-pound packages of bacon and two cartons of eggs, and paid a total of $7.45. Not knowing that she went to the store, I also went to the same store, purchased two 1-pound packages of bacon and three cartons of eggs, and paid a total of $6.45. Now we want to return two 1-pound packages of bacon and two cartons of eggs. How much will be refunded?

66. Finding the Current of a Stream   Pamela requires 3 hours to swim 15 miles downstream on the Illinois River. The return trip upstream takes 5 hours. Find Pamela's average speed in still water. How fast is the current? (Assume that Pamela's speed is the same in each direction.)

67. Pharmacy A doctor's prescription calls for a daily intake containing 40 mg of vitamin C and 30 mg of vitamin D. Your pharmacy stocks two liquids that can be used: one contains 20% vitamin C and 30% vitamin D, the other 40% vitamin C and 20% vitamin D. How many milligrams of each compound should be mixed to fill the prescription?

68. Pharmacy A doctor's prescription calls for the creation of pills that contain 12 units of vitamin B_{12} and 12 units of vitamin E. Your pharmacy stocks two powders that can be used to make these pills: one contains 20% vitamin B_{12} and 30% vitamin E, the other 40% vitamin B_{12} and 20% vitamin E. How many units of each powder should be mixed in each pill?

69. Curve Fitting Find real numbers a, b, and c so that the graph of the function $y = ax^2 + bx + c$ contains the points $(-1, 4)$, $(2, 3)$, and $(0, 1)$.

70. Curve Fitting Find real numbers a, b, and c so that the graph of the function $y = ax^2 + bx + c$ contains the points $(-1, -2)$, $(1, -4)$, and $(2, 4)$.

71. IS-LM Model in Economics In economics, the IS curve is a linear equation that represents all combinations of income Y and interest rates r that maintain an equilibrium in the market for goods in the economy. The LM curve is a linear equation that represents all combinations of income Y and interest rates r that maintain an equilibrium in the market for money in the economy. In an economy, suppose the equilibrium level of income (in millions of dollars) and interest rates satisfy the system of equations

$$\begin{cases} 0.06Y - 5000r = 240 \\ 0.06Y + 6000r = 900 \end{cases}$$

Find the equilibrium level of income and interest rates.

72. IS-LM Model in Economics In economics, the IS curve is a linear equation that represents all combinations of income Y and interest rates r that maintain an equilibrium in the market for goods in the economy. The LM curve is a linear equation that represents all combinations of income Y and interest rates r that maintain an equilibrium in the market for money in the economy. In an economy, suppose the equilibrium level of income (in millions of dollars) and interest rates satisfy the system of equations

$$\begin{cases} 0.05Y - 1000r = 10 \\ 0.05Y + 800r = 100 \end{cases}$$

Find the equilibrium level of income and interest rates.

73. Electricity: Kirchhoff's Rules An application of Kirchhoff's Rules to the circuit shown results in the following system of equations:

$$\begin{cases} I_2 = I_1 + I_3 \\ 5 - 3I_1 - 5I_2 = 0 \\ 10 - 5I_2 - 7I_3 = 0 \end{cases}$$

Find the currents I_1, I_2, and I_3.

SOURCE: *Physics for Scientists & Engineers* 3/E by Serway. © 1990. Reprinted with permission of Brooks/Cole, a division of Thomson Learning.

74. Electricity: Kirchhoff's Rules An application of Kirchhoff's Rules to the circuit shown results in the following system of equations:

$$\begin{cases} I_3 = I_1 + I_2 \\ 8 = 4I_3 + 6I_2 \\ 8I_1 = 4 + 6I_2 \end{cases}$$

Find the currents I_1, I_2, and I_3.

SOURCE: *Physics for Scientists & Engineers* 3/E by Serway. © 1990. Reprinted with permission of Brooks/Cole, a division of Thomson Learning.

75. Theater Revenues A Broadway theater has 500 seats, divided into orchestra, main, and balcony seating. Orchestra seats sell for $50, main seats for $35, and balcony seats for $25. If all the seats are sold, the gross revenue to the theater is $17,100. If all the main and balcony seats are sold, but only half the orchestra seats are sold, the gross revenue is $14,600. How many are there of each kind of seat?

76. Theater Revenues A movie theater charges $8.00 for adults, $4.50 for children, and $6.00 for senior citizens. One day the theater sold 405 tickets and collected $2320 in receipts. There were twice as many children's tickets sold as adult tickets. How many adults, children, and senior citizens went to the theater that day?

77. Nutrition A dietitian wishes a patient to have a meal that has 66 grams of protein, 94.5 grams of carbohydrates, and 910 milligrams of calcium. The hospital food service tells the dietitian that the dinner for today is chicken, corn, and 2%

milk. Each serving of chicken has 30 grams of protein, 35 grams of carbohydrates, and 200 milligrams of calcium. Each serving of corn has 3 grams of protein, 16 grams of carbohydrates, and 10 milligrams of calcium. Each glass of 2% milk has 9 grams of protein, 13 grams of carbohydrates, and 300 milligrams of calcium. How many servings of each food should the dietitian provide for the patient?

78. **Investments** Kelly has $20,000 to invest. As her financial planner, you recommend that she diversify into three investments: Treasury bills that yield 5% simple interest, Treasury bonds that yield 7% simple interest, and corporate bonds that yield 10% simple interest. Kelly wishes to earn $1390 per year in income. Also, Kelly wants her investment in Treasury bills to be $3000 more than her investment in corporate bonds. How much money should Kelly place in each investment?

79. **Prices of Fast Food** One group of customers bought 8 deluxe hamburgers, 6 orders of large fries, and 6 large colas for $26.10. A second group ordered 10 deluxe hamburgers, 6 large fries, and 8 large colas and paid $31.60. Is there sufficient information to determine the price of each food item? If not, construct a table showing the various possibilities. Assume that the hamburgers cost between $1.75 and $2.25, the fries between $0.75 and $1.00, and the colas between $0.60 and $0.90.

80. **Prices of Fast Food** Use the information given in Problem 79. Suppose that a third group purchased 3 deluxe hamburgers, 2 large fries, and 4 large colas for $10.95. Now is there sufficient information to determine the price of each food item? If so, determine each price.

81. **Painting a House** Three painters, Beth, Bill, and Edie, working together, can paint the exterior of a home in 10 hours. Bill and Edie together have painted a similar house in 15 hours. One day, all three worked on this same kind of house for 4 hours, after which Edie left. Beth and Bill required 8 more hours to finish. Assuming no gain or loss in efficiency, how long should it take each person to complete such a job alone?

Discussion and Writing

82. Make up a system of three linear equations containing three variables that has:
    (a) No solution
    (b) Exactly one solution
    (c) Infinitely many solutions
    Give the three systems to a friend to solve and critique.

83. Write a brief paragraph outlining your strategy for solving a system of two linear equations containing two variables.

84. Do you prefer the method of substitution or the method of elimination for solving a system of two linear equations containing two variables? Give reasons.

'Are You Prepared?' Answers

1. $\{1\}$ 2. (a)   (b) $-\dfrac{3}{4}$

[Graph showing a line through points $(0, 3)$ and $(4, 0)$ with y-axis and x-axis, marks at -2, 2, 4 on x-axis and 2, -2 on y-axis]

5.2 Systems of Linear Equations: Matrices

OBJECTIVES   1 Write the Augmented Matrix of a System of Linear Equations
  2 Write the System from the Augmented Matrix
  3 Perform Row Operations on a Matrix
  4 Solve a System of Linear Equations Using Matrices

The systematic approach of the method of elimination for solving a system of linear equations provides another method of solution that involves a simplified notation.

Consider the following system of linear equations:

$$\begin{cases} x + 4y = 14 \\ 3x - 2y = 0 \end{cases}$$

If we choose not to write the symbols used for the variables, we can represent this system as

$$\left[\begin{array}{cc|c} 1 & 4 & 14 \\ 3 & -2 & 0 \end{array} \right]$$

where it is understood that the first column represents the coefficients of the variable x, the second column the coefficients of y, and the third column the constants on the right side of the equal signs. The vertical line serves as a reminder of the equal signs. The large square brackets are used to denote a *matrix* in algebra.

A **matrix** is defined as a rectangular array of numbers,

$$
\begin{array}{c}
\text{Row 1} \\ \text{Row 2} \\ \vdots \\ \text{Row } i \\ \vdots \\ \text{Row } m
\end{array}
\begin{array}{c}
\overset{\text{Column 1}}{} \quad \overset{\text{Column 2}}{} \quad \overset{\text{Column } j}{} \quad \overset{\text{Column } n}{} \\
\left[\begin{array}{ccccccc}
a_{11} & a_{12} & \cdots & a_{1j} & \cdots & a_{1n} \\
a_{21} & a_{22} & \cdots & a_{2j} & \cdots & a_{2n} \\
\vdots & \vdots & & \vdots & & \vdots \\
a_{i1} & a_{i2} & \cdots & a_{ij} & \cdots & a_{in} \\
\vdots & \vdots & & \vdots & & \vdots \\
a_{m1} & a_{m2} & \cdots & a_{mj} & \cdots & a_{mn}
\end{array} \right]
\end{array}
\quad \textbf{(1)}
$$

Each number a_{ij} of the matrix has two indexes: the **row index** i and the **column index** j. The matrix shown in display (1) has m rows and n columns. The numbers a_{ij} are usually referred to as the **entries** of the matrix. For example, a_{23} refers to the entry in the second row, third column.

1 Write the Augmented Matrix of a System of Linear Equations

Now we will use matrix notation to represent a system of linear equations. The matrices used to represent systems of linear equations are called **augmented matrices**. In writing the augmented matrix of a system, the variables of each equation must be on the left side of the equal sign and the constants on the right side. A variable that does not appear in an equation has a coefficient of 0.

EXAMPLE 1   Writing the Augmented Matrix of a System of Linear Equations

Write the augmented matrix of each system of equations.

(a) $\begin{cases} 3x - 4y = -6 & (1) \\ 2x - 3y = -5 & (2) \end{cases}$

(b) $\begin{cases} 2x - y + z = 0 & (1) \\ x + z - 1 = 0 & (2) \\ x + 2y - 8 = 0 & (3) \end{cases}$

Solution   (a) The augmented matrix is

$$\left[\begin{array}{cc|c} 3 & -4 & -6 \\ 2 & -3 & -5 \end{array} \right]$$

(b) Care must be taken that the system be written so that the coefficients of all variables are present (if any variable is missing, its coefficient is 0). Also, all constants must be to the right of the equal sign. We need to rearrange the given system as follows:

$$\begin{cases} 2x - y + z = 0 & (1) \\ x + z - 1 = 0 & (2) \\ x + 2y - 8 = 0 & (3) \end{cases}$$

$$\begin{cases} 2x - y + z = 0 & (1) \\ x + 0 \cdot y + z = 1 & (2) \\ x + 2y + 0 \cdot z = 8 & (3) \end{cases}$$

The augmented matrix is

$$\begin{bmatrix} 2 & -1 & 1 & \bigm| & 0 \\ 1 & 0 & 1 & \bigm| & 1 \\ 1 & 2 & 0 & \bigm| & 8 \end{bmatrix}$$

◀

If we do not include the constants to the right of the equal sign, that is, to the right of the vertical bar in the augmented matrix of a system of equations, the resulting matrix is called the **coefficient matrix** of the system. For the systems discussed in Example 1, the coefficient matrices are

$$\begin{bmatrix} 3 & -4 \\ 2 & -3 \end{bmatrix} \quad \text{and} \quad \begin{bmatrix} 2 & -1 & 1 \\ 1 & 0 & 1 \\ 1 & 2 & 0 \end{bmatrix}$$

 NOW WORK PROBLEM 7.

2 Write the System from the Augmented Matrix

EXAMPLE 2

Writing the System of Linear Equations from the Augmented Matrix

Write the system of linear equations corresponding to each augmented matrix.

(a) $$\begin{bmatrix} 5 & 2 & \bigm| & 13 \\ -3 & 1 & \bigm| & -10 \end{bmatrix}$$

(b) $$\begin{bmatrix} 3 & -1 & -1 & \bigm| & 7 \\ 2 & 0 & 2 & \bigm| & 8 \\ 0 & 1 & 1 & \bigm| & 0 \end{bmatrix}$$

Solution

(a) The matrix has two rows and so represents a system of two equations. The two columns to the left of the vertical bar indicate that the system has two variables. If x and y are used to denote these variables, the system of equations is

$$\begin{cases} 5x + 2y = 13 & (1) \\ -3x + y = -10 & (2) \end{cases}$$

(b) Since the augmented matrix has three rows, it represents a system of three equations. Since there are three columns to the left of the vertical bar, the system contains three variables. If x, y, and z are the three variables, the system of equations is

$$\begin{cases} 3x - y - z = 7 & (1) \\ 2x + 2z = 8 & (2) \\ y + z = 0 & (3) \end{cases}$$

◀

3 Perform Row Operations on a Matrix

Row operations on a matrix are used to solve systems of equations when the system is written as an augmented matrix. There are three basic row operations.

Row Operations

1. Interchange any two rows.
2. Replace a row by a nonzero multiple of that row.
3. Replace a row by the sum of that row and a constant nonzero multiple of some other row.

These three row operations correspond to the three rules given earlier for obtaining an equivalent system of equations. When a row operation is performed on a matrix, the resulting matrix represents a system of equations equivalent to the system represented by the original matrix.

For example, consider the augmented matrix

$$\left[\begin{array}{cc|c} 1 & 2 & 3 \\ 4 & -1 & 2 \end{array}\right]$$

Suppose that we want to apply a row operation to this matrix that results in a matrix whose entry in row 2, column 1 is a 0. The row operation to use is

$$\text{Multiply each entry in row 1 by } -4 \text{ and add the result} \atop \text{to the corresponding entries in row 2.} \tag{2}$$

If we use R_2 to represent the new entries in row 2 and we use r_1 and r_2 to represent the original entries in rows 1 and 2, respectively, then we can represent the row operation in statement (2) by

$$R_2 = -4r_1 + r_2$$

Then

$$\left[\begin{array}{cc|c} 1 & 2 & 3 \\ 4 & -1 & 2 \end{array}\right] \underset{\substack{\uparrow \\ R_2 = -4r_1 + r_2}}{\rightarrow} \left[\begin{array}{cc|c} 1 & 2 & 3 \\ -4(1)+4 & -4(2)+(-1) & -4(3)+2 \end{array}\right] = \left[\begin{array}{cc|c} 1 & 2 & 3 \\ 0 & -9 & -10 \end{array}\right]$$

As desired, we now have the entry 0 in row 2, column 1.

EXAMPLE 3	**Applying a Row Operation to an Augmented Matrix**

Apply the row operation $R_2 = -3r_1 + r_2$ to the augmented matrix

$$\left[\begin{array}{cc|c} 1 & -2 & 2 \\ 3 & -5 & 9 \end{array}\right]$$

Solution The row operation $R_2 = -3r_1 + r_2$ tells us that the entries in row 2 are to be replaced by the entries obtained after multiplying each entry in row 1 by -3 and adding the result to the corresponding entries in row 2.

$$\begin{bmatrix} 1 & -2 & | & 2 \\ 3 & -5 & | & 9 \end{bmatrix} \rightarrow \begin{bmatrix} 1 & -2 & | & 2 \\ -3(1)+3 & (-3)(-2)+(-5) & | & -3(2)+9 \end{bmatrix} = \begin{bmatrix} 1 & -2 & | & 2 \\ 0 & 1 & | & 3 \end{bmatrix}$$

$$R_2 = -3r_1 + r_2$$

✏ NOW WORK PROBLEM **17.**

| EXAMPLE 4 | **Finding a Particular Row Operation** |

Find a row operation that will result in the augmented matrix

$$\begin{bmatrix} 1 & -2 & | & 2 \\ 0 & 1 & | & 3 \end{bmatrix}$$

having a 0 in row 1, column 2.

Solution We want a 0 in row 1, column 2. This result can be accomplished by multiplying row 2 by 2 and adding the result to row 1. That is, we apply the row operation $R_1 = 2r_2 + r_1$.

$$\begin{bmatrix} 1 & -2 & | & 2 \\ 0 & 1 & | & 3 \end{bmatrix} \rightarrow \begin{bmatrix} 2(0)+1 & 2(1)+(-2) & | & 2(3)+2 \\ 0 & 1 & | & 3 \end{bmatrix} = \begin{bmatrix} 1 & 0 & | & 8 \\ 0 & 1 & | & 3 \end{bmatrix}$$

$$R_1 = 2r_2 + r_1$$

A word about the notation that we have introduced. A row operation such as $R_1 = 2r_2 + r_1$ changes the entries in row 1. Note also that for this type of row operation we change the entries in a given row by multiplying the entries in some other row by an appropriate nonzero number and adding the results to the original entries of the row to be changed.

4 **Solve a System of Linear Equations Using Matrices**

To solve a system of linear equations using matrices, we use row operations on the augmented matrix of the system to obtain a matrix that is in *row echelon form*.

A matrix is in **row echelon form** when

1. The entry in row 1, column 1 is a 1, and 0's appear below it.
2. The first nonzero entry in each row after the first row is a 1, 0's appear below it, and it appears to the right of the first nonzero entry in any row above.
3. Any rows that contain all 0's to the left of the vertical bar appear at the bottom.

For example, for a system of three equations containing three variables with a unique solution, the augmented matrix is in row echelon form if it is of the form

$$\begin{bmatrix} 1 & a & b & | & d \\ 0 & 1 & c & | & e \\ 0 & 0 & 1 & | & f \end{bmatrix}$$

where a, b, c, d, e, and f are real numbers. The last row of this augmented matrix states that $z = f$. We can then determine the value of y using back-substitution with $z = f$, since row 2 represents the equation $y + cz = e$. Finally, x is determined using back-substitution again.

Two advantages of solving a system of equations by writing the augmented matrix in row echelon form are the following:

1. The process is algorithmic; that is, it consists of repetitive steps that can be programmed on a computer.

2. The process works on any system of linear equations, no matter how many equations or variables are present.

The next example shows how to write a matrix in row echelon form.

EXAMPLE 5

Solving a System of Linear Equations Using Matrices (Row Echelon Form)

Solve: $\begin{cases} 2x + 2y \quad\quad = 6 & (1) \\ x + y + z = 1 & (2) \\ 3x + 4y - z = 13 & (3) \end{cases}$

Solution

First, we write the augmented matrix that represents this system.

$$\begin{bmatrix} 2 & 2 & 0 & | & 6 \\ 1 & 1 & 1 & | & 1 \\ 3 & 4 & -1 & | & 13 \end{bmatrix}$$

The first step requires getting the entry 1 in row 1, column 1. An interchange of rows 1 and 2 is the easiest way to do this. [Note that this is equivalent to interchanging equations (1) and (2) of the system.]

$$\begin{bmatrix} 1 & 1 & 1 & | & 1 \\ 2 & 2 & 0 & | & 6 \\ 3 & 4 & -1 & | & 13 \end{bmatrix}$$

Next, we want a 0 in row 2, column 1 and a 0 in row 3, column 1. We use the row operations $R_2 = -2r_1 + r_2$ and $R_3 = -3r_1 + r_3$ to accomplish this. Notice that row 1 is unchanged using these row operations. Also, do you see that performing these row operations simultaneously is the same as doing one followed by the other?

$$\begin{bmatrix} 1 & 1 & 1 & | & 1 \\ 2 & 2 & 0 & | & 6 \\ 3 & 4 & -1 & | & 13 \end{bmatrix} \rightarrow \begin{bmatrix} 1 & 1 & 1 & | & 1 \\ 0 & 0 & -2 & | & 4 \\ 0 & 1 & -4 & | & 10 \end{bmatrix}$$

$$\begin{aligned} R_2 &= -2r_1 + r_2 \\ R_3 &= -3r_1 + r_3 \end{aligned}$$

Now we want the entry 1 in row 2, column 2. Interchanging rows 2 and 3 will accomplish this.

$$\begin{bmatrix} 1 & 1 & 1 & | & 1 \\ 0 & 0 & -2 & | & 4 \\ 0 & 1 & -4 & | & 10 \end{bmatrix} \rightarrow \begin{bmatrix} 1 & 1 & 1 & | & 1 \\ 0 & 1 & -4 & | & 10 \\ 0 & 0 & -2 & | & 4 \end{bmatrix}$$

Finally, we want a 1 in row 3, column 3. To obtain it, we use the row operation $R_3 = -\dfrac{1}{2}r_3$. The result is

$$
\begin{bmatrix}
1 & 1 & 1 & \bigm| & 1 \\
0 & 1 & -4 & \bigm| & 10 \\
0 & 0 & -2 & \bigm| & 4
\end{bmatrix}
\longrightarrow
\begin{bmatrix}
1 & 1 & 1 & \bigm| & 1 \\
0 & 1 & -4 & \bigm| & 10 \\
0 & 0 & 1 & \bigm| & -2
\end{bmatrix}
$$

$$R_3 = -\frac{1}{2}r_3$$

This matrix is the row echelon form of the augmented matrix. The third row of this matrix represents the equation $z = -2$. Using $z = -2$, we back-substitute into the equation $y - 4z = 10$ (from the second row) and obtain

$$
\begin{aligned}
y - 4z &= 10 \\
y - 4(-2) &= 10 \qquad z = -2 \\
y &= 2 \qquad \text{Solve for } y.
\end{aligned}
$$

Finally, we back-substitute $y = 2$ and $z = -2$ into the equation $x + y + z = 1$ (from the first row) and obtain

$$
\begin{aligned}
x + y + z &= 1 \\
x + 2 + (-2) &= 1 \qquad y = 2, z = -2 \\
x &= 1 \qquad \text{Solve for } x.
\end{aligned}
$$

The solution of the system is $x = 1$, $y = 2$, $z = -2$.   ◀

The steps that we used to solve the system of linear equations in Example 5 can be summarized as follows:

Matrix Method for Solving a System of Linear Equations (Row Echelon Form)

STEP 1: Write the augmented matrix that represents the system.

STEP 2: Perform row operations that place the entry 1 in row 1, column 1.

STEP 3: Perform row operations that leave the entry 1 in row 1, column 1 unchanged, while causing 0's to appear below it in column 1.

STEP 4: Perform row operations that place the entry 1 in row 2, column 2, but leave the entries in columns to the left unchanged. If it is impossible to place a 1 in row 2, column 2, then proceed to place a 1 in row 2, column 3. Once a 1 is in place, perform row operations to place 0's below it.
[Place any rows that contain only 0's on the left side of the vertical bar, at the bottom of the matrix.]

STEP 5: Now repeat Step 4, placing a 1 in the next row, but one column to the right. Continue until the bottom row or the vertical bar is reached.

STEP 6: The matrix that results is the row echelon form of the augmented matrix. Analyze the system of equations corresponding to it to solve the original system.

In the next example, we solve a system of linear equations using these steps. In addition, we solve the system using a graphing utility.

EXAMPLE 6 | **Solving a System of Linear Equations Using Matrices (Row Echelon Form)**

$$\text{Solve: } \begin{cases} x - y + z = 8 & (1) \\ 2x + 3y - z = -2 & (2) \\ 3x - 2y - 9z = 9 & (3) \end{cases}$$

Algebraic Solution

STEP 1: The augmented matrix of the system is

$$\begin{bmatrix} 1 & -1 & 1 & | & 8 \\ 2 & 3 & -1 & | & -2 \\ 3 & -2 & -9 & | & 9 \end{bmatrix}$$

STEP 2: Because the entry 1 is already present in row 1, column 1, we can go to step 3.

STEP 3: Perform the row operations $R_2 = -2r_1 + r_2$ and $R_3 = -3r_1 + r_3$. Each of these leaves the entry 1 in row 1, column 1 unchanged, while causing 0's to appear under it.

$$\begin{bmatrix} 1 & -1 & 1 & | & 8 \\ 2 & 3 & -1 & | & -2 \\ 3 & -2 & -9 & | & 9 \end{bmatrix} \rightarrow \begin{bmatrix} 1 & -1 & 1 & | & 8 \\ 0 & 5 & -3 & | & -18 \\ 0 & 1 & -12 & | & -15 \end{bmatrix}$$

$$\begin{array}{l} R_2 = -2r_1 + r_2 \\ R_3 = -3r_1 + r_3 \end{array}$$

STEP 4: The easiest way to obtain the entry 1 in row 2, column 2 without altering column 1 is to interchange rows 2 and 3 (another way would be to multiply row 2 by $\frac{1}{5}$, but this introduces fractions).

$$\begin{bmatrix} 1 & -1 & 1 & | & 8 \\ 0 & 1 & -12 & | & -15 \\ 0 & 5 & -3 & | & -18 \end{bmatrix}$$

To get a 0 under the 1 in row 2, column 2, perform the row operation $R_3 = -5r_2 + r_3$.

$$\begin{bmatrix} 1 & -1 & 1 & | & 8 \\ 0 & 1 & -12 & | & -15 \\ 0 & 5 & -3 & | & -18 \end{bmatrix} \rightarrow \begin{bmatrix} 1 & -1 & 1 & | & 8 \\ 0 & 1 & -12 & | & -15 \\ 0 & 0 & 57 & | & 57 \end{bmatrix}$$

$$R_3 = -5r_2 + r_3$$

STEP 5: Continuing, we obtain a 1 in row 3, column 3 by using $R_3 = \frac{1}{57}r_3$.

$$\begin{bmatrix} 1 & -1 & 1 & | & 8 \\ 0 & 1 & -12 & | & -15 \\ 0 & 0 & 57 & | & 57 \end{bmatrix} \rightarrow \begin{bmatrix} 1 & -1 & 1 & | & 8 \\ 0 & 1 & -12 & | & -15 \\ 0 & 0 & 1 & | & 1 \end{bmatrix}$$

$$R_3 = \frac{1}{57}r_3$$

Graphing Solution

The augmented matrix of the system is

$$\begin{bmatrix} 1 & -1 & 1 & | & 8 \\ 2 & 3 & -1 & | & -2 \\ 3 & -2 & -9 & | & 9 \end{bmatrix}$$

We enter this matrix into our graphing utility and name it A. See Figure 7(a). Using the REF (Row Echelon Form) command on matrix A, we obtain the results shown in Figure 7(b). Since the entire matrix does not fit on the screen, we need to scroll right to see the rest of it. See Figure 7(c).

Figure 7

(a)

(b)

(c)

The system of equations represented by the matrix in row echelon form is

STEP 6: The matrix on the right is the row echelon form of the augmented matrix. The system of equations represented by the matrix in row echelon form is

$$\begin{cases} x - y + z = 8 & (1) \\ y - 12z = -15 & (2) \\ z = 1 & (3) \end{cases}$$

Using $z = 1$, we back-substitute to get

$$\begin{cases} x - y + 1 = 8 & (1) \\ y - 12(1) = -15 & (2) \end{cases} \xrightarrow[\text{Simplify.}]{} \begin{cases} x - y = 7 & (1) \\ y = -3 & (2) \end{cases}$$

We get $y = -3$, and back-substituting into $x - y = 7$, we find that $x = 4$. The solution of the system is $x = 4, y = -3, z = 1$.

◄

$$\begin{cases} x - \dfrac{2}{3}y - 3z = 3 & (1) \\ y + \dfrac{15}{13}z = -\dfrac{24}{13} & (2) \\ z = 1 & (3) \end{cases}$$

Using $z = 1$, we back-substitute to get

$$\begin{cases} x - \dfrac{2}{3}y - 3(1) = 3 & (1) \\ y + \dfrac{15}{13}(1) = -\dfrac{24}{13} & (2) \end{cases}$$

$$\begin{cases} x - \dfrac{2}{3}y = 6 & (1) \\ y = -\dfrac{39}{13} = -3 & (2) \end{cases}$$

From the second equation we find that $y = -3$. Back-substituting $y = -3$ into $x - \dfrac{2}{3}y = 6$, we find that $x = 4$. The solution of the system is $x = 4, y = -3, z = 1$.

◄

Notice that the row echelon form of the augmented matrix in the graphing solution differs from the row echelon form in the algebraic solution, yet both matrices provide the same solution! This is because the two solutions used different row operations to obtain the row echelon form. In all likelihood, the two solutions parted ways in Step 4 of the algebraic solution, where we avoided introducing fractions by interchanging rows 2 and 3.

Sometimes it is advantageous to write a matrix in **reduced row echelon form**. In this form, row operations are used to obtain entries that are 0 above (as well as below) the leading 1 in a row. For example, the row echelon form obtained in the algebraic solution to Example 6 is

$$\begin{bmatrix} 1 & -1 & 1 & \bigm| & 8 \\ 0 & 1 & -12 & \bigm| & -15 \\ 0 & 0 & 1 & \bigm| & 1 \end{bmatrix}$$

To write this matrix in reduced row echelon form, we proceed as follows:

$$\begin{bmatrix} 1 & -1 & 1 & \bigm| & 8 \\ 0 & 1 & -12 & \bigm| & -15 \\ 0 & 0 & 1 & \bigm| & 1 \end{bmatrix} \rightarrow \begin{bmatrix} 1 & 0 & -11 & \bigm| & -7 \\ 0 & 1 & -12 & \bigm| & -15 \\ 0 & 0 & 1 & \bigm| & 1 \end{bmatrix} \rightarrow \begin{bmatrix} 1 & 0 & 0 & \bigm| & 4 \\ 0 & 1 & 0 & \bigm| & -3 \\ 0 & 0 & 1 & \bigm| & 1 \end{bmatrix}$$

$$ R_1 = r_2 + r_1 R_1 = 11r_3 + r_1$$
$$R_2 = 12r_3 + r_2$$

The matrix is now written in reduced row echelon form. The advantage of writing the matrix in this form is that the solution to the system, $x = 4, y = -3, z = 1$, is

Figure 8

readily found, without the need to back-substitute. Another advantage will be seen in Section 5.4, where the inverse of a matrix is discussed.

Most graphing utilities also have the ability to put a matrix in reduced row echelon form. Figure 8 shows the reduced row echelon form of the augmented matrix from Example 6 using the RREF command on a TI-84 Plus graphing calculator.

For the remaining examples in this section, we will only provide algebraic solutions to the systems. The reader is encouraged to verify the results using a graphing utility.

NOW WORK PROBLEMS 37 AND 47.

The matrix method for solving a system of linear equations also identifies systems that have infinitely many solutions and systems that are inconsistent. Let's see how.

EXAMPLE 7 | **Solving a Dependent System of Linear Equations Using Matrices**

Solve: $\begin{cases} 6x - y - z = 4 & (1) \\ -12x + 2y + 2z = -8 & (2) \\ 5x + y - z = 3 & (3) \end{cases}$

Solution We start with the augmented matrix of the system and proceed to obtain a 1 in row 1, column 1 with zeros below.

$$\begin{bmatrix} 6 & -1 & -1 & | & 4 \\ -12 & 2 & 2 & | & -8 \\ 5 & 1 & -1 & | & 3 \end{bmatrix} \rightarrow \begin{bmatrix} 1 & -2 & 0 & | & 1 \\ -12 & 2 & 2 & | & -8 \\ 5 & 1 & -1 & | & 3 \end{bmatrix} \rightarrow \begin{bmatrix} 1 & -2 & 0 & | & 1 \\ 0 & -22 & 2 & | & 4 \\ 0 & 11 & -1 & | & -2 \end{bmatrix}$$

$$\uparrow \qquad\qquad\qquad \uparrow$$
$$R_1 = -1r_3 + r_1 \qquad R_2 = 12r_1 + r_2$$
$$R_3 = -5r_1 + r_3$$

Obtaining a 1 in row 2, column 2 without altering column 1 can be accomplished by $R_2 = -\dfrac{1}{22}r_2$ or by $R_3 = \dfrac{1}{11}r_3$ and interchanging rows 2 and 3 or by $R_2 = \dfrac{23}{11}r_3 + r_2$. We shall use the first of these.

$$\begin{bmatrix} 1 & -2 & 0 & | & 1 \\ 0 & -22 & 2 & | & 4 \\ 0 & 11 & -1 & | & -2 \end{bmatrix} \rightarrow \begin{bmatrix} 1 & -2 & 0 & | & 1 \\ 0 & 1 & -\dfrac{1}{11} & | & -\dfrac{2}{11} \\ 0 & 11 & -1 & | & -2 \end{bmatrix} \rightarrow \begin{bmatrix} 1 & -2 & 0 & | & 1 \\ 0 & 1 & -\dfrac{1}{11} & | & -\dfrac{2}{11} \\ 0 & 0 & 0 & | & 0 \end{bmatrix}$$

$$\uparrow \qquad\qquad\qquad\qquad \uparrow$$
$$R_2 = -\dfrac{1}{22}r_2 \qquad\qquad R_3 = -11r_2 + r_3$$

This matrix is in row echelon form. Because the bottom row consists entirely of 0's, the system actually consists of only two equations.

$$\begin{cases} x - 2y = 1 & (1) \\ y - \dfrac{1}{11}z = -\dfrac{2}{11} & (2) \end{cases}$$

To make it easier to write down some of the solutions, we express both x and y in terms of z.

From the second equation, $y = \dfrac{1}{11}z - \dfrac{2}{11}$. Now back-substitute this solution for y into the first equation to get

$$x = 2y + 1 = 2\left(\frac{1}{11}z - \frac{2}{11}\right) + 1 = \frac{2}{11}z + \frac{7}{11}$$

The original system is equivalent to the system

$$\begin{cases} x = \dfrac{2}{11}z + \dfrac{7}{11} & \text{(1)} \\[2mm] y = \dfrac{1}{11}z - \dfrac{2}{11} & \text{(2)} \end{cases}$$

where z can be any real number.

Let's look at the situation. The original system of three equations is equivalent to a system containing two equations. This means that any values of x, y, z that satisfy both

$$x = \frac{2}{11}z + \frac{7}{11} \quad \text{and} \quad y = \frac{1}{11}z - \frac{2}{11}$$

will be solutions. For example, $z = 0$, $x = \dfrac{7}{11}$, $y = -\dfrac{2}{11}$; $z = 1$, $x = \dfrac{9}{11}$, $y = -\dfrac{1}{11}$; and $z = -1$, $x = \dfrac{5}{11}$, $y = -\dfrac{3}{11}$ are some of the solutions of the original system. There are, in fact, infinitely many values of $x, y,$ and z for which the two equations are satisfied. That is, the original system has infinitely many solutions. We will write the solution of the original system as

$$\begin{cases} x = \dfrac{2}{11}z + \dfrac{7}{11} \\[2mm] y = \dfrac{1}{11}z - \dfrac{2}{11} \end{cases}$$

where z can be any real number.   ◄

We can also find the solution by writing the augmented matrix in reduced row echelon form. Starting with the row echelon form, we have

$$\left[\begin{array}{ccc|c} 1 & -2 & 0 & 1 \\[2mm] 0 & 1 & -\dfrac{1}{11} & -\dfrac{2}{11} \\[2mm] 0 & 0 & 0 & 0 \end{array}\right] \xrightarrow[\underset{R_1 = 2r_2 + r_1}{\uparrow}]{} \left[\begin{array}{ccc|c} 1 & 0 & -\dfrac{2}{11} & \dfrac{7}{11} \\[2mm] 0 & 1 & -\dfrac{1}{11} & -\dfrac{2}{11} \\[2mm] 0 & 0 & 0 & 0 \end{array}\right]$$

The matrix on the right is in reduced row echelon form. The corresponding system of equations is

$$\begin{cases} x - \dfrac{2}{11}z = \dfrac{7}{11} & \text{(1)} \\[2mm] y - \dfrac{1}{11}z = -\dfrac{2}{11} & \text{(2)} \end{cases}$$

or, equivalently,

$$\begin{cases} x = \dfrac{2}{11}z + \dfrac{7}{11} & \text{(1)} \\[2mm] y = \dfrac{1}{11}z - \dfrac{2}{11} & \text{(2)} \end{cases}$$

where z can be any real number.

NOW WORK PROBLEM 53.

EXAMPLE 8 **Solving an Inconsistent System of Linear Equations Using Matrices**

Solve: $\begin{cases} x + y + z = 6 \\ 2x - y - z = 3 \\ x + 2y + 2z = 0 \end{cases}$

Solution We proceed as follows, beginning with the augmented matrix.

$$\begin{bmatrix} 1 & 1 & 1 & | & 6 \\ 2 & -1 & -1 & | & 3 \\ 1 & 2 & 2 & | & 0 \end{bmatrix} \rightarrow \begin{bmatrix} 1 & 1 & 1 & | & 6 \\ 0 & -3 & -3 & | & -9 \\ 0 & 1 & 1 & | & -6 \end{bmatrix} \rightarrow \begin{bmatrix} 1 & 1 & 1 & | & 6 \\ 0 & 1 & 1 & | & -6 \\ 0 & -3 & -3 & | & -9 \end{bmatrix} \rightarrow \begin{bmatrix} 1 & 1 & 1 & | & 6 \\ 0 & 1 & 1 & | & -6 \\ 0 & 0 & 0 & | & -27 \end{bmatrix}$$

$$\begin{array}{llll} & R_2 = -2r_1 + r_2 & \text{Interchange rows 2 and 3.} & R_3 = 3r_2 + r_3 \\ & R_3 = -1r_1 + r_3 & & \end{array}$$

This matrix is in row echelon form. The bottom row is equivalent to the equation

$$0x + 0y + 0z = -27$$

which has no solution. Hence, the original system is inconsistent. ◀

NOW WORK PROBLEM 27.

The matrix method is especially effective for systems of equations for which the number of equations and the number of variables are unequal. Here, too, such a system is either inconsistent or consistent. If it is consistent, it will have either exactly one solution or infinitely many solutions.

Let's look at a system of four equations containing three variables.

EXAMPLE 9 **Solving a System of Linear Equations Using Matrices**

Solve: $\begin{cases} x - 2y + z = 0 & \text{(1)} \\ 2x + 2y - 3z = -3 & \text{(2)} \\ y - z = -1 & \text{(3)} \\ -x + 4y + 2z = 13 & \text{(4)} \end{cases}$

Solution We proceed as follows, beginning with the augmented matrix.

$$
\begin{bmatrix}
1 & -2 & 1 & | & 0 \\
2 & 2 & -3 & | & -3 \\
0 & 1 & -1 & | & -1 \\
-1 & 4 & 2 & | & 13
\end{bmatrix}
\rightarrow
\begin{bmatrix}
1 & -2 & 1 & | & 0 \\
0 & 6 & -5 & | & -3 \\
0 & 1 & -1 & | & -1 \\
0 & 2 & 3 & | & 13
\end{bmatrix}
\rightarrow
\begin{bmatrix}
1 & -2 & 1 & | & 0 \\
0 & 1 & -1 & | & -1 \\
0 & 6 & -5 & | & -3 \\
0 & 2 & 3 & | & 13
\end{bmatrix}
$$

$R_2 = -2r_1 + r_2$ Interchange rows 2 and 3.
$R_4 = r_1 + r_4$

$$
\rightarrow
\begin{bmatrix}
1 & -2 & 1 & | & 0 \\
0 & 1 & -1 & | & -1 \\
0 & 0 & 1 & | & 3 \\
0 & 0 & 5 & | & 15
\end{bmatrix}
\rightarrow
\begin{bmatrix}
1 & -2 & 1 & | & 0 \\
0 & 1 & -1 & | & -1 \\
0 & 0 & 1 & | & 3 \\
0 & 0 & 0 & | & 0
\end{bmatrix}
$$

$R_3 = -6r_2 + r_3$ $R_4 = -5r_3 + r_4$
$R_4 = -2r_2 + r_4$

We could stop here, since the matrix is in row echelon form, and back-substitute $z = 3$ to find x and y. Or we can continue to obtain the reduced row echelon form.

$$
\rightarrow
\begin{bmatrix}
1 & 0 & -1 & | & -2 \\
0 & 1 & -1 & | & -1 \\
0 & 0 & 1 & | & 3 \\
0 & 0 & 0 & | & 0
\end{bmatrix}
\rightarrow
\begin{bmatrix}
1 & 0 & 0 & | & 1 \\
0 & 1 & 0 & | & 2 \\
0 & 0 & 1 & | & 3 \\
0 & 0 & 0 & | & 0
\end{bmatrix}
$$

$R_1 = 2r_2 + r_1$ $R_1 = r_3 + r_1$
                   $R_2 = r_3 + r_2$

The matrix is now in reduced row echelon form, and we can see that the solution is $x = 1$, $y = 2$, $z = 3$. ◀

NOW WORK PROBLEM 69.

EXAMPLE 10 **Nutrition**

A dietitian at Cook County Hospital wants a patient to have a meal that has 65 grams of protein, 95 grams of carbohydrates, and 905 milligrams of calcium. The hospital food service tells the dietitian that the dinner for today is chicken a la king, baked potatoes, and 2% milk. Each serving of chicken a la king has 30 grams of protein, 35 grams of carbohydrates, and 200 milligrams of calcium. Each serving of baked potatoes contains 4 grams of protein, 33 grams of carbohydrates, and 10 milligrams of calcium. Each glass of 2% milk contains 9 grams of protein, 13 grams of carbohydrates, and 300 milligrams of calcium. How many servings of each food should the dietitian provide for the patient?

Solution Let c, p, and m represent the number of servings of chicken a la king, baked potatoes, and milk, respectively. The dietitian wants the patient to have 65 grams of protein. Each serving of chicken a la king has 30 grams of protein, so c servings will have $30c$ grams of protein. Each serving of baked potatoes contains 4 grams of protein, so p potatoes will have $4p$ grams of protein. Finally, each glass of milk has 9 grams of protein, so m glasses of milk will have $9m$ grams of protein. The same logic will result in equations for carbohydrates and calcium, and we have the following system of equations:

$$
\begin{cases}
30c + 4p + 9m = 65 & \text{Protein equation} \\
35c + 33p + 13m = 95 & \text{Carbohydrate equation} \\
200c + 10p + 300m = 905 & \text{Calcium equation}
\end{cases}
$$

Figure 9

The augmented matrix of this system is:

$$A = \begin{bmatrix} 30 & 4 & 9 & | & 65 \\ 35 & 33 & 13 & | & 95 \\ 200 & 10 & 300 & | & 905 \end{bmatrix}$$

Then, using the RREF command on a TI-84 Plus graphing calculator, we obtain the matrix in Figure 9.

To meet the dietary requirements, the patient should receive 1.5 servings of chicken a la king, $\frac{1}{2}$ of a baked potato, and 2 glasses of milk. ◀

5.2 Assess Your Understanding

Concepts and Vocabulary

1. An m by n rectangular array of numbers is called a(n) _____.

2. The matrix used to represent a system of linear equations is called a(n) _____ matrix.

3. *True or False*: The augmented matrix of a system of two equations containing three variables has two rows and four columns.

4. *True or False*: The matrix $\begin{bmatrix} 1 & 3 & | & -2 \\ 0 & 1 & | & 5 \\ 0 & 0 & | & 0 \end{bmatrix}$ is in row echelon form.

Skill Building

In Problems 5–16, write the augmented matrix of the given system of equations.

5. $\begin{cases} x - 5y = 5 \\ 4x + 3y = 6 \end{cases}$

6. $\begin{cases} 3x + 4y = 7 \\ 4x - 2y = 5 \end{cases}$

7. $\begin{cases} 2x + 3y - 6 = 0 \\ 4x - 6y + 2 = 0 \end{cases}$

8. $\begin{cases} 9x - y = 0 \\ 3x - y - 4 = 0 \end{cases}$

9. $\begin{cases} 0.01x - 0.03y = 0.06 \\ 0.13x + 0.10y = 0.20 \end{cases}$

10. $\begin{cases} \frac{4}{3}x - \frac{3}{2}y = \frac{3}{4} \\ -\frac{1}{4}x + \frac{1}{3}y = \frac{2}{3} \end{cases}$

11. $\begin{cases} x - y + z = 10 \\ 3x + 3y = 5 \\ x + y + 2z = 2 \end{cases}$

12. $\begin{cases} 5x - y - z = 0 \\ x + y = 5 \\ 2x - 3z = 2 \end{cases}$

13. $\begin{cases} x + y - z = 2 \\ 3x - 2y = 2 \\ 5x + 3y - z = 1 \end{cases}$

14. $\begin{cases} 2x + 3y - 4z = 0 \\ x - 5z + 2 = 0 \\ x + 2y - 3z = -2 \end{cases}$

15. $\begin{cases} x - y - z = 10 \\ 2x + y + 2z = -1 \\ -3x + 4y = 5 \\ 4x - 5y + z = 0 \end{cases}$

16. $\begin{cases} x - y + 2z - w = 5 \\ x + 3y - 4z + 2w = 2 \\ 3x - y - 5z - w = -1 \end{cases}$

In Problems 17–24, perform each row operation on the given augmented matrix.

17. $\begin{bmatrix} 1 & -3 & | & -2 \\ 2 & -5 & | & 5 \end{bmatrix}$ $R_2 = -2r_1 + r_2$

18. $\begin{bmatrix} 1 & -3 & | & -3 \\ 2 & -5 & | & -4 \end{bmatrix}$ $R_2 = -2r_1 + r_2$

19. $\begin{bmatrix} 1 & -3 & 4 & | & 3 \\ 2 & -5 & 6 & | & 6 \\ -3 & 3 & 4 & | & 6 \end{bmatrix}$
(a) $R_2 = -2r_1 + r_2$
(b) $R_3 = 3r_1 + r_3$

20. $\begin{bmatrix} 1 & -3 & 3 & | & -5 \\ 2 & -5 & -3 & | & -5 \\ -3 & -2 & 4 & | & 6 \end{bmatrix}$
(a) $R_2 = -2r_1 + r_2$
(b) $R_3 = 3r_1 + r_3$

21. $\begin{bmatrix} 1 & -3 & 2 & | & -6 \\ 2 & -5 & 3 & | & -4 \\ -3 & -6 & 4 & | & 6 \end{bmatrix}$
(a) $R_2 = -2r_1 + r_2$
(b) $R_3 = 3r_1 + r_3$

22. $\begin{bmatrix} 1 & -3 & -4 & | & -6 \\ 2 & -5 & 6 & | & -6 \\ -3 & 1 & 4 & | & 6 \end{bmatrix}$
(a) $R_2 = -2r_1 + r_2$
(b) $R_3 = 3r_1 + r_3$

23. $\begin{bmatrix} 1 & -3 & 1 & | & -2 \\ 2 & -5 & 6 & | & -2 \\ -3 & 1 & 4 & | & 6 \end{bmatrix}$
(a) $R_2 = -2r_1 + r_2$
(b) $R_3 = 3r_1 + r_3$

24. $\begin{bmatrix} 1 & -3 & -1 & | & 2 \\ 2 & -5 & 2 & | & 6 \\ -3 & -6 & 4 & | & 6 \end{bmatrix}$
(a) $R_2 = -2r_1 + r_2$
(b) $R_3 = 3r_1 + r_3$

In Problems 25–36, the reduced row echelon form of a system of linear equations is given. Write the system of equations corresponding to the given matrix. Use x, y; or x, y, z; or x_1, x_2, x_3, x_4 as variables. Determine whether the system is consistent or inconsistent. If it is consistent, give the solution.

25. $\begin{bmatrix} 1 & 0 & | & 5 \\ 0 & 1 & | & -1 \end{bmatrix}$

26. $\begin{bmatrix} 1 & 0 & | & -4 \\ 0 & 1 & | & 0 \end{bmatrix}$

27. $\begin{bmatrix} 1 & 0 & 0 & | & 1 \\ 0 & 1 & 0 & | & 2 \\ 0 & 0 & 0 & | & 3 \end{bmatrix}$

28. $\begin{bmatrix} 1 & 0 & 0 & | & 0 \\ 0 & 1 & 0 & | & 0 \\ 0 & 0 & 0 & | & 2 \end{bmatrix}$

29. $\begin{bmatrix} 1 & 0 & 2 & | & -1 \\ 0 & 1 & -4 & | & -2 \\ 0 & 0 & 0 & | & 0 \end{bmatrix}$

30. $\begin{bmatrix} 1 & 0 & 4 & | & 4 \\ 0 & 1 & 3 & | & 2 \\ 0 & 0 & 0 & | & 0 \end{bmatrix}$

31. $\begin{bmatrix} 1 & 0 & 0 & 0 & | & 1 \\ 0 & 1 & 0 & 1 & | & 2 \\ 0 & 0 & 1 & 2 & | & 3 \end{bmatrix}$

32. $\begin{bmatrix} 1 & 0 & 0 & 0 & | & 1 \\ 0 & 1 & 0 & 2 & | & 2 \\ 0 & 0 & 1 & 3 & | & 0 \end{bmatrix}$

33. $\begin{bmatrix} 1 & 0 & 0 & 4 & | & 2 \\ 0 & 1 & 1 & 3 & | & 3 \\ 0 & 0 & 0 & 0 & | & 0 \end{bmatrix}$

34. $\begin{bmatrix} 1 & 0 & 0 & 0 & | & 1 \\ 0 & 1 & 0 & 0 & | & 2 \\ 0 & 0 & 1 & 2 & | & 3 \end{bmatrix}$

35. $\begin{bmatrix} 1 & 0 & 0 & 1 & | & -2 \\ 0 & 1 & 0 & 2 & | & 2 \\ 0 & 0 & 1 & -1 & | & 0 \\ 0 & 0 & 0 & 0 & | & 0 \end{bmatrix}$

36. $\begin{bmatrix} 1 & 0 & 0 & 0 & | & 1 \\ 0 & 1 & 0 & 0 & | & 2 \\ 0 & 0 & 1 & 0 & | & 3 \\ 0 & 0 & 0 & 1 & | & 0 \end{bmatrix}$

In Problems 37–72, solve each system of equations using matrices (row operations). If the system has no solution, say that it is inconsistent.

37. $\begin{cases} x + y = 8 \\ x - y = 4 \end{cases}$

38. $\begin{cases} x + 2y = 5 \\ x + y = 3 \end{cases}$

39. $\begin{cases} 2x - 4y = -2 \\ 3x + 2y = 3 \end{cases}$

40. $\begin{cases} 3x + 3y = 3 \\ 4x + 2y = \dfrac{8}{3} \end{cases}$

41. $\begin{cases} x + 2y = 4 \\ 2x + 4y = 8 \end{cases}$

42. $\begin{cases} 3x - y = 7 \\ 9x - 3y = 21 \end{cases}$

43. $\begin{cases} 2x + 3y = 6 \\ x - y = \dfrac{1}{2} \end{cases}$

44. $\begin{cases} \dfrac{1}{2}x + y = -2 \\ x - 2y = 8 \end{cases}$

45. $\begin{cases} 3x - 5y = 3 \\ 15x + 5y = 21 \end{cases}$

46. $\begin{cases} 2x - y = -1 \\ x + \dfrac{1}{2}y = \dfrac{3}{2} \end{cases}$

47. $\begin{cases} x - y = 6 \\ 2x - 3z = 16 \\ 2y + z = 4 \end{cases}$

48. $\begin{cases} 2x + y = -4 \\ -2y + 4z = 0 \\ 3x - 2z = -11 \end{cases}$

49. $\begin{cases} x - 2y + 3z = 7 \\ 2x + y + z = 4 \\ -3x + 2y - 2z = -10 \end{cases}$

50. $\begin{cases} 2x + y - 3z = 0 \\ -2x + 2y + z = -7 \\ 3x - 4y - 3z = 7 \end{cases}$

51. $\begin{cases} 2x - 2y - 2z = 2 \\ 2x + 3y + z = 2 \\ 3x + 2y = 0 \end{cases}$

52. $\begin{cases} 2x - 3y - z = 0 \\ -x + 2y + z = 5 \\ 3x - 4y - z = 1 \end{cases}$

53. $\begin{cases} -x + y + z = -1 \\ -x + 2y - 3z = -4 \\ 3x - 2y - 7z = 0 \end{cases}$

54. $\begin{cases} 2x - 3y - z = 0 \\ 3x + 2y + 2z = 2 \\ x + 5y + 3z = 2 \end{cases}$

55. $\begin{cases} 2x - 2y + 3z = 6 \\ 4x - 3y + 2z = 0 \\ -2x + 3y - 7z = 1 \end{cases}$

56. $\begin{cases} 3x - 2y + 2z = 6 \\ 7x - 3y + 2z = -1 \\ 2x - 3y + 4z = 0 \end{cases}$

57. $\begin{cases} x + y - z = 6 \\ 3x - 2y + z = -5 \\ x + 3y - 2z = 14 \end{cases}$

58. $\begin{cases} x - y + z = -4 \\ 2x - 3y + 4z = -15 \\ 5x + y - 2z = 12 \end{cases}$

59. $\begin{cases} x + 2y - z = -3 \\ 2x - 4y + z = -7 \\ -2x + 2y - 3z = 4 \end{cases}$

60. $\begin{cases} x + 4y - 3z = -8 \\ 3x - y + 3z = 12 \\ x + y + 6z = 1 \end{cases}$

61. $\begin{cases} 3x + y - z = \dfrac{2}{3} \\ 2x - y + z = 1 \\ 4x + 2y = \dfrac{8}{3} \end{cases}$

62. $\begin{cases} x + y = 1 \\ 2x - y + z = 1 \\ x + 2y + z = \dfrac{8}{3} \end{cases}$

63. $\begin{cases} x + y + z + w = 4 \\ 2x - y + z = 0 \\ 3x + 2y + z - w = 6 \\ x - 2y - 2z + 2w = -1 \end{cases}$

64. $\begin{cases} x + y + z + w = 4 \\ -x + 2y + z = 0 \\ 2x + 3y + z - w = 6 \\ -2x + y - 2z + 2w = -1 \end{cases}$

65. $\begin{cases} x + 2y + z = 1 \\ 2x - y + 2z = 2 \\ 3x + y + 3z = 3 \end{cases}$

66. $\begin{cases} x + 2y - z = 3 \\ 2x - y + 2z = 6 \\ x - 3y + 3z = 4 \end{cases}$

67. $\begin{cases} x - y + z = 5 \\ 3x + 2y - 2z = 0 \end{cases}$

68. $\begin{cases} 2x + y - z = 4 \\ -x + y + 3z = 1 \end{cases}$

69. $\begin{cases} 2x + 3y - z = 3 \\ x - y - z = 0 \\ -x + y + z = 0 \\ x + y + 3z = 5 \end{cases}$

70. $\begin{cases} x - 3y + z = 1 \\ 2x - y - 4z = 0 \\ x - 3y + 2z = 1 \\ x - 2y = 5 \end{cases}$

71. $\begin{cases} 4x + y + z - w = 4 \\ x - y + 2z + 3w = 3 \end{cases}$

72. $\begin{cases} -4x + y = 5 \\ 2x - y + z - w = 5 \\ z + w = 4 \end{cases}$

Applications and Extensions

73. Curve Fitting Find the function $y = ax^2 + bx + c$ whose graph contains the points $(1, 2)$, $(-2, -7)$, and $(2, -3)$.

74. Curve Fitting Find the function $y = ax^2 + bx + c$ whose graph contains the points $(1, -1)$, $(3, -1)$, and $(-2, 14)$.

75. Curve Fitting Find the function $f(x) = ax^3 + bx^2 + cx + d$ for which $f(-3) = -112, f(-1) = -2, f(1) = 4,$ and $f(2) = 13.$

76. Curve Fitting Find the function $f(x) = ax^3 + bx^2 + cx + d$ for which $f(-2) = -10, f(-1) = 3, f(1) = 5,$ and $f(3) = 15.$

77. Nutrition A dietitian at Palos Community Hospital wants a patient to have a meal that has 78 grams of protein, 59 grams of carbohydrates, and 75 milligrams of vitamin A. The hospital food service tells the dietitian that the dinner for today is salmon steak, baked eggs, and acorn squash. Each serving of salmon steak has 30 grams of protein, 20 grams of carbohydrates, and 2 milligrams of vitamin A. Each serving of baked eggs contains 15 grams of protein, 2 grams of carbohydrates, and 20 milligrams of vitamin A. Each serving of acorn squash contains 3 grams of protein, 25 grams of carbohydrates, and 32 milligrams of vitamin A. How many servings of each food should the dietitian provide for the patient?

78. Nutrition A dietitian at General Hospital wants a patient to have a meal that has 47 grams of protein, 58 grams of carbohydrates, and 630 milligrams of calcium. The hospital food service tells the dietitian that the dinner for today is pork chops, corn on the cob, and 2% milk. Each serving of pork chops has 23 grams of protein, 0 grams of carbohydrates, and 10 milligrams of calcium. Each serving of corn on the cob contains 3 grams of protein, 16 grams of carbohydrates, and 10 milligrams of calcium. Each glass of 2% milk contains 9 grams of protein, 13 grams of carbohydrates, and 300 milligrams of calcium. How many servings of each food should the dietitian provide for the patient?

79. Financial Planning Carletta has $10,000 to invest. As her financial consultant, you recommend that she invest in Treasury bills that yield 6%, Treasury bonds that yield 7%, and corporate bonds that yield 8%. Carletta wants to have an annual income of $680, and the amount invested in corporate bonds must be half that invested in Treasury bills. Find the amount in each investment.

80. Financial Planning John has $20,000 to invest. As his financial consultant, you recommend that he invest in Treasury bills that yield 5%, Treasury bonds that yield 7%, and corporate bonds that yield 9%. John wants to have an annual income of $1280, and the amount invested in Treasury bills must be two times the amount invested in corporate bonds. Find the amount in each investment.

81. Production To manufacture an automobile requires painting, drying, and polishing. Epsilon Motor Company produces three types of cars: the Delta, the Beta, and the Sigma. Each Delta requires 10 hours for painting, 3 hours for drying, and 2 hours for polishing. A Beta requires 16 hours for painting, 5 hours for drying, and 3 hours for polishing, while a Sigma requires 8 hours for painting, 2 hours for drying, and 1 hour for polishing. If the company has 240 hours for painting, 69 hours for drying, and 41 hours for polishing per month, how many of each type of car are produced?

82. Production A Florida juice company completes the preparation of its products by sterilizing, filling, and labeling bottles. Each case of orange juice requires 9 minutes for sterilizing, 6 minutes for filling, and 1 minute for labeling. Each case of grapefruit juice requires 10 minutes for sterilizing, 4 minutes for filling, and 2 minutes for labeling. Each case of tomato juice requires 12 minutes for sterilizing, 4 minutes for filling, and 1 minute for labeling. If the company runs the sterilizing machine for 398 minutes, the filling machine for 164 minutes, and the labeling machine for 58 minutes, how many cases of each type of juice are prepared?

83. **Electricity: Kirchhoff's Rules** An application of Kirchhoff's Rules to the circuit shown results in the following system of equations:

$$\begin{cases} -4 + 8 - 2I_2 = 0 \\ 8 = 5I_4 + I_1 \\ 4 = 3I_3 + I_1 \\ I_3 + I_4 = I_1 \end{cases}$$

Find the currents I_1, I_2, I_3, and I_4.

SOURCE: Based on Raymond Serway, *Physics*, 3rd ed. (Philadelphia: Saunders, 1990), Prob. 34, p. 790.

84. **Electricity: Kirchhoff's Rules** An application of Kirchhoff's Rules to the circuit shown results in the following system of equations:

$$\begin{cases} I_1 = I_3 + I_2 \\ 24 - 6I_1 - 3I_3 = 0 \\ 12 + 24 - 6I_1 - 6I_2 = 0 \end{cases}$$

Find the currents I_1, I_2, and I_3.

SOURCE: Ibid., Prob. 38, p. 791.

Discussion and Writing

89. Write a brief paragraph or two that outlines your strategy for solving a system of linear equations using matrices.

90. When solving a system of linear equations using matrices, do you prefer to place the augmented matrix in row echelon form or in reduced row echelon form? Give reasons for your choice.

85. **Financial Planning** Three retired couples each require an additional annual income of $2000 per year. As their financial consultant, you recommend that they invest some money in Treasury bills that yield 7%, some money in corporate bonds that yield 9%, and some money in junk bonds that yield 11%. Prepare a table for each couple showing the various ways that their goals can be achieved:
    (a) If the first couple has $20,000 to invest.
    (b) If the second couple has $25,000 to invest.
    (c) If the third couple has $30,000 to invest.
    (d) What advice would you give each couple regarding the amount to invest and the choices available?

    [**Hint:** Higher yields generally carry more risk.]

86. **Financial Planning** A young couple has $25,000 to invest. As their financial consultant, you recommend that they invest some money in Treasury bills that yield 7%, some money in corporate bonds that yield 9%, and some money in junk bonds that yield 11%. Prepare a table showing the various ways that this couple can achieve the following goals:
    (a) The couple wants $1500 per year in income.
    (b) The couple wants $2000 per year in income.
    (c) The couple wants $2500 per year in income.
    (d) What advice would you give this couple regarding the income that they require and the choices available?

    [**Hint:** Higher yields generally carry more risk.]

87. **Pharmacy** A doctor's prescription calls for a daily intake of a supplement containing 40 mg of vitamin C and 30 mg of vitamin D. Your pharmacy stocks three supplements that can be used: one contains 20% vitamin C and 30% vitamin D; a second, 40% vitamin C and 20% vitamin D; and a third, 30% vitamin C and 50% vitamin D. Create a table showing the possible combinations that could be used to fill the prescription.

88. **Pharmacy** A doctor's prescription calls for the creation of pills that contain 12 units of vitamin B_{12} and 12 units of vitamin E. Your pharmacy stocks three powders that can be used to make these pills: one contains 20% vitamin B_{12} and 30% vitamin E; a second, 40% vitamin B_{12} and 20% vitamin E; and a third, 30% vitamin B_{12} and 40% vitamin E. Create a table showing the possible combinations of each powder that could be mixed in each pill.

91. Make up a system of three linear equations containing three variables that has:
    (a) No solution
    (b) Exactly one solution
    (c) Infinitely many solutions
    Give the three systems to a friend to solve and critique.

5.3 Systems of Linear Equations: Determinants

OBJECTIVES 1 Evaluate 2 by 2 Determinants
2 Use Cramer's Rule to Solve a System of Two Equations Containing Two Variables
3 Evaluate 3 by 3 Determinants
4 Use Cramer's Rule to Solve a System of Three Equations Containing Three Variables
5 Know Properties of Determinants

In the preceding section, we described a method of using matrices to solve a system of linear equations. This section deals with yet another method for solving systems of linear equations; however, it can be used only when the number of equations equals the number of variables. Although the method will work for any system (provided that the number of equations equals the number of variables), it is most often used for systems of two equations containing two variables or three equations containing three variables. This method, called *Cramer's Rule*, is based on the concept of a *determinant*.

1 Evaluate 2 by 2 Determinants

If a, b, c, and d are four real numbers, the symbol

$$D = \begin{vmatrix} a & b \\ c & d \end{vmatrix}$$

is called a **2 by 2 determinant**. Its value is the number $ad - bc$; that is,

$$D = \begin{vmatrix} a & b \\ c & d \end{vmatrix} = ad - bc \tag{1}$$

The following device may be helpful for remembering the value of a 2 by 2 determinant:

EXAMPLE 1 **Evaluating a 2 × 2 Determinant**

Evaluate: $\begin{vmatrix} 3 & -2 \\ 6 & 1 \end{vmatrix}$

Algebraic Solution

$$\begin{vmatrix} 3 & -2 \\ 6 & 1 \end{vmatrix} = (3)(1) - (6)(-2)$$

$$= 3 - (-12)$$

$$= 15 \qquad \blacktriangleleft$$

Graphing Solution

First, we enter the matrix whose entries are those of the determinant into the graphing utility and name it A. Using the determinant command, we obtain the result shown in Figure 10.

Figure 10

```
[A]
              [[3 -2]
               [6  1 ]]
det([A])
                     15
```

$\blacktriangleleft$

NOW WORK PROBLEM 7.

2 Use Cramer's Rule to Solve a System of Two Equations Containing Two Variables

Let's now see the role that a 2 by 2 determinant plays in the solution of a system of two equations containing two variables. Consider the system

$$\begin{cases} ax + by = s & (1) \\ cx + dy = t & (2) \end{cases} \qquad \textbf{(2)}$$

We shall use the method of elimination to solve this system.

Provided $d \neq 0$ and $b \neq 0$, this system is equivalent to the system

$$\begin{cases} adx + bdy = sd & (1) \quad \text{Multiply by } d. \\ bcx + bdy = tb & (2) \quad \text{Multiply by } b. \end{cases}$$

On subtracting the second equation from the first equation, we get

$$\begin{cases} (ad - bc)x + 0 \cdot y = sd - tb & (1) \\ bcx + bdy = tb & (2) \end{cases}$$

Now the first equation can be rewritten using determinant notation.

$$\begin{vmatrix} a & b \\ c & d \end{vmatrix} x = \begin{vmatrix} s & b \\ t & d \end{vmatrix}$$

If $D = \begin{vmatrix} a & b \\ c & d \end{vmatrix} = ad - bc \neq 0$, we can solve for x to get

$$x = \frac{\begin{vmatrix} s & b \\ t & d \end{vmatrix}}{\begin{vmatrix} a & b \\ c & d \end{vmatrix}} = \frac{\begin{vmatrix} s & b \\ t & d \end{vmatrix}}{D} \qquad \textbf{(3)}$$

Return now to the original system (2). Provided that $a \neq 0$ and $c \neq 0$, the system is equivalent to

$$\begin{cases} acx + bcy = cs & (1) \quad \text{Multiply by } c. \\ acx + ady = at & (2) \quad \text{Multiply by } a. \end{cases}$$

On subtracting the first equation from the second equation, we get

$$\begin{cases} acx + \quad bcy \quad = \quad cs & (1) \\ 0 \cdot x + (ad - bc)y = at - cs & (2) \end{cases}$$

The second equation can now be rewritten using determinant notation.

$$\begin{vmatrix} a & b \\ c & d \end{vmatrix} y = \begin{vmatrix} a & s \\ c & t \end{vmatrix}$$

If $D = \begin{vmatrix} a & b \\ c & d \end{vmatrix} = ad - bc \neq 0$, we can solve for y to get

$$y = \frac{\begin{vmatrix} a & s \\ c & t \end{vmatrix}}{\begin{vmatrix} a & b \\ c & d \end{vmatrix}} = \frac{\begin{vmatrix} a & s \\ c & t \end{vmatrix}}{D} \qquad (4)$$

Equations (3) and (4) lead us to the following result, called **Cramer's Rule**.

Theorem

Cramer's Rule for Two Equations Containing Two Variables

The solution to the system of equations

$$\begin{cases} ax + by = s & (1) \\ cx + dy = t & (2) \end{cases} \qquad (5)$$

is given by

$$x = \frac{\begin{vmatrix} s & b \\ t & d \end{vmatrix}}{\begin{vmatrix} a & b \\ c & d \end{vmatrix}}, \qquad y = \frac{\begin{vmatrix} a & s \\ c & t \end{vmatrix}}{\begin{vmatrix} a & b \\ c & d \end{vmatrix}} \qquad (6)$$

provided that

$$D = \begin{vmatrix} a & b \\ c & d \end{vmatrix} = ad - bc \neq 0$$

In the derivation given for Cramer's Rule above, we assumed that none of the numbers a, b, c, and d was 0. In Problem 60 you will be asked to complete the proof under the less stringent condition that $D = ad - bc \neq 0$.

Now look carefully at the pattern in Cramer's Rule. The denominator in the solution (6) is the determinant of the coefficients of the variables.

$$\begin{cases} ax + by = s \\ cx + dy = t \end{cases} \qquad D = \begin{vmatrix} a & b \\ c & d \end{vmatrix}$$

In the solution for x, the numerator is the determinant, denoted by D_x, formed by replacing the entries in the first column (the coefficients of x) of D by the constants on the right side of the equal sign.

$$D_x = \begin{vmatrix} s & b \\ t & d \end{vmatrix}$$

In the solution for y, the numerator is the determinant, denoted by D_y, formed by replacing the entries in the second column (the coefficients in y) of D by the constants on the right side of the equal sign.

$$D_y = \begin{vmatrix} a & s \\ c & t \end{vmatrix}$$

Cramer's Rule then states that, if $D \neq 0$,

$$x = \frac{D_x}{D}, \qquad y = \frac{D_y}{D} \tag{7}$$

EXAMPLE 2   **Solving a System of Linear Equations Using Determinants**

Use Cramer's Rule, if applicable, to solve the system

$$\begin{cases} 3x - 2y = 4 & (1) \\ 6x + y = 13 & (2) \end{cases}$$

Algebraic Solution

The determinant D of the coefficients of the variables is

$$D = \begin{vmatrix} 3 & -2 \\ 6 & 1 \end{vmatrix} = (3)(1) - (6)(-2) = 15$$

Because $D \neq 0$, Cramer's Rule (7) can be used.

$$x = \frac{D_x}{D} = \frac{\begin{vmatrix} 4 & -2 \\ 13 & 1 \end{vmatrix}}{15} \qquad y = \frac{D_y}{D} = \frac{\begin{vmatrix} 3 & 4 \\ 6 & 13 \end{vmatrix}}{15}$$

$$= \frac{(4)(1) - (13)(-2)}{15} \qquad = \frac{(3)(13) - (6)(4)}{15}$$

$$= \frac{30}{15} \qquad\qquad = \frac{15}{15}$$

$$= 2 \qquad\qquad\quad = 1$$

Graphing Solution

We enter the coefficient matrix into our graphing utility. Call it A and evaluate $\det[A]$. Since $\det[A] \neq 0$, we can use Cramer's Rule. We enter the matrices D_x and D_y into our graphing utility and call them B and C, respectively. Finally, we find x by calculating $\dfrac{\det[B]}{\det[A]}$ and y by calculating $\dfrac{\det[C]}{\det[A]}$. The results are shown in Figure 11.

Figure 11

The solution is $x = 2$, $y = 1$. ◀

In attempting to use Cramer's Rule, if the determinant D of the coefficients of the variables is found to equal 0 (so that Cramer's Rule is not applicable), then the system either is inconsistent or has infinitely many solutions.

NOW WORK PROBLEM 15.

3 Evaluate 3 by 3 Determinants

To use Cramer's Rule to solve a system of three equations containing three variables, we need to define a 3 by 3 determinant.

A **3 by 3 determinant** is symbolized by

$$\begin{vmatrix} a_{11} & a_{12} & a_{13} \\ a_{21} & a_{22} & a_{23} \\ a_{31} & a_{32} & a_{33} \end{vmatrix} \tag{8}$$

in which $a_{11}, a_{12}, \ldots$, are real numbers.

As with matrices, we use a double subscript to identify an entry by indicating its row and column numbers. For example, the entry a_{23} is in row 2, column 3.

The value of a 3 by 3 determinant may be defined in terms of 2 by 2 determinants by the following formula:

Minus

$$\begin{vmatrix} a_{11} & a_{12} & a_{13} \\ a_{21} & a_{22} & a_{23} \\ a_{31} & a_{32} & a_{33} \end{vmatrix} = a_{11} \begin{vmatrix} a_{22} & a_{23} \\ a_{32} & a_{33} \end{vmatrix} - a_{12} \begin{vmatrix} a_{21} & a_{23} \\ a_{31} & a_{33} \end{vmatrix} + a_{13} \begin{vmatrix} a_{21} & a_{22} \\ a_{31} & a_{32} \end{vmatrix} \tag{9}$$

↑
2 by 2 determinant left after removing row and column containing a_{11}

↑
2 by 2 determinant left after removing row and column containing a_{12}

↑
2 by 2 determinant left after removing row and column containing a_{13}

The 2 by 2 determinants shown in formula (9) are called **minors** of the 3 by 3 determinant. For an n by n determinant, the **minor** M_{ij} of entry a_{ij} is the determinant resulting from removing the ith row and jth column.

EXAMPLE 3 | **Finding Minors of a 3 by 3 Determinant**

For the determinant $A = \begin{vmatrix} 2 & -1 & 3 \\ -2 & 5 & 1 \\ 0 & 6 & -9 \end{vmatrix}$, find: (a) M_{12} (b) M_{23}

Solution (a) M_{12} is the determinant that results from removing the first row and second column from A.

$$A = \begin{vmatrix} 2 & -1 & 3 \\ -2 & 5 & 1 \\ 0 & 6 & -9 \end{vmatrix} \qquad M_{12} = \begin{vmatrix} -2 & 1 \\ 0 & -9 \end{vmatrix} = (-2)(-9) - (0)(1) = 18$$

(b) M_{23} is the determinant that results from removing the second row and third column from A.

$$A = \begin{vmatrix} 2 & -1 & 3 \\ -2 & 5 & 1 \\ 0 & 6 & -9 \end{vmatrix} \qquad M_{23} = \begin{vmatrix} 2 & -1 \\ 0 & 6 \end{vmatrix} = (2)(6) - (0)(-1) = 12$$

◀

Referring back to formula (9), we see that each element a_{ij} is multiplied by its minor, but sometimes this term is added and other times, subtracted. To determine whether to add or subtract a term, we must consider the *cofactor*.

For an n by n determinant A, the **cofactor** of entry a_{ij}, denoted by A_{ij}, is given by

$$A_{ij} = (-1)^{i+j}M_{ij}$$

where M_{ij} is the minor of entry a_{ij}.

The exponent of $(-1)^{i+j}$ is the sum of the row and column of the entry a_{ij}, so if $i + j$ is even, $(-1)^{i+j}$ will equal 1, and if $i + j$ is odd, $(-1)^{i+j}$ will equal -1.

To find the value of a determinant, multiply each entry in any row or column by its cofactor and sum the results. This process is referred to as **expanding across a row or column**. For example, the value of the 3 by 3 determinant in formula (9) was found by expanding across row 1.

If we choose to expand down column 2, we obtain

$$\begin{vmatrix} a_{11} & a_{12} & a_{13} \\ a_{21} & a_{22} & a_{23} \\ a_{31} & a_{32} & a_{33} \end{vmatrix} = (-1)^{1+2}a_{12}\begin{vmatrix} a_{21} & a_{23} \\ a_{31} & a_{33} \end{vmatrix} + (-1)^{2+2}a_{22}\begin{vmatrix} a_{11} & a_{13} \\ a_{31} & a_{33} \end{vmatrix} + (-1)^{3+2}a_{32}\begin{vmatrix} a_{11} & a_{13} \\ a_{21} & a_{23} \end{vmatrix}$$

Expand down column 2.

If we choose to expand across row 3, we obtain

$$\begin{vmatrix} a_{11} & a_{12} & a_{13} \\ a_{21} & a_{22} & a_{23} \\ a_{31} & a_{32} & a_{33} \end{vmatrix} = (-1)^{3+1}a_{31}\begin{vmatrix} a_{12} & a_{13} \\ a_{22} & a_{23} \end{vmatrix} + (-1)^{3+2}a_{32}\begin{vmatrix} a_{11} & a_{13} \\ a_{21} & a_{23} \end{vmatrix} + (-1)^{3+3}a_{33}\begin{vmatrix} a_{11} & a_{12} \\ a_{21} & a_{22} \end{vmatrix}$$

Expand across row 3.

It can be shown that the value of a determinant does not depend on the choice of the row or column used in the expansion. However, expanding across a row or column that has an element equal to 0 reduces the amount of work needed to compute the value of the determinant.

EXAMPLE 4   **Evaluating a 3 × 3 Determinant**

Find the value of the 3 by 3 determinant: $\begin{vmatrix} 3 & 4 & -1 \\ 4 & 6 & 2 \\ 8 & -2 & 3 \end{vmatrix}$

Solution   We choose to expand across row 1.

$$\begin{vmatrix} 3 & 4 & -1 \\ 4 & 6 & 2 \\ 8 & -2 & 3 \end{vmatrix} = (-1)^{1+1}3\begin{vmatrix} 6 & 2 \\ -2 & 3 \end{vmatrix} + (-1)^{1+2}4\begin{vmatrix} 4 & 2 \\ 8 & 3 \end{vmatrix} + (-1)^{1+3}(-1)\begin{vmatrix} 4 & 6 \\ 8 & -2 \end{vmatrix}$$

$$= 3(18 + 4) - 4(12 - 16) + (-1)(-8 - 48)$$

$$= 3(22) - 4(-4) + (-1)(-56)$$

$$= 66 + 16 + 56 = 138$$ ◄

We could also find the value of the 3 by 3 determinant in Example 4 by expanding down column 3.

$$\begin{vmatrix} 3 & 4 & -1 \\ 4 & 6 & 2 \\ 8 & -2 & 3 \end{vmatrix} = (-1)^{1+3}(-1)\begin{vmatrix} 4 & 6 \\ 8 & -2 \end{vmatrix} + (-1)^{2+3}2\begin{vmatrix} 3 & 4 \\ 8 & -2 \end{vmatrix} + (-1)^{3+3}3\begin{vmatrix} 3 & 4 \\ 4 & 6 \end{vmatrix}$$

$$= -1(-8 - 48) - 2(-6 - 32) + 3(18 - 16)$$
$$= 56 + 76 + 6 = 138$$

Evaluating 3×3 determinants on a graphing utility follows the same procedure as evaluating 2×2 determinants.

NOW WORK PROBLEM 11.

4 Use Cramer's Rule to Solve a System of Three Equations Containing Three Variables

Consider the following system of three equations containing three variables.

$$\begin{cases} a_{11}x + a_{12}y + a_{13}z = c_1 \\ a_{21}x + a_{22}y + a_{23}z = c_2 \\ a_{31}x + a_{32}y + a_{33}z = c_3 \end{cases} \tag{10}$$

It can be shown that if the determinant D of the coefficients of the variables is not 0, that is, if

$$D = \begin{vmatrix} a_{11} & a_{12} & a_{13} \\ a_{21} & a_{22} & a_{23} \\ a_{31} & a_{32} & a_{33} \end{vmatrix} \neq 0$$

then the unique solution of system (10) is given by

Cramer's Rule for Three Equations Containing Three Variables

$$x = \frac{D_x}{D} \qquad y = \frac{D_y}{D} \qquad z = \frac{D_z}{D}$$

where

$$D_x = \begin{vmatrix} c_1 & a_{12} & a_{13} \\ c_2 & a_{22} & a_{23} \\ c_3 & a_{32} & a_{33} \end{vmatrix} \qquad D_y = \begin{vmatrix} a_{11} & c_1 & a_{13} \\ a_{21} & c_2 & a_{23} \\ a_{31} & c_3 & a_{33} \end{vmatrix} \qquad D_z = \begin{vmatrix} a_{11} & a_{12} & c_1 \\ a_{21} & a_{22} & c_2 \\ a_{31} & a_{32} & c_3 \end{vmatrix}$$

Do you see the similarity of this pattern and the pattern observed earlier for a system of two equations containing two variables?

EXAMPLE 5

Using Cramer's Rule

Use Cramer's Rule, if applicable, to solve the following system:

$$\begin{cases} 2x + y - z = 3 & (1) \\ -x + 2y + 4z = -3 & (2) \\ x - 2y - 3z = 4 & (3) \end{cases}$$

Solution The value of the determinant D of the coefficients of the variables is

$$D = \begin{vmatrix} 2 & 1 & -1 \\ -1 & 2 & 4 \\ 1 & -2 & -3 \end{vmatrix} = (-1)^{1+1}2 \begin{vmatrix} 2 & 4 \\ -2 & -3 \end{vmatrix} + (-1)^{1+2}1 \begin{vmatrix} -1 & 4 \\ 1 & -3 \end{vmatrix} + (-1)^{1+3}(-1) \begin{vmatrix} -1 & 2 \\ 1 & -2 \end{vmatrix}$$

$$= 2(2) - 1(-1) + (-1)(0)$$

$$= 4 + 1 = 5$$

Because $D \neq 0$, we proceed to find the values of D_x, D_y, and D_z.

$$D_x = \begin{vmatrix} 3 & 1 & -1 \\ -3 & 2 & 4 \\ 4 & -2 & -3 \end{vmatrix} = (-1)^{1+1}3 \begin{vmatrix} 2 & 4 \\ -2 & -3 \end{vmatrix} + (-1)^{1+2}1 \begin{vmatrix} -3 & 4 \\ 4 & -3 \end{vmatrix} + (-1)^{1+3}(-1) \begin{vmatrix} -3 & 2 \\ 4 & -2 \end{vmatrix}$$

$$= 3(2) - 1(-7) + (-1)(-2) = 15$$

$$D_y = \begin{vmatrix} 2 & 3 & -1 \\ -1 & -3 & 4 \\ 1 & 4 & -3 \end{vmatrix} = (-1)^{1+1}2 \begin{vmatrix} -3 & 4 \\ 4 & -3 \end{vmatrix} + (-1)^{1+2}3 \begin{vmatrix} -1 & 4 \\ 1 & -3 \end{vmatrix} + (-1)^{1+3}(-1) \begin{vmatrix} -1 & -3 \\ 1 & 4 \end{vmatrix}$$

$$= 2(-7) - 3(-1) + (-1)(-1)$$

$$= -14 + 3 + 1 = -10$$

$$D_z = \begin{vmatrix} 2 & 1 & 3 \\ -1 & 2 & -3 \\ 1 & -2 & 4 \end{vmatrix} = (-1)^{1+1}2 \begin{vmatrix} 2 & -3 \\ -2 & 4 \end{vmatrix} + (-1)^{1+2}1 \begin{vmatrix} -1 & -3 \\ 1 & 4 \end{vmatrix} + (-1)^{1+3}3 \begin{vmatrix} -1 & 2 \\ 1 & -2 \end{vmatrix}$$

$$= 2(2) - 1(-1) + 3(0) = 5$$

As a result,

$$x = \frac{D_x}{D} = \frac{15}{5} = 3, \qquad y = \frac{D_y}{D} = \frac{-10}{5} = -2, \qquad z = \frac{D_z}{D} = \frac{5}{5} = 1$$

The solution is $x = 3$, $y = -2$, $z = 1$. ◀

If the determinant of the coefficients of the variables of a system of three linear equations containing three variables is 0, then Cramer's Rule is not applicable. In such a case, the system either is inconsistent or has infinitely many solutions.

Solving systems of three equations containing three variables using Cramer's Rule on a graphing utility follows the same procedure as that for solving systems of two equations containing two variables.

NOW WORK PROBLEM 33.

5 Know Properties of Determinants

Determinants have several properties that are sometimes helpful for obtaining their value. We list some of them here.

Theorem The value of a determinant changes sign if any two rows (or any two columns) are interchanged. **(11)**

Proof for 2 by 2 Determinants

$$\begin{vmatrix} a & b \\ c & d \end{vmatrix} = ad - bc \quad \text{and} \quad \begin{vmatrix} c & d \\ a & b \end{vmatrix} = bc - ad = -(ad - bc)$$ ■

| EXAMPLE 6 | **Demonstrating Theorem (11)** |

$$\begin{vmatrix} 3 & 4 \\ 1 & 2 \end{vmatrix} = 6 - 4 = 2 \qquad \begin{vmatrix} 1 & 2 \\ 3 & 4 \end{vmatrix} = 4 - 6 = -2 \qquad \blacktriangleleft$$

Theorem   If all the entries in any row (or any column) equal 0, the value of the determinant is 0.   **(12)**

Proof   Expand across the row (or down the column) containing the 0's.   ∎

Theorem   If any two rows (or any two columns) of a determinant have corresponding entries that are equal, the value of the determinant is 0.   **(13)**

You are asked to prove this result for a 3 by 3 determinant in which the entries in column 1 equal the entries in column 3 in Problem 63.

| EXAMPLE 7 | **Demonstrating Theorem (13)** |

$$\begin{vmatrix} 1 & 2 & 3 \\ 1 & 2 & 3 \\ 4 & 5 & 6 \end{vmatrix} = (-1)^{1+1} 1 \begin{vmatrix} 2 & 3 \\ 5 & 6 \end{vmatrix} + (-1)^{1+2} 2 \begin{vmatrix} 1 & 3 \\ 4 & 6 \end{vmatrix} + (-1)^{1+3} 3 \begin{vmatrix} 1 & 2 \\ 4 & 5 \end{vmatrix}$$

$$= 1(-3) - 2(-6) + 3(-3) = -3 + 12 - 9 = 0 \qquad \blacktriangleleft$$

Theorem   If any row (or any column) of a determinant is multiplied by a nonzero number k, the value of the determinant is also changed by a factor of k.   **(14)**

You are asked to prove this result for a 3 by 3 determinant using row 2 in Problem 62.

| EXAMPLE 8 | **Demonstrating Theorem (14)** |

$$\begin{vmatrix} 1 & 2 \\ 4 & 6 \end{vmatrix} = 6 - 8 = -2$$

$$\begin{vmatrix} k & 2k \\ 4 & 6 \end{vmatrix} = 6k - 8k = -2k = k(-2) = k \begin{vmatrix} 1 & 2 \\ 4 & 6 \end{vmatrix} \qquad \blacktriangleleft$$

Theorem   If the entries of any row (or any column) of a determinant are multiplied by a nonzero number k and the result is added to the corresponding entries of another row (or column), the value of the determinant remains unchanged.   **(15)**

In Problem 64, you are asked to prove this result for a 3 by 3 determinant using rows 1 and 2.

EXAMPLE 9	**Demonstrating Theorem (15)**

$$\begin{vmatrix} 3 & 4 \\ 5 & 2 \end{vmatrix} = -14 \qquad \begin{vmatrix} 3 & 4 \\ 5 & 2 \end{vmatrix} \rightarrow \begin{vmatrix} -7 & 0 \\ 5 & 2 \end{vmatrix} = -14$$

Multiply row 2 by -2 and add to row 1.   ◄

5.3 Assess Your Understanding

Concepts and Vocabulary

1. Cramer's Rule uses _____ to solve a system of linear equations.

2. $D = \begin{vmatrix} a & b \\ c & d \end{vmatrix} = $ _____

3. *True or False:* A 3 by 3 determinant can never equal 0.

4. *True or False:* The value of a determinant remains unchanged if any two rows or any two columns are interchanged.

Skill Building

In Problems 5–14, find the value of each determinant.

5. $\begin{vmatrix} 3 & 1 \\ 4 & 2 \end{vmatrix}$

6. $\begin{vmatrix} 6 & 1 \\ 5 & 2 \end{vmatrix}$

7. $\begin{vmatrix} 6 & 4 \\ -1 & 3 \end{vmatrix}$

8. $\begin{vmatrix} 8 & -3 \\ 4 & 2 \end{vmatrix}$

9. $\begin{vmatrix} -3 & -1 \\ 4 & 2 \end{vmatrix}$

10. $\begin{vmatrix} -4 & 2 \\ -5 & 3 \end{vmatrix}$

11. $\begin{vmatrix} 3 & 4 & 2 \\ 1 & -1 & 5 \\ 1 & 2 & -2 \end{vmatrix}$

12. $\begin{vmatrix} 1 & 3 & -2 \\ 6 & 1 & -5 \\ 8 & 2 & 3 \end{vmatrix}$

13. $\begin{vmatrix} 4 & -1 & 2 \\ 6 & -1 & 0 \\ 1 & -3 & 4 \end{vmatrix}$

14. $\begin{vmatrix} 3 & -9 & 4 \\ 1 & 4 & 0 \\ 8 & -3 & 1 \end{vmatrix}$

In Problems 15–42, solve each system of equations using Cramer's Rule if it is applicable. If Cramer's Rule is not applicable, say so.

15. $\begin{cases} x + y = 8 \\ x - y = 4 \end{cases}$

16. $\begin{cases} x + 2y = 5 \\ x - y = 3 \end{cases}$

17. $\begin{cases} 5x - y = 13 \\ 2x + 3y = 12 \end{cases}$

18. $\begin{cases} x + 3y = 5 \\ 2x - 3y = -8 \end{cases}$

19. $\begin{cases} 3x = 24 \\ x + 2y = 0 \end{cases}$

20. $\begin{cases} 4x + 5y = -3 \\ -2y = -4 \end{cases}$

21. $\begin{cases} 3x - 6y = 24 \\ 5x + 4y = 12 \end{cases}$

22. $\begin{cases} 2x + 4y = 16 \\ 3x - 5y = -9 \end{cases}$

23. $\begin{cases} 3x - 2y = 4 \\ 6x - 4y = 0 \end{cases}$

24. $\begin{cases} -x + 2y = 5 \\ 4x - 8y = 6 \end{cases}$

25. $\begin{cases} 2x - 4y = -2 \\ 3x + 2y = 3 \end{cases}$

26. $\begin{cases} 3x + 3y = 3 \\ 4x + 2y = \dfrac{8}{3} \end{cases}$

27. $\begin{cases} 2x - 3y = -1 \\ 10x + 10y = 5 \end{cases}$

28. $\begin{cases} 3x - 2y = 0 \\ 5x + 10y = 4 \end{cases}$

29. $\begin{cases} 2x + 3y = 6 \\ x - y = \dfrac{1}{2} \end{cases}$

30. $\begin{cases} \dfrac{1}{2}x + y = -2 \\ x - 2y = 8 \end{cases}$

31. $\begin{cases} 3x - 5y = 3 \\ 15x + 5y = 21 \end{cases}$

32. $\begin{cases} 2x - y = -1 \\ x + \dfrac{1}{2}y = \dfrac{3}{2} \end{cases}$

33. $\begin{cases} x + y - z = 6 \\ 3x - 2y + z = -5 \\ x + 3y - 2z = 14 \end{cases}$

34. $\begin{cases} x - y + z = -4 \\ 2x - 3y + 4z = -15 \\ 5x + y - 2z = 12 \end{cases}$

35. $\begin{cases} x + 2y - z = -3 \\ 2x - 4y + z = -7 \\ -2x + 2y - 3z = 4 \end{cases}$

36. $\begin{cases} x + 4y - 3z = -8 \\ 3x - y + 3z = 12 \\ x + y + 6z = 1 \end{cases}$

37. $\begin{cases} x - 2y + 3z = 1 \\ 3x + y - 2z = 0 \\ 2x - 4y + 6z = 2 \end{cases}$

38. $\begin{cases} x - y + 2z = 5 \\ 3x + 2y = 4 \\ -2x + 2y - 4z = -10 \end{cases}$

39. $\begin{cases} x + 2y - z = 0 \\ 2x - 4y + z = 0 \\ -2x + 2y - 3z = 0 \end{cases}$

40. $\begin{cases} x + 4y - 3z = 0 \\ 3x - y + 3z = 0 \\ x + y + 6z = 0 \end{cases}$

41. $\begin{cases} x - 2y + 3z = 0 \\ 3x + y - 2z = 0 \\ 2x - 4y + 6z = 0 \end{cases}$

42. $\begin{cases} x - y + 2z = 0 \\ 3x + 2y = 0 \\ -2x + 2y - 4z = 0 \end{cases}$

In Problems 43–48, solve for x.

43. $\begin{vmatrix} x & x \\ 4 & 3 \end{vmatrix} = 5$ **44.** $\begin{vmatrix} x & 1 \\ 3 & x \end{vmatrix} = -2$ **45.** $\begin{vmatrix} x & 1 & 1 \\ 4 & 3 & 2 \\ -1 & 2 & 5 \end{vmatrix} = 2$ **46.** $\begin{vmatrix} 3 & 2 & 4 \\ 1 & x & 5 \\ 0 & 1 & -2 \end{vmatrix} = 0$ **47.** $\begin{vmatrix} x & 2 & 3 \\ 1 & x & 0 \\ 6 & 1 & -2 \end{vmatrix} = 7$ **48.** $\begin{vmatrix} x & 1 & 2 \\ 1 & x & 3 \\ 0 & 1 & 2 \end{vmatrix} = -4x$

In Problems 49–56, use properties of determinants to find the value of each determinant if it is known that

$$\begin{vmatrix} x & y & z \\ u & v & w \\ 1 & 2 & 3 \end{vmatrix} = 4$$

49. $\begin{vmatrix} 1 & 2 & 3 \\ u & v & w \\ x & y & z \end{vmatrix}$ **50.** $\begin{vmatrix} x & y & z \\ u & v & w \\ 2 & 4 & 6 \end{vmatrix}$ **51.** $\begin{vmatrix} x & y & z \\ -3 & -6 & -9 \\ u & v & w \end{vmatrix}$ **52.** $\begin{vmatrix} 1 & 2 & 3 \\ x-u & y-v & z-w \\ u & v & w \end{vmatrix}$

53. $\begin{vmatrix} 1 & 2 & 3 \\ x-3 & y-6 & z-9 \\ 2u & 2v & 2w \end{vmatrix}$ **54.** $\begin{vmatrix} x & y & z-x \\ u & v & w-u \\ 1 & 2 & 2 \end{vmatrix}$ **55.** $\begin{vmatrix} 1 & 2 & 3 \\ 2x & 2y & 2z \\ u-1 & v-2 & w-3 \end{vmatrix}$ **56.** $\begin{vmatrix} x+3 & y+6 & z+9 \\ 3u-1 & 3v-2 & 3w-3 \\ 1 & 2 & 3 \end{vmatrix}$

Applications and Extensions

57. Geometry: Equation of a Line An equation of the line containing the two points (x_1, y_1) and (x_2, y_2) may be expressed as the determinant

$$\begin{vmatrix} x & y & 1 \\ x_1 & y_1 & 1 \\ x_2 & y_2 & 1 \end{vmatrix} = 0$$

Prove this result by expanding the determinant and comparing the result to the 2-point form of the equation of a line.

58. Geometry: Collinear Points Using the result obtained in Problem 57, show that three distinct points (x_1, y_1), (x_2, y_2), and (x_3, y_3) are collinear (lie on the same line) if and only if

$$\begin{vmatrix} x_1 & y_1 & 1 \\ x_2 & y_2 & 1 \\ x_3 & y_3 & 1 \end{vmatrix} = 0$$

59. Show that $\begin{vmatrix} x^2 & x & 1 \\ y^2 & y & 1 \\ z^2 & z & 1 \end{vmatrix} = (y-z)(x-y)(x-z)$.

60. Complete the proof of Cramer's Rule for two equations containing two variables.

[**Hint:** In system (5), page 538, if $a = 0$, then $b \neq 0$ and $c \neq 0$, since $D = -bc \neq 0$. Now show that equation (6) provides a solution of the system when $a = 0$. There are then three remaining cases: $b = 0$, $c = 0$, and $d = 0$.]

61. Interchange columns 1 and 3 of a 3 by 3 determinant. Show that the value of the new determinant is -1 times the value of the original determinant.

62. Multiply each entry in row 2 of a 3 by 3 determinant by the number k, $k \neq 0$. Show that the value of the new determinant is k times the value of the original determinant.

63. Prove that a 3 by 3 determinant in which the entries in column 1 equal those in column 3 has the value 0.

64. Prove that, if row 2 of a 3 by 3 determinant is multiplied by k, $k \neq 0$, and the result is added to the entries in row 1, there is no change in the value of the determinant.

5.4 Matrix Algebra

OBJECTIVES   1   Find the Sum and Difference of Two Matrices
               2   Find Scalar Multiples of a Matrix
               3   Find the Product of Two Matrices
               4   Find the Inverse of a Matrix
               5   Solve a System of Linear Equations Using Inverse Matrices

In Section 5.2, we defined a matrix as a rectangular array of real numbers and used an augmented matrix to represent a system of linear equations. There is, however, a branch of mathematics, called **linear algebra**, that deals with matrices in such a way that an algebra of matrices is permitted. In this section, we provide a survey of how this **matrix algebra** is developed.

Before getting started, we restate the definition of a matrix.

A **matrix** is defined as a rectangular array of numbers:

$$
\begin{array}{c}
\quad \\
\text{Row 1} \\
\text{Row 2} \\
\vdots \\
\text{Row } i \\
\vdots \\
\text{Row } m
\end{array}
\begin{bmatrix}
a_{11} & a_{12} & \cdots & a_{1j} & \cdots & a_{1n} \\
a_{21} & a_{22} & \cdots & a_{2j} & \cdots & a_{2n} \\
\vdots & \vdots & & \vdots & & \vdots \\
a_{i1} & a_{i2} & \cdots & a_{ij} & \cdots & a_{in} \\
\vdots & \vdots & & \vdots & & \vdots \\
a_{m1} & a_{m2} & \cdots & a_{mj} & \cdots & a_{mn}
\end{bmatrix}
$$

Column 1   Column 2   Column j Column n

Each number a_{ij} of the matrix has two indexes: the **row index** i and the **column index** j. The matrix shown above has m rows and n columns. The numbers a_{ij} are usually referred to as the **entries** of the matrix. For example, a_{23} refers to the entry in the second row, third column.

Let's begin with an example that illustrates how matrices can be used to conveniently represent an array of information.

EXAMPLE 1   **Arranging Data in a Matrix**

In a survey of 900 people, the following information was obtained:

200 males	Thought federal defense spending was too high
150 males	Thought federal defense spending was too low
45 males	Had no opinion
315 females	Thought federal defense spending was too high
125 females	Thought federal defense spending was too low
65 females	Had no opinion

We can arrange these data in a rectangular array as follows:

	Too High	Too Low	No Opinion
Male	200	150	45
Female	315	125	65

or as the matrix

$$
\begin{bmatrix}
200 & 150 & 45 \\
315 & 125 & 65
\end{bmatrix}
$$

This matrix has two rows (representing males and females) and three columns (representing "too high," "too low," and "no opinion"). ◄

The matrix we developed in Example 1 has 2 rows and 3 columns. In general, a matrix with m rows and n columns is called an **m by n matrix**. The matrix we developed in Example 1 is a 2 by 3 matrix and contains $2 \cdot 3 = 6$ entries. An m by n matrix will contain $m \cdot n$ entries.

If an m by n matrix has the same number of rows as columns, that is, if $m = n$, then the matrix is referred to as a **square matrix**.

| EXAMPLE 2 | **Examples of Matrices** |

(a) $\begin{bmatrix} 5 & 0 \\ -6 & 1 \end{bmatrix}$ A 2 by 2 square matrix   (b) $\begin{bmatrix} 1 & 0 & 3 \end{bmatrix}$ A 1 by 3 matrix

(c) $\begin{bmatrix} 6 & -2 & 4 \\ 4 & 3 & 5 \\ 8 & 0 & 1 \end{bmatrix}$ A 3 by 3 square matrix

◀

1 Find the Sum and Difference of Two Matrices

We begin our discussion of matrix algebra by first defining what is meant by two matrices being equal and then defining the operations of addition and subtraction. It is important to note that these definitions require each matrix to have the same number of rows *and* the same number of columns as a pre-requisite for equality and for addition and subtraction.

We usually represent matrices by capital letters, such as A, B, C, and so on.

Two m by n matrices A and B are said to be **equal**, written as

$$A = B$$

provided that each entry a_{ij} in A is equal to the corresponding entry b_{ij} in B.

For example,

$$\begin{bmatrix} 2 & 1 \\ 0.5 & -1 \end{bmatrix} = \begin{bmatrix} \sqrt{4} & 1 \\ \frac{1}{2} & -1 \end{bmatrix} \quad \text{and} \quad \begin{bmatrix} 3 & 2 & 1 \\ 0 & 1 & -2 \end{bmatrix} = \begin{bmatrix} \sqrt{9} & \sqrt{4} & 1 \\ 0 & 1 & \sqrt[3]{-8} \end{bmatrix}$$

$$\begin{bmatrix} 4 & 1 \\ 6 & 1 \end{bmatrix} \neq \begin{bmatrix} 4 & 0 \\ 6 & 1 \end{bmatrix} \quad \text{Because the entries in row 1, column 2 are not equal}$$

$$\begin{bmatrix} 4 & 1 & 2 \\ 6 & 1 & 2 \end{bmatrix} \neq \begin{bmatrix} 4 & 1 & 2 & 3 \\ 6 & 1 & 2 & 4 \end{bmatrix} \quad \text{Because the matrix on the left is 2 by 3 and the matrix on the right is 2 by 4}$$

Suppose that A and B represent two m by n matrices. We define their **sum** $A + B$ to be the m by n matrix formed by adding the corresponding entries a_{ij} of A and b_{ij} of B. The **difference** $A - B$ is defined as the m by n matrix formed by subtracting the entries b_{ij} in B from the corresponding entries a_{ij} in A. Addition and subtraction of matrices are allowed only for matrices having the same number m of rows and the same number n of columns. For example, a 2 by 3 matrix and a 2 by 4 matrix cannot be added or subtracted.

Graphing utilities make the sometimes tedious process of matrix algebra easy. Let's compare how a graphing utility adds and subtracts matrices with doing it algebraically.

| EXAMPLE 3 | **Adding and Subtracting Matrices** |

Suppose that

$$A = \begin{bmatrix} 2 & 4 & 8 & -3 \\ 0 & 1 & 2 & 3 \end{bmatrix} \quad \text{and} \quad B = \begin{bmatrix} -3 & 4 & 0 & 1 \\ 6 & 8 & 2 & 0 \end{bmatrix}$$

Find: (a) $A + B$ (b) $A - B$

Algebraic Solution

(a) $A + B = \begin{bmatrix} 2 & 4 & 8 & -3 \\ 0 & 1 & 2 & 3 \end{bmatrix} + \begin{bmatrix} -3 & 4 & 0 & 1 \\ 6 & 8 & 2 & 0 \end{bmatrix}$

$= \begin{bmatrix} 2 + (-3) & 4 + 4 & 8 + 0 & -3 + 1 \\ 0 + 6 & 1 + 8 & 2 + 2 & 3 + 0 \end{bmatrix}$ *Add correspond-ing entries.*

$= \begin{bmatrix} -1 & 8 & 8 & -2 \\ 6 & 9 & 4 & 3 \end{bmatrix}$

(b) $A - B = \begin{bmatrix} 2 & 4 & 8 & -3 \\ 0 & 1 & 2 & 3 \end{bmatrix} - \begin{bmatrix} -3 & 4 & 0 & 1 \\ 6 & 8 & 2 & 0 \end{bmatrix}$

$= \begin{bmatrix} 2 - (-3) & 4 - 4 & 8 - 0 & -3 - 1 \\ 0 - 6 & 1 - 8 & 2 - 2 & 3 - 0 \end{bmatrix}$ *Subtract corre-sponding entries.*

$= \begin{bmatrix} 5 & 0 & 8 & -4 \\ -6 & -7 & 0 & 3 \end{bmatrix}$

Graphing Solution

Enter the matrices into a graphing utility. Name them $[A]$ and $[B]$. Figure 12 shows the results of adding and subtracting $[A]$ and $[B]$.

Figure 12

```
[A]+[B]
    [[-1  8  8  -2]
     [6   9  4   3]]
[A]-[B]
    [[5   0  8  -4]
     [-6 -7  0   3]]
```

◀ ──── **NOW WORK PROBLEM 7.**

Many of the algebraic properties of sums of real numbers are also true for sums of matrices. Suppose that A, B, and C are m by n matrices. Then matrix addition is **commutative**. That is,

Commutative Property

$$A + B = B + A$$

Matrix addition is also **associative**. That is,

Associative Property

$$(A + B) + C = A + (B + C)$$

Although we shall not prove these results, the proofs, as the following example illustrates, are based on the commutative and associative properties for real numbers.

EXAMPLE 4   **Demonstrating the Commutative Property**

$\begin{bmatrix} 2 & 3 & -1 \\ 4 & 0 & 7 \end{bmatrix} + \begin{bmatrix} -1 & 2 & 1 \\ 5 & -3 & 4 \end{bmatrix} = \begin{bmatrix} 2 + (-1) & 3 + 2 & -1 + 1 \\ 4 + 5 & 0 + (-3) & 7 + 4 \end{bmatrix}$

$= \begin{bmatrix} -1 + 2 & 2 + 3 & 1 + (-1) \\ 5 + 4 & -3 + 0 & 4 + 7 \end{bmatrix}$

$= \begin{bmatrix} -1 & 2 & 1 \\ 5 & -3 & 4 \end{bmatrix} + \begin{bmatrix} 2 & 3 & -1 \\ 4 & 0 & 7 \end{bmatrix}$ ◀

A matrix whose entries are all equal to 0 is called a **zero matrix**. Each of the following matrices is a zero matrix.

$$\begin{bmatrix} 0 & 0 \\ 0 & 0 \end{bmatrix} \text{ 2 by 2 square} \atop \text{zero matrix} \qquad \begin{bmatrix} 0 & 0 & 0 \\ 0 & 0 & 0 \end{bmatrix} \text{ 2 by 3 zero} \atop \text{matrix} \qquad \begin{bmatrix} 0 & 0 & 0 \end{bmatrix} \text{ 1 by 3 zero} \atop \text{matrix}$$

Zero matrices have properties similar to the real number 0. If A is an m by n matrix and 0 is an m by n zero matrix, then

$$A + 0 = A$$

In other words, the zero matrix is the additive identity in matrix algebra.

2 Find Scalar Multiples of a Matrix

We can also multiply a matrix by a real number. If k is a real number and A is an m by n matrix, the matrix kA is the m by n matrix formed by multiplying each entry a_{ij} in A by k. The number k is sometimes referred to as a **scalar**, and the matrix kA is called a **scalar multiple** of A.

EXAMPLE 5 Operations Using Matrices

Suppose that

$$A = \begin{bmatrix} 3 & 1 & 5 \\ -2 & 0 & 6 \end{bmatrix} \qquad B = \begin{bmatrix} 4 & 1 & 0 \\ 8 & 1 & -3 \end{bmatrix} \qquad C = \begin{bmatrix} 9 & 0 \\ -3 & 6 \end{bmatrix}$$

Find: (a) $4A$ (b) $\dfrac{1}{3}C$ (c) $3A - 2B$

Algebraic Solution

(a) $4A = 4\begin{bmatrix} 3 & 1 & 5 \\ -2 & 0 & 6 \end{bmatrix} = \begin{bmatrix} 4 \cdot 3 & 4 \cdot 1 & 4 \cdot 5 \\ 4(-2) & 4 \cdot 0 & 4 \cdot 6 \end{bmatrix} = \begin{bmatrix} 12 & 4 & 20 \\ -8 & 0 & 24 \end{bmatrix}$

(b) $\dfrac{1}{3}C = \dfrac{1}{3}\begin{bmatrix} 9 & 0 \\ -3 & 6 \end{bmatrix} = \begin{bmatrix} \frac{1}{3} \cdot 9 & \frac{1}{3} \cdot 0 \\ \frac{1}{3}(-3) & \frac{1}{3} \cdot 6 \end{bmatrix} = \begin{bmatrix} 3 & 0 \\ -1 & 2 \end{bmatrix}$

(c) $3A - 2B = 3\begin{bmatrix} 3 & 1 & 5 \\ -2 & 0 & 6 \end{bmatrix} - 2\begin{bmatrix} 4 & 1 & 0 \\ 8 & 1 & -3 \end{bmatrix}$

$= \begin{bmatrix} 3 \cdot 3 & 3 \cdot 1 & 3 \cdot 5 \\ 3(-2) & 3 \cdot 0 & 3 \cdot 6 \end{bmatrix} - \begin{bmatrix} 2 \cdot 4 & 2 \cdot 1 & 2 \cdot 0 \\ 2 \cdot 8 & 2 \cdot 1 & 2(-3) \end{bmatrix}$

$= \begin{bmatrix} 9 & 3 & 15 \\ -6 & 0 & 18 \end{bmatrix} - \begin{bmatrix} 8 & 2 & 0 \\ 16 & 2 & -6 \end{bmatrix}$

$= \begin{bmatrix} 9 - 8 & 3 - 2 & 15 - 0 \\ -6 - 16 & 0 - 2 & 18 - (-6) \end{bmatrix}$

$= \begin{bmatrix} 1 & 1 & 15 \\ -22 & -2 & 24 \end{bmatrix}$

Graphing Solution

Enter the matrices $[A]$, $[B]$, and $[C]$ into a graphing utility. Figure 13 shows the required computations.

Figure 13

(a)

(b)

(c)

◀

━ NOW WORK PROBLEM **11.**

We list next some of the algebraic properties of scalar multiplication. Let h and k be real numbers, and let A and B be m by n matrices. Then

Properties of Scalar Multiplication

$$k(hA) = (kh)A$$
$$(k + h)A = kA + hA$$
$$k(A + B) = kA + kB$$

3   Find the Product of Two Matrices

Unlike the straightforward definition for adding two matrices, the definition for multiplying two matrices is not what we might expect. In preparation for this definition, we need the following definitions:

A **row vector** R is a 1 by n matrix

$$R = \begin{bmatrix} r_1 & r_2 & \cdots & r_n \end{bmatrix}$$

A **column vector** C is an n by 1 matrix

$$C = \begin{bmatrix} c_1 \\ c_2 \\ \vdots \\ c_n \end{bmatrix}$$

The **product** RC of R times C is defined as the number

$$RC = \begin{bmatrix} r_1 & r_2 \cdots r_n \end{bmatrix} \begin{bmatrix} c_1 \\ c_2 \\ \vdots \\ c_n \end{bmatrix} = r_1 c_1 + r_2 c_2 + \cdots + r_n c_n$$

Notice that a row vector and a column vector can be multiplied only if they contain the same number of entries.

EXAMPLE 6   **The Product of a Row Vector by a Column Vector**

If $R = \begin{bmatrix} 3 & -5 & 2 \end{bmatrix}$ and $C = \begin{bmatrix} 3 \\ 4 \\ -5 \end{bmatrix}$, then

$$RC = \begin{bmatrix} 3 & -5 & 2 \end{bmatrix} \begin{bmatrix} 3 \\ 4 \\ -5 \end{bmatrix} = 3 \cdot 3 + (-5)4 + 2(-5) = 9 - 20 - 10 = -21$$

◀

Let's look at an application of the product of a row vector by a column vector.

| EXAMPLE 7 | **Using Matrices to Compute Revenue** |

A clothing store sells men's shirts for $40, silk ties for $20, and wool suits for $400. Last month, the store had sales consisting of 100 shirts, 200 ties, and 50 suits. What was the total revenue due to these sales?

Solution We set up a row vector R to represent the prices of each item and a column vector C to represent the corresponding number of items sold. Then

$$
\begin{array}{cc}
\text{Prices} & \text{Number} \\
\text{Shirts Ties Suits} & \text{sold}
\end{array}
$$

$$
R = [40 \quad 20 \quad 400] \qquad C = \begin{bmatrix} 100 \\ 200 \\ 50 \end{bmatrix} \begin{array}{l} \text{Shirts} \\ \text{Ties} \\ \text{Suits} \end{array}
$$

The total revenue obtained is the product RC. That is,

$$
RC = [40 \ 20 \ 400] \begin{bmatrix} 100 \\ 200 \\ 50 \end{bmatrix}
$$

$$
= \underbrace{40 \cdot 100}_{\text{Shirt revenue}} + \underbrace{20 \cdot 200}_{\text{Tie revenue}} + \underbrace{400 \cdot 50}_{\text{Suit revenue}} = \underbrace{\$28{,}000}_{\text{Total revenue}}
$$

◀

The definition for multiplying two matrices is based on the definition of a row vector times a column vector.

Let A denote an m by r matrix, and let B denote an r by n matrix. The **product** AB is defined as the m by n matrix whose entry in row i, column j is the product of the ith row of A and the jth column of B.

The definition of the product AB of two matrices A and B, in this order, requires that the number of columns of A equal the number of rows of B; otherwise, no product is defined.

In Words
To find the product AB, the number of columns in A must equal the number of rows in B.

$$
\begin{array}{ccc}
A & & B \\
m \text{ by } r & & r \text{ by } n
\end{array}
$$

Must be same
for AB to be defined
AB is m by n.

An example will help to clarify the definition.

| EXAMPLE 8 | **Multiplying Two Matrices** |

Find the product AB if

$$
A = \begin{bmatrix} 2 & 4 & -1 \\ 5 & 8 & 0 \end{bmatrix} \quad \text{and} \quad B = \begin{bmatrix} 2 & 5 & 1 & 4 \\ 4 & 8 & 0 & 6 \\ -3 & 1 & -2 & -1 \end{bmatrix}
$$

Solution First, we note that A is 2 by 3 and B is 3 by 4, so the product AB is defined and will be a 2 by 4 matrix.

Algebraic Solution

Suppose that we want the entry in row 2, column 3 of AB. To find it, we find the product of the row vector from row 2 of A and the column vector from column 3 of B.

$$\underset{\text{Row 2 of }A}{[5 \quad 8 \quad 0]} \overset{\text{Column 3 of }B}{\begin{bmatrix} 1 \\ 0 \\ -2 \end{bmatrix}} = 5 \cdot 1 + 8 \cdot 0 + 0(-2) = 5$$

So far, we have

$$AB = \begin{bmatrix} \underline{\quad} & \underline{\quad} & \overset{\overset{\text{Column 3}}{\downarrow}}{5} & \underline{\quad} \end{bmatrix} \quad \leftarrow \text{Row 2}$$

Now, to find the entry in row 1, column 4 of AB, we find the product of row 1 of A and column 4 of B.

$$\underset{\text{Row 1 of }A}{[2 \quad 4 \quad -1]} \overset{\text{Column 4 of }B}{\begin{bmatrix} 4 \\ 6 \\ -1 \end{bmatrix}} = 2 \cdot 4 + 4 \cdot 6 + (-1)(-1) = 33$$

Continuing in this fashion, we find AB.

$$AB = \begin{bmatrix} 2 & 4 & -1 \\ 5 & 8 & 0 \end{bmatrix} \begin{bmatrix} 2 & 5 & 1 & 4 \\ 4 & 8 & 0 & 6 \\ -3 & 1 & -2 & -1 \end{bmatrix}$$

$$= \begin{bmatrix} \text{Row 1 of }A & \text{Row 1 of }A & \text{Row 1 of }A & \text{Row 1 of }A \\ \text{times} & \text{times} & \text{times} & \text{times} \\ \text{column 1 of }B & \text{column 2 of }B & \text{column 3 of }B & \text{column 4 of }B \\ \\ \text{Row 2 of }A & \text{Row 2 of }A & \text{Row 2 of }A & \text{Row 2 of }A \\ \text{times} & \text{times} & \text{times} & \text{times} \\ \text{column 1 of }B & \text{column 2 of }B & \text{column 3 of }B & \text{column 4 of }B \end{bmatrix}$$

$$= \begin{bmatrix} 2 \cdot 2 + 4 \cdot 4 + (-1)(-3) & 2 \cdot 5 + 4 \cdot 8 + (-1)1 & 2 \cdot 1 + 4 \cdot 0 + (-1)(-2) & 33 \text{ (from earlier)} \\ 5 \cdot 2 + 8 \cdot 4 + 0(-3) & 5 \cdot 5 + 8 \cdot 8 + 0 \cdot 1 & 5 \text{ (from earlier)} & 5 \cdot 4 + 8 \cdot 6 + 0(-1) \end{bmatrix}$$

$$= \begin{bmatrix} 23 & 41 & 4 & 33 \\ 42 & 89 & 5 & 68 \end{bmatrix}$$

◀

Graphing Solution

Enter the matrices A and B into a graphing utility. Figure 14 shows the product AB.

Figure 14

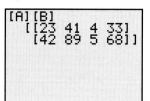

◀

---- **NOW WORK PROBLEM 23.**

Notice that for the matrices given in Example 8 the product BA is not defined, because B is 3 by 4 and A is 2 by 3. Try calculating BA on a graphing utility. What do you notice?

Another result that can occur when multiplying two matrices is illustrated in the next example.[*]

---

[*] For most of the examples that follow, we will multiply matrices algebraically. You should verify each result using a graphing utility.

| EXAMPLE 9 | **Multiplying Two Matrices** |

If

$$A = \begin{bmatrix} 2 & 1 & 3 \\ 1 & -1 & 0 \end{bmatrix} \quad \text{and} \quad B = \begin{bmatrix} 1 & 0 \\ 2 & 1 \\ 3 & 2 \end{bmatrix}$$

find: (a) AB (b) BA

Solution   (a) $AB = \begin{bmatrix} 2 & 1 & 3 \\ 1 & -1 & 0 \end{bmatrix} \begin{bmatrix} 1 & 0 \\ 2 & 1 \\ 3 & 2 \end{bmatrix} = \begin{bmatrix} 13 & 7 \\ -1 & -1 \end{bmatrix}$

   $\underbrace{\qquad}_{2 \text{ by } 3} \quad \underbrace{\qquad}_{3 \text{ by } 2} \qquad \underbrace{\qquad}_{2 \text{ by } 2}$

   (b) $BA = \begin{bmatrix} 1 & 0 \\ 2 & 1 \\ 3 & 2 \end{bmatrix} \begin{bmatrix} 2 & 1 & 3 \\ 1 & -1 & 0 \end{bmatrix} = \begin{bmatrix} 2 & 1 & 3 \\ 5 & 1 & 6 \\ 8 & 1 & 9 \end{bmatrix}$

   $\underbrace{\qquad}_{3 \text{ by } 2} \quad \underbrace{\qquad}_{2 \text{ by } 3} \qquad \underbrace{\qquad}_{3 \text{ by } 3}$ ◀

Notice in Example 9 that AB is 2 by 2 and BA is 3 by 3. It is possible for both AB and BA to be defined, yet be unequal. In fact, even if A and B are both n by n matrices so that AB and BA are each defined and n by n, AB and BA will usually be unequal.

| EXAMPLE 10 | **Multiplying Two Square Matrices** |

If

$$A = \begin{bmatrix} 2 & 1 \\ 0 & 4 \end{bmatrix} \quad \text{and} \quad B = \begin{bmatrix} -3 & 1 \\ 1 & 2 \end{bmatrix}$$

find: (a) AB (b) BA

Solution   (a) $AB = \begin{bmatrix} 2 & 1 \\ 0 & 4 \end{bmatrix} \begin{bmatrix} -3 & 1 \\ 1 & 2 \end{bmatrix} = \begin{bmatrix} -5 & 4 \\ 4 & 8 \end{bmatrix}$

   (b) $BA = \begin{bmatrix} -3 & 1 \\ 1 & 2 \end{bmatrix} \begin{bmatrix} 2 & 1 \\ 0 & 4 \end{bmatrix} = \begin{bmatrix} -6 & 1 \\ 2 & 9 \end{bmatrix}$ ◀

The preceding examples demonstrate that an important property of real numbers, the commutative property of multiplication, is not shared by matrices. In general:

Theorem   Matrix multiplication is not commutative.

✎━━━ **NOW WORK PROBLEMS 13 AND 15.**

Next we give two of the properties of real numbers that are shared by matrices. Assuming that each product and sum is defined, we have the following:

Associative Property

$$A(BC) = (AB)C$$

Distributive Property

$$A(B + C) = AB + AC$$

For an n by n square matrix, the entries located in row i, column i, $1 \le i \le n$, are called the **diagonal entries**. An n by n square matrix whose diagonal entries are 1's, while all other entries are 0's, is called the **identity matrix I_n**. For example,

$$I_2 = \begin{bmatrix} 1 & 0 \\ 0 & 1 \end{bmatrix} \qquad I_3 = \begin{bmatrix} 1 & 0 & 0 \\ 0 & 1 & 0 \\ 0 & 0 & 1 \end{bmatrix}$$

and so on.

EXAMPLE 11   **Multiplication with an Identity Matrix**

Let

$$A = \begin{bmatrix} -1 & 2 & 0 \\ 0 & 1 & 3 \end{bmatrix} \quad \text{and} \quad B = \begin{bmatrix} 3 & 2 \\ 4 & 6 \\ 5 & 2 \end{bmatrix}$$

Find: (a) AI_3 (b) I_2A (c) BI_2

Solution   (a) $AI_3 = \begin{bmatrix} -1 & 2 & 0 \\ 0 & 1 & 3 \end{bmatrix} \begin{bmatrix} 1 & 0 & 0 \\ 0 & 1 & 0 \\ 0 & 0 & 1 \end{bmatrix} = \begin{bmatrix} -1 & 2 & 0 \\ 0 & 1 & 3 \end{bmatrix} = A$

(b) $I_2A = \begin{bmatrix} 1 & 0 \\ 0 & 1 \end{bmatrix} \begin{bmatrix} -1 & 2 & 0 \\ 0 & 1 & 3 \end{bmatrix} = \begin{bmatrix} -1 & 2 & 0 \\ 0 & 1 & 3 \end{bmatrix} = A$

(c) $BI_2 = \begin{bmatrix} 3 & 2 \\ 4 & 6 \\ 5 & 2 \end{bmatrix} \begin{bmatrix} 1 & 0 \\ 0 & 1 \end{bmatrix} = \begin{bmatrix} 3 & 2 \\ 4 & 6 \\ 5 & 2 \end{bmatrix} = B$ ◀

Example 11 demonstrates the following property:

Identity Property

If A is an m by n matrix, then

$$I_m A = A \quad \text{and} \quad AI_n = A$$

If A is an n by n square matrix, then $AI_n = I_n A = A$.

An identity matrix has properties similar to those of the real number 1. In other words, the identity matrix is a multiplicative identity in matrix algebra.

4 Find the Inverse of a Matrix

Let A be a square n by n matrix. If there exists an n by n matrix A^{-1}, read "A inverse," for which

$$AA^{-1} = A^{-1}A = I_n$$

then A^{-1} is called the **inverse** of the matrix A.

As we shall soon see, not every square matrix has an inverse. When a matrix A does have an inverse A^{-1}, then A is said to be **nonsingular**. If a matrix A has no inverse, it is called **singular**.[*]

EXAMPLE 12 **Multiplying a Matrix by Its Inverse**

Show that the inverse of

$$A = \begin{bmatrix} 3 & 1 \\ 2 & 1 \end{bmatrix} \quad \text{is} \quad A^{-1} = \begin{bmatrix} 1 & -1 \\ -2 & 3 \end{bmatrix}$$

Solution   We need to show that $AA^{-1} = A^{-1}A = I_2$.

$$AA^{-1} = \begin{bmatrix} 3 & 1 \\ 2 & 1 \end{bmatrix}\begin{bmatrix} 1 & -1 \\ -2 & 3 \end{bmatrix} = \begin{bmatrix} 1 & 0 \\ 0 & 1 \end{bmatrix} = I_2$$

$$A^{-1}A = \begin{bmatrix} 1 & -1 \\ -2 & 3 \end{bmatrix}\begin{bmatrix} 3 & 1 \\ 2 & 1 \end{bmatrix} = \begin{bmatrix} 1 & 0 \\ 0 & 1 \end{bmatrix} = I_2$$ ◀

We now show one way to find the inverse of

$$A = \begin{bmatrix} 3 & 1 \\ 2 & 1 \end{bmatrix}$$

Suppose that A^{-1} is given by

$$A^{-1} = \begin{bmatrix} x & y \\ z & w \end{bmatrix} \tag{1}$$

where x, y, z, and w are four variables. Based on the definition of an inverse, if A has an inverse, then

$$AA^{-1} = I_2$$

$$\begin{bmatrix} 3 & 1 \\ 2 & 1 \end{bmatrix}\begin{bmatrix} x & y \\ z & w \end{bmatrix} = \begin{bmatrix} 1 & 0 \\ 0 & 1 \end{bmatrix}$$

$$\begin{bmatrix} 3x + z & 3y + w \\ 2x + z & 2y + w \end{bmatrix} = \begin{bmatrix} 1 & 0 \\ 0 & 1 \end{bmatrix}$$

Because corresponding entries must be equal, it follows that this matrix equation is equivalent to four ordinary equations.

$$\begin{cases} 3x + z = 1 \\ 2x + z = 0 \end{cases} \qquad \begin{cases} 3y + w = 0 \\ 2y + w = 1 \end{cases}$$

---

[*]If the determinant of A is zero, then A is singular. (Refer to Section 5.3.)

The augmented matrix of each system is

$$\begin{bmatrix} 3 & 1 & | & 1 \\ 2 & 1 & | & 0 \end{bmatrix} \quad \begin{bmatrix} 3 & 1 & | & 0 \\ 2 & 1 & | & 1 \end{bmatrix} \tag{2}$$

The usual procedure would be to transform each augmented matrix into reduced row echelon form. Notice, though, that the left sides of the augmented matrices are equal, so the same row operations (see Section 5.2) can be used to reduce each one. Thus, we find it more efficient to combine the two augmented matrices (2) into a single matrix, as shown next, and then transform it into reduced row echelon form.

$$\begin{bmatrix} 3 & 1 & | & 1 & 0 \\ 2 & 1 & | & 0 & 1 \end{bmatrix}$$

We attempt to transform the left side into an identity matrix.

$$\begin{bmatrix} 3 & 1 & | & 1 & 0 \\ 2 & 1 & | & 0 & 1 \end{bmatrix} \underset{\substack{\uparrow \\ R_1 = -1r_2 + r_1}}{\rightarrow} \begin{bmatrix} 1 & 0 & | & 1 & -1 \\ 2 & 1 & | & 0 & 1 \end{bmatrix}$$

$$\underset{\substack{\uparrow \\ R_2 = -2r_1 + r_2}}{\rightarrow} \begin{bmatrix} 1 & 0 & | & 1 & -1 \\ 0 & 1 & | & -2 & 3 \end{bmatrix} \tag{3}$$

Matrix (3) is in reduced row echelon form.

Now we reverse the earlier step of combining the two augmented matrices in (2) and write the single matrix (3) as two augmented matrices.

$$\begin{bmatrix} 1 & 0 & | & 1 \\ 0 & 1 & | & -2 \end{bmatrix} \quad \text{and} \quad \begin{bmatrix} 1 & 0 & | & -1 \\ 0 & 1 & | & 3 \end{bmatrix}$$

We conclude from these matrices that $x = 1$, $z = -2$, and $y = -1$, $w = 3$. Substituting these values into matrix (1), we find that

$$A^{-1} = \begin{bmatrix} 1 & -1 \\ -2 & 3 \end{bmatrix}$$

Notice in display (3) that the 2 by 2 matrix to the right of the vertical bar is, in fact, the inverse of A. Also notice that the identity matrix I_2 is the matrix that appears to the left of the vertical bar. These observations and the procedures followed above will work in general.

Procedure for Finding the Inverse of a Nonsingular Matrix

To find the inverse of an n by n nonsingular matrix A, proceed as follows:

STEP 1: Form the matrix $[A | I_n]$.

STEP 2: Transform the matrix $[A | I_n]$ into reduced row echelon form.

STEP 3: The reduced row echelon form of $[A | I_n]$ will contain the identity matrix I_n on the left of the vertical bar; the n by n matrix on the right of the vertical bar is the inverse of A.

In other words, if A is nonsingular, we begin with the matrix $[A | I_n]$ and, after transforming it into reduced row echelon form, we end up with the matrix $[I_n | A^{-1}]$.

Let's look at another example.

EXAMPLE 13   **Finding the Inverse of a Matrix**

The matrix

$$A = \begin{bmatrix} 1 & 1 & 0 \\ -1 & 3 & 4 \\ 0 & 4 & 3 \end{bmatrix}$$

is nonsingular. Find its inverse.

Algebraic Solution

First, we form the matrix

$$[A|I_3] = \begin{bmatrix} 1 & 1 & 0 & | & 1 & 0 & 0 \\ -1 & 3 & 4 & | & 0 & 1 & 0 \\ 0 & 4 & 3 & | & 0 & 0 & 1 \end{bmatrix}$$

Next, we use row operations to transform $[A|I_3]$ into reduced row echelon form.

$$\begin{bmatrix} 1 & 1 & 0 & | & 1 & 0 & 0 \\ -1 & 3 & 4 & | & 0 & 1 & 0 \\ 0 & 4 & 3 & | & 0 & 0 & 1 \end{bmatrix} \rightarrow \begin{bmatrix} 1 & 1 & 0 & | & 1 & 0 & 0 \\ 0 & 4 & 4 & | & 1 & 1 & 0 \\ 0 & 4 & 3 & | & 0 & 0 & 1 \end{bmatrix} \rightarrow \begin{bmatrix} 1 & 1 & 0 & | & 1 & 0 & 0 \\ 0 & 1 & 1 & | & \dfrac{1}{4} & \dfrac{1}{4} & 0 \\ 0 & 4 & 3 & | & 0 & 0 & 1 \end{bmatrix}$$

$$\uparrow \quad R_2 = r_1 + r_2 \qquad\qquad \uparrow \quad R_2 = \frac{1}{4}r_2$$

$$\rightarrow \begin{bmatrix} 1 & 0 & -1 & | & \dfrac{3}{4} & -\dfrac{1}{4} & 0 \\ 0 & 1 & 1 & | & \dfrac{1}{4} & \dfrac{1}{4} & 0 \\ 0 & 0 & -1 & | & -1 & -1 & 1 \end{bmatrix} \rightarrow \begin{bmatrix} 1 & 0 & -1 & | & \dfrac{3}{4} & -\dfrac{1}{4} & 0 \\ 0 & 1 & 1 & | & \dfrac{1}{4} & \dfrac{1}{4} & 0 \\ 0 & 0 & 1 & | & 1 & 1 & -1 \end{bmatrix}$$

$$\uparrow \quad \begin{array}{l} R_1 = -1r_2 + r_1 \\ R_3 = -4r_2 + r_3 \end{array} \qquad\qquad \uparrow \quad R_3 = -1r_3$$

$$\rightarrow \begin{bmatrix} 1 & 0 & 0 & | & \dfrac{7}{4} & \dfrac{3}{4} & -1 \\ 0 & 1 & 0 & | & -\dfrac{3}{4} & -\dfrac{3}{4} & 1 \\ 0 & 0 & 1 & | & 1 & 1 & -1 \end{bmatrix}$$

$$\uparrow \quad \begin{array}{l} R_1 = r_3 + r_1 \\ R_2 = -1r_3 + r_2 \end{array}$$

Graphing Solution

Enter the matrix A into a graphing utility. Figure 15 shows A^{-1}.

Figure 15

```
[A]⁻¹
[[1.75 .75   -1]
 [-.75 -.75  1 ]
 [1    1    -1]]
```

◄

The matrix $[A|I_3]$ is now in reduced row echelon form, and the identity matrix I_3 is on the left of the vertical bar. Hence, the inverse of A is

$$A^{-1} = \begin{bmatrix} \dfrac{7}{4} & \dfrac{3}{4} & -1 \\ -\dfrac{3}{4} & -\dfrac{3}{4} & 1 \\ 1 & 1 & -1 \end{bmatrix}$$

You can (and should) verify that this is the correct inverse by showing that

$$AA^{-1} = A^{-1}A = I_3.$$

◄

NOW WORK PROBLEM 31.

If transforming the matrix $[A|I_n]$ into reduced row echelon form does not result in the identity matrix I_n to the left of the vertical bar, then A is singular and has no inverse. The next example demonstrates such a matrix.

EXAMPLE 14   **Showing That a Matrix Has No Inverse**

Show that the matrix $A = \begin{bmatrix} 4 & 6 \\ 2 & 3 \end{bmatrix}$ has no inverse.

Algebraic Solution

Proceeding as in Example 13, we form the matrix

$$[A|I_2] = \begin{bmatrix} 4 & 6 & | & 1 & 0 \\ 2 & 3 & | & 0 & 1 \end{bmatrix}$$

Then we use row operations to transform $[A|I_2]$ into reduced row echelon form.

$$[A|I_2] = \begin{bmatrix} 4 & 6 & | & 1 & 0 \\ 2 & 3 & | & 0 & 1 \end{bmatrix} \rightarrow \begin{bmatrix} 1 & \dfrac{3}{2} & | & \dfrac{1}{4} & 0 \\ 2 & 3 & | & 0 & 1 \end{bmatrix} \rightarrow \begin{bmatrix} 1 & \dfrac{3}{2} & | & \dfrac{1}{4} & 0 \\ 0 & 0 & | & -\dfrac{1}{2} & 1 \end{bmatrix}$$

$$\uparrow \qquad\qquad\qquad \uparrow$$
$$R_1 = \tfrac{1}{4}r_1 \qquad\qquad R_2 = -2r_1 + r_2$$

The matrix $[A|I_2]$ is sufficiently reduced for us to see that the identity matrix cannot appear to the left of the vertical bar. We conclude that A is singular and so has no inverse. ◄

Graphing Solution

Enter the matrix A. Figure 16 shows the result when we try to find its inverse. The ERRor comes about because A is singular.

Figure 16

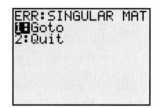

— **Seeing the Concept** —
Compute the determinant of A in Example 14 using a graphing utility. What is the result? Are you surprised?

NOW WORK PROBLEM 59.

---

5   Solve a System of Linear Equations Using Inverse Matrices

Inverse matrices can be used to solve systems of equations in which the number of equations is the same as the number of variables.

EXAMPLE 15   **Using the Inverse Matrix to Solve a System of Linear Equations**

Solve the system of equations: $\begin{cases} x + y & = 3 \\ -x + 3y + 4z & = -3 \\ 4y + 3z & = 2 \end{cases}$

Solution   If we let

$$A = \begin{bmatrix} 1 & 1 & 0 \\ -1 & 3 & 4 \\ 0 & 4 & 3 \end{bmatrix} \qquad X = \begin{bmatrix} x \\ y \\ z \end{bmatrix} \qquad B = \begin{bmatrix} 3 \\ -3 \\ 2 \end{bmatrix}$$

then the original system of equations can be written compactly as the matrix equation

$$AX = B \tag{4}$$

We know from Example 13 that the matrix A has the inverse A^{-1}, so we multiply each side of equation (4) by A^{-1}.

$$AX = B$$
$$A^{-1}(AX) = A^{-1}B \qquad \text{Multiply both sides by } A^{-1}.$$
$$(A^{-1}A)X = A^{-1}B \qquad \text{Associative property of multiplication}$$
$$I_3 X = A^{-1}B \qquad \text{Definition of an inverse matrix}$$
$$X = A^{-1}B \qquad \text{Property of the identity matrix} \tag{5}$$

Now we use (5) to find $X = \begin{bmatrix} x \\ y \\ z \end{bmatrix}$. This can be done either algebraically or graphically.

Algebraic Solution

$$X = \begin{bmatrix} x \\ y \\ z \end{bmatrix} = A^{-1}B = \begin{bmatrix} \dfrac{7}{4} & \dfrac{3}{4} & -1 \\[2mm] -\dfrac{3}{4} & -\dfrac{3}{4} & 1 \\[2mm] 1 & 1 & -1 \end{bmatrix} \begin{bmatrix} 3 \\ -3 \\ 2 \end{bmatrix} = \begin{bmatrix} 1 \\ 2 \\ -2 \end{bmatrix}$$

Example 13

Graphing Solution

Enter the matrices A and B into a graphing utility. Figure 17 shows the solution to the system of equations.

Figure 17

```
[A]-1[B]
            [[1 ]
             [2 ]
             [-2]]
```

So, $x = 1$, $y = 2$, $z = -2$. ◄

The method used in Example 15 to solve a system of equations is particularly useful when it is necessary to solve several systems of equations in which the constants appearing to the right of the equal signs change, while the coefficients of the variables on the left side remain the same. See Problems 39–58 for some illustrations. Be careful; this method can only be used if the inverse exists. If it does not exist, row reduction must be used since the system is either inconsistent or dependent.

➤ **NOW WORK PROBLEM 43.**

HISTORICAL FEATURE

Arthur Cayley
(1821–1895)

Matrices were invented in 1857 by Arthur Cayley (1821–1895) as a way of efficiently computing the result of substituting one linear system into another (see Historical Problem 2). The resulting system had incredible richness, in the sense that a wide variety of mathematical systems could be mimicked by the matrices. Cayley and his friend James J. Sylvester (1814–1897) spent much of the rest of their lives elaborating the theory. The torch was then passed to Georg Frobenius (1849–1917), whose deep investigations established a central place for matrices in modern mathematics. In 1924, rather to the surprise of physicists, it was found that matrices (with complex numbers in them) were exactly the right tool for describing the behavior of atomic systems. Today, matrices are used in a wide variety of applications.

Historical Problems

1. Matrices and Complex Numbers Frobenius emphasized in his research how matrices could be used to mimic other mathematical systems. Here, we mimic the behavior of complex numbers using matrices. Mathematicians call such a relationship an *isomorphism*.

$$\text{Complex number} \longleftrightarrow \text{Matrix}$$

$$a + bi \longleftrightarrow \begin{bmatrix} a & b \\ -b & a \end{bmatrix}$$

Note that the complex number can be read off the top line of the matrix. Thus,

$$2 + 3i \longleftrightarrow \begin{bmatrix} 2 & 3 \\ -3 & 2 \end{bmatrix} \quad \text{and} \quad \begin{bmatrix} 4 & -2 \\ 2 & 4 \end{bmatrix} \longleftrightarrow 4 - 2i$$

(a) Find the matrices corresponding to $2 - 5i$ and $1 + 3i$.
(b) Multiply the two matrices.
(c) Find the corresponding complex number for the matrix found in part (b).
(d) Multiply $2 - 5i$ and $1 + 3i$. The result should be the same as that found in part (c).

The process also works for addition and subtraction. Try it for yourself.

2. Cayley's Definition of Matrix Multiplication Cayley invented matrix multiplication to simplify the following problem:

$$\begin{cases} u = ar + bs \\ v = cr + ds \end{cases} \qquad \begin{cases} x = ku + lv \\ y = mu + nv \end{cases}$$

(a) Find x and y in terms of r and s by substituting u and v from the first system of equations into the second system of equations.
(b) Use the result of part (a) to find the 2 by 2 matrix A in

$$\begin{bmatrix} x \\ y \end{bmatrix} = A \begin{bmatrix} r \\ s \end{bmatrix}$$

(c) Now look at the following way to do it. Write the equations in matrix form.

$$\begin{bmatrix} u \\ v \end{bmatrix} = \begin{bmatrix} a & b \\ c & d \end{bmatrix} \begin{bmatrix} r \\ s \end{bmatrix} \qquad \begin{bmatrix} x \\ y \end{bmatrix} = \begin{bmatrix} k & l \\ m & n \end{bmatrix} \begin{bmatrix} u \\ v \end{bmatrix}$$

So

$$\begin{bmatrix} x \\ y \end{bmatrix} = \begin{bmatrix} k & l \\ m & n \end{bmatrix} \begin{bmatrix} a & b \\ c & d \end{bmatrix} \begin{bmatrix} r \\ s \end{bmatrix}$$

Do you see how Cayley defined matrix multiplication?

5.4 Assess Your Understanding

Concepts and Vocabulary

1. A matrix B, for which $AB = I_n$, the identity matrix, is called the _____ of A.

2. A matrix that has the same number of rows as columns is called a(n) _____ matrix.

3. In the algebra of matrices, the matrix that has properties similar to the number 1 is called the _____ matrix.

4. *True or False:* Every square matrix has an inverse.

5. *True or False:* Matrix multiplication is commutative.

6. *True or False:* Any pair of matrices can be multiplied.

Skill Building

In Problems 7–22, use the following matrices to compute the given expression (a) algebraically and (b) using a graphing utility.

$$A = \begin{bmatrix} 0 & 3 & -5 \\ 1 & 2 & 6 \end{bmatrix} \qquad B = \begin{bmatrix} 4 & 1 & 0 \\ -2 & 3 & -2 \end{bmatrix} \qquad C = \begin{bmatrix} 4 & 1 \\ 6 & 2 \\ -2 & 3 \end{bmatrix}$$

7. $A + B$

8. $A - B$

9. $4A$

10. $-3B$

11. $3A - 2B$

12. $2A + 4B$

13. AC

14. BC

15. CA

16. CB

17. $C(A + B)$

18. $(A + B)C$

19. $AC - 3I_2$

20. $CA + 5I_3$

21. $CA - CB$

22. $AC + BC$

In Problems 23–28, compute each product (a) algebraically and (b) using a graphing utility.

23. $\begin{bmatrix} 2 & -2 \\ 1 & 0 \end{bmatrix} \begin{bmatrix} 2 & 1 & 4 & 6 \\ 3 & -1 & 3 & 2 \end{bmatrix}$

24. $\begin{bmatrix} 4 & 1 \\ 2 & 1 \end{bmatrix} \begin{bmatrix} -6 & 6 & 1 & 0 \\ 2 & 5 & 4 & -1 \end{bmatrix}$

25. $\begin{bmatrix} 1 & 2 & 3 \\ 0 & -1 & 4 \end{bmatrix} \begin{bmatrix} 1 & 2 \\ -1 & 0 \\ 2 & 4 \end{bmatrix}$

26. $\begin{bmatrix} 1 & -1 \\ -3 & 2 \\ 0 & 5 \end{bmatrix} \begin{bmatrix} 2 & 8 & -1 \\ 3 & 6 & 0 \end{bmatrix}$

27. $\begin{bmatrix} 1 & 0 & 1 \\ 2 & 4 & 1 \\ 3 & 6 & 1 \end{bmatrix} \begin{bmatrix} 1 & 3 \\ 6 & 2 \\ 8 & -1 \end{bmatrix}$

28. $\begin{bmatrix} 4 & -2 & 3 \\ 0 & 1 & 2 \\ -1 & 0 & 1 \end{bmatrix} \begin{bmatrix} 2 & 6 \\ 1 & -1 \\ 0 & 2 \end{bmatrix}$

In Problems 29–38, each matrix is nonsingular. Find the inverse of each matrix. Be sure to check your answer using a graphing utility (when possible).

29. $\begin{bmatrix} 2 & 1 \\ 1 & 1 \end{bmatrix}$

30. $\begin{bmatrix} 3 & -1 \\ -2 & 1 \end{bmatrix}$

31. $\begin{bmatrix} 6 & 5 \\ 2 & 2 \end{bmatrix}$

32. $\begin{bmatrix} -4 & 1 \\ 6 & -2 \end{bmatrix}$

33. $\begin{bmatrix} 2 & 1 \\ a & a \end{bmatrix}, \quad a \neq 0$

34. $\begin{bmatrix} b & 3 \\ b & 2 \end{bmatrix}, \quad b \neq 0$

35. $\begin{bmatrix} 1 & -1 & 1 \\ 0 & -2 & 1 \\ -2 & -3 & 0 \end{bmatrix}$

36. $\begin{bmatrix} 1 & 0 & 2 \\ -1 & 2 & 3 \\ 1 & -1 & 0 \end{bmatrix}$

37. $\begin{bmatrix} 1 & 1 & 1 \\ 3 & 2 & -1 \\ 3 & 1 & 2 \end{bmatrix}$

38. $\begin{bmatrix} 3 & 3 & 1 \\ 1 & 2 & 1 \\ 2 & -1 & 1 \end{bmatrix}$

In Problems 39–58, use the inverses found in Problems 29–38 to solve each system of equations algebraically.

39. $\begin{cases} 2x + y = 8 \\ x + y = 5 \end{cases}$

40. $\begin{cases} 3x - y = 8 \\ -2x + y = 4 \end{cases}$

41. $\begin{cases} 2x + y = 0 \\ x + y = 5 \end{cases}$

42. $\begin{cases} 3x - y = 4 \\ -2x + y = 5 \end{cases}$

43. $\begin{cases} 6x + 5y = 7 \\ 2x + 2y = 2 \end{cases}$

44. $\begin{cases} -4x + y = 0 \\ 6x - 2y = 14 \end{cases}$

45. $\begin{cases} 6x + 5y = 13 \\ 2x + 2y = 5 \end{cases}$

46. $\begin{cases} -4x + y = 5 \\ 6x - 2y = -9 \end{cases}$

47. $\begin{cases} 2x + y = -3 \\ ax + ay = -a \end{cases} \quad a \neq 0$

48. $\begin{cases} bx + 3y = 2b + 3 \\ bx + 2y = 2b + 2 \end{cases} \quad b \neq 0$

49. $\begin{cases} 2x + y = \dfrac{7}{a} \\ ax + ay = 5 \end{cases} \quad a \neq 0$

50. $\begin{cases} bx + 3y = 14 \\ bx + 2y = 10 \end{cases} \quad b \neq 0$

51. $\begin{cases} x - y + z = 0 \\ -2y + z = -1 \\ -2x - 3y = -5 \end{cases}$

52. $\begin{cases} x + 2z = 6 \\ -x + 2y + 3z = -5 \\ x - y = 6 \end{cases}$

53. $\begin{cases} x - y + z = 2 \\ -2y + z = 2 \\ -2x - 3y = \dfrac{1}{2} \end{cases}$

54. $\begin{cases} x + 2z = 2 \\ -x + 2y + 3z = -\dfrac{3}{2} \\ x - y = 2 \end{cases}$

55. $\begin{cases} x + y + z = 9 \\ 3x + 2y - z = 8 \\ 3x + y + 2z = 1 \end{cases}$

56. $\begin{cases} 3x + 3y + z = 8 \\ x + 2y + z = 5 \\ 2x - y + z = 4 \end{cases}$

57. $\begin{cases} x + y + z = 2 \\ 3x + 2y - z = \dfrac{7}{3} \\ 3x + y + 2z = \dfrac{10}{3} \end{cases}$

58. $\begin{cases} 3x + 3y + z = 1 \\ x + 2y + z = 0 \\ 2x - y + z = 4 \end{cases}$

In Problems 59–64, show that each matrix has no inverse algebraically. Verify your result using a graphing utility.

59. $\begin{bmatrix} 4 & 2 \\ 2 & 1 \end{bmatrix}$

60. $\begin{bmatrix} -3 & \dfrac{1}{2} \\ 6 & -1 \end{bmatrix}$

61. $\begin{bmatrix} 15 & 3 \\ 10 & 2 \end{bmatrix}$

62. $\begin{bmatrix} -3 & 0 \\ 4 & 0 \end{bmatrix}$

63. $\begin{bmatrix} -3 & 1 & -1 \\ 1 & -4 & -7 \\ 1 & 2 & 5 \end{bmatrix}$

64. $\begin{bmatrix} 1 & 1 & -3 \\ 2 & -4 & 1 \\ -5 & 7 & 1 \end{bmatrix}$

In Problems 65–68, use a graphing utility to find the inverse, if it exists, of each matrix. Round answers to two decimal places.

65. $\begin{bmatrix} 25 & 61 & -12 \\ 18 & -2 & 4 \\ 8 & 35 & 21 \end{bmatrix}$

66. $\begin{bmatrix} 18 & -3 & 4 \\ 6 & -20 & 14 \\ 10 & 25 & -15 \end{bmatrix}$

67. $\begin{bmatrix} 44 & 21 & 18 & 6 \\ -2 & 10 & 15 & 5 \\ 21 & 12 & -12 & 4 \\ -8 & -16 & 4 & 9 \end{bmatrix}$

68. $\begin{bmatrix} 16 & 22 & -3 & 5 \\ 21 & -17 & 4 & 8 \\ 2 & 8 & 27 & 20 \\ 5 & 15 & -3 & -10 \end{bmatrix}$

In Problems 69–72, use the idea behind Example 15 with a graphing utility to solve the following systems of equations. Round answers to two decimal places.

69. $\begin{cases} 25x + 61y - 12z = 10 \\ 18x - 12y + 7y = -9 \\ 3x + 4y - z = 12 \end{cases}$

70. $\begin{cases} 25x + 61y - 12z = 15 \\ 18x - 12y + 7z = -3 \\ 3x + 4y - z = 12 \end{cases}$

71. $\begin{cases} 25x + 61y - 12z = 21 \\ 18x - 12y + 7z = 7 \\ 3x + 4y - z = -2 \end{cases}$

72. $\begin{cases} 25x + 61y - 12z = 25 \\ 18x - 12y + 7z = 10 \\ 3x + 4y - z = -4 \end{cases}$

Applications and Extensions

73. Computing the Cost of Production   The Acme Steel Company is a producer of stainless steel and aluminum containers. On a certain day, the following stainless steel containers were manufactured: 500 with 10-gallon capacity, 350 with 5-gallon capacity, and 400 with 1-gallon capacity. On the same day, the following aluminum containers were manufactured: 700 with 10-gallon capacity, 500 with 5-gallon capacity, and 850 with 1-gallon capacity.

(a) Find a 2 by 3 matrix representing the above data. Find a 3 by 2 matrix to represent the same data.

(b) If the amount of material used in the 10-gallon containers is 15 pounds, the amount used in the 5-gallon containers is 8 pounds, and the amount used in the 1-gallon containers is 3 pounds, find a 3 by 1 matrix representing the amount of material.

(c) Multiply the 2 by 3 matrix found in part (a) and the 3 by 1 matrix found in part (b) to get a 2 by 1 matrix showing the day's usage of material.

(d) If stainless steel costs Acme $0.10 per pound and aluminum costs $0.05 per pound, find a 1 by 2 matrix representing cost.

(e) Multiply the matrices found in parts (c) and (d) to determine the total cost of the day's production.

74. Computing Profit   Rizza Ford has two locations, one in the city and the other in the suburbs. In January, the city loca-

tion sold 400 subcompacts, 250 intermediate-size cars, and 50 SUVs; in February, it sold 350 subcompacts, 100 intermediates, and 30 SUVs. At the suburban location in January, 450 subcompacts, 200 intermediates, and 140 SUVs were sold. In February, the suburban location sold 350 subcompacts, 300 intermediates, and 100 SUVs.

(a) Find 2 by 3 matrices that summarize the sales data for each location for January and February (one matrix for each month).

(b) Use matrix addition to obtain total sales for the two-month period.

(c) The profit on each kind of car is $100 per subcompact, $150 per intermediate, and $200 per SUV. Find a 3 by 1 matrix representing this profit.

(d) Multiply the matrices found in parts (b) and (c) to get a 2 by 1 matrix showing the profit at each location.

75. Consider the 2 by 2 square matrix

$$A = \begin{bmatrix} a & b \\ c & d \end{bmatrix}$$

If $D = ad - bc \neq 0$, show that A is nonsingular and that

$$A^{-1} = \frac{1}{D} \begin{bmatrix} d & -b \\ -c & a \end{bmatrix}$$

Discussion and Writing

76. Create a situation different from any found in the text that can be represented by a matrix.

5.5 Partial Fraction Decomposition

PREPARING FOR THIS SECTION   *Before getting started, review the following:*

- Identity (Section 1.2 p. 101)
- Proper and Improper Rational Functions (Section 3.3, pp. 339–340)

- Factoring Polynomials (Section R.5, pp. 45–51)
- Fundamental Theorem of Algebra (Section 3.7, pp. 381–385)

Now work the 'Are You Prepared?' problems on page 570.

OBJECTIVES   1 Decompose $\dfrac{P}{Q}$, Where Q Has Only Nonrepeated Linear Factors

2 Decompose $\dfrac{P}{Q}$, Where Q Has Repeated Linear Factors

3 Decompose $\dfrac{P}{Q}$, Where Q Has a Nonrepeated Irreducible Quadratic Factor

4 Decompose $\dfrac{P}{Q}$, Where Q Has Repeated Irreducible Quadratic Factors

Consider the problem of adding two rational expressions:

$$\frac{3}{x + 4} \quad \text{and} \quad \frac{2}{x - 3}$$

The result is

$$\frac{3}{x+4} + \frac{2}{x-3} = \frac{3(x-3) + 2(x+4)}{(x+4)(x-3)} = \frac{5x-1}{x^2+x-12}$$

The reverse procedure, of starting with the rational expression $\dfrac{5x-1}{x^2+x-12}$

and writing it as the sum (or difference) of the two simpler fractions $\dfrac{3}{x+4}$ and

$\dfrac{2}{x-3}$, is referred to as **partial fraction decomposition**, and the two simpler fractions are called **partial fractions**. Decomposing a rational expression into a sum of partial fractions is important in solving certain types of calculus problems. This section presents a systematic way to decompose rational expressions.

We begin by recalling that a rational expression is the ratio of two polynomials, say, P and $Q \neq 0$. We assume that P and Q have no common factors. Recall also that a rational expression $\dfrac{P}{Q}$ is called **proper** if the degree of the polynomial in the numerator is less than the degree of the polynomial in the denominator. Otherwise, the rational expression is termed **improper**.

Because any improper rational expression can be reduced by long division to a mixed form consisting of the sum of a polynomial and a proper rational expression, we shall restrict the discussion that follows to proper rational expressions.

The partial fraction decomposition of the rational expression $\dfrac{P}{Q}$ depends on the factors of the denominator Q. Recall (from Section 3.7) that any polynomial whose co-efficients are real numbers can be factored (over the real numbers) into products of linear and/or irreducible quadratic factors. This means that the denominator Q of the rational expression $\dfrac{P}{Q}$ will contain only factors of one or both of the following types:

1. *Linear factors* of the form $x - a$, where a is a real number.

2. *Irreducible quadratic factors* of the form $ax^2 + bx + c$, where a, b, and c are real numbers, $a \neq 0$, and $b^2 - 4ac < 0$ (which guarantees that $ax^2 + bx + c$ cannot be written as the product of two linear factors with real coefficients).

As it turns out, there are four cases to be examined. We begin with the case for which Q has only nonrepeated linear factors.

1 Decompose $\dfrac{P}{Q}$, Where Q Has Only Nonrepeated Linear Factors

Case 1: Q has only nonrepeated linear factors.

Under the assumption that Q has only nonrepeated linear factors, the polynomial Q has the form

$$Q(x) = (x - a_1)(x - a_2) \cdots \cdots (x - a_n)$$

where none of the numbers $a_1, a_2, \ldots, a_n$ is equal. In this case, the partial fraction decomposition of $\dfrac{P}{Q}$ is of the form

$$\frac{P(x)}{Q(x)} = \frac{A_1}{x - a_1} + \frac{A_2}{x - a_2} + \cdots + \frac{A_n}{x - a_n} \tag{1}$$

where the numbers $A_1, A_2, \ldots, A_n$ are to be determined.

We show how to find these numbers in the example that follows.

| EXAMPLE 1 | **Nonrepeated Linear Factors** |

Write the partial fraction decomposition of $\dfrac{x}{x^2 - 5x + 6}$

Solution First, we factor the denominator,

$$x^2 - 5x + 6 = (x - 2)(x - 3)$$

and conclude that the denominator contains only nonrepeated linear factors. Then we decompose the rational expression according to equation (1):

$$\frac{x}{x^2 - 5x + 6} = \frac{A}{x - 2} + \frac{B}{x - 3} \qquad \textbf{(2)}$$

where A and B are to be determined. To find A and B, we clear the fractions by multiplying each side by $(x - 2)(x - 3) = x^2 - 5x + 6$. The result is

$$x = A(x - 3) + B(x - 2) \qquad \textbf{(3)}$$

or

$$x = (A + B)x + (-3A - 2B)$$

This equation is an identity in x. We equate the coefficients of like powers of x to get

$$\begin{cases} 1 = A + B & \text{Equate coefficients of x: 1x = (A + B)x.} \\ 0 = -3A - 2B & \text{Equate coefficients of } x^0, \text{ the constants: } 0x^0 = (-3A - 2B)x^0. \end{cases}$$

This system of two equations containing two variables, A and B, can be solved using whatever method you wish. Solving it, we get

$$A = -2 \qquad B = 3$$

From equation (2), the partial fraction decomposition is

$$\frac{x}{x^2 - 5x + 6} = \frac{-2}{x - 2} + \frac{3}{x - 3}$$

✔ **CHECK:** The decomposition can be checked by adding the rational expressions.

$$\frac{-2}{x - 2} + \frac{3}{x - 3} = \frac{-2(x - 3) + 3(x - 2)}{(x - 2)(x - 3)} = \frac{x}{(x - 2)(x - 3)}$$

$$= \frac{x}{x^2 - 5x + 6} \qquad ◄$$

The numbers to be found in the partial fraction decomposition can sometimes be found more readily by using suitable choices for x (which may include complex numbers) in the identity obtained after fractions have been cleared. In Example 1, the identity after clearing fractions is equation (3):

$$x = A(x - 3) + B(x - 2)$$

If we let $x = 2$ in this expression, the term containing B drops out, leaving $2 = A(-1)$, or $A = -2$. Similarly, if we let $x = 3$, the term containing A drops out, leaving $3 = B$. As before, $A = -2$ and $B = 3$.

NOW WORK PROBLEM 13.

2 Decompose $\dfrac{P}{Q}$ Where Q Has Repeated Linear Factors

Case 2: **Q has repeated linear factors.**

If the polynomial Q has a repeated linear factor, say $(x - a)^n$, $n \geq 2$ an integer, then, in the partial fraction decomposition of $\dfrac{P}{Q}$, we allow for the terms

$$\frac{A_1}{x - a} + \frac{A_2}{(x - a)^2} + \cdots + \frac{A_n}{(x - a)^n}$$

where the numbers $A_1, A_2, \ldots, A_n$ are to be determined.

EXAMPLE 2

Repeated Linear Factors

Write the partial fraction decomposition of $\dfrac{x + 2}{x^3 - 2x^2 + x}$.

Solution First, we factor the denominator,

$$x^3 - 2x^2 + x = x(x^2 - 2x + 1) = x(x - 1)^2$$

and find that the denominator has the nonrepeated linear factor x and the repeated linear factor $(x - 1)^2$. By Case 1, we must allow for the term $\dfrac{A}{x}$ in the decomposition; and, by Case 2, we must allow for the terms $\dfrac{B}{x - 1} + \dfrac{C}{(x - 1)^2}$ in the decomposition. We write

$$\frac{x + 2}{x^3 - 2x^2 + x} = \frac{A}{x} + \frac{B}{x - 1} + \frac{C}{(x - 1)^2} \tag{4}$$

Again, we clear fractions by multiplying each side by $x^3 - 2x^2 + x = x(x - 1)^2$. The result is the identity

$$x + 2 = A(x - 1)^2 + Bx(x - 1) + Cx \tag{5}$$

If we let $x = 0$ in this expression, the terms containing B and C drop out, leaving $2 = A(-1)^2$, or $A = 2$. Similarly, if we let $x = 1$, the terms containing A and B drop out, leaving $3 = C$. Then, equation (5) becomes

$$x + 2 = 2(x - 1)^2 + Bx(x - 1) + 3x$$

Now let $x = 2$ (any choice other than 0 or 1 will work as well). The result is

$$4 = 2(1)^2 + B(2)(1) + 3(2)$$
$$4 = 2 + 2B + 6$$
$$2B = -4$$
$$B = -2$$

We have $A = 2$, $B = -2$, and $C = 3$.

From equation (4), the partial fraction decomposition is

$$\frac{x + 2}{x^3 - 2x^2 + x} = \frac{2}{x} + \frac{-2}{x - 1} + \frac{3}{(x - 1)^2}$$

◀

| EXAMPLE 3 | **Repeated Linear Factors** |

Write the partial fraction decomposition of $\dfrac{x^3 - 8}{x^2(x-1)^3}$.

Solution   The denominator contains the repeated linear factor x^2 and the repeated linear factor $(x-1)^3$. The partial fraction decomposition takes the form

$$\frac{x^3 - 8}{x^2(x-1)^3} = \frac{A}{x} + \frac{B}{x^2} + \frac{C}{x-1} + \frac{D}{(x-1)^2} + \frac{E}{(x-1)^3} \qquad \textbf{(6)}$$

As before, we clear fractions and obtain the identity

$$x^3 - 8 = Ax(x-1)^3 + B(x-1)^3 + Cx^2(x-1)^2 + Dx^2(x-1) + Ex^2 \qquad \textbf{(7)}$$

Let $x = 0$. (Do you see why this choice was made?) Then

$$-8 = B(-1)$$
$$B = 8$$

Now let $x = 1$ in equation (7). Then

$$-7 = E$$

Use $B = 8$ and $E = -7$ in equation (7) and collect like terms.

$$x^3 - 8 = Ax(x-1)^3 + 8(x-1)^3$$
$$+ Cx^2(x-1)^2 + Dx^2(x-1) - 7x^2$$
$$x^3 - 8 - 8(x^3 - 3x^2 + 3x - 1) + 7x^2 = Ax(x-1)^3 + Cx^2(x-1)^2 + Dx^2(x-1)$$
$$-7x^3 + 31x^2 - 24x = x(x-1)[A(x-1)^2 + Cx(x-1) + Dx]$$
$$x(x-1)(-7x + 24) = x(x-1)[A(x-1)^2 + Cx(x-1) + Dx]$$
$$-7x + 24 = A(x-1)^2 + Cx(x-1) + Dx \qquad \textbf{(8)}$$

We now work with equation (8). Let $x = 0$. Then

$$24 = A$$

Now let $x = 1$ in equation (8). Then

$$17 = D$$

Use $A = 24$ and $D = 17$ in equation (8) and collect like terms.

$$-7x + 24 = 24(x-1)^2 + Cx(x-1) + 17x$$

Now let $x = 2$. Then

$$-14 + 24 = 24 + C(2) + 34$$
$$-48 = 2C$$
$$-24 = C$$

We now know all the numbers $A, B, C, D,$ and E, so, from equation (6), we have the decomposition

$$\frac{x^3 - 8}{x^2(x-1)^3} = \frac{24}{x} + \frac{8}{x^2} + \frac{-24}{x-1} + \frac{17}{(x-1)^2} + \frac{-7}{(x-1)^3} \qquad \blacktriangleleft$$

The method employed in Example 3, although somewhat tedious, is still preferable to solving the system of five equations containing five variables that the expansion of equation (6) leads to.

═══ **NOW WORK PROBLEM 19.**

The final two cases involve irreducible quadratic factors. A quadratic factor is irreducible if it cannot be factored into linear factors with real coefficients. A quadratic expression $ax^2 + bx + c$ is irreducible whenever $b^2 - 4ac < 0$. For example, $x^2 + x + 1$ and $x^2 + 4$ are irreducible.

3 Decompose $\dfrac{P}{Q}$, where Q has a Nonrepeated Irreducible Quadratic Factor

Case 3: **Q contains a nonrepeated irreducible quadratic factor.**

If Q contains a nonrepeated irreducible quadratic factor of the form $ax^2 + bx + c$, then, in the partial fraction decomposition of $\dfrac{P}{Q}$, allow for the term

$$\frac{Ax + B}{ax^2 + bx + c}$$

where the numbers A and B are to be determined.

| **EXAMPLE 4** | **Nonrepeated Irreducible Quadratic Factor** |

Write the partial fraction decomposition of $\dfrac{3x - 5}{x^3 - 1}$.

Solution We factor the denominator,

$$x^3 - 1 = (x - 1)(x^2 + x + 1)$$

and find that it has a nonrepeated linear factor $x - 1$ and a nonrepeated irreducible quadratic factor $x^2 + x + 1$. We allow for the term $\dfrac{A}{x - 1}$ by Case 1, and we allow for the term $\dfrac{Bx + C}{x^2 + x + 1}$ by Case 3. We write

$$\frac{3x - 5}{x^3 - 1} = \frac{A}{x - 1} + \frac{Bx + C}{x^2 + x - 1} \tag{9}$$

We clear fractions by multiplying each side of equation (9) by $x^3 - 1 = (x - 1)(x^2 + x + 1)$ to obtain the identity

$$3x - 5 = A(x^2 + x + 1) + (Bx + C)(x - 1) \tag{10}$$

Now let $x = 1$. Then equation (10) gives $-2 = A(3)$, or $A = -\dfrac{2}{3}$. We use this value of A in equation (10) and simplify.

$$3x - 5 = -\frac{2}{3}(x^2 + x + 1) + (Bx + C)(x - 1)$$

$$3(3x - 5) = -2(x^2 + x + 1) + 3(Bx + C)(x - 1) \qquad \text{Multiply each side by 3.}$$

$$9x - 15 = -2x^2 - 2x - 2 + 3(Bx + C)(x - 1)$$

$$2x^2 + 11x - 13 = 3(Bx + C)(x - 1) \qquad \text{Collect terms.}$$

$$(2x + 13)(x - 1) = 3(Bx + C)(x - 1) \qquad \text{Factor the left side.}$$

$$2x + 13 = 3Bx + 3C \qquad \text{Cancel } x - 1 \text{ on each side.}$$

$$2 = 3B \quad \text{and} \quad 13 = 3C \qquad \text{Equate coefficients.}$$

$$B = \frac{2}{3} \qquad\qquad C = \frac{13}{3}$$

From equation (9), we see that

$$\frac{3x - 5}{x^3 - 1} = \frac{-\dfrac{2}{3}}{x - 1} + \frac{\dfrac{2}{3}x + \dfrac{13}{3}}{x^2 + x + 1}$$
◀

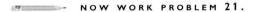 **N O W   W O R K   P R O B L E M   2 1 .**

4 Decompose $\dfrac{P}{Q}$, where Q Has Repeated Irreducible Quadratic Factors

Case 4: Q contains repeated irreducible quadratic factors.

If the polynomial Q contains a repeated irreducible quadratic factor $(ax^2 + bx + c)^n$, $n \geq 2$, n an integer, then, in the partial fraction decomposition of $\dfrac{P}{Q}$, allow for the terms

$$\boxed{\frac{A_1 x + B_1}{ax^2 + bx + c} + \frac{A_2 x + B_2}{(ax^2 + bx + c)^2} + \cdots + \frac{A_n x + B_n}{(ax^2 + bx + c)^n}}$$

where the numbers $A_1, B_1, A_2, B_2, \ldots, A_n, B_n$ are to be determined.

EXAMPLE 5 | **Repeated Irreducible Quadratic Factor**

Write the partial fraction decomposition of $\dfrac{x^3 + x^2}{(x^2 + 4)^2}$.

Solution   The denominator contains the repeated irreducible quadratic factor $(x^2 + 4)^2$, so we write

$$\frac{x^3 + x^2}{(x^2 + 4)^2} = \frac{Ax + B}{x^2 + 4} + \frac{Cx + D}{(x^2 + 4)^2} \qquad \textbf{(11)}$$

We clear fractions to obtain

$$x^3 + x^2 = (Ax + B)(x^2 + 4) + Cx + D$$

Collecting like terms yields

$$x^3 + x^2 = Ax^3 + Bx^2 + (4A + C)x + D + 4B$$

Equating coefficients, we arrive at the system

$$\begin{cases} A = 1 \\ B = 1 \\ 4A + C = 0 \\ D + 4B = 0 \end{cases}$$

The solution is $A = 1$, $B = 1$, $C = -4$, $D = -4$. From equation (11),

$$\frac{x^3 + x^2}{(x^2 + 4)^2} = \frac{x + 1}{x^2 + 4} + \frac{-4x - 4}{(x^2 + 4)^2}$$

◀

NOW WORK PROBLEM 35.

5.5 Assess Your Understanding

'Are You Prepared?'

Answers are given at the end of these exercises. If you get a wrong answer, read the pages listed in red.

1. *True or False:* The equation $(x - 1)^2 - 1 = x(x - 2)$ is an example of an identity. (p. 101)

2. *True or False:* The rational expression $\dfrac{5x^2 - 1}{x^3 + 1}$ is proper. (pp. 339–340)

3. Factor completely: $3x^4 + 6x^3 + 3x^2$. (pp. 50–51)

4. *True or False:* Every polynomial with real numbers as coefficients can be factored into products of linear and/or irreducible quadratic factors. (p. 384)

Skill Building

In Problems 5–12, tell whether the given rational expression is proper or improper. If improper, rewrite it as the sum of a polynomial and a proper rational expression.

5. $\dfrac{x}{x^2 - 1}$

6. $\dfrac{5x + 2}{x^3 - 1}$

7. $\dfrac{x^2 + 5}{x^2 - 4}$

8. $\dfrac{3x^2 - 2}{x^2 - 1}$

9. $\dfrac{5x^3 + 2x - 1}{x^2 - 4}$

10. $\dfrac{3x^4 + x^2 - 2}{x^3 + 8}$

11. $\dfrac{x(x - 1)}{(x + 4)(x - 3)}$

12. $\dfrac{2x(x^2 + 4)}{x^2 + 1}$

In Problems 13–46, write the partial fraction decomposition of each rational expression.

13. $\dfrac{4}{x(x - 1)}$

14. $\dfrac{3x}{(x + 2)(x - 1)}$

15. $\dfrac{1}{x(x^2 + 1)}$

16. $\dfrac{1}{(x + 1)(x^2 + 4)}$

17. $\dfrac{x}{(x - 1)(x - 2)}$

18. $\dfrac{3x}{(x + 2)(x - 4)}$

19. $\dfrac{x^2}{(x - 1)^2(x + 1)}$

20. $\dfrac{x + 1}{x^2(x - 2)}$

21. $\dfrac{1}{x^3 - 8}$

22. $\dfrac{2x + 4}{x^3 - 1}$

23. $\dfrac{x^2}{(x - 1)^2(x + 1)^2}$

24. $\dfrac{x + 1}{x^2(x - 2)^2}$

25. $\dfrac{x - 3}{(x + 2)(x + 1)^2}$

26. $\dfrac{x^2 + x}{(x + 2)(x - 1)^2}$

27. $\dfrac{x + 4}{x^2(x^2 + 4)}$

28. $\dfrac{10x^2 + 2x}{(x - 1)^2(x^2 + 2)}$

29. $\dfrac{x^2 + 2x + 3}{(x + 1)(x^2 + 2x + 4)}$

30. $\dfrac{x^2 - 11x - 18}{x(x^2 + 3x + 3)}$

31. $\dfrac{x}{(3x - 2)(2x + 1)}$

32. $\dfrac{1}{(2x + 3)(4x - 1)}$

33. $\dfrac{x}{x^2 + 2x - 3}$

34. $\dfrac{x^2 - x - 8}{(x + 1)(x^2 + 5x + 6)}$

35. $\dfrac{x^2 + 2x + 3}{(x^2 + 4)^2}$

36. $\dfrac{x^3 + 1}{(x^2 + 16)^2}$

37. $\dfrac{7x + 3}{x^3 - 2x^2 - 3x}$

38. $\dfrac{x^5 + 1}{x^6 - x^4}$

39. $\dfrac{x^2}{x^3 - 4x^2 + 5x - 2}$

40. $\dfrac{x^2 + 1}{x^3 + x^2 - 5x + 3}$

41. $\dfrac{x^3}{(x^2 + 16)^3}$

42. $\dfrac{x^2}{(x^2 + 4)^3}$

43. $\dfrac{4}{2x^2 - 5x - 3}$

44. $\dfrac{4x}{2x^2 + 3x - 2}$

45. $\dfrac{2x + 3}{x^4 - 9x^2}$

46. $\dfrac{x^2 + 9}{x^4 - 2x^2 - 8}$

'Are You Prepared?' Answers

1. True   **2.** True   **3.** $3x^2(x + 1)^2$ **4.** True

5.6 Systems of Linear Inequalities

PREPARING FOR THIS SECTION   *Before getting started, review the following:*

• Solving Inequalities (Section 1.7, pp. 150–159)

• Lines (Section 1.8, pp. 163–176)

• Circles (Section 1.9, pp. 180–185)

• Graphing Techniques: Transformation (Section 2.7, pp. 266–274)

 Now work the 'Are You Prepared?' problems on page 578.

OBJECTIVES   1   Graph an Inequality by Hand

2   Graph an Inequality Using a Graphing Utility

3   Graph a Linear Inequality

4   Graph a System of Linear Inequalities

In Chapter 1, we discussed inequalities in one variable. In this section, we discuss inequalities in two variables.

EXAMPLE 1   **Examples of Inequalities in Two Variables**

(a) $3x + y \leq 6$ (b) $x^2 + y^2 < 4$ (c) $y^2 \leq x$ ◀

1 Graph an Inequality by Hand

An inequality in two variables x and y is **satisfied** by an ordered pair (a, b) if, when x is replaced by a and y by b, a true statement results. The **graph of an inequality in two variables** x and y consists of all points (x, y) whose coordinates satisfy the inequality. Let's look at an example.

EXAMPLE 2   **Graphing an Inequality by Hand**

Graph the linear inequality: $3x + y \leq 6$

Solution   We begin by graphing the equation

$$3x + y = 6$$

formed by replacing (for now) the $\leq$ symbol with an $=$ sign. The graph of the equation is a line. See Figure 18(a). This line is part of the graph of the inequality that we seek because the inequality is nonstrict. (Do you see why? We are seeking points for which $3x + y$ is less than *or equal to* 6.)

Figure 18

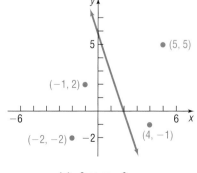

(a) $3x + y = 6$

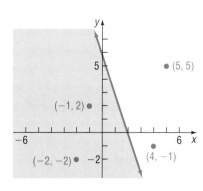

(b) Graph of
$3x + y \leq 6$

Now let's test a few randomly selected points to see whether they belong to the graph of the inequality.

	$3x + y \leq 6$	**Conclusion**
$(4, -1)$	$3(4) + (-1) = 11 > 6$	Does not belong to graph
$(5, 5)$	$3(5) + 5 = 20 > 6$	Does not belong to graph
$(-1, 2)$	$3(-1) + 2 = -1 \leq 6$	Belongs to graph
$(-2, -2)$	$3(-2) + (-2) = -8 \leq 6$	Belongs to graph

Look again at Figure 18(a). Notice that the two points that belong to the graph both lie on the same side of the line, and the two points that do not belong to the graph lie on the opposite side. As it turns out, this is always the case. The graph we seek consists of all points that lie on the same side of the line as $(-1, 2)$ and $(-2, -2)$ and is shown as the shaded region in Figure 18(b). ◀

NOW WORK PROBLEM 15 BY HAND.

NOTE
The strict inequalities are $<$ or $>$. The nonstrict inequalities are $\leq$ or $\geq$. ■

The graph of any inequality in two variables may be obtained in a like way. First, the equation corresponding to the inequality is graphed, using dashes if the inequality is strict and solid marks if it is nonstrict. This graph, in almost every case, will separate the xy-plane into two or more regions. In each region, either all points satisfy the inequality or no points satisfy the inequality. The use of a single test point in each region is all that is required to determine whether the points of that region are part of the graph. The steps to follow are given next.

Steps for Graphing an Inequality by Hand

STEP 1: Replace the inequality symbol by an equal sign and graph the resulting equation. If the inequality is strict, use dashes; if it is nonstrict, use a solid mark. This graph separates the xy-plane into two or more regions.

STEP 2: In each region, select a test point P.
(a) If the coordinates of P satisfy the inequality, then so do all the points in that region. Indicate this by shading the region.
(b) If the coordinates of P do not satisfy the inequality, then none of the points in that region do.

EXAMPLE 3   **Graphing an Inequality by Hand**

Graph: $x^2 + y^2 \leq 4$

Figure 19

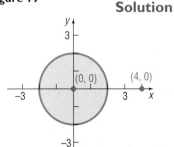

Solution   First, we graph the equation $x^2 + y^2 = 4$, a circle of radius 2, center at the origin. A solid circle will be used because the inequality is not strict. We use two test points, one inside the circle, the other outside.

Inside	$(0, 0)$:	$x^2 + y^2 = 0^2 + 0^2 = 0 \leq 4$	*Belongs to the graph*
Outside	$(4, 0)$:	$x^2 + y^2 = 4^2 + 0^2 = 16 > 4$	*Does not belong to the graph*

All the points inside and on the circle satisfy the inequality. See Figure 19. ◀

NOW WORK PROBLEM 17 BY HAND.

2 Graph an Inequality Using a Graphing Utility

Graphing utilities can also be used to graph inequalities. The steps to follow are given next.

Steps for Graphing an Inequality Using a Graphing Utility

STEP 1: Replace the inequality symbol by an equal sign and graph the resulting equation. This graph separates the xy-plane into two or more regions.

STEP 2: Select a test point P in each region.
(a) Use a graphing utility to determine if the test point P satisfies the inequality. If the test point satisfies the inequality, then so do all the points in this region. Indicate this by using the graphing utility to shade the region.
(b) If the coordinates of P do not satisfy the inequality, then none of the points in that region do.

EXAMPLE 4	**Graphing an Inequality Using a Graphing Utility**

Use a graphing utility to graph $3x + y \leq 6$.

Solution **STEP 1:** We begin by graphing the equation $3x + y = 6$ $(Y_1 = -3x + 6)$. See Figure 20.

Figure 20

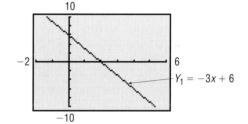

STEP 2: Select a test point in one of the regions and determine whether it satisfies the inequality. To test the point $(-1, 2)$, for example, enter $3(-1) + 2 \leq 6$. See Figure 21(a). The 1 that appears indicates that the statement entered (the inequality) is true. When the point $(5, 5)$ is tested, a 0 appears, indicating that the statement entered is false. So $(-1, 2)$ is a part of the graph of the inequality and $(5, 5)$ is not. We shade the region containing the point $(-1, 2)$ that is below Y_1. Figure 21(b) shows the graph of the inequality on a TI-84 Plus.

Figure 21

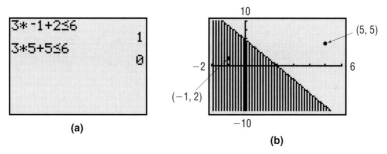

(a)

(b)

◄

NOW WORK PROBLEM **15** USING A GRAPHING UTILITY.

Figure 22

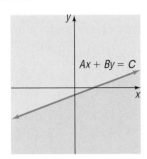

$Ax + By = C$

3   Graph a Linear Inequality

Linear inequalities are inequalities in one of the forms

$$Ax + By < C \qquad Ax + By > C \qquad Ax + By \leq C \qquad Ax + By \geq C$$

where A and B are not both zero.

The graph of the corresponding equation of a linear inequality is a line that separates the xy-plane into two regions, called **half-planes**. See Figure 22.

As shown, $Ax + By = C$ is the equation of the boundary line and it divides the plane into two half-planes: one for which $Ax + By < C$ and the other for which $Ax + By > C$. Because of this, for linear inequalities, only one test point is required.

EXAMPLE 5	**Graphing Linear Inequalities**

Graph: (a) $y < 2$ (b) $y \geq 2x$

Solution   (a)   The graph of the equation $y = 2$ is a horizontal line and is not part of the graph of the inequality. Since $(0, 0)$ satisfies the inequality, the graph consists of the half-plane below the line $y = 2$. See Figure 23.

(b)   The graph of the equation $y = 2x$ is a line and is part of the graph of the inequality. Using $(3, 0)$ as a test point, we find it does not satisfy the inequality $[0 < 2 \cdot 3]$. Points in the half-plane on the opposite side of $(3, 0)$ satisfy the inequality. See Figure 24.

Figure 23

Figure 24

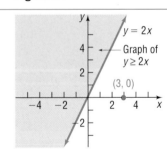

NOW WORK PROBLEM **13**.

4   Graph a System of Linear Inequalities

The **graph of a system of inequalities** in two variables x and y is the set of all points (x, y) that simultaneously satisfy *each* of the inequalities in the system. The graph of a system of inequalities can be obtained by graphing each inequality individually and then determining where, if at all, they intersect.

EXAMPLE 6	**Graphing a System of Linear Inequalities by Hand**

Graph the system: $\begin{cases} x + y \geq 2 \\ 2x - y \leq 4 \end{cases}$

Solution   First, we graph the inequality $x + y \geq 2$ as the shaded region in Figure 25(a). Next, we graph the inequality $2x - y \leq 4$ as the shaded region in Figure 25(b). Now, superimpose the two graphs, as shown in Figure 25(c). The points that are in both shaded regions [the overlapping, darker region in Figure 25(c)] are the solutions we seek to the system, because they simultaneously satisfy each linear inequality.

Figure 25

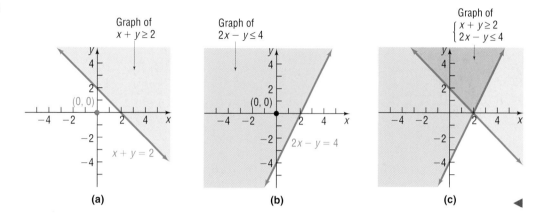

(a)   (b)   (c)

| **EXAMPLE 7** | **Graphing a System of Linear Inequalities Using a Graphing Utility** |

Graph the system: $\begin{cases} x + y \geq 2 \\ 2x - y \leq 4 \end{cases}$

Solution   First, we graph the lines $x + y = 2$ ($Y_1 = -x + 2$) and $2x - y = 4$ ($Y_2 = 2x - 4$). See Figure 26.

Figure 26

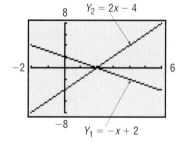

Notice that the graphs divide the viewing window into four regions. We select a test point for each region and determine whether the point makes *both* inequalities true. We choose to test $(0, 0)$, $(2, 3)$, $(4, 0)$, and $(2, -2)$. Figure 27(a) shows that $(2, 3)$ is the only point for which both inequalities are true. We obtain the graph shown in Figure 27(b).

Figure 27

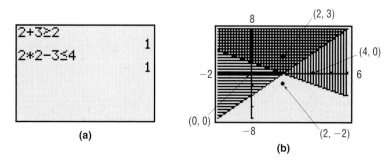

(a)   (b)

Rather than testing four points, we could just test the point $(0, 0)$ on each inequality. For example, $(0, 0)$ does not satisfy $x + y \geq 2$, so we shade above the line $x + y = 2$. In addition, $(0, 0)$ does satisfy $2x - y \leq 4$, so we shade above the line $2x - y = 4$. The intersection of the shaded regions gives us the result presented in Figure 27(b).

NOW WORK PROBLEM **23**.

| **EXAMPLE 8** | **Graphing a System of Linear Inequalities by Hand** |

Graph the system: $\begin{cases} x + y \leq 2 \\ x + y \geq 0 \end{cases}$

Solution   See Figure 28. The overlapping purple-shaded region between the two boundary lines is the graph of the system.

Figure 28

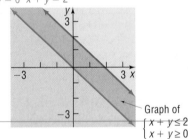

NOW WORK PROBLEM **29**.

| **EXAMPLE 9** | **Graphing a System of Linear Inequalities by Hand** |

Graph the system: $\begin{cases} 2x - y \geq 0 \\ 2x - y \geq 2 \end{cases}$

Solution   See Figure 29. The overlapping purple-shaded region is the graph of the system. Note that the graph of the system is identical to the graph of the single inequality $2x - y \geq 2$.

Figure 29

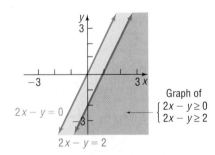

EXAMPLE 10   Graphing a System of Linear Inequalities by Hand

Graph the system: $\begin{cases} x + 2y \le 2 \\ x + 2y \ge 6 \end{cases}$

Solution   See Figure 30. Because no overlapping region results, there are no points in the xy-plane that simultaneously satisfy each inequality. Hence, the system has no solution.

Figure 30

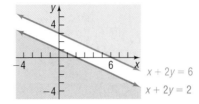

$x + 2y = 6$
$x + 2y = 2$

◀

EXAMPLE 11   Graphing a System of Four Linear Inequalities by Hand

Figure 31

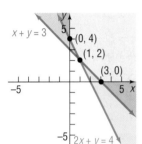

Graph the system: $\begin{cases} x + y \ge 3 \\ 2x + y \ge 4 \\ x \ge 0 \\ y \ge 0 \end{cases}$

Solution   The two inequalities $x \ge 0$ and $y \ge 0$ require the graph of the system to be in quadrant I. We concentrate on the remaining two inequalities. The intersection of the graphs of these two inequalities and quadrant I, shown in light gray in Figure 31, is the graph of the system.   ◀

EXAMPLE 12   Financial Planning

A retired couple has up to $25,000 to invest. As their financial adviser, you recommend that they place at least $15,000 in Treasury bills yielding 6% and at most $5000 in corporate bonds yielding 9%.

(a) Using x to denote the amount of money invested in Treasury bills and y the amount invested in corporate bonds, write a system of linear inequalities that describes the possible amounts of each investment. We shall assume that x and y are in thousands of dollars.

(b) Graph the system.

Figure 32

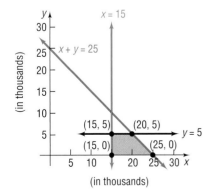

(in thousands)

Solution   (a) The system of linear inequalities is

$\begin{cases} x \ge 0 & \text{x and y are nonnegative variables since they represent} \\ & \text{money invested in thousands of dollars.} \\ y \ge 0 & \\ x + y \le 25 & \text{The total of the two investments, x + y, cannot} \\ & \text{exceed \$25,000.} \\ x \ge 15 & \text{At least \$15,000 in Treasury bills.} \\ y \le 5 & \text{At most \$5000 in corporate bonds.} \end{cases}$

(b) See the shaded region in Figure 32. Note that the inequalities $x \ge 0$ and $y \ge 0$ again require that the graph of the system be in quadrant I.   ◀

The graph of the system of linear inequalities in Figure 32 is said to be **bounded**, because it can be contained within some circle of sufficiently large radius. A graph that cannot be contained in any circle is said to be **unbounded**. For example, the graph of the system of linear inequalities in Figure 31 is unbounded, since it extends indefinitely in a particular direction.

Notice in Figures 31 and 32 that those points belonging to the graph that are also points of intersection of boundary lines have been plotted. Such points are referred to as **vertices** or **corner points** of the graph. The system graphed in Figure 31 has three corner points: $(0, 4)$, $(1, 2)$, and $(3, 0)$. The system graphed in Figure 32 has four corner points: $(15, 0)$, $(25, 0)$, $(20, 5)$, and $(15, 5)$.

These ideas will be used in the next section in developing a method for solving linear programming problems, an important application of linear inequalities.

—— **NOW WORK PROBLEM 37.**

5.6 Assess Your Understanding

'Are You Prepared?'

Answers are given at the end of these exercises. If you get a wrong answer, read the pages listed in red.

1. Solve the inequality: $3x + 4 < 8 - x$. (pp. 154–156)

2. Graph the equation: $3x - 2y = 6$. (pp. 170–172)

3. Graph the equation: $x^2 + y^2 = 9$. (pp. 182–183)

4. Graph the equation: $y = x^2 + 4$. (pp. 266–268)

5. *True or False:* The lines $2x + y = 4$ and $4x + 2y = 0$ are parallel. (pp. 172–174)

6. The graph of $y = (x - 2)^2$ may be obtained by shifting the graph of ____ to the (left/right) a distance of ____ units. (pp. 267–268)

Concepts and Vocabulary

7. An inequality in two variables x and y is ____ by an ordered pair (a, b) if, when x is replaced by a and y by b, a true statement results.

8. The graph of a linear inequality is called a(n) ____.

9. *True or False:* The graph of a linear inequality is a line.

10. *True or False:* The graph of a system of linear inequalities is sometimes unbounded.

Skill Building

In Problems 11–22, graph each inequality (a) by hand and (b) by using a graphing utility.

11. $x \geq 0$

12. $y \geq 0$

13. $x \geq 4$

14. $y \leq 2$

15. $2x + y \geq 6$

16. $3x + 2y \leq 6$

17. $x^2 + y^2 > 1$

18. $x^2 + y^2 \leq 9$

19. $y \leq x^2 - 1$

20. $y > x^2 + 2$

21. $xy \geq 4$

22. $xy \leq 1$

In Problems 23–34, graph each system of inequalities by hand. Verify your results using a graphing utility.

23. $\begin{cases} x + y \leq 2 \\ 2x + y \geq 4 \end{cases}$

24. $\begin{cases} 3x - y \geq 6 \\ x + 2y \leq 2 \end{cases}$

25. $\begin{cases} 2x - y \leq 4 \\ 3x + 2y \geq -6 \end{cases}$

26. $\begin{cases} 4x - 5y \leq 0 \\ 2x - y \geq 2 \end{cases}$

27. $\begin{cases} 2x - 3y \leq 0 \\ 3x + 2y \leq 6 \end{cases}$

28. $\begin{cases} 4x - y \geq 2 \\ x + 2y \geq 2 \end{cases}$

29. $\begin{cases} x - 2y \leq 6 \\ 2x - 4y \geq 0 \end{cases}$

30. $\begin{cases} x + 4y \leq 8 \\ x + 4y \geq 4 \end{cases}$

31. $\begin{cases} 2x + y \geq -2 \\ 2x + y \geq 2 \end{cases}$

32. $\begin{cases} x - 4y \leq 4 \\ x - 4y \geq 0 \end{cases}$

33. $\begin{cases} 2x + 3y \geq 6 \\ 2x + 3y \leq 0 \end{cases}$

34. $\begin{cases} 2x + y \geq 0 \\ 2x + y \geq 2 \end{cases}$

In Problems 35–44, graph each system of linear inequalities. Tell whether the graph is bounded or unbounded, and label the corner points.

35. $\begin{cases} x \geq 0 \\ y \geq 0 \\ 2x + y \leq 6 \\ x + 2y \leq 6 \end{cases}$

36. $\begin{cases} x \geq 0 \\ y \geq 0 \\ x + y \geq 4 \\ 2x + 3y \geq 6 \end{cases}$

37. $\begin{cases} x \geq 0 \\ y \geq 0 \\ x + y \geq 2 \\ 2x + y \geq 4 \end{cases}$

38. $\begin{cases} x \geq 0 \\ y \geq 0 \\ 3x + y \leq 6 \\ 2x + y \leq 2 \end{cases}$

39. $\begin{cases} x \geq 0 \\ y \geq 0 \\ x + y \geq 2 \\ 2x + 3y \leq 12 \\ 3x + y \leq 12 \end{cases}$

40. $\begin{cases} x \geq 0 \\ y \geq 0 \\ x + y \geq 2 \\ x + y \leq 10 \\ 2x + y \leq 3 \end{cases}$ **41.** $\begin{cases} x \geq 0 \\ y \geq 0 \\ x + y \geq 2 \\ x + y \leq 8 \\ 2x + y \leq 10 \end{cases}$ **42.** $\begin{cases} x \geq 0 \\ y \geq 0 \\ x + y \geq 2 \\ x + y \leq 8 \\ x + 2y \geq 1 \end{cases}$ **43.** $\begin{cases} x \geq 0 \\ y \geq 0 \\ x + 2y \geq 1 \\ x + 2y \leq 10 \end{cases}$ **44.** $\begin{cases} x \geq 0 \\ y \geq 0 \\ x + 2y \geq 1 \\ x + 2y \leq 10 \\ x + y \geq 2 \\ x + y \leq 8 \end{cases}$

In Problems 45–48, write a system of linear inequalities that has the given graph.

45.

46.

47.

48.

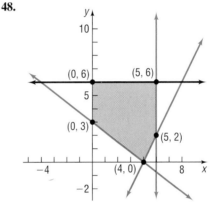

Applications and Extensions

49. Financial Planning A retired couple has up to $50,000 to invest. As their financial adviser, you recommend that they place at least $35,000 in Treasury bills yielding 7% and at most $10,000 in corporate bonds yielding 10%.
  (a) Using x to denote the amount of money invested in Treasury bills and y the amount invested in corporate bonds, write a system of linear inequalities that describes the possible amounts of each investment.
  (b) Graph the system and label the corner points.

50. Manufacturing Trucks Mike's Toy Truck Company manufacturers two models of toy trucks, a standard model and a deluxe model. Each standard model requires 2 hours for painting and 3 hours for detail work; each deluxe model requires 3 hours for painting and 4 hours for detail work. Two painters and three detail workers are employed by the company, and each works 40 hours per week.

  (a) Using x to denote the number of standard model trucks and y to denote the number of deluxe model trucks, write a system of linear inequalities that describes the possible number of each model of truck that can be manufactured in a week.
  (b) Graph the system and label the corner points.

51. Blending Coffee Bill's Coffee House, a store that specializes in coffee, has available 75 pounds of *A* grade coffee and 120 pounds of *B* grade coffee. These will be blended into 1 pound packages as follows: An economy blend that contains 4 ounces of *A* grade coffee and 12 ounces of *B* grade coffee and a superior blend that contains 8 ounces of *A* grade coffee and 8 ounces of *B* grade coffee.
  (a) Using *x* to denote the number of packages of the economy blend and *y* to denote the number of packages of the superior blend, write a system of linear inequalities that describes the possible number of packages of each kind of blend.
  (b) Graph the system and label the corner points.

52. Mixed Nuts Nola's Nuts, a store that specializes in selling nuts, has available 90 pounds of cashews and 120 pounds of peanuts. These are to be mixed in 12-ounce packages as follows: a lower-priced package containing 8 ounces of peanuts and 4 ounces of cashews and a quality package containing 6 ounces of peanuts and 6 ounces of cashews.
  (a) Use *x* to denote the number of lower-priced packages and use *y* to denote the number of quality packages. Write a system of linear inequalities that describes the possible number of each kind of package.
  (b) Graph the system and label the corner points.

53. Transporting Goods A small truck can carry no more than 1600 pounds of cargo nor more than 150 cubic feet of cargo. A printer weighs 20 pounds and occupies 3 cubic feet of space. A microwave oven weighs 30 pounds and occupies 2 cubic feet of space.
  (a) Using *x* to represent the number of microwave ovens and *y* to represent the number of printers, write a system of linear inequalities that describes the number of ovens and printers that can be hauled by the truck.
  (b) Graph the system and label the corner points.

'Are You Prepared?' Answers

1. $\{x | x < 1\}$
2.

3.

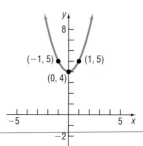

4.

5. True   **6.** $y = x^2$; right; 2

5.7 Linear Programming

OBJECTIVES 1 Set up a Linear Programming Problem
              2 Solve a Linear Programming Problem

Historically, linear programming evolved as a technique for solving problems involving resource allocation of goods and materials for the U.S. Air Force during World War II. Today, linear programming techniques are used to solve a wide variety of problems, such as optimizing airline scheduling and establishing telephone lines. Although most practical linear programming problems involve systems of several hundred linear inequalities containing several hundred variables, we will limit our discussion to problems containing only two variables, because we can solve such problems using graphing techniques.*

*The **simplex method** is a way to solve linear programming problems involving many inequalities and variables. This method was developed by George Dantzig in 1946 and is particularly well suited for computerization. In 1984, Narendra Karmarkar of Bell Laboratories discovered a way of solving large linear programming problems that improves on the simplex method.

1 Set up a Linear Programming Problem

We begin by returning to Example 12 of the previous section.

| EXAMPLE 1 | **Financial Planning** |

A retired couple has up to $25,000 to invest. As their financial adviser, you recommend that they place at least $15,000 in Treasury bills yielding 6% and at most $5000 in corporate bonds yielding 9%. How much money should be placed in each investment so that income is maximized? ◄

The problem given in Example 1 is typical of a *linear programming problem.* The problem requires that a certain linear expression, the income, be maximized. If I represents income, x the amount invested in Treasury bills at 6%, and y the amount invested in corporate bonds at 9%, then

$$I = 0.06x + 0.09y$$

We shall assume, as before, that I, x, and y are in thousands of dollars.

The linear expression $I = 0.06x + 0.09y$ is called the **objective function**. Furthermore, the problem requires that the maximum income be achieved under certain conditions or **constraints**, each of which is a linear inequality involving the variables. (See Example 12 in Section 5.6) The linear programming problem given in Example 1 may be restated as

$$\text{Maximize} \qquad I = 0.06x + 0.09y$$

subject to the conditions that

$$x \geq 0, \qquad y \geq 0$$
$$x + y \leq 25$$
$$x \geq 15$$
$$y \leq 5$$

In general, every linear programming problem has two components:

1. A linear objective function that is to be maximized or minimized.
2. A collection of linear inequalities that must be satisfied simultaneously.

A **linear programming problem** in two variables x and y consists of maximizing (or minimizing) a linear objective function

$$z = Ax + By, \qquad A \text{ and } B \text{ are real numbers, not both } 0$$

subject to certain conditions, or constraints, expressible as linear inequalities in x and y.

2 Solve a Linear Programing Problem

To maximize (or minimize) the quantity $z = Ax + By$, we need to identify points (x, y) that make the expression for z the largest (or smallest) possible. But not all points (x, y) are eligible; only those that also satisfy each linear inequality (constraint) can be used. We refer to each point (x, y) that satisfies the system of linear inequalities (the constraints) as a **feasible point**. In a linear programming problem, we seek the feasible point(s) that maximizes (or minimizes) the objective function.

Let's look again at the linear programming problem in Example 1.

| **EXAMPLE 2** | **Analyzing a Linear Programming Problem** |

Consider the linear programming problem

$$\text{Maximize} \quad I = 0.06x + 0.09y$$

subject to the conditions that

$$x \geq 0, \quad y \geq 0$$
$$x + y \leq 25$$
$$x \geq 15$$
$$y \leq 5$$

Graph the constraints. Then graph the objective function for $I = 0, 0.9, 1.35, 1.65,$ and 1.8.

Solution   Figure 33 shows the graph of the constraints. We superimpose on this graph the graph of the objective function for the given values of I.

Figure 33

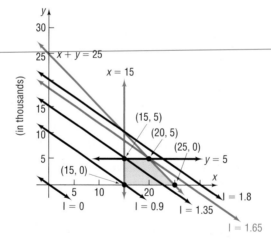

For $I = 0$, the objective function is the line $0 = 0.06x + 0.09y$.

For $I = 0.9$, the objective function is the line $0.9 = 0.06x + 0.09y$.

For $I = 1.35$, the objective function is the line $1.35 = 0.06x + 0.09y$.

For $I = 1.65$, the objective function is the line $1.65 = 0.06x + 0.09y$.

For $I = 1.8$, the objective function is the line $1.8 = 0.06x + 0.09y$.   ◄

A **solution** to a linear programming problem consists of a feasible point that maximizes (or minimizes) the objective function, together with the corresponding value of the objective function.

One condition for a linear programming problem in two variables to have a solution is that the graph of the feasible points be bounded. (Refer to page 577.)

If none of the feasible points maximizes (or minimizes) the objective function, or if there are no feasible points, then the linear programming problem has no solution.

Consider the linear programming problem stated in Example 2, and look again at Figure 33. The feasible points are the points that lie in the shaded region. For example, $(20, 3)$ is a feasible point, as are $(15, 5)$, $(20, 5)$, $(18, 4)$, and so on. To find the solution of the problem requires that we find a feasible point (x, y) that makes $I = 0.06x + 0.09y$ as large as possible. Notice that as I increases in value from $I = 0$ to $I = 0.9$ to $I = 1.35$ to $I = 1.65$ to $I = 1.8$ we obtain a collection of parallel lines. Furthermore, notice that the largest value of I that can be obtained using feasible points is $I = 1.65$, which corresponds to the line $1.65 = 0.06x + 0.09y$. Any larger value of I results in a line that does not pass through any feasible points. Finally, notice that the feasible point that yields $I = 1.65$ is the point $(20, 5)$, a corner point. These observations form the basis of the following result, which we state without proof.

Theorem **Location of the Solution of a Linear Programming Problem**

If a linear programming problem has a solution, it is located at a corner point of the graph of the feasible points.

If a linear programming problem has multiple solutions, at least one of them is located at a corner point of the graph of the feasible points.

In either case, the corresponding value of the objective function is unique.

We shall not consider here linear programming problems that have no solution. As a result, we can outline the procedure for solving a linear programming problem as follows:

Procedure for Solving a Linear Programming Problem

STEP 1: Write an expression for the quantity to be maximized (or minimized). This expression is the objective function.

STEP 2: Write all the constraints as a system of linear inequalities and graph the system.

STEP 3: List the corner points of the graph of the feasible points.

STEP 4: List the corresponding values of the objective function at each corner point. The largest (or smallest) of these is the solution.

EXAMPLE 3 **Solving a Minimum Linear Programming Problem**

Minimize the expression
$$z = 2x + 3y$$
subject to the constraints
$$y \le 5, \quad x \le 6 \quad x + y \ge 2, \quad x \ge 0, \quad y \ge 0$$

Solution The objective function is $z = 2x + 3y$. We seek the smallest value of z that can occur if x and y are solutions of the system of linear inequalities
$$\begin{cases} y \le 5 \\ x \le 6 \\ x + y \ge 2 \\ x \ge 0 \\ y \ge 0 \end{cases}$$

The graph of this system (the set of feasible points) is shown as the shaded region in Figure 34. We have also plotted the corner points. Table 1 lists the corner points and the corresponding values of the objective function. From the table, we can see that the minimum value of z is 4, and it occurs at the point $(2, 0)$.

Figure 34

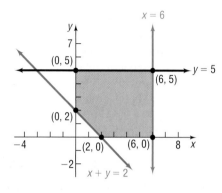

Table 1

Corner Point (x, y)	Value of the Objective Function $z = 2x + 3y$
$(0, 2)$	$z = 2(0) + 3(2) = 6$
$(0, 5)$	$z = 2(0) + 3(5) = 15$
$(6, 5)$	$z = 2(6) + 3(5) = 27$
$(6, 0)$	$z = 2(6) + 3(0) = 12$
$(2, 0)$	$z = 2(2) + 3(0) = 4$

NOW WORK PROBLEMS 5 AND 11.

EXAMPLE 4   **Maximizing Profit**

At the end of every month, after filling orders for its regular customers, a coffee company has some pure Colombian coffee and some special-blend coffee remaining. The practice of the company has been to package a mixture of the two coffees into 1-pound packages as follows: a low-grade mixture containing 4 ounces of Colombian coffee and 12 ounces of special-blend coffee and a high-grade mixture containing 8 ounces of Colombian and 8 ounces of special-blend coffee. A profit of $0.30 per package is made on the low-grade mixture, whereas a profit of $0.40 per package is made on the high-grade mixture. This month, 120 pounds of special-blend coffee and 100 pounds of pure Colombian coffee remain. How many packages of each mixture should be prepared to achieve a maximum profit? Assume that all packages prepared can be sold.

Solution   We begin by assigning symbols for the two variables.

$$x = \text{Number of packages of the low-grade mixture}$$

$$y = \text{Number of packages of the high-grade mixture}$$

If P denotes the profit, then

$$P = \$0.30x + \$0.40y$$

This expression is the objective function. We seek to maximize P subject to certain constraints on x and y. Because x and y represent numbers of packages, the only meaningful values for x and y are nonnegative integers. So we have the two constraints

$$x \geq 0, \qquad y \geq 0 \qquad \text{Nonnegative constraints.}$$

We also have only so much of each type of coffee available. For example, the total amount of Colombian coffee used in the two mixtures cannot exceed 100 pounds, or 1600 ounces. Because we use 4 ounces in each low-grade package and 8 ounces in each high-grade package, we are led to the constraint

$$4x + 8y \leq 1600 \qquad \text{Colombian coffee constraint.}$$

Similarly, the supply of 120 pounds, or 1920 ounces, of special-blend coffee leads to the constraint

$$12x + 8y \leq 1920 \qquad \textit{Special-blend coffee constraint.}$$

The linear programming problem may be stated as

$$\text{Maximize} \qquad P = 0.3x + 0.4y$$

subject to the constraints

$$x \geq 0, \qquad y \geq 0, \qquad 4x + 8y \leq 1600, \qquad 12x + 8y \leq 1920$$

The graph of the constraints (the feasible points) is illustrated in Figure 35. We list the corner points and evaluate the objective function at each. In Table 2, we can see that the maximum profit, $84, is achieved with 40 packages of the low-grade mixture and 180 packages of the high-grade mixture.

Figure 35

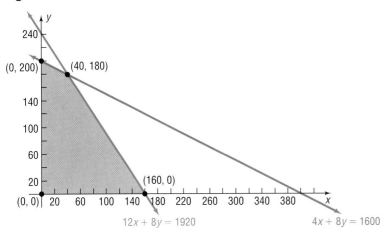

Table 2

Corner Point (x, y)	Value of Profit $P = 0.3x + 0.4y$
(0, 0)	$P = 0$
(0, 200)	$P = 0.3(0) + 0.4(200) = \$80$
(40, 180)	$P = 0.3(40) + 0.4(180) = \$84$
(160, 0)	$P = 0.3(160) + 0.4(0) = \$48$

━━━ **NOW WORK PROBLEM 19.**

5.7 Assess Your Understanding

Concepts and Vocabulary

1. A linear programming problem requires that a linear expression, called the _____ _____, be maximized or minimized.

2. *True or False:* If a linear programming problem has a solution, it is located at a corner point of the graph of the feasible points.

Skill Building

In Problems 3–8, find the maximum and minimum value of the given objective function of a linear programming problem. The figure illustrates the graph of the feasible points.

3. $z = x + y$

6. $z = 10x + y$

4. $z = 2x + 3y$

7. $z = 5x + 7y$

5. $z = x + 10y$

8. $z = 7x + 5y$

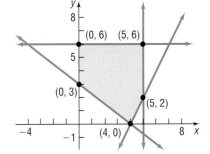

In Problems 9–18, solve each linear programming problem.

9. Maximize $z = 2x + y$ subject to $x \geq 0$, $y \geq 0$, $x + y \leq 6$, $x + y \geq 1$

10. Maximize $z = x + 3y$ subject to $x \geq 0$, $y \geq 0$, $x + y \geq 3$, $x \leq 5$, $y \leq 7$

11. Minimize $z = 2x + 5y$ subject to $x \geq 0$, $y \geq 0$, $x + y \geq 2$, $x \leq 5$, $y \leq 3$

12. Minimize $z = 3x + 4y$ subject to $x \geq 0$, $y \geq 0$, $2x + 3y \geq 6$, $x + y \leq 8$

13. Maximize $z = 3x + 5y$ subject to $x \geq 0$, $y \geq 0$, $x + y \geq 2$, $2x + 3y \leq 12$, $3x + 2y \leq 12$

14. Maximize $z = 5x + 3y$ subject to $x \geq 0$, $y \geq 0$, $x + y \geq 2$, $x + y \leq 8$, $2x + y \leq 10$

15. Minimize $z = 5x + 4y$ subject to $x \geq 0$, $y \geq 0$, $x + y \geq 2$, $2x + 3y \leq 12$, $3x + y \leq 12$

16. Minimize $z = 2x + 3y$ subject to $x \geq 0$, $y \geq 0$, $x + y \geq 3$, $x + y \leq 9$, $x + 3y \geq 6$

17. Maximize $z = 5x + 2y$ subject to $x \geq 0$, $y \geq 0$, $x + y \leq 10$, $2x + y \geq 10$, $x + 2y \geq 10$

18. Maximize $z = 2x + 4y$ subject to $x \geq 0$, $y \geq 0$, $2x + y \geq 4$, $x + y \leq 9$

Applications and Extensions

19. **Maximizing Profit** A manufacturer of skis produces two types: downhill and cross-country. Use the following table to determine how many of each kind of ski should be produced to achieve a maximum profit. What is the maximum profit? What would the maximum profit be if the maximum time available for manufacturing is increased to 48 hours?

	Downhill	Cross-country	Maximum Time Available
Manufacturing time per ski	2 hours	1 hour	40 hours
Finishing time per ski	1 hour	1 hour	32 hours
Profit per ski	$70	$50	

20. **Farm Management** A farmer has 70 acres of land available for planting either soybeans or wheat. The cost of preparing the soil, the workdays required, and the expected profit per acre planted for each type of crop are given in the following table:

	Soybeans	Wheat
Preparation cost per acre	$60	$30
Workdays required per acre	3	4
Profit per acre	$180	$100

The farmer cannot spend more than $1800 in preparation costs nor use more than a total of 120 workdays. How many acres of each crop should be planted to maximize the profit? What is the maximum profit? What is the maximum profit if the farmer is willing to spend no more than $2400 on preparation?

21. **Farm Management** A small farm in Illinois has 100 acres of land available on which to grow corn and soybeans. The following table shows the cultivation cost per acre, the labor cost per acre, and the expected profit per acre. The column on the right shows the amount of money available for each of these expenses. Find the number of acres of each crop that should be planted to maximize profit.

	Soybeans	Corn	Money Available
Cultivation cost per acre	$40	$60	$1800
Labor cost per acre	$60	$60	$2400
Profit per acre	$200	$250	

22. **Dietary Requirements** A certain diet requires at least 60 units of carbohydrates, 45 units of protein, and 30 units of fat each day. Each ounce of Supplement A provides 5 units of carbohydrates, 3 units of protein, and 4 units of fat. Each ounce of Supplement B provides 2 units of carbohydrates, 2 units of protein, and 1 unit of fat. If Supplement A costs $1.50 per ounce and Supplement B costs $1.00 per ounce, how many ounces of each supplement should be taken daily to minimize the cost of the diet?

23. **Production Scheduling** In a factory, machine 1 produces 8-inch pliers at the rate of 60 units per hour and 6-inch pliers at the rate of 70 units per hour. Machine 2 produces 8-inch pliers at the rate of 40 units per hour and 6-inch pliers at the rate of 20 units per hour. It costs $50 per hour to operate machine 1, and machine 2 costs $30 per hour to operate. The production schedule requires that at least 240 units of 8-inch pliers and at least 140 units of 6-inch pliers be produced during each 10-hour day. Which combination of machines will cost the least money to operate?

24. **Farm Management** An owner of a fruit orchard hires a crew of workers to prune at least 25 of his 50 fruit trees. Each newer tree requires one hour to prune, while each older tree needs one-and-a-half hours. The crew contracts to work for at least 30 hours and charge $15 for each newer tree and $20 for each older tree. To minimize his cost, how many of each kind of tree will the orchard owner have pruned? What will be the cost?

25. **Managing a Meat Market** A meat market combines ground beef and ground pork in a single package for meat

loaf. The ground beef is 75% lean (75% beef, 25% fat) and costs the market $0.75 per pound. The ground pork is 60% lean and costs the market $0.45 per pound. The meat loaf must be at least 70% lean. If the market wants to use at least 50 lb of its available pork, but no more than 200 lb of its available ground beef, how much ground beef should be mixed with ground pork so that the cost is minimized?

26. Return on Investment   An investment broker is instructed by her client to invest up to $20,000, some in a junk bond yielding 9% per annum and some in Treasury bills yielding 7% per annum. The client wants to invest at least $8000 in T-bills and no more than $12,000 in the junk bond.

(a) How much should the broker recommend that the client place in each investment to maximize income if the client insists that the amount invested in T-bills must equal or exceed the amount placed in junk bonds?

(b) How much should the broker recommend that the client place in each investment to maximize income if the client insists that the amount invested in T-bills must not exceed the amount placed in junk bonds?

27. Maximizing Profit on Ice Skates   A factory manufactures two kinds of ice skates: racing skates and figure skates. The racing skates require 6 work-hours in the fabrication department, whereas the figure skates require 4 work-hours there. The racing skates require 1 work-hour in the finishing department, whereas the figure skates require 2 work-hours there. The fabricating department has available at most 120 work-hours per day, and the finishing department has no more than 40 work-hours per day available. If the profit on each racing skate is $10 and the profit on each figure skate is $12, how many of each should be manufactured each day to maximize profit? (Assume that all skates made are sold.)

28. Financial Planning   A retired couple has up to $50,000 to place in fixed-income securities. Their financial adviser suggests two securities to them: one is an AAA bond that yields 8% per annum; the other is a certificate of deposit (CD) that yields 4%. After careful consideration of the alternatives, the couple decides to place at most $20,000 in the AAA bond and at least $15,000 in the CD. They also instruct the financial adviser to place at least as much in the CD as in the AAA bond. How should the financial adviser proceed to maximize the return on their investment?

29. Product Design   An entrepreneur is having a design group produce at least six samples of a new kind of fastener that he wants to market. It costs $9.00 to produce each metal fastener and $4.00 to produce each plastic fastener. He wants to have at least two of each version of the fastener and needs to have all the samples 24 hours from now. It takes 4 hours to produce each metal sample and 2 hours to produce each plastic sample. To minimize the cost of the samples, how many of each kind should the entrepreneur order? What will be the cost of the samples?

30. Animal Nutrition   Kevin's dog Amadeus likes two kinds of canned dog food. "Gourmet Dog" costs 40 cents a can and has 20 units of a vitamin complex; the calorie content is 75 calories. "Chow Hound" costs 32 cents a can and has 35 units of vitamins and 50 calories. Kevin likes Amadeus to have at least 1175 units of vitamins a month and at least 2375 calories during the same time period. Kevin has space to store only 60 cans of dog food at a time. How much of each kind of dog food should Kevin buy each month in order to minimize his cost?

31. Airline Revenue   An airline has two classes of service: first class and coach. Management's experience has been that each aircraft should have at least 8 but no more than 16 first-class seats and at least 80 but not more than 120 coach seats.

(a) If management decides that the ratio of first class to coach seats should never exceed 1:12, with how many of each type of seat should an aircraft be configured to maximize revenue?

(b) If management decides that the ratio of first class to coach seats should never exceed 1:8, with how many of each type of seat should an aircraft be configured to maximize revenue?

(c) If you were management, what would you do?

[**Hint:** Assume that the airline charges $C for a coach seat and $F for a first-class seat; $C > 0$, $F > C$.]

32. Minimizing Cost   A farm that specializes in raising frying chickens supplements the regular chicken feed with four vitamins. The owner wants the supplemental food to contain at least 50 units of vitamin I, 90 units of vitamin II, 60 units of vitamin III, and 100 units of vitamin IV per 100 ounces of feed. Two supplements are available: supplement A, which contains 5 units of vitamin I, 25 units of vitamin II, 10 units of vitamin III, and 35 units of vitamin IV per ounce, and supplement B, which contains 25 units of vitamin I, 10 units of vitamin II, 10 units of vitamin III, and 20 units of vitamin IV per ounce. If supplement A costs $0.06 per ounce and supplement B costs $0.08 per ounce, how much of each supplement should the manager of the farm buy to add to each 100 ounces of feed to keep the total cost at a minimum, while still meeting the owner's vitamin specifications?

Discussion and Writing

33. Explain in your own words what a linear programming problem is and how it can be solved.

Chapter Review

Things to Know

Systems of equations (p. 506)

Systems with no solutions are inconsistent.   Systems with a solution are consistent.

Consistent systems of linear equations have either a unique solution or an infinite number of solutions.

Determinants and Cramer's Rule (pp. 536, 538 and 542)

Matrix (pp. 520 and 547)	Rectangular array of numbers, called entries
m by n matrix (p. 547)	Matrix with m rows and n columns
Identity matrix I (p. 555)	Square matrix whose diagonal entries are 1's, while all other entries are 0's
Inverse of a matrix (p. 556)	A^{-1} is the inverse of A if $AA^{-1} = A^{-1}A = I$
Nonsingular matrix (p. 556)	A square matrix that has an inverse

Linear programming (p. 581)

Maximize (or minimize) a linear objective function, $z = Ax + By$, subject to certain conditions, or constraints, expressible as linear inequalities in x and y. A feasible point (x, y) is a point that satisfies the constraints of a linear programming problem.

Location of solution (p. 583)

If a linear programming problem has a solution, it is located at a corner point of the graph of the feasible points. If a linear programming problem has multiple solutions, at least one of them is located at a corner point of the graph of the feasible points. In either case, the corresponding value of the objective function is unique.

Objectives

Section		You should be able to . . .	Review Exercises
5.1	1	Solve systems of equations by substitution (p. 507)	1–14, 87, 88, 91–96
	2	Solve systems of equations by elimination (p. 508)	1–14, 87, 88, 91–96
	3	Identify inconsistent systems of equations containing two variables (p. 510)	9, 10, 13, 84
	4	Express the solution of a system of dependent equations containing two variables (p. 510)	14, 86
	5	Solve systems of three equations containing three variables (p. 511)	15–18, 85, 86, 89
	6	Identify inconsistent systems of equations containing three variables (p. 513)	18
	7	Express the solution of a system of dependent equations containing three variables (p. 514)	17
5.2	1	Write the augmented matrix of a system of linear equations (p. 520)	35–44
	2	Write the system from the augmented matrix (p. 521)	19, 20
	3	Perform row operations on a matrix (p. 522)	35–44
	4	Solve a system of linear equations using matrices (p. 523)	35–44
5.3	1	Evaluate 2 by 2 determinants (p. 536)	45, 46
	2	Use Cramer's Rule to solve a system of two equations containing two variables (p. 537)	51–54
	3	Evaluate 3 by 3 determinants (p. 540)	47–50
	4	Use Cramer's Rule to solve a system of three equations containing three variables (p. 542)	55, 56
	5	Know properties of determinants (p. 543)	57, 58
5.4	1	Find the sum and difference of two matrices (p. 548)	21, 22
	2	Find scalar multiples of a matrix (p. 550)	23, 24
	3	Find the product of two matrices (p. 551)	25–28
	4	Find the inverse of a matrix (p. 556)	29–34
	5	Solve a system of linear equations using inverse matrices (p. 559)	35–37, 39, 40, 43, 44

Review Exercises

In Problems 1–18, solve each system of equations using the method of substitution or the method of elimination. If the system has no solution, say that it is inconsistent. Verify your result using a graphing utility.

1. $\begin{cases} 2x - y = 5 \\ 5x + 2y = 8 \end{cases}$
2. $\begin{cases} 2x + 3y = 2 \\ 7x - y = 3 \end{cases}$
3. $\begin{cases} 3x - 4y = 4 \\ x - 3y = \dfrac{1}{2} \end{cases}$
4. $\begin{cases} 2x + y = 0 \\ 5x - 4y = -\dfrac{13}{2} \end{cases}$

5. $\begin{cases} x - 2y - 4 = 0 \\ 3x + 2y - 4 = 0 \end{cases}$
6. $\begin{cases} x - 3y + 5 = 0 \\ 2x + 3y - 5 = 0 \end{cases}$
7. $\begin{cases} y = 2x - 5 \\ x = 3y + 4 \end{cases}$
8. $\begin{cases} x = 5y + 2 \\ y = 5x + 2 \end{cases}$

9. $\begin{cases} x - 3y + 4 = 0 \\ \dfrac{1}{2}x - \dfrac{3}{2}y + \dfrac{4}{3} = 0 \end{cases}$
10. $\begin{cases} x + \dfrac{1}{4}y = 2 \\ y + 4x + 2 = 0 \end{cases}$
11. $\begin{cases} 2x + 3y - 13 = 0 \\ 3x - 2y = 0 \end{cases}$
12. $\begin{cases} 4x + 5y = 21 \\ 5x + 6y = 42 \end{cases}$

13. $\begin{cases} 3x - 2y = 8 \\ x - \dfrac{2}{3}y = 12 \end{cases}$
14. $\begin{cases} 2x + 5y = 10 \\ 4x + 10y = 20 \end{cases}$
15. $\begin{cases} x + 2y - z = 6 \\ 2x - y + 3z = -13 \\ 3x - 2y + 3z = -16 \end{cases}$

16. $\begin{cases} x + 5y - z = 2 \\ 2x + y + z = 7 \\ x - y + 2z = 11 \end{cases}$
17. $\begin{cases} 2x - 4y + z = -15 \\ x + 2y - 4z = 27 \\ 5x - 6y - 2z = -3 \end{cases}$
18. $\begin{cases} x - 4y + 3z = 15 \\ -3x + y - 5z = -5 \\ -7x - 5y - 9z = 10 \end{cases}$

In Problems 19 and 20, write the system of equations corresponding to the given augmented matrix.

19. $\begin{bmatrix} 3 & 2 & | & 8 \\ 1 & 4 & | & -1 \end{bmatrix}$

20. $\begin{bmatrix} 1 & 2 & 5 & | & -2 \\ 5 & 0 & -3 & | & 8 \\ 2 & -1 & 0 & | & 0 \end{bmatrix}$

In Problems 21–28, use the following matrices to compute each expression. Verify your result using a graphing utility.

$$A = \begin{bmatrix} 1 & 0 \\ 2 & 4 \\ -1 & 2 \end{bmatrix}, \qquad B = \begin{bmatrix} 4 & -3 & 0 \\ 1 & 1 & -2 \end{bmatrix}, \qquad C = \begin{bmatrix} 3 & -4 \\ 1 & 5 \\ 5 & 2 \end{bmatrix}$$

21. $A + C$
22. $A - C$
23. $6A$
24. $-4B$
25. AB
26. BA
27. CB
28. BC

In Problems 29–34, find the inverse, if there is one, of each matrix algebraically. If there is not an inverse, say that the matrix is singular. Verify your result using a graphing utility.

29. $\begin{bmatrix} 4 & 6 \\ 1 & 3 \end{bmatrix}$

30. $\begin{bmatrix} -3 & 2 \\ 1 & -2 \end{bmatrix}$

31. $\begin{bmatrix} 1 & 3 & 3 \\ 1 & 2 & 1 \\ 1 & -1 & 2 \end{bmatrix}$

32. $\begin{bmatrix} 3 & 1 & 2 \\ 3 & 2 & -1 \\ 1 & 1 & 1 \end{bmatrix}$

33. $\begin{bmatrix} 4 & -8 \\ -1 & 2 \end{bmatrix}$

34. $\begin{bmatrix} -3 & 1 \\ -6 & 2 \end{bmatrix}$

In Problems 35–44, solve each system of equations algebraically using matrices. If the system has no solution, say that it is inconsistent. Verify your result using a graphing utility.

35. $\begin{cases} 3x - 2y = 1 \\ 10x + 10y = 5 \end{cases}$

36. $\begin{cases} 3x + 2y = 6 \\ x - y = -\dfrac{1}{2} \end{cases}$

37. $\begin{cases} 5x - 6y - 3z = 6 \\ 4x - 7y - 2z = -3 \\ 3x + y - 7z = 1 \end{cases}$

38. $\begin{cases} 2x + y + z = 5 \\ 4x - y - 3z = 1 \\ 8x + y - z = 5 \end{cases}$

39. $\begin{cases} x - 2z = 1 \\ 2x + 3y = -3 \\ 4x - 3y - 4z = 3 \end{cases}$

40. $\begin{cases} x + 2y - z = 2 \\ 2x - 2y + z = -1 \\ 6x + 4y + 3z = 5 \end{cases}$

41. $\begin{cases} x - y + z = 0 \\ x - y - 5z - 6 = 0 \\ 2x - 2y + z - 1 = 0 \end{cases}$

42. $\begin{cases} 4x - 3y + 5z = 0 \\ 2x + 4y - 3z = 0 \\ 6x + 2y + z = 0 \end{cases}$

43. $\begin{cases} x - y - z - t = 1 \\ 2x + y - z + 2t = 3 \\ x - 2y - 2z - 3t = 0 \\ 3x - 4y + z + 5t = -3 \end{cases}$

44. $\begin{cases} x - 3y + 3z - t = 4 \\ x + 2y - z = -3 \\ x + 3z + 2t = 3 \\ x + y + 5z = 6 \end{cases}$

In Problems 45–50, find the value of each determinant algebraically. Verify your result using a graphing utility.

45. $\begin{vmatrix} 3 & 4 \\ 1 & 3 \end{vmatrix}$

46. $\begin{vmatrix} -4 & 0 \\ 1 & 3 \end{vmatrix}$

47. $\begin{vmatrix} 1 & 4 & 0 \\ -1 & 2 & 6 \\ 4 & 1 & 3 \end{vmatrix}$

48. $\begin{vmatrix} 2 & 3 & 10 \\ 0 & 1 & 5 \\ -1 & 2 & 3 \end{vmatrix}$

49. $\begin{vmatrix} 2 & 1 & -3 \\ 5 & 0 & 1 \\ 2 & 6 & 0 \end{vmatrix}$

50. $\begin{vmatrix} -2 & 1 & 0 \\ 1 & 2 & 3 \\ -1 & 4 & 2 \end{vmatrix}$

In Problems 51–56, use Cramer's Rule, if applicable, to solve each system.

51. $\begin{cases} x - 2y = 4 \\ 3x + 2y = 4 \end{cases}$

52. $\begin{cases} x - 3y = -5 \\ 2x + 3y = 5 \end{cases}$

53. $\begin{cases} 2x + 3y - 13 = 0 \\ 3x - 2y = 0 \end{cases}$

54. $\begin{cases} 3x - 4y - 12 = 0 \\ 5x + 2y + 6 = 0 \end{cases}$

55. $\begin{cases} x + 2y - z = 6 \\ 2x - y + 3z = -13 \\ 3x - 2y + 3z = -16 \end{cases}$

56. $\begin{cases} x - y + z = 8 \\ 2x + 3y - z = -2 \\ 3x - y - 9z = 9 \end{cases}$

In Problems 57 and 58, use properties of determinants to find the value of each determinant if it is known that

$$\begin{vmatrix} x & y \\ a & b \end{vmatrix} = 8$$

57. $\begin{vmatrix} 2x & y \\ 2a & b \end{vmatrix}$

58. $\begin{vmatrix} y & x \\ b & a \end{vmatrix}$

In Problems 59–68, write the partial fraction decomposition of each rational expression.

59. $\dfrac{6}{x(x - 4)}$

60. $\dfrac{x}{(x + 2)(x - 3)}$

61. $\dfrac{x - 4}{x^2(x - 1)}$

62. $\dfrac{2x - 6}{(x - 2)^2(x - 1)}$

63. $\dfrac{x}{(x^2 + 9)(x + 1)}$

64. $\dfrac{3x}{(x - 2)(x^2 + 1)}$

65. $\dfrac{x^3}{(x^2 + 4)^2}$

66. $\dfrac{x^3 + 1}{(x^2 + 16)^2}$

67. $\dfrac{x^2}{(x^2 + 1)(x^2 - 1)}$

68. $\dfrac{4}{(x^2 + 4)(x^2 - 1)}$

In Problems 69–72 graph each inequality (a) by hand and (b) by using a graphing utility.

69. $3x + 4y \le 12$ **70.** $2x - 3y \ge 6$ **71.** $y \le x^2$ **72.** $x \ge y^2$

In Problems 73–78, graph each system of inequalities by hand. Tell whether the graph is bounded or unbounded, and label the corner points.

73. $\begin{cases} -2x + y \le 2 \\ x + y \ge 2 \end{cases}$ **74.** $\begin{cases} x - 2y \le 6 \\ 2x + y \ge 2 \end{cases}$ **75.** $\begin{cases} x \ge 0 \\ y \ge 0 \\ x + y \le 4 \\ 2x + 3y \le 6 \end{cases}$

76. $\begin{cases} x \ge 0 \\ y \ge 0 \\ 3x + y \ge 6 \\ 2x + y \ge 2 \end{cases}$ **77.** $\begin{cases} x \ge 0 \\ y \ge 0 \\ 2x + y \le 8 \\ x + 2y \ge 2 \end{cases}$ **78.** $\begin{cases} x \ge 0 \\ y \ge 0 \\ 3x + y \le 9 \\ 2x + 3y \ge 6 \end{cases}$

In Problems 79–82, solve each linear programming problem.

79. Maximize $z = 3x + 4y$ subject to $x \ge 0, y \ge 0, 3x + 2y \ge 6, x + y \le 8$

80. Maximize $z = 2x + 4y$ subject to $x \ge 0, y \ge 0, x + y \le 6, x \ge 2$

81. Minimize $z = 3x + 5y$ subject to $x \ge 0, y \ge 0, x + y \ge 1, 3x + 2y \le 12, x + 3y \le 12$

82. Minimize $z = 3x + y$ subject to $x \ge 0, y \ge 0, x \le 8, y \le 6, 2x + y \ge 4$

---

83. Find A so that the system of equations has infinitely many solutions.

$$\begin{cases} 2x + 5y = 5 \\ 4x + 10y = A \end{cases}$$

84. Find A so that the system in Problem 83 is inconsistent.

85. Curve Fitting Find the quadratic function $y = ax^2 + bx + c$ that passes through the three points $(0, 1)$, $(1, 0)$, and $(-2, 1)$.

86. Curve Fitting Find the general equation of the circle that passes through the three points $(0, 1)$, $(1, 0)$, and $(-2, 1)$.

[**Hint:** The general equation of a circle is
$x^2 + y^2 + Dx + Ey + F = 0$.]

87. Blending Coffee A coffee distributor is blending a new coffee that will cost $3.90 per pound. It will consist of a blend of $3.00 per pound coffee and $6.00 per pound coffee. What amounts of each type of coffee should be mixed to achieve the desired blend?

[**Hint**: Assume that the weight of the blended coffee is 100 pounds.]

$3.00/lb $3.90/lb $6.00/lb

88. Farming A 1000-acre farm in Illinois is used to grow corn and soy beans. The cost per acre for raising corn is $65, and the cost per acre for soy beans is $45. If $54,325 has been budgeted for costs and all the acreage is to be used, how many acres should be allocated for each crop?

89. Cookie Orders A cookie company makes three kinds of cookies, oatmeal raisin, chocolate chip, and shortbread, packaged in small, medium, and large boxes. The small box contains 1 dozen oatmeal raisin and 1 dozen chocolate chip; the medium box has 2 dozen oatmeal raisin, 1 dozen chocolate chip, and 1 dozen shortbread; the large box contains 2 dozen oatmeal raisin, 2 dozen chocolate chip, and 3 dozen shortbread. If you require exactly 15 dozen oatmeal raisin, 10 dozen chocolate chip, and 11 dozen shortbread, how many of each size box should you buy?

90. Mixed Nuts A store that specializes in selling nuts has available 72 pounds of cashews and 120 pounds of peanuts. These are to be mixed in 12-ounce packages as follows: a lower-priced package containing 8 ounces of peanuts and 4 ounces of cashews and a quality package containing 6 ounces of peanuts and 6 ounces of cashews.

(a) Use x to denote the number of lower-priced packages and use y to denote the number of quality packages. Write a system of linear inequalities that describes the possible number of each kind of package.

(b) Graph the system and label the corner points.

91. Determining the Speed of the Current of the Aguarico River On a recent trip to the Cuyabeno Wildlife Reserve in the Amazon region of Ecuador, a 100-kilometer trip by speedboat was taken down the Aguarico River from Chiritza to the Flotel Orellana. As I watched the Amazon unfold, I wondered how fast the speedboat was going and how fast the current of the white-water Aguarico River was. I timed the trip downstream at 2.5 hours and the return trip at 3 hours. What were the two speeds?

92. Finding the Speed of the Jet Stream On a flight between Midway Airport in Chicago and Ft. Lauderdale, Florida, a Boeing 737 jet maintains an airspeed of 475 miles per hour. If the trip from Chicago to Ft. Lauderdale takes 2 hours, 30 minutes and the return flight takes 2 hours, 50 minutes, what is the speed of the jet stream? (Assume that the speed of the jet stream remains constant at the various altitudes of the plane and that the plane flies with the jet stream one way and against it the other way.)

93. Constant Rate Jobs If Bruce and Bryce work together for 1 hour and 20 minutes, they will finish a certain job. If Bryce and Marty work together for 1 hour and 36 minutes, the same job can be finished. If Marty and Bruce work together, they can complete this job in 2 hours and 40 minutes. How long will it take each of them working alone to finish the job?

94. Maximizing Profit on Figurines A factory manufactures two kinds of ceramic figurines: a dancing girl and a mermaid. Each requires three processes: molding, painting, and glazing. The daily labor available for molding is no more than 90 work-hours, labor available for painting does not exceed 120 work-hours, and labor available for glazing is no more than 60 work-hours. The dancing girl requires 3 work-

hours for molding, 6 work-hours for painting, and 2 work-hours for glazing. The mermaid requires 3 work-hours for molding, 4 work-hours for painting, and 3 work-hours for glazing. If the profit on each figurine is $25 for dancing girls and $30 for mermaids, how many of each should be produced each day to maximize profit? If management decides to produce the number of each figurine that maximizes profit, determine which of these processes has work-hours assigned to it that are not used.

95. Minimizing Production Cost A factory produces gasoline engines and diesel engines. Each week the factory is obligated to deliver at least 20 gasoline engines and at least 15 diesel engines. Due to physical limitations, however, the factory cannot make more than 60 gasoline engines nor more than 40 diesel engines in any given week. Finally, to prevent layoffs, a total of at least 50 engines must be produced. If gasoline engines cost $450 each to produce and diesel engines cost $550 each to produce, how many of each should be produced per week to minimize the cost? What is the excess capacity of the factory; that is, how many of each kind of engine is being produced in excess of the number that the factory is obligated to deliver?

96. Describe four ways of solving a system of three linear equations containing three variables. Which method do you prefer? Why?

Chapter Test

In Problems 1–4, solve each system of equations using the method of substitution or the method of elimination. If the system has no solution, say that it is inconsistent. Verify your result using a graphing utility.

1. $\begin{cases} -2x + y = -7 \\ 4x + 3y = 9 \end{cases}$

2. $\begin{cases} \dfrac{1}{3}x - 2y = 1 \\ 5x - 30y = 18 \end{cases}$

3. $\begin{cases} x - y + 2z = 5 \\ 3x + 4y - z = -2 \\ 5x + 2y + 3z = 8 \end{cases}$

4. $\begin{cases} 3x + 2y - 8z = -3 \\ -x - \dfrac{2}{3}y + z = 1 \\ 6x - 3y + 15z = 8 \end{cases}$

5. Write the augmented matrix corresponding to the system of

equations: $\begin{cases} 4x - 5y + z = 0 \\ -2x - y + 6 = -19 \\ x + 5y - 5z = 10 \end{cases}$

6. Write the system of equations corresponding to the given

augmented matrix: $\begin{bmatrix} 3 & 2 & 4 & | & -6 \\ 1 & 0 & 8 & | & 2 \\ -2 & 1 & 3 & | & -11 \end{bmatrix}$

In Problems 7–10, use the given matrices to compute each expression. Verify your result using a graphing utility.

$$A = \begin{bmatrix} 1 & -1 \\ 0 & -4 \\ 3 & 2 \end{bmatrix} \qquad B = \begin{bmatrix} 1 & -2 & 5 \\ 0 & 3 & 1 \end{bmatrix} \qquad C = \begin{bmatrix} 4 & 6 \\ 1 & -3 \\ -1 & 8 \end{bmatrix}$$

7. $2A + C$

8. $A - 3C$

9. AC

10. BA

In Problems 11 and 12, algebraically find the inverse of each nonsingular matrix. Verify your result using a graphing utility.

11. $A = \begin{bmatrix} 3 & 2 \\ 5 & 4 \end{bmatrix}$

12. $B = \begin{bmatrix} 1 & -1 & 1 \\ 2 & 5 & -1 \\ 2 & 3 & 0 \end{bmatrix}$

In Problems 13–16, solve each system of equations algebraically using matrices. If the system has no solution, say that it is inconsistent. Verify your result using a graphing utility.

13. $\begin{cases} 6x + 3y = 12 \\ 2x - y = -2 \end{cases}$

14. $\begin{cases} x + \dfrac{1}{4}y = 7 \\ 8x + 2y = 56 \end{cases}$

15. $\begin{cases} x + 2y + 4z = -3 \\ 2x + 7y + 15z = -12 \\ 4x + 7y + 13z = -10 \end{cases}$

16. $\begin{cases} 2x + 2y - 3z = 5 \\ x - y + 2z = 8 \\ 3x + 5y - 8z = -2 \end{cases}$

In Problems 17 and 18, find the value of each determinant algebraically. Verify your result using a graphing utility.

17. $\begin{vmatrix} -2 & 5 \\ 3 & 7 \end{vmatrix}$

18. $\begin{vmatrix} 2 & -4 & 6 \\ 1 & 4 & 0 \\ -1 & 2 & -4 \end{vmatrix}$

In Problems 19 and 20, use Cramer's Rule, if possible, to solve each system.

19. $\begin{cases} 4x + 3y = -23 \\ 3x - 5y = 19 \end{cases}$

20. $\begin{cases} 4x - 3y + 2z = 15 \\ -2x + y - 3z = -15 \\ 5x - 5y + 2z = 18 \end{cases}$

In Problems 21 and 22, write the partial fraction decomposition of each rational expression.

21. $\dfrac{3x + 7}{(x + 3)^2}$

22. $\dfrac{4x^2 - 3}{x(x^2 + 3)^2}$

---

23. Graph the system of inequalities. Tell whether the graph is bounded or unbounded, and label all corner points.

$$\begin{cases} x \geq 0 \\ y \geq 0 \\ x + 2y \geq 8 \\ 2x - 3y \geq 2 \end{cases}$$

24. Maximize $z = 5x + 8y$ subject to $x \geq 0$, $2x + y \leq 8$, and $x - 3y \leq -3$.

25. Megan went clothes shopping and bought 2 pairs of flare jeans, 2 camisoles, and 4 t-shirts for $90.00. At the same store, Paige bought one pair of flare jeans and 3 t-shirts for $42.50 while Kara bought 1 pair of flare jeans, 3 camisoles, and 2 t-shirts for $62.00. Determine the price of each clothing item.

Chapter Projects

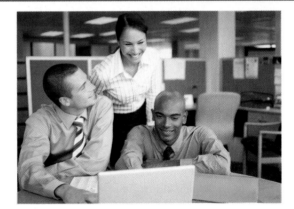

1. **Markov Chains** A **Markov chain** (or process) is one in which future outcomes are determined by a current state. Future outcomes are based on probabilities. The probability of moving to a certain state depends only on the state previously occupied and does not vary with time. An example of

a Markov chain would be the maximum education achieved by children based on the highest education attained by their parents, where the states are (1) earned college degree, (2) high-school diploma only, (3) elementary school only. If p_{ij} is the probability of moving from state i to state j, then the **transition matrix** is the $m \times m$ matrix

$$P = \begin{bmatrix} p_{11} & p_{12} & \cdots & p_{1m} \\ \vdots & \vdots & & \vdots \\ p_{m1} & p_{m2} & \cdots & p_{mm} \end{bmatrix}$$

The table on page 594 represents the probabilities of the highest educational level of children based on the highest educational level of their parents. For example, the table shows that the probability p_{21} is 40% that parents with a high-school education (row 2) will have children with a college education (column 1).
(a) Convert the percentages to decimals.
(b) What is the transition matrix?
(c) Sum across the rows. What do you notice? Why do you think that you obtained this result?

(d) If P is the transition matrix of a Markov chain, then the (i, j)th entry of P^n (nth power of P) gives the probability of passing from state i to state j in n stages. What is the probability that a grandchild of a college graduate is a college graduate?

(e) What is the probability that the grandchild of a high school graduate finishes college?

(f) The row vector $v^{(0)} = [0.267 \ 0.574 \ 0.159]$ represents the proportion of the U.S. population that has college, high school, and elementary school, respectively, as the highest educational level in 2002.* In a Markov chain the probability distribution $v^{(k)}$ after k stages is $v^{(k)} = v^{(0)}P^k$, where P^k is the kth power of the transition matrix. What will be the distribution of highest educational attainment of the grandchildren of the current population?

(g) Calculate $P^3, P^4, P^5, \ldots$. Continue until the matrix does not change. This is called the long-run distribution. What is the long-run distribution of highest educational attainment of the population?

*SOURCE: U.S. Census Bureau.

Highest Educational Level of Parents	Maximum Education That Children Achieve		
	College	High School	Elementary
College	80%	18%	2%
High school	40%	50%	10%
Elementary	20%	60%	20%

The following projects are available at the Instructor's Resource Center (IRC):

2. **Project at Motorola** *Error Control Codings*
3. **Using Matrices to Find the Line of Best Fit**
4. **CBL Experiment**

Cumulative Review

In Problems 1–6, solve each equation algebraically. Verify your result using a graphing utility.

1. $2x^2 - x = 0$
2. $\sqrt{3x + 1} = 4$
3. $2x^3 - 3x^2 - 8x - 3 = 0$
4. $3^x = 9^{x+1}$
5. $\log_3(x - 1) + \log_3(2x + 1) = 2$
6. $3^x = e$

7. Determine whether the function $g(x) = \dfrac{2x^3}{x^4 + 1}$ is even, odd, or neither. Is the graph of g symmetric with respect to the x-axis, y-axis, or origin?

8. Find the center and radius of the circle $x^2 + y^2 - 2x + 4y - 11 = 0$. Graph the circle by hand.

9. Graph $f(x) = 3^{x-2} + 1$ using transformations. What is the domain, range, and horizontal asymptote of f?

10. The function $f(x) = \dfrac{5}{x + 2}$ is one-to-one. Find f^{-1}. Find the domain and the range of f and the domain and the range of f^{-1}.

11. Graph each equation by hand.
    (a) $y = 3x + 6$
    (b) $x^2 + y^2 = 4$
    (c) $y = x^3$
    (d) $y = \dfrac{1}{x}$
    (e) $y = \sqrt{x}$
    (f) $y = e^x$
    (g) $y = \ln x$

12. $f(x) = x^3 - 3x + 5$:
    (a) Using a graphing utility, approximate the zero(s) of f.
    (b) Using a graphing utility, approximate the local maxima and local minima.
    (c) Determine the intervals on which f is increasing.

Analytic Geometry

6

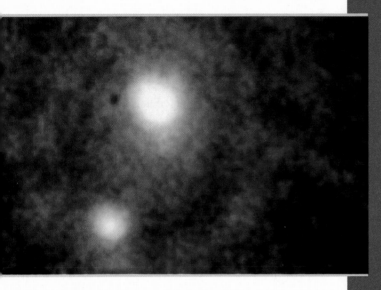

Pluto's Unusual Orbit

Pluto is about 39 times as far from the Sun as Earth is. Its average distance from the Sun is about 3,647,240,000 miles (5,869,660,000 kilometers). Pluto travels around the Sun in an elliptical (oval-shaped) orbit. At some point in its orbit, it comes closer to the Sun than Neptune, usually the second farthest planet. It stays inside Neptune's orbit for about 20 Earth-years. This event occurs every 248 Earth-years, which is about the same number of Earth-years it takes Pluto to travel once around the Sun. Pluto entered Neptune's orbit on January 23, 1979, and remained there until February 11, 1999. Pluto will remain the outermost planet until the year 2227.

SOURCE: Reprinted with permission of The Associated Press.

—*See Chapter Project 1.*

A LOOK BACK In Chapter 1, we introduced rectangular coordinates and showed how geometry problems can be solved algebraically. We defined a circle geometrically and then used the distance formula and rectangular coordinates to obtain an equation for a circle.

A LOOK AHEAD In this chapter we give geometric definitions for the conics and use the distance formula and rectangular coordinates to obtain their equations.

Historically, Apollonius (200 BC) was among the first to study *conics* and discover some of their interesting properties. Today, conics are still studied because of their many uses. *Paraboloids of revolution* (parabolas rotated about their axes of symmetry) are used as signal collectors (the satellite dishes used with radar and cable TV, for example), as solar energy collectors, and as reflectors (telescopes, light projection, and so on). The planets circle the Sun in approximately *elliptical* orbits. Elliptical surfaces can be used to reflect signals such as light and sound from one place to another. And *hyperbolas* can be used to determine the location of lightning strikes.

The Greeks used the methods of Euclidean geometry to study conics. However, we shall use the more powerful methods of analytic geometry, bringing to bear both algebra and geometry, for our study of conics.

The chapter concludes with a section on systems of nonlinear equations and inequalities.

OUTLINE

6.1 Conics

OBJECTIVE 1 Know the Names of the Conics

1 Know the Names of the Conics

The word *conic* derives from the word *cone*, which is a geometric figure that can be constructed in the following way: Let *a* and *g* be two distinct lines that intersect at a point *V*. Keep the line *a* fixed. Now rotate the line *g* about *a* while maintaining the same angle between *a* and *g*. The collection of points swept out (generated) by the line *g* is called a **(right circular) cone**. See Figure 1. The fixed line *a* is called the **axis** of the cone; the point *V* is its **vertex**; the lines that pass through *V* and make the same angle with *a* as *g* are **generators** of the cone. Each generator is a line that lies entirely on the cone. The cone consists of two parts, called **nappes**, that intersect at the vertex.

Figure 1

Axis, *a*

Generators

Vertex, *V*

g

Conics, an abbreviation for **conic sections**, are curves that result from the inter-section of a (right circular) cone and a plane. The conics we shall study arise when the plane does not contain the vertex, as shown in Figure 2. These conics are **circles** when the plane is perpendicular to the axis of the cone and intersects each genera-tor; **ellipses** when the plane is tilted slightly so that it intersects each generator, but intersects only one nappe of the cone; **parabolas** when the plane is tilted farther so that it is parallel to one (and only one) generator and intersects only one nappe of the cone; and **hyperbolas** when the plane intersects both nappes.

If the plane does contain the vertex, the intersection of the plane and the cone is a point, a line, or a pair of intersecting lines. These are usually called **degenerate conics**.

Figure 2

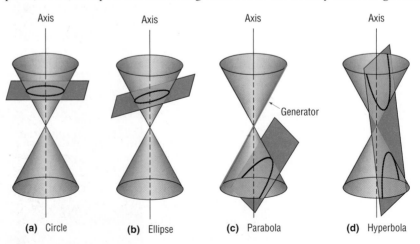

Axis

Axis

Axis

Axis

Generator

(a) Circle   (b) Ellipse   (c) Parabola   (d) Hyperbola

6.2 The Parabola

PREPARING FOR THIS SECTION   *Before getting started, review the following:*

- Distance Formula (Section 1.1, pp. 86–89)
- Symmetry (Section 2.1, pp. 198–200)
- Square Root Method (Section 1.3, pp. 114–115)

- Completing the Square (Section 1.3, pp. 115–116)
- Graphing Techniques: Transformations (Section 2.7, pp. 266–274)

Now work the 'Are You Prepared?' problems on page 605.

OBJECTIVES   1 Work with Parabolas with Vertex at the Origin
2 Work with Parabolas with Vertex at (h, k)
3 Solve Applied Problems Involving Parabolas

We stated earlier (Section 3.1) that the graph of a quadratic function is a parabola. In this section, we begin with a geometric definition of a parabola and use it to obtain an equation.

> A **parabola** is the collection of all points P in the plane that are the same distance from a fixed point F as they are from a fixed line D. The point F is called the **focus** of the parabola, and the line D is its **directrix**. As a result, a parabola is the set of points P for which

$$d(F, P) = d(P, D) \tag{1}$$

Figure 3 shows a parabola. The line through the focus F and perpendicular to the directrix D is called the **axis of symmetry** of the parabola. The point of intersection of the parabola with its axis of symmetry is called the **vertex** V.

Figure 3

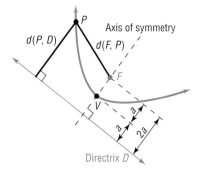

Because the vertex V lies on the parabola, it must satisfy equation (1): $d(F, V) = d(V, D)$. The vertex is midway between the focus and the directrix. We shall let a equal the distance $d(F, V)$ from F to V. Now we are ready to derive an equation for a parabola. To do this, we use a rectangular system of coordinates positioned so that the vertex V, focus F, and directrix D of the parabola are conveniently located.

1 Work with Parabolas with Vertex at the Origin

Figure 4
$y^2 = 4ax$

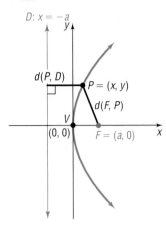

If we choose to locate the vertex V at the origin $(0, 0)$, then we can conveniently position the focus F on either the x-axis or the y-axis. First, we consider the case where the focus F is on the positive x-axis, as shown in Figure 4. Because the distance from F to V is a, the coordinates of F will be $(a, 0)$ with $a > 0$. Similarly, because the distance from V to the directrix D is also a and, because D must be perpendicular to the x-axis (since the x-axis is the axis of symmetry), the equation of the directrix D must be $x = -a$.

Now, if $P = (x, y)$ is any point on the parabola, then P must obey equation (1):

$$d(F, P) = d(P, D)$$

So we have

$$\sqrt{(x - a)^2 + y^2} = |x + a| \qquad \text{Use the Distance Formula.}$$
$$(x - a)^2 + y^2 = (x + a)^2 \qquad \text{Square both sides.}$$
$$x^2 - 2ax + a^2 + y^2 = x^2 + 2ax + a^2 \qquad \text{Remove parentheses.}$$
$$y^2 = 4ax \qquad \text{Simplify.}$$

Theorem

Equation of a Parabola; Vertex at (0, 0), Focus at (a, 0), a > 0

The equation of a parabola with vertex at $(0, 0)$, focus at $(a, 0)$, and directrix $x = -a, a > 0$, is

$$y^2 = 4ax \qquad\qquad (2)$$

EXAMPLE 1 **Finding the Equation of a Parabola and Graphing It**

Find an equation of the parabola with vertex at $(0, 0)$ and focus at $(3, 0)$. Graph the equation.

Solution

Figure 5

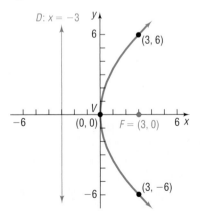

The distance from the vertex $(0, 0)$ to the focus $(3, 0)$ is $a = 3$. Based on equation (2), the equation of this parabola is

$$y^2 = 4ax$$
$$y^2 = 12x \qquad a = 3$$

To graph this parabola, it is helpful to plot the two points on the graph above and below the focus. To locate them, we let $x = 3$. Then

$$y^2 = 12x = 12(3) = 36$$
$$y = \pm 6 \qquad \text{Solve for y.}$$

The points on the parabola above and below the focus are $(3, 6)$ and $(3, -6)$. These points help in graphing the parabola because they determine the "opening." See Figure 5. ◀

In general, the points on a parabola $y^2 = 4ax$ that lie above and below the focus $(a, 0)$ are each at a distance $2a$ from the focus. This follows from the fact that if $x = a$ then $y^2 = 4ax = 4a^2$, so $y = \pm 2a$. The line segment joining these two points is called the **latus rectum**; its length is $4a$.

NOW WORK PROBLEM 19.

EXAMPLE 2

Graphing a Parabola Using a Graphing Utility

Graph the parabola $y^2 = 12x$.

Solution

Figure 6

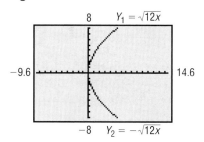

To graph the parabola $y^2 = 12x$, we need to graph the two functions $Y_1 = \sqrt{12x}$ and $Y_2 = -\sqrt{12x}$ on a square screen. Figure 6 shows the graph of $y^2 = 12x$. Notice that the graph fails the vertical line test, so $y^2 = 12x$ is not a function.   ◀

By reversing the steps we used to obtain equation (2), it follows that the graph of an equation of the form of equation (2), $y^2 = 4ax$, is a parabola; its vertex is at $(0, 0)$, its focus is at $(a, 0)$, its directrix is the line $x = -a$, and its axis of symmetry is the x-axis.

For the remainder of this section, the direction "Discuss the equation" will mean to find the vertex, focus, and directrix of the parabola and graph it.

EXAMPLE 3

Discussing the Equation of a Parabola

Discuss the equation: $y^2 = 8x$

Solution

The equation $y^2 = 8x$ is of the form $y^2 = 4ax$, where $4a = 8$ so that $a = 2$. Consequently, the graph of the equation is a parabola with vertex at $(0, 0)$ and focus on the positive x-axis at $(2, 0)$. The directrix is the vertical line $x = -2$. The two points defining the latus rectum are obtained by letting $x = 2$. Then $y^2 = 16$, so $y = \pm 4$. See Figure 7(a) for the graph drawn by hand. Figure 7(b) shows the graph obtained using a graphing utility.

Figure 7

(a)

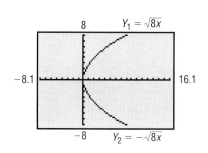

(b)

◀

Recall that we arrived at equation (2) after placing the focus on the positive x-axis. If the focus is placed on the negative x-axis, positive y-axis, or negative y-axis, a different form of the equation for the parabola results. The four forms of the equation of a parabola with vertex at $(0, 0)$ and focus on a coordinate axis a distance a from $(0, 0)$ are given in Table 1, and their graphs are given in Figure 8. Notice that each graph is symmetric with respect to its axis of symmetry.

Table 1

EQUATIONS OF A PARABOLA
VERTEX AT (0, 0); FOCUS ON AN AXIS; $a > 0$

Vertex	Focus	Directrix	Equation	Description
$(0, 0)$	$(a, 0)$	$x = -a$	$y^2 = 4ax$	Parabola, axis of symmetry is the x-axis, opens right
$(0, 0)$	$(-a, 0)$	$x = a$	$y^2 = -4ax$	Parabola, axis of symmetry is the x-axis, opens left
$(0, 0)$	$(0, a)$	$y = -a$	$x^2 = 4ay$	Parabola, axis of symmetry is the y-axis, opens up
$(0, 0)$	$(0, -a)$	$y = a$	$x^2 = -4ay$	Parabola, axis of symmetry is the y-axis, opens down

Figure 8

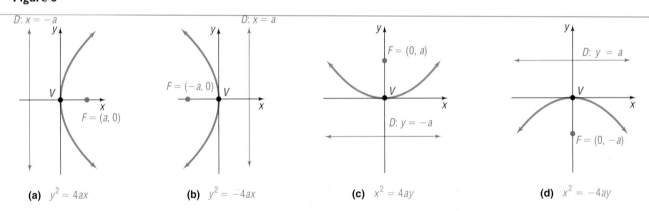

(a) $y^2 = 4ax$　　**(b)** $y^2 = -4ax$　　**(c)** $x^2 = 4ay$　　**(d)** $x^2 = -4ay$

EXAMPLE 4 **Discussing the Equation of a Parabola**

Discuss the equation: $x^2 = -12y$

Solution The equation $x^2 = -12y$ is of the form $x^2 = -4ay$, with $a = 3$. Consequently, the graph of the equation is a parabola with vertex at $(0, 0)$, focus at $(0, -3)$ and directrix the line $y = 3$. The parabola opens down, and its axis of symmetry is the y-axis. To obtain the points defining the latus rectum, let $y = -3$. Then $x^2 = 36$, so $x = \pm 6$. See Figure 9(a) for the graph drawn by hand. Figure 9(b) shows the graph using a graphing utility.

Figure 9

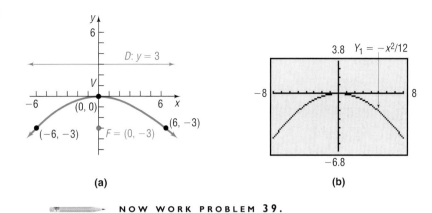

(a)

(b)

NOW WORK PROBLEM **39.**

EXAMPLE 5

Finding the Equation of a Parabola

Figure 10

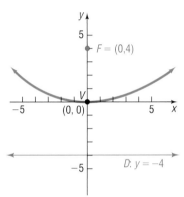

Find the equation of the parabola with focus at $(0, 4)$ and directrix the line $y = -4$. Graph the equation.

Solution   A parabola whose focus is at $(0, 4)$ and whose directrix is the horizontal line $y = -4$ will have its vertex at $(0, 0)$. (Do you see why? The vertex is midway between the focus and the directrix.) Since the focus is on the positive y-axis at $(0, 4)$, the equation of this parabola is of the form $x^2 = 4ay$, with $a = 4$; that is,

$$x^2 = 4ay = 4(4)y = 16y$$
$$\uparrow$$
$$a = 4$$

The points $(8, 4)$ and $(-8, 4)$ determine the latus rectum. Figure 10 shows the graph of $x^2 = 16y$.   ◀

EXAMPLE 6

Finding the Equation of a Parabola

Find the equation of a parabola with vertex at $(0, 0)$ if its axis of symmetry is the x-axis and its graph contains the point $\left(-\dfrac{1}{2}, 2\right)$. Find its focus and directrix, and graph the equation.

Solution   The vertex is at the origin, the axis of symmetry is the x-axis, and the graph contains a point in the second quadrant, so the parabola opens to the left. We see from Table 1 that the form of the equation is

$$y^2 = -4ax$$

Because the point $\left(-\dfrac{1}{2}, 2\right)$ is on the parabola, the coordinates $x = -\dfrac{1}{2}, y = 2$ must satisfy the equation. Substituting $x = -\dfrac{1}{2}$ and $y = 2$ into the equation, we find that

$$4 = -4a\left(-\dfrac{1}{2}\right) \qquad y^2 = -4ax;\ x = -\dfrac{1}{2}, y = 2$$
$$a = 2$$

The equation of the parabola is

$$y^2 = -4(2)x = -8x$$

Figure 11

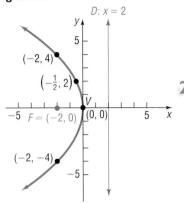

The focus is at $(-2, 0)$ and the directrix is the line $x = 2$. Letting $x = -2$, we find $y^2 = 16$, so $y = \pm 4$. The points $(-2, 4)$ and $(-2, -4)$ define the latus rectum. See Figure 11. ◀

— **NOW WORK PROBLEM 27.**

2 Work with Parabolas with Vertex at (h, k)

If a parabola with vertex at the origin and axis of symmetry along a coordinate axis is shifted horizontally h units and then vertically k units, the result is a parabola with vertex at (h, k) and axis of symmetry parallel to a coordinate axis. The equations of such parabolas have the same forms as those in Table 1, but with x replaced by $x - h$ (the horizontal shift) and y replaced by $y - k$ (the vertical shift). Table 2 gives the forms of the equations of such parabolas. Figures 12(a)–(d) illustrate the graphs for $h > 0, k > 0$.

Table 2

PARABOLAS WITH VERTEX AT (h, k); AXIS OF SYMMETRY PARALLEL TO A COORDINATE AXIS; $a > 0$

Vertex	Focus	Directrix	Equation	Description
(h, k)	$(h + a, k)$	$x = h - a$	$(y - k)^2 = 4a(x - h)$	Parabola, axis of symmetry parallel to x-axis, opens right
(h, k)	$(h - a, k)$	$x = h + a$	$(y - k)^2 = -4a(x - h)$	Parabola, axis of symmetry parallel to x-axis, opens left
(h, k)	$(h, k + a)$	$y = k - a$	$(x - h)^2 = 4a(y - k)$	Parabola, axis of symmetry parallel to y-axis, opens up
(h, k)	$(h, k - a)$	$y = k + a$	$(x - h)^2 = -4a(y - k)$	Parabola, axis of symmetry parallel to y-axis, opens down

Figure 12

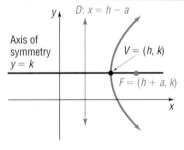

(a) $(y - k)^2 = 4a(x - h)$

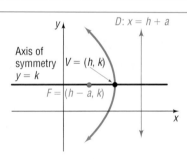

(b) $(y - k)^2 = -4a(x - h)$

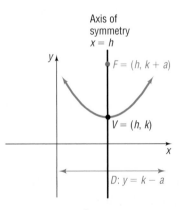

(c) $(x - h)^2 = 4a(y - k)$

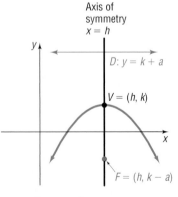

(d) $(x - h)^2 = -4a(y - k)$

EXAMPLE 7   Finding the Equation of a Parabola, Vertex Not at Origin

Figure 13

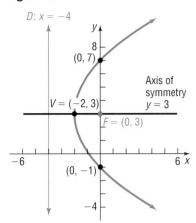

D: x = −4

(0, 7)

V = (−2, 3)   Axis of symmetry y = 3

F = (0, 3)

−6   6 x

(0, −1)

−4

Find an equation of the parabola with vertex at $(-2, 3)$ and focus at $(0, 3)$. Graph the equation.

Solution   The vertex $(-2, 3)$ and focus $(0, 3)$ both lie on the horizontal line $y = 3$ (the axis of symmetry). The distance a from the vertex $(-2, 3)$ to the focus $(0, 3)$ is $a = 2$. Also, because the focus lies to the right of the vertex, we know that the parabola opens to the right. Consequently, the form of the equation is

$$(y - k)^2 = 4a(x - h)$$

where $(h, k) = (-2, 3)$ and $a = 2$. Therefore, the equation is

$$(y - 3)^2 = 4 \cdot 2[x - (-2)]$$
$$(y - 3)^2 = 8(x + 2)$$

If $x = 0$, then $(y - 3)^2 = 16$. Then $y - 3 = \pm 4$, so $y = -1$ or $y = 7$. The points $(0, -1)$ and $(0, 7)$ define the latus rectum; the line $x = -4$ is the directrix. See Figure 13.   ◄

NOW WORK PROBLEM 29.

EXAMPLE 8   Using a Graphing Utility to Graph a Parabola, Vertex Not at Origin

Using a graphing utility, graph the equation $(y - 3)^2 = 8(x + 2)$.

Solution

Figure 14

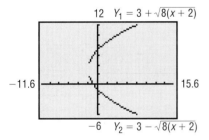

12    $Y_1 = 3 + \sqrt{8(x + 2)}$

−11.6   15.6

−6 $Y_2 = 3 - \sqrt{8(x + 2)}$

First, we must solve the equation for y.

$$(y - 3)^2 = 8(x + 2)$$
$$y - 3 = \pm\sqrt{8(x + 2)} \qquad \text{Use the Square Root Method.}$$
$$y = 3 \pm \sqrt{8(x + 2)} \qquad \text{Add 3 to both sides.}$$

Figure 14 shows the graphs of the equations $Y_1 = 3 + \sqrt{8(x + 2)}$ and $Y_2 = 3 - \sqrt{8(x + 2)}$.   ◄

Polynomial equations define parabolas whenever they involve two variables that are quadratic in one variable and linear in the other. To discuss this type of equation, we first complete the square of the variable that is quadratic.

EXAMPLE 9   Discussing the Equation of a Parabola

Discuss the equation: $x^2 + 4x - 4y = 0$

Solution

Figure 15

Axis of symmetry x = −2   y

4

F = (−2, 0)

−4   4 x

V = (−2, −1)

−3   D: y = −2

To discuss the equation $x^2 + 4x - 4y = 0$, we complete the square involving the variable x.

$$x^2 + 4x - 4y = 0$$
$$x^2 + 4x = 4y \qquad \text{Isolate the terms involving x on the left side.}$$
$$x^2 + 4x + 4 = 4y + 4 \qquad \text{Complete the square on the left side.}$$
$$(x + 2)^2 = 4(y + 1) \qquad \text{Factor.}$$

This equation is of the form $(x - h)^2 = 4a(y - k)$, with $h = -2, k = -1$, and $a = 1$. The graph is a parabola with vertex at $(h, k) = (-2, -1)$ that opens up. The focus is at $(-2, 0)$, and the directrix is the line $y = -2$. See Figure 15.   ◄

NOW WORK PROBLEM 47.

3 Solve Applied Problems Involving Parabolas

Parabolas find their way into many applications. For example, as we discussed in Section 3.1, suspension bridges have cables in the shape of a parabola. Another property of parabolas that is used in applications is their reflecting property.

Suppose that a mirror is shaped like a **paraboloid of revolution**, a surface formed by rotating a parabola about its axis of symmetry. If a light (or any other emitting source) is placed at the focus of the parabola, all the rays emanating from the light will reflect off the mirror in lines parallel to the axis of symmetry. This principle is used in the design of searchlights, flashlights, certain automobile headlights, and other such devices. See Figure 16.

Conversely, suppose that rays of light (or other signals) emanate from a distant source so that they are essentially parallel. When these rays strike the surface of a parabolic mirror whose axis of symmetry is parallel to these rays, they are reflected to a single point at the focus. This principle is used in the design of some solar energy devices, satellite dishes, and the mirrors used in some types of telescopes. See Figure 17.

Figure 16
Searchlight

Figure 17
Telescope

EXAMPLE 10   **Satellite Dish**

A satellite dish is shaped like a paraboloid of revolution. The signals that emanate from a satellite strike the surface of the dish and are reflected to a single point, where the receiver is located. If the dish is 8 feet across at its opening and 3 feet deep at its center, at what position should the receiver be placed?

Solution   Figure 18(a) shows the satellite dish. We draw the parabola used to form the dish on a rectangular coordinate system so that the vertex of the parabola is at the origin and its focus is on the positive y-axis. See Figure 18(b).

Figure 18

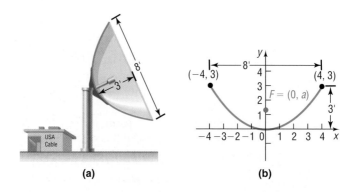

(a)

(b)

The form of the equation of the parabola is

$$x^2 = 4ay$$

and its focus is at $(0, a)$. Since $(4, 3)$ is a point on the graph, we have

$$4^2 = 4a(3)$$

$$a = \frac{4}{3}$$

The receiver should be located $1\frac{1}{3}$ feet from the base of the dish, along its axis of symmetry. ◀

NOW WORK PROBLEM 63.

6.2 Assess Your Understanding

'Are You Prepared?'

Answers are given at the end of these exercises. If you get a wrong answer, read the pages listed in red.

1. The formula for the distance d from $P_1 = (x_1, y_1)$ to $P_2 = (x_2, y_2)$ is $d =$ _____. (pp. 86–89)

2. To complete the square of $x^2 - 4x$, add _____. (pp. 115–116)

3. Use the Square Root Method to find the real solutions of $(x + 4)^2 = 9$. (pp. 114–115)

4. The point that is symmetric with respect to the x-axis to the point $(-2, 5)$ is _____. (pp. 198–200)

5. To graph $y = (x - 3)^2 + 1$, shift the graph of $y = x^2$ to the right _____ units and then _____ 1 unit. (pp. 266–274)

Concepts and Vocabulary

6. A(n) _____ is the collection of all points in the plane such that the distance from each point to a fixed point equals its distance to a fixed line.

7. The surface formed by rotating a parabola about its axis of symmetry is called a _____ _____ _____.

8. *True or False:* The vertex of a parabola is a point on the parabola that also is on its axis of symmetry.

9. *True or False:* If a light is placed at the focus of a parabola, all the rays reflected off the parabola will be parallel to the axis of symmetry.

10. *True or False:* The graph of a quadratic function is a parabola.

Skill Building

In Problems 11–18, the graph of a parabola is given. Match each graph to its equation.

A. $y^2 = 4x$

B. $x^2 = 4y$

C. $y^2 = -4x$

D. $x^2 = -4y$

E. $(y - 1)^2 = 4(x - 1)$

F. $(x + 1)^2 = 4(y + 1)$

G. $(y - 1)^2 = -4(x - 1)$

H. $(x + 1)^2 = -4(y + 1)$

11.

12.

13.

14.

15.

16.

17.

18.

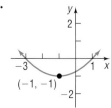

In Problems 19–36, find the equation of the parabola described. Find the two points that define the latus rectum, and graph the equation by hand.

19. Focus at $(4, 0)$; vertex at $(0, 0)$

20. Focus at $(0, 2)$; vertex at $(0, 0)$

21. Focus at $(0, -3)$; vertex at $(0, 0)$

22. Focus at $(-4, 0)$; vertex at $(0, 0)$

23. Focus at $(-2, 0)$; directrix the line $x = 2$

24. Focus at $(0, -1)$; directrix the line $y = 1$

25. Directrix the line $y = -\dfrac{1}{2}$; vertex at $(0, 0)$

26. Directrix the line $x = -\dfrac{1}{2}$; vertex at $(0, 0)$

27. Vertex at $(0, 0)$; axis of symmetry the y-axis; containing the point $(2, 3)$

28. Vertex at $(0, 0)$; axis of symmetry the x-axis; containing the point $(2, 3)$

29. Vertex at $(2, -3)$; focus at $(2, -5)$

30. Vertex at $(4, -2)$; focus at $(6, -2)$

31. Vertex at $(-1, -2)$; focus at $(0, -2)$

32. Vertex at $(3, 0)$; focus at $(3, -2)$

33. Focus at $(-3, 4)$; directrix the line $y = 2$

34. Focus at $(2, 4)$; directrix the line $x = -4$

35. Focus at $(-3, -2)$; directrix the line $x = 1$

36. Focus at $(-4, 4)$; directrix the line $y = -2$

In Problems 37–54, find the vertex, focus, and directrix of each parabola. Graph the equation (a) by hand and (b) by using a graphing utility.

37. $x^2 = 4y$

38. $y^2 = 8x$

39. $y^2 = -16x$

40. $x^2 = -4y$

41. $(y - 2)^2 = 8(x + 1)$

42. $(x + 4)^2 = 16(y + 2)$

43. $(x - 3)^2 = -(y + 1)$

44. $(y + 1)^2 = -4(x - 2)$

45. $(y + 3)^2 = 8(x - 2)$

46. $(x - 2)^2 = 4(y - 3)$

47. $y^2 - 4y + 4x + 4 = 0$

48. $x^2 + 6x - 4y + 1 = 0$

49. $x^2 + 8x = 4y - 8$

50. $y^2 - 2y = 8x - 1$

51. $y^2 + 2y - x = 0$

52. $x^2 - 4x = 2y$

53. $x^2 - 4x = y + 4$

54. $y^2 + 12y = -x + 1$

In Problems 55–62, write an equation for each parabola.

55.

56.

57.

58.

59.

60.

61.

62.
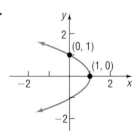

Applications and Extensions

63. **Satellite Dish** A satellite dish is shaped like a paraboloid of revolution. The signals that emanate from a satellite strike the surface of the dish and are reflected to a single point, where the receiver is located. If the dish is 10 feet across at its opening and 4 feet deep at its center, at what position should the receiver be placed?

64. **Constructing a TV Dish** A cable TV receiving dish is in the shape of a paraboloid of revolution. Find the location of the receiver, which is placed at the focus, if the dish is 6 feet across at its opening and 2 feet deep.

65. **Constructing a Flashlight** The reflector of a flashlight is in the shape of a paraboloid of revolution. Its diameter is 4

inches and its depth is 1 inch. How far from the vertex should the light bulb be placed so that the rays will be reflected parallel to the axis?

66. **Constructing a Headlight**   A sealed-beam headlight is in the shape of a paraboloid of revolution. The bulb, which is placed at the focus, is 1 inch from the vertex. If the depth is to be 2 inches, what is the diameter of the headlight at its opening?

67. **Suspension Bridge**   The cables of a suspension bridge are in the shape of a parabola, as shown in the figure. The towers supporting the cable are 600 feet apart and 80 feet high. If the cables touch the road surface midway between the towers, what is the height of the cable at a point 150 feet from the center of the bridge?

68. **Suspension Bridge**   The cables of a suspension bridge are in the shape of a parabola. The towers supporting the cable are 400 feet apart and 100 feet high. If the cables are at a height of 10 feet midway between the towers, what is the height of the cable at a point 50 feet from the center of the bridge?

69. **Searchlight**   A searchlight is shaped like a paraboloid of revolution. If the light source is located 2 feet from the base along the axis of symmetry and the opening is 5 feet across, how deep should the searchlight be?

70. **Searchlight**   A searchlight is shaped like a paraboloid of revolution. If the light source is located 2 feet from the base along the axis of symmetry and the depth of the searchlight is 4 feet, what should the width of the opening be?

71. **Solar Heat**   A mirror is shaped like a paraboloid of revolution and will be used to concentrate the rays of the sun at its focus, creating a heat source. (See the figure.) If the mirror is 20 feet across at its opening and is 6 feet deep, where will the heat source be concentrated?

72. **Reflecting Telescope**   A reflecting telescope contains a mirror shaped like a paraboloid of revolution. If the mirror is 4 inches across at its opening and is 3 feet deep, where will the collected light be concentrated?

73. **Parabolic Arch Bridge**   A bridge is built in the shape of a parabolic arch. The bridge has a span of 120 feet and a maximum height of 25 feet. See the illustration. Choose a suitable rectangular coordinate system and find the height of the arch at distances of 10, 30, and 50 feet from the center.

74. **Parabolic Arch Bridge**   A bridge is to be built in the shape of a parabolic arch and is to have a span of 100 feet. The height of the arch a distance of 40 feet from the center is to be 10 feet. Find the height of the arch at its center.

75. Show that an equation of the form

$$Ax^2 + Ey = 0, \qquad A \neq 0, E \neq 0$$

is the equation of a parabola with vertex at $(0, 0)$ and axis of symmetry the y-axis. Find its focus and directrix.

76. Show that an equation of the form

$$Cy^2 + Dx = 0, \qquad C \neq 0, D \neq 0$$

is the equation of a parabola with vertex at $(0, 0)$ and axis of symmetry the x-axis. Find its focus and directrix.

77. Show that the graph of an equation of the form

$$Ax^2 + Dx + Ey + F = 0, \qquad A \neq 0$$

(a) Is a parabola if $E \neq 0$.
(b) Is a vertical line if $E = 0$ and $D^2 - 4AF = 0$.
(c) Is two vertical lines if $E = 0$ and $D^2 - 4AF > 0$.
(d) Contains no points if $E = 0$ and $D^2 - 4AF < 0$.

78. Show that the graph of an equation of the form

$$Cy^2 + Dx + Ey + F = 0, \qquad C \neq 0$$

(a) Is a parabola if $D \neq 0$.
(b) Is a horizontal line if $D = 0$ and $E^2 - 4CF = 0$.
(c) Is two horizontal lines if $D = 0$ and $E^2 - 4CF > 0$.
(d) Contains no points if $D = 0$ and $E^2 - 4CF < 0$.

'Are You Prepared?' Answers

1. $d = \sqrt{(x_2 - x_1)^2 + (y_2 - y_1)^2}$ **2.** 4   **3.** $x + 4 = \pm 3; \{-7, -1\}$ **4.** $(-2, -5)$ **5.** 3; up

6.3 The Ellipse

PREPARING FOR THIS SECTION *Before getting started, review the following:*

- Distance Formula (Section 1.1, pp. 86–89)
- Completing the Square (Section 1.3, pp. 115–116)
- Intercepts (Section 1.9, p. 183)
- Symmetry (Section 2.1, pp. 198–200)
- Circles (Section 1.9, pp. 180–185)
- Graphing Techniques: Transformations (Section 2.7, pp. 266–274)

Now work the 'Are You Prepared?' problems on page 616.

OBJECTIVES 1 Work with Ellipses with Center at the Origin
2 Work with Ellipses with Center at (h, k)
3 Solve Applied Problems Involving Ellipses

Figure 19

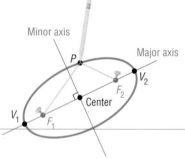

An **ellipse** is the collection of all points in the plane the sum of whose distances from two fixed points, called the **foci**, is a constant.

The definition actually contains within it a physical means for drawing an ellipse. Find a piece of string (the length of this string is the constant referred to in the definition). Then take two thumbtacks (the foci) and stick them on a piece of cardboard so that the distance between them is less than the length of the string. Now attach the ends of the string to the thumbtacks and, using the point of a pencil, pull the string taut. See Figure 19. Keeping the string taut, rotate the pencil around the two thumbtacks. The pencil traces out an ellipse, as shown in Figure 19.

In Figure 19, the foci are labeled F_1 and F_2. The line containing the foci is called the **major axis**. The midpoint of the line segment joining the foci is the **center** of the ellipse. The line through the center and perpendicular to the major axis is the **minor axis**.

The two points of intersection of the ellipse and the major axis are the **vertices**, V_1 and V_2, of the ellipse. The distance from one vertex to the other is the **length of the major axis**. The ellipse is symmetric with respect to its major axis, with respect to its minor axis, and with respect to its center.

Figure 20
$d(F_1, P) + d(F_2, P) = 2a$

1 Work with Ellipses with Center at the Origin

With these ideas in mind, we are now ready to find the equation of an ellipse in a rectangular coordinate system. First, we place the center of the ellipse at the origin. Second, we position the ellipse so that its major axis coincides with a coordinate axis. Suppose that the major axis coincides with the x-axis, as shown in Figure 20. If c is the distance from the center to a focus, then one focus will be at $F_1 = (-c, 0)$ and the other at $F_2 = (c, 0)$. As we shall see, it is convenient to let $2a$ denote the constant distance referred to in the definition. Then, if $P = (x, y)$ is any point on the ellipse, we have

$$d(F_1, P) + d(F_2, P) = 2a \qquad \text{Sum of the distances from } P \text{ to the foci equals a constant, } 2a.$$

$$\sqrt{(x + c)^2 + y^2} + \sqrt{(x - c)^2 + y^2} = 2a \qquad \text{Use the Distance Formula.}$$

$$\sqrt{(x + c)^2 + y^2} = 2a - \sqrt{(x - c)^2 + y^2} \qquad \text{Isolate one radical.}$$

$$(x + c)^2 + y^2 = 4a^2 - 4a\sqrt{(x - c)^2 + y^2} \qquad \text{Square both sides.}$$
$$+ (x - c)^2 + y^2$$

$$x^2 + 2cx + c^2 + y^2 = 4a^2 - 4a\sqrt{(x-c)^2 + y^2}$$ Remove parentheses.
$$+ x^2 - 2cx + c^2 + y^2$$

$$4cx - 4a^2 = -4a\sqrt{(x-c)^2 + y^2}$$ Simplify; isolate the radical.

$$cx - a^2 = -a\sqrt{(x-c)^2 + y^2}$$ Divide each side by 4.

$$(cx - a^2)^2 = a^2[(x-c)^2 + y^2]$$ Square both sides again.

$$c^2x^2 - 2a^2cx + a^4 = a^2(x^2 - 2cx + c^2 + y^2)$$ Remove parentheses.

$$(c^2 - a^2)x^2 - a^2y^2 = a^2c^2 - a^4$$ Rearrange the terms.

$$(a^2 - c^2)x^2 + a^2y^2 = a^2(a^2 - c^2)$$ Multiply each side by -1; factor a^2 on the right side. **(1)**

To obtain points on the ellipse off the x-axis, it must be that $a > c$. To see why, look again at Figure 20.

$$d(F_1, P) + d(F_2, P) > d(F_1, F_2)$$ The sum of the lengths of two sides of a triangle is greater than the length of the third side.

$$2a > 2c$$ $d(F_1, P) + d(F_2, P) = 2a; d(F_1, F_2) = 2c.$

$$a > c$$

Since $a > c$, we also have $a^2 > c^2$, so $a^2 - c^2 > 0$. Let $b^2 = a^2 - c^2, b > 0$. Then $a > b$ and equation (1) can be written as

$$b^2x^2 + a^2y^2 = a^2b^2$$

$$\frac{x^2}{a^2} + \frac{y^2}{b^2} = 1$$ Divide each side by a^2b^2.

Theorem

Equation of an Ellipse; Center at $(0, 0)$; Major Axis along the x-Axis

An equation of the ellipse with center at $(0, 0)$, foci at $(-c, 0)$ and $(c, 0)$, and vertices at $(-a, 0)$ and $(a, 0)$ is

$$\boxed{\frac{x^2}{a^2} + \frac{y^2}{b^2} = 1, \quad \text{where } a > b > 0 \text{ and } b^2 = a^2 - c^2 \quad \textbf{(2)}}$$

The major axis is the x-axis.

As you can verify, the ellipse defined by equation (2) is symmetric with respect to the x-axis, y-axis, and origin.

Because the major axis is the x-axis, we find the vertices of the ellipse defined by equation (2) by letting $y = 0$. The vertices satisfy the equation $\frac{x^2}{a^2} = 1$, the solutions of which are $x = \pm a$. Consequently, the vertices of the ellipse given by equation (2) are $V_1 = (-a, 0)$ and $V_2 = (a, 0)$. The y-intercepts of the ellipse, found by letting $x = 0$, have coordinates $(0, -b)$ and $(0, b)$. These four intercepts, $(a, 0), (-a, 0), (0, b)$, and $(0, -b)$, are used to graph the ellipse. See Figure 21.

Notice in Figure 21 the right triangle formed with the points $(0, 0), (c, 0)$, and $(0, b)$. Because $b^2 = a^2 - c^2$ (or $b^2 + c^2 = a^2$), the distance from the focus at $(c, 0)$ to the point $(0, b)$ is a.

Figure 21

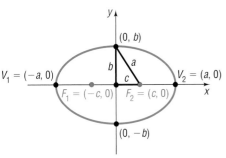

EXAMPLE 1 | **Finding an Equation of an Ellipse**

Find an equation of the ellipse with center at the origin, one focus at $(3, 0)$, and a vertex at $(-4, 0)$. Graph the equation.

Solution

Figure 22

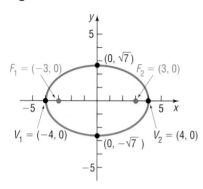

The ellipse has its center at the origin and, since the given focus and vertex lie on the x-axis, the major axis is the x-axis. The distance from the center, $(0, 0)$, to one of the foci, $(3, 0)$, is $c = 3$. The distance from the center, $(0, 0)$, to one of the vertices, $(-4, 0)$, is $a = 4$. From equation (2), it follows that

$$b^2 = a^2 - c^2 = 16 - 9 = 7$$

so an equation of the ellipse is

$$\frac{x^2}{16} + \frac{y^2}{7} = 1$$

Figure 22 shows the graph. ◀

⟶ **NOW WORK PROBLEM 27.**

Notice in Figure 22 how we used the intercepts of the equation to graph the ellipse. Following this practice will make it easier for you to obtain an accurate graph of an ellipse when graphing by hand. It also tells you how to set the initial viewing window when using a graphing utility.

EXAMPLE 2 | **Graphing an Ellipse Using a Graphing Utility**

Use a graphing utility to graph the ellipse $\dfrac{x^2}{16} + \dfrac{y^2}{7} = 1$.

Solution

Figure 23

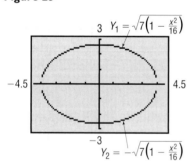

First, we must solve $\dfrac{x^2}{16} + \dfrac{y^2}{7} = 1$ for y.

$$\frac{y^2}{7} = 1 - \frac{x^2}{16} \qquad \text{Subtract } \frac{x^2}{16} \text{ from each side.}$$

$$y^2 = 7\left(1 - \frac{x^2}{16}\right) \qquad \text{Multiply both sides by 7.}$$

$$y = \pm\sqrt{7\left(1 - \frac{x^2}{16}\right)} \qquad \text{Apply the Square Root Method.}$$

Figure 23* shows the graphs of $Y_1 = \sqrt{7\left(1 - \dfrac{x^2}{16}\right)}$ and $Y_2 = -\sqrt{7\left(1 - \dfrac{x^2}{16}\right)}$. ◀

Notice in Figure 23 that we used a square screen. As with circles and parabolas, this is done to avoid a distorted view of the graph.

An equation of the form of equation (2), with $a > b$, is the equation of an ellipse with center at the origin, foci on the x-axis at $(-c, 0)$ and $(c, 0)$, where $c^2 = a^2 - b^2$, and major axis along the x-axis.

---

*The initial viewing window selected was Xmin $= -4$, Xmax $= 4$, Ymin $= -3$, Ymax $= 3$. Then we used the ZOOM-SQUARE option to obtain the window shown.

For the remainder of this section, the direction "Discuss the equation" will mean to find the center, major axis, foci, and vertices of the ellipse and graph it.

EXAMPLE 3 **Discussing the Equation of an Ellipse**

Discuss the equation: $\dfrac{x^2}{25} + \dfrac{y^2}{9} = 1$

Solution The given equation is of the form of equation (2), with $a^2 = 25$ and $b^2 = 9$. The equation is that of an ellipse with center $(0, 0)$ and major axis along the x-axis. The vertices are at $(\pm a, 0) = (\pm 5, 0)$. Because $b^2 = a^2 - c^2$, we find that

$$c^2 = a^2 - b^2 = 25 - 9 = 16$$

The foci are at $(\pm c, 0) = (\pm 4, 0)$. Figure 24(a) shows the graph drawn by hand. Figure 24(b) shows the graph obtained using a graphing utility.

Figure 24

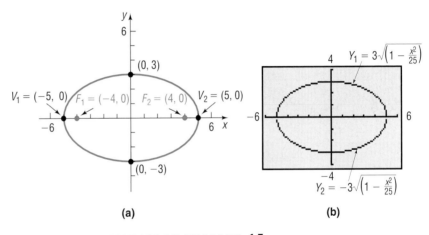

(a)   (b)

NOW WORK PROBLEM **17**.

If the major axis of an ellipse with center at $(0, 0)$ lies on the y-axis, then the foci are at $(0, -c)$ and $(0, c)$. Using the same steps as before, the definition of an ellipse leads to the following result:

Theorem **Equation of an Ellipse; Center at (0, 0); Major Axis along the y-Axis**

An equation of the ellipse with center at $(0, 0)$, foci at $(0, -c)$ and $(0, c)$, and vertices at $(0, -a)$ and $(0, a)$ is

$$\frac{x^2}{b^2} + \frac{y^2}{a^2} = 1, \qquad \text{where } a > b > 0 \text{ and } b^2 = a^2 - c^2 \qquad \text{(3)}$$

The major axis is the y-axis.

Figure 25

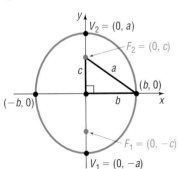

Figure 25 illustrates the graph of such an ellipse. Again, notice the right triangle with the points at $(0, 0)$, $(b, 0)$, and $(0, c)$.

Look closely at equations (2) and (3). Although they may look alike, there is a difference! In equation (2), the larger number, a^2, is in the denominator of the

x^2-term, so the major axis of the ellipse is along the x-axis. In equation (3), the larger number, a^2, is in the denominator of the y^2-term, so the major axis is along the y-axis.

EXAMPLE 4   Discussing the Equation of an Ellipse

Discuss the equation: $9x^2 + y^2 = 9$

Solution   To put the equation in proper form, we divide each side by 9.

$$x^2 + \frac{y^2}{9} = 1$$

The larger number, 9, is in the denominator of the y^2-term so, based on equation (3), this is the equation of an ellipse with center at the origin and major axis along the y-axis. Also, we conclude that $a^2 = 9$, $b^2 = 1$, and $c^2 = a^2 - b^2 = 9 - 1 = 8$. The vertices are at $(0, \pm a) = (0, \pm 3)$, and the foci are at $(0, \pm c) = (0, \pm 2\sqrt{2})$. Figure 26(a) shows the graph drawn by hand. Figure 26(b) shows the graph obtained using a graphing utility.

Figure 26

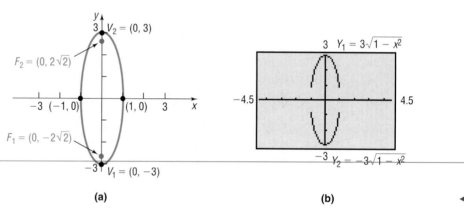

(a)   (b)

◀

NOW WORK PROBLEM 21.

EXAMPLE 5   Finding an Equation of an Ellipse

Find an equation of the ellipse having one focus at $(0, 2)$ and vertices at $(0, -3)$ and $(0, 3)$. Graph the equation by hand.

Figure 27

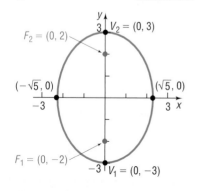

Solution   Because the vertices are at $(0, -3)$ and $(0, 3)$, the center of this ellipse is at their midpoint, the origin. Also, its major axis lies on the y-axis. The distance from the center, $(0, 0)$, to one of the foci, $(0, 2)$, is $c = 2$. The distance from the center, $(0, 0)$, to one of the vertices, $(0, 3)$, is $a = 3$. So $b^2 = a^2 - c^2 = 9 - 4 = 5$. The form of the equation of this ellipse is given by equation (3).

$$\frac{x^2}{b^2} + \frac{y^2}{a^2} = 1$$

$$\frac{x^2}{5} + \frac{y^2}{9} = 1$$

Figure 27 shows the graph.

◀

NOW WORK PROBLEM 29.

The circle may be considered a special kind of ellipse. To see why, let $a = b$ in equation (2) or (3). Then

$$\frac{x^2}{a^2} + \frac{y^2}{a^2} = 1$$
$$x^2 + y^2 = a^2$$

This is the equation of a circle with center at the origin and radius a. The value of c is

$$c^2 = a^2 - b^2 = 0$$

We conclude that the closer the two foci of an ellipse are to the center, the more the ellipse will look like a circle.

2 Work with Ellipses with Center at (h, k)

If an ellipse with center at the origin and major axis coinciding with a coordinate axis is shifted horizontally h units and then vertically k units, the result is an ellipse with center at (h, k) and major axis parallel to a coordinate axis. The equations of such ellipses have the same forms as those given in equations (2) and (3), except that x is replaced by $x - h$ (the horizontal shift) and y is replaced by $y - k$ (the vertical shift). Table 3 gives the forms of the equations of such ellipses and Figure 28 shows their graphs.

Table 3

ELLIPSES WITH CENTER AT (h, k) AND MAJOR AXIS PARALLEL TO A COORDINATE AXIS				
Center	**Major Axis**	**Foci**	**Vertices**	**Equation**
(h, k)	Parallel to x-axis	$(h + c, k)$	$(h + a, k)$	$\dfrac{(x-h)^2}{a^2} + \dfrac{(y-k)^2}{b^2} = 1,$
		$(h - c, k)$	$(h - a, k)$	$a > b$ and $b^2 = a^2 - c^2$
(h, k)	Parallel to y-axis	$(h, k + c)$	$(h, k + a)$	$\dfrac{(x-h)^2}{b^2} + \dfrac{(y-k)^2}{a^2} = 1,$
		$(h, k - c)$	$(h, k - a)$	$a > b$ and $b^2 = a^2 - c^2$

Figure 28

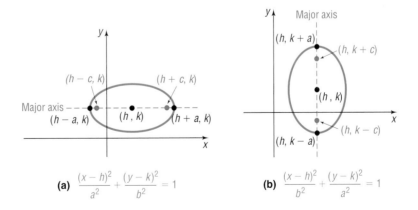

(a) $\dfrac{(x-h)^2}{a^2} + \dfrac{(y-k)^2}{b^2} = 1$ (b) $\dfrac{(x-h)^2}{b^2} + \dfrac{(y-k)^2}{a^2} = 1$

EXAMPLE 6   **Finding an Equation of an Ellipse, Center Not at the Origin**

Find an equation for the ellipse with center at $(2, -3)$, one focus at $(3, -3)$, and one vertex at $(5, -3)$. Graph the equation by hand.

Solution   The center is at $(h, k) = (2, -3)$, so $h = 2$ and $k = -3$. Since the center, focus, and vertex all lie on the line $y = -3$, the major axis is parallel to the x-axis. The distance from the center $(2, -3)$ to a focus $(3, -3)$ is $c = 1$; the distance from the center $(2, -3)$ to a vertex $(5, -3)$ is $a = 3$. Then $b^2 = a^2 - c^2 = 9 - 1 = 8$. The form of the equation is

$$\frac{(x - h)^2}{a^2} + \frac{(y - k)^2}{b^2} = 1, \quad \text{where } h = 2, k = -3, a = 3, b = 2\sqrt{2}$$

$$\frac{(x - 2)^2}{9} + \frac{(y + 3)^2}{8} = 1$$

To graph the equation, we use the center $(h, k) = (2, -3)$ to locate the vertices. The major axis is parallel to the x-axis, so the vertices are $a = 3$ units left and right of the center $(2, -3)$. Therefore, the vertices are

$$V_1 = (2 - 3, -3) = (-1, -3) \quad \text{and} \quad V_2 = (2 + 3, -3) = (5, -3)$$

Figure 29

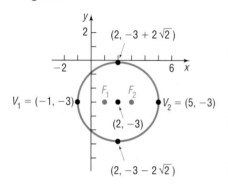

Since $c = 1$ and the major axis is parallel to the x-axis, the foci are 1 unit left and right of the center. Therefore, the foci are

$$F_1 = (2 - 1, -3) = (1, -3) \quad \text{and} \quad F_2 = (2 + 1, -3) = (3, -3)$$

Finally, we use the value of $b = 2\sqrt{2}$ to find the two points above and below the center.

$$\left(2, -3 - 2\sqrt{2}\right) \quad \text{and} \quad \left(2, -3 + 2\sqrt{2}\right)$$

Figure 29 shows the graph. ◀

NOW WORK PROBLEM 55.

| **EXAMPLE 7** | **Using a Graphing Utility to Graph an Ellipse, Center Not at the Origin** |

Using a graphing utility, graph the ellipse: $\dfrac{(x - 2)^2}{9} + \dfrac{(y + 3)^2}{8} = 1$

Solution   First, we must solve the equation $\dfrac{(x - 2)^2}{9} + \dfrac{(y + 3)^2}{8} = 1$ for y.

$$\frac{(y + 3)^2}{8} = 1 - \frac{(x - 2)^2}{9} \qquad \text{Subtract } \frac{(x - 2)^2}{9} \text{ from each side.}$$

$$(y + 3)^2 = 8\left[1 - \frac{(x - 2)^2}{9}\right] \qquad \text{Multiply each side by 8.}$$

$$y + 3 = \pm\sqrt{8\left[1 - \frac{(x - 2)^2}{9}\right]} \qquad \text{Apply the Square Root Method.}$$

$$y = -3 \pm \sqrt{8\left[1 - \frac{(x - 2)^2}{9}\right]} \qquad \text{Subtract 3 from each side.}$$

Figure 30

Figure 30 shows the graphs of $Y_1 = -3 + \sqrt{8\left[1 - \frac{(x - 2)^2}{9}\right]}$ and $Y_2 = -3 - \sqrt{8\left[1 - \frac{(x - 2)^2}{9}\right]}$. ◀

EXAMPLE 8	**Discussing the Equation of an Ellipse**

Discuss the equation: $4x^2 + y^2 - 8x + 4y + 4 = 0$

Solution   We proceed to complete the squares in x and in y.

$$4x^2 + y^2 - 8x + 4y + 4 = 0$$
$$4x^2 - 8x + y^2 + 4y = -4 \qquad \text{Group like variables; place the}$$
$$\text{constant on the right side.}$$
$$4(x^2 - 2x) + (y^2 + 4y) = -4 \qquad \text{Factor out 4 from the first two}$$
$$\text{terms.}$$
$$4(x^2 - 2x + 1) + (y^2 + 4y + 4) = -4 + 4 + 4 \qquad \text{Complete each square.}$$
$$4(x - 1)^2 + (y + 2)^2 = 4 \qquad \text{Factor.}$$
$$(x - 1)^2 + \frac{(y + 2)^2}{4} = 1 \qquad \text{Divide each side by 4.}$$

This is the equation of an ellipse with center at $(1, -2)$ and major axis parallel to the y-axis. Since $a^2 = 4$ and $b^2 = 1$, we have $c^2 = a^2 - b^2 = 4 - 1 = 3$. The vertices are at $(h, k \pm a) = (1, -2 \pm 2)$ or $(1, 0)$ and $(1, -4)$. The foci are at $(h, k \pm c) = (1, -2 \pm \sqrt{3})$ or $(1, -2 - \sqrt{3})$ and $(1, -2 + \sqrt{3})$. Figure 31(a) shows the graph drawn by hand. Figure 31(b) shows the graph obtained using a graphing utility.

Figure 31

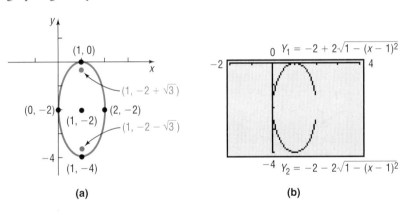

(a)   (b)

NOW WORK PROBLEM **47**.

3   **Solve Applied Problems Involving Ellipses**

Ellipses are found in many applications in science and engineering. For example, the orbits of the planets around the Sun are elliptical, with the Sun's position at a focus. See Figure 32.

Figure 32

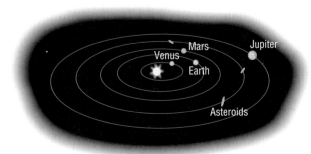

Stone and concrete bridges are often shaped as semielliptical arches. Elliptical gears are used in machinery when a variable rate of motion is required.

Ellipses also have an interesting reflection property. If a source of light (or sound) is placed at one focus, the waves transmitted by the source will reflect off the ellipse and concentrate at the other focus. This is the principle behind *whispering galleries*, which are rooms designed with elliptical ceilings. A person standing at one focus of the ellipse can whisper and be heard by a person standing at the other focus, because all the sound waves that reach the ceiling are reflected to the other person.

| EXAMPLE 9 | **A Whispering Gallery** |

The whispering gallery in the Museum of Science and Industry in Chicago is 47.3 feet long. The distance from the center of the room to the foci is 20.3 feet. Find an equation that describes the shape of the room. How high is the room at its center?

Solution We set up a rectangular coordinate system so that the center of the ellipse is at the origin and the major axis is along the *x*-axis. The equation of the ellipse is

$$\frac{x^2}{a^2} + \frac{y^2}{b^2} = 1$$

Since the length of the room is 47.3 feet, the distance from the center of the room to each vertex (the end of the room) will be $\frac{47.3}{2} = 23.65$ feet; so $a = 23.65$ feet. The distance from the center of the room to each focus is $c = 20.3$ feet. See Figure 33. Since $b^2 = a^2 - c^2$, we find $b^2 = 23.65^2 - 20.3^2 = 147.2325$. An equation that describes the shape of the room is given by

Figure 33

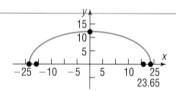

$$\frac{x^2}{23.65^2} + \frac{y^2}{147.2325} = 1$$

The height of the room at its center is $b = \sqrt{147.2325} \approx 12.1$ feet. ◄

6.3 Assess Your Understanding

'Are You Prepared?'

Answers are given at the end of these exercises. If you get a wrong answer, read the pages listed in red.

1. The distance d from $P_1 = (2, -5)$ to $P_2 = (4, -2)$ is $d =$ _____. (pp. 86–89)

2. To complete the square of $x^2 - 3x$, add _____. (p. 115)

3. Find the intercepts of the equation $y^2 = 16 - 4x^2$. (p. 183)

4. The point that is symmetric with respect to the *y*-axis to the point $(-2, 5)$ is _____. (pp. 198–199)

5. To graph $y = (x + 1)^2 - 4$, shift the graph of $y = x^2$ to the (left/right) _____ unit(s) and then (up/down) _____ unit(s). (pp. 266–274)

6. The standard equation of a circle with center at $(2, -3)$ and radius 1 is _____. (pp. 180–185)

Concepts and Vocabulary

7. A(n) _____ is the collection of all points in the plane the sum of whose distances from two fixed points is a constant.

8. For an ellipse, the foci lie on a line called the _____ axis.

9. For the ellipse $\dfrac{x^2}{4} + \dfrac{y^2}{25} = 1$, the vertices are the points _____ and _____.

10. *True or False:* The foci, vertices, and center of an ellipse lie on a line called the axis of symmetry.

11. *True or False:* If the center of an ellipse is at the origin and the foci lie on the y-axis, the ellipse is symmetric with respect to the x-axis, the y-axis, and the origin.

12. *True or False:* A circle is a certain type of ellipse.

Skill Building

In Problems 13–16, the graph of an ellipse is given. Match each graph to its equation.

A. $\dfrac{x^2}{4} + y^2 = 1$

B. $x^2 + \dfrac{y^2}{4} = 1$

C. $\dfrac{x^2}{16} + \dfrac{y^2}{4} = 1$

D. $\dfrac{x^2}{4} + \dfrac{y^2}{16} = 1$

13.

14.

15.

16.

In Problems 17–26, find the vertices and foci of each ellipse. Graph each equation by hand.

17. $\dfrac{x^2}{25} + \dfrac{y^2}{4} = 1$

18. $\dfrac{x^2}{9} + \dfrac{y^2}{4} = 1$

19. $\dfrac{x^2}{9} + \dfrac{y^2}{25} = 1$

20. $x^2 + \dfrac{y^2}{16} = 1$

21. $4x^2 + y^2 = 16$

22. $x^2 + 9y^2 = 18$

23. $4y^2 + x^2 = 8$

24. $4y^2 + 9x^2 = 36$

25. $x^2 + y^2 = 16$

26. $x^2 + y^2 = 4$

In Problems 27–38, find an equation for each ellipse. Graph the equation by hand.

27. Center at $(0,0)$;   focus at $(3,0)$;   vertex at $(5,0)$

28. Center at $(0,0)$;   focus at $(-1,0)$;   vertex at $(3,0)$

29. Center at $(0,0)$;   focus at $(0,-4)$;   vertex at $(0,5)$

30. Center at $(0,0)$;   focus at $(0,1)$;   vertex at $(0,-2)$

31. Foci at $(\pm 2,0)$;   length of the major axis is 6

32. Foci at $(0,\pm 2)$;   length of the major axis is 8

33. Focus at $(-4,0)$;   vertices at $(\pm 5,0)$

34. Focus at $(0,-4)$;   vertices at $(0,\pm 8)$

35. Foci at $(0,\pm 3)$;   x-intercepts are ± 2

36. Vertices at $(\pm 4,0)$;   y-intercepts are ± 1

37. Center at $(0,0)$;   vertex at $(0,4)$; $b = 1$

38. Vertices at $(\pm 5,0)$; $c = 2$

In Problems 39–42, write an equation for each ellipse.

39.

40.

41.

42.
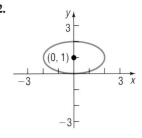

In Problems 43–54, discuss each equation; that is, find the center, foci, and vertices of each ellipse. Graph each equation (a) by hand; and (b) by using a graphing utility.

43. $\dfrac{(x-3)^2}{4} + \dfrac{(y+1)^2}{9} = 1$

44. $\dfrac{(x+4)^2}{9} + \dfrac{(y+2)^2}{4} = 1$

45. $(x+5)^2 + 4(y-4)^2 = 16$

46. $9(x-3)^2 + (y+2)^2 = 18$

47. $x^2 + 4x + 4y^2 - 8y + 4 = 0$

48. $x^2 + 3y^2 - 12y + 9 = 0$

49. $2x^2 + 3y^2 - 8x + 6y + 5 = 0$

50. $4x^2 + 3y^2 + 8x - 6y = 5$

51. $9x^2 + 4y^2 - 18x + 16y - 11 = 0$

52. $x^2 + 9y^2 + 6x - 18y + 9 = 0$

53. $4x^2 + y^2 + 4y = 0$

54. $9x^2 + y^2 - 18x = 0$

In Problems 55–64, find an equation for each ellipse. Graph the equation by hand.

55. Center at $(2, -2)$; vertex at $(7, -2)$; focus at $(4, -2)$

56. Center at $(-3, 1)$; vertex at $(-3, 3)$; focus at $(-3, 0)$

57. Vertices at $(4, 3)$ and $(4, 9)$; focus at $(4, 8)$

58. Foci at $(1, 2)$ and $(-3, 2)$; vertex at $(-4, 2)$

59. Foci at $(5, 1)$ and $(-1, 1)$; length of the major axis is 8

60. Vertices at $(2, 5)$ and $(2, -1)$; $c = 2$

61. Center at $(1, 2)$; focus at $(4, 2)$; contains the point $(1, 3)$

62. Center at $(1, 2)$; focus at $(1, 4)$; contains the point $(2, 2)$

63. Center at $(1, 2)$; vertex at $(4, 2)$; contains the point $(1, 3)$

64. Center at $(1, 2)$; vertex at $(1, 4)$; contains the point $(2, 2)$

In Problems 65–68, graph each function.

[**Hint:** *Notice that each function is half an ellipse.*]

65. $f(x) = \sqrt{16 - 4x^2}$

66. $f(x) = \sqrt{9 - 9x^2}$

67. $f(x) = -\sqrt{64 - 16x^2}$

68. $f(x) = -\sqrt{4 - 4x^2}$

Applications and Extensions

69. Semielliptical Arch Bridge An arch in the shape of the upper half of an ellipse is used to support a bridge that is to span a river 20 meters wide. The center of the arch is 6 meters above the center of the river (see the figure). Write an equation for the ellipse in which the *x*-axis coincides with the water level and the *y*-axis passes through the center of the arch.

70. Semielliptical Arch Bridge The arch of a bridge is a semi-ellipse with a horizontal major axis. The span is 30 feet, and the top of the arch is 10 feet above the major axis. The roadway is horizontal and is 2 feet above the top of the arch. Find the vertical distance from the roadway to the arch at 5-foot intervals along the roadway.

71. Whispering Gallery A hall 100 feet in length is to be designed as a whispering gallery. If the foci are located 25 feet from the center, how high will the ceiling be at the center?

72. Whispering Gallery Jim, standing at one focus of a whispering gallery, is 6 feet from the nearest wall. His friend is standing at the other focus, 100 feet away. What is the length of this whispering gallery? How high is its elliptical ceiling at the center?

73. Semielliptical Arch Bridge A bridge is built in the shape of a semielliptical arch. The bridge has a span of 120 feet and a maximum height of 25 feet. Choose a suitable rectangular coordinate system and find the height of the arch at distances of 10, 30, and 50 feet from the center.

74. Semielliptical Arch Bridge A bridge is to be built in the shape of a semielliptical arch and is to have a span of 100 feet. The height of the arch, at a distance of 40 feet from the center, is to be 10 feet. Find the height of the arch at its center.

75. Semielliptical Arch An arch in the form of half an ellipse is 40 feet wide and 15 feet high at the center. Find the height of the arch at intervals of 10 feet along its width.

76. Semielliptical Arch Bridge An arch for a bridge over a highway is in the form of half an ellipse. The top of the arch is 20 feet above the ground level (the major axis). The highway has four lanes, each 12 feet wide; a center safety strip 8 feet wide; and two side strips, each 4 feet wide. What should the span of the bridge be (the length of its major axis) if the height 28 feet from the center is to be 13 feet?

*In Problems 77–80, use the fact that the orbit of a planet about the Sun is an ellipse, with the Sun at one focus. The **aphelion** of a planet is its greatest distance from the Sun, and the **perihelion** is its shortest distance. The **mean distance** of a planet from the Sun is the length of the semimajor axis of the elliptical orbit. See the illustration.*

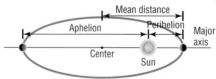

77. Earth The mean distance of Earth from the Sun is 93 million miles. If the aphelion of Earth is 94.5 million miles, what is the perihelion? Write an equation for the orbit of Earth around the Sun.

78. Mars The mean distance of Mars from the Sun is 142 million miles. If the perihelion of Mars is 128.5 million miles, what is the aphelion? Write an equation for the orbit of Mars about the Sun.

79. Jupiter   The aphelion of Jupiter is 507 million miles. If the distance from the Sun to the center of its elliptical orbit is 23.2 million miles, what is the perihelion? What is the mean distance? Write an equation for the orbit of Jupiter around the Sun.

80. Pluto   The perihelion of Pluto is 4551 million miles, and the distance of the Sun from the center of its elliptical orbit is 897.5 million miles. Find the aphelion of Pluto. What is the mean distance of Pluto from the Sun? Write an equation for the orbit of Pluto about the Sun.

81. Racetrack Design   Consult the figure. A racetrack is in the shape of an ellipse, 100 feet long and 50 feet wide. What is the width 10 feet from a vertex?

82. Racetrack Design   A racetrack is in the shape of an ellipse 80 feet long and 40 feet wide. What is the width 10 feet from a vertex?

83. Show that an equation of the form

$$Ax^2 + Cy^2 + F = 0, \qquad A \neq 0, C \neq 0, F \neq 0$$

where A and C are of the same sign and F is of opposite sign,
(a) Is the equation of an ellipse with center at $(0, 0)$ if $A \neq C$.
(b) Is the equation of a circle with center $(0, 0)$ if $A = C$.

84. Show that the graph of an equation of the form

$$Ax^2 + Cy^2 + Dx + Ey + F = 0, \qquad A \neq 0, C \neq 0$$

where A and C are of the same sign,
(a) Is an ellipse if $\dfrac{D^2}{4A} + \dfrac{E^2}{4C} - F$ is the same sign as A.
(b) Is a point if $\dfrac{D^2}{4A} + \dfrac{E^2}{4C} - F = 0$.
(c) Contains no points if $\dfrac{D^2}{4A} + \dfrac{E^2}{4C} - F$ is of opposite sign to A.

Discussion and Writing

85. The **eccentricity** e of an ellipse is defined as the number $\dfrac{c}{a}$, where a and c are the numbers given in equation (2). Because $a > c$, it follows that $e < 1$. Write a brief para- graph about the general shape of each of the following ellipses. Be sure to justify your conclusions.
(a) Eccentricity close to 0   (b) Eccentricity = 0.5
(c) Eccentricity close to 1

'Are You Prepared?' Answers

1. $\sqrt{13}$ **2.** $\dfrac{9}{4}$ **3.** $(-2, 0), (2, 0), (0, -4), (0, 4)$ **4.** $(2, 5)$ **5.** left: 1; down: 4   **6.** $(x - 2)^2 + (y + 3)^2 = 1$

6.4 The Hyperbola

PREPARING FOR THIS SECTION   *Before getting started, review the following:*

- Distance Formula (Section 1.1, pp. 86–89)
- Completing the Square (Section 1.3, p. 115–116)
- Intercepts (Section 1.9, pp. 183)
- Symmetry (Section 2.1, pp. 198–200)

- Asymptotes (Section 3.3, pp. 337–343)
- Graphing Techniques: Transformations (Section 2.7, pp. 266–274)
- Square Root Method (Section 1.3, pp. 114–115)

✎ Now work the 'Are You Prepared?' problems on page 630.

OBJECTIVES   **1** Work with Hyperbolas with Center at the Origin
   2 Find the Asymptotes of a Hyperbola
   3 Work with Hyperbolas with Center at (h, k)
   4 Solve Applied Problems Involving Hyperbolas

A **hyperbola** is the collection of all points in the plane the difference of whose distances from two fixed points, called the **foci**, is a constant.

Figure 34

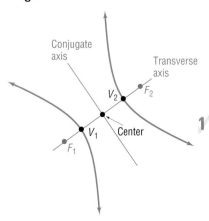

Conjugate axis

Transverse axis

V_2 F_2

V_1 Center

F_1

Figure 34 illustrates a hyperbola with foci F_1 and F_2. The line containing the foci is called the **transverse axis**. The midpoint of the line segment joining the foci is the **center** of the hyperbola. The line through the center and perpendicular to the transverse axis is the **conjugate axis**. The hyperbola consists of two separate curves, called **branches**, that are symmetric with respect to the transverse axis, conjugate axis, and center. The two points of intersection of the hyperbola and the transverse axis are the **vertices**, V_1 and V_2, of the hyperbola.

1 Work with Hyperbolas with Center at the Origin

With these ideas in mind, we are now ready to find the equation of a hyperbola in the rectangular coordinate system. First, we place the center at the origin. Next, we position the hyperbola so that its transverse axis coincides with a coordinate axis. Suppose that the transverse axis coincides with the x-axis, as shown in Figure 35.

If c is the distance from the center to a focus, then one focus will be at $F_1 = (-c, 0)$ and the other at $F_2 = (c, 0)$. Now we let the constant difference of the distances from any point $P = (x, y)$ on the hyperbola to the foci F_1 and F_2 be denoted by $\pm 2a$. (If P is on the right branch, the $+$ sign is used; if P is on the left branch, the $-$ sign is used.) The coordinates of P must satisfy the equation

Figure 35
$d(F_1, P) - d(F_2, P) = \pm 2a$

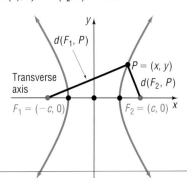

$d(F_1, P)$

Transverse axis

$P = (x, y)$

$d(F_2, P)$

$F_1 = (-c, 0)$ $F_2 = (c, 0)$

$$d(F_1, P) - d(F_2, P) = \pm 2a \qquad \textit{Difference of the distances from P to the foci equals} \pm 2a.$$

$$\sqrt{(x + c)^2 + y^2} - \sqrt{(x - c)^2 + y^2} = \pm 2a \qquad \textit{Use the distance formula.}$$

$$\sqrt{(x + c)^2 + y^2} = \pm 2a + \sqrt{(x - c)^2 + y^2} \qquad \textit{Isolate one radical.}$$

$$(x + c)^2 + y^2 = 4a^2 \pm 4a\sqrt{(x - c)^2 + y^2} \qquad \textit{Square both sides.}$$
$$+ (x - c)^2 + y^2$$

Next we remove the parentheses.

$$x^2 + 2cx + c^2 + y^2 = 4a^2 \pm 4a\sqrt{(x - c)^2 + y^2} + x^2 - 2cx + c^2 + y^2$$

$$4cx - 4a^2 = \pm 4a\sqrt{(x - c)^2 + y^2} \qquad \textit{Simplify; isolate the radical.}$$

$$cx - a^2 = \pm a\sqrt{(x - c)^2 + y^2} \qquad \textit{Divide each side by 4.}$$

$$(cx - a^2)^2 = a^2[(x - c)^2 + y^2] \qquad \textit{Square both sides.}$$

$$c^2x^2 - 2ca^2x + a^4 = a^2(x^2 - 2cx + c^2 + y^2) \qquad \textit{Simplify.}$$

$$c^2x^2 + a^4 = a^2x^2 + a^2c^2 + a^2y^2 \qquad \textit{Remove parentheses and simplify.}$$

$$(c^2 - a^2)x^2 - a^2y^2 = a^2c^2 - a^4 \qquad \textit{Rearrange terms.}$$

$$(c^2 - a^2)x^2 - a^2y^2 = a^2(c^2 - a^2) \qquad \textit{Factor } a^2 \textit{ on the right side.} \qquad \textbf{(1)}$$

To obtain points on the hyperbola off the x-axis, it must be that $a < c$. To see why, look again at Figure 35.

$$d(F_1, P) < d(F_2, P) + d(F_1, F_2) \qquad \textit{Use triangle } F_1PF_2.$$

$$d(F_1, P) - d(F_2, P) < d(F_1, F_2)$$

$$2a < 2c \qquad \textit{P is on the right branch, so } d(F_1, P) - d(F_2, P) = 2a; \textit{ the distance from } F_1 \textit{ to } F_2 \textit{ equals } 2c.$$

$$a < c$$

Since $a < c$, we also have $a^2 < c^2$, so $c^2 - a^2 > 0$. Let $b^2 = c^2 - a^2, b > 0$. Then equation (1) can be written as

$$b^2 x^2 - a^2 y^2 = a^2 b^2$$

$$\frac{x^2}{a^2} - \frac{y^2}{b^2} = 1 \qquad \text{Divide each side by } a^2 b^2.$$

To find the vertices of the hyperbola defined by this equation, let $y = 0$. The vertices satisfy the equation $\frac{x^2}{a^2} = 1$, the solutions of which are $x = \pm a$. Consequently, the vertices of the hyperbola are $V_1 = (-a, 0)$ and $V_2 = (a, 0)$. Notice that the distance from the center $(0, 0)$ to either vertex is a.

Theorem

Equation of a Hyperbola; Center at (0, 0); Transverse Axis along the x-Axis

An equation of the hyperbola with center at $(0, 0)$, foci at $(-c, 0)$ and $(c, 0)$, and vertices at $(-a, 0)$ and $(a, 0)$ is

$$\boxed{\frac{x^2}{a^2} - \frac{y^2}{b^2} = 1, \qquad \text{where } b^2 = c^2 - a^2} \qquad (2)$$

The transverse axis is the x-axis.

Figure 36
$\frac{x^2}{a^2} - \frac{y^2}{b^2} = 1, \quad b^2 = c^2 - a^2$

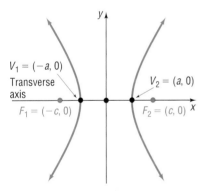

See Figure 36. As you can verify, the hyperbola defined by equation (2) is symmetric with respect to the x-axis, y-axis, and origin. To find the y-intercepts, if any, let $x = 0$ in equation (2). This results in the equation $\frac{y^2}{b^2} = -1$, which has no real solution. We conclude that the hyperbola defined by equation (2) has no y-intercepts. In fact, since $\frac{x^2}{a^2} - 1 = \frac{y^2}{b^2} \geq 0$, it follows that $\frac{x^2}{a^2} \geq 1$. There are no points on the graph for $-a < x < a$.

EXAMPLE 1

Finding and Graphing an Equation of a Hyperbola

Find an equation of the hyperbola with center at the origin, one focus at $(3, 0)$, and one vertex at $(-2, 0)$. Graph the equation by hand.

Solution

The hyperbola has its center at the origin, and the transverse axis coincides with the x-axis. One focus is at $(c, 0) = (3, 0)$, so $c = 3$. One vertex is at $(-a, 0) = (-2, 0)$, so $a = 2$. From equation (2), it follows that $b^2 = c^2 - a^2 = 9 - 4 = 5$, so an equation of the hyperbola is

$$\frac{x^2}{4} - \frac{y^2}{5} = 1$$

To graph a hyperbola, it is helpful to locate and plot other points on the graph. For example, to find the points above and below the foci, we let $x = \pm 3$. Then

$$\frac{x^2}{4} - \frac{y^2}{5} = 1$$

$$\frac{(\pm 3)^2}{4} - \frac{y^2}{5} = 1 \qquad x = \pm 3$$

$$\frac{9}{4} - \frac{y^2}{5} = 1$$

$$\frac{y^2}{5} = \frac{5}{4}$$

$$y^2 = \frac{25}{4}$$

$$y = \pm \frac{5}{2}$$

Figure 37

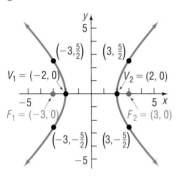

The points above and below the foci are $\left(\pm 3, \frac{5}{2} \right)$ and $\left(\pm 3, -\frac{5}{2} \right)$. These points determine the "opening" of the hyperbola. See Figure 37. ◀

NOW WORK PROBLEM **17**.

EXAMPLE 2   **Using a Graphing Utility to Graph a Hyperbola**

Using a graphing utility, graph the hyperbola: $\dfrac{x^2}{4} - \dfrac{y^2}{5} = 1$

Solution   To graph the hyperbola $\dfrac{x^2}{4} - \dfrac{y^2}{5} = 1$, we need to graph the two functions

Figure 38

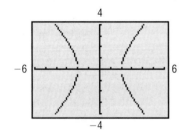

$Y_1 = \sqrt{5} \sqrt{\dfrac{x^2}{4} - 1}$ and $Y_2 = -\sqrt{5} \sqrt{\dfrac{x^2}{4} - 1}$. As with graphing circles, parabolas, and ellipses on a graphing utility, we use a square screen setting so that the graph is not distorted. Figure 38 shows the graph of the hyperbola. ◀

An equation of the form of equation (2) is the equation of a hyperbola with center at the origin, foci on the x-axis at $(-c, 0)$ and $(c, 0)$, where $c^2 = a^2 + b^2$, and transverse axis along the x-axis.

For the remainder of this section, the direction "Discuss the equation" will mean to find the center, transverse axis, vertices, and foci of the hyperbola and graph it.

EXAMPLE 3   **Discussing the Equation of a Hyperbola**

Discuss the equation: $\dfrac{x^2}{16} - \dfrac{y^2}{4} = 1$

Solution   The given equation is of the form of equation (2), with $a^2 = 16$ and $b^2 = 4$. The graph of the equation is a hyperbola with center at $(0, 0)$ and transverse axis along the x-axis. Also, we know that $c^2 = a^2 + b^2 = 16 + 4 = 20$. The vertices are at $(\pm a, 0) = (\pm 4, 0)$, and the foci are at $(\pm c, 0) = (\pm 2\sqrt{5}, 0)$.

To locate the points on the graph above and below the foci, we let $x = \pm 2\sqrt{5}$. Then

$$\frac{x^2}{16} - \frac{y^2}{4} = 1$$

$$\frac{\left(\pm 2\sqrt{5}\right)^2}{16} - \frac{y^2}{4} = 1 \qquad x = \pm 2\sqrt{5}$$

$$\frac{20}{16} - \frac{y^2}{4} = 1$$

$$\frac{5}{4} - \frac{y^2}{4} = 1$$

$$\frac{y^2}{4} = \frac{1}{4}$$

$$y = \pm 1$$

The points above and below the foci are $(\pm 2\sqrt{5}, 1)$ and $(\pm 2\sqrt{5}, -1)$. See Figure 39(a) for the graph drawn by hand. Figure 39(b) shows the graph obtained using a graphing utility.

Figure 39

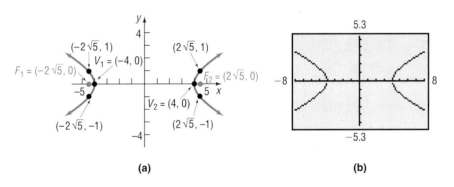

(a)   (b)

The next result gives the form of the equation of a hyperbola with center at the origin and transverse axis along the y-axis.

Theorem

Equation of a Hyperbola; Center at (0, 0); Transverse Axis along the y-Axis

An equation of the hyperbola with center at $(0, 0)$, foci at $(0, -c)$ and $(0, c)$, and vertices at $(0, -a)$ and $(0, a)$ is

$$\boxed{\frac{y^2}{a^2} - \frac{x^2}{b^2} = 1, \qquad \text{where } b^2 = c^2 - a^2} \qquad (3)$$

The transverse axis is the y-axis.

Figure 40
$$\frac{y^2}{a^2} - \frac{x^2}{b^2} = 1, \; b^2 = c^2 - a^2$$

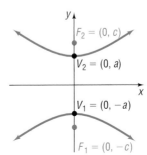

Figure 40 shows the graph of a typical hyperbola defined by equation (3).

An equation of the form of equation (2), $\frac{x^2}{a^2} - \frac{y^2}{b^2} = 1$, is the equation of a hyperbola with center at the origin, foci on the x-axis at $(-c, 0)$ and $(c, 0)$, where $c^2 = a^2 + b^2$, and transverse axis along the x-axis.

An equation of the form of equation (3), $\dfrac{y^2}{a^2} - \dfrac{x^2}{b^2} = 1$, is the equation of a hyperbola with center at the origin, foci on the y-axis at $(0, -c)$ and $(0, c)$, where $c^2 = a^2 + b^2$, and transverse axis along the y-axis.

Notice the difference in the forms of equations (2) and (3). When the y^2-term is subtracted from the x^2-term, the transverse axis is along the x-axis. When the x^2-term is subtracted from the y^2-term, the transverse axis is along the y-axis.

EXAMPLE 4   **Discussing the Equation of a Hyperbola**

Discuss the equation: $y^2 - 4x^2 = 4$

Solution   To put the equation in proper form, we divide each side by 4:

$$\frac{y^2}{4} - x^2 = 1$$

Since the x^2-term is subtracted from the y^2-term, the equation is that of a hyperbola with center at the origin and transverse axis along the y-axis. Also, comparing the above equation to equation (3), we find $a^2 = 4$, $b^2 = 1$, and $c^2 = a^2 + b^2 = 5$. The vertices are at $(0, \pm a) = (0, \pm 2)$, and the foci are at $(0, \pm c) = (0, \pm\sqrt{5})$.

To locate other points on the graph, we let $x = \pm 2$. Then

$$y^2 - 4x^2 = 4$$
$$y^2 - 4(\pm 2)^2 = 4 \qquad x = \pm 2$$
$$y^2 - 16 = 4$$
$$y^2 = 20$$
$$y = \pm 2\sqrt{5}$$

Four other points on the graph are $(\pm 2, 2\sqrt{5})$ and $(\pm 2, -2\sqrt{5})$. See Figure 41(a) for the graph drawn by hand. Figure 41(b) shows the graph obtained using a graphing utility.

Figure 41

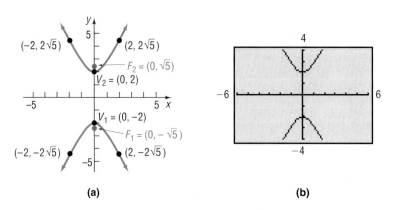

(a)   (b)

EXAMPLE 5   **Finding an Equation of a Hyperbola**

Find an equation of the hyperbola having one vertex at $(0, 2)$ and foci at $(0, -3)$ and $(0, 3)$. Graph the equation by hand.

Solution

Figure 42

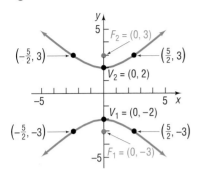

Since the foci are at $(0, -3)$ and $(0, 3)$, the center of the hyperbola is at their midpoint, the origin. Also, the transverse axis is along the y-axis. The given information also reveals that $c = 3$, $a = 2$, and $b^2 = c^2 - a^2 = 9 - 4 = 5$. The form of the equation of the hyperbola is given by equation (3):

$$\frac{y^2}{a^2} - \frac{x^2}{b^2} = 1$$

$$\frac{y^2}{4} - \frac{x^2}{5} = 1$$

Let $y = \pm 3$ to obtain points on the graph across from the foci. See Figure 42.　◀

NOW WORK PROBLEM 19.

Look at the equations of the hyperbolas in Examples 3 and 5. For the hyperbola in Example 3, $a^2 = 16$ and $b^2 = 4$, so $a > b$; for the hyperbola in Example 5, $a^2 = 4$ and $b^2 = 5$, so $a < b$. We conclude that, for hyperbolas, there are no requirements involving the relative sizes of a and b. Contrast this situation to the case of an ellipse, in which the relative sizes of a and b dictate which axis is the major axis. Hyperbolas have another feature to distinguish them from ellipses and parabolas: Hyperbolas have asymptotes.

2　Find the Asymptotes of a Hyperbola

Recall from Section 3.3 that a horizontal or oblique asymptote of a graph is a line with the property that the distance from the line to points on the graph approaches 0 as $x \to -\infty$ or as $x \to \infty$. The asymptotes provide information about the end behavior of the graph of a hyperbola.

Theorem

Asymptotes of a Hyperbola

The hyperbola $\dfrac{x^2}{a^2} - \dfrac{y^2}{b^2} = 1$ has the two oblique asymptotes

$$y = \frac{b}{a}x \quad \text{and} \quad y = -\frac{b}{a}x \tag{4}$$

Proof　We begin by solving for y in the equation of the hyperbola.

$$\frac{x^2}{a^2} - \frac{y^2}{b^2} = 1$$

$$\frac{y^2}{b^2} = \frac{x^2}{a^2} - 1$$

$$y^2 = b^2\left(\frac{x^2}{a^2} - 1\right)$$

Since $x \neq 0$, we can rearrange the right side in the form

$$y^2 = \frac{b^2 x^2}{a^2}\left(1 - \frac{a^2}{x^2}\right)$$

$$y = \pm\frac{bx}{a}\sqrt{1 - \frac{a^2}{x^2}}$$

Now, as $x \to -\infty$ or as $x \to \infty$, the term $\dfrac{a^2}{x^2}$ approaches 0, so the expression under the radical approaches 1. Thus, as $x \to -\infty$ or as $x \to \infty$, the value of y approaches $\pm\dfrac{bx}{a}$; that is, the graph of the hyperbola approaches the lines

$$y = -\frac{b}{a}x \quad \text{and} \quad y = \frac{b}{a}x$$

These lines are oblique asymptotes of the hyperbola. ∎

The asymptotes of a hyperbola are not part of the hyperbola, but they do serve as a guide for graphing a hyperbola. For example, suppose that we want to graph the equation

$$\frac{x^2}{a^2} - \frac{y^2}{b^2} = 1$$

We begin by plotting the vertices $(-a, 0)$ and $(a, 0)$. Then we plot the points $(0, -b)$ and $(0, b)$ and use these four points to construct a rectangle, as shown in Figure 43. The diagonals of this rectangle have slopes $\dfrac{b}{a}$ and $-\dfrac{b}{a}$, and their extensions are the asymptotes $y = \dfrac{b}{a}x$ and $y = -\dfrac{b}{a}x$ of the hyperbola. If we graph the asymptotes, we can use them to establish the "opening" of the hyperbola and avoid plotting other points.

Figure 43
$\dfrac{x^2}{a^2} - \dfrac{y^2}{b^2} = 1$

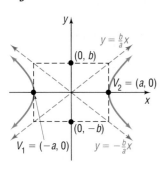

Theorem

Asymptotes of a Hyperbola

The hyperbola $\dfrac{y^2}{a^2} - \dfrac{x^2}{b^2} = 1$ has the two oblique asymptotes

$$y = \frac{a}{b}x \quad \text{and} \quad y = -\frac{a}{b}x \tag{5}$$

You are asked to prove this result in Problem 72.

For the remainder of this section, the direction "Discuss the equation" will mean to find the center, transverse axis, vertices, foci, and asymptotes of the hyperbola and graph it.

EXAMPLE 6

Discussing the Equation of a Hyperbola

Discuss the equation: $9x^2 - 4y^2 = 36$

Solution Divide each side of the equation by 36 to put the equation in proper form.

$$\frac{x^2}{4} - \frac{y^2}{9} = 1$$

We now proceed to analyze the equation. The center of the hyperbola is the origin. Since the x^2-term is first in the equation, we know that the transverse axis is along the x-axis and the vertices and foci will lie on the x-axis. Using equation (2), we find

$a^2 = 4$, $b^2 = 9$, and $c^2 = a^2 + b^2 = 13$. The vertices are $a = 2$ units left and right of the center at $(\pm a, 0) = (\pm 2, 0)$, the foci are $c = \sqrt{13}$ units left and right of the center at $(\pm c, 0) = (\pm\sqrt{13}, 0)$, and the asymptotes have the equations

$$y = \frac{b}{a}x = \frac{3}{2}x \quad \text{and} \quad y = -\frac{b}{a}x = -\frac{3}{2}x$$

To graph the hyperbola by hand, form the rectangle containing the points $(\pm a, 0)$ and $(0, \pm b)$, that is, $(-2, 0)$, $(2, 0)$, $(0, -3)$, and $(0, 3)$. The extensions of the diagonals of this rectangle are the asymptotes. See Figure 44(a) for the graph drawn by hand. Figure 44(b) shows the graph obtained using a graphing utility.

Figure 44

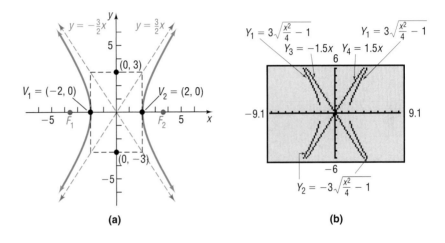

(a)

(b)

NOW WORK PROBLEM 29.

— Seeing the Concept —

Refer to Figure 44(b). Create a TABLE using Y_1 and Y_4 with $x = 10$, 100, 1000, and 10,000. Compare the values of Y_1 and Y_4. Repeat for Y_1 and Y_3, Y_2 and Y_3, and Y_2 and Y_4.

3 Work with Hyperbolas with Center at (h, k)

If a hyperbola with center at the origin and transverse axis coinciding with a coordinate axis is shifted horizontally h units and then vertically k units, the result is a hyperbola with center at (h, k) and transverse axis parallel to a coordinate axis. The equations of such hyperbolas have the same forms as those given in equations (2) and (3), except that x is replaced by $x - h$ (the horizontal shift) and y is replaced by $y - k$ (the vertical shift). Table 4 gives the forms of the equations of such hyperbolas. See Figure 45 for typical graphs.

Table 4

HYPERBOLAS WITH CENTER AT (h, k) AND TRANSVERSE AXIS PARALLEL TO A COORDINATE AXIS					
Center	Transverse Axis	Foci	Vertices	Equation	Asymptotes
(h, k)	Parallel to the x-axis	$(h \pm c, k)$	$(h \pm a, k)$	$\dfrac{(x - h)^2}{a^2} - \dfrac{(y - k)^2}{b^2} = 1$, $\;b^2 = c^2 - a^2$	$y - k = \pm\dfrac{b}{a}(x - h)$
(h, k)	Parallel to the y-axis	$(h, k \pm c)$	$(h, k \pm a)$	$\dfrac{(y - k)^2}{a^2} - \dfrac{(x - h)^2}{b^2} = 1$, $\;b^2 = c^2 - a^2$	$y - k = \pm\dfrac{a}{b}(x - h)$

Figure 45

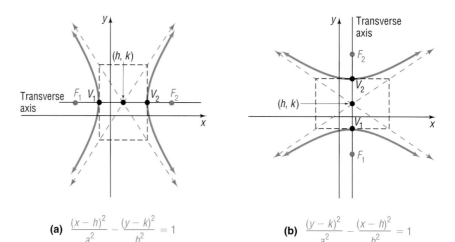

(a) $\dfrac{(x-h)^2}{a^2} - \dfrac{(y-k)^2}{b^2} = 1$

(b) $\dfrac{(y-k)^2}{a^2} - \dfrac{(x-h)^2}{b^2} = 1$

EXAMPLE 7   **Finding an Equation of a Hyperbola, Center Not at the Origin**

Find an equation for the hyperbola with center at $(1, -2)$, one focus at $(4, -2)$, and one vertex at $(3, -2)$. Graph the equation by hand.

Figure 46

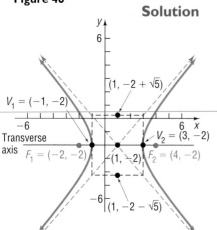

Solution   The center is at $(h, k) = (1, -2)$, so $h = 1$ and $k = -2$. Since the center, focus, and vertex all lie on the line $y = -2$, the transverse axis is parallel to the x-axis. The distance from the center $(1, -2)$ to the focus $(4, -2)$ is $c = 3$; the distance from the center $(1, -2)$ to the vertex $(3, -2)$ is $a = 2$. Thus, $b^2 = c^2 - a^2 = 9 - 4 = 5$. The equation is

$$\frac{(x-h)^2}{a^2} - \frac{(y-k)^2}{b^2} = 1$$

$$\frac{(x-1)^2}{4} - \frac{(y+2)^2}{5} = 1$$

See Figure 46.

◀

NOW WORK PROBLEM 39.

EXAMPLE 8   **Discussing the Equation of a Hyperbola**

Discuss the equation: $-x^2 + 4y^2 - 2x - 16y + 11 = 0$

Solution   We complete the squares in x and in y.

$$-x^2 + 4y^2 - 2x - 16y + 11 = 0$$

$$-(x^2 + 2x) + 4(y^2 - 4y) = -11 \qquad \text{Group terms.}$$

$$-(x^2 + 2x + 1) + 4(y^2 - 4y + 4) = -11 - 1 + 16 \qquad \text{Complete each square.}$$

$$-(x + 1)^2 + 4(y - 2)^2 = 4$$

$$(y - 2)^2 - \frac{(x + 1)^2}{4} = 1 \qquad \text{Divide each side by 4.}$$

This is the equation of a hyperbola with center at $(-1, 2)$ and transverse axis parallel to the y-axis. Also, $a^2 = 1$ and $b^2 = 4$, so $c^2 = a^2 + b^2 = 5$. Since the transverse axis is parallel to the y-axis, the vertices and foci are located a and c units above and below the center, respectively. The vertices are at $(h, k \pm a) = (-1, 2 \pm 1)$, or

$(-1, 1)$ and $(-1, 3)$. The foci are at $(h, k \pm c) = (-1, 2 \pm \sqrt{5})$. The asymptotes are $y - 2 = \dfrac{1}{2}(x + 1)$ and $y - 2 = -\dfrac{1}{2}(x + 1)$. Figure 47(a) shows the graph drawn by hand. Figure 47(b) shows the graph obtained using a graphing utility.

Figure 47

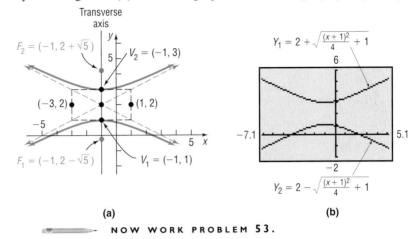

(a)   (b)

NOW WORK PROBLEM **53.**

4   ## Solve Applied Problems Involving Hyperbolas

Look at Figure 48. Suppose that three microphones are located at points O_1, O_2, and O_3 (the foci of the two hyperbolas). In addition, suppose that a gun is fired at S and the microphone at O_1 records the gun shot 1 second after the microphone at O_2. Because sound travels at about 1100 feet per second, we conclude that the microphone at O_1 is 1100 feet farther from the gunshot than O_2. We can model this situation by saying that S lies on the same branch of a hyperbola with foci at O_1 and O_2. (Do you see why? The difference of the distances from S to O_1 and from S to O_2 is the constant 1100.) If the third microphone at O_3 records the gunshot 2 seconds after O_1, then S will lie on a branch of a second hyperbola with foci at O_1 and O_3. In this case, the constant difference will be 2200. The intersection of the two hyperbolas will identify the location of S.

Figure 48

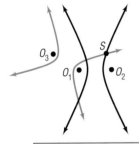

EXAMPLE 9   ### Lightning Strikes

Suppose that two people standing 1 mile apart both see a flash of lightning. After a period of time, the person standing at point A hears the thunder. One second later, the person standing at point B hears the thunder. If the person at B is due west of the person at A and the lightning strike is known to occur due north of the person standing at point A, where did the lightning strike?

Solution   See Figure 49 in which the ordered pair (x, y) represents the location of the lightning strike. We know that sound travels at 1100 feet per second, so the person at point A is 1100 feet closer to the lightning strike than the person at point B. Since the difference

Figure 49

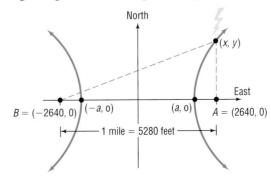

of the distance from (x, y) to A and the distance from (x, y) to B is the constant 1100, the point (x, y) lies on a hyperbola whose foci are at A and B.

An equation of the hyperbola is

$$\frac{x^2}{a^2} - \frac{y^2}{b^2} = 1$$

where $2a = 1100$ so $a = 550$.

Because the distance between the two people is 1 mile (5280 feet) and each person is at a focus of the hyperbola, we have

$$2c = 5280$$

$$c = \frac{5280}{2} = 2640$$

Since $b^2 = c^2 - a^2 = 2640^2 - 550^2 = 6{,}667{,}100$, the equation of the hyperbola that describes the location of the lightning strike is

$$\frac{x^2}{550^2} - \frac{y^2}{6{,}667{,}100} = 1$$

Since the lightning strikes due north of the individual at the point $A = (2640, 0)$, we let $x = 2640$ and solve the resulting equation.

$$\frac{2640^2}{550^2} - \frac{y^2}{6{,}667{,}100} = 1$$

$$-\frac{y^2}{6{,}667{,}100} = -22.04 \qquad \text{Subtract } \frac{2640^2}{550^2} \text{ from both sides.}$$

$$y^2 = 146{,}942{,}884 \qquad \text{Multiply both sides by } -6{,}667{,}100$$

$$y = 12{,}122 \qquad \text{Take the square root of both sides.}$$

✔ **CHECK:** The difference between the distance from $(2640, 12122)$ to the person at the point $B = (-2640, 0)$, and the distance from $(2640, 12122)$ to the person at the point $A = (2640, 0)$, should be 1100. Using the distance formula, we find the difference in the distances is

$$\sqrt{(12{,}122 - 0)^2 + (2640 - (-2640))^2} - \sqrt{(12{,}122 - 0)^2 + (2640 - 2640)^2} = 1100$$

as required.

The lightning strike is 12,122 feet north of the person standing at point A. ◀

NOW WORK PROBLEM 65.

6.4 Assess Your Understanding

'Are You Prepared?'

Answers are given at the end of these exercises. If you get a wrong answer, read the pages listed in red.

1. The distance d from $P_1 = (3, -4)$ to $P_2 = (-2, 1)$ is $d =$ _____. (pp. 86–89)

2. To complete the square of $x^2 + 5x$, add _____. (p. 115)

3. Find the intercepts of the equation $y^2 = 9 + 4x^2$. (p. 183)

4. *True or False:* The equation $y^2 = 9 + x^2$ is symmetric with respect to the x-axis, the y-axis, and the origin. (pp. 198–200)

5. To graph $y = (x - 5)^3 - 4$, shift the graph of $y = x^3$ to the (left/right) _____ unit(s) and then (up/down) _____ unit(s). (pp. 266–274)

6. Find the vertical asymptotes, if any, and the horizontal or oblique asymptotes, if any, of $y = \frac{x^2 - 9}{x^2 - 4}$. (pp. 337–343)

Concepts and Vocabulary

7. A(n) _____ is the collection of points in the plane the difference of whose distances from two fixed points is a constant.

8. For a hyperbola, the foci lie on a line called the _____ _____.

9. The asymptotes of the hyperbola $\dfrac{x^2}{4} - \dfrac{y^2}{9} = 1$ are _____ and _____.

10. *True or False:* The foci of a hyperbola lie on a line called the axis of symmetry.

11. *True or False:* Hyperbolas always have asymptotes.

12. *True or False:* A hyperbola will never intersect its transverse axis.

Skill Building

In Problems 13–16, the graph of a hyperbola is given. Match each graph to its equation.

A. $\dfrac{x^2}{4} - y^2 = 1$ B. $x^2 - \dfrac{y^2}{4} = 1$ C. $\dfrac{y^2}{4} - x^2 = 1$ D. $y^2 - \dfrac{x^2}{4} = 1$

13.

14.

15.

16.
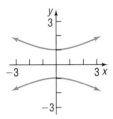

In Problems 17–26, find an equation for the hyperbola described. Graph the equation by hand.

17. Center at $(0, 0)$;   focus at $(3, 0)$;   vertex at $(1, 0)$

18. Center at $(0, 0)$;   focus at $(0, 5)$;   vertex at $(0, 3)$

19. Center at $(0, 0)$;   focus at $(0, -6)$;   vertex at $(0, 4)$

20. Center at $(0, 0)$;   focus at $(-3, 0)$;   vertex at $(2, 0)$

21. Foci at $(-5, 0)$ and $(5, 0)$;   vertex at $(3, 0)$

22. Focus at $(0, 6)$;   vertices at $(0, -2)$ and $(0, 2)$

23. Vertices at $(0, -6)$ and $(0, 6)$;   asymptote the line $y = 2x$

24. Vertices at $(-4, 0)$ and $(4, 0)$;   asymptote the line $y = 2x$

25. Foci at $(-4, 0)$ and $(4, 0)$;   asymptote the line $y = -x$

26. Foci at $(0, -2)$ and $(0, 2)$;   asymptote the line $y = -x$

In Problems 27–34, find the center, transverse axis, vertices, foci, and asymptotes. Graph each equation (a) by hand and (b) by using a graphing utility.

27. $\dfrac{x^2}{25} - \dfrac{y^2}{9} = 1$

28. $\dfrac{y^2}{16} - \dfrac{x^2}{4} = 1$

29. $4x^2 - y^2 = 16$

30. $4y^2 - x^2 = 16$

31. $y^2 - 9x^2 = 9$

32. $x^2 - y^2 = 4$

33. $y^2 - x^2 = 25$

34. $2x^2 - y^2 = 4$

In Problems 35–38, write an equation for each hyperbola.

35.

36.

37.

38.
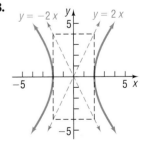

In Problems 39–46, find an equation for the hyperbola described. Graph the equation by hand.

39. Center at $(4, -1)$; focus at $(7, -1)$; vertex at $(6, -1)$

40. Center at $(-3, 1)$; focus at $(-3, 6)$; vertex at $(-3, 4)$

41. Center at $(-3, -4)$; focus at $(-3, -8)$; vertex at $(-3, -2)$

42. Center at $(1, 4)$; focus at $(-2, 4)$; vertex at $(0, 4)$

43. Foci at $(3, 7)$ and $(7, 7)$; vertex at $(6, 7)$

44. Focus at $(-4, 0)$ vertices at $(-4, 4)$ and $(-4, 2)$

45. Vertices at $(-1, -1)$ and $(3, -1)$;
asymptote the line $y + 1 = \dfrac{3}{2}(x - 1)$

46. Vertices at $(1, -3)$ and $(1, 1)$;
asymptote the line $y + 1 = \dfrac{3}{2}(x - 1)$

In Problems 47–60, find the center, transverse axis, vertices, foci, and asymptotes. Graph each equation (a) by hand and (b) by using a graphing utility.

47. $\dfrac{(x - 2)^2}{4} - \dfrac{(y + 3)^2}{9} = 1$

48. $\dfrac{(y + 3)^2}{4} - \dfrac{(x - 2)^2}{9} = 1$

49. $(y - 2)^2 - 4(x + 2)^2 = 4$

50. $(x + 4)^2 - 9(y - 3)^2 = 9$

51. $(x + 1)^2 - (y + 2)^2 = 4$

52. $(y - 3)^2 - (x + 2)^2 = 4$

53. $x^2 - y^2 - 2x - 2y - 1 = 0$

54. $y^2 - x^2 - 4y + 4x - 1 = 0$

55. $y^2 - 4x^2 - 4y - 8x - 4 = 0$

56. $2x^2 - y^2 + 4x + 4y - 4 = 0$

57. $4x^2 - y^2 - 24x - 4y + 16 = 0$

58. $2y^2 - x^2 + 2x + 8y + 3 = 0$

59. $y^2 - 4x^2 - 16x - 2y - 19 = 0$

60. $x^2 - 3y^2 + 8x - 6y + 4 = 0$

In Problems 61–64, graph each function.

[**Hint:** *Notice that each function is half a hyperbola.*]

61. $f(x) = \sqrt{16 + 4x^2}$

62. $f(x) = -\sqrt{9 + 9x^2}$

63. $f(x) = -\sqrt{-25 + x^2}$

64. $f(x) = \sqrt{-1 + x^2}$

Applications and Extensions

65. Fireworks Display Suppose that two people standing 2 miles apart both see the burst from a fireworks display. After a period of time, the first person standing at point A hears the burst. One second later, the second person standing at point B hears the burst. If the display is known to occur due north of the person at point A, where did the fireworks display occur?

66. Lightning Strikes Suppose that two people standing 1 mile apart both see a flash of lightning. After a period of time, the first person standing at point A hears the thunder. Two seconds later, the second person standing at point B hears the thunder. If the lightning strike is known to occur due north of the person standing at point A, where did the lightning strike?

67. Rutherford's Experiment In May 1911, Ernest Rutherford published a paper in *Philosophical Magazine*. In this article, he described the motion of alpha particles as they are shot at a piece of gold foil 0.00004 cm thick. Before conducting this experiment, Rutherford expected that the alpha particles would shoot through the foil just as a bullet would shoot through snow. Instead, a small fraction of the alpha particles bounced off the foil. This led to the conclusion that the nucleus of an atom is dense, while the remainder of the atom is sparse. Only the density of the nucleus could cause the alpha particles to deviate from their path. The figure shows a diagram from Rutherford's paper that indicates that the deflected alpha particles follow the path of one branch of a hyperbola.
(a) Find an equation of the asymptotes under this scenario.
(b) If the vertex of the path of the alpha particles is 10 cm

from the center of the hyperbola, find an equation that describes the path of the particle.

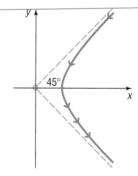

68. An Explosion Two recording devices are set 2400 feet apart, with the device at point A to the west of the device at point B. At a point between the devices, 300 feet from point B, a small amount of explosive is detonated. The recording devices record the time until the sound reaches each. How far directly north of point B should a second explosion be done so that the measured time difference recorded by the devices is the same as that for the first detonation?

69. The **eccentricity** e of a hyperbola is defined as the number $\dfrac{c}{a}$, where a and c are the numbers given in equation (2).
Because $c > a$, it follows that $e > 1$. Describe the general shape of a hyperbola whose eccentricity is close to 1. What is the shape if e is very large?

70. A hyperbola for which $a = b$ is called an **equilateral hyperbola**. Find the eccentricity e of an equilateral hyperbola.
[NOTE: The eccentricity of a hyperbola is defined in Problem 69.]

71. Two hyperbolas that have the same set of asymptotes are called **conjugate**. Show that the hyperbolas

$$\frac{x^2}{4} - y^2 = 1 \quad \text{and} \quad y^2 - \frac{x^2}{4} = 1$$

are conjugate. Graph each hyperbola on the same set of coordinate axes.

72. Prove that the hyperbola

$$\frac{y^2}{a^2} - \frac{x^2}{b^2} = 1$$

has the two oblique asymptotes

$$y = \frac{a}{b}x \quad \text{and} \quad y = -\frac{a}{b}x$$

73. Show that the graph of an equation of the form

$$Ax^2 + Cy^2 + F = 0, \qquad A \neq 0, C \neq 0, F \neq 0$$

where A and C are of opposite sign, is a hyperbola with center at $(0, 0)$.

74. Show that the graph of an equation of the form

$$Ax^2 + Cy^2 + Dx + Ey + F = 0, \qquad A \neq 0, C \neq 0$$

where A and C are of opposite sign,

(a) Is a hyperbola if $\dfrac{D^2}{4A} + \dfrac{E^2}{4C} - F \neq 0$.

(b) Is two intersecting lines if

$$\frac{D^2}{4A} + \frac{E^2}{4C} - F = 0$$

'Are You Prepared?' Answers

1. $5\sqrt{2}$ **2.** $\dfrac{25}{4}$ **3.** $(0, -3), (0, 3)$ **4.** True   **5.** right, 5; down, 4   **6.** Vertical: $x = -2$, $x = 2$; Horizontal: $y = 1$

6.5 Systems of Nonlinear Equations and Inequalities

PREPARING FOR THIS SECTION   *Before getting started, review the following:*

- Lines (Section 1.8, pp. 163–176)
- Quadratic Functions (Section 3.1, pp. 298–305)
- Systems of Linear Inequalities (Section 5.6, pp. 572–579)
- Circles (Section 1.9, pp. 180–185)
- Systems of Linear Equations (Section 5.1, pp. 504–515)

Now work the 'Are You Prepared?' problems on page 641.

OBJECTIVES   1   Solve a System of Nonlinear Equations Using Substitution
             2   Solve a System of Nonlinear Equations Using Elimination
             3   Graph a System of Nonlinear Inequalities

1   Solve a System of Nonlinear Equations Using Substitution

In Section 5.1 we observed that the solution to a system of linear equations could be found geometrically by determining the point(s) of intersection (if any) of the equations in the system. Similarly, when solving systems of nonlinear equations, the solution(s) also represents the point(s) of intersection (if any) of the graphs of the equations.

There is no general methodology for solving a system of nonlinear equations. At times substitution is best; other times, elimination is best; and sometimes neither of these methods works. Experience and a certain degree of imagination are your allies here.

Before we begin, two comments are in order.

1. If the system contains two variables and if the equations in the system are easy to graph, then graph them. By graphing each equation in the system, you can get an idea of how many solutions a system has and approximately where they are located.

2. Extraneous solutions can creep in when solving nonlinear systems, so it is imperative that all apparent solutions be checked.

| EXAMPLE 1 | **Solving a System of Nonlinear Equations** |

Solve the following system of equations:

$$\begin{cases} 3x - y = -2 & \text{(1)} \quad \text{A line} \\ 2x^2 - y = 0 & \text{(2)} \quad \text{A parabola} \end{cases}$$

Algebraic Solution Using Substitution

First, we notice that the system contains two variables and that we know how to graph each equation by hand. See Figure 50. The system apparently has two solutions.

Figure 50

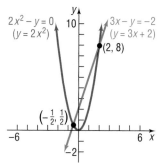

We will use substitution to solve the system. Equation (1) is easily solved for y.

$$3x - y = -2 \qquad \text{(1)}$$
$$y = 3x + 2$$

We substitute this expression for y in equation (2). The result is an equation containing just the variable x, which we can then solve.

$$2x^2 - y = 0 \qquad \text{(2)}$$
$$2x^2 - (3x + 2) = 0 \qquad \text{Substitute } 3x + 2 \text{ for } y$$
$$2x^2 - 3x - 2 = 0 \qquad \text{Remove parentheses.}$$
$$(2x + 1)(x - 2) = 0 \qquad \text{Factor.}$$
$$2x + 1 = 0 \quad \text{or} \quad x - 2 = 0 \qquad \text{Apply the Zero-Product Property.}$$
$$x = -\frac{1}{2} \quad \text{or} \quad x = 2$$

Using these values for x in $y = 3x + 2$, we find

$$y = 3\left(-\frac{1}{2}\right) + 2 = \frac{1}{2} \quad \text{or} \quad y = 3(2) + 2 = 8$$

The apparent solutions are $x = -\frac{1}{2}, y = \frac{1}{2}$ and $x = 2, y = 8$.

✔ **CHECK:** For $x = -\frac{1}{2}, y = \frac{1}{2}$:

$$\begin{cases} 3\left(-\frac{1}{2}\right) - \frac{1}{2} = -\frac{3}{2} - \frac{1}{2} = -2 & \text{(1)} \\ 2\left(-\frac{1}{2}\right)^2 - \frac{1}{2} = 2\left(\frac{1}{4}\right) - \frac{1}{2} = 0 & \text{(2)} \end{cases}$$

For $x = 2, \quad y = 8$:

$$\begin{cases} 3(2) - 8 = 6 - 8 = -2 & \text{(1)} \\ 2(2)^2 - 8 = 2(4) - 8 = 0 & \text{(2)} \end{cases}$$

Each solution checks. Now we know that the graphs in Figure 50 intersect at $\left(-\frac{1}{2}, \frac{1}{2}\right)$ and at $(2, 8)$.

Graphing Solution

We use a graphing utility to graph $Y_1 = 3x + 2$ and $Y_2 = 2x^2$. From Figure 51 we see that the system apparently has two solutions. Using INTERSECT, the solutions to the system of equations are $(-0.5, 0.5)$ and $(2, 8)$.

Figure 51

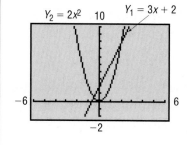

◀

◀

NOW WORK PROBLEM 17 USING SUBSTITUTION.

2 Solve a System of Nonlinear Equations Using Elimination

Our next example illustrates how the method of elimination works for nonlinear systems.

EXAMPLE 2

Solving a System of Nonlinear Equations

Solve: $\begin{cases} x^2 + y^2 = 13 & (1) \quad \text{A circle} \\ x^2 - y = 7 & (2) \quad \text{A parabola} \end{cases}$

Algebraic Solution Using Elimination

First, we graph each equation, as shown in Figure 52. Based on the graph, we expect four solutions. By subtracting equation (2) from equation (1), the variable x can be eliminated.

$$\begin{cases} x^2 + y^2 = 13 \\ x^2 - y = 7 \end{cases} \quad \text{Subtract}$$
$$\overline{\qquad y^2 + y = 6}$$

This quadratic equation in y can be solved by factoring.

$$y^2 + y - 6 = 0$$
$$(y + 3)(y - 2) = 0$$
$$y = -3 \quad \text{or} \quad y = 2$$

We use these values for y in equation (2) to find x.

   If $y = 2$, then $x^2 = y + 7 = 9$, so $x = 3$ or -3.

   If $y = -3$, then $x^2 = y + 7 = 4$, so $x = 2$ or -2.

We have four solutions: $x = 3, y = 2$; $x = -3, y = 2$; $x = 2, y = -3$; and $x = -2, y = -3$.

   You should verify that, in fact, these four solutions also satisfy equation (1), so all four are solutions of the system. The four points, $(3, 2), (-3, 2), (2, -3),$ and $(-2, -3)$, are the points of intersection of the graphs. Look again at Figure 52.

Graphing Solution

We use a graphing utility to graph $x^2 + y^2 = 13$ and $x^2 - y = 7$. (Remember that to graph $x^2 + y^2 = 13$ requires two functions, $Y_1 = \sqrt{13 - x^2}$ and $Y_2 = -\sqrt{13 - x^2}$, and a square screen.) From Figure 53 we see that the system apparently has four solutions. Using INTERSECT, the solutions to the system of equations are $(-3, 2), (3, 2), (-2, -3),$ and $(2, -3)$.

Figure 53

Figure 52

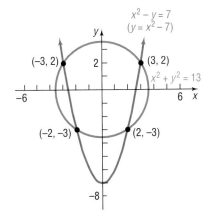

| EXAMPLE 3 | Solving a System of Nonlinear Equations |

Solve:
$$\begin{cases} x^2 + x + y^2 - 3y + 2 = 0 & (1) \\ x + 1 + \dfrac{y^2 - y}{x} = 0 & (2) \end{cases}$$

Algebraic Solution Using Elimination

Since it is not straightforward how to graph the equations in the system, we proceed directly to use the method of elimination.

First, we multiply equation (2) by x to eliminate the fraction. The result is an equivalent system because x cannot be 0. [Look at equation (2) to see why.]

$$\begin{cases} x^2 + x + y^2 - 3y + 2 = 0 & (1) \\ x^2 + x + y^2 - y = 0 & (2) \end{cases}$$

Now subtract equation (2) from equation (1) to eliminate x. The result is

$$-2y + 2 = 0$$
$$y = 1 \qquad \text{Solve for } y.$$

To find x, we back-substitute $y = 1$ in equation (1):

$$x^2 + x + y^2 - 3y + 2 = 0 \quad (1)$$
$$x^2 + x + 1 - 3 + 2 = 0 \quad \text{Substitute } y = 1 \text{ in (1).}$$
$$x^2 + x = 0 \qquad \text{Simplify.}$$
$$x(x + 1) = 0 \qquad \text{Factor.}$$
$$x = 0 \quad \text{or} \quad x = -1 \qquad \text{Apply the Zero-Product Property.}$$

Because x cannot be 0, the value $x = 0$ is extraneous, and we discard it. The solution is $x = -1$, $y = 1$.

✔ **CHECK:** We now check $x = -1$, $y = 1$:

$$\begin{cases} (-1)^2 + (-1) + 1^2 - 3(1) + 2 = 1 - 1 + 1 - 3 + 2 = 0 & (1) \\ -1 + 1 + \dfrac{1^2 - 1}{-1} = 0 + \dfrac{0}{-1} = 0 & (2) \end{cases}$$

◄

Graphing Solution

First, we multiply equation (2) by x to eliminate the fraction. The result is an equivalent system because x cannot be 0 [look at equation (2) to see why]:

$$\begin{cases} x^2 + x + y^2 - 3y + 2 = 0 & (1) \\ x^2 + x + y^2 - y = 0 & (2) \end{cases}$$

We need to solve each equation for y. First, we solve equation (1) for y:

$$x^2 + x + y^2 - 3y + 2 = 0$$
$$y^2 - 3y = -x^2 - x - 2 \qquad \text{Rearrange so that terms involving } y \text{ are on left side.}$$
$$y^2 - 3y + \frac{9}{4} = -x^2 - x - 2 + \frac{9}{4} \qquad \text{Complete the square involving } y.$$
$$\left(y - \frac{3}{2}\right)^2 = -x^2 - x + \frac{1}{4}$$
$$y - \frac{3}{2} = \pm\sqrt{-x^2 - x + \frac{1}{4}} \qquad \text{Square Root Method}$$
$$y = \frac{3}{2} \pm \sqrt{-x^2 - x + \frac{1}{4}} \qquad \text{Solve for } y.$$

Now we solve equation (2) for y:

$$x^2 + x + y^2 - y = 0$$
$$y^2 - y = -x^2 - x \qquad \text{Rearrange so that terms involving } y \text{ are on left side.}$$
$$y^2 - y + \frac{1}{4} = -x^2 - x + \frac{1}{4} \qquad \text{Complete the square involving } y.$$
$$\left(y - \frac{1}{2}\right)^2 = -x^2 - x + \frac{1}{4}$$
$$y - \frac{1}{2} = \pm\sqrt{-x^2 - x + \frac{1}{4}} \qquad \text{Square Root Method}$$
$$y = \frac{1}{2} \pm \sqrt{-x^2 - x + \frac{1}{4}} \qquad \text{Solve for } y.$$

Now graph each equation using a graphing utility. See Figure 54.

Using INTERSECT, the points of intersection are $(-1, 1)$ and $(0, 1)$. Since $x \neq 0$ [look back at the original equation (2)], the graph of Y_3 has a hole at the point $(0, 1)$ and Y_4 has a hole at $(0, 0)$. The value $x = 0$ is extraneous, and we discard it. The only solution is $x = -1$ and $y = 1$.

Figure 54

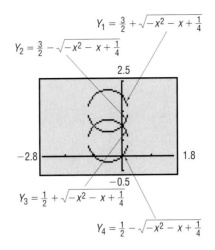

$Y_1 = \frac{3}{2} + \sqrt{-x^2 - x + \frac{1}{4}}$

$Y_2 = \frac{3}{2} - \sqrt{-x^2 - x + \frac{1}{4}}$

$Y_3 = \frac{1}{2} + \sqrt{-x^2 - x + \frac{1}{4}}$

$Y_4 = \frac{1}{2} - \sqrt{-x^2 - x + \frac{1}{4}}$

NOW WORK PROBLEMS 31 AND 51.

| **EXAMPLE 4** | **Solving a System of Nonlinear Equations** |

Solve: $\begin{cases} x^2 - y^2 = 4 & (1) \quad \text{A hyperbola} \\ y = x^2 & (2) \quad \text{A parabola} \end{cases}$

Algebraic Solution

Either substitution or elimination can be used here. We use substitution and replace x^2 by y in equation (1). The result is

$$y - y^2 = 4$$
$$y^2 - y + 4 = 0$$

This is a quadratic equation whose discriminant is $(-1)^2 - 4 \cdot 1 \cdot 4 = 1 - 4 \cdot 4 = -15 < 0$. The equation has no real solutions, so the system is inconsistent. The graphs of these two equations do not intersect. See Figure 55.

Figure 55

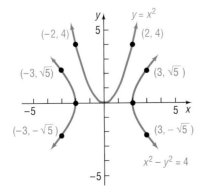

Graphing Solution

We graph $Y_1 = x^2$ and $x^2 - y^2 = 4$ in Figure 56. You will need to graph $x^2 - y^2 = 4$ as two functions:

$$Y_2 = \sqrt{x^2 - 4} \quad \text{and} \quad Y_3 = -\sqrt{x^2 - 4}$$

From Figure 56 we see that the graphs of these two equations do not intersect. The system is inconsistent.

Figure 56

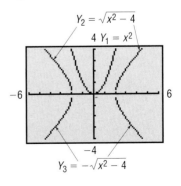

$Y_2 = \sqrt{x^2 - 4}$

$Y_1 = x^2$

$Y_3 = -\sqrt{x^2 - 4}$

| **EXAMPLE 5** | **Solving a System of Nonlinear Equations** |

Solve: $\begin{cases} 3xy - 2y^2 = -2 & (1) \\ 9x^2 + 4y^2 = 10 & (2) \end{cases}$

Algebraic Solution

We multiply equation (1) by 2 and add the result to equation (2) to eliminate the y^2 terms.

$\begin{cases} 6xy - 4y^2 = -4 & (1) \\ 9x^2 + 4y^2 = 10 & (2) \quad \text{Add.} \end{cases}$

$9x^2 + 6xy = 6$

$3x^2 + 2xy = 2 \qquad \text{Divide each side by 3.}$

Since $x \neq 0$ (do you see why?), we can solve for y in this equation to get

$$y = \frac{2 - 3x^2}{2x}, \qquad x \neq 0 \qquad (3)$$

Now substitute for y in equation (2) of the system.

$9x^2 + 4y^2 = 10 \qquad (2)$

$9x^2 + 4\left(\frac{2 - 3x^2}{2x}\right)^2 = 10 \qquad \text{Substitute } y = \frac{2 - 3x^2}{2x} \text{ in (2).}$

$9x^2 + \frac{4 - 12x^2 + 9x^4}{x^2} = 10$

$9x^4 + 4 - 12x^2 + 9x^4 = 10x^2 \qquad \text{Multiply both sides by } x^2.$

$18x^4 - 22x^2 + 4 = 0 \qquad \text{Subtract } 10x^2 \text{ from both sides.}$

$9x^4 - 11x^2 + 2 = 0 \qquad \text{Divide both sides by 2.}$

This quadratic equation (in x^2) can be factored:

$(9x^2 - 2)(x^2 - 1) = 0$

$9x^2 - 2 = 0 \qquad \text{or} \qquad x^2 - 1 = 0$

$x^2 = \frac{2}{9} \qquad\qquad x^2 = 1$

$x = \pm\sqrt{\frac{2}{9}} = \pm\frac{\sqrt{2}}{3} \qquad x = \pm 1$

To find y, we use equation (3):

If $x = \frac{\sqrt{2}}{3}$: $y = \frac{2 - 3x^2}{2x} = \frac{2 - \dfrac{2}{3}}{2\left(\dfrac{\sqrt{2}}{3}\right)} = \frac{4}{2\sqrt{2}} = \sqrt{2}$

If $x = -\frac{\sqrt{2}}{3}$: $y = \frac{2 - 3x^2}{2x} = \frac{2 - \dfrac{2}{3}}{2\left(-\dfrac{\sqrt{2}}{3}\right)} = \frac{4}{-2\sqrt{2}} = -\sqrt{2}$

If $x = 1$: $y = \frac{2 - 3x^2}{2x} = \frac{2 - 3(1)^2}{2} = -\frac{1}{2}$

If $x = -1$: $y = \frac{2 - 3x^2}{2x} = \frac{2 - 3(-1)^2}{-2} = \frac{1}{2}$

The system has four solutions. Check them for yourself.

Graphing Solution

To graph $3xy - 2y^2 = -2$, we need to solve for y. In this instance, it is easier to view the equation as a quadratic equation in the variable y.

$3xy - 2y^2 = -2$

$2y^2 - 3xy - 2 = 0 \qquad \text{Place in standard form.}$

$y = \frac{-(-3x) \pm \sqrt{(-3x)^2 - 4(2)(-2)}}{2(2)}$

Use the quadratic formula with $a = 2$, $b = -3x$, $c = -2$.

$y = \frac{3x \pm \sqrt{9x^2 + 16}}{4} \qquad \text{Simplify.}$

Using a graphing utility, we graph

$Y_1 = \frac{3x + \sqrt{9x^2 + 16}}{4}$ and $Y_2 = \frac{3x - \sqrt{9x^2 + 16}}{4}$.

From (2), we graph $Y_3 = \frac{\sqrt{10 - 9x^2}}{2}$ and $Y_4 = \frac{-\sqrt{10 - 9x^2}}{2}$. See Figure 57.

Figure 57

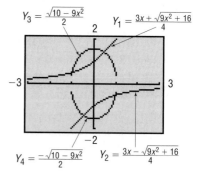

Using INTERSECT, the solutions to the system of equations are $(-1, 0.5)$, $(0.47, 1.41)$, $(1, -0.5)$, and $(-0.47, -1.41)$, each rounded to two decimal places.

◄

══════ NOW WORK PROBLEM **49**.

EXAMPLE 6	**Running a Long Distance Race**

In a 50-mile race, the winner crosses the finish line 1 mile ahead of the second-place runner and 4 miles ahead of the third-place runner. Assuming that each runner maintains a constant speed throughout the race, by how many miles does the second-place runner beat the third-place runner?

Solution   Let v_1, v_2, and v_3 denote the speeds of the first-, second-, and third-place runners, respectively. Let t_1 and t_2 denote the times (in hours) required for the first-place runner and second-place runner to finish the race. Then we have the system of equations

$$\begin{cases} 50 = v_1 t_1 & (1) \quad \text{First-place runner goes 50 miles in } t_1. \\ 49 = v_2 t_1 & (2) \quad \text{Second-place runner goes 49 miles in } t_1. \\ 46 = v_3 t_1 & (3) \quad \text{Third-place runner goes 46 miles in } t_1. \\ 50 = v_2 t_2 & (4) \quad \text{Second-place runner goes 50 miles in } t_2. \end{cases}$$

We seek the distance of the third-place runner from the finish at time t_2. At time t_2, the third-place runner has gone a distance of $v_3 t_2$ miles, so the distance remaining is $50 - v_3 t_2$. Now

$$50 - v_3 t_2 = 50 - v_3 \left(t_1 \cdot \frac{t_2}{t_1} \right)$$

$$= 50 - (v_3 t_1) \cdot \frac{t_2}{t_1}$$

$$= 50 - 46 \cdot \frac{\dfrac{50}{v_2}}{\dfrac{50}{v_1}} \qquad \left\{ \begin{array}{l} \text{From (3), } v_3 t_1 = 46 \\[1ex] \text{From (4), } t_2 = \dfrac{50}{v_2} \\[1ex] \text{From (1), } t_1 = \dfrac{50}{v_1} \end{array} \right.$$

$$= 50 - 46 \cdot \frac{v_1}{v_2}$$

$$= 50 - 46 \cdot \frac{50}{49} \qquad \text{Form the quotient of (1) and (2).}$$

$$\approx 3.06 \text{ miles} \qquad\qquad\qquad\qquad\qquad \blacktriangleleft$$

3   Graph a System of Nonlinear Inequalities

In Section 5.6, we graphed systems of linear inequalities. We now discuss systems of nonlinear inequalities.

Remember, the graph of a system of inequalities in two variables x and y is the set of all points (x, y) that simultaneously satisfy each of the inequalities in the system. The graph of a system of inequalities can be obtained by graphing each inequality individually and then determining where, if at all, they intersect.

| EXAMPLE 7 | **Graphing a System of Nonlinear Inequalities** |

Graph the region below the graph of $x + y = 2$ and above the graph of $y = x^2 - 4$ by graphing the system:

$$\begin{cases} y \geq x^2 - 4 \\ x + y \leq 2 \end{cases}$$

Label all points of intersection.

Solution　Figure 58 shows the graph of the region above the graph of the parabola $y = x^2 - 4$ and below the graph of the line $x + y = 2$. The points of intersection are found by solving the system of equations

Figure 58

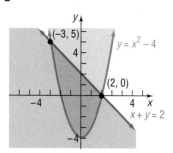

$$\begin{cases} y = x^2 - 4 \\ x + y = 2 \end{cases}$$

Using substitution, we find

$$x + (x^2 - 4) = 2$$

$$x^2 + x - 6 = 0$$

$$(x + 3)(x - 2) = 0$$

$$x = -3 \qquad x = 2$$

The two points of intersection are $(-3, 5)$ and $(2, 0)$.　◀

✏️　**N O W　W O R K　P R O B L E M　7 3 .**

HISTORICAL FEATURE

In the beginning of this section, we said that imagination and experience are important in solving systems of nonlinear equations. Indeed, these kinds of problems lead into some of the deepest and most difficult parts of modern mathematics. Look again at the graphs in Examples 1 and 2 of this section (Figures 50 and 52). We see that Example 1 has two solutions, and Example 2 has four solutions. We might conjecture that the number of solutions is equal to the product of the degrees of the equations involved.

This conjecture was indeed made by Etienne Bezout (1730–1783), but working out the details took about 150 years. It turns out that, to arrive at the correct number of intersections, we must count not only the complex number intersections, but also those intersections that, in a certain sense, lie at infinity. For example, a parabola and a line lying on the axis of the parabola intersect at the vertex and at infinity. This topic is part of the study of algebraic geometry.

Historical Problem

A papyrus dating back to 1950 BC contains the following problem: "A given surface area of 100 units of area shall be represented as the sum of two squares whose sides are to each other as $1 : \dfrac{3}{4}$."

Solve for the sides by solving the system of equations

$$\begin{cases} x^2 + y^2 = 100 \\ x = \dfrac{3}{4}y \end{cases}$$

6.5 Assess Your Understanding

'Are You Prepared?'

Answers are given at the end of these exercises. If you get a wrong answer, read the pages listed in red.

1. Graph the equation: $y = 3x + 2$. (pp. 170–171)
2. Graph the equation: $y = x^2 - 4$. (pp. 298–301)
3. Graph the inequality: $2x + 3y < 6$. (pp. 572–574)

4. Graph the system of linear inequalities: $\begin{cases} y \leq -2x + 7 \\ 4x + 4y \leq 20 \\ x \geq 0 \\ y \geq 0 \end{cases}$
   (pp. 575–578)

5. Solve by substitution: $\begin{cases} y = 2x - 3 \\ 2x - 3y = -7 \end{cases}$
   (pp. 507–508)

6. Solve by elimination: $\begin{cases} 2x + 5y = 7 \\ -4x - 2y = 10 \end{cases}$
   (pp. 508–509)

Skill Building

In Problems 7–26, graph each equation of the system. Then solve the system to find the points of intersection.

7. $\begin{cases} y = x^2 + 1 \\ y = x + 1 \end{cases}$

8. $\begin{cases} y = x^2 + 1 \\ y = 4x + 1 \end{cases}$

9. $\begin{cases} y = \sqrt{36 - x^2} \\ y = 8 - x \end{cases}$

10. $\begin{cases} y = \sqrt{4 - x^2} \\ y = 2x + 4 \end{cases}$

11. $\begin{cases} y = \sqrt{x} \\ y = 2 - x \end{cases}$

12. $\begin{cases} y = \sqrt{x} \\ y = 6 - x \end{cases}$

13. $\begin{cases} x = 2y \\ x = y^2 - 2y \end{cases}$

14. $\begin{cases} y = x - 1 \\ y = x^2 - 6x + 9 \end{cases}$

15. $\begin{cases} x^2 + y^2 = 4 \\ x^2 + 2x + y^2 = 0 \end{cases}$

16. $\begin{cases} x^2 + y^2 = 8 \\ x^2 + y^2 + 4y = 0 \end{cases}$

17. $\begin{cases} y = 3x - 5 \\ x^2 + y^2 = 5 \end{cases}$

18. $\begin{cases} x^2 + y^2 = 10 \\ y = x + 2 \end{cases}$

19. $\begin{cases} x^2 + y^2 = 4 \\ y^2 - x = 4 \end{cases}$

20. $\begin{cases} x^2 + y^2 = 16 \\ x^2 - 2y = 8 \end{cases}$

21. $\begin{cases} xy = 4 \\ x^2 + y^2 = 8 \end{cases}$

22. $\begin{cases} x^2 = y \\ xy = 1 \end{cases}$

23. $\begin{cases} x^2 + y^2 = 4 \\ y = x^2 - 9 \end{cases}$

24. $\begin{cases} xy = 1 \\ y = 2x + 1 \end{cases}$

25. $\begin{cases} y = x^2 - 4 \\ y = 6x - 13 \end{cases}$

26. $\begin{cases} x^2 + y^2 = 10 \\ xy = 3 \end{cases}$

In Problems 27–56, solve each system. Use any method you wish.

27. $\begin{cases} 2x^2 + y^2 = 18 \\ xy = 4 \end{cases}$

28. $\begin{cases} x^2 - y^2 = 21 \\ x + y = 7 \end{cases}$

29. $\begin{cases} y = 2x + 1 \\ 2x^2 + y^2 = 1 \end{cases}$

30. $\begin{cases} x^2 - 4y^2 = 16 \\ 2y - x = 2 \end{cases}$

31. $\begin{cases} x + y + 1 = 0 \\ x^2 + y^2 + 6y - x = -5 \end{cases}$

32. $\begin{cases} 2x^2 - xy + y^2 = 8 \\ xy = 4 \end{cases}$

33. $\begin{cases} 4x^2 - 3xy + 9y^2 = 15 \\ 2x + 3y = 5 \end{cases}$

34. $\begin{cases} 2y^2 - 3xy + 6y + 2x + 4 = 0 \\ 2x - 3y + 4 = 0 \end{cases}$

35. $\begin{cases} x^2 - 4y^2 + 7 = 0 \\ 3x^2 + y^2 = 31 \end{cases}$

36. $\begin{cases} 3x^2 - 2y^2 + 5 = 0 \\ 2x^2 - y^2 + 2 = 0 \end{cases}$

37. $\begin{cases} 7x^2 - 3y^2 + 5 = 0 \\ 3x^2 + 5y^2 = 12 \end{cases}$

38. $\begin{cases} x^2 - 3y^2 + 1 = 0 \\ 2x^2 - 7y^2 + 5 = 0 \end{cases}$

39. $\begin{cases} x^2 + 2xy = 10 \\ 3x^2 - xy = 2 \end{cases}$

40. $\begin{cases} 5xy + 13y^2 + 36 = 0 \\ xy + 7y^2 = 6 \end{cases}$

41. $\begin{cases} 2x^2 + y^2 = 2 \\ x^2 - 2y^2 + 8 = 0 \end{cases}$

42. $\begin{cases} y^2 - x^2 + 4 = 0 \\ 2x^2 + 3y^2 = 6 \end{cases}$

43. $\begin{cases} x^2 + 2y^2 = 16 \\ 4x^2 - y^2 = 24 \end{cases}$

44. $\begin{cases} 4x^2 + 3y^2 = 4 \\ 2x^2 - 6y^2 = -3 \end{cases}$

45. $\begin{cases} \dfrac{5}{x^2} - \dfrac{2}{y^2} + 3 = 0 \\ \dfrac{3}{x^2} + \dfrac{1}{y^2} = 7 \end{cases}$

46. $\begin{cases} \dfrac{2}{x^2} - \dfrac{3}{y^2} + 1 = 0 \\ \dfrac{6}{x^2} - \dfrac{7}{y^2} + 2 = 0 \end{cases}$

47. $\begin{cases} \dfrac{1}{x^4} + \dfrac{6}{y^4} = 6 \\ \dfrac{2}{x^4} - \dfrac{2}{y^4} = 19 \end{cases}$

48. $\begin{cases} \dfrac{1}{x^4} - \dfrac{1}{y^4} = 1 \\ \dfrac{1}{x^4} + \dfrac{1}{y^4} = 4 \end{cases}$

49. $\begin{cases} x^2 - 3xy + 2y^2 = 0 \\ x^2 + xy = 6 \end{cases}$

50. $\begin{cases} x^2 - xy - 2y^2 = 0 \\ xy + x + 6 = 0 \end{cases}$

51. $\begin{cases} y^2 + y + x^2 - x - 2 = 0 \\ y + 1 + \dfrac{x - 2}{y} = 0 \end{cases}$

52. $\begin{cases} x^3 - 2x^2 + y^2 + 3y - 4 = 0 \\ x - 2 + \dfrac{y^2 - y}{x^2} = 0 \end{cases}$

53. $\begin{cases} \log_x y = 3 \\ \log_x(4y) = 5 \end{cases}$

54. $\begin{cases} \log_x(2y) = 3 \\ \log_x(4y) = 2 \end{cases}$

55. $\begin{cases} \ln x = 4 \ln y \\ \log_3 x = 2 + 2 \log_3 y \end{cases}$

56. $\begin{cases} \ln x = 5 \ln y \\ \log_2 x = 3 + 2 \log_2 y \end{cases}$

In Problems 57–62, graph each equation and find the point(s) of intersection, if any.

57. The line $x + 2y = 0$ and
the circle $(x - 1)^2 + (y - 1)^2 = 5$

58. The line $x + 2y + 6 = 0$ and
the circle $(x + 1)^2 + (y + 1)^2 = 5$

59. The circle $(x - 1)^2 + (y + 2)^2 = 4$
and the parabola $y^2 + 4y - x + 1 = 0$

60. The circle $(x + 2)^2 + (y - 1)^2 = 4$
and the parabola $y^2 - 2y - x - 5 = 0$

61. The graph of $y = \dfrac{4}{x - 3}$ and
the circle $x^2 - 6x + y^2 + 1 = 0$

62. The graph of $y = \dfrac{4}{x + 2}$ and
the circle $x^2 + 4x + y^2 - 4 = 0$

In Problems 63–70, use a graphing utility to solve each system of equations. Express the solution(s) rounded to two decimal places.

63. $\begin{cases} y = x^{2/3} \\ y = e^{-x} \end{cases}$

64. $\begin{cases} y = x^{3/2} \\ y = e^{-x} \end{cases}$

65. $\begin{cases} x^2 + y^3 = 2 \\ x^3 y = 4 \end{cases}$

66. $\begin{cases} x^3 + y^2 = 2 \\ x^2 y = 4 \end{cases}$

67. $\begin{cases} x^4 + y^4 = 12 \\ xy^2 = 2 \end{cases}$

68. $\begin{cases} x^4 + y^4 = 6 \\ xy = 1 \end{cases}$

69. $\begin{cases} xy = 2 \\ y = \ln x \end{cases}$

70. $\begin{cases} x^2 + y^2 = 4 \\ y = \ln x \end{cases}$

In Problems 71–78, graph each system of inequalities by hand.

71. $\begin{cases} x^2 + y^2 \le 9 \\ x + y \ge 3 \end{cases}$

72. $\begin{cases} x^2 + y^2 \ge 9 \\ x + y \le 3 \end{cases}$

73. $\begin{cases} y \ge x^2 - 4 \\ y \le x - 2 \end{cases}$

74. $\begin{cases} y^2 \le x \\ y \ge x \end{cases}$

75. $\begin{cases} x^2 + y^2 \le 16 \\ y \ge x^2 - 4 \end{cases}$

76. $\begin{cases} x^2 + y^2 \le 25 \\ y \le x^2 - 5 \end{cases}$

77. $\begin{cases} xy \ge 4 \\ y \ge x^2 + 1 \end{cases}$

78. $\begin{cases} y + x^2 \le 1 \\ y \ge x^2 - 1 \end{cases}$

Applications and Extensions

79. The difference of two numbers is 2 and the sum of their squares is 10. Find the numbers.

80. The sum of two numbers is 7 and the difference of their squares is 21. Find the numbers.

81. The product of two numbers is 4 and the sum of their squares is 8. Find the numbers.

82. The product of two numbers is 10 and the difference of their squares is 21. Find the numbers.

83. The difference of two numbers is the same as their product, and the sum of their reciprocals is 5. Find the numbers.

84. The sum of two numbers is the same as their product, and the difference of their reciprocals is 3. Find the numbers.

85. The ratio of a to b is $\dfrac{2}{3}$. The sum of a and b is 10. What is the ratio of $a + b$ to $b - a$?

86. The ratio of a to b is $4:3$. The sum of a and b is 14. What is the ratio of $a - b$ to $a + b$?

87. Geometry The perimeter of a rectangle is 16 inches and its area is 15 square inches. What are its dimensions?

88. Geometry An area of 52 square feet is to be enclosed by two squares whose sides are in the ratio of $2:3$. Find the sides of the squares.

89. Geometry Two circles have circumferences that add up to 12π centimeters and areas that add up to 20π square centimeters. Find the radius of each circle.

90. Geometry The altitude of an isosceles triangle drawn to its base is 3 centimeters, and its perimeter is 18 centimeters. Find the length of its base.

91. The Tortoise and the Hare In a 21-meter race between a tortoise and a hare, the tortoise leaves 9 minutes before the hare. The hare, by running at an average speed of 0.5 meter per hour faster than the tortoise, crosses the finish line 3 minutes before the tortoise. What are the average speeds of the tortoise and the hare?

92. Running a Race In a 1-mile race, the winner crosses the finish line 10 feet ahead of the second-place runner and 20 feet ahead of the third-place runner. Assuming that each runner maintains a constant speed throughout the race, by how many feet does the second-place runner beat the third-place runner?

93. Constructing a Box A rectangular piece of cardboard, whose area is 216 square centimeters, is made into an open box by cutting a 2-centimeter square from each corner and turning up the sides. See the figure. If the box is to have a volume of 224 cubic centimeters, what size cardboard should you start with?

94. Constructing a Cylindrical Tube A rectangular piece of cardboard, whose area is 216 square centimeters, is made into a cylindrical tube by joining together two sides of the rectangle. See the figure. If the tube is to have a volume of 224 cubic centimeters, what size cardboard should you start with?

95. Fencing A farmer has 300 feet of fence available to enclose a 4500-square-foot region in the shape of adjoining squares, with sides of length x and y. See the figure. Find x and y.

96. Bending Wire A wire 60 feet long is cut into two pieces. Is it possible to bend one piece into the shape of a square and the other into the shape of a circle so that the total area enclosed by the two pieces is 100 square feet? If this is possible, find the length of the side of the square and the radius of the circle.

97. Geometry Find formulas for the length l and width w of a rectangle in terms of its area A and perimeter P.

98. Geometry Find formulas for the base b and one of the equal sides l of an isosceles triangle in terms of its altitude h and perimeter P.

99. Descartes's Method of Equal Roots Descartes's method for finding tangents depends on the idea that, for many graphs, the tangent line at a given point is the *unique* line that intersects the graph at that point only. We will apply his method to find an equation of the tangent line to the parabola $y = x^2$ at the point $(2, 4)$. See the figure.

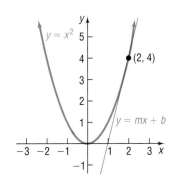

First, we know that the equation of the tangent line must be in the form $y = mx + b$. Using the fact that the point $(2, 4)$ is on the line, we can solve for b in terms of m and get the equation $y = mx + (4 - 2m)$. Now we want $(2, 4)$ to be the *unique* solution to the system

$$\begin{cases} y = x^2 \\ y = mx + 4 - 2m \end{cases}$$

From this system, we get $x^2 - mx + (2m - 4) = 0$. By using the quadratic formula, we get

$$x = \frac{m \pm \sqrt{m^2 - 4(2m - 4)}}{2}$$

To obtain a unique solution for x, the two roots must be equal; in other words, the discriminant $m^2 - 4(2m - 4)$ must be 0. Complete the work to get m, and write an equation of the tangent line.

In Problems 100–106, use Descartes's method from Problem 99 to find the equation of the line tangent to each graph at the given point.

100. $x^2 + y^2 = 10$; at $(1, 3)$

101. $y = x^2 + 2$; at $(1, 3)$

102. $x^2 + y = 5$; at $(-2, 1)$

103. $2x^2 + 3y^2 = 14$; at $(1, 2)$

104. $3x^2 + y^2 = 7$; at $(-1, 2)$

105. $x^2 - y^2 = 3$; at $(2, 1)$

106. $2y^2 - x^2 = 14$; at $(2, 3)$

---

107. If r_1 and r_2 are two solutions of a quadratic equation $ax^2 + bx + c = 0$, then it can be shown that

$$r_1 + r_2 = -\frac{b}{a} \quad \text{and} \quad r_1 r_2 = \frac{c}{a}$$

Solve this system of equations for r_1 and r_2.

Discussion and Writing

108. A circle and a line intersect at most twice. A circle and a parabola intersect at most four times. Deduce that a circle and the graph of a polynomial of degree 3 intersect at most six times. What do you conjecture about a polynomial of degree 4? What about a polynomial of degree n? Can you explain your conclusions using an algebraic argument?

109. Suppose that you are the manager of a sheet metal shop. A customer asks you to manufacture 10,000 boxes, each box being open on top. The boxes are required to have a square base and a 9-cubic-foot capacity. You construct the boxes by cutting out a square from each corner of a square piece of sheet metal and folding along the edges.

(a) What are the dimensions of the square to be cut if the area of the square piece of sheet metal is 100 square feet?

(b) Could you make the box using a smaller piece of sheet metal? Make a list of the dimensions of the box for various pieces of sheet metal.

'Are You Prepared?' Answers

1.

2

3.

4.

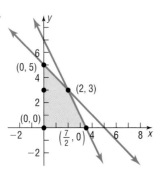

5. $(4, 5)$

6. $(-4, 3)$

Chapter Review

Things to Know

Equations

Parabola	See Tables 1 and 2 (pp. 600 and 602).
Ellipse	See Table 3 (p. 613).
Hyperbola	See Table 4 (p. 627).

Definitions

Parabola (p. 597)	Set of points P in the plane for which $d(F, P) = d(P, D)$, where F is the focus and D is the directrix
Ellipse (p. 608)	Set of points P in the plane, the sum of whose distances from two fixed points (the foci) is a constant
Hyperbola (p. 619)	Set of points P in the plane, the difference of whose distances from two fixed points (the foci) is a constant

Objectives

Section		You should be able to . . .	Review Exercises
6.1	1	Know the names of the conics (p. 596)	1–36
6.2	1	Work with parabolas with vertex at the origin (p. 598)	1, 2, 21, 24
	2	Work with parabolas with vertex at (h, k) (p. 602)	7, 11, 12, 17, 18, 27, 30
	3	Solve applied problems involving parabolas (p. 604)	55, 56
6.3	1	Work with ellipses with center at the origin (p. 608)	5, 6, 10, 22, 25
	2	Work with ellipses with center at (h, k) (p. 613)	14–16, 19, 28, 31
	3	Solve applied problems involving ellipses (p. 615)	57, 58
6.4	1	Work with hyperbolas with center at the origin (p. 620)	3, 4, 8, 9, 23, 26
	2	Find the asymptotes of a hyperbola (p. 625)	3, 4, 8, 9
	3	Work with hyperbolas with center at (h, k) (p. 627)	13, 20, 29, 32–36
	4	Solve applied problems involving hyperbolas (p. 629)	59
6.5	1	Solve a system of nonlinear equations using substitution (p. 633)	37–46
	2	Solve a system of nonlinear equations using elimination (p. 635)	37–46
	3	Graph a system of nonlinear inequalities (p. 639)	47–50

Review Exercises

In Problems 1–20, identify each equation. If it is a parabola, gives its vertex, focus, and directrix; if it is an ellipse, give its center, vertices, and foci; if it is a hyperbola, give its center, vertices, foci, and asymptotes.

1. $y^2 = -16x$

2. $16x^2 = y$

3. $\dfrac{x^2}{25} - y^2 = 1$

4. $\dfrac{y^2}{25} - x^2 = 1$

5. $\dfrac{y^2}{25} + \dfrac{x^2}{16} = 1$

6. $\dfrac{x^2}{9} + \dfrac{y^2}{16} = 1$

7. $x^2 + 4y = 4$

8. $3y^2 - x^2 = 9$

9. $4x^2 - y^2 = 8$

10. $9x^2 + 4y^2 = 36$

11. $x^2 - 4x = 2y$

12. $2y^2 - 4y = x - 2$

13. $y^2 - 4y - 4x^2 + 8x = 4$

14. $4x^2 + y^2 + 8x - 4y + 4 = 0$

15. $4x^2 + 9y^2 - 16x - 18y = 11$

16. $4x^2 + 9y^2 - 16x + 18y = 11$

17. $4x^2 - 16x + 16y + 32 = 0$

18. $4y^2 + 3x - 16y + 19 = 0$

19. $9x^2 + 4y^2 - 18x + 8y = 23$

20. $x^2 - y^2 - 2x - 2y = 1$

In Problems 21–36, obtain an equation of the conic described. Graph the equation by hand.

21. Parabola; focus at $(-2, 0)$; directrix the line $x = 2$

22. Ellipse; center at $(0, 0)$; focus at $(0, 3)$; vertex at $(0, 5)$

23. Hyperbola; center at $(0, 0)$; focus at $(0, 4)$; vertex at $(0, -2)$

24. Parabola; vertex at $(0, 0)$; directrix the line $y = -3$

25. Ellipse; foci at $(-3, 0)$ and $(3, 0)$; vertex at $(4, 0)$

26. Hyperbola; vertices at $(-2, 0)$ and $(2, 0)$; focus at $(4, 0)$

27. Parabola; vertex at $(2, -3)$; focus at $(2, -4)$

28. Ellipse; center at $(-1, 2)$; focus at $(0, 2)$; vertex at $(2, 2)$

29. Hyperbola; center at $(-2, -3)$; focus at $(-4, -3)$; vertex at $(-3, -3)$

30. Parabola; focus at $(3, 6)$; directrix the line $y = 8$

31. Ellipse; foci at $(-4, 2)$ and $(-4, 8)$; vertex at $(-4, 10)$

32. Hyperbola; vertices at $(-3, 3)$ and $(5, 3)$; focus at $(7, 3)$

33. Center at $(-1, 2)$; $a = 3$; $c = 4$; transverse axis parallel to the x-axis

34. Center at $(4, -2)$; $a = 1$; $c = 4$; transverse axis parallel to the y-axis

35. Vertices at $(0, 1)$ and $(6, 1)$; asymptote the line $3y + 2x = 9$

36. Vertices at $(4, 0)$ and $(4, 4)$; asymptote the line $y + 2x = 10$

In Problems 37–46, solve each system of equations algebraically. Verify your result using a graphing utility.

37. $\begin{cases} 2x + y + 3 = 0 \\ x^2 + y^2 = 5 \end{cases}$

38. $\begin{cases} x^2 + y^2 = 16 \\ 2x - y^2 = -8 \end{cases}$

39. $\begin{cases} 2xy + y^2 = 10 \\ 3y^2 - xy = 2 \end{cases}$

40. $\begin{cases} 3x^2 - y^2 = 1 \\ 7x^2 - 2y^2 - 5 = 0 \end{cases}$

41. $\begin{cases} x^2 + y^2 = 6y \\ x^2 = 3y \end{cases}$

42. $\begin{cases} 2x^2 + y^2 = 9 \\ x^2 + y^2 = 9 \end{cases}$

43. $\begin{cases} 3x^2 + 4xy + 5y^2 = 8 \\ x^2 + 3xy + 2y^2 = 0 \end{cases}$

44. $\begin{cases} 3x^2 + 2xy - 2y^2 = 6 \\ xy - 2y^2 + 4 = 0 \end{cases}$

45. $\begin{cases} x^2 - 3x + y^2 + y = -2 \\ \dfrac{x^2 - x}{y} + y + 1 = 0 \end{cases}$

46. $\begin{cases} x^2 + x + y^2 = y + 2 \\ x + 1 = \dfrac{2 - y}{x} \end{cases}$

In Problems 47–50, graph each system of inequalities.

47. $\begin{cases} x^2 + y^2 \le 16 \\ x + y \ge 2 \end{cases}$

48. $\begin{cases} y^2 \le x - 1 \\ x - y \le 3 \end{cases}$

49. $\begin{cases} y \le x^2 \\ xy \le 4 \end{cases}$

50. $\begin{cases} x^2 + y^2 \ge 1 \\ x^2 + y^2 \le 4 \end{cases}$

51. Find an equation of the hyperbola whose foci are the vertices of the ellipse $4x^2 + 9y^2 = 36$ and whose vertices are the foci of this ellipse.

52. Find an equation of the ellipse whose foci are the vertices of the hyperbola $x^2 - 4y^2 = 16$ and whose vertices are the foci of this hyperbola.

53. Describe the collection of points in a plane so that the distance from each point to the point $(3, 0)$ is three-fourths of its distance from the line $x = \dfrac{16}{3}$.

54. Describe the collection of points in a plane so that the distance from each point to the point $(5, 0)$ is five-fourths of its distance from the line $x = \dfrac{16}{5}$.

55. Mirrors A mirror is shaped like a paraboloid of revolution. If a light source is located 1 foot from the base along the axis of symmetry and the opening is 2 feet across, how deep should the mirror be?

56. Parabolic Arch Bridge A bridge is built in the shape of a parabolic arch. The bridge has a span of 60 feet and a maximum height of 20 feet. Find the height of the arch at distances of 5, 10, and 20 feet from the center.

57. Semi-elliptical Arch Bridge A bridge is built in the shape of a semi-elliptical arch. The bridge has a span of 60 feet and

a maximum height of 20 feet. Find the height of the arch at distances of 5, 10, and 20 feet from the center.

58. Whispering Galleries The figure shows the specifications for an elliptical ceiling in a hall designed to be a whispering gallery. Where are the foci located in the hall?

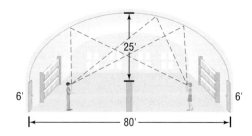

59. Calibrating Instruments In a test of their recording devices, a team of seismologists positioned two of the devices 2000 feet apart, with the device at point A to the west of the device at point B. At a point between the devices and 200 feet from point B, a small amount of explosive was detonated and a note made of the time at which the sound reached each device. A second explosion is to be carried out at a point directly north of point B. How far north should the site of the second explosion be chosen so that the measured time difference recorded by the devices for the second detonation is the same as that recorded for the first detonation?

Chapter Test

In Problems 1–3, identify each equation. If it is a parabola, give its vertex, focus, and directrix; if an ellipse, give its center, vertices, and foci; if a hyperbola, give its center, vertices, foci, and asymptotes.

1. $\dfrac{(x + 1)^2}{4} - \dfrac{y^2}{9} = 1$

2. $8y = (x - 1)^2 - 4$

3. $2x^2 + 3y^2 + 4x - 6y = 13$

In Problems 4–6, obtain an equation of the conic described; graph the equation by hand.

4. parabola: focus $(-1, 4.5)$, vertex $(-1, 3)$

5. ellipse: center $(0, 0)$, vertex $(0, -4)$, focus $(0, 3)$

6. hyperbola: center $(2, 2)$, vertex $(2, 4)$, contains the point $\left(2 + \sqrt{10}, 5\right)$

In Problems 7–8, solve each system of equations algebraically.

7. $\begin{cases} 3x^2 + y^2 = 12 \\ \quad\ \ y^2 = 9x \end{cases}$

8. $\begin{cases} 2y^2 - 3x^2 = 5 \\ \ \ y \ - \ x = 1 \end{cases}$

9. Graph the system of inequalities:

$$\begin{cases} x^2 + \ y^2 \le 100 \\ 4x \ - 3y \ \ge 0 \end{cases}$$

10. A parabolic reflector (paraboloid of revolution) is used by TV crews at football games to pick up the referee's announcements, quarterback signals, and so on. A microphone is placed at the focus of the parabola. If a certain reflector is 4 ft. wide and 1.5 ft deep, where should the microphone be placed?

Chapter Projects

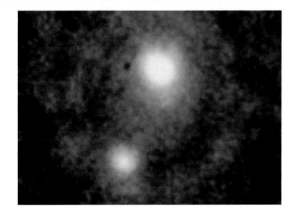

1. **The Orbits of Neptune and Pluto**   The orbit of a planet about the Sun is an ellipse, with the Sun at one focus. The **aphelion** of a planet is its greatest distance from the Sun and the **perihelion** is its shortest distance. The **mean distance** of a planet from the Sun is the length of the semi-major axis of the elliptical orbit. See the illustration.

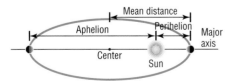

(a) The aphelion of Neptune is 4532.2×10^6 km and its perihelion is 4458.0×10^6 km. Find the equation for the orbit of Neptune around the Sun.

(b) The aphelion of Pluto is 7381.2×10^6 km and its perihelion is 4445.8×10^6 km. Find the equation for the orbit of Pluto around the Sun.

(c) Graph the orbits of Pluto and Neptune on a graphing utility. Knowing that the orbits of the planets intersect, what is wrong with the graphs you obtained?

(d) The graphs of the orbits drawn in part (c) have the same center, so their foci lie in different locations. To see an accurate representation, the location of the Sun (a focus) needs to be the same for both graphs. This can be accomplished by shifting Pluto's orbit to the left. The shift amount is equal to Pluto's distance from the center [in the graph in part (c)] to the Sun minus Neptune's distance from the center to the Sun. Find the new equation representing the orbit of Pluto.

(e) Graph the equation for the orbit of Pluto found in part (d) along with the equation of the orbit of Neptune. Do you see that Pluto's orbit is sometimes inside Neptune's?

(f) Find the point(s) of intersection of the two orbits.

(g) Do you think two planets ever collide?

The following project can be found on the Instructor's Resource Center (IRC):

2. **Constructing a Bridge Over the East River**

Cumulative Review

1. For $f(x) = -3x^2 + 5x - 2$, find
$$\frac{f(x + h) - f(x)}{h}, \qquad h \neq 0$$

2. (a) Find the domain and range of $y = 3^x + 2$.
   (b) Find the inverse of $y = 3^x + 2$ and state its domain and range.

3. Solve the equation $9x^4 + 33x^3 - 71x^2 - 57x - 10 = 0$.

4. For what numbers x is $6 - x \geq x^2$?

5. Find an equation for each of the following graphs:

   (a) Line:

   (b) Circle:

   (c) Ellipse:

   (d) Parabola:

   (e) Hyperbola:

   (f) Exponential:

6. If $f(x) = \log_4(x - 2)$:
   (a) Solve $f(x) = 2$.
   (b) Solve $f(x) \leq 2$.

Sequences; Induction; the Binomial Theorem

A LOOK BACK, A LOOK AHEAD This chapter may be divided into three independent parts: Sections 7.1–7.3, Section 7.4, and Section 7.5.

In Chapter 2, we defined a function and its domain, which was usually some set of real numbers. In Sections 7.1–7.3, we discuss sequences, which are functions whose domain is the set of positive integers.

Throughout this text, where it seemed appropriate, we have given proofs of many of the results. In Section 7.4, a technique for proving theorems involving natural numbers is discussed.

In Chapter R, we gave formulas for expanding $(x + a)^2$ and $(x + a)^3$. In Section 7.5, we discuss the Binomial Theorem, a formula for the expansion of $(x + a)^n$, where n is a positive integer.

The topics introduced in this chapter are covered in more detail in courses titled *Discrete Mathematics*. Applications of these topics can be found in the fields of computer science, engineering, business and economics, the social sciences, and the physical and biological sciences.

The Future of the World Population

WASHINGTON—World population is projected to increase 46 percent by 2050, with most of the growth occurring in the less industrialized areas of the globe.

Projected growth in population by continent, 2003–2050: North America 41.8%; Latin America and the Caribbean 46.2%; Oceania 55.6%; Europe −8.8%; Asia 39.8%; Africa 118.8%.

Note: The United Nations classifies the countries of Latin America and the Caribbean, Asia, Oceania, and Africa as less industrialized with the exception of Australia, New Zealand, and Japan.

Africa's population could soar by more than 1 billion over the next half-century, further straining food and water supplies and social services in areas already struggling, according to a new report.

The latest edition of the "World Population Data Sheet" estimates the global population will rise 46 percent between now and 2050 to about 9 billion, a level also predicted by the United Nations and other groups.

European nations, more industrialized and prosperous, are expected to lose population because of falling birth rates and low immigration.

The U.S. population is expected to grow 45 percent to 422 million in 2050, paced by a stable birth rate and high levels of immigration.

But most of the world's growth will be in developing nations. India's population is estimated to grow 52 percent to 1.6 billion by 2050, when it will surpass China as the world's largest country.

Africa is predicted to more than double in population to 1.9 billion by midcentury.

SOURCE: *The Houston Chronicle* (Houston, TX), July 23, 2003, p. 12.

—*See Chapter Project 1.*

OUTLINE

7.1 Sequences

PREPARING FOR THIS SECTION *Before getting started, review the following concept:*

• Functions (Section 2.2, pp. 204–213) • Compound Interest (Section 4.7, pp. 463–470)

Now work the 'Are You Prepared?' problems on page 659

OBJECTIVES 1 Write the First Several Terms of a Sequence
2 Write the Terms of a Sequence Defined by a Recursive Formula
3 Use Summation Notation
4 Find the Sum of a Sequence Algebraically and Using a Graphing Utility
5 Solve Annuity and Amortization Problems

A **sequence** is a function whose domain is the set of positive integers.

Because a sequence is a function, it will have a graph. In Figure 1(a), we have the graph of the function $f(x) = \dfrac{1}{x}$, $x > 0$. If all the points on this graph were removed except those whose x-coordinates are positive integers, that is, if all points were removed except $(1, 1)$, $\left(2, \dfrac{1}{2}\right)$, $\left(3, \dfrac{1}{3}\right)$, and so on, the remaining points would be the graph of the sequence $f(n) = \dfrac{1}{n}$, as shown in Figure 1(b). Notice that we use n to represent the independent variable in a sequence. This serves to remind us that n is a positive integer.

Figure 1

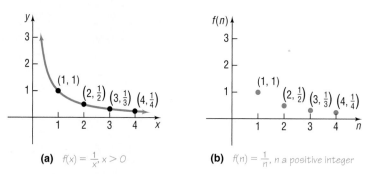

(a) $f(x) = \frac{1}{x}, x > 0$ **(b)** $f(n) = \frac{1}{n}, n$ a positive integer

1 Write the First Several Terms of a Sequence

A sequence is usually represented by listing its values in order. For example, the sequence whose graph is given in Figure 1(b) might be represented as

$$f(1), f(2), f(3), f(4), \ldots \quad \text{or} \quad 1, \frac{1}{2}, \frac{1}{3}, \frac{1}{4}, \ldots$$

The list never ends, as the ellipsis indicates. The numbers in this ordered list are called the **terms** of the sequence.

In dealing with sequences, we usually use subscripted letters, such as a_1, to represent the first term, a_2 for the second term, a_3 for the third term, and so on. For the sequence $f(n) = \dfrac{1}{n}$, we write

$$a_1 = f(1) = 1, \quad a_2 = f(2) = \frac{1}{2}, \quad a_3 = f(3) = \frac{1}{3}, \quad a_4 = f(4) = \frac{1}{4}, \ldots, \quad a_n = f(n) = \frac{1}{n}, \ldots$$

In other words, we usually do not use the traditional function notation $f(n)$ for sequences. For this particular sequence, we have a rule for the nth term, which is $a_n = \dfrac{1}{n}$, so it is easy to find any term of the sequence.

When a formula for the nth term (sometimes called the **general term**) of a sequence is known, rather than write out the terms of the sequence, we usually represent the entire sequence by placing braces around the formula for the nth term.

For example, the sequence whose nth term is $b_n = \left(\dfrac{1}{2}\right)^n$ may be represented as

$$\{b_n\} = \left\{\left(\frac{1}{2}\right)^n\right\}$$

or by

$$b_1 = \frac{1}{2}, \quad b_2 = \frac{1}{4}, \quad b_3 = \frac{1}{8}, \ldots, \quad b_n = \left(\frac{1}{2}\right)^n, \ldots$$

EXAMPLE 1 **Writing the First Several Terms of a Sequence**

Write down the first six terms of the following sequence and graph it.

$$\{a_n\} = \left\{\frac{n-1}{n}\right\}$$

Algebraic Solution

The first six terms of the sequence are

$$a_1 = 0, \quad a_2 = \frac{1}{2}, \quad a_3 = \frac{2}{3},$$

$$a_4 = \frac{3}{4}, \quad a_5 = \frac{4}{5}, \quad a_6 = \frac{5}{6}$$

See Figure 2 for the graph.

Figure 2

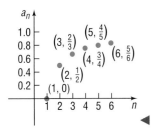

Graphing Solution

Figure 3 shows the sequence generated on a TI-84 Plus graphing calculator. We can see the first few terms of the sequence on the screen. You need to press the right arrow key to scroll right to see the remaining terms of the sequence.

We could also obtain the terms of the sequence using the TABLE feature. First, put the graphing utility in SEQuence mode. Using $Y=$, enter the formula for the sequence into the graphing utility. See Figure 4. Set up the table with TblStart $= 1$ and ΔTbl $= 1$. See Table 1. Finally, we can graph the sequence. See Figure 5. Notice that the first term of the sequence is not visible since it lies on the x-axis. TRACEing the graph will allow you to determine the terms of the sequence.

Figure 3

Figure 4

Plot1 Plot2 Plot3
nMin=1
\u$(n)$⊟$(n-1)/n$
u$(n$Min$)$⊟$\{0\}$
\v$(n)=$
v$(n$Min$)=$
\w$(n)=$
w$(n$Min$)=$

Table 1

n	u(n)
1	0
2	.5
3	.66667
4	.75
5	.8
6	.83333
7	.85714

u(n)⊟$(n-1)/n$

Figure 5

NOW WORK PROBLEM **19**.

We will usually provide solutions done by hand. The reader is encouraged to verify solutions using a graphing utility.

Figure 6

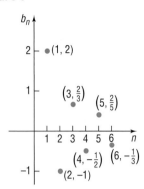

| EXAMPLE 2 | **Writing the First Several Terms of a Sequence** |

Write down the first six terms of the following sequence and graph it.

$$\{b_n\} = \left\{(-1)^{n+1}\left(\frac{2}{n}\right)\right\}$$

Solution The first six terms of the sequence are

$$b_1 = 2, \quad b_2 = -1, \quad b_3 = \frac{2}{3}, \quad b_4 = -\frac{1}{2}, \quad b_5 = \frac{2}{5}, \quad b_6 = -\frac{1}{3}$$

See Figure 6 for the graph. ◀

Notice in the sequence $\{b_n\}$ in Example 2 that the signs of the terms **alternate**. When this occurs, we use factors such as $(-1)^{n+1}$, which equals 1 if n is odd and -1 if n is even, or $(-1)^n$, which equals -1 if n is odd and 1 if n is even.

Figure 7

| EXAMPLE 3 | **Writing the First Several Terms of a Sequence** |

Write down the first six terms of the following sequence and graph it.

$$\{c_n\} = \left\{\begin{array}{ll} n & \text{if } n \text{ is even} \\ \dfrac{1}{n} & \text{if } n \text{ is odd} \end{array}\right\}$$

Solution The first six terms of the sequence are

$$c_1 = 1, \quad c_2 = 2, \quad c_3 = \frac{1}{3}, \quad c_4 = 4, \quad c_5 = \frac{1}{5}, \quad c_6 = 6$$

See Figure 7 for the graph. ◀

NOW WORK PROBLEM **21.**

Sometimes a sequence is indicated by an observed pattern in the first few terms that makes it possible to infer the makeup of the nth term. In the example that follows, a sufficient number of terms of the sequence is given so that a natural choice for the nth term is suggested.

| EXAMPLE 4 | **Determining a Sequence from a Pattern** |

(a) $e, \dfrac{e^2}{2}, \dfrac{e^3}{3}, \dfrac{e^4}{4}, \ldots \qquad a_n = \dfrac{e^n}{n}$

(b) $1, \dfrac{1}{3}, \dfrac{1}{9}, \dfrac{1}{27}, \ldots \qquad b_n = \dfrac{1}{3^{n-1}}$

(c) $1, 3, 5, 7, \ldots \qquad c_n = 2n - 1$

(d) $1, 4, 9, 16, 25, \ldots \qquad d_n = n^2$

(e) $1, -\dfrac{1}{2}, \dfrac{1}{3}, -\dfrac{1}{4}, \dfrac{1}{5}, \ldots \qquad e_n = (-1)^{n+1}\left(\dfrac{1}{n}\right)$

◀

NOW WORK PROBLEM **29.**

The Factorial Symbol

If $n \geq 0$ is an integer, the **factorial symbol** $n!$ is defined as follows:

$$0! = 1 \qquad 1! = 1$$
$$n! = n(n-1) \cdot \ldots \cdot 3 \cdot 2 \cdot 1 \qquad \text{if } n \geq 2$$

For example, $2! = 2 \cdot 1 = 2$, $3! = 3 \cdot 2 \cdot 1 = 6$, $4! = 4 \cdot 3 \cdot 2 \cdot 1 = 24$, and so on. Table 2 lists the values of $n!$ for $0 \leq n \leq 6$.

Table 2

n	0	1	2	3	4	5	6
$n!$	1	1	2	6	24	120	720

Because

$$n! = \underbrace{n(n-1)(n-2) \cdot \ldots \cdot 3 \cdot 2 \cdot 1}_{(n-1)!}$$

we can use the formula

$$n! = n(n-1)!$$

to find successive factorials. For example, because $6! = 720$, we have

$$7! = 7 \cdot 6! = 7(720) = 5040$$

and

$$8! = 8 \cdot 7! = 8(5040) = 40{,}320$$

NOTE
Your calculator has a factorial key. Use it to see how fast factorials increase in value. Find the value of 69!. What happens when you try to find 70!? In fact, 70! is larger than 10^{100} (a **googol**), the largest number most calculators can display.   ∎

2 Write the Terms of a Sequence Defined by a Recursive Formula

A second way of defining a sequence is to assign a value to the first (or the first few) term(s) and specify the nth term by a formula or equation that involves one or more of the terms preceding it. Sequences defined this way are said to be defined **recursively**, and the rule or formula is called a **recursive formula**.

EXAMPLE 5   **Writing the Terms of a Recursively Defined Sequence**

Write down the first five terms of the following recursively defined sequence.

$$s_1 = 1, \qquad s_n = n s_{n-1}$$

Algebraic Solution

The first term is given as $s_1 = 1$. To get the second term, we use $n = 2$ in the formula $s_n = n s_{n-1}$ to get $s_2 = 2s_1 = 2 \cdot 1 = 2$. To get the third term, we use $n = 3$ in the formula to get $s_3 = 3s_2 = 3 \cdot 2 = 6$. To get a new term requires that we know the value of the preceding term. The first five terms are

$$s_1 = 1$$
$$s_2 = 2 \cdot 1 = 2$$
$$s_3 = 3 \cdot 2 = 6$$
$$s_4 = 4 \cdot 6 = 24$$
$$s_5 = 5 \cdot 24 = 120$$

Do you recognize this sequence?
$s_n = n!$ ◀

Graphing Solution

First, put the graphing utility into SEQuence mode. Using $Y =$, enter the recursive formula into the graphing utility. See Figure 8(a). Next, set up the viewing window to generate the desired sequence. Finally, graph the recursion relation and use TRACE to determine the terms in the sequence. See Figure 8(b). For example, we see that the fourth term of the sequence is 24. Table 3 also shows the terms of the sequence.

Figure 8

(a)

Table 3

(b)   ◀

EXAMPLE 6	**Writing the Terms of a Recursively Defined Sequence**

Write down the first five terms of the following recursively defined sequence.

$$u_1 = 1, \quad u_2 = 1, \quad u_{n+2} = u_n + u_{n+1}$$

Solution   We are given the first two terms. To get the third term requires that we know each of the previous two terms. That is,

$$u_1 = 1$$
$$u_2 = 1$$
$$u_3 = u_1 + u_2 = 1 + 1 = 2$$
$$u_4 = u_2 + u_3 = 1 + 2 = 3$$
$$u_5 = u_3 + u_4 = 2 + 3 = 5$$

◄

The sequence defined in Example 6 is called a **Fibonacci sequence**, and the terms of this sequence are called **Fibonacci numbers**. These numbers appear in a wide variety of applications (see Problems 91–94).

✎ **NOW WORK PROBLEMS 37 AND 45.**

3 Use Summation Notation

It is often important to be able to find the sum of the first n terms of a sequence $\{a_n\}$, that is,

$$a_1 + a_2 + a_3 + \cdots + a_n$$

Rather than write down all these terms, we introduce a more concise way to express the sum, called **summation notation**. Using summation notation, we would write the sum as

$$a_1 + a_2 + a_3 + \cdots + a_n = \sum_{k=1}^{n} a_k$$

The symbol Σ (the Greek letter sigma, which is an S in our alphabet) is simply an instruction to sum, or add up, the terms. The integer k is called the **index** of the sum; it tells you where to start the sum and where to end it. The expression

$$\sum_{k=1}^{n} a_k$$

is an instruction to add the terms a_k of the sequence $\{a_n\}$ starting with $k = 1$ and ending with $k = n$. We read the expression as "the sum of a_k from $k = 1$ to $k = n$."

EXAMPLE 7	**Expanding Summation Notation**

Write out each sum.

(a) $\displaystyle\sum_{k=1}^{n} \frac{1}{k}$ (b) $\displaystyle\sum_{k=1}^{n} k!$

Solution   (a) $\displaystyle\sum_{k=1}^{n} \frac{1}{k} = 1 + \frac{1}{2} + \frac{1}{3} + \cdots + \frac{1}{n}$ (b) $\displaystyle\sum_{k=1}^{n} k! = 1! + 2! + \cdots + n!$

◄

EXAMPLE 8	**Writing a Sum in Summation Notation**

Express each sum using summation notation.

(a) $1^2 + 2^2 + 3^2 + \cdots + 9^2$ (b) $1 + \dfrac{1}{2} + \dfrac{1}{4} + \dfrac{1}{8} + \cdots + \dfrac{1}{2^{n-1}}$

Solution   (a) The sum $1^2 + 2^2 + 3^2 + \cdots + 9^2$ has 9 terms, each of the form k^2, and starts at $k = 1$ and ends at $k = 9$:

$$1^2 + 2^2 + 3^2 + \cdots + 9^2 = \sum_{k=1}^{9} k^2$$

(b) The sum

$$1 + \frac{1}{2} + \frac{1}{4} + \frac{1}{8} + \cdots + \frac{1}{2^{n-1}}$$

has n terms, each of the form $\dfrac{1}{2^{k-1}}$, and starts at $k = 1$ and ends at $k = n$:

$$1 + \frac{1}{2} + \frac{1}{4} + \frac{1}{8} + \cdots + \frac{1}{2^{n-1}} = \sum_{k=1}^{n} \frac{1}{2^{k-1}} \qquad \blacktriangleleft$$

The index of summation need not always begin at 1 or end at n; for example, we could have expressed the sum in Example 8(b) as

$$\sum_{k=0}^{n-1} \frac{1}{2^k} = 1 + \frac{1}{2} + \frac{1}{4} + \cdots + \frac{1}{2^{n-1}}$$

Letters other than k may be used as the index. For example,

$$\sum_{j=1}^{n} j! \quad \text{and} \quad \sum_{i=1}^{n} i!$$

each represent the same sum as the one given in Example 7(b).

NOW WORK PROBLEMS 53 AND 63.

4   Find the Sum of a Sequence Algebraically and Using a Graphing Utility

Next we list some properties of sequences using summation notation. These properties are useful for adding the terms of a sequence algebraically.

Theorem   **Properties of Sequences**

If $\{a_n\}$ and $\{b_n\}$ are two sequences and c is a real number, then:

$$\sum_{k=1}^{n} c = \underbrace{c + c + \cdots + c}_{n \text{ terms}} = cn \tag{1}$$

$$\sum_{k=1}^{n} (ca_k) = ca_1 + ca_2 + \cdots + ca_n = c(a_1 + a_2 + \cdots + a_n) = c\sum_{k=1}^{n} a_k \tag{2}$$

$$\sum_{k=1}^{n} (a_k + b_k) = \sum_{k=1}^{n} a_k + \sum_{k=1}^{n} b_k \tag{3}$$

$$\sum_{k=1}^{n} (a_k - b_k) = \sum_{k=1}^{n} a_k - \sum_{k=1}^{n} b_k \tag{4}$$

$$\sum_{k=1}^{n} a_k = \sum_{k=1}^{j} a_k + \sum_{k=j+1}^{n} a_k, \quad \text{where } 0 < j < n \tag{5}$$

$$\sum_{k=1}^{n} k = 1 + 2 + 3 + \cdots + n = \frac{n(n+1)}{2} \tag{6}$$

$$\sum_{k=1}^{n} k^2 = 1^2 + 2^2 + 3^2 + \cdots + n^2 = \frac{n(n+1)(2n+1)}{6} \tag{7}$$

$$\sum_{k=1}^{n} k^3 = 1^3 + 2^3 + 3^3 + \cdots + n^3 = \left[\frac{n(n+1)}{2}\right]^2 \tag{8}$$

We shall not prove these properties. The proofs of (1) through (5) are based on properties of real numbers; the proofs of (7) and (8) require mathematical induction, which is discussed in Section 7.4. See Problem 97 for a derivation of (6).

EXAMPLE 9	**Finding the Sum of a Sequence**

Find the sum of each sequence.

(a) $\sum_{k=1}^{5} (3k)$ (b) $\sum_{k=1}^{3} (k^3 + 1)$ (c) $\sum_{k=1}^{4} (k^2 - 7k + 2)$

Algebraic Solution

(a) $\sum_{k=1}^{5} (3k) = 3 \sum_{k=1}^{5} k$ Property (2)

$= 3 \left(\frac{5(5 + 1)}{2} \right)$ Property (6)

$= 3(15)$

$= 45$

(b) $\sum_{k=1}^{3} (k^3 + 1) = \sum_{k=1}^{3} k^3 + \sum_{k=1}^{3} 1$ Property (3)

$= \left(\frac{3(3 + 1)}{2} \right)^2 + 1(3)$ Properties (1) and (8)

$= 36 + 3$

$= 39$

(c) $\sum_{k=1}^{4} (k^2 - 7k + 2) = \sum_{k=1}^{4} k^2 - \sum_{k=1}^{4} (7k) + \sum_{k=1}^{4} 2$ Properties (3), (4)

$= \sum_{k=1}^{4} k^2 - 7 \sum_{k=1}^{4} k + \sum_{k=1}^{4} 2$ Property (2)

$= \frac{4(4 + 1)(2 \cdot 4 + 1)}{6} - 7 \left(\frac{4(4 + 1)}{2} \right) + 2(4)$ Properties (1), (6), (7)

$= 30 - 70 + 8$

$= -32$ ◄

Graphing Solution

(a) Figure 9 shows the solution using a TI-84 Plus graphing calculator.

Figure 9

So, $\sum_{k=1}^{5} (3k) = 45$

(b) Figure 10 shows the solution using a TI-84 Plus graphing calculator.

Figure 10

So, $\sum_{k=1}^{3} (k^3 + 1) = 39$

(c) Figure 11 shows the solution using a TI-84 Plus graphing calculator.

Figure 11

So, $\sum_{k=1}^{4} (k^2 - 7k + 2) = -32$ ◄

━━━ **NOW WORK PROBLEM 75.**

5 Solve Annuity and Amortization Problems

In Section 4.7 we developed the compound interest formula, which gives the future value when a fixed amount of money is deposited in an account that pays interest compounded periodically. Often, though, money is invested in small amounts at periodic intervals. An **annuity** is a sequence of equal periodic deposits. The periodic deposits may be made annually, quarterly, monthly, or daily.

When deposits are made at the same time that the interest is credited, the annuity is called **ordinary**. We will only deal with ordinary annuities here. The **amount of an annuity** is the sum of all deposits made plus all interest paid.

Suppose that the initial amount deposited in an annuity is M, the periodic deposit is P, and the per annum rate of interest is $r\%$ (expressed as a decimal) compounded N times per year. The periodic deposit is made at the same time that the interest is credited, so N deposits are made per year. The amount A_n of the annuity after n deposits will equal A_{n-1}, the amount of the annuity after $n-1$ deposits, plus the interest earned on this amount, plus P, the periodic deposit. That is,

$$A_n = A_{n-1} + \frac{r}{N}A_{n-1} + P = \left(1 + \frac{r}{N}\right)A_{n-1} + P$$

Amount   Amount Interest   Periodic
after in previous earned   deposit
n deposits period

We have established the following result:

Theorem

Annuity Formula

If $A_0 = M$ represents the initial amount deposited in an annuity that earns $r\%$ per annum compounded N times per year, and if P is the periodic deposit made at each payment period, then the amount A_n of the annuity after n deposits is given by the recursive sequence

$$A_0 = M, \qquad A_n = \left(1 + \frac{r}{N}\right)A_{n-1} + P, \qquad n \geq 1 \qquad \textbf{(9)}$$

Formula (9) may be explained as follows: the money in the account initially, A_0, is M; the money in the account after $n-1$ payments, A_{n-1} earns interest $\dfrac{r}{N}$ during the nth period; so when the periodic payment of P dollars is added, the amount after n payments, A_n, is obtained.

EXAMPLE 10

Saving for Spring Break

A trip to Cancun during spring break will cost $450 and full payment is due March 2. To have the money, a student, on September 1, deposits $100 in a savings account that pays 4% per annum compounded monthly. On the first of each month, the student deposits $50 in this account.

(a) Find a recursive sequence that explains how much is in the account after n months.

(b) Use the TABLE feature to list the amounts of the annuity for the first 6 months.

(c) After the deposit on March 1 is made, is there enough in the account to pay for the Cancun trip?

(d) If the student deposits $60 each month, will there be enough for the trip after the March 1 deposit?

Solution

(a) The initial amount deposited in the account is $A_0 = \$100$. The monthly deposit is $P = \$50$, and the per annum rate of interest is $r = 0.04$ compounded $N = 12$

Table 4

n	$u(n)$
0	100
1	150.33
2	200.83
3	251.5
4	302.34
5	353.35
6	404.53

$u(n)\boxminus(1+.04/12)...$

Table 5

n	$u(n)$
0	100
1	160.33
2	220.87
3	281.6
4	342.54
5	403.68
6	465.03

$u(n)\boxminus(1+.04/12)...$

times per year. The amount A_n in the account after n monthly deposits is given by the recursive sequence

$$A_0 = 100, \qquad A_n = \left(1 + \frac{r}{N}\right)A_{n-1} + P = \left(1 + \frac{0.04}{12}\right)A_{n-1} + 50$$

(b) In SEQuence mode on a TI-84 Plus, enter the sequence $\{A_n\}$ and create Table 4. On September 1 ($n = 0$), there is $100 in the account. After the first payment on October 1, the value of the account is $150.33. After the second payment on November 1, the value of the account is $200.83. After the third payment on December 1, the value of the account is $251.50, and so on.

(c) On March 1 ($n = 6$), there is only $404.53, not enough to pay for the trip to Cancun.

(d) If the periodic deposit, P, is $60, then on March 1, there is $465.03 in the account, enough for the trip. See Table 5. ◀

Recursive sequences can also be used to compute information about loans. When equal periodic payments are made to pay off a loan, the loan is said to be **amortized**.

Theorem

Amortization Formula

If B is borrowed at an interest rate of $r\%$ (expressed as a decimal) per annum compounded monthly, the balance A_n due after n monthly payments of P is given by the recursive sequence

$$A_0 = B, \qquad A_n = \left(1 + \frac{r}{12}\right)A_{n-1} - P, \qquad n \geq 1 \qquad \textbf{(10)}$$

Formula (10) may be explained as follows: The initial loan balance is B. The balance due A_n after n payments will equal the balance due previously, A_{n-1}, plus the interest charged on that amount, reduced by the periodic payment P.

EXAMPLE 11 Mortgage Payments

Table 6

n	$u(n)$
0	180000
1	179852
2	179704
3	179555
4	179405
5	179254
6	179102

$u(n)\boxminus(1+.07/12)...$

Table 7

n	$u(n)$
52	171067
53	170868
54	170667
55	170465
56	170262
57	170057
58	169852

$u(n)\boxminus(1+.07/12)...$

John and Wanda borrowed $180,000 at 7% per annum compounded monthly for 30 years to purchase a home. Their monthly payment is determined to be $1197.54.

(a) Find a recursive formula that represents their balance after each payment of $1197.54 has been made.

(b) Determine their balance after the first payment is made.

(c) When will their balance be below $170,000?

Solution

(a) We use formula (10) with $A_0 = 180,000$, $r = 0.07$, and $P = \$1197.54$. Then

$$A_0 = 180,000 \qquad A_n = \left(1 + \frac{0.07}{12}\right)A_{n-1} - 1197.54$$

(b) In SEQuence mode on a TI-84 Plus, enter the sequence $\{A_n\}$ and create Table 6. After the first payment is made, the balance is $A_1 = \$179,852$.

(c) Scroll down until the balance is below $170,000. See Table 7. After the fifty-eighth payment is made ($n = 58$), the balance is below $170,000. ◀

NOW WORK PROBLEM 83.

7.1 Assess Your Understanding

'Are You Prepared?'

Answers are given at the end of these exercises. If you get a wrong answer, read the pages listed in red.

1. For the function $f(x) = \dfrac{x-1}{x}$, find $f(2)$ and $f(3)$.
(pp. 209–211)

2. *True or False:* A function is a relation between two sets D and R so that each element x in the first set D is related to exactly one element y in the second set R. (pp. 204–209)

3. If $1000 is invested at 4% per annum compounded semi-annually, how much is in the account after 2 years? (pp. 463–467)

4. How much do you need to invest now at 5% per annum compounded monthly so that in 1 year you will have $10,000? (pp. 463–467)

Concepts and Vocabulary

5. A(n) _____ is a function whose domain is the set of positive integers.

6. For the sequence $\{s_n\} = \{4n - 1\}$, the first term is $s_1 =$ _____ and the fourth term is $s_4 =$ _____.

7. $\displaystyle\sum_{k=1}^{4} (2k) =$ _____.

8. *True or False:* Sequences are sometimes defined recursively.

9. *True or False:* A sequence is a function.

10. *True or False:* $\displaystyle\sum_{k=1}^{2} k = 3$

Skill Building

In Problems 11–16, evaluate each factorial expression. Verify your results using a graphing utility.

11. $10!$

12. $9!$

13. $\dfrac{9!}{6!}$

14. $\dfrac{12!}{10!}$

15. $\dfrac{3!\,7!}{4!}$

16. $\dfrac{5!\,8!}{3!}$

In Problems 17–28, write down the first five terms of each sequence.

17. $\{n\}$

18. $\{n^2 + 1\}$

19. $\left\{\dfrac{n}{n+2}\right\}$

20. $\left\{\dfrac{2n+1}{2n}\right\}$

21. $\{(-1)^{n+1} n^2\}$

22. $\left\{(-1)^{n-1}\left(\dfrac{n}{2n-1}\right)\right\}$

23. $\left\{\dfrac{2^n}{3^n + 1}\right\}$

24. $\left\{\left(\dfrac{4}{3}\right)^n\right\}$

25. $\left\{\dfrac{(-1)^n}{(n+1)(n+2)}\right\}$

26. $\left\{\dfrac{3^n}{n}\right\}$

27. $\left\{\dfrac{n}{e^n}\right\}$

28. $\left\{\dfrac{n^2}{2^n}\right\}$

In Problems 29–36, the given pattern continues. Write down the nth term of each sequence suggested by the pattern.

29. $\dfrac{1}{2}, \dfrac{2}{3}, \dfrac{3}{4}, \dfrac{4}{5}, \dots$

30. $\dfrac{1}{1\cdot 2}, \dfrac{1}{2\cdot 3}, \dfrac{1}{3\cdot 4}, \dfrac{1}{4\cdot 5}, \dots$

31. $1, \dfrac{1}{2}, \dfrac{1}{4}, \dfrac{1}{8}, \dots$

32. $\dfrac{2}{3}, \dfrac{4}{9}, \dfrac{8}{27}, \dfrac{16}{81}, \dots$

33. $1, -1, 1, -1, 1, -1, \dots$

34. $1, \dfrac{1}{2}, 3, \dfrac{1}{4}, 5, \dfrac{1}{6}, 7, \dfrac{1}{8}, \dots$

35. $1, -2, 3, -4, 5, -6, \dots$

36. $2, -4, 6, -8, 10, \dots$

In Problems 37–50, a sequence is defined recursively. Write the first five terms.

37. $a_1 = 2; \quad a_n = 3 + a_{n-1}$

38. $a_1 = 3; \quad a_n = 4 - a_{n-1}$

39. $a_1 = -2; \quad a_n = n + a_{n-1}$

40. $a_1 = 1; \quad a_n = n - a_{n-1}$

41. $a_1 = 5; \quad a_n = 2a_{n-1}$

42. $a_1 = 2; \quad a_n = -a_{n-1}$

43. $a_1 = 3; \quad a_n = \dfrac{a_{n-1}}{n}$

44. $a_1 = -2; \quad a_n = n + 3a_{n-1}$

45. $a_1 = 1; \quad a_2 = 2; \quad a_n = a_{n-1} \cdot a_{n-2}$

46. $a_1 = -1; \quad a_2 = 1; \quad a_n = a_{n-2} + n a_{n-1}$

47. $a_1 = A; \quad a_n = a_{n-1} + d$

48. $a_1 = A; \quad a_n = r a_{n-1}, \quad r \neq 0$

49. $a_1 = \sqrt{2}; \quad a_n = \sqrt{2 + a_{n-1}}$

50. $a_1 = \sqrt{2}; \quad a_n = \sqrt{\dfrac{a_{n-1}}{2}}$

In Problems 51–60, write out each sum.

51. $\displaystyle\sum_{k=1}^{n} (k + 2)$

52. $\displaystyle\sum_{k=1}^{n} (2k + 1)$

53. $\displaystyle\sum_{k=1}^{n} \dfrac{k^2}{2}$

54. $\displaystyle\sum_{k=1}^{n} (k + 1)^2$

55. $\displaystyle\sum_{k=0}^{n} \dfrac{1}{3^k}$

56. $\sum_{k=0}^{n} \left(\frac{3}{2}\right)^{k}$ **57.** $\sum_{k=0}^{n-1} \frac{1}{3^{k+1}}$ **58.** $\sum_{k=0}^{n-1} (2k+1)$ **59.** $\sum_{k=2}^{n} (-1)^{k} \ln k$ **60.** $\sum_{k=3}^{n} (-1)^{k+1} 2^{k}$

In Problems 61–70, express each sum using summation notation.

61. $1 + 2 + 3 + \cdots + 20$

62. $1^{3} + 2^{3} + 3^{3} + \cdots + 8^{3}$

63. $\frac{1}{2} + \frac{2}{3} + \frac{3}{4} + \cdots + \frac{13}{13+1}$

64. $1 + 3 + 5 + 7 + \cdots + [2(12) - 1]$

65. $1 - \frac{1}{3} + \frac{1}{9} - \frac{1}{27} + \cdots + (-1)^{6}\left(\frac{1}{3^{6}}\right)$

66. $\frac{2}{3} - \frac{4}{9} + \frac{8}{27} - \cdots + (-1)^{11+1}\left(\frac{2}{3}\right)^{11}$

67. $3 + \frac{3^{2}}{2} + \frac{3^{3}}{3} + \cdots + \frac{3^{n}}{n}$

68. $\frac{1}{e} + \frac{2}{e^{2}} + \frac{3}{e^{3}} + \cdots + \frac{n}{e^{n}}$

69. $a + (a + d) + (a + 2d) + \cdots + (a + nd)$

70. $a + ar + ar^{2} + \cdots + ar^{n-1}$

In Problems 71–82, find the sum of each sequence (a) algebraically and (b) using a graphing utility.

71. $\sum_{k=1}^{10} 5$ **72.** $\sum_{k=1}^{20} 8$ **73.** $\sum_{k=1}^{6} k$ **74.** $\sum_{k=1}^{4} (-k)$

75. $\sum_{k=1}^{5} (5k + 3)$ **76.** $\sum_{k=1}^{6} (3k - 7)$ **77.** $\sum_{k=1}^{3} (k^{2} + 4)$ **78.** $\sum_{k=0}^{4} (k^{2} - 4)$

79. $\sum_{k=1}^{6} (-1)^{k} 2^{k}$ **80.** $\sum_{k=1}^{4} (-1)^{k} 3^{k}$ **81.** $\sum_{k=1}^{4} (k^{3} - 1)$ **82.** $\sum_{k=0}^{3} (k^{3} + 2)$

Applications and Extensions

83. Credit Card Debt John has a balance of $3000 on his Discover card that charges 1% interest per month on any unpaid balance. John can afford to pay $100 toward the balance each month. His balance each month after making a $100 payment is given by the recursively defined sequence

$$B_{0} = \$3000, \qquad B_{n} = 1.01 B_{n-1} - 100$$

(a) Determine John's balance after making the first payment. That is, determine B_{1}.
(b) Using a graphing utility, determine when John's balance will be below $2000. How many payments of $100 have been made?
(c) Using a graphing utility, determine when John will pay off the balance. What is the total of all the payments?
(d) What was John's interest expense?

84. Car Loans Phil bought a car by taking out a loan for $18,500 at 0.5% interest per month. Phil's normal monthly payment is $434.47 per month, but he decides that he can afford to pay $100 extra toward the balance each month. His balance each month is given by the recursively defined sequence

$$B_{0} = \$18,500, \qquad B_{n} = 1.005 B_{n-1} - 534.47$$

(a) Determine Phil's balance after making the first payment. That is, determine B_{1}.
(b) Using a graphing utility, determine when Phil's balance will be below $10,000. How many payments of $534.47 have been made?
(c) Using a graphing utility, determine when Phil will pay off the balance. What is the total of all the payments?
(d) What was Phil's interest expense?

85. Trout Population A pond currently has 2000 trout in it. A fish hatchery decides to add an additional 20 trout each month. In addition, it is known that the trout population is growing 3% per month. The size of the population after n months is given by the recursively defined sequence

$$p_{0} = 2000, \qquad p_{n} = 1.03 p_{n-1} + 20$$

(a) How many trout are in the pond at the end of the second month? That is, what is p_{2}?
(b) Using a graphing utility, determine how long it will be before the trout population reaches 5000.

86. Environmental Control The Environmental Protection Agency (EPA) determines that Maple Lake has 250 tons of pollutants as a result of industrial waste and that 10% of the pollutant present is neutralized by solar oxidation every year. The EPA imposes new pollution control laws that result in 15 tons of new pollutant entering the lake each year. The amount of pollutant in the lake at the end of each year is given by the recursively defined sequence

$$p_{0} = 250, \qquad p_{n} = 0.9 p_{n-1} + 15$$

(a) Determine the amount of pollutant in the lake at the end of the second year. That is, determine p_{2}.
(b) Using a graphing utility, provide pollutant amounts for the next 20 years.
(c) What is the equilibrium level of pollution in Maple Lake? That is, what is $\lim\limits_{n \to \infty} p_{n}$?

87. Roth IRA On January 1, 1999, Bob decided to place $500 at the end of each quarter into a Roth Individual Retirement Account.

(a) Find a recursive formula that represents Bob's balance at the end of each quarter if the rate of return is assumed to be 8% per annum compounded quarterly.

(b) How long will it be before the value of the account exceeds $100,000?

(c) What will be the value of the account in 25 years when Bob retires?

88. Education IRA On January 1, 1999, John's parents decided to place $45 at the end of each month into an Education IRA.

(a) Find a recursive formula that represents the balance at the end of each month if the rate of return is assumed to be 6% per annum compounded monthly.

(b) How long will it be before the value of the account exceeds $4000?

(c) What will be the value of the account in 16 years when John goes to college?

89. Home Loan Bill and Laura borrowed $150,000 at 6% per annum compounded monthly for 30 years to purchase a home. Their monthly payment is determined to be $899.33.

(a) Find a recursive formula for their balance after each monthly payment has been made.

(b) Determine Bill and Laura's balance after the first payment.

(c) Using a graphing utility, create a table showing Bill and Laura's balance after each monthly payment.

(d) Using a graphing utility, determine when Bill and Laura's balance will be below $140,000.

(e) Using a graphing utility, determine when Bill and Laura will pay off the balance.

(f) Determine Bill and Laura's interest expense when the loan is paid.

(g) Suppose that Bill and Laura decide to pay an additional $100 each month on their loan. Answer parts (a) to (f) under this scenario.

(h) Is it worthwhile for Bill and Laura to pay the additional $100? Explain.

90. Home Loan Jodi and Jeff borrowed $120,000 at 6.5% per annum compounded monthly for 30 years to purchase a home. Their monthly payment is determined to be $758.48.

(a) Find a recursive formula for their balance after each monthly payment has been made.

(b) Determine Jodi and Jeff's balance after the first payment.

(c) Using a graphing utility, create a table showing Jodi and Jeff's balance after each monthly payment.

(d) Using a graphing utility, determine when Jodi and Jeff's balance will be below $100,000.

(e) Using a graphing utility, determine when Jodi and Jeff will pay off the balance.

(f) Determine Jodi and Jeff's interest expense when the loan is paid.

(g) Suppose that Jodi and Jeff decide to pay an additional $100 each month on their loan. Answer parts (a) to (f) under this scenario.

(h) Is it worthwhile for Jodi and Jeff to pay the additional $100? Explain.

91. Growth of a Rabbit Colony A colony of rabbits begins with one pair of mature rabbits, which will produce a pair of offspring (one male, one female) each month. Assume that all rabbits mature in 1 month and produce a pair of offspring (one male, one female) after 2 months. If no rabbits ever die, how many pairs of mature rabbits are there after 7 months?
[**Hint:** A Fibonacci sequence models this colony. Do you see why?]

1 mature pair

1 mature pair

2 mature pairs

3 mature pairs

92. Fibonacci Sequence Let

$$u_n = \frac{\left(1 + \sqrt{5}\right)^n - \left(1 - \sqrt{5}\right)^n}{2^n \sqrt{5}}$$

define the nth term of a sequence.

(a) Show that $u_1 = 1$ and $u_2 = 1$.

(b) Show that $u_{n+2} = u_{n+1} + u_n$.

(c) Draw the conclusion that $\{u_n\}$ is a Fibonacci sequence.

93. Pascal's Triangle Divide the triangular array shown (called Pascal's triangle) using diagonal lines as indicated. Find the sum of the numbers in each of these diagonal rows. Do you recognize this sequence?

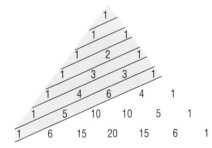

94. Fibonacci Sequence Use the result of Problem 92 to do the following problems:

(a) Write the first 10 terms of the Fibonacci sequence.

(b) Compute the ratio $\dfrac{u_{n+1}}{u_n}$ for the first 10 terms.

(c) As n gets large, what number does the ratio approach? This number is referred to as the **golden ratio**. Rectangles whose sides are in this ratio were considered pleasing to the eye by the Greeks. For example, the facade of the Parthenon was constructed using the golden ratio.

(d) Compute the ratio $\dfrac{u_n}{u_{n+1}}$ for the first 10 terms.

(e) As n gets large, what number does the ratio approach? This number is also referred to as the **golden ratio**. This ratio is believed to have been used in the construction of the Great Pyramid in Egypt. The ratio equals the sum of the areas of the four face triangles divided by the total surface area of the Great Pyramid.

95. Approximating e^x   In calculus, it can be shown that

$$e^x = \sum_{k=0}^{\infty} \frac{x^k}{k!}$$

We can approximate the value of e^x for any x using the following sum

$$e^x \approx \sum_{k=0}^{n} \frac{x^k}{k!}$$

for some n.

(a) Approximate $e^{1.3}$ with $n = 4$.
(b) Approximate $e^{1.3}$ with $n = 7$.
(c) Use a calculator to approximate the value of $e^{1.3}$.
(d) Using trial and error along with a graphing utility's SEQuence mode, determine the value of n required to approximate $e^{1.3}$ correct to eight decimal places.

96. Approximating e^x   Refer to Problem 95.

(a) Approximate $e^{-2.4}$ with $n = 3$.
(b) Approximate $e^{-2.4}$ with $n = 6$.

(c) Use a calculator to approximate the value of $e^{-2.4}$.
(d) Using trial and error along with a graphing utility's SEQuence mode, determine the value of n required to approximate $e^{-2.4}$ correct to eight decimal places.

97. Show that

$$1 + 2 + \cdots + (n - 1) + n = \frac{n(n + 1)}{2}$$

[Hint: Let

$$S = 1 + 2 + \cdots + (n - 1) + n$$
$$S = n + (n - 1) + (n - 2) + \cdots + 1$$

Add these equations. Then

$$2S = \underbrace{[1 + n] + [2 + (n - 1)] + \cdots + [n + 1]}_{n \text{ terms in brackets}}$$

Now complete the derivation.**]**

Discussion and Writing

98. Investigate various applications that lead to a Fibonacci sequence, such as art, architecture, or financial markets. Write an essay on these applications.

'Are You Prepared?' Answers

1. $f(2) = \dfrac{1}{2}; f(3) = \dfrac{2}{3}$ **2.** True   **3.** $1082.43   **4.** $9513.28

7.2 Arithmetic Sequences

OBJECTIVES   1 Determine If a Sequence Is Arithmetic
2 Find a Formula for an Arithmetic Sequence
3 Find the Sum of an Arithmetic Sequence

1 Determine If a Sequence Is Arithmetic

When the difference between successive terms of a sequence is always the same number, the sequence is called **arithmetic**. An **arithmetic sequence**[*] may be defined recursively as $a_1 = a$, $a_n - a_{n-1} = d$, or as

$$a_1 = a, \qquad a_n = a_{n-1} + d \tag{1}$$

where $a = a_1$ and d are real numbers. The number a is the first term, and the number d is called the **common difference**.

The terms of an arithmetic sequence with first term a and common difference d follow the pattern

$$a, \quad a + d, \quad a + 2d, \quad a + 3d, \ldots$$

[*]Sometimes called an **arithmetic progression**.

| EXAMPLE 1 | **Determining If a Sequence Is Arithmetic** |

The sequence
$$4, \quad 7, \quad 10, \quad 13, \ldots$$
is arithmetic since the difference of successive terms is 3. The first term is 4, and the common difference is 3.   ◀

| EXAMPLE 2 | **Determining If a Sequence Is Arithmetic** |

Show that the following sequence is arithmetic. Find the first term and the common difference.
$$\{s_n\} = \{3n + 5\}$$

Solution   The first term is $s_1 = 3 \cdot 1 + 5 = 8$. The nth and $(n - 1)$st terms of the sequence $\{s_n\}$ are
$$s_n = 3n + 5 \quad \text{and} \quad s_{n-1} = 3(n - 1) + 5 = 3n + 2$$

Their difference is
$$s_n - s_{n-1} = (3n + 5) - (3n + 2) = 5 - 2 = 3$$

Since the difference of two successive terms is constant, the sequence is arithmetic and the common difference is 3.   ◀

| EXAMPLE 3 | **Determining If a Sequence Is Arithmetic** |

Show that the sequence $\{t_n\} = \{4 - n\}$ is arithmetic. Find the first term and the common difference.

Solution   The first term is $t_1 = 4 - 1 = 3$. The nth and $(n - 1)$st terms are
$$t_n = 4 - n \quad \text{and} \quad t_{n-1} = 4 - (n - 1) = 5 - n$$

Their difference is
$$t_n - t_{n-1} = (4 - n) - (5 - n) = 4 - 5 = -1$$

Since the difference of two successive terms is constant, $\{t_n\}$ is an arithmetic sequence whose common difference is -1.   ◀

NOW WORK PROBLEM 5.

2   Find a Formula for an Arithmetic Sequence

Suppose that a is the first term of an arithmetic sequence whose common difference is d. We seek a formula for the nth term, a_n. To see the pattern, we write down the first few terms.

$$a_1 = a$$
$$a_2 = a_1 + d = a + 1 \cdot d$$
$$a_3 = a_2 + d = (a + d) + d = a + 2 \cdot d$$
$$a_4 = a_3 + d = (a + 2 \cdot d) + d = a + 3 \cdot d$$
$$a_5 = a_4 + d = (a + 3 \cdot d) + d = a + 4 \cdot d$$
$$\vdots$$
$$a_n = a_{n-1} + d = [a + (n - 2)d] + d = a + (n - 1)d$$

We are led to the following result:

Theorem | *n*th **Term of an Arithmetic Sequence**

For an arithmetic sequence $\{a_n\}$ whose first term is a and whose common difference is d, the nth term is determined by the formula

$$a_n = a + (n - 1)d \qquad \textbf{(2)}$$

EXAMPLE 4 | **Finding a Particular Term of an Arithmetic Sequence**

Find the thirteenth term of the arithmetic sequence: $2, 6, 10, 14, 18, \ldots$

Solution | The first term of this arithmetic sequence is $a = 2$, and the common difference is 4. By formula (2), the nth term is

$$a_n = 2 + (n - 1)4$$

Hence, the thirteenth term is

$$a_{13} = 2 + 12 \cdot 4 = 50 \qquad \blacktriangleleft$$

─── **Exploration** ───

Use a graphing utility to find the thirteenth term of the sequence given in Example 4. Use it to find the twentieth and fiftieth terms.

EXAMPLE 5 | **Finding a Recursive Formula for an Arithmetic Sequence**

The eighth term of an arithmetic sequence is 75, and the twentieth term is 39. Find the first term and the common difference. Give a recursive formula for the sequence. What is the nth term of the sequence?

Solution | By formula (2), we know that $a_n = a + (n - 1)d$. As a result,

$$\begin{cases} a_8 = a + 7d = 75 \\ a_{20} = a + 19d = 39 \end{cases}$$

This is a system of two linear equations containing two variables, a and d, which we can solve by elimination. Subtracting the second equation from the first equation, we get

$$-12d = 36$$
$$d = -3$$

With $d = -3$, we use $a + 7d = 75$ and find that $a = 75 - 7d = 75 - 7(-3) = 96$. The first term is $a = 96$, and the common difference is $d = -3$.

Using formula (1), a recursive formula for this sequence is

$$a_1 = 96, \qquad a_n = a_{n-1} - 3$$

Using formula (2), a formula for the nth term of the sequence $\{a_n\}$ is

$$a_n = a + (n - 1)d = 96 + (n - 1)(-3) = 99 - 3n \qquad \blacktriangleleft$$

─── **Exploration** ───

Graph the recursive formula from Example 5, $a_1 = 96$, $a_n = a_{n-1} - 3$, using a graphing utility. Conclude that the graph of the recursive formula behaves like the graph of a linear function. How is d, the common difference, related to m, the slope of a line?

NOW WORK PROBLEMS 21 AND 27.

3   Find the Sum of an Arithmetic Sequence

The next result gives a formula for finding the sum of the first n terms of an arithmetic sequence.

Theorem

Sum of n Terms of an Arithmetic Sequence

Let $\{a_n\}$ be an arithmetic sequence with first term a and common difference d. The sum S_n of the first n terms of $\{a_n\}$ is

$$S_n = \frac{n}{2}[2a + (n-1)d] = \frac{n}{2}(a + a_n) \qquad\qquad \textbf{(3)}$$

Proof

$$
\begin{aligned}
S_n &= a_1 + a_2 + a_3 + \cdots + a_n && \text{Sum of first } n \text{ terms}\\
&= a + (a + d) + (a + 2d) + \cdots + [a + (n-1)d] && \text{Formula (2)}\\
&= \underbrace{(a + a + \cdots + a)}_{n \text{ terms}} + [d + 2d + \cdots + (n-1)d] && \text{Rearrange terms}\\
&= na + d[1 + 2 + \cdots + (n-1)]\\
&= na + d\left[\frac{(n-1)n}{2}\right] && \text{Property 6, Section 7.1}\\
&= na + \frac{n}{2}(n-1)d\\
&= \frac{n}{2}[2a + (n-1)d] && \text{Factor out } \frac{n}{2} && \textbf{(4)}\\
&= \frac{n}{2}[a + a + (n-1)d]\\
&= \frac{n}{2}(a + a_n) && \text{Formula (2)} && \textbf{(5)}
\end{aligned}
$$

∎

Formula (3) provides two ways to find the sum of the first n terms of an arithmetic sequence. Notice that formula (4) involves the first term and common difference, whereas formula (5) involves the first term and the nth term. Use whichever form is easier.

| EXAMPLE 6 | Finding the Sum of n Terms of an Arithmetic Sequence |

Find the sum S_n of the first n terms of the sequence $\{3n + 5\}$; that is, find

$$8 + 11 + 14 + \cdots + (3n + 5)$$

Solution

The sequence $\{3n + 5\}$ is an arithmetic sequence with first term $a = 8$ and the nth term $(3n + 5)$. To find the sum S_n, we use formula (5).

$$S_n = \frac{n}{2}(a + a_n) = \frac{n}{2}[8 + (3n + 5)] = \frac{n}{2}(3n + 13) \qquad \blacktriangleleft$$

◾━━━ **NOW WORK PROBLEM 35.**

| **EXAMPLE 7** | **Using a Graphing Utility to Find the Sum of 20 Terms of an Arithmetic Sequence** |

Figure 12

Use a graphing utility to find the sum of the first 20 terms of the sequence $\{9.5n + 2.6\}$.

Solution Figure 12 shows the results obtained using a TI-84 Plus graphing calculator.

The sum of the first 20 terms of the sequence $\{9.5n + 2.6\}$ is 2047. ◄

➤ **WORK EXAMPLE 7 USING FORMULA (3).**

➤ **NOW WORK PROBLEM 43.**

| **EXAMPLE 8** | **Creating a Floor Design** |

A ceramic tile floor is designed in the shape of a trapezoid 20 feet wide at the base and 10 feet wide at the top. See Figure 13. The tiles, 12 inches by 12 inches, are to be placed so that each successive row contains one less tile than the preceding row. How many tiles will be required?

Figure 13

Solution The bottom row requires 20 tiles and the top row, 10 tiles. Since each successive row requires one less tile, the total number of tiles required is

$$S = 20 + 19 + 18 + \cdots + 11 + 10$$

This is the sum of an arithmetic sequence; the common difference is -1. The number of terms to be added is $n = 11$, with the first term $a = 20$ and the last term $a_{11} = 10$. The sum S is

$$S = \frac{n}{2}(a + a_{11}) = \frac{11}{2}(20 + 10) = 165$$

In all, 165 tiles will be required. ◄

7.2 Assess Your Understanding

Concepts and Vocabulary

1. In a(n) _____ sequence, the difference between successive terms is a constant.

2. *True or False:* In an arithmetic sequence the sum of the first and last terms equals twice the sum of all the terms.

Skill Building

In Problems 3–12, show that each sequence is arithmetic. Find the common difference and write out the first four terms.

3. $\{n + 4\}$ **4.** $\{n - 5\}$ **5.** $\{2n - 5\}$ **6.** $\{3n + 1\}$ **7.** $\{6 - 2n\}$

8. $\{4 - 2n\}$ **9.** $\left\{\dfrac{1}{2} - \dfrac{1}{3}n\right\}$ **10.** $\left\{\dfrac{2}{3} + \dfrac{n}{4}\right\}$ **11.** $\{\ln 3^n\}$ **12.** $\{e^{\ln n}\}$

In Problems 13–20, find the nth term of the arithmetic sequence whose initial term a and common difference d are given. What is the fifth term?

13. $a = 2$; $d = 3$ **14.** $a = -2$; $d = 4$ **15.** $a = 5$; $d = -3$ **16.** $a = 6$; $d = -2$

17. $a = 0$; $d = \dfrac{1}{2}$ **18.** $a = 1$; $d = -\dfrac{1}{3}$ **19.** $a = \sqrt{2}$; $d = \sqrt{2}$ **20.** $a = 0$; $d = \pi$

In Problems 21–26, find the indicated term in each arithmetic sequence.

21. 12th term of $2, 4, 6, \ldots$ **22.** 8th term of $-1, 1, 3, \ldots$

23. 10th term of $1, -2, -5, \ldots$ **24.** 9th term of $5, 0, -5, \ldots$

25. 8th term of $a, a + b, a + 2b, \ldots$ **26.** 7th term of $2\sqrt{5}, 4\sqrt{5}, 6\sqrt{5}, \ldots$

In Problems 27–34, find the first term and the common difference of the arithmetic sequence described. Give a recursive formula for the sequence. Find a formula for the nth term.

27. 8th term is 8; 20th term is 44   **28.** 4th term is 3; 20th term is 35   **29.** 9th term is -5; 15th term is 31

30. 8th term is 4; 18th term is -96 **31.** 15th term is 0; 40th term is -50 **32.** 5th term is -2; 13th term is 30

33. 14th term is -1; 18th term is -9 **34.** 12th term is 4; 18th term is 28

In Problems 35–42, find each sum.

35. $1 + 3 + 5 + \cdots + (2n - 1)$ **36.** $2 + 4 + 6 + \cdots + 2n$ **37.** $7 + 12 + 17 + \cdots + (2 + 5n)$

38. $-1 + 3 + 7 + \cdots + (4n - 5)$ **39.** $2 + 4 + 6 + \cdots + 70$ **40.** $1 + 3 + 5 + \cdots + 59$

41. $5 + 9 + 13 + \cdots + 49$ **42.** $2 + 5 + 8 + \cdots + 41$

For Problems 43–48, use a graphing utility to find the sum of each sequence.

43. $\{3.45n + 4.12\}$, $n = 20$ **44.** $\{2.67n - 1.23\}$, $n = 25$ **45.** $2.8 + 5.2 + 7.6 + \cdots + 36.4$

46. $5.4 + 7.3 + 9.2 + \cdots + 32$ **47.** $4.9 + 7.48 + 10.06 + \cdots + 66.82$ **48.** $3.71 + 6.9 + 10.09 + \cdots + 80.27$

Applications and Extensions

49. Find x so that $x + 3, 2x + 1$, and $5x + 2$ are consecutive terms of an arithmetic sequence.

50. Find x so that $2x, 3x + 2$, and $5x + 3$ are consecutive terms of an arithmetic sequence.

51. Drury Lane Theater The Drury Lane Theater has 25 seats in the first row and 30 rows in all. Each successive row contains one additional seat. How many seats are in the theater?

52. Football Stadium The corner section of a football stadium has 15 seats in the first row and 40 rows in all. Each successive row contains two additional seats. How many seats are in this section?

53. Creating a Mosaic A mosaic is designed in the shape of an equilateral triangle, 20 feet on each side. Each tile in the mosaic is in the shape of an equilateral triangle, 12 inches to a side. The tiles are to alternate in color as shown in the illustration. How many tiles of each color will be required?

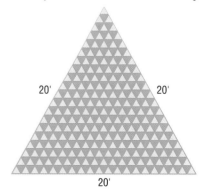

54. Constructing a Brick Staircase A brick staircase has a total of 30 steps. The bottom step requires 100 bricks. Each successive step requires two less bricks than the prior step.
(a) How many bricks are required for the top step?
(b) How many bricks are required to build the staircase?

55. Stadium Construction How many rows are in the corner section of a stadium containing 2040 seats if the first row has 10 seats and each successive row has 4 additional seats?

56. Salary Suppose that you just received a job offer with a starting salary of $35,000 per year and a guaranteed raise of $1400 per year. How many years will it take before your aggregate salary is $280,000?

[**Hint:** Your aggregate salary after 2 years is $35,000 + ($35,000 + $1400).]

Discussion and Writing

57. Make up an arithmetic sequence. Give it to a friend and ask for its twentieth term.

58. Describe the similarities and differences between arithmetic sequences and linear functions.

7.3 Geometric Sequences; Geometric Series

OBJECTIVES 1 Determine If a Sequence Is Geometric
2 Find a Formula for a Geometric Sequence
3 Find the Sum of a Geometric Sequence
4 Find the Sum of a Geometric Series

1 Determine If a Sequence Is Geometric

When the ratio of successive terms of a sequence is always the same nonzero number, the sequence is called **geometric**. A **geometric sequence**[*] may be defined recursively as $a_1 = a$, $\dfrac{a_n}{a_{n-1}} = r$, or as

$$a_1 = a, \qquad a_n = ra_{n-1} \qquad \text{(1)}$$

where $a_1 = a$ and $r \neq 0$ are real numbers. The number a is the first term, and the nonzero number r is called the **common ratio**.

The terms of a geometric sequence with first term a and common ratio r follow the pattern

$$a, \quad ar, \quad ar^2, \quad ar^3, \ldots$$

EXAMPLE 1 **Determining If a Sequence Is Geometric**

The sequence

$$2, \ 6, \ 18, \ 54, \ 162, \ldots$$

is geometric since the ratio of successive terms is $3 \left(\dfrac{6}{2} = \dfrac{18}{6} = \dfrac{54}{18} = \cdots = 3 \right)$. The first term is 2, and the common ratio is 3. ◄

EXAMPLE 2 **Determining If a Sequence Is Geometric**

Show that the following sequence is geometric.

$$\{s_n\} = 2^{-n}$$

Find the first term and the common ratio.

[*]Sometimes called a **geometric progression**.

Solution The first term is $s_1 = 2^{-1} = \dfrac{1}{2}$. The nth and $(n-1)$st terms of the sequence $\{s_n\}$ are

$$s_n = 2^{-n} \quad \text{and} \quad s_{n-1} = 2^{-(n-1)}$$

Their ratio is

$$\frac{s_n}{s_{n-1}} = \frac{2^{-n}}{2^{-(n-1)}} = 2^{-n+(n-1)} = 2^{-1} = \frac{1}{2}$$

Because the ratio of successive terms is a nonzero constant, the sequence $\{s_n\}$ is geometric with common ratio $\dfrac{1}{2}$. ◄

EXAMPLE 3 **Determining If a Sequence Is Geometric**

Show that the following sequence is geometric.
$$\{t_n\} = \{4^n\}$$
Find the first term and the common ratio.

Solution The first term is $t_1 = 4^1 = 4$. The nth and $(n-1)$st terms are

$$t_n = 4^n \quad \text{and} \quad t_{n-1} = 4^{n-1}$$

Their ratio is

$$\frac{t_n}{t_{n-1}} = \frac{4^n}{4^{n-1}} = 4^{n-(n-1)} = 4$$

The sequence, $\{t_n\}$, is a geometric sequence with common ratio 4. ◄

NOW WORK PROBLEM 7.

2 Find a Formula for a Geometric Sequence

Suppose that a is the first term of a geometric sequence with common ratio $r \neq 0$. We seek a formula for the nth term a_n. To see the pattern, we write down the first few terms:

$$a_1 = a = a \cdot 1 = ar^0$$
$$a_2 = ra_1 = ar^1$$
$$a_3 = ra_2 = r(ar) = ar^2$$
$$a_4 = ra_3 = r(ar^2) = ar^3$$
$$a_5 = ra_4 = r(ar^3) = ar^4$$
$$\vdots$$
$$a_n = ra_{n-1} = r(ar^{n-2}) = ar^{n-1}$$

We are led to the following result:

Theorem

*n*th Term of a Geometric Sequence

For a geometric sequence $\{a_n\}$ whose first term is a and whose common ratio is r, the *n*th term is determined by the formula

$$a_n = ar^{n-1}, \qquad r \neq 0 \qquad \qquad \textbf{(2)}$$

EXAMPLE 4

Finding a Particular Term of a Geometric Sequence

(a) Find the ninth term of the geometric sequence: $10,\ 9,\ \dfrac{81}{10},\ \dfrac{729}{100}\cdots$

(b) Find a recursive formula for this sequence.

Solution

(a) The first term of this geometric sequence is $a = 10$ and the common ratio is $\dfrac{9}{10}$.

— **Exploration** —

Use a graphing utility to find the ninth term of the sequence given in Example 4. Use it to find the twentieth and fiftieth terms. Now use a graphing utility to graph the recursive formula found in Example 4(b). Conclude that the graph of the recursive formula behaves like the graph of an exponential function. How is r, the common ratio, related to a, the base of the exponential function $y = a^x$?

$\left(\text{Use } \dfrac{9}{10}, \text{ or } \dfrac{\frac{81}{10}}{9} = \dfrac{9}{10}, \text{ or any two successive terms.}\right)$ By formula (2), the *n*th term is

$$a_n = 10\left(\frac{9}{10}\right)^{n-1}$$

The ninth term is

$$a_9 = 10\left(\frac{9}{10}\right)^{9-1} = 10\left(\frac{9}{10}\right)^8 \approx 4.3046721$$

(b) The first term in the sequence is 10 and the common ratio is $r = \dfrac{9}{10}$. Using formula (1), the recursive formula is $a_1 = 10,\ a_n = \dfrac{9}{10}a_{n-1}$. ◀

✏ **NOW WORK PROBLEMS 29 AND 37.**

3 ## Find the Sum of a Geometric Sequence

The next result gives us a formula for finding the sum of the first n terms of a geometric sequence.

Theorem

Sum of *n* Terms of a Geometric Sequence

Let $\{a_n\}$ be a geometric sequence with first term a and common ratio r, where $r \neq 0, r \neq 1$. The sum S_n of the first n terms of $\{a_n\}$ is

$$S_n = a \cdot \frac{1 - r^n}{1 - r}, \qquad r \neq 0, 1 \qquad \qquad \textbf{(3)}$$

Proof The sum S_n of the first n terms of $\{a_n\} = \{ar^{n-1}\}$ is

$$S_n = a + ar + \cdots + ar^{n-1} \qquad \qquad \textbf{(4)}$$

Multiply each side by r to obtain

$$rS_n = ar + ar^2 + \cdots + ar^n \qquad \qquad \textbf{(5)}$$

Now, subtract (5) from (4). The result is

$$S_n - rS_n = a - ar^n$$
$$(1 - r)S_n = a(1 - r^n)$$

Since $r \neq 1$, we can solve for S_n.

$$S_n = a \cdot \frac{1 - r^n}{1 - r}$$

∎

| **EXAMPLE 5** | **Finding the Sum of n Terms of a Geometric Sequence** |

Find the sum S_n of the first n terms of the sequence $\left\{ \left(\frac{1}{2}\right)^n \right\}$; that is, find

$$\frac{1}{2} + \frac{1}{4} + \frac{1}{8} + \cdots + \left(\frac{1}{2}\right)^n$$

Solution The sequence $\left\{ \left(\frac{1}{2}\right)^n \right\}$ is a geometric sequence with $a = \frac{1}{2}$ and $r = \frac{1}{2}$. The sum S_n that we seek is the sum of the first n terms of the sequence, so we use formula (3) to get

$$S_n = \sum_{k=1}^{n} \left(\frac{1}{2}\right)^k = \frac{1}{2} + \frac{1}{4} + \frac{1}{8} + \cdots + \left(\frac{1}{2}\right)^n$$

$$= \frac{1}{2} \left[\frac{1 - \left(\frac{1}{2}\right)^n}{1 - \frac{1}{2}} \right] \qquad \text{Formula (3)}$$

$$= \frac{1}{2} \left[\frac{1 - \left(\frac{1}{2}\right)^n}{\frac{1}{2}} \right]$$

$$= 1 - \left(\frac{1}{2}\right)^n$$

◀

━━━ **NOW WORK PROBLEM 43.**

| **EXAMPLE 6** | **Using a Graphing Utility to Find the Sum of a Geometric Sequence** |

Use a graphing utility to find the sum of the first 15 terms of the sequence $\left\{ \left(\frac{1}{3}\right)^n \right\}$; that is, find

$$\frac{1}{3} + \frac{1}{9} + \frac{1}{27} + \cdots + \left(\frac{1}{3}\right)^{15}$$

Figure 14

Solution Figure 14 shows the result obtained using a TI-84 Plus graphing calculator. The sum of the first 15 terms of the sequence $\left\{ \left(\frac{1}{3}\right)^n \right\}$ is 0.4999999652. ◀

━━━ **NOW WORK PROBLEM 49.**

4 Find the Sum of a Geometric Series

An infinite sum of the form

$$a + ar + ar^2 + \cdots + ar^{n-1} + \cdots$$

with first term a and common ratio r, is called an **infinite geometric series** and is denoted by

$$\sum_{k=1}^{\infty} ar^{k-1}$$

Based on formula (3), the sum S_n of the first n terms of a geometric series is

$$S_n = a \cdot \frac{1 - r^n}{1 - r} = \frac{a}{1 - r} - \frac{ar^n}{1 - r} \tag{6}$$

If this finite sum S_n approaches a number L as $n \to \infty$, then we say the infinite geometric series $\sum_{k=1}^{\infty} ar^{k-1}$ **converges**. We call L the **sum of the infinite geometric series**, and we write

$$L = \sum_{k=1}^{\infty} ar^{k-1}$$

If a series does not converge, it is called a **divergent series**.

Theorem

Sum of an Infinite Geometric Series

If $|r| < 1$, the sum of the infinite geometric series $\sum_{k=1}^{\infty} ar^{k-1}$ is

$$\sum_{k=1}^{\infty} ar^{k-1} = \frac{a}{1 - r} \tag{7}$$

Intuitive Proof Since $|r| < 1$, it follows that $|r^n|$ approaches 0 as $n \to \infty$. Then, based on formula (6), the sum S_n approaches $\dfrac{a}{1 - r}$ as $n \to \infty$. ∎

EXAMPLE 7

Finding the Sum of a Geometric Series

Find the sum of the geometric series: $\quad 2 + \dfrac{4}{3} + \dfrac{8}{9} + \cdots$

Solution

The first term is $a = 2$, and the common ratio is

$$r = \frac{\frac{4}{3}}{2} = \frac{4}{6} = \frac{2}{3}$$

Since $|r| < 1$, we use formula (7) to find that

$$2 + \frac{4}{3} + \frac{8}{9} + \cdots = \frac{2}{1 - \dfrac{2}{3}} = 6$$

— Exploration —

Use a graphing utility to graph $U_n = 2\left(\dfrac{2}{3}\right)^{n-1}$ in sequence mode. TRACE the graph for large values of n. What happens to the value of U_n as n increases without bound? What can you conclude about $\sum_{n=1}^{\infty} 2\left(\dfrac{2}{3}\right)^{n-1}$?

NOW WORK PROBLEM 55.

| EXAMPLE 8 | **Repeating Decimals** |

Show that the repeating decimal $0.999\ldots$ equals 1.

Solution $0.999\ldots = \dfrac{9}{10} + \dfrac{9}{100} + \dfrac{9}{1000} + \cdots$

The decimal $0.999\ldots$ is a geometric series with first term $\dfrac{9}{10}$ and common ratio $\dfrac{1}{10}$. Using formula (7), we find

$$0.999\ldots = \frac{\dfrac{9}{10}}{1 - \dfrac{1}{10}} = \frac{\dfrac{9}{10}}{\dfrac{9}{10}} = 1$$

◀

| EXAMPLE 9 | **Pendulum Swings** |

Figure 15

Initially, a pendulum swings through an arc of 18 inches. See Figure 15. On each successive swing, the length of the arc is 0.98 of the previous length.

(a) What is the length of the arc of the 10th swing?

(b) On which swing is the length of the arc first less than 12 inches?

(c) After 15 swings, what total distance will the pendulum have swung?

(d) When it stops, what total distance will the pendulum have swung?

Solution   (a) The length of the first swing is 18 inches.

The length of the second swing is $0.98(18)$ inches.

The length of the third swing is $0.98(0.98)(18) = 0.98^2(18)$ inches.

The length of the arc of the 10th swing is

$$(0.98)^9(18) \approx 15.007 \text{ inches}$$

(b) The length of the arc of the nth swing is $(0.98)^{n-1}(18)$. For this to be exactly 12 inches requires that

$$(0.98)^{n-1}(18) = 12$$

$$(0.98)^{n-1} = \frac{12}{18} = \frac{2}{3} \qquad \text{\textit{Divide both sides by 18.}}$$

$$n - 1 = \log_{0.98}\left(\frac{2}{3}\right) \qquad \text{\textit{Express as a logarithm.}}$$

$$n = 1 + \frac{\ln\left(\dfrac{2}{3}\right)}{\ln 0.98} \approx 1 + 20.07 = 21.07 \qquad \text{\textit{Solve for n; use the Change of Base Formula.}}$$

The length of the arc of the pendulum exceeds 12 inches on the 21st swing and is first less than 12 inches on the 22nd swing.

(c) After 15 swings, the pendulum will have swung the following total distance L:

$$L = \underset{\text{1st}}{18} + \underset{\text{2nd}}{0.98(18)} + \underset{\text{3rd}}{(0.98)^2(18)} + \underset{\text{4th}}{(0.98)^3(18)} + \cdots + \underset{\text{15th.}}{(0.98)^{14}(18)}$$

This is the sum of a geometric sequence. The common ratio is 0.98; the first term is 18. The sum has 15 terms, so

$$L = 18 \cdot \frac{1 - 0.98^{15}}{1 - 0.98} \approx 18(13.07) \approx 235.3 \text{ inches}$$

The pendulum will have swung through 235.3 inches after 15 swings.

(d) When the pendulum stops, it will have swung the following total distance T:

$$T = 18 + 0.98(18) + (0.98)^2(18) + (0.98)^3(18) + \cdots$$

This is the sum of a geometric series. The common ratio is $r = 0.98$; the first term is $a = 18$. The sum is

$$T = \frac{a}{1 - r} = \frac{18}{1 - 0.98} = 900$$

The pendulum will have swung a total of 900 inches when it finally stops. ◀

NOW WORK PROBLEM 69.

HISTORICAL FEATURE

Fibonacci

Sequences are among the oldest objects of mathematical investigation, having been studied for over 3500 years. After the initial steps, however, little progress was made until about 1600.

Arithmetic and geometric sequences appear in the Rhind papyrus, a mathematical text containing 85 problems copied around 1650 BC by the Egyptian scribe Ahmes from an earlier work (see Historical Problem 1). Fibonacci (AD 1220) wrote about problems similar to those found in the Rhind papyrus, leading one to suspect that Fibonacci may have had material available that is now lost. This material would have been in the non-Euclidean Greek tradition of Heron (about AD 75) and Diophantus (about AD 250). One problem, again modified slightly, is still with us in the familiar puzzle rhyme "As I was going to St. Ives ..." (see Historical Problem 2).

The Rhind papyrus indicates that the Egyptians knew how to add up the terms of an arithmetic or geometric sequence, as did the Babylonians. The rule for summing up a geometric sequence is found in Euclid's *Elements* (Book IX, 35, 36), where, like all Euclid's algebra, it is presented in a geometric form.

Investigations of other kinds of sequences began in the 1500s, when algebra became sufficiently developed to handle the more complicated problems. The development of calculus in the 1600s added a powerful new tool, especially for finding the sum of infinite series, and the subject continues to flourish today.

Historical Problems

1. *Arithmetic sequence problem from the Rhind papyrus (statement modified slightly for clarity)* One hundred loaves of bread are to be divided among five people so that the amounts that they receive form an arithmetic sequence. The first two together receive one-seventh of what the last three receive. How many loaves does each receive?
   [*Partial answer:* First person receives $1\frac{2}{3}$ loaves.]

2. The following old English children's rhyme resembles one of the Rhind papyrus problems.

   As I was going to St. Ives
   I met a man with seven wives

   Each wife had seven sacks
   Each sack had seven cats
   Each cat had seven kits [kittens]
   Kits, cats, sacks, wives
   How many were going to St. Ives?

   (a) Assuming that the speaker and the cat fanciers met by traveling in opposite directions, what is the answer?
   (b) How many kittens are being transported?
   (c) Kits, cats, sacks, wives; how many?
   [**Hint:** It is easier to include the man, find the sum with the formula, and then subtract 1 for the man.]

7.3 Assess Your Understanding

Concepts and Vocabulary

1. In a(n) _____ sequence, the ratio of successive terms is a constant.

2. If $|r| < 1$, the sum of the geometric series $\sum_{k=1}^{\infty} ar^{k-1}$ is _____.

3. *True or False:* A geometric sequence may be defined recursively.

4. *True or False:* In a geometric sequence, the common ratio is always a positive number.

Skill Building

In Problems 5–14, show that each sequence is geometric. Find the common ratio and write out the first four terms.

5. $\{3^n\}$

6. $\{(-5)^n\}$

7. $\left\{-3\left(\frac{1}{2}\right)^n\right\}$

8. $\left\{\left(\frac{5}{2}\right)^n\right\}$

9. $\left\{\frac{2^{n-1}}{4}\right\}$

10. $\left\{\frac{3^n}{9}\right\}$

11. $\{2^{n/3}\}$

12. $\{3^{2n}\}$

13. $\left\{\frac{3^{n-1}}{2^n}\right\}$

14. $\left\{\frac{2^n}{3^{n-1}}\right\}$

In Problems 15–28, determine whether the given sequence is arithmetic, geometric, or neither. If the sequence is arithmetic, find the common difference; if it is geometric, find the common ratio.

15. $\{n + 2\}$

16. $\{2n - 5\}$

17. $\{4n^2\}$

18. $\{5n^2 + 1\}$

19. $\left\{3 - \frac{2}{3}n\right\}$

20. $\left\{8 - \frac{3}{4}n\right\}$

21. $1, 3, 6, 10, \ldots$

22. $2, 4, 6, 8, \ldots$

23. $\left\{\left(\frac{2}{3}\right)^n\right\}$

24. $\left\{\left(\frac{5}{4}\right)^n\right\}$

25. $-1, -2, -4, -8, \ldots$ **26.** $1, 1, 2, 3, 5, 8, \ldots$

27. $\{3^{n/2}\}$

28. $\{(-1)^n\}$

In Problems 29–36, find the fifth term and the nth term of the geometric sequence whose initial term a and common ratio r are given.

29. $a = 2;\quad r = 3$

30. $a = -2;\quad r = 4$

31. $a = 5;\quad r = -1$

32. $a = 6;\quad r = -2$

33. $a = 0;\quad r = \frac{1}{2}$

34. $a = 1;\quad r = -\frac{1}{3}$

35. $a = \sqrt{2};\quad r = \sqrt{2}$

36. $a = 0;\quad r = \frac{1}{\pi}$

In Problems 37–42, find the indicated term of each geometric sequence.

37. 7th term of $1, \frac{1}{2}, \frac{1}{4}, \ldots$

38. 8th term of $1, 3, 9, \ldots$

39. 9th term of $1, -1, 1, \ldots$

40. 10th term of $-1, 2, -4, \ldots$

41. 8th term of $0.4, 0.04, 0.004, \ldots$

42. 7th term of $0.1, 1.0, 10.0, \ldots$

In Problems 43–48, find each sum.

43. $\frac{1}{4} + \frac{2}{4} + \frac{2^2}{4} + \frac{2^3}{4} + \cdots + \frac{2^{n-1}}{4}$

44. $\frac{3}{9} + \frac{3^2}{9} + \frac{3^3}{9} + \cdots + \frac{3^n}{9}$

45. $\sum_{k=1}^{n} \left(\frac{2}{3}\right)^k$

46. $\sum_{k=1}^{n} 4 \cdot 3^{k-1}$

47. $-1 - 2 - 4 - 8 - \cdots - (2^{n-1})$

48. $2 + \frac{6}{5} + \frac{18}{25} + \cdots + 2\left(\frac{3}{5}\right)^{n-1}$

For Problems 49–54, use a graphing utility to find the sum of each geometric sequence.

49. $\frac{1}{4} + \frac{2}{4} + \frac{2^2}{4} + \frac{2^3}{4} + \cdots + \frac{2^{14}}{4}$

50. $\frac{3}{9} + \frac{3^2}{9} + \frac{3^3}{9} + \cdots + \frac{3^{15}}{9}$

51. $\sum_{n=1}^{15} \left(\frac{2}{3}\right)^n$

52. $\sum_{n=1}^{15} 4 \cdot 3^{n-1}$

53. $-1 - 2 - 4 - 8 - \cdots - 2^{14}$

54. $2 + \frac{6}{5} + \frac{18}{25} + \cdots + 2\left(\frac{3}{5}\right)^{15}$

In Problems 55–64, find the sum of each infinite geometric series.

55. $1 + \frac{1}{3} + \frac{1}{9} + \cdots$

56. $2 + \frac{4}{3} + \frac{8}{9} + \cdots$

57. $8 + 4 + 2 + \cdots$

58. $6 + 2 + \frac{2}{3} + \cdots$

59. $2 - \frac{1}{2} + \frac{1}{8} - \frac{1}{32} + \cdots$

60. $1 - \frac{3}{4} + \frac{9}{16} - \frac{27}{64} + \cdots$

61. $\sum_{k=1}^{\infty} 5\left(\frac{1}{4}\right)^{k-1}$

62. $\sum_{k=1}^{\infty} 8\left(\frac{1}{3}\right)^{k-1}$

63. $\sum_{k=1}^{\infty} 6\left(-\frac{2}{3}\right)^{k-1}$

64. $\sum_{k=1}^{\infty} 4\left(-\frac{1}{2}\right)^{k-1}$

Applications and Extensions

65. Find x so that $x, x + 2$, and $x + 3$ are consecutive terms of a geometric sequence.

66. Find x so that $x - 1, x$, and $x + 2$ are consecutive terms of a geometric sequence.

67. Salary Increases   Suppose that you have just been hired at an annual salary of $18,000 and expect to receive annual increases of 5%. What will your salary be when you begin your fifth year?

68. Equipment Depreciation   A new piece of equipment cost a company $15,000. Each year, for tax purposes, the company depreciates the value by 15%. What value should the company give the equipment after 5 years?

69. Pendulum Swings   Initially, a pendulum swings through an arc of 2 feet. On each successive swing, the length of the arc is 0.9 of the previous length.

(a) What is the length of the arc of the 10th swing?

(b) On which swing is the length of the arc first less than 1 foot?

(c) After 15 swings, what total length will the pendulum have swung?

(d) When it stops, what total length will the pendulum have swung?

70. Bouncing Balls A ball is dropped from a height of 30 feet. Each time it strikes the ground, it bounces up to 0.8 of the previous height.

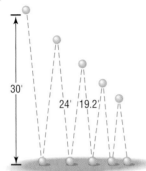

(a) What height will the ball bounce up to after it strikes the ground for the third time?

(b) How high will it bounce after it strikes the ground for the nth time?

(c) How many times does the ball need to strike the ground before its bounce is less than 6 inches?

(d) What total distance does the ball travel before it stops bouncing?

71. Grains of Wheat on a Chess Board In an old fable, a commoner who had saved the king's life was told he could ask the king for any just reward. Being a shrewd man, the commoner said, "A simple wish, sire. Place one grain of wheat on the first square of a chessboard, two grains on the second square, four grains on the third square, continuing until you have filled the board. This is all I seek." Compute the total number of grains needed to do this to see why the request, seemingly simple, could not be granted. (A chessboard consists of $8 \times 8 = 64$ squares.)

Discussion and Writing

77. A Rich Man's Promise A rich man promises to give you $1000 on September 1, 2001. Each day thereafter he will give you $\dfrac{9}{10}$ of what he gave you the previous day. What is the first date on which the amount you receive is less than 1¢? How much have you received when this happens?

72. Look at the figure below. What fraction of the square is eventually shaded if the indicated shading process continues indefinitely?

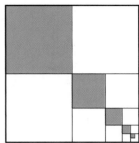

73. Multiplier Suppose that, throughout the U.S. economy, individuals spend 90% of every additional dollar that they earn. Economists would say that an individual's **marginal propensity to consume** is 0.90. For example, if Jane earns an additional dollar, she will spend $0.9(1) = \$0.90$ of it. The individual that earns $0.90 (from Jane) will spend 90% of it or $0.81. This process of spending continues and results in an infinite geometric series as follows:

$$1, 0.90, 0.90^2, 0.90^3, 0.90^4, \ldots.$$

The sum of this infinite geometric series is called the **multiplier**. What is the multiplier if individuals spend 90% of every additional dollar that they earn?

74. Multiplier Refer to Problem 73. Suppose that the marginal propensity to consume throughout the U.S. economy is 0.95. What is the multiplier for the U.S. economy?

75. Stock Price One method of pricing a stock is to discount the stream of future dividends of the stock. Suppose that a stock pays $P per year in dividends and, historically, the dividend has been increased i% per year. If you desire an annual rate of return of r%, this method of pricing a stock states that the price that you should pay is the present value of an infinite stream of payments:

$$\text{Price} = P + P\frac{1+i}{1+r} + P\left(\frac{1+i}{1+r}\right)^2 + P\left(\frac{1+i}{1+r}\right)^3 + \cdots$$

The price of the stock is the sum of an infinite geometric series. Suppose that a stock pays an annual dividend of $4.00 and, historically, the dividend has been increased 3% per year. You desire an annual rate of return of 9%. What is the most you should pay for the stock?

76. Stock Price Refer to Problem 75. Suppose that a stock pays an annual dividend of $2.50 and, historically, the dividend has increased 4% per year. You desire an annual rate of return of 11%. What is the most that you should pay for the stock?

78. Critical Thinking You are interviewing for a job and receive two offers:

   A: $20,000 to start, with guaranteed annual increases of 6% for the first 5 years

   B: $22,000 to start, with guaranteed annual increases of 3% for the first 5 years

Which offer is best if your goal is to be making as much as possible after 5 years? Which is best if your goal is to make as much money as possible over the contract (5 years)?

79. Critical Thinking Which of the following choices, *A* or *B*, results in more money?

 A: To receive $1000 on day 1, $999 on day 2, $998 on day 3, with the process to end after 1000 days

 B: To receive $1 on day 1, $2 on day 2, $4 on day 3, for 19 days

80. Critical Thinking You have just signed a 7-year professional football league contract with a beginning salary of $2,000,000 per year. Management gives you the following options with regard to your salary over the 7 years.

 1. A bonus of $100,000 each year
 2. An annual increase of 4.5% per year beginning after 1 year
 3. An annual increase of $95,000 per year beginning after 1 year

Which option provides the most money over the 7-year period? Which the least? Which would you choose? Why?

81. Can a sequence be both arithmetic and geometric? Give reasons for your answer.

82. Make up a geometric sequence. Give it to a friend and ask for its 20th term.

83. Make up two infinite geometric series, one that has a sum and one that does not. Give them to a friend and ask for the sum of each series.

84. Describe the similarities and differences between geometric sequences and exponential functions.

7.4 Mathematical Induction

OBJECTIVE 1 Prove Statements Using Mathematical Induction

1 Prove Statements Using Mathematical Induction

Mathematical induction is a method for proving that statements involving natural numbers are true for all natural numbers.*

For example, the statement "$2n$ is always an even integer" can be proved true for all natural numbers n by using mathematical induction. Also, the statement "the sum of the first n positive odd integers equals n^2," that is,

$$1 + 3 + 5 + \cdots + (2n - 1) = n^2 \tag{1}$$

can be proved for all natural numbers n by using mathematical induction.

Before stating the method of mathematical induction, let's try to gain a sense of the power of the method. We shall use the statement in equation (1) for this purpose by restating it for various values of $n = 1, 2, 3, \ldots$.

$n = 1$	The sum of the first positive odd integer is 1^2; $1 = 1^2$.
$n = 2$	The sum of the first 2 positive odd integers is 2^2; $1 + 3 = 4 = 2^2$.
$n = 3$	The sum of the first 3 positive odd integers is 3^2; $1 + 3 + 5 = 9 = 3^2$.
$n = 4$	The sum of the first 4 positive odd integers is 4^2; $1 + 3 + 5 + 7 = 16 = 4^2$.

Although from this pattern we might conjecture that statement (1) is true for any choice of n, can we really be sure that it does not fail for some choice of n? The method of proof by mathematical induction will, in fact, prove that the statement is true for all n.

*Recall that the natural numbers are the numbers 1, 2, 3, 4, ... In other words, the terms *natural numbers* and *positive integers* are synonymous.

Theorem

The Principle of Mathematical Induction

Suppose that the following two conditions are satisfied with regard to a statement about natural numbers:

CONDITION I: The statement is true for the natural number 1.

CONDITION II: If the statement is true for some natural number k, it is also true for the next natural number $k + 1$.

Then the statement is true for all natural numbers.

We shall not prove this principle. However, we can provide a physical interpretation that will help us to see why the principle works. Think of a collection of natural numbers obeying a statement as a collection of infinitely many dominoes. See Figure 16.

Figure 16

Now, suppose that we are told two facts:

1. The first domino is pushed over.

2. If one domino falls over, say the kth domino, then so will the next one, the $(k + 1)$st domino.

Is it safe to conclude that *all* the dominoes fall over? The answer is yes, because if the first one falls (Condition I), then the second one does also (by Condition II); and if the second one falls, then so does the third (by Condition II); and so on.

Now let's prove some statements about natural numbers using mathematical induction.

EXAMPLE 1	**Using Mathematical Induction**

Show that the following statement is true for all natural numbers n.

$$1 + 3 + 5 + \cdots + (2n - 1) = n^2 \tag{2}$$

Solution

We need to show first that statement (2) holds for $n = 1$. Because $1 = 1^2$, statement (2) is true for $n = 1$. Condition I holds.

Next, we need to show that Condition II holds. Suppose that we know for some k that

$$1 + 3 + \cdots + (2k - 1) = k^2 \tag{3}$$

We wish to show that, based on equation (3), statement (2) holds for $k + 1$. We look at the sum of the first $k + 1$ positive odd integers to determine whether this sum equals $(k + 1)^2$.

$$1 + 3 + \cdots + (2k - 1) + [2(k + 1) - 1] = \underbrace{[1 + 3 + \cdots + (2k - 1)]}_{= \, k^2 \text{ by equation (3)}} + (2k + 1)$$

$$= k^2 + (2k + 1)$$
$$= k^2 + 2k + 1 = (k + 1)^2$$

Conditions I and II are satisfied; by the Principle of Mathematical Induction, statement (2) is true for all natural numbers n. ◀

EXAMPLE 2	**Using Mathematical Induction**

Show that the following statement is true for all natural numbers n.

$$2^n > n$$

Solution First, we show that the statement $2^n > n$ holds when $n = 1$. Because $2^1 = 2 > 1$, the inequality is true for $n = 1$. Condition I holds.

Next, we assume, for some natural number k, that $2^k > k$. We wish to show that the formula holds for $k + 1$; that is, we wish to show that $2^{k+1} > k + 1$. Now

$$2^{k+1} = 2 \cdot 2^k > 2 \cdot k = k + k \geq k + 1$$

We know that $2^k > k$. $\quad\quad k \geq 1$

If $2^k > k$, then $2^{k+1} > k + 1$, so Condition II of the Principle of Mathematical Induction is satisfied. The statement $2^n > n$ is true for all natural numbers n. ◄

| **EXAMPLE 3** | **Using Mathematical Induction** |

Show that the following formula is true for all natural numbers n.

$$1 + 2 + 3 + \cdots + n = \frac{n(n + 1)}{2} \quad\quad \textbf{(4)}$$

Solution First, we show that formula (4) is true when $n = 1$. Because

$$\frac{1(1 + 1)}{2} = \frac{1(2)}{2} = 1$$

Condition I of the Principle of Mathematical Induction holds.

Next, we assume that formula (4) holds for some k, and we determine whether the formula then holds for $k + 1$. We assume that

$$1 + 2 + 3 + \cdots + k = \frac{k(k + 1)}{2} \quad \text{for some } k \quad\quad \textbf{(5)}$$

Now we need to show that

$$1 + 2 + 3 + \cdots + k + (k + 1) = \frac{(k + 1)(k + 1 + 1)}{2} = \frac{(k + 1)(k + 2)}{2}$$

We do this as follows:

$$1 + 2 + 3 + \cdots + k + (k + 1) = \underbrace{[1 + 2 + 3 + \cdots + k]}_{= \frac{k(k + 1)}{2} \quad \text{by equation (5)}} + (k + 1)$$

$$= \frac{k(k + 1)}{2} + (k + 1)$$

$$= \frac{k^2 + k + 2k + 2}{2}$$

$$= \frac{k^2 + 3k + 2}{2} = \frac{(k + 1)(k + 2)}{2}$$

Condition II also holds. As a result, formula (4) is true for all natural numbers n. ◄

NOW WORK PROBLEM 1.

| **EXAMPLE 4** | **Using Mathematical Induction** |

Show that $3^n - 1$ is divisible by 2 for all natural numbers n.

Solution First, we show that the statement is true when $n = 1$. Because $3^1 - 1 = 3 - 1 = 2$ is divisible by 2, the statement is true when $n = 1$. Condition I is satisfied.

Next, we assume that the statement holds for some k, and we determine whether the statement then holds for $k + 1$. We assume that $3^k - 1$ is divisible by 2 for some k. We need to show that $3^{k+1} - 1$ is divisible by 2. Now

$$3^{k+1} - 1 = 3^{k+1} - 3^k + 3^k - 1 \qquad \text{Subtract and add } 3^k.$$
$$= 3^k(3 - 1) + (3^k - 1) = 3^k \cdot 2 + (3^k - 1)$$

Because $3^k \cdot 2$ is divisible by 2 and $3^k - 1$ is divisible by 2, it follows that $3^k \cdot 2 + (3^k - 1) = 3^{k+1} - 1$ is divisible by 2. Condition II is also satisfied. As a result, the statement "$3^n - 1$ is divisible by 2" is true for all natural numbers n. ◀

WARNING The conclusion that a statement involving natural numbers is true for all natural numbers is made only after both Conditions I and II of the Principle of Mathematical Induction have been satisfied. Problem 27 demonstrates a statement for which only Condition I holds, but the statement is not true for all natural numbers. Problem 28 demonstrates a statement for which only Condition II holds, but the statement is not true for any natural number. ■

7.4 Assess Your Understanding

Skill Building

In Problems 1–26, use the Principle of Mathematical Induction to show that the given statement is true for all natural numbers n.

1. $2 + 4 + 6 + \cdots + 2n = n(n + 1)$

2. $1 + 5 + 9 + \cdots + (4n - 3) = n(2n - 1)$

3. $3 + 4 + 5 + \cdots + (n + 2) = \dfrac{1}{2}n(n + 5)$

4. $3 + 5 + 7 + \cdots + (2n + 1) = n(n + 2)$

5. $2 + 5 + 8 + \cdots + (3n - 1) = \dfrac{1}{2}n(3n + 1)$

6. $1 + 4 + 7 + \cdots + (3n - 2) = \dfrac{1}{2}n(3n - 1)$

7. $1 + 2 + 2^2 + \cdots + 2^{n-1} = 2^n - 1$

8. $1 + 3 + 3^2 + \cdots + 3^{n-1} = \dfrac{1}{2}(3^n - 1)$

9. $1 + 4 + 4^2 + \cdots + 4^{n-1} = \dfrac{1}{3}(4^n - 1)$

10. $1 + 5 + 5^2 + \cdots + 5^{n-1} = \dfrac{1}{4}(5^n - 1)$

11. $\dfrac{1}{1 \cdot 2} + \dfrac{1}{2 \cdot 3} + \dfrac{1}{3 \cdot 4} + \cdots + \dfrac{1}{n(n + 1)} = \dfrac{n}{n + 1}$

12. $\dfrac{1}{1 \cdot 3} + \dfrac{1}{3 \cdot 5} + \dfrac{1}{5 \cdot 7} + \cdots + \dfrac{1}{(2n - 1)(2n + 1)} = \dfrac{n}{2n + 1}$

13. $1^2 + 2^2 + 3^2 + \cdots + n^2 = \dfrac{1}{6}n(n + 1)(2n + 1)$

14. $1^3 + 2^3 + 3^3 + \cdots + n^3 = \dfrac{1}{4}n^2(n + 1)^2$

15. $4 + 3 + 2 + \cdots + (5 - n) = \dfrac{1}{2}n(9 - n)$

16. $-2 - 3 - 4 - \cdots - (n + 1) = -\dfrac{1}{2}n(n + 3)$

17. $1 \cdot 2 + 2 \cdot 3 + 3 \cdot 4 + \cdots + n(n + 1) = \dfrac{1}{3}n(n + 1)(n + 2)$

18. $1 \cdot 2 + 3 \cdot 4 + 5 \cdot 6 + \cdots + (2n - 1)(2n) = \dfrac{1}{3}n(n + 1)(4n - 1)$

19. $n^2 + n$ is divisible by 2.

20. $n^3 + 2n$ is divisible by 3.

21. $n^2 - n + 2$ is divisible by 2.

22. $n(n + 1)(n + 2)$ is divisible by 6.

23. If $x > 1$, then $x^n > 1$.

24. If $0 < x < 1$, then $0 < x^n < 1$.

25. $a - b$ is a factor of $a^n - b^n$.

26. $a + b$ is a factor of $a^{2n+1} + b^{2n+1}$.

[**Hint:** $a^{k+1} - b^{k+1} = a(a^k - b^k) + b^k(a - b)$]

Applications and Extensions

27. Show that the statement "$n^2 - n + 41$ is a prime number" is true for $n = 1$, but is not true for $n = 41$.

28. Show that the formula

$$2 + 4 + 6 + \cdots + 2n = n^2 + n + 2$$

obeys Condition II of the Principle of Mathematical Induction. That is, show that if the formula is true for some k it is also true for $k + 1$. Then show that the formula is false for $n = 1$ (or for any other choice of n).

29. Use mathematical induction to prove that if $r \neq 1$ then

$$a + ar + ar^2 + \cdots + ar^{n-1} = a\frac{1 - r^n}{1 - r}$$

30. Use mathematical induction to prove that

$$a + (a + d) + (a + 2d)$$
$$+ \cdots + [a + (n - 1)d] = na + d\frac{n(n - 1)}{2}$$

31. Extended Principle of Mathematical Induction The Extended Principle of Mathematical Induction states that if Conditions I and II hold, that is,

(I)   A statement is true for a natural number j.

(II) If the statement is true for some natural number $k \geq j$, then it is also true for the next natural number $k + 1$.

then the statement is true for all natural numbers $\geq j$.

Use the Extended Principle of Mathematical Induction to show that the number of diagonals in a convex polygon of n

sides is $\dfrac{1}{2}n(n - 3)$.

[**Hint:** Begin by showing that the result is true when $n = 4$ (Condition I).]

32. Geometry Use the Extended Principle of Mathematical Induction to show that the sum of the interior angles of a convex polygon of n sides equals $(n - 2) \cdot 180°$.

Discussion and Writing

33. How would you explain the Principle of Mathematical Induction to a friend?

7.5 The Binomial Theorem

OBJECTIVES **1** Evaluate $\dbinom{n}{j}$

2 Use the Binomial Theorem

Formulas have been given for expanding $(x + a)^n$ for $n = 2$ and $n = 3$. The *Binomial Theorem*[*] is a formula for the expansion of $(x + a)^n$ for any positive integer n. If $n = 1, 2, 3,$ and 4, the expansion of $(x + a)^n$ is straightforward.

$(x + a)^1 = x + a$ Two terms, beginning with x^1 and ending with a^1

$(x + a)^2 = x^2 + 2ax + a^2$ Three terms, beginning with x^2 and ending with a^2

$(x + a)^3 = x^3 + 3ax^2 + 3a^2x + a^3$ Four terms, beginning with x^3 and ending with a^3

$(x + a)^4 = x^4 + 4ax^3 + 6a^2x^2 + 4a^3x + a^4$ Five terms, beginning with x^4 and ending with a^4

Notice that each expansion of $(x + a)^n$ begins with x^n and ends with a^n. As you read from left to right, the powers of x are decreasing by one, while the powers of a are increasing by one. Also, the number of terms that appears equals $n + 1$. Notice, too, that the degree of each monomial in the expansion equals n. For example, in the

[*]The name *binomial* is derived from the fact that $x + a$ is a binomial; that is, it contains two terms.

expansion of $(x + a)^3$, each monomial $(x^3, 3ax^2, 3a^2x, a^3)$ is of degree 3. As a result, we might conjecture that the expansion of $(x + a)^n$ would look like this:

$$(x + a)^n = x^n + \underline{\quad} ax^{n-1} + \underline{\quad} a^2 x^{n-2} + \cdots + \underline{\quad} a^{n-1}x + a^n$$

where the blanks are numbers to be found. This is, in fact, the case, as we shall see shortly.

Before we can fill in the blanks, we need to introduce the symbol $\binom{n}{j}$.

1 **Evaluate** $\binom{n}{j}$

We define the symbol $\binom{n}{j}$, read "n taken j at a time," as follows:

NOTE
On a graphing calculator, the symbol $\binom{n}{j}$ may be denoted by the key $\boxed{nCr}$. ∎

If j and n are integers with $0 \leq j \leq n$, the symbol $\binom{n}{j}$ is defined as

$$\binom{n}{j} = \frac{n!}{j!(n-j)!} \qquad \textbf{(1)}$$

EXAMPLE 1   **Evaluating** $\binom{n}{j}$

Find:

(a) $\binom{3}{1}$ (b) $\binom{4}{2}$ (c) $\binom{8}{7}$ (d) $\binom{65}{15}$

Solution   (a) $\binom{3}{1} = \frac{3!}{1!(3-1)!} = \frac{3!}{1! \, 2!} = \frac{3 \cdot 2 \cdot 1}{1(2 \cdot 1)} = \frac{6}{2} = 3$

(b) $\binom{4}{2} = \frac{4!}{2!(4-2)!} = \frac{4!}{2! \, 2!} = \frac{4 \cdot 3 \cdot 2 \cdot 1}{(2 \cdot 1)(2 \cdot 1)} = \frac{24}{4} = 6$

Figure 17

```
65 nCr 15
   2.073746998E14
```

(c) $\binom{8}{7} = \frac{8!}{7!(8-7)!} = \frac{8!}{7! \, 1!} = \frac{8 \cdot 7!}{7! \cdot 1!} = \frac{8}{1} = 8$

$\uparrow$
$8! = 8 \cdot 7!$

(d) Figure 17 shows the solution using a TI-84 Plus graphing calculator. So,

$$\binom{65}{15} \approx 2.073746998 \times 10^{14}$$

◀

✏ ── **NOW WORK PROBLEM 5.**

Four useful formulas involving the symbol $\binom{n}{j}$ are

$$\binom{n}{0} = 1 \qquad \binom{n}{1} = n \qquad \binom{n}{n-1} = n \qquad \binom{n}{n} = 1$$

Proof $\dbinom{n}{0} = \dfrac{n!}{0!(n-0)!} = \dfrac{n!}{0!\,n!} = \dfrac{1}{1} = 1$

$\dbinom{n}{1} = \dfrac{n!}{1!(n-1)!} = \dfrac{n!}{(n-1)!} = \dfrac{n\cancel{(n-1)!}}{\cancel{(n-1)!}} = n$

You are asked to show the remaining two formulas in Problem 45. ∎

Suppose that we arrange the various values of the symbol $\dbinom{n}{j}$ in a triangular display, as shown next and in Figure 18.

$$\binom{0}{0}$$

$$\binom{1}{0}\ \binom{1}{1}$$

$$\binom{2}{0}\ \binom{2}{1}\ \binom{2}{2}$$

$$\binom{3}{0}\ \binom{3}{1}\ \binom{3}{2}\ \binom{3}{3}$$

$$\binom{4}{0}\ \binom{4}{1}\ \binom{4}{2}\ \binom{4}{3}\ \binom{4}{4}$$

$$\binom{5}{0}\ \binom{5}{1}\ \binom{5}{2}\ \binom{5}{3}\ \binom{5}{4}\ \binom{5}{5}$$

Figure 18
Pascal triangle

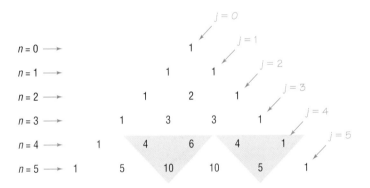

This display is called the **Pascal triangle**, named after Blaise Pascal (1623–1662), a French mathematician.

The Pascal triangle has 1's down the sides. To get any other entry, add the two nearest entries in the row above it. The shaded triangles in Figure 18 illustrate this feature of the Pascal triangle. Based on this feature, the row corresponding to $n = 6$ is found as follows:

$$n = 5 \rightarrow \quad 1 \quad 5 \quad 10 \quad 10 \quad 5 \quad 1$$
$$n = 6 \rightarrow \quad 1 \quad 6 \quad 15 \quad 20 \quad 15 \quad 6 \quad 1$$

Later we shall prove that this addition always works (see the theorem on page 686).

Although the Pascal triangle provides an interesting and organized display of the symbol $\dbinom{n}{j}$, in practice it is not all that helpful. For example, if you wanted to know the value of $\dbinom{12}{5}$, you would need to produce 13 rows of the triangle before seeing the answer. It is much faster to use the definition (1).

2 Use the Binomial Theorem

Now we are ready to state the **Binomial Theorem**.

Theorem

Binomial Theorem

Let x and a be real numbers. For any positive integer n, we have

$$(x + a)^n = \binom{n}{0}x^n + \binom{n}{1}ax^{n-1} + \cdots + \binom{n}{j}a^jx^{n-j} + \cdots + \binom{n}{n}a^n$$

$$= \sum_{j=0}^{n}\binom{n}{j}x^{n-j}a^j \qquad\qquad (2)$$

Now you know why we needed to introduce the symbol $\binom{n}{j}$; these symbols are the numerical coefficients that appear in the expansion of $(x + a)^n$. Because of this, the symbol $\binom{n}{j}$ is called a **binomial coefficient**.

EXAMPLE 2 | **Expanding a Binomial**

Use the Binomial Theorem to expand $(x + 2)^5$.

Solution In the Binomial Theorem, let $a = 2$ and $n = 5$. Then

$$(x + 2)^5 = \binom{5}{0}x^5 + \binom{5}{1}2x^4 + \binom{5}{2}2^2x^3 + \binom{5}{3}2^3x^2 + \binom{5}{4}2^4x + \binom{5}{5}2^5$$

↑
Use equation (2)

$$= 1 \cdot x^5 + 5 \cdot 2x^4 + 10 \cdot 4x^3 + 10 \cdot 8x^2 + 5 \cdot 16x + 1 \cdot 32$$

↑
Use row $n = 5$ of the Pascal triangle or formula (1) for $\binom{n}{j}$.

$$= x^5 + 10x^4 + 40x^3 + 80x^2 + 80x + 32 \qquad \blacktriangleleft$$

EXAMPLE 3 | **Expanding a Binomial**

Expand $(2y - 3)^4$ using the Binomial Theorem.

Solution First, we rewrite the expression $(2y - 3)^4$ as $[2y + (-3)]^4$. Now we use the Binomial Theorem with $n = 4$, $x = 2y$, and $a = -3$.

$$[2y + (-3)]^4 = \binom{4}{0}(2y)^4 + \binom{4}{1}(-3)(2y)^3 + \binom{4}{2}(-3)^2(2y)^2$$

$$+ \binom{4}{3}(-3)^3(2y) + \binom{4}{4}(-3)^4$$

$$= 1 \cdot 16y^4 + 4(-3)8y^3 + 6 \cdot 9 \cdot 4y^2 + 4(-27)2y + 1 \cdot 81$$

↑
Use row $n = 4$ of the Pascal triangle or formula (1) for $\binom{n}{j}$.

$$= 16y^4 - 96y^3 + 216y^2 - 216y + 81$$

In this expansion, note that the signs alternate due to the fact that $a = -3 < 0$. $\blacktriangleleft$

NOW WORK PROBLEM 21.

| **EXAMPLE 4** | **Finding a Particular Coefficient in a Binomial Expansion** |

Find the coefficient of y^8 in the expansion of $(2y + 3)^{10}$.

Solution   We write out the expansion using the Binomial Theorem.

$$(2y + 3)^{10} = \binom{10}{0}(2y)^{10} + \binom{10}{1}(2y)^9(3)^1 + \binom{10}{2}(2y)^8(3)^2 + \binom{10}{3}(2y)^7(3)^3$$

$$+ \binom{10}{4}(2y)^6(3)^4 + \cdots + \binom{10}{9}(2y)(3)^9 + \binom{10}{10}(3)^{10}$$

From the third term in the expansion, the coefficient of y^8 is

$$\binom{10}{2}(2)^8(3)^2 = \frac{10!}{2!\,8!} \cdot 2^8 \cdot 9 = \frac{10 \cdot 9 \cdot 8!}{2 \cdot 8!} \cdot 2^8 \cdot 9 = 103{,}680 \qquad \blacktriangleleft$$

As this solution demonstrates, we can use the Binomial Theorem to find a particular term in an expansion without writing the entire expansion.

Based on the expansion of $(x + a)^n$, the term containing x^j is

$$\boxed{\binom{n}{n-j}a^{n-j}x^j} \qquad \qquad \textbf{(3)}$$

For example, we can solve Example 4 by using formula (3) with $n = 10$, $a = 3$, $x = 2y$, and $j = 8$. Then the term containing y^8 is

$$\binom{10}{10-8}3^{10-8}(2y)^8 = \binom{10}{2} \cdot 3^2 \cdot 2^8 \cdot y^8 = \frac{10!}{2!\,8!} \cdot 9 \cdot 2^8 y^8$$

$$= \frac{10 \cdot 9 \cdot 8!}{2 \cdot 8!} \cdot 9 \cdot 2^8 y^8 = 103{,}680 y^8$$

| **EXAMPLE 5** | **Finding a Particular Term in a Binomial Expansion** |

Find the sixth term in the expansion of $(x + 2)^9$.

Solution A   We expand using the Binomial Theorem until the sixth term is reached.

$$(x + 2)^9 = \binom{9}{0}x^9 + \binom{9}{1}x^8 \cdot 2 + \binom{9}{2}x^7 \cdot 2^2 + \binom{9}{3}x^6 \cdot 2^3 + \binom{9}{4}x^5 \cdot 2^4$$

$$+ \binom{9}{5}x^4 \cdot 2^5 + \cdots$$

The sixth term is

$$\binom{9}{5}x^4 \cdot 2^5 = \frac{9!}{5!\,4!} \cdot x^4 \cdot 32 = 4032x^4$$

Solution B   The sixth term in the expansion of $(x + 2)^9$, which has 10 terms total, contains x^4. (Do you see why?) By formula (3), the sixth term is

$$\binom{9}{9-4}2^{9-4}x^4 = \binom{9}{5}2^5x^4 = \frac{9!}{5!\,4!} \cdot 32x^4 = 4032x^4 \qquad \blacktriangleleft$$

NOW WORK PROBLEMS **29** AND **35**.

Next we show that the *triangular addition* feature of the Pascal triangle illustrated in Figure 18 always works.

Theorem If n and j are integers with $1 \le j \le n$, then

$$\binom{n}{j-1} + \binom{n}{j} = \binom{n+1}{j} \qquad \textbf{(4)}$$

Proof

$$\binom{n}{j-1} + \binom{n}{j} = \frac{n!}{(j-1)![n-(j-1)]!} + \frac{n!}{j!(n-j)!}$$

$$= \frac{n!}{(j-1)!(n-j+1)!} + \frac{n!}{j!(n-j)!}$$

$$= \frac{jn!}{j(j-1)!(n-j+1)!} + \frac{(n-j+1)n!}{j!(n-j+1)(n-j)!}$$

Multiply the first term by $\dfrac{j}{j}$ and the second term by $\dfrac{n-j+1}{n-j+1}$. Now the denominators are equal.

$$= \frac{jn!}{j!(n-j+1)!} + \frac{(n-j+1)n!}{j!(n-j+1)!}$$

$$= \frac{jn! + (n-j+1)n!}{j!(n-j+1)!}$$

$$= \frac{n!(j+n-j+1)}{j!(n-j+1)!}$$

$$= \frac{n!(n+1)}{j!(n-j+1)!} = \frac{(n+1)!}{j![(n+1)-j]!} = \binom{n+1}{j} \qquad \blacksquare$$

HISTORICAL FEATURE

Omar Khayyám
(1050–1123)

The case $n = 2$ of the Binomial Theorem, $(a + b)^2$, was known to Euclid in 300 BC, but the general law seems to have been discovered by the Persian mathematician and astronomer Omar Khayyám (1050–1123), who is also well known as the author of the *Rubáiyát*, a collection of four-line poems making observations on the human condition. Omar Khayyám did not state the Binomial Theorem explicitly, but he claimed to have a method for extracting third, fourth, fifth roots, and so on. A little study shows that one must know the Binomial Theorem to create such a method.

The heart of the Binomial Theorem is the formula for the numerical coefficients, and, as we saw, they can be written out in a symmetric triangular form. The Pascal triangle appears first in the books of Yang Hui (about 1270) and Chu Shih-chieh (1303). Pascal's name is attached to the triangle because of the many applications he made of it, especially to counting and probability. In establishing these results, he was one of the earliest users of mathematical induction.

Many people worked on the proof of the Binomial Theorem, which was finally completed for all n (including complex numbers) by Niels Abel (1802–1829).

7.5 Assess Your Understanding

Concepts and Vocabulary

1. The _____ _____ is a triangular display of the binomial coefficients.

2. $\binom{6}{2} =$ _____

3. *True of False:* $\binom{n}{j} = \dfrac{j!}{(n-j)!\, n!}$

4. The _____ _____ can be used to expand expressions like $(2x + 3)^6$.

Skill Building

In Problems 5–16, evaluate each expression.

5. $\dbinom{5}{3}$ **6.** $\dbinom{7}{3}$ **7.** $\dbinom{7}{5}$ **8.** $\dbinom{9}{7}$

9. $\dbinom{50}{49}$ **10.** $\dbinom{100}{98}$ **11.** $\dbinom{1000}{1000}$ **12.** $\dbinom{1000}{0}$

13. $\dbinom{55}{23}$ **14.** $\dbinom{60}{20}$ **15.** $\dbinom{47}{25}$ **16.** $\dbinom{37}{19}$

In Problems 17–28, expand each expression using the Binomial Theorem.

17. $(x + 1)^5$ **18.** $(x - 1)^5$ **19.** $(x - 2)^6$ **20.** $(x + 3)^5$

21. $(3x + 1)^4$ **22.** $(2x + 3)^5$ **23.** $(x^2 + y^2)^5$ **24.** $(x^2 - y^2)^6$

25. $\left(\sqrt{x} + \sqrt{2}\right)^6$ **26.** $\left(\sqrt{x} - \sqrt{3}\right)^4$ **27.** $(ax + by)^5$ **28.** $(ax - by)^4$

In Problems 29–42, use the Binomial Theorem to find the indicated coefficient or term.

29. The coefficient of x^6 in the expansion of $(x + 3)^{10}$

30. The coefficient of x^3 in the expansion of $(x - 3)^{10}$

31. The coefficient of x^7 in the expansion of $(2x - 1)^{12}$

32. The coefficient of x^3 in the expansion of $(2x + 1)^{12}$

33. The coefficient of x^7 in the expansion of $(2x + 3)^9$

34. The coefficient of x^2 in the expansion of $(2x - 3)^9$

35. The fifth term in the expansion of $(x + 3)^7$

36. The third term in the expansion of $(x - 3)^7$

37. The third term in the expansion of $(3x - 2)^9$

38. The sixth term in the expansion of $(3x + 2)^8$

39. The coefficient of x^0 in the expansion of $\left(x^2 + \dfrac{1}{x}\right)^{12}$

40. The coefficient of x^0 in the expansion of $\left(x - \dfrac{1}{x^2}\right)^9$

41. The coefficient of x^4 in the expansion of $\left(x - \dfrac{2}{\sqrt{x}}\right)^{10}$

42. The coefficient of x^2 in the expansion of $\left(\sqrt{x} + \dfrac{3}{\sqrt{x}}\right)^8$

Applications and Extensions

43. Use the Binomial Theorem to find the numerical value of $(1.001)^5$ correct to five decimal places.
   [Hint: $(1.001)^5 = (1 + 10^{-3})^5]$

44. Use the Binomial Theorem to find the numerical value of $(0.998)^6$ correct to five decimal places.

45. Show that $\dbinom{n}{n - 1} = n$ and $\dbinom{n}{n} = 1$.

46. Show that if n and j are integers with $0 \le j \le n$ then

$$\binom{n}{j} = \binom{n}{n - j}$$

   Conclude that the Pascal triangle is symmetric with respect to a vertical line drawn from the topmost entry.

47. If n is a positive integer, show that

$$\binom{n}{0} + \binom{n}{1} + \cdots + \binom{n}{n} = 2^n$$

   [Hint: $2^n = (1 + 1)^n$; now use the Binomial Theorem.]

48. If n is a positive integer, show that

$$\binom{n}{0} - \binom{n}{1} + \binom{n}{2} - \cdots + (-1)^n \binom{n}{n} = 0$$

49. $\dbinom{5}{0}\left(\dfrac{1}{4}\right)^5 + \dbinom{5}{1}\left(\dfrac{1}{4}\right)^4\left(\dfrac{3}{4}\right) + \dbinom{5}{2}\left(\dfrac{1}{4}\right)^3\left(\dfrac{3}{4}\right)^2$

$+ \dbinom{5}{3}\left(\dfrac{1}{4}\right)^2\left(\dfrac{3}{4}\right)^3 + \dbinom{5}{4}\left(\dfrac{1}{4}\right)\left(\dfrac{3}{4}\right)^4 + \dbinom{5}{5}\left(\dfrac{3}{4}\right)^5 = ?$

50. Stirling's Formula   An approximation for $n!$, when n is large, is given by

$$n! \approx \sqrt{2n\pi}\left(\dfrac{n}{e}\right)^n\left(1 + \dfrac{1}{12n - 1}\right)$$

   Calculate 12!, 20!, and 25! on your calculator. Then use Stirling's formula to approximate 12!, 20!, and 25!.

Chapter Review

Things to Know

Sequence (p. 650)   A function whose domain is the set of positive integers.

Factorials (p. 652) $0! = 1, 1! = 1, n! = n(n - 1) \cdot \ldots \cdot 3 \cdot 2 \cdot 1$ if $n \geq 2$

Arithmetic sequence $a_1 = a, a_n = a_{n-1} + d$, where a = first term, d = common difference
(pp. 662 and 664) $a_n = a + (n - 1)d$

Sum of the first n terms $S_n = \dfrac{n}{2}[2a + (n - 1)d] = \dfrac{n}{2}(a + a_n)$
of an arithmetic sequence (p. 665)

Geometric sequence $a_1 = a, \quad a_n = ra_{n-1}$, where a = first term, r = common ratio
(pp. 668 and 670) $a_n = ar^{n-1}, \quad r \neq 0$

Sum of the first n terms of $S_n = a\dfrac{1 - r^n}{1 - r}, \quad r \neq 0, 1$
a geometric sequence (p. 670)

Infinite geometric series (p. 672) $a + ar + \cdots + ar^{n-1} + \cdots = \displaystyle\sum_{k=1}^{\infty} ar^{k-1}$

Sum of an infinite $\displaystyle\sum_{k=1}^{\infty} ar^{k-1} = \dfrac{a}{1 - r}, \quad |r| < 1$
geometric series (p. 672)

Principle of Mathematical   Suppose the following two conditions are satisfied.
Induction (p. 678)   Condition I: The statement is true for the natural number 1.
    Condition II: If the statement is true for some natural number k, it is also true for $k + 1$.
    Then the statement is true for all natural numbers n.

Binomial coefficient (p. 682) $\dbinom{n}{j} = \dfrac{n!}{j!(n - j)!}$

Pascal triangle (p. 683)   See Figure 18.

Binomial Theorem (p. 684) $(x + a)^n = \dbinom{n}{0}x^n + \dbinom{n}{1}ax^{n-1} + \cdots + \dbinom{n}{j}a^jx^{n-j} + \cdots + \dbinom{n}{n}a^n$

Objectives

Section		You should be able to . . .	Review Exercises
7.1	1	Write the first several terms of a sequence (p. 650)	1–4
	2	Write the terms of a sequence defined by a recursive formula (p. 653)	5–8
	3	Use summation notation (p. 654)	9–12
	4	Find the sum of a sequence algebraically and using a graphing utility (p. 655)	25–30
	5	Solve annuity and amortization problems (p. 656)	66, 67
7.2	1	Determine if a sequence is arithmetic (p. 662)	13–24
	2	Find a formula for an arithmetic sequence (p. 663)	31, 32, 35, 37–40, 63, 64
	3	Find the sum of an arithmetic sequence (p. 665)	13, 14, 19, 20, 63, 64
7.3	1	Determine if a sequence is geometric (p. 668)	13–24
	2	Find a formula for a geometric sequence (p. 669)	17, 18, 21, 22, 33–36
	3	Find the sum of a geometric sequence (p. 670)	17, 18, 21, 22, 65(a)-(c)
	4	Find the sum of a geometric series (p. 672)	41–46, 65(d)
7.4	1	Prove statements using mathematical induction (p. 677)	47–52
7.5	1	Evaluate $\dbinom{n}{j}$ (p. 682)	53–54
	2	Use the Binomial Theorem (p. 684)	55–62

Review Exercises

In Problems 1–8, write down the first five terms of each sequence.

1. $\left\{(-1)^n\left(\dfrac{n+3}{n+2}\right)\right\}$ **2.** $\{(-1)^{n+1}(2n+3)\}$ **3.** $\left\{\dfrac{2^n}{n^2}\right\}$ **4.** $\left\{\dfrac{e^n}{n}\right\}$

5. $a_1 = 3;\ \ a_n = \dfrac{2}{3}a_{n-1}$ **6.** $a_1 = 4;\ \ a_n = -\dfrac{1}{4}a_{n-1}$ **7.** $a_1 = 2;\ \ a_n = 2 - a_{n-1}$ **8.** $a_1 = -3;\ \ a_n = 4 + a_{n-1}$

In Problems 9 and 10, write out each sum.

9. $\displaystyle\sum_{k=1}^{4}(4k+2)$ **10.** $\displaystyle\sum_{k=1}^{3}(3-k^2)$

In Problems 11 and 12, express each sum using summation notation.

11. $1 - \dfrac{1}{2} + \dfrac{1}{3} - \dfrac{1}{4} + \cdots + \dfrac{1}{13}$ **12.** $2 + \dfrac{2^2}{3} + \dfrac{2^3}{3^2} + \cdots + \dfrac{2^{n+1}}{3^n}$

In Problems 13–24, determine whether the given sequence is arithmetic, geometric, or neither. If the sequence is arithmetic, find the common difference and the sum of the first n terms. If the sequence is geometric, find the common ratio and the sum of the first n terms.

13. $\{n+5\}$ **14.** $\{4n+3\}$ **15.** $\{2n^3\}$ **16.** $\{2n^2-1\}$

17. $\{2^{3n}\}$ **18.** $\{3^{2n}\}$ **19.** $0,4,8,12,\ldots$ **20.** $1,-3,-7,-11,\ldots$

21. $3,\dfrac{3}{2},\dfrac{3}{4},\dfrac{3}{8},\dfrac{3}{16},\cdots$ **22.** $5,-\dfrac{5}{3},\dfrac{5}{9},-\dfrac{5}{27},\dfrac{5}{81},\cdots$ **23.** $\dfrac{2}{3},\dfrac{3}{4},\dfrac{4}{5},\dfrac{5}{6},\cdots$ **24.** $\dfrac{3}{2},\dfrac{5}{4},\dfrac{7}{6},\dfrac{9}{8},\dfrac{11}{10},\cdots$

In Problems 25–30, evaluate each sum (a) algebraically and (b) using a graphing utility.

25. $\displaystyle\sum_{k=1}^{5}(k^2+12)$ **26.** $\displaystyle\sum_{k=1}^{3}(k+2)^2$ **27.** $\displaystyle\sum_{k=1}^{10}(3k-9)$

28. $\displaystyle\sum_{k=1}^{9}(-2k+8)$ **29.** $\displaystyle\sum_{k=1}^{7}\left(\dfrac{1}{3}\right)^k$ **30.** $\displaystyle\sum_{k=1}^{10}(-2)^k$

In Problems 31–36, find the indicated term in each sequence.
[Hint: Find the general term first.]

31. 9th term of $3,7,11,15,\ldots$ **32.** 8th term of $1,-1,-3,-5,\ldots$ **33.** 11th term of $1,\dfrac{1}{10},\dfrac{1}{100},\ldots$

34. 11th term of $1,2,4,8,\ldots$ **35.** 9th term of $\sqrt{2},2\sqrt{2},3\sqrt{2},\ldots$ **36.** 9th term of $\sqrt{2},2,2^{3/2},\ldots$

In Problems 37–40, find a general formula for each arithmetic sequence.

37. 7th term is 31; 20th term is 96   **38.** 8th term is -20; 17th term is -47

39. 10th term is 0; 18th term is 8   **40.** 12th term is 30; 22nd term is 50

In Problems 41–46, find the sum of each infinite geometric series.

41. $3 + 1 + \dfrac{1}{3} + \dfrac{1}{9} + \cdots$ **42.** $2 + 1 + \dfrac{1}{2} + \dfrac{1}{4} + \cdots$ **43.** $2 - 1 + \dfrac{1}{2} - \dfrac{1}{4} + \cdots$

44. $6 - 4 + \dfrac{8}{3} - \dfrac{16}{9} + \cdots$ **45.** $\displaystyle\sum_{k=1}^{\infty}4\left(\dfrac{1}{2}\right)^{k-1}$ **46.** $\displaystyle\sum_{k=1}^{\infty}3\left(-\dfrac{3}{4}\right)^{k-1}$

In Problems 47–52, use the Principle of Mathematical Induction to show that the given statement is true for all natural numbers.

47. $3 + 6 + 9 + \cdots + 3n = \dfrac{3n}{2}(n+1)$ **48.** $2 + 6 + 10 + \cdots + (4n-2) = 2n^2$

49. $2 + 6 + 18 + \cdots + 2\cdot3^{n-1} = 3^n - 1$ **50.** $3 + 6 + 12 + \cdots + 3\cdot2^{n-1} = 3(2^n - 1)$

51. $1^2 + 4^2 + 7^2 + \cdots + (3n-2)^2 = \dfrac{1}{2}n(6n^2 - 3n - 1)$ **52.** $1\cdot3 + 2\cdot4 + 3\cdot5 + \cdots + n(n+2) = \dfrac{n}{6}(n+1)(2n+7)$

In Problems 53 and 54, evaluate each binomial coefficient.

53. $\dbinom{5}{2}$

54. $\dbinom{8}{6}$

In Problems 55–58, expand each expression using the Binomial Theorem.

55. $(x + 2)^5$
56. $(x - 3)^4$
57. $(2x + 3)^5$
58. $(3x - 4)^4$

---

59. Find the coefficient of x^7 in the expansion of $(x + 2)^9$.

60. Find the coefficient of x^3 in the expansion of $(x - 3)^8$.

61. Find the coefficient of x^2 in the expansion of $(2x + 1)^7$.

62. Find the coefficient of x^6 in the expansion of $(2x + 1)^8$.

63. Constructing a Brick Staircase A brick staircase has a total of 25 steps. The bottom step requires 80 bricks. Each successive step requires three less bricks than the prior step.
(a) How many bricks are required for the top step?
(b) How many bricks are required to build the staircase?

64. Creating a Floor Design A mosaic tile floor is designed in the shape of a trapezoid 30 feet wide at the base and 15 feet wide at the top. The tiles, 12 inches by 12 inches, are to be placed so that each successive row contains one less tile than the row below. How many tiles will be required?

[**Hint:** Refer to Figure 13, page 666.]

65. Bouncing Balls A ball is dropped from a height of 20 feet. Each time it strikes the ground, it bounces up to three-quarters of the previous height.
(a) What height will the ball bounce up to after it strikes the ground for the third time?
(b) How high will it bounce after it strikes the ground for the nth time?
(c) How many times does the ball need to strike the ground before its bounce is less than 6 inches?
(d) What total distance does the ball travel before it stops bouncing?

66. Home Loan Mike and Yola borrowed $190,000 at 6.75% per annum compounded monthly for 30 years to purchase a home. Their monthly payment is determined to be $1232.34.
(a) Find a recursive formula for the balance after each monthly payment has been made.

(b) Determine the balance after the first payment.
(c) Using a graphing utility, create a table showing the balance after each monthly payment.
(d) Using a graphing utility, determine when the balance will be below $100,000.
(e) Using a graphing utility, determine when Mike and Yola will pay off the balance.
(f) Determine their interest expense when the loan is paid.
(g) Suppose that Mike and Yola decide to pay an additional $100 each month on the loan. Answer parts (a) to (f) under this scenario.

67. Credit Card Debt Beth just charged $5000 on a VISA card that charges 1.5% interest per month on any unpaid balance. She can afford to pay $100 toward the balance each month. Her balance at the beginning of each month after making a $100 payment is given by the recursively defined sequence

$$b = \$5000, \quad b_n = 1.015b_{n-1} - 100$$

(a) Determine Beth's balance after making the first payment; that is, determine b_2, the balance at the beginning of the second month.
(b) Using a graphing utility, graph the recursively defined sequence.
(c) Using a graphing utility, determine when Beth's balance will be below $4000. How many payments of $100 have been made?
(d) Using a graphing utility, determine the number of payments it will take until Beth pays off the balance. What is the total of all the payments?
(e) What was Beth's interest expense?

Chapter Test

In Problems 1 and 2, write down the first five terms of each sequence.

1. $\left\{ \dfrac{n^2 - 1}{n + 8} \right\}$

2. $a_1 = 4, a_n = 3a_{n-1} + 2$

In Problems 3 and 4, write out each sum. Evaluate each sum.

3. $\displaystyle\sum_{k=1}^{3} (-1)^{k+1} \left(\dfrac{k + 1}{k^2} \right)$

4. $\displaystyle\sum_{k=1}^{4} \left[\left(\dfrac{2}{3} \right)^k - k \right]$

5. Write the following sum using summation notation.

$$-\dfrac{2}{5} + \dfrac{3}{6} - \dfrac{4}{7} + \cdots + \dfrac{11}{14}$$

6. Find the sum using a graphing utility.

$$\displaystyle\sum_{k=1}^{100} (-1)^k (k^2)$$

In Problems 7–12, determine whether the given sequence is arithmetic, geometric, or neither. If the sequence is arithmetic, find the common difference and the sum of the first n terms. If the sequence is geometric, find the common ratio and the sum of the first n terms.

7. $6, 12, 36, 144, \ldots$

8. $\left\{ -\dfrac{1}{2} \cdot 4^n \right\}$

9. $-2, -10, -18, -26, \ldots$

10. $\left\{ -\dfrac{n}{2} + 7 \right\}$

11. $25, 10, 4, \dfrac{8}{5}, \ldots$

12. $\left\{ \dfrac{2n - 3}{2n + 1} \right\}$

---

13. Find the sum of the infinite geometric series
$256 - 64 + 16 - 4 + \cdots$

14. Expand $(3m + 2)^5$ using the Binomial Theorem.

15. Use the Principle of Mathematical Induction to show that the given statement is true for all natural numbers.
$$\left(1 + \dfrac{1}{1}\right)\left(1 + \dfrac{1}{2}\right)\left(1 + \dfrac{1}{3}\right) \cdots \left(1 + \dfrac{1}{n}\right) = n + 1$$

16. A 2004 Dodge Durango sold for \$31,000. If the vehicle loses 15% of its value each year, how much will it be worth after 10 years?

17. A weightlifter begins his routine by benching 100 pounds and increases the weight by 30 pounds for each set. If he does 10 repetitions in each set, what is the total weight lifted after 5 sets?

Chapter Projects

1. **Population Growth** The size of the population of the United States essentially depends on its current population, the birth and death rates of the population, and immigration. Suppose that b represents the birth rate of the U.S. population and d represents its death rate. Then $r = b - d$ represents the growth rate of the population, where r varies from year to year. The U.S. population after n years can be modeled using the recursive function

$$p_n = (1 + r)p_{n-1} + I$$

where I represents net immigration into the United States.

(a) Using data from the National Center for Health Statistics *www.fedstats.gov*, determine the birth and death rates for all races for the most recent year that data are available. Birth rates are given as the number of live births per 1000 population, while death rates are given as the number of deaths per 100,000 population. Each must be computed as the number of births (deaths) per individual. For example, in 1990, the birth rate was 16.7 per 1000 and the death rate was 863.8 per 100,000, so

$$b = \frac{16.7}{1000} = 0.0167, \text{ while } d = \frac{863.8}{100,000} = 0.008638.$$

Next, using data from the Immigration and Naturalization Service *www.fedstats.gov*, determine the net immigration to the United States for the same year used to obtain b and d in part (a).

(b) Determine the value of r, the growth rate of the population.

(c) Find a recursive formula for the population of the United States.

(d) Use the recursive formula to predict the population of the United States in the following year. In other words, if data are available up to the year 2003, predict the U.S. population in 2004.

(e) Compare your prediction to actual data.

(f) Do you think the recursive formula found in part (c) will be useful in predicting future populations? Why or why not?

The following projects are available at the Instructor's Resource Center (IRC):

2. **Project at Motorola** *Digital Wireless Communication*

3. **Economics**

4. **Standardized Tests**

Cumulative Review

1. Find all the solutions, real and complex, of the equation

$$|x^2| = 9$$

2. (a) Graph the circle $x^2 + y^2 = 100$ and the parabola $y = 3x^2$

   (b) Solve the system of equations: $\begin{cases} x^2 + y^2 = 100 \\ \quad\ y = 3x^2 \end{cases}$

   (c) Where do the circle and the parabola intersect?

3. Solve the equation $2e^x = 5$.

4. Find an equation of the line with slope 5 and x-intercept 2.

5. Find the general equation of the circle whose center is the point $(-1, 2)$ if $(3, 5)$ is a point on the circle.

6. $f(x) = \dfrac{3x}{x - 2}$, $g(x) = 2x + 1$

   Find:
   (a) $(f \circ g)(2)$
   (b) $(g \circ f)(4)$
   (c) $(f \circ g)(x)$
   (d) The domain of $(f \circ g)(x)$
   (e) $(g \circ f)(x)$
   (f) The domain of $(g \circ f)(x)$
   (g) The function g^{-1} and its domain
   (h) The function f^{-1} and its domain

7. Find the equation of an ellipse with center at the origin, a focus at $(0, 3)$, and a vertex at $(0, 4)$.

8. Find the equation of a parabola with vertex at $(-1, 2)$ and focus at $(-1, 3)$.

Counting and Probability

8

The Two-children Problem

PROBLEM: A woman and a man (unrelated) each have two children. At least one of the woman's children is a boy and the man's older child is a boy. Do the chances that the woman has two boys equal the chances that the man has two boys?

The above problem was posed to Marilyn vos Savant in her column *Ask Marilyn*. Her original answer, based on theoretical probabilities, was that the chances that the woman has two boys are 1 in 3 and the chances that the man has two boys are 1 in 2. This is found by looking at the sample space of two-child families: BB, BG, GB, GG. In the case of the man, we know that his older child is a boy and the sample space reduces to BB and BG. Hence the probability that he has two boys is 1 out of 2. In the case of the woman, since we only know that she has at least one boy, the sample space reduces to BB, BG, and GB. Thus, her chances of having two boys is 1 out of 3.

The answer about the woman's chances created quite a bit of controversy, resulting in many letters that challenged the correctness of her answer (*Parade*, July 27, 1997). Marilyn proposed that readers with exactly two children and at least one boy write in and tell the sex of both their children.

Reprinted with permission from PARADE and Marilyn vos Savant, copyright © 1997.

—*See Chapter Project 1.*

8.1 Sets and Counting

OBJECTIVES 1 Find All the Subsets of a Set
2 Find the Intersection and Union of Sets
3 Find the Complement of a Set
4 Count the Number of Elements in a Set

A **set** is a well-defined collection of distinct objects. The objects of a set are called its **elements**. By **well-defined**, we mean that there is a rule that enables us to determine whether a given object is an element of the set. If a set has no elements, it is called the **empty set**, or **null set**, and is denoted by the symbol ∅.

Because the elements of a set are distinct, we never repeat elements. For example, we would never write {1, 2, 3, 2}; the correct listing is {1, 2, 3}. Because a set is a collection, the order in which the elements are listed is immaterial. {1, 2, 3}, {1, 3, 2}, {2, 1, 3}, and so on, all represent the same set.

EXAMPLE 1 **Writing the Elements of a Set**

Write the set consisting of the possible results (outcomes) from tossing a coin twice. Use H for *heads* and T for *tails*.

Solution In tossing a coin twice, we can get heads each time, HH; or heads the first time and tails the second, HT; or tails the first time and heads the second, TH; or tails each time, TT. Because no other possibilities exist, the set of outcomes is

$$\{HH, HT, TH, TT\} \qquad ◄$$

Find All the Subsets of a Set

We now look at ways that two sets can be compared, beginning with set equality.

If two sets A and B have precisely the same elements, we say that A and B are **equal** and write $A = B$.

If each element of a set A is also an element of a set B, we say that A is a **subset** of B and write $A \subseteq B$.

If $A \subseteq B$ and $A \neq B$, then we say that A is a **proper subset** of B and write $A \subset B$.

If $A \subseteq B$, every element in set A is also in set B, but B may or may not have additional elements. If $A \subset B$, every element in A is also in B, and B has at least one element not found in A.

Finally, we agree that the empty set is a subset of every set; that is,

$$\varnothing \subseteq A, \qquad \text{for any set } A$$

EXAMPLE 2 **Finding All the Subsets of a Set**

Write down all the subsets of the set $\{a, b, c\}$.

Solution To organize our work, we write down all the subsets with no elements, then those with one element, then those with two elements, and finally those with three elements. These will give us all the subsets. Do you see why?

0 Elements	1 Element	2 Elements	3 Elements
$\emptyset$	$\{a\}, \{b\}, \{c\}$	$\{a, b\}, \{b, c\}, \{a, c\}$	$\{a, b, c\}$	◄

NOW WORK PROBLEM 25.

2 Find the Intersection and Union of Sets

If A and B are sets, the **intersection** of A with B, denoted $A \cap B$, is the set consisting of elements that belong to both A and B. The **union** of A with B, denoted $A \cup B$, is the set consisting of elements that belong to either A or B, or both.

EXAMPLE 3   **Finding the Intersection and Union of Sets**

Let $A = \{1, 3, 5, 8\}$, $B = \{3, 5, 7\}$, and $C = \{2, 4, 6, 8\}$. Find:

(a) $A \cap B$ (b) $A \cup B$ (c) $B \cap (A \cup C)$

Solution
(a) $A \cap B = \{1, 3, 5, 8\} \cap \{3, 5, 7\} = \{3, 5\}$
(b) $A \cup B = \{1, 3, 5, 8\} \cup \{3, 5, 7\} = \{1, 3, 5, 7, 8\}$
(c) $B \cap (A \cup C) = \{3, 5, 7\} \cap [\{1, 3, 5, 8\} \cup \{2, 4, 6, 8\}]$

$$= \{3, 5, 7\} \cap \{1, 2, 3, 4, 5, 6, 8\} = \{3, 5\}$$ ◄

NOW WORK PROBLEM 9.

3 Find the Complement of a Set

Usually, in working with sets, we designate a **universal set** U, the set consisting of all the elements that we wish to consider. Once a universal set has been designated, we can consider elements of the universal set not found in a given set.

If A is a set, the **complement** of A, denoted $\overline{A}$, is the set consisting of all the elements in the universal set that are not in A.[*]

EXAMPLE 4   **Finding the Complement of a Set**

If the universal set is $U = \{1, 2, 3, 4, 5, 6, 7, 8, 9\}$ and if $A = \{1, 3, 5, 7, 9\}$, then $\overline{A} = \{2, 4, 6, 8\}$. ◄

It follows that $A \cup \overline{A} = U$ and $A \cap \overline{A} = \emptyset$. Do you see why?

NOW WORK PROBLEM 17.

It is often helpful to draw pictures of sets. Such pictures, called **Venn diagrams**, represent sets as circles enclosed in a rectangle, which represents the universal set. Such diagrams often help us to visualize various relationships among sets. See Figure 1.

Figure 1

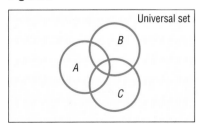

[*]Some books use the notation A' for the complement of A.

If we know that $A \subset B$, we might use the Venn diagram in Figure 2(a). If we know that A and B have no elements in common, that is, if $A \cap B = \varnothing$, we might use the Venn diagram in Figure 2(b). The sets A and B in Figure 2(b) are said to be **disjoint**.

Figure 2

(a) $A \subset B$
proper subset

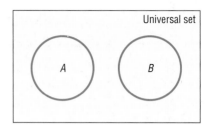

(b) $A \cap B = \varnothing$
disjoint sets

Figures 3(a), 3(b), and 3(c) use Venn diagrams to illustrate the definitions of intersection, union, and complement, respectively.

Figure 3

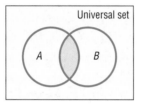

(a) $A \cap B$
intersection

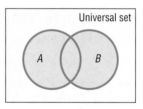

(b) $A \cup B$
union

(c) $\bar{A}$
complement

4 Count the Number of Elements in a Set

As you count the number of students in a classroom or the number of pennies in your pocket, what you are really doing is matching, on a one-to-one basis, each object to be counted with the set of counting numbers $1, 2, 3, \ldots, n$, for some number n. If a set A matched up in this fashion with the set $\{1, 2, \ldots, 25\}$, you would conclude that there are 25 elements in the set A. We use the notation $n(A) = 25$ to indicate that there are 25 elements in the set A.

Because the empty set has no elements, we write

$$n(\varnothing) = 0$$

If the number of elements in a set is a nonnegative integer, we say that the set is **finite**. Otherwise, it is **infinite**. We shall concern ourselves only with finite sets.

Look again at Example 2. A set with 3 elements has $2^3 = 8$ subsets. This result can be generalized.

If A is a set with n elements, then A has 2^n subsets.

For example, the set $\{a, b, c, d, e\}$ has $2^5 = 32$ subsets.

| EXAMPLE 5 | **Analyzing Survey Data** |

In a survey of 100 college students, 35 were registered in College Algebra, 52 were registered in Computer Science I, and 18 were registered in both courses.

(a) How many students were registered in College Algebra or Computer Science I?

(b) How many were registered in neither course?

Solution   (a) First, let A = set of students in College Algebra

B = set of students in Computer Science I

Then the given information tells us that

$$n(A) = 35 \qquad n(B) = 52 \qquad n(A \cap B) = 18$$

Figure 4

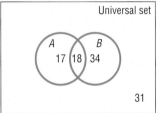

Refer to Figure 4. Since $n(A \cap B) = 18$, we know that the common part of the circles representing set A and set B has 18 elements. In addition, we know that the remaining portion of the circle representing set A will have $35 - 18 = 17$ elements. Similarly, we know that the remaining portion of the circle representing set B has $52 - 18 = 34$ elements. We conclude that $17 + 18 + 34 = 69$ students were registered in College Algebra or Computer Science I.

(b) Since 100 students were surveyed, it follows that $100 - 69 = 31$ were registered in neither course. ◀

NOW WORK PROBLEMS 33 AND 39.

The solution to Example 5 contains the basis for a general counting formula. If we count the elements in each of two sets A and B, we necessarily count twice any elements that are in both A and B, that is, those elements in $A \cap B$. To count correctly the elements that are in A or B, that is, to find $n(A \cup B)$, we need to subtract those in $A \cap B$ from $n(A) + n(B)$.

Theorem   **Counting Formula**

If A and B are finite sets, then

$$n(A \cup B) = n(A) + n(B) - n(A \cap B) \tag{1}$$

Refer back to Example 5. Using (1), we have

$$n(A \cup B) = n(A) + n(B) - n(A \cap B)$$
$$= 35 + 52 - 18$$
$$= 69$$

There are 69 students registered in College Algebra or Computer Science I.

A special case of the counting formula (1) occurs if A and B have no elements in common. In this case, $A \cap B = \varnothing$, so $n(A \cap B) = 0$.

Theorem   **Addition Principle of Counting**

If two sets A and B have no elements in common, that is,

$$\text{if } A \cap B = \varnothing, \text{ then } n(A \cup B) = n(A) + n(B) \tag{2}$$

We can generalize formula (2).

Theorem

General Addition Principle of Counting

If, for n sets $A_1, A_2, \ldots, A_n$, no two have elements in common, then

$$n(A_1 \cup A_2 \cup \cdots \cup A_n) = n(A_1) + n(A_2) + \cdots + n(A_n) \qquad \textbf{(3)}$$

EXAMPLE 6

Counting

As of June, 2002, federal agencies employed 93,445 full-time personnel authorized to make arrests and to carry firearms. Table 1 lists the type of law-enforcement officer and the corresponding number of full-time officers. No officer is classified as more than one type of officer.

Table 1

Type of Officer	Number of Full-time Federal Officers
Criminal (investigation/enforcement)	37,208
Police response and patrol	20,955
Corrections	16,915
Noncriminal (investigation/inspection)	12,801
Court operations	4,090
Security/protection	1,320
Other	156

SOURCE: Bureau of Justice Statistics

(a) How many full-time law-enforcement officers in the United States federal government were criminal officers or corrections officers?

(b) How many full-time law-enforcement officers in the United States federal government were criminal officers, corrections officers, or noncriminal officers?

Solution

Let A represent the set of criminal officers, B represent the set of corrections officers, and C represent the set of noncriminal officers. No two of the sets A, B, and C have elements in common since a single officer cannot be classified as more than one type of officer. Then

$$n(A) = 37{,}208 \qquad n(B) = 16{,}915 \qquad n(C) = 12{,}801$$

(a) Using formula (2), we have

$$n(A \cup B) = n(A) + n(B) = 37{,}208 + 16{,}915 = 54{,}123$$

There were 54,123 officers that were criminal or corrections officers.

(b) Using formula (3), we have

$$n(A \cup B \cup C) = n(A) + n(B) + n(C) = 37{,}208 + 16{,}915 + 12{,}801 = 66{,}924$$

There were 66,924 officers that were criminal, corrections, or noncriminal officers. ◀

NOW WORK PROBLEM **43**.

8.1 Assess Your Understanding

Concepts and Vocabulary

1. The _____ of A and B consists of all elements in either A or B or both.

2. The _____ of A with B consists of all elements in both A and B.

3. *True or False:* The intersection of two sets is always a subset of their union.

4. *True or False:* If A is a set, the complement of A is the set of all the elements in the universal set that are not in A.

Skill Building

In Problems 5–14, use $A = \{1, 3, 5, 7, 9\}$, $B = \{1, 5, 6, 7\}$, and $C = \{1, 2, 4, 6, 8, 9\}$ to find each set.

5. $A \cup B$

6. $A \cup C$

7. $A \cap B$

8. $A \cap C$

9. $(A \cup B) \cap C$

10. $(A \cap C) \cup (B \cap C)$

11. $(A \cap B) \cup C$

12. $(A \cup B) \cup C$

13. $(A \cup C) \cap (B \cup C)$

14. $(A \cap B) \cap C$

In Problems 15–24, use U = universal set = $\{0, 1, 2, 3, 4, 5, 6, 7, 8, 9\}$, $A = \{1, 3, 4, 5, 9\}$, $B = \{2, 4, 6, 7, 8\}$, and $C = \{1, 3, 4, 6\}$ to find each set.

15. $\overline{A}$

16. $\overline{C}$

17. $\overline{A \cap B}$

18. $\overline{B \cup C}$

19. $\overline{A} \cup \overline{B}$

20. $\overline{B} \cap \overline{C}$

21. $\overline{A \cap C}$

22. $\overline{B \cup C}$

23. $\overline{A \cup B \cup C}$

24. $\overline{A \cap B \cap C}$

25. Write down all the subsets of $\{a, b, c, d\}$.

26. Write down all the subsets of $\{a, b, c, d, e\}$.

27. If $n(A) = 15$, $n(B) = 20$, and $n(A \cap B) = 10$, find $n(A \cup B)$.

28. If $n(A) = 30$, $n(B) = 40$, and $n(A \cup B) = 45$, find $n(A \cap B)$.

29. If $n(A \cup B) = 50$, $n(A \cap B) = 10$, and $n(B) = 20$, find $n(A)$.

30. If $n(A \cup B) = 60$, $n(A \cap B) = 40$, and $n(A) = n(B)$, find $n(A)$.

In Problems 31–38, use the information given in the figure.

31. How many are in set A?

32. How many are in set B?

33. How many are in A or B?

34. How many are in A and B?

35. How many are in A but not C?

36. How many are not in A?

37. How many are in A and B and C?

38. How many are in A or B or C?

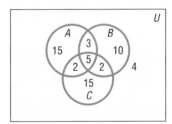

Applications and Extensions

39. **Analyzing Survey Data** In a consumer survey of 500 people, 200 indicated that they would be buying a major appliance within the next month, 150 indicated that they would buy a car, and 25 said that they would purchase both a major appliance and a car. How many will purchase neither? How many will purchase only a car?

40. **Analyzing Survey Data** In a student survey, 200 indicated that they would attend Summer Session I and 150 indicated Summer Session II. If 75 students plan to attend both summer sessions and 275 indicated that they would attend neither session, how many students participated in the survey?

41. **Analyzing Survey Data** In a survey of 100 investors in the stock market,

    50 owned shares in IBM

    40 owned shares in AT&T

    45 owned shares in GE

    20 owned shares in both IBM and GE

    15 owned shares in both AT&T and GE

    20 owned shares in both IBM and AT&T

    5 owned shares in all three

    (a) How many of the investors surveyed did not have shares in any of the three companies?

    (b) How many owned just IBM shares?

    (c) How many owned just GE shares?

    (d) How many owned neither IBM nor GE?

    (e) How many owned either IBM or AT&T but no GE?

42. **Classifying Blood Types** Human blood is classified as either Rh+ or Rh−. Blood is also classified by type: A, if it contains an A antigen; B, if it contains a B antigen; AB, if it contains both A and B antigens; and O, if it contains neither antigen. Draw a Venn diagram illustrating the various blood types. Based on this classification, how many different kinds of blood are there?

43. The following data represent the marital status of males 18 years old and older in 2002.

Marital Status	Number (in thousands)
Married	61,212
Widowed	2,632
Divorced	8,659
Never married	28,107

SOURCE: Current Population Survey

(a) Determine the number of males 18 years old and older who are widowed or divorced.
(b) Determine the number of males 18 years old and older who are married, widowed, or divorced.

44. The following data represent the marital status of females 18 years old and older in 2002.

Marital Status	Number (in thousands)
Married	62,037
Widowed	11,404
Divorced	12,236
Never married	23,036

SOURCE: Current Population Survey

(a) Determine the number of females 18 years old and older who are widowed or divorced.
(b) Determine the number of females 18 years old and older who are married, widowed, or divorced.

Discussion and Writing

45. Make up a problem different from any found in the text that requires the addition principle of counting to solve. Give it to a friend to solve and critique.

46. Investigate the notion of counting as it relates to infinite sets. Write an essay on your findings.

8.2 Permutations and Combinations

PREPARING FOR THIS SECTION   *Before getting started, review the following:*

• Factorial (Section 7.1, pp. 652–653).

Now work the 'Are You Prepared?' problems on page 708.

OBJECTIVES 1 Solve Counting Problems Using the Multiplication Principle
2 Solve Counting Problems Using Permutations
3 Solve Counting Problems Using Combinations
4 Solve Counting Problems Using Permutations Involving *n* Nondistinct Objects

1 Solve Counting Problems Using the Multiplication Principle

Counting plays a major role in many diverse areas, such as probability, statistics, and computer science; counting techniques are a part of a branch of mathematics called **combinatorics**. In this section we shall look at special types of counting problems and develop general formulas for solving them.

We begin with an example that will demonstrate a general counting principle.

EXAMPLE 1   **Counting the Number of Possible Meals**

The fixed-price dinner at Mabenka Restaurant provides the following choices:

Appetizer: soup or salad
Entree: baked chicken, broiled beef patty, baby beef liver, or roast beef au jus
Dessert: ice cream or cheese cake

How many different meals can be ordered?

Solution   Ordering such a meal requires three separate decisions:

Choose an Appetizer	**Choose an Entree**	**Choose a Dessert**
2 choices	4 choices	2 choices

Look at the **tree diagram** in Figure 5. We see that, for each choice of appetizer, there are 4 choices of entrees. And for each of these $2 \cdot 4 = 8$ choices, there are 2 choices for dessert. A total of

$$2 \cdot 4 \cdot 2 = 16$$

different meals can be ordered.

Figure 5

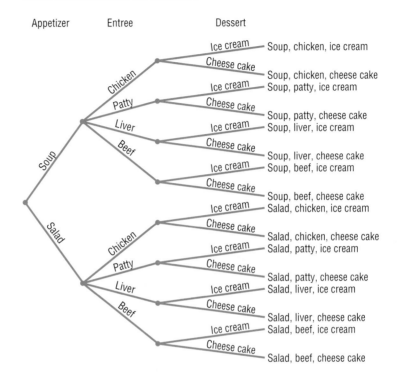

Appetizer   Entree   Dessert

Ice cream — Soup, chicken, ice cream
Cheese cake — Soup, chicken, cheese cake
Ice cream — Soup, patty, ice cream
Cheese cake — Soup, patty, cheese cake
Ice cream — Soup, liver, ice cream
Cheese cake — Soup, liver, cheese cake
Ice cream — Soup, beef, ice cream
Cheese cake — Soup, beef, cheese cake
Ice cream — Salad, chicken, ice cream
Cheese cake — Salad, chicken, cheese cake
Ice cream — Salad, patty, ice cream
Cheese cake — Salad, patty, cheese cake
Ice cream — Salad, liver, ice cream
Cheese cake — Salad, liver, cheese cake
Ice cream — Salad, beef, ice cream
Cheese cake — Salad, beef, cheese cake

Theorem

Multiplication Principle of Counting

If a task consists of a sequence of choices in which there are p selections for the first choice, q selections for the second choice, r selections for the third choice, and so on, then the task of making these selections can be done in

$$p \cdot q \cdot r \cdot \ldots$$

different ways.

EXAMPLE 2   **Forming Codes**

How many two-symbol codewords can be formed if the first symbol is an uppercase letter and the second symbol is a digit?

Solution   It sometimes helps to begin by listing some of the possibilities. The code consists of an uppercase letter followed by a digit, so some possibilities are A1, A2, B3, X0, and so on. The task consists of making two selections: the first selection requires choosing an uppercase letter (26 choices) and the second task requires choosing a digit (10 choices). By the Multiplication Principle, there are

$$26 \cdot 10 = 260$$

different codewords of the type described.   ◀

NOW WORK PROBLEM 31.

2 Solve Counting Problems Using Permutations

We begin with a definition.

A **permutation** is an ordered arrangement of r objects chosen from n objects.

We discuss three types of permutations:

1. The n objects are distinct (different), and repetition is allowed in the selection of r of them. [Distinct, with repetition]

2. The n objects are distinct (different), and repetition is not allowed in the selection of r of them, where $r \leq n$. [Distinct, without repetition]

3. The n objects are not distinct, and we use all of them in the arrangement. [Not distinct]

We take up the first two types here and deal with the third type at the end of this section.

The first type of permutation is handled using the Multiplication Principle.

EXAMPLE 3 | **Counting Airport Codes [Permutation: Distinct, with Repetition]**

The International Airline Transportation Association (IATA) assigns three-letter codes to represent airport locations. For example, the airport code for Ft. Lauderdale, Florida, is FLL. Notice that repetition is allowed in forming this code. How many airport codes are possible?

Solution   We are choosing 3 letters from 26 letters and arranging them in order. In the ordered arrangement a letter may be repeated. This is an example of a permutation with repetition in which 3 objects are chosen from 26 distinct objects.

The task of counting the number of such arrangements consists of making three selections. Each selection requires choosing a letter of the alphabet (26 choices). By the Multiplication Principle, there are

$$26 \cdot 26 \cdot 26 = 17{,}576$$

different airport codes.   ◀

The solution given to Example 3 can be generalized.

Theorem   **Permutations: Distinct Objects with Repetition**

The number of ordered arrangements of r objects chosen from n objects, in which the n objects are distinct and repetition is allowed, is n^r.

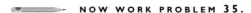 **NOW WORK PROBLEM 35.**

We begin the discussion of permutations in which the objects are distinct and repetition is not allowed with an example.

EXAMPLE 4 | **Forming Codes**
[Permutation: Distinct, without Repetition]

Suppose that we wish to establish a three-letter code using any of the 26 uppercase letters of the alphabet, but we require that no letter be used more than once. How many different three-letter codes are there?

Solution Some of the possibilities are: ABC, ABD, ABZ, ACB, CBA, and so on. The task consists of making three selections. The first selection requires choosing from 26 letters. Because no letter can be used more than once, the second selection requires choosing from 25 letters. The third selection requires choosing from 24 letters. (Do you see why?) By the Multiplication Principle, there are

$$26 \cdot 25 \cdot 24 = 15,600$$

different three-letter codes with no letter repeated. ◄

For the second type of permutation, we introduce the following notation.

The notation $P(n, r)$ represents the number of ordered arrangements of r objects chosen from n distinct objects, where $r \leq n$ and repetition is not allowed.

For example, the question posed in Example 4 asks for the number of ways that the 26 letters of the alphabet can be arranged in order using three nonrepeated letters. The answer is

$$P(26, 3) = 26 \cdot 25 \cdot 24 = 15,600$$

| EXAMPLE 5 | **Lining Up People** |

In how many ways can 5 people be lined up?

Solution The 5 people are distinct. Once a person is in line, that person will not be repeated elsewhere in the line; and, in lining up people, order is important. We have a permutation of 5 objects taken 5 at a time. We can line up 5 people in

$$P(5, 5) = \underbrace{5 \cdot 4 \cdot 3 \cdot 2 \cdot 1}_{5 \text{ factors}} = 120 \text{ ways}$$

◄

 NOW WORK PROBLEM **37**.

To arrive at a formula for $P(n, r)$, we note that the task of obtaining an ordered arrangement of n objects in which only $r \leq n$ of them are used, without repeating any of them, requires making r selections. For the first selection, there are n choices; for the second selection, there are $n - 1$ choices; for the third selection, there are $n - 2$ choices; ...; for the rth selection, there are $n - (r - 1)$ choices. By the Multiplication Principle, we have

$$
\begin{array}{cccc}
\text{1st} & \text{2nd} & \text{3rd} & \text{rth} \\
\end{array}
$$
$$P(n, r) = n \cdot (n - 1) \cdot (n - 2) \cdot \cdots \cdot [n - (r - 1)]$$
$$= n \cdot (n - 1) \cdot (n - 2) \cdot \cdots \cdot (n - r + 1)$$

This formula for $P(n, r)$ can be compactly written using factorial notation.[*]

$$P(n, r) = n \cdot (n - 1) \cdot (n - 2) \cdot \cdots \cdot (n - r + 1)$$
$$= n \cdot (n - 1) \cdot (n - 2) \cdot \cdots \cdot (n - r + 1) \cdot \frac{(n - r) \cdot \cdots \cdot 3 \cdot 2 \cdot 1}{(n - r) \cdot \cdots \cdot 3 \cdot 2 \cdot 1} = \frac{n!}{(n - r)!}$$

[*]Recall that $0! = 1$, $1! = 1$, $2! = 2 \cdot 1, \ldots, n! = n(n - 1) \cdot \cdots \cdot 3 \cdot 2 \cdot 1$.

Theorem

Permutations of *r* Objects Chosen from *n* Distinct Objects without Repetition

The number of arrangements of *n* objects using $r \leq n$ of them, in which

1. the *n* objects are distinct,
2. once an object is used it cannot be repeated, and
3. order is important,

is given by the formula

$$P(n, r) = \frac{n!}{(n - r)!} \qquad \textbf{(1)}$$

| EXAMPLE 6 | **Computing Permutations** |

Evaluate: (a) $P(7, 3)$ (b) $P(6, 1)$ (c) $P(52, 5)$

Solution We shall work parts (a) and (b) in two ways.

(a) $P(7, 3) = \underbrace{7 \cdot 6 \cdot 5}_{3\ factors} = 210$

or

$$P(7, 3) = \frac{7!}{(7 - 3)!} = \frac{7!}{4!} = \frac{7 \cdot 6 \cdot 5 \cdot 4\!\!\!/}{4\!\!\!/} = 210$$

(b) $P(6, 1) = \underbrace{6}_{1\ factor} = 6$

Figure 6

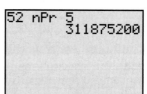

or

$$P(6, 1) = \frac{6!}{(6 - 1)!} = \frac{6!}{5!} = \frac{6 \cdot 5\!\!\!/}{5\!\!\!/} = 6$$

(c) Figure 6 shows the solution using a TI-84 Plus graphing calculator. So,
$$P(52, 5) = 311{,}875{,}200 \qquad \blacktriangleleft$$

✏ **NOW WORK PROBLEM 7.**

| EXAMPLE 7 | **The Birthday Problem** |

All we know about Shannon, Patrick, and Ryan is that they have different birthdays. If we listed all the possible ways this could occur, how many would there be? Assume that there are 365 days in a year.

Solution This is an example of a permutation in which 3 birthdays are selected from a possible 365 days, and no birthday may repeat itself. The number of ways that this can occur is

$$P(365, 3) = \frac{365!}{(365 - 3)!} = \frac{365 \cdot 364 \cdot 363 \cdot 362!}{362!} = 365 \cdot 364 \cdot 363 = 48,228,180$$

There are 48,228,180 ways in a group of three people that each has a different birthday.   ◀

NOW WORK PROBLEM 53.

3   Solve Counting Problems Using Combinations

In a permutation, order is important. For example, the arrangements ABC, CAB, BAC, . . . are considered different arrangements of the letters A, B, and C. In many situations, though, order is unimportant. For example, in the card game of poker, the order in which the cards are received does not matter; it is the *combination* of the cards that matters.

> A **combination** is an arrangement, without regard to order, of r objects selected from n distinct objects without repetition, where $r \le n$. The notation $C(n, r)$ represents the number of combinations of n distinct objects using r of them.

EXAMPLE 8   **Listing Combinations**

List all the combinations of the 4 objects a, b, c, d taken 2 at a time. What is $C(4, 2)$?

Solution   One combination of a, b, c, d taken 2 at a time is

$$ab$$

We exclude ba from the list because order is not important in a combination (this means that we do not distinguish ab from ba). The list of all such combinations (convince yourself of this) is

$$ab, \quad ac, \quad ad, \quad bc, \quad bd, \quad cd$$

so

$$C(4, 2) = 6 \qquad ◀$$

We can find a formula for $C(n, r)$ by noting that the only difference between a permutation of type 2 (distinct, without repetition) and a combination is that we disregard order in combinations. To determine $C(n, r)$, we need only eliminate from the formula for $P(n, r)$ the number of permutations that were simply rearrangements of a given set of r objects. This can be determined from the formula for $P(n, r)$ by calculating $P(r, r) = r!$. So, if we divide $P(n, r)$ by $r!$, we will have the desired formula for $C(n, r)$:

$$C(n, r) = \frac{P(n, r)}{r!} = \underset{\substack{\uparrow \\ \text{Use formula (1).}}}{\frac{\dfrac{n!}{(n - r)!}}{r!}} = \frac{n!}{(n - r)! \, r!}$$

We have proved the following result:

Theorem

Number of Combinations of *n* Distinct Objects Taken *r* at a Time

The number of arrangements of *n* objects using $r \leq n$ of them, in which

1. the *n* objects are distinct,
2. once an object is used, it cannot be repeated, and
3. order is not important,

is given by the formula

$$C(n, r) = \frac{n!}{(n - r)!r!} \tag{2}$$

Based on formula (2), we discover that the symbol $C(n, r)$ and the symbol $\binom{n}{r}$ for the binomial coefficients are, in fact, the same. The Pascal triangle (see Section 7.5) can be used to find the value of $C(n, r)$. However, because it is more practical and convenient, we will use formula (2) instead.

EXAMPLE 9 **Using Formula (2)**

Use formula (2) to find the value of each expression.

(a) $C(3, 1)$ (b) $C(6, 3)$ (c) $C(n, n)$ (d) $C(n, 0)$ (e) $C(52, 5)$

Solution

(a) $C(3, 1) = \dfrac{3!}{(3 - 1)!1!} = \dfrac{3!}{2!1!} = \dfrac{3 \cdot 2 \cdot 1}{2 \cdot 1 \cdot 1} = 3$

(b) $C(6, 3) = \dfrac{6!}{(6 - 3)!3!} = \dfrac{6 \cdot 5 \cdot 4 \cdot 3!}{3! \cdot 3!} = \dfrac{6 \cdot 5 \cdot 4}{6} = 20$

Figure 7

52 nCr 5
2598960

(c) $C(n, n) = \dfrac{n!}{(n - n)!n!} = \dfrac{n!}{0!n!} = \dfrac{1}{1} = 1$

(d) $C(n, 0) = \dfrac{n!}{(n - 0)!0!} = \dfrac{n!}{n!0!} = \dfrac{1}{1} = 1$

(e) Figure 7 shows the solution using a TI-84 Plus graphing calculator. So,

$$C(52, 5) = 2,598,960 \qquad \blacktriangleleft$$

NOW WORK PROBLEM 15.

EXAMPLE 10 **Forming Committees**

How many different committees of 3 people can be formed from a pool of 7 people?

Solution

The 7 people are distinct. More important, though, is the observation that the order of being selected for a committee is not significant. The problem asks for the number of combinations of 7 objects taken 3 at a time.

$$C(7, 3) = \frac{7!}{4!3!} = \frac{7 \cdot 6 \cdot 5 \cdot 4!}{4!3!} = \frac{7 \cdot 6 \cdot 5}{6} = 35 \qquad \blacktriangleleft$$

EXAMPLE 11 **Forming Committees**

In how many ways can a committee consisting of 2 faculty members and 3 students be formed if 6 faculty members and 10 students are eligible to serve on the committee?

Solution The problem can be separated into two parts: the number of ways that the faculty members can be chosen, $C(6, 2)$, and the number of ways that the student members can be chosen, $C(10, 3)$. By the Multiplication Principle, the committee can be formed in

$$C(6, 2) \cdot C(10, 3) = \frac{6!}{4!2!} \cdot \frac{10!}{7!3!} = \frac{6 \cdot 5 \cdot \cancel{4!}}{\cancel{4!}2!} \cdot \frac{10 \cdot 9 \cdot 8 \cdot \cancel{7!}}{\cancel{7!}3!}$$

$$= \frac{30}{2} \cdot \frac{720}{6} = 1800 \text{ ways} \qquad \blacktriangleleft$$

NOW WORK PROBLEM **55**.

4 Solve Counting Problems Using Permutations Involving *n* Nondistinct Objects

We begin with an example.

EXAMPLE 12 **Forming Different Words**

How many different words (real or imaginary) can be formed using all the letters in the word REARRANGE?

Solution Each word formed will have 9 letters: 3 R's, 2 A's, 2 E's, 1 N, and 1 G. To construct each word, we need to fill in 9 positions with the 9 letters:

$$\overline{1} \quad \overline{2} \quad \overline{3} \quad \overline{4} \quad \overline{5} \quad \overline{6} \quad \overline{7} \quad \overline{8} \quad \overline{9}$$

The process of forming a word consists of five tasks:

Task 1: Choose the positions for the 3 R's.
Task 2: Choose the positions for the 2 A's.
Task 3: Choose the positions for the 2 E's.
Task 4: Choose the position for the 1 N.
Task 5: Choose the position for the 1 G.

Task 1 can be done in $C(9, 3)$ ways. There then remain 6 positions to be filled, so Task 2 can be done in $C(6, 2)$ ways. There remain 4 positions to be filled, so Task 3 can be done in $C(4, 2)$ ways. There remain 2 positions to be filled, so Task 4 can be done in $C(2, 1)$ ways. The last position can be filled in $C(1, 1)$ way. Using the Multiplication Principle, the number of possible words that can be formed is

$$C(9, 3) \cdot C(6, 2) \cdot C(4, 2) \cdot C(2, 1) \cdot C(1, 1) = \frac{9!}{3! \cdot \cancel{6!}} \cdot \frac{\cancel{6!}}{2! \cdot \cancel{4!}} \cdot \frac{\cancel{4!}}{2! \cdot \cancel{2!}} \cdot \frac{\cancel{2!}}{1! \cdot \cancel{1!}} \cdot \frac{\cancel{1!}}{0! \cdot 1!}$$

$$= \frac{9!}{3! \cdot 2! \cdot 2! \cdot 1! \cdot 1!} \qquad \blacktriangleleft$$

The form of the answer to Example 12 is suggestive of a general result. Had the letters in REARRANGE each been different, there would have been $P(9, 9) = 9!$ possible words formed. This is the numerator of the answer. The presence of 3 R's, 2 A's, and 2 E's reduces the number of different words, as the entries in the denominator illustrate. We are led to the following result:

Theorem

Permutations Involving n Objects That Are Not Distinct

The number of permutations of n objects of which n_1 are of one kind, n_2 are of a second kind, ..., and n_k are of a kth kind is given by

$$\frac{n!}{n_1! \cdot n_2! \cdot \cdots \cdot n_k!} \quad \text{(3)}$$

where $n = n_1 + n_2 + \cdots + n_k$.

EXAMPLE 13 | **Arranging Flags**

How many different vertical arrangements are there of 8 flags if 4 are white, 3 are blue, and 1 is red?

Solution

We seek the number of permutations of 8 objects, of which 4 are of one kind, 3 are of a second kind, and 1 is of a third kind. Using formula (3), we find that there are

$$\frac{8!}{4! \cdot 3! \cdot 1!} = \frac{8 \cdot 7 \cdot 6 \cdot 5 \cdot 4!}{4! \cdot 3! \cdot 1!} = 280 \text{ different arrangements}$$

◀

━━━━━━ **NOW WORK PROBLEM 57.**

8.2 Assess Your Understanding

'Are You Prepared?'

Answers are given at the end of these exercises. If you get a wrong answer, read the pages listed in red.

1. $0! = $ _____; $1! = $ _____. (p. 652)

2. *True or False:* $n! = \dfrac{(n + 1)!}{n}$. (pp. 652–653)

Concepts and Vocabulary

3. A(n) _____ is an ordered arrangement of r objects chosen from n objects.

4. A(n) _____ is an arrangement of r objects chosen from n distinct objects, without repetition and without regard to order.

5. *True or False:* In a combination problem, order is not important.

6. *True or False:* In some permutation problems, once an object is used, it cannot be repeated.

Skill Building

In Problems 7–14, find the value of each permutation.

7. $P(6, 2)$ **8.** $P(7, 2)$ **9.** $P(4, 4)$ **10.** $P(8, 8)$

11. $P(7, 0)$ **12.** $P(9, 0)$ **13.** $P(8, 4)$ **14.** $P(8, 3)$

In Problems 15–22, use formula (2) to find the value of each combination.

15. $C(8, 2)$ **16.** $C(8, 6)$ **17.** $C(7, 4)$ **18.** $C(6, 2)$

19. $C(15, 15)$ **20.** $C(18, 1)$ **21.** $C(26, 13)$ **22.** $C(18, 9)$

Applications and Extensions

23. List all the ordered arrangements of 5 objects a, b, c, d, and e choosing 3 at a time without repetition. What is $P(5, 3)$?

24. List all the ordered arrangements of 5 objects a, b, c, d, and e choosing 2 at a time without repetition. What is $P(5, 2)$?

25. List all the ordered arrangements of 4 objects 1, 2, 3, and 4 choosing 3 at a time without repetition. What is $P(4, 3)$?

26. List all the ordered arrangements of 6 objects 1, 2, 3, 4, 5, and 6 choosing 3 at a time without repetition. What is $P(6, 3)$?

27. List all the combinations of 5 objects a, b, c, d, and e taken 3 at a time. What is $C(5, 3)$?

28. List all the combinations of 5 objects a, b, c, d, and e taken 2 at a time. What is $C(5, 2)$?

29. List all the combinations of 4 objects 1, 2, 3, and 4 taken 3 at a time. What is $C(4, 3)$?

30. List all the combinations of 6 objects 1, 2, 3, 4, 5, and 6 taken 3 at a time. What is $C(6, 3)$?

31. Shirts and Ties A man has 5 shirts and 3 ties. How many different shirt and tie arrangements can he wear?

32. Blouses and Skirts A woman has 3 blouses and 5 skirts. How many different outfits can she wear?

33. Forming Codes How many two-letter codes can be formed using the letters A, B, C, and D? Repeated letters are allowed.

34. Forming Codes How many two-letter codes can be formed using the letters A, B, C, D, and E? Repeated letters are allowed.

35. Forming Numbers How many three-digit numbers can be formed using the digits 0 and 1? Repeated digits are allowed.

36. Forming Numbers How many three-digit numbers can be formed using the digits 0, 1, 2, 3, 4, 5, 6, 7, 8, and 9? Repeated digits are allowed.

37. Lining People Up In how many ways can 4 people be lined up?

38. Stacking Boxes In how many ways can 5 different boxes be stacked?

39. Forming Codes How many different three-letter codes are there if only the letters A, B, C, D, and E can be used and no letter can be used more than once?

40. Forming Codes How many different four-letter codes are there if only the letters A, B, C, D, E, and F can be used and no letter can be used more than once?

41. Stocks on the NYSE Companies whose stocks are listed on the New York Stock Exchange (NYSE) have their company name represented by either 1, 2, or 3 letters (repetition of letters is allowed). What is the maximum number of companies that can be listed on the NYSE?

42. Stocks on the NASDAQ Companies whose stocks are listed on the NASDAQ stock exchange have their company name represented by either 4 or 5 letters (repetition of letters is allowed). What is the maximum number of companies that can be listed on the NASDAQ?

43. Establishing Committees In how many ways can a committee of 4 students be formed from a pool of 7 students?

44. Establishing Committees In how many ways can a committee of 3 professors be formed from a department having 8 professors?

45. Possible Answers on a True/False Test How many arrangements of answers are possible for a true/false test with 10 questions?

46. Possible Answers on a Multiple-choice Test How many arrangements of answers are possible in a multiple-choice test with 5 questions, each of which has 4 possible answers?

47. Four-digit Numbers How many four-digit numbers can be formed using the digits 0, 1, 2, 3, 4, 5, 6, 7, 8, and 9 if the first digit cannot be 0? Repeated digits are allowed.

48. Five-digit Numbers How many five-digit numbers can be formed using the digits 0, 1, 2, 3, 4, 5, 6, 7, 8, and 9 if the first digit cannot be 0 or 1? Repeated digits are allowed.

49. Arranging Books Five different mathematics books are to be arranged on a student's desk. How many arrangements are possible?

50. Forming License Plate Numbers How many different license plate numbers can be made using 2 letters followed by 4 digits selected from the digits 0 through 9, if
(a) letters and digits may be repeated?
(b) letters may be repeated, but digits may not be repeated?
(c) neither letters nor digits may be repeated?

51. Stock Portfolios As a financial planner, you are asked to select one stock each from the following groups: 8 DOW stocks, 15 NASDAQ stocks, and 4 global stocks. How many different portfolios are possible?

52. Combination Locks   A combination lock displays 50 numbers. To open it, you turn to a number, then rotate clockwise to a second number, and then counterclockwise to the third number.

(a) How many different lock combinations are there?

(b) Comment on the description of such a lock as a *combination* lock.

53. Birthday Problem   In how many ways can 2 people each have different birthdays? Assume that there are 365 days in a year.

54. Birthday Problem   In how many ways can 5 people each have different birthdays? Assume that there are 365 days in a year.

55. Forming a Committee   A student dance committee is to be formed consisting of 2 boys and 3 girls. If the membership is to be chosen from 4 boys and 8 girls, how many different committees are possible?

56. Forming a Committee   The student relations committee of a college consists of 2 administrators, 3 faculty members, and 5 students. Four administrators, 8 faculty members, and 20 students are eligible to serve. How many different committees are possible?

57. Forming Words   How many different 9-letter words (real or imaginary) can be formed from the letters in the word ECONOMICS?

58. Forming Words   How many different 11-letter words (real or imaginary) can be formed from the letters in the word MATHEMATICS?

59. Selecting Objects   An urn contains 7 white balls and 3 red balls. Three balls are selected. In how many ways can the 3 balls be drawn from the total of 10 balls:

(a) If 2 balls are white and 1 is red?

(b) If all 3 balls are white?

(c) If all 3 balls are red?

60. Selecting Objects   An urn contains 15 red balls and 10 white balls. Five balls are selected. In how many ways can the 5 balls be drawn from the total of 25 balls:

(a) If all 5 balls are red?

(b) If 3 balls are red and 2 are white?

(c) If at least 4 are red balls?

61. Senate Committees   The U.S. Senate has 100 members. Suppose that it is desired to place each senator on exactly 1 of 7 possible committees. The first committee has 22 members, the second has 13, the third has 10, the fourth has 5, the fifth has 16, and the sixth and seventh have 17 apiece. In how many ways can these committees be formed?

62. Football Teams   A defensive football squad consists of 25 players. Of these, 10 are linemen, 10 are linebackers, and 5 are safeties. How many different teams of 5 linemen, 3 linebackers, and 3 safeties can be formed?

63. Baseball   In the American Baseball League, a designated hitter may be used. How many batting orders is it possible for a manager to use? (There are 9 regular players on a team.)

64. Baseball   In the National Baseball League, the pitcher usually bats ninth. If this is the case, how many batting orders is it possible for a manager to use?

65. Baseball Teams   A baseball team has 15 members. Four of the players are pitchers, and the remaining 11 members can play any position. How many different teams of 9 players can be formed?

66. World Series   In the World Series the American League team (A) and the National League team (N) play until one team wins four games. If the sequence of winners is designated by letters (for example, $NAAAA$ means that the National League team won the first game and the American League won the next four), how many different sequences are possible?

67. Basketball Teams   A basketball team has 6 players who play guard (2 of 5 starting positions). How many different teams are possible, assuming that the remaining 3 positions are filled and it is not possible to distinguish a left guard from a right guard?

68. Basketball Teams   On a basketball team of 12 players, 2 only play center, 3 only play guard, and the rest play forward (5 players on a team: 2 forwards, 2 guards, and 1 center). How many different teams are possible, assuming that it is not possible to distinguish left and right guards and left and right forwards?

Discussion and Writing

69. Create a problem different from any found in the text that requires the Multiplication Principle to solve. Give it to a friend to solve and critique.

70. Create a problem different from any found in the text that requires a permutation to solve. Give it to a friend to solve and critique.

71. Create a problem different from any found in the text that requires a combination to solve. Give it to a friend to solve and critique.

72. Explain the difference between a permutation and a combination. Give an example to illustrate your explanation.

'Are You Prepared?' Answers

1. $1; 1$ **2.** False

8.3 Probability

> **OBJECTIVES** 1 Construct Probability Models
> 2 Compute Probabilities of Equally Likely Outcomes
> 3 Use the Addition Rule to Find Probabilities
> 4 Use the Complement Rule to Find Probabilities

Probability is an area of mathematics that deals with experiments that yield random results, yet admit a certain regularity. Such experiments do not always produce the same result or outcome, so the result of any one observation is not predictable. However, the results of the experiment over a long period do produce regular patterns that enable us to predict with remarkable accuracy.

EXAMPLE 1   **Tossing a Fair Coin**

In tossing a fair coin, we know that the outcome is either a head or a tail. On any particular throw, we cannot predict what will happen, but, if we toss the coin many times, we observe that the number of times that a head comes up is approximately equal to the number of times that we get a tail. It seems reasonable, therefore, to assign a probability of $\frac{1}{2}$ that a head comes up and a probability of $\frac{1}{2}$ that a tail comes up.   ◀

---

1   Construct Probability Models

The discussion in Example 1 constitutes the construction of a **probability model** for the experiment of tossing a fair coin once. A probability model has two components: a sample space and an assignment of probabilities. A **sample space** S is a set whose elements represent all the possibilities that can occur as a result of the experiment. Each element of S is called an **outcome**. To each outcome, we assign a number, called the **probability** of that outcome, which has two properties:

1. The probability assigned to each outcome is nonnegative.

2. The sum of all the probabilities equals 1.

If a probability model has the sample space

$$S = \{e_1, e_2, \ldots, e_n\}$$

where $e_1, e_2, \ldots, e_n$ are the possible outcomes, and if $P(e_1), P(e_2), \ldots, P(e_n)$ denote the respective probabilities of these outcomes, then

$$P(e_1) \geq 0, P(e_2) \geq 0, \ldots, P(e_n) \geq 0 \qquad \textbf{(1)}$$

$$\sum_{i=1}^{n} P(e_i) = P(e_1) + P(e_2) + \cdots + P(e_n) = 1 \qquad \textbf{(2)}$$

EXAMPLE 2

Determining Probability Models

In a bag of M&Ms, the candies are colored red, green, blue, brown, yellow, and orange. Suppose that a candy is drawn from the bag and the color is recorded. The sample space of this experiment is {red, green, blue, brown, yellow, orange}. Determine which of the following are probability models.

(a) Outcome	Probability		(b) Outcome	Probability
{red}	0.3		{red}	0.1
{green}	0.15		{green}	0.1
{blue}	0		{blue}	0.1
{brown}	0.15		{brown}	0.4
{yellow}	0.2		{yellow}	0.2
{orange}	0.2		{orange}	0.3

(c) Outcome	Probability		(d) Outcome	Probability
{red}	0.3		{red}	0
{green}	−0.3		{green}	0
{blue}	0.2		{blue}	0
{brown}	0.4		{brown}	0
{yellow}	0.2		{yellow}	1
{orange}	0.2		{orange}	0

Solution

(a) This model is a probability model since all the outcomes have probabilities that are nonnegative and the sum of the probabilities is 1.

(b) This model is not a probability model because the sum of the probabilities is not 1.

(c) This model is not a probability model because $P(\text{green})$ is less than 0. Recall, all probabilities must be nonnegative.

(d) This model is a probability model because all the outcomes have probabilities that are nonnegative, and the sum of the probabilities is 1. Notice that $P(\text{yellow}) = 1$, meaning that this outcome will occur with 100% certainty each time that the experiment is repeated. This means that the bag of M&Ms has only yellow candies. ◄

NOW WORK PROBLEM 7.

EXAMPLE 3 | **Constructing a Probability Model**

An experiment consists of rolling a fair die once. A die is a cube with each face having either 1, 2, 3, 4, 5, or 6 dots on it. See Figure 8. Construct a probability model for this experiment.

Figure 8

Solution   A sample space S consists of all the possibilities that can occur. Because rolling the die will result in one of six faces showing, the sample space S consists of

$$S = \{1, 2, 3, 4, 5, 6\}$$

Because the die is fair, one face is no more likely to occur than another. As a result, our assignment of probabilities is

$$P(1) = \frac{1}{6} \qquad P(2) = \frac{1}{6}$$

$$P(3) = \frac{1}{6} \qquad P(4) = \frac{1}{6}$$

$$P(5) = \frac{1}{6} \qquad P(6) = \frac{1}{6}$$ ◀

Now suppose that a die is loaded (weighted) so that the probability assignments are

$$P(1) = 0, \quad P(2) = 0, \quad P(3) = \frac{1}{3}, \quad P(4) = \frac{2}{3}, \quad P(5) = 0, \quad P(6) = 0$$

This assignment would be made if the die were loaded so that only a 3 or 4 could occur and the 4 is twice as likely as the 3 to occur. This assignment is consistent with the definition, since each assignment is nonnegative and the sum of all the probability assignments equals 1.

NOW WORK PROBLEM **23.**

EXAMPLE 4 | **Constructing a Probability Model**

An experiment consists of tossing a coin. The coin is weighted so that heads (H) is three times as likely to occur as tails (T). Construct a probability model for this experiment.

Solution   The sample space S is $S = \{H, T\}$. If x denotes the probability that a tail occurs, then

$$P(T) = x \quad \text{and} \quad P(H) = 3x$$

Since the sum of the probabilities of the possible outcomes must equal 1, we have

$$P(T) + P(H) = x + 3x = 1$$
$$4x = 1$$
$$x = \frac{1}{4}$$

We assign the probabilities

$$P(T) = \frac{1}{4} \qquad P(H) = \frac{3}{4}$$ ◀

NOW WORK PROBLEM **27.**

In working with probability models, the term **event** is used to describe a set of possible outcomes of the experiment. An event E is some subset of the sample space S. The **probability of an event** E, $E \neq \emptyset$, denoted by $P(E)$, is defined as the sum of the probabilities of the outcomes in E. We can also think of the probability of an event E as the likelihood that the event E occurs. If $E = \emptyset$, then $P(E) = 0$; if $E = S$, then $P(E) = P(S) = 1$.

2 Compute Probabilities of Equally Likely Outcomes

When the same probability is assigned to each outcome of the sample space, the experiment is said to have **equally likely outcomes**.

Theorem

Probability for Equally Likely Outcomes

If an experiment has n equally likely outcomes and if the number of ways that an event E can occur is m, then the probability of E is

$$P(E) = \frac{\text{Number of ways that } E \text{ can occur}}{\text{Number of all logical possibilities}} = \frac{m}{n} \qquad (3)$$

If S is the sample space of this experiment, then

$$P(E) = \frac{n(E)}{n(S)} \qquad (4)$$

| EXAMPLE 5 | Calculating Probabilities of Events Involving Equally Likely Outcomes |

Calculate the probability that in a 3-child family there are 2 boys and 1 girl. Assume equally likely outcomes.

Solution

Figure 9

1st child 2nd child 3rd child

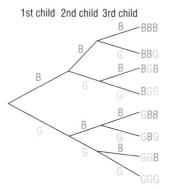

We begin by constructing a tree diagram to help in listing the possible outcomes of the experiment. See Figure 9, where B stands for boy and G for girl. The sample space S of this experiment is

$$S = \{\text{BBB, BBG, BGB, BGG, GBB, GBG, GGB, GGG}\}$$

so $n(S) = 8$.

We wish to know the probability of the event E: "having two boys and one girl." From Figure 9, we conclude that $E = \{\text{BBG, BGB, GBB}\}$, so $n(E) = 3$. Since the outcomes are equally likely, the probability of E is

$$P(E) = \frac{n(E)}{n(S)} = \frac{3}{8}$$

◀

NOW WORK PROBLEM 37.

So far, we have calculated probabilities of single events. We will now compute probabilities of multiple events, called **compound probabilities**.

| EXAMPLE 6 | **Computing Compound Probabilities** |

Consider the experiment of rolling a single fair die. Let E represent the event "roll an odd number," and let F represent the event "roll a 1 or 2."

(a) Write the event E and F.
(b) Write the event E or F.
(c) Compute $P(E)$ and $P(F)$.
(d) Compute $P(E \cap F)$.
(e) Compute $P(E \cup F)$.

Solution The sample space S of the experiment is $\{1, 2, 3, 4, 5, 6\}$, so $n(S) = 6$. Since the die is fair, the outcomes are equally likely. The event E: "roll an odd number" is $\{1, 3, 5\}$, and the event F: "roll a 1 or 2" is $\{1, 2\}$, so $n(E) = 3$ and $n(F) = 2$.

(a) The word *and* in probability means the intersection of two events. The event E and F is

$$E \cap F = \{1, 3, 5\} \cap \{1, 2\} = \{1\} \qquad n(E \cap F) = 1$$

(b) The word *or* in probability means the union of the two events. The event E or F is

$$E \cup F = \{1, 3, 5\} \cup \{1, 2\} = \{1, 2, 3, 5\} \qquad n(E \cup F) = 4$$

(c) We use formula (4).

$$P(E) = \frac{n(E)}{n(S)} = \frac{3}{6} = \frac{1}{2} \qquad P(F) = \frac{n(F)}{n(S)} = \frac{2}{6} = \frac{1}{3}$$

(d) $P(E \cap F) = \dfrac{n(E \cap F)}{n(S)} = \dfrac{1}{6}$

(e) $P(E \cup F) = \dfrac{n(E \cup F)}{n(S)} = \dfrac{4}{6} = \dfrac{2}{3}$ ◀

3 Use the Addition Rule to Find Probabilities

The **Addition Rule** can be used to find the probability of the union of two events.

Theorem **Addition Rule**

For any two events E and F,

$$P(E \cup F) = P(E) + P(F) - P(E \cap F) \qquad (5)$$

For example, we can use the Addition Rule to find $P(E \cup F)$ in Example 6(e). Then

$$P(E \cup F) = P(E) + P(F) - P(E \cap F) = \frac{1}{2} + \frac{1}{3} - \frac{1}{6} = \frac{3}{6} + \frac{2}{6} - \frac{1}{6} = \frac{4}{6} = \frac{2}{3}$$

as before.

EXAMPLE 7	**Computing Probabilities of Compound Events Using the Addition Rule**

If $P(E) = 0.2$, $P(F) = 0.3$, and $P(E \cap F) = 0.1$, find the probability of E or F; that is, find $P(E \cup F)$.

Solution   We use the Addition Rule, formula (5).

$$\text{Probability of } E \text{ or } F = P(E \cup F) = P(E) + P(F) - P(E \cap F)$$
$$= 0.2 + 0.3 - 0.1 = 0.4 \quad \blacktriangleleft$$

A Venn diagram can sometimes be used to obtain probabilities. To construct a Venn diagram representing the information in Example 7, we draw two sets E and F. We begin with the fact that $P(E \cap F) = 0.1$. See Figure 10(a). Then, since $P(E) = 0.2$ and $P(F) = 0.3$, we fill in E with $0.2 - 0.1 = 0.1$ and F with $0.3 - 0.1 = 0.2$. See Figure 10(b). Since $P(S) = 1$, we complete the diagram by inserting $1 - (0.1 + 0.1 + 0.2) = 0.6$ outside the circles. See Figure 10(c). Now it is easy to see, for example, that the probability of F, but not E, is 0.2. Also, the probability of neither E nor F is 0.6.

Figure 10

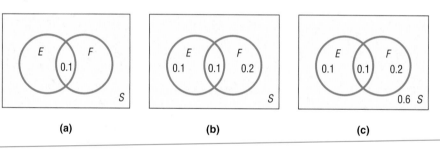

(a)   (b)   (c)

NOW WORK PROBLEM 45.

If events E and F are disjoint so that $E \cap F = \varnothing$, we say they are **mutually exclusive**. In this case, $P(E \cap F) = 0$, and the Addition Rule takes the following form:

Theorem	**Mutually Exclusive Events**

If E and F are **mutually exclusive events**, then

$$P(E \cup F) = P(E) + P(F) \qquad \textbf{(6)}$$

EXAMPLE 8	**Computing Compound Probabilities of Mutually Exclusive Events**

If $P(E) = 0.4$ and $P(F) = 0.25$, and E and F are mutually exclusive, find $P(E \cup F)$.

Solution   Since E and F are mutually exclusive, we use formula (6).

$$P(E \cup F) = P(E) + P(F) = 0.4 + 0.25 = 0.65 \quad \blacktriangleleft$$

NOW WORK PROBLEM 47.

4 Use the Complement Rule to Find Probabilities

Recall, if A is a set, the complement of A, denoted $\overline{A}$, is the set of all elements in the universal set U not in A. We similarly define the complement of an event.

Complement of an Event

Let S denote the sample space of an experiment, and let E denote an event. The **complement of E**, denoted $\overline{E}$, is the set of all outcomes in the sample space S that are not outcomes in the event E.

The complement of an event E, that is, $\overline{E}$, in a sample space S has the following two properties:

$$E \cap \overline{E} = \varnothing \qquad E \cup \overline{E} = S$$

Since E and $\overline{E}$ are mutually exclusive, it follows from (6) that

$$P(E \cup \overline{E}) = P(S) = 1 \qquad P(E) + P(\overline{E}) = 1 \qquad P(\overline{E}) = 1 - P(E)$$

We have the following result:

Theorem

Computing Probabilities of Complementary Events

If E represents any event and $\overline{E}$ represents the complement of E, then

$$P(\overline{E}) = 1 - P(E) \qquad \qquad \textbf{(7)}$$

EXAMPLE 9

Computing Probabilities Using Complements

On the local news the weather reporter stated that the probability of rain tomorrow is 40%. What is the probability that it will not rain?

Solution   The complement of the event "rain" is "no rain."

$$P(\text{no rain}) = 1 - P(\text{rain}) = 1 - 0.4 = 0.6$$

There is a 60% chance of no rain tomorrow.   ◀

NOW WORK PROBLEM **51**.

EXAMPLE 10

Birthday Problem

What is the probability that in a group of 10 people at least 2 people have the same birthday? Assume that there are 365 days in a year.

Solution   We assume that a person is as likely to be born on one day as another, so we have equally likely outcomes.

We first determine the number of outcomes in the sample space S. There are 365 possibilities for each person's birthday. Since there are 10 people in the group, there are 365^{10} possibilities for the birthdays. [For one person in the group, there are 365 days on which his or her birthday can fall; for two people, there are $(365)(365) = 365^2$ pairs of days; and, in general, using the Multiplication Principle, for n people there are 365^n possibilities.] So

$$n(S) = 365^{10}$$

We wish to find the probability of the event E: "at least two people have the same birthday." It is difficult to count the elements in this set; it is much easier to count the elements of the complementary event $\overline{E}$: "no two people have the same birthday."

We find $n(\overline{E})$ as follows: Choose one person at random. There are 365 possibilities for his or her birthday. Choose a second person. There are 364 possibilities for this birthday, if no two people are to have the same birthday. Choose a third person. There are 363 possibilities left for this birthday. Finally, we arrive at the tenth person. There are 356 possibilities left for this birthday. By the Multiplication Principle, the total number of possibilities is

$$n(\overline{E}) = 365 \cdot 364 \cdot 363 \cdots \cdot 356$$

The probability of the event $\overline{E}$ is

$$P(\overline{E}) = \frac{n(\overline{E})}{n(S)} = \frac{365 \cdot 364 \cdot 363 \cdots \cdot 356}{365^{10}} \approx 0.883$$

The probability of two or more people in a group of 10 people having the same birthday is then

$$P(E) = 1 - P(\overline{E}) \approx 1 - 0.883 = 0.117 \qquad \blacktriangleleft$$

The birthday problem can be solved for any group size. The following table gives the probabilities for two or more people having the same birthday for various group sizes. Notice that the probability is greater than $\frac{1}{2}$ for any group of 23 or more people.

	Number of People															
	5	10	15	20	21	22	23	24	25	30	40	50	60	70	80	90
Probability That Two or More Have the Same Birthday	0.027	0.117	0.253	0.411	0.444	0.476	0.507	0.538	0.569	0.706	0.891	0.970	0.994	0.99916	0.99991	0.99999

NOW WORK PROBLEM **69.**

HISTORICAL FEATURE

Blaise Pascal (1623–1662)

Set theory, counting, and probability first took form as a systematic theory in an exchange of letters (1654) between Pierre de Fermat (1601–1665) and Blaise Pascal (1623–1662). They discussed the problem of how to divide the stakes in a game that is interrupted before completion, knowing how many points each player needs to win. Fermat solved the problem by listing all possibilities and counting the favorable ones, whereas Pascal made use of the triangle that now bears his name. As mentioned in the text, the entries in Pascal's triangle are equivalent to $C(n, r)$. This recognition of the role of $C(n, r)$ in counting is the foundation of all further developments.

The first book on probability, the work of Christiaan Huygens (1629–1695), appeared in 1657. In it, the notion of mathematical expectation is explored. This allows the calculation of the profit or loss that a gambler might expect, knowing the probabilities involved in the game (see the Historical Problems that follow).

Although Girolamo Cardano (1501–1576) wrote a treatise on probability, it was not published until 1663 in Cardano's collected works, and this was too late to have any effect on the development of the theory.

In 1713, the posthumously published *Ars Conjectandi* of Jakob Bernoulli (1654–1705) gave the theory the form it would have until 1900. Recently, both combinatorics (counting) and probability have undergone rapid development due to the use of computers.

A final comment about notation. The notations $C(n, r)$ and $P(n, r)$ are variants of a form of notation developed in England after 1830. The notation $\dbinom{n}{r}$ for $C(n, r)$ goes back to Leonhard Euler (1707–1783), but is now losing ground because it has no clearly related symbolism of the same type for permutations. The set symbols $\cup$ and $\cap$ were introduced by Giuseppe Peano (1858–1932) in 1888 in a slightly different context. The inclusion symbol $\subset$ was introduced by E. Schroeder (1841–1902) about 1890. The treatment of set theory in the text is due to George Boole (1815–1864), who wrote $A + B$ for $A \cup B$ and AB for $A \cap B$ (statisticians still use AB for $A \cap B$).

Historical Problems

1. *The Problem Discussed by Fermat and Pascal*   A game between two equally skilled players, A and B, is interrupted when A needs 2 points to win and B needs 3 points. In what proportion would the stakes be divided?

   (a) *Fermat's solution*   List all possible outcomes that can occur as a result of four more plays. The probabilities for A to win and B to win then determine how the stakes should be divided.

   (b) *Pascal's solution*   Use combinations to determine the number of ways that the 2 points needed for A to win could occur in four plays. Then use combinations to determine the number of ways that the 3 points needed for B to win could occur. This is trickier than it looks, since A can win with 2 points in either two plays, three plays, or four plays. Compute the probabilities and compare with the results in part (a).

2. *Huygen's Mathematical Expectation*   In a game with n possible outcomes with probabilities $p_1, p_2, \ldots, p_n$, suppose that the net winnings are $w_1, w_2, \ldots, w_n$, respectively. Then the mathematical expectation is

$$E = p_1 w_1 + p_2 w_2 + \cdots + p_n w_n$$

   The number E represents the profit or loss per game in the long run. The following problems are a modification of those of Huygens.

   (a) A fair die is tossed. A gambler wins \$3 if he throws a 6 and \$6 if he throws a 5. What is his expectation?
   [**Hint:** $w_1 = w_2 = w_3 = w_4 = 0$]

   (b) A gambler plays the same game as in part (a), but now the gambler must pay \$1 to play. This means that $w_5 = \$5$, $w_6 = \$2$, and $w_1 = w_2 = w_3 = w_4 = -\$1$. What is the expectation?

8.3 Assess Your Understanding

Concepts and Vocabulary

1. When the same probability is assigned to each outcome of a sample space, the experiment is said to have _____ _____ outcomes.

2. The _____ of an event E is the set of all outcomes in the sample space S that are not outcomes in the event E.

3. *True or False:* The probability of an event can never equal 0.

4. *True or False:* In a probability model, the sum of all probabilities is 1.

Skill Building

5. In a probability model, which of the following numbers could be the probability of an outcome?

$$0, \quad 0.01, \quad 0.35, \quad -0.4, \quad 1, \quad 1.4$$

6. In a probability model, which of the following numbers could be the probability of an outcome?

$$1.5, \quad \frac{1}{2}, \quad \frac{3}{4}, \quad \frac{2}{3}, \quad 0, \quad -\frac{1}{4}$$

7. Determine whether the following is a probability model.

Outcome	Probability
{1}	0.2
{2}	0.3
{3}	0.1
{4}	0.4

8. Determine whether the following is a probability model.

Outcome	Probability
{Jim}	0.4
{Bob}	0.3
{Faye}	0.1
{Patricia}	0.2

9. Determine whether the following is a probability model.

Outcome	Probability
{Linda}	0.3
{Jean}	0.2
{Grant}	0.1
{Ron}	0.3

10. Determine whether the following is a probability model.

Outcome	Probability
{Lanny}	0.3
{Joanne}	0.2
{Nelson}	0.1
{Rich}	0.5
{Judy}	−0.1

In Problems 11–16, construct a probability model for each experiment.

11. Tossing a fair coin twice

12. Tossing two fair coins once

13. Tossing two fair coins, then a fair die

14. Tossing a fair coin, a fair die, and then a fair coin

15. Tossing three fair coins once

16. Tossing one fair coin three times

In Problems 17–22, use the following spinners to construct a probability model for each experiment.

Spinner I   Spinner II   Spinner III

17. Spin spinner I, then spinner II. What is the probability of getting a 2 or a 4, followed by Red?

18. Spin spinner III, then spinner II. What is the probability of getting Forward, followed by Yellow or Green?

19. Spin spinner I, then II, then III. What is the probability of getting a 1, followed by Red or Green, followed by Backward?

20. Spin spinner II, then I, then III. What is the probability of getting Yellow, followed by a 2 or a 4, followed by Forward?

21. Spin spinner I twice, then spinner II. What is the probability of getting a 2, followed by a 2 or a 4, followed by Red or Green?

22. Spin spinner III, then spinner I twice. What is the probability of getting Forward, followed by a 1 or a 3, followed by a 2 or a 4?

In Problems 23–26, consider the experiment of tossing a coin twice. The table lists six possible assignments of probabilities for this experiment. Using this table, answer the following questions.

	Sample Space			
Assignments	**HH**	**HT**	**TH**	**TT**
A	$\frac{1}{4}$	$\frac{1}{4}$	$\frac{1}{4}$	$\frac{1}{4}$
B	0	0	0	1
C	$\frac{3}{16}$	$\frac{5}{16}$	$\frac{5}{16}$	$\frac{3}{16}$
D	$\frac{1}{2}$	$\frac{1}{2}$	$-\frac{1}{2}$	$\frac{1}{2}$
E	$\frac{1}{4}$	$\frac{1}{4}$	$\frac{1}{4}$	$\frac{1}{8}$
F	$\frac{1}{9}$	$\frac{2}{9}$	$\frac{2}{9}$	$\frac{4}{9}$

23. Which of the assignments of probabilities are consistent with the definition of a probability model?

24. Which of the assignments of probabilities should be used if the coin is known to be fair?

25. Which of the assignments of probabilities should be used if the coin is known to always come up tails?

26. Which of the assignments of probabilities should be used if tails is twice as likely as heads to occur?

27. Assigning Probabilities A coin is weighted so that heads is four times as likely as tails to occur. What probability should we assign to heads? to tails?

28. Assigning Probabilities A coin is weighted so that tails is twice as likely as heads to occur. What probability should we assign to heads? to tails?

29. Assigning Probabilities A die is weighted so that an odd-numbered face is twice as likely to occur as an even-numbered face. What probability should we assign to each face?

30. Assigning Probabilities A die is weighted so that a six cannot appear. The other faces occur with the same probability. What probability should we assign to each face?

For Problems 31–34, let the sample space be S = {1, 2, 3, 4, 5, 6, 7, 8, 9, 10}. Suppose that the outcomes are equally likely.

31. Compute the probability of the event $E = \{1, 2, 3\}$.

32. Compute the probability of the event $F = \{3, 5, 9, 10\}$.

33. Compute the probability of the event *E*: "an even number."

34. Compute the probability of the event *F*: "an odd number."

For Problems 35 and 36, an urn contains 5 white marbles, 10 green marbles, 8 yellow marbles, and 7 black marbles.

35. If one marble is selected, determine the probability that it is white.

36. If one marble is selected, determine the probability that it is black.

In Problems 37–40, assume equally likely outcomes.

37. Determine the probability of having 3 boys in a 3-child family.

38. Determine the probability of having 3 girls in a 3-child family.

39. Determine the probability of having 1 girl and 3 boys in a 4-child family.

40. Determine the probability of having 2 girls and 2 boys in a 4-child family.

For Problems 41–44, two fair dice are rolled.

41. Determine the probability that the sum of the two dice is 7.

42. Determine the probability that the sum of the two dice is 11.

43. Determine the probability that the sum of the two dice is 3.

44. Determine the probability that the sum of the two dice is 12.

In Problems 45–48, find the probability of the indicated event if P(A) = 0.25 and P(B) = 0.45.

45. $P(A \cup B)$ if $P(A \cap B) = 0.15$

46. $P(A \cap B)$ if $P(A \cup B) = 0.6$

47. $P(A \cup B)$ if A, B are mutually exclusive

48. $P(A \cap B)$ if A, B are mutually exclusive

49. If $P(A) = 0.60$, $P(A \cup B) = 0.85$, and $P(A \cap B) = 0.05$, find $P(B)$.

50. If $P(B) = 0.30$, $P(A \cup B) = 0.65$, and $P(A \cap B) = 0.15$, find $P(A)$.

51. According to the Federal Bureau of Investigation, in 2002 there was a 26.5% probability of theft from a motor vehicle. If a victim of theft is randomly selected, what is the probability that he or she was not a victim of theft from a motor vehicle?

52. According to the Federal Bureau of Investigation, in 2002 there was a 3.9% probability of theft involving a bicycle. If a victim of theft is randomly selected, what is the probability that he or she was not a victim of bicycle theft?

53. In Chicago, there is a 30% probability that Memorial Day will have a high temperature in the 70s. What is the probability that next Memorial Day will not have a high temperature in the 70s in Chicago?

54. In Chicago, there is a 4% probability that Memorial Day will have a low temperature in the 30s. What is the probability that next Memorial Day will not have a low temperature in the 30s in Chicago?

For Problems 55–58, a golf ball is selected at random from a container. If the container has 9 white balls, 8 green balls, and 3 orange balls, find the probability of each event.

55. The golf ball is white or green.

56. The golf ball is white or orange.

57. The golf ball is not white.

58. The golf ball is not green.

59. On the "Price is Right" there is a game in which a bag is filled with 3 strike chips and 5 numbers. Let's say that the numbers in the bag are 0, 1, 3, 6, and 9. What is the probability of selecting a strike chip or the number 1?

60. Another game on the "Price is Right" requires the contestant to spin a wheel with numbers 5, 10, 15, 20, ..., 100. What is the probability that the contestant spins 100 or 30?

Problems 61–64, are based on a consumer survey of annual incomes in 100 households. The following table gives the data.

Income	$0–9999	$10,000–19,999	$20,000–29,999	$30,000–39,999	$40,000 or more
Number of households	5	35	30	20	10

61. What is the probability that a household has an annual income of $30,000 or more?

62. What is the probability that a household has an annual income between $10,000 and $29,999, inclusive?

63. What is the probability that a household has an annual income of less than $20,000?

64. What is the probability that a household has an annual income of $20,000 or more?

65. **Surveys** In a survey about the number of TV sets in a house, the following probability table was constructed:

Number of TV sets	0	1	2	3	4 or more
Probability	0.05	0.24	0.33	0.21	0.17

Find the probability of a house having:
(a) 1 or 2 TV sets
(b) 1 or more TV sets
(c) 3 or fewer TV sets
(d) 3 or more TV sets
(e) Less than 2 TV sets
(f) Less than 1 TV set
(g) 1, 2, or 3 TV sets
(h) 2 or more TV sets

66. **Checkout Lines** Through observation, it has been determined that the probability for a given number of people waiting in line at the "5 items or less" checkout register of a supermarket is as follows:

Number waiting in line	0	1	2	3	4 or more
Probability	0.10	0.15	0.20	0.24	0.31

Find the probability of:
(a) At most 2 people in line
(b) At least 2 people in line
(c) At least 1 person in line

67. In a certain Algebra and Trigonometry class, there are 18 freshmen and 15 sophomores. Of the 18 freshmen, 10 are male, and of the 15 sophomores, 8 are male. Find the probability that a randomly selected student is:
(a) A freshman or female
(b) A sophomore or male

68. The faculty of the mathematics department at Joliet Junior College is composed of 4 females and 9 males. Of the 4 females, 2 are under the age of 40, and of the males 3 are under age 40. Find the probability that a randomly selected faculty member is:
(a) Female or under age 40
(b) Male or over age 40

69. Birthday Problem What is the probability that at least 2 people have the same birthday in a group of 12 people? Assume that there are 365 days in a year.

70. Birthday Problem What is the probability that at least 2 people have the same birthday in a group of 35 people? Assume that there are 365 days in a year.

71. Winning a Lottery In a certain lottery, there are ten balls, numbered 1, 2, 3, 4, 5, 6, 7, 8, 9, 10. Of these, five are drawn in order. If you pick five numbers that match those drawn in the correct order, you win $1,000,000. What is the probability of winning such a lottery?

Chapter Review

Things to Know

Set (p. 694) Well-defined collection of distinct objects, called elements

Null set (p. 694)	$\varnothing$	Set that has no elements
Equality (p. 694)	$A = B$	A and B have the same elements.
Subset (p. 694)	$A \subseteq B$	Each element of A is also an element of B.
Intersection (p. 695)	$A \cap B$	Set consisting of elements that belong to both A and B.
Union (p. 695)	$A \cup B$	Set consisting of elements that belong to either A or B, or both.
Universal set (p. 695)	U	Set consisting of all the elements that we wish to consider.
Complement (p. 695)	$\overline{A}$	Set consisting of elements of the universal set that are not in A.
Finite set (p. 696)		The number of elements in the set is a nonnegative integer.
Infinite set (p. 696)		A set that is not finite.

Counting formula (p. 697) $n(A \cup B) = n(A) + n(B) - n(A \cap B)$

Addition Principle (p. 697) If $A \cap B = \varnothing$, then $n(A \cup B) = n(A) + n(B)$

Multiplication Principle (p. 701) If a task consists of a sequence of choices in which there are p selections for the first choice, q selections for the second choice, and so on, then the task of making these selections can be done in $p \cdot q \cdot \ldots$ different ways.

Permutation (p. 702) An ordered arrangement of r objects chosen from n objects.

Permutation: Distinct, with repetition (p. 702) n^r
The n objects are distinct (different), and repetition is allowed in the selection of r of them.

Permutation: Distinct, without repetition (p. 704) $P(n, r) = n(n - 1) \cdot \cdots \cdot [n - (r - 1)] = \dfrac{n!}{(n - r)!}$
The n objects are distinct (different), and repetition is not allowed in the selection of r of them where $r \leq n$.

Combination (p. 706) $C(n, r) = \dfrac{P(n, r)}{r!} = \dfrac{n!}{(n - r)! r!}$
An arrangement, without regard to order, of r objects selected from n distinct objects where $r \leq n$.

Permutation: Not distinct, with repetition (p. 708)	$\dfrac{n!}{n_1!n_2!\cdots n_k!}$ The number of permutations of n objects of which n_1 are of one kind, n_2 are of a second kind, ..., and n_k are of a kth kind, where $n = n_1 + n_2 + \cdots + n_k$
Sample space (p. 711)	Set whose elements represent all the logical possibilities that can occur as a result of an experiment.
Probability (p. 711)	A nonnegative number assigned to each outcome of a sample space; the sum of all the probabilities of the outcomes equals 1.
Equally likely outcomes (p. 714)	$P(E) = \dfrac{n(E)}{n(S)}$ The same probability is assigned to each outcome.
Addition Rule (p. 715)	$P(E \cup F) = P(E) + P(F) - P(E \cap F)$
Complement of an event (p. 717)	$P(\overline{E}) = 1 - P(E)$

Objectives

Section		You should be able to . . .	Review Exercises
8.1	1	Find all the subsets of a set (p. 694)	1, 2
	2	Find the intersection and union of sets (p. 695)	3–6
	3	Find the complement of a set (p. 695)	7–10
	4	Count the number of elements in a set (p. 696)	11–18
8.2	1	Solve counting problems using the Multiplication Principle (p. 700)	23–26, 32–36
	2	Solve counting problems using permutations (p. 702)	19, 20, 27, 28, 41(a)
	3	Solve counting problems using combinations (p. 705)	21, 22, 29–31, 39–40
	4	Solve counting problems using permutations involving n nondistinct objects (p. 707)	37, 38
8.3	1	Construct probability models (p. 711)	41(b)
	2	Compute probabilities of equally likely outcomes (p. 714)	41(b), 42(a), 43(a), 44–47
	3	Use the Addition Rule to find probabilities (p. 715)	48
	4	Use the Complement Rule to find probabilities (p. 717)	41(c), 42(b), 43(b), 44

Review Exercises

1. Write down all the subsets of the set {Dave, Joanne, Erica}.   **2.** Write down all the subsets of the set {Green, Blue, Red}.

In Problems 3–10, use U = universal set = {1, 2, 3, 4, 5, 6, 7, 8, 9}, A = {1, 3, 5, 7}, B = {3, 5, 6, 7, 8}, and C = {2, 3, 7, 8, 9} to find each set.

3. $A \cup B$ **4.** $B \cup C$ **5.** $A \cap C$ **6.** $A \cap B$

7. $\overline{A} \cup \overline{B}$ **8.** $\overline{B} \cap \overline{C}$ **9.** $\overline{B \cap C}$ **10.** $\overline{A \cup B}$

11. If $n(A) = 8, n(B) = 12$, and $n(A \cap B) = 3$, find $n(A \cup B)$.   **12.** If $n(A) = 12, n(A \cup B) = 30$, and $n(A \cap B) = 6$, find $n(B)$.

In Problems 13–18, use the information supplied in the figure.

13. How many are in A?

14. How many are in A or B?

15. How many are in A and C?

16. How many are not in B?

17. How many are in neither A nor C?

18. How many are in B but not in C?

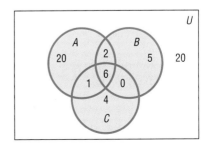

In Problems 19–22, compute the given expression.

19. $P(8, 3)$ **20.** $P(7, 3)$ **21.** $C(8, 3)$ **22.** $C(7, 3)$

23. A clothing store sells pure wool and polyester–wool suits. Each suit comes in 3 colors and 10 sizes. How many suits are required for a complete assortment?

24. In connecting a certain electrical device, 5 wires are to be connected to 5 different terminals. How many different wirings are possible if 1 wire is connected to each terminal?

25. Baseball On a given day, the American Baseball League schedules 7 games. How many different outcomes are possible, assuming that each game is played to completion?

26. Baseball On a given day, the National Baseball League schedules 6 games. How many different outcomes are possible, assuming that each game is played to completion?

27. If 4 people enter a bus having 9 vacant seats, in how many ways can they be seated?

28. How many different arrangements are there of the letters in the word ROSE?

29. In how many ways can a squad of 4 relay runners be chosen from a track team of 8 runners?

30. A professor has 10 similar problems to put on a test with 3 problems. How many different tests can she design?

31. Baseball In how many ways can 2 teams from 14 teams in the American League be chosen without regard to which team is at home?

32. Arranging Books on a Shelf There are 5 different French books and 5 different Spanish books. How many ways are there to arrange them on a shelf if:
(a) Books of the same language must be grouped together, French on the left, Spanish on the right?
(b) French and Spanish books must alternate in the grouping, beginning with a French book?

33. Telephone Numbers Using the digits 0, 1, 2, . . . , 9, how many 7-digit numbers can be formed if the first digit cannot be 0 or 9 and if the last digit is greater than or equal to 2 and less than or equal to 3? Repeated digits are allowed.

34. Home Choices A contractor constructs homes with 5 different choices of exterior finish, 3 different roof arrangements, and 4 different window designs. How many different types of homes can be built?

35. License Plate Possibilities A license plate consists of 1 letter, excluding O and I, followed by a 4-digit number that cannot have a 0 in the lead position. How many different plates are possible?

36. Using the digits 0 and 1, how many different numbers consisting of 8 digits can be formed?

37. Forming Different Words How many different words, real or imaginary, can be formed using all the letters in the word MISSING?

38. Arranging Flags How many different vertical arrangements are there of 10 flags if 4 are white, 3 are blue, 2 are green, and 1 is red?

39. Forming Committees A group of 9 people is going to be formed into committees of 4, 3, and 2 people. How many committees can be formed if:
(a) A person can serve on any number of committees?
(b) No person can serve on more than one committee?

40. Forming Committees A group consists of 5 men and 8 women. A committee of 4 is to be formed from this group, and policy dictates that at least 1 woman be on this committee.
(a) How many committees can be formed that contain exactly 1 man?
(b) How many committees can be formed that contain exactly 2 women?
(c) How many committees can be formed that contain at least 1 man?

41. Birthday Problem For this problem, assume that a year has 365 days.
(a) How many ways can 18 people have different birthdays?
(b) What is the probability that nobody has the same birthday in a group of 18 people?
(c) What is the probability in a group of 18 people that at least 2 people have the same birthday?

42. Death Rates According to the U.S. National Center for Health Statistics, 29% of all deaths in 2001 were due to heart disease.
(a) What is the probability that a randomly selected death in 2001 was due to heart disease?
(b) What is the probability that a randomly selected death in 2001 was not due to heart disease?

43. Unemployment According to the U.S. Bureau of Labor Statistics, 5.8% of the U.S. labor force was unemployed in 2002.
(a) What is the probability that a randomly selected member of the U.S. labor force was unemployed in 2002?
(b) What is the probability that a randomly selected member of the U.S. labor force was not unemployed in 2002?

44. From a box containing three 40-watt bulbs, six 60-watt bulbs, and eleven 75-watt bulbs, a bulb is drawn at random. What is the probability that the bulb is 40 watts? What is the probability that it is not a 75-watt bulb?

45. You have four $1 bills, three $5 bills, and two $10 bills in your wallet. If you pick a bill at random, what is the probability that it will be a $1 bill?

46. Each of the letters in the word ROSE is written on an index card and the cards are then shuffled. What is the probability that, when the cards are dealt out, they spell the word ROSE?

47. Each of the numbers, $1, 2, \ldots, 100$ is written on an index card and the cards are then shuffled. If a card is selected at random, what is the probability that the number on the card is divisible by 5? What is the probability that the card selected is either a 1 or names a prime number?

48. At the Milex tune-up and brake repair shop, the manager has found that a car will require a tune-up with a probabili-

ty of 0.6, a brake job with a probability of 0.1, and both with a probability of 0.02.
(a) What is the probability that a car requires either a tune-up or a brake job?
(b) What is the probability that a car requires a tune-up but not a brake job?
(c) What is the probability that a car requires neither a tune-up nor a brake job?

Chapter Test

In Problems 1–6, use U = Universal set = $\{0, 1, 2, 3, 4, 5, 6, 7, 8, 9\}$, A = $\{0, 1, 4, 9\}$, B = $\{2, 4, 6, 8\}$, and C = $\{1, 3, 5, 7, 9\}$ to find each set.

1. $A \cap B$

2. $A \cup C$

3. $(A \cup B) \cap C$

4. $\overline{A \cup B}$

5. $\overline{C}$

6. $A \cap \overline{C}$

In Problems 7–10, a survey of 70 college freshmen asked whether students planned to take biology, chemistry, or physics during their first year. Use the diagram to answer each question.

7. How many of the surveyed students plan to take physics during their first year?

8. How many of the surveyed students do not plan to take biology, chemistry, or physics during their first year?

9. How many of the surveyed students plan to take only biology and chemistry during their first year?

10. How many of the surveyed students plan to take physics or chemistry during their first year?

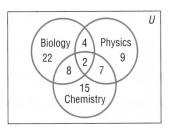

In Problems 11–13, compute the given expression.

11. $7!$

12. $P(10, 6)$

13. $C(11, 5)$

14. M&M's® offers customers the opportunity to create their own color mix of candy. There are 21 colors to choose from, and customers are allowed to select up to 6 different colors. How many different color mixes are possible, assuming that no color is selected more than once?

15. How many distinct 8-letter words (real or imaginary) can be formed from the letters in the word REDEEMED?

16. In horse racing, an exacta bet requires the bettor to pick the first two horses in the exact order. If there are 8 horses in a race, in how many ways could you make an exacta bet?

17. On February 20, 2004, the Ohio Bureau of Motor Vehicles unveiled the state's new license plate format. The plate consists of three letters (A–Z) followed by 4 digits (0–9). Assume that all letters and digits may be used except that the third letter cannot be O, I, or Z. If repetitions are allowed, how many different plates are possible?

18. Kiersten applies for admission to the University of Southern California (USC) and Florida State University (FSU). She estimates that she has a 60% chance of being admitted to USC, a 70% chance of being admitted to FSU, and a 35% chance of being admitted to both universities.
(a) What is the probability that she will be admitted to either USC or FSU?

(b) What is the probability that she will not be admitted to FSU?

19. A cooler contains 8 bottles of Pepsi, 5 bottles of Coke, 4 bottles of Mountain Dew, and 3 bottles of IBC.
(a) What is the probability that a bottle chosen at random is Coke?
(b) What is the probability that a bottle chosen at random is either Pepsi or IBC?

20. A study on the age distribution of a community college gave the following table:

Age	17 and under	18–20	21–24	25–34	35–64	65 and over
Probability	0.03	???	0.23	0.29	0.25	0.01

What must be the probability that a randomly selected student at the college is between 18 and 20 years old?

21. Powerball is a multi-state lottery where 5 white balls from a drum with 53 balls and 1 red ball from a drum with 42 red balls are selected. For a $1 ticket, players get one chance at winning the jackpot by matching all 6 numbers. What is the probability of selecting the winning numbers on a $1 play?

22. If you roll 1 die five times, what is the probability that you obtain exactly 2 fours?

Chapter Projects

1. **Simulation** In the Winter 1998 edition of *Eightysomething!*, Mike Koehler uses simulation to calculate the following probabilities: "A woman and man (unrelated) each have two children. At least one of the woman's children is a boy, and the man's older child is a boy. Do the chances that the woman has two boys equal the chances that the man has two boys?" Perform a simulation to answer the question.

The following projects are available at the Instructor's Resource Center (IRC):

2. **Project at Motorola** *Probability of Error Digital Wireless Comunications.*

3. **Surveys**

4. **Law of Large Numbers**

Cumulative Review

1. Solve $3x^2 - 2x = -1$.

2. Graph $f(x) = x^2 + 4x - 5$ by determining whether the graph opens up or down and by finding the vertex, axis of symmetry, and intercepts.

3. Graph $f(x) = 2(x + 1)^2 - 4$ using transformations.

4. Solve $|x - 4| \leq 0.01$.

5. Find the complex zeros of
$$f(x) = 5x^4 - 9x^3 - 7x^2 - 31x - 6.$$

6. Graph $g(x) = 3^{x-1} + 5$ using transformations. Determine the domain, the range, and horizontal asymptote of g.

7. What is the exact value of $\log_3 9$?

8. Solve $\log_2(3x - 2) + \log_2 x = 4$.

9. Solve the system:
$$\begin{cases} x - 2y + z = 15 \\ 3x + y - 3z = -8 \\ -2x + 4y - z = -27 \end{cases}$$

10. What is the 33rd term in the sequence $-3, 1, 5, 9, \ldots$? What is the sum of the first 20 terms?

ANSWERS

CHAPTER R Review

Historical Problems (page 14)

1. (a) 1,20 **(b)** 2,50 **2. (a)** $\dfrac{7}{3} = 2.333\ldots$ **(b)** $\dfrac{39}{8} = 4.875$ **(c)** $\dfrac{84,823}{27,000} = 3.141592592\ldots$

R.1 Assess Your Understanding (page 14)

1. rational **2.** 31 **3.** distributive **4.** $5(x+3) = 6$ **5.** T **6.** F **7.** F **8.** T **9. (a)** $\{2,5\}$ **(b)** $\{-6,2,5\}$ **(c)** $\left\{-6,\dfrac{1}{2},-1.333\ldots,2,5\right\}$
(d) $\{\pi\}$ **(e)** $\left\{-6,\dfrac{1}{2},-1.333\ldots,\pi,2,5\right\}$ **11. (a)** $\{1\}$ **(b)** $\{0,1\}$ **(c)** $\left\{0,1,\dfrac{1}{2},\dfrac{1}{3},\dfrac{1}{4}\right\}$ **(d)** None **(e)** $\left\{0,1,\dfrac{1}{2},\dfrac{1}{3},\dfrac{1}{4}\right\}$ **13. (a)** None **(b)** None
(c) None **(d)** $\left\{\sqrt{2},\pi,\sqrt{2}+1,\pi+\dfrac{1}{2}\right\}$ **(e)** $\left\{\sqrt{2},\pi,\sqrt{2}+1,\pi+\dfrac{1}{2}\right\}$ **15. (a)** 18.953 **(b)** 18.952 **17. (a)** 28.653 **(b)** 28.653

19. (a) 0.063 **(b)** 0.062 **21. (a)** 9.999 **(b)** 9.998 **23. (a)** 0.429 **(b)** 0.428 **25. (a)** 34.733 **(b)** 34.733 **27.** $3+2 = 5$ **29.** $x+2 = 3 \cdot 4$

31. $3 \cdot y = 1 + 2$ **33.** $x - 2 = 6$ **35.** $\dfrac{x}{2} = 6$ **37.** 7 **39.** 6 **41.** 1 **43.** $\dfrac{13}{3}$ **45.** 11 **47.** 11 **49.** -4 **51.** 1 **53.** 6 **55.** $\dfrac{2}{7}$ **57.** $\dfrac{4}{45}$ **59.** $\dfrac{23}{20}$
61. $\dfrac{79}{30}$ **63.** $\dfrac{13}{36}$ **65.** $-\dfrac{16}{45}$ **67.** $\dfrac{1}{60}$ **69.** $\dfrac{15}{22}$ **71.** $6x+24$ **73.** $x^2 - 4x$ **75.** $x^2 + 6x + 8$ **77.** $x^2 - x - 2$ **79.** $x^2 - 10x + 16$ **81.** $x^2 - 4$
83. $2x + 3x = (2+3)x = 5x$ **85.** $2(3 \cdot 4) = 2 \cdot 12 = 24; (2 \cdot 3) \cdot (2 \cdot 4) = 6 \cdot 8 = 48$ **87.** No; $2 - 3 \neq 3 - 2$ **89.** No; $\dfrac{2}{3} \neq \dfrac{3}{2}$

91. Symmetric Property **93.** No; no **95.** 1

R.2 Assess Your Understanding (page 25)

1. variable **2.** origin **3.** strict **4.** base; exponent or power **5.** 1.2345678×10^3 **6.** T **7.** F **8.** F **9.** F **10.** F
11. **13.** $>$ **15.** $>$ **17.** $>$ **19.** $=$ **21.** $<$ **23.** $x > 0$ **25.** $x < 2$ **27.** $x \leq 1$ **29.**

31. **33.** 1 **35.** 2 **37.** 6 **39.** 4 **41.** -28 **43.** $\dfrac{4}{5}$ **45.** 0 **47.** 1 **49.** 5 **51.** 1 **53.** 22 **55.** 2 **57.** $x = 0$ **59.** $x = 3$

61. None **63.** $x = 1, x = 0, x = -1$ **65.** $\{x | x \neq 5\}$ **67.** $\{x | x \neq -4\}$ **69.** $0°C$ **71.** $25°C$ **73.** 16 **75.** $\dfrac{1}{16}$ **77.** $\dfrac{1}{9}$ **79.** 9 **81.** 5 **83.** 4 **85.** $64x^6$
87. $\dfrac{x^4}{y^2}$ **89.** $\dfrac{x}{y}$ **91.** $-\dfrac{8x^3z}{9y}$ **93.** $\dfrac{16x^2}{9y^2}$ **95.** -4 **97.** 5 **99.** 4 **101.** 2 **103.** $\sqrt{5}$ **105.** $\dfrac{1}{2}$ **107.** 10; 0 **109.** 81 **111.** 304,006.671 **113.** 0.004

115. 481.890 **117.** 0.000 **119.** 4.542×10^2 **121.** 1.3×10^{-2} **123.** 3.2155×10^4 **125.** 4.23×10^{-4} **127.** 61,500 **129.** 0.001214 **131.** 110,000,000

133. 0.081 **135.** $A = lw$ **137.** $C = \pi d$ **139.** $A = \dfrac{\sqrt{3}}{4}x^2$ **141.** $V = \dfrac{4}{3}\pi r^3$ **143.** $V = x^3$ **145. (a)** \$6000 **(b)** \$8000

147. (a) $2 \leq 5$ **(b)** $6 > 5$ **149. (a)** Yes **(b)** No **151.** 400,000,000 m **153.** 0.0000005 m **155.** 5×10^{-4} in. **157.** 5.865696×10^{12} mi

159. No; $\dfrac{1}{3}$ is larger; 0.000333... **161.** No

R.3 Assess Your Understanding (page 32)

1. right; hypotenuse **2.** $A = \dfrac{1}{2}bh$ **3.** $C = 2\pi r$ **4.** T **5.** T **6.** F **7.** 13 **9.** 26 **11.** 25 **13.** Right triangle; 5 **15.** Not a right triangle

17. Right triangle; 25 **19.** Not a right triangle **21.** 8 in.^{2} **23.** 4 in.^{2} **25.** $A = 25\pi$ m^2; $C = 10\pi$ m **27.** $V = 224$ ft^3; $S = 232$ ft^2

29. $V = \dfrac{256}{3}\pi$ cm^3; $S = 64\pi$ cm^2 **31.** $V = 648\pi$ in.^{3}; $S = 306\pi$ in.^{2} **33.** π square units **35.** 2π square units **37.** About 16.8 ft **39.** 64 ft^2

41. $24 + 2\pi \approx 30.28$ ft^2; $16 + 2\pi \approx 22.28$ ft **43.** About 5.477 mi **45.** From 100 ft: 12.247 mi; from 150 ft: 15.000 mi

R.4 Assess Your Understanding (page 43)

1. 4; 3 **2.** $x^2 - 16$ **3.** $x^3 - 8$ **4.** F **5.** T **6.** F **7.** Monomial; variable: x; coefficient: 2; degree: 3 **9.** Not a monomial
11. Monomial; variables: x, y; coefficient: -2; degree: 3 **13.** Not a monomial **15.** Not a monomial **17.** Yes; 2 **19.** Yes; 0 **21.** No **23.** Yes; 3
25. No **27.** $x^2 + 7x + 2$ **29.** $x^3 - 4x^2 + 9x + 7$ **31.** $6x^5 + 5x^4 + 3x^2 + x$ **33.** $7x^2 - x - 7$ **35.** $-2x^3 + 18x^2 - 18$ **37.** $2x^2 - 4x + 6$
39. $15y^2 - 27y + 30$ **41.** $x^3 + x^2 - 4x$ **43.** $-8x^5 - 10x^2$ **45.** $x^3 + 3x^2 - 2x - 4$ **47.** $x^2 + 6x + 8$ **49.** $2x^2 + 9x + 10$ **51.** $x^2 - 2x - 8$
53. $x^2 - 5x + 6$ **55.** $2x^2 - x - 6$ **57.** $-2x^2 + 11x - 12$ **59.** $2x^2 + 8x + 8$ **61.** $x^2 - xy - 2y^2$ **63.** $-6x^2 - 13xy - 6y^2$ **65.** $x^2 - 49$
67. $4x^2 - 9$ **69.** $x^2 + 8x + 16$ **71.** $x^2 - 8x + 16$ **73.** $9x^2 - 16$ **75.** $4x^2 - 12x + 9$ **77.** $x^2 - y^2$ **79.** $9x^2 - y^2$ **81.** $x^2 + 2xy + y^2$
83. $x^2 - 4xy + 4y^2$ **85.** $x^3 - 6x^2 + 12x - 8$ **87.** $8x^3 + 12x^2 + 6x + 1$ **89.** $4x^2 - 11x + 23$; remainder -45 **91.** $4x - 3$; remainder $x + 1$

93. $5x^2 - 13$; remainder $x + 27$ **95.** $2x^2$; remainder $-x^2 + x + 1$ **97.** $x^2 - 2x + \dfrac{1}{2}$; remainder $\dfrac{5}{2}x + \dfrac{1}{2}$ **99.** $-4x^2 - 3x - 3$; remainder -7
101. $x^2 - x - 1$; remainder $2x + 2$ **103.** $x^2 + ax + a^2$; remainder 0

R.5 Assess Your Understanding (page 52)

1. $3x(x - 2)(x + 2)$ **2.** Prime **3.** T **4.** F **5.** $3(x + 2)$ **7.** $a(x^2 + 1)$ **9.** $x(x^2 + x + 1)$ **11.** $2x(x - 1)$ **13.** $3xy(x - 2y + 4)$
15. $(x + 1)(x - 1)$ **17.** $(2x + 1)(2x - 1)$ **19.** $(x + 4)(x - 4)$ **21.** $(5x + 2)(5x - 2)$ **23.** $(x + 1)^2$ **25.** $(x + 2)^2$ **27.** $(x - 5)^2$
29. $(2x + 1)^2$ **31.** $(4x + 1)^2$ **33.** $(x - 3)(x^2 + 3x + 9)$ **35.** $(x + 3)(x^2 - 3x + 9)$ **37.** $(2x + 3)(4x^2 - 6x + 9)$ **39.** $(x + 2)(x + 3)$
41. $(x + 6)(x + 1)$ **43.** $(x + 5)(x + 2)$ **45.** $(x - 8)(x - 2)$ **47.** $(x - 8)(x + 1)$ **49.** $(x + 8)(x - 1)$ **51.** $(x + 2)(2x + 3)$
53. $(x - 2)(2x + 1)$ **55.** $(2x + 3)(3x + 2)$ **57.** $(3x + 1)(x + 1)$ **59.** $(z + 1)(2z + 3)$ **61.** $(x + 2)(3x - 4)$ **63.** $(x - 2)(3x + 4)$
65. $(x + 4)(3x + 2)$ **67.** $(x + 4)(3x - 2)$ **69.** $(x + 6)(x - 6)$ **71.** $2(1 + 2x)(1 - 2x)$ **73.** $(x + 2)(x + 5)$ **75.** $(x - 7)(x - 3)$
77. $4(x^2 - 2x + 8)$ **79.** Prime **81.** $-(x - 5)(x + 3)$ **83.** $3(x + 2)(x - 6)$ **85.** $y^2(y + 5)(y + 6)$ **87.** $(2x + 3)^2$ **89.** $2(3x + 1)(x + 1)$
91. $(x - 3)(x + 3)(x^2 + 9)$ **93.** $(x - 1)^2(x^2 + x + 1)^2$ **95.** $x^5(x - 1)(x + 1)$ **97.** $(4x + 3)^2$ **99.** $-(4x - 5)(4x + 1)$
101. $(2y - 5)(2y - 3)$ **103.** $-(3x - 1)(3x + 1)(x^2 + 1)$ **105.** $(x + 3)(x - 6)$ **107.** $(x + 2)(x - 3)$ **109.** $(3x - 5)(9x^2 - 3x + 7)$
111. $(x + 5)(3x + 11)$ **113.** $(x - 1)(x + 1)(x + 2)$ **115.** $(x - 1)(x + 1)(x^2 - x + 1)$ **117.** $2(3x + 4)(9x + 13)$ **119.** $2x(3x + 5)$
121. $5(x + 3)(x - 2)^2(x + 1)$ **123.** $3(4x - 3)(4x - 1)$ **125.** $6(3x - 5)(2x + 1)^2(5x - 4)$ **127.** The possibilities are
$(x \pm 1)(x \pm 4) = x^2 \pm 5x + 4$ or $(x \pm 2)(x \pm 2) = x^2 \pm 4x + 4$, none of which equals $x^2 + 4$.

R.6 Assess Your Understanding (page 56)

1. quotient; divisor; remainder **2.** $-3\overline{)2\ 0\ -5\ 1}$ **3.** T **4.** T **5.** $x^2 + x + 4$; remainder 12 **7.** $3x^2 + 11x + 32$; remainder 99
9. $x^4 - 3x^3 + 5x^2 - 15x + 46$; remainder -138 **11.** $4x^5 + 4x^4 + x^3 + x^2 + 2x + 2$; remainder 7 **13.** $0.1x^2 - 0.11x + 0.321$; remainder -0.3531
15. $x^4 + x^3 + x^2 + x + 1$; remainder 0 **17.** No **19.** Yes **21.** Yes **23.** No **25.** Yes **27.** -9

R.7 Assess Your Understanding (page 66)

1. lowest terms **2.** least common multiple **3.** T **4.** F **5.** $\dfrac{3}{x - 3}$ **7.** $\dfrac{x}{3}$ **9.** $\dfrac{4x}{2x - 1}$ **11.** $\dfrac{y + 5}{2(y + 1)}$ **13.** $\dfrac{x + 5}{x - 1}$ **15.** $-(x + 7)$

17. $\dfrac{3}{5x(x - 2)}$ **19.** $\dfrac{2x}{x + 4}$ **21.** $\dfrac{8}{3x}$ **23.** $\dfrac{x - 3}{x + 7}$ **25.** $\dfrac{4x}{(x - 2)(x - 3)}$ **27.** $\dfrac{4}{5(x - 1)}$ **29.** $-\dfrac{(x - 4)^2}{4x}$ **31.** $\dfrac{(x + 3)^2}{(x - 3)^2}$ **33.** $\dfrac{(x - 4)(x + 3)}{(x - 1)(2x + 1)}$

35. $\dfrac{x + 5}{2}$ **37.** $\dfrac{(x - 2)(x + 2)}{2x - 3}$ **39.** $\dfrac{3x - 2}{x - 3}$ **41.** $\dfrac{x + 9}{2x - 1}$ **43.** $\dfrac{4 - x}{x - 2}$ **45.** $\dfrac{2(x + 5)}{(x - 1)(x + 2)}$ **47.** $\dfrac{3x^2 - 2x - 3}{(x + 1)(x - 1)}$ **49.** $\dfrac{-(11x + 2)}{(x + 2)(x - 2)}$

51. $\dfrac{2(x^2 - 2)}{x(x - 2)(x + 2)}$ **53.** $(x - 2)(x + 2)(x + 1)$ **55.** $x(x - 1)(x + 1)$ **57.** $x^3(2x - 1)^2$ **59.** $x(x - 1)^2(x + 1)(x^2 + x + 1)$

61. $\dfrac{5x}{(x - 6)(x - 1)(x + 4)}$ **63.** $\dfrac{2(2x^2 + 5x - 2)}{(x - 2)(x + 2)(x + 3)}$ **65.** $\dfrac{5x + 1}{(x - 1)^2(x + 1)^2}$ **67.** $\dfrac{-x^2 + 3x + 13}{(x - 2)(x + 1)(x + 4)}$ **69.** $\dfrac{x^3 - 2x^2 + 4x + 3}{x^2(x + 1)(x - 1)}$

71. $\dfrac{-1}{x(x + h)}$ **73.** $\dfrac{x + 1}{x - 1}$ **75.** $\dfrac{(x - 1)(x + 1)}{2x(2x + 1)}$ **77.** $\dfrac{2(5x - 1)}{(x - 2)(x + 1)^2}$ **79.** $\dfrac{-2x(x^2 - 2)}{(x + 2)(x^2 - x - 3)}$ **81.** $\dfrac{-1}{x - 1}$ **83.** $\dfrac{19}{(3x - 5)^2}$

85. $\dfrac{x^2 - 1}{(x^2 + 1)^2}$ **87.** $\dfrac{x(3x + 2)}{(3x + 1)^2}$ **89.** $-\dfrac{3x^2 + 8x - 3}{(x^2 + 1)^2}$ **91.** $f = \dfrac{R_1 \cdot R_2}{(n - 1)(R_1 + R_2)}$; $\dfrac{2}{15}$ m

R.8 Assess Your Understanding (page 74)

3. index **4.** T **5.** cube root **6.** F **7.** 3 **9.** -2 **11.** $2\sqrt{2}$ **13.** $-2x\sqrt[3]{x}$ **15.** x^3y^2 **17.** x^2y **19.** $6\sqrt{x}$ **21.** $6x\sqrt{x}$ **23.** $15\sqrt[3]{3}$ **25.** $12\sqrt{3}$

27. $7\sqrt{2}$ **29.** $\sqrt{2}$ **31.** $2\sqrt{3}$ **33.** $-\sqrt[3]{2}$ **35.** $x - 2\sqrt{x} + 1$ **37.** $(2x - 1)\sqrt[3]{2x}$ **39.** $(2x - 15)\sqrt{2x}$ **41.** $-(x + 5y)\sqrt[3]{2xy}$ **43.** $\dfrac{\sqrt{2}}{2}$

45. $-\dfrac{\sqrt{15}}{5}$ **47.** $\dfrac{(5 + \sqrt{2})\sqrt{3}}{23}$ **49.** $\dfrac{19 + 8\sqrt{5}}{41}$ **51.** $\dfrac{5\sqrt[3]{4}}{2}$ **53.** $\dfrac{2x + h - 2\sqrt{x}\sqrt{x + h}}{h}$ **55.** 4 **57.** -3 **59.** 64 **61.** $\dfrac{1}{27}$ **63.** $\dfrac{27\sqrt{2}}{32}$

65. $\dfrac{27\sqrt{2}}{32}$ **67.** $x^{7/12}$ **69.** xy^2 **71.** $x^{4/3}y^{5/3}$ **73.** $\dfrac{8x^{3/2}}{y^{1/4}}$ **75.** $\dfrac{3x + 2}{(1 + x)^{1/2}}$ **77.** $\dfrac{x(3x^2 + 2)}{(x^2 + 1)^{1/2}}$ **79.** $\dfrac{22x + 5}{10\sqrt{x - 5}\sqrt{4x + 3}}$ **81.** $\dfrac{2 + x}{2(1 + x)^{3/2}}$

83. $\dfrac{4 - x}{(x + 4)^{3/2}}$ **85.** $\dfrac{1}{x^2(x^2 - 1)^{1/2}}$ **87.** $\dfrac{1 - 3x^2}{2\sqrt{x}(1 + x^2)^2}$ **89.** $\dfrac{1}{2}(5x + 2)(x + 1)^{1/2}$ **91.** $2x^{1/2}(3x - 4)(x + 1)$ **93.** $(x^2 + 4)^{1/3}(11x^2 + 12)$

95. $(3x + 5)^{1/3}(2x + 3)^{1/2}(17x + 27)$ **97.** $\dfrac{3(x + 2)}{2x^{1/2}}$ **99.** 1.41 **101.** 1.59 **103.** 4.89 **105.** 2.15 **107. (a)** 15,660.4 gal **(b)** 390.7 gal

109. $2\sqrt{2}\,\pi \approx 8.89$ sec **111.** $\dfrac{\pi\sqrt{3}}{6} \approx 0.91$ sec

Review Exercises (page 78)

1. (a) None **(b)** $\{-10\}$ **(c)** $\left\{-10, 0.65, 1.343434\ldots, \dfrac{1}{9}\right\}$ **(d)** $\{\sqrt{7}\}$ **(e)** $\left\{-10, 0.65, 1.343434\ldots, \sqrt{7}, \dfrac{1}{9}\right\}$ **3. (a)** 782.57 **(b)** 782.56 **5. (a)** 9.62

(b) 9.62 **7.** 14 **9.** $\dfrac{3}{4}$ **11.** 3 **13.** $4x - 12$ **15.** **17.** 5 **19.** -10 **21.** -49 **23.** 5 **25.** $\{x \mid x \neq 6\}$ **27.** $4\sqrt{2}$ **29.** $-2\sqrt[3]{2}$

31. $2\sqrt{2}$ **33.** $\dfrac{1}{x^5y}$ **35.** $\dfrac{125}{x^2y}$ **37.** 5.0625 **39.** Coefficients: $3, 4, -2, 0, 5, -12$; degree: 5 **41.** $2x^4 - 2x^3 + x^2 + 5x + 3$ **43.** $6x^2 - 7x - 5$

45. $16x^2 - 1$ **47.** $x^3 - 7x - 6$ **49.** $9x^2 + 6xy + y^2$ **51.** $x^3 + 6x^2y + 12xy^2 + 8y^3$ **53.** $3x^2 + 8x + 25$; remainder 79

55. $-3x^2 + 4$; remainder -2 **57.** $x^4 - x^3 + x^2 - x + 1$; remainder 0 **59.** $(x + 7)(x - 2)$ **61.** $(3x + 2)(2x - 3)$ **63.** $3(x + 2)(x - 7)$

65. $(2x + 1)(4x^2 - 2x + 1)$ **67.** $(2x + 3)(x - 1)(x + 1)$ **69.** $(5x - 2)(5x + 2)$ **71.** Prime **73.** $(x + 4)^2$ **75.** $\dfrac{2x + 7}{x - 2}$ **77.** $\dfrac{3(3x - 1)}{(x + 3)(3x + 1)}$

79. $\dfrac{4x}{(x + 1)(x - 1)}$ **81.** $\dfrac{x^2 + 17x + 2}{(x - 2)(x + 2)^2}$ **83.** $\dfrac{x - 1}{x - 3}$ **85.** $\dfrac{3x}{5y^2}$ **87.** $3xy^4\sqrt[3]{x}$ **89.** $\dfrac{4\sqrt{5}}{5}$ **91.** $-2 - 2\sqrt{2}$ **93.** $-\dfrac{3 + \sqrt{5}}{2}$ **95.** $\dfrac{2(1 + x^2)}{(2 + x^2)^{1/2}}$

97. $\dfrac{x(3x + 16)}{2(x + 4)^{3/2}}$ **99.** $\dfrac{1}{x^2\sqrt{x^2 - 1}}$ **101.** $(x^2 + 4)^{1/3}(11x^2 + 12)$ **103.** 2.81421906×10^8 **105.** Yes **107.** $0.35 per share **109.** 216 ft^2; 84 ft

111. Yes; about 229.2 mi

Chapter Test (page 81)

1. (a) $\{6\}$ **(b)** $\{-5, 0, 6\}$ **(c)** $\left\{-5, 0, \dfrac{4}{3}, 1.5, 2.333\ldots, 6\right\}$ **(d)** $\left\{\dfrac{\pi}{4}, \sqrt{3}\right\}$ **(e)** $\left\{-5, 0, \dfrac{\pi}{4}, \dfrac{4}{3}, 1.5, \sqrt{3}, 2.3\ldots, 6\right\}$ **2. (a)** -2.54 **(b)** -2.53

3. (a) 3.14 **(b)** 3.14 **4.** 8 **5.** $\dfrac{2}{3}$ **6.** 3 **7.** $\dfrac{4}{27}$ **8.** $4x^3 - 19x^2 - 19x + 6$ **9.** $4a^2 - 1$ **10.** $3y^2 - 13yz - 10z^2$ **11.** $\dfrac{3x + 7}{(x + 2)(x - 7)}$

12. $2x^3 - x^2 - 5x + 12$ **13.** $9m^2 + 42mn + 49n^2$ **14.** $\dfrac{2x^2 + 10x + 16}{x + 3}$ **15.** $8b^3 + 1$ **16.** $\dfrac{x}{6}, x \neq -2, 0$

17. $x^2 - 7x + 12, x \neq -3, -2$ **18.** $\sqrt{2xy}\,(4x - 8y)$ **19.** $-\dfrac{a^2b^3}{2c^2}$ **20.** $\dfrac{y^2}{x^2z^4}$ **21.** $\dfrac{x^2 + x + 2}{x^2 - 1}$ **22.** $\dfrac{x^{5/6}}{y^{1/2}}$ **23.** $\dfrac{9}{16}$, undefined; $\dfrac{16}{9}$; $x = 0$

24. $3\sqrt{2} + 3$ **25** $(z - 7)(z + 3)$ **26.** Prime **27.** $(4w + 5)(16w^2 - 20w + 25)$ **28.** $c^2(4c + 3)(4c - 3)$ **29.** $5a(a - 3b)^2$

30. $(4x - 5)(2x + 3)$ **31.** $(y - 4)(3y^2 + 2)$ **32. (a)** $3x^2 - 8x + 3$; remainder 2 **(b)** $3x^2 - 8x + 3$; remainder 2

33. (a) $(x + 32)(x - 20)$ **(b)** 1440 ft^2 **34.** 20 ft **35.** 1.41 m^3

C H A P T E R   1   Graphs, Equations, and Inequalities

1.1 Assess Your Understanding *(page 97)*

5. abscissa; ordinate   **6.** quadrants   **7.** midpoint   **8.** F   **9.** F   **10.** T

11. (a) Quadrant II   **(b)** Positive x-axis

(c) Quadrant III   **(d)** Quadrant I

(e) Negative y-axis   **(f)** Quadrant IV

13. The points will be on a vertical line that is 2 units to the right of the y-axis.

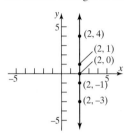

15. $(-1, 4)$; Quadrant II   **17.** $(3, 1)$; Quadrant I   **19.** $X\min = -11, X\max = 5, X\mathrm{scl} = 1, Y\min = -3, Y\max = 6, Y\mathrm{scl} = 1$ **21.** $X\min = -30,$
$X\max = 50, X\mathrm{scl} = 10, Y\min = -90, Y\max = 50, Y\mathrm{scl} = 10$ **23.** $X\min = -10, X\max = 110, X\mathrm{scl} = 10, Y\min = -10, Y\max = 160, Y\mathrm{scl} = 10$
25. $X\min = -6, X\max = 6, X\mathrm{scl} = 2, Y\min = -4, Y\max = 4, Y\mathrm{scl} = 2$ **27.** $X\min = -6, X\max = 6, X\mathrm{scl} = 2, Y\min = -1, Y\max = 3, Y\mathrm{scl} = 1$
29. $X\min = 3, X\max = 9, X\mathrm{scl} = 1, Y\min = 2, Y\max = 10, Y\mathrm{scl} = 2$ **31.** $\sqrt{5}$ **33.** $\sqrt{10}$ **35.** $2\sqrt{17}$ **37.** $\sqrt{85}$ **39.** $\sqrt{53}$ **41.** $\sqrt{6.89} \approx 2.62$
43. $\sqrt{a^2 + b^2}$ **45.** $4\sqrt{10}$ **47.** $2\sqrt{65}$

49. $d(A, B) = \sqrt{13}$
$d(B, C) = \sqrt{13}$
$d(A, C) = \sqrt{26}$
$(\sqrt{13})^2 + (\sqrt{13})^2 = (\sqrt{26})^2$

Area $= \dfrac{13}{2}$ square units

51. $d(A, B) = \sqrt{130}$
$d(B, C) = \sqrt{26}$
$d(A, C) = 2\sqrt{26}$
$(\sqrt{26})^2 + (2\sqrt{26})^2 = (\sqrt{130})^2$

Area $= 26$ square units

53. $d(A, B) = 4$
$d(A, C) = 5$
$d(B, C) = \sqrt{41}$
$4^2 + 5^2 = 16 + 25 = (\sqrt{41})^2$

Area $= 10$ square units

55. $(4, 0)$ **57.** $\left(\dfrac{3}{2}, 1\right)$ **59.** $(5, -1)$ **61.** $(1.05, 0.7)$ **63.** $\left(\dfrac{a}{2}, \dfrac{b}{2}\right)$ **65.** $(0, 0)$ is on the graph.   **67.** $(0, 3)$ is on the graph.

69. $(0, 2)$ and $(\sqrt{2}, \sqrt{2})$ are on the graph.   **71.** $(-1, 0), (1, 0)$ **73.** $\left(-\dfrac{\pi}{2}, 0\right), (0, 1), \left(\dfrac{\pi}{2}, 0\right)$ **75.** $(0, 0)$ **77.** $(-4, 0), (-1, 0), (0, -3), (4, 0)$

79.

81.

83.

85.

$y = -x^2 + 4$, points $(0, 4)$, $(-2, 0)$, $(2, 0)$

87.

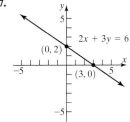

$2x + 3y = 6$, points $(0, 2)$, $(3, 0)$

89.

$(0, 9)$, $(-2, 0)$, $(2, 0)$, $9x^2 + 4y = 36$

91. x-intercept: 6.5
y-intercept: -13

93. x-intercepts: $-2.74, 2.74$
y-intercept: -15

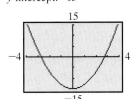

95. x-intercept: 14.33
y-intercept: -21.5

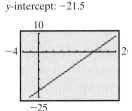

97. x-intercepts: $-2.72, 2.72$
y-intercept: 12.33

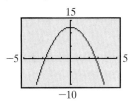

99. $\sqrt{17}; 2\sqrt{5}; \sqrt{29}$ **101.** $d(P_1, P_2) = 6; d(P_2, P_3) = 4; d(P_1, P_3) = 2\sqrt{13}$; right triangle   **103.** $d(P_1, P_2) = 2\sqrt{17}; d(P_2, P_3) = \sqrt{34}$; $d(P_1, P_3) = \sqrt{34}$; isosceles right triangle   **105.** $90\sqrt{2} \approx 127.28$ ft   **107. (a)** $(90, 0), (90, 90), (0, 90)$ **(b)** $5\sqrt{2161} \approx 232.43$ ft **(c)** $30\sqrt{149} \approx 366.20$ ft   **109.** $d = 50t$ **111. (a)** $y = \sqrt{x^2}$ and $y = |x|$ have the same graph.   **(b)** $\sqrt{x^2} = |x|$ **(c)** $x \geq 0$ for $y = (\sqrt{x})^2$, while x can be any real number for $y = x$.   **(d)** $y \geq 0$ for $y = \sqrt{x^2}$

1.2 Assess Your Understanding *(page 109)*

3. equivalent   **4.** identity   **5.** linear; first-degree   **6.** F   **7.** T   **8.** F   **9.** $\{3\}$ **11.** $\{-5\}$ **13.** $\left\{\dfrac{3}{2}\right\}$ **15.** $\left\{\dfrac{5}{4}\right\}$ **17.** $-2.21, 0.54, 1.68$

19. $-1.55, 1.15$ **21.** $-1.12, 0.36$ **23.** $-2.69, -0.49, 1.51$ **25.** $-2.86, -1.34, 0.20, 1.00$ **27.** No real solutions   **29.** $\{-2\}$

31. $\{3\}$ **33.** $\{-1\}$ **35.** $\{-2\}$ **37.** $\{-18\}$ **39.** $\{-4\}$ **41.** $\left\{-\dfrac{3}{4}\right\}$ **43.** $\{-20\}$ **45.** $\{-2\}$ **47.** $\{0.5\}$ **49.** $\left\{\dfrac{29}{10}\right\}$ **51.** $\{2\}$ **53.** $\{8\}$ **55.** $\{2\}$

57. $\{-1\}$ **59.** $\{3\}$ **61.** No solution   **63.** No solution   **65.** $\{-6\}$ **67.** $\{34\}$ **69.** $\left\{-\dfrac{20}{39}\right\}$ **71.** $\{-1\}$ **73.** $\left\{-\dfrac{11}{6}\right\}$ **75.** $\{-6\}$ **77.** $-\dfrac{2}{5}$

79. $2a + 3b = 6$ **81.** $x = \dfrac{b + c}{a}$ **83.** $x = \dfrac{abc}{a + b}$ **85.** $x = a^2$ **87.** $a = 3$ **89.** $R = \dfrac{R_1 R_2}{R_1 + R_2}$ **91.** $R = \dfrac{mv^2}{F}$ **93.** $r = \dfrac{S - a}{S}$

95. \$11,500 will be invested in bonds and \$8500 in CDs.   **97.** Scott will receive \$400,000, Alice \$300,000, and Tricia \$200,000.   **99.** The regular hourly rate is \$8.50.   **101.** Brooke needs a score of 85.   **103.** The original price was \$147,058.82; purchasing the model saves \$22,058.82. **105.** The bookstore paid \$70 for the book.   **107.** There were 400 adults.   **109.** The length is 19 ft; the width is 11 ft.   **111.** In obtaining step (7) we divided by $x - 2$. Since $x = 2$ from step (1), we actually divided by 0.

Historical Problems *(page 121)*

1. The area of each shaded square is 9, so the larger square will have area $85 + 4(9) = 121$. The area of the larger square is also given by the expression $(x + 6)^2$, so $(x + 6)^2 = 121$. Taking the positive square root of each side, $x + 6 = 11$ or $x = 5$.

2. Let $z = -6$, so $z^2 + 12z - 85 = -121$. We get the equation $u^2 - 121 = 0$ or $u^2 = 121$. Thus $u = \pm 11$, so $x = \pm 11 - 6$. $x = -17$ or $x = 5$.

3.

$$\left(x + \frac{b}{2a}\right)^2 = \left(\frac{\sqrt{b^2 - 4ac}}{2a}\right)^2$$

$$\left(x + \frac{b}{2a}\right)^2 - \left(\frac{\sqrt{b^2 - 4ac}}{2a}\right)^2 = 0$$

$$\left(x + \frac{b}{2a} - \frac{\sqrt{b^2 - 4ac}}{2a}\right)\left(x + \frac{b}{2a} + \frac{\sqrt{b^2 - 4ac}}{2a}\right) = 0$$

$$\left(x + \frac{b - \sqrt{b^2 - 4ac}}{2a}\right)\left(x + \frac{b + \sqrt{b^2 - 4ac}}{2a}\right) = 0$$

$$x = \frac{-b + \sqrt{b^2 - 4ac}}{2a} \text{ or } x = \frac{-b - \sqrt{b^2 - 4ac}}{2a}$$

1.3 Assess Your Understanding *(page 121)*

5. repeated; multiplicity two **6.** discriminant; negative **7.** second **8.** T **9.** F **10.** T **11.** $\{0, 9\}$ **13.** $\{-5, 5\}$

15. $\{-3, 2\}$ **17.** $\left\{-\frac{1}{2}, 3\right\}$ **19.** $\{-4, 4\}$ **21.** $\{-6, -2\}$ **23.** $\left\{\frac{3}{2}\right\}$ **25.** $\left\{-\frac{5}{2}, 3\right\}$ **27.** $\left\{-\frac{3}{4}, 2\right\}$ **29.** $\{-5, 5\}$ **31.** $\{-1, 3\}$ **33.** $\{-3, 0\}$

35. 16 **37.** $\frac{1}{16}$ **39.** $\{-7, 3\}$ **41.** $\left\{-\frac{1}{4}, \frac{3}{4}\right\}$ **43.** $\left\{\frac{-1 - \sqrt{7}}{6}, \frac{-1 + \sqrt{7}}{6}\right\}$ **45.** $\{2 - \sqrt{2}, 2 + \sqrt{2}\}$ **47.** $\{2 - \sqrt{5}, 2 + \sqrt{5}\}$ **49.** $\left\{1, \frac{3}{2}\right\}$

51. No real solution **53.** $\left\{\frac{-1 - \sqrt{5}}{4}, \frac{-1 + \sqrt{5}}{4}\right\}$ **55.** $\left\{\frac{9 - \sqrt{113}}{8}, \frac{9 + \sqrt{113}}{8}\right\}$ **57.** $\left\{\frac{1}{3}\right\}$ **59.** $\left\{-\frac{2}{3}, 1\right\}$ **61.** $\left\{\frac{1 - \sqrt{33}}{8}, \frac{1 + \sqrt{33}}{8}\right\}$

63. $\{-\sqrt{5}, \sqrt{5}\}$ **65.** $\left\{\frac{1}{4}\right\}$ **67.** $\left\{-\frac{3}{5}, \frac{5}{2}\right\}$ **69.** $\left\{-\frac{1}{2}, \frac{2}{3}\right\}$ **71.** $\left\{\frac{-\sqrt{2} + 2}{2}, \frac{-\sqrt{2} - 2}{2}\right\}$ **73.** $\left\{\frac{-1 - \sqrt{17}}{2}, \frac{-1 + \sqrt{17}}{2}\right\}$

75. No real solution **77.** Repeated real solution **79.** Two unequal real solutions **81.** The dimensions are 11 ft by 13 ft. **83.** The dimensions are 5 m by 8 m. **85.** The dimensions should be 4 ft by 4 ft. **87. (a)** The ball strikes the ground after 6 sec. **(b)** The ball passes the top of the building on its way down after 5 sec. **89.** The dimensions should be 11.55 cm by 6.55 cm by 3 cm. **91.** The border will be 2.71 ft wide. **93.** The border will be 2.56 ft wide. **95.** 36 consecutive integers must be added.

97. $\dfrac{-b + \sqrt{b^2 - 4ac}}{2a} + \dfrac{-b - \sqrt{b^2 - 4ac}}{2a} = \dfrac{-2b}{2a} = -\dfrac{b}{a}$ **99.** $k = \dfrac{1}{2}$ or $k = -\dfrac{1}{2}$

101. $ax^2 + bx + c = 0, x = \dfrac{-b \pm \sqrt{b^2 - 4ac}}{2a}; ax^2 - bx + c = 0, x = \dfrac{b \pm \sqrt{(-b)^2 - 4ac}}{2a} = \dfrac{b \pm \sqrt{b^2 - 4ac}}{2a} = -\dfrac{-b \pm \sqrt{b^2 - 4ac}}{2a}$

1.4 Assess Your Understanding *(page 131)*

4. real; imaginary; imaginary unit **5.** $-2i, 2i$ **6.** F **7.** T **8.** F **9.** $8 + 5i$ **11.** $-7 + 6i$ **13.** $-6 - 11i$ **15.** $6 - 18i$ **17.** $6 + 4i$ **19.** $10 - 5i$

21. 37 **23.** $\frac{6}{5} + \frac{8}{5}i$ **25.** $1 - 2i$ **27.** $\frac{5}{2} - \frac{7}{2}i$ **29.** $-\frac{1}{2} + \frac{\sqrt{3}}{2}i$ **31.** $2i$ **33.** $-i$ **35.** i **37.** -6 **39.** $-10i$ **41.** $-2 + 2i$ **43.** 0 **45.** 0 **47.** $2i$

49. $5i$ **51.** $5i$ **53.** $\{-2i, 2i\}$ **55.** $\{-4, 4\}$ **57.** $\{3 - 2i, 3 + 2i\}$ **59.** $\{3 - i, 3 + i\}$ **61.** $\left\{\frac{1}{4} - \frac{1}{4}i, \frac{1}{4} + \frac{1}{4}i\right\}$ **63.** $\left\{-\frac{1}{5} - \frac{2}{5}i, -\frac{1}{5} + \frac{2}{5}i\right\}$

65. $\left\{-\frac{1}{2} - \frac{\sqrt{3}}{2}i, -\frac{1}{2} + \frac{\sqrt{3}}{2}i\right\}$ **67.** $\{2, -1 - \sqrt{3}i, -1 + \sqrt{3}i\}$ **69.** $\{-2, 2, -2i, 2i\}$ **71.** $\{-3i, -2i, 2i, 3i\}$ **73.** Two complex solutions

75. Two unequal real solutions **77.** A repeated real solution **79.** $2 - 3i$ **81.** 6 **83.** 25

85. $z + \bar{z} = (a + bi) + (a - bi) = 2a; z - \bar{z} = (a + bi) - (a - bi) = 2bi$

87. $\overline{z + w} = \overline{(a + bi) + (c + di)} = \overline{(a + c) + (b + d)i} = (a + c) - (b + d)i = (a - bi) + (c - di) = \bar{z} + \bar{w}$

1.5 Assess Your Understanding *(page 139)*

6. extraneous **7.** quadratic in form **8.** T **9.** $\{9\}$ **11.** $\{22\}$ **13.** $\{1\}$ **15.** No real solution **17.** $\{-13\}$ **19.** $\{4\}$ **21.** $\{-1\}$ **23.** $\{0, 64\}$

25. $\{3\}$ **27.** $\{2\}$ **29.** $\left\{-\frac{8}{5}\right\}$ **31.** $\{8\}$ **33.** $\{-1, 3\}$ **35.** $\{1, 5\}$ **37.** $\{1\}$ **39.** $\{5\}$ **41.** $\{2\}$ **43.** $\{-4, 4\}$ **45.** $\{0, 3\}$ **47.** $\{-2, 2\}$

49. $\{-2, -1, 1, 2\}$ **51.** $\{-1, 1\}$ **53.** $\{-2, 1\}$ **55.** $\{-6, -5\}$ **57.** $\left\{-\frac{1}{3}\right\}$ **59.** $\left\{-\frac{3}{2}, 2\right\}$ **61.** $\{0, 16\}$ **63.** $\{16\}$ **65.** $\{1\}$

67. $\left\{\left(\frac{9 - \sqrt{17}}{8}\right)^4, \left(\frac{9 + \sqrt{17}}{8}\right)^4\right\}$ **69.** $\{\sqrt{2}, \sqrt{3}\}$ **71.** $\left\{-2, -\frac{1}{2}\right\}$ **73.** $\left\{-\frac{3}{2}, \frac{1}{3}\right\}$ **75.** $\left\{-\frac{1}{8}, 27\right\}$ **77.** $\left\{-2, -\frac{4}{5}\right\}$ **79.** $\{-6, 6\}$ **81.** $\{-4, 1\}$

83. $\left\{-1, \frac{3}{2}\right\}$ **85.** $\{-4, 4\}$ **87.** $\left\{-\frac{1}{2}, \frac{1}{2}\right\}$ **89.** $\left\{-\frac{27}{2}, \frac{27}{2}\right\}$ **91.** $\left\{-\frac{36}{5}, \frac{24}{5}\right\}$ **93.** No real solution **95.** $\{-3, 3\}$ **97.** $\{-1, 3\}$ **99.** $\{-2, -1, 0, 1\}$

101. $\{-3, 0, 3\}$ **103.** $\left\{0, \frac{3}{4}\right\}$ **105.** $\{-5, 0, 4\}$ **107.** $\{-1, 1\}$ **109.** $\{-2, 2, 3\}$ **111.** $\left\{-2, \frac{1}{2}, 2\right\}$ **113.** $\left\{\frac{2}{5}\right\}$ **115.** $\{0.34, 11.66\}$ **117.** $\{-1.03, 1.03\}$

119. $\{-1.85, 0.17\}$ **121.** $\left\{\frac{3}{2}, 5\right\}$ **123.** $(2, 2); (2, -4)$ **125.** $(0, 0); (8, 0)$ **127.** The distance is approximately 229.94 ft.

1.6 Assess Your Understanding *(page 147)*

1. mathematical modeling **2.** interest **3.** uniform motion **4.** F **5.** T **6.** $100 - x$ **7.** $A = \pi r^2; r = $ radius, $A = $ area **9.** $A = s^2; A = $ area, $s = $ length of a side **11.** $F = ma; F = $ force, $m = $ mass, $a = $ acceleration **13.** $W = Fd; W = $ work, $F = $ force, $d = $ distance **15.** $C = 150x$; $C = $ total variable cost, $x = $ number of dishwashers **17.** Invest \$31,250 in bonds and \$18,750 in CDs. **19.** \$11,600 was loaned out at 8%. **21.** Mix 75 lb of Earl Grey tea with 25 lb of Orange Pekoe tea. **23.** Mix 40 lb of cashews with the peanuts. **25.** The speed of the current is 2.286 mi/hr. **27.** The speed of the current is 5 mi/hr. **29.** Working together, it takes 12 min. **31. (a)** The dimensions are 10 ft by 5 ft. **(b)** The area is 50 sq ft. **(c)** The dimensions would be 7.5 ft by 7.5 ft. **(d)** The area would be 56.25 sq ft. **33.** The defensive back catches up to the tight end at the tight end's 45 yd line.   **35.** Add $\frac{2}{3}$ gal of water.   **37.** Evaporate 10.67 oz of water.   **39.** 40 g of 12-karat gold should be mixed with 20 g of pure gold.

41. Mike passes Dan $\frac{1}{3}$ mi from the start, 2 min from the time Mike started to race.   **43.** Start the auxiliary pump at 9:45 AM

45. The tub will fill in 1 hr.   **47.** 175 boxes were ordered.   **49.** Lewis would beat Burke by 16.75 m.   **53.** The average speed is 49.5 mi/hr.

1.7 Assess Your Understanding *(page 159)*

3. negative   **4.** closed interval   **5.** multiplication properties   **6.** T   **7.** T   **8.** T   **9.** F   **10.** T   **11.** $[0, 2]; 0 \le x \le 2$ **13.** $(-1, 2); -1 < x < 2$
15. $[0, 3); 0 \le x < 3$ **17.(a)** $6 < 8$ **(b)** $-2 < 0$ **(c)** $9 < 15$ **(d)** $-6 > -10$ **19.(a)** $7 > 0$ **(b)** $-1 > -8$ **(c)** $12 > -9$ **(d)** $-8 < 6$
21.(a) $2x + 4 < 5$ **(b)** $2x - 4 < -3$ **(c)** $6x + 3 < 6$ **(d)** $-4x - 2 > -4$
23. $[0, 4]$ **25.** $[4, 6)$ **27.** $[4, \infty)$ **29.** $(-\infty, -4)$

31. $2 \le x \le 5$ **33.** $-3 < x < -2$ **35.** $x \ge 4$ **37.** $x < -3$

39. $<$ **41.** $>$ **43.** $\ge$ **45.** $<$ **47.** $\le$ **49.** $<$ **51.** $\ge$
53. $\{x | x < 4\}; (-\infty, 4)$ **55.** $\{x | x \ge -1\}; [-1, \infty)$ **57.** $\{x | x > 3\}; (3, \infty)$ **59.** $\{x | x \ge 2\}; [2, \infty)$

61. $\{x | x > -7\}; (-7, \infty)$ **63.** $\left\{x \middle| x \le \frac{2}{3}\right\}; \left(-\infty, \frac{2}{3}\right]$ **65.** $\{x | x < -20\}; (-\infty, -20)$ **67.** $\left\{x \middle| x \ge \frac{4}{3}\right\}; \left[\frac{4}{3}, \infty\right)$

69. $\{x | 3 \le x \le 5\}; [3, 5]$ **71.** $\left\{x \middle| \frac{2}{3} \le x \le 3\right\}; \left[\frac{2}{3}, 3\right]$ **73.** $\left\{x \middle| -\frac{11}{2} < x < \frac{1}{2}\right\}; \left(-\frac{11}{2}, \frac{1}{2}\right)$

75. $\{x | -6 < x < 0\}; (-6, 0)$ **77.** $\{x | x < -5\}; (-\infty, -5)$ **79.** $\{x | x \ge -1\}; [-1, \infty)$ **81.** $\left\{x \middle| \frac{1}{2} \le x < \frac{5}{4}\right\}; \left[\frac{1}{2}, \frac{5}{4}\right)$

83. $\{x | -6 < x < 6\}; (-6, 6)$ **85.** $\{x | x < -4 \text{ or } x > 4\};$ **87.** $\{x | -4 < x < 4\}; (-4, 4)$ **89.** $\{x | x < -4 \text{ or } x > 4\};$
                                    $(-\infty, -4) \text{ or } (4, \infty)$ $(-\infty, -4) \text{ or } (4, \infty)$

91. $\{x | 1 < x < 3\}; (1, 3)$ **93.** $\left\{t \middle| -\frac{2}{3} \le t \le 2\right\}; \left[-\frac{2}{3}, 2\right]$ **95.** $\{x | x \le 1 \text{ or } x \ge 5\};$ **97.** $\left\{x \middle| -1 < x < \frac{3}{2}\right\}; \left(-1, \frac{3}{2}\right)$
                                                                                 $(-\infty, 1] \text{ or } [5, \infty)$

99. $\{x | x < -1 \text{ or } x > 2\};$ **101.** No real solution; $\varnothing$ **103.** $\{x | x > 5\}; (5, \infty)$ **105.** $\{x | x > 3\}; (3, \infty)$
       $(-\infty, -1) \text{ or } (2, \infty)$

107. $\left| x - 2 \right| < \frac{1}{2}; \left\{x \middle| \frac{3}{2} < x < \frac{5}{2}\right\}$ **109.** $|x + 3| > 2; \{x | x < -5 \text{ or } x > -1\}$ **111.** $21 < \text{age} < 30$

113. $|x - 98.6| \ge 1.5; \{x | x \le 97.1 \text{ or } x \ge 100.1\}$ **115. (a)** Male ≥ 75.6 **(b)** Female ≥ 80.4 **(c)** A female can expect to live at least 4.8 years longer.
117. The agent's commission ranges from $45,000 to $95,000, inclusive. As a percent of selling price, the commission ranges from 5% to approximately 8.6%, inclusive.   **119.** The amount withheld varies from $76.35 to $101.35, inclusive.   **121.** The usage varies from approximately 675.41 to 2500.91 kilowatt-hours, inclusive.   **123.** The dealer's cost varies from $7457.63 to $7857.14, inclusive.   **125.** You need at least a 74 on the last test.   **127.** The amount of gasoline ranged from 12 to 20 gal, inclusive.

129. $\dfrac{a+b}{2} - a = \dfrac{a+b-2a}{2} = \dfrac{b-a}{2} > 0$; therefore, $a < \dfrac{a+b}{2}$ **131.** $(\sqrt{ab})^2 - a^2 = ab - a^2 = a(b-a) > 0$; thus, $(\sqrt{ab})^2 > a^2$ and $\sqrt{ab} > a.\ b^2 - (\sqrt{ab})^2 = b^2 - ab = b(b-a) > 0$; thus $b^2 > (\sqrt{ab})^2$ and $b > \sqrt{ab}$.

1.8 Assess Your Understanding *(page 176)*

1. undefined; zero   **2.** $m_1 = m_2$; y-intercepts; $m_1 = -\dfrac{1}{m_2}$ **3.** $y = b$; y-intercept   **4.** T   **5.** F   **6.** F

7. (a) $\dfrac{1}{2}$ **(b)** If x increases by 2 units, y will increase by 1 unit.   **9. (a)** $-\dfrac{1}{3}$ **(b)** If x increases by 3 units, y will decrease by 1 unit.

11. Slope $= -\dfrac{3}{2}$ **13.** Slope $= -\dfrac{1}{2}$ **15.** Slope $= 0$ **17.** Slope undefined

19.   **21.**   **23.**   **25.**

27. $(2,6)$; $(3,10)$; $(4,14)$ **29.** $(4,-7)$; $(6,-10)$; $(8,-13)$ **31.** $(-1,-5)$; $(0,-7)$; $(1,-9)$ **33.** $x - 2y = 0$ or $y = \dfrac{1}{2}x$

35. $x + y = 2$ or $y = -x + 2$ **37.** $2x - y = 3$ or $y = 2x - 3$ **39.** $x + 2y = 5$ or $y = -\dfrac{1}{2}x + \dfrac{5}{2}$ **41.** $3x - y = -9$ or $y = 3x + 9$

43. $2x + 3y = -1$ or $y = -\dfrac{2}{3}x - \dfrac{1}{3}$ **45.** $x - 2y = -5$ or $y = \dfrac{1}{2}x + \dfrac{5}{2}$ **47.** $3x + y = 3$ or $y = -3x + 3$ **49.** $x - 2y = 2$ or $y = \dfrac{1}{2}x - 1$

51. $x = 2$; no slope–intercept form   **53.** $2x - y = -4$ or $y = 2x + 4$ **55.** $2x - y = 0$ or $y = 2x$ **57.** $x = 4$; no slope–intercept form

59. $2x + y = 0$ or $y = -2x$ **61.** $x - 2y = -3$ or $y = \dfrac{1}{2}x + \dfrac{3}{2}$ **63.** $y = 4$

65. Slope $= 2$; y-intercept $= 3$ **67.** $y = 2x - 2$; slope $= 2$; y-intercept $= -2$

69. Slope $= \frac{1}{2}$; y-intercept $= 2$

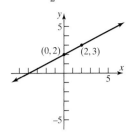

71. $y = -\frac{1}{2}x + 2$; slope $= -\frac{1}{2}$; y-intercept $= 2$

73. $y = \frac{2}{3}x - 2$; slope $= \frac{2}{3}$; y-intercept $= -2$

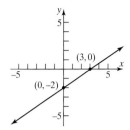

75. $y = -x + 1$; slope $= -1$; y-intercept $= 1$

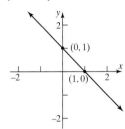

77. Slope undefined; no y-intercept

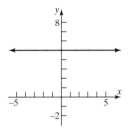

79. Slope $= 0$; y-intercept $= 5$

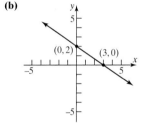

81. $y = x$; slope $= 1$; y-intercept $= 0$

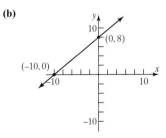

83. $y = \frac{3}{2}x$; slope $= \frac{3}{2}$; y-intercept $= 0$

85. (a) x-intercept: 3; y-intercept: 2

(b)

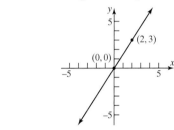

87. (a) x-intercept: -10; y-intercept: 8

(b)

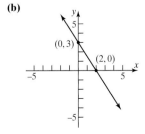

89. (a) x-intercept: 3; y-intercept: $\frac{21}{2}$

(b)

91. (a) x-intercept: 2; y-intercept: 3

(b)

93. (a) x-intercept: 5; y-intercept: -2

(b)

95. $y = 0$ **97.** (b)   **99.** (d)   **101.** $x - y = -2$ or $y = x + 2$ **103.** $x + 3y = 3$ or $y = -\dfrac{1}{3}x + 1$ **105.** $C = 0.07x + 29$; $36.70; $45.10

107. $C = 0.53x + 1{,}070{,}000$

109. (a) $C = 0.08275x + 7.58, 0 \le x \le 400$ **(b)**

(c) $15.86   **(d)** $32.41

(e) Each additional kW-h used adds $0.08275 to the bill.

111. $°C = \dfrac{5}{9}(°F - 32)$; approximately 21°C   **113. (a)** $A = \dfrac{1}{5}(x - 100{,}000) + 40{,}000$ **(b)** $80{,}000

(c) Each additional box sold requires an additional $0.20 in advertising.

115. All have the same slope, 2; the lines are parallel.

117. (b), (c), (e), (g)
119. (c)
125. No; no
127. They are the same line.
129. No

1.9 Assess Your Understanding *(page 185)*

1. F   **2.** radius   **3.** T   **4.** F   **5.** Center $(2, 1)$; radius 2; $(x - 2)^2 + (y - 1)^2 = 4$ **7.** Center $\left(\dfrac{5}{2}, 2\right)$; radius $\dfrac{3}{2}$; $\left(x - \dfrac{5}{2}\right)^2 + (y - 2)^2 = \dfrac{9}{4}$

9. $x^2 + y^2 = 4$;
$x^2 + y^2 - 4 = 0$

11. $x^2 + (y - 2)^2 = 4$;
$x^2 + y^2 - 4y = 0$

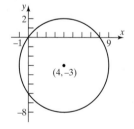

13. $(x - 4)^2 + (y + 3)^2 = 25$;
$x^2 + y^2 - 8x + 6y = 0$

15. $(x + 2)^2 + (y - 1)^2 = 16$;
$x^2 + y^2 + 4x - 2y - 11 = 0$

17. $\left(x - \dfrac{1}{2}\right)^2 + y^2 = \dfrac{1}{4}$;
$x^2 + y^2 - x = 0$

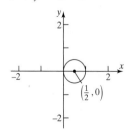

19. (a) $(h, k) = (0, 0); r = 2$

(b)

(c) $(\pm 2, 0); (0, \pm 2)$

21. (a) $(h, k) = (3, 0); r = 2$ **23. (a)** $(h, k) = (1, 2); r = 3$ **25. (a)** $(h, k) = (-2, 2); r = 3$

(b)   **(b)**   **(b)**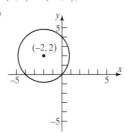

(c) $(1, 0); (5, 0)$ **(c)** $(1 \pm \sqrt{5}, 0); (0, -2 \pm 2\sqrt{2})$ **(c)** $(-2 \pm \sqrt{5}, 0); (0, 2 \pm \sqrt{5})$

27. (a) $(h, k) = \left(\dfrac{1}{2}, -1\right); r = \dfrac{1}{2}$ **29. (a)** $(h, k) = (3, -2); r = 5$

(b)   **(b)**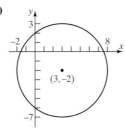

(c) $(0, -1)$ **(c)** $(3 \pm \sqrt{21}, 0); (0, -6), (0, 2)$

31. $x^2 + y^2 - 13 = 0$ **33.** $x^2 + y^2 - 4x - 6y + 4 = 0$ **35.** $x^2 + y^2 + 2x - 6y + 5 = 0$ **37. (c)**   **39. (b)**   **41.** $(x + 3)^2 + (y - 1)^2 = 16$
43. $(x - 2)^2 + (y - 2)^2 = 9$ **45.** $x^2 + y^2 + 2x + 4y - 4168.16 = 0$ **47.** $\sqrt{2}x + 4y - 9\sqrt{2} = 0$ **49.** $(1, 0)$ **51.** $y = 2$
53. (b), (e), (g)

Review Exercises *(page 190)*

1. $\{-18\}$ **3.** $\{6\}$ **5.** $\left\{\dfrac{1}{5}\right\}$ **7.** $\{6\}$ **9.** No real solution   **11.** $\left\{\dfrac{11}{8}\right\}$ **13.** $\left\{-2, \dfrac{3}{2}\right\}$ **15.** $\left\{\dfrac{1 - \sqrt{13}}{4}, \dfrac{1 + \sqrt{13}}{4}\right\}$ **17.** $\{-3, 3\}$

19. No real solution   **21.** $\{-2, -1, 1, 2\}$ **23.** $\{2\}$ **25.** $\left\{\dfrac{13}{2}\right\}$ **27.** $\left\{\dfrac{\sqrt{5}}{2}\right\}$ **29.** $\left\{\dfrac{9}{4}\right\}$ **31.** $\left\{-1, \dfrac{1}{2}\right\}$ **33.** $\left\{\dfrac{m}{1 - n}, \dfrac{m}{1 + n}\right\}$ **35.** $\left\{-\dfrac{9b}{5a}, \dfrac{2b}{a}\right\}$

37. $\left\{-\dfrac{9}{5}\right\}$ **39.** $\{-5, 2\}$ **41.** $\left\{-\dfrac{5}{3}, 3\right\}$ **43.** $\left\{0, \dfrac{3}{2}\right\}$ **45.** $\left\{-\dfrac{5}{2}, -2, 2\right\}$ **47.** $\{3\}$ **49.** $\{-1, 5\}$ **51.** $-2.49, 0.66, 1.83$ **53.** $-1.14, 1.64$

55. $\{x | x \geq 14\}; [14, \infty)$ **57.** $\left\{x \Big| -\dfrac{31}{2} \leq x \leq \dfrac{33}{2}\right\}; \left[-\dfrac{31}{2}, \dfrac{33}{2}\right]$ **59.** $\{x | -23 < x < -7\}; (-23, -7)$

61. $\left\{x \Big| -\dfrac{3}{2} < x < -\dfrac{7}{6}\right\}; \left(-\dfrac{3}{2}, -\dfrac{7}{6}\right)$ **63.** $\{x | x \leq -2 \text{ or } x \geq 7\}; (-\infty, -2] \text{ or } [7, \infty)$

65. $\left\{x \Big| 0 \leq x \leq \dfrac{4}{3}\right\}; \left[0, \dfrac{4}{3}\right]$ **67.** $\left\{x \Big| x < -1 \text{ or } x > \dfrac{7}{3}\right\}; (-\infty, -1) \text{ or } \left(\dfrac{7}{3}, \infty\right)$

69. 9 **71.** $\dfrac{4}{9}$ **73.** $4 + 7i$ **75.** $-3 + 2i$ **77.** $\dfrac{9}{10} - \dfrac{3}{10}i$ **79.** -1 **81.** $-46 + 9i$ **83.** $\left\{-\dfrac{1}{2} - \dfrac{\sqrt{3}}{2}i, -\dfrac{1}{2} + \dfrac{\sqrt{3}}{2}i\right\}$ **85.** $\left\{\dfrac{-1 - \sqrt{17}}{4}, \dfrac{-1 + \sqrt{17}}{4}\right\}$

87. $\left\{\dfrac{1}{2} - \dfrac{\sqrt{11}}{2}i, \dfrac{1}{2} + \dfrac{\sqrt{11}}{2}i\right\}$ **89.** $\left\{\dfrac{1}{2} - \dfrac{\sqrt{23}}{2}i, \dfrac{1}{2} + \dfrac{\sqrt{23}}{2}i\right\}$ **91. (a)** $2\sqrt{5}$ **(b)** $(2, 1)$ **(c)** $\dfrac{1}{2}$ **(d)** For each run of 2, there is a rise of 1.

93. (a) 5   **(b)** $\left(-\dfrac{1}{2}, 1\right)$ **(c)** $-\dfrac{4}{3}$ **(d)** For each run of 3, there is a rise of −4.   **95. (a)** 12   **(b)** $(4, 2)$ **(c)** undefined   **(d)** No change in x

97.

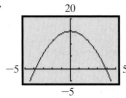

99. $2x + y = 5$ or $y = -2x + 5$

101. $x = -3$; no slope–intercept form

103. $x + 5y = -10$ or $y = -\dfrac{1}{5}x - 2$

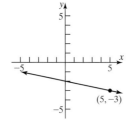

105. $2x - 3y = -19$ or $y = \dfrac{2}{3}x + \dfrac{19}{3}$

107. $-x + y = -7$ or $y = x - 7$

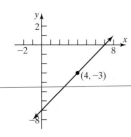

109. x-intercept: 9; y-intercept: 6

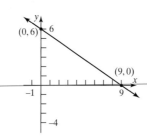

111. x-intercept: 20; y-intercept: 4

113.

115.

117.

119. $(x + 2)^2 + (y - 3)^2 = 16$
121. $(x + 1)^2 + (y + 2)^2 = 1$

123. Center $(1, -2)$; radius $= 3$

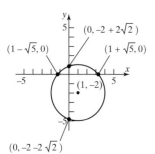

125. Center $(1, -2)$; Radius $= \sqrt{5}$

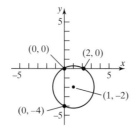

127. $d(A, B) = \sqrt{13}; d(B, C) = \sqrt{13}$ **129.** $m_{AB} = -1; m_{BC} = -1$ **131.** Center $(1, -2)$; radius $= 4\sqrt{2}; x^2 + y^2 - 2x + 4y - 27 = 0$
133. $p = 2l + 2w$ **135.** The interest is \$630.   **137.** The storm is 3300 ft away.   **139.** The search plane can go as far as 616 mi.
141. The helicopter will reach the life raft in a little less than 1 hr 35 min.   **143.** The Metra commuter averages 30 mi/hr; the Amtrak averages
80 mi/hr.   **145.** It takes Clarissa 10 days by herself.   **147.** Mix 90 cm³ of 15% HCl to obtain 150 cm³ of 25% HCl.   **149.** Add 256 oz of water.
151. 5 cm and 12 cm   **153.** The freight train is 190.67 ft long.   **155.** It will take the smaller pump 2 hr.   **157.** 36 seniors went on the trip; each paid
\$13.40.   **159. (a)** No **(b)** Todd wins again. **(c)** Todd wins by $\dfrac{1}{4}$ m. **(d)** Todd should line up 5.26 m behind the start line. **(e)** Yes

Chapter Test *(page 194)*

1. (a) $d = 10$ **(b)** $M = (1, 1)$ **(c)** $(x - 1)^2 + (y - 1)^2 = 25$ **(d)**

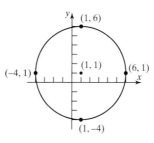

2. $\left\{ -\dfrac{3}{2}, -1 \right\}$ **3.** $\{2\}$ **4.** $\left\{ -\dfrac{3}{2}, 2 \right\}$

5. $\{2\}$ **6.** $\{0, 3\}$ **7.** $\{-2, 2, -i\sqrt{2}, i\sqrt{2}\}$

8. $\{2 - i\sqrt{3}, 2 + i\sqrt{3}\}$

9. $\{3 - 2\sqrt{3}, 3 + 2\sqrt{3}\}$

10.

11.

12.

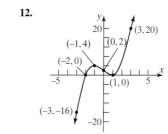

13. (a) $m = \dfrac{5}{3}$ **(b)** The x-intercept is 2. The y-intercept is $-\dfrac{10}{3}$.   **(c)**

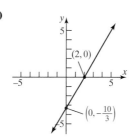

(d) $y = -\dfrac{3}{5}x + \dfrac{22}{5}$ **(e)** $y = \dfrac{5}{3}x + \dfrac{20}{3}$

14. $-1, 0.5,$ and 1 **15.** $-2.50,$ and 2.50 **16.** $-2.46, -0.24,$ and 1.70 **17.** $\left\{ x \mid x < -\dfrac{11}{2} \right\}; \left(-\infty, -\dfrac{11}{2} \right)$

18. $\{x \mid x \le -5 \text{ or } x \ge 2\}; (-\infty, -5] \cup [2, \infty)$

19. $\{x \mid -1 \le x < 2\}; [-1, 2)$

20. $2 - 25i$ **21.** $14 + 83i$ **22.** $\dfrac{7}{34} + \dfrac{11}{34}i$ **23.** About 204.63 minutes (3.41 hours)   **24.** 23.75 pounds of banana chips   **25.** The sale price is \$159.50.
26. Glenn will make \$100 in interest after 3 months.

CHAPTER 2   Functions and Their Graphs

2.1 Assess Your Understanding *(page 202)*

3. *y*-axis **4.** origin **5.** T **6.** T

7.

9.

11.

13.

15.

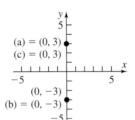

17. Intercepts: $(-1, 0)$, $(1, 0)$; symmetry: x-axis, y-axis, and origin

19. Intercepts: $\left(\dfrac{-\pi}{2}, 0\right)$, $(0, 1)$, $\left(\dfrac{\pi}{2}\right)$; symmetry: y-axis

21. Intercepts: $(0, 0)$; symmetry: x-axis

23. Intercepts: $(-2, 0)$, $(0, 0)$, $(2, 0)$; symmetry: origin

25.

27.

29.

31.

33. (a) $(0, 0)$
 (b) Symmetric with respect to the origin
39. (a) $(0, 0)$, $(-2\sqrt{2}, 0)$, $(2\sqrt{2}, 0)$
 (b) Symmetric with respect to the origin

35. (a) $(0, -9)$, $(-3, 0)$, $(3, 0)$
 (b) Symmetric with respect to the y-axis
41. (a) $(-3, 0)$, $(3, 0)$, $(0, -3)$, $(0, 3)$
 (b) Symmetric with respect to the y-axis, the x-axis, and the origin

37. (a) $(0, -2)$, $(0, 2)$, $(-4, 0)$
 (b) Symmetric with respect to the x-axis
43. (a) $(0, -4)$, $(-1, 0)$, $(4, 0)$
 (b) No symmetry

45. (a) $(0, -4)$, $(-4, 0)$, $(4, 0)$
 (b) Symmetric with respect to the y-axis
47. (a) $(-3, 0)$, $(3, 0)$
 (b) Symmetric with respect to the origin
49. $(-1, -2)$ **51.** 4

2.2 Assess Your Understanding *(page 216)*

5. independent; dependent **6.** range **7.** $[0, 5]$ **8.** $\neq$; f; g **9.** $(g - f)(x)$ **10.** F **11.** T **12.** T **13.** F **14.** F
15. Function; Domain: {Elvis, Colleen, Kaleigh, Marissa}, Range: {January 8, March 15, September 17} **17.** Not a function
19. Not a function **21.** Function; Domain: $\{1, 2, 3, 4\}$; Range: $\{3\}$ **23.** Not a function **25.** Function; Domain: $\{-2, -1, 0, 1\}$, Range: $\{0, 1, 4\}$
27. Function **29.** Function **31.** Not a function **33.** Not a function **35.** Function **37.** Not a function **39. (a)** -4 **(b)** 1 **(c)** -3
 (d) $3x^2 - 2x - 4$ **(e)** $-3x^2 - 2x + 4$ **(f)** $3x^2 + 8x + 1$ **(g)** $12x^2 + 4x - 4$ **(h)** $3x^2 + 6xh + 3h^2 + 2x + 2h - 4$
41. (a) 0 **(b)** $\dfrac{1}{2}$ **(c)** $-\dfrac{1}{2}$ **(d)** $\dfrac{-x}{x^2 + 1}$ **(e)** $\dfrac{-x}{x^2 + 1}$ **(f)** $\dfrac{x + 1}{x^2 + 2x + 2}$ **(g)** $\dfrac{2x}{4x^2 + 1}$ **(h)** $\dfrac{x + h}{x^2 + 2xh + h^2 + 1}$
43. (a) 4 **(b)** 5 **(c)** 5 **(d)** $|x| + 4$ **(e)** $-|x| - 4$ **(f)** $|x + 1| + 4$ **(g)** $2|x| + 4$ **(h)** $|x + h| + 4$
45. (a) $-\dfrac{1}{5}$ **(b)** $-\dfrac{3}{2}$ **(c)** $\dfrac{1}{8}$ **(d)** $\dfrac{-2x + 1}{-3x - 5}$ **(e)** $\dfrac{-2x - 1}{3x - 5}$ **(f)** $\dfrac{2x + 3}{3x - 2}$ **(g)** $\dfrac{4x + 1}{6x - 5}$ **(h)** $\dfrac{2x + 2h + 1}{3x + 3h - 5}$
47. All real numbers **49.** All real numbers **51.** $\{x | x \neq -4, x \neq 4\}$ **53.** $\{x | x \neq 0\}$ **55.** $\{x | x \geq 4\}$ **57.** $\{x | x > 9\}$ **59.** $\{x | x > 1\}$
61. (a) $(f + g)(x) = 5x + 1$; All real numbers **(b)** $(f - g)(x) = x + 7$; All real numbers
 (c) $(f \cdot g)(x) = 6x^2 - x - 12$; All real numbers **(d)** $\left(\dfrac{f}{g}\right)(x) = \dfrac{3x + 4}{2x - 3}$; $\left\{x \middle| x \neq \dfrac{3}{2}\right\}$

63. (a) $(f + g)(x) = 2x^2 + x - 1$; All real numbers   **(b)** $(f - g)(x) = -2x^2 + x - 1$; All real numbers

(c) $(f \cdot g)(x) = 2x^3 - 2x^2$; All real numbers   **(d)** $\left(\dfrac{f}{g}\right)(x) = \dfrac{x - 1}{2x^2}$; $\{x | x \neq 0\}$

65. (a) $(f + g)(x) = \sqrt{x} + 3x - 5$; $\{x | x \geq 0\}$ **(b)** $(f - g)(x) = \sqrt{x} - 3x + 5$; $\{x | x \geq 0\}$

(c) $(f \cdot g)(x) = 3x\sqrt{x} - 5\sqrt{x}$; $\{x | x \geq 0\}$ **(d)** $\left(\dfrac{f}{g}\right)(x) = \dfrac{\sqrt{x}}{3x - 5}$; $\left\{x \middle| x \geq 0, x \neq \dfrac{5}{3}\right\}$

67. (a) $(f + g)(x) = 1 + \dfrac{2}{x}$; $\{x | x \neq 0\}$ **(b)** $(f - g)(x) = 1$; $\{x | x \neq 0\}$ **(c)** $(f \cdot g)(x) = \dfrac{1}{x} + \dfrac{1}{x^2}$; $\{x | x \neq 0\}$

(d) $\left(\dfrac{f}{g}\right)(x) = x + 1$; $\{x | x \neq 0\}$ **69. (a)** $(f + g)(x) = \dfrac{6x + 3}{3x - 2}$; $\left\{x \middle| x \neq \dfrac{2}{3}\right\}$ **(b)** $(f - g)(x) = \dfrac{-2x + 3}{3x - 2}$; $\left\{x \middle| x \neq \dfrac{2}{3}\right\}$

(c) $(f \cdot g)(x) = \dfrac{8x^2 + 12x}{(3x - 2)^2}$; $\left\{x \middle| x \neq \dfrac{2}{3}\right\}$ **(d)** $\left(\dfrac{f}{g}\right)(x) = \dfrac{2x + 3}{4x}$; $\left\{x \middle| x \neq 0, x \neq \dfrac{2}{3}\right\}$ **71.** $g(x) = 5 - \dfrac{7}{2}x$ **73.** 4   **75.** $2x + h - 1$

77. $3x^2 + 3xh + h^2$ **79.** $A = -\dfrac{7}{2}$ **81.** $A = -4$ **83.** $A = 8$; undefined at $x = 3$ **85.** $A(x) = \dfrac{1}{2}x^2$ **87.** $G(x) = 10x$

89. (a) 15.1 m, 14.07 m, 12.94 m, 11.72 m   **(b)** 1.01 sec, 1.43 sec, 1.75 sec   **(c)** 2.02 sec   **91. (a)** \$222   **(b)** \$225   **(c)** \$220   **(d)** \$230

93. $R(x) = \dfrac{L(x)}{P(x)}$ **95.** $H(x) = P(x) \cdot I(x)$ **97.** Only $h(x) = 2x$

2.3 Assess Your Understanding *(page 223)*

3. vertical   **4.** $5; -3$ **5.** $a = -2$ **6.** F   **7.** F   **8.** T

9. (a) $f(0) = 3; f(-6) = -3$ **(b)** $f(6) = 0; f(11) = 1$ **(c)** Positive   **(d)** Negative   **(e)** $-3, 6,$ and 10 **(f)** $-3 < x < 6; 10 < x \leq 11$

(g) $\{x | -6 \leq x \leq 11\}$ **(h)** $\{y | -3 \leq y \leq 4\}$ **(i)** $-3, 6, 10$ **(j)** 3   **(k)** 3 times   **(l)** Once   **(m)** $0, 4$ **(n)** $-5, 8$

11. Not a function   **13.** Function   **(a)** Domain: $\{x | -\pi \leq x \leq \pi\}$; Range: $\{y | -1 \leq y \leq 1\}$ **(b)** $\left(-\dfrac{\pi}{2}, 0\right), \left(\dfrac{\pi}{2}, 0\right), (0, 1)$ **(c)** y-axis

15. Not a function   **17.** Function   **(a)** Domain: $\{x | x > 0\}$; Range: all real numbers   **(b)** $(1, 0)$ **(c)** None

19. Function   **(a)** Domain: all real numbers; Range: $\{y | y \leq 2\}$ **(b)** $(-3, 0), (3, 0), (0, 2)$ **(c)** y-axis

21. Function   **(a)** Domain: all real numbers; Range: $\{y | y \geq -3\}$ **(b)** $(1, 0), (3, 0), (0, 9)$ **(c)** None

23. (a) Yes   **(b)** $f(-2) = 9; (-2, 9)$ **(c)** $0, \dfrac{1}{2}; (0, -1), \left(\dfrac{1}{2}, -1\right)$ **(d)** All real numbers   **(e)** $-\dfrac{1}{2}, 1$ **(f)** -1

25. (a) No   **(b)** $f(4) = -3; (4, -3)$ **(c)** $14; (14, 2)$ **(d)** $\{x | x \neq 6\}$ **(e)** -2 **(f)** $-\dfrac{1}{3}$ **27. (a)** Yes   **(b)** $f(2) = \dfrac{8}{17}; \left(2, \dfrac{8}{17}\right)$

(c) $-1, 1; (-1, 1), (1, 1)$ **(d)** All real numbers   **(e)** 0 **(f)** 0

29. (a) About 81.07 ft   **(b)** About 129.59 ft

(c) About 26.63 ft   **(d)** About 528.13 ft   **31. (a)**   **(b)**   **(c)** 600 mi/hr

(e)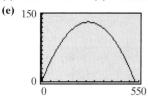

(f) 115.07 ft and 413.05 ft

(g) 275 ft; maximum height shown in the table is 131.8 ft

(h) 264 ft

35. There is at most one y-intercept.   **37. (a)** III   **(b)** IV   **(c)** I   **(d)** V   **(e)** II

39.

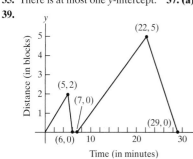

41. (a) 2 hr elapsed during which Kevin was between 0 and 3 mi from home.

(b) 0.5 hr elapsed during which Kevin was 3 mi from home.

(c) 0.3 hr elapsed during which Kevin was between 0 and 3 mi from home.

(d) 0.2 hr elapsed during which Kevin was 0 mi from home.

(e) 0.9 hr elapsed during which Kevin was between 0 and 2.8 mi from home.

(f) 0.3 hr elapsed during which Kevin was 2.8 mi from home.

(g) 1.1 hr elapsed during which Kevin was between 0 and 2.8 mi from home.

(h) 3 mi   **(i)** 2 times

43. No points whose x-coordinate is 5 or whose y-coordinate is 0 can be on the graph.

2.4 Assess Your Understanding (page 236)

6. increasing **7.** even; odd **8.** T **9.** T **10.** F **11.** Yes **13.** No **15.** $(-8, -2)$; $(0, 2)$; $(5, \infty)$ **17.** Yes; 10 **19.** $-2, 2$; 6, 10

21. (a) $(-2, 0), (0, 3), (2, 0)$ **(b)** Domain: $\{x | -4 \le x \le 4\}$ or $[-4, 4]$; Range: $\{y | 0 \le y \le 3\}$ or $[0, 3]$ **(c)** Increasing on $(-2, 0)$ and $(2, 4)$; Decreasing on $(-4, -2)$ and $(0, 2)$ **(d)** Even **23. (a)** $(0, 1)$ **(b)** Domain: all real numbers; Range: $\{y | y > 0\}$ or $(0, \infty)$ **(c)** Increasing on $(-\infty, \infty)$ **(d)** Neither **25. (a)** $(-\pi, 0), (0, 0), (\pi, 0)$

(b) Domain: $\{x | -\pi \le x \le \pi\}$ or $[-\pi, \pi]$; Range: $\{y | -1 \le y \le 1\}$ or $[-1, 1]$ **(c)** Increasing on $\left(-\dfrac{\pi}{2}, \dfrac{\pi}{2}\right)$; Decreasing on $\left(-\pi, -\dfrac{\pi}{2}\right)$

and $\left(\dfrac{\pi}{2}, \pi\right)$ **(d)** Odd **27. (a)** $\left(0, \dfrac{1}{2}\right), \left(\dfrac{1}{2}, 0\right), \left(\dfrac{5}{2}, 0\right)$ **(b)** Domain: $\{x | -3 \le x \le 3\}$ or $[-3, 3]$; Range: $\{y | -1 \le y \le 2\}$ or $[-1, 2]$

(c) Increasing on $(2, 3)$; Decreasing on $(-1, 1)$; Constant on $(-3, -1)$ and $(1, 2)$ **(d)** Neither **29. (a)** $0; 3$ **(b)** $-2, 2; 0, 0$

31. (a) $\dfrac{\pi}{2}; 1$ **(b)** $-\dfrac{\pi}{2}; -1$ **33.** Odd **35.** Even **37.** Odd **39.** Neither **41.** Even **43.** Odd

45.

Increasing: $(-2, -1), (1, 2)$
Decreasing: $(-1, 1)$
Local maximum: $(-1, 4)$
Local minimum: $(1, 0)$

47.

Increasing: $(-2, -0.77), (0.77, 2)$
Decreasing: $(-0.77, 0.77)$
Local maximum: $(-0.77, 0.19)$
Local minimum: $(0.77, -0.19)$

49.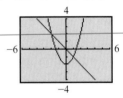

Increasing: $(-3.77, 1.77)$
Decreasing: $(-6, -3.77), (1.77, 4)$
Local maximum: $(1.77, -1.91)$
Local minimum: $(-3.77, -18.89)$

51.

Increasing: $(-1.87, 0), (0.97, 2)$
Decreasing: $(-3, -1.87), (0, 0.97)$
Local maximum: $(0, 3)$
Local minima: $(-1.87, 0.95), (0.97, 2.65)$

53. (a) -4 **(b)** -8 **(c)** -10 **55. (a)** 17 **(b)** -1 **(c)** 11

57. (a) 5 **(b)** 5; the slope of the line joining $(1, f(1))$ and $(3, f(3))$ is 5. **(c)** $y = 5x - 2$ **(d)**

59. (a) $x - 2$ **(b)** 1; the slope of the line joining $(-2, g(-2))$ and $(1, g(1))$ is -1. **(c)** $y = -x$ **(d)**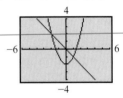

61. (a) x
(b) 4; the slope of the line joining $(2, h(2))$ and $(4, h(4))$ is 4.
(c) $y = 4x - 8$
(d)

63. (a) $V(x) = x(24 - 2x)^2$
(b) 972 in.³ **(c)** 160 in.³
(d) V is largest when $x = 4$.

65. (a)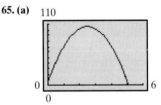
(b) 2.5 sec
(c) 106 ft

67. (a)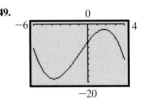
(b) 10 riding lawn mowers/hr
(c) $239/mower

69. (a), (b), (e)

(c) 1120 dollars/bicycle
(d) For each additional bicycle sold between 0 and 25 bicycles, total revenue increases, on average, by $1120.
(f) 75 dollars/bicycle
(g) For each additional bicycle sold between 190 and 223 bicycles, total revenue increases, on average, by $75.
(h) The average rate of change is decreasing as the number of bicycles increase.

71. (a) 1 **(b)** 0.5 **(c)** 0.1 **(d)** 0.01 **(e)** 0.001

(f)

$y = x^2$ $y = x$

$y = 0.5x$
$y = 0.1x$
$y = 0.01x,$
$y = 0.001x$

(g) They are getting closer to the tangent line at $(0, 0)$.

(h) They are getting closer to 0.

73. (a) 2 **(b)** 2; 2; 2; 2 **(c)** $y = 2x + 5$

(d)

75. (a) $2x + h + 2$ **(b)** 4.5; 4.1; 4.01; 4

(c) $y = 4.01x - 1.01$

(d)

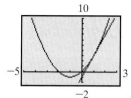

77. (a) $4x + 2h - 3$ **(b)** 2; 1.2; 1.02; 1

(c) $y = 1.02x - 1.02$

(d)

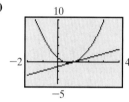

79. (a) $-\dfrac{1}{(x + h)x}$ **(b)** $-\dfrac{2}{3}; -\dfrac{10}{11}; -\dfrac{100}{101}; -1$

(c) $y = -\dfrac{100}{101}x + \dfrac{201}{101}$

(d)

83. At most one

85. Yes; the function $f(x) = 0$ is both even and odd.

2.5 Assess Your Understanding *(page 249)*

5. slope; y-intercept **6.** scatter diagram **7.** $y = kx$ **8.** T **9.** T **10.** T

11.

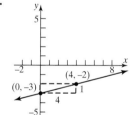

; 2

13.

; −3

15.

; $\dfrac{1}{4}$

17.

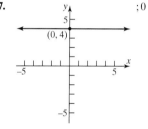

; 0

19. Linear relation **21.** Linear relation **23.** Nonlinear relation

25. (a)

(b) Answers will vary. Using $(4, 6)$ and $(8, 14)$: $y = 2x - 2$

(c)

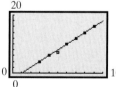

(d) $y = 2.0357x - 2.3571$ **(e)**

27. (a)

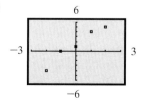

(b) Answers will vary. Using $(-2, -4)$ and $(1, 4)$: $y = \dfrac{8}{3}x + \dfrac{4}{3}$

(c)

(d) $y = 2.2x + 1.2$ **(e)**

29. (a)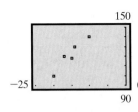
150
−25
90
0

(b) Answers will vary. Using $(-20, 100)$ and $(-15, 118)$: $y = \dfrac{18}{5}x + 172$

(c)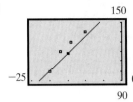
150
−25
90
0

(d) $y = 3.8613x + 180.2920$ **(e)**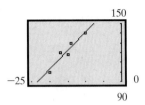
150
−25
90
0

31. (a) \$45 **(b)** 180 mi **(c)** 259 mi   **33. (a)** \$778.22 **(b)** 2006 **(c)** 2012   **35. (a)** $p = \$16, q = 600$ T-shirts **(b)** $0 < p < \$16$

37. (a) $x = 5000$ **(b)** $x > 5000$

39. (a) $V(x) = -1000x + 3000$

(b)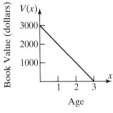

(c) \$1000 **(d)** After 1 year

41. (a) $C(x) = 90x + 1800$

(b)

(c) \$3060 **(d)** 22 bicycles

43. (a) $C(x) = 0.07x + 29$ **(b)** \$36.70; \$45.10

45. $p(b) = 0.00649B; \$941.05$

47. $R(g) = 1.95g; \$20.48$ **49.** 5 gallons

51. (a)
27,000
19,000   29,000
18,000

(b) $C(I) = 0.9241I + 479.6584$

(c) If disposable income increases by \$1, consumption increases by \$0.92.

(d) \$27,048 **(e)** \$28,590

53. (a)
14
0
0   175

(b) $L(G) = 0.0261G + 7.8738$

(c) If gestation period increases by 1 day, then life expectancy increases by about 0.0261 years.

(d) About 10.2 years

55. (a) No

(b)

(c) $D = -1.3355p + 86.1974$

(d) If the price increases \$1, the quantity sold per day decreases by about 1.34 pairs of jeans.

(e) $D(p) = -1.3355p + 86.1974$

(f) $\{p|p > 0\}$

(g) $D(28) \approx 48.80$; about 49 pairs

57.

The data do not follow a linear pattern.

59. When the y-intercept is 0. Yes, if the slope is 0.

61. No linear relation

2.6 Assess Your Understanding *(page 262)*

4. less **5.** piecewise-defined **6.** T **7.** F **8.** F

9. C **11.** E **13.** B **15.** F

17.

19.

21.

23.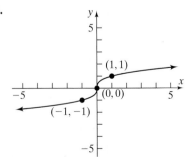

25. (a) 4 **(b)** 2 **(c)** 5   **27. (a)** 2 **(b)** 3 **(c)** −4

29. (a) All real numbers

(b) $(0, 1)$

(c)

(d) $\{y | y \neq 0\}; (-\infty, 0)$ or $(0, \infty)$

31. (a) All real numbers

(b) $(0, 3)$

(c)
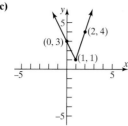

(d) $\{y | y \geq 1\}; [1, \infty)$

33. (a) $\{x | x \geq -2\}; [-2, \infty)$

(b) $(0, 3), (2, 0)$

(c)

(d) $\{y | y < 4, y = 5\}; (-\infty, 4)$ and $\{5\}$

35. (a) All real numbers

(b) $(-1, 0), (0, 0)$

(c)

(d) All real numbers

37. (a) $\{x | x \geq -2\}; [-2, \infty)$

(b) $(0, 1)$

(c)
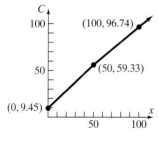

(d) $\{y | y > 0\}; (0, \infty)$

39. (a) All real numbers

(b) $(x, 0)$ for $0 \leq x < 1$

(c)

(d) Set of even integers

41. $f(x) = \begin{cases} -x & \text{if } -1 \leq x \leq 0 \\ \frac{1}{2}x & \text{if } 0 < x \leq 2 \end{cases}$ (Other answers are possible.)

43. $f(x) = \begin{cases} -x & \text{if } x \leq 0 \\ -x + 2 & \text{if } 0 < x \leq 2 \end{cases}$ (Other answers are possible.)

45. (a) \$35 **(b)** \$61 **(c)** \$35.40

47. (a) \$59.33 **(b)** \$396.04

(c) $C = \begin{cases} 0.99755x + 9.45 & \text{if } 0 \leq x \leq 50 \\ 0.74825x + 21.915 & \text{if } x > 50 \end{cases}$

(d)
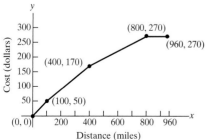

49. For schedule X: $f(x) = \begin{cases} 0.10x & \text{if} & 0 < x \leq 7150 \\ 715 + 0.15(x - 7150) & \text{if} & 7150 < x \leq 29{,}050 \\ 4000 + 0.25(x - 29{,}050) & \text{if} & 29{,}050 < x \leq 70{,}350 \\ 14{,}325 + 0.28(x - 70{,}350) & \text{if} & 70{,}350 < x \leq 146{,}750 \\ 35{,}717 + 0.33(x - 146{,}750) & \text{if} & 146{,}750 < x \leq 319{,}100 \\ 92{,}592.50 + 0.35(x - 319{,}100) & \text{if} & x > 319{,}100 \end{cases}$

51. (a)

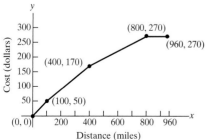

(b) $C = 50 + 0.4(x - 100)$

(c) $C = 170 + 0.25(x - 400)$

53. $f(x) = \begin{cases} x & \text{if} & 0 \le x < 10 \\ 10 & \text{if} & 10 \le x < 500 \\ 30 & \text{if} & 500 \le x < 1000 \\ 50 & \text{if} & 1000 \le x < 1500 \\ 70 & \text{if} & 1500 \le x \end{cases}$

55. (a) 10°C **(b)** 4°C **(c)** −3°C **(d)** −4°C **(e)** The wind chill is equal to the air temperature. **(f)** At wind speed greater than 20 m/sec, the wind chill factor depends only on the air temperature. **57.** Each graph is that of $y = x^2$, but shifted vertically. If $y = x^2 + k, k > 0$, the shift is up k units; if $y = x^2 - k, k > 0$, the shift is down k units. **59.** Each graph is that of $y = |x|$, but either compressed or stretched. If $y = k|x|$ and $k > 1$, the graph is stretched vertically; if $y = k|x|, 0 < k < 1$, the graph is compressed vertically. **61.** The graph of $y = f(-x)$ is the reflection about the y-axis of the graph of $y = f(x)$. **63.** They are all ∪-shaped and open upward. All three go through the points $(-1, 1), (0, 0)$ and $(1, 1)$. As the exponent increases, the steepness of the curve increases (except near $x = 0$).

2.7 Assess Your Understanding *(page 274)*

1. horizontal; right **2.** y **3.** −5; −2; 2 **4.** T **5.** F **6.** T **7.** B **9.** H **11.** I **13.** L **15.** F **17.** G **19.** $y = (x - 4)^3$ **21.** $y = x^3 + 4$
23. $y = -x^3$ **25.** $y = 4x^3$ **27.** (1) $y = \sqrt{x} + 2$; (2) $y = -(\sqrt{x} + 2)$; (3) $y = -(\sqrt{-x} + 2)$
29. (1) $y = -\sqrt{x}$; (2) $y = -\sqrt{x} + 2$; (3) $y = -\sqrt{x + 3} + 2$ **31.** (c) **33.** (c)

35.

37.

39.

41.

43.

45.

47.

49.

51.

53.

55.

57.

59.

61.

63.

65. (a) $F(x) = f(x) + 3$

(b) $G(x) = f(x + 2)$

(c) $P(x) = -f(x)$

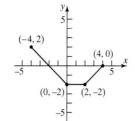

(d) $H(x) = f(x + 1) - 2$

(e) $Q(x) = \frac{1}{2}f(x)$

(f) $g(x) = f(-x)$

(g) $h(x) = f(2x)$

67. (a) $F(x) = f(x) + 3$

(b) $G(x) = f(x + 2)$

(c) $P(x) = -f(x)$

(d) $H(x) = f(x + 1) - 2$

(e) $Q(x) = \frac{1}{2}f(x)$

(f) $g(x) = f(-x)$

(g) $h(x) = f(2x)$

69. (a) -7 and 1 **(b)** -3 and 5 **(c)** -5 and 3 **(d)** -3 and 5 **71. (a)** $(-3, 3)$ **(b)** $(4, 10)$ **(c)** Decreasing on $(-1, 5)$ **(d)** Decreasing on $(-5, 1)$

73. (a)

(b)

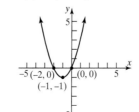

75. $f(x) = (x + 1)^2 - 1$

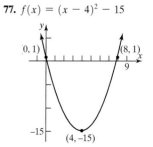

77. $f(x) = (x - 4)^2 - 15$

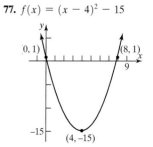

79. $f(x) = \left(x + \dfrac{1}{2}\right)^2 + \dfrac{3}{4}$

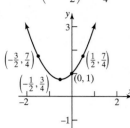

81. $f(x) = 2(x - 3)^2 + 1$

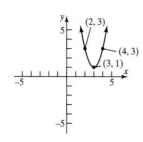

83. $f(x) = -3(x + 2)^2 - 5$

85.

87.

89. (a)

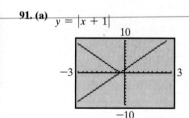

(b) 10% tax

(c) Y_1 is the graph of $p(x)$ shifted down vertically 10,000 units.
Y_2 is the graph of $p(x)$ vertically compressed by a factor of 0.9.

(d) 10% tax

91. (a) $y = |x + 1|$

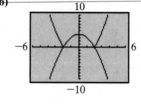

$y = x + 1$

(b)

(c)

(d) Any part of the graph of $y = f(x)$ that lies below the x-axis is reflected about the x-axis to obtain the graph of $y = |f(x)|$.

2.8 Assess Your Understanding *(page 282)*

1. $V(r) = 2\pi r^3$

3. (a) $R(x) = -\dfrac{1}{6}x^2 + 100x$

(b) \$13,333.33

(c) 15,000

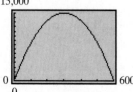

(d) 300; \$15,000 **(e)** \$50

5. (a) $R(x) = -\dfrac{1}{5}x^2 + 20x$

(b) \$255

(c) 500

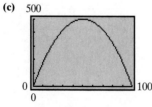

(d) 50; \$500 **(e)** \$10

7. (a) $A(x) = -x^2 + 200x$

(b) $0 < x < 200$

(c) A is largest when $x = 100$ yd.

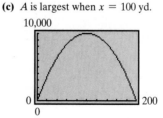

9. (a) $d(x) = \sqrt{x^4 - 15x^2 + 64}$

(b) $d(0) = 8$

(c) $d(1) = \sqrt{50} \approx 7.07$

(d)

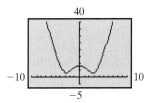

(e) d is smallest when $x \approx -2.74$

11. (a) $d(x) = \sqrt{x^2 - x + 1}$

(b)

(c) d is smallest when $x = \dfrac{1}{2}$.

13. $A(x) = \dfrac{1}{2}x^4$

15. (a) $A(x) = x(16 - x^2)$

(b) Domain: $\{x \mid 0 < x < 4\}$

(c) The area is largest when $x \approx 2.31$.

17. (a) $A(x) = 4x\sqrt{4 - x^2}$ **(b)** $p(x) = 4x + 4\sqrt{4 - x^2}$ **(c)** A is largest when $x \approx 1.41$.   **(d)** p is largest when $x \approx 1.41$.

19. (a) $A(x) = x^2 + \dfrac{25 - 20x + 4x^2}{\pi}$

(b) Domain: $\{x \mid 0 < x < 2.5\}$

(c) A is smallest when $x \approx 1.40$ m.

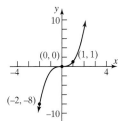

21. (a) $C(x) = x$ **(b)** $A(x) = \dfrac{x^2}{4\pi}$

23. (a) $A(r) = 2r^2$ **(b)** $p(r) = 6r$

25. $A(x) = \left(\dfrac{\pi}{3} - \dfrac{\sqrt{3}}{4}\right)x^2$

27. (a) $d(t) = \sqrt{2500t^2 - 360t + 13}$

(b) d is smallest when $t \approx 0.07$.

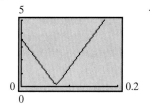

29. $V(r) = \dfrac{\pi H(R - r)r^2}{R}$ **31. (a)** $T(x) = \dfrac{12 - x}{5} + \dfrac{\sqrt{x^2 + 4}}{3}$ **(b)** $\{x \mid 0 \le x \le 12\}$ **(c)** 3.09 hr   **(d)** 3.55 hr

Review Exercises *(page 288)*

1. x-axis   **3.** x-axis, y-axis, origin   **5.** y-axis   **7.** No symmetry

9.

11. Function; domain $\{-1, 2, 4\}$, range $\{0, 3\}$

13. (a) 2 **(b)** -2 **(c)** $-\dfrac{3x}{x^2 - 1}$ **(d)** $-\dfrac{3x}{x^2 - 1}$ **(e)** $\dfrac{3(x - 2)}{x^2 - 4x + 3}$ **(f)** $\dfrac{6x}{4x^2 - 1}$

15. (a) 0 **(b)** 0 **(c)** $\sqrt{x^2 - 4}$ **(d)** $-\sqrt{x^2 - 4}$ **(e)** $\sqrt{x^2 - 4x}$ **(f)** $2\sqrt{x^2 - 1}$

17. (a) 0 **(b)** 0 **(c)** $\dfrac{x^2 - 4}{x^2}$ **(d)** $-\dfrac{x^2 - 4}{x^2}$ **(e)** $\dfrac{x(x - 4)}{(x - 2)^2}$ **(f)** $\dfrac{x^2 - 1}{x^2}$

19. $\{x \mid x \ne -3, x \ne 3\}$

21. $\{x \mid x \le 2\}$ **23.** $\{x \mid x > 0\}$ **25.** $\{x \mid x \ne -3, x \ne 1\}$

27. $(f + g)(x) = 2x + 3$; Domain: all real numbers

$(f - g)(x) = -4x + 1$; Domain: all real numbers

$(f \cdot g)(x) = -3x^2 + 5x + 2$; Domain: all real numbers

$\left(\dfrac{f}{g}\right)(x) = \dfrac{2 - x}{3x + 1}$; Domain: $\left\{x \mid x \ne -\dfrac{1}{3}\right\}$

29. $(f + g)(x) = 3x^2 + 4x + 1$; Domain: all real numbers

$(f - g)(x) = 3x^2 - 2x + 1$; Domain: all real numbers

$(f \cdot g)(x) = 9x^3 + 3x^2 + 3x$; Domain: all real numbers

$\left(\dfrac{f}{g}\right)(x) = \dfrac{3x^2 + x + 1}{3x}$; Domain: $\{x \mid x \ne 0\}$

31. $(f + g)(x) = \dfrac{x^2 + 2x - 1}{x(x - 1)}$; Domain: $\{x | x \neq 0, 1\}$

$(f - g)(x) = \dfrac{x^2 + 1}{x(x - 1)}$; Domain: $\{x | x \neq 0, 1\}$

$(f \cdot g)(x) = \dfrac{x + 1}{x(x - 1)}$; Domain: $\{x | x \neq 0, 1\}$

$\left(\dfrac{f}{g}\right)(x) = \dfrac{x(x + 1)}{x - 1}$; Domain: $\{x | x \neq 0, 1\}$

33. $-4x + 1 - 2h$

35. (a) Domain: $\{x | -4 \leq x \leq 3\}$; Range: $\{y | -3 \leq y \leq 3\}$ (b) $(0, 0)$ (c) -1 (d) -4 (e) $\{x | 0 < x \leq 3\}$

(f)   (g)   (h)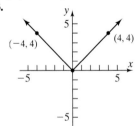

37. (a) Domain: $\{x | -4 \leq x \leq 4\}$ or $[-4, 4]$; Range: $\{y | -3 \leq y \leq 1\}$ or $[-3, 1]$

(b) Increasing on $(-4, -1)$ and $(3, 4)$; Decreasing on $(-1, 3)$

(c) Local maxima are 1 and 0 and occur at $x = -1$ and $x = 4$;
Local minima are -2 and -3 and occur at $x = -4$ and $x = 3$

(d) No symmetry

(e) Neither

(f) x-intercepts: $-2, 0, 4$; y-intercept: 0

39. Odd
41. Even
43. Neither
45. Odd

47.

Local maximum: $(-0.91, 4.04)$
Local minimum: $(0.91, -2.04)$
Increasing: $(-3, -0.91)$; $(0.91, 3)$
Decreasing: $(-0.91, 0.91)$

49.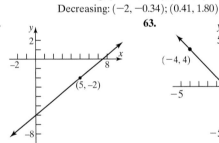

Local maximum: $(0.41, 1.53)$
Local minimum: $(-0.34, 0.54)$; $(1.80, -3.56)$
Increasing: $(-0.34, 0.41)$; $(1.80, 3)$
Decreasing: $(-2, -0.34)$; $(0.41, 1.80)$

51. (a) 23   (b) 7   (c) 47
53. -5
55. $-4x - 5$
57. (b)

59.

61.

63.

65.

Intercepts: $(-4, 0)$, $(4, 0)$, $(0, -4)$
Domain: all real numbers
Range: $\{y | y \geq -4\}$ or $[-4, \infty)$

67.

Intercept: $(0, 0)$
Domain: all real numbers
Range: $\{y | y \leq 0\}$ or $(-\infty, 0]$

69.

Intercept: $(1, 0)$
Domain: $\{x | x \geq 1\}$ or $[1, \infty)$
Range: $\{y | y \geq 0\}$ or $[0, \infty)$

71.

Intercepts: $(0, 1)$, $(1, 0)$
Domain: $\{x | x \leq 1\}$ or $(-\infty, 1]$
Range: $\{y | y \geq 0\}$ or $[0, \infty)$

73.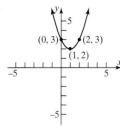

Intercept: $(0, 3)$

Domain: all real numbers

Range: $\{y | y \geq 2\}$ or $[2, \infty)$

75.

Intercepts: $(0, -2)$,

$\left(1 - \dfrac{\sqrt[3]{9}}{3}, 0\right)$ or about $(0.3, 0)$

Domain: all real numbers

Range: all real numbers

77. (a) $\{x | x > -2\}; (-2, \infty)$

(b) $(0, 0)$

(c)

(d) $\{y | y > -6\}; (-6, \infty)$

79. (a) $\{x | x \geq -4\}; [-4, \infty)$

(b) $(0, 1)$

(c)

(d) $\{y | y \geq -4\}; [-4, \infty)$

81. $f(x) = -2x + 3$ **83.** $A = 11$ **85.** $T(h) = -0.0025h + 30, 0 \leq x \leq 10,000$

87. $V(S) = \dfrac{S}{6}\sqrt{\dfrac{S}{\pi}}$; If the surface area doubles, the volume increases by a factor of $2\sqrt{2}$.

89. $S(x) = kx(36 - x^2)^{3/2}$; Domain: $\{x | 0 < x < 6\}$

91. (a) Yes

(b)

(c) $G = 0.964x + 0.072$ **(d)** As high school GPA increases 1 point, college GPA increases by 0.964 point.

(e) $G(x) = 0.964x + 0.072$ **(f)** $\{x | 0 \leq = x \leq = 4\}$

(g) 3.19

93. $R(g) = 2.14g; \$23.97$ **95.** 2 by 8 ft   **97. (a)** 63   **(b)** $\$151.90$

99. $(2, 3)$ **101. (a)** $A(x) = 2x^2 + \dfrac{40}{x}$ **(b)** 42 ft^2 **(c)** 28 ft^2

(d)

Chapter Test *(page 293)*

1. x-axis only   **2. (a)** Function; domain: $\{2, 4, 6, 8\}$; range: $\{5, 6, 7, 8\}$ **(b)** Not a function   **(c)** Not a function   **(d)** Function; domain: all real numbers; range: $\{y | y \geq 2\}$ **3.** Domain: $\left\{x | x \leq \dfrac{4}{5}\right\}; f(-1) = 3$ **4.** Domain: $\{x | x \neq -2\}; g(-1) = 1$ **5.** Domain: $\{x | x \neq -9, x \neq 4\}; h(-1) = \dfrac{1}{8}$

6. (a) Domain: $\{x | x -5 \leq x \leq 5\}$; range: $\{y | -3 \leq y \leq 3\}$ **(b)** $(0, 2), (-2, 0)$, and $(2, 0)$ **(c)** $f(1) = 3$ **(d)** $x = -5$ and $x = 3$

(e) $\{x | -5 \leq x < -2 \text{ or } 2 < x \leq 5\}; [-5, -2) \cup (2, 5]$

7. Local maxima: $f(-0.85) \approx -0.86$

$f(2.35) \approx 15.55$

Local minima: $f(0) = -2$

The function is increasing on the intervals $(-5, -0.85)$ and

$(0, 2.35)$ and decreasing on the intervals $(-0.85, 0)$ and $(2.35, 5)$.

8. (a)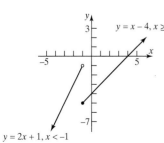

(b) $(0, -4), (4, 0)$

(c) $g(-5) = -9$

(d) $g(2) = -2$

9. $\dfrac{f(x) - f(3)}{x - 3} = 3x + 7, x \neq 3$

10. (a) $(f - g) = 2x^2 - 3x + 3$

(b) $(f \cdot g) = 6x^3 - 4x^2 + 3x - 2$

(c) $f(x + h) - f(x) = 4xh + 2h^2$

11. (a)

(b)

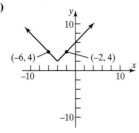

12. (a) Set B is more linear.   **(b)** $y = 2.02x - 5.33$

13. (a) 8.67% occurring in 1997 ($x \approx 5$)

    (b) The model predicts that the interest rate will be -10.343%. This is not reasonable.

14. (a) $V(x) = \dfrac{x^2}{8} - \dfrac{5x}{4} + \dfrac{\pi x^2}{64}$ **(b)** 1297.61 ft³

Cumulative Review *(page 295)*

1. $\left\{\dfrac{4}{5}\right\}$ **2.** $\{3, 4\}$ **3.** $\left\{-\dfrac{1}{3}, 2\right\}$ **4.** $\left\{-\dfrac{1}{2}\right\}$ **5.** No real solution   **6.** $\{-7\}$ **7.** $\{-31\}$ **8.** $\left\{\dfrac{1}{3}, 1\right\}$ **9.** $\left\{\dfrac{1 - \sqrt{15}i}{4}, \dfrac{1 + \sqrt{15}i}{4}\right\}$

10. $\{x \mid 1 < x < 4\}$ **11.**

12.

13.

14.

15. (a) Domain: $\{x \mid -4 \le x \le 4\}$; Range: $\{y \mid -1 \le y \le 3\}$ **(b)** $(-1, 0), (0, -1), (1, 0)$ **(c)** y-axis   **(d)** 1   **(e)** -4 and 4   **(f)** $\{x \mid -1 \le x < 1\}$

(g)

(h)

(i)

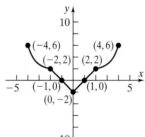

(j) even
(k) $(0, 4)$
(l) $(-4, 0)$
(m) f has a local minimum of -1 at $x = 0$
(n) 1

16. $5\sqrt{2}$ **17.** $(-2, -1)$ and $(2, 3)$ are on the graph.   **18.** x-intercepts: $-5, \dfrac{1}{3}$; y-intercept: -5 **19.** $-1.10, 0.26, 1.48, 2.36$

20. $y = -\dfrac{1}{2}x + \dfrac{13}{2}$

21. Yes, a function
22. (a) -3
    (b) $x^2 - 4x - 2$
    (c) $x^2 + 4x + 1$
    (d) $-x^2 + 4x - 1$
    (e) $x^2 - 3$
    (f) $2x + h - 4$

23. $\{z \mid z \ne -1, z \ne 7\}$
24. Yes, a function
25. (a) No
    (b) $-1; (-2, -1)$ is on the graph.
    (c) $-8; (-8, 2)$ is on the graph.

C H A P T E R   3   Polynomial and Rational Functions

3.1 Assess Your Understanding *(page 311)*

5. parabola   **6.** axis or axis of symmetry   **7.** $-\dfrac{b}{2a}$ **8.** T   **9.** T   **10.** T

11. *C*   **13.** *F*   **15.** *G*   **17.** *H*

19.

21.

23.

25.
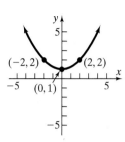

27. $f(x) = (x + 2)^2 - 2$

29. $f(x) = 2(x - 1)^2 - 1$

31. $f(x) = -(x + 1)^2 + 1$

33. $f(x) = \dfrac{1}{2}(x + 1)^2 - \dfrac{3}{2}$

35.

37.

39.

41.
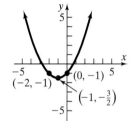

Domain: $(-\infty, \infty)$
Range $[-1, \infty)$
Decreasing on $(-\infty, -1)$
Increasing on $(-1, \infty)$

Domain: $(-\infty, \infty)$
Range $(-\infty, 9]$
Increasing on $(-\infty, -3)$
Decreasing on $(-3, \infty)$

Domain: $(-\infty, \infty)$
Range $[-8, \infty)$
Decreasing on $(-\infty, 2)$
Increasing on $(2, \infty)$

Domain: $(-\infty, \infty)$
Range $[-9, \infty)$
Decreasing on $(-\infty, -1)$
Increasing on $(-1, \infty)$

43.

45.

47.

49.

Domain: $(-\infty, \infty)$

Range $[0, \infty)$

Decreasing on $(-\infty, -1)$

Increasing on $(-1, \infty)$

Domain: $(-\infty, \infty)$

Range $\left[\dfrac{15}{8}, \infty\right)$

Decreasing on $\left(-\infty, \dfrac{1}{4}\right)$

Increasing on $\left(\dfrac{1}{4}, \infty\right)$

Domain: $(-\infty, \infty)$

Range $\left(-\infty, -\dfrac{5}{2}\right]$

Decreasing on $\left(\dfrac{1}{2}, \infty\right)$

Increasing on $\left(-\infty, \dfrac{1}{2}\right)$

Domain: $(-\infty, \infty)$

Range $[-1, \infty)$

Decreasing on $(-\infty, -1)$

Increasing on $(-1, \infty)$

51.

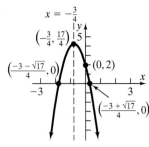

$$x = -\frac{3}{4}$$

$\left(-\frac{3}{4}, \frac{17}{4}\right)$

$\left(\frac{-3-\sqrt{17}}{4}, 0\right)$ $(0, 2)$

$\left(\frac{-3+\sqrt{17}}{4}, 0\right)$

Domain: $(-\infty, \infty)$

Range $\left(-\infty, \frac{17}{4}\right]$

Decreasing on $\left(-\frac{3}{4}, \infty\right)$

Increasing on $\left(-\infty, -\frac{3}{4}\right)$

53. $f(x) = (x + 1)^2 - 2 = x^2 + 2x - 1$
55. $f(x) = -(x + 3)^2 + 5 = -x^2 - 6x - 4$
57. $f(x) = 2(x - 1)^2 - 3 = 2x^2 - 4x - 1$
59. Minimum value; -18 **61.** Minimum value; -21
63. Maximum value; 21
65. Maximum value; 13   **67.** $a = 6, b = 0, c = 2$
69. (a) $a = 1: f(x) = (x + 3)(x - 1) = x^2 + 2x - 3$
  $a = 2: f(x) = 2(x + 3)(x - 1) = 2x^2 + 4x - 6$
  $a = -2: f(x) = -2(x + 3)(x - 1) = -2x^2 - 4x + 6$
  $a = 5: f(x) = 5(x + 3)(x - 1) = 5x^2 + 10x - 15$
  (b) The value of a does not affect the x-intercepts
  but it changes the y-intercept by a factor of a.

  (c) The value of a does not affect the axis of symmetry.
  It is $x = -1$ for all values of a.

(d) The value of a does not affect the
  x-coordinate of the vertex. However,
  the y-coordinate of the vertex is
  multiplied by a.
(e) The midpoint of the x-intercepts is
  the x-coordinate of the vertex.

71. $500; $1,000,000   **73. (a)** $R(x) = -\frac{1}{6}x^2 + 100x$ **(b)** $13,333.33   **(c)** 300; $15,000   **(d)** $50   **75. (a)** $R(x) = -\frac{1}{5}x^2 + 20x$ **(b)** $255

(c) 50; $500   **(d)** $10   **77. (a)** $A(w) = -w^2 + 200w$ **(b)** A is largest when $w = 100$ yd.   **(c)** 10,000 sq yd   **79.** 2,000,000 m²

81. (a) $\frac{625}{16} \approx 39$ ft

(b) 219.5 ft

(c) 170 ft

(d) 220

(e) When the height is 100 ft, the
projectile is 135.7 ft from the cliff.

83. 18.75 m

85. 3 in.

87. $\frac{750}{\pi}$ by 375 m

89. $x = \frac{a}{2}$

91. $\frac{38}{3}$

93. $\frac{248}{3}$

95. 25 square units

97. (a) $56,600; 3685 hunters

(b) 4000

0   120
0
Increasing

99. (a) 1795   **(b)** 28 years old

(c) 5000

0   90
0

(d) The number of victims initially
decreases, then begins to increase.

101. (a) Quadratic, $a < 0$

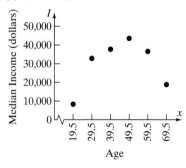

(b) $I(x) = -42.67x^2 + 3998.27x - 51,873.20$
(c) 46.9 yr old
(d) $41,788
(e) 50,000

0   80
0

103. (a) Quadratic, $a < 0$

(b) $h(x) = -0.0037x^2 + 1.0318x + 5.6667$
(c) 139.4 ft
(d) 77.6 ft
(e) 80

0   220
0

3.2 Assess Your Understanding *(page 330)*

5. smooth; continuous **6.** zero or root **7.** touches **8.** T **9.** F **10.** F **11.** Yes; degree 3 **13.** Yes; degree 2

15. No; x is raised to the -1 power. **17.** No; x is raised to the $\frac{3}{2}$ power. **19.** Yes; degree 4 **21.** Yes; degree 4

23.   **25.**   **27.**   **29.**

31.   **33.**   **35.**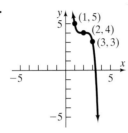

37. $f(x) = x^3 - 3x^2 - x + 3$ for $a = 1$

39. $f(x) = x^3 - x^2 - 12x$ for $a = 1$

41. $f(x) = x^4 - 15x^2 + 10x + 24$ for $a = 1$

43. $f(x) = x^3 - 5x^2 + 3x + 9$ for $a = 1$

45. (a) 7, multiplicity 1; -3, multiplicity 2 **(b)** graph touches the x-axis at -3 and crosses it at 7 **(c)** $y = 3x^3$

47. (a) 2, multiplicity 3 **(b)** graph crosses the x-axis at 2 **(c)** $y = 4x^5$

49. (a) $-\frac{1}{2}$, multiplicity 2 **(b)** graph touches the x-axis at $-\frac{1}{2}$ **(c)** $y = -2x^6$

51. (a) 5, multiplicity 3; -4, multiplicity 2 **(b)** graph touches the x-axis at -4 and crosses it at 5 **(c)** $y = x^5$

53. (a) no real zeros **(b)** graph neither crosses nor touches the x-axis **(c)** $y = 3x^6$

55. (a) 0, multiplicity 2; $-\sqrt{2}$, $\sqrt{2}$, multiplicity 1 **(b)** graph touches the x-axis at 0 and crosses at $-\sqrt{2}$ and $\sqrt{2}$ **(c)** $y = -2x^4$

57. (a) Degree 2; $y = x^2$

(b) x-intercept: 1; y-intercept: 1

(c) 1: Touches

(d)

(e) Local minimum at $(1, 0)$

(f)

(g) Domain: $(-\infty, \infty)$
Range: $[0, \infty)$

(h) Increasing on $(1, \infty)$
Decreasing on $(-\infty, 1)$

59. (a) Degree 3; $y = x^3$

(b) x-intercepts: 0, 3; y-intercept: 0

(c) 0: Touches; 3: Crosses

(d)

(e) Local minimum at $(2, -4)$
Local maximum at $(0, 0)$

(f)

(g) Domain: $(-\infty, \infty)$
Range: $(-\infty, \infty)$

(h) Increasing on $(-\infty, 0)$ and $(2, \infty)$
Decreasing on $(0, 2)$

61. (a) Degree 3; $y = x^3$

(b) x-intercepts: -4, 2; y-intercept: 16

(c) -4: Crosses; 2: Touches

(d)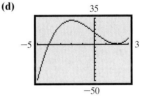

(e) Local maximum at $(-2, 32)$
Local minimum at $(2, 0)$

(f)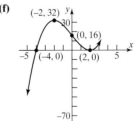

(g) Domain: $(-\infty, \infty)$
Range: $(-\infty, \infty)$

(h) Increasing on $(-\infty, -2)$ and $(2, \infty)$
Decreasing on $(-2, 2)$

63. (a) Degree 4; $y = -2x^4$
(b) x-intercepts: $-2, 2$
y-intercept: 32
(c) -2: Crosses; 2: Crosses
(d)

(e) Local maximum at $(-1, 54)$
(f)

(g) Domain: $(-\infty, \infty)$
Range: $(-\infty, 54]$
(h) Increasing on $(-\infty, -1)$
Decreasing on $(-1, \infty)$

65. (a) Degree 3; $y = x^3$
(b) x-intercepts: $-4, -1, 2$
y-intercept: -8
(c) $-4, -1, 2$: Crosses
(d)

(e) Local maximum at $(-2.73, 10.39)$
Local minimum at $(0.73, -10.39)$
(f)
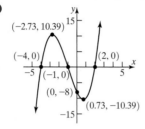
(g) Domain: $(-\infty, \infty)$
Range: $(-\infty, \infty)$
(h) Increasing on $(-\infty, -2.73)$ and
$(0.73, \infty)$
Decreasing on $(-2.73, 0.73)$

67. (a) Degree 3; $y = -x^3$
(b) x-intercepts: $-2, 0, 2$
y-intercept: 0
(c) $-2, 0, 2$: Crosses
(d)

(e) Local minimum at $(-1.15, -3.08)$
Local maximum at $(1.15, 3.08)$
(f)
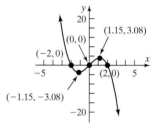
(g) Domain: $(-\infty, \infty)$
Range: $(-\infty, \infty)$
(h) Increasing on $(-1.15, 1.15)$
Decreasing on $(-\infty, -1.15)$
and $(1.15, \infty)$

69. (a) Degree 4; $y = x^4$
(b) x-intercepts: $-2, 0, 2$
y-intercept: 0
(c) $-2, 2$: Crosses; 0: Touches
(d)

(e) Local minima at $(-1.41, -4)$,
$(1.41, -4)$; Local maximum at $(0, 0)$
(f)
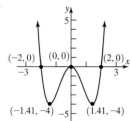
(g) Domain: $(-\infty, \infty)$
Range: $[-4, \infty)$
(h) Increasing on $(-1.41, 0)$ and $(1.41, \infty)$
Decreasing on $(-\infty, -1.41)$ and $(0, 1.41)$

71. (a) Degree 4; $y = x^4$
(b) x-intercepts: $-1, 2$
y-intercept: 4
(c) $-1, 2$: Touches
(d)

(e) Local minima at $(-1, 0), (2, 0)$
Local maximum at $(0.5, 5.06)$
(f)
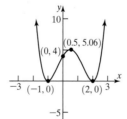
(g) Domain: $(-\infty, \infty)$
Range: $[0, \infty)$
(h) Increasing on $(-1, 0.5)$ and $(2, \infty)$
Decreasing on $(-\infty, -1)$ and $(0.5, 2)$

73. (a) Degree 4; $y = x^4$
(b) x-intercepts: $-1, 0, 3$
y-intercept: 0
(c) $-1, 3$: Crosses; 0: Touches
(d)

(e) Local minima at $(-0.69, -0.54)$,
$(2.19, -12.39)$; Local maximum at $(0, 0)$
(f)
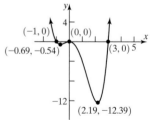
(g) Domain: $(-\infty, \infty)$
Range: $(-12.39, \infty)$
(h) Increasing on $(-0.69, 0)$ and $(2.19, \infty)$
Decreasing on $(-\infty, -0.69)$ and $(0, 2.19)$

75. (a) Degree 4; $y = x^4$

  (b) x-intercepts: $-2, 4$; y-intercept: 64

  (c) $-2, 4$: Touches

  (d)

  (e) Local minima at $(-2, 0)$, $(4, 0)$
  Local maximum at $(1, 81)$

  (f)

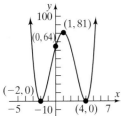

  (g) Domain: $(-\infty, \infty)$; Range: $(0, \infty)$

  (h) Increasing on $(-2, 1)$ and $(4, \infty)$
  Decreasing on $(-\infty, -2)$ and $(1, 4)$

77. (a) Degree 5; $y = x^5$

  (b) x-intercepts: $0, 2$; y-intercept: 0

  (c) 0: Touches; 2: Crosses

  (d)

  (e) Local maximum at $(0, 0)$
  Local minimum at $(1.48, -5.91)$

  (f)

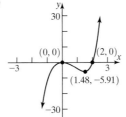

  (g) Domain: $(-\infty, \infty)$; Range: $(-\infty, \infty)$

  (h) Increasing on $(-\infty, 0)$ and $(1.48, \infty)$
  Decreasing on $(0, 1.48)$

79. (a) Degree 5; $y = -x^5$

  (b) x-intercepts: $-1, 0, 1$; y-intercept: 0

  (c) $-1, 0$: Touches; 1: Crosses

  (d)

  (e) Local minima at $(-1, 0)$, $(0, 0)$
  Local maxima at $(-0.54, 0.10)$, $(0.74, 0.43)$

  (f)

  (g) Domain: $(-\infty, \infty)$; Range: $(-\infty, \infty)$

  (h) Increasing on $(-1, -0.54)$ and $(0, 0.74)$
  Decreasing on $(-\infty, -1)$, $(-0.54, 0)$, and $(0.74, \infty)$

81. (a) Degree 3; $y = x^3$

  (b)

  (c) x-intercepts: $-1.26, -0.20, 1.26$

  y-intercept: -0.31752

  (d)

  (e) Local maximum at $(-0.80, 0.57)$
  Local minimum at $(0.66, -0.99)$

  (f)

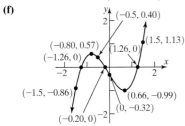

  (g) Domain: $(-\infty, \infty)$; Range: $(-\infty, \infty)$

  (h) Increasing on $(-\infty, -0.80)$ and $(0.66, \infty)$
  Decreasing on $(-0.80, 0.66)$

83. (a) Degree 3; $y = x^3$

  (b)

  (c) x-intercepts: $-3.56, 0.50$

  y-intercept: 0.89

  (d)

  (e) Local maximum at $(-2.21, 9.91)$
  Local minimum at $(0.50, 0)$

  (f)

  (g) Domain: $(-\infty, \infty)$; Range: $(-\infty, \infty)$

  (h) Increasing on $(-\infty, -2.21)$ and $(0.50, \infty)$
  Decreasing on $(-2.21, 0.50)$

85. (a) Degree 4; $y = x^4$

  (b)

  (c) x-intercepts: $-1.5, -0.5, 0.5, 1.5$

  y-intercept: 0.5625

  (d)

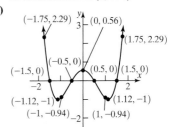

  (e) Local minima at $(-1.12, -1)$, $(1.12, -1)$
  Local maximum at $(0, 0.56)$

  (f)

  (g) Domain: $(-\infty, \infty)$; Range: $[-1, \infty]$

  (h) Increasing on $(-1.12, 0)$ and $(1.12, \infty)$
  Decreasing on $(-\infty, -1.12)$ and $(0, 1.12)$

87. (a) Degree 4; $y = 2x^4$

(b)

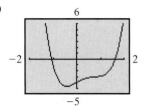

(c) x-intercepts: $-1.07, 1.62$
y-intercept: -4

(d)

(e) Local minimum at $(-0.42, -4.64)$

(f)

(g) Domain: $(-\infty, \infty)$
Range: $[-4.64, \infty)$

(h) Increasing on $(-0.42, \infty)$
Decreasing on $(-\infty, -0.42)$

89. (a) Degree 5; $y = -2x^5$

(b)

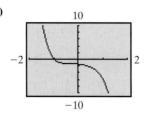

(c) x-intercept: -0.98
y-intercept: $-\sqrt{2}$

(d)

(e) None

(f)

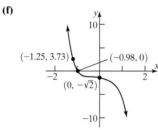

(g) Domain: $(-\infty, \infty)$
Range: $(-\infty, \infty)$

(h) Decreasing on $(-\infty, \infty)$

91. Answers will vary. One possibility is
$f(x) = x(x - 1)(x - 2)$.

93. Answers will vary. One possibility
is $f(x) = -\dfrac{1}{2}(x + 1)(x - 1)(x - 2)$.

95. (a) Cubic, $a > 0$

(b) $7000 per car
(c) $20,000 per car
(d) $C(x) = 0.2156x^3 - 2.3473x^2 + 14.3275x + 10.2238$
(e)

(f) About $171,000
(g) Fixed costs of about $10,200

97. (a) Cubic, $a > 0$

(b) $T(x) = 1.2582x^3 - 28.6146x^2 + 139.5808x + 1453.2098$
(c)

(d) About 1,738,000 thefts

99. No; yes   **103.** $f(x) = \text{int}(x)$; $g(x) = |x|$

3.3 Assess Your Understanding (page 343)

5. $y = 1$ **6.** $x = -1$ **7.** proper   **8.** False   **9.** True   **10.** True   **11.** All real numbers except 3; $\{x \mid x \ne 3\}$

13. All real numbers except 2 and -4; $\{x \mid x \ne 2, x \ne -4\}$ **15.** All real numbers except $-\dfrac{1}{2}$ and 3; $\left\{x \mid x \ne -\dfrac{1}{2}, x \ne 3\right\}$

17. All real numbers except 2; $\{x \mid x \ne 2\}$ **19.** All real numbers   **21.** All real numbers except -3 and 3; $\{x \mid x \ne -3, x \ne 3\}$
23. (a) Domain: $\{x \mid x \ne 2\}$; Range: $\{y \mid y \ne 1\}$ **(b)** $(0, 0)$ **(c)** $y = 1$ **(d)** $x = 2$ **(e)** None
25. (a) Domain: $\{x \mid x \ne 0\}$; Range: all real numbers   **(b)** $(-1, 0), (1, 0)$ **(c)** None   **(d)** $x = 0$ **(e)** $y = 2x$
27. (a) Domain: $\{x \mid x \ne -2, x \ne 2\}$; Range: $\{y \mid y \le 0, y > 1\}$ **(b)** $(0, 0)$ **(c)** $y = 1$ **(d)** $x = -2, x = 2$ **(e)** None

29.

31.

33.

35.

37.

39.

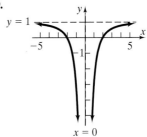

41. Horizontal asymptote: $y = 3$; vertical asymptote: $x = -4$

43. Vertical asymptote: $x = 2$; oblique asymptote: $y = x + 5$ **45.** Horizontal asymptote: $y = 0$; vertical asymptotes: $x = 1, x = -1$

47. Horizontal asymptote: $y = 0$; vertical asymptote: $x = 0$ **49.** Oblique asymptote: $y = 3x$; vertical asymptote: $x = 0$

51. Oblique asymptote: $y = -(x + 1)$; vertical asymptote: $x = 0$ **53. (a)** 9.8208 m/sec^2 **(b)** 9.8195 m/sec^2 **(c)** 9.7936 m/sec^2 **(d)** h-axis

3.4 Concepts and Vocabulary (*page 355*)

3. in lowest terms   **4.** F   **5.** F   **6.** T

7. 1. Domain: $\{x \mid x \neq 0, x \neq -4\}$

    2. $R(x)$ is in lowest terms.

    3. x-intercept: -1; no y-intercept

    4. No symmetry

    5. Vertical asymptotes: $x = 0, x = -4$

    6. Horizontal asymptote: $y = 0$, intersected at $(-1, 0)$

7.

8.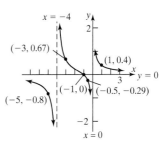

9. 1. Domain: $\{x \mid x \neq -2\}$

    2. $R(x) = \dfrac{3(x + 1)}{2(x + 2)}$

    3. x-intercept: -1; y-intercept: $\dfrac{3}{4}$

    4. No symmetry

    5. Vertical asymptote: $x = -2$

    6. Horizontal asymptote: $y = \dfrac{3}{2}$, not intersected

7.

8.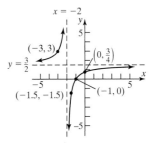

11. 1. Domain: $\{x \mid x \neq -2, x \neq 2\}$

    2. $R(x) = \dfrac{3}{(x + 2)(x - 2)}$

    3. No x-intercept; y-intercept: $-\dfrac{3}{4}$

    4. Symmetric with respect to the y-axis

    5. Vertical asymptotes: $x = 2, x = -2$

    6. Horizontal asymptote: $y = 0$, not intersected

7.

8.

13. 1. Domain: $\{x | x \neq -1, x \neq 1\}$

2. $P(x) = \dfrac{x^4 + x^2 + 1}{(x + 1)(x - 1)}$

3. No x-intercept; y-intercept: -1

4. Symmetric with respect to the y-axis

5. Vertical asymptotes: $x = -1, x = 1$

6. No horizontal or oblique asymptotes

7.

8.

15. 1. Domain: $\{x | x \neq -3, x \neq 3\}$

2. $H(x) = \dfrac{(x - 1)(x^2 + x + 1)}{(x + 3)(x - 3)}$

3. x-intercept: 1; y-intercept: $\dfrac{1}{9}$

4. No symmetry

5. Vertical asymptotes: $x = 3, x = -3$

6. Oblique asymptote: $y = x$, intersected at $\left(\dfrac{1}{9}, \dfrac{1}{9}\right)$

7.

8.

17. 1. Domain: $\{x \neq -3, x \neq 2\}$

2. $R(x) = \dfrac{x^2}{(x + 3)(x - 2)}$

3. Intercept: $(0, 0)$

4. No symmetry

5. Vertical asymptotes: $x = 2, x = -3$

6. Horizontal asymptote: $y = 1$, intersected at $(6, 1)$

7.

8.

19. 1. Domain: $\{x | x \neq -2, x \neq 2\}$

2. $G(x) = \dfrac{x}{(x + 2)(x - 2)}$

3. Intercept: $(0, 0)$

4. Symmetry with respect to the origin

5. Vertical asymptotes: $x = -2, x = 2$

6. Horizontal asymptote: $y = 0$, intersected at $(0, 0)$

7.

8.

21. 1. Domain: $\{x | x \neq 1, x \neq -2, x \neq 2\}$

2. $R(x) = \dfrac{3}{(x - 1)(x - 2)(x + 2)}$

3. No x-intercept; y-intercept: $\dfrac{3}{4}$

4. No symmetry

5. Vertical asymptotes: $x = -2, x = 1, x = 2$

6. Horizontal asymptote: $y = 0$, not intersected

7.

8.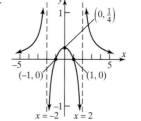

23. 1. Domain: $\{x | x \neq -2, x \neq 2\}$

2. $H(x) = \dfrac{4(x + 1)(x - 1)}{(x^2 + 4)(x + 2)(x - 2)}$

3. x-intercepts: $-1, 1$; y-intercept: $\dfrac{1}{4}$

4. Symmetry with respect to the y-axis

5. Vertical asymptotes: $x = -2, x = 2$

6. Horizontal asymptote: $y = 0$, intersected at $(-1, 0)$ and $(1, 0)$

25. 1. Domain: $\{x|x \neq -2\}$

2. $F(x) = \dfrac{(x-4)(x+1)}{x+2}$

3. x-intercepts: $-1, 4$; y-intercept: -2

4. No symmetry

5. Vertical asymptote: $x = -2$

6. Oblique asymptote: $y = x - 5$, not intersected

7.

8.

27. 1. Domain: $\{x|x \neq 4\}$

2. $R(x) = \dfrac{(x+4)(x-3)}{x-4}$

3. x-intercepts: $-4, 3$; y-intercept: 3

4. No symmetry

5. Vertical asymptote: $x = 4$

6. Oblique asymptote: $y = x + 5$, not intersected

7.

8.

29. 1. Domain: $\{x|x \neq -2\}$

2. $F(x) = \dfrac{(x+4)(x-3)}{x+2}$

3. x-intercepts: $-4, 3$; y-intercept: -6

4. No symmetry

5. Vertical asymptote: $x = -2$

6. Oblique asymptote: $y = x - 1$, not intersected

7.

8.

31. 1. Domain: $\{x|x \neq -3\}$

2. $R(x)$ is in lowest terms.

3. x-intercepts: $0, 1$; y-intercept: 0

4. No symmetry

5. Vertical asymptote: $x = -3$

6. Horizontal asymptote: $y = 1$, not intersected

7.

8.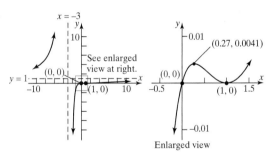

33. 1. Domain: $\{x|x \neq -2, x \neq 3\}$

2. $R(x) = \dfrac{x+4}{x+2}$

3. x-intercept: -4; y-intercept: 2

4. No symmetry

5. Vertical asymptote: $x = -2$; hole at $\left(3, \dfrac{7}{5}\right)$

6. Horizontal asymptote: $y = 1$, not intersected

7.

8.

35. 1. Domain: $\left\{x \middle| x \neq \dfrac{3}{2}, x \neq 2\right\}$

2. $R(x) = \dfrac{3x + 1}{x - 2}$

3. x-intercept: $-\dfrac{1}{3}$; y-intercept: $-\dfrac{1}{2}$

4. No symmetry

5. Vertical asymptote: $x = 2$; hole at $\left(\dfrac{3}{2}, -11\right)$

6. Horizontal asymptote: $y = 3$, not intersected

7.

8.

37. 1. Domain: $\{x | x \neq -3\}$

2. $R(x) = x + 2$

3. x-intercept: -2; y-intercept: 2

4. No symmetry

5. Vertical asymptote: none; hole at $(-3, -1)$

6. Oblique asymptote: $y = x + 2$

7.

8.

39. 1. Domain: $\{x | x \neq 0\}$

2. $f(x) = \dfrac{x^2 + 1}{x}$

3. No x-intercepts; no y-intercept

4. Symmetric with respect to the origin

5. Vertical asymptote: $x = 0$

6. Oblique asymptote: $y = x$, not intersected

7.

8.

41. 1. Domain: $\{x | x \neq 0\}$

2. $f(x) = \dfrac{x^3 + 1}{x} = \dfrac{(x + 1)(x^2 - x + 1)}{x}$

3. x-intercept: -1; no y-intercept

4. No symmetry

5. Vertical asymptote: $x = 0$

6. No horizontal or oblique asymptotes

7.

8.

43. 1. Domain: $\{x | x \neq 0\}$

2. $f(x) = \dfrac{x^4 + 1}{x^3}$

3. No x-intercepts; no y-intercept

4. Symmetric with respect to the origin

5. Vertical asymptote: $x = 0$

6. Oblique asymptote: $y = x$, not intersected

7.

8.

45. Minimum value: 2.00 at $x = 1.00$

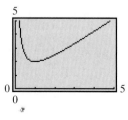

47. Minimum value: 1.89 at $x = 0.79$

49. Minimum value: 1.75 at $x = 1.32$

51. One possibility: $R(x) = \dfrac{x^2}{x^2 - 4}$ **53.** One possibility: $R(x) = \dfrac{(x - 1)(x - 3)(x^2 + \frac{4}{3})}{(x + 1)^2(x - 2)^2}$

55. (a) t-axis; $C(t) \to 0$

57. (a) $\overline{C}(x) = \dfrac{0.2x^3 - 2.3x^2 + 14.3x + 10.2}{x}$

59. (a) $S(x) = 2x^2 + \dfrac{40{,}000}{x}$

(b)
0.4

0   12
0

(b) \$9400

(c) $\approx$ \$10,933

(d)
60

0   20
0

(b) 10,000

0   60
0

(c) 0.71 hr after injection

(e) 6

(f) \$9400

(c) 2784.95 sq in.

(d) 21.54 in. $\times$ 21.54 in. $\times$ 21.54 in.

(e) To minimize the cost of material needed for construction

61. (a) $C(x) = 0.12\pi r^2 + \dfrac{40}{r}$

63. (a) $D(p) = \dfrac{429}{p}$ **(b)** 143   **65.** 450 cm^3 **67.** 124.76 pounds

69. $V = \pi r^2 h$ **71.** $\sqrt[3]{6} \approx 1.82$ in.

73. 900 foot-lb   **75.** 384 psi

77. No. Each function is a quotient of polynomials, but it is not written in lowest terms. Each function is undefined for $x = 1$; each graph has a hole at $x = 1$.

(b)
60

0   10
0

The cost is smallest when $r \approx 3.76$ cm.

3.5 Assess Your Understanding *(page 365)*

2. T   **3.** $\{x|-2 < x < 5\}; (-2, 5)$ **5.** $\{x|x < 0 \text{ or } x > 4\}; (-\infty, 0) \text{ or } (4, \infty)$ **7.** $\{x|-3 < x < 3\}; (-3, 3)$

9. $\{x|x \le 4 \text{ or } x \ge 3\}; (-\infty, -4] \text{ or } [3, \infty)$ **11.** $\left\{x\left|-\dfrac{1}{2} < x < 3\right.\right\}; \left(-\dfrac{1}{2}, 3\right)$ **13.** $\{x|x < -1 \text{ or } x > 8\}; (-\infty, -1) \text{ or } (8, \infty)$

15. No real solution   **17.** $\left\{x\left|x \le -\dfrac{2}{3} \text{ or } x \ge \dfrac{3}{2}\right.\right\}; \left(-\infty, -\dfrac{2}{3}\right] \text{ or } \left[\dfrac{3}{2}, \infty\right)$ **19.** $\{x|x > 1\}; (1, \infty)$ **21.** $\{x|x < 1 \text{ or } 2 < x < 3\}; (-\infty, 1) \text{ or } (2, 3)$

23. $\{x|-1 \le x \le 0 \text{ or } x \ge 3\}; [-1, 0] \text{ or } [3, \infty)$ **25.** $\{x|x < -1 \text{ or } x > 1\}; (-\infty, -1) \text{ or } (1, \infty)$ **27.** $\{x|x > 1\}; (1, \infty)$

29. $\{x|x < -1 \text{ or } x > 1\}; (-\infty, -1) \text{ or } (1, \infty)$ **31.** $\{x|-1 < x < 8\}; (-1, 8)$ **33.** $\{x|x < -2 \text{ or } x > 2\}; (-\infty, -2) \text{ or } (2, \infty)$

35. $\{x|x < -1 \text{ or } x > 1\}; (-\infty, -1) \text{ or } (1, \infty)$ **37.** $\{x|x < -1 \text{ or } 0 < x < 1\}; (-\infty, -1) \text{ or } (0, 1)$ **39.** $\{x|x < -1 \text{ or } x > 1\}; (-\infty, -1) \text{ or } (1, \infty)$

41. $\left\{x\left|x < -\dfrac{2}{3} \text{ or } 0 < x < \dfrac{3}{2}\right.\right\}; \left(-\infty, -\dfrac{2}{3}\right) \text{ or } \left(0, \dfrac{3}{2}\right)$ **43.** $\{x|x < 2\}; (-\infty, 2)$ **45.** $\{x|-2 < x \le 9\}; (-2, 9]$

47. $\{x|x < 2 \text{ or } 3 < x < 5\}; (-\infty, 2) \text{ or } (3, 5)$ **49.** $\{x|x < -3 \text{ or } -1 < x < 1 \text{ or } x > 2\}; (-\infty, -3) \text{ or } (-1, 1) \text{ or } (2, \infty)$

51. $\{x|x < -5 \text{ or } -4 < x < -3 \text{ or } x > 1\}; (-\infty, -5) \text{ or } (-4, -3) \text{ or } (1, \infty)$ **53.** $\left\{x\left|x \le -\dfrac{1}{2} \text{ or } 1 \le x < 4\right.\right\}; \left(-\infty, -\dfrac{1}{2}\right] \text{ or } [1, 4)$

55. $\left\{x\left|\dfrac{-3 - \sqrt{13}}{2} < x < -3 \text{ or } x > \dfrac{-3 + \sqrt{13}}{2}\right.\right\}; \left(\dfrac{-3 - \sqrt{13}}{2}, -3\right) \text{ or } \left(\dfrac{-3 + \sqrt{13}}{2}, \infty\right)$ **57.** $\{x|x > 4\}; (4, \infty)$

59. $\{x|x \le -4 \text{ or } x \ge 4\}; (-\infty, -4] \text{ or } [4, \infty)$ **61.** $\{x|x < -4 \text{ or } x \ge 2\}; (-\infty, -4) \text{ or } [2, \infty)$

63. (a) The ball is more than 96 feet above the ground from time t between 2 and 3 seconds, $2 < t < 3$.

(b)
100

0   5
0

(c) 100 ft

(d) 2.5 sec

65. (a) For a profit of at least \$50, between 8 and 32 watches must be sold, $8 \le x \le 32$.   **(b)** 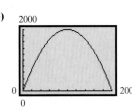   **(c)** \$2000   **(d)** 100   **(e)** 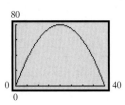   **(f)** \$80   **(g)** 20

67. Chevy can produce at most 8 Cavaliers in an hour, assuming that cars cannot be partially completed in an hour.

Historical Problems (page 378)

1.
$$\left(x - \frac{b}{3}\right)^3 + b\left(x - \frac{b}{3}\right)^2 + c\left(x - \frac{b}{3}\right) + d = 0$$

$$x^3 - bx^2 + \frac{b^2 x}{3} - \frac{b^3}{27} + bx^2 - \frac{2b^2 x}{3} + \frac{b^3}{9} + cx - \frac{bc}{3} + d = 0$$

$$x^3 + \left(c - \frac{b^2}{3}\right)x + \left(\frac{2b^3}{27} - \frac{bc}{3} + d\right) = 0$$

Let $p = c - \dfrac{b^2}{3}$ and $q = \dfrac{2b^3}{27} - \dfrac{bc}{3} + d$. Then $x^3 + px + q = 0$.

2.
$$(H + K)^3 + p(H + K) + q = 0$$
$$H^3 + 3H^2 K + 3HK^2 + K^3 + pH + pK + q = 0$$
Let $3HK = -p$.
$$H^3 - pH - pK + K^3 + pH + pK + q = 0$$
$$H^3 + K^3 = -q$$

3.
$$3HK = -p$$
$$K = -\frac{p}{3H}$$
$$H^3 + \left(-\frac{p}{3H}\right)^3 = -q$$
$$H^3 - \frac{p^3}{27H^3} = -q$$
$$27H^6 - p^3 = -27qH^3$$
$$27H^6 + 27qH^3 - p^3 = 0$$
$$H^3 = \frac{-27q \pm \sqrt{(27q)^2 - 4(27)(-p^3)}}{2 \cdot 27}$$
$$H^3 = \frac{-q}{2} \pm \sqrt{\frac{27^2 q^2}{2^2(27^2)} + \frac{4(27)p^3}{2^2(27^2)}}$$
$$H^3 = \frac{-q}{2} \pm \sqrt{\frac{q^2}{4} + \frac{p^3}{27}}$$
$$H = \sqrt[3]{\frac{-q}{2} + \sqrt{\frac{q^2}{4} + \frac{p^3}{27}}}$$

4. $\;H^3 + K^3 = -q$
$$K^3 = -q - H^3$$
$$K^3 = -q - \left[\frac{-q}{2} + \sqrt{\frac{q^2}{4} + \frac{p^3}{27}}\right]$$
$$K^3 = \frac{-q}{2} - \sqrt{\frac{q^2}{4} + \frac{p^3}{27}}$$
$$K = \sqrt[3]{\frac{-q}{2} - \sqrt{\frac{q^2}{4} + \frac{p^3}{27}}}$$

Choose the positive root for now.

5. $x = H + K$
$$x = \sqrt[3]{\frac{-q}{2} + \sqrt{\frac{q^2}{4} + \frac{p^3}{27}}} + \sqrt[3]{\frac{-q}{2} - \sqrt{\frac{q^2}{4} + \frac{p^3}{27}}}$$ (Note that if we had used the negative root in 3 the result would be the same.)

6. $x = 3$ **7.** $x = 2$ **8.** $x = 2$

3.6 Assess Your Understanding (page 378)

5. Remainder; dividend   **6.** $f(c)$ **7.** -4 **8.** F   **9.** F   **10.** T   **11.** No; $f(3) = 61$ **13.** No; $f(1) = 2$
15. Yes; $f(x) = (x + 2)(3x^5 - 6x^4 + 12x^3 - 22x^2 + 44x - 88)$ **17.** Yes; $f(x) = (x - 4)(4x^5 + 16x^4 + x + 4)$

19. No; $f\left(-\dfrac{1}{2}\right) = -\dfrac{7}{4}$ **21.** $4; \pm 1, \pm\dfrac{1}{3}$ **23.** $5; \pm 1, \pm 3$ **25.** $3; \pm 1, \pm 2, \pm\dfrac{1}{4}, \pm\dfrac{1}{2}$ **27.** $4; \pm 1, \pm 2, \pm\dfrac{1}{3}, \pm\dfrac{2}{3}$ **29.** $5; \pm 1, \pm 2, \pm 4, \pm\dfrac{1}{2}$

31. $4; \pm 1, \pm 2, \pm\dfrac{1}{6}, \pm\dfrac{1}{3}, \pm\dfrac{1}{2}, \pm\dfrac{2}{3}$

33. -1 and 1 **35.** -12 and 12 **37.** -10 and 10

39. $-3, -1, 2; f(x) = (x + 3)(x + 1)(x - 2)$ **41.** $\frac{1}{2}, 3, 3; f(x) = (2x - 1)(x - 3)^2$ **43.** $-\frac{1}{3}; f(x) = (3x + 1)(x^2 + x + 1)$

45. $3, \dfrac{5 + \sqrt{17}}{2}, \dfrac{5 - \sqrt{17}}{2}; f(x) = (x - 3)\left(x - \left(\dfrac{5 + \sqrt{17}}{2}\right)\right)\left(x - \left(\dfrac{5 - \sqrt{17}}{2}\right)\right)$

47. $-2, -1, 1, 1; f(x) = (x + 2)(x + 1)(x - 1)^2$ **49.** $-5, -3, -\dfrac{3}{2}, 1; f(x) = (x + 5)(x + 3)(2x + 3)(x - 1)$

51. $-2, -\dfrac{3}{2}, 1, 4; f(x) = (x + 2)(2x + 3)(x - 1)(x - 4)$ **53.** $-\dfrac{1}{2}, \dfrac{1}{2}; f(x) = (2x + 1)(2x - 1)(x^2 + 2)$

55. $\dfrac{\sqrt{2}}{2}, -\dfrac{\sqrt{2}}{2}, 2; f(x) = (x - 2)(2x - \sqrt{2})(2x + \sqrt{2})\left(x^2 + \dfrac{1}{2}\right)$ **57.** $-5.9, -0.3, 3$ **59.** $-3.8, 4.5$ **61.** $-43.5, 1, 23$ **63.** $\{-1, 2\}$

65. $\left\{\dfrac{2}{3}, -1 + \sqrt{2}, -1 - \sqrt{2}\right\}$ **67.** $\left\{\dfrac{1}{3}, \sqrt{5}, -\sqrt{5}\right\}$ **69.** $\{-3, -2\}$ **71.** $\left\{-\dfrac{1}{3}\right\}$ **73.** $f(0) = -1; f(1) = 10;$ Zero: 0.22

75. $f(-5) = -58; f(-4) = 2;$ Zero: -4.05 **77.** $f(1.4) = -0.17536; f(1.5) = 1.40625;$ Zero: 1.41 **79.** ≈ 27 Cavaliers   **81.** $k = 5$ **83.** -7

85. 5 **87.** 7 in.   **89.** If $f(x) = x^n - c^n$, then $f(c) = c^n - c^n = 0$, so $x - c$ is a factor of f.

91. All the potential rational zeros are integers. Hence, r is either an integer or is not a rational zero (and is therefore irrational).

3.7 Assess Your Understanding (page 385)

3. one   **4.** $3 - 4i$ **5.** T   **6.** F   **7.** $4 + i$ **9.** $-i, 1 - i$ **11.** $-i, -2i$ **13.** $-i$ **15.** $2 - i, -3 + i$

17. $f(x) = x^4 - 14x^3 + 77x^2 - 200x + 208; a = 1$ **19.** $f(x) = x^5 - 4x^4 + 7x^3 - 8x^2 + 6x - 4; a = 1$

21. $f(x) = x^4 - 6x^3 + 10x^2 - 6x + 9; a = 1$ **23.** $-2i, 4$ **25.** $2i, -3, \dfrac{1}{2}$ **27.** $3 + 2i, -2, 5$ **29.** $4i, -\sqrt{11}, \sqrt{11}, -\dfrac{2}{3}$

31. $1, -\dfrac{1}{2} - \dfrac{\sqrt{3}}{2}i, -\dfrac{1}{2} + \dfrac{\sqrt{3}}{2}i; f(x) = (x - 1)\left(x + \dfrac{1}{2} + \dfrac{\sqrt{3}}{2}i\right)\left(x + \dfrac{1}{2} - \dfrac{\sqrt{3}}{2}i\right)$

33. $2, 3 - 2i, 3 + 2i; f(x) = (x - 2)(x - 3 + 2i)(x - 3 - 2i)$ **35.** $-i, i, -2i, 2i; f(x) = (x + i)(x - i)(x + 2i)(x - 2i)$

37. $-5i, 5i, -3, 1; f(x) = (x + 5i)(x - 5i)(x + 3)(x - 1)$ **39.** $-4, \dfrac{1}{3}, 2 - 3i, 2 + 3i; f(x) = 3(x + 4)\left(x - \dfrac{1}{3}\right)(x - 2 + 3i)(x - 2 - 3i)$

41. Zeros that are complex numbers must occur in conjugate pairs; or a polynomial with real coefficients of odd degree must have at least one real zero.
43. If the remaining zero were a complex number, then its conjugate would also be a zero, creating a polynomial of degree 5.

Review Exercises (page 388)

1.

3.

5.

7.

9.

11.

13.

$x = -\frac{1}{3}$

$\left(-\frac{1}{3}, \frac{1}{2}\right)$ (0, 1)

15.

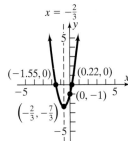

$x = -\frac{2}{3}$

(−1.55, 0)   (0.22, 0)

(0, −1)

$\left(-\frac{2}{3}, -\frac{7}{3}\right)$

17. Minimum value; 1   **19.** Maximum value; 12   **21.** Maximum value; 16   **23.** Polynomial of degree 5   **25.** Not a polynomial

27.

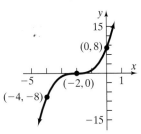

(0, 8)

(−2, 0)

(−4, −8)

29.

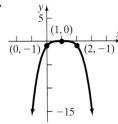

(1, 0)

(0, −1)   (2, −1)

31.

(−2, 4)   (0, 4)

(−1, 2)

33. (a) *x*-intercepts: −4, −2, 0; *y*-intercept: 0   **(b)** −4, −2, 0: Crosses   **(c)** $y = x^3$

(d)

(e) 2; (−3.15, 3.08), (−0.85, −3.08)   **(g)** Domain: $(-\infty, \infty)$

(f)

(−3.15, 3.08)

(−4, 0)   (−2, 0)   (0, 0)

(−0.85, −3.08)

Range: $(-\infty, \infty)$

(h) Increasing on $(-\infty, -3.15)$ and $(-0.85, \infty)$

Decreasing on $(-3.15, -0.85)$

35. (a) *x*-intercepts: −4, 2; *y*-intercept: 16
(b) −4: Crosses; 2: Touches
(c) $y = x^3$

(d)

(e) 2; (−2, 32), (2, 0)
(f)

(−2, 32)   (0, 16)

(−4, 0)

(2, 0)

(g) Domain: $(-\infty, \infty)$
Range: $(-\infty, \infty)$
(h) Increasing on $(-\infty, -2)$ and $(2, \infty)$
Decreasing on $(-2, 2)$

37. $f(x) = x^3 - 4x^2 = x^2(x - 4)$
(a) *x*-intercepts: 0, 4; *y*-intercept: 0
(b) 0: Touches; 4: Crosses
(c) $y = x^3$
(d)

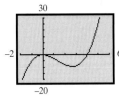

(e) 2; (0, 0), (2.67, −9.48)
(f)

(0, 0)   (4, 0)

(2.67, −9.48)

(g) Domain: $(-\infty, \infty)$
Range: $(-\infty, \infty)$
(h) Increasing on $(-\infty, 0)$ and $(2.67, \infty)$
Decreasing on $(0, 2.67)$

39. (a) *x*-intercepts: −3, −1, 1;
y-intercept: 3
(b) −3, −1: Crosses; 1: Touches
(c) $y = x^4$
(d)

(e) 3; (−2.28, −9.91), (−0.22, 3.23), (1, 0)
(f)

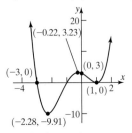

(−0.22, 3.23)

(−3, 0)   (0, 3)

(1, 0)

(−2.28, −9.91)

(g) Domain: $(-\infty, \infty)$
Range: $(-\infty, \infty)$
(h) Increasing on $(-2.28, -0.22)$ and $(1, \infty)$
Decreasing on $(-\infty, -2.28)$ and $(-0.22, 1)$

41. Domain: $\{x \mid x \neq -3, x \neq 3\}$; Horizontal asymptote: $y = 0$; Vertical asymptotes: $x = -3, x = 3$
43. Domain: $\{x \mid x \neq -2\}$; Horizontal asymptote: $y = 1$; Vertical asymptote: $x = -2$

45. 1. Domain: $\{x \mid x \neq 0\}$
  2. $R(x)$ is in lowest terms.
  3. x-intercept: 3; no y-intercept
  4. No symmetry
  5. Vertical asymptote: $x = 0$
  6. Horizontal asymptote: $y = 2$, not intersected

7.

8.

47. 1. Domain: $\{x \mid x \neq 0, x \neq 2\}$
  2. $H(x)$ is in lowest terms
  3. x-intercept: -2; no y-intercept
  4. No symmetry
  5. Vertical asymptotes: $x = 0, x = 2$
  6. Horizontal asymptote: $y = 0$, intersected at $(-2, 0)$

7.

8.

49. 1. Domain: $\{x \mid x \neq -2, x \neq 3\}$
  2. $R(x) = \dfrac{(x + 3)(x - 2)}{(x - 3)(x + 2)}$
  3. x-intercepts: $-3, 2$; y-intercept: 1
  4. No symmetry
  5. Vertical asymptotes: $x = -2, x = 3$
  6. Horizontal asymptote: $y = 1$, intersected at $(0, 1)$

7.

8.

51. 1. Domain: $\{x \mid x \neq -2, x \neq 2\}$
  2. $F(x) = \dfrac{x^3}{(x + 2)(x - 2)}$
  3. Intercept: $(0, 0)$
  4. Symmetric with respect to the origin
  5. Vertical asymptotes: $x = -2, x = 2$
  6. Oblique asymptote: $y = x$, intersected at $(0, 0)$

7.

8.

53. 1. Domain: $\{x \mid x \neq 1\}$
  2. $R(x)$ is in lowest terms
  3. Intercept: $(0, 0)$
  4. No symmetry
  5. Vertical asymptote: $x = 1$
  6. No oblique or horizontal asymptote

7.

8.

55. 1. Domain: $\{x | x \neq -1, x \neq 2\}$

2. $G(x) = \dfrac{x+2}{x+1}$

3. x-intercept: -2; y-intercept: 2

4. No symmetry

5. Vertical asymptote: $x = -1$; hole at $\left(2, \dfrac{4}{3}\right)$

6. Horizontal asymptote: $y = 1$, not intersected

7.

8.

57. $\left\{x \middle| -4 < x < \dfrac{3}{2}\right\}; \left(-4, \dfrac{3}{2}\right)$ **59.** $\{x | -3 < x \le 3\}; (-3, 3]$ **61.** $\{x | x < 1 \text{ or } x > 2\}; (-\infty, 1) \text{ or } (2, \infty)$

63. $\{x | 1 < x < 2 \text{ or } x > 3\}; (1, 2) \text{ or } (3, \infty)$ **65.** $\{x | x < -4 \text{ or } 2 < x < 4 \text{ or } x > 6\}; (-\infty, -4) \text{ or } (2, 4) \text{ or } (6, \infty)$

67. $R = 10$; g is not a factor of f. **69.** $R = 0$; g is a factor of f. **71.** $f(4) = 47{,}105$ **73.** $8; \pm\dfrac{1}{2}, \pm 1, \pm\dfrac{3}{2}, \pm 3$ **75.** $-2, 1, 4$ **77.** $\dfrac{1}{2}$, multiplicity 2; -2

79. 2, multiplicity 2 **81.** $-2.5, 3.1, 5.32$ **83.** $-11.3, -0.6, 4, 9.33$ **85.** $\{-3, 2\}$ **87.** $\left\{-3, -1, -\dfrac{1}{2}, 1\right\}$

89. -5 and 5 **91.** $-\dfrac{37}{2}$ and $\dfrac{37}{2}$

93. $f(0) = -1, f(1) = 1; 0.85$ **95.** $f(0) = -1; f(1) = 1; 0.94$ **97.** $4 - i; f(x) = x^3 - 14x^2 + 65x - 102$

99. $-i, 1 - i; f(x) = x^4 - 2x^3 + 3x^2 - 2x + 2$ **101.** $\left\{-\dfrac{1}{2} - \dfrac{\sqrt{3}}{2}i, -\dfrac{1}{2} + \dfrac{\sqrt{3}}{2}i\right\}$

103. $\left\{\dfrac{-1 - \sqrt{17}}{4}, \dfrac{-1 + \sqrt{17}}{4}\right\}$ **105.** $\left\{\dfrac{1}{2} - \dfrac{\sqrt{11}}{2}i, \dfrac{1}{2} + \dfrac{\sqrt{11}}{2}i\right\}$

107. $\left\{\dfrac{1}{2} - \dfrac{\sqrt{23}}{2}i, \dfrac{1}{2} + \dfrac{\sqrt{23}}{2}i\right\}$ **109.** $\{-\sqrt{2}, \sqrt{2}, -2i, 2i\}$ **111.** $\{-3, 2\}$ **113.** $\left\{\dfrac{1}{3}, 1, -i, i\right\}$ **115.** $(2, 2)$ **117.** $4{,}166{,}666.7 \text{ m}^2$

119. The side with the semicircles should be $\dfrac{50}{\pi}$ ft; the other side should be 25 ft. **121. (a)** 63 clubs **(b)** \$151.90

123. 199.9 pounds

125. **(a)** Quadratic, $a < 0$. **(b)** $R(A) = -7.760A^2 + 411.875A + 942.721$

(c) About \$26.5 thousand

(d) \$6408 thousand

(e)

127. (a) $A(r) = 2\pi r^2 + \dfrac{500}{r}$ **(b)** 223.22 cm^2 **(c)** 257.08 cm^2 **(d)** A is smallest when $r \approx 3.41$ cm.

Chapter Test *(page 392)*

1.

2. (a) The leading coefficient (i.e., the coefficient on x^2) is positive so the graph will open up. Thus, the graph has a minimum.

(b) $(2, -8)$

(c) $x = 2$

(d) $(0, 4), \left(\dfrac{6 - 2\sqrt{6}}{3}, 0\right), \left(\dfrac{6 + 2\sqrt{6}}{3}, 0\right).$

(e)

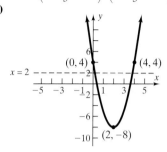

3. (a) 3

(b) Every zero of g lies between -15 and 15.

(c) $\dfrac{p}{q}$: $\pm\dfrac{1}{2}, \pm 1, \pm\dfrac{3}{2}, \pm\dfrac{5}{2}, \pm 3, \pm 5, \pm\dfrac{15}{2}, \pm 15$

(d) $-5, -\dfrac{1}{2}, 3; g(x) = (x + 5)(2x + 1)(x - 3)$

4. $4, -5i, 5i$

5. $\left\{1, \dfrac{5 - \sqrt{61}}{6}, \dfrac{5 + \sqrt{61}}{6}\right\}$

6. Domain: $\{x | x \neq -10, x \neq 4\}$
Asymptotes: $x = -10, y = 2$

7. Domain: $\{x | x \neq -1\}$;
Asymptotes: $x = -1, y = x + 1$

8.

9. Answers may vary. One possibility is
$f(x) = x^4 - 4x^3 - 2x^2 + 20x$

10. Answers may vary. One possibility is $r(x) = \dfrac{2(x - 9)(x - 1)}{(x - 4)(x - 9)}$

11. $f(0) = 8; f(4) = -36$
Since $f(0) = 8 > 0$ and $f(4) = -36 < 0$, the Intermediate Value Theorem guarantees that there is at least one real zero between 0 and 4.

12. $\{x | -\infty < x \leq -3 \text{ or } 2 \leq x < \infty\}$, or $(-\infty, -3] \cup [2, \infty)$

13. $\{x | x < 3 \text{ or } x > 8\}$, or $(-\infty, 3) \cup (8, \infty)$

14. (a) $(0, -4), (-2, 0), (1, 0)$

(d) Because the power function has an odd degree, we expect the ends of the graph to go in opposite directions. The leading coefficient is negative, so the left side of the graph will go up and the right side will go down.

Reading the graph from left to right, we would expect to see it decreasing and cross the x-axis at $x = -2$. Somewhere between $x = -2$ and $x = 1$ the graph turns so that it is increasing, touches the x-axis at $x = 1$, turns around and decreases from that point on.

(b) Touches at 1 and crosses at -2.   **(c)** $y = -2x^3$

(e) $(-1, -8)$, and $(1, 0)$.

(f)

15. (a) $A(x) = -0.604x^2 + 6.038x + 25.350$

(b) The model predicts that the average 2004 home game attendance for the St. Louis Cardinals will be 30,768.

Cumulative Review *(page 393)*

1. $\sqrt{26}$

2. $\{x|x \le 0 \text{ or } x \ge 1\}$ or $(-\infty, 0] \cup [1, \infty)$

3. $\{x|-1 < x < 4\}$ or $(-1, 4)$

4. $f(x) = -3x + 1$

5. $y = 2x - 1$

6.

7. Not a function; 3 has two images **8.** $\{0, 2, 4\}$ **9.** $\left\{x \middle| x \ge \dfrac{3}{2}\right\}; \left[\dfrac{3}{2}, \infty\right)$ **10.** Center: $(-2, 1)$; Radius: 3

11. x-intercepts: $-3, 0, 3$
y-intercept: 0
Symmetric with respect to the origin

12. $y = -\dfrac{2}{3}x + \dfrac{17}{3}$

13. Not a function; it fails the Vertical Line Test.

14. (a) 22 **(b)** $x^2 - 5x - 2$ **(c)** $-x^2 - 5x + 2$ **(d)** $9x^2 + 15x - 2$ **(e)** $2x + h + 5$

15. (a) $\{x|x \ne 1\}$ **(b)** No, $(2, 7)$ is on the graph. **(c)** 4; $(3, 4)$ is on the graph. **(d)** $\dfrac{7}{4}$; $\left(\dfrac{7}{4}, 9\right)$ is on the graph.

16.

17.

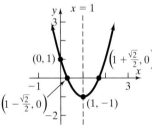

18. $x + 4$; $m_{\text{sec}} = 6$

19. (a) x-intercepts: $-5, -1, 5$; y-intercept: -3
(b) No symmetry
(c) Neither
(d) Increasing: $(-\infty, -3)$ and $(2, \infty)$
Decreasing: $(-3, 2)$
(e) Local maximum is 5 and occurs at $x = -3$.
(f) Local minimum is -6 and occurs at $x = 2$.

20. Odd **21. (a)** Domain: $\{x|-3 < x\}$ or $(-3, \infty)$
(b) x-intercept: $-\dfrac{1}{2}$; y-intercept: 1
(c)

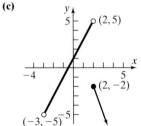

(d) Range: $\{y|y < 5\}$ or $(-\infty, 5)$

22.

23. (a) $(f + g)(x) = x^2 - 9x - 6$; domain: all real numbers
(b) $\left(\dfrac{f}{g}\right)(x) = \dfrac{x^2 - 5x + 1}{-4x - 7}$; domain: $\left\{x \middle| x \ne -\dfrac{7}{4}\right\}$

24. (a) $R(x) = -\dfrac{1}{10}x^2 + 150x$ **(b)** $\$14,000$ **(c)** 750; $\$56,250$ **(d)** $\$75$

C H A P T E R   4   Exponential and Logarithmic Functions

4.1 Assess Your Understanding (page 401)

4. $(g \circ f)(x)$ 5. F   6. F   7. (a) -1 (b) -1 (c) 8   (d) 0   (e) 8   (f) -7

9. (a) 4   (b) 5   (c) -1 (d) -2 11. (a) 98   (b) 49   (c) 4   (d) 4   13. (a) 97   (b) $-\dfrac{163}{2}$

(c) 1   (d) $-\dfrac{3}{2}$ 15. (a) $2\sqrt{2}$ (b) $2\sqrt{2}$ (c) 1   (d) 0   17. (a) $\dfrac{1}{17}$ (b) $\dfrac{1}{5}$ (c) 1   (d) $\dfrac{1}{2}$

19. (a) $\dfrac{3}{\sqrt[3]{4}+1}$ (b) 1   (c) $\dfrac{6}{5}$ (d) 0   21. $\{x | x \neq 0, x \neq 2\}$ 23. $\{x | x \neq -4, x \neq 0\}$ 25. $\left\{x \middle| x \geq -\dfrac{3}{2}\right\}$ 27. $\{x | x \geq 1\}$

29. (a) $(f \circ g)(x) = 6x + 3$; All real numbers   (b) $(g \circ f)(x) = 6x + 9$; All real numbers   (c) $(f \circ f)(x) = 4x + 9$; All real numbers
(d) $(g \circ g)(x) = 9x$; All real numbers   31. (a) $(f \circ g)(x) = 3x^2 + 1$; All real numbers   (b) $(g \circ f)(x) = 9x^2 + 6x + 1$; All real numbers
(c) $(f \circ f)(x) = 9x + 4$; All real numbers   (d) $(g \circ g)(x) = x^4$; All real numbers   33. (a) $(f \circ g)(x) = x^4 + 8x^2 + 16$; All real numbers
(b) $(g \circ f)(x) = x^4 + 4$; All real numbers   (c) $(f \circ f)(x) = x^4$; All real numbers   (d) $(g \circ g)(x) = x^4 + 8x^2 + 20$; All real numbers

35. (a) $(f \circ g)(x) = \dfrac{3x}{2 - x}$; $\{x | x \neq 0, x \neq 2\}$ (b) $(g \circ f)(x) = \dfrac{2(x - 1)}{3}$; $\{x | x \neq 1\}$ (c) $(f \circ f)(x) = \dfrac{3(x - 1)}{4 - x}$; $\{x | x \neq 1, x \neq 4\}$

(d) $(g \circ g)(x) = x$; $\{x | x \neq 0\}$ 37. (a) $(f \circ g)(x) = \dfrac{4}{4 + x}$; $\{x | x \neq -4, x \neq 0\}$ (b) $(g \circ f)(x) = \dfrac{-4(x - 1)}{x}$; $\{x | x \neq 0, x \neq 1\}$

(c) $(f \circ f)(x) = x$; $\{x | x \neq 1\}$ (d) $(g \circ g)(x) = x$; $\{x | x \neq 0\}$ 39. (a) $(f \circ g)(x) = \sqrt{2x + 3}$; $\left\{x \middle| x \geq -\dfrac{3}{2}\right\}$ (b) $(g \circ f)(x) = 2\sqrt{x} + 3$; $\{x | x \geq 0\}$

(c) $(f \circ f)(x) = \sqrt[4]{x}$; $\{x | x \geq 0\}$ (d) $(g \circ g)(x) = 4x + 9$; All real numbers   41. (a) $(f \circ g)(x) = x$; $\{x | x \geq 1\}$ (b) $(g \circ f)(x) = |x|$; All real numbers
(c) $(f \circ f)(x) = x^4 + 2x^2 + 2$; All real numbers   (d) $(g \circ g)(x) = \sqrt{\sqrt{x - 1} - 1}$; $\{x | x \geq 2\}$ 43. (a) $(f \circ g)(x) = acx + ad + b$; All real numbers
(b) $(g \circ f)(x) = acx + bc + d$; All real numbers   (c) $(f \circ f)(x) = a^2x + ab + b$; All real numbers   (d) $(g \circ g)(x) = c^2x + cd + d$; All real numbers

45. $(f \circ g)(x) = f(g(x)) = f\left(\dfrac{1}{2}x\right) = 2\left(\dfrac{1}{2}x\right) = x$; $(g \circ f)(x) = g(f(x)) = g(2x) = \dfrac{1}{2}(2x) = x$ 47. $(f \circ g)(x) = f(g(x)) = f(\sqrt[3]{x}) = (\sqrt[3]{x})^3 = x$;

$(g \circ f)(x) = g(f(x)) = g(x^3) = \sqrt[3]{x^3} = x$ 49. $(f \circ g)(x) = f(g(x)) = f\left(\dfrac{1}{2}(x + 6)\right) = 2\left[\dfrac{1}{2}(x + 6)\right] - 6 = x + 6 - 6 = x$;

$(g \circ f)(x) = g(f(x)) = g(2x - 6) = \dfrac{1}{2}(2x - 6 + 6) = x$ 51. $(f \circ g)(x) = f(g(x)) = f\left(\dfrac{1}{a}(x - b)\right) = a\left[\dfrac{1}{a}(x - b)\right] + b = x$;

$(g \circ f)(x) = g(f(x)) = g(ax + b) = \dfrac{1}{a}(ax + b - b) = x$ 53. $f(x) = x^4$; $g(x) = 2x + 3$ (Other answers are possible.)   55. $f(x) = \sqrt{x}$;

$g(x) = x^2 + 1$ (Other answers are possible.)   57. $f(x) = |x|$; $g(x) = 2x + 1$ (Other answers are possible.)   59. $(f \circ g)(x) = 11$; $(g \circ f)(x) = 2$

61. $-3, 3$ 63. $S(t) = \dfrac{16}{9}\pi t^6$ 65. $C(t) = 15,000 + 800,000t - 40,000t^2$ 67. $C(p) = \dfrac{2\sqrt{100 - p}}{25} + 600, 0 \leq p \leq 100$ 69. $V(r) = 2\pi r^3$

71. (a) $f(x) = 0.857118x$ (b) $g(x) = 128.6054x$ (c) $g(f(x)) = g(0.857118x) = 110.23x$ (d) 110,230.00 yen

73. f is an odd function, so $f(-x) = -f(x)$. g is an even function, so $g(-x) = g(x)$. Then $(f \circ g)(-x) = f(g(-x)) = f(g(x)) = (f \circ g)(x)$. So $f \circ g$
is even. Also, $(g \circ f)(-x) = g(f(-x)) = g(-f(x)) = g(f(x)) = (g \circ f)(x)$, so $g \circ f$ is even.

4.2 Assess Your Understanding (page 415)

4. One-to-one   5. $y = x$ 6. $[4, \infty)$ 7. F   8. T   9. One-to-one   11. Not one-to-one   13. Not one-to-one   15. One-to-one   17. One-to-one
19. Not one-to-one   21. One-to-one

23.
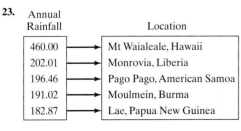

Domain: $\{460.00, 202.01, 196.46, 191.02; 182.87\}$

Range: $\{$Mt. Waialeale, Monrovia, Pago Pago, Moulmein, Lae$\}$

25.

Monthly Cost of Life Insurance		Age
$7.09	$\rightarrow$	30
$8.40	$\rightarrow$	40
$11.29	$\rightarrow$	45

Domain: $\{\$7.09, \$8.04, \$11.29\}$

Range: $\{30, 40, 45\}$

27. $\{(5, -3), (9, -2), (2, -1), (11, 0), (-5, 1)\}$
Domain: $\{5, 9, 2, 11, -5\}$
Range: $\{-3, -2, -1, 0, 1\}$

29. $\{(1, -2), (2, -3), (0, -10), (9, 1), (4, 2)\}$
Domain: $\{1, 2, 0, 9, 4\}$
Range: $\{-2, -3, -10, 1, 2\}$

31.

33.

35.

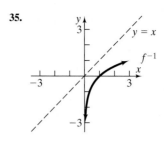

37. $f(g(x)) = f\left(\dfrac{1}{3}(x-4)\right) = 3\left[\dfrac{1}{3}(x-4)\right] + 4$

$\qquad = (x-4) + 4 = x;$

$\quad g(f(x)) = g(3x+4) = \dfrac{1}{3}[(3x+4) - 4] = \dfrac{1}{3}(3x) = x$

39. $f(g(x)) = fx\left(\dfrac{x}{4} + 2\right) = 4\left[\dfrac{x}{4} + 2\right] - 8 = (x+8) - 8 = x;$

$\quad g(f(x)) = g(4x-8) = \dfrac{4x-8}{4} + 2 = (x-2) + 2 = x$

41. $f(g(x)) = f(\sqrt[3]{x+8}) = (\sqrt[3]{x+8})^3 - 8 = (x+8) - 8 = x;$

$\quad g(f(x)) = g(x^3 - 8) = \sqrt[3]{(x^3-8)+8} = \sqrt[3]{x^3} = x$

43. $f(g(x)) = f\left(\dfrac{1}{x}\right) = \dfrac{1}{\left(\dfrac{1}{x}\right)} = x; g(f(x)) = g\left(\dfrac{1}{x}\right) = \dfrac{1}{\left(\dfrac{1}{x}\right)} = x$

45. $f(g(x)) = f\left(\dfrac{4x-3}{2-x}\right) = \dfrac{2\left(\dfrac{4x-3}{2-x}\right) + 3}{\dfrac{4x-3}{2-x} + 4}$

$\qquad = \dfrac{2(4x-3) + 3(2-x)}{4x-3 + 4(2-x)} = \dfrac{5x}{5} = x;$

$\quad g(f(x)) = g\left(\dfrac{2x+3}{x+4}\right) = \dfrac{4\left(\dfrac{2x+3}{x+4}\right) - 3}{2 - \dfrac{2x+3}{x+4}}$

$\qquad = \dfrac{4(2x+3) - 3(x+4)}{2(x+4) - (2x+3)} = \dfrac{5x}{5} = x$

47. $f^{-1}(x) = \dfrac{1}{3}x$

$\quad f(f^{-1}(x)) = f\left(\dfrac{1}{3}x\right) = 3\left(\dfrac{1}{3}x\right) = x$

$\quad f^{-1}(f(x)) = f^{-1}(3x) = \dfrac{1}{3}(3x) = x$

Domain f = Range f^{-1} = All real numbers

Range f = Domain f^{-1} = All real numbers

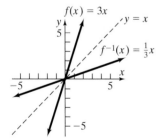

49. $f^{-1}(x) = \dfrac{x}{4} - \dfrac{1}{2}$

$\quad f(f^{-1}(x)) = f\left(\dfrac{x}{4} - \dfrac{1}{2}\right) = 4\left(\dfrac{x}{4} - \dfrac{1}{2}\right) + 2 = (x-2) + 2 = x$

$\quad f^{-1}(f(x)) = f^{-1}(4x+2) = \dfrac{4x+2}{4} - \dfrac{1}{2} = \left(x + \dfrac{1}{2}\right) - \dfrac{1}{2} = x$

Domain f = Range f^{-1} = All real numbers

Range f = Domain f^{-1} = All real numbers

51. $f^{-1}(x) = \sqrt[3]{x + 1}$

$f(f^{-1}(x)) = f(\sqrt[3]{x + 1}) = (\sqrt[3]{x + 1})^3 - 1 = x$

$f^{-1}(f(x)) = f^{-1}(x^3 - 1) = \sqrt[3]{(x^3 - 1) + 1} = x$

Domain f = Range f^{-1} = All real numbers

Range f = Domain f^{-1} = All real numbers

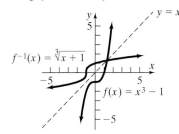

53. $f^{-1}(x) = \sqrt{x - 4}, x \geq 4$

$f(f^{-1}(x)) = f(\sqrt{x - 4}) = (\sqrt{x - 4})^2 + 4 = x$

$f^{-1}(f(x)) = f^{-1}(x^2 + 4) = \sqrt{(x^2 + 4) - 4} = \sqrt{x^2} = x, x \geq 0$

Domain f = Range f^{-1} = $\{x | x \geq 0\}$ or $[0, \infty)$

Range f = Domain f^{-1} = $\{x | x \geq 4\}$ or $[4, \infty)$

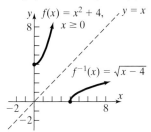

55. $f^{-1}(x) = \dfrac{4}{x}$

$f(f^{-1}(x)) = f\left(\dfrac{4}{x}\right) = \dfrac{4}{\left(\dfrac{4}{x}\right)} = x$

$f^{-1}(f(x)) = f^{-1}\left(\dfrac{4}{x}\right) = \dfrac{4}{\left(\dfrac{4}{x}\right)} = x$

Domain f = Range f^{-1} = All real numbers except 0
Range f = Domain f^{-1} = All real numbers except 0

57. $f^{-1}(x) = \dfrac{2x + 1}{x}$

$f(f^{-1}(x)) = f\left(\dfrac{2x + 1}{x}\right) = \dfrac{1}{\dfrac{2x + 1}{x} - 2} = \dfrac{x}{(2x + 1) - 2x} = x$

$f^{-1}(f(x)) = f^{-1}\left(\dfrac{1}{x - 2}\right) = \dfrac{2\left(\dfrac{1}{x - 2}\right) + 1}{\dfrac{1}{x - 2}} = \dfrac{2 + (x - 2)}{1} = x$

Domain f = Range f^{-1} = All real numbers except 2
Range f = Domain f^{-1} = All real numbers except 0

59. $f^{-1}(x) = \dfrac{2 - 3x}{x}$

$f(f^{-1}(x)) = f\left(\dfrac{2 - 3x}{x}\right) = \dfrac{2}{3 + \dfrac{2 - 3x}{x}} = \dfrac{2x}{3x + 2 - 3x} = \dfrac{2x}{2} = x$

$f^{-1}(f(x)) = f^{-1}\left(\dfrac{2}{3 + x}\right) = \dfrac{2 - 3\left(\dfrac{2}{3 + x}\right)}{\dfrac{2}{3 + x}} = \dfrac{2(3 + x) - 3 \cdot 2}{2} = \dfrac{2x}{2} = x$

Domain f = All real numbers except -3
Range f = Domain f^{-1} = All real numbers except 0

61. $f^{-1}(x) = \dfrac{-2x}{x-3}$

$$f(f^{-1}(x)) = f\left(\dfrac{-2x}{x-3}\right) = \dfrac{3\left(\dfrac{-2x}{x-3}\right)}{\dfrac{-2x}{x-3}+2} = \dfrac{3(-2x)}{-2x+2(x-3)} = \dfrac{-6x}{-6} = x$$

$$f^{-1}(f(x)) = f^{-1}\left(\dfrac{3x}{x+2}\right) = \dfrac{-2\left(\dfrac{3x}{x+2}\right)}{\dfrac{3x}{x+2}-3} = \dfrac{-2(3x)}{3x-3(x+2)} = \dfrac{-6x}{-6} = x$$

Domain f = All real numbers except -2

Range f = Domain f^{-1} = All real numbers except 3

63. $f^{-1}(x) = \dfrac{x}{3x-2}$

$$f(f^{-1}(x)) = f\left(\dfrac{x}{3x-2}\right) = \dfrac{2\left(\dfrac{x}{3x-2}\right)}{3\left(\dfrac{x}{3x-2}\right)-1} = \dfrac{2x}{3x-(3x-2)} = \dfrac{2x}{2} = x$$

$$f^{-1}(f(x)) = f^{-1}\left(\dfrac{2x}{3x-1}\right) = \dfrac{\dfrac{2x}{3x-1}}{3\left(\dfrac{2x}{3x-1}\right)-2} = \dfrac{2x}{6x-2(3x-1)} = \dfrac{2x}{2} = x$$

Domain f = All real numbers except $\dfrac{1}{3}$

Range f = Domain f^{-1} = All real numbers except $\dfrac{2}{3}$

65. $f^{-1}(x) = \dfrac{3x+4}{2x-3}$

$$f(f^{-1}(x)) = f\left(\dfrac{3x+4}{2x-3}\right) = \dfrac{3\left(\dfrac{3x+4}{2x-3}\right)+4}{2\left(\dfrac{3x+4}{2x-3}\right)-3} = \dfrac{3(3x+4)+4(2x-3)}{2(3x+4)-3(2x-3)}$$

$$= \dfrac{17x}{17} = x$$

$$f^{-1}(f(x)) = f^{-1}\left(\dfrac{3x+4}{2x-3}\right) = \dfrac{3\left(\dfrac{3x+4}{2x-3}\right)+4}{2\left(\dfrac{3x+4}{2x-3}\right)-3} = \dfrac{3(3x+4)+4(2x-3)}{2(3x+4)-3(2x-3)}$$

$$= \dfrac{17x}{17} = x$$

Domain f = All real numbers except $\dfrac{3}{2}$

Range f = Domain f^{-1} = All real numbers except $\dfrac{3}{2}$

67. $f^{-1}(x) = \dfrac{-2x + 3}{x - 2}$

$$f(f^{-1}(x)) = f\left(\dfrac{-2x + 3}{x - 2}\right) = \dfrac{2\left(\dfrac{-2x + 3}{x - 2}\right) + 3}{\dfrac{-2x + 3}{x - 2} + 2}$$

$$= \dfrac{2(-2x + 3) + 3(x - 2)}{-2x + 3 + 2(x - 2)} = \dfrac{-x}{-1} = x$$

$$f^{-1}(f(x)) = f^{-1}\left(\dfrac{2x + 3}{x + 2}\right) = \dfrac{2\left(\dfrac{-2x + 3}{x - 2}\right) + 3}{\dfrac{-2x + 3}{x - 2} + 2}$$

$$= \dfrac{2(-2x + 3) + 3(x - 2)}{-2x + 3 + 2(x - 2)} = \dfrac{-x}{-1} = x$$

Domain f = All real numbers except -2
Range f = Domain f^{-1} = All real numbers except 2

69. $f^{-1}(x) = \dfrac{2}{\sqrt{1 - 2x}}$

$$f(f^{-1}(x)) = f\left(\dfrac{2}{\sqrt{x - 2x}}\right) = \dfrac{\dfrac{4}{1 - 2x} - 4}{2 \cdot \dfrac{4}{1 - 2x}} = \dfrac{4 - 4(1 - 2x)}{2 \cdot 4} = \dfrac{8x}{8} = x$$

$$f^{-1}(f(x)) = f^{-1}\left(\dfrac{x^2 - 4}{2x^2}\right) = \dfrac{2}{\sqrt{1 - 2\left(\dfrac{x^2 - 4}{2x^2}\right)}} = \dfrac{2}{\sqrt{\dfrac{4}{x^2}}} = \sqrt{x^2}$$

$$= x, \text{ since } x > 0.$$

Domain f = $\{x | x > 0\}$ or $(0, \infty)$

Range f = Domain f^{-1} = $\left\{x \,\middle|\, x < \dfrac{1}{2}\right\}$ or $\left(-\infty, \dfrac{1}{2}\right)$

71. (a) 0 **(b)** 2 **(c)** 0 **(d)** 1 **73.** $f^{-1}(x) = \dfrac{1}{m}(x - b), m \neq 0$ **75.** Quadrant I **77.** Possible answer: $f(x) = |x|, x \geq 0$, is one-to-one; $f^{-1}(x) = x, x \geq 0$

79. (a) $C(H) = \dfrac{H + 10.53}{2.15}$ **(b)** 16.99 in. **81.** $x(p) = \dfrac{300 - p}{50}, p \leq 300$ **83.** $f^{-1}(x) = \dfrac{-dx + b}{cx - a}; f = f^{-1}$ if $a = -d$

87. Yes, if the domain is $\{x | x \geq 0\}$. **89.** On the line $y = x$; no; no.

4.3 Assess Your Understanding *(page 430)*

6. $\left(-1, \dfrac{1}{a}\right), (0, 1), (1, a)$ **7.** 1 **8.** 4 **9.** F **10.** F **11. (a)** 11.212 **(b)** 11.587 **(c)** 11.664 **(d)** 11.665 **13. (a)** 8.815 **(b)** 8.821 **(c)** 8.824

(d) 8.825 **15. (a)** 21.217 **(b)** 22.217 **(c)** 22.440 **(d)** 22.459 **17.** 3.320 **19.** 0.427 **21.** Not exponential **23.** Exponential; $a = 4$
25. Exponential; $a = 2$ **27.** Not exponential **29.** B **31.** D **33.** A **35.** E

37. **39.** **41.** **43.**

Domain: All real numbers
Range: $\{y | y > 1\}$ or $(1, \infty)$
Horizontal asymptote: $y = 1$

Domain: All real numbers
Range: $\{y | y > -2\}$ or $(-2, \infty)$
Horizontal asymptote: $y = -2$

Domain: All real numbers
Range: $\{y | y > 2\}$ or $(2, \infty)$
Horizontal asymptote: $y = 2$

Domain: All real numbers
Range: $\{y | y > 2\}$ or $(2, \infty)$
Horizontal asymptote: $y = 2$

45. **47.** **49.** **51.**

Domain: All real numbers
Range: $\{y | y > 0\}$ or $(0, \infty)$
Horizontal asymptote: $y = 0$

Domain: All real numbers
Range: $\{y | y > 0\}$ or $(0, \infty)$
Horizontal asymptote: $y = 0$

Domain: All real numbers
Range: $\{y | y < 5\}$ or $(-\infty, 5)$
Horizontal asymptote: $y = 5$

Domain: All real numbers
Range: $\{y | y < 2\}$ or $(-\infty, 2)$
Horizontal asymptote: $y = 2$

53. $\left\{\dfrac{1}{2}\right\}$ **55.** $\{-\sqrt{2}, 0, \sqrt{2}\}$ **57.** $\left\{1 - \dfrac{\sqrt{6}}{3}, 1 + \dfrac{\sqrt{6}}{3}\right\}$ **59.** $\{0\}$ **61.** $\{4\}$ **63.** $\left\{\dfrac{3}{2}\right\}$ **65.** $\{1, 2\}$ **67.** $\dfrac{1}{49}$ **69.** $\dfrac{1}{4}$

71. $f(x) = 3^x$ **73.** $f(x) = -6^x$ **75. (a)** 74% **(b)** 47% **77. (a)** \$12,123 **(b)** \$6443 **79.** 3.35 mg; 0.45 mg

81. (a) 0.63 **(b)** 0.98 **(c)** 1

(d)

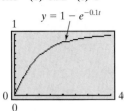

$y = 1 - e^{-0.1t}$

(e) About 7 min

83. (a) 0.0516 **(b)** 0.0888
85.(a) 70.95% **(b)** 72.62% **(c)** 100%

87. (a) 5.41 amp, 7.59 amp, 10.38 amp **(b)** 12 amp **(d)** 3.34 amp, 5.31 amp, 9.44 amp **(e)** 24 amp

(c), (f)

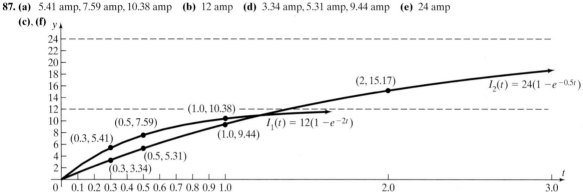

89. $n = 4: 2.7083; n = 6: 2.7181;$
$n = 8: 2.7182788; n = 10: 2.7182818$

91. $\dfrac{f(x+h) - f(x)}{h} = \dfrac{a^{x+h} - a^x}{h} = \dfrac{a^x a^h - a^x}{h} = \dfrac{a^x(a^h - 1)}{h}$

93. $f(-x) = a^{-x} = \dfrac{1}{a^x} = \dfrac{1}{f(x)}$

95. (a) $f(-x) = \dfrac{1}{2}(e^{-x} - e^{-(-x)}) = \dfrac{1}{2}(e^{-x} - e^x)$

$= -\dfrac{1}{2}(e^x - e^{-x}) = -f(x)$

97. $f(1) = 5, f(2) = 17, f(3) = 257, f(4) = 65{,}537,$

$f(5) = 4{,}294{,}967{,}297 = 641 \times 6{,}700{,}417$

(b)

$y = \frac{1}{2}(e^x - e^{-x})$

4.4 Assess Your Understanding *(page 444)*

4. $\{x | x > 0\}$ or $(0, \infty)$ **5.** $\left(\dfrac{1}{a}, -1\right), (1, 0), (a, 1)$ **6.** 1 **7.** F **8.** T **9.** $2 = \log_3 9$ **11.** $2 = \log_a 1.6$ **13.** $2 = \log_{1.1} M$ **15.** $x = \log_2 7.2$

17. $\sqrt{2} = \log_x \pi$ **19.** $x = \ln 8$ **21.** $2^3 = 8$ **23.** $a^6 = 3$ **25.** $3^x = 2$ **27.** $2^{1.3} = M$ **29.** $(\sqrt{2})^x = \pi$ **31.** $e^x = 4$ **33.** 0 **35.** 2 **37.** −4 **39.** $\dfrac{1}{2}$

41. 4 **43.** $\dfrac{1}{2}$ **45.** $\{x | x > 3\}; (3, \infty)$ **47.** All real numbers except 0; $\{x | x \neq 0\}$ **49.** $\{x | x > 0\}; (0, \infty)$ **51.** $\{x | x > -1\}; (-1, \infty)$

53. $\{x | x < -1 \text{ or } x > 0\}; (-\infty, -1) \text{ or } (0, \infty)$ **55.** $\{x | x \geq 1\}; [1, \infty)$ **57.** 0.511 **59.** 30.099 **61.** $\sqrt{2}$

63.

65.

67. B **69.** D **71.** A **73.** E

75.
Domain: $(-4, \infty)$
Range: $(-\infty, \infty)$
Vertical asymptote: $x = -4$

77.
Domain: $(0, \infty)$
Range: $(-\infty, \infty)$
Vertical asymptote: $x = 0$

79.
Domain: $(0, \infty)$
Range: $(-\infty, \infty)$
Vertical asymptote: $x = 0$

81.
Domain: $(0, \infty)$
Range: $(-\infty, \infty)$
Vertical asymptote: $x = 0$

83.
Domain: $(4, \infty)$
Range: $(-\infty, \infty)$
Vertical asymptote: $x = 4$

85.
Domain: $(0, \infty)$
Range: $(-\infty, \infty)$
Vertical asymptote: $x = 0$

87.
Domain: $(0, \infty)$
Range: $(-\infty, \infty)$
Vertical asymptote: $x = 0$

89.
Domain: $(-2, \infty)$
Range: $(-\infty, \infty)$
Vertical asymptote: $x = -2$

91. $\{9\}$ **93.** $\left\{\dfrac{7}{2}\right\}$ **95.** $\{2\}$ **97.** $\{5\}$ **99.** $\{3\}$ **101.** $\{2\}$ **103.** $\left\{\dfrac{\ln 10}{3}\right\}$ **105.** $\left\{\dfrac{\ln 8 - 5}{2}\right\}$ **107.** $\{-2\sqrt{2}, 2\sqrt{2}\}$ **109.** $\{-1\}$

111. (a)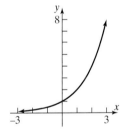
Domain: $(-\infty, \infty)$
Range: $(0, \infty)$
Horizontal asymptote: $y = 0$

(b) $f^{-1}(x) = \log_2 x$

(c)
Domain of f^{-1} = Range of $f = (0, \infty)$
Range of f^{-1} = Domain of $f = (-\infty, \infty)$
Vertical asymptote of f^{-1}: $x = 0$

113. (a)
Domain: $(-\infty, \infty)$
Range: $(0, \infty)$
Horizontal asymptote: $y = 0$

(b) $f^{-1}(x) = \log_2 x + 3$

(c)
Domain of f^{-1} = Range of $f = (0, \infty)$
Range of f^{-1} = Domain of $f = (-\infty, \infty)$
Vertical asymptote of f^{-1}: $x = 0$

115. (a) 1 **(b)** 2 **(c)** 3 **(d)** It increases. **(e)** 0.000316 **(f)** 3.981×10^{-8} **117. (a)** 5.97 km **(b)** 0.90 km **119. (a)** 6.93 min **(b)** 16.09 min
(c) No, since $F(t)$ can never equal 1 **121.** $h \approx 2.29$, so the time between injections is about 2 hr, 17 min

123. 0.2695 sec
0.8959 sec

125. 50 decibels (dB)
127. 110 dB
129. 8.1
131. (a) $k = 20.07$ **(b)** 91% **(c)** 0.175 **(d)** 0.08
133. Because $y = \log_1 x$ means $1^y = 1 = x$, which cannot be true for $x \neq 1$

4.5 Assess Your Understanding (page 455)

1. Sum **2.** 7 **3.** $r \log_a M$ **4.** F **5.** F **6.** T **7.** 71 **9.** -4 **11.** 7 **13.** 1 **15.** 1 **17.** 2 **19.** $\dfrac{5}{4}$ **21.** 4 **23.** $a + b$ **25.** $b - a$ **27.** $3a$

29. $\dfrac{1}{5}(a + b)$ **31.** $2 + \log_5 x$ **33.** $3 \log_2 z$ **35.** $1 + \ln x$ **37.** $\ln x + x$ **39.** $2 \log_a u + 3 \log_a v$ **41.** $2 \ln x + \dfrac{1}{2}\ln(1 - x)$

43. $3 \log_2 x - \log_2(x - 3)$ **45.** $\log x + \log(x + 2) - 2 \log(x + 3)$ **47.** $\dfrac{1}{3}\ln(x - 2) + \dfrac{1}{3}\ln(x + 1) - \dfrac{2}{3}\ln(x + 4)$

49. $\ln 5 + \ln x + \dfrac{1}{2}\ln(1 + 3x) - 3 \ln(x - 4)$ **51.** $\log_5 u^3 v^4$ **53.** $-\dfrac{5}{2}\log_3 x$ **55.** $\log_4\left[\dfrac{x - 1}{(x + 1)^4}\right]$ **57.** $-2 \ln(x - 1)$ **59.** $\log_2[x(3x - 2)^4]$

61. $\log_a\left(\dfrac{25x^6}{\sqrt{2x + 3}}\right)$ **63.** $\log_2\left[\dfrac{(x + 1)^2}{(x + 3)(x - 1)}\right]$ **65.** 2.771 **67.** -3.880 **69.** 5.615 **71.** 0.874

73. $y = \dfrac{\log x}{\log 4}$ **75.** $y = \dfrac{\log(x + 2)}{\log 2}$ **77.** $y = \dfrac{\log(x + 1)}{\log(x - 1)}$

79. $y = Cx$ **81.** $y = Cx(x + 1)$ **83.** $y = Ce^{3x}$ **85.** $y = Ce^{-4x} + 3$ **87.** $y = \dfrac{\sqrt[3]{C}(2x + 1)^{1/6}}{(x + 4)^{1/9}}$ **89.** 3 **91.** 1

93. $\log_a(x + \sqrt{x^2 - 1}) + \log_a(x - \sqrt{x^2 - 1}) = \log_a[(x + \sqrt{x^2 - 1})(x - \sqrt{x^2 - 1})] = \log_a[x^2 - (x^2 - 1)] = \log_a 1 = 0$

95. $\ln(1 + e^{2x}) = \ln[e^{2x}(e^{-2x} + 1)] = \ln e^{2x} + \ln(e^{-2x} + 1) = 2x + \ln(1 + e^{-2x})$

97. $y = f(x) = \log_a x; a^y = x$ implies $a^y = \left(\dfrac{1}{a}\right)^{-y} = x$, so $-y = \log_{1/a} x = -f(x)$. **99.** $f(x) = \log_a x; f\left(\dfrac{1}{x}\right) = \log_a \dfrac{1}{x} = \log_a 1 - \log_a x = -f(x)$

101. $\log_a \dfrac{M}{N} = \log_a(M \cdot N^{-1}) = \log_a M + \log_a N^{-1} = \log_a M - \log_a N$, since $a^{\log_a N^{-1}} = N^{-1}$ implies $a^{-\log_a N^{-1}} = N$; i.e., $\log_a N = -\log_a N^{-1}$.

4.6 Assess Your Understanding (page 461)

5. 6 **7.** 16 **9.** 8 **11.** 3 **13.** 5 **15.** $-1 + \sqrt{1 + e^4} \approx 6.456$ **17.** $\dfrac{\ln 3}{\ln 2} \approx 1.585$ **19.** 0 **21.** $\dfrac{\ln 10}{\ln 2} \approx 3.322$ **23.** $-\dfrac{\ln 1.2}{\ln 8} \approx -0.088$

25. $\dfrac{\ln 3}{2 \ln 3 + \ln 4} \approx 0.307$ **27.** $\dfrac{\ln 7}{\ln 0.6 + \ln 7} \approx 1.356$ **29.** 0 **31.** $\dfrac{\ln \pi}{1 + \ln \pi} \approx 0.534$ **33.** $\dfrac{\ln 1.6}{3 \ln 2} \approx 0.226$

35. $\dfrac{9}{2}$ **37.** 2 **39.** 1 **41.** 16 **43.** $-1, \dfrac{2}{3}$ **45.** 0 **47.** $\ln(2 + \sqrt{5}) \approx 1.444$ **49.** $10^{\frac{1}{\log 5} + \frac{1}{\log 3}}$ **51.** 2.79 **53.** -0.57 **55.** -0.70 **57.** 0.57

59. 0.39, 1.00 **61.** 1.32 **63.** 1.31 **65. (a)** Around the middle of the year 2006 **(b)** In the beginning of the year 2021
67. (a) After 2.4 years **(b)** After 6.5 years **(c)** After 10 years

4.7 Assess Your Understanding *(page 470)*

3. $108.29 **5.** $609.50 **7.** $697.09 **9.** $12.46 **11.** $125.23 **13.** $88.72 **15.** $860.72 **17.** $554.09 **19.** $59.71 **21.** $361.93 **23.** 5.35%

25. 26% **27.** $6\frac{1}{4}$% compounded annually **29.** 9% compounded monthly **31.** 104.32 mo (about 8.7 yr); 103.97 mo (about 8.66 yr)

33. 61.02 mo; 60.82 mo **35.** 15.27 yr or 15 yr, 3 mo **37.** $104,335 **39.** $12,910.62 **41.** About $30.17 per share or $3017 **43.** 9.35%

45. Not quite. Jim will have $1057.60. The second bank gives a better deal, since Jim will have $1060.62 after 1 yr.

47. Will has $11,632.73; Henry has $10,947.89. **49. (a)** Interest is $30,000 **(b)** Interest is $38,613.59

(c) Interest is $37,752.73. Simple interest at 12% is best. **51. (a)** $1364.62 **(b)** $1353.35 **53.** $4631.93

55. (a) 6.1 yr **(b)** 18.45 yr **(c)** $mP = P\left(1 + \dfrac{r}{n}\right)^{nt}$

$$m = \left(1 + \frac{r}{n}\right)^{nt}$$

$$\ln m = \ln\left(1 + \frac{r}{n}\right)^{nt} = nt \ln\left(1 + \frac{r}{n}\right)$$

$$t = \frac{\ln m}{n \ln\left(1 + \dfrac{r}{n}\right)}$$

4.8 Assess Your Understanding *(page 482)*

1. (a) 500 insects
(b) $0.02 = 2\%$
(c)

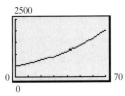

(d) About 611 insects
(e) After about 23.5 days
(f) After about 34.7 days

3. (a) $-0.0244 = -2.44\%$
(b)

(c) About 391.7 g
(d) After about 9.1 yr
(e) 28.4 y

5. 5832; 3.9 days
7. 25,198
9. 9.797 g

11. (a) 9727 yr ago
(b)

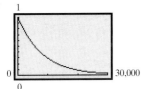

(c) 5600 yr

13. (a) 5:18 PM
(b)

(c) About 14.3 min
(d) The temperature of the pizza approaches 70°F.

15. (a) 18.63°C; 25.1°C
(b)

17. 7.34 kg; 76.6 h **19.** 26.6 days

21. (a) 90% **(b)** 12.86%
(c)

(d) 85.77% **(e)** 1996
(f) About 5.6 yr

23. (a) 1000 g **(b)** 43.9% **(c)** 30 g
(d)

(e) 616.6 g
(f) After 9.85 h
(g) About 7.9

25. (a) 9.23×10^{-3}, or about 0 **(b)** 0.81, or about 1
(c) 5.01, or about 5 **(d)** 57.91°, 43.99°, 30.07°

27. (a) In 1984, 91.8% of households did not own a personal computer

(b)

(c) 70.6% **(d)** During 2007

29. (a)

(b) 0.78 or 78%
(c) 50 people

4.9 Assess Your Understanding *(page 490)*

1. (a)

(b) $y = 0.0903(1.3384)^x$
(c) $N(t) = 0.0903e^{0.2915t}$
(d)

(e) 0.69 **(f)** After about 7.26 h

3. (a)

(b) $y = 100.326(0.8769)^x$
(c) $A = 100.326e^{-0.1314t}$
(d)

(e) 5.3 weeks **(f)** 0.14 g
(g) After about 12.3 weeks

5. (a)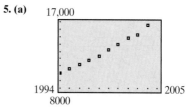

(b) $y = 2.2908 \times 10^{-44}(1.056554737)^x$
(c) 5.66%
(d) \$44,230.54
(e) In 2023

7. (a)

(b) $y = 32,741.02 - 6070.96 \ln x$

(c)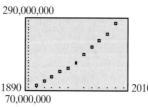

(d) Approximately 168 computers

9. (a)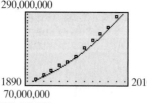

(b) $y = \dfrac{799,475,916.5}{1 + 1.56344 \times 10^{14}e^{-0.0160x}}$

(c)

(d) 799,475,917
(e) Approximately 279,809,184 **(f)** 2011

11. (a)

(b)
$$y = \dfrac{14,471,245.24}{1 + 3.860 \times 10^{20}\, e^{-0.0246x}}$$

(c)

(d) 14,471,245
(e) Approximately 12,811,429

Review Exercises *(page 496)*

1. (a) -26 **(b)** -241 **(c)** 16 **(d)** -1 **3. (a)** $\sqrt{11}$ **(b)** 1 **(c)** $\sqrt{\sqrt{6} + 2}$ **(d)** 19 **5. (a)** e^4 **(b)** $3e^{-2} - 2$ **(c)** e^{e^4} **(d)** -17
7. $(f \circ g)(x) = 1 - 3x$, all real numbers; $(g \circ f)(x) = 7 - 3x$, all real numbers;
$(f \circ f)(x) = x$, all real numbers; $(g \circ g)(x) = 9x + 4$, all real numbers

9. $(f \circ g)(x) = 27x^2 + 3|x| + 1$, all real numbers; $(g \circ f)(x) = 3|3x^2 + x + 1|$, all real numbers;
$(f \circ f)(x) = 3(3x^2 + x + 1)^2 + 3x^2 + x + 2$, all real numbers; $(g \circ g)(x) = 9|x|$, all real numbers

11. $(f \circ g)(x) = \dfrac{1 + x}{1 - x}$, $\{x | x \neq 0, x \neq 1\}$; $(g \circ f)(x) = \dfrac{x - 1}{x + 1}$, $\{x | x \neq -1, x \neq 1\}$; $(f \circ f)(x) = x$, $\{x | x \neq 1\}$; $(g \circ g)(x) = x$, $\{x | x \neq 0\}$

13. (a) One-to-one   **(b)** $\{(2, 1), (5, 3), (8, 5), (10, 6)\}$

15.

17. $f^{-1}(x) = \dfrac{2x + 3}{5x - 2}$

$$f(f^{-1}(x)) = \dfrac{2\left(\dfrac{2x + 3}{5x - 2}\right) + 3}{5\left(\dfrac{2x + 3}{5x - 2}\right) - 2} = x$$

$$f^{-1}(f(x)) = \dfrac{2\left(\dfrac{2x + 3}{5x - 2}\right) + 3}{5\left(\dfrac{2x + 3}{5x - 2}\right) - 2} = x$$

Domain of f = Range of f^{-1} = all real numbers except $\dfrac{2}{5}$

Range of f = Domain of f^{-1} = all real numbers except $\dfrac{2}{5}$

19. $f^{-1}(x) = \dfrac{x + 1}{x}$

$$f(f^{-1}(x)) = \dfrac{1}{\dfrac{x + 1}{x} - 1} = x$$

$$f^{-1}(f(x)) = \dfrac{\dfrac{1}{x - 1} + 1}{\dfrac{1}{x - 1}} = x$$

Domain of f = Range of f^{-1} = all real numbers except 1
Range of f = Domain of f^{-1} = all real numbers except 0

21. $f^{-1}(x) = \dfrac{27}{x^3}$

$$f(f^{-1}(x)) = \dfrac{3}{\left(\dfrac{27}{x^3}\right)^{1/3}} = x$$

$$f^{-1}(f(x)) = \dfrac{27}{\left(\dfrac{3}{x^{1/3}}\right)^3} = x$$

Domain of f = Range of f^{-1} = all real numbers except 0
Range of f = Domain of f^{-1} = all real numbers except 0

23. (a) 81   **(b)** 2   **(c)** $\dfrac{1}{9}$ **(d)** -3 **25.** $\log_5 z = 2$ **27.** $5^{13} = u$ **29.** $\left\{x \middle| x > \dfrac{2}{3}\right\}$; $\left(\dfrac{2}{3}, \infty\right)$ **31.** $\{x | x < 1 \text{ or } x > 2\}$; $(-\infty, 1) \text{ or } (2, \infty)$

33. -3 **35.** $\sqrt{2}$ **37.** 0.4   **39.** $\log_3 u + 2 \log_3 v - \log_3 w$ **41.** $2 \log x + \dfrac{1}{2} \log(x^3 + 1)$ **43.** $\ln x + \dfrac{1}{3} \ln(x^2 + 1) - \ln(x - 3)$

45. $\dfrac{25}{4} \log_4 x$ **47.** $-2 \ln(x + 1)$ **49.** $\log\left(\dfrac{4x^3}{[(x + 3)(x - 2)]^{1/2}}\right)$ **51.** 2.124

53.

55.

Domain: $(-\infty, \infty)$
Range: $(0, \infty)$
Horizontal asymptote: $y = 0$

57.

Domain: $(-\infty, \infty)$
Range: $(0, \infty)$
Horizontal asymptote: $y = 0$

59.

Domain: $(-\infty, \infty)$
Range: $(-\infty, 1)$
Horizontal asymptote: $y = 1$

61.

Domain: $(0, \infty)$
Range: $(-\infty, \infty)$
Vertical asymptote: $x = 0$

63.

Domain: $(-\infty, \infty)$
Range: $(-\infty, 3)$
Horizontal asymptote: $y = 3$

65. $\left\{\dfrac{1}{4}\right\}$ **67.** $\left\{\dfrac{-1-\sqrt{3}}{2}, \dfrac{-1+\sqrt{3}}{2}\right\}$ **69.** $\left\{\dfrac{1}{4}\right\}$ **71.** $\left\{\dfrac{2\ln 3}{\ln 5 - \ln 3} \approx 4.301\right\}$ **73.** $\left\{\dfrac{12}{5}\right\}$ **75.** $\{83\}$ **77.** $\left\{\dfrac{1}{2}, -3\right\}$ **79.** $\{-1\}$

81. $\{1 - \ln 5 \approx -0.609\}$ **83.** $\left\{\dfrac{\ln 3}{3\ln 2 - 2\ln 3} \approx -9.327\right\}$ **85.** 3229.5 m **87. (a)** 37.3 W **(b)** 6.9 dB **89. (a)** 9.85 yr **(b)** 4.27 yr

91. \$41,668.97 **93.** 24,203 yr ago **95.** 6,835,600,129

97. (a) 0.3

(b) 0.8

(c)

(d) In 2023

99. (a)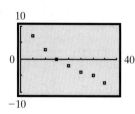

(b) Wind chill $= 18.921 - 7.096 \ln$ (wind speed)

(c)

(d) Approximately $-3°$F

Chapter Test *(page 500)*

1. (a) $f \circ g = \dfrac{2x + 7}{2x + 3}$; Domain: $\left\{x \,\middle|\, x \neq -\dfrac{3}{2}\right\}$ **(b)** $(g \circ f)(-2) = 5$ **(c)** $(f \circ g)(-2) = -3$

2. (a) The function is not one-to-one. **(b)** The function is one-to-one.

3. $f^{-1}(x) = \dfrac{2 + 5x}{3x}$

	Domain	Range
f	$x \neq \dfrac{5}{3}$	$y \neq 0$
f^{-1}	$x \neq 0$	$y \neq \dfrac{5}{3}$

4. The point $(-5, 3)$ must be on the graph of f^{-1}. **5.** $x = 5$ **6.** $b = 4$ **7.** $x = 625$ **8.** $e^3 + 2 \approx 22.086$ **9.** $\log 20 \approx 1.301$

10. $\log_3 21 = \dfrac{\ln 21}{\ln 3} \approx 2.771$

11. $\ln 133 \approx 4.890$

12. Domain: $\{x \,|\, -\infty < x < \infty\}$
Range: $\{y \,|\, y > -2\}$
Asymptote: $y = -2$

13. Domain: $\{x \,|\, x > 2\}$
Range: $\{y \,|\, -\infty < y < \infty\}$
Asymptote: $x = 2$

14. $x = 1$ **15.** $x = 91$ **16.** $x = -\ln 2 \approx -0.693$ **17.** $\dfrac{1 - \sqrt{13}}{2} \approx -1.303$, $\dfrac{1 + \sqrt{13}}{2} \approx 2.303$ **18.** $x = \dfrac{3\ln 7}{1 - \ln 7} \approx -6.172$

19. $x = 2\sqrt{6} \approx 4.899$ **20.** $2 + 3\log_2 x - \log_2 (x - 6) - \log_2 (x + 3)$ **21.** About 250.39 days **22. (a)** \$35,298 **(b)** \$21,409

23. (a) About 83 decibles **(b)** The pain threshold will be exceeded if 31,623 people shouted at the same time.

24. $y = \dfrac{213}{1 + 205.86e^{-0.3564t}}$; 197 million U.S. residents

Cumulative Review *(page 501)*

1. Yes; no   **2. (a)** 10   **(b)** $2x^2 + 3x + 1$ **(c)** $2x^2 + 4xh + 2h^2 - 3x - 3h + 1$ **3.** $\left(\dfrac{1}{2}, \dfrac{\sqrt{3}}{2}\right)$ is on the graph   **4.** -26

5.

6. (a)

7. $f(x) = 2(x - 4)^2 - 8 = 2x^2 - 16x + 24$

8.

(b) $\{x | -\infty < x < \infty\}$

9. $f(g(x)) = \dfrac{4}{(x - 3)^2} + 2$; domain: $\{x | x \neq 3\}$; 3

10. (a) Zeros: $-4, -\dfrac{1}{4}, 2$

(b) x-intercepts: $-4, -\dfrac{1}{4}, 2$; y-intercept: -8

(c) Local maximum of 60.75 occurs at $x = -2.5$;
Local minimum of -25 occurs at $x = 1$

(d)

11. (a), (c)

Domain g = Range $g^{-1} = (-\infty, \infty)$
Range g = Domain $g^{-1} = (2, \infty)$

(b) $g^{-1}(x) = \log_3(x - 2)$

12. $-\dfrac{3}{2}$ **13.** 2   **14. (a)** -1 **(b)** $\{x | x > -1\}$ or $(-1, \infty)$ **(c)** 25   **15. (a)**

(b) Answers will vary.

C H A P T E R   5   Systems of Equations and Inequalities

5.1 Assess Your Understanding *(page 515)*

3. inconsistent   **4.** consistent   **5.** F   **6.** T

7. $\begin{cases} 2(2) - (-1) = 5 \\ 5(2) + 2(-1) = 8 \end{cases}$ **9.** $\begin{cases} 3(2) - 4\left(\dfrac{1}{2}\right) = 4 \\ \dfrac{1}{2}(2) - 3\left(\dfrac{1}{2}\right) = -\dfrac{1}{2} \end{cases}$ **11.** $\begin{cases} 4 - 1 = 3 \\ \dfrac{1}{2}(4) + 1 = 3 \end{cases}$ **13.** $\begin{cases} 3(1) + 3(-1) + 2(2) = 4 \\ 1 - (-1) - 2 = 0 \\ 2(-1) - 3(2) = -8 \end{cases}$ **15.** $\begin{cases} 3(2) + 3(-2) + 2(2) = 4 \\ 2 - 3(-2) + 2 = 10 \\ 5(2) - 2(-2) - 3(2) = 8 \end{cases}$

17. $x = 6, y = 2$ **19.** $x = 3, y = 2$ **21.** $x = 8, y = -4$ **23.** $x = \dfrac{1}{3}, y = -\dfrac{1}{6}$ **25.** Inconsistent   **27.** $x = 1, y = 2$

29. $x = 4 - 2y$, y is any real number   **31.** $x = 1, y = 1$ **33.** $x = \dfrac{3}{2}, y = 1$ **35.** $x = 4, y = 3$ **37.** $x = \dfrac{4}{3}, y = \dfrac{1}{5}$ **39.** $x = \dfrac{1}{5}, y = \dfrac{1}{3}$

41. $x = 8, y = 2, z = 0$ **43.** $x = 2, y = -1, z = 1$ **45.** Inconsistent   **47.** $x = 5z - 2, y = 4z - 3$, where z is any real number   **49.** Inconsistent

51. $x = 1, y = 3, z = -2$ **53.** $x = -3, y = \dfrac{1}{2}, z = 1$ **55.** Length 30 ft; width 15 ft   **57.** Cheeseburger $1.55; shake $0.85   **59.** 22.5 lb

61. Average wind speed 25 mph; average airspeed 175 mph   **65.** \$5.56   **67.** Mix 50 mg of first compound with 75 mg of second.

69. $a = \dfrac{4}{3}, b = -\dfrac{5}{3}, c = 1$ **71.** $y = 9000, r = 0.06$ **73.** $I_1 = \dfrac{10}{71}, I_2 = \dfrac{65}{71}, I_3 = \dfrac{55}{71}$ **75.** 100 orchestra, 210 main, and 190 balcony seats

77. 1.5 chicken, 1 corn, 2 milk   **79.** If $x = $ price of hamburgers, $y = $ price of fries, $z = $ price of colas, then $x = 2.75 - z, y = \dfrac{41}{60} + \dfrac{1}{3}z$, $\$0.60 \le z \le \$0.90$.

There is not sufficient information:

x	\$2.13	\$2.01	\$1.86
y	\$0.89	\$0.93	\$0.98
z	\$0.62	\$0.74	\$0.89

81. It will take Beth 30 hr, Bill 24 hr, and Edie 40 hr.

5.2 Assess Your Understanding (page 532)

1. matrix   **2.** augmented   **3.** T   **4.** T

5. $\begin{bmatrix} 1 & -5 & | & 5 \\ 4 & 3 & | & 6 \end{bmatrix}$ **7.** $\begin{bmatrix} 2 & 3 & | & 6 \\ 4 & -6 & | & -2 \end{bmatrix}$ **9.** $\begin{bmatrix} 0.01 & -0.03 & | & 0.06 \\ 0.13 & 0.10 & | & 0.20 \end{bmatrix}$ **11.** $\begin{bmatrix} 1 & -1 & 1 & | & 10 \\ 3 & 3 & 0 & | & 5 \\ 1 & 1 & 2 & | & 2 \end{bmatrix}$ **13.** $\begin{bmatrix} 1 & 1 & -1 & | & 2 \\ 3 & -2 & 0 & | & 2 \\ 5 & 3 & -1 & | & 1 \end{bmatrix}$

15. $\begin{bmatrix} 1 & -1 & -1 & | & 10 \\ 2 & 1 & 2 & | & -1 \\ -3 & 4 & 0 & | & 5 \\ 4 & -5 & 1 & | & 0 \end{bmatrix}$ **17.** $\begin{bmatrix} 1 & -3 & | & -2 \\ 0 & 1 & | & 9 \end{bmatrix}$ **19. (a)** $\begin{bmatrix} 1 & -3 & 4 & | & 3 \\ 0 & 1 & -2 & | & 0 \\ -3 & 3 & 4 & | & 6 \end{bmatrix}$ **(b)** $\begin{bmatrix} 1 & -3 & 4 & | & 3 \\ 2 & -5 & 6 & | & 6 \\ 0 & -6 & 16 & | & 15 \end{bmatrix}$

21. (a) $\begin{bmatrix} 1 & -3 & 2 & | & -6 \\ 0 & 1 & -1 & | & 8 \\ -3 & -6 & 4 & | & 6 \end{bmatrix}$ **(b)** $\begin{bmatrix} 1 & -3 & 2 & | & -6 \\ 2 & -5 & 3 & | & -4 \\ 0 & -15 & 10 & | & -12 \end{bmatrix}$ **23. (a)** $\begin{bmatrix} 1 & -3 & 1 & | & -2 \\ 0 & 1 & 4 & | & 2 \\ -3 & 1 & 4 & | & 6 \end{bmatrix}$ **(b)** $\begin{bmatrix} 1 & -3 & 1 & | & -2 \\ 2 & -5 & 6 & | & -2 \\ 0 & -8 & 7 & | & 0 \end{bmatrix}$

25. $\begin{cases} x = 5 \\ y = -1 \end{cases}$
Consistent; $x = 5, y = -1$

27. $\begin{cases} x = 1 \\ y = 2 \\ 0 = 3 \end{cases}$
Inconsistent

29. $\begin{cases} x + 2z = -1 \\ y - 4z = -2 \\ 0 = 0 \end{cases}$
Consistent;
$x = -1 - 2z,$
$y = -2 + 4z,$
z is any real number

31. $\begin{cases} x_1 = 1 \\ x_2 + x_4 = 2 \\ x_3 + 2x_4 = 3 \end{cases}$
Consistent;
$x_1 = 1, x_2 = 2 - x_4,$
$x_3 = 3 - 2x_4,$
x_4 is any real number

33. $\begin{cases} x_1 + 4x_4 = 2 \\ x_2 + x_3 + 3x_4 = 3 \\ 0 = 0 \end{cases}$
Consistent;
$x_1 = 2 - 4x_4,$
$x_2 = 3 - x_3 - 3x_4,$
x_3, x_4 are any real numbers

35. $\begin{cases} x_1 + x_4 = -2 \\ x_2 + 2x_4 = 2 \\ x_3 - x_4 = 0 \end{cases}$
Consistent;
$x_1 = -2 - x_4,$
$x_2 = 2 - 2x_4,$
$x_3 = x_4,$
x_4 is any real number

37. $x = 6, y = 2$ **39.** $x = \dfrac{1}{2}, y = \dfrac{3}{4}$ **41.** $x = 4 - 2y, y$ is any real number   **43.** $x = \dfrac{3}{2}, y = 1$ **45.** $x = \dfrac{4}{3}, y = \dfrac{1}{5}$ **47.** $x = 8, y = 2, z = 0$

49. $x = 2, y = -1, z = 1$ **51.** Inconsistent   **53.** $x = 5z - 2, y = 4z - 3$, where z is any real number   **55.** Inconsistent   **57.** $x = 1, y = 3, z = -2$

59. $x = -3, y = \dfrac{1}{2}, z = 1$ **61.** $x = \dfrac{1}{3}, y = \dfrac{2}{3}, z = 1$ **63.** $x = 1, y = 2, z = 0, w = 1$ **65.** $y = 0, z = 1 - x, x$ is any real number

67. $x = 2, y = z - 3, z$ is any real number   **69.** $x = \dfrac{13}{9}, y = \dfrac{7}{18}, z = \dfrac{19}{18}$ **71.** $x = \dfrac{7}{5} - \dfrac{3}{5}z - \dfrac{2}{5}w, y = -\dfrac{8}{5} + \dfrac{7}{5}z + \dfrac{13}{5}w$, where z and w are any real

numbers   **73.** $y = -2x^2 + x + 3$ **75.** $f(x) = 3x^3 - 4x^2 + 5$ **77.** 1.5 salmon steak, 2 baked eggs, 1 acorn squash   **79.** \$4000 in Treasury bills,

\$4000 in Treasury bonds, \$2000 in corporate bonds   **81.** 8 Deltas, 5 Betas, 10 Sigmas   **83.** $I_1 = \dfrac{44}{23}, I_2 = 2, I_3 = \dfrac{16}{23}, I_4 = \dfrac{28}{23}$

85. (a)

Amount Invested At		
7%	9%	11%
0	10,000	10,000
1000	8000	11,000
2000	6000	12,000
3000	4000	13,000
4000	2000	14,000
5000	0	15,000

(b)

Amount Invested At		
7%	9%	11%
12,500	12,500	0
14,500	8500	2000
16,500	4500	4000
18,750	0	6250

(c) All the money invested at 7% provides \$2100, more than what is required.

87.

First Liquid	Second Liquid	Third Liquid
50 mg	75 mg	0 mg
36 mg	76 mg	8 mg
22 mg	77 mg	16 mg
8 mg	78 mg	24 mg

5.3 Assess Your Understanding *(page 545)*

1. determinants **2.** $ad - bc$ **3.** F **4.** F **5.** 2 **7.** 22 **9.** -2 **11.** 10 **13.** -26 **15.** $x = 6, y = 2$ **17.** $x = 3, y = 2$ **19.** $x = 8, y = -4$

21. $x = 4, y = -2$ **23.** Not applicable **25.** $x = \dfrac{1}{2}, y = \dfrac{3}{4}$ **27.** $x = \dfrac{1}{10}, y = \dfrac{2}{5}$ **29.** $x = \dfrac{3}{2}, y = 1$

31. $x = \dfrac{4}{3}, y = \dfrac{1}{5}$ **33.** $x = 1, y = 3, z = -2$ **35.** $x = -3, y = \dfrac{1}{2}, z = 1$ **37.** Not applicable **39.** $x = 0, y = 0, z = 0$

41. Not applicable **43.** -5 **45.** $\dfrac{13}{11}$ **47.** 0 or -9 **49.** -4 **51.** 12 **53.** 8 **55.** 8

57. $(y_1 - y_2)x - (x_1 - x_2)y + (x_1 y_2 - x_2 y_1) = 0$

$$(y_1 - y_2)x + (x_2 - x_1)y = x_2 y_1 - x_1 y_2$$
$$(x_2 - x_1)y - (x_2 - x_1)y_1 = (y_2 - y_1)x + x_2 y_1 - x_1 y_2 - (x_2 - x_1)y_1$$
$$(x_2 - x_1)(y - y_1) = (y_2 - y_1)x - (y_2 - y_1)x_1$$
$$y - y_1 = \frac{y_2 - y_1}{x_2 - x_1}(x - x_1)$$

59. $\begin{vmatrix} x^2 & x & 1 \\ y^2 & y & 1 \\ z^2 & z & 1 \end{vmatrix} = x^2 \begin{vmatrix} y & 1 \\ z & 1 \end{vmatrix} - x \begin{vmatrix} y^2 & 1 \\ z^2 & 1 \end{vmatrix} + \begin{vmatrix} y^2 & y \\ z^2 & z \end{vmatrix}$

$$= x^2(y - z) - x(y^2 - z^2) + yz(y - z)$$
$$= (y - z)[x^2 - x(y + z) + yz]$$
$$= (y - z)[(x^2 - xy) - (xz - yz)]$$
$$= (y - z)[x(x - y) - z(x - y)]$$
$$= (y - z)(x - y)(x - z)$$

61. $\begin{vmatrix} a_{13} & a_{12} & a_{11} \\ a_{23} & a_{22} & a_{21} \\ a_{33} & a_{32} & a_{31} \end{vmatrix} = a_{13}(a_{22}a_{31} - a_{32}a_{21}) - a_{12}(a_{23}a_{31} - a_{33}a_{21}) + a_{11}(a_{23}a_{32} - a_{33}a_{22})$

$$= -[a_{11}(a_{22}a_{33} - a_{32}a_{23}) - a_{12}(a_{21}a_{33} - a_{31}a_{23}) + a_{13}(a_{21}a_{32} - a_{31}a_{22})] = - \begin{vmatrix} a_{11} & a_{12} & a_{13} \\ a_{21} & a_{22} & a_{23} \\ a_{31} & a_{32} & a_{33} \end{vmatrix}$$

63. $\begin{vmatrix} a_{11} & a_{12} & a_{11} \\ a_{21} & a_{22} & a_{21} \\ a_{31} & a_{32} & a_{31} \end{vmatrix} = a_{11}(a_{22}a_{31} - a_{32}a_{21}) - a_{12}(a_{21}a_{31} - a_{31}a_{21}) + a_{11}(a_{21}a_{32} - a_{31}a_{22})$

$$= a_{11}a_{22}a_{31} - a_{11}a_{32}a_{21} - a_{12}(0) + a_{11}a_{21}a_{32} - a_{11}a_{31}a_{22} = 0$$

Historical Problems *(page 561)*

1. (a) $2 - 5i \longleftrightarrow \begin{bmatrix} 2 & -5 \\ 5 & 2 \end{bmatrix}, 1 + 3i \longleftrightarrow \begin{bmatrix} 1 & 3 \\ -3 & 1 \end{bmatrix}$ **(b)** $\begin{bmatrix} 2 & -5 \\ 5 & 2 \end{bmatrix}\begin{bmatrix} 1 & 3 \\ -3 & 1 \end{bmatrix} = \begin{bmatrix} 17 & 1 \\ -1 & 17 \end{bmatrix}$ **(c)** $17 + i$ **(d)** $17 + i$

2. (a) $x = k(ar + bs) + l(cr + ds) = r(ka + lc) + s(kb + ld)$ **(b)** $A = \begin{bmatrix} ka + lc & kb + ld \\ ma + nc & mb + nd \end{bmatrix}$
$y = m(ar + bs) + n(cr + ds) = r(ma + nc) + s(mb + nd)$

5.4 Assess Your Understanding *(page 561)*

1. inverse **2.** square **3.** identity **4.** F **5.** F **6.** F

7. $\begin{bmatrix} 4 & 4 & -5 \\ -1 & 5 & 4 \end{bmatrix}$ **9.** $\begin{bmatrix} 0 & 12 & -20 \\ 4 & 8 & 24 \end{bmatrix}$ **11.** $\begin{bmatrix} -8 & 7 & -15 \\ 7 & 0 & 22 \end{bmatrix}$

13. $\begin{bmatrix} 28 & -9 \\ 4 & 23 \end{bmatrix}$ **15.** $\begin{bmatrix} 1 & 14 & -14 \\ 2 & 22 & -18 \\ 3 & 0 & 28 \end{bmatrix}$ **17.** $\begin{bmatrix} 15 & 21 & -16 \\ 22 & 34 & -22 \\ -11 & 7 & 22 \end{bmatrix}$ **19.** $\begin{bmatrix} 25 & -9 \\ 4 & 20 \end{bmatrix}$

21. $\begin{bmatrix} -13 & 7 & -12 \\ -18 & 10 & -14 \\ 17 & -7 & 34 \end{bmatrix}$ **23.** $\begin{bmatrix} -2 & 4 & 2 & 8 \\ 2 & 1 & 4 & 6 \end{bmatrix}$ **25.** $\begin{bmatrix} 5 & 14 \\ 9 & 16 \end{bmatrix}$ **27.** $\begin{bmatrix} 9 & 2 \\ 34 & 13 \\ 47 & 20 \end{bmatrix}$ **29.** $\begin{bmatrix} 1 & -1 \\ -1 & 2 \end{bmatrix}$ **31.** $\begin{bmatrix} 1 & -\frac{5}{2} \\ -1 & 3 \end{bmatrix}$ **33.** $\begin{bmatrix} 1 & -\frac{1}{a} \\ -1 & \frac{2}{a} \end{bmatrix}$

35. $\begin{bmatrix} 3 & -3 & 1 \\ -2 & 2 & -1 \\ -4 & 5 & -2 \end{bmatrix}$ **37.** $\begin{bmatrix} -\frac{5}{7} & \frac{1}{7} & \frac{3}{7} \\ \frac{9}{7} & \frac{1}{7} & -\frac{4}{7} \\ \frac{3}{7} & -\frac{2}{7} & \frac{1}{7} \end{bmatrix}$ **39.** $x = 3, y = 2$ **41.** $x = -5, y = 10$ **43.** $x = 2, y = -1$ **45.** $x = \frac{1}{2}, y = 2$

47. $x = -2, y = 1$ **49.** $x = \frac{2}{a}, y = \frac{3}{a}$ **51.** $x = -2, y = 3, z = 5$ **53.** $x = \frac{1}{2}, y = -\frac{1}{2}, z = 1$ **55.** $x = -\frac{34}{7}, y = \frac{85}{7}, z = \frac{12}{7}$

57. $x = \frac{1}{3}, y = 1, z = \frac{2}{3}$

59. $\begin{bmatrix} 4 & 2 & | & 1 & 0 \\ 2 & 1 & | & 0 & 1 \end{bmatrix} \rightarrow \begin{bmatrix} 1 & \frac{1}{2} & | & \frac{1}{4} & 0 \\ 2 & 1 & | & 0 & 1 \end{bmatrix} \rightarrow \begin{bmatrix} 1 & \frac{1}{2} & | & \frac{1}{4} & 0 \\ 0 & 0 & | & -\frac{1}{2} & 1 \end{bmatrix}$ **61.** $\begin{bmatrix} 15 & 3 & | & 1 & 0 \\ 10 & 2 & | & 0 & 1 \end{bmatrix} \rightarrow \begin{bmatrix} 1 & \frac{1}{5} & | & \frac{1}{15} & 0 \\ 10 & 2 & | & 0 & 1 \end{bmatrix} \rightarrow \begin{bmatrix} 1 & \frac{1}{5} & | & \frac{1}{15} & 0 \\ 0 & 0 & | & -\frac{2}{3} & 1 \end{bmatrix}$

63. $\begin{bmatrix} -3 & 1 & -1 & | & 1 & 0 & 0 \\ 1 & -4 & -7 & | & 0 & 1 & 0 \\ 1 & 2 & 5 & | & 0 & 0 & 1 \end{bmatrix} \rightarrow \begin{bmatrix} 1 & 2 & 5 & | & 0 & 0 & 1 \\ 1 & -4 & -7 & | & 0 & 1 & 0 \\ -3 & 1 & -1 & | & 1 & 0 & 0 \end{bmatrix} \rightarrow \begin{bmatrix} 1 & 2 & 5 & | & 0 & 0 & 1 \\ 0 & -6 & -12 & | & 0 & 1 & -1 \\ 0 & 7 & 14 & | & 1 & 0 & 3 \end{bmatrix} \rightarrow \begin{bmatrix} 1 & 2 & 5 & | & 0 & 0 & 1 \\ 0 & 1 & 2 & | & 0 & -\frac{1}{6} & \frac{1}{6} \\ 0 & 1 & 2 & | & \frac{1}{7} & 0 & \frac{3}{7} \end{bmatrix}$

$\rightarrow \begin{bmatrix} 1 & 2 & 5 & | & 0 & 0 & 1 \\ 0 & 1 & 2 & | & 0 & -\frac{1}{6} & \frac{1}{6} \\ 0 & 0 & 0 & | & \frac{1}{7} & \frac{1}{6} & \frac{11}{42} \end{bmatrix}$

65. $\begin{bmatrix} 0.01 & 0.05 & -0.01 \\ 0.01 & -0.02 & 0.01 \\ -0.02 & 0.01 & 0.03 \end{bmatrix}$ **67.** $\begin{bmatrix} 0.02 & -0.04 & -0.01 & 0.01 \\ -0.02 & 0.05 & 0.03 & -0.03 \\ 0.02 & 0.01 & -0.04 & 0.00 \\ -0.02 & 0.06 & 0.07 & 0.06 \end{bmatrix}$ **69.** $x = 4.57, y = -6.44, z = -24.07$ **71.** $x = -1.19, y = 2.46, z = 8.27$

73. (a) $\begin{bmatrix} 500 & 350 & 400 \\ 700 & 500 & 850 \end{bmatrix}; \begin{bmatrix} 500 & 700 \\ 350 & 500 \\ 400 & 850 \end{bmatrix}$ **(b)** $\begin{bmatrix} 15 \\ 8 \\ 3 \end{bmatrix}$ **(c)** $\begin{bmatrix} 11{,}500 \\ 17{,}050 \end{bmatrix}$ **(d)** $\begin{bmatrix} 0.10 & 0.05 \end{bmatrix}$ **(e)** $2002.50

75. If $D = ad - bc \neq 0$, then $a \neq 0$ and $d \neq 0$, or $b \neq 0$ and $c \neq 0$. Assuming the former then,

$$\begin{bmatrix} a & b & | & 1 & 0 \\ c & d & | & 0 & 1 \end{bmatrix} \rightarrow \begin{bmatrix} 1 & \frac{b}{a} & | & \frac{1}{a} & 0 \\ c & d & | & 0 & 1 \end{bmatrix} \rightarrow \begin{bmatrix} 1 & \frac{b}{a} & | & \frac{1}{a} & 0 \\ 0 & \frac{D}{a} & | & -\frac{c}{a} & 1 \end{bmatrix} \rightarrow \begin{bmatrix} 1 & \frac{b}{a} & | & \frac{1}{a} & 0 \\ 0 & 1 & | & -\frac{c}{D} & \frac{a}{D} \end{bmatrix} \rightarrow \begin{bmatrix} 1 & 0 & | & \frac{d}{D} & -\frac{b}{D} \\ 0 & 1 & | & -\frac{c}{D} & \frac{a}{D} \end{bmatrix}$$

$$R_1 = \frac{1}{a}r_1 \qquad R_2 = -cr_1 + r_2 \qquad R_2 = \frac{a}{D}r_2 \qquad R_1 = -\frac{b}{a}r_2 + r_1$$

5.5 Assess Your Understanding *(page 570)*

5. Proper **7.** Improper; $1 + \dfrac{9}{x^2 - 4}$ **9.** Improper; $5x + \dfrac{22x - 1}{x^2 - 4}$ **11.** Improper; $1 + \dfrac{-2(x - 6)}{(x + 4)(x - 3)}$ **13.** $\dfrac{-4}{x} + \dfrac{4}{x - 1}$ **15.** $\dfrac{1}{x} + \dfrac{-x}{x^2 + 1}$

17. $\dfrac{-1}{x - 1} + \dfrac{2}{x - 2}$ **19.** $\dfrac{\frac{1}{4}}{x + 1} + \dfrac{\frac{3}{4}}{x - 1} + \dfrac{\frac{1}{2}}{(x - 1)^2}$ **21.** $\dfrac{\frac{1}{12}}{x - 2} + \dfrac{-\frac{1}{12}(x + 4)}{x^2 + 2x + 4}$ **23.** $\dfrac{\frac{1}{4}}{x - 1} + \dfrac{\frac{1}{4}}{(x - 1)^2} + \dfrac{-\frac{1}{4}}{x + 1} + \dfrac{\frac{1}{4}}{(x + 1)^2}$

25. $\dfrac{-5}{x + 2} + \dfrac{5}{x + 1} + \dfrac{-4}{(x + 1)^2}$ **27.** $\dfrac{\frac{1}{4}}{x} + \dfrac{1}{x^2} + \dfrac{-\frac{1}{4}(x + 4)}{x^2 + 4}$ **29.** $\dfrac{\frac{2}{3}}{x + 1} + \dfrac{\frac{1}{3}(x + 1)}{x^2 + 2x + 4}$ **31.** $\dfrac{\frac{2}{7}}{3x - 2} + \dfrac{\frac{1}{7}}{2x + 1}$ **33.** $\dfrac{\frac{3}{4}}{x + 3} + \dfrac{\frac{1}{4}}{x - 1}$

35. $\dfrac{1}{x^2 + 4} + \dfrac{2x - 1}{(x^2 + 4)^2}$ **37.** $\dfrac{-1}{x} + \dfrac{2}{x - 3} + \dfrac{-1}{x + 1}$ **39.** $\dfrac{4}{x - 2} + \dfrac{-3}{x - 1} + \dfrac{-1}{(x - 1)^2}$ **41.** $\dfrac{x}{(x^2 + 16)^2} + \dfrac{-16x}{(x^2 + 16)^3}$

43. $\dfrac{-\dfrac{8}{7}}{2x + 1} + \dfrac{\dfrac{4}{7}}{x - 3}$ **45.** $\dfrac{-\dfrac{2}{9}}{x} + \dfrac{-\dfrac{1}{3}}{x^2} + \dfrac{\dfrac{1}{6}}{x - 3} + \dfrac{\dfrac{1}{18}}{x + 3}$

5.6 Assess Your Understanding (page 578)

7. satisfied   **8.** half-plane   **9.** F   **10.** T

11.

$x \geq 0$

13.

$x \geq 4$

15.

$2x + y \geq 6$

17.

$x^2 + y^2 > 1$

19.

$y \leq x^2 - 1$

21.

$xy \geq 4$

23.
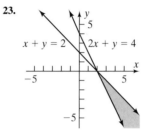
$x + y = 2$ $2x + y = 4$

25.

$3x + 2y = -6$ $2x - y = 4$

27.
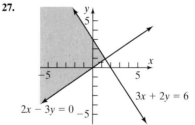
$2x - 3y = 0$ $3x + 2y = 6$

29.
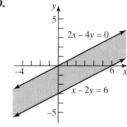
$2x - 4y = 0$ $x - 2y = 6$

31.

$2x + y = -2$ $2x + y = 2$

33. No solution
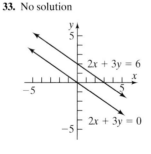
$2x + 3y = 6$ $2x + 3y = 0$

35. Bounded; corner points
$(0, 0), (3, 0), (2, 2), (0, 3)$
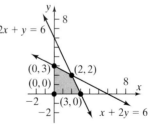
$2x + y = 6$ $(0, 3)$ $(2, 2)$ $(0, 0)$ $(3, 0)$ $x + 2y = 6$

37. Unbounded; corner points
$(2, 0), (0, 4)$
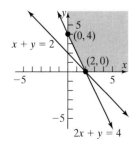
$x + y = 2$ $(0, 4)$ $(2, 0)$ $2x + y = 4$

39. Bounded; corner points $(2, 0), (4, 0),$
$\left(\dfrac{24}{7}, \dfrac{12}{7}\right), (0, 4), (0, 2)$
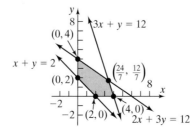
$3x + y = 12$ $(0, 4)$ $x + y = 2$ $\left(\dfrac{24}{7}, \dfrac{12}{7}\right)$ $(0, 2)$ $(2, 0)$ $(4, 0)$ $2x + 3y = 12$

41. Bounded; corner points $(2, 0), (5, 0),$
$(2, 6), (0, 8), (0, 2)$
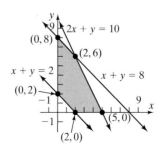
$2x + y = 10$ $(0, 8)$ $(2, 6)$ $x + y = 2$ $x + y = 8$ $(0, 2)$ $(5, 0)$ $(2, 0)$

43. Bounded; corner points

$(1, 0), (10, 0), (0, 5), \left(0, \dfrac{1}{2}\right)$

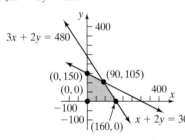

45. $\begin{cases} x \le 4 \\ x + y \le 6 \\ x \ge 0 \\ y \ge 0 \end{cases}$

47. $\begin{cases} x \le 20 \\ y \ge 15 \\ x + y \le 50 \\ x - y \le 0 \\ x \ge 0 \end{cases}$

49. (a) $\begin{cases} x + y \le 50{,}000 \\ x \ge 35{,}000 \\ y \le 10{,}000 \\ x \ge 0 \\ y \ge 0 \end{cases}$

(b)

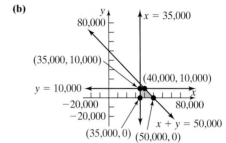

51. (a) $\begin{cases} x \ge 0 \\ y \ge 0 \\ x + 2y \le 300 \\ 3x + 2y \le 480 \end{cases}$

(b)

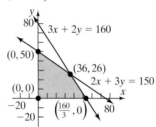

53. (a) $\begin{cases} 3x + 2y \le 160 \\ 2x + 3y \le 150 \\ x \ge 0 \\ y \ge 0 \end{cases}$

(b)

5.7 Assess Your Understanding *(page 585)*

1. objective function   **2.** T   **3.** Maximum value is 11; minimum value is 3.   **5.** Maximum value is 65; minimum value is 4.
7. Maximum value is 67; minimum value is 20.   **9.** The maximum value of z is 12, and it occurs at the point $(6, 0)$.
11. The minimum value of z is 4, and it occurs at the point $(2, 0)$.   **13.** The maximum value of z is 20, and it occurs at the point $(0, 4)$.
15. The minimum value of z is 8, and it occurs at the point $(0, 2)$.   **17.** The maximum value of z is 50, and it occurs at the point $(10, 0)$.
19. 8 downhill, 24 cross-country; $1760; $1920   **21.** 30 acres of soybeans and 10 acres of corn   **23.** $\dfrac{1}{2}$ hr on machine 1; $5\dfrac{1}{4}$ hr on machine 2
25. 100 lb of ground beef and 50 lb of pork   **27.** 10 racing skates, 15 figure skates   **29.** 2 metal samples, 4 plastic samples; $34
31. (a) 10 first class, 120 coach   **(b)** 15 first class, 120 coach

Review Exercises *(page 589)*

1. $x = 2, y = -1$ **3.** $x = 2, y = \dfrac{1}{2}$ **5.** $x = 2, y = -1$ **7.** $x = \dfrac{11}{5}, y = -\dfrac{3}{5}$ **9.** Inconsistent   **11.** $x = 2, y = 3$

13. Inconsistent   **15.** $x = -1, y = 2, z = -3$ **17.** $x = \dfrac{7}{4}z + \dfrac{39}{4}, y = \dfrac{9}{8}z + \dfrac{69}{8}, z$ is any real number.   **19.** $\begin{cases} 3x + 2y = 8 \\ x + 4y = -1 \end{cases}$

21. $\begin{bmatrix} 4 & -4 \\ 3 & 9 \\ 4 & 4 \end{bmatrix}$ **23.** $\begin{bmatrix} 6 & 0 \\ 12 & 24 \\ -6 & 12 \end{bmatrix}$ **25.** $\begin{bmatrix} 4 & -3 & 0 \\ 12 & -2 & -8 \\ -2 & 5 & -4 \end{bmatrix}$ **27.** $\begin{bmatrix} 8 & -13 & 8 \\ 9 & 2 & -10 \\ 22 & -13 & -4 \end{bmatrix}$ **29.** $\begin{bmatrix} \dfrac{1}{2} & -1 \\ -\dfrac{1}{6} & \dfrac{2}{3} \end{bmatrix}$ **31.** $\begin{bmatrix} -\dfrac{5}{7} & \dfrac{9}{7} & \dfrac{3}{7} \\ \dfrac{1}{7} & \dfrac{1}{7} & -\dfrac{2}{7} \\ \dfrac{3}{7} & -\dfrac{4}{7} & \dfrac{1}{7} \end{bmatrix}$ **33.** Singular

35. $x = \dfrac{2}{5}, y = \dfrac{1}{10}$ **37.** $x = 9, y = \dfrac{13}{3}, z = \dfrac{13}{3}$ **39.** $x = -\dfrac{1}{2}, y = -\dfrac{2}{3}, z = -\dfrac{3}{4}$ **41.** $z = -1, x = y + 1, y$ is any real number.

43. $x = 4, y = 2, z = 3, t = -2$ **45.** 5   **47.** 108   **49.** -100 **51.** $x = 2, y = -1$ **53.** $x = 2, y = 3$ **55.** $x = -1, y = 2, z = -3$

57. 16   **59.** $\dfrac{-\dfrac{3}{2}}{x} + \dfrac{\dfrac{3}{2}}{x - 4}$ **61.** $\dfrac{-3}{x - 1} + \dfrac{3}{x} + \dfrac{4}{x^2}$ **63.** $\dfrac{-\dfrac{1}{10}}{x + 1} + \dfrac{\dfrac{1}{10}x + \dfrac{9}{10}}{x^2 + 9}$ **65.** $\dfrac{x}{x^2 + 4} + \dfrac{-4x}{(x^2 + 4)^2}$ **67.** $\dfrac{\dfrac{1}{2}}{x^2 + 1} + \dfrac{\dfrac{1}{4}}{x - 1} + \dfrac{-\dfrac{1}{4}}{x + 1}$

69. (a)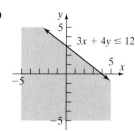

$3x + 4y \leq 12$

(b)

71. (a)

$y = x^2$

(b)

73. Unbounded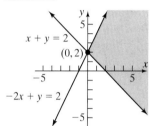

$x + y = 2$

$(0, 2)$

$-2x + y = 2$

75. Bounded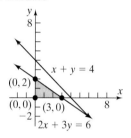

$x + y = 4$

$(0, 2)$

$(0, 0)$ $(3, 0)$

$2x + 3y = 6$

77. Bounded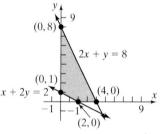

$(0, 8)$

$2x + y = 8$

$(0, 1)$

$x + 2y = 2$ $(4, 0)$

$(2, 0)$

79. The maximum value is 32 when $x = 0$ and $y = 8$. **81.** The minimum value is 3 when $x = 1$ and $y = 0$. **83.** 10 **85.** $y = -\frac{1}{3}x^2 - \frac{2}{3}x + 1$

87. 70 pounds of \$3 coffee and 30 pounds of \$6 coffee **89.** 1 small, 5 medium, 2 large **91.** Speedboat: 36.67 km/hr; Aguarico River: 3.33 km/hr

93. Bruce: 4 hours; Bryce: 2 hours; Marty: 8 hours **95.** 35 gasoline engines, 15 diesel engines; 15 gasoline engines, 0 diesel engines

Chapter Test *(page 592)*

1. $x = 3, y = -1$ **2.** Inconsistent. **3.** $x = -z + \frac{18}{7}, y = z - \frac{17}{7}$ where z can be any real number. **4.** $x = \frac{1}{3}, y = -2, z = 0$

5. $\begin{bmatrix} 1 & -5 & 1 & | & 0 \\ -2 & -1 & 0 & | & -25 \\ 1 & 5 & -5 & | & 10 \end{bmatrix}$ **6.** $\begin{cases} 3x + 2y + 4z = -6 \\ 1x + 0y + 8z = 2 \\ -2x + 1y + 3z = -11 \end{cases}$ or $\begin{cases} 3x + 2y + 4z = -6 \\ x + 8z = 2 \\ -2x + y + 3z = -11 \end{cases}$ **7.** $\begin{bmatrix} 6 & 4 \\ 1 & -11 \\ 5 & 12 \end{bmatrix}$ **8.** $\begin{bmatrix} -11 & -19 \\ -3 & 5 \\ 6 & -22 \end{bmatrix}$

9. The operation cannot be performed. **10.** $\begin{bmatrix} 16 & 17 \\ 3 & -10 \end{bmatrix}$ **11.** $\begin{bmatrix} 2 & -1 \\ -\frac{5}{2} & \frac{3}{2} \end{bmatrix}$ **12.** $B^{-1} = \begin{bmatrix} 3 & 3 & -4 \\ -2 & -2 & 3 \\ -4 & -5 & 7 \end{bmatrix}$ **13.** $x = \frac{1}{2}, y = 3$

14. The system is dependent and therefore has an infinite number of solutions. Any ordered pair satisfying the equation

$x = \frac{1}{4}y - 7$ or $y = -4x + 28$, is a solution to the system. **15.** $x = 1, y = -2, z = 0$ **16.** Inconsistent **17.** -29 **18.** -12 **19.** $x = -2, y = -5$

20. $x = 1, y = -1, z = 4$ **21.** $\frac{3}{x + 3} + \frac{-2}{(x + 3)^2}$ **22.** $\frac{-\frac{1}{3}}{x} + \frac{\frac{1}{3}x}{(x^2 + 3)} + \frac{5x}{(x^2 + 3)^2}$

23. The graph is unbounded. The corner points are $(4, 2)$ and $(8, 0)$

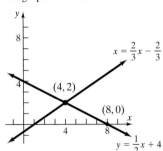

24. The maximum value of z is 64, and it occurs at the point $(0, 8)$.

25. USC received 48 first place votes and Oklahoma received 11 first place votes.

26. Flare jeans cost $24.50, camisoles cost $8.50, and t-shirts cost $6.00.

Cumulative Review *(page 594)*

1. $\left\{0, \dfrac{1}{2}\right\}$ **2.** $\{5\}$ **3.** $\left\{-1, -\dfrac{1}{2}, 3\right\}$ **4.** $\{-2\}$ **5.** $\left\{\dfrac{5}{2}\right\}$ **6.** $\left\{\dfrac{1}{\ln 3}\right\}$ **7.** Odd; symmetric with respect to origin

8. Center: $(1, -2)$; radius $= 4$

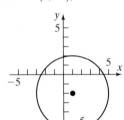

9. Domain: all real numbers
Range: $\{y|y > 1\}$
Horizontal asymptote: $y = 1$

10. $f^{-1}(x) = \dfrac{5}{x} - 2$

Domain of f: $\{x|x \neq -2\}$
Range of f: $\{y|y \neq 0\}$
Domain of f^{-1}: $\{x|x \neq 0\}$
Range of f^{-1}: $\{y|y \neq -2\}$

11. (a)

(b)

(c)

(d)

(e)

(f)

(g)

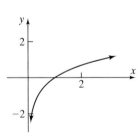

12. (a) -2.28

   (b) Local maximum of 7 at $x = -1$
        Local minimum of 3 at $x = 1$

   (c) $(-\infty, -1), (1, \infty)$

C H A P T E R 6   Conics

6.2 Assess Your Understanding *(page 605)*

6. parabola **7.** paraboloid of revolution **8.** T **9.** T **10.** T **11.** B **13.** E **15.** H **17.** C

19. $y^2 = 16x$

21. $x^2 = -12y$

23. $y^2 = -8x$

25. $x^2 = 2y$

27. $x^2 = \dfrac{4}{3}y$

29. $(x - 2)^2 = -8(y + 3)$

31. $(y + 2)^2 = 4(x + 1)$

33. $(x + 3)^2 = 4(y - 3)$

35. $(y + 2)^2 = -8(x + 1)$

37. Vertex: $(0, 0)$; Focus: $(0, 1)$;
Directrix: $y = -1$

(a)

(b)

39. Vertex: $(0, 0)$; Focus: $(-4, 0)$;
Directrix: $x = 4$

(a)

(b)

41. Vertex: $(-1, 2)$; Focus: $(1, 2)$;
Directrix: $x = -3$

(a)

(b)

43. Vertex: $(3, -1)$; Focus: $\left(3, -\dfrac{5}{4}\right)$;
Directrix: $y = -\dfrac{3}{4}$

(a)

(b)

45. Vertex: $(2, -3)$; Focus: $(4, -3)$;
Directrix: $x = 0$

(a)

(b)

47. Vertex: $(0, 2)$; Focus: $(-1, 2)$;
Directrix: $x = 1$

(a)

(b)

49. Vertex: $(-4, -2)$; Focus: $(-4, -1)$;

Directrix: $y = -3$

(a)

51. Vertex: $(-1, -1)$; Focus: $\left(-\dfrac{3}{4}, -1\right)$;

Directrix: $x = -\dfrac{5}{4}$

(a)

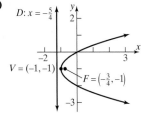

53. Vertex: $(2, -8)$; Focus: $\left(2, -\dfrac{31}{4}\right)$;

Directrix: $y = -\dfrac{33}{4}$

(b)

(b)

(b)

57. $(y - 1)^2 = -(x - 2)$ **59.** $x^2 = 4(y - 1)$ **61.** $y^2 = \dfrac{1}{2}(x + 2)$

63. 1.5625 ft from the base of the dish, along the axis of symmetry   **65.** 1 in. from the vertex   **67.** 20 ft   **69.** 0.78125 ft
71. 4.17 ft from the base along the axis of symmetry   **73.** 24.31 ft, 18.75 ft, 7.64 ft

75. $Ax^2 + Ey = 0, A \neq 0, E \neq 0$ This is the equation of a parabola with vertex at $(0, 0)$ and axis of symmetry the y-axis.

$Ax^2 = -Ey$ The focus is $\left(0, -\dfrac{E}{4A}\right)$; the directrix is the line $y = \dfrac{E}{4A}$. The parabola opens up if $-\dfrac{E}{A} > 0$

$x^2 = -\dfrac{E}{A}y$ and down if $-\dfrac{E}{A} < 0$.

77. $Ax^2 + Dx + Ey + F = 0, A \neq 0$ **(a)** If $E \neq 0$, then the equation may be written as

$Ax^2 + Dx = -Ey - F$

$$\left(x + \frac{D}{2A}\right)^2 = \frac{E}{A}\left(y - \frac{D^2 - 4AF}{4AE}\right)$$

$x^2 + \dfrac{D}{A}x = -\dfrac{E}{A}y - \dfrac{F}{A}$ This is the equation of a parabola with vertex at

$\left(x + \dfrac{D}{2A}\right)^2 = -\dfrac{E}{A}y - \dfrac{F}{A} + \dfrac{D^2}{4A^2}$

$\left(-\dfrac{D}{2A}, \dfrac{D^2 - 4AF}{4AE}\right)$ and axis of symmetry parallel to the y-axis.

$\left(x + \dfrac{D}{2A}\right)^2 = -\dfrac{E}{A}y + \dfrac{D^2 - 4AF}{4A^2}$ **(b)–(d)** If $E = 0$, the graph of the equation contains no points if

$D^2 - 4AF < 0$, is a single vertical line if $D^2 - 4AF = 0$, and is two vertical lines if $D^2 - 4AF > 0$.

6.3 Assess Your Vocabulary *(page 616)*

7. ellipse   **8.** major   **9.** $(0, -5); (0, 5)$ **10.** F   **11.** T   **12.** T   **13.** C   **15.** B

17. Vertices: $(-5, 0), (5, 0)$

Foci: $(-\sqrt{21}, 0), (\sqrt{21}, 0)$

19. Vertices: $(0, -5), (0, 5)$

Foci: $(0, -4), (0, 4)$

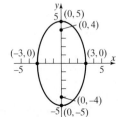

21. $\dfrac{x^2}{4} + \dfrac{y^2}{16} = 1$

Vertices: $(0, -4), (0, 4)$
Foci: $(0, -2\sqrt{3}), (0, 2\sqrt{3})$

23. $\dfrac{x^2}{8} + \dfrac{y^2}{2} = 1$

Vertices: $(-2\sqrt{2}, 0), (2\sqrt{2}, 0)$
Foci: $(-\sqrt{6}, 0), (\sqrt{6}, 0)$

25. $\dfrac{x^2}{16} + \dfrac{y^2}{16} = 1$

Vertices: $(-4, 0), (4, 0),$
$(0, -4), (0, 4)$
Focus: $(0, 0)$

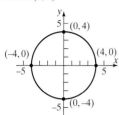

27. $\dfrac{x^2}{25} + \dfrac{y^2}{16} = 1$

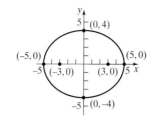

29. $\dfrac{x^2}{9} + \dfrac{y^2}{25} = 1$

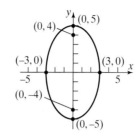

31. $\dfrac{x^2}{9} + \dfrac{y^2}{5} = 1$

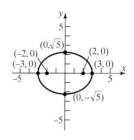

33. $\dfrac{x^2}{25} + \dfrac{y^2}{9} = 1$

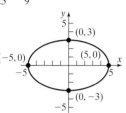

35. $\dfrac{x^2}{4} + \dfrac{y^2}{13} = 1$

37. $x^2 + \dfrac{y^2}{16} = 1$

39. $\dfrac{(x+1)^2}{4} + (y-1)^2 = 1$

41. $(x-1)^2 + \dfrac{y^2}{4} = 1$

43. Center: $(3, -1)$; Vertices: $(3, -4), (3, 2)$

Foci: $(3, -1-\sqrt{5}), (3, -1+\sqrt{5})$

(a)

(b)

45. $\dfrac{(x+5)^2}{16} + \dfrac{(y-4)^2}{4} = 1$

Foci: $(-5-2\sqrt{3}, 4), (-5+2\sqrt{3}, 4)$

(a)

(b)

47. $\dfrac{(x+2)^2}{4} + (y-1)^2 = 1$

Center: $(-2, 1)$; Vertices: $(-4, 1), (0, 1)$

Foci: $(-2-\sqrt{3}, 1), (-2+\sqrt{3}, 1)$

(a)

(b)

49. $\dfrac{(x-2)^2}{3} + \dfrac{(y+1)^2}{2} = 1$

Center: $(2, -1)$; Vertices: $(2-\sqrt{3}, -1),$
Foci: $(2+\sqrt{3}, -1)$; Foci: $(1, -1), (3, -1)$

(a)

(b)

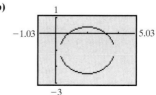

51 $\dfrac{(x-1)^2}{4} + \dfrac{(y+2)^2}{9} = 1$

Center: $(1,-2)$; Vertices: $(1,-5)$, $(1,1)$
Foci: $(1,-2-\sqrt{5})$, $(1,-2+\sqrt{5})$

(a)

53. $x^2 + \dfrac{(y+2)^2}{4} = 1$

Center: $(0,-2)$; Vertices: $(0,-4)$, $(0,0)$
Foci: $(0,-2-\sqrt{3})$, $(0,-2+\sqrt{3})$

a)

55. $\dfrac{(x-2)^2}{25} + \dfrac{(y+2)^2}{21} = 1$

(b)

(b)

57.

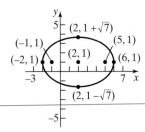

59. $\dfrac{(x-2)^2}{16} + \dfrac{(y-1)^2}{7} = 1$

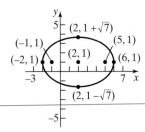

61. $\dfrac{(x-1)^2}{10} + (y-2)^2 = 1$

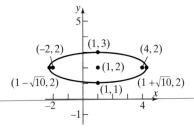

63. $\dfrac{(x-1)^2}{9} + (y-2)^2 = 1$

65.

67.

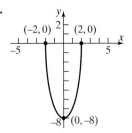

69. $\dfrac{x^2}{100} + \dfrac{y^2}{36} = 1$ **71.** 43.3 ft   **73.** 24.65 ft, 21.65 ft, 13.82 ft   **75.** 0 ft, 12.99 ft, 15 ft, 12.99 ft, 0 ft   **77.** 91.5 million mi; $\dfrac{x^2}{(93)^2} + \dfrac{y^2}{8646.75} = 1$

79. perihelion: 460.6 million mi; mean distance: 483.8 million mi; $\dfrac{x^2}{(483.8)^2} + \dfrac{y^2}{233,524.2} = 1$ **81.** 30 ft

83. (a) $Ax^2 + Cy^2 + F = 0$ If A and C are of the same sign and F is of opposite sign, then the equation takes the form

$$Ax^2 + Cy^2 = -F$$

$$\dfrac{x^2}{\left(-\dfrac{F}{A}\right)} + \dfrac{y^2}{\left(-\dfrac{F}{C}\right)} = 1, \text{ where } -\dfrac{F}{A} \text{ and } -\dfrac{F}{C} \text{ are positive. This is the equation of an ellipse}$$

with center at $(0,0)$.

(b) If $A = C$, the equation may be written as $x^2 + y^2 = -\dfrac{F}{A}$.

This is the equation of a circle with center at $(0,0)$ and radius equal to $\sqrt{-\dfrac{F}{A}}$.

6.4 Assess Your Understanding *(page 630)*

7. hyperbola   **8.** transverse axis   **9.** $3x = -2y, 3x = 2y$ **10.** F   **11.** T   **12.** F   **13.** B   **15.** A

17. $x^2 - \dfrac{y^2}{8} = 1$

19. $\dfrac{y^2}{16} - \dfrac{x^2}{20} = 1$

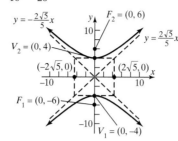

21. $\dfrac{x^2}{9} - \dfrac{y^2}{16} = 1$

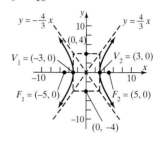

23. $\dfrac{y^2}{36} - \dfrac{x^2}{9} = 1$

25. $\dfrac{x^2}{8} - \dfrac{y^2}{8} = 1$

27. $\dfrac{x^2}{25} - \dfrac{y^2}{9} = 1$

Center: $(0, 0)$

Transverse axis: x-axis

Vertices: $(-5, 0), (5, 0)$

Foci: $(-\sqrt{34}, 0), (\sqrt{34}, 0)$

Asymptotes: $y = \pm\dfrac{3}{5}x$

(a)

(b)

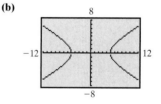

29. $\dfrac{x^2}{4} - \dfrac{y^2}{16} = 1$

Center: $(0, 0)$

Transverse axis: x-axis

Vertices: $(-2, 0), (2, 0)$

Foci: $(-2\sqrt{5}, 0), (2\sqrt{5}, 0)$

Asymptotes: $y = \pm2x$

(a)

(b)

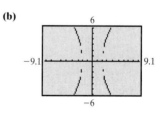

31. $\dfrac{y^2}{9} - x^2 = 1$

Center: $(0, 0)$

Transverse axis: y-axis

Vertices: $(0, -3), (0, 3)$

Foci: $(0, -\sqrt{10}), (0, \sqrt{10})$

Asymptotes: $y = \pm3x$

(a)

(b)

33. $\dfrac{y^2}{25} - \dfrac{x^2}{25} = 1$

Center: $(0,0)$

Transverse axis: y-axis

Vertices: $(0,-5)$, $(0,5)$

Foci: $(0,-5\sqrt{2})$, $(0,5\sqrt{2})$

Asymptotes: $y = \pm x$

(a)

(b)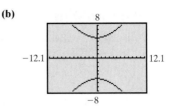

35. $x^2 - y^2 = 1$ **37.** $\dfrac{y^2}{36} - \dfrac{x^2}{9} = 1$

39. $\dfrac{(x-4)^2}{4} - \dfrac{(y+1)^2}{5} = 1$

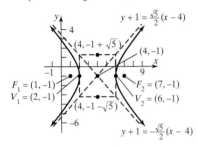

41. $\dfrac{(y+4)^2}{4} - \dfrac{(x+3)^2}{12} = 1$

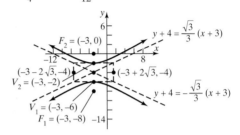

43. $(x-5)^2 - \dfrac{(y-7)^2}{3} = 1$

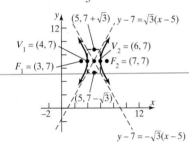

45. $\dfrac{(x-1)^2}{4} - \dfrac{(y+1)^2}{9} = 1$

47. $\dfrac{(x-2)^2}{4} - \dfrac{(y+3)^2}{9} = 1$

Center: $(2,-3)$

Transverse axis: Parallel to x-axis

Vertices: $(0,-3)$, $(4,-3)$

Foci: $(2-\sqrt{13},-3)$, $(2+\sqrt{13},-3)$

Asymptotes: $y + 3 = \pm\dfrac{3}{2}(x-2)$

(a)

(b)

49. $\dfrac{(y-2)^2}{4} - (x+2)^2 = 1$

Center: $(-2,2)$

Transverse axis: Parallel to y-axis

Vertices: $(-2,0)$, $(-2,4)$

Foci: $(-2,2-\sqrt{5})$, $(-2,2+\sqrt{5})$

Asymptotes: $y - 2 = \pm 2(x+2)$

(a)

(b)

51. $\dfrac{(x+1)^2}{4} - \dfrac{(y+2)^2}{4} = 1$

Center: $(-1, -2)$

Transverse axis: Parallel to x-axis

Vertices: $(-3, -2), (1, -2)$

Foci: $(-1 - 2\sqrt{2}, -2), (-1 + 2\sqrt{2}, -2)$

Asymptotes: $y + 2 = \pm(x + 1)0$

(a)

(b)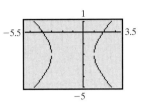

53. $(x-1)^2 - (y+1)^2 = 1$

Center: $(1, -1)$

Transverse axis: Parallel to x-axis

Vertices: $(0, -1), (2, -1)$

Foci: $(1 - \sqrt{2}, -1), (1 + \sqrt{2}, -1)$

Asymptotes: $y + 1 = \pm(x - 1)$

(a)

(b)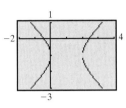

55. $\dfrac{(y-2)^2}{4} - (x+1)^2 = 1$

Center: $(-1, 2)$

Transverse axis: Parallel to y-axis

Vertices: $(-1, 0), (-1, 4)$

Foci: $(-1, 2 - \sqrt{5}), (-1, 2 + \sqrt{5})$

Asymptotes: $y - 2 = \pm 2(x + 1)$

(a)

(b)

57. $\dfrac{(x-3)^2}{4} - \dfrac{(y+2)^2}{16} = 1$

Center: $(3, -2)$

Transverse axis: Parallel to x-axis

Vertices: $(1, -2), (5, -2)$

Foci: $(3 - 2\sqrt{5}, -2), (3 + 2\sqrt{5}, -2)$

Asymptotes: $y + 2 = \pm 2(x - 3)$

(a)

(b)

59. $\dfrac{(y-1)^2}{4} - (x+2)^2 = 1$

Center: $(-2, 1)$

Transverse axis: Parallel to y-axis

Vertices: $(-2, -1), (-2, 3)$

Foci: $(-2, 1 - \sqrt{5}), (-2, 1 + \sqrt{5})$

Asymptotes: $y - 1 = \pm 2(x + 2)$

(a)

(b)

61.

63.

65. The fireworks display is 50,138 feet north of the person at point A.

67. (a) $y = \pm x$ **(b)** $\dfrac{x^2}{100} - \dfrac{y^2}{100} = 1$

71. $\dfrac{x^2}{4} - y^2 = 1$: asymptotes $y = \pm\dfrac{1}{2}x$; $y^2 - \dfrac{x^2}{4} = 1$: asymptotes $y = \pm\dfrac{1}{2}x$

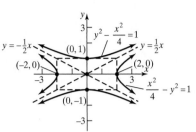

73. $Ax^2 + Cy^2 + F = 0$ If A and C are of opposite sign and $F \neq 0$, this equation may be written as $\dfrac{x^2}{\left(-\dfrac{F}{A}\right)} + \dfrac{y^2}{\left(-\dfrac{F}{C}\right)} = 1$,

$\qquad\qquad Ax^2 + Cy^2 = -F$ where $-\dfrac{F}{A}$ and $-\dfrac{F}{C}$ are opposite in sign. This is the equation of a hyperbola with center $(0, 0)$.

The transverse axis is the x-axis if $-\dfrac{F}{A} > 0$; the transverse axis is the y-axis if $-\dfrac{F}{A} < 0$.

Historical Problem *(page 640)*

$x - 6$ units, $y = 8$ units

6.5 Assess Your Understanding *(page 641)*

7.

9.

11.

13.

15.

17.

19.

21.

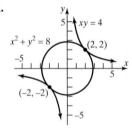

23. No points of intersection

25.

27. $x = 1, y = 4; x = -1, y = -4; x = 2\sqrt{2}, y = \sqrt{2}; x = -2\sqrt{2}, y = -\sqrt{2}$

29. $x = 0, y = 1; x = -\dfrac{2}{3}, y = -\dfrac{1}{3}$ **31.** $x = 0, y = -1; x = \dfrac{5}{2}, y = -\dfrac{7}{2}$

33. $x = 2, y = \dfrac{1}{3}; x = \dfrac{1}{2}, y = \dfrac{4}{3}$ **35.** $x = 3, y = 2; x = 3, y = -2; x = -3, y = 2; x = -3, y = -2$

37. $x = \dfrac{1}{2},\ y = \dfrac{3}{2};\ x = \dfrac{1}{2},\ y = -\dfrac{3}{2};\ x = -\dfrac{1}{2},\ y = \dfrac{3}{2};\ x = -\dfrac{1}{2},\ y = -\dfrac{3}{2}$ **39.** $x = \sqrt{2},\ y = 2\sqrt{2};\ x = -\sqrt{2},\ y = -2\sqrt{2}$

41. No real solution exists.   **43** $x = \dfrac{8}{3},\ y = \dfrac{2\sqrt{10}}{3};\ x = -\dfrac{8}{3},\ y = \dfrac{2\sqrt{10}}{3};\ x = \dfrac{8}{3},\ y = -\dfrac{2\sqrt{10}}{3};\ x = -\dfrac{8}{3},\ y = -\dfrac{2\sqrt{10}}{3}$

45. $x = 1,\ y = \dfrac{1}{2};\ x = -1,\ y = \dfrac{1}{2};\ x = 1,\ y = -\dfrac{1}{2};\ x = -1,\ y = -\dfrac{1}{2}$ **47.** No real solution exists.

49. $x = \sqrt{3},\ y = \sqrt{3};\ x = -\sqrt{3},\ y = -\sqrt{3};\ x = 2,\ y = 1;\ x = -2,\ y = -1$ **51.** $x = 0,\ y = -2;\ x = 0,\ y = 1;\ x = 2,\ y = -1$

53. $x = 2,\ y = 8$ **55.** $x = 81,\ y = 3$

57.

59.

61.

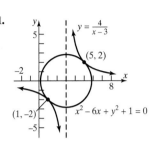

63. $x = 0.48,\ y = 0.62$ **65.** $x = -1.65,\ y = -0.89$ **67.** $x = 0.58,\ y = 1.86;\ x = 1.81,\ y = 1.05;\ x = 0.58,\ y = -1.86;\ x = 1.81,\ y = -1.05$

69. $x = 2.35,\ y = 0.85$

71.

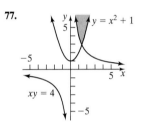

73.

75.

77.

79. 3 and 1; -3 and -1 **81.** 2 and 2; -2 and -2 **83.** $\dfrac{1}{2}$ and $\dfrac{1}{3}$ **85.** 5   **87.** 5 in. by 3 in.   **89.** 2 cm and 4 cm   **91.** tortoise: 7 m/hr, hare: $7\dfrac{1}{2}$ m/hr

93. 12 cm by 18 cm   **95.** $x = 60$ ft; $y = 30$ ft   **97.** $l = \dfrac{P + \sqrt{P^2 - 16A}}{4};\ w = \dfrac{P - \sqrt{P^2 - 16A}}{4}$ **99.** $y = 4x - 4$ **101.** $y = 2x + 1$

103. $y = -\dfrac{1}{3}x + \dfrac{7}{3}$ **105.** $y = 2x - 3$ **107.** $r_1 = \dfrac{-b + \sqrt{b^2 - 4ac}}{2a};\ r_2 = \dfrac{-b - \sqrt{b^2 - 4ac}}{2a}$ **109. (a)** 4.274 ft by 4.274 ft or 0.093 ft by 0.093 ft

Review Exercises *(page 645)*

1. Parabola; vertex $(0, 0)$, focus $(-4, 0)$, directrix $x = 4$

3. Hyperbola; center $(0, 0)$, vertices $(5, 0)$ and $(-5, 0)$, foci $(\sqrt{26}, 0)$ and $(-\sqrt{26}, 0)$, asymptotes $y = \dfrac{1}{5}x$ and $y = -\dfrac{1}{5}x$

5. Ellipse; center $(0, 0)$, vertices $(0, 5)$ and $(0, -5)$, foci $(0, 3)$ and $(0, -3)$

7. $x^2 = -4(y - 1)$: Parabola; vertex $(0, 1)$, focus $(0, 0)$, directrix $y = 2$

9. $\dfrac{x^2}{2} - \dfrac{y^2}{8} = 1$: Hyperbola; center $(0, 0)$, vertices $(\sqrt{2}, 0)$ and $(-\sqrt{2}, 0)$, foci $(\sqrt{10}, 0)$ and $(-\sqrt{10}, 0)$, asymptotes $y = 2x$ and $y = -2x$

11. $(x - 2)^2 = 2(y + 2)$: Parabola; vertex $(2, -2)$, focus $\left(2, -\dfrac{3}{2}\right)$, directrix $y = -\dfrac{5}{2}$

13. $\dfrac{(y - 2)^2}{4} - (x - 1)^2 = 1$: Hyperbola; center $(1, 2)$, vertices $(1, 4)$ and $(1, 0)$, foci $(1, 2 + \sqrt{5})$ and $(1, 2 - \sqrt{5})$,
asymptotes $y - 2 = \pm 2(x - 1)$

15. $\dfrac{(x - 2)^2}{9} + \dfrac{(y - 1)^2}{4} = 1$: Ellipse; center $(2, 1)$, vertices $(5, 1)$ and $(-1, 1)$, foci $(2 + \sqrt{5}, 1)$ and $(2 - \sqrt{5}, 1)$

17. $(x - 2)^2 = -4(y + 1)$: Parabola; vertex $(2, -1)$, focus $(2, -2)$, directrix $y = 0$

19. $\dfrac{(x - 1)^2}{4} + \dfrac{(y + 1)^2}{9} = 1$: Ellipse; center $(1, -1)$, vertices $(1, 2)$ and $(1, -4)$, foci $(1, -1 + \sqrt{5})$ and $(1, -1 - \sqrt{5})$

21. $y^2 = -8x$

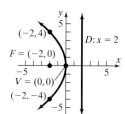

23. $\dfrac{y^2}{4} - \dfrac{x^2}{12} = 1$

25. $\dfrac{x^2}{16} + \dfrac{y^2}{7} = 1$

27. $(x-2)^2 = -4(y+3)$

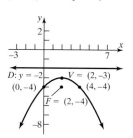

29. $(x+2)^2 - \dfrac{(y+3)^2}{3} = 1$

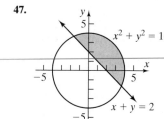

31. $\dfrac{(x+4)^2}{16} + \dfrac{(y-5)^2}{25} = 1$

33. $\dfrac{(x+1)^2}{9} - \dfrac{(y-2)^2}{7} = 1$

35. $\dfrac{(x-3)^2}{9} - \dfrac{(y-1)^2}{4} = 1$

37. $x = -\dfrac{2}{5}, y = -\dfrac{11}{5}; x = -2, y = 1$ **39.** $x = 2\sqrt{2}, y = \sqrt{2}; x = -2\sqrt{2}, y = -\sqrt{2}$ **41.** $x = 0, y = 0; x = -3, y = 3; x = 3, y = 3$

43. $x = \sqrt{2}, y = -\sqrt{2}; x = -\sqrt{2}, y = \sqrt{2}; x = \dfrac{4}{3}\sqrt{2}, y = -\dfrac{2}{3}\sqrt{2}; x = -\dfrac{4}{3}\sqrt{2}, y = \dfrac{2}{3}\sqrt{2}$ **45.** $x = 1, y = -1$

47.

49.

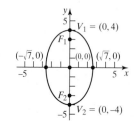

51. $\dfrac{x^2}{5} - \dfrac{y^2}{4} = 1$ **53.** The ellipse $\dfrac{x^2}{16} + \dfrac{y^2}{7} = 1$ **55.** $\dfrac{1}{4}$ ft or 3 in.   **57.** 19.72 ft, 18.86 ft, 14.91 ft   **59.** 450 ft

Chapter Test *(page 647)*

1. Hyperbola; center: $(-1, 0)$; vertices: $(-3, 0)$ and $(1, 0)$; foci: $(-1 - \sqrt{13}, 0)$ and $(-1 + \sqrt{13}, 0)$; asymptotes: $y = -\dfrac{3}{2}(x+1)$ and $y = \dfrac{3}{2}(x+1)$

2. Parabola; vertex: $\left(1, -\dfrac{1}{2}\right)$; focus: $\left(1, \dfrac{3}{2}\right)$; directrix: $y = -\dfrac{5}{2}$

3. Ellipse; center: $(-1, 1)$; foci: $(-1 - \sqrt{3}, 1)$ and $(-1 + \sqrt{3}, 1)$; vertices: $(-4, 1)$ and $(2, 1)$

4. $(x+1)^2 = 6(y-3)$

5. $\dfrac{x^2}{7} + \dfrac{y^2}{16} = 1$

6. $\dfrac{(y-2)^2}{4} - \dfrac{(x-2)^2}{8} = 1$

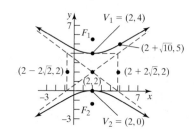

7. $(1, -3)$ and $(1, 3)$

8. $(3, 4)$ and $(1, 2)$

9.

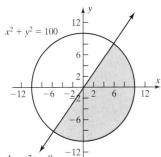

10. The microphone should be located $\dfrac{2}{3}$ feet from the base of the reflector, along its axis of symmetry.

Cumulative Review *(page 648)*

1. $-6x + 5 - 3h$ **2. (a)** Domain: $(-\infty, \infty)$; Range: $(2, \infty)$ **(b)** $y = \log_3(x - 2)$; Domain: $(2, \infty)$; Range: $(-\infty, \infty)$

3. $x = 2$ or $x = -\dfrac{1}{3}$ or $x = -5$ **4.** $-3 \le x \le 2$ or $[-3, 2]$

5. (a) $y = 2x - 2$ **(b)** $(x - 2)^2 + y^2 = 4$ **(c)** $\dfrac{x^2}{9} + \dfrac{y^2}{4} = 1$ **(d)** $y = 2(x - 1)^2$ **(e)** $\dfrac{y^2}{1} - \dfrac{x^2}{3} = 1$ **(f)** $y = 4^x$

6. (a) 18 **(b)** $(2, 18]$

C H A P T E R   7   Sequences; Induction; the Binomial Theorem

7.1 Assess Your Understanding *(page 659)*

5. sequence **6.** $3; 15$ **7.** 20 **8.** T **9.** T **10.** T **11.** $3{,}628{,}800$ **13.** 504 **15.** 1260 **17.** $1, 2, 3, 4, 5$ **19.** $\dfrac{1}{3}, \dfrac{1}{2}, \dfrac{3}{5}, \dfrac{2}{3}, \dfrac{5}{7}$ **21.** $1, -4, 9, -16, 25$

23. $\dfrac{1}{2}, \dfrac{2}{5}, \dfrac{2}{7}, \dfrac{8}{41}, \dfrac{8}{61}$ **25.** $-\dfrac{1}{6}, \dfrac{1}{12}, -\dfrac{1}{20}, \dfrac{1}{30}, -\dfrac{1}{42}$ **27.** $\dfrac{1}{e}, \dfrac{2}{e^2}, \dfrac{3}{e^3}, \dfrac{4}{e^4}, \dfrac{5}{e^5}$ **29.** $a_n = \dfrac{n}{n + 1}$ **31.** $a_n = \dfrac{1}{2^{n-1}}$ **33.** $a_n = (-1)^{n+1}$ **35.** $a_n = (-1)^{n+1}n$

37. $a_1 = 2, a_2 = 5, a_3 = 8, a_4 = 11, a_5 = 14$ **39.** $a_1 = -2, a_2 = 0, a_3 = 3, a_4 = 7, a_5 = 12$ **41.** $a_1 = 5, a_2 = 10, a_3 = 20, a_4 = 40, a_5 = 80$

43. $a_1 = 3, a_2 = \dfrac{3}{2}, a_3 = \dfrac{1}{2}, a_4 = \dfrac{1}{8}, a_5 = \dfrac{1}{40}$ **45.** $a_1 = 1, a_2 = 2, a_3 = 2, a_4 = 4, a_5 = 8$

47. $a_1 = A, a_2 = A + d, a_3 = A + 2d, a_4 = A + 3d, a_5 = A + 4d$

49. $a_1 = \sqrt{2}, a_2 = \sqrt{2 + \sqrt{2}}, a_3 = \sqrt{2 + \sqrt{2 + \sqrt{2}}}, a_4 = \sqrt{2 + \sqrt{2 + \sqrt{2 + \sqrt{2}}}}, a_5 = \sqrt{2 + \sqrt{2 + \sqrt{2 + \sqrt{2 + \sqrt{2}}}}}$

51. $3 + 4 + \cdots + (n + 2$ **53.** $\dfrac{1}{2} + 2 + \dfrac{9}{2} + \cdots + \dfrac{n^2}{2}$ **55.** $1 + \dfrac{1}{3} + \dfrac{1}{9} + \cdots + \dfrac{1}{3^n}$ **57.** $\dfrac{1}{3} + \dfrac{1}{9} + \cdots + \dfrac{1}{3^n}$

59. $\ln 2 - \ln 3 + \ln 4 - \cdots + (-1)^n \ln n$ **61.** $\displaystyle\sum_{k=1}^{20} k$ **63.** $\displaystyle\sum_{k=1}^{13} \dfrac{k}{k + 1}$ **65.** $\displaystyle\sum_{k=0}^{6} (-1)^k \left(\dfrac{1}{3^k}\right)$ **67.** $\displaystyle\sum_{k=1}^{n} \dfrac{3^k}{k}$ **69.** $\displaystyle\sum_{k=0}^{n} (a + kd)$ or $\displaystyle\sum_{k=1}^{n+1} [a + (k - 1)d]$

71. 50 **73.** 21 **75.** 90 **77.** 26 **79.** 42 **81.** 96 **83. (a)** $\$2930$ **(b)** 14 payments have been made. **(c)** 36 payments; $\$3584.62$ **(d)** $\$584.62$

85. (a) 2162 **(b)** After 26 months

87. (a) $a_0 = 0, a_n = (1.02)a_{n-1} + 500$ **(b)** After 82 quarters **(c)** $\$156{,}116.15$

89. (a) $a_0 = 150{,}000, a_n = (1.005)a_{n-1} - 899.33$

   (b) \$149,850.67   **(c)**

   (d) After 58 payments or 4 years and 10 months later
   (e) After 359 payments of \$899.33, plus last payment
      of \$895.10
   (f) \$173,754.57

   (g) (a) $a_0 = 150{,}000, a_n = (1.005)a_{n-1} - 999.33$

      (b) \$149,750.67   (c)

   (d) After 37 payments or 3 years and 1 month later
   (e) After 278 payments of \$999.33, plus last payment
      of $\$353.69(1.005) = \$355.46$
   (f) \$128,169.20

91. 21   **93.** A Fibonacci sequence   **95. (a)** 3.630170833   **(b)** 3.669060828   **(c)** 3.669296668   **(d)** 12

97. $2S = \underbrace{(1 + n) + (1 + n) + \cdots + (n + 1)}_{n \text{ terms}}$

$2S = n(n + 1)$

$S = \dfrac{1}{2}n(n + 1)$

7.2 Assess Your Understanding *(page 666)*

1. arithmetic   **2.** F   **3.** $t_n - t_{n-1} = (n + 4) - [(n - 1) + 4] = n + 4 - (n + 3) = n + 4 - n - 3 = 1$, a constant; $d = 1; 5, 6, 7, 8$

5. $t_n - t_{n-1} = (2n - 5) - [2(n - 1) - 5] = 2n - 5 - (2n - 2 - 5) = 2n - 5 - (2n - 7) = 2n - 5 - 2n + 7 = 2$, a constant; $d = 2; -3, -1, 1, 3$

7. $t_n - t_{n-1} = (6 - 2n) - [6 - 2(n - 1)] = 6 - 2n - (6 - 2n + 2) = 6 - 2n - (8 - 2n) = 6 - 2n - 8 + 2n = -2$, a constant; $d = -2; 4, 2, 0, -2$

9. $t_n - t_{n-1} = \left(\dfrac{1}{2} - \dfrac{1}{3}n\right) - \left[\dfrac{1}{2} - \dfrac{1}{3}(n - 1)\right] = \dfrac{1}{2} - \dfrac{1}{3}n - \left(\dfrac{1}{2} - \dfrac{1}{3}n + \dfrac{1}{3}\right) = \dfrac{1}{2} - \dfrac{1}{3}n - \left(\dfrac{5}{6} - \dfrac{1}{3}n\right) = \dfrac{1}{2} - \dfrac{1}{3}n - \dfrac{5}{6} + \dfrac{1}{3}n = -\dfrac{1}{3}$, a constant; $d = -\dfrac{1}{3}; \dfrac{1}{6}, -\dfrac{1}{6}, -\dfrac{1}{2}, -\dfrac{5}{6}$

11. $t_n - t_{n-1} = \ln 3^n - \ln 3^{n-1} = n\ln 3 - (n - 1)\ln 3 = n\ln 3 - (n\ln 3 - \ln 3) = n\ln 3 - n\ln 3 + \ln 3 = \ln 3$, a constant; $d = \ln 3; \ln 3, 2\ln 3, 3\ln 3, 4\ln 3$

13. $a_n = 3n - 1; a_5 = 14$ **15.** $a_n = 8 - 3n; a_5 = -7$ **17.** $a_n = \dfrac{1}{2}(n - 1); a_5 = 2$ **19.** $a_n = \sqrt{2}n; a_5 = 5\sqrt{2}$ **21.** $a_{12} = 24$ **23.** $a_{10} = -26$

25. $a_8 = a + 7b$ **27.** $a_1 = -13; d = 3; a_n = a_{n-1} + 3; a_n = -16 + 3n$ **29.** $a_1 = -53; d = 6; a_n = a_{n-1} + 6; a_n = -59 + 6n$ **31.** $a_1 = 28; d = -2;$ $a_n = a_{n-1} - 2; a_n = 30 - 2n$ **33.** $a_1 = 25; d = -2; a_n = a_{n-1} - 2; a_n = 27 - 2n$ **35.** n^2 **37.** $\dfrac{n}{2}(9 + 5n)$ **39.** 1260   **41.** 324

43.
```
sum(seq(3.45n+4.
12,n,1,20,1))
          806.9
```
45.
```
sum(seq(2.4n+.4,
n,1,15,1))
          294
```
47.
```
sum(seq(2.58n+2.
32,n,1,25,1))
          896.5
```
49. $-\dfrac{3}{2}$

51. 1185 seats   **53.** 210 of beige and 190 blue   **55.** 30 rows

Historical Problems *(page 674)*

1. $1\dfrac{2}{3}$ loaves, $10\dfrac{5}{6}$ loaves, 20 loaves, $29\dfrac{1}{6}$ loaves, $38\dfrac{1}{3}$ loaves   **2. (a)** 1   **(b)** 2401   **(c)** 2800

7.3 Assess Your Understanding (page 708)

1. geometric **2.** $\dfrac{a}{1-r}$ **3.** T **4.** F **5.** $\dfrac{S_n}{S_{n-1}} = \dfrac{3^n}{3^{n-1}} = 3^{n-(n-1)} = 3^1 = 3$, a nonzero constant; $r = 3$; 3, 9, 27, 81

7. $\dfrac{S_n}{S_{n-1}} = \dfrac{-3\left(\frac{1}{2}\right)^n}{-3\left(\frac{1}{2}\right)^{n-1}} = \left(\dfrac{1}{2}\right)^{n-(n-1)} = \left(\dfrac{1}{2}\right)^1 = \dfrac{1}{2}$, a nonzero constant; $r = \dfrac{1}{2}$; $-\dfrac{3}{2}, -\dfrac{3}{4}, -\dfrac{3}{8}, -\dfrac{3}{16}$

9 $\dfrac{S_n}{S_{n-1}} = \dfrac{\frac{2^{n-1}}{4}}{\frac{2^{(n-1)-1}}{4}} = 2^{(n-1)-[(n-1)-1]} = 2^1 = 2$, a nonzero constant; $r = 2$; $\dfrac{1}{4}, \dfrac{1}{2}, 1, 2$

11. $\dfrac{S_n}{S_{n-1}} = \dfrac{2^{n/3}}{2^{(n-1)/3}} = 2^{n/3-[(n-1)/3]} = 2^{1/3}$, a nonzero constant; $r = 2^{1/3}$; $2^{1/3}, 2^{2/3}, 2, 2^{4/3}$

13. $\dfrac{S_n}{S_{n-1}} = \dfrac{\frac{3^{n-1}}{2^n}}{\frac{3^{(n-1)-1}}{2^{n-1}}} = \dfrac{3^{n-1}}{2^n} \cdot \dfrac{2^{n-1}}{3^{n-2}} = \dfrac{3^{n-1} \cdot 2^{n-1}}{2^n \cdot 3^{n-2}} = 3^{n-1-(n-2)} \cdot 2^{n-1-n} = 3^1 \cdot 2^{-1} = \dfrac{3}{2}$, a nonzero constant; $r = \dfrac{3}{2}$; $\dfrac{1}{2}, \dfrac{3}{4}, \dfrac{9}{8}, \dfrac{27}{16}$

15. Arithmetic; $d = 1$ **17.** Neither **19.** Arithmetic; $d = -\dfrac{2}{3}$ **21.** Neither **23.** Geometric; $r = \dfrac{2}{3}$ **25.** Geometric; $r = 2$

27. Geometric; $r = 3^{1/2}$ **29.** $a_5 = 162$; $a_n = 2 \cdot 3^{n-1}$ **31.** $a_5 = 5$; $a_n = 5 \cdot (-1)^{n-1}$ **33.** $a_5 = 0$; $a_n = 0$ **35.** $a_5 = 4\sqrt{2}$; $a_n = (\sqrt{2})^n$

37. $a_7 = \dfrac{1}{64}$ **39.** $a_9 = 1$ **41.** $a_8 = 0.00000004$ **43.** $-\dfrac{1}{4}(1 - 2^n)$ **45.** $2\left[1 - \left(\dfrac{2}{3}\right)^n\right]$ **47.** $1 - 2^n$

49. `(1/4)sum(seq(2^n` `,n,0,14,1))` `8191.75` **51.** `sum(seq((2/3)^n,` `n,1,15,1))` `1.995432683` **53.** `-1sum(seq(2^n,n,` `0,14,1))` `-32767`

55. $\dfrac{3}{2}$ **57.** 16 **59.** $\dfrac{8}{5}$ **61.** $\dfrac{20}{3}$ **63.** $\dfrac{18}{5}$ **65.** -4 **67.** \$21,879.11 **69. (a)** 0.775 ft **(b)** 8th **(c)** 15.88 ft **(d)** 20 ft **71.** 1.845×10^{19} **73.** 10

75. \$72.67 per share **77.** December 20, 2001 (111 days); \$9999.92 **79.** Option B results in more money (\$524,287 versus \$500,500).

81. Yes. A constant sequence is both arithmetic and geometric. For example, 3, 3, 3, ... is an arithmetic sequence with $a_1 = 3$ and $d = 0$ and is a geometric sequence with $a_1 = 3$ and $r = 1$.

7.4 Assess Your Understanding (page 680)

1. (I) $n = 1$: $2(1) = 2$ and $1(1 + 1) = 2$
  (II) If $2 + 4 + 6 + \cdots + 2k = k(k + 1)$, then $2 + 4 + 6 + \cdots + 2k + 2(k + 1) = (2 + 4 + 6 + \cdots + 2k) + 2(k + 1)$
   $= k(k + 1) + 2(k + 1) = k^2 + 3k + 2 = (k + 1)(k + 2) = (k + 1)[(k + 1) + 1]$.

3. (I) $n = 1$: $1 + 2 = 3$ and $\dfrac{1}{2}(1)(1 + 5) = \dfrac{1}{2}(6) = 3$

  (II) If $3 + 4 + 5 + \cdots + (k + 2) = \dfrac{1}{2}k(k + 5)$, then $3 + 4 + 5 + \cdots + (k + 2) + [(k + 1) + 2]$

   $= [3 + 4 + 5 + \cdots + (k + 2)] + (k + 3) = \dfrac{1}{2}k(k + 5) + k + 3 = \dfrac{1}{2}(k^2 + 7k + 6) = \dfrac{1}{2}(k + 1)(k + 6)$

   $= \dfrac{1}{2}(k + 1)[(k + 1) + 5]$.

5. (I) $n = 1$: $3(1) - 1 = 2$ and $\dfrac{1}{2}(1)[3(1) + 1] = \dfrac{1}{2}(4) = 2$

  (II) If $2 + 5 + 8 + \cdots + (3k - 1) = \dfrac{1}{2}k(3k + 1)$, then $2 + 5 + 8 + \cdots + (3k - 1) + [3(k + 1) - 1]$

   $= [2 + 5 + 8 + \cdots + (3k - 1)] + (3k + 2) = \dfrac{1}{2}k(3k + 1) + (3k + 2) = \dfrac{1}{2}(3k^2 + 7k + 4) = \dfrac{1}{2}(k + 1)(3k + 4)$

   $= \dfrac{1}{2}(k + 1)[3(k + 1) + 1]$.

7. (I) $n = 1: 2^{1-1} = 1$ and $2^1 - 1 = 1$

(II) If $1 + 2 + 2^2 + \cdots + 2^{k-1} = 2^k - 1$, then $1 + 2 + 2^2 + \cdots + 2^{k-1} + 2^{(k+1)-1} = (1 + 2 + 2^2 + \cdots + 2^{k-1}) + 2^k$
$= 2^k - 1 + 2^k = 2(2^k) - 1 = 2^{k+1} - 1$.

9. (I) $n = 1: 4^{1-1} = 1$ and $\frac{1}{3}(4^1 - 1) = \frac{1}{3}(3) = 1$

(II) If $1 + 4 + 4^2 + \cdots + 4^{k-1} = \frac{1}{3}(4^k - 1)$, then $1 + 4 + 4^2 + \cdots + 4^{k-1} + 4^{(k+1)-1} = (1 + 4 + 4^2 + \cdots + 4^{k-1}) + 4^k$

$= \frac{1}{3}(4^k - 1) + 4^k = \frac{1}{3}[4^k - 1 + 3(4^k)] = \frac{1}{3}[4(4^k) - 1] = \frac{1}{3}(4^{k+1} - 1)$.

11. (I) $n = 1: \dfrac{1}{1 \cdot 2} = \dfrac{1}{2}$ and $\dfrac{1}{1 + 1} = \dfrac{1}{2}$

(II) If $\dfrac{1}{1 \cdot 2} + \dfrac{1}{2 \cdot 3} + \dfrac{1}{3 \cdot 4} + \cdots + \dfrac{1}{k(k + 1)} = \dfrac{k}{k + 1}$, then $\dfrac{1}{1 \cdot 2} + \dfrac{1}{2 \cdot 3} + \dfrac{1}{3 \cdot 4} + \cdots + \dfrac{1}{k(k + 1)} + \dfrac{1}{(k + 1)[(k + 1) + 1]}$

$= \left[\dfrac{1}{1 \cdot 2} + \dfrac{1}{2 \cdot 3} + \dfrac{1}{3 \cdot 4} + \cdots + \dfrac{1}{k(k + 1)}\right] + \dfrac{1}{(k + 1)(k + 2)} = \dfrac{k}{k + 1} + \dfrac{1}{(k + 1)(k + 2)} = \dfrac{k(k + 2) + 1}{(k + 1)(k + 2)}$

$= \dfrac{k^2 + 2k + 1}{(k + 1)(k + 2)} = \dfrac{(k + 1)^2}{(k + 1)(k + 2)} = \dfrac{k + 1}{k + 2} = \dfrac{k + 1}{(k + 1) + 1}$.

13. (I) $n = 1: 1^2 = 1$ and $\frac{1}{6} \cdot 1 \cdot 2 \cdot 3 = 1$

(II) If $1^2 + 2^2 + 3^2 + \cdots + k^2 = \frac{1}{6}k(k + 1)(2k + 1)$, then $1^2 + 2^2 + 3^2 + \cdots + k^2 + (k + 1)^2$

$= (1^2 + 2^2 + 3^2 + \cdots + k^2) + (k + 1)^2 = \frac{1}{6}k(k + 1)(2k + 1) + (k + 1)^2 = \frac{1}{6}(2k^3 + 9k^2 + 13k + 6)$

$= \frac{1}{6}(k + 1)(k + 2)(2k + 3) = \frac{1}{6}(k + 1)[(k + 1) + 1][2(k + 1) + 1]$.

15. (I) $n = 1: 5 - 1 = 4$ and $\frac{1}{2}(1)(9 - 1) = \frac{1}{2} \cdot 8 = 4$

(II) If $4 + 3 + 2 + \cdots + (5 - k) = \frac{1}{2}k(9 - k)$, then $4 + 3 + 2 + \cdots + (5 - k) + [5 - (k + 1)]$

$= [4 + 3 + 2 + \cdots + (5 - k)] + 4 - k = \frac{1}{2}k(9 - k) + 4 - k = \frac{1}{2}(9k - k^2 + 8 - 2k) = \frac{1}{2}(-k^2 + 7k + 8)$

$= \frac{1}{2}(k + 1)(8 - k) = \frac{1}{2}(k + 1)[9 - (k + 1)]$.

17. (I) $n = 1: 1 \cdot (1 + 1) = 2$ and $\frac{1}{3} \cdot 1 \cdot 2 \cdot 3 = 2$

(II) If $1 \cdot 2 + 2 \cdot 3 + 3 \cdot 4 + \cdots + k(k + 1) = \frac{1}{3}k(k + 1)(k + 2)$, then $1 \cdot 2 + 2 \cdot 3 + 3 \cdot 4 + \cdots + k(k + 1)$

$+ (k + 1)[(k + 1) + 1] = [1 \cdot 2 + 2 \cdot 3 + 3 \cdot 4 + \cdots + k(k + 1)] + (k + 1)(k + 2)$

$= \frac{1}{3}k(k + 1)(k + 2) + \frac{1}{3} \cdot 3 (k + 1)(k + 2) = \frac{1}{3}(k + 1)(k + 2)(k + 3) = \frac{1}{3}(k + 1)[(k + 1) + 1][(k + 1) + 2]$.

19. (I) $n = 1: 1^2 + 1 = 2$, which is divisible by 2.

(II) If $k^2 + k$ is divisible by 2, then $(k + 1)^2 + (k + 1) = k^2 + 2k + 1 + k + 1 = (k^2 + k) + 2k + 2$. Since $k^2 + k$ is divisible by 2 and $2k + 2$ is divisible by 2, $(k + 1)^2 + (k + 1)$ is divisible by 2.

21. (I) $n = 1: 1^2 - 1 + 2 = 2$ which is divisible by 2.

(II) If $k^2 - k + 2$ is divisible by 2, then $(k + 1)^2 - (k + 1) + 2 = k^2 + 2k + 1 - k - 1 + 2 = (k^2 - k + 2) + 2k$.
Since $k^2 - k + 2$ is divisible by 2 and $2k$ is divisible by 2, $(k + 1)^2 - (k + 1) + 2$ is divisible by 2.

23. (I) $n = 1$: If $x > 1$, then $x^1 = x > 1$.

(II) Assume, for an arbitrary natural number k, that if $x > 1$ then $x^k > 1$. Multiply both sides of the inequality $x^k > 1$ by x. If $x > 1$, then $x^{k+1} > x > 1$.

25. (I) $n = 1: a - b$ is a factor of $a^1 - b^1 = a - b$.

(II) If $a - b$ is a factor of $a^k - b^k$, then $a^{k+1} - b^{k+1} = a(a^k - b^k) + b^k(a - b)$.
Since $a - b$ is a factor of $a^k - b^k$ and $a - b$ is a factor of $a - b$, then $a - b$ is a factor of $a^{k+1} - b^{k+1}$.

27. $n = 1: 1^2 - 1 + 41 = 41$ which is a prime number.
$n = 41: 41^2 - 41 + 41 = 1681 = 41^2$, which is not prime.

29. (I) $n = 1$: $ar^{1-1} = a \cdot 1 = a$ and $a \cdot \dfrac{1 - r^1}{1 - r} = a$, because $r \neq 1$.

(II) If $a + ar + ar^2 + \cdots + ar^{k-1} = a \cdot \dfrac{1 - r^k}{1 - r}$, then $a + ar + ar^2 + \cdots + ar^{k-1} + ar^{(k+1)-1} = (a + ar + ar^2 + \cdots + ar^{k-1}) + ar^k$

$= a \cdot \dfrac{1 - r^k}{1 - r} + ar^k = \dfrac{a(1 - r^k) + ar^k(1 - r)}{1 - r} = \dfrac{a - ar^k + ar^k - ar^{k+1}}{1 - r} = a \cdot \dfrac{1 - r^{k+1}}{1 - r}$.

31. (I) $n = 4$: The number of diagonals in a convex polygon of 4 sides is 2 and $\dfrac{1}{2} \cdot 4 \cdot (4 - 3) = 2$.

(II) If the number of diagonals in a convex polygon of k sides is $\dfrac{1}{2} k(k - 3)$ then that of $(k + 1)$ sides is increased by

$(k + 1) - 2 = k - 1$. Thus, the number of diagonals in a convex polygon of $(k + 1)$ sides is $\dfrac{1}{2} k(k - 3) + (k - 1)$

$= \dfrac{1}{2}[k^2 - 3k + 2k - 2] = \dfrac{1}{2}[k^2 - k - 2] = \dfrac{1}{2}(k + 1)(k - 2) = \dfrac{1}{2}(k + 1)[(k + 1) - 3]$.

7.5 Assess Your Understanding *(page 686)*

1. Pascal triangle   **2.** 15   **3.** False   **4.** Binomial Theorem   **5.** 10   **7.** 21   **9.** 50   **11.** 1   **13.** $\approx 1.8664 \times 10^{15}$ **15.** $\approx 1.4834 \times 10^{13}$
17. $x^5 + 5x^4 + 10x^3 + 10x^2 + 5x + 1$ **19.** $x^6 - 12x^5 + 60x^4 - 160x^3 + 240x^2 - 192x + 64$ **21.** $81x^4 + 108x^3 + 54x^2 + 12x + 1$
23. $x^{10} + 5x^8 y^2 + 10x^6 y^4 + 10x^4 y^6 + 5x^2 y^8 + y^{10}$ **25.** $x^3 + 6\sqrt{2}x^{5/2} + 30x^2 + 40\sqrt{2}x^{3/2} + 60x + 24\sqrt{2}x^{1/2} + 8$
27. $a^5 x^5 + 5a^4 bx^4 y + 10a^3 b^2 x^3 y^2 + 10a^2 b^3 x^2 y^3 + 5ab^4 xy^4 + b^5 y^5$ **29.** 17,010   **31.** $-101,376$ **33.** 41,472   **35.** $2835x^3$ **37.** $314,928x^7$
39. 495   **41.** 3360   **43.** 1.00501
45. $\dbinom{n}{n - 1} = \dfrac{n!}{(n - 1)![n - (n - 1)]!} = \dfrac{n!}{(n - 1)!1!} = \dfrac{n \cdot (n - 1)!}{(n - 1)!} = n;\ \dbinom{n}{n} = \dfrac{n!}{n!(n - n)!} = \dfrac{n!}{n!0!} = \dfrac{n!}{n!} = 1$

47. $2^n = (1 + 1)^n = \dbinom{n}{0}1^n + \dbinom{n}{1}(1)^{n-1}(1) + \cdots + \dbinom{n}{n}1^n = \dbinom{n}{0} + \dbinom{n}{1} + \cdots + \dbinom{n}{n}$

49. 1

Review Exercises *(page 689)*

1. $-\dfrac{4}{3}, \dfrac{5}{4}, -\dfrac{6}{5}, \dfrac{7}{6}, -\dfrac{8}{7}$ **3.** $2, 1, \dfrac{8}{9}, 1, \dfrac{32}{25}$ **5.** $3, 2, \dfrac{4}{3}, \dfrac{8}{9}, \dfrac{16}{27}$ **7.** $2, 0, 2, 0, 2$ **9.** $6 + 10 + 14 + 18 = 48$ **11.** $\displaystyle\sum_{k=1}^{13} (-1)^{k+1} \dfrac{1}{k}$

13. Arithmetic; $d = 1$; $S_n = \dfrac{n}{2}(n + 11)$ **15.** Neither

17. Geometric; $r = 8$; $S_n = \dfrac{8}{7}(8^n - 1)$ **19.** Arithmetic; $d = 4$; $S_n = 2n(n - 1)$ **21.** Geometric; $r = \dfrac{1}{2}$; $S_n = 6\left[1 - \left(\dfrac{1}{2}\right)^n\right]$

23. Neither   **25.** 115   **27.** 75   **29.** $\dfrac{1093}{2187} \approx 0.49977$ **31.** 35   **33.** $\dfrac{1}{10^{10}}$ **35.** $9\sqrt{2}$ **37.** $a_n = 5n - 4$ **39.** $a_n = n - 10$ **41.** $\dfrac{9}{2}$ **43.** $\dfrac{4}{3}$ **45.** 8
47. (I) $n = 1$: $3 \cdot 1 = 3$ and $\dfrac{3 \cdot 1}{2}(1 + 1) = 3$

(II) If $3 + 6 + 9 + \cdots + 3k = \dfrac{3k}{2}(k + 1)$, then $3 + 6 + 9 + \cdots + 3k + 3(k + 1) = (3 + 6 + 9 + \cdots + 3k) + (3k + 3)$

$= \dfrac{3k}{2}(k + 1) + (3k + 3) = \dfrac{3k^2}{2} + \dfrac{3k}{2} + \dfrac{6k}{2} + \dfrac{6}{2} = \dfrac{3}{2}(k^2 + 3k + 2) = \dfrac{3}{2}(k + 1)(k + 2) = \dfrac{3(k + 1)}{2}[(k + 1) + 1]$ **49.**   (I)

$n = 1$: $2 \cdot 3^{1-1} = 2$ and $3^1 - 1 = 2$

(II) If $2 + 6 + 18 + \cdots + 2 \cdot 3^{k-1} = 3^k - 1$, then $2 + 6 + 18 + \cdots + 2 \cdot 3^{k-1} + 2 \cdot 3^{(k+1)-1} = (2 + 6 + 18 + \cdots + 2 \cdot 3^{k-1}) + 2 \cdot 3^k$
$= 3^k - 1 + 2 \cdot 3^k = 3 \cdot 3^k - 1 = 3^{k+1} - 1$.

51. (I) $n = 1$: $(3 \cdot 1 - 2)^2 = 1$ and $\dfrac{1}{2} \cdot 1 \cdot [6(1)^2 - 3(1) - 1] = 1$

(II) If $1^2 + 4^2 + 7^2 + \cdots + (3k - 2)^2 = \dfrac{1}{2}k(6k^2 - 3k - 1)$, then $1^2 + 4^2 + 7^2 + \cdots + (3k - 2)^2 + [3(k + 1) - 2]^2$

$= [1^2 + 4^2 + 7^2 + \cdots + (3k - 2)^2] + (3k + 1)^2 = \dfrac{1}{2}k(6k^2 - 3k - 1) + (3k + 1)^2 = \dfrac{1}{2}(6k^3 - 3k^2 - k) + (9k^2 + 6k + 1)$

$= \dfrac{1}{2}(6k^3 + 15k^2 + 11k + 2) = \dfrac{1}{2}(k + 1)(6k^2 + 9k + 2) = \dfrac{1}{2}(k + 1)[6(k + 1)^2 - 3(k + 1) - 1]$.

53. 10 **55.** $x^5 + 10x^4 + 40x^3 + 80x^2 + 80x + 32$ **57.** $32x^5 + 240x^4 + 720x^3 + 1080x^2 + 810x + 243$ **59.** 144 **61.** 84

63. (a) 8 **(b)** 1100 **65. (a)** $20\left(\dfrac{3}{4}\right)^3 = \dfrac{135}{16}$ ft **(b)** $20\left(\dfrac{3}{4}\right)^n$ ft **(c)** 13 times **(d)** 140 ft

67. (a) \$4975

(b)

(c) At the beginning of the thirty-third month the balance is less than \$4000. At this time, 32 payments of \$100 each have been made.

(d) 93 payments of \$100 plus 1 payment of \$11.18; \$9311.18

(e) \$4311.18

Chapter Test *(page 690)*

1. $0, \dfrac{3}{10}, \dfrac{8}{11}, \dfrac{5}{4}, \dfrac{24}{13}$ **2.** 4, 14, 44, 134, 404 **3.** $2 - \dfrac{3}{4} + \dfrac{4}{9} = \dfrac{61}{36}$ **4.** $-\dfrac{1}{3} - \dfrac{14}{9} - \dfrac{73}{27} - \dfrac{308}{81} = -\dfrac{680}{81}$ **5.** $\displaystyle\sum_{k=1}^{10} (-1)^k \left(\dfrac{k+1}{k+4}\right)$

6. 5050 **7.** Neither **8.** Geometric; $r = 4$; $S_n = \dfrac{2}{3}(1 - 4^n)$ **9.** Arithmetic; $d = -8$; $S_n = n(2 - 4n)$

10. Arithmetic; $d = -\dfrac{1}{2}$; $Sn = \dfrac{n}{4}(27 - n)$ **11.** Geometric; $r = \dfrac{2}{5}$; $Sn = \dfrac{125}{3}\left(1 - \left(\dfrac{2}{5}\right)^n\right)$

12. Neither **13.** $\dfrac{1024}{5}$ **14.** $243m^5 + 810m^4 + 1080m^3 + 720m^2 + 240m + 32$

15. First we show that the statement holds for $n = 1$.

$$\left(1 + \dfrac{1}{1}\right) = 1 + 1 = 2$$

The equality is true for $n = 1$ so Condition I holds.
Next we assume that

$$\left(1 + \dfrac{1}{1}\right)\left(1 + \dfrac{1}{2}\right)\left(1 + \dfrac{1}{3}\right)...\left(1 + \dfrac{1}{n}\right) = n + 1 \text{ is true for some } k, \text{ and}$$

we determine whether the formula then holds for $k + 1$. We assume that

$$\left(1 + \dfrac{1}{1}\right)\left(1 + \dfrac{1}{2}\right)\left(1 + \dfrac{1}{3}\right)...\left(1 + \dfrac{1}{k}\right) = k + 1$$

Now we need to show that

$$\left(1 + \dfrac{1}{1}\right)\left(1 + \dfrac{1}{2}\right)\left(1 + \dfrac{1}{3}\right)...\left(1 + \dfrac{1}{k}\right)\left(1 + \dfrac{1}{k+1}\right) = (k + 1) + 1 = k + 2$$

We do this as follows:
$$\left(1 + \dfrac{1}{1}\right)\left(1 + \dfrac{1}{2}\right)\left(1 + \dfrac{1}{3}\right)...\left(1 + \dfrac{1}{k}\right)\left(1 + \dfrac{1}{k+1}\right) =$$
$$\left[\left(1 + \dfrac{1}{1}\right)\left(1 + \dfrac{1}{2}\right)\left(1 + \dfrac{1}{3}\right)...\left(1 + \dfrac{1}{k}\right)\right]\left(1 + \dfrac{1}{k+1}\right) = (k + 1)\left(1 + \dfrac{1}{k+1}\right) \text{ (induction assumption)}$$
$$= (k + 1)\cdot 1 + (k + 1)\cdot\dfrac{1}{k+1}$$
$$= k + 1 + 1$$
$$= k + 2$$

Condition II also holds. Thus, formula holds true for all natural numbers.

17. The weightlifter will have lifted a total of 8000 pounds after 5 sets.

Cumulative Review Exercises *(page 692)*

1. $-3, 3, -3i, 3i$

2. (a)

(b) $\left(\sqrt{\dfrac{-1 + \sqrt{3601}}{18}}, \dfrac{-1 + \sqrt{3601}}{6} \right), \left(-\sqrt{\dfrac{-1 + \sqrt{3601}}{18}}, \dfrac{-1 + \sqrt{3601}}{6} \right)$

(c) The circle and the parabola intersect at

$\left(\sqrt{\dfrac{-1 + \sqrt{3601}}{18}}, \dfrac{-1 + \sqrt{3601}}{6} \right), \left(-\sqrt{\dfrac{-1 + \sqrt{3601}}{18}}, \dfrac{-1 + \sqrt{3601}}{6} \right).$

3. $\left\{ \ln\left(\dfrac{5}{2}\right) \right\}$ **4.** $y = 5x - 10$ **5.** $x^2 + y^2 + 2x - 4y - 20 = 0$ **6. (a)** 5 **(b)** 13 **(c)** $\dfrac{6x + 3}{2x - 1}$ **(d)** $\left\{ x \,\middle|\, x \neq \dfrac{1}{2} \right\}$ **(e)** $\dfrac{7x - 2}{x - 2}$ **(f)** $\{x | x \neq 2\}$

(g) $g^{-1}(x) = \dfrac{1}{2}(x - 1)$; all reals **(h)** $f^{-1}(x) = \dfrac{2x}{x - 3}$; $\{x | x \neq 3\}$ **7.** $\dfrac{x^2}{7} + \dfrac{y^2}{16} = 1$ **8.** $(x + 1)^2 = 4(y - 2)$

C H A P T E R   8   Counting and Probability

8.1 Assess Your Understanding *(page 699)*

1. union **2.** intersection **3.** T **4.** T **5.** $\{1, 3, 5, 6, 7, 9\}$ **7.** $\{1, 5, 7\}$ **9.** $\{1, 6, 9\}$ **11.** $\{1, 2, 4, 5, 6, 7, 8, 9\}$ **13.** $\{1, 2, 4, 5, 6, 7, 8, 9\}$
15. $\{0, 2, 6, 7, 8\}$ **17.** $\{0, 1, 2, 3, 5, 6, 7, 8, 9\}$ **19.** $\{0, 1, 2, 3, 5, 6, 7, 8, 9\}$ **21.** $\{0, 1, 2, 3, 4, 6, 7, 8\}$ **23.** $\{0\}$ **25.** $\varnothing, \{a\}, \{b\}, \{c\}, \{d\}, \{a, b\},$
$\{a, c\}, \{a, d\}, \{b, c\}, \{b, d\}, \{c, d\}, \{a, b, c\}, \{b, c, d\}, \{a, c, d\}, \{a, b, d\}, \{a, b, c, d\}$ **27.** 25 **29.** 40 **31.** 25 **33.** 37 **35.** 18 **37.** 5 **39.** $175; 125$
41. (a) 15 **(b)** 15 **(c)** 15 **(d)** 25 **(e)** 40 **43. (a)** 11,291 thousand **(b)** 72,503 thousand

8.2 Assess Your Understanding *(page 708)*

3. permutation **4.** combination **5.** T **6.** T **7.** 30 **9.** 24 **11.** 1 **13.** 1680 **15.** 28 **17.** 35 **19.** 1 **21.** 10,400,600 **23.** $\{abc, abd, abe, acb,$
$acd, ace, adb, adc, ade, aeb, aec, aed, bac, bad, bae, bca, bcd, bce, bda, bdc, bde, bea, bec, bed, cab, cad, cae, cba, cbd, cbe, cda, cdb, cde, cea, ceb, ced,$
$dab, dac, dae, dba, dbc, dbe, dca, dcb, dce, dea, deb, dec, eab, eac, ead, eba, ebc, ebd, eca, ecb, ecd, eda, edb, edc\}; 60$ **25.** $\{123, 124, 132, 134, 142, 143,$
$213, 214, 231, 234, 241, 243, 312, 314, 321, 324, 341, 342, 412, 413, 421, 423, 431, 432\}; 24$ **27.** $\{abc, abd, abe, acd, ace, ade, bcd, bce, bde, cde\}; 10$
29. $\{123, 124, 134, 234\}; 4$ **31.** 15 **33.** 16 **35.** 8 **37.** 24 **39.** 60 **41.** 18,278 **43.** 35 **45.** 1024 **47.** 9000 **49.** 120 **51.** 480 **53.** 132,860
55. 336 **57.** 90,720 **59. (a)** 63 **(b)** 35 **(c)** 1 **61.** 1.157×10^{76} **63.** 362,880 **65.** 660 **67.** 15

Historical Problems *(page 719)*

1. (a) $\{AAAA, AAAB, AABA, AABB, ABAA, ABAB, ABBA, ABBB, BAAA, BAAB, BABA, BABB, BBAA, BBAB, BBBA, BBBB\}$

(b) $P(A \text{ wins}) = \dfrac{C(4, 2) + C(4, 3) + C(4, 4)}{2^4} = \dfrac{6 + 4 + 1}{16} = \dfrac{11}{16}; P(B \text{ wins}) = \dfrac{C(4, 3) + C(4, 4)}{2^4} = \dfrac{4 + 1}{16} = \dfrac{5}{16}$

2. (a) $\$\dfrac{3}{2} = \$1.50$ **(b)** $\$\dfrac{1}{2} = \$0.50$

8.3 Assess Your Understanding *(page 719)*

1. equally likely **2.** complement **3.** F **4.** T **5.** $0, 0.01, 0.35, 1$ **7.** Probability model **9.** Not a probability model

11. $S = \{HH, HT, TH, TT\}; P(HH) = \dfrac{1}{4}, P(HT) = \dfrac{1}{4}, P(TH) = \dfrac{1}{4}, P(TT) = \dfrac{1}{4}$

13. $S = \{HH1, HH2, HH3, HH4, HH5, HH6, HT1, HT2, HT3, HT4, HT5, HT6, TH1, TH2, TH3, TH4, TH5, TH6, TT1, TT2, TT3, TT4, TT5, TT6\};$
each outcome has the probability of $\dfrac{1}{24}$.

15. $S = \{HHH, HHT, HTH, HTT, THH, THT, TTH, TTT\};$ each outcome has the probability of $\dfrac{1}{8}$.

17. $S = \{1 \text{ Yellow}, 1 \text{ Red}, 1 \text{ Green}, 2 \text{ Yellow}, 2 \text{ Red}, 2 \text{ Green}, 3 \text{ Yellow}, 3 \text{ Red}, 3 \text{ Green}, 4 \text{ Yellow}, 4 \text{ Red}, 4 \text{ Green}\};$ each outcome has the
probability of $\dfrac{1}{12}$; thus, $P(2 \text{ Red}) + P(4 \text{ Red}) = \dfrac{1}{12} + \dfrac{1}{12} = \dfrac{1}{6}$.

19. $S = \{1$ Yellow Forward, 1 Yellow Backward, 1 Red Forward, 1 Red Backward, 1 Green Forward, 1 Green Backward, 2 Yellow Forward,

2 Yellow Backward, 2 Red Forward, 2 Red Backward, 2 Green Forward, 2 Green Backward, 3 Yellow Forward, 3 Yellow Backward,

3 Red Forward, 3 Red Backward, 3 Green Forward, 3 Green Backward, 4 Yellow Forward, 4 Yellow Backward, 4 Red Forward,

4 Red Backward, 4 Green Forward, 4 Green Backward$\}$; each outcome has the probability of $\dfrac{1}{24}$; thus,

$P(1$ Red Backward$) + P(1$ Green Backward$) = \dfrac{1}{24} + \dfrac{1}{24} = \dfrac{1}{12}$.

21. $S = \{11$ Red, 11 Yellow, 11 Green, 12 Red, 12 Yellow, 12 Green, 13 Red, 13 Yellow, 13 Green, 14 Red, 14 Yellow, 14 Green, 21 Red,

21 Yellow, 21 Green, 22 Red, 22 Yellow, 22 Green, 23 Red, 23 Yellow, 23 Green, 24 Red, 24 Yellow, 24 Green, 31 Red, 31 Yellow,

31 Green, 32 Red, 32 Yellow, 32 Green, 33 Red, 33 Yellow, 33 Green, 34 Red, 34 Yellow, 34 Green, 41 Red, 41 Yellow, 41 Green, 42 Red,

42 Yellow, 42 Green, 43 Red, 43 Yellow, 43 Green, 44 Red, 44 Yellow, 44 Green$\}$; each outcome has the probability of $\dfrac{1}{48}$;

thus, $E = \{22$ Red, 22 Green, 24 Red, 24 Green$\}$; $P(E) = \dfrac{n(E)}{n(S)} = \dfrac{4}{48} = \dfrac{1}{12}$.

23. A, B, C, F **25.** B **27.** $P(\text{H}) = \dfrac{4}{5}$; $P(\text{T}) = \dfrac{1}{5}$ **29.** $P(1) = P(3) = P(5) = \dfrac{2}{9}$; $P(2) = P(4) = P(6) = \dfrac{1}{9}$ **31.** $\dfrac{3}{10}$ **33.** $\dfrac{1}{2}$ **35.** $\dfrac{1}{6}$ **37.** $\dfrac{1}{8}$

39. $\dfrac{1}{4}$ **41.** $\dfrac{1}{6}$ **43.** $\dfrac{1}{18}$ **45.** 0.55 **47.** 0.70 **49.** 0.30 **51.** 0.735 **53.** 0.7 **55.** $\dfrac{17}{20}$ **57.** $\dfrac{11}{20}$ **59.** $\dfrac{1}{2}$ **61.** $\dfrac{3}{10}$ **63.** $\dfrac{2}{5}$

65. (a) 0.57 **(b)** 0.95 **(c)** 0.83 **(d)** 0.38 **(e)** 0.29 **(f)** 0.05 **(g)** 0.78 **(h)** 0.71 **67. (a)** $\dfrac{25}{33}$ **(b)** $\dfrac{25}{33}$ **69.** 0.167 **71.** 0.000033069

Review Exercises *(page 723)*

1. $\varnothing$, {Dave}, {Joanne}, {Erica}, {Dave, Joanne}, {Dave, Erica}, {Joanne, Erica}, {Dave, Joanne, Erica}   **3.** $\{1, 3, 5, 6, 7, 8\}$ **5.** $\{3, 7\}$

7. $\{1, 2, 4, 6, 8, 9\}$ **9.** $\{1, 2, 4, 5, 6, 9\}$ **11.** 17 **13.** 29 **15.** 7 **17.** 25 **19.** 336 **21.** 56 **23.** 60 **25.** 128 **27.** 3024 **29.** 1680 **31.** 91

33. 1,600,000 **35.** 216,000 **37.** 1260 **39. (a)** 381,024 **(b)** 1260 **41. (a)** $8.634628387 \times 10^{45}$ **(b)** 0.6531 **(c)** 0.3469 **43. (a)** 0.058 **(b)** 0.942

45. $\dfrac{4}{9}$ **47.** 0.2; 0.26

Chapter Test *(page 725)*

1. $\{4\}$ **2.** $\{0, 1, 3, 4, 5, 7, 9\}$ **3.** $\{1, 9\}$ **4.** $\{3, 5, 7\}$ **5.** $\{0, 2, 4, 6, 8\}$ **6.** $\{1, 2, 3, 5, 6, 7, 8, 9\}$ **7.** $n(\text{physics}) = 22$

8. $n(\text{none of the three}) = 3$ **9.** $n(\text{only biology } and \text{ chemistry}) = 8$ **10.** $n(\text{physics } or \text{ chemistry}) = 45$ **11.** 5040 **12.** 151,200

13. 462 **14.** There are 54,264 ways to choose 6 colors from the 21 available colors. **15.** There are 840 distinct arrangementss of the letters in the word REDEEMED. **16.** There are 56 different exacta bets for an 8-horse race. **17.** There are 155,480,000 possible license plates using the new format. **18. (a)** 0.95 **(b)** 0.30 **19. (a)** 0.25 **(b)** 0.55 **20.** 0.19

21. $P(\text{win on \$1 play}) = \dfrac{1}{120,526,770}$ **22.** $P(\text{exactly 2 fours}) = \dfrac{1250}{7776} \approx 0.1608.$
≈ 0.0000000083

Cumulative Review *(page 726)*

1. $\left\{\dfrac{1}{3} - \dfrac{\sqrt{2}}{3}i, \dfrac{1}{3} + \dfrac{\sqrt{2}}{3}i\right\}$ **2.**   **3.**   **4.** $\{x \mid 3.99 \le x \le 4.01\}$ or $[3.99, 4.01]$

5. $\left\{-\dfrac{1}{2} + \dfrac{\sqrt{7}}{2}i, -\dfrac{1}{2} - \dfrac{\sqrt{7}}{2}i, -\dfrac{1}{5}, 3\right\}$

6.   **7.** 2 **8.** $\left\{\dfrac{8}{3}\right\}$ **9.** $x = 2, y = -5, z = 3$ **10.** 125; 700

Domain: all real numbers
Range: $\{y \mid y > 5\}$
Horizontal asymptote: $y = 5$

Index

CONICS

Parabola

  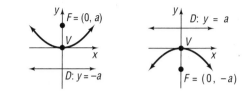

$$y^2 = 4ax \qquad y^2 = -4ax \qquad x^2 = 4ay \qquad x^2 = -4ay$$

Ellipse

 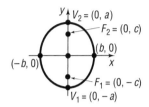

$$\frac{x^2}{a^2} + \frac{y^2}{b^2} = 1, \quad a > b, \quad c^2 = a^2 - b^2 \qquad\qquad \frac{x^2}{b^2} + \frac{y^2}{a^2} = 1, \quad a > b, \quad c^2 = a^2 - b^2$$

Hyperbola

$$\frac{x^2}{a^2} - \frac{y^2}{b^2} = 1, \quad c^2 = a^2 + b^2 \qquad\qquad \frac{y^2}{a^2} - \frac{x^2}{b^2} = 1, \quad c^2 = a^2 + b^2$$

$$\text{Asymptotes:} \quad y = \frac{b}{a}x, \quad y = -\frac{b}{a}x \qquad\qquad \text{Asymptotes:} \quad y = \frac{a}{b}x, \quad y = -\frac{a}{b}x$$

PROPERTIES OF LOGARITHMS

$$\log_a(MN) = \log_a M + \log_a N$$

$$\log_a\left(\frac{M}{N}\right) = \log_a M - \log_a N$$

$$\log_a M^r = r \log_a M$$

$$\log_a M = \frac{\log M}{\log a} = \frac{\ln M}{\ln a}$$

PERMUTATIONS/COMBINATIONS

$$0! = 1 \qquad 1! = 1$$

$$n! = n(n-1) \cdot \ldots \cdot (3)(2)(1)$$

$$P(n, r) = \frac{n!}{(n-r)!}$$

$$C(n, r) = \binom{n}{r} = \frac{n!}{(n-r)!r!}$$

BINOMIAL THEOREM

$$(a + b)^n = a^n + \binom{n}{1}ba^{n-1} + \binom{n}{2}b^2 a^{n-2}$$
$$+ \cdots + \binom{n}{n-1}b^{n-1}a + b^n$$

ARITHMETIC SEQUENCE

$$a + (a + d) + (a + 2d) + \cdots + [a + (n-1)d]$$
$$= \frac{n}{2}[2a + (n-1)d]$$

GEOMETRIC SEQUENCE

$$a + ar + ar^2 + \cdots + ar^{n-1} = a \cdot \frac{1 - r^n}{1 - r}$$

GEOMETRIC SERIES

$$\text{If } |r| < 1, \ a + ar + ar^2 + \cdots = \sum_{k=1}^{\infty} ar^{k-1}$$
$$= \frac{a}{1 - r}$$

FORMULAS/EQUATIONS

Distance Formula If $P_1 = (x_1, y_1)$ and $P_2 = (x_2, y_2)$, the distance from P_1 to P_2 is

$$d(P_1, P_2) = \sqrt{(x_2 - x_1)^2 + (y_2 - y_1)^2}$$

Standard Equation of a Circle The standard equation of a circle of radius r with center at (h, k) is

$$(x - h)^2 + (y - k)^2 = r^2$$

Slope Formula The slope m of the line containing the points $P_1 = (x_1, y_1)$ and $P_2 = (x_2, y_2)$ is

$$m = \frac{y_2 - y_1}{x_2 - x_1} \quad \text{if } x_1 \neq x_2$$

$$m \text{ is undefined} \quad \text{if } x_1 = x_2$$

Point–Slope Equation of a Line The equation of a line with slope m containing the point (x_1, y_1) is

$$y - y_1 = m(x - x_1)$$

Slope–Intercept Equation of a Line The equation of a line with slope m and y-intercept b is

$$y = mx + b$$

Quadratic Formula The solutions of the equation $ax^2 + bx + c = 0, a \neq 0$, are

$$x = \frac{-b \pm \sqrt{b^2 - 4ac}}{2a}$$

If $b^2 - 4ac > 0$, there are two unequal real solutions.
If $b^2 - 4ac = 0$, there is a repeated real solution.
If $b^2 - 4ac < 0$, there are two complex solutions that are not real.

GEOMETRY FORMULAS

Circle

r = Radius, A = Area, C = Circumference
$A = \pi r^2 \qquad C = 2\pi r$

Triangle

b = Base, h = Altitude (Height), A = area
$A = \frac{1}{2}bh$

Rectangle

l = Length, w = Width, A = area, P = perimeter
$A = lw \qquad P = 2l + 2w$

Rectangular Box

l = Length, w = Width, h = Height, V = Volume, S = Surface area
$V = lwh \qquad S = 2lw + 2lh + 2wh$

Sphere

r = Radius, V = Volume, S = Surface area
$V = \frac{4}{3}\pi r^3 \qquad S = 4\pi r^2$

Right Circular Cylinder

r = Radius, h = Height, V = Volume, S = Surface area
$V = \pi r^2 h \qquad S = 2\pi r^2 + 2\pi rh$